Yong Shi   Geert Di
Jack Dongarra   Pet

Computational Science – ICCS 2007

7th International Conference
Beijing, China, May 27 - 30, 2007
Proceedings, Part I

 Springer

Volume Editors

Yong Shi
Graduate University of the Chinese Academy of Science
Beijing 100080, China
E-mail: yshi@gucas.as.cn

Geert Dick van Albada
Peter M.A. Sloot
University of Amsterdam, Section Computational Science
1098 SJ Amsterdam, The Netherlands
E-mail: {dick, sloot}@science.uva.nl

Jack Dongarra
University of Tennessee, Computer Science Department
Knoxville, TN 37996-3450, USA
E-mail: dongarra@cs.utk.edu

Library of Congress Control Number: 2007927049

CR Subject Classification (1998): F, D, G, H,I.1, I.3. I.6, J, K.3 C.2-3

LNCS Sublibrary: SL 1 – Theoretical Computer Science and General Issues

ISSN   0302-9743
ISBN-10   3-540-72583-0 Springer Berlin Heidelberg New York
ISBN-13   978-3-540-72583-1 Springer Berlin Heidelberg New York

Springer is a part of Springer Science+Business Media

springer.com

© Springer-Verlag Berlin Heidelberg 2007
Printed in Germany

Typesetting: Camera-ready by author, data conversion by Scientific Publishing Services, Chennai, India
Printed on acid-free paper   SPIN: 12065691   06/3180   5 4 3 2 1 0

Preface

The Seventh International Conference on Computational Science (ICCS 2007) was held in Beijing, China, May 27-30, 2007. This was the continuation of previous conferences in the series: ICCS 2006 in Reading, UK; ICCS 2005 in Atlanta, Georgia, USA; ICCS 2004 in Krakow, Poland; ICCS 2003 held simultaneously at two locations in, Melbourne, Australia and St. Petersburg, Russia; ICCS 2002 in Amsterdam, The Netherlands; and ICCS 2001 in San Francisco, California, USA. Since the first conference in San Francisco, the ICCS series has become a major platform to promote the development of Computational Science. The theme of ICCS 2007 was "Advancing Science and Society through Computation." It aimed to bring together researchers and scientists from mathematics and computer science as basic computing disciplines, researchers from various application areas who are pioneering the advanced application of computational methods to sciences such as physics, chemistry, life sciences, and engineering, arts and humanitarian fields, along with software developers and vendors, to discuss problems and solutions in the area, to identify new issues, and to shape future directions for research, as well as to help industrial users apply various advanced computational techniques.

During the opening of ICCS 2007, Siwei Cheng (Vice-Chairman of the Standing Committee of the National People's Congress of the People's Republic of China and the Dean of the School of Management of the Graduate University of the Chinese Academy of Sciences) presented the welcome speech on behalf of the Local Organizing Committee, after which Hector Ruiz (President and CEO, AMD) made remarks on behalf of international computing industries in China. Seven keynote lectures were delivered by Vassil Alexandrov (Advanced Computing and Emerging Technologies, University of Reading, UK) - Efficient Scalable Algorithms for Large-Scale Computations; Hans Petter Langtangen (Simula Research Laboratory, Lysaker, Norway) - Computational Modelling of Huge Tsunamis from Asteroid Impacts; Jiawei Han (Department of Computer Science, University of Illinois at Urbana-Champaign, USA) - Research Frontiers in Advanced Data Mining Technologies and Applications; Ru-qian Lu (Institute of Mathematics, Chinese Academy of Sciences) - Knowledge Engineering and Knowledge Ware; Alessandro Vespignani (School of Informatics, Indiana University, USA) -Computational Epidemiology and Emergent Disease Forecast; David Keyes (Department of Applied Physics and Applied Mathematics, Columbia University) - Scalable Solver Infrastructure for Computational Science and Engineering; and Yves Robert (Ecole Normale Suprieure de Lyon , France) - Think Before Coding: Static Strategies (and Dynamic Execution) for Clusters and Grids. We would like to express our thanks to all of the invited and keynote speakers for their inspiring talks. In addition to the plenary sessions, the conference included 14 parallel oral sessions and 4 poster sessions. This year, we received more than 2,400 submissions for all tracks combined, out of which 716 were accepted.

This includes 529 oral papers, 97 short papers, and 89 poster papers, spread over 35 workshops and a main track. For the main track we had 91 papers (80 oral papers and 11 short papers) in the proceedings, out of 360 submissions. We had some 930 people doing reviews for the conference, with 118 for the main track. Almost all papers received three reviews. The accepted papers are from more than 43 different countries and 48 different Internet top-level domains.

The papers cover a large volume of topics in computational science and related areas, from multiscale physics to wireless networks, and from graph theory to tools for program development.

We would like to thank all workshop organizers and the Program Committee for the excellent work in maintaining the conference's standing for high-quality papers. We would like to express our gratitude to staff and graduates of the Chinese Academy of Sciences Research Center on Data Technology and Knowledge Economy and the Institute of Policy and Management for their hard work in support of ICCS 2007. We would like to thank the Local Organizing Committee and Local Arrangements Committee for their persistent and enthusiastic work towards the success of ICCS 2007. We owe special thanks to our sponsors, AMD, Springer; University of Nebraska at Omaha, USA and Graduate University of Chinese Academy of Sciences, for their generous support.

ICCS 2007 was organized by the Chinese Academy of Sciences Research Center on Data Technology and Knowledge Economy, with support from the Section Computational Science at the Universiteit van Amsterdam and Innovative Computing Laboratory at the University of Tennessee, in cooperation with the Society for Industrial and Applied Mathematics (SIAM), the International Association for Mathematics and Computers in Simulation (IMACS), the Chinese Society for Management Modernization (CSMM), and the Chinese Society of Optimization, Overall Planning and Economical Mathematics (CSOOPEM).

May 2007   Yong Shi

Organization

ICCS 2007 was organized by the Chinese Academy of Sciences Research Center on Data Technology and Knowledge Economy, with support from the Section Computational Science at the Universiteit van Amsterdam and Innovative Computing Laboratory at the University of Tennessee, in cooperation with the Society for Industrial and Applied Mathematics (SIAM), the International Association for Mathematics and Computers in Simulation (IMACS), and the Chinese Society for Management Modernization (CSMM).

Conference Chairs

Conference Chair - Yong Shi (Chinese Academy of Sciences, China/University of Nebraska at Omaha USA)
Program Chair - Dick van Albada (Universiteit van Amsterdam, The Netherlands)
ICCS Series Overall Scientific Co-chair - Jack Dongarra (University of Tennessee, USA)
ICCS Series Overall Scientific Chair - Peter M.A. Sloot (Universiteit van Amsterdam, The Netherlands)

Local Organizing Committee

Weimin Zheng (Tsinghua University, Beijing, China) – Chair
Hesham Ali (University of Nebraska at Omaha, USA)
Chongfu Huang (Beijing Normal University, Beijing, China)
Masato Koda (University of Tsukuba, Japan)
Heeseok Lee (Korea Advanced Institute of Science and Technology, Korea)
Zengliang Liu (Beijing University of Science and Technology, Beijing, China)
Jen Tang (Purdue University, USA)
Shouyang Wang (Academy of Mathematics and System Science, Chinese Academy of Sciences, Beijing, China)
Weixuan Xu (Institute of Policy and Management, Chinese Academy of Sciences, Beijing, China)
Yong Xue (Institute of Remote Sensing Applications, Chinese Academy of Sciences, Beijing, China)
Ning Zhong (Maebashi Institute of Technology, USA)
Hai Zhuge (Institute of Computing Technology, Chinese Academy of Sciences, Beijing, China)

Local Arrangements Committee

Weixuan Xu, Chair
Yong Shi, Co-chair of events
Benfu Lu, Co-chair of publicity
Hongjin Yang, Secretary
Jianping Li, Member
Ying Liu, Member
Jing He, Member
Siliang Chen, Member
Guanxiong Jiang, Member
Nan Xiao, Member
Zujin Deng, Member

Sponsoring Institutions

AMD
Springer
World Scientific Publlishing
University of Nebraska at Omaha, USA
Graduate University of Chinese Academy of Sciences
Institute of Policy and Management, Chinese Academy of Sciences Universiteit
van Amsterdam

Program Committee

J.H. Abawajy, Deakin University, Australia
D. Abramson, Monash University, Australia
V. Alexandrov, University of Reading, UK
I. Altintas, San Diego Supercomputer Center, UCSD
M. Antolovich, Charles Sturt University, Australia
E. Araujo, Universidade Federal de Campina Grande, Brazil
M.A. Baker, University of Reading, UK
B. Balis, Krakow University of Science and Technology, Poland
A. Benoit, LIP, ENS Lyon, France
I. Bethke, University of Amsterdam, The Netherlands
J.A.R. Blais, University of Calgary, Canada
I. Brandic, University of Vienna, Austria
J. Broeckhove, Universiteit Antwerpen, Belgium
M. Bubak, AGH University of Science and Technology, Poland
K. Bubendorfer, Victoria University of Wellington, Australia
B. Cantalupo, DATAMAT S.P.A, Italy
J. Chen Swinburne, University of Technology, Australia
O. Corcho, University of Manchester, UK
J.C. Cunha, Univ. Nova de Lisboa, Portugal

M. Mascagni, Florida State University, USA
V. Maxville, Curtin Business School, Australia
A.S. McGough, London e-Science Centre, UK
E.D. Moreno, UEA-BENq, Manaus, Brazil
J.T. Moscicki, Cern, Switzerland
S. Naqvi, CoreGRID Network of Excellence, France
P.O.A. Navaux, Universidade Federal do Rio Grande do Sul, Brazil
Z. Nemeth, Computer and Automation Research Institute, Hungarian Academy
of Science, Hungary
J. Ni, University of Iowa, USA
G. Norman, Joint Institute for High Temperatures of RAS, Russia
B. Ó Nualláin, University of Amsterdam, The Netherlands
C.W. Oosterlee, Centrum voor Wiskunde en Informatica, CWI, The Netherlands
S. Orlando, Università Ca' Foscari, Venice, Italy
M. Paprzycki, IBS PAN and SWPS, Poland
M. Parashar, Rutgers University, USA
L.M. Patnaik, Indian Institute of Science, India
C.P. Pautasso, ETH Zürich, Switzerland
R. Perrott, Queen's University, Belfast, UK
V. Prasanna, University of Southern California, USA
T. Priol, IRISA, France
M.R. Radecki, Krakow University of Science and Technology, Poland
M. Ram, C-DAC Bangalore Centre, India
A. Rendell, Australian National University, Australia
P. Rhodes, University of Mississippi, USA
M. Riedel, Research Centre Juelich, Germany
D. Rodríguez García, University of Alcalá, Spain
K. Rycerz, Krakow University of Science and Technology, Poland
R. Santinelli, CERN, Switzerland
J. Schneider, Technische Universität Berlin, Germany
B. Schulze, LNCC, Brazil
J. Seo, The University of Manchester, UK
Y. Shi, Chinese Academy of Sciences, Beijing, China
D. Shires, U.S. Army Research Laboratory, USA
A.E. Solomonides, University of the West of England, Bristol, UK
V. Stankovski, University of Ljubljana, Slovenia
H. Stockinger, Swiss Institute of Bioinformatics, Switzerland
A. Streit, Forschungszentrum Jülich, Germany
H. Sun, Beihang University, China
R. Tadeusiewicz, AGH University of Science and Technology, Poland
J. Tang, Purdue University USA
M. Taufer, University of Texas El Paso, USA
C. Tedeschi, LIP-ENS Lyon, France
A. Thandavan, ACET Center, University of Reading, UK
A. Tirado-Ramos, University of Amsterdam, The Netherlands

P. Tvrdik, Czech Technical University Prague, Czech Republic
G.D. van Albada, Universiteit van Amsterdam, The Netherlands
F. van Lingen, California Institute of Technology, USA
J. Vigo-Aguiar, University of Salamanca, Spain
D.W. Walker, Cardiff University, UK
C.L. Wang, University of Hong Kong, China
A.L. Wendelborn, University of Adelaide, Australia
Y. Xue, Chinese Academy of Sciences, China
L.T. Yang, St. Francis Xavier University, Canada
C.T. Yang, Tunghai University, Taichung, Taiwan
J. Yu, The University of Melbourne, Australia
Y. Zheng, Zhejiang University, China
W. Zheng, Tsinghua University, Beijing, China
L. Zhu, University of Florida, USA
A. Zomaya, The University of Sydney, Australia
E.V. Zudilova-Seinstra, University of Amsterdam, The Netherlands

Reviewers

J.H. Abawajy
D. Abramson
A. Abran
P. Adriaans
W. Ahn
R. Akbani
K. Akkaya
R. Albert
M. Aldinucci
V.N. Alexandrov
B. Alidaee
I. Altintas
K. Altmanninger
S. Aluru
S. Ambroszkiewicz
L. Anido
K. Anjyo
C. Anthes
M. Antolovich
S. Antoniotti
G. Antoniu
H. Arabnia
E. Araujo
E. Ardeleanu
J. Aroba
J. Astalos

B. Autin
M. Babik
G. Bai
E. Baker
M.A. Baker
S. Balfe
B. Balis
W. Banzhaf
D. Bastola
S. Battiato
M. Baumgarten
M. Baumgartner
P. Beckaert
A. Belloum
O. Belmonte
A. Belyaev
A. Benoit
G. Bergantz
J. Bernsdorf
J. Berthold
I. Bethke
I. Bhana
R. Bhowmik
M. Bickelhaupt
J. Bin Shyan
J. Birkett

J.A.R. Blais
A. Bode
B. Boghosian
S. Bolboaca
C. Bothorel
A. Bouteiller
I. Brandic
S. Branford
S.J. Branford
R. Braungarten
R. Briggs
J. Broeckhove
W. Bronsvoort
A. Bruce
C. Brugha
Y. Bu
K. Bubendorfer
I. Budinska
G. Buemi
B. Bui
H.J. Bungartz
A. Byrski
M. Cai
Y. Cai
Y.Q. Cai
Z.Y. Cai

B. Cantalupo
K. Cao
M. Cao
F. Capkovic
A. Cepulkauskas
K. Cetnarowicz
Y. Chai
P. Chan
G.-L. Chang
S.C. Chang
W.A. Chaovalitwongse
P.K. Chattaraj
C.-K. Chen
E. Chen
G.Q. Chen
G.X. Chen
J. Chen
J. Chen
J.J. Chen
K. Chen
Q.S. Chen
W. Chen
Y. Chen
Y.Y. Chen
Z. Chen
G. Cheng
X.Z. Cheng
S. Chiu
K.E. Cho
Y.-Y. Cho
B. Choi
J.K. Choi
D. Choinski
D.P. Chong
B. Chopard
M. Chover
I. Chung
M. Ciglan
B. Cogan
G. Cong
J. Corander
J.C. Corchado
O. Corcho
J. Cornil
H. Cota de Freitas

E. Coutinho
J.J. Cuadrado-Gallego
Y.F. Cui
J.C. Cunha
V. Curcin
A. Curioni
R. da Rosa Righi
S. Dalai
M. Daneva
S. Date
P. Dazzi
S. de Marchi
V. Debelov
E. Deelman
J. Della Dora
Y. Demazeau
Y. Demchenko
H. Deng
X.T. Deng
Y. Deng
M. Mat Deris
F. Desprez
M. Dewar
T. Dhaene
Z.R. Di
G. di Biasi
A. Diaz Guilera
P. Didier
I.T. Dimov
L. Ding
G.D. Dobrowolski
T. Dokken
J.J. Dolado
W. Dong
Y.-L. Dong
J. Dongarra
F. Donno
C. Douglas
G.J. Garcke
R.P. Mundani
R. Drezewski
D. Du
B. Duan
J.F. Dufourd
H. Dun

C. Earley
P. Edmond
T. Eitrich
A. El Rhalibi
T. Ernst
V. Ervin
D. Estrin
L. Eyraud-Dubois
J. Falcou
H. Fang
Y. Fang
X. Fei
Y. Fei
R. Feng
M. Fernandez
K. Fisher
C. Fittschen
G. Fox
F. Freitas
T. Friesz
K. Fuerlinger
M. Fujimoto
T. Fujinami
W. Funika
T. Furumura
A. Galvez
L.J. Gao
X.S. Gao
J.E. Garcia
H.J. Gardner
M. Garre
G. Garsva
F. Gava
G. Geethakumari
M. Geimer
J. Geiser
J.-P. Gelas
A. Gerbessiotis
M. Gerndt
S. Gimelshein
S.G. Girdzijauskas
S. Girtelschmid
Z. Gj
C. Glasner
A. Goderis

D. Godoy
J. Golebiowski
S. Gopalakrishnan
Y. Gorbachev
A.M. Goscinski
M. Govindaraju
E. Grabska
G.A. Gravvanis
C.H. Grelck
D.J. Groen
L. Gross
P. Gruer
A. Grzech
J.F. Gu
Y. Guang Xue
T. Gubala
V. Guevara-Masis
C.H. Guo
X. Guo
Z.Q. Guo
L. Guohui
C. Gupta
I. Gutman
A. Haffegee
K. Han
M. Hardt
A. Hasson
J. He
J. He
K. He
T. He
J. He
M.R. Head
P. Heinzlreiter
H. Chojnacki
J. Heo
S. Hirokawa
G. Hliniak
L. Hluchy
T.B. Ho
A. Hoekstra
W. Hoffmann
A. Hoheisel
J. Hong
Z. Hong

D. Horvath
F. Hu
L. Hu
X. Hu
X.H. Hu
Z. Hu
K. Hua
H.W. Huang
K.-Y. Huang
L. Huang
L. Huang
M.S. Huang
S. Huang
T. Huang
W. Huang
Y. Huang
Z. Huang
Z. Huang
B. Huber
E. Hubo
J. Hulliger
M. Hultell
M. Humphrey
P. Hurtado
J. Huysmans
T. Ida
A. Iglesias
K. Iqbal
D. Ireland
N. Ishizawa
I. Lukovits
R. Jamieson
J.K. Jan
P. Janderka
M. Jankowski
L. Jäntschi
S.J.K. Jensen
N.J. Jeon
T.H. Jeon
T. Jeong
H. Ji
X. Ji
D.Y. Jia
C. Jiang
H. Jiang

M.J. Jiang
P. Jiang
W. Jiang
Y. Jiang
H. Jin
J. Jin
L. Jingling
G.-S. Jo
D. Johnson
J. Johnstone
J.J. Jung
K. Juszczyszyn
J.A. Kaandorp
M. Kabelac
B. Kadlec
R. Kakkar
C. Kameyama
B.D. Kandhai
S. Kandl
K. Kang
S. Kato
S. Kawata
T. Kegl
W.A. Kelly
J. Kennedy
G. Khan
J.B. Kido
C.H. Kim
D.S. Kim
D.W. Kim
H. Kim
J.G. Kim
J.H. Kim
M. Kim
T.H. Kim
T.W. Kim
P. Kiprof
R. Kirner
M. Kisiel-Dorohinicki
J. Kitowski
C.R. Kleijn
M. Kluge
A. Knüpfer
I.S. Ko
Y. Ko

R. Kobler
B. Koblitz
G.A. Kochenberger
M. Koda
T. Koeckerbauer
M. Koehler
I. Kolingerova
V. Korkhov
T. Korkmaz
L. Kotulski
G. Kou
J. Kozlak
M. Krafczyk
D. Kranzlmüller
B. Kryza
V.V. Krzhizhanovskaya
M. Kunze
D. Kurzyniec
E. Kusmierek
S. Kwang
Y. Kwok
F. Kyriakopoulos
H. Labiod
A. Lagana
H. Lai
S. Lai
Z. Lan
G. Le Mahec
B.G. Lee
C. Lee
H.K. Lee
J. Lee
J. Lee
J.H. Lee
S. Lee
S.Y. Lee
V. Lee
Y.H. Lee
L. Lefevre
L. Lei
F. Lelj
A. Lesar
D. Lesthaeghe
Z. Levnajic
A. Lewis

A. Li
D. Li
D. Li
E. Li
J. Li
J. Li
J.P. Li
M. Li
P. Li
X. Li
X.M. Li
X.S. Li
Y. Li
Y. Li
J. Liang
L. Liang
W.K. Liao
X.F. Liao
G.G. Lim
H.W. Lim
S. Lim
A. Lin
I.C. Lin
I-C. Lin
Y. Lin
Z. Lin
P. Lingras
C.Y. Liu
D. Liu
D.S. Liu
E.L. Liu
F. Liu
G. Liu
H.L. Liu
J. Liu
J.C. Liu
R. Liu
S.Y. Liu
W.B. Liu
X. Liu
Y. Liu
Y. Liu
Y. Liu
Y. Liu
Y.J. Liu

Y.Z. Liu
Z.J. Liu
S.-C. Lo
R. Loogen
B. López
A. López García de
    Lomana
F. Loulergue
G. Lu
J. Lu
J.H. Lu
M. Lu
P. Lu
S. Lu
X. Lu
Y.C. Lu
C. Lursinsap
L. Ma
M. Ma
T. Ma
A. Macedo
N. Maillard
M. Malawski
S. Maniccam
S.S. Manna
Z.M. Mao
M. Mascagni
E. Mathias
R.C. Maurya
V. Maxville
A.S. McGough
R. Mckay
T.-G. MCKenzie
K. Meenal
R. Mehrotra
M. Meneghin
F. Meng
M.F.J. Meng
E. Merkevicius
M. Metzger
Z. Michalewicz
J. Michopoulos
J.-C. Mignot
R. mikusauskas
H.Y. Ming

G. Miranda Valladares
M. Mirua
G.P. Miscione
C. Miyaji
A. Miyoshi
J. Monterde
E.D. Moreno
G. Morra
J.T. Moscicki
H. Moshkovich
V.M. Moskaliova
G. Mounie
C. Mu
A. Muraru
H. Na
K. Nakajima
Y. Nakamori
S. Naqvi
S. Naqvi
R. Narayanan
A. Narjess
A. Nasri
P. Navaux
P.O.A. Navaux
M. Negoita
Z. Nemeth
L. Neumann
N.T. Nguyen
J. Ni
Q. Ni
K. Nie
G. Nikishkov
V. Nitica
W. Nocon
A. Noel
G. Norman
B. Ó Nualláin
N. O'Boyle
J.T. Oden
Y. Ohsawa
H. Okuda
D.L. Olson
C.W. Oosterlee
V. Oravec
S. Orlando

F.R. Ornellas
A. Ortiz
S. Ouyang
T. Owens
S. Oyama
B. Ozisikyilmaz
A. Padmanabhan
Z. Pan
Y. Papegay
M. Paprzycki
M. Parashar
K. Park
M. Park
S. Park
S.K. Pati
M. Pauley
C.P. Pautasso
B. Payne
T.C. Peachey
S. Pelagatti
F.L. Peng
Q. Peng
Y. Peng
N. Petford
A.D. Pimentel
W.A.P. Pinheiro
J. Pisharath
G. Pitel
D. Plemenos
S. Pllana
S. Ploux
A. Podoleanu
M. Polak
D. Prabu
B.B. Prahalada Rao
V. Prasanna
P. Praxmarer
V.B. Priezzhev
T. Priol
T. Prokosch
G. Pucciani
D. Puja
P. Puschner
L. Qi
D. Qin

H. Qin
K. Qin
R.X. Qin
X. Qin
G. Qiu
X. Qiu
J.Q. Quinqueton
M.R. Radecki
S. Radhakrishnan
S. Radharkrishnan
M. Ram
S. Ramakrishnan
P.R. Ramasami
P. Ramsamy
K.R. Rao
N. Ratnakar
T. Recio
K. Regenauer-Lieb
R. Rejas
F.Y. Ren
A. Rendell
P. Rhodes
J. Ribelles
M. Riedel
R. Rioboo
Y. Robert
G.J. Rodgers
A.S. Rodionov
D. Rodríguez García
C. Rodriguez Leon
F. Rogier
G. Rojek
L.L. Rong
H. Ronghuai
H. Rosmanith
F.-X. Roux
R.K. Roy
U. Rüde
M. Ruiz
T. Ruofeng
K. Rycerz
M. Ryoke
F. Safaei
T. Saito
V. Sakalauskas

F. Wang
F.L. Wang
H. Wang
H.G. Wang
H.W. Wang
J. Wang
J. Wang
J. Wang
J. Wang
J.H. Wang
K. Wang
L. Wang
M. Wang
M.Z. Wang
Q. Wang
Q.Q. Wang
S.P. Wang
T.K. Wang
W. Wang
W.D. Wang
X. Wang
X.J. Wang
Y. Wang
Y.Q. Wang
Z. Wang
Z.T. Wang
A. Wei
G.X. Wei
Y.-M. Wei
X. Weimin
D. Weiskopf
B. Wen
A.L. Wendelborn
I. Wenzel
A. Wibisono
A.P. Wierzbicki
R. Wismüller
F. Wolf
C. Wu
C. Wu
F. Wu
G. Wu
J.N. Wu
X. Wu
X.D. Wu

Y. Wu
Z. Wu
B. Wylie
M. Xavier Py
Y.M. Xi
H. Xia
H.X. Xia
Z.R. Xiao
C.F. Xie
J. Xie
Q.W. Xie
H. Xing
H.L. Xing
J. Xing
K. Xing
L. Xiong
M. Xiong
S. Xiong
Y.Q. Xiong
C. Xu
C.-H. Xu
J. Xu
M.W. Xu
Y. Xu
G. Xue
Y. Xue
Z. Xue
A. Yacizi
B. Yan
N. Yan
N. Yan
W. Yan
H. Yanami
C.T. Yang
F.P. Yang
J.M. Yang
K. Yang
L.T. Yang
L.T. Yang
P. Yang
X. Yang
Z. Yang
W. Yanwen
S. Yarasi
D.K.Y. Yau

P.-W. Yau
M.J. Ye
G. Yen
R. Yi
Z. Yi
J.G. Yim
L. Yin
W. Yin
Y. Ying
S. Yoo
T. Yoshino
W. Youmei
Y.K. Young-Kyu Han
J. Yu
J. Yu
L. Yu
Z. Yu
Z. Yu
W. Yu Lung
X.Y. Yuan
W. Yue
Z.Q. Yue
D. Yuen
T. Yuizono
J. Zambreno
P. Zarzycki
M.A. Zatevakhin
S. Zeng
A. Zhang
C. Zhang
D. Zhang
D.L. Zhang
D.Z. Zhang
G. Zhang
H. Zhang
H.R. Zhang
H.W. Zhang
J. Zhang
J.J. Zhang
L.L. Zhang
M. Zhang
N. Zhang
P. Zhang
P.Z. Zhang
Q. Zhang

S. Zhang	Z. Zhao	L.G. Zhou
W. Zhang	L. Zhen	X.J. Zhou
W. Zhang	B. Zheng	X.L. Zhou
Y.G. Zhang	G. Zheng	Y.T. Zhou
Y.X. Zhang	W. Zheng	H.H. Zhu
Z. Zhang	Y. Zheng	H.L. Zhu
Z.W. Zhang	W. Zhenghong	L. Zhu
C. Zhao	P. Zhigeng	X.Z. Zhu
H. Zhao	W. Zhihai	Z. Zhu
H.K. Zhao	Y. Zhixia	M. Zhu.
H.P. Zhao	A. Zhmakin	J. Zivkovic
J. Zhao	C. Zhong	A. Zomaya
M.H. Zhao	X. Zhong	E.V. Zudilova-Seinstra
W. Zhao	K.J. Zhou	

Workshop Organizers

Sixth International Workshop on Computer Graphics and Geometric Modelling

A. Iglesias, University of Cantabria, Spain

Fifth International Workshop on Computer Algebra Systems and Applications

A. Iglesias, University of Cantabria, Spain,
A. Galvez, University of Cantabria, Spain

PAPP 2007 - Practical Aspects of High-Level Parallel Programming (4th International Workshop)

A. Benoit, ENS Lyon, France
F. Loulerge, LIFO, Orlans, France

International Workshop on Collective Intelligence for Semantic and Knowledge Grid (CISKGrid 2007)

N.T. Nguyen, Wroclaw University of Technology, Poland
J.J. Jung, INRIA Rhône-Alpes, France
K. Juszczyszyn, Wroclaw University of Technology, Poland

Simulation of Multiphysics Multiscale Systems, 4th International Workshop

V.V. Krzhizhanovskaya, Section Computational Science, University of
    Amsterdam, The Netherlands
A.G. Hoekstra, Section Computational Science, University of Amsterdam,
    The Netherlands

S. Sun, Clemson University, USA
J. Geiser, Humboldt University of Berlin, Germany

2nd Workshop on Computational Chemistry and Its Applications (2nd CCA)

P.R. Ramasami, University of Mauritius

Efficient Data Management for HPC Simulation Applications

R.-P. Mundani, Technische Universität München, Germany
J. Abawajy, Deakin University, Australia
M. Mat Deris, Tun Hussein Onn College University of Technology, Malaysia

Real Time Systems and Adaptive Applications (RTSAA-2007)

J. Hong, Soongsil University, South Korea
T. Kuo, National Taiwan University, Taiwan

The International Workshop on Teaching Computational Science (WTCS 2007)

L. Qi, Department of Information and Technology, Central China Normal
    University, China
W. Yanwen, Department of Information and Technology, Central China Normal
    University, China
W. Zhenghong, East China Normal University, School of Information Science
    and Technology, China

GeoComputation

Y. Xue, IRSA, China

Risk Analysis

C.F. Huang, Beijing Normal University, China

Advanced Computational Approaches and IT Techniques in Bioinformatics

M.A. Pauley, University of Nebraska at Omaha, USA
H.A. Ali, University of Nebraska at Omaha, USA

Workshop on Computational Finance and Business Intelligence

Y. Shi, Chinese Acedemy of Scienes, China
S.Y. Wang, Academy of Mathematical and System Sciences, Chinese Academy
    of Sciences, China
X.T. Deng, Department of Computer Science, City University of Hong Kong,
    China

Collaborative and Cooperative Environments

C. Anthes, Institute of Graphics and Parallel Processing, JKU, Austria
V.N. Alexandrov, ACET Centre, The University of Reading, UK
D. Kranzlmüller, Institute of Graphics and Parallel Processing, JKU, Austria
J. Volkert, Institute of Graphics and Parallel Processing, JKU, Austria

Tools for Program Development and Analysis in Computational Science

A. Knüpfer, ZIH, TU Dresden, Germany
A. Bode, TU Munich, Germany
D. Kranzlmüller, Institute of Graphics and Parallel Processing, JKU, Austria
J. Tao, CAPP, University of Karlsruhe, Germany
R. Wissmüller FB12, BSVS, University of Siegen, Germany
J. Volkert, Institute of Graphics and Parallel Processing, JKU, Austria

Workshop on Mining Text, Semi-structured, Web or Multimedia Data (WMTSWMD 2007)

G. Kou, Thomson Corporation, R&D, USA
Y. Peng, Omnium Worldwide, Inc., USA
J.P. Li, Institute of Policy and Management, Chinese Academy of Sciences, China

2007 International Workshop on Graph Theory, Algorithms and Its Applications in Computer Science (IWGA 2007)

M. Li, Dalian University of Technology, China

2nd International Workshop on Workflow Systems in e-Science (WSES 2007)

Z. Zhao, University of Amsterdam, The Netherlands
A. Belloum, University of Amsterdam, The Netherlands

2nd International Workshop on Internet Computing in Science and Engineering (ICSE 2007)

J. Ni, The University of Iowa, USA

Workshop on Evolutionary Algorithms and Evolvable Systems (EAES 2007)

B. Zheng, College of Computer Science, South-Central University for
    Nationalities, Wuhan, China
Y. Li, State Key Lab. of Software Engineering, Wuhan University, Wuhan, China
J. Wang, College of Computer Science, South-Central University for
    Nationalities, Wuhan, China
L. Ding, State Key Lab. of Software Engineering, Wuhan University, Wuhan,
    China

Wireless and Mobile Systems 2007 (WMS 2007)

H. Choo, Sungkyunkwan University, South Korea

**WAFTS: WAvelets, FracTals, Short-Range Phenomena —
Computational Aspects and Applications**

C. Cattani, University of Salerno, Italy
C. Toma, Polythecnica, Bucharest, Romania

Dynamic Data-Driven Application Systems - DDDAS 2007

F. Darema, National Science Foundation, USA

**The Seventh International Workshop on Meta-synthesis and
Complex Systems (MCS 2007)**

X.J. Tang, Academy of Mathematics and Systems Science, Chinese Academy of
    Sciences, China
J.F. Gu, Institute of Systems Science, Chinese Academy of Sciences, China
Y. Nakamori, Japan Advanced Institute of Science and Technology, Japan
H.C. Wang, Shanghai Jiaotong University, China

**The 1st International Workshop on Computational Methods in
Energy Economics**

L. Yu, City University of Hong Kong, China
J. Li, Chinese Academy of Sciences, China
D. Qin, Guangdong Provincial Development and Reform Commission, China

High-Performance Data Mining

Y. Liu, Data Technology and Knowledge Economy Research Center, Chinese
    Academy of Sciences, China
A. Choudhary, Electrical and Computer Engineering Department, Northwestern
    University, USA
S. Chiu, Department of Computer Science, College of Engineering, Idaho State
    University, USA

Computational Linguistics in Human–Computer Interaction

H. Ji, Sungkyunkwan University, South Korea
Y. Seo, Chungbuk National University, South Korea
H. Choo, Sungkyunkwan University, South Korea

Intelligent Agents in Computing Systems

K. Cetnarowicz, Department of Computer Science, AGH University of Science
    and Technology, Poland
R. Schaefer, Department of Computer Science, AGH University of Science and
    Technology, Poland

Networks: Theory and Applications

B. Tadic, Jozef Stefan Institute, Ljubljana, Slovenia
S. Thurner, COSY, Medical University Vienna, Austria

Workshop on Computational Science in Software Engineering

D. Rodrguez, University of Alcala, Spain
J.J. Cuadrado-Gallego, University of Alcala, Spain

International Workshop on Advances in Computational Geomechanics and Geophysics (IACGG 2007)

H.L. Xing, The University of Queensland and ACcESS Major National Research
    Facility, Australia
J.H. Wang, Shanghai Jiao Tong University, China

2nd International Workshop on Evolution Toward Next-Generation Internet (ENGI)

Y. Cui, Tsinghua University, China

Parallel Monte Carlo Algorithms for Diverse Applications in a Distributed Setting

V.N. Alexandrov, ACET Centre, The University of Reading, UK

The 2007 Workshop on Scientific Computing in Electronics Engineering (WSCEE 2007)

Y. Li, National Chiao Tung University, Taiwan

High-Performance Networked Media and Services 2007 (HiNMS 2007)

I.S. Ko, Dongguk University, South Korea
Y.J. Na, Honam University, South Korea

Table of Contents – Part I

Table of Contents – Part II

Table of Contents – Part III

Table of Contents – Part IV

A Composite Finite Element-Finite Difference Model Applied to Turbulence Modelling

Lale Balas and Asu İnan

Department of Civil Engineering, Faculty of Engineering and Architecture
Gazi University, 06570 Ankara, Turkey
lalebal@gazi.edu.tr,
asuinan@gazi.edu.tr

Abstract. Turbulence has been modeled by a two equation k-ω turbulence model to investigate the wind induced circulation patterns in coastal waters. Predictions of the model have been compared by the predictions of two equation k-ε turbulence model. Kinetic energy of turbulence is k, dissipation rate of turbulence is ε, and frequency of turbulence is ω. In the three dimensional modeling of turbulence by k-ε model and by k-ω model, a composite finite element-finite difference method has been used. The governing equations are solved by the Galerkin Weighted Residual Method in the vertical plane and by finite difference approximations in the horizontal plane. The water depths in coastal waters are divided into the same number of layers following the bottom topography. Therefore, the vertical layer thickness is proportional to the local water depth. It has been seen that two equation k-ω turbulence model leads to better predictions compared to k-ε model in the prediction of wind induced circulation in coastal waters.

Keywords: Finite difference, finite element, modeling, turbulence, coastal.

1 Introduction

There are different applications of turbulence models in the modeling studies of coastal transport processes. Some of the models use a constant eddy viscosity for the whole flow field, whose value is found from experimental or from trial and error calculations to match the observations to the problem considered. In some of the models, variations in the vertical eddy viscosity are described in algebraic forms. Algebraic or zero equation turbulence models invariably utilize the Boussinesq assumption. In these models mixing length distribution is rather problem dependent and therefore models lack universality. Further problems arise, because the eddy viscosity and diffusivity vanish whenever the mean velocity gradient is zero. To overcome these limitations, turbulence models were developed which accounted for transport or history effects of turbulence quantities by solving differential transport equations for them. In one-equation turbulence models, for velocity scale, the most meaningful scale is $k^{0.5}$, where k is the kinetic energy of the turbulent motion per unit mass[1]. In one-equation models, it is difficult to determine the length scale distribution. Therefore the trend has been to move on to two-equation models

Y. Shi et al. (Eds.): ICCS 2007, Part I, LNCS 4487, pp. 1 – 8, 2007.

which determine the length scale from a transport equation. One of the two equation models is k-ε turbulence model in which the length scale is obtained from the transport equation of dissipation rate of the kinetic energy ε [2],[3]. The other two equation model is k-ω turbulence model that includes two equations for the turbulent kinetic energy k and for the specific turbulent dissipation rate or the turbulent frequency ω [4].

2 Theory

The implicit baroclinic three dimensional numerical model (HYDROTAM-3), has been improved by a two equation k-ω turbulence model. Developed model is capable of computing water levels and water particle velocity distributions in three principal directions by solving the Navier-Stokes equations.   The governing hydrodynamic equations in the three dimensional cartesian coordinate system with the z-axis vertically upwards, are [5],[6],[7],[8]:

$$\frac{\partial u}{\partial x}+\frac{\partial v}{\partial y}+\frac{\partial w}{\partial z}=0 \tag{1}$$

$$\frac{\partial u}{\partial t}+u\frac{\partial u}{\partial x}+v\frac{\partial u}{\partial y}+w\frac{\partial u}{\partial z}=fv-\frac{1}{\rho_o}\frac{\partial p}{\partial x}+2\frac{\partial}{\partial x}(v_h\frac{\partial u}{\partial x})+\frac{\partial}{\partial y}(v_h(\frac{\partial u}{\partial y}+\frac{\partial v}{\partial x}))+\frac{\partial}{\partial z}(v_z(\frac{\partial u}{\partial z}+\frac{\partial w}{\partial x})) \tag{2}$$

$$\frac{\partial v}{\partial t}+u\frac{\partial v}{\partial x}+v\frac{\partial v}{\partial y}+w\frac{\partial v}{\partial z}=-fu-\frac{1}{\rho_o}\frac{\partial p}{\partial y}+2\frac{\partial}{\partial y}(v_h\frac{\partial v}{\partial y})+\frac{\partial}{\partial x}(v_h(\frac{\partial v}{\partial x}+\frac{\partial u}{\partial y}))+\frac{\partial}{\partial z}(v_z(\frac{\partial v}{\partial z}+\frac{\partial w}{\partial y})) \tag{3}$$

$$\frac{\partial w}{\partial t}+u\frac{\partial w}{\partial x}+v\frac{\partial w}{\partial y}+w\frac{\partial w}{\partial z}=-\frac{1}{\rho_o}\frac{\partial p}{\partial z}-g+\frac{\partial}{\partial y}(v_h(\frac{\partial w}{\partial y}+\frac{\partial v}{\partial z}))+\frac{\partial}{\partial x}(v_h(\frac{\partial w}{\partial x}+\frac{\partial u}{\partial z}))+\frac{\partial}{\partial z}(v_z\frac{\partial w}{\partial z}) \tag{4}$$

where, x,y:horizontal coordinates, z:vertical coordinate, t:time, u,v,w:velocity components in x,y,z directions at any grid locations in space, v_z:eddy viscosity coefficients in z direction, v_h:horizontal eddy viscosity coefficient, f:corriolis coefficient, ρ(x,y,z,t):water density, g:gravitational acceleration, p:pressure.
   As the turbulence model, firstly, modified k-ω turbulence model is used. Model includes two equations for the turbulent kinetic energy k and for the specific turbulent dissipation rate or the turbulent frequency ω. Equations of k-ω turbulence model are given by the followings.

$$\frac{dk}{dt}=\frac{\partial}{\partial z}\left[\sigma^*v_z\frac{\partial k}{\partial z}\right]+P+\frac{\partial}{\partial x}\left[\sigma^*v_h\frac{\partial k}{\partial x}\right]+\frac{\partial}{\partial y}\left[\sigma^*v_h\frac{\partial k}{\partial y}\right]-\beta^*\varpi k \tag{5}$$

$$\frac{d\varpi}{dt}=\frac{\partial}{\partial z}\left[\sigma^*v_z\frac{\partial\varpi}{\partial z}\right]+\alpha\frac{\varpi}{k}P+\frac{\partial}{\partial x}\left[\sigma^*v_h\frac{\partial\varpi}{\partial x}\right]+\frac{\partial}{\partial y}\left[\sigma^*v_h\frac{\partial\varpi}{\partial y}\right]-\beta\varpi^2 \tag{6}$$

The stress production of the kinetic energy P, and eddy viscosity v_z are defined by;

$$P=v_h\left[2\left(\frac{\partial u}{\partial x}\right)^2+2\left(\frac{\partial v}{\partial y}\right)^2+\left(\frac{\partial u}{\partial y}+\frac{\partial v}{\partial x}\right)^2\right]+v_z\left[\left(\frac{\partial u}{\partial z}\right)^2+\left(\frac{\partial v}{\partial z}\right)^2\right]; \quad v_z=\frac{k}{\varpi} \quad (7)$$

At high Reynolds Numbers(R_T), the constants are used as; $\alpha=5/9$, $\beta=3/40$, $\beta^*=9/100,\sigma=1/2$ and $\sigma^*=1/2$. Whereas at lower Reynolds numbers they are calculated as;

$$\alpha^*=\frac{1/40+R_T/6}{1+R_T/6}; \alpha=\frac{5}{9}\frac{1/10+R_T/2.7}{1+R_T/2.7}(\alpha^*)^{-1}; R_T=\frac{k}{\varpi v}; \beta^*=\frac{9}{100}\frac{5/18+(R_T/8)^4}{1+(R_T/8)^4} \quad (8)$$

Secondly, as the turbulence model a two equation k-ε model has been applied. Equations of k-ε turbulence model are given by the followings.

$$\frac{\partial k}{\partial t}+u\frac{\partial k}{\partial x}+v\frac{\partial k}{\partial y}+w\frac{\partial k}{\partial z}=\frac{\partial}{\partial z}\left(\frac{v_z}{\sigma_k}\frac{\partial k}{\partial z}\right)+P-\varepsilon+\frac{\partial}{\partial x}\left(v_h\frac{\partial k}{\partial x}\right)+\frac{\partial}{\partial y}\left(v_h\frac{\partial k}{\partial y}\right) \quad (9)$$

$$\frac{\partial \varepsilon}{\partial t}+u\frac{\partial \varepsilon}{\partial x}+v\frac{\partial \varepsilon}{\partial y}+w\frac{\partial \varepsilon}{\partial z}=\frac{\partial}{\partial z}\left(\frac{v_z}{\sigma_\varepsilon}\frac{\partial \varepsilon}{\partial z}\right)+C_{1\varepsilon}P\frac{\varepsilon}{k}-C_{2\varepsilon}\frac{\varepsilon^2}{k}+\frac{\partial}{\partial x}\left(v_h\frac{\partial \varepsilon}{\partial x}\right)+\frac{\partial}{\partial y}\left(v_h\frac{\partial \varepsilon}{\partial y}\right) \quad (10)$$

where, k:Kinetic energy, ε:Rate of dissipation of kinetic energy, P: Stress production of the kinetic energy. The following universal k-ε turbulence model empirical constants are used and the vertical eddy viscosity is calculated by:

$$v_z=C_\mu\frac{k^2}{\varepsilon}; \quad C_\mu=0.09, \sigma_\varepsilon=1.3, C_{1\varepsilon}=1.44, C_{2\varepsilon}=1.92. \quad (11)$$

Some other turbulence models have also been widely applied in three dimensional numerical modeling of wind induced currents such as one equation turbulence model and mixing length models. They are also used in the developed model HYROTAM-3, however it is seen that two equation turbulence models give better predictions compared to the others.

3 Solution Method

Solution method is a composite finite difference-finite element method. Equations are solved numerically by approximating the horizontal gradient terms using a staggered finite difference scheme (Fig.1a). In the vertical plane however, the Galerkin Method of finite elements is utilized. Water depths are divided into the same number of layers following the bottom topography (Fig.1b). At all nodal points, the ratio of the length (thickness) of each element (layer) to the total depth is constant. The mesh size may be varied in the horizontal plane. By following the finite element approach, all the variables at any point over the depth are written in terms of the discrete values of these variables at the vertical nodal points by using linear shape functions.

$$\tilde{G}=N_1 G_1^k + N_2 G_2^k \ ; \ N_1 = \frac{z_2-z}{l_k} \ ; \ N_2 = \frac{z-z_1}{l_k} \ ; l_k = z_2 - z_1 \qquad (12)$$

where $\tilde{G}$:shape function; G: any of the variables, k: element number; N_1,N_2: linear interpolation functions; l_k:length of the k'th element; z_1,z_2:beginning and end elevations of the element k; z: transformed variable that changes from z_1 to z_2 in an element.

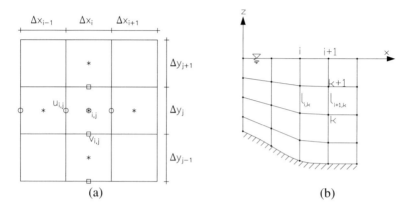

(a)   (b)

Fig. 1. a) Horizontal staggered finite difference scheme, ○: longitudinal horizontal velocity, u; □: lateral horizontal velocity, v; *: all other variables b) Finite element scheme throughout the water depth

After the application of the Galerkin Method, any derivative terms with respect to horizontal coordinates appearing in the equations are replaced by their central finite difference approximations. The system of nonlinear equations is solved by the Crank Nicholson Method which has second order accuracy in time. Some of the finite difference approximations are given in the following equations.

$$\left(\frac{\partial l}{\partial x}\right)_{i,j} = \frac{(\Delta x_i + \Delta x_{i+1})(l_{i,j} - l_{i-1,j})}{(\Delta x_i + \Delta x_{i-1})\left(\Delta x_i + \frac{\Delta x_{i+1} + \Delta x_{i-1}}{2}\right)} + \frac{(\Delta x_i + \Delta x_{i-1})(l_{i+1,j} - l_{i,j})}{(\Delta x_i + \Delta x_{i+1})\left(\Delta x_i + \frac{\Delta x_{i+1} + \Delta x_{i-1}}{2}\right)} \qquad (13)$$

$$\left(\frac{\partial l}{\partial y}\right)_{i,j} = \frac{(\Delta y_j + \Delta y_{j+1})(l_{i,j} - l_{i,j-1})}{(\Delta y_j + \Delta y_{j-1})\left(\Delta y_j + \frac{\Delta y_{j-1} + \Delta y_{j+1}}{2}\right)} + \frac{(\Delta y_j + \Delta y_{j-1})(l_{i,j+1} - l_{i,j})}{(\Delta y_j + \Delta y_{j+1})\left(\Delta y_j + \frac{\Delta y_{j-1} + \Delta y_{j+1}}{2}\right)} \qquad (14)$$

$$\left(\frac{\partial C}{\partial x}\right)_{i,j} = \frac{(\Delta x_i + \Delta x_{i+1})(C_{i,j} - C_{i-1,j})}{(\Delta x_i + \Delta x_{i-1})\left(\Delta x_i + \frac{\Delta x_{i+1} + \Delta x_{i-1}}{2}\right)} + \frac{(\Delta x_i + \Delta x_{i-1})(C_{i+1,j} - C_{i,j})}{(\Delta x_i + \Delta x_{i+1})\left(\Delta x_i + \frac{\Delta x_{i+1} + \Delta x_{i-1}}{2}\right)} \qquad (15)$$

$$\left(\frac{\partial C}{\partial y}\right)_{i,j}=\frac{\left(\Delta y_j+\Delta y_{j+1}\right)\left(C_{i,j}-C_{i,j-1}\right)}{\left(\Delta y_j+\Delta y_{j-1}\right)\left(\Delta y_j+\frac{\Delta y_{j-1}+\Delta y_{j+1}}{2}\right)}+\frac{\left(\Delta y_j+\Delta y_{j-1}\right)\left(C_{i,j+1}-C_{i,j}\right)}{\left(\Delta y_j+\Delta y_{j+1}\right)\left(\Delta y_j+\frac{\Delta y_{j-1}+\Delta y_{j+1}}{2}\right)} \tag{16}$$

$$\left(\frac{\partial^2 C}{\partial y^2}\right)_{i,j}=2\left(\frac{C_{i,j-1}}{\frac{\left(\Delta y_j+\Delta y_{j-1}\right)}{2}\left(\Delta y_j+\frac{\Delta y_{j-1}+\Delta y_{j+1}}{2}\right)}-\frac{C_{i,j}}{\frac{\left(\Delta y_j+\Delta y_{j-1}\right)}{2}\frac{\left(\Delta y_j+\Delta y_{j+1}\right)}{2}}\right.$$
$$\left.+\frac{C_{i,j+1}}{\frac{\left(\Delta y_j+\Delta y_{j+1}\right)}{2}\left(\Delta y_j+\frac{\Delta y_{j+1}+\Delta y_{j-1}}{2}\right)}\right) \tag{17}$$

$$\left(\frac{\partial^2 C}{\partial x^2}\right)_{i,j}=2\left(\frac{C_{i-1,j}}{\frac{\left(\Delta x_{i-1}+\Delta x_i\right)}{2}\left(\Delta x_i+\frac{\Delta x_{i-1}+\Delta x_{i+1}}{2}\right)}-\frac{C_{i,j}}{\frac{\left(\Delta x_{i-1}+\Delta x_i\right)}{2}\frac{\left(\Delta x_{i+1}+\Delta x_i\right)}{2}}\right.$$
$$\left.+\frac{C_{i+1,j}}{\frac{\left(\Delta x_{i+1}+\Delta x_i\right)}{2}\left(\Delta x_i+\frac{\Delta x_{i-1}+\Delta x_{i+1}}{2}\right)}\right) \tag{18}$$

$$u_{i+\frac{1}{2},j}=\frac{u_{i-1,j}\left(\Delta x_i\right)^2}{4\Delta x_{i-1}\left(\Delta x_i+\Delta x_{i-1}\right)}+\frac{u_{i,j}\left(\frac{\Delta x_i}{2}+\Delta x_{i-1}\right)}{2\Delta x_{i-1}}+\frac{u_{i+1,j}\left(\frac{\Delta x_i}{2}+\Delta x_{i-1}\right)}{2\left(\Delta x_i+\Delta x_{i-1}\right)} \tag{19}$$

$$v_{i,j+\frac{1}{2}}=\frac{v_{i,j-1}\left(\Delta y_j\right)^2}{4\Delta y_{j-1}\left(\Delta y_j+\Delta y_{j-1}\right)}+\frac{v_{i,j}\left(\frac{\Delta y_j}{2}+\Delta y_{j-1}\right)}{2\Delta y_{j-1}}+\frac{v_{i,j+1}\left(\frac{\Delta y_j}{2}+\Delta y_{j-1}\right)}{2\left(\Delta y_j+\Delta y_{j-1}\right)} \tag{20}$$

4   Model Applications

Simulated velocity profiles by using k-ε turbulence model, k-ω turbulence model have been compared with the experimental results of wind driven turbulent flow of an homogeneous fluid conducted by Tsanis and Leutheusser [9]. Laboratory basin had a length of 2.4 m., a width of 0.72 m. and depth of H=0.05 meters. The Reynolds

Number, $R_s=\dfrac{u_s H\rho}{\mu}$ was 3000 (u_s is the surface velocity, H is the depth of the

flow, ρ is the density of water and μ is the dynamic viscosity). The velocity profiles obtained by using k-ε turbulence model and k-ω turbulence model are compared with the measurements in Fig.2a and vertical eddy viscosity distributions are given in Fig.2b.

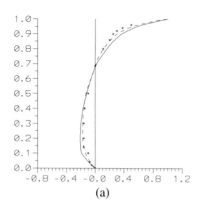 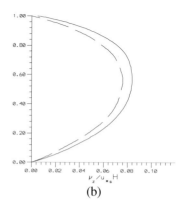

(a)   (b)

Fig. 2. a)Velocity profiles, b) Distribution of vertical eddy viscosity (solid line: k-ε turbulence model, dashed line: k-ω turbulence model, *: experimental data)

The root mean square error of the nondimensional horizontal velocity predicted by the k-ε turbulence model is 0.08, whereas it drops to 0.02 in the predictions by using k-ω turbulence model. This basically due to a better estimation of vertical distribution of vertical eddy viscosity by k-ω turbulence model.

Developed three dimensional numerical model (HYROTAM-3) has been implemented to Bay of Fethiye located at the Mediterranean coast of Turkey. Water depths in the Bay are plotted in Fig.3a. The grid system used has a square mesh size of 100x100 m. Wind characteristics are obtained from the measurements of the meteorological station in Fethiye for the period of 1980-2002. The wind analysis shows that the critical wind direction for wind speeds more than 7 m/s, is WNW-WSW direction. Some field measurements have been performed in the area. The current pattern over the area is observed by tracking drogues, which are moved by currents at the water depths of 1 m., 5 m and 10 m.. At Station I and at Station II shown in Fig.3a, continuous velocity measurements throughout water depth, at Station III water level measurements were taken for 27 days. In the application measurement period has been simulated and model is forced by the recorded wind as shown in Fig. 3b. No significant density stratification was recorded at the site. Therefore water density is taken as a constant. A horizontal grid spacing of Δx=Δy=100 m. is used. Horizontal eddy viscosities are calculated by the sub-grid scale turbulence model and the vertical eddy viscosity is calculated by k-ε turbulence model and also by k-ω turbulence model. The sea bottom is treated as a rigid boundary. Model predictions are in good agreement with the measurements. Simulated velocity profiles over the depth at the end of 4 days are compared with the measurements taken at Station I and Station II and are shown in Fig.4. At Station I, the root mean square error of the horizontal velocity is 0.19 cm/s in the predictions by k-ε turbulence model and it is 0.11cm/s in the predictions by k-ω turbulence model. At Station II, the root mean square error of the horizontal velocity is 0.16 cm/s in the predictions by k-ε turbulence model and it is 0.09cm/s in the predictions by k-ω turbulence model.

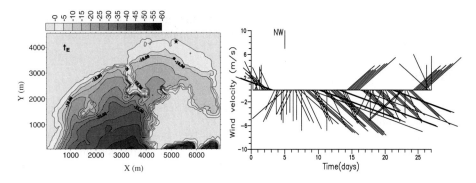

Fig. 3. a)Water depths(m) of Fethiye Bay, +:Station I, •:Station II,∗ :Station III. b) Wind speeds and directions during the measurement period.

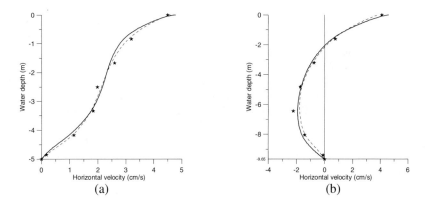

Fig. 4. Simulated velocity profiles over the depth at the end of 4 days; solid line: k-ε turbulence model, dashed line: k-ω turbulence model, *: experimental data, a) at Station I, b) at Station II

5 Conclusions

From the two equation turbulence models, k-ε model and k-ω model have been used in three dimensional modeling of coastal flows. The main source of coastal turbulence production is the surface current shear stress generated by the wind action. In the numerical solution a composite finite element-finite difference method has been applied. Governing equations are solved by the Galerkin Weighted Residual Method in the vertical plane and by finite difference approximations in the horizontal plane on a staggered scheme. Generally, two-equation turbulence models give improved estimations compared to other turbulence models. In the comparisons of model predictions with both the experimental and field measurements, it is seen that the two equation k-ω turbulence model predictions are better than the predictions of two equation k-ε turbulence model. This is basically due to the better parameterizations of the non-linear processes in the formulations leading a more reliable and numerically rather easy to handle vertical eddy viscosity distribution in the k-ω turbulence model.

Acknowledgment. The author wishes to thank the anonymous referees for their careful reading of the manuscript and their fruitful comments and suggestions.

References

1. Li, Z., Davies, A.G.: Turbulence Closure Modelling of Sediment Transport Beneath Large Waves. Continental Shelf Research (2001) 243-262
2. Bonnet-Verdier,C., Angot P., Fraunie, P., Coantic, M.: Three Dimensional Modelling of Coastal Circulations with Different k-ε Closures. Journal of Marine Systems (2006) 321-339
3. Baumert, H., Peters, H.: Turbulence Closure, Steady State, and Collapse into Waves. Journal of Physical Oceanography 34 (2004) 505-512
4. Neary, V.S., Sotiropoulos, F., Odgaard, A.J.: Three Dimensional Numerical Model of Lateral Intake Inflows. Journal of Hyraulic Engineering 125 (1999) 126-140
5. Balas,L., Özhan, E.: An Implicit Three Dimensional Numerical Model to Simulate Transport Processes in Coastal Water Bodies, International Journal for Numerical Methods in Fluids 34 (2000) 307-339
6. Balas, L., Özhan, E.: Three Dimensional Modelling of Stratified Coastal Waters, Estuarine, Coastal and Shelf Science 56 (2002) 75-87
7. Balas, L.: Simulation of Pollutant Transport in Marmaris Bay. China Ocean Engineering, Nanjing Hydraulics Research Institute (NHRI) 15 (2001) 565-578
8. Balas, L., Özhan, E: A Baroclinic Three Dimensional Numerical Model Applied to Coastal Lagoons. Lecture Notes in Computer Science 2658 (2003) 205-212
9. Tsanis,K.I., Leutheusser, H.J.:The Structure of Turbulent Shear-Induced Countercurrent Flow, Journal of Fluid Mechanics 189 (1998) 531-552

Vortex Identification in the Wall Region of Turbulent Channel Flow

Giancarlo Alfonsi[1] and Leonardo Primavera[2]

[1] Dipartimento di Difesa del Suolo, Università della Calabria
Via P. Bucci 42b, 87036 Rende (Cosenza), Italy
alfonsi@dds.unical.it
[2] Dipartimento di Fisica, Università della Calabria
Via P. Bucci 33b, 87036 Rende (Cosenza), Italy
lprimavera@fis.unical.it

Abstract. Four widely-used techniques for vortex detection in turbulent flows are investigated and compared. The flow of a viscous incompressible fluid in a plane channel is simulated numerically by means of a parallel computational code based on a mixed spectral-finite difference algorithm for the numerical integration of the Navier-Stokes equations. The *DNS* approach (Direct Numerical Simulation) is followed in the calculations, performed at friction Reynolds number $Re_\tau = 180$. A database representing the turbulent statistically steady state of the velocity field through 10 viscous time units is assembled and the different vortex-identification techniques are applied to the database. It is shown that the method of the "imaginary part of the complex eigenvalue pair of the velocity-gradient tensor" gives the best results in identifying hairpin-like vortical structures in the wall region of turbulent channel flow.

Keywords: turbulence, direct numerical simulation, wall-bounded flows, vortex-eduction methods.

1 Introduction

Organized vortical structures in wall-bounded flows have been investigated by several authors. One of the first contributions to the issue of the presence of vortices in the wall region of turbulent shear flows is due to Theodorsen [1] who introduced the *hairpin* vortex model. Robinson [2] confirmed the existence of non-symmetric *arch-* and *quasi-streamwise* vortices on the basis of the evaluation of *DNS* results. Studies involving the dynamics of hairpin vortices in the boundary layer have been performed by Perry & Chong [3], Acarlar & Smith [4,5], Smith *et al.* [6], Haidari & Smith [7] and Singer & Joslin [8]. On these bases, a picture of vortex generation and reciprocal interaction in the boundary layer emerges in which processes of interaction of existing vortices with wall-layer fluid involve viscous-inviscid interaction, generation of new vorticity, redistribution of existing vorticity, vortex stretching near the wall and vortex relaxation in the outer region. The process of evolution of a hairpin vortex involves the development of vortex legs in regions of increasing shear with intensification of vorticity in the legs themselves. The leg of a vortex – considered in isolation – may

Y. Shi et al. (Eds.): ICCS 2007, Part I, LNCS 4487, pp. 9–16, 2007.
© Springer-Verlag Berlin Heidelberg 2007

appear as a quasi-streamwise vortex near the wall. The head of a vortex instead, rises through the shear flow, entering regions of decreasing shear. As a consequence, the vorticity in the vortex head diminishes.

In spite of the remarkable amount of scientific work accomplished in this field, still there are no definite conclusions on the character of the phenomena occurring in the wall region of wall-bounded turbulent flows. Modern techniques for the numerical integration of the Navier-Stokes equations (advanced numerical methods and high-performance computing) have the ability of remarkably increasing the amount of data gathered during a research of computational nature, bringing to the condition of managing large amounts of data. A typical turbulent-flow database includes all three components of the fluid velocity in all points of a three-dimensional domain, evaluated for an adequate number of time steps of the turbulent statistically steady state. Mathematically-founded methods for the identification of vortical structures from a turbulent-flow database have been introduced by: *i)* Perry & Chong [9], based on the complex eigenvalues of the velocity-gradient tensor; *ii)* Hunt *et al.* [10] and Zhong *et al.* [11], based on the second invariant of the velocity-gradient tensor; *iii)* Zhou *et al.* [12], based on the imaginary part of the complex eigenvalue pair of the velocity-gradient tensor; *iv)* Jeong & Hussain [13], based on the analysis of the Hessian of the pressure. These techniques for vortex eduction are extensively used in turbulence research but no work exists in which their ability in vortex detection is systematically compared. In the present work the capability of vortex identification of the four techniques outlined above, is analyzed.

2 Vortex-Identification Techniques

2.1 Complex Eigenvalues of the Velocity-Gradient Tensor (Method A)

Perry & Chong [9] proposed a definition of a vortex as a region of space where the rate-of-deformation tensor has complex eigenvalues. By considering the system of the Navier-Stokes equations, an arbitrary point O can be chosen in the flow field and a Taylor series expansion of each velocity component can be performed in terms of the space coordinates, with the origin in O. The first-order pointwise linear approximation at point O is:

$$u_i = \dot{x}_i = A_i + A_{ij}x_j \tag{1}$$

and if O is located at a critical point, the zero-order terms A_i are equal to zero, being $A_{ij} = \partial u_i / \partial x_j$ the velocity-gradient tensor (rate-of-deformation tensor, $A = \nabla u$). In the case of incompressible flow, the characteristic equation of A_{ij} becomes:

$$\lambda^3 + Q\lambda + R = 0 \tag{2}$$

where Q and R are invariants of the velocity-gradient tensor (the other invariant $P = 0$ by continuity). Complex eigenvalues of the velocity-gradient tensor occur when the discriminant of A_{ij}, $D > 0$. According to this method, whether or not a region of vorticity appears as a vortex depends on its environment, i.e. on the local rate-of-strain field induced by the motions outside of the region of interest.

2.2 Second Invariant of the Velocity-Gradient Tensor (Method B)

Hunt *et al.* [10] and Zhong *et al.* [11] devised another criterion, in defining a eddy zone a region characterized by positive values of the second invariant Q of the velocity-gradient tensor. The velocity-gradient tensor can be split into symmetric and antisymmetric parts:

$$A_{ij} = S_{ij} + W_{ij} \qquad (3)$$

being S_{ij} the rate-of-strain tensor (corresponding to the pure irrotational motion) and W_{ij} the rate-of-rotation tensor (corresponding to the pure rotational motion). The second invariant of A_{ij} can be written as:

$$Q = \left(W_{ij}W_{ij} - S_{ij}S_{ij} \right)/2 \qquad (4)$$

where the first term on the right-hand side of (4) is proportional to the enstrophy density and the second term is proportional to the rate of dissipation of kinetic energy.

2.3 Imaginary Part of the Complex Eigenvalue Pair of the Velocity-Gradient Tensor (Method C)

Zhou *et al.* [12] adopted the criterion of identifying vortices by visualizing isosurfaces of the imaginary part of the complex eigenvalue pair of the velocity-gradient tensor (actually the square of). By considering equation (2) and defining the quantities:

$$J = \left(-\frac{R}{2} + \sqrt{\frac{R^2}{4} + \frac{Q^3}{27}} \right)^{\frac{1}{3}}, \quad K = -\left(+\frac{R}{2} + \sqrt{\frac{R^2}{4} + \frac{Q^3}{27}} \right)^{\frac{1}{3}} \qquad (5)$$

one has:

$$\lambda_1 = J + K, \quad \lambda_2 = -\frac{J+K}{2} + \frac{J-K}{2}\sqrt{-3}, \quad \lambda_3 = -\frac{J+K}{2} - \frac{J-K}{2}\sqrt{-3}. \qquad (6)$$

The method of visualizing isosurfaces (of the square) of the imaginary part of the complex eigenvalue pair of the velocity-gradient tensor is frame independent and due to the fact that the eigenvalue is complex only in regions of local circular or spiralling streamline, it automatically eliminates regions having vorticity but no local spiralling motion, such as shear layers.

2.4 Analysis of the Hessian of the Pressure (Method D)

Jeong & Hussain [13] proposed a definition of a vortex by reasoning on the issue of the pressure minimum, as follows "... a vortex core is a connected region characterized by two negative eigenvalues of the tensor $B = S^2 + W^2$ ...", where S

and W are the symmetric and antisymmetric parts of the velocity-gradient tensor. The gradient of the Navier-Stokes equation is considered and decomposed into symmetric and antisymmetric parts. By considering the symmetric part (the antisymmetric portion is the vorticity-transport equation), one has:

$$\frac{DS_{ij}}{Dt} - v\frac{\partial}{\partial x_k \partial x_k}S_{ij} + B_{ij} = -\frac{1}{\rho}\frac{\partial p}{\partial x_i \partial x_j} \tag{7}$$

where:

$$B_{ij} = S_{ik}S_{kj} + W_{ik}W_{kj} . \tag{8}$$

The existence of a local pressure minimum requires two positive eigenvalues for the Hessian tensor ($\partial p/\partial x_i \partial x_j$). By neglecting the contribution of the first two terms on the left-hand side of equation (7), only tensor (8) is considered to determine the existence of a local pressure minimum due to a vortical motion, i.e. the presence of two negative eigenvalues of B. The tensor B is symmetric by construction, all its eigenvalues are real and can be ordered ($\lambda_1 \geq \lambda_2 \geq \lambda_3$). According to this method a vortex is defined as a connected region of the flow with the requirement that the intermediate eigenvalue of B, $\lambda_2 < 0$.

3 Numerical Simulations

The numerical simulations are performed with a parallel computational code based on a mixed spectral-finite difference technique. The unsteady Navier-Stokes equation (besides continuity) for incompressible fluids with constant properties in three dimensions and non-dimensional conservative form, is considered ($i \& j = 1,2,3$):

$$\frac{\partial u_i}{\partial t} + \frac{\partial}{\partial x_j}(u_i u_j) = -\frac{\partial p}{\partial x_i} + \frac{1}{Re_\tau}\frac{\partial^2 u_i}{\partial x_j \partial x_j} \tag{9}$$

where $u_i(u,v,w)$ are the velocity components in the cartesian coordinate system $x_i(x,y,z)$. Equation (9) is nondimensionalized by the channel half-width h for lenghts, wall shear velocity u_τ for velocities, ρu_τ^2 for pressure and h/u_τ for time, being $Re_\tau = (u_\tau h/v)$ the friction Reynolds number. The fields are admitted to be periodic in the streamwise (x) and spanwise (z) directions, and equation (9) are Fourier transformed accordingly. The nonlinear terms in the momentum equation are evaluated pseudospectrally by anti-transforming the velocities back in physical space to perform the products (*FFTs* are used). In order to have a better spatial resolution near the walls, a grid-stretching law of hyperbolic-tangent type is introduced for the grid points along y, the direction orthogonal to the walls. For time advancement, a third-order Runge-Kutta algorithm is implemented and time marching is executed with the fractional-step method. No-slip boundary conditions at the walls and cyclic conditions in the streamwise and spanwise directions are applied to the velocity. More detailed descriptions of the numerical scheme, of its reliability and of the performance obtained on the parallel computers that have been used, can be found in Alfonsi *et al.*

[14] and Passoni *et al.* [15,16,17]. The characteristic parameters of the numerical simulations are the following. Computing domain: $L_x^+ = 1131$, $L_y^+ = 360$, $L_z^+ = 565$ (wall units). Computational grid: $N_x = 96$, $N_y = 129$, $N_z = 64$. Grid spacing: $\Delta x^+ = 11.8$, $\Delta y_{center}^+ = 4.4$, $\Delta y_{wall}^+ = 0.87$, $\Delta z^+ = 8.8$ (wall units). It can be verified that there are 6 grid points in the y direction within the viscous sublayer ($y^+ \leq 5$). After the insertion of appropriate initial conditions, the initial transient of the flow in the channel is simulated, the turbulent statistically steady state is reached and calculated for a time $t = 10\delta/u_\tau$ ($t^+ = 1800$). 20000 time steps are calculated with a temporal resolution of $\Delta t = 5 \times 10^{-4} \delta/u_\tau$ ($\Delta t^+ = 0.09$). In Figure 1, the computed turbulence intensities (in wall units) of present work are compared with the results of Moser *et al.* [18] at $Re_\tau = 180$. The agreement between the present results and the results of Moser *et al.* [18] (obtained with a fully spectral code) is rather satisfactory.

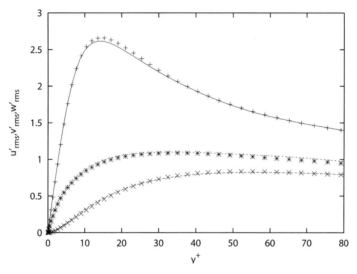

Fig. 1. *Rms* values of the velocity fluctuations normalized by the friction velocity in wall coordinates. Present work: (—) $\left(u'_{rms}\right)$, (---) $\left(v'_{rms}\right)$, (···) $\left(w'_{rms}\right)$. Data from Moser *et al.* [18]: (+) $\left(u'_{rms}\right)$, (×) $\left(v'_{rms}\right)$, (*) $\left(w'_{rms}\right)$.

4   Results

In Figures 2*a-b* the vortical structure that occurs at $t^+ = 1065.6$ as detected with method *A*, is represented and visualized from two different points of view (isosurfaces corresponding to the 5% of the maximum value are used in all representations). The top and side views show a structure not very well corresponding to a hairpin vortex. Portions of head and legs are visible, being the neck almost missing. The visualizations of the flow structure shows in practice a sketch of an hairpin, that can be completed only by intuition. Of the four methods examined, method *A* gives the less satisfactory representation of the flow structure at the bottom wall of the computational domain at $t^+ = 1065.6$.

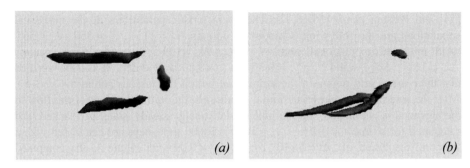

Fig. 2. Method *A*. Representation of hairpin vortex: *a)* top view; *b)* side view.

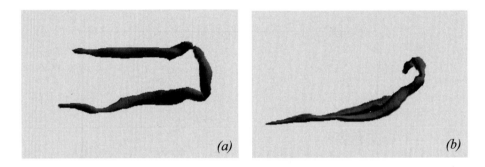

Fig. 3. Method *B*. Representation of hairpin vortex: *a)* top view; *b)* side view.

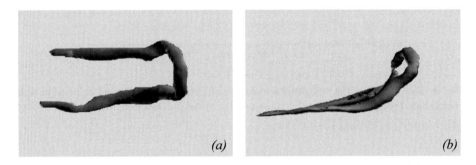

Fig. 4. Method *C*. Representation of hairpin vortex: *a)* top view; *b)* side view.

In Figures 3*a-b* the flow structure at as educted with method *B*, is represented. A hairpin-like vortical structure more complete with respect to the former case is visible. The head of the vortex is almost complete and well defined. Of the two legs, one is longer than the other and both are longer with respect to those of method *A*. In turn, a portion of the vortex neck is missing, as can be seen from Figure 3*b*. Figures 4*a-b* show the vortex structure extracted with method *C*. The Figures show a complete and well-defined hairpin vortex, with legs, neck and head clearly represented and no missing parts anywhere. Of the four eduction techniques tested, this is the best

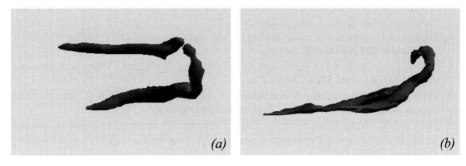

(a)   *(b)*

Fig. 5. Method *D*. Representation of hairpin vortex: *a)* top view; *b)* side view.

representation that can be obtained. From Figure 4*a* it can be noted that the hairpin exhibits two legs of length comparable to those shown in Figure 3*a*. Figure 4*b* shows a hairpin neck remarkably more thick with respect to that of Figure 3*b*. The results obtained with the use of method *D* are shown in Figures 5*a-b*. Also in this case a not complete hairpin vortex structure appears. Figure 5*b* shows that a portion of the neck of the vortex is missing.

5   Concluding Remarks

The case of the flow of a viscous incompressible fluid in a plane channel is simulated numerically at $Re_{\tau} = 180$ and a turbulent-flow database is assembled. Four different criteria for vortex eduction are applied to the database and compared, showing that: *i)* the method of the "complex eigenvalues of the velocity-gradient tensor" gives the less satisfactory results in terms of vortex representation; *ii)* the methods of the "second invariant of the velocity-gradient tensor" and that based on the "analysis of the Hessian of the pressure" gives intermediate results, in the sense that a hairpin-like vortical structure actually appears, otherwise with missing parts or not optimal definitions; *iii)* the best results are obtained by using the method of the "imaginary part of the complex eigenvalue pair of the velocity-gradient tensor".

References

1. Theodorsen, T.: Mechanism of turbulence. In *Proc. 2nd Midwestern Mechanics Conf.* (1952) 1
2. Robinson, S.K.: Coherent motions in the turbulent boundary layer. *Annu. Rev. Fluid Mech.* **23** (1991) 601
3. Perry, A.E., Chong, M.S.: On the mechanism of wall turbulence. *J. Fluid Mech.* **119** (1982) 173
4. Acarlar, M.S., Smith, C.R.: A study of hairpin vortices in a laminar boundary layer. Part 1. Hairpin vortices generated by a hemisphere protuberance. *J. Fluid Mech.* **175** (1987) 1
5. Acarlar, M.S., Smith, C.R.: A study of hairpin vortices in a laminar boundary layer. Part 2. Hairpin vortices generated by fluid injection. *J. Fluid Mech.* **175** (1987) 43

6. Smith, C.R., Walker, J.D.A., Haidari A.H., Soburn U.: On the dynamics of near-wall turbulence. *Phil. Trans. R. Soc. A* **336** (1991) 131

7. Haidari, A.H., Smith, C.R.: The generation and regeneration of single hairpin vortices. *J. Fluid Mech.* **277** (1994) 135

8. Singer, B.A., Joslin R.D.: Metamorphosis of a hairpin vortex into a young turbulent spot. *Phys. Fluids* **6** (1994) 3724

9. Perry, A.E., Chong, M.S.: A description of eddying motions and flow patterns using critical-point concepts. *Annu. Rev. Fluid Mech.* **19** (1987) 125

10. Hunt, J.C.R., Wray, A.A., Moin, P.: Eddies, streams and convergence zones in turbulent flows. In *Proc. Center Turbulence Research 1988 Summer Prog.* NASA Ames/Stanford University (1988) 193

11. Zhong, J., Huang, T.S., Adrian, R.J.: Extracting 3D vortices in turbulent fluid flow. *IEEE Trans. Patt. Anal. Machine Intell.* **20** (1998) 193

12. Zhou, J., Adrian, R.J., Balachandar, S., Kendall, T.M.: Mechanisms for generating coherent packets of hairpin vortices in channel flow. *J. Fluid Mech.* **387** (1999) 353

13. Jeong, J., Hussain, F.: On the definition of a vortex. *J. Fluid Mech.* **285** (1995) 69

14. Alfonsi, G., Passoni, G., Pancaldo, L., Zampaglione D.: A spectral-finite difference solution of the Navier-Stokes equations in three dimensions. *Int. J. Num. Meth. Fluids* **28** (1998) 129

15. Passoni, G., Alfonsi, G., Tula, G., Cardu, U.: A wavenumber parallel computational code for the numerical integration of the Navier-Stokes equations. *Parall. Comp.* **25** (1999) 593

16. Passoni, G., Cremonesi, P., Alfonsi, G.: Analysis and implementation of a parallelization strategy on a Navier-Stokes solver for shear flow simulations. *Parall. Comp.* **27** (2001) 1665

17. Passoni, G., Alfonsi, G., Galbiati, M.: Analysis of hybrid algorithms for the Navier-Stokes equations with respect to hydrodynamic stability theory. *Int. J. Num. Meth. Fluids* **38** (2002) 1069

18. Moser, R.D., Kim, J., Mansour, N.N.: Direct numerical simulation of turbulent channel flow up to $Re_\tau = 590$. *Phys. Fluids* **11** (1999) 943

Numerical Solution of a Two-Class LWR Traffic Flow Model by High-Resolution Central-Upwind Scheme

Jianzhong Chen[1], Zhongke Shi[1], and Yanmei Hu[2]

[1] College of Automation, Northwestern Polytechnical University, Xi'an, Shaanxi
710072, P.R. China
jzhchen1976@126.com,zkeshi@nwpu.edu.cn
[2] College of Science, Chang'an University, Xi'an, Shaanxi 710064, P.R. China
yanmeihu@126.com

Abstract. A high-resolution semi-discrete central-upwind scheme for solving a two class Lighthill-Whitham-Richards (LWR) traffic flow model is investigated in this paper. This scheme is based on combining a fourth-order central weighted essentially nonoscillatory (CWENO) reconstruction with semi-discrete central-upwind numerical flux. The CWENO re construction is chosen to improve the accuracy and guarantee the non-oscillatory behavior of the present method. The strong stability preserving Runge-Kutta method is used for time integration. The resulting method is applied to simulating several tests such as mixture of the two traffic flows. The simulated results illustrate the effectiveness of the present method.

Keywords: Traffic flow model, central-upwind scheme, CWENO reconstruction.

1   Introduction

Continuum traffic flow models are of great practical importance in many applications such as traffic simulation, traffic control, and, etc. The Lighthill-Whitham-Richards (LWR) model proposed independently by Lighthill and Whitham [1] and Richards [2] is the forerunner for all other continuum traffic flow models. In recent years an amount of research was done in implementing and extending the LWR model. Zhang [3] and Jiang et al. [4] proposed higher-order continuum models. Wong and Wong [5] presented a multi-class LWR traffic flow model(MCLWR model). For numerical method, the Lax-Friedrichs scheme was used to solve the MCLWR model in [5]. The Lax-Friedrichs scheme is only first-order accurate and yields a relatively poor resolution due to the excessive numerical dissipation. Recently, the Godunov scheme was also employed to solve the LWR model [6] and higher-order model [7]. However, the Godunov scheme needs to use exact or approximate Riemann solvers, which make the scheme complicated and time-consuming. Zhang, et al. [8] pointed out that the scalar LWR model and those higher-order continuum models proposed so far contain hyperbolic partial differential equations. One important feature of this type equation is that it

Y. Shi et al. (Eds.): ICCS 2007, Part I, LNCS 4487, pp. 17–24, 2007.

admits both smooth and discontinuous solutions such as shocks. However, the lower order numerical methods may produce smeared solutions near discontinuities due to excessive numerical viscosity. The high-order scheme can provide the satisfactory resolution. Moreover, the problems in which solutions contain rich smooth region structures can be resolved by the high-order scheme on a relatively small number of grid points. To embody traffic phenomena described by traffic flow model completely and resolve discontinuities well, a high-resolution shock-capturing numerical method is required. A recent application of the weighted essentially non-oscillatory (WENO) scheme can be found in [8,9].

In this paper we study another type shock-capturing scheme, the so-called high-resolution semi-discrete central-upwind schemes originally introduced in [10], which have attracted considerable attention more recently. These schemes enjoy the advantages of high-resolution central schemes. They are free of Riemann solvers, require no characteristic decompositions and retain high resolution similar to the upwind results. At the same time, they have an upwind nature. These features make the semi-discrete central-upwind schemes a universal, efficient and robust tool for a wide variety of applications. With regard to its application to traffic flow problems, we have not yet seen any research works. In this work the semi-discrete central-upwind scheme combined with fourth-order central WENO (CWENO) reconstruction [11] is applied to a two class LWR traffic flow model.

This paper is organized as follows. Section 2 presents the two class LWR traffic flow model. In section 3 we describe our numerical method. Numerical simulations are carried out in section 4. The conclusions are given in section 5.

2   The Tow-Class Model

The MCLWR model [5] describes the characteristics of traffic flow of M classes of road users with different speed choice behaviors in response to the same density when traveling on a highway section. There are some difficulties to compute the eigenvalues and prove the hyperbolicity of the model for $M > 3$. In this paper, we consider the two-class($M = 2$) LWR traffic flow model, which can be written in conservation form as

$$\mathbf{u}_t + \mathbf{f}(\mathbf{u})_x = 0 \ , \tag{1}$$

where $\mathbf{u}$ is the vector of conserved variables and $\mathbf{f}(\mathbf{u})$ is the vector of fluxes. These are given respectively by

$$\mathbf{u} = \begin{bmatrix} \rho_1 \\ \rho_2 \end{bmatrix} , \mathbf{f}(\mathbf{u}) = \begin{bmatrix} \rho_1 u_1(\rho) \\ \rho_2 u_2(\rho) \end{bmatrix} \ ,$$

where ρ_1 and ρ_2 are the densities for Class 1 and Class 2 traffic, respectively, $\rho = \rho_1 + \rho_2$ is the total density, and $u_1(\rho)$ and $u_2(\rho)$ are the velocity-density relationships. The two eigenvalues of the Jacobian are

$$\lambda_{1,2} = (u_1(\rho) + \rho_1 u_1'(\rho) + u_2(\rho) + \rho_2 u_2'(\rho) \pm \sqrt{\Delta})/2 \ , \tag{2}$$

where

$$\Delta = ((u_1(\rho) + \rho_1 u_1'(\rho)) - (u_2(\rho) + \rho_2 u_2'(\rho)))^2 + 4\rho_1 \rho_2 u_1'(\rho) u_2'(\rho) \ . \quad (3)$$

Since $\Delta \geq 0$ and $\lambda_{1,2}$ are real, the model is hyperbolic.

3 Numerical Scheme

For simplicity, let us consider a uniform grid, $x_\alpha = \alpha \Delta x, t^n = n\Delta t$, where Δx and Δt are the uniform spatial and time step, respectively. The cell average in the spatial cell $I_j = [x_{j-1/2}, x_{j+1/2}]$ at time $t = t^n$ is denoted by $\mathbf{u}_j^n(t) = \frac{1}{\Delta x} \int_{I_j} \mathbf{u}(x, t^n) \, dx$. Starting with the given cell averages $\{\mathbf{u}_j^n\}$, a piecewise polynomial interpolant is reconstructed

$$\tilde{\mathbf{u}}(x) = \sum_j p_j^n(x) \chi_j(x) \ . \quad (4)$$

Here χ_j is the characteristic function of the interval I_j and $p_j^n(x)$ is a polynomial of a suitable degree. Different semi-discrete central-upwind schemes will be characteristic of different reconstructions. Given such a reconstruction, the point-values of $\tilde{\mathbf{u}}$ at the interface points $\{x_{j+1/2}\}$ are denoted by $\mathbf{u}_{j+1/2}^+ = p_{j+1}^n(x_{j+1/2}, t^n)$ and $\mathbf{u}_{j+1/2}^- = p_j^n(x_{j+1/2}, t^n)$. The discontinuities of the construction (4) at the cell interfaces propagate with right- and left-sided local speeds, which can be estimated by

$$a_{j+1/2}^+ = \max\left\{\lambda_N\left(\frac{\partial \mathbf{f}}{\partial \mathbf{u}}\left(\mathbf{u}_{j+1/2}^-\right)\right), \lambda_N\left(\frac{\partial \mathbf{f}}{\partial \mathbf{u}}\left(\mathbf{u}_{j+1/2}^+\right)\right), 0\right\}$$

$$a_{j+1/2}^- = \min\left\{\lambda_1\left(\frac{\partial \mathbf{f}}{\partial \mathbf{u}}\left(\mathbf{u}_{j+1/2}^-\right)\right), \lambda_1\left(\frac{\partial \mathbf{f}}{\partial \mathbf{u}}\left(\mathbf{u}_{j+1/2}^+\right)\right), 0\right\} \ . \quad (5)$$

Here $\lambda_1, \cdots, \lambda_N$ denote the N eigenvalues of $\partial \mathbf{f}/\partial \mathbf{u}$. The semi-discrete central-upwind scheme for the spatial discretization of equation (1) can be given by (see [10] for the detailed derivation)

$$\frac{d}{dt}\mathbf{u}_j(t) = -\frac{H_{j+1/2}(t) - H_{j-1/2}(t)}{\Delta x} \ , \quad (6)$$

where the numerical fluxes $H_{j+1/2}$ is

$$H_{j+1/2}(t) = \frac{a_{j+1/2}^+ \mathbf{f}(\mathbf{u}_{j+1/2}^-) - a_{j+1/2}^- \mathbf{f}(\mathbf{u}_{j+1/2}^+)}{a_{j+1/2}^+ - a_{j+1/2}^-}$$

$$+ \frac{a_{j+1/2}^+ a_{j+1/2}^-}{a_{j+1/2}^+ - a_{j+1/2}^-}\left[\mathbf{u}_{j+1/2}^+ - \mathbf{u}_{j+1/2}^-\right] \ . \quad (7)$$

Note that different semi-discrete central-upwind schemes are typical of different reconstructions. The accuracy of the semi-discrete scheme (6)-(7) depends

on the accuracy of the reconstruction (4). One can use the second order piecewise linear reconstruction, the third-order piecewise quadratic reconstruction,
highly accurate essentially non-oscillatory (ENO) reconstruction, highly accurate WENO reconstruction or highly accurate CWENO reconstruction. In this
work, we have used an fourth-order CWENO reconstruction proposed in [11] to
compute the point values $\mathbf{u}_{j+1/2}^{\pm}$. To simplify notations, the superscript n will be
omitted below. In each cell I_j, the reconstruction, $p_j(x)$, is a convex combination
of three quadratic polynomials, $q_l(x), l = j - 1, j, j + 1$,

$$p_j(x) = \omega_{j-1}^j q_{j-1}(x) + \omega_j^j q_j(x) + \omega_{j+1}^j q_{j+1}(x) \ , \tag{8}$$

where ω_l^j are the weights which satisfy $\omega_l^j \geq 0$ and $\sum_{l=j-1}^{j+1} \omega_l^j = 1$. The quadratic
polynomials $q_l(x), l = j - 1, j, j + 1$, are given by

$$q_l(x) = \tilde{\mathbf{u}}_l + \tilde{\mathbf{u}}_l'(x - x_l) + \tilde{\mathbf{u}}_l''(x - x_l)^2, l = j - 1, j, j + 1 \ . \tag{9}$$

Here $\tilde{\mathbf{u}}_l'' = \frac{\mathbf{u}_{l+1} - 2\mathbf{u}_l + \mathbf{u}_{l-1}}{\Delta x^2}, \tilde{\mathbf{u}}_l' = \frac{\mathbf{u}_{l+1} + \mathbf{u}_{l-1}}{2\Delta x}$ and $\tilde{\mathbf{u}}_l = \mathbf{u}_l - \frac{\tilde{\mathbf{u}}_l''}{24}$. The weights ω_l^j are
defined as

$$\omega_l^j = \frac{\alpha_l^j}{\alpha_{j-1}^j + \alpha_j^j + \alpha_{j+1}^j}, \alpha_l^j = \frac{C_l}{(\epsilon + IS_l^j)^2}, l = j - 1, j, j + 1 \ , \tag{10}$$

where $C_{j-1} = C_{j+1} = 3/16$, $C_j = 5/8$. The constant ϵ is used to prevent the
denominators from becoming zero and is taken as $\epsilon = 10^{-6}$. The smoothness
indicators, IS_l^j, are calculated by

$$IS_{j-1}^j = \frac{13}{12}(\mathbf{u}_{j-2} - 2\mathbf{u}_{j-1} + \mathbf{u}_j)^2 + \frac{1}{4}(\mathbf{u}_{j-2} - 4\mathbf{u}_{j-1} + 3\mathbf{u}_j)^2 \ ,$$

$$IS_j^j = \frac{13}{12}(\mathbf{u}_{j-1} - 2\mathbf{u}_j + \mathbf{u}_{j+1})^2 + \frac{1}{4}(\mathbf{u}_{j-1} - \mathbf{u}_{j+1})^2 \ ,$$

$$IS_{j+1}^j = \frac{13}{12}(\mathbf{u}_j - 2\mathbf{u}_{j+1} + \mathbf{u}_{j+2})^2 + \frac{1}{4}(3\mathbf{u}_j - 4\mathbf{u}_{j+1} + \mathbf{u}_{j+2})^2 \ . \tag{11}$$

The time discretization of the semi-discrete scheme is achieved by third-order
strong stability preserving Runge-Kutta solver [12].

4   Numerical Examples

In this section, we choose several numerical examples presented in [9] as out test
case. The results demonstrate the performance of the present method for the
two-class LWR traffic flow model. In all examples, the following velocity-density
relationships are chose:

$$u_1(\rho) = u_{1f}(1 - \rho/\rho_m), u_2(\rho) = u_{2f}(1 - \rho/\rho_m) \ , \tag{12}$$

where ρ_m is maximal density and u_{1f} and u_{2f} are the free flow velocity for Class
1 and Class 2 traffic, respectively. Moreover, the variables of space, time, density

and velocity are scaled by L, T, ρ_m and u_f, where L is the length of the road, T is computational time and $u_f = \max(u_{1f}, u_{2f})$. A variable is also non-dimensional if it is not followed by its unit.

Example 1: Mixture of the two traffic flows.
The computational parameters are $L = 6000$m, $T = 400$s, $\Delta x = 60$m, $\Delta t = 0.4$s, $u_{1f} = 14$m/s and $u_{2f} = 20$m/s. The initial data is taken as the following Riemann problem:

$$\mathbf{u}(x,0) = \begin{cases} (0,0.4) , & x < 0.1 , \\ (0.4,0) , & x > 0.1 . \end{cases} \tag{13}$$

In this test Class 2 traffic will mix in Class 1 traffic, which causes the increase of total density. Its solution contains a shock, a constant region and a rarefaction. The total density computed by the presented method is shown in Fig. 1. To illustrate the advantage of using high-order schemes, the Godunov scheme with the Rusanov approximate Riemann solver [13,14] is also adopted to compute the same problem using the same parameters. Here and below, this scheme is abbreviated to GR. The scheme presented in this paper is abbreviated to CP4. The result computed by GR scheme is presented in Fig. 2. This comparison demonstrates the clear advantage of SD4 scheme over GR scheme. The SD4 scheme has the higher shock resolution and smaller numerical dissipation.

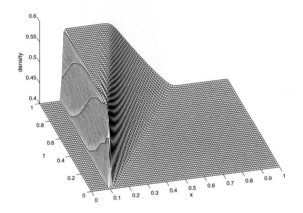

Fig. 1. Example 1: The total density. The solution is computed with CP4 scheme.

Example 2: Separation of the two traffic flows.
The parameters are $L = 8000$m, $T = 400$s, $\Delta x = 80$m, $\Delta t = 0.4$s, $u_{1f} = 10$m/s and $u_{2f} = 20$m/s. The Riemann initial data is used:

$$\mathbf{u}(x,0) = \begin{cases} (0.2,0) , & x < 0.1 , \\ (0,0.2) , & x > 0.1 . \end{cases} \tag{14}$$

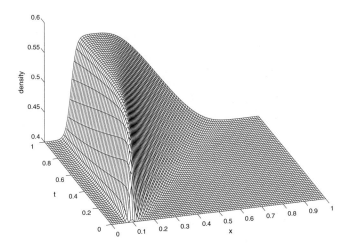

Fig. 2. Example 1: The total density. The solution is computed with GR scheme.

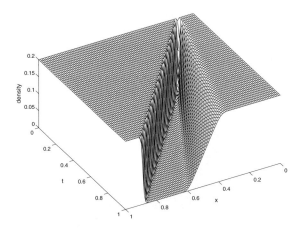

Fig. 3. Example 2: The total density. The solution is computed with CP4 scheme.

Note that $u_2 = 16\text{m/s} > u_{1f}$ and thus Class 1 drivers can not keep up with Class 2 drivers. A vacuum region is formed between Class 1 and Class 2 traffic. This test has solution consisting of a right shock, a constant region and a left rarefaction. Figs. 3 and 4 show the results obtained with CP4 and GR scheme, respectively. It can be seen that discontinuities are well resolved by CP4 scheme.

Example 3: A close following of the two traffic flows.
The parameters are $L = 4000\text{m}$, $T = 240\text{s}$, $\Delta x = 80\text{m}$, $\Delta t = 0.4\text{s}$, $u_{1f} = 14\text{m/s}$ and $u_{2f} = 20\text{m/s}$. The Riemann initial data is used:

$$\mathbf{u}(x,0) = \begin{cases} (0.2, 0) , & x < 0.1 , \\ (0, 0.44) , & x > 0.1 . \end{cases} \tag{15}$$

The high resolution properties of CP4 scheme are illustrated in Fig. 5.

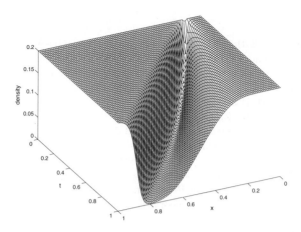

Fig. 4. Example 2: The total density. The solution is computed with GR scheme.

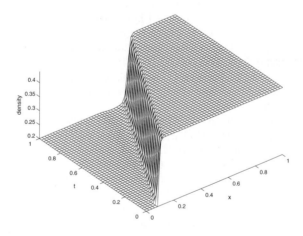

Fig. 5. Example 3: The total density. The solution is computed with CP4 scheme.

5   Conclusions

As an attempt to simulate traffic flow by high-resolution finite difference schemes, we have applied the semi-discrete central-upwind scheme to a two class LWR traffic flow model in this paper. The numerical results demonstrate that the semi-discrete central-upwind scheme resolve the shock and rarefaction waves well. This universal, efficient and high-resolution scheme will be implemented and applied to higher-order continuum models and multi-class models to simulate traffic flow in our future work.

References

1. Lighthill, M. J., Whitham, G. B.: On kinematic waves (II)-A theory of traffic flow on long crowed roads. Proc. R. Sco. London, Ser. A **22** (1955) 317-345
2. Richards, P. I.: Shock waves on the highway. Oper. Res. **4** (1956) 42-51
3. Zhang, H. M.: A non-equilibrium traffic model devoid of gas-like behavior. Transportation Research B **36** (2002) 275-290
4. Jiang, R., Wu, Q. S., Zhu, Z. J.: A new continuum model for traffic flow and numerical tests. Transportation Research B **36** (2002) 405-419
5. Wong, G. C. K., Wong, S. C.: A multi-class traffic flow model-an extension of LWR model with heterogeneous drivers. Transportation Research A **36** (2002) 827-841
6. Lebacque, J. P.: The Godunov scheme and what it means for first order traffic flow models. In: Lesort, J. B. (eds.): Proceedings of the 13th International Symposium on Transportation and Traffic Theory. Elsevier Science Ltd., Lyon France (1996) 647-677
7. Zhang, H. M.: A finite difference approximation of a non-equilibrium traffic flow model. Transportation Research B **35** (2001) 337-365
8. Zhang, M. P., Shu, C.-W., Wong, G. C. K., Wong, S. C.: A weighted essentially non-oscillatory numerical scheme for a multi-class Lighthill-Whitham-Richards traffic flow model. Journal of Computational Physics **191** (2003) 639-659
9. Zhang, P., Liu, R. X., Dai, S. Q.: Theoretical analysis and numerical simulation on a two-phase traffic flow LWR model. Journal of university of science and technology of China **35** (2005) 1-11
10. Kurganov, A., Noelle, S., Petrova, G.: Semi-discrete central-upwind schemes for hyperbolic conservation laws and Hamilton-Jacobi equations. SIAM J. Sci. Comput. **23** (2001) 707-740
11. Levy, D., Puppo, G., Russo, G.: Central WENO schemes for hyperbolic systems of conservation laws. Math. Model. Numer. Anal. **33** (1999) 547-571
12. Gottlieb, S., Shu, C.-W., Tadmor, E.: Strong stability preserving high order time discretization methods. SIAM Rev. **43** (2001) 89-112
13. Toro, E. F.: Riemann Solvers and Numerical Methods for Fluid Dynamics. Springer-Verlag, Berlin Heidelberg New York (1997)
14. Rusanov, V. V.: Calculation of interaction of non-steady shock waves with obstacles. J. Comput. Math. Phys. **1** (1961) 267-279

User-Controllable GPGPU-Based
Target-Driven Smoke Simulation

Jihyun Ryu[1] and Sanghun Park[2,*]

[1] Dept. of Applied Mathematics, Sejong University, Seoul 143-747, Republic of Korea
ajhryu@sogang.ac.kr
[2] Dept. of Multimedia, Dongguk University, Seoul 100-715, Republic of Korea
mshpark@dongguk.edu

Abstract. The simulation of fluid phenomena, such as smoke, water, fire, has developed rapidly in computer games, special effects, and animation. The various physics-based methods can result in high quality images. However, the simulation speed is also important issue for consideration in applications. This paper describes an efficient method for controlling smoke simulation, running entirely on the GPU. Our interest is in how to reach given user-defined target smoke states in real time. Given an initial smoke state, we propose to simulate the smoke towards a special target state. This control is made by adding special external force terms to the standard flow equation.

Keywords: GPGPU, Navier-Stokes equations, interactive simulation.

1 Introduction

The modeling of natural phenomena such as smoke, fire, and liquid has received considerable attention from the computer graphics industry. This is especially true for visual smoke models, which have many applications in the creation of special effects and interactive games. It is important to produce highly realistic results as well as simulate effects in real time. This becomes a more challenging if the produced animation can be controlled by users. Recently computer graphics researchers have created simulations for controlling fluids. Treuille et al. [2] introduced a method to control fluid flows to obtain the target shapes. However, this method is too slow in shape controlled flow simulation. In fact, in real-time applications such as computer games, the simulation speed is more important than image quality. This paper presents a method of controlling interactive fluids in real time using the GPU (Graphics Processing Unit). It is based on the results of Fattal et al. [1], and our goal is to perform all the steps on the GPU. This technique can interactively create target shapes using computer-generated smoke.

2 Mathematical Background

To simulate the behavior of fluid, we must have a mathematical representation of the state of the fluid at any given time. The greatest quantity to represent

---

* Corresponding author.

Y. Shi et al. (Eds.): ICCS 2007, Part I, LNCS 4487, pp. 25–29, 2007.

is the velocity of the fluid. But the fluid's velocity varies in both time and space, so we represent this as a vector field. The key to fluid simulation is to take steps in time and determine the velocity field at each time step. We can achieve this by solving special equations. In physics, we assume an incompressible and homogeneous fluid for fluid simulation. This means that density is constant in both time and space. Under these assumptions, the state of the fluid over time can be described using the Navier-Stokes equations for incompressible flow: $\frac{\partial u}{\partial t} = -u \cdot \nabla u - \nabla p + \nu \nabla^2 u + F$, $\nabla \cdot u = 0$ where $u(x, t)$ is the velocity vector of the position vector x at time t, $p(x, t)$ is the pressure, ν is the kinematic viscosity, and F represents any external forces that act on the fluid. In our case, we may simulate the non-viscosity fluid and therefore solve the Euler Equation with $\nu = 0$;

$$\frac{\partial u}{\partial t} = -u \cdot \nabla u - \nabla p + F, \qquad \nabla \cdot u = 0 \tag{1}$$

Let $\rho = \rho(x, t)$ be the density scalar field at position x and time t. In order to describe the transport of the smoke along the fluid's velocity fields, we solve the additional equation;

$$\frac{\partial \rho}{\partial t} = -u \cdot \nabla \rho \tag{2}$$

Moreover, the external forces term provides an important means of control over the smoke simulation. In the result of Fattal et al. [1], the special external term $F(\rho, \rho^*)$ depends on the smoke density ρ with the target density ρ^*. This result has the same direction as the gradient vector of ρ^*. In addition, a "normalized" gradient can be used by $F(\rho, \rho^*) \propto \frac{\nabla \rho^*}{\rho^*}$. The blurring filter of ρ^* must have sufficiently large support, since the target density ρ^* is constant and $\nabla \rho^* = 0$. In order to ensure $\nabla \rho^* \neq 0$, the blurred version of ρ^*, denoted by $\tilde{\rho}^*$, can be used. The force is $F(\rho, \rho^*) = \tilde{\rho} \frac{\nabla \tilde{\rho}^*}{\tilde{\rho}^*} \equiv F_{df}$ where F_{df} is named the "driving force". In summary, we have two modified equations for the controlled fluid simulations. The first is the advection equation for density; the second is the momentum equation using the driving force.

$$\frac{\partial \rho}{\partial t} = -u \cdot \nabla \rho, \qquad \frac{\partial u}{\partial t} = -u \cdot \nabla u + \nabla p + F_{df} \tag{3}$$

In addition, another external force, denoted by F_{ui}, for user interaction, can be applied to fluid by clicking and dragging with the mouse. The force is computed from the direction and length of the mouse drag.

3   Implementation Details

The goal of this research has been to improve user interaction with computer-generated smoke. We implemented GPGPU (General-Purpose computation on GPUs) based techniques to simulate dynamic smoke that can be described by PDEs. Algorithm 1 shows a method to process the user-controllable interactive target-driven simulation. Fattal et al. [1] introduced offline target-driven

---

Algorithm 1. User-controllable interactive target-driven simulation algorithm

---

1: Load a target key-frame, initialize target density field ρ^*
2: **while** (1) **do**
3:    Setup external force F_{ui} and density field ρ_{ui} // from user interaction
4:    $\rho \leftarrow \tilde{\rho}, \quad u \leftarrow u + F_{ui}$ // apply gaussian filter and add external force
5:    **if** (target-driven) **then**
6:       $F_{df} \leftarrow \frac{\nabla \tilde{\rho^*}}{\rho^*}\rho, \quad u \leftarrow u + F_{df}$ // add driving force
7:    **end if**
8:    $v \leftarrow \nabla \times u, \ u \leftarrow u + v\Delta t$ // vorticity
9:    $\frac{\partial u}{\partial t} = -u \cdot \nabla u$ // advection of velocity
10:    $\nabla^2 p = \nabla \cdot u, \ u \leftarrow u - \nabla p$ // projection
11:    $\rho \leftarrow \rho + \rho_{ui}$ // add density
12:    $\frac{\partial \rho}{\partial t} = -u \cdot \nabla \rho$ // advection of density
13:    draw(ρ)
14: **end while**

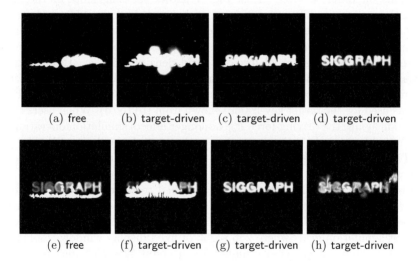

(a) free   (b) target-driven   (c) target-driven   (d) target-driven

(e) free   (f) target-driven   (g) target-driven   (h) target-driven

Fig. 1. User-controllable interactive simulation snapshots

animation on CPUs, our algorithm is based on this technique. In free simulation mode, users can generate external force F_{ui} and density field ρ_{ui} from mouse movement. The users can change the current simulation mode to **target-driven** by pressing a keyboard button, and the driving force F_{df} is applied to current velocity field u. This results in driving the current density field ρ to a pre-defined target image ρ^*. Gaussian filtering for the blurring operation can be implemented in a fragment shader, since recent shading languages support nested loops in shader programs. All of the operations in the algorithm are implemented efficiently as fragment shaders to maximize the degree of rendering speed enhancement; it is possible for users to control smoke simulation interactively.

Table 1. Simulation speeds (frames per sec) where resolution of display windows is 256×256 and n_{ji} is the number of Jacobian iteration. The results were evaluated on a general purpose PC, equipped with a 3.0 GHz Intel Pentium D processor, 1 GB of main memory, and a graphics card with an NVIDIA GeForce 7900 GT processor and 256 MB of memory.

n_{ji}	Grid of simulation fields		
	128×128	256×256	512×512
50	59.64	15.03	4.60
150	58.01	14.99	3.99
300	29.62	11.99	3.32

To verify the effectiveness of our proposed GPGPU-based system, in smoke simulation, we present timing performances for solving the Navier-Stokes equations on different sized grids. Fig. 1 shows the snapshots of the smoke simulation sequence. The resolution of each image is 256×256, and starting simulation mode is free. The system allows users to change the current simulation mode to either target-driven or free. In (a), users can interactively generate smoke by calculating external force F_{ui} and density fields ρ_{ui} through mouse movement in free mode. In selecting target-driven mode, current smoke starts moving to a pre-defined target key-frame (in this example, "SIGGRAPH" image was used for the target key-frame) shown from (b) to (d). When users choose free mode again, the driving force F_{df} no longer affects simulation operation and users are able to apply their interaction to the current simulation field shown in (e). The status of target-driven mode re-activated without any user interaction is shown in (f) and (g). We can see that the image quality of (g) is almost identical to that of (d). When the mode is switched to target-driven after running in free mode for a few of seconds, the situation where smoke gathers around the target key-frame is shown in (h). Table 1 shows the simulation speeds on different simulation grids where 256×256 of fixed display window resolution is used. The system can run smoke simulation at about $12 \sim 15$ frames per second, where both the grid of simulation fields and the resolution of display windows are 256×256.

4   Concluding Remarks

Interactive applications, such as computer games, demand realism, but cannot afford to sacrifice speed to achieve this requirement. We developed user-controllable smoke simulation techniques based on GPGPU computations with these requirements as consideration. This permits the user to select and change the behavior of features as the simulation progresses. Users can create and modify density and flow elements of Navier-Stokes simulation, through a graphic user interface. This system also provides an interactive approach to simulate the smoke towards a special target state.

Acknowledgements

This work was supported by the Korea Research Foundation Grant funded by the Korean Government (MOEHRD) (KRF-2006-331-D00497) and Seoul R&BD Program (10672).

References

1. Fattal, R., Lischinski, D.: Target-driven smoke animation. Transactions on Graphics, Vol. 23, No. 3. ACM (2004) 441–448
2. Treuille, A., McNamara, A., Popovic, Z., Stam, J.: Keyframe control of smoke simulation. Transactions on Graphics, Vol. 22, No. 3. ACM (2003) 716–723

Variable Relaxation Solve
for Nonlinear Thermal Conduction*

Jin Chen

Princeton Plasma Physics Laboratory, Princeton, NJ, USA
jchen@pppl.gov

Abstract. Efficient and robust nonlinear solvers, based on Variable Relaxation, is developed to solve nonlinear anisotropic thermal conduction
arising from fusion plasma simulations. By adding first and/or second
order time derivatives to the system, this type of methods advances corresponding time-dependent nonlinear systems to steady state, which is
the solution to be sought. In this process, only the stiffness matrix itself
is involved so that the numerical complexity and errors can be greatly
reduced. In fact, this work is an extension of implementing efficient linear
solvers for fusion simulation on Cray X1E.

Two schemes are derived in this work, first and second order Variable Relaxations. Four factors are observed to be critical for efficiency
and preservation of solution's symmetric structure arising from periodic
boundary condition: mesh scales, initialization, variable time step, and
nonlinear stiffness matrix computation. First finer mesh scale should be
taken in strong transport direction; Next the system is carefully initialized by the solution with linear conductivity; Third, time step and relaxation factor are vertex-based varied and optimized at each time step;
Finally, the nonlinear stiffness matrix is updated by just scaling corresponding linear one with the vector generated from nonlinear thermal
conductivity.

1 Introduction

In plasma physics modeling[1], the steady state of nonlinear anisotropic thermal
conduction can be modeled by the following nonlinear elliptic equation

$$\frac{\partial}{\partial x}(\kappa_x \frac{\partial T}{\partial x}) + \frac{\partial}{\partial y}(\kappa_y \frac{\partial T}{\partial y}) = s \tag{1}$$

on a 2D rectangular domain ABCD: $[0, L_x] \times [0, L_y]$ with four vertexes at A$(0,0)$,
B$(0, L_x)$, C(L_x, L_y), and D$(0, L_y)$. $L_x < L_y$. The coordinate is given in Cartesian
(x, y) system. The magnetic field is directed in the y direction, and accordingly
we can set $\kappa_x = 1$ and κ_y as an nonlinear function of the temperature T, parallel
to magnetic field line. Therefore we can omit κ_x and denote κ_y by $\kappa_\parallel$ to make
its meaning more clear. The periodic boundary condition is set on edges AD and

---

* This work is supported by DOE contract DE-AC02-76CH03073.

Y. Shi et al. (Eds.): ICCS 2007, Part I, LNCS 4487, pp. 30–37, 2007.
© Springer-Verlag Berlin Heidelberg 2007

BC, and Dirichlet boundary conditions are set on edges AB and CD. This setup allows us to separate the effects of grid misalignment from the boundary effects. The upper boundary, CD, represent the material surface where the temperature is low, and the boundary condition there is $T_{CD} = 1$. At the lower boundary, AB, the inflow boundary condition is $T_{AB}(x) = 10 + 40e^{(-|x-L_x/2|)}$.

Finite element discretization[2] generates the following nonlinear system

$$(S_{xx} + S_{yy}(T))T = Ms. \tag{2}$$

M is the mass matrix. S_{xx} and $S_{yy}(T)$ are the stiffness matrices contributed by operator $\frac{\partial^2 T}{\partial x^2}$ and $\frac{\partial}{\partial y}(\kappa_\parallel \frac{\partial T}{\partial y})$, respectively. T is the temperature profile to be solved. When $\kappa_\parallel$ is linear, $S_{yy}(T)$ reduced to $\kappa_\parallel S_{yy}$. Newton-Krylov method can be used to solve system (2). But usually it is quite expensive to update Jacobian at each iteration. Although the Jacobian-free variation[3][4] is more efficient, information of the Jacobian is still needed to form the preconditioner and preconditioning is expensive.

In this work we present an alternative way, Variable Relaxation[5], to solve the nonlinear system (1). This is a class of iterative methods which solve the elliptic equations by adding first and/or second order time derivative terms to eq.(1) to convert it to nonlinear parabolic or hyperbolic equation and then marching the system to steady state. In this marching process, only the nonlinear stiffness matrix $S_{yy}(T)$ itself is involved and needs to be updated regularly.

We have been using this type of idea on Cray X1E to design efficient linear elliptic solvers for M3D code[6]. Although It takes longer to converge, each iteration is much cheaper than other iterative solvers[7] so that it still wins on vector architecture machines.

The nonlinear iteration can be completed in two steps:

Step 1: solve eq.(1) with linear conductivity $10^0 \le \kappa_\parallel \le 10^9$.
Step 2: solve eq.(1) with nonlinear conductivity $\kappa_\parallel = T^{5/2}$.

The solution from "**Step 1**" is used as an initial guess for "**Step 2**". Experiments will show that this is a very powerful strategy to accelerate convergence. We will also demonstrate how to choose artificial time step from CFL condition and relaxation factor from dispersion relation to achieve optimization. An efficient way to generate the stiffness matrix is also to be discussed in order to preserve the symmetry structure of the solution as a result of periodic boundary condition.

2   First Order Relaxation and Numerical Schemes

The so called first order relaxation is obtained by adding a first order time derivative term to eq. (1)

$$\frac{\partial u}{\partial t} = \frac{\partial^2 T}{\partial x^2} + \frac{\partial}{\partial y}(\kappa_\parallel \frac{\partial T}{\partial y}). \tag{3}$$

Discretizing it in temporal direction by finite difference and spatial directions as in system (2), we have

$$(\frac{1}{\delta t}M - \theta S_{non})T^{k+1} = [\frac{1}{\delta t}M + (1-\theta)S_{non})]T^k - Ms. \tag{4}$$

$0 \le \theta \le 1$. When $\theta = 0$, the system is fully explicit; when $\theta = 1$, the system is fully implicit; when $\theta = \frac{1}{2}$, the system is stable and has smallest truncation error as well. $S_{non} = S_{xx} + \bar{S}_{yy}(T)$. δt is the artificial time step which should be chosen to be small enough to make the scheme stable and big enough to allow the system approach steady state quickly. According to CFL condition, δt is related to mesh scales δx in x direction and δy in y direction by

$$\delta t = \frac{1}{2}\frac{1}{\frac{1}{\delta x^2} + \kappa_{\|}\frac{1}{\delta y^2}} = \frac{\delta x \delta y}{4}\frac{2}{\frac{\delta y}{\delta x} + \kappa_{\|}\frac{\delta x}{\delta y}} \equiv \frac{\delta x \delta y}{4}\bar{\delta t}. \tag{5}$$

Obviously, when $\kappa_{\|} = 1$, $\bar{\delta t}$ is symmetric in $(\delta x, \delta y)$ and gets maximized at $\delta x = \delta y$. More can be derived if we different $\bar{\delta t}$ with respect to δx and δy

$$\frac{\partial \bar{\delta t}}{\partial \delta x} = -2\frac{-\frac{\delta y}{\delta x^2} + \kappa_{\|}\frac{1}{\delta y}}{(\frac{\delta y}{\delta x} + \kappa_{\|}\frac{\delta x}{\delta y})^2} = 2\frac{1}{\delta y}\frac{\frac{\delta y^2}{\delta x^2} - \kappa_{\|}}{(\frac{\delta y}{\delta x} + \kappa_{\|}\frac{\delta x}{\delta y})^2},$$

$$\frac{\partial \bar{\delta t}}{\partial \delta y} = -2\frac{\frac{1}{\delta x} - \kappa_{\|}\frac{\delta x}{\delta y^2}}{(\frac{\delta y}{\delta x} + \kappa_{\|}\frac{\delta x}{\delta y})^2} = 2\frac{\delta x}{\delta y^2}\frac{\kappa_{\|} - \frac{\delta y^2}{\delta x^2}}{(\frac{\delta y}{\delta x} + \kappa_{\|}\frac{\delta x}{\delta y})^2}.$$

When $\kappa_{\|} > 1$, most likely we will have $\frac{\partial \bar{\delta t}}{\partial \delta x} < 0$ and $\frac{\partial \bar{\delta t}}{\partial \delta y} > 0$. This suggests that δx should be taken as large as possible, while δy as small as possible.

The convergence of scheme (4) can be analyzed in the following way. Given the form of transient solution of eq.(3) as $\tilde{u} = e^{-\gamma t}\sin\frac{m\pi x}{L_x}\sin\frac{n\pi y}{L_y}$, the operator $\frac{\partial^2}{\partial x^2} + \frac{\partial}{\partial y}(\kappa_{\|}\frac{\partial}{\partial y})$ has eigenvalues $\lambda_{mn} = \pi^2(\frac{m^2}{L_x^2} + \kappa_{\|}\frac{n^2}{L_y^2})$. m and n are the mode numbers in x and y directions, respectively. Then the decaying rate is $-\lambda_{11}$ and the corresponding decaying time can be found by

$$t = \frac{1}{\lambda_{11}} = \frac{1}{\pi^2}\frac{1}{\frac{1}{L_x^2} + \kappa_{\|}\frac{1}{L_y^2}}.$$

The number of iterations needed for convergence can be predicted by

$$N_{its} \equiv \frac{t}{\delta t} = \frac{2}{\pi^2}\frac{\frac{1}{\delta x^2} + \kappa_{\|}\frac{1}{\delta y^2}}{\frac{1}{L_x^2} + \kappa_{\|}\frac{1}{L_y^2}} = \frac{2}{\pi^2}\frac{\frac{N_x^2}{L_x^2} + \kappa_{\|}\frac{N_y^2}{L_y^2}}{\frac{1}{L_x^2} + \kappa_{\|}\frac{1}{L_y^2}}.$$

When $\kappa_{\|} \to \infty$

$$N_{its} \to \frac{2}{\pi^2}N_y^2 \approx \frac{1}{5}\frac{N_y}{N_x}(N_x N_y) \equiv c(N_x N_y).$$

$(N_x N_y)$ is the number of unknowns. After some experiments, we found the optimized coefficient should be $c = 0.64$ for the problem we are studying. Also from the following expression we found the number of iterations increases as $\kappa_{\|}$ gets larger

$$\frac{dN_{its}}{d\kappa_{\|}} = \frac{2}{\pi^2}\frac{(N_y^2 - N_x^2)}{(\frac{L_y}{L_x} + \kappa_{\|}\frac{L_x}{L_y})^2} > 0$$

as long as $\delta y \leq \delta x$.

3   Second Order Relaxation and Numerical Schemes

Besides the addition of the first order derivative term in eq. (3), the second order relaxation is obtained by adding a relaxation factor, τ, and a second order time derivative term to eq. (1)

$$\frac{\partial^2 u}{\partial t^2} + \frac{2}{\tau}\frac{\partial u}{\partial t} = \frac{\partial^2 T}{\partial x^2} + \frac{\partial}{\partial y}(\kappa_{\|}\frac{\partial T}{\partial y}). \tag{6}$$

Again it can be discretized and rearranged as

$$[(1 + \tfrac{\delta t}{\tau})M - \theta S_{non}]T^{k+1} = \\ -(1 - \tfrac{\delta t}{\tau})MT^{k-1} + [2M + \delta t^2(1 - \theta)S_{non}]T^k - \delta t^2 M s. \tag{7}$$

The CFL condition can be expressed as $\delta t^2(\frac{1}{\delta_x^2} + \kappa_{\|}\frac{1}{\delta_y^2}) \leq 1$. Therefore,

$$\delta t \leq \frac{1}{\sqrt{\frac{1}{\delta x^2} + \kappa_{\|}\frac{1}{\delta_y^2}}} = \frac{\sqrt{\delta x \delta y}}{\sqrt{\frac{\delta y}{\delta x} + \kappa_{\|}\frac{\delta x}{\delta y}}} = \frac{\sqrt{\delta x \delta y}}{\sqrt{2}}\frac{\sqrt{2}}{\sqrt{\frac{\delta y}{\delta x} + \kappa_{\|}\frac{\delta x}{\delta y}}}. \tag{8}$$

The relaxation factor can be found again by looking for the transient solution of eq.(6). The decay rates satisfy $\gamma^2 - \frac{2}{\tau}\gamma + \lambda^{mn} = 0$, or $\gamma = \frac{1}{\tau} \pm (\frac{1}{\tau^2} - \lambda^{mn})^{1/2}$. For optimal damping, we choose $\tau^2 = \frac{1}{\lambda^{11}} = 1/[(\frac{L_y^2}{L_x^2} + \kappa_{\|})\frac{\pi^2}{L_y^2}]$, i.e.,

$$\tau = \frac{1}{\pi\sqrt{\frac{1}{L_x^2} + \kappa_{\|}\frac{1}{L_y^2}}} \tau = \frac{\sqrt{L_x L_y}}{\sqrt{2\pi}}\frac{\sqrt{2}}{\sqrt{\frac{L_y}{L_x} + \kappa_{\|}\frac{L_x}{L_y}}} \tag{9}$$

and the number of iterations for convergence can be predicted by

$$N_{its} \equiv \frac{\tau}{\delta t}$$

$$= \frac{1}{\pi}\frac{\sqrt{\frac{1}{\delta x^2} + \kappa_{\|}\frac{1}{\delta_y^2}}}{\sqrt{\frac{1}{L_x^2} + \kappa_{\|}\frac{1}{L_y^2}}} = \frac{1}{\pi}\frac{\sqrt{\frac{N_x^2}{L_x^2} + \kappa_{\|}\frac{N_y^2}{L_y^2}}}{\sqrt{\frac{1}{L_x^2} + \kappa_{\|}\frac{1}{L_y^2}}} = \frac{1}{\pi}\frac{\sqrt{\frac{25}{9}N_x^2 + \kappa_{\|}N_y^2}}{\sqrt{\frac{25}{9} + \kappa_{\|}}} = \frac{1}{\pi}\frac{\sqrt{\frac{25}{9}\frac{N_x}{N_y} + \kappa_{\|}\frac{N_y}{N_x}}}{\sqrt{\frac{25}{9} + \kappa_{\|}}}\sqrt{N_x N_y}$$

When $\kappa_{\|} \to \infty$

$$N_{its} \to (\frac{1}{\pi}\sqrt{\frac{N_y}{N_x}})\sqrt{N_x N_y} \equiv c\sqrt{N_x N_y}.$$

Experiments show that the optimal coefficient would be $c = 0.6$. The number of iteration increases as the conductivity $\kappa_\parallel$ increases. This can be understood from the following expression.

$$\frac{dN_{its}}{d\kappa_\parallel} = \frac{1}{\pi} \frac{\sqrt{3N_y^2 + \kappa_\parallel N_y^2} - \sqrt{3N_x^2 + \kappa_\parallel N_y^2}}{3 + \kappa_\parallel} > 0.$$

4 Variable Relaxations

When $\kappa_\parallel$ is an nonlinear function of T, $\kappa_\parallel$ changes as T_{ij}^k changes at every vertex ij and every time step k. Therefore, time step and relaxation factor changes as well. This is why the name "Variable" is given. From now on, schemes (4) is called VR(4), scheme (7) is called VR(7), and $\kappa_\parallel$ is rewritten as κ_{ij}^k in nonlinear case. From the analysis given in the previous two sections, we have

$$\delta t_{ij}^k = \frac{1}{2} \frac{1}{\frac{1}{\delta x^2} + \kappa_{ij}^k \frac{1}{\delta y^2}} = \frac{\delta x \delta y}{4} \frac{2}{\frac{\delta y}{\delta x} + \kappa_{ij}^k \frac{\delta x}{\delta y}} \tag{10}$$

for VR(4) and

$$\delta t_{ij}^k \leq \frac{1}{\sqrt{\frac{1}{\delta x^2} + \kappa_{ij}^k \frac{1}{\delta_y^2}}} = \frac{\sqrt{\delta x \delta y}}{\sqrt{\frac{\delta y}{\delta x} + \kappa_{ij}^k \frac{\delta x}{\delta y}}} = \frac{\sqrt{\delta x \delta y}}{\sqrt{2}} \frac{\sqrt{2}}{\sqrt{\frac{\delta y}{\delta x} + \kappa_{ij}^k \frac{\delta x}{\delta y}}} \tag{11}$$

$$\tau_{ij}^k = \frac{\sqrt{L_x L_y}}{\sqrt{2\pi}} \frac{\sqrt{2}}{\sqrt{\frac{L_y}{L_x} + \kappa_{ij}^k \frac{L_x}{L_y}}} \tag{12}$$

for VR(7).

5 Numerical Issues

In practical application due to nonuniform meshes and nonlinearity of the problem, δt and the damping factor τ are modified by scaling factors t_{scale} and τ_{scale}. The optimal δt and τ in both cases can be found by tuning these two parameters. This is summarized in the following table:

VR(4) for linear problem	VR(7) for linear problem
$\delta t = \frac{\delta x \delta y}{4} \frac{2}{\frac{\delta y}{\delta x} + \kappa_\parallel \frac{\delta x}{\delta y}} \cdot t_{scale}$	$\delta t = \frac{\sqrt{\delta x \delta y}}{\sqrt{2}} \frac{\sqrt{2}}{\sqrt{\frac{\delta y}{\delta x} + \kappa_\parallel \frac{\delta x}{\delta y}}} \cdot t_{scale}$ $\tau = \frac{\sqrt{L_x L_y}}{\sqrt{2\pi}} \frac{\sqrt{2}}{\sqrt{\frac{L_y}{L_x} + \kappa_\parallel \frac{L_x}{L_y}}} \cdot \tau_{scale}$
VR(4) for nonlinear problem	VR(7) for nonlinear problem
$\delta t_{ij}^k = \frac{\delta x \delta y}{4} \frac{2}{\frac{\delta y}{\delta x} + \kappa_{ij}^k \frac{\delta x}{\delta y}} \cdot t_{scale}$	$\delta t_{ij}^k = \frac{\sqrt{\delta x \delta y}}{\sqrt{2}} \frac{\sqrt{2}}{\sqrt{\frac{\delta y}{\delta x} + \kappa_{ij}^k \frac{\delta x}{\delta y}}} \cdot t_{scale}$ $\tau_{ij}^k = \frac{\sqrt{L_x L_y}}{\sqrt{2\pi}} \frac{\sqrt{2}}{\sqrt{\frac{L_y}{L_x} + \kappa_{ij}^k \frac{L_x}{L_y}}} \cdot \tau_{scale}$

$\frac{\delta x \delta y}{4}$ is the stability criterion for VR(4) when $\kappa_\| = 1$. $\frac{2}{\frac{\delta y}{\delta x}+\kappa_\| \frac{\delta x}{\delta y}}$ or $\frac{2}{\frac{\delta y}{\delta x}+\kappa_{ij}^k \frac{\delta x}{\delta y}}$ is the extra term if $\kappa_\|$ is larger than one or nonlinear. $\frac{\sqrt{\delta x \delta y}}{\sqrt{2}}$ is the stability criterion for VR(7) when $\kappa_\| = 1$. $\frac{\sqrt{2}}{\sqrt{\frac{\delta y}{\delta x}+\kappa_\| \frac{\delta x}{\delta y}}}$ or $\frac{\sqrt{2}}{\sqrt{\frac{\delta y}{\delta x}+\kappa_{ij}^k \frac{\delta x}{\delta y}}}$ is the extra term if $\kappa_\|$ is larger than one or nonlinear. For the relaxation factor τ, $\frac{\sqrt{L_x L_y}}{\sqrt{2\pi}}$ is the criterion for VR(7) when $\kappa_\| = 1$ and $\frac{\sqrt{2}}{\sqrt{\frac{L_y}{L_x}+\kappa_\| \frac{L_x}{L_y}}}$ or $\frac{\sqrt{2}}{\sqrt{\frac{L_y}{L_x}+\kappa_{ij}^k \frac{L_x}{L_y}}}$ is the extra term when $\kappa_\|$ is larger than one or nonlinear.

δx and δy are chosen based on the guidelines discussed in the previous sections so that as an example we have $N_x = (16-1)*2+1$ is 3 times less than $N_y = (51-1)*2+1$. N_x and N_y are the number of corresponding grid points in x and y directions. In this case VR(4) converged in 29708 number of iterations at optimal $t_{scale} = 0.174$; while VR(7) converged in 1308 number of iterations at optimal $t_{scale} = 0.41$, $\tau_{scale} = 0.87$. From here we can say that VR(7) is more than 20 times faster than VR(4). Hence from now on we will only use VR(7). Although iteration numbers seems to be large, each iteration is very cheap even compared to JFNK which requires preconditioning.

Next let's study the impact of initializing on convergence. As mentioned before, the nonlinear process can be initialized by the solution from the linear system with constant $\kappa_\|$. Given the linear solution with different size of $\kappa_\|$, the number of iterations for the nonlinear system to reach steady state is given in the following table. We found as long as the linear solution has $\kappa_\| \geq 2$, the nonlinear convergence doesn't have much difference. It only diverges when a guess has $\kappa_\| = 1$.

$\kappa_\|$	1	2	3	4	5	6	7,8,9,10[1] $\sim 10^9$
N_{its}	diverge	1313	1310	1309	1309	1309	1308

The marching process is even accelerated by varying δt and τ at each vertex ij and every time step k. We found the iteration won't even converge if uniform δt and τ are used.

Finally we give an efficient approach to update the nonlinear stiffness matrix $S_{yy}(T)$ at each time step. The numerical integration has to be carefully chosen in order to keep the symmetric structure as a result of periodic boundary condition. Generally

$$S_{yy}(T) = -\int\int \kappa_\| \frac{\partial N_i}{\partial y} \frac{\partial N_j}{\partial y} d\sigma$$

where N_i and N_j are the ith and jth base functions in finite element space. On each triangle, assuming n is the index running through all of the collocation points, then one way to formulate $S_{yy}(T)$ at kth time step would be

$$S_{yy}^{ij}(T) = \sum_n w(n)\kappa^k(n)\frac{\partial N_i}{\partial y}(n)\frac{\partial N_j}{\partial y}(n)J(n)$$

where $w(n)$, $\kappa^k(n)$, and $J(n)$ are the corresponding weight, conductivity, and Jacobian at nth point. $\frac{\partial N_i}{\partial y}(n)$ and $\frac{\partial N_j}{\partial y}(n)$ are also valued at these points as well. As a function of T, $\kappa^k(n)$ can be found by

$$\sum_l (T_l^k)^{5/2} N_l(n) \quad \text{or} \quad \sum_l [T_l^k N_l(n)]^{5/2}$$

where l is the index running through all of the vertexes on each triangle. But experiments show that the symmetric structure is destroyed by the above two formulations. Then we worked out the following formula

$$S_{yy}^{ij}(T) = \kappa_{ij}^k \sum_n w_n \frac{\partial N_i}{\partial y}(n) \frac{\partial N_j}{\partial y}(n) J(n)$$

which leads to

$$S_{non} = S_{xx} + B^k S_{yy}$$

where B^k is a vector with component $B_{ij} = \kappa_{ij}^k$ at each vertex given by ij. Therefore, we conclude that the nonlinear stiffness matrix S_{yy} can be updated by just scaling the linear stiffness matrix S_{yy} using nonlinear vector B. This approach not only saves computation complexity, but also preserves the symmetric structure of the periodic solution. The nonlinear solution is shown in Fig. 1 again in (x,y) coordinate system. The linear initial guess with $\kappa_\parallel = 2 \times 10^4$ given in the left plot is applied.

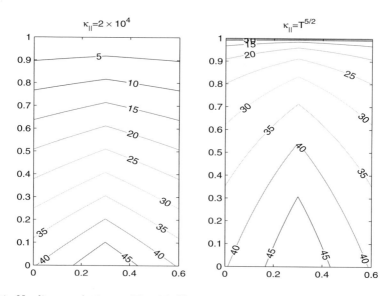

Fig. 1. Nonlinear solution at Nx=31, Ny=101, $t_{scale} = 0.41, \tau_{scale} = 0.87$. VR(7) is stable when $t_{scale} \leq 0.41$; VR(4) is stable when $t_{scale} \leq 0.174$.

6   Conclusions

As an extension of developing efficient linear elliptic solvers for fusion simulation on Cray X1E, nonlinear solver, based on Variable Relaxation, is constructed by by adding first and/or second order time derivative to the nonlinear elliptic equation and marching the resulting time-dependent PDEs to steady state. Instead of Jacobian, Only the stiffness matrix itself is involved and needs to be updated at each iteration.

Two schemes has been given, first and second order Variable Relaxations. four numerical issues has been fully discussed: The mesh scale ratio, nonlinear process initialization, variable time step and relaxation factor, efficient calculation of the nonlinear stiffness matrix. In summary, the mesh needs to be finer in direction with strong conductivity; convergence can be sped up by using the solution from corresponding linear system as an initial guess; time step and relaxation factor has to be varied at each grid point and every time step as well; only the nonlinear vector, used to update the nonlinear stiffness matrix, needs to be updated regularly. Therefore, the only computation consists of renewing δt_{ij}^k, τ_{ij}^k, and B^k at each iteration, and apparently these approaches give an efficient and robust algorithm to solve nonlinear systems.

References

1. W Park et al, Nonlinear simulation studies of tokamaks and STs, Nucl. Fusion **43** (2003) 483.
2. J Chen, S C Jardin, H R Strauss, *Solving Anisotropic Transport Equation on Misaligned Grids*, LNCS 3516, pp. 1076-1079 (2005).
3. D A Knoll, D E Keyes, Jacobian-free Newton-Krylov methods: a survey of approaches and applications, J comp. Phys. 193(2004) 357-397.
4. A Ern, V Giovangigli, D E Keyes, M D Smooke, Towards polyalgorithmic linear system solvers for nonlinear elliptic problems, SIAM J Sci. Comput. 15(1994) 681-703.
5. Y T Feng, On the discrete dynamic nature of the conjugate gradient method, J comp. Phys. 211(2006) 91-98.
6. J Chen, J Breslau, G Fu, S Jardin, W Park, *New Applications of Dynamic Relaxation in Advanced Scientific Computing*, proceedings of ISICS'06 Conference held at Dalian, China, Aug 15-18, 2006.
7. Y Saad, *Iterative Methods for Sparse Linearsystems*, PWS Publishing Company, (1996).

A Moving Boundary Wave Run-Up Model

Asu İnan[1] and Lale Balas[2]

Department of Civil Engineering, Faculty of Engineering and Architecture,
Gazi University, 06570 Ankara Turkey
asuinan@gazi.edu.tr, lalebal@gazi.edu.tr

Abstract. A numerical model has been developed for the simulation long wave propagation and run-up accounting for the bottom friction. Shallow water continuity and momentum equations are solved numerically by a two time step finite difference method. The upwind/downwind method is applied to the nonlinear advection terms and the continuity equation. Effects of damping and bottom friction on the computations are investigated. Since the equations lose their validity when waves break, wave breaking has been checked at every step of the computations. A point can be either wet or dry at different time levels. A moving boundary description and staggered grid are used to overcome the difficulty of determining wet and dry points. Equations are solved by the finite difference approximations of second and third order accuracy. Furthermore, space filters are applied to prevent parasitic short wave oscillations.

Keywords: Finite difference, long wave, run-up, moving boundary, filters.

1 Introduction

Nonbreaking long waves induced by tsunami, tide or storm cause catastrophic damages on the coasts because of high run-up levels. The numerical simulation of long wave propagation is an important tool in the damage control of catastrophic long waves. Estimation of the boundary between wet and dry points is a difficult problem in the simulation of wave run-up. During the simulations a computational point can be either wet or dry at different time levels. Therefore a moving boundary description is necessary.

Two dimensional nonlinear shallow water equations including bed shear stress were numerically solved by some researchers [1],[2]. Lynett et al. (2002) proposed a moving boundary technique to calculate wave run-up and run-down with depth-integrated equations [3]. An eddy viscosity model was inserted in the numerical model to account breaking phenomena. Kanoglu (2004) focused on initial value problem of the nonlinear evolution and run-up and run-down of long waves on sloping beaches for different initial wave forms [4]. Shermeneva and Shugan (2006) calculated the run-up of long gravity waves on a sloping beach using a high-order Boussinesq model [5]. Wei et al. (2006) simulated the long wave run-up under nonbreaking and breaking conditions with two dimensional well-balanced finite volume methods [6].

Y. Shi et al. (Eds.): ICCS 2007, Part I, LNCS 4487, pp. 38–45, 2007.
© Springer-Verlag Berlin Heidelberg 2007

In this study, nonlinear long wave equations have been solved numerically to simulate the wave run-up by upwind-downwind method. The effects wave breaking, wetting-drying of boundary points, friction and nonlinear terms on wave run-up, have been investigated.

2 Theory

Continuity equation and equation of motion for long waves are given below where x is the propagation direction.

$$\frac{\partial u}{\partial t} + u\frac{\partial u}{\partial x} = -g\frac{\partial \eta}{\partial x} - \frac{ru|u|}{D} \tag{1}$$

$$\frac{\partial \eta}{\partial t} = -\frac{\partial(Du)}{\partial x} \tag{2}$$

where, u, r, D, H, η are horizontal velocity, bed friction factor, total depth, water depth and water elevation above still water level, respectively. Total depth is the sum of water depth and water elevation above the still water level.

Two-time level numerical scheme has been used for the solution of the system. Upwind/downwind method has been applied to the nonlinear (advective) terms. The following algorithm is used to check the wet and dry points. Total depth has a positive value at a wet point and it is zero on the boundary[7].

$$\begin{array}{ll} 0.5(D_{j-1}+D_j) \geq 0 & \text{wet for point } D_j \\ 0.5(D_{j-1}+D_j) < 0 & \text{dry for point } D_j \end{array} \tag{3}$$

At every time level, wet and dry points are controlled and the validity of equation of motion has been provided. The equation of motion is modelled as given below;

$$\frac{u_j^{t+1}-u_j^t}{\Delta t}+u_p\frac{\left(u_j^t-u_{j-1}^t\right)}{\Delta x}+u_n\frac{\left(u_{j+1}^t-u_j^t\right)}{\Delta x}=-g\frac{\left(\eta_j^t-\eta_{j-1}^t\right)}{\Delta x}+\frac{ru_j^t|u_j^t|}{0.5\left(D_j^t+D_{j-1}^t\right)} \tag{4}$$

where, the u_p and u_n values are defined as;

$$u_p=\frac{\left(u_j^t+|u_j^t|\right)}{2}; \quad u_n=\frac{\left(u_j^t-|u_j^t|\right)}{2} \tag{5}$$

Upwind/downwind approach has been applied to the continuity equation. Following scheme is applied to overcome the stability problem at the boundaries for wet and dry regions.

$$\frac{\eta_j^{t+1}-\eta_j^t}{\Delta t}=\frac{u_{pj1}D_j^t+u_{nj1}D_{j+1}^t-u_{pj}D_{j-1}^t-u_{nj}D_j^t}{\Delta x} \tag{6}$$

where, $u_{pj1}, u_{pj}, u_{nj1}, u_{nj}$ are calculated as given below.

$$u_{pj1}=\frac{\left(u_{j+1}^{t+1}+\left|u_{j+1}^{t+1}\right|\right)}{2} \; ; \; u_{nj1}=\frac{\left(u_{j+1}^{t+1}-\left|u_{j+1}^{t+1}\right|\right)}{2} \; ; \; u_{pj}=\frac{\left(u_{j}^{t+1}+\left|u_{j}^{t+1}\right|\right)}{2} \; ; \; u_{nj}=\frac{\left(u_{j}^{t+1}-\left|u_{j}^{t+1}\right|\right)}{2} \tag{7}$$

A staggered grid is used for spatial derivatives at wet and dry boundaries as shown in Fig. 1.

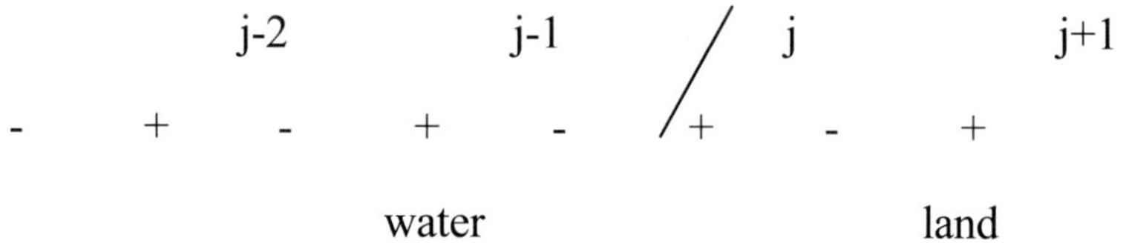

Fig. 1. Staggered grid

The intersection of wet and dry areas is a single point. The following extrapolation has been applied to the first dry point [8].

$$\left(\eta_j^{t+1}\right)_{ext}=2\eta_{j-1}^{t+1}-\eta_{j-2}^{t+1} \; ; \quad \left(u_j^{t+1}\right)_{ext}=2u_{j-1}^{t+1}-u_{j-2}^{t+1} \tag{8}$$

When waves break, the continuity equation and equation of motion lose their validity.

To check the breaking, a dimensionless parameter B_r is used [9].

$$B_r=\frac{\eta_0 w^2}{g\alpha^2} \tag{9}$$

where, η_0 is the initial wave amplitude, w is angular frequency, g is gravitational acceleration, α is bottom slope. If $B_r<1$, nonbreaking waves occur, otherwise the waves break while climbing up the sloping beach [2].

3 Applications of the Model

A sinusodial wave is assumed to enter from the left side of the channel with a water elevation function given below,

$$\eta=\eta_0\sin\left(\frac{2\Pi t}{T_p}\right) \tag{10}$$

where T_p is the wave period.

The application channel has a length of 2km. The bottom slope is linear. The water depth of the channel begins with 10 m at the left side of the channel and ends with -10m at the right side. Wave period is 1000 sec. The positive value of the depth descibes a wet region in the run-up computations. The horizontal grid size is 5m in the propagation direction and the time step is 0.5 sec. [10].

Computations begin with describing a sinusodial signal at the left side of the channel. Then, the total depth is computed at each time level related according to the changes of water level with time. The boundary is described with j_m index. If the water level at point j_m, is greater than the water depth, then, the point j_{m+1} is considered in the calculations. After this control, velocity is calculated. Velocity at point j_{m+1} is determined with the linear extrapolation and a similar approach is applied to the water elevation computation.

Computed water level changes at different levels of wave run-up and run- down is shown in Fig. 2. Bottom friction is neglected in this application. Curves 1 and 2 are the initial wave run-up curves, and curves 3 and 4 are the curves of wave run-down. Curve 5 shows the initiation of run- up of the second wave.

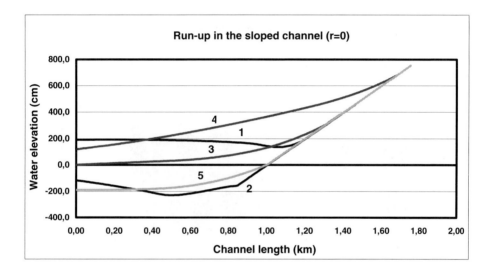

Fig. 2. Wave Run- up in the sloped channel

In the second application, bottom friction, nonlinear terms and damping effects are included in the computations while simulating the wave propagation along a channel with a positive slope. The length of the channel is 100km. Water depth varies linearly from 50m to 5m. Long wave propagates from deep water to shallow water with a wave period of $T_p=5$ min. The time step is 0.1sec and the grid size is 25m. Simulated wave amplitude and velocity distributions throughout the channel are given in Fig.3.

Distributions of water elevation and velocity show a significant difference. Wave amplitudes change slightly along the channel, but velocities change significantly as shown in Fig.3. Damping is more donimant on wave amplitudes compared to velocities. Strong nonlinear interactions occur in shallow waters and they cause stability problems. To test whether the waves with small periods appeared at the beginning and at the end of the main wave are the results of a physical phenemenon

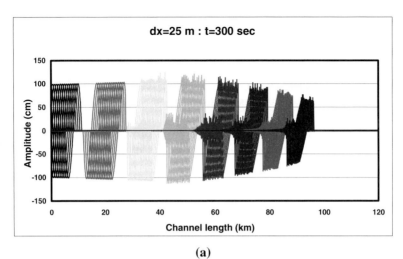

(a)

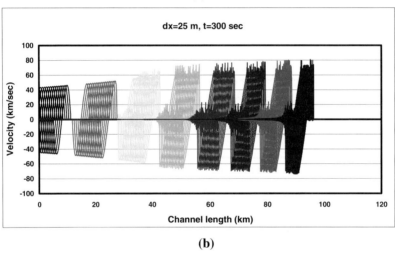

(b)

Fig. 3. a) Wave amplitude distribution b) Velocity distribution

or not, a finer resolution in the spatial domain has been used in the numerical computations. Therefore grid size is reduced to 5m and the simulated wave amplitude changes are depicted in Fig. 4. Short wave oscillation decreases as resolution increases, so it can be concluded that short wave oscillations are the results of coarse spatial resolution.

The main sources of nonlinear effects are the advective terms. Because of stability problem, higher order accurate upwind approximations have been applied to first order derivatives . Three points are used in the solution scheme as;

$$u\frac{\partial u}{\partial x} \approx u_p \frac{\left(3u_j^t - 4u_{j-1}^t + u_{j-2}^t\right)}{2h} + u_n \frac{\left(-u_{j+2}^t + 4u_{j+1}^t - 3u_j^t\right)}{2h} + \theta\left(h^3\right) \quad (11)$$

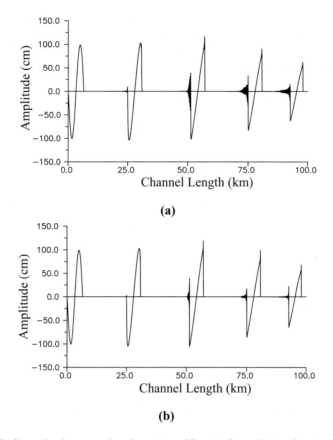

Fig. 4. Second order approximations a) Δx=25m, Δt=0.1sec b) Δx=5m, Δt=0.1sec

Higher order spatial approaches are used for the calculations of water elevations. The mesh point j is the point where velocity u_j is calculated. Water elevation is calculated at Δx/2 distances away from point j at two sides. Higher order equations are obtained by combining the equations at points located at Δx/2 and 3Δx/2 distances and considering the second order Taylor series. The new formulation of the first derivative of water elevation is given below.

$$\frac{\partial \eta}{\partial x}=\left[27\left(\eta_j-\eta_{j-1}\right)-\left(\eta_{j+1}-\eta_{j-2}\right)\right]/24\Delta x+\theta\left(h^4\right) \tag{12}$$

A similar approach is used for the spatial derivative of velocity in the continuity equation. However, water elevation is computed at the midpoint in velocity computations and therefore spatial index is shifted to the right and point j+1 is inserted instead of point j in the numerical calculations.

$$\frac{\partial(Hu)}{\partial x}=\left\{\begin{matrix}27\left[u_{j+1}\left(h_j+h_{j+1}\right)/2-u_j\left(h_j+h_{j-1}\right)/2\right]-\\ \left[u_{j+2}\left(h_{j+2}+h_{j+1}\right)/2-u_{j-1}\left(h_{j-2}+h_{j-1}\right)/2\right]\end{matrix}\right\}/24\Delta x+\theta\left(h^4\right) \tag{13}$$

Parasitic short wave oscillations are prevented with the application of higher order spatial and time resolutions, but application of simple space filters is easier to prevent the formation of parasitic short waves. Space filters are applied only to the computed velocities. The filtered new form of velocity is given in the equation (14).

$$u_n(j) = u_j^{t+1}(1-\alpha^*) + 0.25(u_{j-1}^{t+1} + 2u_j^{t+1} + u_{j+1}^{t+1})\alpha^* \tag{14}$$

where, α^* is a filter parameter and taken as 0.005. The results obtained after the application of higher order approaches and filter technique are shown in Fig. 5.

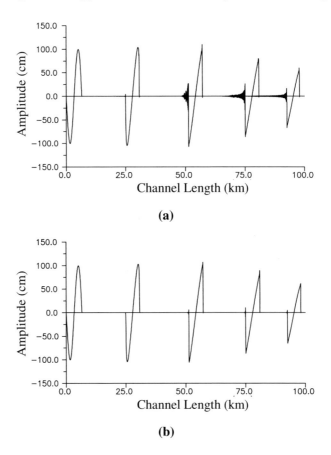

Fig. 5. a) Third order approximations b) Third order approximations and filter technique ($\Delta x=25m$, $\Delta t=0.1sec$)

4 Conclusions

Motion and continuity equations of long wave run-up are solved numerically by a two time level approach. The effects of nonlinear terms, damping, bottom friction and wet/dry boundaries have been investigated. Application of upwind/downwind approximations to continuity equation gives stable results at the boundaries of wet and dry regions. Parasitic short wave oscillations are eliminated by increasing the resolutions in space and in time. Application of higher order accurate approximations increases the accuracy of the solution. It is observed that filter technique is a reliable tool for preventing parasitic short wave oscillations in the solutions obtained with lower order accurate approximations. The proposed numerical model successfully simulates the long wave run-up.

References

1. Hubbard, M.E., Dodd, N.:A 2D Numerical Model of Wave Runup and Overtopping. Coastal Engineering 47 (2002) 1-26.
2. Kowalik, Z.: Basic Relations between Tsunami Calculations and Their Physics. Science of Tsunami Hazard.19 (2001) 99-115.
3. Lynett, P. J., Wu, T-R., Liu P. L-F.:Modeling Wave Runup with Depth-Integrated Equations. Coastal Engineering, 46 (2002) 89-107.
4. Kanoglu, U.: Nonlinear Evolution and Runup- Rundown of Long Waves over a Sloping Beach. Journal of Fluid Mechanics, 513 (2004) 363-372.
5. Shermeneva, M.A., Shugan, I.V.: Calculating the wave runup on a low-sloping beach using a higher- order Boussinesq Model. Technical Physics Letter, 32 (2006) 64-66.
6. Wei, Y., Mao, X.Z., Cheung, K.F.: Well- Balanced Finite- Volume Model for Long- Wave Runup. Journal of Waterway, Port, Coastal and Ocean Engineering, ASCE, 132 (2006) 114-124.
7. Flather,R.A., Heaps, N.S. :Tidal computations for Morecamble Bay, Geophys. J. Roy. Astr. 42 (1975) 423-436.
8. Sielecki, A., Wurtele, M.G.: The numerical integration of the nonlinear shallow-water equations with sloping boundaries. J. Computational Physics 6 (1970) 219-236.
9. Voltzinger, N.E., Klevanny, K.A., Pelinovsky, E.N.: Long-Wave Dynamics of the Coastal Zone. Gidrometizdat., Leningrad (1989)
10. 10.İnan, A., Temiz, M., Balas, L.: Simulation of Long Wave Run-up, Proceedings of the V. National Conference on Turkish Coast'04, Adana, Turkey, 2 (2004) 891-898

Enabling Very-Large Scale Earthquake Simulations on Parallel Machines

Yifeng Cui[1], Reagan Moore[1], Kim Olsen[2], Amit Chourasia[1], Philip Maechling[4],
Bernard Minster[3], Steven Day[2], Yuanfang Hu[1], Jing Zhu[1], Amitava Majumdar[1],
and Thomas Jordan[4]

[1] San Diego Supercomputer Center, 9500 Gilman Drive, La Jolla, CA 92093-0505 USA
{yfcui,moore,yhu,jzhu,majumdar}@sdsc.edu
[2] San Diego State University, 5500 Campanile Drive, San Diego, CA 92182 USA
{kolsen,steven.day}@geology.sdsu.edu
[3] Scripps Institution of Oceanography, 9500 Gilman Drive, La Jolla, CA 92024 USA
jbminster@ucsd.edu
[4] University of Southern California, Southern California Earthquake Center, Los Angeles,
CA 90089 USA
{maechlin,tjordan}@usc.edu

Abstract. The Southern California Earthquake Center initiated a major large-scale earthquake simulation called TeraShake. The simulations propagated seismic waves across a domain of 600x300x80 km at 200 meter resolution, some of the largest and most detailed earthquake simulations of the southern San Andreas fault. The output from a single simulation may be as large as 47 terabytes of data and 400,000 files. The execution of these large simulations requires high levels of expertise and resource coordination. We describe how we performed single-processor optimization of the application, optimization of the I/O handling, and the optimization of execution initialization. We also look at the challenges presented by run-time data archive management and visualization. The improvements made to the application as it was recently scaled up to 40k BlueGene processors have created a community code that can be used by the wider SCEC community to perform large scale earthquake simulations.

Keywords: parallel computing, scalability, earthquake simulation, data management, visualization, TeraShake.

1 Introduction

The southern portion of the San Andreas Fault, between Cajon Creek and Bombay Beach in the state of California in the United States has not seen a major event since 1690, and has accumulated a slip deficit of 5-6 meters [13]. The potential for this portion of the fault to rupture in an earthquake as large as magnitude 7.7 is a major component of seismic hazard in southern California and northern Mexico. To gain insights into how intensely the region will shake during such an event, the Southern California Earthquake Center (SCEC) initiated a major large-scale earthquake

Y. Shi et al. (Eds.): ICCS 2007, Part I, LNCS 4487, pp. 46–53, 2007.

simulation in 2004, called TeraShake [8][11]. The simulations used a 3,000 by 1,500 by 400 mesh, dividing the volume into 1.8 billion cubes with a spatial resolution of 200 meters and a maximum frequency of 0.5 hertz, some of the largest and most detailed earthquake simulations of this region. TeraShake runs used up to 2048 processors on the NSF funded TeraGrid [12], and in some cases produced 47 TB of time-varying volumetric data outputs for a single run. The outputs were then registered in a digital library, managed by San Diego Supercomputer Center's Storage Resource Broker (SRB) [10], with a second copy archived into SDSC's High Performance Storage System.

The TeraShake-2 simulations added a physics-based dynamic rupture component to the simulation, which was run at a very high 100 meter resolution, to create the earthquake source description for the San Andreas Fault. This is more physically realistic than the kinematic source description previously used in TeraShake-1. The resulting seismic wave propagation gave a more realistic picture of the strong ground motions that may occur in the event of such an earthquake, which can be especially intense in sediment-filled basins such as the Los Angeles area.

In this paper, we look at challenges we faced porting the application, optimizing the application performance, optimizing the I/O handling, and optimizing the run initialization (Section 2 and 3). Section 4 discusses challenges for large-scale executions. We also discuss the challenges for data archive and management (Section 5), as well as the expertise required for analyzing the results (Section 6). Final section briefly discusses the results of our simulations and draws some conclusions.

2   Challenges for Porting and Optimization

To compute the propagation of the seismic waves that travel along complex paths from a fault rupture across an entire domain, the anelastic wave model (AWM), developed by Kim Olsen et al. [2][4][6][7][8], was picked as the primary model for the SCEC TeraShake simulation. The AWM uses a structured 3D grid with fourth-order staggered-grid finite differences for velocity and stress. One of the significant advantages of the code is the use of Perfectly Matched Layers absorbing boundary conditions on the sides and bottom of the grid, and a zero-stress free surface boundary condition at the top [7]. The AWM code is written in Fortran 90. Message passing is done with MPI using domain decomposition. I/O is done using MPI-I/O so that all processors write velocity output to a single file. The code was extensively validated for a wide range of problems, from simple point sources in a half-space to dipping propagating faults in 3D crustal models [7].

The computational challenges in porting the AWM code to the TeraGrid were two fold. First we identified and fixed the bugs related to Message Passing Interface (MPI) and MPI-IO that caused the code to hang on the target platforms. We found that the original design of the MPI-IO data type in the code that represents count blocks was defined at each time step, which caused a memory leak problem. Our improved version of the code defined indexed data type once only at the initialization phase, and effectively set new views by each task of a file group to obtain efficient MPI-IO performance. For MPI-IO optimization, we modified the collective writes

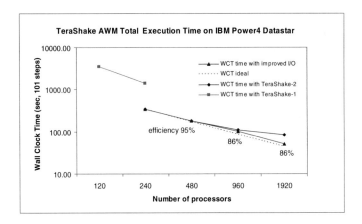

Fig. 1. Strong scaling of AWM with a parallel efficiency of 86% on 1920 processors

from using an individual file pointer to using an explicit offset, which not only made large output writing possible, but also greatly improved the I/O performance.

The second effort was to enhance and integrate features necessary for large-scale simulations. Previous simulations of the Southern California region with the AWM code were only tested up to a 592 x 592 x 592 mesh. For the TeraShake case with 1.8 billion mesh points, new problems emerged in managing memory requirements for problem initialization. The code was enhanced from 32-bit to 64-bit for managing 1.8 billion mesh points. To improve the code performance, we profiled the execution time of each part of the code, and identified its performance bottleneck. We optimized cache performance and reduced instruction counts. Some of the very time-consuming functions were in-lined, which immediately saved more than 50% of the initialization time. The reduction of the required memory size and tuning of data operations were necessary steps to scale the code up to the TeraShake scale.

As part of the TeraShake-2 effort, we integrated a new dynamic rupture component into the AWM. This new feature models slip dynamically on the fault surface to generate a more realistic source than the kinematic source description used in TeraShake-1.

The TeraShake simulation poses significant challenges for I/O handling. In the heavy-I/O case, the I/O takes 46% of the total elapsed time on 240 processors of the 10 teraflops TeraGrid Power4 p655 Datastar at SDSC, and the performance saturates quickly as the number of processors increases to more than a few hundred. To improve the disk write performance, we carefully calculated the runtime memory utilization of writes, and accumulated output data in a memory buffer until it reached an optimized size before writing the data to disk.. This optimization through best tradeoff between I/O performance and memory overhead alone reduced the I/O time by a factor of 10 or more, resulting in a very small fraction of the surface velocity write time compared to the total elapsed time.

The simulation algorithm showed very good scaling as a function of the number of processors. The integrated AWM code scales up to 2,048 processors on Datastar. Figure 1 illustrates the significant improvement of scaling after I/O tuning. The figure also shows the improvement of single CPU performance using machine-specific aggressive optimization flags. The overall performance optimization of the code forms the basis for a parallel efficiency of 96% on 40,960 Blue Gene/L processors at IBM TJ Watson, the latest achievement for petascale earthquake simulation.

3   Challenges for Initialization

AWM initialization presented a significant challenge as we scaled up to TeraShake problem size. Originally the AWM didn't separate the mesh generator processing from the finite difference solver. This made it difficult to scale the code up to a large problem size. While allocated memory is about 1 GB per processor for the finite difference solver, the mesh generation processing performed during the initialization stage required much more memory. Tests performed on the 32-way 256 GB memory Datastar p690 used around 230 GB of memory for initialization. Note that while TeraShake-1 used an extended kinematic source defined at 18,886 points, TeraShake-2 used dynamic sources which were defined at 89,095 points. Memory required per processor using the dynamic source exceeds 4 GB, far beyond the limit of the memory available per processor on both target machines TeraGrid IA-64 and Datastar.

To reduce the memory requirements, we deallocated the arrays not being actively used, and reused existing arrays. More importantly, we separated the source and mesh initialization step from the main production run, so that a pre-processing step is performed to prepare sub-domain velocity model and source partition. With this strategy, the production run only reads in source and mesh input data needed by each processor. This means the production runs with dynamic sources required only the memory size associated with the point source, which reduced the memory requirement by a factor of 8.

To improve disk read performance, we optimized the code by reading data in bulk. We aggressively read data, with data retrieval going beyond the disk attached to the local processor. We calculated the actual location of the data and then assigned the read to the corresponding processors, which improved the disk read performance by a factor of 10.

The final production initialization for TeraShake-1 used a 3-D crustal structure based on the SCEC Community Velocity Model Version 3.0. The source model is based on that inferred for the 2002 Denali Earthquake (M7.9), and some modifications were made in order to apply it to the southern San Andreas fault [8].

4   Challenges for Executions

The large TeraShake simulations were expected to take multiple days to complete. As we prepared the code for use in this simulation, we recognized that the foremost needs were the capabilities of checkpoint and restart which were not available in the original code. We integrated and validated these capabilities, partly prepared by Bernard Minster's group at Scripps Institution of Oceanography. Subsequently, we added more checkpoints/restart features for the initialization partition, as well as for the dynamic rupture mode. To prepare for post-processing visualizations, we separated the writes of volume velocity data output from writes of velocity surface data output. The latter was output at each time step. To track and verify the integrity of the simulation data collections, we generated MD5 checksums in parallel at each processor, for each mesh sub-array in core memory. The parallelized MD5 approach substantially decreased the time needed to checksum several Terabytes of data.

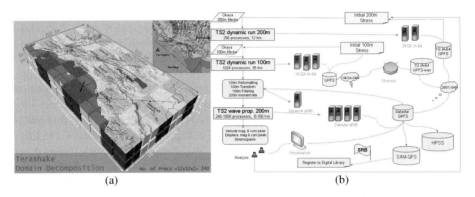

(a)   (b)

Fig. 2. (a) TeraShake domain decomposition on SDSC IBM Power4 DataStar p655 nodes. The rectangle in red on the top right inset shows the simulation region that is 600km long and 300km wide. Domain decomposition of the region onto 240 processors is shown in the center. (b) SCEC seismic wave simulation data flow.

The TeraShake runs required a powerful computational infrastructure as well as an efficient and large scale data handling system. We used multiple TeraGrid computers for the production runs at different stages of the project. The data-intensive TeraShake1.2 simulation, which generated 47 TB volume outputs, used 18,000 CPU hours on 240 Datastar processors (Fig. 2a). The optimal processor configuration was a trade-off between computational and I/O demands. Volume data was generated at each 10th to 100th time step for the run. Surface data were archived for every time step. Checkpoint files were created at each 1000th step in case restarts were required due to reconfigurations or eventual run failures. The model computed 22,728 time steps of 0.011 second duration for the first 250 seconds of the earthquake scenario.

The TeraShake-2 dynamic rupture simulations used a mesh size of 2992 x 800 x 400 cells at 100m resolution, after the appropriate dynamic parameters were determined from several coarser-grid simulations with 200 m cells. Figure 2b shows the execution and dataflow between compute resources and archiving storage system. The 200m resolution runs were conducted on 256 TeraGrid IA-64 processors at SDSC.   The 100m resolution runs were conducted on 1024 TeraGrid IA-64 processors at the National Center for Supercomputing Applications. The TeraShake-2 wave propagation runs were executed on up to 2000 processors of Datastar, determined to be the most efficient available processor. The simulation output data were written to the DataStar GPFS parallel disk cache, archived on the Sun Sam-QFS file system, and registered into the SCEC Community Digital Library supported by the SDSC SRB.

5   Challenges for Data Archive and Management

The data management was highly constrained by the massive scale of the simulation. The output from the seismic wave propagation was migrated onto both a Sun Sam-QFS file system and the IBM High Performance Storage System (HPSS) archive as the run progressed – and moving it fast enough at sustained data transfer rate over 120 MB/sec to keep up with the 10 terabytes per day of simulation output.

The TeraShake simulations have generated hundreds of terabytes of output and more than one million files, with 90,000 - 120,000 files per simulation. Each simulation is organized as a separate sub-collection in the SRB data grid. The sub-collections are published through the SCEC community digital library. The files are labeled with metadata attributes which defined the time steps in the simulation, the velocity component, the size of the file, the creation date, the grid spacing, and the number of cells, etc [5]. All files registered into the data grid can be accessed by their logical file name, independently of whether the data were on parallel file system GPFS, Sam-QFS, or the HPSS archive. General properties of the simulation such as the source characterization are associated as metadata for the simulation collection. Integrity information is associated with each file (MD5 checksum) as well as existence of replicas. Since even tape archives are subject to data corruption, selected files are replicated onto either multiple storage media or multiple storage systems.

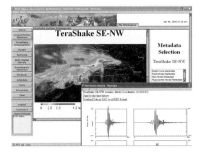   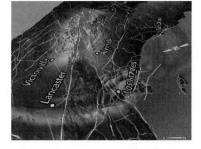

Fig. 3. User interaction with the TeraShake Surface Seismograms portlet at the SCECLib Portal

Fig. 4. TeraShake Surface rendering of displacement magnitude with topographic deformation

The SCEC digital library includes the digital entities (simulation output, observational data, and visualizations), metadata about each digital entity, and services that can be used to access and display selected data sets. The services have been integrated through the SCEC portal into seismic-oriented interaction environments [5][9]. A researcher can then select an earthquake simulation scenario and select a location on the surface, by pointing and clicking over the interactive cumulative peak velocity map, or interact with the full service resolution data amounting to one terabyte (Fig. 3).

6   Challenges for Analysis of Results

Verification of the simulation progress at runtime and thereafter seismological assessment of data computed was a major concern for the success of the TeraShake project. Visualization techniques helped solve this problem by rendering the output data during the simulation run. Animations of these renderings were instantly made available for analysis. SDSC's volume rendering tool Vista, based on the Scalable Visualization Toolkit (SVT), was used for visualizations. Vista employs ray casting for performing volumetric rendering. Surface data have been visualized with different variables (velocities and displacements) and data ranges in multiple modes. The resulting animations have proven valuable not only to domain scientists but also to a broader audience by providing an intuitive way to understand the TeraShake simulation results. Visualizations alone have consumed more than 40,000 CPU hours on Datastar and IA-64 at SDSC. Upwards of 100 visualization runs were performed, with each run utilizing 8 to 256 processors in a distributed manner. The results have produced over 130,000 images [1] (Fig. 4 shows an example).

Scientists want to conduct hands on analysis in an attempt to gain a better understanding of output data. The size of TeraShake data poses a significant problem for accessibility and analysis. We developed a web front end where scientists can download the data and are able to create custom visualizations over the web directly from surface data. The portal uses LAMP (Linux, Apache, MySQL, PHP) and Java technology for web middle-ware and on the back-end compute side relies on specialized programs to fetch data from the archive, visualize, composite, annotate and make it available to client browser.

7   Summary

The TeraShake simulation was one of the early projects at SCEC targeting capability computing, and the code accuracy has been extensively verified for anelastic wave propagation and dynamic fault rupture [3]. A major result of the simulation was the identification of the critical role a sedimentary waveguide along the southern border of the San Bernardino and San Gabriel Mountains has in channeling seismic energy into the heavily populated San Gabriel and Los Angeles basin areas. The simulations have considerable implications for seismic hazards in southern California and northern Mexico.

The TeraShake simulations demonstrated that optimization and enhancement of major applications codes are essential for using large resources (number of processors, number of CPU-hours, terabytes of data produced). TeraShake also showed that multiple types of resources are needed for large problems: initialization, run-time execution, analysis resources, and long-term data collection management.

The improvements made to the TeraShake AWM have created a community code that can be used by the wider SCEC community to perform large scale earthquake simulations. The TeraShake code is already being integrated for use in other SCEC Projects such as the SCEC Earthworks Science Gateway.

SCEC has identified a PetaShake platform for petascale simulations of dynamic ruptures and ground motions with outer/inner scale ratios as high as $10^{4.5}$. Excellent scalability of TeraShake AWM shown on 40k BG/L processors has demonstrated an important step toward petascale earthquake rupture and wave propagation computing.

Acknowledgments. SDSC's computational collaboration effort was supported through the NSF-funded SDSC Strategic Applications Collaborations programs (award SCI 0438741). The TeraShake Digital Library effort is funded by the National Science Foundation Information Technology Research program through the award EAR 0122464. SDSC's visualization effort was supported through the NSF-funded SDSC core program. This project was also funded through NSF Grant EAR-0122464: The SCEC Community Modeling Environment (SCEC/CME): An Information Infrastructure for System-Level Earthquake Research.

References

1. Chourasia. A., Cutchin. S. M., Olsen. K.B., Minster. B., Day. S., Cui. Y., Maechling. P., Moore. R., Jordan. T.: Insights gained through visualization for large earthquake simulations, submitted to Computer Graphics and Application Journal (2006).
2. Cui, Y., Olsen, K., Hu, Y., Day, S., Dalguer, L., Minster, B., Moore, R., Zhu, J., Maechling, P., Jordan, T.: Optimization and Scalability of A Large-Scale Earthquake Simulation Application. Eos, Trans, AGU 87(52), Fall Meet. Suppl. (2006), Abstract S41C-1351.
3. Dalguer, L.A. and Day, S.: Staggered-grid split-node method for spontaneous rupture simulation, J. Geophys. Res. (2007), accepted.
4. Day, S.M. and Bradley, C.: Memory-efficient simulation of an-elastic wave propagation, Bull. Seis. Soc. Am. 91 (2001) 520-531
5. Faerman, M., Moore, R., Cui, Y., Hu, Y., Zhu, J., Minister, B. and Maechling, P.: Managing Large Scale Data for Earthquake Simulations, Journal of Grid Computing, Springer-Verlag DOI 10.1007/s10723-007-9072-x (2007)
6. Olsen, Kim B.: Simulation of three-dimensional wave propagation in the Salt Lake Basin, Ph.D. thesis, The University of Utah, (1994)
7. Olsen, K.B., Day, S.M., and Bradley, C.R.: Estimation of Q for long-period (>2 s) waves in the Los Angeles Basin, Bull. Seis. Soc. Am. 93 (2003) 627-638
8. Olsen, K., Day, S.M., Minster, J.B., Cui, Y., Chourasia, A., Faerman, M., Moore, R., Maechling, P. and Jordan, T.: Strong Shaking in Los Angeles Expected from Southern San Andreas Earthquake, Geophysical Research Letters, Vol 33 (2006), 1-4
9. Olsen, K., Zhu, J. and Talley, J.: Dynamic User Interface for Cross-plot, Filtering and Upload/Download of Time Series Data. Eos, Trans, AGU 87(52), Fall Meet. Suppl. (2006), Abstract IN51B-0814
10. Moore, R., Rajasekar, A., Wan, M.: Data Grids, Digital Libraries and Persistent Archives: An Integrated Approach to Publishing, Sharing and Archiving Data. Special Issue of the Proceedings of the IEEE on Grid Computing, Vol. 93, No.3 (2005) 578-588
11. SCEC/CME Web Site: http://www.scec.org/cme.
12. TeraGrid Website: http://teragrid.org/about/
13. Weldon, R., K. Scharer, T. Furnal and Biasi, G.: Wrightwood and the earthquake cycle: What a long recurrence record tells us about how faults work, Geol. Seismol. Am. Today, 14 (2004) 4-10

Fast Insolation Computation in Large Territories

Siham Tabik[1], Jesús M. Vías[2], Emilio L. Zapata[1], and Luis F. Romero[1]

[1] Depto de Arquitectura de Computadores, campus de Teatinos, 29080 Málaga, Spain
[2] Depto de Geografía, campus de Teatinos, 29080 Málaga, Spain
siham@ac.uma.es

Abstract. This paper presents a new model for the computation of the insolation in large territories. The main novelty of this work consists in that it reduces significatively the number of necessary arithmetic operations, and simplifies the calculations introducing errors that are inferior to the ones imposed by the climate in a territory. In particular, a faster algorithm to compute the horizon at all points of a terrain is provided. This algorithm is also useful for shading and visibility applications. Moreover, parallel computing is introduced into this model using grid computing technology to extend its application to very large territories.

Keywords: Insolation, horizon computation, very large territories, Globus.

1 Introduction

The knowledge of the amount of incoming solar energy at diverse geographic locations is of a paramount importance in several fields, in solar energy utilization, agriculture, forestry, meteorology, environmental assessment and ecology. However, most solar radiation models that take into account the topographic heterogeneity and climatological features, like solar analyst [1] and r.sun model [2] have a large computational cost and high storage requirements, which become unapproachable for large territories. This paper presents a fast and accurate model that computes the insolation in large terrains considering all the involved factors, i.e., the localization and the horizons of all the points of the territory, the localization of the sun and the meteorological factors. We demonstrate that a great part of calculations can be reduced substantially by searching equivalences, until reducing it to the minimum possible. In addition, an implementation in grid computing have been developed to allow the extension of this model to very large territories, of the order of millions of kilometers. We describe the performance results of the application of the model to Andalucia territory.

2 The Sky

The sun passes from many parts of the sky. In each one of these localization, the incoming radiations into a horizontal surface have different orientation (or azimuth) α and inclination (or altitude) Θ. If in addition, each point of the

Y. Shi et al. (Eds.): ICCS 2007, Part I, LNCS 4487, pp. 54–61, 2007.

territory has different orientation *head* and inclination *tilt*, a priori, it would be necessary to know the incidence in each point of the territory, at each instant of the year.

However, the sun passes many times form the same zone of the sky. Our first simplification consists in supposing that the considered territory is horizontal and construct a celestial map of hemispherical shape that covers the horizon. We discretize this map into cells, and determine the quantity of radiation received by each point of the territory from each one of these cells. According to the spherical coordinates, the discretization results in $N_s = 90 \times 360$ celestial cells that can be represented by a matrix of 90 lines and 360 columns. Each coefficient of this matrix correspond to a celestial cell with different coordinates α and Θ.

The calculation of the position of the sun can be performed using the equations of the earth movement with respect to the sun [3].

Fig. 1. Insolation of one year of exposition in a point of the horizon (black) displayed from the left to the right, in the west, south, east and north. The dark and light colors show low and high values of insolation respectively.

Performing these calculations, each cell of the obtained matrix of results accumulates the received radiation, i.e., the number of hours the sun has emitted energy from that cell. As illustration, Fig. 1 shows the insolation one year of exposition. The set of N_s cells or celestial sectors are represented with the symbol Δ.

The error produced in the discretization of the sky is lower than the committed by nature due to cloudiness. This error can be quantified in order to determine an appropriate dimension of celestial sectors.

3   The Ground

A terrain is most often represented with a digital elevation map (DEM), that can be discretized into cells. Typically, a very large territory contains millions of points, however, many of these N points are similar. In particular, each localization P has a given orientation $head_i$ and inclination $tilt_i$ with respect to the horizontal but with a very small error. We can consider that there exist only $N_f = 90 \times 360$ possible orientations in our territory. Consequently, there exist only N_s different cells, in which we want to compute the insolation. Let's consider Ω the set of those N_f possible orientations. If there would be no mountains in the horizon, all points with the same orientation will receive the same quantity of radiation, that we call $E_{full}(head_i, tilt_i)$.

For each one of the N_s different cases we compute the insolation received from the corresponding N_s celestial sectors Δ, in which the angle of incidence is taken into account, this can be computed with only $5 \times N_f$ trigonometrical operations for each element ø of Ω. In total, we obtain $5 \times N_f \times N_s$ operations.

Up to here, all the calculations are performed with a low computational cost.

4   The Horizon

Each point of the total N points of a territory suffers from a partial concealment of the sun due to the orography that surrounds it (from the orto to the astronomical decline). If we consider that the orography is different in each point of the territory, the computation becomes very complicated. We will have $N \times N_s$ possible cases, where N is a very large number, of the order of millions. In addition, an accurate computation of solar occultation requires an extraordinary amount of arithmetical operations as discussed in next section.

4.1   The Calculation of Shaded Zones

In principle, to determine the shadowed regions for each position of the sun we should first compute the trajectory of the sun rays for each terrestrial cell. However, the execution of such calculation, from this perspective, that we could call *celestial*, implies a large redundant calculation. Herein, we have chosen a different vision of the problem. In particular, we compute the visibility of the celestial cells, through which sometimes the sun has passed, from any cell of the territory. In other words, we are interested in knowing the profile or the horizon of a territory.

Tackling the problem from this perspective, that we could call *terrestrial*, is absolutely equivalent, besides, this has some advantages that we reveals in next sections.

Generally, the horizon of each point must analyzes the position of the rest of points of the grid (i.e., the territory), with a computational complexity $O(N)$. The computation of the horizon of all these points is $O(N^2)$, which implies an impractical computational cost, i.e., a N-body problem.

However, there exist several algorithms that simplify the calculations. For example the approximative method of Cabral, Max and Springermeyer (CMS) [4] divides the horizon in s sectors, and in each sector it determines the elevation of the horizon considering solely the points of the terrain that are in the central line of the sector. This algorithm has a computational cost $O(s(N^{1.5}))$. Stewart [5] proposed a more precise algorithm that computes the horizon in all the sector with a cost $O(sN(\log^2 N + s))$. This method is faster than the CMS, and the quality of the image generated is better especially for very large terrains, i.e., more than 100.000 points.

4.2   Stewart Model and the Horizon Computation

Stewart [5] divides the horizon of each point P in s sectors $\sigma_{i,p}$ $(1 \leq i \leq s)$ that form an angle $\frac{2\pi}{s}$ radians each one.

This algorithm consists in performing for each sector i, four important steps:

1. compute the new coordinates of all the points in the coordinate system $(a^\perp, b^\perp, z)$, where a and b are two horizontal vectors in the directions $\frac{2\pi}{s} i$ and $\frac{2\pi}{s}(i+1)$ respectively, and sort all the points by $a^\perp$ and $b^\perp$ for each direction
2. compute the horizons using the last updated hull structure
3. update the hull structure including the information of the considered point
4. insert the horizon of the considered point in the original structure

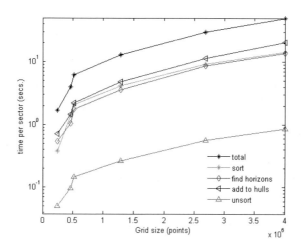

Fig. 2. Times by sector in logarithmic scale of the four main steps of Stewart's algorithm

Figure 2 shows the time per sector of each step of Stewart's algorithm, for several grid sizes. As it can be seen from this figure, the time increases enormously as the size of the grid increases. This means that the calculation of the horizon of Andalucia DEM that has 100000×50000 cells of precision 10×10 m^2 will suppose very long execution time. However, dividing this DEM into subgrids of 2000×2000 cells requires only $256MB$, which is reasonable from a computational point of view but this produces an undesired edge effect as explained in next section.

4.3 Our Horizon Algorithm

In this work we propose an optimized algorithm that in addition to compute the horizon of large terrains in a reasonable time, it eliminates the edge effect due to the division into regular grids. First, to eliminate the edge effect we apply a two-partition technique that consists in considering the horizon of a given point as the overlapping of its horizons in the grid A and B, both of size 10×10 points. Each point P belongs to a grid in the division A (black grid in Fig. 3)

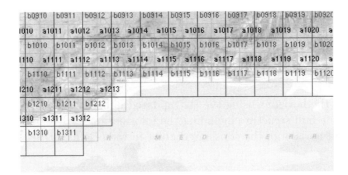

Fig. 3. Two-partition technique. Grid A (black color) and grid B (grey color).

and an other grid in the division B (grey grid in Fig. 3). Each single grid $A_{i,j}$ is considered as four sub-grids: $A_{i,j}^{nw}$, $A_{i,j}^{ne}$, $A_{i,j}^{se}$ and $A_{i,j}^{sw}$, where each of which belongs to four B-grids (see Fig. 3) in such a way that:

$$A_{i,j}^{nw} = B_{i,j}^{se}, \ A_{i,j}^{ne} = B_{i,j+1}^{sw}, \ A_{i,j}^{se} = B_{i+1,j+2}^{nw} \text{ and } A_{i,j}^{sw} = B_{i+1,j}^{ne}$$

The horizons are computed only in the opposite direction of the position of each one of the four sub-grids inside the grid, for example, a sub-grid nw computes its horizon only in the direction se. This reduces the time of step 2 in the algorithm by half.

To compute the horizons of more than 10 km, we have used a low precision grid that we call C grid, of dimension 2000×1000 points, which covers Andalucia with steps of 200 meters.

As stated before, Stewart's algorithm computes the horizon in a main loop by sector. The time taken in computing the horizons of Andalucia using Stewart's algorithm is about (518 grids × 64 sectors × 50 seconds) equivalent to 19 days. However, taking into account the regularity of the grid and applying some optimizations, this time can be significantly reduced.

- Our first optimization consists in reusing the sorting of sector i by $b^{\perp}$ for sorting sector $i+1$ by $a^{\perp}$. Thus, we reduce the computational time of step 1 to approximately half the time. This optimization can be applied to all sectors except the first one.
- The second optimization consists is that in a regular grid, there exist an easier way to sort the points by $a^{\perp}$ and $b^{\perp}$ without the need to compute its coordinates in that coordinate reference. It can be done by just counting the number of points behind the running line, using very simple algebraic operations. As the position number is directly computed, it is not necessary to use any sorting algorithm. In practise, the time spent in the optimized phase 1 is NULL.
- Our third optimization is the use of the two-partition technique presented here, so that we do not need to perform the 2nd phase for every sector, but only for the half of them, as the horizon for the other half is obtained from the overlapped grid B (and vice versa).

The calculated horizon by the optimized algorithm is stored in a matrix μ of dimension 1×360.

As a result of the proposed optimizations, the expected computational time is reduced to less than 11 days. The improvement in runtime when applying the optimizations 1,2,3 and 4 is evident as shown in Fig. 4.

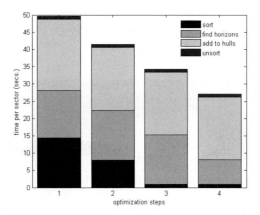

Fig. 4. The improvement in times applying the optimizations 1,2,3 and 4 for a grid of size 2000×2000

4.4   Parallel Computing

An important quality of the proposed horizon algorithm is that it does not include any data dependencies between sectors, which means that parallelism can be easily applied. By using dual core technology, the execution time for any grid is reduced by half (about 18 second per grid), and just a week for the whole territory.

Although it was unnecessary to exploit more parallelism in Andalucia project, we have employed the Grid technology and Globus [6], to reduce even more the computational time, and to extend the possibilities to tackle very large territories. With this technology, and using the computational power of a laboratory with 32 dual-core processors, the horizons for Andalucia has been computed in only 5 hours. Where the grids A and B are of dimensions 268 and 250 respectively.

5   The Accumulated Sky

As mentioned in section 3, each element ø of the set Ω watches the sky from the same perspective, but with different horizons in each case. This means that each element ø of Ω (or a point P of the territory) have a horizon μ_i, with $\mu_i \in \Lambda$. Being N the total number of elements of Λ; as there are so many horizons as points in the territory we obtain $N \times 32400$ operations.

But, for each element ø(*head,tilt*) of Ω, there exist only 90 possible heights for the obstacles that can have in each azimuth α. The received insolation by a given point P from the celestial cells column k of azimuth $\alpha(k)$ could be calculated adding the energies received by all the celestial cells of that column from the zenith (90°) to the correspondent obstacle $\mu_i(k)$. Let's define $E_{real}(\alpha_i, \theta_i, k, \mu_i(k))$ as the insolation in the element ø. The resulting matrix is of dimension 32400×32400, and stores the values of E_{real}.

In order to compute the yearly insolation matrix, we have to redefine our representation of the celestial map, modifying it in such a way that each celestial cell stores the received energy in the whole year, adding the values of the column that goes from the zenith to that cell.

Fig. 5. The insolation of one year computed using the accumulative representation of the celestial matrix

Let's $\sum \Delta$ be the new accumulative celestial matrix, of dimension 32400×32400. Our definitive model for the computation of one year insolation in large territories can be resumed as follows:

– For each minutes of the year ($365 \times 24 \times 60$ minutes)
  • Compute the position of the sun. The resulting celestial matrix is of dimension 90×360)

– For each possible orientation in the territory
  • Compute E_{full} received from all N_s celestial cells in absence of horizons
  • Compute the matrix E_{real} of one year accumulated insolation in each celestial cell

– For each point P of the territory
  • Determine its coordinates (i, j)
  • Compute its horizon μ_i(a matrix of dimension 1×360 points)
  • For each column k of the celestial matrix
    * Compute the accumulate insolation taking into account its corresponding horizon $\mu_{i,k}$
    If the terrestrial cell is horizontal:

$$E(i) = E_{full}(azm_i, elv_i) \qquad (1)$$

    If the terrestrial cell is a mountainous site:

$$E(i) = \sum_{azm_k=0}^{360} E_{real}(head, tilt, azm_k, \mu_{i,k}) \qquad (2)$$

For the analysis of the attenuation of solar radiation passing through clouds, we have extrapolated the behavior of cloudiness in the past to the future. This hypothesis is true if we neglige the effect of global change.

The calculation of the insolation taking into account the cloudiness for a grid of dimension 2000×2000 using our model takes 25 seconds.

6   Conclusions

From a computational point of view, this paper has proposed an efficient model to compute the insolation in large and very large territories. In particular, an algorithm for horizon computation about two times faster that the fastest algorithm provided in literature has been provided. In addition, a parallel implementation of the horizon algorithm in grid computing has been carried out to make possible the study of very large territories, which is useful for different applications, including shading and visibility ones.

Acknowledgment

This work was supported by the Spanish Ministry of Education and Science through grants TIC2003-06623.

References

1. Fu, P., and Rich, P.M.: The solar analyst 1.0 User Manual. WWW document, http://www.fs.fed.us/informs/solaranalyst/solar_analyst_users_guide.pdf, 2000.
2. Marcel, S., Jaroslav, H.: A New GIS-based Solar Radiation Model and Its Application to Photovoltaic Assessments. Transaction in GIS, **8** (2004) 175-190.
3. Page, J., K.: Prediction of Solar Radiation on inclined surfaces. (1899) Springer.
4. Cabral, B., Max, N., Springmeyer, R.: Bidirectional reflection functions from surface bump maps. Computer Graphics, Proceedings of SIGGRAPH'87, **21** (1987) 273-281.
5. James, A., Stewart.: Fast Horizon Computation at all Points of a Terrain with Visibility and Shading Applications. IEEE Trans. on Visualization and Computer Graphics, **4** (1998) 82–93.
6. http://www.globus.org/toolkit/

Non-equilibrium Thermodynamics, Thermomechanics, Geodynamics

Klaus Regenauer-Lieb[1,2], Bruce Hobbs[1], Alison Ord[1], and Dave A. Yuen[3]

[1] CSIRO Exploration and Mining, PO Box 1130, Bentley, WA 6102, Australia
[2] School of Earth and Geographical Sciences, The University of Western Australia, Perth, WA 6009, Australia
[3] Dept. of Geology and Geophysics and Minnesota Supercomputing Institute, Univ. of Minnesota, Minneapolis, MN 55455-0219, U.S.A
{Alison.Ord,Bruce.Hobbs,klaus.regenauer-lieb}@csiro.au,
daveyuen@gmail.com

Abstract. The subject of non-equilibrium thermodynamics is now quite old, dating back to Prigogine, Biot and Truesdell in the 1950's and 1960's. It has had a resurgence in the physical sciences in the past five years largely due to a consolidation of ideas on thermodynamics as a whole and to advances in computer speeds. Non-equilibrium thermodynamics has now advanced to a stage where it forms an umbrella approach to understanding and modelling coupled phenomena in Earth Sciences. Currently, approaches are pioneered independently in geodynamics, seismology, material sciences (solid mechanics), atmospheric and marine sciences (fluid dynamics) and the chemical sciences. We present a first attempt at consolidating some ideas and show a simple example with potential significance for geodynamics, structural geology and seismology.

Keywords: Thermodynamics, Geodynamics, Structural Geology, Slow Earthquakes.

1 Introduction

In this contribution we advance new approaches for the problem of deformation of geological solids and attempt to develop a common framework for the different approaches [1-4]. The application of thermodynamics to the plastic deformation of solids requires special caution in a number of respects [5]. First, plastic deformation is essentially irreversible. It is hence important to define a reference state from which deviations define a driving force. Second, plastic deformation depends sensitively on shear stresses, and it leads to shape changes and not just volume changes. While volumetric strain is easily defined as a state variable, the definition of a non-volumetric strain as a state variable is not straight forward. Third, in plasticity, the number of relevant material parameters may far exceed the number of macroscopically observable relations; the use of energy, volume, etc., as "state variables" then may become meaningless.

Y. Shi et al. (Eds.): ICCS 2007, Part I, LNCS 4487, pp. 62–69, 2007.

In the light of these difficulties it is obvious that thermodynamic approaches to continuum mechanics have, in the past, been replaced by constitutive approaches. These have up to now neglected the basic principles of thermodynamics. However, recently, significant progress has been made for understanding the plastic deformation of soils where all three cautions can be addressed rigorously [6]. Continuum mechanics can be recast and verified on the basis of thermodynamics for isothermal deformation. For soils the influence of the state variable temperature, is thought to be negligible. It is hence possible, to formulate, to first order, a theory of plastic deformation without local gradients in temperature, which must arise naturally through the process of dissipation.

1.1 Isothermal Thermomechanics

Thermomechanics has been pushed forward as a robust theory in the 70s [7] continuing earlier breakthroughs in thermodynamics of chemical systems. However, since then the topic of thermomechanics of continua was not considered forefront of continuum mechanics. Recently, thermomechanics has been rediscovered [8]. Modern thermomechanical approaches assume that there is no heat flow, thus allowing a very concise treatment of the basic underlying principles. Thermomechanics addresses only the second law of thermodynamics as an overarching constraint to deformation. This solves the reference state for irreversible deformation by constraints on energy.

The strain energy rate density is defined as the double dot product (scalar product) of the applied stress σ and the strain rate tensor thus the local power (work rate) of a reference volume [9].

$$\sigma : \dot{\varepsilon} = \dot{\Psi} + \Phi .\tag{1}$$

where Ψ is the Helmholtz free energy function and Φ is the rate of dissipation per unit volume. In the isothermal case the Clausius-Duhem Inequality (second law of thermodynamics) collapses to:

$$\Phi \geq 0 .\tag{2}$$

The Clausius-Duhem Inequality can hence be understood as the isothermal thermal-mechanical stability condition. Note, that the Clausius-Duhem Inequality has a broader meaning in that not only the mechanical dissipation is positive but that the thermal conduction is positive. Combining eqns. (1) and (2) it follows that the thermomechanical equilibrium state is constrained by the condition that the rate of change of the Helmholtz free energy is smaller than the local strain energy rate density [5].

$$\dot{\Psi} \leq \sigma : \dot{\varepsilon} .\tag{3}$$

The equality holds for isentropic (elastic) deformation while plastic deformation causes dissipation.

The second caution of a definition of a strain measure as a differential function of the free energy is solved by expanding the local strain rate energy density into dissipative and non-dissipative microstrains α_k assuming a summation over k of the various microstructural processes

$$\boldsymbol{\sigma} : \dot{\boldsymbol{\varepsilon}} = \frac{\partial \Psi}{\partial \boldsymbol{\varepsilon}} : \dot{\boldsymbol{\varepsilon}} + \frac{\partial \Psi}{\partial \boldsymbol{\alpha}^k} : \dot{\boldsymbol{\alpha}}^k + \frac{\partial \Phi}{\partial \dot{\boldsymbol{\alpha}}^k} : \dot{\boldsymbol{\alpha}}^k .$$ (4)

In the same framework the Cauchy stress tensor can be defined as a partial derivative of the Helmholtz free energy.

$$\boldsymbol{\sigma} \equiv \frac{\partial \Psi}{\partial \boldsymbol{\varepsilon}} ;$$ (5)

and the small strain tensor as the partial derivative of the Gibbs free energy:

$$\boldsymbol{\varepsilon} \equiv \frac{\partial G}{\partial \boldsymbol{\sigma}} .$$ (6)

The partial derivatives of the Helmholtz free energy (equation 4) and the dissipation potential over their microstrains and microstrain rates, respectively, define the recoverable elastic small strain measure and the dissipative small strain measure, thus giving the familiar additive elasto-dissipative strain rate decomposition [10].

$$\dot{\boldsymbol{\varepsilon}} = \frac{\partial^2 G}{\partial \boldsymbol{\sigma}^2} : \dot{\boldsymbol{\sigma}} + \frac{\partial^2 G}{\partial \boldsymbol{\sigma} \partial \boldsymbol{\alpha}^k} : \dot{\boldsymbol{\alpha}}^k = \dot{\boldsymbol{\varepsilon}}^{elastic} + \dot{\boldsymbol{\varepsilon}}^{diss} .$$ (7)

The third caution put forward by Kocks et al. [5] is the caution that the number of necessary material parameters for interpreting the strongly nonlinear system can exceed the number of available equations. For isothermal approximations (e.g. soil) this does not appear to be the case [6]. For the more general application to non-isothermal cases it is a challenge to modern computational approaches. Nowadays, such complex non-linear systems can readily be analyzed by computational methods.

1.2 Non-isothermal Thermomechanics

Temperature needs to be added as a state-variable, therefore we have to consider a full thermodynamic approach where heat Q is allowed to flow into the reference volume. Integrating the strain energy rate density in equation (1) with respect to time, we obtain the mechanical work W done in the volume element. The first law of thermodynamics states that in a closed system the increment of specific internal energy δU of the reference volume is the sum of the increment of heat δQ flowing into the solid and the increment of mechanical work δW done on the solid.

$$\delta U = \delta Q + \delta W .$$ (8)

The internal energy is also defined by:

$$U \equiv \Psi(T, \alpha_j) + sT ,$$ (9)

where the internal energy is now written in terms of entropy s and α_j are state variables other than temperature T. The entropy is defined by:

$$s \equiv -\frac{\partial \Psi}{\partial T} \, , \tag{10}$$

The first law is often expressed in irreversible thermodynamics in terms of the equivalent time derivatives rather than using virtual variations. Taking the material time derivative (in our notation D/Dt) of the entropy we obtain

$$\frac{Ds}{Dt} = -\frac{\partial^2 \Psi}{\partial T^2} \frac{DT}{Dt} - \frac{\partial^2 \Psi}{\partial T \partial \alpha_j} \frac{D\alpha_j}{Dt} \, , \tag{11}$$

where the specific heat c_α is defined as:

$$c_\alpha \equiv -T \frac{\partial^2 \Psi}{\partial T^2} \, . \tag{12}$$

We switch from a unit volume based to a unit mass based framework ($\tilde{\Psi} = \Psi / \rho$), the entropy flux is then related to the heat by:

$$\rho T \frac{Ds}{Dt} = \rho c_\alpha \frac{DT}{Dt} - \rho T \frac{\partial^2 \tilde{\Psi}}{\partial T \partial \alpha_j} \frac{D\alpha_j}{Dt} \, . \tag{13}$$

Considering equations (8-10) and rearranging terms we can now write the heat equation as [11].

$$\rho c_\alpha \frac{DT}{Dt} = \left(\boldsymbol{\sigma} : \dot{\boldsymbol{\varepsilon}} - \rho \frac{\partial \tilde{\Psi}}{\partial \alpha_j} \frac{D\alpha_j}{Dt} \right) + \rho T \frac{\partial^2 \tilde{\Psi}}{\partial T \partial \alpha_j} \frac{D\alpha_j}{Dt} + q_i \, , \tag{14}$$

where the first term described by the bracket on the right side is the mechanical dissipation potential Φ defined in equation (1), the second term is the isentropic thermal-mechanical coupling term and q_i is the heat flux.

This equation is a generalized thermodynamic formulation without specification of the mode of mechanical dissipative and stored microstrain-processes, as well as the particular mechanism heat is transferred. It describes the importance of two important thermal-mechanical feedback effects within the energy equation. The first bracketed term on the right side, the dissipated energy, is also known as the shear heating term, and the second term is the important feedback given by the stored energy. While this equation is generally applicable we will now go on and illustrate the application of the method in a simple example for a specific case study, simplifying what we believe to be the most essential case in geodynamics.

1.3 Simplified Thermodynamic Approach for Geodynamics

In terms of non-dissipative processes producing elastic strain we only consider thermal expansion and elastic deformation:

$$\dot{\boldsymbol{\varepsilon}}^{elastic} = \left(\frac{1+v}{E} \frac{D\tilde{\sigma}'}{Dt} + \frac{v}{E} \frac{Dp}{Dt} + \alpha \frac{DT}{Dt} \delta_{ij} \right). \tag{15}$$

where E is Young's modulus , v is Poisson ratio and and α is the coefficient of thermal expansion. $\tilde{\sigma}'$ is the objective co-rotational stress tensor and δ_{ij} is the Kronecker delta. For the dissipative processes in equation (7) we assume only two basic micro-mechanisms causing macroscopic dissipative strains:

$$\dot{\boldsymbol{\varepsilon}}^{diss} = \left(\dot{\boldsymbol{\varepsilon}}^{pl} \frac{\sigma'}{2\tau} \right)_{plastic} + \left(A\sigma' J_2^{n-1} \exp\left(-\frac{Q}{RT} \right) \right)_{creep}. \tag{16}$$

The plastic yield stress is τ and is here assumed to be linearly dependent on pressure; A and n are power-law material constants, Q is the activation enthalpy and R is the universal gas constant. J_2 is second invariant of the deviatoric stress tensor σ'.

The crucial step for a computational approach producing patterns out of nonlinear thermodynamic feedback processes is common to all *ab-initio* style calculations. Here, the method is applied to homogenized microstrain processes in a continuum approach upscaled for geodynamics. The key is to solve the triplet of equilibrium equations, *continuity*, *momentum* and *heat* in a fully coupled way. Without coupling there is no pattern development. The computational strategy, the convergence requirements, and the time stepping routine are laid out in different paper [12]. Here we will just point out the two key feedback loops appearing in the form of the energy equation (14) which now reads:

$$\rho c_P \frac{DT}{Dt} = \boldsymbol{\sigma} : \dot{\boldsymbol{\varepsilon}}^{diss} + \alpha T \frac{Dp}{Dt} + \rho c_P \kappa \nabla^2 T. \tag{17}$$

Discussing this specific simplification of equation (13) from back to front we now only consider for q_i the phonon part of heat conduction with thermal diffusivity κ , c_P is the specific heat at constant pressure. For the isentropic feedback term we only consider the state variable pressure p including the thermal-elastic feedback loop and in the first term on the right we consider that all dissipated work is released as heat. This is of course an extreme mathematical idealization of a real material. For most materials the shear heating efficiency appears to be around 85% and 95% for large strain [13]. The remaining 5-15% clearly require incorporation of additional feedback loops in the expansion of the second term of equation (14), thereby diminishing the energy available for shear heating feedback by the same amount [1, 3]. This feedback is important for very shallow crustal conditions. We only described deeper crustal conditions. A surprising dynamics, not recorded in classical quasi-static continuum mechanical approaches, is based on the physics of two feedback loops.

2 Feedback Loops

With this formulation we have encapsulated two basic thermodynamics feedback processes into the energy equation, based on the two state variables temperature and pressure, these are: 1) Thermal expansion feedback dependent on elasto-plastic (semi-brittle, *p*-dependent) displacements; 2) Shear heating feedback dependent on elasto-visco-plastic (ductile, *T*-dependent) displacements [14].

3 Model

The model setup is a very basic one. We consider an isothermal quartz-feldspar composite slab with random thermal perturbations and extend it by a factor of 4 from 3.3 x 12 km to 13.2 x 3 km. This is considered to be a generic slice of a continental crust, where the individual *p-T* conditions correspond to a particular depth. Since the pressure and temperature increases in the crust with depth, the pressure sensitive brittle feedback mechanism is suppressed for higher starting model temperatures. Random temperature perturbations with a maximum amplitude of 70 K are assumed at the start of the extension and they diffuse during deformation and self-organize into a characteristic pattern controlled by the two feedback loops. Owing to the pure shear boundary conditions, the feldspar and quartz layers are expected to assume a shape shown in Figure 1, either through plastic deformation or viscous creep or both, depending on the base temperature/pressure assumed at the start of the model run. No internal structure is expected in the absence of thermal-mechanical feedback.

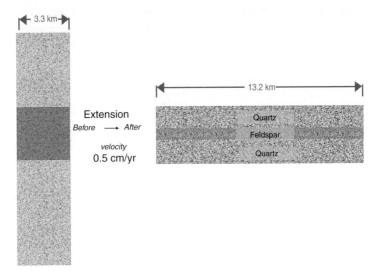

Fig. 1. Pure shear extension of an isothermal composite piece of continental crust. Without thermal-mechanical feedback random thermal perturbation have no effect and both feldspar and quartz are extended homogenously as shown.

4 Results

The resulting patterns deviate significantly from the expected patterns for a standard continuum mechanics solution. Figure 2 clearly shows three significantly different mechanical regimes.

We have shown here an example of the new approach originally designed for geodynamic simulations [14]. The rich spatial patterns developing out of an originally unpatterned state are typical for thermodynamic feedback. The 500K case, in particular, is accompanied by a dynamic pattern of creep/plastic slip events, which could throw a light on the dynamic events underlying the source region of major earthquakes such as the recent Sumatra event [15].

430 Kelvin Extension

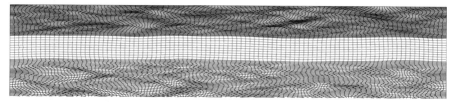

500 Kelvin Extension

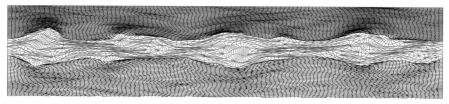

530 Kelvin Extension

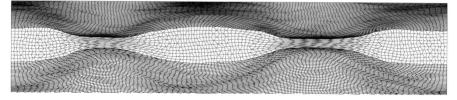

Fig. 2. Only the final pattern after extension is shown. With feedback random temperature perturbation have a significant effect on the deformation. Three mechanical regimes are identified. At 430K brittle, thermal-expansion feedback dominates, at 530K ductile shear-heating feedback dominates. Mixed-mode brittle/ductile feedback produces the richest pattern at around 500K.

5 Summary and Outlook

Computer sciences are now allowing us to tackle problems that have previously been out of reach of exact analytical approaches. This revolutionizes new non-equilibrium thermodynamic approaches for coupling chemistry and mechanics (both fluid and solid). We have shown here a promising research path with the potential to bridge seismology, geodynamics, geomechanics and materials sciences in the near future.

Acknowledgments. *Pmd*CRC*, CSIRO, UWA , DAY acknowledges support from ITR and CSEDI programs of National Science Foundation.

References

1. Y. Ben-Zion, V. Lyakhovsky: Analysis of Aftershocks in a Lithospheric Model with Seismogenic Zone Governed by Damage Rheology, Geophysical Journal International 165, 197 (2006).
2. I. F. Collins: The concept of stored plastic work or frozen elastic energy in soil mechnics, Geotechnique 55, 373 (2005).
3. D. Bercovici, Y. Ricard: Energetics of a two-phase model of lithospheric damage, shear localization and plate-boundary formation, Geophysical Journal International 152, 581 (2003).
4. B. Hobbs, K. Regenauer-Lieb, A. Ord: The thermodynamics of folding in the middle to lower crust, Geology 35, 175 (2007).
5. U. F. Kocks, A. S. Argon, M. F. Ashby, *Thermodynamics and kinetics of slip* (Pergamon Press, Oxford, 1975), pp. 293.
6. I. F. Collins: Elastic/plastic models for soils and sands, International Journal of Mechanical Sciences 47, 493 (2005).
7. H. Ziegler, *An Introducton to thermomechanics* (North-Holland Publishing Company, Amsterdam, ed. 2nd Edition, 1983), pp. 358.
8. I. F. Collins, G. T. Houlsby: Application of thermomechanical principles to the modelling of geotechnical materials, Proceedings - Royal Society of London, A 453 1964, 1975 (1997).
9. I. F. Collins, P. A. Kelly: A thermomechanical analysis of a family of soil models, Geotechnique 52, 507 (2002).
10. S. Nemat-Nasser: Decomposition of strain measures and their rates in finite deformation elasto-plasticity, International Journal of Solids and Structures 15, 155 (1979).
11. A. Chrysochoos, R. Peyroux: Modelisation numerique des couplages en thermomecanique des solides, Revue europeene des element finis 6, 673 (1997).
12. K. Regenauer-Lieb, D. A. Yuen: Positive feedback of interacting ductile faults from coupling of equation of state, rheology and thermal-mechanics, Physics of Earth and Planetary Interiors 142, 113 (2004).
13. A. Chrysochoos, O. Maisonneuve, G. Martin, H. Caumon, J. C. Chezeaux: Plastic and dissipated work and stored energy, Nuclear Engineering and Design 114, 323 (1989).
14. K. Regenauer-Lieb, R. Weinberg, G. Rosenbaum: The effect of energy feedbacks on continental strength, Nature 442, 67 (2006).
15. K. Regenauer-Lieb, D. Yuen: Quartz Rheology and short time-scale crustal instabilities, Pure and Applied Geophysics 163 (1915-1932, 2006).

A Finite Element Model for Epidermal Wound Healing

F.J. Vermolen[1] and J.A. Adam[2]

[1] Delft University of Technology, Delft Institute of Applied Mathematics,
Mekelweg 4, 2628 CD Delft, The Netherlands
[2] Old Dominion University, Department of Mathematics and Statistics,
Norfolk, VA 23529-0077, USA
f.j.vermolen@tudelft.nl

Abstract. A finite element model for epidermal wound healing is proposed. The model takes into account the sequential steps of angiogenesis (neo-vascularization) and wound contraction (the actual healing of a wound). An innovation in the present study is the combination of both partially overlapping processes, yielding novel insights into the process of wound healing. The models consist of nonlinearly coupled diffusion-reaction equations, in which transport of oxygen, growth factors and epidermal cells and mitosis are taken into account.

Keywords: finite elements, wound contraction, neo-vascularization.

1 Introduction

Models for bone regeneration and wound healing often rely on experiments on animals. When a wound occurs, blood vessels are cut and blood enters the wound. Due to coagulation of blood inside the wound, the wound is temporarily closed and as a result the blood vessels adjacent to the wound are also closed. In due course contaminants will be removed from the wounded area and the blood vessel network will be restored, but initially due to insufficient blood supply, there will be a low concentration of nutrients which are necessary for cell division and wound healing. Wound healing, if it occurs, proceeds by a combination of several processes: chemotaxis (movement of cells induced by a concentration gradient), neo-vascularization, synthesis of extracellular matrix proteins, and scar modeling. Previous models incorporate cell mitosis, cell proliferation, cell death, capillary formation, oxygen supply and growth factor generation, including studies by Sherratt *et al.* [1], Fillion *et al.* [2], Maggelakis [3] and Gaffney *et al.* [4], to mention just a few. A recent work devoted to mathematical biology has been written by Murray [5], in which the issue of wound healing is also treated. The wound healing process can roughly be divided into the following partially overlapping consecutive stages:

1. Formation of a blood cloth on the wound to prevent undesired chemicals from entering the tissue of the organism (blood clothing/inflammatory phase);

Y. Shi et al. (Eds.): ICCS 2007, Part I, LNCS 4487, pp. 70–77, 2007.

2. Formation of a network of tiny arteries (capillaries) for blood flow to supply the necessary nutrients for wound healing;
3. Division and growth of epidermal cells (mitosis), taking place during the actual healing of the wound (remodeling phase).

A good supply of nutrients and constituents is necessary for the process of cell division and cellular growth. For this purpose tiny capillaries are formed during the process of angiogenesis. Some models for capillary network formation have been proposed by Gaffney et al. [4] and Maggelakis [6].

Wound contraction is modeled by Sherratt & Murray [1] who consider cell division and growth factor generation simultaneously for healing of epidermal wounds. Their model consists of a system of reaction-diffusion equations. Among many others an alternative model, based on an active layer at the wound edge, for wound contraction was proposed by Adam et al. [7] and for more references, see [8]. Until now, the conditions for wound healing were only analyzed for geometries where only one spatial co-ordinate could be used. As far as we know, in all the mathematical studies mentioned before, either neo-vascularization or wound contraction is modeled. Hence, these processes are considered to be sequential. However, according to the medical literature [9], these sequential processes partially overlap. Therefore, this study aims at the creation of a model in which the stages of neo-vascularization and wound contraction are treated as sequential, but partially overlapping processes. Hence, the key innovations in the present study are the following: Finite Element solutions for arbitrary wound geometries; and a combination of models for angiogenesis and wound contraction as partially overlapping consecutive processes.

The present paper is organized as follows. First, an existing model for angiogenesis (neo-vascularization) with an extension is described. This is followed by a treatment of a model for wound contraction and a proposition of how to combine these models. We continue with a description of the numerical method to solve the resulting nonlinearly coupled system of partial differential equations and present some numerical experiments, followed by some concluding remarks.

2   The Mathematical Model

In this section a model for epidermal wound healing is presented. The model consists of neo-vascularization and wound contraction and the construction of the model relies on a combination of the ideas developed by Maggelakis [6], Gaffney et al. [4] and Sherratt and Murray [1].

2.1   Neo-vascularization

Maggelakis constructed this model on angiogenesis in 2004 [6]. It is assumed that the tips of the capillaries act as the only sources for oxygen supply. The domain of computation is assumed to be divided into the wound region Ω_w and the undamaged region Ω_u, that is $\Omega = \Omega_w \cup \Omega_u \cup (\overline{\Omega}_w \cap \overline{\Omega}_u)$ and Ω_w is embedded within Ω_u. The closures of Ω_w and Ω_u respectively are denoted by $\overline{\Omega}_w$ and $\overline{\Omega}_u$.

Let n and c_o respectively denote the capillary density and oxygen concentration and let them be functions of time t and space within the domain of computation Ω; then a mass balance results into the following partial differential equation (PDE):

$$\frac{\partial c_o}{\partial t} = D_o \text{ div grad } c_o + \lambda_n n - \lambda_o c_o, \text{ for } \mathbf{x} \in \Omega_w,$$

$$\text{subject to } c_o(\mathbf{x}, t) = c_i \text{ for } \mathbf{x} \in \overline{\Omega}_w \cap \overline{\Omega}_u. \tag{1}$$

The initial oxygen concentration is assumed to be zero in Ω_w. The above equation is based on the assumption that the oxygen supply and oxygen consumption depend linearly on the capillary density and oxygen concentration respectively. When the oxygen level is low, macrophages appear at the wound site. The macrophages release MDGF's (macrophage derived growth factors) which are hormones that enhance the growth of blood vessels and collagen deposition and hence help to restore the capillaries that provide the skin with the necessary nutrients and oxygen for cell division needed for wound contraction. An assumption in the model is that macrophages are produced if the oxygen level is below a threshold value, say c_θ. The production rate, Q, is assumed to depend linearly on the lack of oxygen, that is

$$Q = Q(c_0) = \begin{cases} 1 - \dfrac{c_0}{c_\theta}, & \text{if } c_0 < c_\theta, \\ 0, & \text{if } c_0 \geq c_\theta. \end{cases} \tag{2}$$

The mass balance of MDGF's results into the following PDE's in the wound region Ω_w and out of the wound region Ω_u:

$$\frac{\partial c_m}{\partial t} = D_m \text{ div grad } c_m + \lambda_m Q(c_o) - \lambda c_m, \text{ for } \mathbf{x} \in \Omega_w,$$

$$\frac{\partial c_m}{\partial t} = D_m \text{ div grad } c_m - \lambda_c c_m, \text{ for } \mathbf{x} \in \Omega_u. \tag{3}$$

The initial MDGF concentration, denoted by c_m, is assumed to be zero in the entire domain of computation Ω and a homogeneous Neumann boundary condition is used. The capillary density is assumed to grow as a result of the MDGF's in a logistic manner, that is

$$\frac{\partial n}{\partial t} = D_n \text{ div grad } n + \mu c_m n(1 - \frac{n}{n_m}), \text{ for } \mathbf{x} \in \Omega, \tag{4}$$

where n_m denotes the maximum capillary density. The capillary density is assumed to satisfy the following initial condition

$$n(\mathbf{x}, 0) = \begin{cases} 0, & \text{for } \mathbf{x} \in \Omega_w, \\ n_m, & \text{for } \mathbf{x} \in \Omega_u. \end{cases} \tag{5}$$

A homogeneous Neumann boundary condition is used for n. We assume the capillary tips to migrate via a random walk. This random walk was not incorporated into Maggelakis' model but it was extended with a bias in Gaffney's model.

Further, Gaffney *et al.* [4] distinguish between the capillaries and the capillary tips. The bias is neglected in this paper but it will be dealt with in the future work to be presented in the talk. But it is reasonable since away from the wound the capillary network is assumed to be undamaged. Further, Maggelakis sets in a nonzero artificial starting value for the capillary density to have the capillary density to increase up to the equilibrium value. Analytic solutions are considered in [6] for the above equations under the assumptions that the diffusion of oxygen and growth factors is instantaneous and that $D_n = 0$. The induced growth of MDGF's due to a lack of oxygen can be classified as a negative feedback mechanism. In the presentation we will extend this model with model hypotheses due to Gaffney *et al.* [4] who incorporate the interaction of the capillary tip density with the endothelial cell density and use a biased random walk resulting into nonlinear cross diffusion.

2.2   Wound Contraction

The mechanism for wound contraction is cell division and growth (mitosis). This mechanism is triggered by a complicated system of growth factors. In the present model we use the simplification that wound contraction is influenced by one generic growth factor only. The growth factor concentration influences the production of epidermal cells. Following Sherratt and Murray [1], we assume that the main source of growth factors is the epidermal cells. Growth factors diffuse through the tissue region and the concentration decays due to reactions with other chemicals present in the tissue. Further, the epidermal cells need the supply of nutrients by the capillaries in order to be able to grow and divide. A constant blood flow through the capillaries is assumed, hence, the mitotic rate can be related to capillary density n. The mitotic rate is assumed to depend on the capillary density in a nonlinear fashion by a function $\phi_p(y)$, given by

$$\phi_p(y) = \frac{y^p}{(1-y)^p + y^p},\tag{6}$$

where p is assumed to be positive. Herewith, the expression of Sherratt and Murray where the accumulation of the epidermal cells is determined by proliferation (diffusive transport), mitosis and cell death, is adjusted to

$$\frac{\partial m}{\partial t} = D_1 \text{ div grad } m + \phi_p(\frac{n}{n_m})s(c)m(2 - \frac{m}{m_0}) - km, \qquad \mathbf{x} \in \Omega_w,\tag{7}$$

subject to $m(\mathbf{x}, t) = \hat{m}$, for $\mathbf{x} \in \overline{\Omega}_w \cap \overline{\Omega}_u$.

in order to depend on the capillary density. In the above equation $m = m(\mathbf{x}, t)$ and $c = c(\mathbf{x}, t)$ respectively denote the epidermal cell density and growth factor concentration. The function $s = s(c)$ is a nonlinear function of the growth concentration describing the mitotic rate, see [5, 1]. For the growth factor accumulation a similar relationship due to diffusive transport, production and decay is obtained with a similar adaptation for the dependence of the capillary density:

$$\frac{\partial c}{\partial t} = D_2 \text{ div grad } c + \phi_p(\frac{n}{n_m})f(m) - \lambda c, \qquad \mathbf{x} \in \Omega_w.\tag{8}$$

subject to $c(\mathbf{x}, t) = \hat{c}$, for $\mathbf{x} \in \overline{\Omega}_w \cap \overline{\Omega}_u$.

As initial conditions for m and c we have $m(\mathbf{x}, 0) = 0$ and $c(\mathbf{x}, 0) = 0$. In the above equation $f(m)$ denotes a nonlinear relation for the growth factor regeneration. In the model due to Sherratt and Murray, two different types of growth factors are considered: 1. activators; and 2. inhibitors, both with their characteristic functions for s and f. We assume that within the wound initially $m = c = 0$ and at the wound edge, $m = m_0$ and $c = c_0$. Hence we want the initial state to be unstable and the boundary conditions to be stable, so that the functions m and c converge to the values of the boundary conditions m_0 and c_0 as $t \to \infty$. In other words, the unwounded state is stable with respect to small perturbations, whereas the wounded state is unstable. Note that the capillary density from the neo-vascularization model described in the previous subsection is needed. Furthermore, $\lim_{p \to \infty} \phi_p(y) = u_{1/2}(y)$, which is in the class of Heaviside functions with the jump at $y = 1/2$. The behavior is motivated by the intuitively obvious notion that the mitotic rate and production rate of the growth factors increase with the capillary density. The rest of the equations remains the same. The present coupling between the two models is an innovation and will be evaluated during the presentation. Further, the model will be extended to include the interaction of the capillary density with the endothelial cell density, which is the building block of the capillaries. A further future implementation is the feedback from the contraction model to the neo-vascularization model.

3   Numerical Method

The numerical method is explained for the wound contraction model since this model contains most of the difficulties. To solve the system of partial differential equations a finite element strategy is used. In the presentation the solution and a discussion of the entire set of model equations will be considered. In this section we present the solution method for our preliminary results consisting of the solution of the equations of Sherratt and Murray. Piecewise linear basis functions are used for the triangular elements with a Newton-Cotes integration, which makes the mass matrix diagonal. For the time integration method, we use the following IMEX method:

$$
\begin{aligned}
M\frac{\underline{m}^{k+1}}{\Delta t} &= M\frac{\underline{m}^k}{\Delta t} + D_1 S \underline{m}^{k+1} + M_1(\underline{m}^k, \underline{c}^k)\underline{m}^{k+1}, \\
M\frac{\underline{c}^{k+1}}{\Delta t} &= M\frac{\underline{c}^k}{\Delta t} + D_2 S \underline{c}^{k+1} - \lambda M \underline{c}^{k+1} + \underline{b}^k.
\end{aligned}
\tag{9}
$$

Here M and S respectively denote the mass and stiffness matrices. By the use of the Newton-Cotes integration rule, the element mass matrix for M_1 becomes for an element Ω_ε surrounded by nodes i, j and p:

$$
M_1^{\Omega_\varepsilon} = \frac{|\Delta|}{6} \operatorname{diag}(s(c_\alpha^k)(2 - n_\alpha^k)), \text{ for } \alpha \in \{i, j, p\},
\tag{10}
$$

where $\frac{|\Delta|}{2}$ represents the area of the triangular element. The α-th component of the element vector for $\underline{b}$ becomes

$$b_\alpha^{\Omega_\varepsilon} = \frac{|\Delta|}{6}(f(n_\alpha^k)), \text{ for } \alpha \in \{i, j, p\}. \tag{11}$$

These functions have been implemented within the package SEPRAN.

4   Preliminary Results

Since the thickness of human skin is in the order of a millimeter, the model can be considered as two-dimensional. First, we show some finite element results for a circular wound with radius 1 of the neo-vascularization model. Our simulations indicate that the oxygen and growth factor concentrations reach the steady-state solution rapidly. This confirms the assumption under which Maggelakis found the analytic solutions. The capillary tip density exhibits a transient behavior. Due to diffusion, the tip density decreases for a while outside the wound region, whereas it increases at all stages within the wound region. As time proceeds, it converges towards its equilibrium value. The results are not shown in the present paper.

Subsequently, we consider the actual wound contraction time, where we assume the neo-vascularization process to have taken place entirely, as a function of the aspect ratio of rectangular and elliptic wounds. The aspect ratio is defined by the ratio of the length and the width for a rectangular wound and by the ratio of the longer and shorter axis for elliptic wounds. Following Sherratt & Murray [1] the wound edge is identified by the level curve $n = 0.8$. In Figure 1 we show the healing time as a function of the aspect ratio for ellipsoidal and circular wounds. It can be seen that the width of the wound is the most important factor to determine the healing time. However, the relation is not linear, the decrease of the healing time as a function of the aspect ratio is much less than for the linear case.

Subsequently, when the models are combined, the full system of PDE's is solved, and the influence of the oxygen production on the healing rate is shown in Figure 2. For larger values of λ_o wound healing slows down. This is caused by the decrease of the MDGF growth factor concentration as more oxygen is supplied. The decrease of MDGF growth factors, decreases the speed of capillary formation and hence the supply of nutrients needed for cell division in the wound contraction process deteriorates. This results into a delay of mitosis, and hence the wound contraction slows down. Finally, the influence of the capillary density on the healing time is investigated. The influence is quantified by the power p in the function $\phi_p(y)$. For $0 < p < 1$ the influence is large for $n/n_m < 0.5$ and small elsewhere. For $p > 1$, the behavior is opposite. The calculations, not shown here, result into a strange behavior for $0 < p \le 1$ that the wound heals from the edge at the early stages and starts healing from the wound center at the later stages. This behavior is not realistic. Note that $p = 1$ corresponds to a linear coupling. For $p > 1$ the wound only heals from the edge, which is more realistic

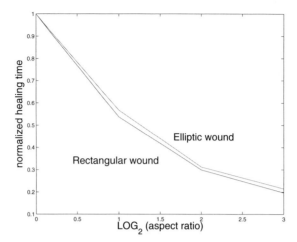

Fig. 1. Healing times as a function of the aspect ratio of rectangular and ellipsoidal wounds with the same area

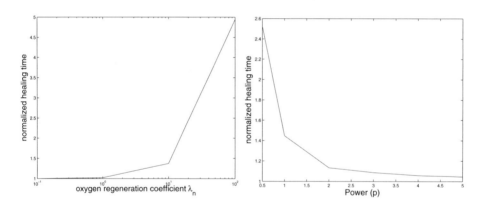

Fig. 2. Circular wound:Left: The influence of oxygen production on the healing time ($p = 5$); Right: The influence of the power p in ϕ on the healing time ($\lambda_o = 0.1$)

from an intuitive point of view. Therefore, the expectation is that $p > 1$ reflects the right behavior. For the input values we used the same values as Sherratt and Murray [1] and Maggelakis [6].

5   Conclusions

Some preliminary results for a wound healing model have been presented. An innovation in this paper is the assumption to allow overlapping between the processes of neo-vascularization and wound contraction. During the presentation further extensions to the neo-vascularization model will be discussed and the coupling between the two models will be analyzed in more detail. In this paper the nonlinear impact of the aspect ratio of the wound on the healing time is also demonstrated. In future work (to be presented in the talk) the model will be extended with tip-sprout interaction (anastomosis) and a biased random walk, resulting into cross diffusion.

References

1. Sherratt, J.A., Murray, J.D.: *Mathematical analysis of a basic model for epidermal wound healing*, J. Math. Biol., **29** (1991), 389–404.
2. Filion, J., Popel, A.P.: *A reaction diffusion model of basic fibroblast growth factor interactions with cell surface receptors*, Annals of Biomed. Engng., **32(5)** (2004), 645–663.
3. Maggelakis, S.A.: *A mathematical model for tissue replacement during epidermal wound healing*, Appl. Math. Modell., **27(3)** (2003), 189–196.
4. Gaffney, E.A., Pugh, K., Maini, P.K.: *Investigating a simple model for cutaneous wound healing angiogenesis*, J. Math. Biol,, *45 (4)* (2002), 337–374.
5. Murray, J.D.: *Mathematical biology II: spatial models and biomedical applications*, Springer-Verlag, New York, 2004.
6. Maggelakis, S.A.: *Modeling the role of angiogenesis in epidermal wound healing*, Discr. and Cont. Sys., *4* (2004), 267–273.
7. Adam, J.A.: *A simplified model of wound healing (with particular reference to the critical size defect)*, Math. and Comput. Modell., *30* (1999), 23–32.
8. Vermolen, F.J., van Rossum, W.G., Javierre, E., Adam, J.A.: *Modeling of self-healing of skin tissue, in: Self-healing materials*, Springer-Verlag, to appear, 2007.
9. Stadelman, W.K., Digenis, A.G., Tobin, G.R.: *Physiology and healing dynamics of chronic cutaneous wounds*, The American Journal of Surgery, *176(2)* (1997), 265–385.

Predicting Binding Sites of Hepatitis C Virus Complexes Using Residue Binding Propensity and Sequence Entropy

Guang-Zheng Zhang*, Chirag Nepal, and Kyungsook Han**

School of Computer Science & Engineering, Inha Univ., Incheon 402751, Korea
aokunzhang@126.com, chirag@inhaian.net, khan@inha.ac.kr

Abstract. Hepatitis C virus (HCV) remains a dangerous health problem for the reason that the mechanism of hepatocyte infection is still unknown. Hence, much attention has been put on the problem of interaction between HCV and human proteins. However, the research is still standing at the beginning point due to the lack of structure information of HCV and human proteins. We extracted the most representative set of 18 complexes all known HCV protein complexes involving human proteins, and computed the binding propensity of each residue and sequence entropy of each HCV protein. analyzed the most representative set of 18 complexes. Using a radial basis function neural network (RBFNN), we predicted binding sites with an overall sensitivity of 77%. The approach will help understand the interaction between HCV and human proteins.

Keywords: hepatitis C virus, radial basis function neural networks, binding site prediction, binding propensity, sequence entropy.

1 Introduction

Since the discovery of hepatitis C virus (HCV) in 1986, much attention has been focused on the research of interactions between HCV and human proteins. However, identifying the patterns of interactions between these proteins is hindered by a lack of structure information on these proteins. Hence, identification of the binding sites in HCV proteins, one of the most important median steps towards the interaction between HCV and human complex, has been developed for the purpose of understanding the HCV interaction mechanism [1].

However, the exact location of the binding sites within complexes, which can not always be addressed with X-ray crystallography, NMR or other experimental approaches, poses a problem. As an extension of our previous work [2], we have developed a radial basis function neural network (RBFNN) model for predicting the binding sites of HCV complexes. The RBFNN model is used to capture the relations between HCV primary sequence features and the known binding sites.

---

* This work is done while G.-Z. Zhang is on leave from the Red Star Institution, Hefei 230031, Anhui, China.
** Corresponding author.

Y. Shi et al. (Eds.): ICCS 2007, Part I, LNCS 4487, pp. 78–85, 2007.

Table 1. Dataset of non-redundant HCV complexes with multiple chains

PDB code	Chains	Description	Sequence length	[♮]Number of known binding residues
1CSJ	A/B	Crystal str. of the RNA polymerase of HCV	531/531	32/33
1NHU	A/B	HCV RNA Polymerase complex with analogue inhibitor	578/578	12/12
[♭]1A1R	A/B	HCV NS3 protease domain: NS4A peptide complex	198/198	43/53
	C/D		23/23	
1C2P	A/B	Hepatitis C virus NS5B RNA-Dependent ...	576/576	14/14
1HEI	A/B	Str. of the hepatitis C virus RNA Helicase domain	451/451	38/196
[♭]1N64	H/L	Immunodominant antigenic site on HCV protein ...	220/218	56/117
	P		16	
1NHV	A/B	HCV RNA Polymerase in complex with ...	578/578	12/12
1ZH1	A/B	Hepatitis C virus NS5B RNA-Dependent genotype 2a	163/163	24/19
2AWZ	A/B	HCV NS5b RNA Polymerase in complex inhibitor (5h)	580/580	20/17
2AX1	A/B	HCV NS5b RNA Polymerase in complex inhibitor (5ee)	580/580	20/18
2D3U	A/B	X-ray crystal str of HCV RNA polymerase in complex	570/570	11/11
2D41	A/B	X-ray crystal str of HCV RNA dependent complex	570/570	12/12
2GC8	A/B	Str. of HCV NS5b Polymerase	578/578	7/8
2HWH	A/B	HCV NS5B allosteric inhibitor complex	578/576	14/14
2I1R	A/B	Novel Thiazolones as HCV NS5B Polymerase	576/576	14/13
[♭]1RTL	A/B	Str. of HCV NS3 protease domain	200/200	58/45
	C/D		23/23	
[♭]1NS3	A/B	Str. of HCV protease (BK strain)	186/186	64/66
	C/D		14/14	
[♭]2A4R	A/C	HCV NS3 Protease Domain	200/200	58/84
	B/D		23/23	

[♭] To compute the numbers of binding sites, we considered the 2 longest chains in each protein complex, for the reason that chains C and D of 1A1R, 1RTL, 1NS3, chain P of 1N64, and chains B and D of 2A4R are very short sequences, and have limited influence on binding site identification.

[♮] This column gives the numbers of binding sites (identified by ASA method) in the 2 longest chains of HCV complexes.

Conformational parameters such as sequence entropy, residue binding propensity, evolutionary distance, secondary structure, as well as residue biological-physical properties, are coupled as input feature vectors for the RBFNN model for locating binding residues. The model was trained and tested using a dataset of HCV complexes with multiple chains, which were retrieved from the protein data bank (PDB) [3]. Simulation results indicate that our proposed method can successfully predict binding sites within HCV complexes.

2   Materials

We extracted 64 HCV complexes with known structures from PDB, and removed the proteins with identity of 90% or higher using the program PSI-BLAST [4]. We then selected the protein complexes with multiple chains as the representative set of HCV protein complexes. 18 structurally non-redundant HCV complexes with multiple chains were left for our study of binding site prediction, as shown in Table 1.

The accessible surface area (ASA) method [5] was adopted to compute binding residues within the HCV complexes. Specifically, in a given HCV protein sequence, we calculated the ASA for each residue both in the unbound molecule

Table 2. Binding residues in the HCV complexes

HCV ID	A	R	N	D	C	Q	E	G	H	I	L	K	M	F	P	S	T	W	Y	V	SUM
1CSJ	4	4	4	4	2	2	6		2	2	2	4			3	8	4		4	2	65
1NHU		6	2	4		2	2				4	2								2	24
1A1R	8	8		2	3	6	4	5	2	5	7			2	9	6	14	2	3	10	96
1C2P		6	2	6		2	2		2		4	2								2	28
1HEI	16	11	4	17	3	10	18	21	9	6	15	14	6	5	16	13	25	1	4	20	234
1N64	9	7	10	7	1	9	11	4		3	7	18	2	6	14	26	19	4	8	8	173
1NHV		6	2	4		2	2				4	2								2	24
1ZH1	5	3	2		4	2	1		2			2		4	5	4	4	2	3		43
2AWZ		3		2		4	3	9			5	5	1	1	2	2					37
2AX1	1	4		2		4	2	8		1	5	5	1	1	2	2					38
2D3U		4	2	4		2	2				4	2								2	22
1NS3	11	10	1	6	5	11	4	10	2	4	13	5		2	4	9	13	2	4	14	130
2D41		6	2	4		2	2				4	2								2	24
2GC8	2							1			2	6				2	2				15
2HWH		5	2	6		2	2		2		4	3								2	28
1RTL	12	8		4	3	6	4	5	2	6	7			2	11	5	13	2	3	10	103
2I1R		5	2	6		2	2		2		4	2								2	27
2A4R	13	14	2	9	3	9	6	8	2	6	7	3	2		14	8	17	1	6	12	142

and in the complex using the program GETAREA 1.1 [6] with a probe sphere of radius 1.4 Å. An amino acid is defined to be a *binding residue* if its ASA in the complex is less than that in the monomer by at least 1 Å² [7]. In this way, we computed the number of binding residues for each HCV complex (Table 2).

3   Methods

Due to the simple topological structure and powerful nonlinear mapping ability, the radial basis function neural network (RBFNN) model has been successfully adopted to solve nonlinear classification problems with multi-parameters and multi-models, including protein inter-residue contact map prediction [8, 2]. We adopted the RBFNN model, described in our previous study [2], to address the binding sites prediction problem.

3.1   Sequence Entropy

The importance of different residue for maintaining structure stability and particular function of a protein can usually be inferred from the alignment of that protein and its homologies. These information could provide some useful guidelines for analyzing and predicting HCV-human protein binding sites. In the present study, we used the normalized sequence entropy score [11] to measure the variability at a particular position of a given HCV protein sequence. Specifically, this sequence entropy is defined by the following equation:

$$S_{column(j)} = -\sum_{k=1}^{20} P_i ln P_i \times \frac{1}{lnK} \qquad (1)$$

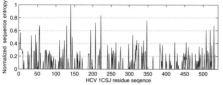
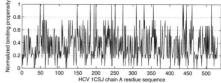

(a) Normalized sequence entropy   (b) Normalized residue binding propensity

Fig. 1. Distributions of sequence entropy and residue binding propensity of chain A of HCV protein 1CSJ. (a) Normalized sequence entropy of residues along chain A of 1CSJ, (b) Normalized residue propensity of chain A of 1CSJ.

where $P(i) = \frac{n_i}{N}$ is the fractional frequency of type i residue in the jth column, $N = \sum_{i=1}^{20}(n_i)$ is the total occurrences of all residues in the column. A low value of sequence entropy, $S_{column(j)}$, implies that position j has been subjected to relatively higher evolutionary conservation than other positions in the same alignment having a higher sequence entropy.

Considering HCV protein 1CSJ as an example, its 14 homologous sequences (PDB ID: 2D3U, 2BRK, 1QUV, 1NHU, 2AWZ, 1OS5, 2I1R, 2HAI, 1GX5, 1GX6, 1C2P, 1NB4, 1YVE, and 1YUY) were retrieved from the PDB by using the program BLASTP [12]. Then, its sequence entropy was calculated from the alignment of these homologies by the help of Bioinformatics Toolbox of MATLAB 7.1, which is illustrated in Fig.1(a).

3.2   Binding Propensity

Several previous studies demonstrated that some critical amino acids, such as structural or functional conserved residues, contribute significantly to protein binding segments, as well as to protein-protein interaction [13, 14, 15]. This suggests that some particular residues may have higher propensities than others in protein-protein binding interactions, and these propensities can in some degree be used to improve binding site prediction. The frequencies of residues in the binding sites of HCV complexes can be used to derive the residue propensities. The overall propensities of the 20 amino acids were computed based on the occurrences of each residue, as shown in Table 2. Briefly, for a particular type residue R_i, the binding propensity $P_{bind}(R_i)$ is defined as:

$$P_{bind}(R_i) = \frac{f_{bind}^{R_i}}{f_{whole}^{R_i}} \tag{2}$$

$$f_{bind}^{R_i} = \frac{O_{bind}(R_i)}{\sum_{i=1}^{20} O_{bind}(R_i)} \tag{3}$$

$$f_{whole}^{R_i} = \frac{N(R_i)}{\sum_{i=1}^{20} N(R_i)} \tag{4}$$

Table 3. Binding propensities of residue

Residue Type	Num. of Residues	$f_{whole}^{r_i}$	Num. of Binding Residues	$f_{bind}^{r_i}$	Binding Propensity $P_{bind}(r_i)$
A	1361	5.01	81	6.51	1.29
R	945	3.49	110	8.84	2.53
N	475	1.76	37	2.97	1.69
D	691	2.55	87	6.99	2.74
C	526	1.94	24	1.93	0.99
Q	501	1.85	77	6.18	3.34
E	659	2.43	73	5.86	2.41
G	989	3.65	71	5.70	1.56
H	656	2.42	27	2.17	0.89
I	356	1.31	33	2.65	2.02
L	1481	5.47	98	7.87	1.44
K	797	2.94	77	6.18	2.10
M	257	0.95	12	0.96	1.02
F	390	1.44	23	1.85	1.28
P	832	3.07	80	6.42	2.09
S	1304	4.82	85	6.83	1.42
T	1121	4.14	111	8.92	2.15
W	236	0.87	14	1.12	1.29
Y	605	2.24	35	2.81	1.26
V	1130	4.18	90	7.22	1.73
SUM	27065	100	1245	100	

where $f_{bind}^{R_i}$ is the fraction of residue R_i in the binding sites, $O_{bind}(R_i)$ is the number of observed occurrence of residue R_i, $N(R_i)$ denotes the number of r_i residue in the whole HCV set. Residues with a higher propensity indicate that the residue occurs more frequently in the binding surface in comparison with other residues with lower propensity. Table 3 displays the detailed binding propensities of 20 different types of residue. For the purpose of encoding the binding propensity to the input feature vector, we normalized all the computed propensities. In this manner, all the 20 residues' propensities lie in the area of $[0, 1]$. Detailed distribution of residue binding propensity along the chain A of HCV 1CSJ was demonstrated in Fig.1(b).

3.3   Encoding Scheme

Conformational features such as sequence entropy, binding propensity, residue type, residue classification, and secondary structure, were integrated to construct the input feature vector of our RBFNN model for each residue. Specifically, sequence entropy was represented by a 5-bit vector, i.e., 00001 denotes the sequence entropy which value lies in the region of $[0, 0.2)$, 00010 denotes the sequence entropy in the range of $[0.2, 0.4)$, $\cdots\cdots$, 10000 denotes the region of $[0.8, 1]$. Furthermore, binding propensity is also represented by a 5-bit vector, which is similar to that of sequence entropy. The detailed encoding scheme of the other three conformational attributes, residue type, residue classification and secondary structure, can be found in our previous study [2]. Hence, a given residue is represented by a 39-bit (5+5+20+6+3) vector. Moreover, when we attempt to determine whether a particular residue is in a binding site, the sliding windows method [16,17] was adopted to construct an input feature vector of our RBFNN predictor, to consider the contribution of sequence adjacent residues.

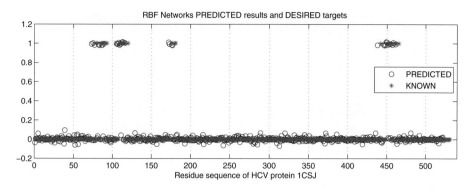

Fig. 2. Predicted and known targets of chain_A within HCV protein sequence 1CSJ, blue 'o' denotes the predicted binding sites, while red '*' is the known binding sites

Table 4. Detailed predicted results of the proposed RBFNN model

PDB code	Correctly predicted numbers		Incorrectly predicted numbers		Specificity	Sensitivity
	T_p	F_p	T_n	F_n		
1CSJ	47	959	38	18	3.81%	76.17%
1NHU	18	1106	26	6	2.30%	77.51%
2AX1	29	1104	18	9	1.60%	78.03%
Average					3%	77%

4   Results and Discussion

Here we take the HCV protein sequence, 1CSJ, as an example to illustrate the result. The predicted scores of 531 amino acid residues along the chain A of 1CSJ are shown in Fig.2. The predicted scores of most residues lie in the area of [-0.15, 0.15], and are labeled as non-binding sites. Roughly, the predicted binding residues have scores in the range of [0.97, 1.05], and all the predicted binding sites were located in the known 4 binding segments successfully. Fig.3 presents the predicted and known binding sites for HCV 1NHU chain A and 2AX1 chain A. The results indicate that our proposed method and RBFNN predictor can catch the binding sites successfully.

To evaluate the performance of the RBFNN model, 3 randomly selected HCV complexes, 1CSJ, 1NHU, and 2AX1, were selected as a test set. As shown in Table 4, the sensitivity and specificity of the model are 77% and 3%, respectively. Moreover, our RBFNN model successfully predicted 94 binding sites out of the 127 binding sites from the 3 testing protein complexes, with an overall prediction accuracy of 74%. Although the sensitivity is relatively high, the specificity remains to be improved. The residue properties, which were not used in this study, such as positive charge, negative charge, polar, and hydrophobic, can be encoded

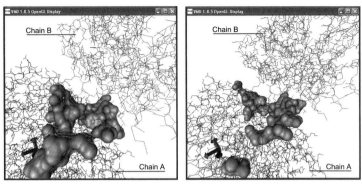

(a) Known binding sites of 1NHU (b) Predicted biding sites of 1NHU

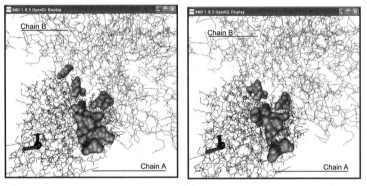

(c) Known binding sites of 2AX1 (d) Predicted binding sites of 2AX1

Fig. 3. Comparison of predicted results and known binding area within chain A of HCV proteins 1NHU and 2AX1

for more accurate prediction. However, prediction with accuracy 74% of binding sites only with residue binding propensity and sequence entropy indicates this method is very powerful and promising for narrowing down the search space.

5   Conclusion

We have developed a method for predicting binding sites in HCV complexes using the statistical residue conformational property (binding propensity) and protein evolutionary information (sequence entropy). Experimental results demonstrate that the method can accurately predict binding residues in HCV complexes. In our next work, we will focus on the problem of identifying interaction areas between HCV and human proteins, and the results of this study will help identify interaction segments between HCV and human proteins.

Acknowledgments. This work was supported in part by the Korea Research Foundation Grant funded by the Korean Government (KRF-2006-D00038) and in part by KOSEF through the Systems Bio-Dynamics Research Center.

References

1. Drummer, H.E., Wilson, K.A. and Poumbourios, P.: Identification of the Hepatitis C Virus E2 Glycoprotein Binding Site on the Large Extracellular Loop of CD81. Journal of Virology, **76(21)** (2002), 11143–11147

2. Zhang, G.Z., Huang, D.S.: Prediction of inter-residue contacts map based on genetic algorithm optimized radial basis function neural network and binary input encoding scheme. Journal of Computer Aided Molecular Design, **18** (2004), 797–810

3. Berman, H.M., Westbrook, J., Feng, Z. and et al.: The Protein Data Bank. Nucleic Acids Research, **28** (2000), 235-242.

4. Altschul, S.F., Madden, T.L., Schaffer, A.A. and et al.: Gapped BLAST and PSI-BLAST: a new generation of protein database search programs. Nucleic Acids Research, **25** (1997), 3389–3402

5. Reš, I. and Lichtarge, O.; Character and evolution of protein-protein interfaces. Physical Biology, **2** (2005), S36–S43

6. Fraczkiewicz, R. and Braun, W.: Exact and efficient analytical calculation of the accessible surface areas and their gradients for macromolecules. Journal of Computational Chemistry, **19** (1998), 319–333

7. Valdar, W.S.J. and Thornton, J.M.: Conservation Helps to Identify Biologically Relevant Crystal Contacts. J. Mol. Biol., **313** (2001), 399–416

8. Fariselli, P., and Casadio, R.: Neural network based prediction of residue contacts in protein. Protein Engineering, **12** (1999), 15–21

9. Jores, R., Alzari, P.M, Meo, T.: Resolution of hypervariable regions in T-cell receptor β chains by a modified Wu-Kabat index of amino acid diversity. Proc. Natl. Acad. Sci. USA, **87** (1990), 9138–9142

10. Schneider, T.D.: Information content of individual genetic sequences. Journal of Theoretical Biology, **189** (1997), 427–441

11. Guharoy, M. and Chakrabarti, P.: Conservation and relative importance of residues across protein-protein interfaces. Proc. Natl. Acad. Sci. USA, **102** (2005), 15447–15452

12. Schäffer, A.A., Aravind, L., Madden, T.L., Shavirin, S. and et al.: Improving the accuracy of PSI-BLAST protein database searches with composition-based statistics and other refinements. Nucleic Acids Research, **29** (2001), 2994-3005.

13. Hu, Z.J., Ma, B.Y., Wolfson, H. and Nussinov, R.: Conservation of Polar Residues as Hot Spots at Protein Interfaces. PROTEINS: Structure, Function, and Genetics, **39** (2000), 331–342

14. Ma, B., Elkayam, T., Wolfson, H. and Nussinov, R.: Protein-protein interactions: Structurally conserved residues distinguish between binding sites and exposed protein surfaces. Proc. Natl. Acad. Sci. USA, **100(10)** (2003), 5772–5777

15. De-Vries, S.J. and Bonvin, A.M.J.J.: Intramolecular surface contacts contain information about protein-protein interface regions. Bioinformatics, **22(17)** (2006), 2094–2098

16. Rost, B. and Sander, C.: Prediction of Protein Secondary Structure at Better than 70% Accuracy. J. Mol. Biol., **232(1)** (1993), 584–599

17. Kim, K. and Park, H.: Protein secondary structure prediction based on an improved support vector machines approach. Protein Engineering, **16(8)** (2003), 553–560

Use of Parallel Simulated Annealing for Computational Modeling of Human Head Conductivity

Adnan Salman[1], Allen Malony[1], Sergei Turovets[1], and Don Tucker[2]

[1] NeuroInformatics Center, 5219 University of Oregon, Eugene, OR 97403, USA
[2] Electrical Geodesic, Inc., 1600 Millrace Dr, Eugene, OR 97403, USA

Abstract. We present a parallel computational environment used to determine conductivity properties of human head tissues when the effects of skull inhomogeneities are modeled. The environment employs a parallel simulated annealing algorithm to overcome poor convergence rates of the simplex method for larger numbers of head tissues required for accurate modeling of electromagnetic dynamics of brain function. To properly account for skull inhomogeneities, parcellation of skull parts is necessary. The multi-level parallel simulated annealing algorithm is described and performance results presented. Significant improvements in both convergence rate and speedup are achieved. The simulated annealing algorithm was successful in extracting conductivity values for up to thirteen head tissues without showing computational deficiency.

1 Introduction

Accurate knowledge of the geometry of human head tissues and their conductivities plays a fundamental role in high-resolution neuroimaging. This knowledge is necessary and essential to create computational models of the electromagnetic characteristics of the human head, allowing precise monitoring of brain dynamics in both space and time. Dense-array electroencephalography (EEG) up to 256 sensor channels can be projected to cortex locations using computational head models (this is known as the *inverse* or *source mapping* problem), but tradeoffs concerning simulation complexity and solution accuracy and speed will determine model precision. While tissue geometry is observable through structural MRI or CT measures, the regional conductivities of the human head are largely unknown. Because the skull is the most resistive tissue, the lack of accurate skull conductivity estimates is particularly problematic given also the developmental variations in the human skull from infancy through adolescence. Without an accurate forward model specifying the volume conduction from cortex to scalp, even advanced inverse efforts cannot achieve precision with EEG data as the error of source localization due to the conductivity uncertainty may reach a few centimeters [1].

Our group has developed a high-performance modeling framework for simulation of human head electromagnetics based on finite difference methods (FDM) with realistic head shape. In our 2005 ICCS paper [2], we combined FDM modeling with a parameterized electrical impedance tomography (EIT) measurement procedure and a downhill simplex algorithm to extract three- and four-tissue conductivities in simulations with reasonable accuracy. We parallelized the simplex search using a multi-start technique and good speedup was achieved. However, two factors argue for a new approach. First,

Y. Shi et al. (Eds.): ICCS 2007, Part I, LNCS 4487, pp. 86–93, 2007.

Fig. 1. The Geodesic Sensor Net, current injection between selected electrode pairs and simultaneous acquisition of the return potentials from the sensor array

experimental studies report that skull is anisotropic and highly heterogeneous, and can not be modeled as a uniform tissue[6]. Second, if we increased the number of segmented tissues to more accurately model skull inhomogeneities, the viability of the simplex method diminishes quickly. Six tissues are enough to reduce the probability of convergence to less than 10% and of finding optimal solutions to much less.

This paper reports on our work replacing the simplex method with a simulated annealing algorithm that can discover conductivities up to thirteen segmented tissues. We have parallelized the conductivity search problem and evaluated its convergence and scalability attributes on the SDSC DataStar machine. We have also prototyped an exponential parallelization strategy and tested its scalability potential. The sections that follow provide more background on the problem domain, describe the computational design, and present our performance results. The paper concludes with future work.

2   Methods

To non-invasively determine head tissue conductivities, a tomographic-based search procedure must be used to optimally match EEG data measured from a subject's head to simulated solutions of the electromagnetic response. Small currents are injected into the head and electrical response measured at dense-array sensors on the scalp (see Figure 1). Given these measurements, we can search for conductivity solutions using a FDM model of the subject's head. Once a set of conductivities is chosen, "forward" calculations simulate the electromagnetic effects when current is injected at the chosen locations. The simulated electrical potentials are computed and compared to those measured. Based on the error, the "inverse" calculation attempts to improve the next choice of conductivity parameters. Together, the forward and inverse calculations define the (indirect) conductivity modeling problem. A complete formal description of the forward and inverse problems can be found in our early ICCS paper [2]. Here we provide only a brief review.

2.1   Forward and Inverse Problems

The electrical forward problem can be stated as follows: given the positions and magnitudes of current sources, as well as geometry and electrical conductivity of the head volume Ω calculate the distribution of the electrical potential on the surface of the head (scalp) Γ_Ω. Mathematically, it means solving the linear Poisson equation [3]:

$$\nabla \cdot \sigma(x,y,z)\nabla\phi(x,y,z) = S, \tag{1}$$

in Ω with no-flux Neumann boundary conditions on the scalp:

$$\sigma(\nabla\phi) \cdot n = 0 \quad , \tag{2}$$

on Γ_Ω. Here n is the normal to Γ_Ω, $\sigma = \sigma_{ij}(x,y,z)$ is an inhomogeneous tensor of the head tissues conductivity and S is the source current. We have built a FD forward problem solver for Eq. (1) and (2) based on the multi-component alternating directions implicit (ADI) algorithm [4]. It is a generalization of the classic ADI algorithm [5].

The inverse problem for the electrical imaging modality has the general tomographic structure. From the assumed distribution of the head tissue conductivities, σ_{ij}, and the given injection current configuration, S, it is possible to predict the set of potential measurement values, ϕ^p , given a forward model F (Eq. (1), (2)), as the nonlinear functional [5]:

$$\phi^p = F(\sigma_{ij}(x,y,z)). \tag{3}$$

Then an appropriate objective function is defined, which describes the difference between the measured, V, and predicted data, ϕ^p, and a search for the global minimum is undertaken using advanced nonlinear optimization algorithms. In this paper, we used the simple least square error norm:

$$E = \left(\frac{1}{N}\sum_{i=1}^{N}(\phi_i^p - V_i)^2\right)^{1/2}, \tag{4}$$

where N is a total number of the measuring electrodes. With the constraints imposed by the segmented MRI data, one needs to know only the average regional conductivities of a few tissues, for example, scalp, skull, cerebrospinal fluid (CSF), and brain, which significantly reduces the dimensionality of the parameter space in the inverse search.

In our earlier work, to solve the nonlinear optimization problem in Eq. (4), we employed the downhill simplex method of Nelder and Mead. Our observation was that simplex search performs well when the number of parameters is few (three or four parameters), while it fails completely when the number of parameters is larger than six. To avoid the local minima in the simplex search, we used a statistical approach. The inverse procedure was repeated for hundreds sets of conductivity guesses from appropriate physiological intervals, and then the solutions closest to the global minimum solutions were selected using a simple error threshold criteria $E < E_{threshold}$.

2.2   Skull Inhomogeneities

The human skull includes eight cranial bones and fourteen facial bones in addition to the sutures. Figure 2 shows how the head bones are parcellated based on anatomical features. It is expected that anatomically different parts of the skull have different conductivities values and experiments bear this out [6]. For instance, conductivity of trilayer

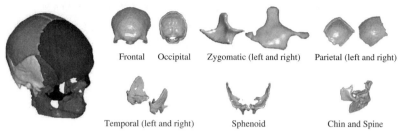

Frontal   Occipital   Zygomatic (left and right)   Parietal (left and right)

Temporal (left and right)   Sphenoid   Chin and Spine

Fig. 2. Anatomically parcellated-skull into 11 major bones

bones increases linearly with thickness, the parts with absence of cancellous bones are less conductive, and sutures are highly conductive. Thus, it is important to characterize the skull inhomogeneities as much as possible in the conductivity modeling.

However, increasing the number of modeled tissues also increases search and computational complexity. For this reason, it has been impossible to date to determine the tradeoffs of tissue dimensionality, simulation time-to-solution, and conductivity model accuracy. We know that the simplex method is not viable beyond six tissues. When skull parts are separately modeled, a more powerful technique is required. The results here are based on the simulated annealing algorithm [7]. For evaluation purposes, we computed scalp potentials for thirteen preset tissue conductivities and tested how well the model performed as the number of tissues varies. To address the increased computational demands, we parallelized the conductivity search based on simulated annealing. The computational design is described in the next section.

3 Computational Design

With higher resolution and more detailed parcellation of the skull, the computational requirements of the forward and inverse calculation increase significantly. Beyond thirteen search parameters and $1mm^3$ MRI/CT resolution, the computation quickly becomes impractical. Certainly, to pursue higher dimensionality, it is clear we must replace the simplex method used in our earlier work. We chose the advanced simulated annealing algorithm [7] which has been shown to be more robust for optimization across complex multi-variate search spaces. This algorithm has allowed us to extend the conductivity modeling to study the impact of skull inhomogeneities on the conductivity modeling convergence and performance.

The simulated annealing method consists of three nested loops:

Temperature cooling: The outer loop controls the temperature cooling, the temperature is reduced by a factor of r after executing N_t search-radius loops.

Search radius: The intermediate loop, the search radius loop, executes the inner loop N_s times before the maximum step length adjusted. The maximum step lengths are adjusted such that approximately half of the moves are accepted.

Control point: The inner loop considers new moves in all directions by perturbing the control point in all direction; the perturbation is constraint between 0 and the maximum step length in each direction.

```
input : Initial Temprature T₀ and initial Point X₀
output: X_optimal
T = T₀, X = X₀, F = Cost(X₀)
while T>0 do
    for i ← 0 to N_t do
        for j ← 0 to Ns do
            for k ← 0 to N do
                X_k ← purturb(X, k)
                F_k ← Cost(X_k)
                ΔF ← F_k − F
                if ΔF < 0 then accept X_k and update X_optimal
                else accept X_k with probability ∝ exp(ΔF/T)
            end
        end
        adjust-maximum-step-length
    end
    check-termination-condition
    reduce temparature T = rT
end
                                                    (a)
```

```
input : Initial Temprature T₀ and initial Point X₀
output: X_optimal
T = T₀, X = X₀, F = Cost(X₀)
N_task = N_s/Number of tasks
while T>0 do
    for i ← 0 to N_t do
        X_best = X, F_bset = F
        for j ← 0 to N_task do
            for k ← 0 to N do
                X_k ← purturb(X, k)
                F_k ← Cost(X_k)
                ΔF ← F_k − F
                if ΔF < 0 then accept X_k and update X_best
                else accept X_k with probability ∝ exp(ΔF/T)
            end
        end
        MPI communication
        adjust-maximum-step-length
    end
    Master task gets each task best point
    Update X_optimal
    check-termination-condition
    Master task Brodcast X_optimal
    X = X_optimal, F = F_optimal
    reduce temparature T = rT
end
                                                    (b)
```

Fig. 3. Pseudo code for simulated annealing algorithm: a) Serial search about the control point with parallel forward calculations, and b) Parallelization along the search radius with parallel forward calculations

Transitions that lower the cost function are always accepted, while transitions that raise the cost function are accepted with probability based on the temperature and the size (cost) of the move (Metropolis criteria). The acceptance of the uphill moves allows the system to overcome local minima and make the algorithm insensitive to the initial starting values. The simulated annealing converges when the minimum become stable that does not change for more than epsilon after several temperature reduction loop iterations. To reduce the number of forward calculation, we added a second termination condition when the value of the cost function becomes less than some tolerance value. The complete algorithm is given in Figure 3(a).

The inner loop spends most of the time in solving the forward problem, since the forward solver is the highest cost component of conductivity modeling. In our earlier work, we parallelized the forward solver using OpenMP and achieved speedups of 5.8 on eight processors and 8.2 on sixteen processors on a 16-processor IBM p690 system (see [2]). Here we also parallelized the simulated annealing algorithm along the search-radius, modifying the parallel MPI-based methods described in [7] for our problem domain. Figure 4(a) shows a high-level view of the parallel simulated annealing approach, with the intermediate loop distributed across several nodes (1 task per node), each running parallelized forward calculations. All tasks start with the same initial values, but with different random generator seeds. Each task perform a random search around the control point by perturbing the control point in all direction. The perturbation is constrained to be within the *maximum step length*. At the end of the search radius loop, the master task gathers every task best solution and updates the optimal solution and all nodes communicate to adjust the maximum step length. At the end of the temperature reduction loop all tasks updates the control point with optimal solution. The complete algorithm is given in Figure 3(b).

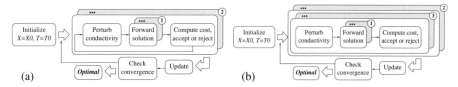

Fig. 4. Alternative parallel simulated annealing algorithms: a) Two-level parallelism – parallel execution occurs on the level of the intermediate loop, where each task runs the parallel forward calculation, and b) Three-level parallelism – in addition to the parallel forward calculation, each task computes the inner loop in parallel

Most of the performance numbers reported in the next section are for the parallel simulated annealing algorithm in Figure 4(a). Choosing twelve tasks and 16-way forward solves allows the amount of parallel execution to reach 192 processors. There does not appear to be much benefit in increasing the number of tasks beyond twelve and the OpenMP performance flattens beyond sixteen processors. To create greater potential parallelism, we decided to enumerate all possible perturbation paths around the control point, increasing the number of candidate points from N to 2^N, and breaking the serialization introduced in the former algorithm when the next perturbation point depends on the acceptance or rejection of the prior point. This new algorithm (shown graphically in Figure 4(b)) effectively multiplies the degree of parallelism by $2^N/N$ with a potential performance gain of N. The numbers in Figure 4 for each case indicate the different types of parallelism. Interestingly, the nature by which this algorithm was created guarantees it to produce at least as optimal results as the former, but it requires significantly more resources to achieve greater performance benefits.

4   Computational Results

We conducted a series of experiments to test both the convergence properties and the performance of the conductivity modeling based on parallel simulated annealing. All experiments were performed on the San Diego Supercomputing Centere DataStar system, a cluster of shared-memory nodes consisting of 16-processor IBM p690 and 8-processor IBM p655 machines. All results presented below were performed using 8-way OpenMP tasks running forward calculations each on a separate p655 node.

As mentioned earlier, we preset thirteen tissue conductivity values (eleven skull parts, scalp, brain) and ran experiments to test the conductivity model accuracy on fewer tissue numbers. We started with eleven tissue to verify convergence to acceptable values. The simulated annealing search starts with initial random conductivities selected from the biomedical ranges and stops when one of three criteria is met as described in the computational design. Our results verify the ability of simulated annealing to extract eleven tissues with good accuracy and precision. Figure 5(right) shows the dynamics of the 11-tissue inverse search convergence, giving the temperature cooling, the cost function, and one tissue's conductivity. This calculation was done on a single 8-processor p655 node in our lab and took approximately 31 hours to complete.

Having verified convergence for a large number of tissues, we decided to limit the number of tissues to five (three skull parts, scalp, brain) to test the performance

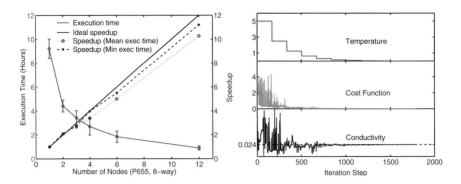

Fig. 5. The Speedup, the mean and the standard deviation of the execusion time of the two-level parallel simulated annealing solver. Each point corresponds to three experimental runs using 5 tissues (left). Simulated annealing dynamics, the temperature cooling scheduale, cost function, and the retrieved conductivity of frontal bone (right).

properties of the parallel simulated annealing algorithm. This will allow us also to contrast performance with the earlier simplex outcome. The execution time and performance speedup for a $2mm^3$ size problem from one to twelve tasks (each task run on a 8-processor p655 node) is shown in Figure 5(left). The speedup is almost linear with the number of nodes. Three experiments were run on each data point to show the performance variation due to the random number generator sequence.

Even though we have excellent speedup, the degree of parallelism is limited. The second version significantly increases parallelism by generating N random numbers, one for perturbing the control point in each direction, and generated all possible points by enumerating the perturbation in all directions. Then the cost function at these points is evaluated in parallel. After computing all possible paths, the simulated annealing criteria is applied. In theory this parallelism can speedup the computation by a factor of N. In addition, we can get further speedup by selecting the best point from all points that was computed and not only from those points on a simulated annealing path. This speedup is due to speeding the convergence.

For verification purposes, we compared this new algorithm with the former for a problem where the conductivities of three tissues are found. We used a single task and two processors for the forward calculation. Thus, in the former algorithm, the parallelism degree is two. In the new algorithm, the parallelism is of degree sixteen (2^3 perturbed points by two processors in the forward calculation), allowing two 8-processor p655 nodes to be used, with a speedup potential of three (the number of tissues). Our experimental results show an overall performance improvement of 1.98. The best point selection gives a convergence speedup of 8%, while the inner loop parallelism produces a speedup of 1.77. We believe better speedup can be achieved by eliminating communication overheads. For larger numbers of tissues, the number of processors needed to realize the potential parallelism in the new algorithm increases by a power of two. When this resource scaling is unavailable, the method can be throttled to use a smaller degree of perturbation fanout.

5   Conclusion

We provided an efficient computational environment that made the reconstruction of the human head tissues conductivities possible. The simulated annealing algorithm proved to be stable in extracting up to 13 tissues without any sign of failure. Remarkably, using the parallel simulated annealing algorithm, we were able to extract five tissues in 40 minutes, compared to over three days using our previous simplex methods. More importantly, the new algorithms are very robust. In all the testing for this paper, the simulated annealing runs never failed to converge.

There are still improvements to come. The potential parallelism in our new algorithm begs for large-scale testing. For instance, with eleven tissues, over 2048-way parallelism is possible. We intend to perform experiments on IBM BG/L and Cray XTE machines soon. This level of performance is important because our research group at the Neuroinformatics Center will begin creating a head model database for several demographic populations in the next year.

References

1. Huiskamp, G., Vroejenstijn, M., van Dijk, R., Wieneke, G., Huffe-len, A.C.: The need for correct realistic geometry in the inverse EEG problem. IEEE Transactions on Biomedical Engineering **46** (1999) 121–1287
2. Salman, A., Turovets, S., Malony, A., Eriksen, J., Tucker, D.: Computational Modeling of Human Head Conductivity. International Conference on Computational Science **3514**, (2005) 631–638
3. Gulrajani, R.M.: Bioelectricity and Biomagnetism. John Wiley & Sons, New York (1998)
4. Abrashin, V.N., Egorov, A.A., Zhadaeva, N.G.: On the Convergence Rate of Additive Iterative Methods. Differential Equations **37** (2001) 867–879
5. Hielscher, A.H., Klose, A.D., Hanson, K.M.: Gradient Based Iterative Image Reconstruction Scheme for Time-Resolved Optical Tomography. IEEE Transactions on Medical Imaging **18**(1999) 262–271
6. Law, S.: Thickness and resistivity variations over the upper surface of the human skull. Brain Topography **6** (1993) 99–109
7. Higginson, J.S., Neptune, R.R., Anderson, F.C.: Simulated parallel annealing within a neighborhood for optimization of biomechanical systems. Journal Biomechanics **38** (2005) 1938–1942

Mining Molecular Structure Data for the Patterns of Interactions Between Protein and RNA

Kyungsook Han[*] and Chirag Nepal

School of Computer Science and Engineering, Inha University
Inchon, South Korea
khan@inha.ac.kr

Abstract. Mining useful information from a large amount of biological data is becoming important, but most data mining research in bioinformatics is limited to molecular sequence data. We have developed a set of algorithms for analyzing hydrogen bond and van der Waals interactions between protein and RNA. Analysis of the most representative set of protein-RNA complexes revealed several interesting observations: (1) in both hydrogen bond and van der Waals interactions, arginine has the highest interaction propensity, whereas cytosine has the lowest interaction propensity; (2) side chain contacts are more frequent than main chain contacts in amino acids, whereas backbone contacts are more frequent than base contacts in nucleotides; (3) amino acids, in which side chain contacts are dominant, reveal more diverse interaction propensities than nucleotides; and (4) valine rarely binds to any nucleotide. The interaction patterns found in this study should prove useful for determining binding sites in protein-RNA complexes.

1 Introduction

As the full genome sequences of several organisms are known, discovering useful information from the huge, noisy biological data has become an interesting problem. Many research of data mining in bioinformatics are limited to biological sequence data of macromolecules such as DNA, RNA and protein. Biological sequences are easy to analyze due to their sequential nature and have many well-developed algorithms since they can be treated as strings of characters. The structure of a molecule is much more complex than its sequence but plays an important role since it determines the biological function of the molecule. However, there have been few attempts to mine the structures of molecules, because much less is known about the structures than the sequences and there is no readily usable algorithm to handle structural data. However, an increasing number of the three-dimensional structures of molecules is known recently and continues to increase in public databases. The sequence of a molecule is a string of four bases and twenty amino acids, respectively, and is relatively simple to handle. The three-dimensional structure data of a molecule contains the x, y, and z coordinate values of all the atoms contained in the molecules and the information on embedded structure elements. Therefore, for the same molecule, its structure data is much larger and more complex than its sequence data.

---

[*] Corresponding author.

Y. Shi et al. (Eds.): ICCS 2007, Part I, LNCS 4487, pp. 94–101, 2007.
© Springer-Verlag Berlin Heidelberg 2007

Over the past years a variety of problems concerned with protein-DNA complexes have been investigated (Deng et al., 1999; Luscombe et al., 2001), but protein-RNA complexes have received much less attention despite their importance. In contrast to the regular helical structure of DNA, RNA molecules form complex secondary and tertiary structures consisting of stems, loops, and pseudoknots. The structural elements arranged into three-dimensional space are often recognized by specific proteins. RNA structures display hydrogen bonding, electrostatic, and hydrophobic groups that can interact with small molecules to form specific contacts. However, it is not clear how proteins interact with RNA with specificity.

Since our previous study on hydrogen-bonding interactions between protein and RNA (Kim et al., 2003), the structures of many protein-RNA complexes have been determined. In this study we attempt a more rigorous study of the interactions by considering van der Waals interactions as well as hydrogen-bonding interactions in a new, extensive dataset of protein-RNA complexes. The primary focus of this work is to find how proteins selectively bind specific sites of RNA molecules in both types of interactions. The rest of this paper presents a set of algorithms for analyzing hydrogen-bonding and van der Waals interactions and the interaction patterns discovered from a new dataset of 45 protein-RNA complexes.

2 Identifying Hydrogen Bonds and van der Waals Contacts

2.1 Datasets of Protein-RNA Complexes

The protein-RNA complex structures were obtained from the PDB database (Berman et al. 2000). The complexes that were determined by X-ray crystallography with a resolution of 3.0 Å or better were selected. As of January 2006, there were 312 protein-RNA complexes in PDB and the number of complexes with a resolution of 3.0 Å or better was 207. We used PSI-BLAST (Altschul et al. 1997) for similarity search on each of the protein and RNA sequences in these 207 protein-RNA complexes, in order to eliminate the equivalent amino acids or nucleotides in the homologous protein or RNA structures. 45 out of 207 protein-RNA complexes were left as the representative, nonhomologous complexes after running the PSI-BLAST program with an E value of 0.001 and an identity value of 90% or below. The final dataset of 45 protein-RNA complexes was used in our study (Table 1).

Table 1. The list of 45 protein-RNA complexes in the data set

PDB ID									
1A9N	1AQ3	1ASY	1B7F	1B23	1C9S	1DFU	1DI2	1DK1	1DUL
1E7K	1EC6	1EUY	1F7U	1F8V	1FFY	1G59	1H3E	1H4Q	1HC8
1I6U	1JBR	1JID	1K8W	1M8Y	1MJI	1MZP	1R3E	1RC7	1RPU
1SI3	1TFW	1TTT	1WSU	1XOK	1YTY	1YVP	2ASB	2AZ0	2B3J
2BGG	2BTE	2BX2	2F8K	2FMT					

2.2 Hydrogen Bonds and van der Waals Contacts

The number of hydrogen bonds (H-bonds) between amino acids and nucleotides in protein-RNA complexes was calculated using CLEAN and HBPLUS version 3.15

(McDonald and Thornton 1994). The positions of hydrogen atoms (H) were inferred from the surrounding atoms, as hydrogen atoms are invisible in purely X-ray-derived structures. H-bonds were identified by finding all proximal atom pairs between H-bond donors (D) and acceptors (A) that satisfy the following geometric criteria: contacts with maximum D-A distance of 2.7 Å, maximum H-A distance of 2.5 Å, and minimum D-H-A and H-A-AA angles set to 90°, where AA is an acceptor antecedent. All protein-RNA bonds were extracted from the HBPLUS output files.

The criteria considered for the van der Waals interactions are: contacts with maximum D-A distance of 3.9 Å, maximum H-A distance of 2.5 Å, and minimum D-H-A and H-A-AA angles set to 90°. The interactions that are considered as H-bonds are not included in van der Waals interactions. The definitions of H-bonds and van der Waals interactions are slightly different from those used by Jones et al. (2001).

2.3 Binding Propensity

Counting the number of H-bonds does not yield precise interaction propensities because it does not take into account the number of residues in the complex and on the surface of the complex. For example, amino acid A may have a weak propensity, even though it may be involved in many H-bonds or may occur very frequently. Therefore, we employ a function for determining propensities that considers H-bond numbers, number of amino acids on the surface, nucleotide numbers and explosion values. The propensity function is based on that reported by Moodie et al. (1996), but we modified their function to determine the interaction propensities of amino acids and nucleotides pair on the surface of a complex. Amino acids are considered to be on the surface if their relative accessibility exceeds 5% according to the NACCESS program (Hubbard and Thornton 1993).

The interaction propensity P_{ab} between amino acid a and nucleotide b is defined by equation (1), where N_{ab} is the number of residues of amino acid a hydrogen bonding to nucleotide b, $\sum N_{ij}$ is the total number of amino acids hydrogen bonding to any nucleotide, N_a is the number of residues of amino acid a, $\sum N_i$ is the total number of amino acid residues, N_b is the number of nucleotide b, $\sum N_j$ is the total number of nucleotides, and the numbers refer to residues on the surface. The numerator $N_{ab}/\sum N_{ij}$ represents the ratio of the co-occurrences of amino acid a binding with nucleotide b, to the total number of all amino acids binding to any nucleotide on the surface. The term $N_a/\sum N_i$ of the denominator represents the ratio of the frequency of amino acid a to that of all amino acids on the surface, and the second term $N_b/\sum N_j$ represents the ratio of the frequency of nucleotide b to that of all nucleotides on the surface.

$$P_{ab} = \frac{\dfrac{N_{ab}}{\sum N_{ij}}}{\dfrac{N_a \cdot N_b}{\sum N_i \cdot \sum N_j}} \tag{1}$$

The interaction propensity is computed as the proportion of a particular amino acid binding to a particular nucleotide on the surface divided by the proportion of each on the surface. Therefore, the propensity represents the frequency of co-occurrence of amino acids and nucleotides in protein-RNA complexes, for every pair of amino acids and nucleotides. A propensity greater than 1 indicates that the given amino acid occurs more frequently in combination with a given nucleotide on the surface, whereas a propensity less than 1 indicates that the amino acid occurs less frequently.

2.4   Structure Elements of Protein and RNA

An RNA nucleotide consists of sugar, phosphoric acid, and base. A base consists of a fixed number of atoms that provide important clues for extracting base pairs and classifying them. Base pairs are formed by hydrogen bonding between the atoms of bases. For example, the canonical A-U pair has two hydrogen bonds between N1 of adenine (A) and N3 of uracil (U), and between N6 of A and O4 of U. Thus, we can establish definite rules for hydrogen bonding between fixed atoms, which we call base pair rules. Base pairs can be classified into canonical base pairs (Watson-crick base pairs) and non-canonical base pairs. We consider base pairs of 28 types (Tinoco 1993) comprising of both canonical and non-canonical base pairs. Base pair rules are used for extracting base pairs and secondary and tertiary structural elements of RNA.

In this study, the secondary structure elements of protein are categorized into four types: helix (α-helix, 3/10 helix and π-helix), sheet (ß-ladder and ß-bridge), turn (hydrogen-bonded turn), and others (bend and other structures). These secondary structure elements are assigned using the DSSP program (Kabsch and Sander 1983). The secondary structure elements of RNA are categorized into two types: paired and unpaired. If at least one H-bond exists between the base atoms of two nucleotides, these nucleotides are considered to be paired. If not, they are considered to be unpaired. If an H-bond exists between other parts of two nucleotides than the base part, the two nucleotides are considered to be unpaired.

2.5   Algorithms for Extracting Secondary and Tertiary Structures of RNA

The first part of the algorithm extracts the secondary and tertiary structures of the RNA. The second part generates a visual representation of the RNA structure by integrating this information with knowledge of the coordinates of the nucleotides.

1. From the PDB files extract H-bonds that link the base of one nucleotide and that of another, and classify the H-bonds into the 28 types of base pairs using the base pair rules; they are then recorded in the Base Pair List.
2. Extract the RNA sequence data from the PDB and records it in RNA-SEQ. This contains information on all nucleotides in the RNA, which is used again in Step 4.
3. Derive the three-dimensional coordinates of the nucleotides. We define the average coordinate value of all the atoms that make up a nucleotide as the three-dimensional coordinate value of the nucleotide.
4. Integrate the sequence data in RNA-SEQ with the base pair data in the Base Pair List. All the nucleotides in RNA-SEQ are matched to the nucleotides in the Base Pair List to determine the bonding relationships between the nucleotides.
5. Integrate the three dimensional coordinates of the nucleotides obtained in Step 3 with the RNA structure data obtained in Step 4 and visualize the RNA structure.

2.6   Algorithms for Analyzing Protein-RNA Complexes

This algorithm analyzes the H-bonds and van der Waals contacts and generates the patterns of binding between protein and RNA.

1. Extract hydrogen bonds from the structure data of the protein-RNA complexes and assign secondary structure elements to each atom of the proteins.
2. Extract only hydrogen bonds that link RNA bases to get base pairs. Information on the base pairs obtained in this step, together with the information about the secondary structure elements of the protein obtained in Step 1, is used to identify the interaction pattern between protein and RNA at the secondary level.
3. Classify the amino acids into main chain and side chain. This can provide information on which type tends to form hydrogen bonds with nucleotides.
4. Show the analysis results between protein and RNA at various structure levels.

3   Results and Discussion

The algorithms were implemented in a web-based program called PRI-Modeler (http://wilab.inha.ac.kr/primodeler/) using Visual C#. PRI-Modeler is executable within a web browser on any Windows systems. Given one or more PDB files as input, PRI-Modeler produces several files containing H-bonding or van der Waals interactions between amino acids and nucleotides, secondary and tertiary structures of RNA, and interaction patterns of protein-RNA complexes.

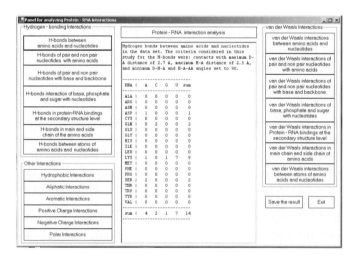

Fig. 1. Analysis of the interactions between protein and RNA using PRI-Modeler

Fig. 1 shows an exemplary user interface of PRI-Modeler. It consists of twenty menu buttons and one text panel. When the user specifies the types of interactions to analyze using the menu buttons (such as H-bonding interactions, van der Waals interactions, hydrophobic interactions, aliphatic interactions, aromatic interactions, positively and negatively charged interactions and polar interactions), PRI-Modeler analyzes the specified interactions and displays the analysis results in the text panel. The rest of this section presents some analysis results of 45 protein-RNA complexes.

Table 2 shows the hydrogen bonds and van der Waals interactions between amino acids and nucleotides, and Fig. 2 shows their interaction propensities. There are a total

318 H-bonds and 1056 van der Waals contacts in the data set. Interestingly, both in the H-bonding and van der Waals interactions, arginine (Arg) showed the highest interaction propensity, whereas cytosine (C) showed the lowest interaction propensity (Fig. 2). Amino acids show more diverse nature (H-bonding interaction propensity in the range [0-2.44], and van der Waals interaction propensity in the range [0.04-3.16]) than nucleotides ([0.74-1.12] and [0.67-1.03], respectively). This indicates that amino acids are more distinguishable than nucleotides by their interaction propensities. Amino acids have a main chain in common, and nucleotides have a backbone in common. In protein, both interactions are more frequently observed in the side chain (average 73%) than in the main chain (average 27%). In contrast, the backbone part of RNA has more interactions (average 62% in both interactions) than in the base part (average 38%). Amino acids, in which side chain contacts are dominant, naturally reveal more diverse interaction propensities than nucleotides. One of the reasons that backbone contacts are more frequent than base contacts in nucleotides is that the backbone part has more atoms and highly electronegative atoms than the base part, which makes the backbone more favorable to form hydrogen bonds than the base.

Many amino acids show a strong preference for specific nucleotides with which they form stable H-bonds. For example, amino acids that contain polar atoms (e.g., arginine and lysine) have a strong tendency to form H-bonds. In H-bonding interactions, Arg-U, Gln-C, Arg-G, and Ser-A are the binding pairs with high propensity. In van der Waals interactions, Arg-U, Arg-G, and Asn-U are the binding pairs with high propensity. Valine rarely binds to any nucleotide.

Table 2. Hydrogen bonds and van der Waals interactions between amino acids and nucleotides

Amino acids	Frequency	hydrogen bonds					van der Waals interactions				
		A	C	G	U	Total	A	C	G	U	Total
Glu	3390	4	4	19	0	27	10	17	34	6	67
Lys	2866	9	8	8	16	41	39	55	50	26	170
Arg	2728	10	9	24	20	63	47	79	95	64	285
Leu	2270	0	0	0	1	1	5	8	0	1	14
Gly	2096	2	7	2	3	14	7	9	19	4	39
Asp	1960	8	1	10	8	27	1	14	26	8	49
Thr	1838	4	7	7	12	30	17	13	25	17	72
Ala	1764	0	1	0	2	3	2	8	6	8	24
Val	1688	1	0	0	0	1	0	3	0	0	3
Ser	1640	10	12	7	5	34	22	23	18	8	71
Pro	1628	3	0	0	0	3	1	4	2	0	7
Asn	1416	6	4	5	3	18	14	7	24	32	77
Gln	1290	2	13	3	4	22	11	12	28	19	70
Ile	1206	1	0	0	1	2	10	2	0	4	16
Tyr	1122	4	7	4	8	23	5	20	0	12	37
Phe	1032	0	0	0	0	0	1	0	12	3	16
His	870	0	5	4	0	9	6	8	5	2	21
Met	474	0	0	0	0	0	3	1	4	0	8
Trp	382	0	0	0	0	0	0	4	1	1	6
Cys	172	0	0	0	0	0	2	0	2	0	4
Total	31832	64	78	93	83	318	203	287	351	215	1056

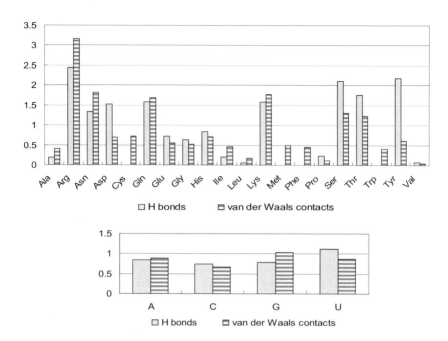

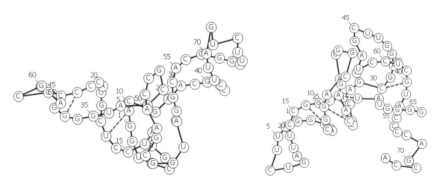

Fig. 2. Average interaction propensities of amino acids (top) and nucleotides (bottom) for the hydrogen bonding and van der Waals interactions

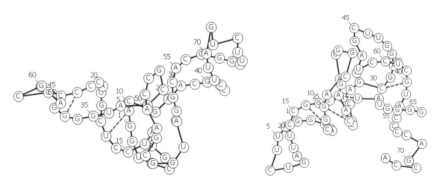

Fig. 3. RNA structure extracted from a protein-RNA complex (PDB ID: 1EUY). H-bonds are shown in red dotted lines (left) and van der Waals contacts are shown green dotted lines (right).

Fig. 3 shows the RNA structures extracted from a protein-RNA complex (PDB ID: 1EUY). They are three-dimensional (3D) drawings, and can be rotated as well as zoomed in/out to get a clear perspective view. With programs like Rasmol (http://www.umass.edu/microbio/rasmol/) and MOLSCRIPT (Per 1991), one cannot easily obtain information about each nucleotide and the binding relations between nucleotides, because the programs represent the structures at the atomic level. On the other hand, PRI-Modeler shows not only the 3D configuration of RNA but also the bonding relations and types of base pairs between nucleotides.

4   Conclusion

We have analyzed 318 hydrogen bonds and 1056 van der Waals contacts in the most representative set of 45 protein-RNA complexes. The interaction propensity function we developed for this analysis indicates the frequency of co-occurrences of amino acids and nucleotides in the protein-RNA complexes for every combination of amino acids and nucleotides. This interaction propensity function is more refined than others since our primary focus for the analysis is RNA as well as protein. Amino acids, in which side chain contacts are much more dominant than main chain contacts, reveal more diverse interaction propensities than nucleotides, in which backbone contacts are dominant. Since our analysis is restricted to protein-RNA complexes that were crystallized and examined by X-ray crystallography, the analysis results may give rise to a high proportion of strong hydrogen bonding interactions. Our long-term goal of this study is to predict the structure of RNA binding protein, and we plan to extend this study to predict binding sites of virus proteins.

Acknowledgment. This work was supported by Inha University Research Grant.

References

1. Altschul, S.F., Madden, T.L., Schaffer, A.A., Zhang, J., Zhang, Z., Miller, W., Lipman, D.J.: Gapped BLAST and PSI-BLAST: a new generation of protein database search programs. Nucleic Acids Res. **25** (1997) 3389-3402
2. Berman, H.M., Westbrook, J., Feng, Z., Gilliland, G., Bhat, T.N., Weissig, H., Shindyalov, I.N., Bourne, P.E.: The Protein Data Bank. Nucleic Acids Res. **28** (2000) 235-242
3. Deng, Y., Glimm, J., Wang, Y., Korobka, A., Eisenberg, M., Grollman, A.P.: Prediction of Protein Binding to DNA in the Presence of Water-Mediated Hydrogen Bonds. Journal of Molecular Modeling **5** (1999) 125-133
4. Hubbard, S.J., Thornton, J.M.: NACCESS, Computer Program, Department of Biochemistry and Molecular Biology, (1993) University College, London
5. Jones, S., Daley, D.T.A., Luscombe, N.M., Berman, H.M., Thornton, J.M.: Protein-RNA interactions: a structural analysis. Nucleic Acids Res. **29** (2001) 943-954
6. Kabsch, W., Sander, C.: Dictionary of protein secondary structures: Pattern recognition of hydrogen bonded and geometrical features. Biopolymers **22** (1983) 2577-2637
7. Kim, H., Jeong, E., Lee, S.-W., Han, K.: Computational Analysis of Hydrogen Bonds in Protein-RNA Complexes for Interaction Patterns. FEBS Letters **552** (2003) 231-239
8. Luscombe, N.M., Laskowski, R.A., Thornton, J.M.: Amino acid-base interactions: a three dimensional analysis of protein-DNA interactions at an atomic level. Nucleic Acids Res. **29** (2001) 2860-2874
9. McDonald, I.K., Thornton, J.M.: Satisfying hydrogen bonding potential in proteins. J. Mol. Biol. **238** (1994) 777-793
10. Moodie, S.L., Mitchell, J., Thornton, J.: Protein recognition of adenylate: An example of a fuzzy recognition template. J. Mol. Biol. **263** (1996) 468-500
11. Per, J.K.: MOLSCRIPT: a program to produce both detailed and schematic plots of protein structures. Journal of Applied Crystallography **24** (1991) 946-950
12. Tinoco, Jr.: The RNA World. In: Gesteland, R.F., Atkins, J.F. (eds), Cold Spring Harbor Laboratory Press. (1993) 603-607

Detecting Periodically Expression in Unevenly Spaced Microarray Time Series

Jun Xian[1], Jinping Wang[2], and Dao-Qing Dai[1,3]

[1] Department of Mathematics, Sun Yat-Sen (Zhongshan) University, Guangzhou, 510275,
China
[2] Department of Mathematics, Ningbo University, Ningbo, Zhejiang, 315211, China
[3] Center for Computer Vision, Sun Yat-Sen (Zhongshan) University, Guangzhou, 510275,
China
xianjun@mail.sysu.edu.cn, stsddq@mail.sysu.edu.cn

Abstract. Spectral estimation has important applications to microarray time series analysis. For unevenly sampled data, a common spectral estimation technique is to use the Lomb-Scargle algorithm. In this paper, we introduce a new reconstruction algorithm and singular spectrum analysis (SSA) method to deal with unevenly sampled microarray time series. The new reconstruction method is based on signal reconstruction technique in aliased shift-invariant signal spaces and a direct implemental algorithm is developed based on the B-spline basis. We experiments on simulated noisy signals and gene expression profiles show different effects for our designed three methods. The three methods are based on our presented reconstruction algorithm, SSA, classical FFT periodogram method and Lomb-Scargle periodogram method.

Keywords: Spectral estimation; periodically expressed genes; unevenly sampled data; the Lomb-Scargle method; signal reconstruction; B-splines.

1 Introduction

Spectral estimation has been a classical research topic in signal processing communities. Many approaches have been proposed in the past decades, including the modified periodogram, autoregressive (AR) model, the MUSIC algorithm and the multitaper method [1,2]. Although all these algorithms have their own advantages, they are all developed based on a basic assumption: the input signal is evenly sampled. However, in many real-world applications, the data can be unevenly sampled. For example, in DAN microarray gene expression experiments, a time series may be obtained with different time sampling intervals [3-5]. Furthermore, an evenly sampled time series may contain missing values due to corruption or absence of some expression measurements. A time series with missing values can be considered as one with unevenly data samples in general.

Lu et al. [6] proposed a periodic-normal mixture (PNM) model to fit transcription profiles of periodically expressed genes in cell cycle microarray experiments and obtained some nice results. Ruf [7] is one of the first to treat evenly sampled gene expression time series with missing values as unevenly sampled data for spectral

Y. Shi et al. (Eds.): ICCS 2007, Part I, LNCS 4487, pp. 102–110, 2007.
© Springer-Verlag Berlin Heidelberg 2007

analysis using the Lomb-Scargle periodogram. Recently, Bohn et al. [8] have used the Lomb-Scargle periodogram in their attempt to detect rhythmic components in the circadian cycle of the Crassulacean acid metabolism plants. The Lomb-Scargle periodogram was originally developed for analysis of noisy unevenly sampled data from astronomical observations.

Although spectral analysis can be applied directly to the original data, noise and outliers would degrade the final results. In DNA microarray analysis, a major challenge is to effectively dissociate actual gene expression values from noise. As a robust model-free technique in time series analysis, SSA can be used to extract as much reliable information as possible from short and noisy time series without prior knowledge of the dynamics underlying. It has been proven to be a powerful tool for processing many types of time series in geophysics, economics, medicine and other sciences [9]. SSA method was originally developed for analysis of noisy unevenly sampled data from geophysical observations. We will introduce SSA method into DAN microarray gene expression analysis. To the best of our knowledge, we firstly introduce SSA into DAN microarray gene expression data analysis.

In this paper, we propose a new spectral estimation technique for unevenly sampled data. Our method models the signal in the aliased shift-invariant signal space that is a generalization of shift-invariant signal space, for which many theories and algorithms are available [10-22]. The proposed algorithm is also flexible in adjusting the order of the B-spline basis based on the number of sampling points. We experiments on simulated noisy signals and gene expression profiles show different effects for our designed three methods. The three methods are based on our presented reconstruction algorithm, SSA, classical FFT periodogram method and Lomb-Scargle periodogram method.

2 Mathematical Theory and Algorithm

2.1 Reconstruction Algorithm of Signal in Aliased Shift-Invariant Signal Spaces

In the following, we first review existing work on signal analysis in the shift-invariant signal space.

The space of a bandlimited signal is introduced, which is inspired by Shannon's signal sampling and reconstruction theorem:

$$V(\text{sinc}) = \{ f(x) = \sum_{k \in Z} c_k \text{sinc}(x-k) : (c_k) \in \ell^2 \} \tag{1}$$

where $\text{sinc}(x) = \dfrac{\sin(x)}{x}$.

Since the sinc function has an infinite support and slow decay, it is not convenient in real applications. Xian and Lin [21] found a good decay function that can replace the sinc basis function, but the new function still has an infinite support. To replace the sinc function with a general function ϕ, we can introduce a signal space that is called the shift-invariant (also called time-invariant) signal space [10, 13-14,17-18]. Its aliased version is defined by

$$V_L(\phi) = \{f : f(x) = \sum_{k \in Z} c_k \phi(x - Lk) : (c_k) \in \ell^2\} \tag{2}$$

where $L > 0$ is a constant.

Signal reconstructing in the shift-invariant space is an active research area and there are many mathematical theories and computer algorithms on the topic [10-22]. When the signal $f \in V_L(\phi)$, we hope to reconstruct signal f from sampled value $\{f(x_i)\}$, where $\{x_i\}$ is the sampling point set.

In fact, the well-known autoregressive (AR) model can be regarded as a special case of signal reconstruction in the above signal space. Yeung et al. [23] have presented an application of the AR model to microarray gene expression time series analysis.

In order to avoid introducing spurious periodic components caused by basis function ϕ with infinite support, for example, the *sinc* function, here we only consider basis functions with compact support. The B-spline function is a good choice and is widely adopted in the wavelet reconstruction theory. We use supp ϕ to indicate the support of basis function ϕ. Assume that supp $\phi \subset [-\Omega, \Omega]$ and $f(x) \in V_L(\phi)$ that is defined in finite interval $[A_1, A_2]$, then f can be determined completely by the coefficients $\{c_k\}$ for $k \in (\frac{A_1 - \Omega}{L}, \frac{A_2 + \Omega}{L}) \cap Z$ with

$$f(x) = \sum_{k = \frac{A_1 - \Omega}{L} + 1}^{\frac{A_2 + \Omega}{L} - 1} c_k \phi(x - Lk) \tag{3}$$

where supp $\phi = \overline{\{x : \phi(x) \neq 0\}}$.

The coefficients, $\{c_i\}$, can be calculated according to following steps:

(1) Given sampling points $x_1, \cdots, x_J \in [A_1, A_2]$ and corresponding discrete function value $y = (y_1, \cdots, y_J)$. Assume that $J \geq \frac{A_2 - A_1 + 2\Omega}{L} - 1$ and that the truncated matrix T defined below is invertible. Computing matrix: $U = (U_{jk})$, $T = (T_{kl})$, where

$$U_{jk} = \phi(x_j - Lk), T_{kl} = \sum_{j=1}^{J} \overline{\phi(x_j - Lk)} \phi(x_j - Ll),$$

$$j = 1, \cdots, J, \ k, l = \frac{A_1 - \Omega}{L} + 1, \cdots, \frac{A_2 + \Omega}{L} - 1.$$

(2) Compute $c = T^{-1} b$ according to $b = \overline{U} y$, where $\overline{U}$ denotes complex conjugate of U and T^{-1} is the inverse of T.

Remark

(1) The construction of matrix T has the advantage that it is a positive operator on $\ell^2(Z)$ [24].

(2) In the case of B-spline, that matrix T is invertible under the condition that

$$\sup_j(x_{j+1} - x_j) \le \frac{L(A_2 - A_1)}{A_2 - A_1 + 2\Omega - L} \tag{4}$$

That is, the Schoenberg-Whitney Theorem implies that T is invertible [25].

(3) Since the numerical solution of (c_k) could be sensitive to particular sets of intersample spacings, we can approximate the inverse of the singular or ill-conditioned T by its pseudoinverse using singular valued decomposition (SVD).

2.2 Lomb-Scargle(L-S) Periodogram Method

In the following, we briefly give an expression of Lomb-Scargle periodogram method. For a times series $X(t_i)$, where $i = 1, 2, \cdots, N_0$, the periodogram as a function of the frequence ω is defined as

$$P_{xx}(\omega) = \frac{1}{2\sigma^2} \left\{ \frac{\left[\sum_{j=1}^{N_0} X(t_j) \cos \omega(t_j - \tau) \right]^2}{\sum_{j=1}^{N_0} \cos^2 \omega(t_j - \tau)} + \frac{\left[\sum_{j=1}^{N_0} X(t_j) \sin \omega(t_j - \tau) \right]^2}{\sum_{j=1}^{N_0} \sin^2 \omega(t_j - \tau)} \right\} \tag{5}$$

where τ is defined by the equation

$$\tan(2\omega\tau) = (\sum_{j=1}^{N_0} \sin 2\omega t_j) / (\sum_{j=1}^{N_0} \cos 2\omega t_j)$$

and $\sigma^2 = \frac{1}{N_0 - 1} \sum_{i=1}^{N_0} (X(t_i))^2$.

The above Lomb-Scargle periodogram method gives superior results on unevenly sampled data with classical FFT method: it weights the data on a "per point" basis instead of on a "per time interval" basis. Lomb proved that his periodogram is the same as the classical periodogram (see [26]) in the case of equally spaced data. But since it assumes there is a single stationary sinusoid wave that has infinite support in Lomb-Scargle method, it may introduce some spurious periodic components for finite data. Also due to the effect of noise in the given data, it may produce inaccurate estimation results.

2.3 Singular Spectrum Analysis (SSA) Method

Although spectral analysis can be applied directly to the original data, noise and outliers would degrade the final results. In DNA microarray analysis, a major challenge is to effectively dissociate actual gene expression values from noise.

As a robust model-free technique in time series analysis, the SSA can be used to extract as much reliable information as possible from short and noisy time series without prior knowledge of the dynamics underlying [9]. The main idea of the SSA is to perform singular value decomposition (SVD) of the so-called trajectory matrix obtained from the original time series with subsequent reconstruction of the series. It is aimed at an expansion of the original time series into a sum of a small number of independent and additive components that typically can be interpreted as trend components, oscillatory components and noise components. These properties make SSA a useful technique for gene expression data processing. The SSA decomposition into three parts is indeed related to the separation of the spectral range into three regions: low frequencies corresponding to the trend, the higher frequencies describing the oscillatory component, and the residual as noise. The Fourier analysis demonstrates that the power spectrum of the expression profiles are dominated by low frequency components. Therefore, SSA could be used to extracts the trend curve that represents the dominative component in order to reduce the effect of noise, which in fact is a process of data fitting. The detail of SSA can be found in [9].

2.4 Fisher's Test

To search for periodic gene expression, a formal test should be carried out to determine a peak in periodogram whether or not is significant or not. Fisher [7] proposed a test statistic and derived the null distribution of the Fisher's G-statistic for finite samples, and this is what will be utilized next. In the context of microarray gene expression data, the observed significance value or p-value for the hypothesis testing of the periodicity of a fixed gene g, using G-statistic as the test statistic, can be obtained by explicit expression. The G-statistic value calculated from the Fisher's G-statistic. So Fisher's Test gives a p-value that allows to test whether a gene behaves like a purely random process or whether the maximum peak in the periodogram is significant. For details on the G test statistic, its null distribution and the percentage points of the G test statistics, please refer to [27-29].

3 Experimental Results

3.1 Design of Experiment

In our numerical test, we choose B-spline of order N as the basis function ϕ in our reconstruction algorithm. It is obvious that $\operatorname{supp} \phi \subset [-\frac{N+1}{2}, \frac{N+1}{2}]$.

L-S method can directly based on unevenly sampling data, but it may be deduced by missing data and noise. Though SSA can be used to extract as much reliable information as possible from short and noisy time series without prior knowledge of the dynamics underlying, SSA are based on evenly sampling data. The classical periodogram method also has the disadvantage. So we need a reconstruction data from original raw unevenly signal for SSA and the classical periodogram method. We adopt our presented reconstruction algorithm in Section 2.1.

So experiments are designed with the following three cases:

I.   Raw data with missing points →L-S Method

II.   Raw data with missing points→Our Reconstruction Algorithm→Classical Periodogram Method

III. Raw data with missing points→Our Reconstruction Algorithm→SSA→ Classical Periodogram Method

3.2 Simulated Data

First, we use simulated data to test the proposed method with the three cases. For a gene g and expression level observed at time t_i, we denote the time series by $Y_g(t_i)$ for $i = 1, \cdots, N$ and $g = 1, \cdots, G$.

$$Y_g(t_i) = \beta \cos(\frac{t_i \pi}{12}) + \varepsilon_g(t_i) \qquad (6)$$

In the case of the simulated data, the length of a whole ideal time series (genes) is evenly 48 points. That is, $t_i = i(i = 1, \cdots, 48)$.

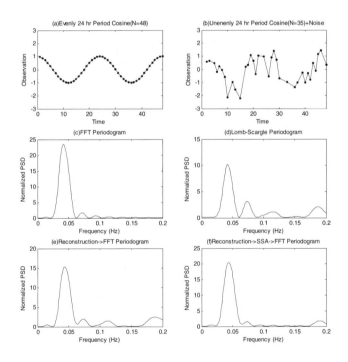

Fig. 1. Comparison of spectral estimation for simulated data: (a) simulated cosine signal with even sampling (N=48), (b) simulated noisy cosine signal with uneven sampling, (c) the periodogram of the simulated signal in (a) obtained using the Fourier transform, and the periodograms of the signal in (c) obtained using the three methods mentioned in Section 3.1 are shown in (d), (e) and (f), respectively

First, we test a gene that presented in (6). We give a classical periodogram of 24 hours period cosine with evenly 48 sampling points as a standard contrastive figure.

Let FFT periodogram based on evenly sampling points be a standard periodogram (See Figure 1(c)). Method III produces fewer and smaller false peaks than Method II in Figure 1(e), (f). Method II produces fewer and smaller false peaks than Method I in in Figure 1(d), (e). The peaks are highest among the three methods and closest the standard (ture) peak value with Method III (see Figure 1(f)).

3.3 Genes Data

The gene expression for S. Pombe in [30] were gathered from three experiments. Hence the data sets are called cdc25, elutA and elutB. In our test, we only apply our methods to deal with the gene expression data from the cdc25 experiment.

From the Table 1, we observe that Method III can find more periodic genes than Method I and Method II under the same FDR level threshold. The reasons are the reconstruction algorithm repairing the missing data and SSA reducing noise. Because the missing rate of cdc25 data is lower, the effect of Method I is almost as same as Method II. We believe that the effect of Method II is better than Method I in a high missing rate data.

Table 1. Data sets analyzed with different methods

Cell type	Experiment	Number of periodic expression		
		Method I	Method II	Method III
S. Pombe (Oliven et al., [30])	cdc25	420	411	493

Notation: The number of periodic genes that are statistically significant for a FDR level of q=0.03.

4   Conclusions

In this paper, we have proposed a new spectral estimation algorithm based on a signal reconstruction technique in an unevenly sampled space. The new algorithm is flexible since the order of the B-spline basis function can be adjusted. Experiments on simulated signals and real gene expression data show that Method III (Reconstruction→SSA→Classical Periodogram Method) is more effective in identifying periodically expressed genes. Due to the limitation of eight pages, we omit some simulated tests and real data tests.

Acknowledgements

This project is supported in part by NSF of China (60175031, 10231040, 60575004), Grants from the Ministry of Education of China (NCET-04-0791), NSF of Guang Dong and Grants from Sun Yat-Sen University 10471123, the Mathematical Tianyuan Foundation of China NSF (10526036) and China Postdoctoral Science Foundation

(20060391063), Natural Science Foundation of Zhejiang Province (Y606093) and Scientific Research Fund of Zhejiang Provincial Education Department（20051760）.

References

1. Hayes, M. H., Statistical digital signal processing and modeling. John Wiley and Sons, Inc. (1996)
2. Kay, S.M., Marple Jr, S.L.: Spectrum analysis-A modern perspective. Proceeding of IEEE. 69(11) (1981) 1380-1418
3. Chu, S., DeRisi, J., et al. : The transcriptional program of sporulation in budding yeast. Science. 282(1998) 699-705
4. Eisen, M.B., Spellman, P.T., Brown PO, Botstein D: Cluster analysis and display of genome-wide expression patterns. PNAS. 95(1998) 14863-14868
5. Spellman, T.S., Sherlock, G., et al: Comprehensive identification of cell cycle-regulated genes of the yeast Saccharomyces cerevisia by microarray hybridization. Mol. Biol. Cell. 9(1998)3273-3297.
6. Lu, X., Zhang, W., Qin, Z.H., Kwast, K.E., Liu, J.S.: Statistical resynchronization and Bayesian detection of periodically expressed genes. Nucleic Acids Research. 32(2)(2004) 447-455
7. Ruf, T.: The Lomb-Scargle Periodogram in biological rhythm research: Analysis of incomplete and unequally spaced time-series. Biological Rhythm Research. 30(1999) 178-201
8. Bohn, A., Hinderlich, S., Hutt, M.T., Kaiser, F., Luttge, U.: Identification of rhythmic subsystems in the circadian cycle of crassulacean acid metabolism under thermoperiodic perturbations. Biol. Chem. 384(2003)721-728
9. Golyandina N., Nekrutkin V., Zhigljavsky A., Analysis of time series structure: SSA and related techniques, Chapman & Hall/CRC, 2001.
10. Benedetto, J.J.: Irregular sampling and frames, in: C. K. Chui (Ed.), wavelets: A Tutorial in Theory and Applications (1992)445-507
11. Chen, W., Han, B., Jia, R.Q.: Estimate of aliasing error for non-smooth signals prefiltered by quasi-projections into shift-invariant spaces. IEEE Trans. Signal. Processing. 53(5) (2005) 1927-1933
12. Chen, W., Han, B., Jia, R.Q.: On simple oversampled A/D conversion in shift-invariant spaces. IEEE Trans. Inform. Theory. 51(2) (2005) 648-657
13. Chui, C.K.: An introduction to Wavelet, Academic Press, New York 1992
14. Daubechies, I.: Ten Lectures on Wavelets, SIAM 1992
15. Ericsson, S., Grip, N.: An analysis method for sampling in shift-invariant spaces. Int. J. Wavelet. Multi. Information. Processing. 3 (3) (2005) 301-319
16. Goh, S. S., Ong, I. G. H: Reconstruction of bandlimited signals from irregular samples. Signal. Processing. 46(3) (1995)315-329
17. Jia, R.Q.: Shift-invariant spaces and linear operator equations. Israel Math. J. 103(1998)259-288
18. Lei, J.J, Jia, R.Q., Cheney, E.W.: Approximation from shift-invariant spaces by integral operators. SIAM J. Math. Anal. 28(2) (1997) 481-498
19. Liu, Y.: Irregular sampling for spline wavelet subspaces. IEEE Trans. Information Theory. 42(2) (1996) 623 – 627
20. Liu, Y., Walter, G.G.: Irregular sampling in wavelet subspaces. J. Fourier. Anal. Appl. 2(2) (1995)181-189

21. Xian, J., Lin, W.: Sampling and reconstruction in time-warped spaces and their applications. Appl. Math. Comput. 157(2004)153-173
22. Xian, J., Luo, S.P., Lin, W.: Weighted sampling and signal reconstruction in spline subspaces. Signal. Processing. 86(2)(2006)331-340
23. Yeung, L.K., Szeto, L.K., Liew, A.W.C., Yan, H.: Dominant spectral component analysis for transcriptional regulations using microarray time-series data. Bioinformatics. 20(5)(2004)742-749
24. Golub, G.H., Van Loan, C.F.: Matrix Computations, 3rd ed., Johns Hopkins University Press, Baltimore, MD, (1996)
25. Schumarker, L.L.: Spline functions: Basic Theory, Wiley-Interscience, New York (1981)
26. Lomb, N.R.: Least-squares frequency analysis of unequally spaced data. Astrophys Space Science. 39(1976)447-462
27. Fisher, R.A.: Tests of significance in Harmonic Analysis. Proceedings of the Royal Society of London, Series A. 125(1929)54-59.
28. Davis, H.T.: The Analysis of Economic Time Series. Principia Press, Bloomington, Indiana; (1941)
29. Priestley, M.B.: Spectral Analysis and Time Series. Academic Press: San Diego. (1981)
30. Oliva, A., Rosebrock, A., Ferrezuelo, F., et al.: The cell cycle-regulated genes of Schizosaccharomyces pombe. Plos Biology. 3 (7)(2005)1239-1260

Creating Individual Based Models of the Plankton Ecosystem

Wes Hinsley[1], Tony Field[1], and John Woods[2]

[1] Imperial College Department of Computing, London, UK
w.hinsley@imperial.ac.uk
[2] Imperial College Department of Earth Science and Engineering, London, UK

Abstract. The Virtual Ecology Workbench (VEW) is a suite of utilities for creating, executing and analysing biological models of the ocean. At its core is a mathematical language allowing individual plankters to be modelled using equations from laboratory experiments. The language uses conventional mathematical assignments and seven plankton-specific functions. A model consists of a number of different plankton species, each with different behaviour. The compiler produces Java classes which when executed perform a timestep-based, agent-based simulation. Each agent is a Lagrangian Ensemble agent [13] representing a dynamic number of individuals, (a sub-population), that follow the same trajectory. The agents are simulated in a virtual water column that can be anchored, or can drift with ocean currents. This paper shows how the language allows biological oceanographers to create models without the need of conventional programming, the benefits of this approach and some examples of the type of scientific experiments made possible.

1 Introduction

There are two main methods for modelling plankton. Population-based modelling involves computing the size of a population of plankters from statistics and rules which apply to whole populations. Such models, first established by Lotka and Volterra [8], are computationally cheap and have thus been until recently the predominant method [1,4]. Alternatively individual-based modelling aims to describe the behaviour of an individual plankter and allow the properties of the population to emerge by integration. Popova et al used a population-based model to show the ocean ecosystem exhibits chaotic behaviour [11], whereas Woods et al with a comparable individual-based model produced stable results [14]. Lomnicki has argued that intra-population variation, unique to individual-based modelling, is the reason the two approaches may give different results. However since individual-based models are harder to code and computationally more expensive, population-based modelling has remained the preferred approach.

The Virtual Ecology Workbench (VEW) is designed to address the difficulties in creating individual-based models. The core of the VEW is the Planktonica language and the compiler which translates equations familiar to a biological oceanographer into Java, thus giving the oceanographer a familiar interface language with which to build models.

Y. Shi et al. (Eds.): ICCS 2007, Part I, LNCS 4487, pp. 111–118, 2007.

2   Creating Models

2.1   Functional Groups and States

A model consists of any number of functional groups which the user creates. A functional group is defined as a set of plankters that behave similarly. They may vary dynamically in terms of their position, volume, mass and any other user-defined property. They may also have parameters that are constant for all plankters of that type. The user creates a set of functions for each functional group. Each function describes some aspect of a plankter's behaviour in response to its biological properties and the properties of its local (ambient) environment. A function can have any number of rules which can be thought of as mathematical statements. These rules may be assignments to variables or calls to one of seven special-purpose functions, which may be executed conditionally or unconditionally.

Two important points must be noted. Firstly when writing functions and rules, the user is considering a single plankter and not a population, nor an agent, even though the simulation is agent-based. Secondly rules are executed each timestep and the user must ensure that the units of time in their rules are appropriate.

A functional group has at least one state, possibly many. These are usually used to represent stages in a life cycle. A plankter exists in one of its states at any time; in each state a subset of its functions are switched on and the remainder switched off.

2.2   The Water Column

The simulation is conducted in a virtual water column stratified into one-metre layers. The user can create any number of chemicals and for each, a concentration variable is automatically created in each layer. Chemicals can have pigmentation properties and when a plankter contains a quantity of such a pigment, the transfer of light and temperature through the water containing that plankter will be affected (called bio-optic feedback).

The physical properties of each layer are generated by a built in physics module, which calculates the irradiance (based on Morel [9]), temperature, salinity and density of water throughout the column. A separate layer structure offering higher resolution in the top metre is used, since this is where the irradiance and temperature change most rapidly with respect to depth.

The physics module also calculates the depth of the turbocline. Above this depth, the water can be assumed to be mixed in each half-hour timestep. The chemicals in solution above the turboclnie are therefore mixed each timestep. Below the turbocline, laminar flow may be assumed. Note that the plankters are not automatically mixed as part of the physics, to allow the user to make some plankters more buoyant than others.

2.3   Variables and Constants

The user can create three types of variable. The first is a 'biological parameter' which is a constant for all plankters of a functional group. The second is a 'biological property' which defines a property of an individual at a given time. It can be assigned a new value in each timestep. Some biological properties are created automatically such as a plankter's depth and internal pool for each chemical the user has created. It also inherits a read-only 'incoming pool' for each chemical which provides the amount of chemical gained in the previous timestep by uptake from the environment, and ingestion of other plankters. Thirdly the user can for convenience create intermediate variables for breaking up a complex equation into parts, or for sharing an intermediate value between other rules or functions.

Other variables may be read by the user but not written. These include ambient physical properties (temperature and irradiance for example), ambient chemical concentrations for each chemical the user has created, and ambient biological properties - the local concentration of any functional group. Additionally, the timestep size in hours and the depth of the turbocline can be read. The latter is needed in order to write motion equations, since it is necessary to know whether the plankter is above or below the turbocline.

2.4   Special-Purpose Functions

Along with basic assignments, there are seven special-purpose functions defined below which facilitate interactions between the plankter and its environment.

Uptake(chemical, amount) and Release(chemical, amount). The uptake call is used when the plankter attempts to absorb chemical from its ambient environment. Requests are proportionally issued over each layer that the plankter visits on the journey between the current timestep and the next, scaled by how long it spends in each layer (or part thereof). In cases where the amount of chemical available is insufficient (since many plankters may try to uptake in the same layer), all the requests will be proportionally scaled down so that they sum to the available amount. The actual amount gained by an individual will be available in its 'chemical-gained' biological property for that chemical. The release call performs the opposite; the specified chemical is released into the environment proportionally in each layer the plankter moves through.

Divide(x) and Create(x, state, [assignments]). These two functions model cell division and reproduction respectively. The two are treated separately because cell division offers a useful optimisation, namely that having divided into two, the two parts will be identical. When the user writes divide(x), they are stating that the plankter should divide into 'x' indentical parts. Since cell divisions among diatoms are extremely common in Spring, creating new a plankter agent for each division would be costly on memory. Instead the sub-population size of the agent is multiplied by 'x'.

The create function by contrast creates a new agent with its own trajectory. The agent represents a number of plankters with the same biological properties, but they may differ from those of the parent. The number of offspring is defined by 'x'; this is the number of individuals that the new agent will represent. The state of the offspring is defined by 'state' and a list of assignments may be provided to initialise the biological properties of the offspring.

Change(state) and Pchange(p,state). Two methods of changing state are provided. The 'change' function causes an unconditional change from the current state to a newly specified one. When the user writes the 'pchange' function they are stating there is a probability 'p' that the plankter should change into its new state. Planktonica interprets this as splitting the agent into two, one representing the proportion 'p' that did change and the remainder are left in their original state.

Ingest(foods, rates, thresholds). The final function call is for plankters that perform ingestion. Any number of foods may be selected, where a food is defined by its functional group and stage. Each food may be ingested at a different 'rate', in individuals per second, provided that the concentration of prey is above a specified 'threshold'. More specifically, the type of food is specified by 'species' and stage; species will be defined in the following section.

Like the uptake call, requests are made in each layer that the plankter will pass through in the next timestep. If there is insufficient food of any of the types, the request for that food-type will be scaled down. The sub-population sizes of the prey that were ingested will be reduced accordingly and the chemicals in the pools of the ingested plankters will be transferred to the chemical-gain biological property of the predator. A further system variable is available called 'ingested', which returns the number of individuals of each food-type that were ingested.

3   Further Specification

3.1   Species

Having created the functional groups and chemicals the user creates at least one species of each functional group. A species is a parameterisation of a functional group in which a value is given for each of the biological parameters for that functional group. A common example is to create a number of size-based variations of a functional group so that the species all exhibit the same behaviour, but with different size-dependent rates. Having defined the species the user can now specify the foods, rates and thresholds for each of the ingestion calls they may have made in their model.

3.2   Scenario

The scenario defines the input data for the simulation. This includes the starting time and duration of the simulation and the track that the water column takes.

The column may be anchored at a location or it may drift in response to ocean currents, supplied by OCCAM [10]. Data is created for each timestep, providing the astronomical and climatological values required by the physics code. This data can be taken from Bunker [2] or the ERA40 data set [3]. The physical and chemical profiles of the water column also must be initialised. For some chemicals, data is available from the Levitus World Atlas [6].

3.3   Agent Initialisation and Management

Initial distributions of plankters are then configured; any number of sets of plankters may be distributed at a given density with a given set of initial values for their biological properties and an initial state. Agent management rules can then be specified which allow the user to control the compromise between demographic noise and computational cost. Two rule types are available. The first allows splitting of the largest agents (that is, agents with the largest sub-population size), should the number of agents of a given species and stage fall below a given limit. The second allows merging of the smallest agents if the number of agents exceeds a given limit. Both of these can be specified to apply to each one-metre layer of the column, or to the column as a whole.

3.4   Logging Options

The output of the simulation can include water column properties such as the total number of plankters of each species and stage, or field data can be recorded layer-by-layer for all or part of the column, such as concentrations of plankters, concentrations of chemical (in solution or particulate form), and physical properties such as temperature and irradiance. Audit trails, which are values of the biological properties of individual plankters and their ambient environmental properties may be plotted. These are unique to individual-based modelling, and not easily observable in nature. Additionally, demographic statistics can be plotted which show how many plankters changed state due to a particular function and the location of those plankters in the column. Finally, for debugging purposes the local variables the user created can be logged.

4   Compilation and Execution

A single XML file stores the model and all the associated specification information. The compiler then produces Java classes from the specification file, for each functional group and chemical. These are then compiled by the standard Java compiler along with a set of kernel classes which drive the simulation. The climate data specified in the scenario is extracted from the available datasets and written to binary files. All the files are then packaged into a single executable JAR file.

Execution can be done silently, or alternatively a utility called 'LiveSim' is provided for running the simulation step by step and inspecting every property of the system visually. This is useful for debugging models and as an interactive teaching tool to explain the behaviour of plankters. The VEW contains a documentation tool which writes all the information necessary to recreate the model in an indexed HTML format.

5   Applications and Example Results

The modelling language was designed with reference to the WB model [13], an individual based containing Nitrogen, Phytoplankton, Zooplankton and Detritus (NPZD). The VEW has been recently used to develop the Lagrangian Ensemble Recruitment Model (LERM), which includes diatoms, copepods and squid larvae in an environment with nitrogen and silicate [12].

Below are some of the typical plots available using the VEW's own Analyser package, taken from a simulation of the WB model [13]. Figure 1(a) shows the yearly cycle of diatoms and copepods. The diatoms bloom in Spring until they run out of nutrient. The copepods then grow by feeding on the diatoms until they can reproduce. The parent copepods then die of old age and the next generation will have to wait until the following year until there is enough food for them to grow to reproductive size.

Figure 1(b) is an example plot of field data. It shows the nitrogen concentration varying over depth and time, with the turbocline marked. Through Summer nitrogen near the surface is depleted by the diatoms, which are nitrogen-limited. In Autumn, the turbocline descends and nitrogen from deeper water is mixed above the turbocline. Copepods excrete nitrogen and dead diatoms, dead copepods and copepod pellets remineralise nitrogen.

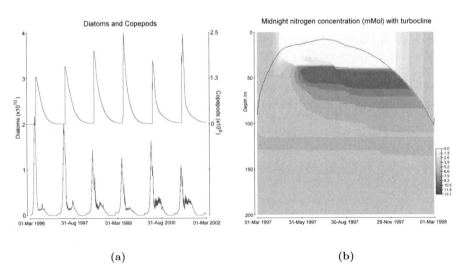

(a)   (b)

Fig. 1. (a): Diatom and Copepod yearly cycles. (b): Nitrogen throughout the column at midnight with turbocline.

Figure 2 shows the unique characteristic of individual-based modelling. A single diatom's energy pool is plotted along with its depth. The more intense the diatom's ambient irradiance, the higher its energy gain by photosynthesis will be, whereas in relatively dark water, or at night, respiration losses outweigh energy gains. Note also the effect of the turbocline on the depth of the plankter; below the turbocline it sinks, whereas above the turbocline it is randomly displaced.

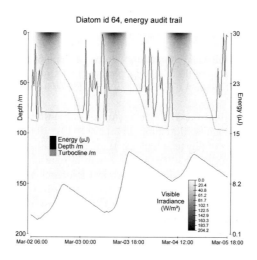

Fig. 2. Energy of an individual diatom

6   Conclusion

The Virtual Ecology Workbench with the Planktonica language at its core offers the best known method for creating individual-based plankton models. The equations for the WB model above can be shown succinctly in a few pages of equations, whereas conventional methods were only able to produce raw computing source code. The latter gave rise to inherent problems of maintainability, requiring a mediator between the biological modeller and the simulation code, as few biological modellers are skilled programmers.

While it is necessary for the modeller to familiarise themselves with the special function calls and to take the necessary care in ensuring their equations are correct and input correctly to the system, the amount of pure computing skill required has been significantly reduced.

While individual-based models offer advantages over population-based models [7,14], the difficulty in constructing such simulations has limited research into the precise nature of the them. The VEW now offers a convenient method for building these simulations, and investigating more thoroughly the issues regarding these two approaches to modelling plankton.

References

1. Anderson, T.R., Pondaven, P.: Non-redfield carbon and nitrogen cycling in the Sargasso Sea: pelagic imbalances and export flux. Deep-Sea Research I **50** (2003) 573–591
2. The Bunker Climate Atlas of the North Atlantic Ocean. http://dss.ucar.edu/datasets/ds209.2/
3. The ERA-40 Re-analysis Dataset. http://www.ecmwf.int/research/era/
4. Fasham M.J.R., Ducklow H.W., McKelvie S.M.: A nitrogen-based model of plankton dynamics in the oceanic mixed layer. Journal of Marine Research **48** (1990) 591–639
5. Hinsley W.: Planktonica: A System for Doing Biological Oceanography by Computer. PhD Thesis, Imperial College Department of Computing. (2005)
6. The Levitus World Ocean Atlas. http://www.cdc.noaa.gov/cdc/data.nodc.woa98.html
7. Lomnicki, A.: Individual-based models and the individual-based approach to population ecology. Ecological Modelling **115** (1999) 191–198
8. Lotka, A.J.: Elements of physical biology. Baltimore: Williams & Wilkins co. (1925)
9. Morel, A.: Optical modelling of the upper ocean in relation to its biogenous matter content (case 1 water). Journal of Geophysical Research **93(c9)**, 10749
10. The OCCAM global ocean model. http://www.noc.soton.ac.uk/JRD/OCCAM/
11. Popova E.E., Fasham M.J.R., Osipov A.V.O., Ryabchenko V.A.: Chaotic behaviour of an ocean ecosystem model under seasonal external forcing. Journal of Plankton Research **19:10** (1997) 1495–1515
12. Sinerchia M.: Testing Fisheries Recruitment Models. Ph.D in preparation, Imperial College Department of Earth Sciences and Engineering. (2006)
13. Woods, J.D.: The Lagrangian Ensemble metamodel for simulation plankton ecosystems. Progress in Oceanography **78** (2005) 84–159
14. Woods, J.D., Perilli, A., Barkmann, W.: Stability and predictability of a virtual plankton ecosystem created with an individual-based model. Progress in Oceanography **67** (2005) 43-83

A Hybrid Agent-Based Model of Chemotaxis

Zaiyi Guo and Joc Cing Tay

Evolutionary and Complex Systems Program
School of Computer Engineering, Nanyang Technological University, Nanyang Avenue,
Singapore 639798
asjctay@ntu.edu.sg

Abstract. Models for systems biology commonly adopt Differential Equations or Agent-Based modeling approaches for simulating the processes as a whole. These choices need not necessarily be mutually exclusive. We propose a hybrid agent-based approach where biological cells are modeled as individuals (agents) while chemical molecules are kept as quantities. This hybridization in entity representation entails a combined modeling strategy with agent-based behavioral rules and differential equations, thereby balancing requirements of modeling extensibility with computational tractability. We demonstrate the efficacy of this approach with a realistic model of chemotaxis based on receptor kinetics involving an assay of 10^3 cells and 1.2×10^6 molecules. The simulation is efficient and the results are agreeable with laboratory experiments.

Keywords: Hybrid Models, Multi-Agents, Chemotaxis, Receptor Kinetics.

1 Introduction

A large part of *systems biology* involves mathematical models constructed to understand the biological system (such as the human immune system) as a *whole* [1] from integrating knowledge obtained from laboratory experiments on *sub-parts* of the system. To simulate these *system level* dynamics efficiently, such models necessarily assume compartmentalized homogeneity and spatial independence [2, 3]. An alternative approach that has been gaining popularity are Multi-Agent (or MA) models, which explicitly model biological cells and molecules as individual agents [4-7]. As the real biological systems comprise millions to trillions of cells, MA models inevitably face issues of computational tractability and model scalability [8]. Despite these drawbacks, MA models are appealing for their advantages in phenomenological encoding of biological cell behaviors as agent rules [8]. Attempts have been made to overcome the computational tractability; such as choosing an intermediate level of abstraction [6] and applying the model to a distributed or parallel computational framework [9].

Though MA models generally deal with individuals instead of quantities, we believe that the DE modeling techniques for characterizing aggregated dynamics can be applied to MA models as well, giving rise to a hybrid modeling approach. In this paper, we propose a hybrid approach for designing MA system models where cells remain modeled as individual agents and molecules kept as quantities. This approach is motivated by the morphological distinction between cells and molecules, where

Y. Shi et al. (Eds.): ICCS 2007, Part I, LNCS 4487, pp. 119–127, 2007.

molecules are much smaller (in the order of nanometers vs. cells in the order of micrometers), more abundant, have less complex behaviors and are well-characterized by physical laws (such as mass action law and Fick's diffusion law) as compared to cells. We demonstrate the efficacy of this approach with a model of chemotaxis. Our experimental design mimics the Under-Agarose Assay [10]: an *in vitro* experiment designed to study the chemotactic behavior of cells. Simulation results are found to be comparable with laboratory experiments.

2 Integrating Agents and Quantities

We consider the hybridization on two levels: globally in the environment, and locally with a cell (on its surface *and* within its body).

The environment is where biological cells and molecules reside and interact. The environment handles the storage and dissemination of spatial proximity information of cells and molecules. Consider the neutrophils (a type of white blood cell) that are attracted to the site of an injury as a result of communicated information in the form of inflammatory signals. The spread of these inflammatory signals, which is essentially based on molecular diffusion, is an environment level event. Together with this, another exogenous event involves blood flow (within the blood vessels) that forces the neutrophils and other related immune cells and molecules to recirculate in the human body. And then, upon entering the damaged tissue, their movement becomes more governed by the process of chemotaxis. Therefore, the environment is not merely a passive 'holder' to place the agents, but an entity with its own activities and special properties (e.g., blood flows). How agents behave is greatly influenced by its environment, hence the modeling of the environment is vital to agent-based system models of microbiological phenomena. We propose a three-layer environment, namely, 1) An *Agent Holder Layer*, where all agents (individuals) reside, 2) A *Molecule Space Layer*, where all molecule distributions are recorded, 3) A *Flow Field Layer*, where background flows, such as blood flows and lymph flows are modeled. Fig. 1a illustrates the three layers as composable two-dimensional planes.

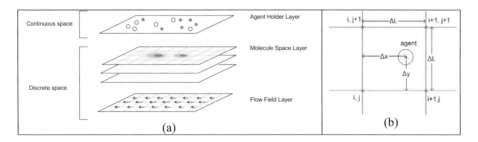

Fig. 1. (a) Three-layer structure of environment (b) Grid point interpolation by agent

The layers are either continuous or discrete in space. In continuous space layers, real-valued coordinates are used. In the discrete space layers, values are only stored at discrete 'grid points', hence values at real-valued coordinates may require interpolation. Of the three layers, only the *Agent Holder Layer* is continuous. The distance between two neighbouring grid points ΔL, for a discrete space determines its *granularity*. Each grid point has a corresponding real-valued coordinate in the continuous space of the Agent Holder Layer, and each agent coordinate falls within a grid cell. The correspondence in spatial location enables spatial proximity-based inter-layer communication.

The *Molecule Space Layer* consists of sub-layers, each representing the spatial distribution of a certain type of molecule. We let $Q_{i,j}$ denote the number of molecules surrounding the grid point (i,j). In our model, the net change in molecule quantity over the entire spatial distribution is a combined result of 1) *Molecular Production*, where cells synthesize and release cytokines into its neighbourhood according to specific agent rules, 2) *Molecular Diffusion*, where molecules spread due to Brownian movement based on Fick's diffusion law, 3) *Fluid Mechanics*, in which molecules are 'pushed' by the local fluid flow within certain compartments such as the blood vessels and lymph duct. This is modeled by sampling the local velocity of the flow from the *Flow Field Layer*, and computing the quantity shift along the flow velocity by redistributing the molecular quantity to neighbouring grid points, and 4) *Molecular Consumption / Degradation*, where certain molecules are 'consumed' by binding to cell receptors. Molecules also decay through various pathways (e.g., instability or enzymatic reactions), which is quantitatively specified by the decay rate constants (λ) or half-life values ($t_{1/2}$).

The *Flow Field Layer* characterizes the background flow in the environment. It represents the systematic transport superimposed onto the random movements of molecules and agents. The flow velocity $\mathbf{F}_{i,j}$ is specified for each grid point (i,j), The Flow Field Layer is sampled by the Molecule Space Layer for computing the systematic shift of molecules, as stated in point (3) above.

The agents in the Agent Holder Layer move not only passively via systematic transport due to existence of flow, but also actively in response to the local molecular concentration and gradient, exhibiting *chemokinesis* and *chemotaxis*. Formally, we specify the displacement of an agent $\Delta \mathbf{p}$ within time Δt as a function of the local flow velocity $\mathbf{F_p}$, molecular concentration $C_{m,p}$ and gradient $\mathbf{G_{m,p}}$ for a set of molecules $\mathbf{m}$. Consider an agent in Fig. 1b, whose nearest grid points are (i,j), $(i,j+1)$, $(i+1,j)$, and $(i+1,j+1)$. The local concentration of molecule m at the agent's position obtained by bilinear interpolation is:

$$C_{m,p} = \left(Q_{i,j}v_x v_y + Q_{i+1,j}w_x v_y + Q_{i,j+1}v_x w_y + Q_{i+1,j+1}w_x w_y\right)/\Delta L^2$$
$$\text{with } w_x = \Delta x/\Delta L, v_x = 1 - \Delta x/\Delta L, \quad w_y = \Delta y/\Delta L, v_y = 1 - \Delta y/\Delta L \tag{1}$$

The local gradient $\mathbf{G}_{m,p} = \nabla C_{m,p}$ is a vector with components:

$$G_x = \left(\left(Q_{i+1,j} - Q_{i,j}\right)v_y + \left(Q_{i+1,j+1} - Q_{i,j+1}\right)w_y\right)/\Delta L^3$$
$$G_y = \left(\left(Q_{i,j+1} - Q_{i,j}\right)v_x + \left(Q_{i+1,j+1} - Q_{i+1,j}\right)w_x\right)/\Delta L^3 \tag{2}$$

The local flow velocity is:

$$\mathbf{F_p} = \mathbf{F}_{i,j}v_x v_y + \mathbf{F}_{i+1,j}w_x v_y + \mathbf{F}_{i,j+1}v_x w_y + \mathbf{F}_{i+1,j+1}w_x w_y \tag{3}$$

The functional form of $\Delta \mathbf{p}/\Delta t$ is determined by the agent's ruleset based on its biological properties: some types of cells are highly motile and sensitive to molecular stimuli while others are rather fixed.

We have described the hybrid modeling of the environment. We next introduce the interaction rule building scheme for biological cells.

Biological cells are modeled at the individual level as *agents*, and their behaviors are specified by a set of rules. Most MA models use phenomenological if-then rules [7, 11]. In reality, cell behaviors emerge from complex signaling pathways of molecular interactions and chemical reactions [6]. Therefore we believe the incorporation of molecular interactions (through surface receptors) into agent rules is necessary for a more quantitatively accurate MA model. As a first step, we develop a model for cellular locomotion through receptor kinetics [12]. We assume an agent's behaviour is governed by the following events: 1) *Receptor-ligand binding*, where free ligand molecules bind to receptors on the cell surface, 2) *Receptor internalization* and *ligand consumption*, where the receptor-ligand complex are internalized into the cytosol, and ligand is 'used up'/digested by other biological processes, and 3) *Receptor recycling*, where the previously internalized receptors are returned to the cell surface, available for binding.

Table 1. The individual reaction rules and corresponding quantity change equations

Individual Reaction	Quantity Change	
Receptor-ligand binding $$L + R \xrightarrow{k_b} LR$$	$$\frac{d[L]}{dt} = -K_b[L][R]$$	(4)
Receptor internalization and ligand consumption $$LR \xrightarrow{k_i} R*$$	$$\frac{d[R]}{dt} = -K_b[L][R] - K_r[R^*]$$	(5)
Receptor recycling $$R* \xrightarrow{k_r} R$$	$$\frac{d[LR]}{dt} = K_b[L][R] - K_i[LR]$$	(6)
	$$\frac{d[R^*]}{dt} = K_i[LR] - K_r[R^*]$$	(7)

The leftmost panel in Table 1 gives the molecular reactions of the three events underlying receptor kinetics. L and R denote individual ligand and receptor molecules respectively. The labels on the arrows characterize the probability of reaction per unit time. The right panel gives the corresponding differential equations, which are the macroscopic quantity-level changes that emerge from the individual reactions and are derived through mass action law. The rate constants K_x ($x \in \{b,i,r\}$) are proportional to their microscopic counterpart k_x and denote the number of reactions per unit time per reactant(s). By convention, [•] represents the *concentration* of the molecules in the whole space, but here we will define it as *quantities* of molecules *in the proximity of*, or *within* the agent. Readers may find these equations familiar, however, we emphasize that instead of a direct overall description of the molecular kinetics, these equations are *local* to an agent.

Table 2. Agent Rule: Receptor-Ligand Interaction

If Alive Then

      Sample the environment: get current local ligand quantity $[L]$:

      Get the values of individually stored quantities $[R], [LR]$, and $[R*]$

      Compute $d[L], d[R], d[LR], d[R*]$ using equations **(4)** - **(7)**

      Update local quantities: $[X] = [X] + d[X]$ ($X \in \{R, L, LR, R*\}$)

End If

Table 2 expresses the agent rule that utilizes these equations for updating the local molecule quantities $[L]$, $[R]$, $[LR]$, and $[R*]$. The rule we specify here is quite flat as there is only one proposition and all the equations appear in the action part. However, it is possible to specify more complex situations that an agent can be in and the different responses it could make. We incorporate this receptor-ligand interaction in a hybrid model of chemotaxis in the next Section.

3 Chemotaxis

Chemotaxis refers to the directed cell movement towards or away from a chemical source (chemoattractant or chemorepellent) by sensing and following the *chemical gradient* [13]. Through chemotaxis, a sperm cell finds the ovum, and a white blood cell finds the place of injury or inflammation [13]. Much of the research effort has been devoted to discover the mechanisms that transduce the extracellular chemical signals into cell locomotion [14-16], including molecular events such as G-protein activation, lipid remodeling, protein kinase activation, calcium elevation [13, 15] and cellular events such as cell polarization, pseudopod extension, excitation and adaptation [13, 16]. However, these molecular details are seldom captured quantitatively into models for simulation. Keller and Segel used Partial-DEs to model chemotaxis [17]: both cells and molecules are treated as quantities, hence the intracellular molecule kinetics are largely ignored. Our model differs in that chemotactic behaviors are described at an individual level instead of a population level. Though the model we present here is simplified, the hybrid approach makes it possible to incorporate more complex intracellular signaling pathways for future extensions. We next describe how an agent's displacement is computed based on our model of chemotaxis.

We define the overall displacement of an agent (or cell) as the vector sum of Brownian, chemotactic and systematic displacement due to external fluid flows. In this experiment, we define chemotactic response to be proportional to the difference in the receptor occupancy of the front and rear of the cell [18]. The receptor-ligand interactions have been described in Equation (4) to (7). Let r and V represent the radius and the volume of an cell, and let $[R]$, $[LR]$, $[R*]$ represent the quantities of the free surface receptors, occupied surface receptors and the internalized receptors respectively. The amount of ligand available at the front and the rear of the cell is approximated by:

$$[L]_{front} = (C+|\mathbf{G}|r)V/2 \ , \ [L]_{rear} = (C-|\mathbf{G}|r)V/2 \tag{8}$$

The change in newly bound receptors (LR) in time Δt at the front and rear are:

$$\Delta[LR]_{front} = K_b[L]_{front}[R]_{front}\Delta t \ , \ \Delta[LR]_{rear} = K_b[L]_{rear}[R]_{rear}\Delta t \tag{9}$$

and we assume $[R]_{front} = [R]_{rear} = [R]/2$. The difference in newly bound receptors between the cell front and the rear in time Δt is therefore

$$\Delta[LR]_{diff} = K_b([L]_{front} - [L]_{rear})[R]\Delta t/2 \tag{10}$$

Assuming linear dependency of the agent's displacement on $\Delta[LR]_{diff}$ gives rise to

$$\Delta\mathbf{p} = k\Delta[LR]_{diff}\,\mathbf{G}/|\mathbf{G}| \tag{11}$$

where k is a proportionality parameter. The larger the value of k, the more sensitive is the cell is to the change in receptor occupancy at its front and rear.

We conduct an *in silico* simulation of the Under-Agarose Assay, which was designed to investigate the *in vitro* chemotactic behavior of lymphoblastic cells [10]. Fig. 2 illustrates the assay and a sample result. The tissue culture dish is filled with Agarose (gel) and two wells are incised, where one is filled with cells and the other with a chemoattractant (lysophosphatidylcholine). The cells are observed to migrate towards the chemoattractant well by squeezing under the gel (Fig. 2).

Fig. 2. Under-Agarose Assay (Figures adopted from [10] and [19])

Table 3. Model Parameters for Chemotaxis Experiment

Parameter	Value	Description
Δt	1 min	Time-step
ΔL	20 μm	Granularity of the site grid
D	10 μm²min⁻¹	Diffusion constant of molecule L
V	50 μm²	Volume of cell
r	4 μm	Radius of cell
l	2 μm min⁻¹	Random motility of cell: max step length per unit time
R_{tot}	50000	The total amount of receptor molecule R per cell
K_r	0.5 min⁻¹	The recycle rate of internalized receptor R^*
K_b	2×10⁻⁶ min⁻¹	The association rate of L and R
K_i	0.3 min⁻¹	The internalization rate of occupied receptor LR

To mimic the Under-Agarose Assay, we construct a virtual environment as a 3mm by 2mm enclosed two-dimensional space (Fig. 3, left). The cell and attractant wells are both 250μm in radius and are placed 1mm apart (Fig. 3, left, red circle). Initially, the cell well is filled with 1000 cells, and the attractant well is filled with attractant molecules at a concentration of 2500 particles per 20μm×20μm area (for a total of about 1.2×10^6 molecules). Table 3 summarizes the parameter values for the simulation.

The simulation model is implemented in Java. The simulation is run on a Windows XP Professional system with 2.4 GHz Pentium 4 CPU and 1GB of Memory. One single run of the simulation takes approximately 3 minutes. The video capture of the simulation can be downloaded freely from http://www.ntu.edu.sg/home/asjctay/data/chemotaxis.zip (7.56 MB in size).

Fig. 3 shows the simulation results with k set to 5 and 30. The left figures show the migration pattern at time step t=5000. Both exhibit the biased random walk of the cells towards the attractant, and with larger k values, the spread accelerates. Due to ligand consumption, the space occupied by the cells is at a lower concentration, resulting in a sharp gradient at the edge of the diffusing molecular population, which promotes the movement of cells. The right figures show the movement of the front (x_{max}), rear (x_{min}), and mean (x_{avg}) positions of the molecular migration shape over time. L_x and L_y denotes the bounding width and length of this shape along the x and y axes, and L_x/L_y measures its distortion. Interestingly, though the spread towards the source of the attractant is evident, the average shift along the x axis is very small,

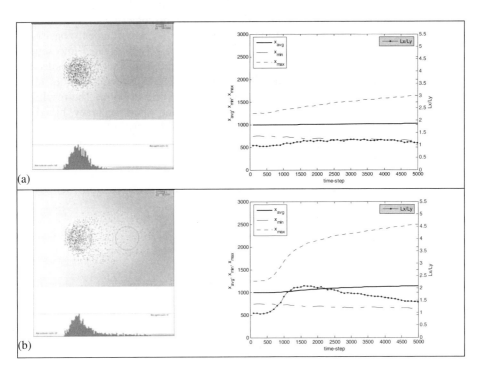

Fig. 3. Simulation results of Chemotaxis at time steps (a) k=5 and (b) k=30

as a large amount of the cells are still within the cell well and exhibit no chemotactic movement. This is likely due to low concentration created by the front-end cells, hence the rear-end cells are blocked from sensing the distant gradient. For $k =30$ the distortion maximizes at t=1500, showing that cells are attracted along the x axis. The distortion then slowly decreases, indicating increasing spread along the y axis due to gradient created at the edge of the migration shape. The cells spread rather than move with a narrow front, hence our results differ from the illustration from [10] but closer to the photo in [19] (see Fig. 2). Our simulation results show that the spread is caused by the distortion of the gradient landscape due to both diffusion and interactions between cells and molecules. The efficacy of the hybrid approach to construct agent-based models for biological phenomena has been demonstrated with a model of chemotaxis verified qualitatively with laboratory experimentation.

4 Conclusions

We present a hybrid agent-based modeling approach, where biological cells are modeled as individual agents and molecules as quantities, thereby balancing requirements of modeling extensibility (i.e. to handle complex cell behaviors) with computational tractability. The hybridization of the quantity-based macroscopic modeling (for molecules) and the individual-based microscopic modeling (for cells) exists at both the environment and individual levels. The environment consists of three layers: an Agent Holder Layer, a Molecule Space Layer (divided into several sub-layers), and a Field Layer. Agents (individual cells) interact with molecules (quantities) based on spatial proximity. The model allows agent behaviors to be specified by differential equations, if-then logic, or a hybrid of both. The result is a modeling approach that provides extensibility and increased accuracy based on available information. We demonstrate the efficacy of this hybrid approach with a realistic model of chemotaxis based on receptor kinetics for an Under-Agarose assay of 10^3 cells and 1.2×10^6 molecules. The experiment successfully and efficiently reproduces a "biased random walk" as reported in the research literature, with results showing strong agreement with migration patterns observed from laboratory experiments.

References

1. Kitano, H.: System Biology: A Brief Overview. Science 295 (2002) 1662-1664
2. Perelson, A.S., Nelson, P.W.: Mathematical Analysis of HIV-1 Dynamics in Vivo. SIAM Review 41 (1999) 3-44
3. Stekel, D.J., Parker, C.E., Nowak, M.A.: A Model of Lymphocyte Recirculation. Immunology Today 18 (1997) 216-221
4. Casal, A., Sumen, C., Reddy, T.E., Alber, M.S., Lee, P.P.: Agent-Based Modeling of the Context Dependency in T Cell Recognition. Journal of Theoretical Biology 236 (2005) 376-391
5. Webb, K., White, T. Cell Modeling Using Agent-Based Formalisms. In: AAMAS'04, New York, USA (2004)

6. Khan, S., Makkena, R., McGeary, F., Decker, K., Grills, W., Schmidt, C. A Multi-Agent System for the Quantitative Simulation of Biological Networks. In: AAMAS 03, Melbourne, Australia (2003) 385-392

7. Guo, Z., Han, H.K., Tay, J.C. Sufficiency Verification of HIV-1 Pathogenesis Based on Multi-Agent Simulation. In: Proceedings of the Genetic and Evolutionary Conference 2005, Vol. I. Washington D.C. (2005) 305-312

8. Guo, Z., Tay, J.C. A Comparative Study of Modeling Strategies of Immune System Dynamics under HIV-1 Infection. Lecture Notes in Computer Science, Vol. 3627. Springer-Verlag, Banff, Albeta, Canada (2005) 220-233

9. Logan, B., Theodoropoulos, G.: The Distributed Simulation of Multi-Agent Systems. Proceedings of the IEEE 89 (2001) 174-185

10. Heit, B., Kubes, P.: Measuring Chemotaxis and Chemokinesis: The under-Agarose Cell Migration Assay. Sci. STKE 170 (2003) pl5.

11. Jacob, C., Litorco, J., Lee, L. Immunity through Swarms: Agent-Based Simulations of the Human Immune System. In: Proceedings of 3rd International Conference on Artificial Immune Systems, Vol. 3239. Springer-Verlag GmbH, Catania, Sicily, Italy (2004) 400-412

12. Zigmond, S.H., Sullivan, J.L., Lauffenburger, D.A.: Kinetic Analysis of Chemotactic Peptide Receptor Modulation. The Journal of Cell Biology 92 (1982) 34-43

13. Eisenbach, M., Lengeler, J.W., Varon, M., Gutnick, D., Meili, R., Firtel, R.A., Segall, J., E., Omann, G.M., Tamada, A., Murakami, F.: Chemotaxis. Imperial College Press, London (2004).

14. Mitchison, T.J., Cramer, L.P.: Actin-Based Cell Motility and Cell Locomotion. Cell 84 (1996) 371-379

15. Firtel, R.A., Chung, C.Y.: The Molecular Genetics of Chemotaxis: Sensing and Responding to Chemoattractant Gradients. BioEssays 22 (2000) 603-615

16. Sanchez-Madrid, F., Del Pozo, M.A.: Leukocyte Polarization in Cell Migration and Immune Interactions. The EMBO Journal 18 (1999) 501-511

17. Keller, E.F., Segel, L.A.: Model for Chemotaxis. Journal of Theoretical Biology 30 (1971) 225-234

18. Devreotes, P.N., Zigmond, S.H.: Chemotaxis in Eukaryotic Cells: A Focus on Leukocytes and Dictyostelium. Ann. Rev. Cell. Biol. 4 (1988) 649-686

19. Hoffman, R.D., Klingerman, M., Sundt, T.M., Anderson, N.D., Shin, H.S.: Stereospecific Chemoattraction of Lymphoblastic Cells by Gradients of Lysophosphatidylcholine. Proc. Natl. Acad. Sci. USA 79 (1982) 3285-3289

Block-Based Approach to Solving Linear Systems

Sunil R. Tiyyagura and Uwe Küster

High Performance Computing Center Stuttgart,
University of Stuttgart,
Nobelstrasse 19, 70569 Stuttgart, Germany
`sunil,kuester@hlrs.de`

Abstract. This paper addresses the efficiency issues in solving large sparse linear systems parallely on scalar and vector architectures. Linear systems arise in numerous applications that need to solve PDEs on complex domains. The major time consuming part of large scale implicit Finite Element (FE) or Finite Volume (FV) simulation is solving the assembled global system of equations. First, the performance of widely used public domain solvers which target performance on scalar machines is analyzed on a typical vector machine. Then, a newly developed parallel sparse iterative solver (Block-based Linear Iterative Solver – BLIS) targeting performance on both scalar and vector systems is introduced and the time needed for solving linear systems is compared on different architectures. Finally, the reasons behind the scaling behaviour of parallel iterative solvers is analysed.

Keywords: Sparse linear algebra, Scalability, Block algorithms, Indirect memory addressing.

1 Introduction

The most powerful supercomputers today are an agglomeration of thousands of scalar processors connected with innovative interconnects. The major concern with this developing trend is the scalability of communication intensive applications which need global synchronization at many points, for example a conjugate gradient solver. Vector architectures seem to be a better alternative for such applications by providing a cluster of very powerful SMP (Symmetric Multiprocessing) nodes capable of processing a huge amount of computation and thereby reducing the overhead for synchronization. The rapidly increasing gap between sustained and peak performance of the high performance computing architectures [1] is another concern facing computational scientists today. The sustained computation to communication ratio and the computation to memory bandwidth ratio are much better for vector architectures when compared to clusters of commodity processors [2]. This emphasizes the need to look towards vector computing as a future alternative for certain class of applications.

In this paper we focus on the performance of parallel linear iterative solver on both scalar and vector machines. The work described here was done on the basis

Y. Shi et al. (Eds.): ICCS 2007, Part I, LNCS 4487, pp. 128–135, 2007.

of the research finite element program *Computer Aided Research Analysis Tool* (CCARAT), that is jointly developed and maintained at the Institute of Structural Mechanics of the University of Stuttgart and the Chair of Computational Mechanics at the Technical University of Munich. The research code CCARAT is a multipurpose finite element program covering a wide range of applications in computational mechanics, like e.g. multi-field and multi-scale problems, structural and fluid dynamics, shape and topology optimization, material modeling and finite element technology. The code is parallelized using MPI and runs on a variety of platforms.

The major time consuming portions of a finite element simulation are calculating the local element contributions to the globally assembled matrix and solving the assembled global system of equations. As much as 90% of the time in a very large scale simulation can be spent in the linear solver, specially if the problem to be solved is ill-conditioned. While the time taken in element calculation scales linearly with the size of the problem, often the time in the sparse solver does not. Major reason being the kind of preconditioning needed for a successful solution. In Sect. 2 of this paper, the performance of public domain solvers is analysed on vector architecture. In Sect. 3, a newly developed parallel iterative solver (Block-based Linear Iterative solver – BLIS) targeting performance on both the architectures is introduced. Sect. 4 discusses performance and scaling results of BLIS.

2   Public Domain Solvers

Most public domain solvers like AZTEC [3], PETSc, Trilinos [4], etc. do not perform on vector architecture as well as they do on superscalar architectures. The main reason being their design considerations that primarily target performance on superscalar architectures thereby neglecting the following performance critical features of vector systems.

2.1   Average Vector Length

This is an important metric that has a huge effect on performance. In sparse linear algebra, the matrix object is sparse where as the vectors are still dense. So, any operations involving only vectors, like the dot product, result in a good average vector length as the innermost vectorized loop runs over long vectors. The problem is with operations involving the sparse matrix object like the matrix vector product (MVP) which is a key kernel of Krylov subspace methods. The data structure used to store the sparse matrix plays an important role in the performance of such operators. Naturally resulting row based data structures used in the above mentioned solvers result in low average vector length which is further problem dependent. This leads to partially empty pipeline processing and hence hinders performance of a critical operator on vector machines.

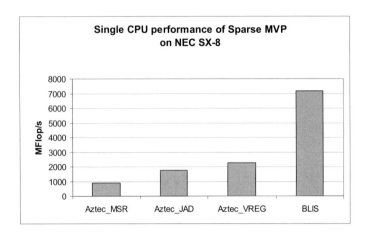

Fig. 1. Single CPU performance of Sparse MVP on NEC SX-8

A well known solution to this problem is to use pseudo diagonal data structure to store the sparse matrix [5]. We tried to introduce similar functionality into AZTEC. Figure 1 shows the single CPU performance in Sparse MVP on the NEC SX-8 with the native row based (Modified Sparse Row - MSR) and the introduced pseudo diagonal based (JAgged Diagonal - JAD) matrix storage formats. Using vector registers to reduce memory operations for loading and storing the result vector further improves the performance of JAD based sparse MVP to 2.2 GFlop/s. Further optimizations result in a maximum performance of 20% vector peak (which is 16 GFlop/s) for sparse MVP on NEC SX-8 [6].

2.2   Indirect Memory Addressing

The performance of sparse MVP on vector as well as on superscalar architectures is not limited by memory bandwidth, but by latencies. Due to the sparse storage, the vector to be multiplied in a sparse MVP is accessed randomly (non-strided access). Hence, the performance of this operation completely depends on the implementation and performance of the "vector gather" assembly instruction on vector machines. Though the memory bandwidth and byte/flop ratios of a vector architecture are in general far more superior than any superscalar architecture, superscalar architectures have the advantage of cache re-usage for this operation. But, the cost of accessing the main memory is so high on superscalar machines that without special optimizations [7], sparse MVP performs at around 5% peak.

It is interesting in this context to look into architectures which combine the advantages of vector pipelining with memory caching, like the CRAY X1. Sparse MVP with pseudo diagonal format on this machine performs at about 14% peak [8]. This is comparable to the performance achieved for point-based JAD algorithm on the NEC SX-8. An upper bound of around 30% peak is estimated for most sparse matrix computations on the CRAY X1.

3   Block-Based Linear Iterative Solver (BLIS)

In the sparse MVP kernel discussed so far, the major hurdle to performance is not memory bandwidth but the latencies involved due to indirect memory addressing. Block based computations exploit the fact that many FE problems typically have more than one physical variable to be solved per grid point. Thus, small blocks can be formed by grouping the equations at each grid point. Operating on such dense blocks considerably reduces the amount of indirect addressing required for sparse MVP [6]. This improves the performance of the kernel dramatically on vector machines [9] and also remarkably on superscalar architectures [10,11]. BLIS uses this approach primarily to reduce the penalty incurred due to indirect memory access.

3.1   Available Functionality

Presently, BLIS is working with finite element applications that have 4 unknowns to be solved per grid point. JAD sparse storage format is used to store the dense blocks. This assures sufficient average vector length for operations done using the sparse matrix object (Preconditioning, Sparse MVP). The single CPU performance of sparse MVP, Fig. 1, with a matrix consisting of 4x4 dense blocks is around 7.2 GFlop/s (about 45% vector peak) on the NEC SX-8.

BLIS is based on MPI and includes well known Krylov subspace methods such as BiCGSTAB and GMRES. Block scaling, block Jacobi and block ILU(0) on subdomains are the available matrix preconditioner methods. Exchange of halos in sparse MVP can be done using MPI non-blocking or MPI persistent communication.

3.2   Future Work

BLIS presently works with applications that have 4 unknowns to be evaluated per grid point. This will further be extended to handle any number of unknowns. Blocking functionality will be provided in the solver in order to free the users from preparing blocked matrices. This makes adaptation of the library to an application easy. Scheduling communication via coloring and reducing global synchronization at different places in Krylov subspace algorithms has to be extensively looked into for further improving scaling of the solver [12].

4   Performance

This section explains the performance of BLIS on both scalar and vector architectures. The machines tested are a cluster of NEC SX-8 SMPs and a cluster of Intel 3.2 GHz Xeon EM64T processors. The network interconnect available on NEC SX-8 is a proprietary multi-stage crossbar called IXS and on the Xeon cluster a Gigabit ethernet. Vendor tuned MPI library is used on the SX-8 and Voltaire MPI library on the Xeon cluster to run the following example.

Table 1. Different discretizations of the introduced numerical example

Discretization	No.of elements	No. of nodes	No. of unknowns
1	33750	37760	151040
2	81200	88347	353388
3	157500	168584	674336
4	270000	285820	1143280
5	538612	563589	2254356
6	911250	946680	3786720

4.1   Numerical Example

In this example the laminar, unsteady 3-dimensional flow around a cylinder with a square cross-section is examined. The setup was introduced as a benchmark example by the DFG Priority Research Program "Flow Simulation on High Performance Computers" to compare different solution approaches of the Navier-Stokes equations[13]. The fluid is assumed to be incompressible Newtonian with a kinematic viscosity $\nu = 10^{-3}\ m^2/s$ and a density of $\rho = 1.0\ kg/m^3$. The rigid cylinder (cross-section: 0.1 m x 0.1m) is placed in a 2.5 m long channel with a square cross-section of 0.41 m by 0.41 m. On one side a parabolic inflow condition with the mean velocity $u_m = 2.25\ m/s$ is applied. No-slip boundary conditions are assumed on the four sides of the channel and on the cylinder.

4.2   BLIS Scaling on Vector Machine

The scaling behaviour of the solver on NEC SX-8 was tested for the above mentioned numerical example using stabilized 3D hexahedral fluid elements in CCARAT. Table 1 lists all the six discretizations of the example used.

Figure 3 plots strong scaling of BLIS upto 128 processors using BiCGSTAB algorithm along with block Jacobi preconditioning on the NEC SX-8. All problems were run for 5 time steps where each non-linear time step needs about 3-5 newton iterations for convergence. The number of iterations needed for convergence in BLIS for each newton step varies largely between 200-2000 depending on the problem size (number of equations). The plots show the drop in sustained floating point performance of BLIS from above 6 GFlop/s to 3 GFlop/s

Fig. 2. Velocity in the midplane of the channel

depending on the number of processors used for each problem size. The right plot of Fig. 3 explains the reason for this drop in performance in terms of drop in computation to communication ratio in BLIS. It has to be noted that major part of the communication with the increase in processor count is spent in MPI global reduction call which needs global syncronization.

Figure 4 plots the same data for increasing problem size per CPU. Each curve represents performance using particular number of CPUs with varying problem size. Here the effect of global synchronization on scaling can be clearly seen. As the processor count increases, the performance curves climb slowly till the performance saturates. This behaviour can be directly attributed to the time spent in communication which is clear from the right plot. These plots are hence important as they accentuate the problem with Krylov subspace algorithms where large problem sizes are needed to sustain high performance on large processor counts. This is a drawback for certain class of applications where the demand for HPC (High Performance Computing) is due to the largely transient nature of the problem. For instance, in some Fluid-Structure interaction examples, the problem size is not too large but thousands of time steps are necessary to simulate the transient effects.

4.3   Performance Comparison on Scalar and Vector Machines

Here, we compare the performance of the linear solver on both scalar and vector machines. A relatively smaller discretization (353K unknowns) was used. Elapsed time (seconds) in linear iterative solver for 5 time steps is listed on the y-axis of the plot in Fig. 5. Fastest solution method (BiCGSTAB) and preconditioning (ILU on the Xeon cluster and block Jacobi on NEC SX-8) was used on the respective machines.

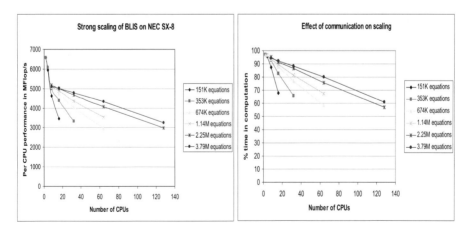

Fig. 3. Strong scaling of BLIS (left) and computation to communication ratio in BLIS on NEC SX-8 (right)

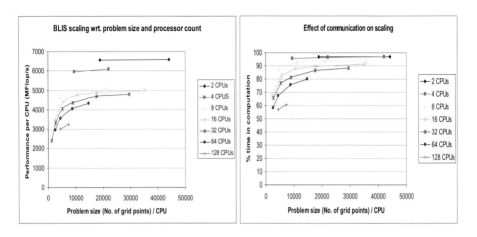

Fig. 4. Scaling of BLIS wrt. problem size on NEC SX-8 (left) Computation to communication ratio in BLIS on NEC SX-8 (right)

The performance of BLIS is compared to AZTEC on both the machines. BLIS is about 20% faster than AZTEC on the Xeon cluster due to better convergence and block based approach. BLIS is about 10 times faster than AZTEC on the NEC SX-8 for lower processor counts when the problem size per processor provides sufficient average vector length. This is the main reason behind developing a new linear solver. The solution of the linear system is about 4-5 times faster on the NEC SX-8 than on the Xeon cluster for lower processor counts. As the number of processors increases, the performance of BLIS on NEC SX-8 drops due to decrease in computation to communication ratio and also drop in average vector length. Contrarily, this is an advantage (better cache utilization) on the Xeon cluster due to a smaller problem size on each processor. This can be noticed from the super linear speedup on the Xeon cluster.

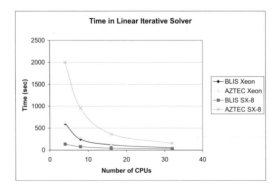

Fig. 5. Elapsed time (seconds) in linear solver

5   Summary

The reasons behind the dismal performance of most of the public domain sparse iterative solvers on vector machines were analyzed. We then introduced the Block-based Linear Iterative Solver (BLIS) which is currently under development targeting performance on all architectures. Results show an order of magnitude performance improvement over other public domain libraries on the tested vector system. A moderate performance improvement is also seen on the scalar machines.

References

1. Oliker, L., Canning, A., Carter, J., Shalf, J., Ethier, S.: Scientific computations on modern parallel vector systems. In: Proceedings of the ACM/IEEE Supercomputing Conference (SC 2004), Pittsburgh, USA (2004)
2. Rabenseifner, R., Tiyyagura, S.R., Müller, M.: Network bandwidth measurements and ratio analysis with the hpc challenge benchmark suite (hpcc). In Martino, B.D., Mueller, D.K., Dongarra, J., eds.: Proceedings of the 12th European PVM/MPI Users' Group Meeting (EURO PVM/MPI 2005). LNCS 3666, Sorrento, Italy, Springer (2005) 368–378
3. Tuminaro, R.S., Heroux, M., Hutchinson, S.A., Shadid, J.N.: Aztec user's guide: Version 2.1. Technical Report SAND99-8801J, Sandia National Laboratories (1999)
4. Heroux, M.A., Willenbring, J.M.: Trilinos users guide. Technical Report SAND2003-2952, Sandia National Laboratories (2003)
5. Saad, Y.: Iterative Methods for Sparse Linear Systems, Second Edition. SIAM, Philadelphia, PA (2003)
6. Tiyyagura, S.R., Küster, U., Borowski, S.: Performance improvement of sparse matrix vector product on vector machines. In Alexandrov, V., van Albada, D., Sloot, P., Dongarra, J., eds.: Proceedings of the Sixth International Conference on Computational Science (ICCS 2006). LNCS 3991, Reading, UK, Springer (2006)
7. Im, E.J., Yelick, K.A., Vuduc, R.: Sparsity: An optimization framework for sparse matrix kernels. International Journal of High Performance Computing Applications **(1)18** (2004) 135–158
8. Agarwal, P., et al.: ORNL Cray X1 evaluation status report. Technical Report LBNL-55302, Lawrence Berkeley National Laboratory (May 1, 2004)
9. Nakajima, K.: Parallel iterative solvers of geofem with selective blocking preconditioning for nonlinear contact problems on the earth simulator. GeoFEM 2003-005, RIST/Tokyo (2003)
10. Jones, M.T., Plassmann, P.E.: Blocksolve95 users manual: Scalable library software for the parallel solution of sparse linear systems. Technical Report ANL-95/48, Argonne National Laboratory (1995)
11. Tuminaro, R.S., Shadid, J.N., Hutchinson, S.A.: Parallel sparse matrix vector multiply software for matrices with data locality. Concurrency: Practice and Experience **(3)10** (1998) 229–247
12. Demmel, J., Heath, M., van der Vorst, H.: Parallel numerical linear algebra. Acta Numerica **2** (1993) 53–62
13. Schäfer, M., Turek, S.: Benchmark Computations of Laminar Flow Around a Cylinder. Notes on Numerical Fluid Mechanics **52** (1996) 547–566

Numerical Tests with Gauss-Type Nested Implicit Runge-Kutta Formulas*

Gennady Yu. Kulikov and Sergey K. Shindin

School of Computational and Applied Mathematics, University of the Witwatersrand,
Private Bag 3, Wits 2050, Johannesburg, South Africa
gkulikov@cam.wits.ac.za, sshindin@cam.wits.ac.za

Abstract. In this paper we conduct a detailed numerical analysis of the Gauss-type Nested Implicit Runge-Kutta formulas of order 4, introduced by Kulikov and Shindin in [4]. These methods possess many important practical properties such as high order, good stability, symmetry and so on. They are also conjugate to a symplectic method of order 6 at least. All of these make them efficient for solving many nonstiff and stiff ordinary differential equations (including Hamiltonian and reversible systems). On the other hand, Nested Implicit Runge-Kutta formulas have only explicit internal stages, in the sense that they are easily reduced to a single equation of the same dimension as the source problem. This means that such Runge-Kutta schemes admit a cheap implementation in practice. Here, we check the above-mentioned properties numerically. Different strategies of error estimation are also examined with the purpose of finding an effective one.

1 Introduction

In this paper we deal with numerical solution of ordinary differential equations (ODE's) of the form

$$x'(t) = g(t, x(t)), \quad t \in [t_0, t_{end}], \quad x(t_0) = x^0 \tag{1}$$

where $x(t) \in \mathbb{R}^n$ and $g : D \subset \mathbb{R}^{n+1} \to \mathbb{R}^n$ is a sufficiently smooth function. Problem (1) is assumed to have a unique solution $x(t)$ on the interval $[t_0, t_{end}]$.

An *l-stage Nested Implicit Runge-Kutta* (NIRK) formula applied to ODE (1) reads

$$x_{kj}^2 = a_{j1}^2 x_k + a_{j2}^2 x_{k+1} + \tau_k \left(d_{j1}^2 g(t_k, x_k) + d_{j2}^2 g(t_{k+1}, x_{k+1}) \right), \tag{2a}$$

$$
\begin{aligned}
x_{kj}^i = a_{j1}^i x_k + a_{j2}^i x_{k+1} + \tau_k \left(d_{j1}^i g(t_k, x_k) + d_{j2}^i g(t_{k+1}, x_{k+1}) \right) \\
+ \tau_k \sum_{m=1}^{i-1} d_{j,m+2}^i g(t_{km}^{i-1}, x_{km}^{i-1}), \quad i = 3, 4, \dots, l, \quad j = 1, 2, \dots, i,
\end{aligned}
\tag{2b}
$$

---

* This work was supported in part by the National Research Foundation of South Africa under grant No. FA2004033000016.

Y. Shi et al. (Eds.): ICCS 2007, Part I, LNCS 4487, pp. 136–143, 2007.

$$x_{k+1} = x_k + \tau_k \sum_{i=1}^{l} b_i g(t_{ki}^l, x_{ki}^l). \tag{2c}$$

where $x_0 = x^0$, $t_{kj}^i = t_k + \tau_k c_j^i$ and τ_k is a step size. It is also required that $a_{j1}^i + a_{j2}^i = 1$ and $c_j^i = a_{j2}^i + \sum_{m=1}^{i+1} d_{jm}^i$. We stress that method (2) is an RK formula and its Butcher tableau is given by

$$
\begin{array}{c|ccccccc}
0 & 0 & 0 & 0 & \cdots & 0 & 0 & 0 \\
c^2 & d_1^2 & 0 & 0 & \cdots & 0 & a^2 b^T & d_2^2 \\
c^3 & d_1^3 & D^3 & 0 & \cdots & 0 & a^3 b^T & d_2^3 \\
c^4 & d_1^4 & 0 & D^4 & \cdots & 0 & a^4 b^T & d_2^4 \\
\vdots & \vdots & \vdots & \vdots & \ddots & \vdots & \vdots & \vdots \\
c^l & d_1^l & 0 & 0 & \cdots & D^l & a^l b^T & d_2^l \\
1 & 0 & 0 & 0 & \cdots & 0 & b^T & 0 \\
\hline
 & 0 & 0 & 0 & \cdots & 0 & b^T & 0
\end{array}
\tag{3}
$$

where

$$
D^i = \begin{pmatrix} d_{12}^i & \cdots & d_{1,i+1}^i \\ \vdots & \ddots & \vdots \\ d_{i2}^i & \cdots & d_{i,i+1}^i \end{pmatrix}, \quad
d_1^i = \begin{pmatrix} d_{11}^i \\ \vdots \\ d_{i1}^i \end{pmatrix}, \quad
d_2^i = \begin{pmatrix} d_{12}^i \\ \vdots \\ d_{i2}^i \end{pmatrix},
$$

$$
a^i = \begin{pmatrix} a_{12}^i \\ \vdots \\ a_{i2}^i \end{pmatrix}, \quad
c^i = \begin{pmatrix} c_1^i \\ \vdots \\ c_i^i \end{pmatrix}, \quad
b = \begin{pmatrix} b_1 \\ \vdots \\ b_l \end{pmatrix}.
$$

It is quite clear that NIRK method (2) is easily reduced to a single equation of dimension n. This is of great advantage in practice. We refer the reader to [4] and [6] for more information about the reason to design NIRK formulas. Our intention here is to examine numerically the Gauss-type NIRK methods of order 4 introduced in the cited papers. Especially, we are interested to test adaptive formulas of this type to find a proper computational technique for practical use.

Finally, we want to point out that the first research on the topic under discussion was done by van Bokhoven [1], who introduced cheap RK methods termed Implicit Endpoint Quadrature (IEQ) formulas. It is obvious that NIRK methods are a particular case of IEQ formulas. Nevertheless, van Bokhoven made some mistakes in his paper and failed to find good practical methods, in the sense of the properties mentioned above (see [6] for explanation). Thus, we concentrate here on the NIRK methods designed by Kulikov and Shindin.

2   Adaptive Gauss-Type NIRK Methods

The Gauss-type NIRK methods of orders 2, 4 and 6 have been developed and investigated theoretically in [4] and [6]. So we intend to discuss different error estimations which are suitable for the NIRK methods in Section 2. We do our analysis for the methods of order 4. However, the results obtained can be extended to the Gauss-type NIRK formulas of order 6. Our research covers both Embedded Methods Approach (EMA) and Richardson Extrapolation Technique (RET) as well. We will see below that methods (2) admit one more idea for error estimation, termed Embedded Stages Approach (ESA).

We emphasize that van Bokhoven [1] failed to construct RK schemes with built-in error estimation which are effective to treat stiff ODE's (see [6] for details). Therefore we apply EMA to solve this problem for Gauss-type NIRK methods of order 4. In other words, we deal with embedded RK formulas of the form

$$
\begin{array}{c|ccccc}
0 & 0 & 0 & 0 & 0 \\[2mm]
c_1^2 & \dfrac{6(c_1^2+\theta)-5}{12} & \dfrac{1-\theta}{2} & \dfrac{1-\theta}{2} & \dfrac{6(c_1^2+\theta)-7}{12} \\[3mm]
1-c_1^2 & \dfrac{7-6(c_1^2+\theta)}{12} & \dfrac{\theta}{2} & \dfrac{\theta}{2} & \dfrac{5-6(c_1^2+\theta)}{12} \\[3mm]
1 & 0 & \dfrac{1}{2} & \dfrac{1}{2} & 0 \\[3mm]
\hline
& 0 & \dfrac{1}{2} & \dfrac{1}{2} & 0 \\[3mm]
\hline
& \dfrac{1}{2} & -\dfrac{1}{2} & -\dfrac{1}{2} & \dfrac{1}{2}
\end{array}
\tag{4}
$$

where $c_1^2 = (3 - \sqrt{3})/6$ and θ is a real free parameter. We recall that NIRK method (4) is of stage order 3 when $\theta = 1/2 + 2\sqrt{3}/9$ and of stage order 2 otherwise (see [4]). In this paper we are going to examine numerically the following five different error estimations introduced in [6]:

Embedded Methods Error Estimation (EMEE) is taken to be

$$
le_{k+1} = \frac{\tau_k}{2}\Big(g(t_k, x_k) - g(t_{k1}^2, x_{k1}^2) - g(t_{k2}^2, x_{k2}^2) + g(t_{k+1}, x_{k+1})\Big).
\tag{5}
$$

Estimate (5) is cheap and of order 3. It is based on Trapezoidal Rule. Unfortunately, one of the RK schemes in the embedded pair (4) is not A-stable. Hence, EMEE can be inefficient for some stiff ODE's. That is why we offer an improved error estimation based on Shampine's idea in the above-cited paper.

Modified Embedded Methods Error Estimation (MEMEE) uses the formula

$$
Q_1(\tau_k J)\tilde{le}_{k+1} = le_{k+1}
\tag{6}
$$

where $Q_1(z) = (1 - z/4)^3$ and J means the Jacobi matrix evaluated at the point (t_k, x_k). The last formula implies that the improved error estimate is obtained by solving linear system (6). We stress that MEMEE is not expensive in practice because it means three solutions of linear systems with the coefficient matrix $I - \tau_k J/4$. Anyway, the latter matrix is computed and decomposed to advance a step of method (4) (see [4] and [5] for more detail). Eventually, MEMEE is suitable for integration of stiff problems (see [6]).

One more idea of error estimation discovered in the above-mentioned paper and referred to as *Embedded Stages Error Estimation* (ESEE) uses the fact that calculation of stage values is explicit and, hence, very cheap in NIRK methods. We know that method (4) is of stage order 3 when $\theta = 1/2 + 2\sqrt{3}/9$. It is of stage order 2 for other θ's. Therefore we choose another $\hat{\theta} \neq 1/2 + 2\sqrt{3}/9$ to calculate the stage values of the different order. When taking the difference we come to the error estimate

$$\hat{le}_{k+1} = \frac{\tau_k}{8} \Big(g(t_k, x_k) - g(t_{k1}^2, x_{k1}^2) - g(t_{k2}^2, x_{k2}^2) + g(t_{k+1}, x_{k+1}) \Big). \qquad (7)$$

ESEE is of order 3 and used in a stepsize selection algorithm in the usual way. Again, error estimate (7) can be inefficient to treat stiff ODE's.

Hopefully, a modification is possible to make ESEE suitable for stiff problems. For them, it is recommended to apply *Modified Embedded Stages Error Estimation* (MESEE) presented by the formula

$$Q_2(\tau J)\bar{le}_{k+1} = \frac{\tau_k}{8} \Big(g(t_k, x_k) - g(t_{k1}^2, x_{k1}^2) - g(t_{k2}^2, x_{k2}^2) + g(t_{k+1}, x_{k+1}) \Big) \qquad (8)$$

where $Q_2(z) = 1 - z/4$. It is not difficult to check that error estimate (8) is limited for any step size. It is also cheap because of the reasons given above. We point out that the last formula is superior to error estimation (6) in terms of CPU time since only one extra solution is implemented in MESEE.

For the sake of completeness, we include *Richardson Extrapolation Error Estimation* (REEE) applied to the Gauss-type NIRK method of order 4 when $\theta = 1/2 + 2\sqrt{3}/9$ in our testing below. Details of REEE can be found in [3], for example. We stress that the NIRK method of order 4 and with REEE showed sufficiently good numerical results when applied to the two-dimensional Brusselator with diffusion and the periodic boundary conditions. We compared it with the usual Gauss method of order 4 which uses the same error estimation and stepsize selection strategy (see [4]).

3   Test Problems

In the experiments below we want to check how our NIRK methods perform for problems of different sorts. We cover Hamiltonian systems and the usual nonstiff and stiff ODE's. We also include some ODE's with known exact solutions in order to check the quality of numerical solutions obtained. This allows us to exhibit practical properties of the adaptive NIRK methods introduced in [4] and [6]. We are interested to find the methods which compute good numerical solutions for the minimum CPU time. To implement our plan we take the following test problems:

The Kepler problem is given by

$$x_1''(t) = \frac{-x_1(t)}{\big(x_1^2(t) + x_2^2(t)\big)^{3/2}}, \quad x_2''(t) = \frac{-x_2(t)}{\big(x_1^2(t) + x_2^2(t)\big)^{3/2}} \qquad (9)$$

where $x_1(0) = 1 - e$, $x_1'(0) = 0$, $x_2(0) = 0$, $x_2'(0) = \sqrt{1+e}/\sqrt{1-e}$ and $e = 0.2$. Problem (9) has two first integrals: the total energy $H(x'(t), x(t))$ and the angular momentum $L(x'(t), x(t))$ where $x(t) = (x_1(t), x_2(t))$. Its exact solution is also well-known. Thus, the Kepler problem is considered to be a good example of Hamiltonian systems and is often used to test numerical methods suitable for such sort of problems (see, for instance, [2]).

The simple problem is presented by

$$x_1'(t) = 2t x_2^{1/5}(t) x_4(t), \quad x_2'(t) = 10t \exp\left(5(x_3(t) - 1)\right) x_4(t), \qquad (10a)$$

$$x_3'(t) = 2t x_4(t), \quad x_4'(t) = -2t \ln(x_1(t)), \quad t \in [0, 5] \qquad (10b)$$

where $x(0) = (1, 1, 1, 1)^T$. It possesses the exact solution

$$x_1(t) = \exp(\sin t^2), \quad x_2(t) = \exp(5 \sin t^2), \quad x_3(t) = \sin t^2 + 1, \quad x_4(t) = \cos t^2,$$

which will be used to check the quality of numerical solutions obtained in our experiments.

The restricted three body problem is

$$x_1''(t) = x_1(t) + 2x_2'(t) - \mu_1 \frac{x_1(t) + \mu_2}{y_1(t)} - \mu_2 \frac{x_1(t) - \mu_1}{y_2(t)}, \qquad (11a)$$

$$x_2''(t) = x_2(t) - 2x_1'(t) - \mu_1 \frac{x_2(t)}{y_1(t)} - \mu_2 \frac{x_2(t)}{y_2(t)}, \qquad (11b)$$

$$y_1(t) = \left((x_1(t) + \mu_2)^2 + x_2^2(t)\right)^{3/2}, \quad y_2(t) = \left((x_1(t) - \mu_1)^2 + x_2^2(t)\right)^{3/2} \quad (11c)$$

where $t \in [0, T]$, $T = 17.065216560157962558891$, $\mu_1 = 1 - \mu_2$ and $\mu_2 = 0.012277471$. The initial values of problem (11) are: $x_1(0) = 0.994$, $x_1'(0) = 0$, $x_2(0) = 0$, $x_2'(0) = -2.00158510637908252240$. Problem (11) has no analytic solution, but it is still useful to gain experience concerning the quality of numerical solutions because its solution-path is periodic. We can monitor the error at the end point T to verify the quality.

A pattern of stiff ODE's is presented by *Van der Pol's equation*

$$x_1'(t) = x_2(t), \quad x_2'(x) = \mu^2\left((1 - x_1^2(t)) x_2(t) - x_1(t)\right), \quad t \in [0, 2] \qquad (12)$$

where $x(0) = (2, 0)^T$ and $\mu = 1000$. Problem (12) is considered to be very stiff.

4   Numerical Experiments

We start this section with numerical experiments on the Kepler problem. We stress that the standard stepsize selection is not suitable for integration of Hamiltonian and reversible problems because it ruins nice properties of specially designed numerical methods (see Section VIII in [2]). So we just apply the fixed-stepsize version of method (4) when $\theta = 1/2 + 2\sqrt{3}/9$ to see how it works for long time integration of problem (9).

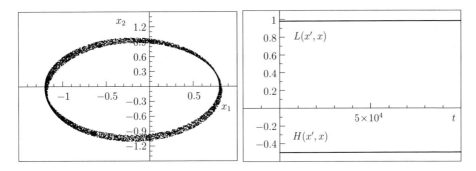

Fig. 1. The numerical solution of the Kepler problem (the left-hand graph) and its first integrals (the right-hand graph) computed by the fixed-stepsize Gauss-type NIRK method of order 4

Table 1. The global errors of the adaptive Gauss-type NIRK method of order 4 with different error estimation techniques applied to the simple problem

Error Tolerance	Error Estimation Technique				
	EMEE	ESEE	MEMEE	MESEE	REEE
10^{-1}	$3.139 \cdot 10^{+1}$	$1.199 \cdot 10^{+2}$	$3.646 \cdot 10^{+1}$	$1.479 \cdot 10^{+2}$	$2.534 \cdot 10^{+2}$
$5 \cdot 10^{-2}$	$1.876 \cdot 10^{+1}$	$6.355 \cdot 10^{+1}$	$1.409 \cdot 10^{+1}$	$1.557 \cdot 10^{+2}$	$1.750 \cdot 10^{+2}$
10^{-2}	$1.766 \cdot 10^{+0}$	$1.779 \cdot 10^{+1}$	$1.575 \cdot 10^{+0}$	$1.720 \cdot 10^{+1}$	$2.450 \cdot 10^{+1}$
$5 \cdot 10^{-3}$	$7.694 \cdot 10^{-1}$	$4.566 \cdot 10^{+0}$	$6.969 \cdot 10^{-1}$	$4.605 \cdot 10^{+0}$	$1.183 \cdot 10^{+1}$
10^{-3}	$7.740 \cdot 10^{-2}$	$5.502 \cdot 10^{-1}$	$7.568 \cdot 10^{-2}$	$5.624 \cdot 10^{-1}$	$3.207 \cdot 10^{+0}$
$5 \cdot 10^{-4}$	$2.501 \cdot 10^{-2}$	$2.055 \cdot 10^{-1}$	$2.582 \cdot 10^{-2}$	$2.096 \cdot 10^{-1}$	$1.311 \cdot 10^{+0}$
10^{-4}	$2.051 \cdot 10^{-3}$	$1.731 \cdot 10^{-2}$	$1.839 \cdot 10^{-3}$	$1.706 \cdot 10^{-2}$	$1.547 \cdot 10^{-1}$
$5 \cdot 10^{-5}$	$6.609 \cdot 10^{-4}$	$5.961 \cdot 10^{-3}$	$6.650 \cdot 10^{-4}$	$5.944 \cdot 10^{-3}$	$6.716 \cdot 10^{-2}$
10^{-5}	$7.155 \cdot 10^{-5}$	$4.976 \cdot 10^{-4}$	$6.533 \cdot 10^{-5}$	$5.005 \cdot 10^{-4}$	$1.385 \cdot 10^{-2}$

Figure 1 shows the behaviour of the numerical solution and the first integrals calculated by our NIRK method with the step size $\tau_k = 0.1$ on the interval $[0, 10^{+5}]$. We observe that this result corresponds well to the exact solution and the first integrals of the Kepler problem (see [2, p. 9]). Thus, method (4) is suitable for the numerical integration of Hamiltonian and reversible problems in practice. We recall that the method under discussion is conjugate to a symplectic method of order 6 at least (see [6]).

Now we come to tests with the adaptive Gauss-type NIRK methods. For that, we perform a number of numerical integrations of problems (10) and (11) by the method (4) when $\theta = 1/2 + 2\sqrt{3}/9$ and with the error estimation strategies mentioned in Section 2. To implement the Gauss-type NIRK method, we apply a modified Newton iteration with nontrivial predictor, as explained in [4] or [5]. We perform two iteration steps per each grid point and display the global

Table 2. CPU time (in sec.) of the adaptive Gauss-type NIRK method of order 4 with different error estimation techniques applied to the simple problem

Error Tolerance	Error Estimation Technique				
	EMEE	ESEE	MEMEE	MESEE	REEE
10^{-1}	$0.375 \cdot 10^{+0}$	$0.235 \cdot 10^{+0}$	$0.328 \cdot 10^{+0}$	$0.141 \cdot 10^{+0}$	$0.328 \cdot 10^{+0}$
$5 \cdot 10^{-2}$	$0.328 \cdot 10^{+0}$	$0.266 \cdot 10^{+0}$	$0.391 \cdot 10^{+0}$	$0.235 \cdot 10^{+0}$	$0.391 \cdot 10^{+0}$
10^{-2}	$0.547 \cdot 10^{+0}$	$0.422 \cdot 10^{+0}$	$0.531 \cdot 10^{+0}$	$0.391 \cdot 10^{+0}$	$0.453 \cdot 10^{+0}$
$5 \cdot 10^{-3}$	$0.579 \cdot 10^{+0}$	$0.437 \cdot 10^{+0}$	$0.657 \cdot 10^{+0}$	$0.422 \cdot 10^{+0}$	$0.469 \cdot 10^{+0}$
10^{-3}	$0.938 \cdot 10^{+0}$	$0.625 \cdot 10^{+0}$	$1.047 \cdot 10^{+0}$	$0.609 \cdot 10^{+0}$	$0.500 \cdot 10^{+0}$
$5 \cdot 10^{-4}$	$1.204 \cdot 10^{+0}$	$0.781 \cdot 10^{+0}$	$1.219 \cdot 10^{+0}$	$0.828 \cdot 10^{+0}$	$0.547 \cdot 10^{+0}$
10^{-4}	$1.844 \cdot 10^{+0}$	$1.328 \cdot 10^{+0}$	$1.875 \cdot 10^{+0}$	$1.343 \cdot 10^{+0}$	$0.672 \cdot 10^{+0}$
$5 \cdot 10^{-5}$	$2.391 \cdot 10^{+0}$	$1.500 \cdot 10^{+0}$	$2.422 \cdot 10^{+0}$	$1.687 \cdot 10^{+0}$	$0.719 \cdot 10^{+0}$
10^{-5}	$3.984 \cdot 10^{+0}$	$2.672 \cdot 10^{+0}$	$3.985 \cdot 10^{+0}$	$2.656 \cdot 10^{+0}$	$1.031 \cdot 10^{+0}$

Table 3. The errors evaluated at the point T for the adaptive Gauss-type NIRK method of order 4 with different error estimation techniques applied to the restricted three body problem

Error Tolerance	Error Estimation Technique				
	EMEE	ESEE	MEMEE	MESEE	REEE
10^{-1}	$1.988 \cdot 10^{+0}$	$8.816 \cdot 10^{+1}$	$1.347 \cdot 10^{+0}$	$3.474 \cdot 10^{+0}$	$1.832 \cdot 10^{+0}$
$5 \cdot 10^{-2}$	$1.900 \cdot 10^{+0}$	$2.163 \cdot 10^{+0}$	$1.813 \cdot 10^{+0}$	$1.823 \cdot 10^{+0}$	$1.915 \cdot 10^{+0}$
10^{-2}	$9.913 \cdot 10^{-1}$	$1.945 \cdot 10^{+0}$	$8.998 \cdot 10^{-1}$	$1.898 \cdot 10^{+0}$	$1.435 \cdot 10^{+0}$
$5 \cdot 10^{-3}$	$4.124 \cdot 10^{-1}$	$1.580 \cdot 10^{+0}$	$6.277 \cdot 10^{-1}$	$1.536 \cdot 10^{+0}$	$1.330 \cdot 10^{+0}$
10^{-3}	$1.485 \cdot 10^{-1}$	$4.496 \cdot 10^{-1}$	$1.695 \cdot 10^{-1}$	$4.768 \cdot 10^{-1}$	$8.324 \cdot 10^{-1}$
$5 \cdot 10^{-4}$	$7.058 \cdot 10^{-2}$	$2.721 \cdot 10^{-1}$	$8.218 \cdot 10^{-2}$	$2.955 \cdot 10^{-1}$	$1.821 \cdot 10^{-1}$
10^{-4}	$1.092 \cdot 10^{-2}$	$5.593 \cdot 10^{-2}$	$1.180 \cdot 10^{-2}$	$5.825 \cdot 10^{-2}$	$3.470 \cdot 10^{-2}$
$5 \cdot 10^{-5}$	$4.336 \cdot 10^{-3}$	$2.471 \cdot 10^{-2}$	$4.610 \cdot 10^{-3}$	$2.557 \cdot 10^{-2}$	$4.986 \cdot 10^{-3}$
10^{-5}	$5.306 \cdot 10^{-4}$	$3.224 \cdot 10^{-3}$	$5.530 \cdot 10^{-4}$	$3.282 \cdot 10^{-3}$	$1.925 \cdot 10^{-3}$

errors calculated by the exact solutions of these test problems in the sup-norm in Tables 1 and 3. The CPU time of these integrations is shown in Tables 2 and 4, respectively. The CPU time for numerical solutions of Van der Pol's equation is presented in Table 5.

Tables 1–5 show clearly that ESEE and MESEE produce numerical solutions of good quality for less CPU time. Certainly, REEE works faster when applied to problems of small size because this estimate is of order 5. The remaining estimates are of order 3. On the other hand, REEE is very expensive for large-scale problems (see Table 1 in [6]). In conclusion, we consider that the error estimation based on ESA is quite promising and deserves further study. We intend to develop such error estimation techniques for the Gauss-type NIRK method of order 6. In future it is also planned to supply our NIRK schemes with a facility to provide Automatic Global Error Control.

Table 4. CPU time (in sec.) of the adaptive Gauss-type NIRK method of order 4 with different error estimation techniques applied to the restricted three body problem

Error Tolerance	Error Estimation Technique				
	EMEE	ESEE	MEMEE	MESEE	REEE
10^{-1}	$0.156 \cdot 10^{+0}$	$0.062 \cdot 10^{+0}$	$0.094 \cdot 10^{+0}$	$0.047 \cdot 10^{+0}$	$0.187 \cdot 10^{+0}$
$5 \cdot 10^{-2}$	$0.140 \cdot 10^{+0}$	$0.125 \cdot 10^{+0}$	$0.125 \cdot 10^{+0}$	$0.078 \cdot 10^{+0}$	$0.235 \cdot 10^{+0}$
10^{-2}	$0.204 \cdot 10^{+0}$	$0.125 \cdot 10^{+0}$	$0.203 \cdot 10^{+0}$	$0.109 \cdot 10^{+0}$	$0.281 \cdot 10^{+0}$
$5 \cdot 10^{-3}$	$0.266 \cdot 10^{+0}$	$0.156 \cdot 10^{+0}$	$0.219 \cdot 10^{+0}$	$0.156 \cdot 10^{+0}$	$0.281 \cdot 10^{+0}$
10^{-3}	$0.359 \cdot 10^{+0}$	$0.235 \cdot 10^{+0}$	$0.391 \cdot 10^{+0}$	$0.250 \cdot 10^{+0}$	$0.360 \cdot 10^{+0}$
$5 \cdot 10^{-4}$	$0.468 \cdot 10^{+0}$	$0.312 \cdot 10^{+0}$	$0.453 \cdot 10^{+0}$	$0.328 \cdot 10^{+0}$	$0.375 \cdot 10^{+0}$
10^{-4}	$0.672 \cdot 10^{+0}$	$0.484 \cdot 10^{+0}$	$0.657 \cdot 10^{+0}$	$0.515 \cdot 10^{+0}$	$0.438 \cdot 10^{+0}$
$5 \cdot 10^{-5}$	$0.797 \cdot 10^{+0}$	$0.531 \cdot 10^{+0}$	$0.813 \cdot 10^{+0}$	$0.625 \cdot 10^{+0}$	$0.532 \cdot 10^{+0}$
10^{-5}	$1.250 \cdot 10^{+0}$	$0.906 \cdot 10^{+0}$	$1.266 \cdot 10^{+0}$	$0.891 \cdot 10^{+0}$	$0.703 \cdot 10^{+0}$

Table 5. CPU time (in sec.) of the adaptive Gauss-type NIRK method of order 4 with different error estimation techniques applied to Van der Pol's equation

Error Tolerance	Error Estimation Technique				
	EMEE	ESEE	MEMEE	MESEE	REEE
10^{-1}	$3.219 \cdot 10^{+0}$	$2.406 \cdot 10^{+0}$	$3.593 \cdot 10^{+0}$	$1.734 \cdot 10^{+0}$	$1.796 \cdot 10^{+0}$
$5 \cdot 10^{-2}$	$3.703 \cdot 10^{+0}$	$2.859 \cdot 10^{+0}$	$2.687 \cdot 10^{+0}$	$2.047 \cdot 10^{+0}$	$1.547 \cdot 10^{+0}$
10^{-2}	$5.516 \cdot 10^{+0}$	$3.984 \cdot 10^{+0}$	$6.234 \cdot 10^{+0}$	$2.937 \cdot 10^{+0}$	$1.641 \cdot 10^{+0}$
$5 \cdot 10^{-3}$	$6.531 \cdot 10^{+0}$	$4.500 \cdot 10^{+0}$	$5.250 \cdot 10^{+0}$	$3.516 \cdot 10^{+0}$	$2.625 \cdot 10^{+0}$
10^{-3}	$1.053 \cdot 10^{+1}$	$6.922 \cdot 10^{+0}$	$8.563 \cdot 10^{+0}$	$5.562 \cdot 10^{+0}$	$2.828 \cdot 10^{+0}$
$5 \cdot 10^{-4}$	$1.276 \cdot 10^{+1}$	$8.578 \cdot 10^{+0}$	$1.053 \cdot 10^{+1}$	$6.985 \cdot 10^{+0}$	$2.937 \cdot 10^{+0}$
10^{-4}	$2.090 \cdot 10^{+1}$	$1.379 \cdot 10^{+1}$	$1.806 \cdot 10^{+1}$	$1.168 \cdot 10^{+1}$	$3.797 \cdot 10^{+0}$
$5 \cdot 10^{-5}$	$2.646 \cdot 10^{+1}$	$1.736 \cdot 10^{+1}$	$2.285 \cdot 10^{+1}$	$1.479 \cdot 10^{+1}$	$4.422 \cdot 10^{+0}$
10^{-5}	$4.325 \cdot 10^{+1}$	$2.782 \cdot 10^{+1}$	$3.731 \cdot 10^{+1}$	$2.457 \cdot 10^{+1}$	$6.000 \cdot 10^{+0}$

References

1. Van Bokhoven, W.M.G.: Efficient higher order implicit one-step methods for integration of stiff differential equations. BIT. **20** (1980) 34–43
2. Hairer, E., Lubich, C., Wanner, G.: Geometric numerical integration: Structure preserving algorithms for ordinary differential equations, Springer-Verlag, Berlin, 2002
3. Hairer, E., Nørsett, S.P., Wanner, G.: Solving ordinary differential equations I: Nonstiff problems. Springer-Verlag, Berlin, 1993
4. Kulikov, G.Yu., Shindin, S.K.: On a family of cheap symmetric one-step methods of order four. In: Vassil N. Alexandrov et al (eds.): Computational Science — ICCS 2006. 6th International Conference, Reading, UK, May 28–31, 2006. Proceedings, Part I. Lecture Notes in Computer Science. **3991** (2006) 781–785
5. Kulikov, G.Yu., Merkulov, A.I., Shindin, S.K.: Asymptotic error estimate for general Newton-type methods and its application to differential equations. Russ. J. Numer. Anal. Math. Model. **22** (2007) (to appear)
6. Kulikov, G.Yu., Shindin, S.K.: Adaptive nested implicit Runge-Kutta formulas of the Gauss type (in preparation)

An Efficient Implementation of the Thomas-Algorithm for Block Penta-diagonal Systems on Vector Computers

Katharina Benkert[1] and Rudolf Fischer[2]

[1] High Performance Computing Center Stuttgart (HLRS), University of Stuttgart,
70569 Stuttgart, Germany
benkert@hlrs.de
[2] NEC High Performance Computing Europe GmbH, Prinzenallee 11,
40549 Duesseldorf, Germany
rfischer@hpce.nec.com

Abstract. In simulations of supernovae, linear systems of equations with a block penta-diagonal matrix possessing small, dense matrix blocks occur. For an efficient solution, a compact multiplication scheme based on a restructured version of the Thomas algorithm and specifically adapted routines for LU factorization as well as forward and backward substitution are presented. On a NEC SX-8 vector system, runtime could be decreased between 35% and 54% for block sizes varying from 20 to 85 compared to the original code with BLAS and LAPACK routines.

Keywords: Thomas algorithm, vector architecture.

1 Introduction

Neutrino transport and neutrino interactions in dense matter play a crucial role in stellar core collapse, supernova explosions and neutron star formation. The multidimensional neutrino radiation hydrodynamics code PROMETHEUS / VERTEX [1] discretizes the angular moment equations of the Boltzmann equation giving rise to a non-linear algebraic system. It is solved by a Newton Raphson procedure, which in turn requires the solution of multiple block-penta-diagonal linear systems with small, dense matrix blocks in each step. This is achieved by the Thomas algorithm and takes a major part of the overall computing time. Since the code already performs well on vector computers, this kind of architecture has been the focus of the current work.

The Thomas algorithm [2,3] is a simplified form of Gaussian elimination without pivoting, as originally applied to tridiagonal systems. Bieniasz [4] gives a comprehensive overview of the numerous adaptations for special cases and mutations of tridiagonal systems, the extensions to cyclic tridiagonal systems and the transfer to block tridiagonal matrices. However, an efficient implementation for block penta-diagonal systems has not yet been considered in the literature. An additional challenge consists in solving systems with relatively small matrices

Y. Shi et al. (Eds.): ICCS 2007, Part I, LNCS 4487, pp. 144–151, 2007.

as used in the Thomas algorithm. Optimizations of LAPACK routines usually target large matrices as necessitated for example for the HPL benchmark [5].

The remainder of the paper is organized as follows: after introducing the Thomas algorithm and the concept of vector architectures, section 2 presents the compact multiplication scheme. Section 3 explains implementation issues for the newly introduced scheme and the LU decomposition and states the results obtained, followed by the summary of the paper in section 4.

1.1   Thomas Algorithm

Consider a linear system of equations consisting of a block penta-diagonal (BPD) matrix with n blocks of size $k \times k$ in each column resp. row, a solution vector $\underline{x}$ and a right hand side (RHS) $\underline{f}$. The vectors $\underline{x} = (\underline{x}_1^T \underline{x}_2^T \ldots \underline{x}_n^T)^T$ and $\underline{f} = (\underline{f}_1^T \underline{f}_2^T \ldots \underline{f}_n^T)^T$ are each of dimension $k \cdot n$. The BPD system is defined by

$$A_i \underline{x}_{i-2} + B_i \underline{x}_{i-1} + C_i \underline{x}_i + D_i \underline{x}_{i+1} + E_i \underline{x}_{i+2} = \underline{f}_i, \quad 1 \le i \le n, \tag{1}$$

by setting, for ease of notation, $A_1 = B_1 = A_2 = E_{n-1} = D_n = E_n = 0$, and implementing $\underline{x}$ and $\underline{f}$ as $(\underline{x}_{-1}^T \underline{x}_0^T \ldots \underline{x}_n^T)^T$ and $(\underline{f}_{-1}^T \underline{f}_0^T \ldots \underline{f}_n^T)^T$.

Eliminating the sub-diagonal matrix blocks A_i and B_i and inverting the diagonal matrix blocks C_i would result in the following system:

$$\begin{aligned}
\underline{x}_i + Y_i \, \underline{x}_{i+1} + Z_i \, \underline{x}_{i+2} &= \underline{r}_i, \quad 1 \le i \le n-2, \\
\underline{x}_{n-1} + Y_{n-1} \underline{x}_n &= \underline{r}_{n-1}, \\
\underline{x}_n &= \underline{r}_n.
\end{aligned} \tag{2}$$

In the Thomas algorithm, the new components Y_i, Z_i and $\underline{r}_i$ are computed by substituting $\underline{x}_{i-2}$ and $\underline{x}_{i-1}$ in (1) using the appropriate equations of (2) and comparing coefficients. This results in

$$\left. \begin{aligned}
Y_i &= G_i^{-1} (D_i - K_i Z_{i-1}) \\
Z_i &= G_i^{-1} E_i \\
\underline{r}_i &= G_i^{-1} (\underline{f}_i - A_i \underline{r}_{i-2} - K_i \underline{r}_{i-1})
\end{aligned} \right\} \quad i = 1, n, \tag{3}$$

where

$$\left. \begin{aligned}
K_i &= B_i - A_i Y_{i-2} \\
G_i &= C_i - K_i Y_{i-1} - A_i Z_{i-2}
\end{aligned} \right\} \quad i = 1, n, \tag{4}$$

assuming again, for ease of notation, $Y_{-1} = Z_{-1} = Y_0 = Z_0 = 0$. Then the solution is computed using backward substitution, i.e.

$$\begin{aligned}
\underline{x}_n &= \underline{r}_n, \\
\underline{x}_{n-1} &= \underline{r}_{n-1} - Y_{n-1} \, \underline{x}_n, \\
\underline{x}_i &= \underline{r}_i \quad - Y_i \, \underline{x}_{i+1} - Z_i \underline{x}_{i+2}, \quad i = n-2, -1, 1.
\end{aligned} \tag{5}$$

Since in practice, the inversion of G_i in (3) is replaced by a LU-decomposition $G_i = L_i U_i$, it's crucial to understand that solving a BPD system includes two levels of Gaussian elimination (GE) and backward substitution (BS): one for the whole system and one for each block row. The GE of the whole system (GE_S) requires the calculation of K_i and G_i shown in (4). GE and BS are then applied to each block row (GE_R / BS_R) computing Y_i, Z_i and $\underline{r}_i$ in (3). Finally, backward substitution (5) is applied to the whole system (BS_S) to obtain the solution $\underline{x}$.

1.2   Vector Architecture

Two different kinds of computer architectures are available today: scalar and vector. They use different approaches to tackle the common problem, memory latencies and memory bandwidth. Vector architectures [6,7] use SIMD instructions in particular also for memory access to hide memory latencies. This requires large, independent data sets for efficient calculation. The connection to main memory is considerably faster than for scalar architectures. Since there are no caches, data are directly transfered to and from main memory. Temporary results are stored in vector registers (VRs) of hardware vector length VL, which is 256 on the NEC SX-8 [8,9]. As not all instructions are vectorizable, scalar ones are integrated into vector architectures. Because of their slow execution compared with commodity processors their use should be avoided by improving the vectorization ratio of the code.

2   Reordering the Computational Steps

For operations on small matrix blocks, memory traffic is the limiting factor for total performance. Our approach involves reordering the steps in (3) and (4) as

$$
\begin{aligned}
K_i &= B_i - A_i\,Y_{i-2} & G_i &= G_i' - K_i\,Y_{i-1} & Y_i &= G_i^{-1}\cdot H_i \\
G_i' &= C_i - A_i\,Z_{i-2} & H_i &= D_i - K_i\,Z_{i-1} & Z_i &= G_i^{-1}\cdot\quad E_i \qquad (6) \\
\underline{r}_i' &= \underline{f}_i - A_i\,\underline{r}_{i-2} & \underline{r}_i'' &= \underline{r}_i' - K_i\,\underline{r}_{i-1} & \underline{r}_i &= G_i^{-1}\cdot\qquad \underline{r}_i'',
\end{aligned}
$$

where the following spaces can be shared: B_i and K_i; C_i, G_i' and G_i; D_i, H_i and Y_i; E_i and Z_i, as well as $\underline{f}_i$, $\underline{r}_i'$, $\underline{r}_i''$ and $\underline{r}_i$. The improvements of this rearrangement are valuable for the GE of the whole system as well as for the solution of the block rows. First, during GE_S, A_i and K_i are loaded only k times from main memory instead of $3k$ times as is the case with a straight forward implementation. Second, by storing H_i, E_i and $\underline{r}_i''$ contiguously in memory, the inverse of G is applied simultaneously to a combined matrix of size $k \times (2k+1)$ during GE_R / BS_R. Third, by supplying a specifically adapted own version of LAPACK's xGETRF (factorization) and xGETRS (forward and backward substitution) routines [10], memory traffic is further reduced. By combining factorization and forward substitution, L_i is applied to the RHS vectors during the elimination process and therefore not reloaded from main memory.

3   Implementation Details and Numerical Results

In this section, the performance of our new approach presented in section 2 is compared to the standard approach which uses LAPACK's xGETRF and xGETRS routines to compute the LU-factorization and apply forward / backward substitution and the BLAS routines xGEMM and xGEMV [11,12] for matrix-matrix products and matrix-vector products, respectively. All computations were carried out on a NEC SX-8 vector computer using the proprietary Fortran90 compiler version 2.0 / Rev. 340 [13].

After presenting the results for the compact multiplication scheme used for GE_S, different implementations of GE_R and BS_R are analyzed. Since the number of RHS vectors is $2k + 1$ and therefore depends on the block size k, a small, a medium and a large test case with block sizes of $k = 20$, 55 and 85 are selected representatively. After that, the performance of the solution process for one block row as well as for the whole system (1) is described.

3.1   Applying Gaussian Elimination to the Whole System

The first two systems of (6) are of the form

$$
\begin{aligned}
D &= B + A \cdot C, \\
G &= E + A \cdot F, \\
\underline{z} &= \underline{v} + A \cdot \underline{w}.
\end{aligned}
\tag{7}
$$

Using BLAS, each line is split into two operations, one copy operation, e.g. $D = B$, and one call to xGEMM or xGEMV. Instead, if A should only "once" be loaded from main memory, the system (7) has to be treated as a single entity and xGEMM or xGEMV can not be used any longer since neither the matrices nor the vectors are stored contiguously in memory.

Our implementation listed below is a compact multiplication scheme. It contains two parts, the first one calculating the first column of D and G as well as the whole vector $\underline{z}$, the second computing the remaining columns of D and G. The loop in i is stripmined to work on vectors smaller or equal to the hardware vector length VL.

```
do is = 1, k, VL
   ie = min( is+VL-1, k )

! start with vector z and first columns of D and G
   save v(is:ie), B(is:ie,1) and E(is:ie,1) to VRs vx, vm1 and vm2
   do l = 1, k
      store w(l), C(1,1) and F(1,1) to scalars val1, val2 and val3
      multiply A(is:ie,l) with val1, val2 and val3 and save
         results to VRs vxt, vm1t and vm2t
      add VRs vxt(is:ie), vm1t(is:ie) and vm2t(is:ie) to VRs
         vx(is:ie), vm1(is:ie) and vm2(is:ie)
   end do
   store VRs vx, vm1 and vm2 to z(is:ie), D(is:ie,1) and G(is:ie,1)

! do rest of columns of D and G
   do j = 2, k
```

```
    save B(is:ie,j) and E(is:ie,j) to VRs vm1 and vm2
    do l = 1, k
        store C(l,j) and F(l,j) to scalars val1 and val2
        multiply A(is:ie,l) with val1 and val2 and save results
            to VRs vm1t and vm2t
        add VRs vm1t(is:ie) and vm2t(is:ie) to VRs vm1(is:ie) and
            vm2(is:ie)
    end do
    store VRs vm1 and vm2 to D(is:ie,j) and G(is:ie,j)
  end do
end do
```

Tbl. 1 compares the performance of the BLAS code to the introduced compact multiplication scheme. The latter leads to an average decrease of $11 - 31\%$ in execution time. It is considerably faster for small and medium block sizes, whereas for large block sizes, its performance is approached by BLAS.

Table 1. Execution time of different implementations solving (7) for 150000 calls

$k =$	20	30	40	50	60	70	80	90
LAPACK [s]	2.23	4.65	7.68	11.97	16.92	22.49	28.44	37.99
comp. mult. [s]	1.54	3.25	5.93	9.07	13.19	18.49	25.28	33.71
decr. in runtime [%]	31.1	30.1	22.9	24.2	22.1	17.8	11.1	11.3

3.2 Applying Gaussian Elimination and Backward Substitution to One Block Row

Gaussian Elimination. The search of the pivot element and exchange of matrix rows is not very costly. Measurements on the NEC SX-8 indicated that the search for the maximum value in the pivot column using Fortran's intrinsic function *maxval* and then looping until the correct index has been found is generally faster than a search over the whole pivot column.

The performance of the different versions of GE is depicted in figure 1, 2 and 3. The simplest implementation of row-oriented GE with immediate update [14] for a given step j consists of just two loops in i (column index) and l (row index) updating the remaining parts of the matrix and RHS vectors. The loop in l over the remaining rows is the longer one and therefore the innermost. This loop order is switched by the compiler which naturally degrades performance. The next variant "val + UR(4)", introducing a temporary scalar for elements of the pivot row, still with the compiler's switching of loops, leads to an unrolling of depth 4 in l causing strided memory access. The compiler directive select(vector) on the SX-8, normally used for parallel constructs, provides a simple means to

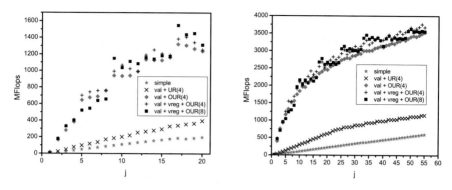

Fig. 1. Performance for GE_R for $k = 20$ **Fig. 2.** Performance for GE_R for $k = 55$

enforce the correct order of the loops. This version "val + OUR(4)" with a temporary scalar now has a compiler-generated outer unrolling in i with depth 4. The remaining two variants "val + vreg + OUR(4)" and "val + vreg + OUR(8)" use VRs for the pivot row, a temporary scalar for elements of the pivot column as well as outer unrolling in i with depths 4 or 8. As can be seen from these figures, the correct loop order is essential for good performance. Using temporary scalars for elements of the pivot column, a VR for elements of the pivot row and outer unrolling in i of depth 8 leads essentially to the highest MFlop rates on the tested vector system.

Backward Substitution. The results for the different versions of row-wise backward substitution explained below are shown in figure 4, 5 and 6. The simplest version for a given step j also consists of just two loops. The obvious first improvement, introducing a temporary scalar for elements of the "pivot row" leads to compiler-generated inner unrolling with depth 9993 and is therefore not shown. The second variant "val + OUR(4)" imposes outer unrolling with depth 4. The remaining variants "val + vreg + OUR(8)", "val + vreg + OUR(16)" and "val + vreg + OUR(4/16)" use one VR for elements of the pivot row, temporary scalars for elements of the pivot column as well as outer unrolling of different depths. Apparently, the fastest version uses a VR for elements of the pivot row, temporary scalars for elements of the pivot column and explicitly coded, yet compiler-generated, outer unrolling of depth 16 as long as possible and of depth 4 for the remaining columns.

Performance Results. Using the optimal implementation for GE_R / BS_R, runtimes on the NEC SX-8 may be reduced by the following factors (for 50000 calls) compared to the LAPACK version: by 61.5% from 5.04s to 1.94s for $k = 20$, by 43.3% from 23.98s to 13.59s for $k = 55$ and by 41.2% from 54.54s to 32.08s for $k = 85$.

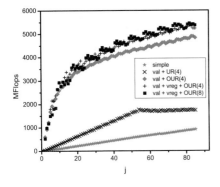

Fig. 3. Performance for GE_R for $k = 85$

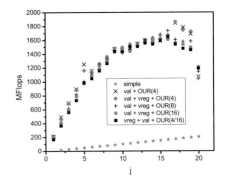

Fig. 4. Performance for BS_R for $k = 20$

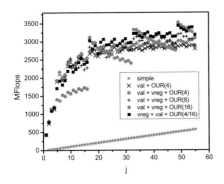

Fig. 5. Performance for BS_R for $k = 55$

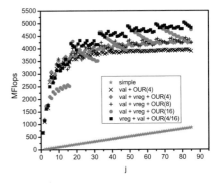

Fig. 6. Performance for BS_R for $k = 85$

3.3   Solution of a Sample BPD System

The compact multiplication scheme and the improvements of the solution of
the block rows are integrated into a new BPD solver. Its execution times are
compared to a traditional BLAS/LAPACK solver in table 2 for 100 systems
with block sizes $k = 20, 55$ and 85 and $n = 500$.

Except for the first version using bandwise storage, the diagonals are stored
as five separate vectors of matrix blocks. If H_i, Z_i and $\underline{r}_i$ in (6) are stored
contiguously in memory, only one call to xGETRS is needed instead of three.

Table 2. Execution times for BPD solver

	$k =$	20	55	85
BLAS/LAPACK + bandw. stor. [s]		11.63	36.79	79.70
BLAS + 3 LAPACK calls [s]		8.33	40.22	85.26
BLAS + 1 LAPACK call ($\star$) [s]		6.43	33.80	76.51
comp. mult. + new solver ($\star\star$) [s]		3.79	23.10	55.10
decr. in runtime between ($\star$) and ($\star\star$) [%]		54.5	42.6	35.4

4   Summary

The solution of a linear system of equations with a BPD matrix with block sizes ranging from 20 to 85 was investigated. A substitute for xGEMM and xGEMV based on a restructured version of the Thomas algorithm as well as specifically adapted versions of LAPACK's xGETRF and xGETRS routines were implemented. Best results were obtained with a compact multiplication routine for the global system and a combined factorization and forward substitution scheme for each block row. The latter uses temporary scalars for elements of the pivot column, a vector register for elements of the pivot row and outer unrolling of depth 8 during Gaussian elimination and of depth 16 and 4 during backward substitution. For a NEC SX-8 vector system a decrease in total runtime between 35% and 54% for test cases of block size 20, 55 and 85 compared to the original code using BLAS and LAPACK was achieved.

An efficient implementation for scalar architectures will be the subject of further research.

References

1. Rampp, M., Janka, H.T.: Radiation hydrodynamics with neutrinos. Astron. Astrophys. **396** (2002) 361–392
2. Thomas, L.H.: Elliptic problems in linear difference equations over a network. Watson Sci. Comput. Lab. Rept., Columbia University, New York (1949)
3. Bruce, G.H., Peaceman, D.W., Jr. Rachford, H.H., Rice, J.D.: Calculations of unsteady-state gas flow through porous media. Petrol. Trans. AIME **198** (1953) 79–92
4. Bieniasz, L.K.: Extension of the Thomas algorithm to a class of algebraic linear equation systems involving quasi-block-tridiagonal matrices with isolated block-pentadiagonal rows, assuming variable block dimensions. Computing **67** (2001) 269–284
5. Dongarra, J.: Performance of various computers using standard linear equations software. Technical Report CS-89-85, University of Tennessee (1989)
6. Kitagawa, K., Tagaya, S., Hagihara, Y., Kanoh, Y.: A hardware overview of SX-6 and SX-7 supercomputer. NEC Res. & Develop. **44** (2003) 2–7
7. Haan, O.: Vektorrechner: Architektur - Programmierung - Anwendung. Saur, München (1993)
8. Joseph, E., Snell, A., Willard, C.G.: NEC launches next-generation vector supercomputer: The SX-8. IDC # 4290 (2004) 1–15 White Paper.
9. HLRS. http://www.hlrs.de/hw-access/platforms/sx8/ (2006)
10. Anderson, E., Blackford, L.S., Sorensen, D., eds.: LAPACK User's Guide. Society for Industrial & Applied Mathematics (2000)
11. Dongarra, J.J., Croz, J.D., Hammarling, S., Hanson, R.J.: An extended set of Fortran basic linear algebra subprograms. ACM Trans. Math. Soft. **14** (1988) 1–17
12. Dongarra, J.J., Croz, J.D., Hammarling, S., Duff, I.S.: A set of level 3 basic linear algebra subprograms. ACM Trans. Math. Soft. **16** (1990) 1–17
13. NEC Corporation: SUPER-UX Fortran90/SX Programmer's Guide. Revision 1.2. (2005)
14. Ortega, J.M.: Introduction to parallel and vector solution of linear systems. Frontiers of computer science. Plenum Press, New York (1988)

Compatibility of Scalapack with the Discrete Wavelet Transform

Liesner Acevedo[1,2], Victor M. Garcia[1], and Antonio M. Vidal[1]

[1] Departamento de Sistemas Informáticos y Computación
Universidad Politécnica de Valencia
Camino de Vera s/n, 46022 Valencia, España
{lacevedo, vmgarcia, avidal}@dsic.upv.es
[2] Departamento de Técnicas de Programación
Universidad de las Ciencias Informaticas
Carretera a San Antonio de los Baños Km 2 s/n, La Habana, Cuba

Abstract. Parallelization of the Discrete Wavelet Transform (DWT) oriented to linear algebra applications, specially the solution of large linear systems, is studied in this paper. We propose that for parallel applications using linear algebra and wavelets, it can be advantageous to use directly the two-dimensional block cyclic distribution (2DBC), used in ScaLAPACK [1]. Although the parallel computation of the DWT may be faster with other data distributions, the efficiency of the 2DBC distribution for some linear algebra applications can make it more appropriate. We have tested this approach implementing a preconditioner for iterative solution of linear systems, which uses wavelets and was proposed in [2]. The results using the 2DBC Distribution are better than those obtained with a popular parallel data distribution for computing wavelets, the "Communication-Efficient" scheme proposed in [3].

Keywords: Discrete Wavelet Transform, Parallel Linear Algebra, Scalapack.

1 Introduction

The Discrete Wavelet transform is still a very recent field, so that it is the subject of active research. The number of applications which can benefit from this tool increases nearly every day, such as time series analysis, signal analysis, signal denoising, signal and image compression, fractals, and many more.

The main motivation of this work has been the parallelization of linear algebra algorithms which use wavelets at any stage. There are many papers describing different techniques for the parallelization of the DWT; however, the algorithms described in these papers put the emphasis in computing the DWT as fast as possible; therefore the data distribution among the processors is chosen to minimize the computing time needed to obtain the transformed matrix.

In linear algebra algorithms, the matrix obtained after applying the DWT may have to be used for some calculation far more expensive than the DWT

Y. Shi et al. (Eds.): ICCS 2007, Part I, LNCS 4487, pp. 152–159, 2007.

in itself, such as solution of linear systems. The standard parallel library for linear systems with dense matrices is ScaLAPACK. However, in order to use the ScaLAPACK functions the matrix must be distributed using the bidimensional block cyclic distribution (2DBC distribution), the ScaLAPACK standard. If the matrix is not distributed in this way, one must face either a redistribution of the matrix, (with a high communication cost) or a heavy load imbalance (if no redistribution is done).

We propose to compute the DWT directly over the matrix distributed with the 2DBC distribution. As an application study, we have applied it to the parallelization of a multilevel preconditioner proposed by T. Chan and K. Chen in [2], based on the DWT and the Schur complement method.

The paper starts by giving a small review of the DWT and its application to solution of linear systems. In that section, the multilevel preconditioner is described. Then starts the main part of paper, where the results about parallel DWT are presented. Finally, the code implementing the parallel version of the preconditioner is described, and some numerical results are given.

2   Discrete Wavelet Transform

The theoretical background of the DWT is based on the concept of Multiresolution analysis and is already quite well known. However, these concepts are not useful when considering the implementation details and the parallelization, so that we will refer to the interested reader to the following references: [4,5].

The DWT applied over a signal or vector is obtained through digital filters, a lowpass filter G determined by the coefficients $\{g_i\}_{i=1}^{D}$ and a highpass filter H $\{h_i\}_{i=1}^{D}$.

Not all pairs of highpass and lowpass filters are valid for a DWT; however, there are many such pairs. This means that the DWT is not a single, unique operation, since it depends on the filters. The orthogonal wavelets are the most popular group of wavelets. In this case, to obtain a reversible DWT the filter coefficients must satisfy $h_i = (-1)^i g_i$. The length D of the filter is called the "order" of the wavelet. For simplicity, we shall concentrate only on orthogonal wavelets, and, among these, only on the wavelets depending on short filters, since these will minimize the communications in the parallel DWT.

A single stage of the DWT (or a one-level DWT) over a vector v of length $n = 2^k$ is performed, first, convolving both filters with the vector v, obtaining two sequences of length n. Next, every even element of these sequences is discarded, and the resulting sequences (of length $\frac{n}{2}$) are assembled together.

These two steps are equivalent to the multiplication of the data vector by the matrix $W_n = \begin{pmatrix} H \\ G \end{pmatrix}$, where H and G (of dimension $\frac{n}{2} \times n$) are

$$H = \begin{pmatrix} h_1 & h_2 & ... & h_D & 0 & \cdots\cdots\cdots\cdots & 0 \\ 0 & 0 & h_1 & h_2 & ... h_D & 0 & \cdots\cdots & 0 \\ 0 & \cdots\cdots & \ddots & & \ddots & \ddots & \vdots & 0 \\ h_3 & h_4 & \cdots h_D & 0 & \cdots & & 0 & h_1 & h_2 \end{pmatrix} \tag{1}$$

The matrix G has the same structure. This product would result in

$$W_n \cdot v = \begin{pmatrix} H \cdot v \\ G \cdot v \end{pmatrix} = \begin{pmatrix} d_1 \\ \vdots \\ d_{n/2} \\ s_1 \\ \vdots \\ s_{n/2} \end{pmatrix} \tag{2}$$

where the d_i are the high frequency (or "detail") components and the s_i are the low frequency (or "smooth") components. If the wavelet is orthogonal, then the matrix W_n must be orthogonal as well. The orthogonal wavelets with shorter filters are the Haar wavelet, with coefficients $\left(g_1 = \frac{\sqrt{2}}{2}, g_2 = \frac{\sqrt{2}}{2} \right)$

The L-Level DWT (or, simply, the DWT) is obtained applying L times the one-level DWT over the low frequency part of the data vector.

The two-dimensional wavelet transform is applied to a matrix by applying 1-D transforms along each dimension. So, a one-level 2-D DWT applied over a matrix $T_0 \in \Re^{n,n}$ can be written as $\tilde{T}_0 = W_n T_0 W_n^T$, with W_n defined as above. In this case, we obtain the following partition:

$$\tilde{T}_0 = W_n T_0 W_n^T = \begin{pmatrix} A_1 & B_1 \\ C_1 & T_1 \end{pmatrix}, \tag{3}$$

where $T_1 = GT_0G^T$, $A_1 = HT_0H^T$, $B_1 = HT_0G^T$ and $C_1 = GT_0H^T$.

If, as usual, we want to apply more than one DWT level, at least two different possibilities appear. The so-called "Standard" DWT form of a matrix is obtained applying the L-level DWT over the columns, and then applying the L-level DWT over the rows. The Non-Standard form [6] is obtained applying a one level DWT over the rows, and one over the columns (obtaining a partition as in (3). Then, another one-level DWT by rows $(W_{n/2}^T)$ and another DWT by columns $(W_{n/2})$ are applied over the low-frequency submatrix T_1, and this is repeated L times, always over the low frequency submatrix.

When working with matrices, the choice usually is the Non-Standard form or some slightly different version, such as the level-by level form, proposed in [2]. Therefore, all the considerations from now on will be referred to the Non-Standard version.

2.1   Wavelets and Linear Systems

There are several research lines, oriented towards efficient solution of linear systems using wavelets [6,7]. One of these is based on the decomposition shown in equation (3). The main algorithm of interest for us is the recursive Wavelet-Schur preconditioner proposed by Chan and Chen in [2].

It is not possible for us, due to space reasons, to give an appropriate description of the algorithm. Therefore we will only give a brief outline; all the details can be found in [2].

This algorithm uses the block decomposition (3) produced by the DWT. To build a preconditioner, the submatrices A_1, B_1 and C_1 are approximated by "band approximations" (all the elements except those on the desired band are set to zero), and the Schur complement method is used to solve, approximately, a linear system with this new matrix. Combining this technique with standard iterative methods (GMRES, Richardson)[8] a multilevel algorithm for solution of linear systems is built.

The implementation of this algorithm requires the computation of several DWT levels of a matrix, the computation of matrix-vector products of DWT-transformed matrices, and, in the inner level, needs the solution of a linear system through LU decomposition, where the coefficient matrix is again a DWT-transformed matrix.

2.2   Parallel DWT; Data Distributions

The parallel DWT of a matrix is relatively simple to implement [3]. The most critical issue is the data distribution to avoid communications. The scheme with seemingly better properties is the "Communication-efficient" (C-E) DWT proposed by Nielsen and Hegland [3], in which the matrix is distributed by block of contiguous columns.

On the other hand (as mentioned above), if the matrix should be used as coefficient matrix for solution of linear systems, it would be better to use a 2DBC distribution, the used as basis of the Scalapack library. The ScaLAPACK data distribution is described in detail in [1]. Here we only show an example. Let us consider the bidimensional grid of processors of Figure 1.

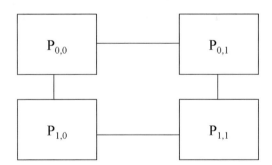

Fig. 1. Bidimensional processor mesh

Figure 2 shows how a matrix would be distributed over the processor mesh in Fig.1. The distribution for C-E DWT is displayed on the left, while the figure on the right shows a possible two dimensional cyclic distribution, suitable for use with Scalapack.

We will examine now which are the communications needed for the computation of a single wavelet level, using both distributions. Let us start with the C-E distribution.

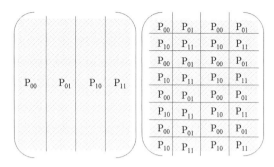

Fig. 2. Matrix on the left, distributed by columns (for C-E DWT); on the right, a possible two-dimensional cyclic distribution

Using basic results from the Nielsen-Hegland paper [3], we have that with the column distribution every processor will need, for each row, $D-2$ elements from the processor holding the columns on the right. These elements are needed to compute the row-oriented 1D DWT. (Since in this distribution each processor holds complete columns, there is no need of communications for the column-oriented 1D DWT). The only exception is the last processor (P_{11} in the figure), which must receive these elements form the first processor (P_{00} in the Figure).

Therefore, considering all rows, each processor will need $N \cdot (D-2)$ elements, and the total number of words to be transferred is $P \cdot N \cdot (D-2)$, where P is the number of processors.

Using a 2DBC distribution, the communications are somewhat more involved. Let us assume first that the matrix is distributed over a square processor mesh of diameter $\sqrt{P}$, and that the block size, N_b divides exactly the dimension of the matrix n. So,if $k = \frac{N}{N_b}$, then the total number of blocks in the matrix is k^2.

Let us consider now a single block, located, for the example, in the processor P_{00}. Using the same result from [3] mentioned above, we obtain that to complete the DWT of that block, the processor P_{00} should receive $N_b \cdot (D-2)$ elements from the processor P_{01} (needed for the row-oriented DWT) and another $N_b \cdot (D-2)$ elements from the processor P_{10} (for the column-oriented DWT). If we consider the last row of blocks, these should be receiving the same amount of data, but from the processor(s) holding the first row of blocks, like in the CE distribution. If we consider the last column of blocks, it would happen exactly the same.

It turns out that every block (no matter in which processor are they located) requires a communication of $2N_b \cdot (D-2)$ words . Therefore, for a single DWT level, the total number of words to be transferred shall be $k^2 \cdot 2N_b \cdot (D-2) = 2N \cdot k \cdot (D-2)$. Clearly, for "normal" values of P and N_b, more communications are needed with the 2DBC distribution.

Of course, if we were interested only in computing the DWT, it would not be sensible to consider a Scalapack-like distribution, since it is simpler and more efficient to compute the DWT with the communication-efficient strategy. Nevertheless, the situation changes when, as in this case, some linear algebra tasks must be performed in parallel with the transformed matrices.

3   Example of Application: Parallel Implementation of the Wavelet-Schur Preconditioner

The Wavelet-Schur preconditioner, described briefly above, is a good example of algorithm which might benefit of the use of the ScaLAPACK library. It needs to compute several matrix-vector products and, above all, it needs to solve a dense linear system.

To test the performance of the data distributions considered, we have written a parallel linear system iterative solver, based on GMRES and preconditioned with the Wavelet-Schur preconditioner.

The GMRES method for dense matrices in a 2DBC distribution is implemented easily using the PBLAS and ScaLAPACK functions for the matrix-vector product, saxpy update, dot product and 2-norm: **pdgemv**, **pdaxpy**, **pddot** and **pdnorm2**. In the coarsest level where the direct method is used, the LU decomposition of ScaLAPACK **pdgesv** is used.

The band approximations of the A, B and C matrices have been chosen as diagonal matrices.

The linear systems used to evaluate the parallel algorithm proposed were generated with the following matrices:

$$a_{i,j} = \begin{cases} 2 & if \ i = j \\ \frac{1}{|i-j|} & otherwise \end{cases} \tag{4}$$

For simplicity, the sizes of the matrix have been chosen as powers of two: $768, 1152, \cdots, 4024$. All the tests were carried out in a cluster with 20 2GHz Intel Xeon biprocessors, each one with 1 Gbyte of RAM, disposed in a 4x5 mesh with 2D torus topology and interconnected through a SCI network. The processor grids used have sizes 2×2, 3×3 and 4×4 and the block size was 32.

3.1   Numerical Results

The table 1 and 2 shows the CPU times of the algorithm in sequential and parallel versions, using 1 and 2 wavelet levels in the proposed processor meshes with the 2DBC Distribution.

The same algorithm was implemented as well computing the parallel DWT with the CE distribution. It would have been interesting to apply directly the algorithm for solution of linear systems to the CE distribution, but the ScaLA-PACK routine for the LU decomposition request that the matrix is distributed using square blocks, so that it cannot be used with the distribution by blocks of columns used for the communication-efficient DWT. Therefore, the matrix had to be redistributed to a 2DBC distribution after computing the DWT (The redistribution was done using the ScaLAPACK redistribution subroutine **pdgemr2d**).

So, the difference between execution times for both distributions can be assigned entirely to the redistribution of the matrix form the CE (column) distribution to a 2DBC distribution. The results in that case can be seen in Tables 3 and 4.

Table 1. CPU time (sec.) for the algorithm with 1 wavelet level, Scalapack distribution

N	768	1152	1536	1920	2304	2688	3072	3456	3840	4224
1 Processor	0.15	0.36	0.68	1.14	1.76	2.45	3.41	4.76	6.19	7.93
2 × 2 Processors	0.13	0.29	0.40	0.62	0.80	1.07	1.40	1.80	2.29	2.76
3 × 3 Processors	0.08	0.14	0.20	0.29	0.42	0.52	0.69	0.93	1.17	1.42
4 × 4 Processors	0.05	0.10	0.12	0.18	0.25	0.32	0.43	0.56	0.72	0.85

Table 2. CPU time (sec.) for the algorithm with 2 wavelet level, Scalapack distribution

N	768	1152	1536	1920	2304	2688	3072	3456	3840	4224
1 Processor	0.13	0.26	0.47	0.76	1.13	1.65	1.96	2.80	3.37	4.22
2 × 2 Processors	0.12	0.23	0.30	0.47	0.64	0.78	0.98	1.17	1.37	1.63
3 × 3 Processors	0.07	0.12	0.14	0.21	0.29	0.39	0.43	0.61	0.66	0.79
4 × 4 Processors	0.05	0.08	0.09	0.13	0.17	0.23	0.27	0.35	0.40	0.46

Table 3. CPU time (sec.) for the algorithm with 1 wavelet level, Communication-efficient distribution

N	768	1152	1536	1920	2304	2688	3072	3456	3840	4224
1 Processor	0.15	0.36	0.68	1.14	1.76	2.45	3.41	4.76	6.19	7.93
2 × 2 Processors	0.22	0.49	0.70	1.08	1.50	2.03	2.62	3.30	4.14	4.98
3 × 3 Processors	0.25	0.38	0.58	0.87	1.13	1.47	1.92	2.44	3.07	3.66
4 × 4 Processors	0.26	0.40	0.52	0.77	0.98	1.33	1.66	2.13	2.64	3.15

Table 4. CPU time (sec.) for the algorithm with 2 wavelet level, Communication-efficient distribution

N	768	1152	1536	1920	2304	2688	3072	3456	3840	4224
1 Processor	0.13	0.26	0.47	0.76	1.13	1.65	1.96	2.80	3.37	4.22
2 × 2 Processors	0.14	0.27	0.38	0.59	0.81	1.04	1.31	1.54	1.83	2.21
3 × 3 Processors	0.18	0.25	0.31	0.42	0.56	0.70	0.85	1.07	1.20	1.43
4 × 4 Processors	0.27	0.34	0.41	0.52	0.53	0.60	0.69	0.86	0.99	1.15

It is quite clear (and obvious) that the times using directly the ScaLAPACK distribution are significantly better than the sequential times and the times obtained with the C-E distribution. Considering that the parallel computation of the DWT is faster with the CE distribution than with the 2DBC distribution, this means that the time needed to redistribute the matrix (from the CE distribution to the 2DBC distribution) is quite large. The efficiencies obtained are not very good in any case, due to the high communications requirements of the LU decomposition and the matrix-vector products. However, the efficiency improves when the number of processors and the size of the problem increases; in the case of the 2DBC distribution, the improvement is more pronounced.

4   Conclusions

In this paper we have considered the computation of the parallel DWT of a matrix, and specially to its application in numerical linear algebra applications.

We have shown that the 2DBC distribution is an interesting distribution for computing the parallel DWT, since it would allow the direct use of ScaLAPACK routines. As an example, the Wavelet-Schur preconditioner has been parallelized with good results with respect to the sequential versions, and with respect to the same algorithm using the C-E DWT proposed by Nielsen and Hegland.

Acknowledgement

This work has been supported by Spanish MCYT and FEDER under Grant TIC2003-08238-C02-02.

References

1. L.S. Blackford et al., Software, Enviroment, Tools. ScaLAPACK Users Guide, SIAM (1997).
2. T.F. Chan, K. Chen, On Two Variants of an Algebraic Wavelet Preconditioner, SIAM J. Sci. Comput.,24 (2002),260–283
3. O.M. Nielsen, M.A. Hegland, A Scalable Parallel 2D Wavelet Transform Algorithm. REPORT TR-CS-97-21, Australian National University, 1997. http://cs.anu.edu.au/techreports.
4. I. Daubechies, Ten Lectures on Wavelets, CBMS-NSF Regional Conference Series In Applied Mathematics, Vol. 61. Society for Industrial and Applied Mathematics Philadelphia, 1992.
5. M. V. Wickerhauser, Adapted Wavelet Analysis from Theory to software. A.K. Peters, IEEE Press, 1994.
6. G. Beylkin, R. Coifman, V. Rokhlin, Fast Wavelet Transforms and Numerical algorithms I, Commun. Pure Appl. Math., XLIV (1991), pp. 141-183.
7. T.F. Chan, W.P. Tang, W.L. Wan, Wavelet sparse approximate inverse preconditioners, BIT, 37 (1997), pp. 644-660.
8. R. Barrett, M. Berry, T. Chan, J. Demmel, J. Donato, J. Dongarra, V. Eijkhout, R. Pozo, C. Romine, H. Vorst, Templates for the Solution of Linear Systems: Building Blocks for Iterative Methods, Society for Industrial and Applied Mathematics, Philadelphia, 1993.

A Model for Representing Topological Relations Between Simple Concave Regions

Jihong OuYang[1,2], Qian Fu[1,2], and Dayou Liu[1,2]

[1] College of Computer Science and Technology, Jilin University,
Changchun 130012, China
[2] Key Laboratory of Symbolic Computation and Knowledge Engineering of Ministry
of Education, Jilin University, Changchun 130012, China
ouyangjihong@yahoo.com.cn, fuqian2008@yahoo.com.cn

Abstract. At present, qualitative spatial reasoning has become the hot
issues in many research fields. The most popular models of spatial topological relations are Region Connection Calculus (RCC) and 9-inter-
section model. However, there are few contributions on topological
relations of concave regions in which the representative model is Cohn's
RCC23. There are some limitations of RCC23 especially in practical applications due to its less expressiveness. In order to construct a more
expressive model of topological relations between concave regions, this
paper completed the following works: 9-intersection matrix is extended
to 16-intersection matrix, and RCC23 is refined to RCC62 based on 16-
intersection matrix. More relations can be distinguished in RCC62, which
is more expressive than RCC23. In order to further reason about relations in RCC62, the Conceptual Neighborhood Graph (CNG) and the
Closest Topological Relation Graph (CTRG) of RCC62 are given.

Keywords: topological relation, simple concave region, 9-intersection
model, 16-intersection matrix, RCC23.

1   Introduction

The development of formal models on spatial relations is an essential topic in spatial reasoning, geographic information system (GIS) and computer vision which
got much attention from researchers in relative fields during recent years [1], [2],
[3]. Particularly, significant progresses have been made in models on topological
spatial relations of regions. Topology is perhaps the most basic aspect of space.
It is clear that topology must form a fundamental aspect of qualitative spatial
reasoning since topology certainly can only make qualitative distinctions [3].

At present, most of the models on topological spatial relations adopt either
logical or algebraic methods, among which the most typical models are Region
Connection Calculus (RCC) [4] and intersection models [5]. Moreover, there are
many improved methods proposed by other researchers. Most of the existing
models mainly concentrate on topological relations between convex regions [4],
[5], but less works have been made on topological relations between concave

Y. Shi et al. (Eds.): ICCS 2007, Part I, LNCS 4487, pp. 160–167, 2007.

regions, in which one of the typical models is Cohn's RCC23 [6], which twenty-three topological relations between two simple concave regions are defined based on two primitive of connection and convex hull. However, there are limitations of RCC23 especially in spatial query on account of its less expressiveness. Sometimes, the querying results would be inexact or the querying request may be failed. Thus, it is essential to establish more expressive models for topological relations between concave regions.

This paper make an extension of 9-intersection to 16-intersection matrix, and based on 16-intersection matrix, RCC23 [6] is refined to a set of sixty-two relations, thus derived RCC62 which is more expressive than RCC23. In order to further research the reasoning of RCC62, the Conceptual Neighborhood Graph (CNG) [7] and the Closest Topological Relation Graph (CTRG) [8] of RCC62 are given.

2   Background

2.1   RCC8 and RCC23

Randell et al. [4] proposed the RCC theory based on a calculus of individuals developed by Clarke. The basic part of RCC theory assumes a primitive dyadic relation: $C(x, y)$ read as 'x connects with y' which is defined on regions. Using the relation $C(x, y)$, a basic set of dyadic relations are defined: DC, EC, PO, TPP, NTPP, TPPi, NTPPi, EQ. To reason about spatial change, the Conceptual Neighborhood Graph (CNG) [7] is used. Fig. 1(left) shows the basic relations and also the CNG of RCC8 which indicates the continuous transitions among the RCC8 relations.

Cohn et al. [6] defined three predicates (INSIDE, P-INSIDE, and OUTSIDE) based on the primitive of convex hull to test for a region being inside, partly inside and outside another. Each of these relations is asymmetric so they have inverses, denoted by INSIDEi, P-INSIDEi and OUTSIDEi. The definitions naturally give rise to a set of twenty-three JEPD relations, which we call RCC23, as illustrated in Fig. 1 (right).

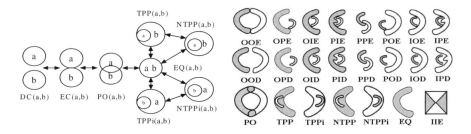

Fig. 1. (left) 2D illustrations of the relations of RCC 8 and their continuous transitions (CNG) (right) The pictorial representation of the relations of RCC23

2.2   Intersection Models

Egenhofer et al. [5] presented the 9-intersection model for binary topological relations between crisp regions by comparing the interiors (A°), boundaries (∂A) and exteriors (A^-) of the two regions. Considering the binary values empty and non-empty for these intersections, one can distinguish 512 binary topological relationships. Eight topological relations can be realized between two regions embedded in $\mathbf{IR}^2$. The 9-intersection matrix is shown in Fig. 2 (left). In order to represent the continuous change of spatial relations, Egenhofer presented the Closest Topological Relation Graph (CTRG) [7] based on the definition of topological distance. Fig. 2 (right) shows the CTRG of eight binary relations.

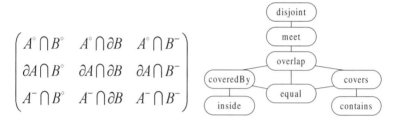

$$\begin{pmatrix} A^\circ \cap B^\circ & A^\circ \cap \partial B & A^\circ \cap B^- \\ \partial A \cap B^\circ & \partial A \cap \partial B & \partial A \cap B^- \\ A^- \cap B^\circ & A^- \cap \partial B & A^- \cap B^- \end{pmatrix}$$

Fig. 2.   (left) The 9-intersection matrix (right) The CTRG of eight binary relations

3   RCC62

An extension of 9-intersection matrix to 16-intersection matrix is made, then RCC23 is refined by adding 39 new relations based on 16-intersection matrix, thus sixty-two topological relations between two simple concave regions are derived, which we call RCC62. The CNG and the CTRG of RCC62 are also given.

3.1   16-Intersection Matrix

Some basic definitions used in this paper are given below.

Definition 1. *The convex hull of region x, denoted by* $\mathtt{conv}(x)$*, is the smallest convex region of which x is a part.* [4]

Definition 2. *The inside of concave region x, denoted by* $\mathtt{inside}(x)$*, is the difference of the convex hull of region x and the region itself.*[4]

$$\mathtt{inside}(x) \equiv_{def} \iota y [\forall z [C(z, y) \leftrightarrow \exists w [\mathtt{INSIDE}(w, x) \wedge C(z, w)]]] \qquad (1)$$

Definition 3. *The outside of concave region x, denoted by* $\mathtt{outside}(x)$*, is the complement part of the convex hull of region x.*[4]

$$\mathtt{outside}(x) \equiv_{def} \iota y [\forall z [C(z, y) \leftrightarrow \exists w [\mathtt{OUTSIDE}(w, x) \wedge C(z, w)]]] \qquad (2)$$

Definition 4. *x is a concavity of region y, denoted by* Concavity(x, y), *when predicate* Concavity(x, y) *is true.*[9]

$$\text{Concavity}(x, y) \equiv_{def} \text{MAX\_P}(x, \text{inside}(y)) \tag{3}$$

Definition 5. *Region x is a simple concave region, if it is concave and have only one concavity.*

Definition 6. *The topological relation between two concave regions a and b, is characterized by comparing a's outside (a_0), boundary (a_1), interior (a_2), inside (a_3) with b's outside (b_0), boundary (b_1), interior (b_2), inside (b_3), thus concisely represented as a 4×4 matrix, called the 16-intersection matrix which denoted by*

$$R(a, b) = \begin{pmatrix} a_0 \cap b_0 & a_1 \cap b_0 & a_2 \cap b_0 & a_3 \cap b_0 \\ a_0 \cap b_1 & a_1 \cap b_1 & a_2 \cap b_1 & a_3 \cap b_1 \\ a_0 \cap b_2 & a_1 \cap b_2 & a_2 \cap b_2 & a_3 \cap b_2 \\ a_0 \cap b_3 & a_1 \cap b_3 & a_2 \cap b_3 & a_3 \cap b_3 \end{pmatrix} \tag{4}$$

If the intersection of a_i and b_j is non-empty, then the value of $a_i \cap b_j$ is 1, otherwise the value of $a_i \cap b_j$ is 0. Different combinations in the intersection matrix can represent different topological relations. Note that since a_3 is a virtual area component, no boundary is defined between a_3 and a_0. Thus if a_3 has a non-empty intersection with b_1, then it has also to intersect with b_2. Similarly, if b_3 has a non-empty intersection with a_1, then it has also to intersect with a_2.

As illustrated in Fig. 3 (left), the topological relation between two simple concave regions a and b is represented by their 16-intersection matrix.

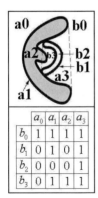

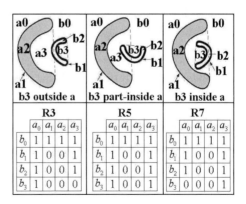

Fig. 3. (left) The topological relation between two simple concave region a, b and their 16-intersection matrix. (right) 16-intersection matrix for distinguishing b_3 (the bay of the island) outsides, partially-insides or insides island a.

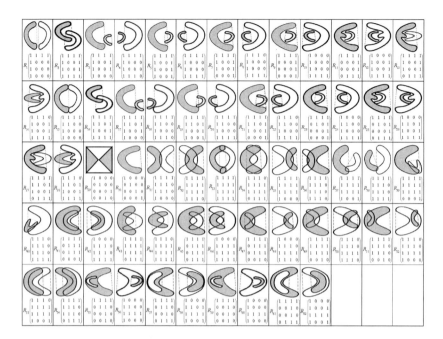

Fig. 4. A pictorial representation of the basic relations of RCC62 and their corresponding 16-intersection matrixes

3.2   Basic Relations of RCC62

The definition of 16-intersection matrix is useful for distinguish topological spatial relations of simple concave regions such as the bay of a small island b (i.e. b_3) is outside, partially-inside, or inside a large island a, as shown in Fig. 3 (right). These distinctions cannot be made using RCC23 since they represent the same relation OPD.

In RCC23, the concave regions are considered as a whole part, while in RCC62 it is decomposed into four sub-parts (i.e. a_0, a_1, a_2, a_3), thus based on 16-intersection matrix more topological relations can be derived between two concave regions.

Here relation OPD in RCC23 is refined to three new relations (i.e. R_3, R_5, R_7) by considering that b_3 intersects only with a_0, both with a_0, a_3, and only with a_3. In the same way, RCC23 is refined based on 16-intersection matrix, and totally sixty-two topological relations between two simple concave regions are defined, which we call RCC62. The pictorial representation of the base relations of RCC62 and their corresponding 16-intersection matrixes are shown in Fig. 4.

3.3   Relationship Between RCC23 and RCC62

Since RCC62 is derived from RCC23, there is a natural connection between them. The relationship between RCC23 and RCC62 is given in Fig. 5. Basic

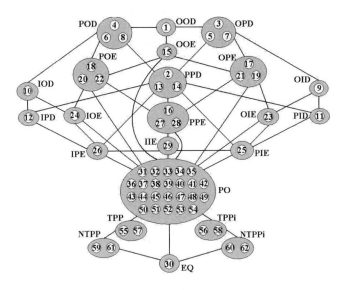

Fig. 5. Basic relations of RCC23 (denoted by the solid ovals labeled with letters), and their corresponding refinements to RCC62 (denoted by the hollow circles labeled with numbers), e.g. relation POD in RCC23 are refined to three new relations (i.e. R_4, R_6, R_8) in RCC62

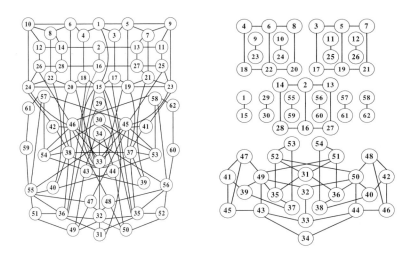

Fig. 6. (left) The CNG of RCC62, in which a node for each relation and an arc between two relations represents a smooth transition that can transform a relation to the other and vice versa. (right) The CTRG of RCC62, in which a node for each relation and an arc for each pair of matrices at minimum topological distance that measured in terms of the number of different values in the corresponding matrices.

relations of RCC23 are represented by the solid ovals labeled with letters, while relations in RCC62 are described by the hollow circles labeled with numbers. Take an example, relation POD in RCC23 are refined to three new relations (i.e. R_4, R_6, R_8) in RCC62. The arc between two basic relations of RCC23 denotes a smooth transition between them. As shown in Fig. 5, there are more relations can be distinguished in RCC62, thus it is more expressive than RCC23.

3.4   CNG and CTRG of RCC62

It is necessary to represent the change of spatial relations in spatial reasoning. In qualitative reasoning we assume that change is continuous. The primary methods to represent the continuous change of spatial relations are the Conceptual Neighborhood Graph (CNG) [7] and the Closest Topological Relation Graph (CTRG) [8]. We give both the CNG and CTRG of RCC62, as shown in Fig. 6 (left) and Fig. 6 (right) respectively.

4   Conclusions

Topological relation is the most basic relation of spatial objects which is one of the elementary aspects of spatial reasoning. The most popular models of topological relation are RCC and intersection models. However, few contributions are available for topological relations between concave regions, in which one of the typical models is Cohn's RCC23. Nevertheless, RCC23 has some limitations in the practical fields especially for spatial query. Therefore it is essential to establish a more expressive model of topological relations between concave regions.

In order to represent the topological relations between simple concave regions in 2D space, an extension of 9-intersection matrix is made to 16-intersection matrix, and RCC23 is refined by adding 39 relations based on 16-intersection method, thus derived RCC62, in which more relations can be distinguished, and is more expressive than RCC23. The CNG and the CTRG of RCC62 which are bases for the further reasoning of RCC62 are also given.

Acknowledgments

This research was supported by NSFC Major Research Program No. 60496321, National Natural Science Foundation of China under Grant No. 60573073, the National High-Tech Research and Development Plan of China under Grant No. 2006AA10Z245 and 2006AA10A309 the Major Program of Science and Technology Development Plan of Jilin Province under Grant No. 20020303, the Science and Technology Development Plan of Jilin Province under Grant No. 20030523, European Commission under grant No. TH/Asia Link/010 (111084).

References

1. Jihong, Ou. Yang., Dayou Liu., He Hu., Boyu Chen.: A hybrid spatial reasoning approach based on fuzzy sets. Journal of Jilin University (Science Edition) 42(4) (2004) 565–569

2. Qiangyuan Yu., Dayou Liu., Jihong Ou. Yang.: Topological relations model of fuzzy regions based on interval valued fuzzy sets. Chinese Journal of Electronics 33(1) (2005) 186–189
3. Cohn, A.G., Hazarika, S.M.: Qualitative spatial representation and reasoning: An overview. Fundamenta Informaticae 46(1–2) (2001) 1–29
4. Randell, D.A., Cui, Z., Cohn, A.G.: A spatial logic based on regions and connection. In: Proc. of the 3rd Int. Conf. on Principles of Knowledge Representation and Reasoning. Morgan Kaufmann, Sanmateo (1992) 165–176
5. Egenhofer, M.J., Herring, J.R.: Categorizing binary topological relationships between regions, lines and points in geographic database. University of Maine, Orono, Maine, Dept. of Surveying Engineering, Technical Report (1992)
6. Cohn, A.G., Bennett, B., Gooday, J., Gotts, N.M.: Qualitative spatial representation and reasoning with the region connection calculus. GeoInformatica 1(1) (1997) 1–44
7. Freksa, C.: Temporal reasoning based on semi-intervals. Artificial Intelligence 54 (1992) 199–227
8. Egenhofer, M.J., Al-Taha, K.: Reasoning about gradual changes of topological relationships. In: Frank, A., Campari, I., Formentini, U. (eds.): Theories and models of spatio-temporal reasoning in geographic space. Lecture Notes in Computer Science, Vol. 693 Springer-Verlag, Berlin. (1992) 196–219
9. Cohn, A.G.: A hierarchical representation of qualitative shape based on connection and convexity. In: Proc. of COSIT'95. Semmering (1995)

Speech Emotion Recognition Based on a Fusion of All-Class and Pairwise-Class Feature Selection

Jia Liu[1], Chun Chen[1], Jiajun Bu[1], Mingyu You[1], and Jianhua Tao[2]

[1] College of Computer Science, Zhejiang University, Hangzhou, P.R. China, 310027
{liujia,chenc,bjj,roseyoumy}@zju.edu.cn
[2] National Laboratory of Pattern Recognition, Institute of Automation,
Chinese Academy of Sciences, Beijing, P.R. China, 100080
jhtao@nlpr.ia.ac.cn

Abstract. Traditionally in speech emotion recognition, feature selection(FS) is implemented by considering the features from all classes jointly. In this paper, a hybrid system based on all-class FS and pairwise-class FS is proposed to improve speech emotion classification performance. Besides a subset of features obtained from an all-class structure, FS is performed on the available data from each pair of classes. All these subsets are used in their corresponding K-nearest-neighbors(KNN) or Support Vector Machine(SVM) classifiers and the posterior probabilities of the multi-classifiers are fused hierarchically. The experiment results demonstrate that compared with the classical method based on all-class FS and the pairwise method based on pairwise-class FS, the proposed approach achieves 3.2%-8.4% relative improvement on the average F1-measure in speaker-independent emotion recognition.

1 Introduction

Emotion recognition is one of the latest challenges in intelligent human-machine communication. It has broad applications in areas such as customer service, medical analysis and entertainment, etc. As a major indicator of human affective states, speech plays an important role in detecting emotions[1]. Speech emotion recognition can be formulated as three basic steps: extracting acoustic features from speech signals, selecting a feature subset and detecting emotions with various classifiers. Among them, feature selection(FS) is a significant step since it can decrease computational complexity and prediction error by avoiding the 'curse of dimensionality'[2].

In traditional speech emotion recognition, features from all classes are analyzed simultaneously to make a lower-dimensional set which is subsequently used in training/testing an appropriate classifier. Chuang and Wu[3] adopted Principal Component Analysis(PCA) to reduce dimensionality of features for classifying seven basic emotions. Ververidis et al.[4] used Sequential Forward Selection(SFS) to choose five best features from speech signals for recognizing five emotional states. All of them considered the data from all classes together. But intuitively, the features useful to distinguish the emotion 'Angry'

Y. Shi et al. (Eds.): ICCS 2007, Part I, LNCS 4487, pp. 168–175, 2007.
© Springer-Verlag Berlin Heidelberg 2007

from 'Happy' may differ from those features distinguishing 'Angry' from 'Sad'. In other words, the feature subset produced by all-class FS is not optimal in discriminating between a specific pair of classes. In contrast to the all-class method, a pairwise-class FS has been proposed for musical instrument classification[5] and face recognition[6]. They suggested choosing the most efficient feature subset for each pair of classes. Every subset was dealt with a one-against-one classifier correspondingly and then a 'majority voting' was applied on multi-classifiers to get a final decision. It was reported that compared with the classification based on all-class FS, a relatively high recognition accuracy was achieved by the pairwise method. However, little work of speech emotion recognition has focused on the pairwise framework.

In this paper, we analyze the data by all-class FS and pairwise-class FS respectively. Based on the two views, a hybrid speech emotion recognition system is developed. In the experiments, KNN and SVM are both employed to demonstrate the performance of the novel system.

2   Emotion Corpus and Acoustic Features

The speech corpus contains six emotions: *Neutral, Anger, Fear, Happiness, Sadness* and *Surprise*. Three hundred sentences, each expressed in the six categories, are recorded from four Chinese native speakers(two females and two males). In our experiments, speaker-independent emotion recognition of each gender was investigated. On each gender, 80% of utterances were randomly selected for training, and the remaining 20% for testing. The entire experiment was repeated ten times with different training and testing data sets, and the results were averaged. The classification performance was evaluated in terms of the F1-measure, which is defined as $F1 = 2 \times precision \times recall \div (recall + precision)$. The recall is the portion of the correct categories that were assigned, while the precision measures the portion of the assigned categories that were correct.

In the field of speech emotion recognition, many acoustic features have been presented in the literature[4,7]. We extracted 16 formant frequency features and 48 prosodic features from signals since formant frequency features are widely used in speech processing applications and prosody is believed to be the primary indicator of human affective states. The detailed features are indexed as follows:

01.-04.   *max, min, mean, variance* of the first formant
05.-08.   *max, min, mean, variance* of the second formant
09.-12.   *max, min, mean, variance* of the third formant
13.-16.   *max, min, mean, variance* of the fourth formant
17.-20.   *max, min, mean, median value* of pitch
21.-22.   *mean, median value* of pitch rises
23.-25.   *max, mean, median duration* of pitch rises
26.-27.   *mean, median value* of pitch falls
28.-30.   *max, mean, median duration* of pitch falls
31.-32.   *mean, median value* of plateaux at pitch maxima[4]
33.-35.   *max, mean, median duration* of plateaux at pitch maxima

36.-37.	*mean, median value* of plateaux at pitch minima
38.-40.	*max, mean, median duration* of plateaux at pitch minima
41.-44.	*max, min, mean, median value* of energy
45.-46.	*mean, median value* of energy rises
47.-49.	*max, mean, median duration* of energy rises
50.-51.	*mean, median value* of energy falls
52.-54.	*max, mean, median duration* of energy falls
55.-56.	*mean, median value* of plateaux at energy maxima
57.-59.	*max, mean, median duration* of plateaux at energy maxima
60.-61.	*mean, median value* of plateaux at energy minima
62.-64.	*max, mean, median duration* of plateaux at energy minima

3   Experiment Design

3.1   System Overview

The novel system of speech emotion recognition takes account of the most relevant features from all-class data and the optimal feature subsets from all possible pairs of classes. Mainly, the system is composed of two phases: training(Fig. 1) and testing(Fig. 2). Each phase can be divided into two components: an all-class structure and a pairwise-class structure.

In the off-line training phase, the all-class structure analyzes the data from six classes jointly and produces a selected feature subset. It outputs a classification model based on the subset with its label information and creates a feature index table which is an array of selected feature indices. On the other hand, a parallel pairwise-class structure is implemented. For each pair of classes, feature selection

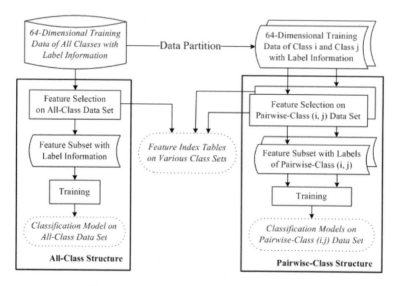

Fig. 1. The framework of the training phase

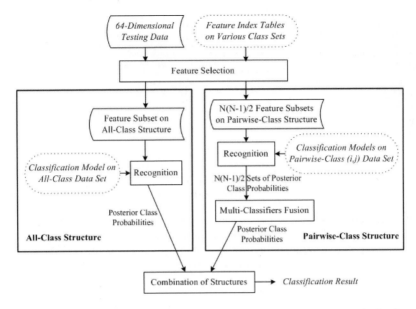

Fig. 2. The framework of the testing phase. The rounded rectangles with dot lines are the acquisition of the training phase.

and corresponding model training are performed. Supposing N is the number of distinct classes, the pairwise-class structure makes $N(N-1)/2$ feature index tables and $N(N-1)/2$ classification models for later testing.

While testing with the 64-dimensional data, FS runs $N(N-1)/2+1$ times according to the different feature index tables. Afterwards, the subset selected by the all-class feature index table is recognized by the model obtained from the all-class structure of training. In this process, an estimate of posterior probabilities rather than only predicting a class label, is required because of a final combination with the pairwise-class structure. Moreover, in the pairwise-class structure, $N(N-1)/2$ one-against-one classifiers implement recognition on their corresponding feature subsets side by side and output $N(N-1)/2$ sets of posterior probabilities. These sets are fused by a multi-classifiers combining strategy to attain the whole posterior probabilities of the pairwise-class structure. Ultimately, two sets of posterior probabilities from the all-class structure and the pairwise-class structure are combined to determine the classification result.

3.2   Feature Subset Selection

SFS, the first algorithm[8] proposed for feature selection, is simple and empirically successful. It starts with an empty subset of features and performs a hill-climbing deterministic search. In each iteration, the feature not yet selected is individually incorporated in the subset to calculate a criterion. Then the feature which yields the best criterion value is included in the new subset. This iteration will not be stopped until no improvement of the criterion value is achieved.

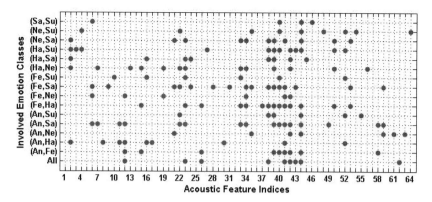

Fig. 3. Indices of 16 feature subsets selected by SFS with KNN on female utterances. X-axis is the indices of all 64 features(see details in Section 2), and Y-axis represents the involved classes. The corresponding words of abbreviated letters are as follows: All-All six emotions, An-Anger, Fe-Fear, Ha-Happiness, Ne-Neutral, Sa-Sadness, Su-Surprise.

The criterion employed to guide the search was the average classification F1-measure by the selected features. The F1-measure was calculated by a 5-fold cross-validation with certain classifier. Fig.3 shows the array elements of some feature index tables selected by SFS with KNN on female utterances. These subsets were quite distinct since they were selected on different involved classes. The SFS with KNN on male and SFS with SVM have also been performed, yet the results are omitted.

3.3   Classifiers and Probability Estimates

In pattern recognition, KNN is the simplest algorithm only based on memory. Given a query sample x, it finds K closest neighbors of x in training nodes by a distance metric(e.g. Euclidean distance in our study) and predicts the class label using majority voting by the labels of the K nodes. In contrast to other statistical classifiers, KNN needs no model to fit. This property simplifies the structures of the training phase by no model training, thus the training phase with KNN classifier only requires selecting features. As a preprocessing of experiments, the value of K was determined by a 5-fold cross-validation on 64-dimensional training data from all classes. It was conducted on the points $K : [10, 15, ..., 70, 75]$. At each K, the average F1-measure on validation data was obtained from the iterations of the cross-validation. The experiment results in Fig.4(a) show that $K = 35$ performs best on each gender.

Another powerful tool 'SVM', has been originally proposed to classify samples within two classes. It maps training samples of two classes into a higher-dimensional space through a kernel function and seeks an optimal separating hyperplane in this new space to maximize its distance from the closest training point. While testing, a query point is categorized according to the distance between the point and the hyperplane. We employed binary SVM with a radial

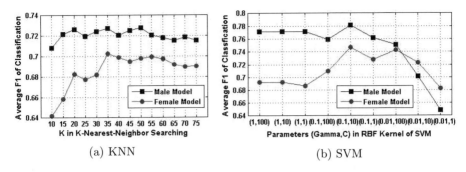

Fig. 4. Average classification performance of KNN/SVM on different values of parameters

basis function(RBF) kernel, and for the all-class structure, we extended binary SVM with the 'one-against-one' rule which has offered empirically good performance in multi-class pattern recognition[9]. The classification error rate of a binary RBF-SVM is largely dependent on a kernel parameter γ and a regularization parameter C. We selected them by another 5-fold cross-validation based on the 64-dimensional training data over all classes. The average F1-measures of the experiments on $(\gamma, C) : [1, 0.1, 0.01] \times [100, 10, 1]$ are shown in the Fig.4(b) and we chose $(0.1, 10)$ as the fix value of (γ, C) in our experiments.

For the sake of probability estimates, probabilistic outputs of KNN and SVM were required rather than label prediction. So a well-known estimate of KNN which treats each neighbor equally was adopted[10]: for a query point x, the posterior probability Pi of class i equals the fraction of the K nearest neighbors that belong to class i, i.e. $Pi = Ki/K$ where Ki denotes the number of the nodes belonging to class i in the K neighbors. For the binary SVM, the Platt's posterior probability estimate[11] by a sigmoid function was calculated as follows: given the training examples labelled by $(1, -1)$, the SVM computes a decision function f such that $sign(f(x))$ is used to predict the label of any query sample x. The posterior probability of x belonging to class '1' is: $P(y = 1|x) \approx P_{A,B}(x) \equiv (1 + exp(Af(x) + B))^{-1}$, where A and B are estimated by minimizing the negative log-likelihood function using the validation data set.

3.4 Fusion Strategy

Once the 15 binary classifiers in the pairwise-class structure have been implemented, the next important task involves the fusion of these individual results. Many multi-classifier combining strategies have been proposed in recognition applications, such as majority voting, sum-rule, product-rule and belief function, etc. The sum-rule, where the continuous outputs like posterior probabilities are summed up, empirically perform well in multi-classifier problems[12]. We adopted it in the fusion of all binary classifiers and denoted the outputs from sum-rule by $P = (p_1, p_2, ..., p_6)$. For the later combination with the probabilities from the all-class structure, an output square sum normalization[13] of P was required to solve the problem of output incomparability. The normalization result was denoted by P', satisfying $\sum_{i=1}^{6} p'_i = 1, 0 \leq p'_i \leq 1$. Supposing the posterior probabilities from the all-class structure are $Q = (q_1, q_2, ..., q_6)$, where $\sum_{i=1}^{6} q_i = 1, 0 \leq q_i \leq 1$, the Q and P' were also combined by sum-rule.

4   Performance Comparison

In order to evaluate the system performance, the classical method based on all-class FS and the pairwise method based on pairwise-class FS were investigated, i.e. the all-class structure and the pairwise-class structure were considered separately to do the speech emotion classification. The average F1-measure of three methods on each emotion is shown below. With KNN on female model(Fig.5(a)), it is observed the F1-measure of the pairwise method is better than those of the all-class method on every emotion. But on male model(Fig.5(b)), these two methods achieve comparable performance. Besides, the comparisons using SVM in Fig.5(c)(d) show that the method based on all-class FS performs better than the pairwise method in a majority of emotions on both female and male models. This is contrary to the previous researches in [5,6] and the results by KNN. Nevertheless, our proposal is more stable and it performs substantially better in almost all cases. Compared with the all-class method, our approach using KNN reaches 7.6% improvement on female and 4.4% on male in the average F1-measure of six emotions. And it increases by 3.2%(female) and 3.6%(male)

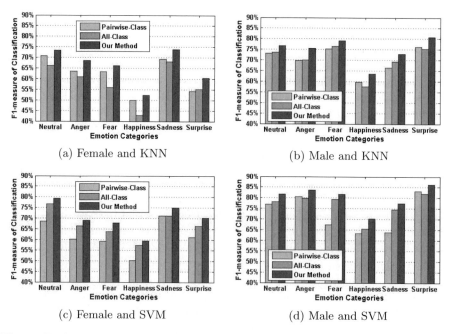

Fig. 5. Performance comparison of All-Class(the method based on all-class FS), Pairwise-Class(the method based on pairwise-class FS) and Our Approach

with SVM. The relative improvements to the pairwise method are 3.9%(female) and 4.7%(male) using KNN, 8.4%(female) and 7.7%(male) with SVM.

5   Conclusions and Future Work

This paper proposes a novel speech emotion recognition system based on a fusion of all-class FS and pairwise-class FS. It is a combination of the classical approach which only produces a feature subset in view of all-class data and the pairwise method selecting features on each pair of classes. Experiment results show that our proposal makes relative improvement in almost all cases, compared with the classical method and the pairwise method by the KNN/SVM classifier. In this paper, we have only investigated SFS, KNN/SVM classifiers and audio features. More effort will be paid on the generalization of the hybrid system and multi-modal emotion recognition will be included in the future work.

References

1. A.Mehrabian: Communication without words. Psychology Today, **2(4)** (1968) 53-56
2. R.Bellman: Adaptive Control Processes: A Guided Tour. Princeton University Press, (1961)
3. Z.J.Chuang, C.H.Wu: Emotion Recognition Using Acoustic Features and Textual Content. Proc. International Conference on Multimedia and Expo, **1** (2004) 53-56
4. D.Ververidis, C.Kotropoulos, I.Pitas: Automatic Emotional Speech Classification. Proc. International Conference on Acoustics, Speech, and Signal Processing, **1** (2004) 593-596
5. S.Essid, G.Richard, B.David: Musical Instrument Recognition Based on Class Pairwise Feature Selection. Proc. International Conference on Music Information Retrieval, (2003) 560-567
6. G.Guo, H.Zhang, S.Li: Pairwise Face Recognition. Proc. International Conference on Computer Vision, **2** (2001) 282-287
7. B.Schuller, G.Rigoll, M.Lang: Hidden Markov Model-based Speech Emotion Recognition. Proc. International Conference on Acoustics, Speech, and Signal Processing, **2** (2003) 1-4
8. I.Guyon, A.Elisseeff: An Introduction to Variables and Feature Selection. Machine Learning Research, (2003) 1157-1182
9. N.Cristianini, J.Shawe-Taylor: An Introduction to Support Vector Machines. Cambridge University Press, (2000)
10. K.Fukunaga, L.Hostetler: K-nearest-neighbor Bayes-risk Estimation. IEEE Transactions on Information Theory, **21(3)** (1975) 285-293
11. J.Platt: Probabilistic Outputs for Support Vector Machines and Comparison to Regularized Likelihood Methods. MIT Press, (2000)
12. J.Kittler, M.Hatef, R.Duin, J.Matas: On Combining Classifiers. IEEE Transactions on Pattern Analysis and Machine Intelligence, **20(3)** (1998) 226-239
13. Y.S.Huang, C.Y.Suen: A Method of Combining Multiple Classifiers- A Neural Network Approach. Proc. the 12th IAPR International Conference on Pattern Recognition, **2** (1994) 473-475

Regularized Knowledge-Based Kernel Machine

Olutayo O. Oladunni and Theodore B. Trafalis

School of Industrial Engineering, The University of Oklahoma
202 West Boyd, CEC 124 Norman, OK 73019 USA
{tayo, ttrafalis}@ou.edu

Abstract. This paper presents a knowledge-based kernel classification model for binary classification of sets or objects with prior knowledge. The prior knowledge is in the form of multiple polyhedral sets belonging to one or two classes, and it is introduced as additional constraints into a regularized knowledge-based optimization problem. The resulting formulation leads to a least squares problem that can be solved using matrix or iterative methods. To evaluate the model, the experimental laminar & turbulent flow data and the Reynolds number equation used as prior knowledge were used to train and test the proposed model.

1 Introduction

In data mining applications, incorporation of prior knowledge is usually not considered because most algorithms do not have the adequate means for incorporating explicitly such types of constraints. In the case of Support Vector Machines (SVMs) model [1], Fung et al. [2], [3] developed explicit formulations for incorporating prior knowledge in the form of multiple polyhedral sets belonging to one or more categories into a linear and nonlinear classification model.

The main motivation of this paper is to develop a nonlinear kernel approach based on the Tikhonov regularization scheme for knowledge-based classification discrimination. The proposed model problem is the kernel-based formulation of the multi-classification model in [4] for a two-class classification problem. The feature of the proposed model is that it leads to a least squares problem that can be solved by solving a linear system of equations that reduces computational time. In contrast, Fung et al. [2], [3] solve a linear programming optimization problem.

In this work, a combination of the prior knowledge sets together with the concept of the least squares SVM (LS-SVM) [5] and regularization based LS-SVM [6] results into a linear system of equations. Our approach is different in the sense that, the proposed minimization problem is an unconstrained optimization problem resulting into a smaller linear system of equations than those proposed in [5], [6]. By expressing the normal vector w as a linear combination of the data points, it turns out that in the least squares knowledge based formulation the prior knowledge(s) for nonlinear classification is the precise implication of the prior knowledge(s) for linear classification. This differs from the knowledge based model by Fung et al. [3], where the prior knowledge(s) for nonlinear classification is not the precise implication of the prior knowledge(s) for linear classification. In their formulation, the knowledge constraints are

Y. Shi et al. (Eds.): ICCS 2007, Part I, LNCS 4487, pp. 176–183, 2007.
© Springer-Verlag Berlin Heidelberg 2007

kernelized so as to fit a standard [7] linear programming formulation for nonlinear kernel classification.

The regularized knowledge-based kernel classification model can be considered as a least squares knowledge based nonlinear kernel formulation of Fung et al. [3]. Note that the resulting regularized knowledge-based model problem minimizes the classical regularization objective function with strict L_2 norm functions. This norm guarantees the differentiability of the objective function, and as result a linear system of equations is derived. In the knowledge based model by Fung et al. [3], the differentiability property is lost because in its model problem the functions are L_1 norm functions, which further restrain the solution of the model problem and also requires an LP solver to obtain a solution.

Benefits of the regularized knowledge-based kernel classification model includes: the reduction of a classification problem to a smaller linear system of equation, the ability to provide fast solutions which require no special solvers, the ability to provide explicit solution in terms of the given data and prior knowledge sets, and the ability to provide effective and robust classifiers as a result of the bounding planes within a cluster of points (margin increase).

2   Prior Knowledge in Two-Class Classification

Suppose that in addition to the points belonging to a class, there is prior information belonging to one or two categories. The knowledge sets [2] in an n dimensional space are given in the form of a polyhedral set determined by the set of linear equalities and linear inequalities. The polyhedral knowledge set $\{x \in R^n \mid Bx \leq b\}$ or $\{x \in R^n \mid \overline{B}x = \overline{b}\}$ should lie in the halfspace $\{x \in R^n \mid x^T w \geq \gamma+1\}$ where $B \in R^{g_u \times n}$, $\overline{B} \in R^{g_{\overline{u}} \times n}$, $b \in R^{g_u}$ and $\overline{b} \in R^{g_{\overline{u}}}$ are the prior information belonging to class 1. g_u or $g_{\overline{u}}$ is the number of prior knowledge (equality or inequality) constraints in class 1. d_v or $d_{\overline{v}}$ is the number of prior knowledge (equality or inequality) constraints in class 2. Therefore, the following implications must hold for a given (w, γ), where w is a normal vector and γ is the location of the optimal separating plane relative to the origin:

$$\left\{ Bx \leq b \Rightarrow x^T w \geq \gamma+1 \right\} \text{ or } \left\{ \overline{B}x = \overline{b} \Rightarrow x^T w \geq \gamma+1 \right\}. \tag{2.1}$$

Using the nonhomogeneous Farkas theorem of the alternative [8] or, using duality in linear programming (LP) [9], equations (2.1) can be transformed into a set of knowledge constraints [2].

$$\left\{ B^T u + w = 0, \ b^T u + \gamma+1 \leq 0, \ u \geq 0 \right\} \text{ or } \left\{ \overline{B}^T \overline{u} + w = 0, \ \overline{b}^T \overline{u} + \gamma+1 \leq 0 \right\}. \tag{2.2}$$

To formulate the nonlinear counterpart of the linear classification, the primal variable w is replaced by its equivalent dual representation $w = \overline{A}^T Y \alpha$, where α is the vector of dual variables, $\overline{A} = [A^{(1)T} \ A^{(2)T}]^T$ whose rows are the points belonging to classes 1 and 2, and Y is a diagonal matrix whose diagonals are +1 for points in class 1 and -1 for points in class 2.

Let $A \in R^{m \times n}$ *and* $B \in R^{n \times k}$. *The* **kernel** $K(A, B)$ *maps* $R^{m \times n} \times R^{n \times k}$ *into* $R^{m \times k}$. The behavior of a kernel matrix strongly relies on Mercer's condition [10], [11] for symmetric kernel functions, i.e., a kernel matrix is a positive semidefinite (PSD) matrix.

Using the dual representation of w and applying the kernel definition, a more complex classifier can be determined, and the nonlinear optimal separating hyperplane is given as:

$$K(x^T, \overline{A}^T)Y\alpha = \gamma. \quad (2.3)$$

The implications for a given (A, Y, α, γ) now becomes the following [3]:

$$\left\{ Bx \leq b \Rightarrow x^T \overline{A}^T Y\alpha \geq \gamma + 1 \right\} \text{ or } \left\{ \overline{B}x = \overline{b} \Rightarrow x^T \overline{A}^T Y\alpha \geq \gamma + 1 \right\}, \quad (2.4)$$

and the prior knowledge constraints can be rewritten as:

$$\left\{ B^T u + \overline{A}^T Y\alpha = 0, \ b^T u + \gamma + 1 \leq 0, \ u \geq 0 \right\} \text{ or } \left\{ \overline{B}^T \overline{u} + \overline{A}^T Y\alpha = 0, \ \overline{b}^T \overline{u} + \gamma + 1 \leq 0 \right\}. \quad (2.5)$$

Notice that there is no kernel present in the prior knowledge, and as a result equation (2.4) is the precise implications of equations (2.1). In the subsequent section, when the prior knowledge constraint is incorporated into a classification model, the kernel will be employed to obtain nonlinear classifiers.

3 Regularized Knowledge-Based Kernel Classification Machine

Consider a problem of classifying data sets with prior knowledge in R^n that are represented by a data matrix $A^{(i)} \in R^{m_i \times n}$, where $i = 1, 2$, and knowledge sets $\{x \mid B^{(1)} x \leq b^{(1)}\}$ or $\{x \mid \overline{B}^{(1)} x = \overline{b}^{(1)}\}$ belonging to class 1, $\{x \mid B^{(2)} x \leq b^{(2)}\}$ or $\{x \mid \overline{B}^{(2)} x = \overline{b}^{(2)}\}$ belonging to class 2. Let $A^{(1)}$ be a $m_1 \times n$ matrix whose rows are points in class 1, and m_1 is the number of data in class 1. Let $A^{(2)}$ be a $m_2 \times n$ matrix whose rows are points in class 2, and m_2 is the number of data in class 2. This problem can be modeled through the following optimization problem:

$$\min_{\substack{\alpha, \gamma, a, c}} f_{NT_R KKM}(\alpha, \gamma, a, c) = \begin{bmatrix} \dfrac{\lambda}{2}\|\alpha\|^2 + \dfrac{1}{2}\|YKY\alpha - \gamma Ye - e\|^2 + \\[2mm] \dfrac{1}{2}\left[\left\|B_u^T a + I_u \overline{A}^T Y\alpha\right\|^2 + \left\|B_v^T c - I_v \overline{A}^T Y\alpha\right\|^2\right] + \\[2mm] \dfrac{1}{2}\left[\left\|B_{bu}^T a + \gamma e_u + e_u\right\|^2 + \left\|B_{bv}^T c - \gamma e_v + e_v\right\|^2\right] \end{bmatrix}. \quad (3.1)$$

Where α is the vector of dual variables; γ is the location of the optimal separating plane; λ is a regularization parameter; $a = [u^T, \overline{u}^T]^T$ is a vector of all multipliers referring to class 1, and $c = [v^T, \overline{v}^T]^T$ is a vector of all multipliers referring to class 2. $K = K(\overline{A}, \overline{A}^T)$ is a kernel matrix; matrix $\overline{A} = [A^{(1)T} \ A^{(2)T}]^T$ whose rows are the points belonging to classes 1 & 2 points respectively; Y is a diagonal matrix whose diagonals are +1 for points in class 1 and -1 for points in class 2; and vector $e = [e^{(1)T} \ e^{(2)T}]^T$

is a vector of ones. Matrices B_u & B_v are diagonal block matrices whose diagonals contain knowledge sets belonging to class 1 & 2. The diagonals of B_u are $B^{(1)} \in R^{g_u \times n}$, $\bar{B}^{(1)} \in R^{g_{\bar{u}} \times n}$ and of B_v are $B^{(2)} \in R^{d_v \times n}$, $\bar{B}^{(2)} \in R^{d_{\bar{v}} \times n}$. Matrices B_{bu} & B_{bv} are created where the diagonals of B_{bu} are $b^{(1)} \in R^{g_u \times 1}$, $\bar{b}^{(1)} \in R^{g_{\bar{u}} \times 1}$ and of B_{bv} are $b^{(2)} \in R^{d_v \times 1}$, $\bar{b}^{(2)} \in R^{d_{\bar{v}} \times 1}$. Matrices $I_u = [I_{(u)}^T \ \bar{I}_{(u)}^T]^T$ & $I_v = [I_{(v)}^T \ \bar{I}_{(v)}^T]^T$ are block matrices, where $I_{(u)}$, $\bar{I}_{(u)}$, $I_{(v)}$, $\bar{I}_{(v)} \in R^{n \times n}$ are identity matrices. Vectors e_u & e_v consist of entries $e_1, \bar{e}_1 \in R$ & $e_2, \bar{e}_2 \in R$, where $e_1, \bar{e}_1, e_2, \bar{e}_2$ are equal to one. Vectors e_u, e_v are vectors of ones, where each entry corresponds to a vector $b^{(1)}, \bar{b}^{(1)}$ ($b^{(2)}, \bar{b}^{(2)}$) respectively.

The (α, γ) taken from a solution of (3.1) generates the nonlinear separating surface (2.3). Problem (3.1) was formulated using the concept of penalty functions [9], and it is called the nonlinear Tikhonov regularization [12] knowledge-based kernel machine classification model (NT$_R$KKM). It is a binary classification formulation of the multi-classification model in [4] to accommodate nonlinearly separable patterns with knowledge sets, and explicit solutions in dual space in terms of the given data can be derived. Problem (3.1) is slightly different from the knowledge based model by Fung et al. [4], which is a linear programming formulation. The difference lies in the selection of the norm distance and the squared error with regularization term (least squares).

Below is the explicit solution to NT$_R$KKM in terms of the given data:

$$\gamma = \bar{h}(d\alpha - t). \tag{3.2}$$

$$\bar{M}\alpha = \bar{z} \Rightarrow \alpha = \bar{M}^{-1}\bar{z}, \text{ where } \alpha \in R^m. \tag{3.3}$$

Note that $\bar{h}$, d, t, $\bar{M}$, $\bar{z}$ are defined as follows.

$$\bar{h} = \left[e^T YYe + e_u^T\left[e_u - B_u^T UB_{bu}e_u\right] + e_v^T\left[e_v - B_v^T VB_{bv}e_v\right]\right]^{-1}. \tag{3.4}$$

$$\begin{aligned} d &= \left[e^T Y\bar{K} + e_u^T B_{bu}^T UK_{Bu}^T Y + e_v^T B_{bv}^T VK_{Bv}^T Y\right] \\ t &= \left[e^T Ye + e_u^T\left[e_u - B_{bu}^T UB_{bu}e_u\right] - e_v^T\left[e_v - B_{bv}^T VB_{bv}e_v\right]\right] \end{aligned}. \tag{3.5}$$

$$\begin{aligned} &U = \left(B_u B_u^T + B_{bu}B_{bu}^T\right)^{-1}, \ V = \left(B_v B_v^T + B_{bv}B_{bv}^T\right)^{-1}, \ \bar{K} = YKY \\ &K_u = K(\bar{A}I_u^T, I_u \bar{A}^T), \ K_v = K(\bar{A}I_v^T, I_v \bar{A}^T), \ K_{Bu} = K(\bar{A}I_u^T, B_u^T), \ K_{Bv} = K(\bar{A}I_v^T, B_v^T) \end{aligned} \tag{3.6}$$

$$\bar{M} = \begin{bmatrix} \lambda I + \bar{K}\left[\bar{K} - Y\bar{h}ed\right] + \\ Y\left[K_u Y - K_{Bu}U\left[K_{Bu}^T Y + B_{bu}\bar{h}e_u d\right]\right] + \\ Y\left[K_v Y - K_{Bv}V\left[K_{Bv}^T Y + B_{bv}\bar{h}e_v d\right]\right] \end{bmatrix}, \ \bar{z} = \begin{bmatrix} \bar{K}^T\left[e - Y\bar{h}et\right] + \\ YK_{Bu}UB_{bu}\left[e_u - \bar{h}e_u t\right] - \\ YK_{Bv}VB_{bv}\left[e_v + \bar{h}e_v t\right] \end{bmatrix}. \tag{3.7}$$

Minimizing the L$_2$ norm of α guarantees a solution for a positive tradeoff constant. It is evident from matrix $\bar{M}$ that only the diagonals change with any change in the tradeoff constant, implying we can always get a diagonally dominant matrix $\bar{M}$ which can ensure a solution. However, if the first term in $\bar{M}$ is replaced with a kernel matrix, then for any constant, all elements in matrix $\bar{M}$ also increase. Therefore, we cannot guarantee a diagonally dominant matrix $\bar{M}$ and hence we cannot always guarantee a solution when minimizing the square norm of the linear combination of w in dual space.

Assume that U, V and $\bar{M}$ are invertible matrices. Solution (3.3) provides an estimate of the dual variables. Such a linear system is easier to solve than a quadratic or linear programming formulation. Methods for solving the system include matrix decomposition methods, or iterative based methods [9], [13]. Its solution involves the inversion of an $m \times m$ dimensional matrix.

The decision function for classifying a point x is given by:

$$D(x) = sign\left[K(x^T, \bar{A}^T)Y\alpha - \gamma \right] = \begin{cases} +1, & \text{if point } (x) \text{ is in class } A^{(1)} \\ -1, & \text{if point } (x) \text{ is in class } A^{(2)} \end{cases}. \qquad (3.8)$$

4 Laminar/Turbulent Flow Pattern Data Set

The fluid flow data uses flow rate, density, viscosity, borehole diameter, drill collar diameter and mud type to delineate the flow pattern (laminar, +1; turbulent, -1) of the model. There are 92 data-points and 5 attributes, 46 instances for laminar flow pattern and 46 instances turbulent flow pattern [14], [15], [16]. The attributes are as follows: ρ, density of fluid (lbm/gal – continuous variable); q, flow rate (gal/min – continuous variable); $(d_2 + d_1)$, summation of borehole and drill collar (OD) diameter (in – continuous variable); μ_p, plastic viscosity (cp – continuous variable); Mud type, water based mud (1) oil based mud (2) (categorical variable).

Prior Knowledge: In addition to the fluid flow data, the Reynold's equation [17] for a Bingham plastic model is used as prior knowledge to develop a knowledge based classification model. As additional constraints, the prior knowledge will represent the transition equations which delineates laminar from turbulent flows. Since the flow pattern data are scaled by taking the natural logarithm of each instance, the prior knowledge needs to be scaled by also considering a natural logarithm transformation of the equations. Below is the equivalent logarithmic transformation for Reynold's equation [17]:

$$\Rightarrow \begin{cases} \ln(\rho) + \ln(q) - \ln(d_2 + d_1) - \ln(\mu_p) \leq 1.9156 - \varepsilon \to x^T w - \gamma \geq +1 \text{ (Laminar)} \\ \ln(\rho) + \ln(q) - \ln(d_2 + d_1) - \ln(\mu_p) \geq 1.9156 + \varepsilon \to x^T w - \gamma \leq -1 \text{ (Turbulent)} \end{cases}, \qquad (4.1)$$

where ε is a deviation factor, a small fraction perturbing the critical Reynold's number. The dataset where divided up as follows; 50% of the whole data is used as training data, while the remaining 50% was used as testing data.

5 Computational Results

In this section, the results of the analyzed data sets and prior sets described in section 4, are presented and discussed. The NT_RKKM model is used to train the data sets with prior knowledge. To demonstrate the uniqueness of the formulations, a comparison between the models above was conducted. The comparisons were made by evaluating a performance parameter (misclassification error) defined below:

$$\beta = 1 - \left(\frac{Total\ number\ of\ correctly\ classified\ points}{Total\ number\ of\ observed\ points} \right). \qquad (5.1)$$

β represents the overall misclassification error rate, i.e., the fraction of misclassified points for a given data set. For 100% classification, $\beta = 0$. The tradeoff constant considered is within the interval $0 - 100$, and the deviation factor $\varepsilon = 0.01$. Further comparisons were made between the NT_RKKM model, LT_RKSVM [16] model problem, the traditional SVM [1], [14], and the mixed integer programming kernel based classifiers (MIPKC$_{ps}$ & MIPKC$_p$, [15]). The polynomial kernel, $k(x_i, x) = (x_i^T x + 1)^p$, where p is the degree of the polynomial ($p = 2$) was used.

Results of the fluid flow pattern data with prior knowledge information trained on the NT_RKKM model, and compared with the LT_RKSVM, SVM, MIPKC$_{ps}$, & MIPKC$_p$ model, are shown in Tables 1 & 2. It should be noted that β (error rate) in the Tables are defined by (5.1), and the computing (cpu) time is measured in seconds. All computations and experiments were performed using MATLAB [18] for NT_RKKM, LT_RKSVM & SVM model [19], and CPLEX OPL [20] for the mixed integer programming kernel based models.

Table 1. Sample and Average random sample validation test error rate for NT_RKKM on Fluid Flow Pattern Data (varying tradeoff, λ)

Statistics	Tradeoff	Test 1	Test 2	Test 3	Test 4	Average
Error	0	-	-	-	-	-
Cpu time		-	-	-	-	-
Error	1	0.0000	0.0278	0.0000	0.0000	0.0070
Cpu time		0.0200	0.0100	0.0200	0.0100	0.0150
Error	5	0.0000	0.0000	0.0000	0.0000	**0.0000**
Cpu time		0.0400	0.0100	0.0100	0.0200	0.0200
Error	10	0.0000	0.0000	0.0000	0.0000	**0.0000**
Cpu time		0.0200	0.0800	0.0200	0.0100	0.0325
Error	50	0.0000	0.0000	0.0000	0.0278	0.0070
Cpu time		0.0100	0.0100	0.1310	0.0700	0.0553
Error	100	0.0000	0.0000	0.0556	0.0278	0.0209
Cpu time		0.0200	0.0100	0.0100	0.0200	0.0150

Table 2. Average error, accuracy & cpu. time of NT_RKKM, LT_RKSVM, MIPKC & SVM models

Average	NT$_R$KKM	LT$_R$KSVM	MIPKCps	MIPKCp	SVM	SVM [14]
Error	0.0000	0.0139	0.0139	0.0348	0.0209	0.0500
Accuracy	1.0000	0.9861	0.9861	0.9652	0.9792	0.9500
Cpu time	0.0200	0.0100	49.2975	83.9400	0.4250	0.4600

Tables 1 & 2 contain results for the NT_RKKM, LT_RKSVM, MIPKC$_{ps}$, MIPKC$_p$, & SVM model on the fluid flow pattern classification data. The models in comparison generally report promising error rates. The NT_RKKM model reports the smallest error (0). This means that the testing data set was correctly classified (100% classification).

The error rate (0.0139) for the model with prior knowledge (LT$_R$KSVM) performs in the same capacity as the data driven mixed integer programming model (MIPKC$_{ps}$). The computation time of both the NT$_R$KKM and LT$_R$KSVM model are comparable, and both clearly outperform the other learning models. The SVM models reports the next best computation time and the MIPKC model reports the worst time.

The fast solution of the LT$_R$KSVM model is due the formulation of the problem that is analyzed analytically in the primal (input) space, i.e., the dimension n of the classification problem is equal to the number of attributes (variables) of the data set; in our case n = 5. The fast solution of the NT$_R$KKM model is due to the small number of data in the test set (m = 46 test observations) and when possible, the avoidance of iterative methods in computing its solution. The next best computation time to the NT$_R$KKM model is the one of the SVM model. Its fast solution is also due to the small number of test observations (m = 46 test observations). Note that in this case iterative methods are employed to find its solution. The MIPKC$_{ps}$, MIPKC$_p$ models require more time because their formulation is based on integer and mixed integer programming techniques which generally perform more computations in obtaining a solution than its linear counterparts such as SVM, LT$_R$KSVM and NT$_R$KKM.

6 Conclusion

In this paper, a binary classification model called the nonlinear classification Tikhonov regularization knowledge-based kernel machine (NT$_R$KKM) is described. This model problem is an unconstrained optimization problem for discriminating between two disjoint sets. The proposed model can be considered as a least squares formulation of Fung et al. [3] knowledge-based nonlinear kernel model. The model was applied to the laminar/turbulent fluid flow data. Comparisons were made with the linear and nonlinear counterpart formulations, and best statistics were obtained by performing training on the NT$_R$KKM model. The NT$_R$KKM model can be applied to determine the flow patterns of fluid flow with different rheology. The fluid model in this paper is a non-Newtonian fluid, pseudoplastic model (Bingham plastic) with two flow patterns (laminar & turbulent flows). The same concept can be applied to a Newtonian or non-Newtonian fluid with three flow patterns (laminar, transition and turbulent).

References

1. Vapnik, V.: The Nature of Statistical Learning Theory. New-York: Springer-Verlag (1995)
2. Fung, G., Mangasarian, O.L., Shavlik, J.W.: Knowledge-based Support Vector Machine Classifiers. Neural Information Processing Systems 2002 (NIPS 2002), Vancouver, BC, December 10-12, (2002)
3. Fung, G., Mangasarian, O.L., Shavlik, J.W.: Knowledge-based Nonlinear Kernel Classifiers, in Manfred Warmuth and Bernhard Scholkopf (eds), Proceedings Conference on Learning Theory (COLT/Kernel 03) and Workshop on Kernel Machines, Washington, D.C., August 24 - 27, (2003) 102-113

4. Oladunni, O.O., Trafalis, T.B., Papavassiliou, D.V.: Knowledge-based multiclass support vector machines applied to vertical two-phase flow, ICCS 2006, Lecture notes in Computer Science, (V.N. Alexandrov et al., eds.), Part I, LNCS 3991. Springer-Verlag, Berlin Heidelberg (2006) 188-195
5. Suykens, J.A.K. and Vandewalle, J.: Least Squares Support Vector Machine Classifiers. Neural Processing Letters, 9, (1999b) 293-300
6. Pelckmans, K., Suykens, J.A.K., De Moor, B.: Morozov, Ivanov and Tikhonov Regularization based LS-SVMs. In Proceedings of the International Conference On Neural Information Processing (ICONIP 2004), Calcutta, India, Nov. 22-25, (2004)
7. Mangasarian, O.L.: Generalized Support Vector Machines. Advances in Large Margin Classifiers, (A. Smola, P. Bartlett, B. Schölkopf, and D. Schuurmans, eds.), MIT Press, Cambridge, MA (2000) ftp://ftp.cs.wisc.edu/pub/dmi/tech-reports/99-14.ps
8. Mangasarian, O.L.: Nonlinear Programming. Philadelphia, PA: SIAM (1994)
9. Bazaraa, M.S., Sherali, H.D., Shetty, C.M.: Nonlinear Programming – Theory and Algorithms. John Wiley & Sons, Inc. (1993)
10. Burges, C.J.C.: A Tutorial on Support Vector Machines for Pattern Classification. Data Mining and Knowledge Discovery, 2(2): 121-167, (1998)
11. Cristianini, N., Shawe-Taylor, J.: Support Vector Machines and other Kernel-based Learning Methods; Cambridge University Press, (2000)
12. Tikhonov, A.N., Arsenin, V.Y.: Solution of Ill-Posed Problems. Winston, Washington D.C., (1977)
13. Lewis, J. M., Lakshmivarahan, S., Dhall, S.: Dynamic Data Assimilation. Cambridge University Press, (2006)
14. Trafalis, T.B.; Oladunni, O.: Single Phase Fluid Flow Classification via Neural Networks & Support Vector Machine. Intelligent Engineering Systems Through Artificial Neural Networks, (C.H. Dagli, A.L. Buczak, D. L. Enke, M.J. Embrechts, and O. Ersoy, eds.), ASME Press, 14, (2004) 427-432
15. Oladunni, O.; Trafalis, T.B.: Mixed-Integer Programming Kernel-Based Classification, WSEAS Transactions on Computers, Issue 7, Vol. 4: 671-678, July (2005)
16. Oladunni, O.; Trafalis, T.B.: Single Phase Laminar & Turbulent Flow Classification in Annulus via a Knowledge-Based Linear Model. Intelligent Engineering Systems Through Artificial Neural Networks, (C.H. Dagli, A.L. Buczak, D. L. Enke, M.J. Embrechts, and O. Ersoy, eds.), ASME Press, 16, (2006) 599-604
17. Bourgoyne, A.T. Jr., Chenevert, M.E., Millheim, K.K., Young, F.S. Jr.: Applied Drilling Engineering, SPE Textbook Series Vol. 2, Richardson, TX, p. 155, (1991)
18. MATLAB User's Guide.: The Math-Works, Inc., Natwick, MA 01760, 1994-2003. http://www.mathworks.com
19. Gunn, S.T.: Support Vector Machine for Classification and Regression, Technical Report, Department of Electronic and Computer Science, University of Southampton, (1998)
20. ILOG OPL STUDIO 3.7.: Language Manual, ILOG, S.A. Gentily, France and ILOG, Inc., Mountain View California, USA. http://www.ilog.com/products/oplstudio/, (2003)

Three-Phase Inverse Design Stefan Problem

Damian Słota

Institute of Mathematics
Silesian University of Technology
Kaszubska 23, 44-100 Gliwice, Poland
d.slota@polsl.pl

Abstract. The method of the convective heat transfer coefficient identification in a three-phase inverse design Stefan problem is presented in
this paper. The convective heat transfer coefficient will be sought in the
form of a continuous function, non-linearly dependent on the parameters sought. A genetic algorithm was used to determine these parameters. The direct Stefan problem was solved via a generalized alternating
phase truncation method. Stability of the whole algorithm was ensured
by applying the Tikhonov regularization technique.

Keywords: Inverse Stefan problem, solidification, genetic algorithm,
Tikhonov regularization.

1 Introduction

A majority of available studies refer to the one- or two-phase inverse Stefan problem [1,2,3,4,8,11,16,18], whereas studies regarding the three-phase inverse Stefan
problem are scarce [5,6,7,15]. In paper [15], three tasks are identified, successively
in a liquid, solid and mush zone, on the basis of which the sought convective heat
transfer coefficient is then determined. The method described in papers [6,7] consists of minimization of a functional, whose value is the norm of difference between
the given positions of phase-change front and the positions reconstructed based
on the selected function describing the convective heat transfer coefficient. In paper [5], a solution is found in a linear combination form of functions satisfying the
heat conduction equation. The coefficients of the combination are determined by
the least square method to minimize the maximal defect in the initial-boundary
data, thus making it possible to find the temperature distribution, the heat flux,
or the convective heat transfer coefficient for the boundary.

The method of the convective heat transfer coefficient identification in a three-
phase inverse design Stefan problem is presented in this paper. The convective
heat transfer coefficient will be sought in the form of a continuous function, non-
linearly dependent on the parameters sought. A genetic algorithm was used to determine these parameters. The direct Stefan problem was solved via a generalized
alternating phase truncation method [9,14]. Stability of the whole algorithm was
ensured by applying the Tikhonov regularization technique [10,17]. The featured
example calculations show a very good approximation of the exact solution.

Y. Shi et al. (Eds.): ICCS 2007, Part I, LNCS 4487, pp. 184–191, 2007.

2   Three-Phase Problem

Now, we are going to describe an algorithm for the solution of a three-phase inverse design Stefan problem. Let the boundary of the domain $D = [0, b] \times [0, t^*] \subset \mathbb{R}^2$ be divided into seven parts (Figure 1), where the boundary and initial conditions are given, and let the D domain be divided into three subdomains D_1, D_2 and D_3 ($D = D_1 \cup D_2 \cup D_3$). Let $\Gamma_{1,2}$ will designate the common boundary of domains D_1 and D_2, whilst $\Gamma_{2,3}$ will mean the common boundary of domains D_2 and D_3. Let us assume that boundary $\Gamma_{k,k+1}$ is described by function $x = \xi_{k,k+1}(t)$.

For the given partial position of interfaces $\Gamma_{1,2}$ and $\Gamma_{2,3}$, we will determine function $\alpha(t)$ defined on boundaries Γ_{2k} ($k = 1, 2, 3$) and temperature distributions T_k, which inside domains D_k ($k = 1, 2, 3$) fulfil the heat conduction equation:

$$c_k \, \varrho_k \, \frac{\partial T_k}{\partial t}(x, t) = \frac{1}{x} \frac{\partial}{\partial x} \left(\lambda_k \, x \, \frac{\partial T_k}{\partial x}(x, t) \right), \tag{1}$$

on boundary Γ_0, they fulfil the initial condition:

$$T_1(x, 0) = T_0, \tag{2}$$

on boundaries Γ_{1k} ($k = 1, 2, 3$), they fulfil the second-kind homogeneous condition:

$$\frac{\partial T_k}{\partial x}(x, t) = 0, \tag{3}$$

on boundaries Γ_{2k} ($k = 1, 2, 3$), they fulfil the third-kind condition:

$$-\lambda_k \frac{\partial T_k}{\partial x}(x, t) = \alpha(t) \left(T_k(x, t) - T_\infty \right), \tag{4}$$

whereas on moving interfaces $\Gamma_{1,2}$ and $\Gamma_{2,3}$, they fulfil the condition of temperature continuity and the Stefan condition ($k = 1, 2$):

$$T_k \big(\xi_{k,k+1}(t), t \big) = T_{k+1} \big(\xi_{k,k+1}(t), t \big) = T^*_{k,k+1}, \tag{5}$$

$$L_{k,k+1} \, \varrho_{k+1} \, \frac{d\xi_{k,k+1}(t)}{dt} = -\lambda_k \frac{\partial T_k(x, t)}{\partial x} \bigg|_{\Gamma_{k,k+1}} + \lambda_{k+1} \frac{\partial T_{k+1}(x, t)}{\partial x} \bigg|_{\Gamma_{k,k+1}}, \tag{6}$$

where c_k, ϱ_k and λ_k are: the specific heat, the mass density and the thermal conductivity in respective phases, α is the convective heat transfer coefficient, T_0 is the initial temperature, T_∞ is the ambient temperature, $T^*_{k,k+1}$ is the phase change temperature, $L_{k,k+1}$ is the latent heat of fusion, and t and x refer to time and spatial location, respectively.

Function $\alpha(t)$, describing the convective heat transfer coefficient, will be sought in the form of a function dependent (linearly or non-linearly) on n parameters:

$$\alpha(t) = \alpha(t; \alpha_1, \alpha_2, \dots, \alpha_n). \tag{7}$$

Let V mean a set of all functions in form (7). In real processes, function $\alpha(t)$ does not have an arbitrary value. Therefore, the problem of minimization with

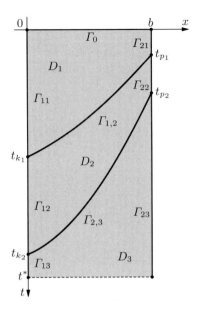

Fig. 1. Domain of the three-phase problem

constraints has some practical importance. Let V_c mean a set of those functions from set V, for which $\alpha_i \in [\alpha_i^l, \alpha_i^u]$, $\alpha_i^l < \alpha_i^u$, $\alpha_i^l, \alpha_i^u \in \mathbb{R}$, for $i = 1, 2, \ldots, n$.

For the given function $\alpha(t) \in V_c$, the problem (1)–(6) becomes a direct Stefan problem, whose solution makes it possible to find the positions of interfaces $\xi_{1,2}(t)$ and $\xi_{2,3}(t)$ corresponding to function $\alpha(t)$. By using the interface positions found, $\xi_{k,k+1}(t)$, and the given positions $\xi_{k,k+1}^*(t)$ ($k = 1, 2$), we can build a functional which will define the error of an approximate solution:

$$J\big(\alpha(t)\big) = \sum_{k=1}^{2} \left\| \xi_{k,k+1}(t) - \xi_{k,k+1}^*(t) \right\|^2 + \gamma \left\| \alpha(t) \right\|^2, \tag{8}$$

where γ is the regularization parameter. The discrepancy principle proposed by Morozov will be used to determine the regularization parameter [17, 10]. The above norms denote the norms in a space of square-integrable functions in the interval $(0, t^*)$. When the exact position of the interface is given only in selected points (hereafter called the "control points"), the first norm present in functional (8) will be calculated from the following dependence:

$$\left\| \xi_{k,k+1}(t) - \xi_{k,k+1}^*(t) \right\|^2 = \sum_{i=1}^{M} \left[A_i \left(\xi_{k,k+1;i} - \xi_{k,k+1;i}^* \right)^2 \right], \tag{9}$$

where A_i are coefficients dependent on the chosen numerical integration method, M is the number of control points, and $\xi_{k,k+1;i}^* = \xi_{k,k+1}^*(t_i)$ and $\xi_{k,k+1;i} = \xi_{k,k+1}(t_i)$ are the given and calculated points respectively, describing the interfaces positions.

3   Genetic Algorithm

For the representation of the vector of decision variables, a chromosome was used in the form of a vector of real numbers (real number representation) [12, 13]. The tournament selection and elitist model were applied in the algorithm. This selection is carried out so that two chromosomes are drawn and the one with better fitness, goes to a new generation. There are as many draws as individuals that the new generation is supposed to include. In the elitist model the best individual of the previous generation is saved and, if all individuals in the current generation are worse, the worst of them is replaced with the saved best individual from the previous population.

As the crossover operator, arithmetical crossover was applied, where as a result of crossing of two chromosomes, their linear combinations are obtained:

$$\boldsymbol{\alpha}^{1'} = r\,\boldsymbol{\alpha}^1 + (1-r)\,\boldsymbol{\alpha}^2, \qquad \boldsymbol{\alpha}^{2'} = r\,\boldsymbol{\alpha}^2 + (1-r)\,\boldsymbol{\alpha}^1, \tag{10}$$

where parameter r is a random number with a uniform distribution from the domain $[0, 1]$.

In the calculations, a nonuniform mutation operator was used as well. During mutation, the α_i gene is transformed according to the equation:

$$\alpha_i' = \begin{cases} \alpha_i + \Delta(\tau, \alpha_i^u - \alpha_i), \\ \alpha_i - \Delta(\tau, \alpha_i - \alpha_i^l), \end{cases} \tag{11}$$

and a decision is taken at random which from the above formulas should be applied, where:

$$\Delta(\tau, x) = x\left(1 - r^{(1-\frac{\tau}{N})d}\right), \tag{12}$$

and r is a random number with a uniform distribution from the domain $[0, 1]$, τ is the current generation number, N is the maximum number of generations and d is a constant parameter (in the calculations, $d = 2$ was assumed).

In calculations parameters used for the genetic algorithm are as follows: population size $n_{pop} = 70$, number of generations $N = 500$, crossover probability $p_c = 0.7$ and mutation probability $p_m = 0.1$.

4   Numerical Example

Now we will present an example illustrating the application of the method discussed. An axisymmetric problem is considered in the example, where: $b = 0.06$ $[m]$, $\lambda_1 = 54$ $[W/(m\,K)]$, $\lambda_2 = 42$ $[W/(m\,K)]$, $\lambda_3 = 30$ $[W/(m\,K)]$, $c_1 = 840$ $[J/(kg\,K)]$, $c_2 = 755$ $[J/(kg\,K)]$, $c_3 = 670$ $[J/(kg\,K)]$, $\varrho_1 = 7000$ $[kg/m^3]$, $\varrho_2 = 7250$ $[kg/m^3]$, $\varrho_3 = 7500$ $[kg/m^3]$, $L_{1,2} = 217600$ $[J/kg]$, $L_{2,3} = 54400$ $[J/kg]$, $T_{1,2}^* = 1773$ $[K]$, $T_{2,3}^* = 1718$ $[K]$, $T_\infty = 303$ $[K]$ and $T_0 = 1803$ $[K]$.

Function $\alpha(t)$ is sought as an exponential function (Figure 2):

$$\alpha(t) = \alpha_1 \exp\left(\frac{t}{t_2}\,\ln\frac{\alpha_2}{\alpha_1}\right). \tag{13}$$

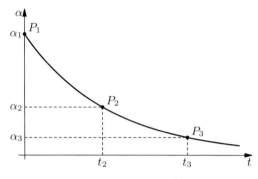

Fig. 2. Function $\alpha(t)$

The parameters describing the exact form of function $\alpha(t)$ are:

$$\alpha_1 = 1200, \quad \alpha_2 = 600, \quad t_2 = 90.$$

Set V_c is defined in the following way:

$$V_c = \left\{ \alpha(t) \in V : \alpha_1 \in [1000, 1400], \; \alpha_2 \in [400, 700], \; t_2 \in [50, 150] \right\}. \quad (14)$$

In the alternating phase truncation method, the finite-difference method was used, the calculations having been made on a grid of discretization intervals equal $\Delta t = 0.1$ and $\Delta x = b/500$. A change (reasonable) of the grid density did not significantly affect the results obtained.

The giving of two points: $P_1(0, \alpha_1)$ and $P_2(t_2, \alpha_2)$ (see Figure 2) explicitly determines the exponential function in form (13), however, the same function can be determined through defining points $P_1(0, \alpha_1)$ and $P_3(t_3, \alpha_3)$. Therefore, the problem of reconstructing function $\alpha(t)$ in form (13) based on three parameters: $(\alpha_1, \alpha_2, t_2)$, has infinitely many solutions. For this reason, the accuracy of solution in this case will be determined for the entire function $\alpha(t)$, and not separately for each of the sought parameters . Thus, the relative percentage error will be calculated from the following relation:

$$e_\alpha = \left(\int\limits_0^{t^*} \left(\alpha_e(t) - \alpha_a(t) \right)^2 dt \right)^{1/2} \cdot \left(\int\limits_0^{t^*} \left(\alpha_e(t) \right)^2 dt \right)^{-1/2} \cdot 100\%, \quad (15)$$

where $\alpha_e(t)$ is the exact value of function $\alpha(t)$, and $\alpha_a(t)$ is an approximate value of function $\alpha(t)$.

The calculations were made for an accurate moving interface position and for a position disturbed with a pseudorandom error of normal distribution. Results for 1%, 2% and 5% disturbance are presented in the paper. Also, the influence of the number of control points, i.e. the number of points where the interface position is known (including the addends in sum (9)), has been examined. The results of interface control carried out every 0.1, 0.2, 0.5 and 1 seconds are presented below. They correspond to a situation where $M = 2093, 1047, 419, 210$.

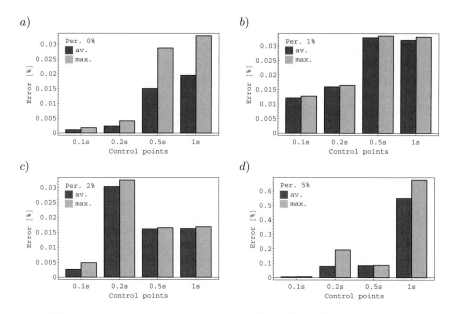

Fig. 3. Average and maximum errors of function $\alpha(t)$ reconstruction

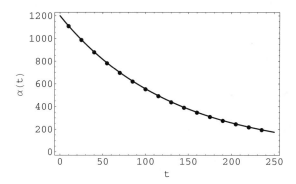

Fig. 4. Exact (solid line) and approximate (dot line) values of function $\alpha(t)$ for interface control every one second and for perturbation equal to 5%

In each case, calculations were carried out for ten different initial settings of a pseudorandom numbers generator.

Figure 3 presents the errors with which function $\alpha(t)$ was reconstructed for different disturbance values and a different number of control points. The mean error value for ten activations of the algorithm and the obtained maximum error value are shown in the figure. It should be noted that where the input data were given without disturbance, the convective heat transfer coefficient was reconstructed with minimal errors resulting from the chosen algorithm termination point, whereas for the disturbed input data, the result's error is much lower than the error at the beginning. For the highest disturbance equal to 5% and for

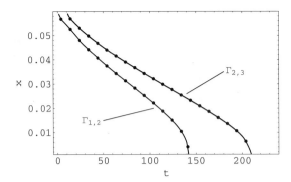

Fig. 5. Exact (solid lines) and reconstructed (dot lines) positions of interfaces $\Gamma_{1,2}$ and $\Gamma_{2,3}$ for control every one second and for perturbation equal to 5%

interface controls every 1 second, the maximum error equals to 0.68%. With a larger amount of control points, errors become significantly lower. An exception is the result obtained for a 2% disturbance and for interface control carried out every second, where the errors are slightly higher than those obtained for a smaller number of control points.

Figure 4 shows the exact and approximate values of function $\alpha(t)$ for interface control every one second and for perturbations equal to 5%. Figure 5 presents the exact and reconstructed positions of interfaces $\Gamma_{1,2}$ and $\Gamma_{2,3}$ for control carried out every one second and for perturbation equal to 5%. In the remaining cases, all curves were reconstructed equally well.

5   Conclusion

This paper discussed the identification of the convective heat transfer coefficient in a three-phase inverse design Stefan problem. The problem consists in the reconstruction of the function which describes the convective heat transfer coefficient, where the position of the moving interfaces of the phase change are well-known. The convective heat transfer coefficient is sought in the form of a continuous function, non-linearly dependent on the parameters sought. A genetic algorithm was used to determine these parameters. The direct Stefan problem was solved via a generalized alternating phase truncation method. Stability of the whole algorithm was ensured by applying the Tikhonov regularization technique.

The calculations made show stability of the proposed algorithm in terms of the input data errors and the number of control points. The application of genetic algorithms yields better results than the classical nonderivative optimization methods (e.g. the Nelder-Mead method). A comparison of the results for the two-phase problem is included in paper [16]. In that case, the scatter of the results obtained is also much smaller.

In the future we are going to use the presented algorithm for the solution of the inverse Stefan problem in which measured temperatures are given at some points of the domain.

References

1. Ang, D.D., Dinh, A.P.N., Thanh, D.N.: Regularization of an inverse two-phase Stefan problem. Nonlinear Anal. **34** (1998) 719–731
2. Briozzo, A.C., Natale, M.F., Tarzia, D.A.: Determination of unknown thermal coefficients for Storm's-type materials through a phase-change process. Int. J. Nonlinear Mech. **34** (1999) 329–340
3. Goldman, N.L.: Inverse Stefan problem. Kluwer, Dordrecht (1997)
4. Grzymkowski, R., Słota, D.: One-phase inverse Stefan problems solved by Adomian decomposition method. Comput. Math. Appl. **51** (2006) 33–40
5. Grzymkowski, R., Słota, D.: Multi-phase inverse Stefan problems solved by approximation method. In: Wyrzykowski, R. et al. (eds.): Parallel Processing and Applied Mathematics. LNCS 2328, Springer-Verlag, Berlin (2002) 679–686
6. Grzymkowski, R., Słota, D.: Numerical calculations of the heat-transfer coefficient during solidification of alloys. In: Sarler, B. et al. (eds.): Moving Boundaries VI. Wit Press, Southampton (2001) 41–50
7. Grzymkowski, R., Słota, D.: Numerical method for multi-phase inverse Stefan design problems. Arch. Metall. Mater. **51** (2006) 161–172
8. Jochum, P.: The numerical solution of the inverse Stefan problem. Numer. Math. **34** (1980) 411–429
9. Kapusta, A., Mochnacki, B.: The analysis of heat transfer processes in the cylindrical radial continuous casting volume. Bull. Pol. Acad. Sci. Tech. Sci. **36** (1988) 309–320
10. Kurpisz, K., Nowak, A.J.: Inverse thermal problems. CMP, Southampton (1995)
11. Liu, J., Guerrier, B.: A comparative study of domain embedding methods for regularized solutions of inverse Stefan problems. Int. J. Numer. Methods Engrg. **40** (1997) 3579–3600
12. Michalewicz, Z.: Genetic algorithms + data structures = evolution programs. Springer-Verlag, Berlin (1996)
13. Osyczka, A.: Evolutionary algorithms for single and multicriteria design optimization. Physica-Verlag, Heidelberg (2002)
14. Rogers, J.C.W., Berger, A.E., Ciment, M.: The alternating phase truncation method for numerical solution of a Stefan problem. SIAM J. Numer. Anal. **16** (1979) 563–587
15. Slodička, M., De Schepper, H.: Determination of the heat-transfer coefficient during soldification of alloys. Comput. Methods Appl. Mech. Engrg. **194** (2005) 491–498
16. Słota, D.: Solving the inverse Stefan design problem using genetic algorithms. Inverse Probl. Sci. Eng. (in review)
17. Tikhonov, A.N., Arsenin, V.Y.: Solution of ill-posed problems. Wiley & Sons, New York (1977)
18. Zabaras, N., Kang, S.: On the solution of an ill-posed design solidification problem using minimization techniques in finite- and infinite-dimensional function space. Int. J. Numer. Methods Engrg. **36** (1993) 3973–3990

Semi-supervised Clustering Using Incomplete Prior Knowledge

Chao Wang, Weijun Chen, Peipei Yin, and Jianmin Wang

School of Software, Tsinghua University, Beijing 100084, P.R. China
chao-wang05@mails.thu.edu.cn

Abstract. Clustering algorithms incorporated with prior knowledge have been widely studied and many nice results were shown in recent years. However, most existing algorithms implicitly assume that the prior information is complete, typically specified in the form of labeled objects with each category. These methods decay and behave unstably when the labeled classes are incomplete. In this paper a new type of prior knowledge which bases on partially labeled data is proposed. Then we develop two novel semi-supervised clustering algorithms to face this new challenge. An empirical study performed on benchmark dataset shows that our proposed algorithms produce better results with limited labeled examples comparing with existing baselines.

Keywords: semi-supervised clustering, seeded clustering, clustering with prior knowledge.

1 Introduction

Semi-supervised clustering algorithms have recently been studied by many researchers with considerable interests. Although better results have been shown by some of the improved clustering algorithms, most of these methods depend heavily on the completeness of the prior knowledge. This means that all classes in the dataset need at least one labeled object which is also called complete seeding. However, in practical domains, usually only partial prior knowledge is provided that some unseeded categories exist. In such a clustering task, the existing algorithms such as Seeded-KMeans (Basu et al., 2002) tend to generate unsatisfied partitions to the data set because they need complete seeding knowledge.

In this paper we present two novel semi-supervised clustering algorithms (FS-KMeans and SS-KMeans) which can take use of the partial prior knowledge to estimate the rest parts of the seed set. Our experiments results show that the new algorithms can not only significantly boost the performance of semi-supervised clustering but also behave more stably on benchmark datasets.

The organization of this paper is as follows. In Section2 two novel semi-supervised clustering algorithms are proposed to solve the new challenge. Section3 presents a comparative study and discusses experimental results on two benchmark datasets, followed by concluding remarks in section4.

Y. Shi et al. (Eds.): ICCS 2007, Part I, LNCS 4487, pp. 192–195, 2007.

2   Algorithms

2.1   Seeded-KMeans

Seeded-KMeans (Basu et al., 2002) is a semi-supervised variant of KMeans, where initial background knowledge, provided in the form of labeled data points, is used in the clustering process. Thus, rather than initializing the KMeans from K random points, the mean of the lth cluster is initialized with the mean of the lth partition of the seed set . Then it repeats the point-assignment and recomputing means until convergence. Notice that the labels of the seeds could be changed in the assignment step. However, with a view to an incomplete seeding problem, we assume that S only contains data points from L classes ($L < K$). So $L < K$ classes have no labeled instances given already for the clustering task. As a result, the initial steps of the Seeded-KMeans and the Constrained-KMeans have to be adapted to this now problem.

2.2   Tow Incomplete-Seeding KMeans Algorithms

Noticing that the centroids of some clusters with seeds should reflect the distributions of the other clusters without seeds, we can utilize this information to choose better initial centers for unseeded clusters. Two more sophisticated algorithms we developed are given bellow.

Table 1. Algorithm:Farthest-Seeded-KMeans

Input Set of data points $X = \{x_1, x_2, \ldots, x_N\}, x_i \in R^d$, number of clusters , set $S_g = \cup_{l=1}^L S_l$ of initial seeds provided.

Output Disjoint K partitions $\{X\}_{l=1}^K$ of such that KMeans objective function is optimized

Method   1. Generating Step

    (a) for the clusters with seeds, $\mu_h^{(0)} \leftarrow \frac{1}{S_h} \sum_{x \in S_h} x$, for $h = 1, \ldots, L$;

    (b) choose $K-L$ data points farthest from any cluster center as new centroids $\mu_h^{(0)}$, for $h = L+1, L+2, \ldots, K$;

  2. Iterating Step

    Repeat until *convergence*

    (a) *assignment*: assign each data point to the cluster h^* (i.e. set $X_{h^*}^{(t+1)}$) where
$$h^* = argmin_{h \in (1,\ldots,K)} \|x - \mu_h^{(t)}\|^2$$

    (b) *update*: recompute the means as $\mu_h^{t+1} \leftarrow \frac{1}{|X_h^{t+1}|} \sum_{x \in X_h^{t+1}} X$;

    (c) $t \leftarrow t + 1$;

In the Farthest-Seeded-KMeans (FS-KMeans), the clustering process is divided into two steps which are the same as original KMeans. However, in the seeds generation, the partial seed set is used to produce initial centroids for the rest clusters. Then the point farthest from any cluster center at present is chosen as the centroid for the cluster. We could also use some other criterions such as choosing the points with the largest standard deviation for one particular attribute. After repeating this for times, the initial centroids for the rest clusters are produced. The algorithm is presented in detail above.

Another algorithm we proposed, Splitting-Seeded-KMeans (SS-KMeans)is based on such an idea: to obtain clusters, split the set of all points into two clusters, select one of these clusters to split, and so on, until clusters have been produced. However, we have some seeds for clusters already, so we firstly generate a partitioning of the dataset, and then split these clusters to produce new clusters. In our experiments, we choose the cluster with the largest SSE to split.The details of SS-KMeans are given below.

Table 2. Algorithm:Splitting-Seeded-KMeans

Input Set of data points $X = \{x_1, x_2, \ldots, x_N\}, x_i \in R^d$, number of clusters , set $S_g = \cup_{l=1}^{L} S_l$ of initial seeds provided.

Output Disjoint K partitions $\{X\}_{l=1}^{K}$ of such that KMeans objective function is optimized

Method   1. Initialization a list Δ of clusters to contain the cluster consisting all points;
  2. Pre-Clustering Step
     (a) Initialization by seeds:$\mu_h^{(0)} \leftarrow \frac{1}{S_h} \sum_{x \in S_h} x$, for $h = 1, \ldots, L$;
     (b) Generate L clusters using KMeans clustering with $\mu_h^{(0)}$;
     (c) Update the list Δ;
  3. Splitting Step
     Repeat follow steps *until* K clusters are generated:
     (a) choose a cluster from list Δ;
     (b) split the cluster into two partitions by KMeans with $K = 2$;
     (c) Update the list Δ;
  4. Refining Step
     (a) Initialize centroids with the means of clusters in list Δ;
     (b) Conduct KMeans on the whole dataset.

---

3   Experiments

The four clustering algorithms - FS-KMeans, SS-KMeans, Seeded-KMeans and random KMeans - are compared on high-dimensional text datasets (subsets of CMU 20-Newsgroups), with varying seeding, using mutual information as evaluation measure. In Seeded-KMeans, FS-KMeans and SS-KMeans, the seeds were selected from the dataset according to the corresponding seed fraction which vary from 0.1 to 1 in steps of 0.1. The four algorithms were compared with the unseeded categories increased from fully seeded to completely unseeded.

Table 3 summarizes the results of our experiments on Newsgroups data set when comparing FS-KMeans and SS-KMeans to Random KMeans and Seeded-KMeans. The number of unseeded categories varied from 0 (i.e. complete seeding) to 5 (i.e. no category had seed set). Clearly, both FS-KMeans and SS-KMeans outperforms their unsupervised and semi-supervised learning counterparts when the prior knowledge is incomplete in the form of partially seeded. However, the FS-KMeans fluctuates in a range because when the number of unseeded categorizes increases, because FS-KMeans has a tendency to choose outliers as the candidate initial centers because it picks the farthest point rather than points in a dense region. Comparatively, SS-KMeans did not only result good MI measures but was also more stable than the other methods even when all the categories were unseeded. This is mainly owed to the Refine Step in the SS-KMeans which guarantees to find a minimum to the objective function.

Table 3. Performance comparisonon 20-Newsgroups dataset with incomplete seeding

unseeded categories	0	1	2	3	4	5
Random KMeans	0.516	0.499	0.508	0.501	0.496	0.497
Seeded-KMeans	0.588	0.581	0.577	0.571	0.563	0.551
FS-KMeans	0.588	0.586	0.582	0.579	0.569	0.563
SS-KMeans	0.615	0.613	0.608	0.605	0.601	0.579

4   Conclusion

We have presented two novel algorithms FS-KMeans and SS-KMeans which effectively utilize the partial prior knowledge provided as incomplete seed set. Experimental results on benchmark datasets show that: (a) the two novel algorithms can estimate the underlying seeds more reasonably than the original seeded KMeans. (b) SS-KMeans does not only produce superior performance but also behave stably when the unseeded categorizes increase.

References

1. Basu, S., Banerjee, A., & Mooney, R. J.:Semi-supervised clustering by seeding. Proceedings of 19th International Conference on Machine Learning (ICML-2002) 19-26
2. Bilmes, J.: A gentle tutorial on the EM algorithm and its application to parameter estimation for Gaussian mixture and hidden Markov models. Tech. rep. (1997) ICSI-TR-97-021, ICSI.
3. Bilenko, M., Basu, S., & Mooney, R. J.: Integrating constraints and metric learning in semi-supervised clustering. Proceedings of 21st International Conference on Machine Learning (ICML-2004)
4. Wagstaff, K., Cardie, C., Rogers, S., & Schroedl, S.: Constrained K-Means clustering with background knowledge. Proceedings of 18th International Conference on Machine Learning (ICML-2001) 577-584

Distributed Reasoning with Fuzzy Description Logics

Jianjiang Lu[1], Yanhui Li[2], Bo Zhou[1], Dazhou Kang[2], and Yafei Zhang[1]

[1] Institute of Command Automation, PLA University of Science and Technology,
Nanjing 210007, China
[2] Department of Computer Science and Engineering, Southeast University,
Nanjing 210096, P.R. China
jjlu@seu.edu.cn

Abstract. By the development of Semantic Web, increasing demands for vague and distributed information representation have triggered a mass of theoretical and applied researches of fuzzy and distributed ontologies, whose main logical infrastructures are fuzzy and distributed description logics. However, current solutions are proposed respectively on one of these two aspects. By integrating $\mathcal{E}$-connection into fuzzy description logics, this paper proposes a novel logical approach to couple both fuzzy and distributed features within description logics. The main contribution of these paper is to propose a discrete tableau algorithm to achieve reasoning within this new logical system.

Keywords: ontologies; $\mathcal{E}$-connection; discrete tableau algorithm.

1 Introduction

The Semantic Web stands for the idea of a future Web, in which information is given well-defined meaning, better enabling intelligent Web information processing [1]. In the Semantic Web, ontology is a crucial knowledge representation model to express a shared understanding of information between users and machines, and description logics (DLs for short) are often named as the logic infrastructure of ontologies [2]. Along with the evolvement from current Web to the Semantic Web, the management of ill-structured, ill-defined or imprecise information plays a more and more important role in applications of the Semantic Web, such as document retrieval [3], search engine [4] and query refinement [5]. This trend calls for ontologies with capability to deal with uncertainty. However, classical DLs are two-value-based languages. The need for expressing uncertainty in the Semantic Web has triggered extending classical DLs with fuzzy capabilities, yielding Fuzzy DLs (FDLs for short) [6,7,8,9]. Meanwhile, Working with multiple distributed ontologies brings a growing body of work in distributed research of description logic. In the distribution extension of classical DLs, Cuenca Grau et al integrated the $\mathcal{E}$-connections formalism [10] into OWL in a compact and natural way by defining "links" that stand for the inter-ontology relations [11]. Their extension is largely based on reasoning technique in classical DLs with general TBoxes.

   The main difficulty in achieving similar distributed extension within FDLs and combining fuzzy and distributed features within DLs is that reasoning with general TBox in FDLs is still a hard problem. In this year, we propose a discrete tableau algorithm

Y. Shi et al. (Eds.): ICCS 2007, Part I, LNCS 4487, pp. 196–203, 2007.

to solve this problem [12], that can be considered as a base technique to achieve distributed reasoning in FDLs. In this paper, we will extend our discretization algorithm in distributed case and combine $\mathcal{E}$-connections to propose a distributed extension of FDLs (here we focus on $\mathcal{FSHIN}$ [8], a complex FDL with inverse role, role hierarchy and unqualified number restriction.) and a corresponding tableau reasoning algorithm within this extension, hence achieve distributed reasoning within multiple FDL KBs.

2 $\mathcal{E}$-Connection Between Fuzzy Description Logics

2.1   Fuzzy Links Between Two Knowledge Bases

Let $\mathcal{K}_1$ and $\mathcal{K}_2$ be two $\mathcal{FSHIN}$ KBs, $\mathcal{I}_1 = \langle \Delta^{\mathcal{I}_1}, \cdot^{\mathcal{I}_1} \rangle$ and $\mathcal{I}_2 = \langle \Delta^{\mathcal{I}_2}, \cdot^{\mathcal{I}_2} \rangle$ be their fuzzy interpretations. E_{12} is a set of fuzzy links (denoted E_{12} and F_{12}) that connect these two $\mathcal{K}_1$ and $\mathcal{K}_2$. We define $\mathcal{I}_{12} = \langle \Delta^{\mathcal{I}_{12}}, \cdot^{\mathcal{I}_{12}} \rangle$ as the fuzzy interpretation of E_{12}, where $\Delta^{\mathcal{I}_{12}} = \Delta^{\mathcal{I}_1} \times \Delta^{\mathcal{I}_2}$ and for any $E_{12} \in E_{12}$, $\cdot^{\mathcal{I}_{12}}$ interprets it as a membership function: $\Delta^{\mathcal{I}_1} \times \Delta^{\mathcal{I}_2} \to [0,1]$. And to describe constraints among fuzzy links, we propose fuzzy link axioms: $E_{12} \sqsubseteq F_{12}$, where E_{12} and F_{12} are fuzzy links. A interpretation $\mathcal{I}_{12}$ satisfies the above fuzzy link axioms, iff for any $d \in \Delta^{\mathcal{I}_1}$ and any $d' \in \Delta^{\mathcal{I}_2}$, $E_{12}^{\mathcal{I}_{12}}(d, d') \leq F_{12}^{\mathcal{I}_{12}}(d, d')$. An LBox L_{12} is a finite set of fuzzy link axioms, $\mathcal{I}_{12}$ satisfies L_{12}, iff it satisfies every axiom in L_{12}.

These two FDL KBs $\mathcal{K}_1$ and $\mathcal{K}_2$ and their LBox L_{12} construct a simple Combined Distributed FDL KB $\Sigma = (\mathcal{K}_1, \mathcal{K}_2, L_{12})$. By introducing fuzzy link, we allow two new concepts $\mathcal{K}_1$: $\exists E_{12}.C_2$ and $\forall E_{12}.C_2$ in $\mathcal{K}_1$, where E_{12} is a fuzzy link in $\mathcal{E}_{12}$ and C_2 is a fuzzy concept in $\mathcal{K}_2$. These two concepts are considered as normal fuzzy concepts in $\mathcal{K}_1$, hence they can appear in TBox and ABox of $\mathcal{K}_1$.

For example, let $\mathcal{K}_1$ and $\mathcal{K}_2$ be two KBs about animal and person respectively. Dog_1 and $Person_2$ are fuzzy concepts in $\mathcal{K}_1$ and $\mathcal{K}_2$, and $lovewith_{12}$ is a fuzzy link in L_{12}. By using fuzzy links, we can define a new fuzzy concept $Friendlydog_1$ in $\mathcal{K}_1$'s TBox:

$$Friendlydog_1 \equiv \exists lovewith_{12}.Person_2 \sqcap Dog_1 \tag{1}$$

2.2   Combined Distributed Fuzzy Description Logic Knowledge Bases

In above subsection, we discuss the fuzzy links between two FDL KBs and give a simple example of Combined Distributed FDL (CDFDL for short) KBs. Now we will give a general definition of it.

Definition 1. *a CDFDL KB is a pair* $\Sigma = (\mathcal{K}_S, L_S)$, *where* $\mathcal{K}_S$ *is a set of FDL KBs:* $\mathcal{K}_S = \{\mathcal{K}_1, \ldots, \mathcal{K}_m\}$, *and* L_S *is a set of LBoxes that connect any two knowledge bases in* $\mathcal{K}_S$: $L_S = \{L_{ij} | 1 \leq i, j \leq m \text{ and } i \neq j\}$. *For any fuzzy concept* C_j *in* $\mathcal{K}_j$ *and any fuzzy link* E_{ij} *in* L_{ij} *and , the following expressions are also considered as fuzzy concepts in* $\mathcal{K}_i$: $\exists E_{ij}.C_j$ *and* $\forall E_{ij}.C_j$.

An interpretation of a CDFDL KB is a pair $\mathcal{I} = (\{\mathcal{I}_i\}, \{\mathcal{I}_{ij}\})$, where $\mathcal{I}_i$ is an interpretation of $\mathcal{K}_i$ and correspondingly $\mathcal{I}_{ij}$ is an interpretation of L_{ij}. For any fuzzy concept C_j (role R_j) in $\mathcal{K}_j$, $C_j^{\mathcal{I}} = C_j^{\mathcal{I}_j}$ $(R_j^{\mathcal{I}} = R_j^{\mathcal{I}_j})$; for any fuzzy link E_{ij} in $\mathcal{I}_{ij}$,

$E_{ij}{}^{\mathcal{I}} = E_{ij}{}^{\mathcal{I}_{ij}}$; for any individual a_j in $\mathcal{K}_j$, $a_j{}^{\mathcal{I}} = a_j{}^{\mathcal{I}_j}$; and for $\exists E_{ij}.C_j$ and $\forall E_{ij}.C_j$, their interpretation are inductively defined as:

$$\exists E_{ij}.C_j{}^{\mathcal{I}}(d) = \sup{}_{d' \in \Delta^{\mathcal{I}_i}} \{\min(E_{ij}{}^{\mathcal{I}_{ij}}(d, d'), C_j{}^{\mathcal{I}_j}(d'))\} \qquad (2)$$

$$\forall E_{ij}.C_j{}^{\mathcal{I}}(d) = \inf{}_{d' \in \Delta^{\mathcal{I}_i}} \{\max(1 - E_{ij}{}^{\mathcal{I}_{ij}}(d, d'), C_j{}^{\mathcal{I}_j}(d'))\}$$

An interpretation $\mathcal{I}$ is a model of $\Sigma = (\mathcal{K}_S, L_S)$, iff $\mathcal{I}$ satisfies every $\mathcal{K}_i$ in $\mathcal{K}_S$ and every L_{ij} in L_S. In this paper, we will propose a discrete tableau algorithm to decide satisfiability of CDFDL KBs Σ, which is based on the semantical discretization technique discussed in the following section.

3   Semantical Discretization

In this section, we will propose a novel semantical discretization technique to achieve such translations: if a CDFDL KB has a fuzzy model, we use the discretization to translate it into a special model, in which any value of membership degree functions belongs to a given discrete degree set S and its cardinality $|S|$ is polynomial of the sum of the cardinality $|\mathcal{A}_i|$ of the ABox $\mathcal{A}_i$ in every KB $\mathcal{K}_i$. And we call it a discrete model within S.

The main issue in semantical discretization is to decide the discrete degree set S. Let us now proceed formally in the creation of S. Given $\Sigma = (\mathcal{K}_S = \{\mathcal{K}_1, \ldots, \mathcal{K}_m\}, L_S = \{L_{ij} | 1 \leq i, j \leq m, i \neq j\})$, and $\mathcal{K}_i = \langle \mathcal{T}_i, \mathcal{R}_i, \mathcal{A}_i \rangle$. Let N_d be the set of degrees appearing in any ABox: $N_d = \{n | \alpha \bowtie n \in \mathcal{A}_i, 1 \leq i \leq m\}$. From N_d, we define the degree closure $N_d^* = \{0, 0.5, 1\} \cup N_d \cup \{n | 1 - n \in N_d\}$ and order degrees in ascending order: $N_d^* = \{n_0, n_1, \ldots, n_s\}$, where for any $0 \leq i \leq s$, $n_i < n_{i+1}$. For any two back-to-back elements $n_i, n_{i+1} \in N_d^*$, we insert their median $m_{i+1} = (n_i + n_{i+1})/2$ to get $S = \{n_0, m_1, n_1, \ldots, n_{s-1}, m_s, n_s\}$. We call S a discrete degree set w.r.t Σ. Obviously for any $1 \leq i \leq s$, $m_i + m_{s+1-i} = 1$ and $n_{i-1} < m_i < n_i$. Note that:

$$|S| = 2s + 1 = O(|N_d|) = O(\sum_{i=1}^{m} |\mathcal{A}_i|). \qquad (3)$$

Lemma 1. *For any $\mathcal{K}_i = \langle \mathcal{T}_i, \mathcal{R}_i, \mathcal{A}_i \rangle$ and any discrete degree set S w.r.t Σ, if $\mathcal{K}_i$ has a fuzzy model, it has a discrete model within S.*

The proof of this lemma is an extension of the proof in FDL cases [12]. Meanwhile, to verify the soundness of our discretization, we have the following lemma.

Lemma 2. *For any L_{ij} and any discrete degree set S w.r.t Σ, if L_{ij} has a fuzzy model, it has a discrete model within S.*

Since a discrete model is also a fuzzy model of Σ, we get the following theorem to guarantee the equivalence between existence of fuzzy models and discrete models.

Theorem 1. *For any $\Sigma = (\mathcal{K}_S = \{\mathcal{K}_1 \ldots \mathcal{K}_m\}, L_S = \{L_{ij}\})$ and any discrete degree set S w.r.t Σ, Σ has a fuzzy model iff it has a discrete model within S.*

4   Discrete Tableau Algorithm

Before expressing discrete tableau algorithms formally, here we introduce some notations. It will be assumed that the concepts appearing in tableau algorithms are written in NNF [13]. The set of subconcepts of a concept C is denoted as $\mathrm{sub}(C)$. For a CDFDL KB Σ, we define $\mathrm{sub}(\mathcal{K}_i)$ as the union of all $\mathrm{sub}(C)$, for any concept C appears in $\mathcal{K}_i$. And we use the symbols $\rhd$ and $\lhd$ as two placeholders for the inequalities $\geq, >$ and $\leq$, $<$, and the symbols $\bowtie^-$, $\rhd^-$ and $\lhd^-$ to denote their reflections, for example, $\geq$ and $\leq$ are reflections to each other. Finally we define $\langle \bowtie, n \rangle$ as a degree pair. Two degree pairs are called conjugated, iff they satisfy the following conditions (see table 1).

Table 1. Conjugated pairs

	$\langle <, m \rangle$	$\langle \leq, m \rangle$
$\langle \geq, n \rangle$	$n \geq m$	$n > m$
$\langle >, n \rangle$	$\neg \exists n_1 \in S$ with $n < n_1 < m$	$n \geq m$

Now we define the discrete tableau for Σ. Let $\mathrm{R}_{\mathcal{K}_i}$ and $\mathrm{O}_{\mathcal{K}_i}$ be the sets of roles and individuals appearing in $\mathcal{K}_i$. A discrete tableau T for Σ within a degree set S is a pair: $\mathrm{T} = \langle \{\mathrm{T}_i\}, \{\mathcal{E}_{ij}\} \rangle$, $\mathrm{T}_i = \langle \mathcal{O}_i, \mathcal{L}_i, \mathcal{E}_i, \mathcal{V}_i \rangle$, $1 \leq i, j \leq m$ and $i \neq j$, where

- $\mathcal{O}_i$: a nonempty set of nodes;
- $\mathcal{L}_i : \mathcal{O}_i \to 2^{M_i}$, $M_i = \mathrm{sub}(\mathcal{K}_i) \times \{\geq, >, \leq, <\} \times S$;
- $\mathcal{E}_i : \mathrm{R}_{\mathcal{K}_i} \to 2^{Q_i}$, $Q_i = \{\mathcal{O}_i \times \mathcal{O}_i\} \times \{\geq, >, \leq, <\} \times S$;
- $\mathcal{V}_i : \mathrm{O}_{\mathcal{K}_i} \to \mathcal{O}_i$, maps any individual into a corresponding node in $\mathcal{O}_i$.
- $\mathcal{E}_{ij} : \mathrm{E}_{ij} \to 2^{Q_{ij}}$, $Q_{ij} = \{\mathcal{O}_i \times \mathcal{O}_j\} \times \{\geq, >, \leq, <\} \times S$;

Any T_i has a forest-like structure, which is a collection of trees that correspond to individuals in the ABox $\mathcal{A}_i$. Every tree consists of nodes standing for the individuals, and edges representing the relations between two nodes (individuals). Each node d is labelled with a set $\mathcal{L}(d)$ of degree triples: $\langle C, \bowtie, n \rangle$, which denotes the membership degree of d being an instance of $C \bowtie n$. A pair of triple $\langle C, \bowtie, n \rangle$ and $\langle C, \bowtie^-, m \rangle$ are conjugated if $\langle \bowtie, n \rangle$ and $\langle \bowtie^-, m \rangle$ are conjugated. In any T_i, for any $d, d' \in \mathcal{O}_i$, $a, b \in \mathrm{O}_{\mathcal{K}_i}$, $C, D \in \mathrm{sub}(\mathcal{K}_i)$ and $R \in \mathrm{R}_{\mathcal{K}_i}$, the following conditions must hold:

1. There does not exist two conjugated degree triples in $\mathcal{L}_i(d)$;
2. There does not exist mistake triples: $\langle \perp, \geq, n \rangle$ $(n > 0)$, $\langle \top, \leq, n \rangle$ $(n < 1)$, $\langle \perp, > , n \rangle$, $\langle \top, <, n \rangle$, $\langle C, >, 1 \rangle$ and $\langle C, <, 0 \rangle$ in $\mathcal{L}_i(d)$;
3. If $C \sqsubseteq D \in \mathcal{T}$, then there must be some $n \in S$ with $\langle C, \leq, n \rangle$ and $\langle D, \geq, n \rangle$ in $\mathcal{L}_i(d)$;
4. If $\langle C, \bowtie, n \rangle \in \mathcal{L}_i(d)$, then $\langle \mathrm{nnf}(\neg C), \bowtie^-, 1 - n \rangle \in \mathcal{L}_i(d)$;
5. If $\langle C \sqcap D, \rhd, n \rangle \in \mathcal{L}_i(d)$, then $\langle C, \rhd, n \rangle$ and $\langle D, \rhd, n \rangle \in \mathcal{L}_i(d)$;
6. If $\langle C \sqcap D, \lhd, n \rangle \in \mathcal{L}_i(d)$, then $\langle C, \lhd, n \rangle$ or $\langle D, \lhd, n \rangle \in \mathcal{L}_i(d)$;
7. If $\langle C \sqcup D, \rhd, n \rangle \in \mathcal{L}_i(d)$, then $\langle C, \rhd, n \rangle$ or $\langle D, \rhd, n \rangle \in \mathcal{L}_i(d)$;
8. If $\langle C \sqcup D, \lhd, n \rangle \in \mathcal{L}_i(d)$, then $\langle C, \lhd, n \rangle$ and $\langle D, \lhd, n \rangle \in \mathcal{L}_i(d)$;

9. If $\langle \forall R.C, \rhd, n\rangle \in \mathcal{L}_i(d), \langle\langle d, d'\rangle, \rhd', m\rangle \in \mathcal{E}_i(R)$, and $\langle \rhd', m\rangle$ is conjugated with $\langle \rhd^-, 1-n\rangle$, then $\langle C, \rhd, n\rangle \in \mathcal{L}_i(d')$;

10. If $\langle \forall R.C, \lhd, n\rangle \in \mathcal{L}_i(d)$, then there must be a node $d' \in \mathcal{O}_i$ with $\langle\langle d, d'\rangle, \lhd^-, 1-n\rangle \in \mathcal{E}_i(R)$ and $\langle C, \lhd, n\rangle \in \mathcal{L}_i(d')$;

11. If $\langle \exists R.C, \rhd, n\rangle \in \mathcal{L}_i(d)$, then there must be a node $d' \in \mathcal{O}_i$ with $\langle\langle d, d'\rangle, \rhd, n\rangle \in \mathcal{E}_i(R)$ and $\langle C, \rhd, n\rangle \in \mathcal{L}_i(d')$;

12. If $\langle \exists R.C, \lhd, n\rangle \in \mathcal{L}_i(d)$, $\langle\langle d, d'\rangle, \rhd', m\rangle \in \mathcal{E}_i(R)$, and $\langle \rhd', m\rangle$ is conjugated with $\langle \lhd, n\rangle$, then $\langle C, \lhd, n\rangle \in \mathcal{L}_i(d')$;

13. If $\langle \forall P.C, \rhd, n\rangle \in \mathcal{L}_i(d)$, $\langle\langle d, d'\rangle, \rhd', m\rangle \in \mathcal{E}_i(R)$ for some $R \sqsubseteq^* P$ with Trans (R)=true and $\langle \rhd', m\rangle$ is conjugated with $\langle \rhd^-, 1-n\rangle$, then $\langle \forall R.C, \rhd, n\rangle \in \mathcal{L}_i(d')$;

14. If $\langle \exists P.C, \lhd, n\rangle \in \mathcal{L}_i(d)$, $\langle\langle d, d'\rangle, \rhd', m\rangle \in \mathcal{E}_i(R)$ for some $R \sqsubseteq^* P$ with Trans (R)=true and $\langle \rhd', m\rangle$ is conjugated with $\langle \lhd, n\rangle$, then $\langle \exists R.C, \lhd, n\rangle \in \mathcal{L}_i(d')$;

15. If $\langle \geq pR, \rhd, n\rangle \in \mathcal{L}_i(d)$, then $|\{d' | \langle\langle d, d'\rangle, \rhd, n\rangle \in \mathcal{E}_i(R)|\} \geq p$;

16. If $\langle \geq pR, \lhd, n\rangle \in \mathcal{L}_i(d)$, then $|\{d' | \langle\langle d, d'\rangle, \rhd', m\rangle \in \mathcal{E}_i(R)\}| < p$, where $\langle \rhd', m\rangle$ is conjugated with $\langle \lhd, n\rangle$;

17. If $\langle \leq pR, \rhd, n\rangle \in \mathcal{L}_i(d)$, then $|\{d' | \langle\langle d, d'\rangle, \rhd', m\rangle \in \mathcal{E}_i(R)\}| < p+1$, where $\langle \rhd', m\rangle$ is conjugated with $\langle \rhd^-, 1-n\rangle$;

18. If $\langle \leq pR, \lhd, n\rangle \in \mathcal{L}_i(d)$, $|\{d' | \langle\langle d, d'\rangle, \lhd^-, 1-n\rangle \in \mathcal{E}_i(R)\}| \geq p+1$;

19. If $\langle\langle d, d'\rangle, \bowtie, n\rangle \in \mathcal{E}_i(R)$, then $\langle\langle d', d\rangle, \bowtie, n\rangle \in \mathcal{E}_i(\mathrm{Inv}(R))$;

20. If $\langle\langle d, d'\rangle, \rhd, n\rangle \in \mathcal{E}_i(R)$ and $R \sqsubseteq^* P$, then $\langle\langle d, d'\rangle, \rhd, n\rangle \in \mathcal{E}_i(P)$;

21. If $a : C \bowtie n \in \mathcal{A}_i$, then $\langle C, \bowtie, n\rangle \in \mathcal{L}_i(\mathcal{V}_i(a))$;

22. If $\langle a, b\rangle : R \bowtie n \in \mathcal{A}_i$, then $\langle\langle \mathcal{V}_i(a), \mathcal{V}_i(b)\rangle, \bowtie, n\rangle \in \mathcal{E}_i(R)$;

23. If $a \neq b \in \mathcal{A}_i$, then $\mathcal{V}_i(a) \neq \mathcal{V}_i(b)$.

From conditions 1-2, the discrete tableau contains no clash. Condition 3 deals with general TBoxes: for any $C \sqsubseteq D \in \mathcal{T}$, we adopt a direct extension of reasoning technique in DLs: since any membership degree value in the discrete models belongs to S, for any node d, we guess $d : C$=n and $d : D$=m, for some $n, m \in S$ and $n \leq m$. Then we add $\langle C, \leq, n\rangle$ and $\langle D, \geq, n\rangle$ in $\mathcal{L}(d)$. Conditions 4-20 are necessary for the soundness of discrete tableaus. Conditions 21-23 ensure the correctness of individual mapping function $\mathcal{V}()$.

Additionally, we add some constraints to deal with fuzzy links. For any $d \in \mathcal{O}_i$, $d' \in \mathcal{O}_j$, $E_{ij}, F_{ij} \in L_{ij}$ and $C_j \in \mathrm{sub}(\mathcal{K}_j)$, the following conditions must hold:

24. If $\langle\langle d, d'\rangle, \rhd, n\rangle \in \mathcal{E}_{ij}(E_{ij})$ and $E_{ij} \sqsubseteq F_{ij} \in L_{ij}$, then $\langle\langle d, d'\rangle, \rhd, n\rangle \in \mathcal{E}_{ij}(F_{ij})$;

25. If $\langle \forall E_{ij}.C_j, \rhd, n\rangle \in \mathcal{L}_i(d), \langle\langle d, d'\rangle, \rhd', m\rangle \in \mathcal{E}_{ij}(E_{ij})$, and $\langle \rhd', m\rangle$ is conjugated with $\langle \rhd^-, 1-n\rangle$, then $\langle C, \rhd, n\rangle \in \mathcal{L}_j(d')$;

26. If $\langle \forall E_{ij}.C_j, \lhd, n\rangle \in \mathcal{L}_i(d)$, then there must be a node $d' \in \mathcal{O}_j$ with $\langle\langle d, d'\rangle, \lhd^-, 1-n\rangle \in \mathcal{E}_{ij}(E_{ij})$ and $\langle C, \lhd, n\rangle \in \mathcal{L}_j(d')$;

27. If $\langle \exists E_{ij}.C_j, \rhd, n\rangle \in \mathcal{L}_i(d)$, then there must be a node $d' \in \mathcal{O}_j$ with $\langle\langle d, d'\rangle, \rhd, n\rangle \in \mathcal{E}_{ij}(E_{ij})$ and $\langle C, \rhd, n\rangle \in \mathcal{L}_j(d')$;

28. If $\langle \exists E_{ij}.C_j, \lhd, n\rangle \in \mathcal{L}_i(d)$, $\langle\langle d, d'\rangle, \rhd', m\rangle \in \mathcal{E}_{ij}(E_{ij})$, and $\langle \rhd', m\rangle$ is conjugated with $\langle \lhd, n\rangle$, then $\langle C, \lhd, n\rangle \in \mathcal{L}_j(d')$;

Condition 24 guarantees that tableau satisfies the restriction of LBoxes. Conditions 25-28 are distributed extensions of classical conditions to deal with $\forall$ and $\exists$ restriction.

Theorem 2. *For any $\Sigma = (\mathcal{K}_S = \{\mathcal{K}_1 \ldots \mathcal{K}_m\}, L_S = \{L_{ij}\})$ and any discrete degree set S w.r.t Σ, Σ has a discrete model within S iff it has a discrete tableau T within S.*

From theorem 1 and 2, an algorithm that constructs a discrete tableau of Σ within S can be considered as a decision procedure for the satisfiability of Σ. The discrete tableau algorithm works on a completion forest F_Σ with a set $S^{\neq}$ to denote "$\neq$" relation between nodes and a tag function $W()$: for any node x, $W(x)$ denotes that x is an individual in the $W(x)$-th KB. When $W(x) = W(y) = i$, x is labelled with $\mathcal{L}_i(x) \subseteq M_i = \mathrm{sub}(\mathcal{K}_i) \times \{\geq, >, \leq, <\} \times S$; and the edge $\langle x, y \rangle$ is labelled $\mathcal{L}_i(\langle x, y \rangle) = \{\langle R, \bowtie, n \rangle\}$, for some $R \in \mathrm{R}_{\mathcal{K}_i}$ and $n \in S$. When $W(x) = i \neq W(y) = j$, the edge $\langle x, y \rangle$ is labelled $\mathcal{L}_{ij}(\langle x, y \rangle) = \{\langle E, \bowtie, n \rangle\}$, for some $E \in \mathrm{E}_{ij}$ and $n \in S$. The tableau algorithm initializes $F_\mathcal{K}$ to contain a root node x_a for each individual a in any $\mathrm{O}_{\mathcal{K}_i}$, sets $W(x_a) = i$ and labels x_a with $\mathcal{L}_i(x_a) = \{\langle C, \bowtie, n \rangle | a : C \bowtie n \in \mathcal{A}_i\}$. Moreover, for any pair $\langle x_a, x_b \rangle$, $\mathcal{L}_i(x_a, x_b) = \{\langle R, \bowtie, n \rangle | \langle a, b \rangle : R \bowtie n \in \mathcal{A}_i\}$, and for any $a \neq b \in \mathcal{A}_i$, we add $\langle x_a, x_b \rangle \in S^{\neq}$. The algorithm expands the forest F_Σ either by extending $\mathcal{L}_i(x)$ for the current node x or by adding new leaf node y with expansion rules in table 2.

In table 2, we adopt a optimized way to reduce "$\lhd$ rules": for any triple $\langle C, \lhd, n \rangle \in \mathcal{L}_i(x)$ with "$\lhd$", we use $\neg^{\bowtie}$ rules to add its equivalence $\langle \mathrm{nnf}(C), \lhd^-, 1 - n \rangle$ to $\mathcal{L}_i(x)$, and then deal it with $\rhd$ rules. Edges are added when expanding $\langle \exists G.C, \rhd, n \rangle$, $\langle \geq pG, \rhd, n \rangle$ in $\mathcal{L}_i(x)$, where G can be a fuzzy role or fuzzy link. A node y is called an G-successor of another node x and x is called a G-predecessor of y, if $\langle G, \bowtie, n \rangle \in \mathcal{L}_{i(ij)}(\langle x, y \rangle)$. Ancestor is the transitive closure of predecessor. And for any two connected nodes x and y, we define $D_G(x, y) = \{\langle \bowtie, n \rangle | P \sqsubseteq^* G, \langle P, \bowtie, n \rangle \in \mathcal{L}_{i(ij)}(\langle x, y \rangle)$ or $\langle \mathrm{Inv}(P), \bowtie, n \rangle \in \mathcal{L}_{i(ij)}(\langle y, x \rangle)\}$. If $D_G(x, y) \neq \emptyset$, y is called a R-neighbor of x. As inverse role and number restriction are allowed in $\mathcal{SHIN}$, we make use of pairwise blocking technique [14] to ensure the termination and correctness of our tableau algorithm: a node x is directly blocked by its ancestor y iff (1) x is not a root node; (2) x and y have predecessors x' and y', such that $\mathcal{L}_i(x) = \mathcal{L}_i(y)$ and $\mathcal{L}_i(x') = \mathcal{L}_i(y')$ and $\mathcal{L}_{i(ij)}(\langle y', y \rangle) = \mathcal{L}_{i(ij)}(\langle x', x \rangle)$. A node x is indirectly blocked if its predecessor is blocked. A node x is blocked iff it is either directly or indirectly blocked.

A completion forest $F_\mathcal{K}$ is said to contain a clash, if for a node x in $F_\mathcal{K}$ with $W(x) = i$, (1)$\mathcal{L}_i(x)$ contains two conjugated triples, or a mistake triple (see condition 2 in discrete tableau restriction); or (2) $\langle \geq pR, \lhd, n \rangle$ or $\langle \leq (p-1)R, \lhd^-, 1-n \rangle \in \mathcal{L}_i(x)$, and there are p nodes $y_1, y_2, \ldots y_p$ in F_Σ: for any $1 \leq k \leq p$, $\langle R, \rhd_k, m_k \rangle \in \mathcal{L}_i(\langle x, y_k \rangle)$, $\langle \rhd_k, m_k \rangle$ is conjugated with $\langle \lhd, n \rangle$ and for any two nodes y_k and y_q, $\langle y_k, y_q \rangle \in S^{\neq}$. A completion forest F_Σ is clash-free if it does not contain a clash, and it is complete if none of the expansion rules are applicable. From pairwise blocking technique, the worst-case complexity of our tableau algorithm is 2NEXPTIME [15]. And the soundness and completeness of our tableau algorithm are guaranteed by the following theorem.

Table 2. Expansion rules of discrete Tableau

Rule name Description
Assume $W(x) = i$

KB rule: if $C \sqsubseteq D \in \mathcal{T}_i$ and there is no n with $\langle C, \leq, n \rangle$ and $\langle D, \geq, n \rangle$ in $\mathcal{L}_i(x)$; then $\overline{\mathcal{L}_i(x)} \rightarrow \mathcal{L}_i(x) \cup \{\langle C, \leq, n \rangle \langle D, \geq, n \rangle\}$ for some $n \in S$.

The following rules are applied to nodes x which is not indirectly blocked.

$\neg^{\bowtie}$ rule: if $\langle C, \bowtie, n \rangle \in \mathcal{L}_i(x)$ and $\langle \mathrm{nnf}(\neg C), \bowtie^-, n \rangle \notin \mathcal{L}_i(x)$; then $\mathcal{L}_i(x) \rightarrow \mathcal{L}_i(x) \cup \{\langle \mathrm{nnf}(\neg C), \bowtie^-, n \rangle\}$.

$\sqcap^{\triangleright}$ rule: if $\langle C \sqcap D, \triangleright, n \rangle \in \mathcal{L}_i(x)$, and $\langle C, \triangleright, n \rangle$ or $\langle D, \triangleright, n \rangle \notin \mathcal{L}_i(x)$; then $\mathcal{L}_i(x) \rightarrow \mathcal{L}_i(x) \cup \{\langle C, \triangleright, n \rangle, \langle D, \triangleright, n \rangle\}$.

$\sqcup^{\triangleright}$ rule: if $\langle C \sqcup D, \triangleright, n \rangle \in \mathcal{L}_i(x)$, and $\langle C, \triangleright, n \rangle, \langle D, \triangleright, n \rangle \notin \mathcal{L}_i(x)$ then $\mathcal{L}_i(x) \rightarrow \mathcal{L}_i(x) \cup \{T\}$, for some $T \in \{\langle C, \triangleright, n \rangle, \langle D, \triangleright, n \rangle\}$

$\forall^{\triangleright}$ rule: if $\langle \forall R.C, \triangleright, n \rangle \in \mathcal{L}_i(x)$, $R \in \mathrm{R}_{\mathcal{K}_i}$, there is a R-neighbor y of x with $\langle \triangleright', m \rangle \in D_R(x, y)$, which is conjugated with $\langle \triangleright^-, 1 - n \rangle$, and $\langle C, \triangleright, n \rangle \notin \mathcal{L}_i(y)$; then $\mathcal{L}_i(y) \rightarrow \mathcal{L}_i(y) \cup \{\langle C, \triangleright, n \rangle\}$.

$\forall^{L\triangleright}$ rule: if $\langle \forall R.C, \triangleright, n \rangle \in \mathcal{L}_i(x)$, $R \in \mathrm{E}_{ij}$, there is a R-neighbor y of x with $\langle \triangleright', m \rangle \in$, $D_R(x, y)$ which is conjugated with $\langle \triangleright^-, 1 - n \rangle$, and $\langle C, \triangleright, n \rangle \notin \mathcal{L}_j(y)$; then $\mathcal{L}_j(y) \rightarrow \mathcal{L}_j(y) \cup \{\langle C, \triangleright, n \rangle\}$.

$\forall^{+\triangleright}$ rule: if $\langle \forall P.C, \triangleright, n \rangle \in \mathcal{L}_i(x)$, $R \in \mathrm{R}_{\mathcal{K}_i}$, there is a R-neighbor y of x with $R \sqsubseteq^* P$, $\mathrm{Trans}(R) = \mathrm{True}$ and $\langle \triangleright', m \rangle \in D_R(x, y)$, $\langle \triangleright', m \rangle$ is conjugated with $\langle \triangleright^-, 1 - n \rangle$ and $\langle \forall R.C, \triangleright, n \rangle \notin \mathcal{L}_i(y)$; then $\mathcal{L}(y)_i \rightarrow \mathcal{L}(y)_i \cup \{\langle \forall R.C, \triangleright, n \rangle\}$.

$\leq p^{\triangleright}$ rule: if $\langle \leq pR, \triangleright, n \in \mathcal{L}_i(x)$; there is $p + 1$ R-successors $y_1, y_2, \ldots, y_{p+1}$ of x with $\langle R, \triangleright_i, m_i \rangle \in \mathcal{L}_i(\langle x, y_i \rangle)$ and $\langle \triangleright_i, m_i \rangle$ is conjugated with $\langle \triangleleft^-, 1 - n \rangle$ for any $1 \leq i \leq p + 1$; and $\langle y_i, y_j \rangle \notin S^{\neq}$ for some $1 \leq i < j \leq p + 1$ then merge two nodes y_i and y_j into one : $\mathcal{L}(y_i) \rightarrow \overline{\mathcal{L}(y_i)} \cup \mathcal{L}(y_j)$; $\forall x, \mathcal{L}(y_i, x) \rightarrow \mathcal{L}(y_i, x) \cup \mathcal{L}(y_j, x), \langle y_j, x \rangle \in S^{\neq}$, add $\langle y_i, x \rangle$ in $S^{\neq}$

The following rules are applied to nodes x which is not blocked.

$\exists^{\triangleright}$ rule: if $\langle \exists R.C, \triangleright, n \rangle \in \mathcal{L}_i(x)$; $R \in \mathrm{R}_{\mathcal{K}_i}$; there is not a R-neighbor y of x with $\langle \triangleright, n \rangle \in D_R(x, y)$ and $\langle C, \triangleright, n \rangle \in \mathcal{L}_i(y)$. then add a new node z with $W(z) = i, \langle R, \triangleright, n \rangle \in \mathcal{L}_i(\langle x, z \rangle)$ and $\langle C, \triangleright, n \rangle \in \mathcal{L}_i(z)$.

$\exists^{L\triangleright}$ rule: if $\langle \exists R.C, \triangleright, n \rangle \in \mathcal{L}_i(x)$; $R \in \mathrm{E}_{ij}$; there is not a R-neighbor y of x with $\langle \triangleright, n \rangle \in D_R(x, y)$ and $\langle C, \triangleright, n \rangle \in \mathcal{L}_j(y)$. then add a new node z with $W(z) = j, \langle R, \triangleright, n \rangle \in \mathcal{L}_{ij}(\langle x, z \rangle)$ and $\langle C, \triangleright, n \rangle \in \mathcal{L}_j(z)$.

$\geq pR^{\triangleright}$ rule: if $\langle \geq pR, \triangleright, n \rangle \in \mathcal{L}_i(x)$, there are not p R-neighbors $y_1, y_2, \ldots, y_p$ of x with $\langle R, \triangleright, n \rangle \in \mathcal{L}_i(\langle x, y_i \rangle)$ and for any $i \neq j$, $\langle y_i, y_j \rangle \in S^{\neq}$. then add p new nodes $z_1, z_2, \ldots, z_p$ with $\langle R, \triangleright, n \rangle \in \mathcal{L}_i(\langle x, z_i \rangle)$, and for any two node z_i and z_j add $\langle z_i, z_j \rangle$ in $S^{\neq}$.

Theorem 3. *For any $\Sigma = (\mathcal{K}_S = \{\mathcal{K}_1 \ldots \mathcal{K}_m\}, L_S = \{L_{ij}\})$ and any discrete degree set S w.r.t Σ has a discrete tableau within S iff the tableau algorithm can construct a complete and clash-free completion forest.*

5   Conclusion

By integrating $\mathcal{E}$-connection into FDLs, this paper proposes a novel logical approach to couple both fuzzy and distributed features within DLs. To achieve reasoning support within this new logical form CDFDL, we extend our semantical discretization in distributed case and design a discrete tableau reasoning algorithm. Our work can be considered as a logical foundation to support reasoning with multiple distributed fuzzy ontologies.

References

1. Berners-Lee, T., Hendler, J., Lassila, O.: The semantic web. Scientific American **284** (2001) 34–43
2. Horrocks, I., Patel-Schneider, P.: Reducing owl entailment to description logic satisfiability. In: Proceeedings of the International Workshop on Description Logics (DL-05). (2003) 1–8
3. Parry, D. Fuzzy Logic and the Semantic Web. In: A fuzzy ontology for medical document retrieval. Elsevier Science, Oxford, UK (2006)
4. Widyantoro, D.H., Yen, J.: A fuzzy ontology-based abstract search engine and its user studies. In: FUZZ-IEEE. (2001) 1291–1294
5. Widyantoro, D., Yen, J.: Using fuzzy ontology for query refinement in a personalized abstract search engine. In: Proceedings of Joint 9th IFSA World Congress and 20th NAFIPS International Conference, Vancouver, Canada (2001)
6. Straccia, U.: A fuzzy description logic. In: Proceedings of AAAI-98, 15th National Conference on Artificial Intelligence, Madison, Wisconsin (1998) 594–599
7. Stoilos, G., Stamou, G., Tzouvaras, V., Pan, J., Horrocks, I.: Fuzzy owl: Uncertainty and the semantic web. In: Proceedings of International Workshop of OWL: Experiences and Directions, Galway (2005)
8. Stoilos, G., Stamou, G., Tzouvaras, V., Pan, J., Horrocks, I.: The fuzzy description logic shin. In: Proceedings of International Workshop of OWL: Experiences and Directions, Galway (2005)
9. Stoilos, G., Stamou, G., Tzouvaras, V., Pan, J., Horrock, I.: A Fuzzy Description Logic for Multimedia Knowledge Representation. In: Proc. of the International Workshop on Multimedia and the Semantic Web. (2005)
10. Kutz, O., Lutz, C., Wolter, F., Zakharyaschev, M.: E-connections of abstract description systems. Artificial Intelligence **156** (2004) 1–73
11. Cuenca Grau, B., Parsia, B., Sirin, E.: Working with multiple ontologies on the semantic web. In: Proceedings of the 3thrd International Semantic Web Conference. (2004)
12. Li, Y.H., Xu, B.W., Lu, J.J., Kang, D.Z.: Discrete tableaus for fshi. In: Proceedings of 2006 International Workshop on Description Logics - DL2006, The Lake District of the UK (2006)
13. Baader, F., Sattler, U.: An overview of tableau algorithms for description logics. Studia Logica **69** (2001) 5–40
14. Horrocks, I., Sattler, U.: A description logic with transitive and inverse roles and role hierarchies. Journal of Logic and Computation **9** (1999) 385–410
15. Horrocks, I., Sattler, U., Tobies, S.: Practical reasoning for expressive description logics. In: Proceedings of of LPAR99. (1999)

Effective Pattern Similarity Match for Multidimensional Sequence Data Sets

Seok-Lyong Lee[1,*] and Deok-Hwan Kim[2,**]

[1] School of Industrial and Information Engineering,
Hankuk University of Foreign Studies, 449-701, Korea
sllee@hufs.ac.kr
[2] School of Electronics and Electrical Engineering,
Inha University, 402-751, Korea
deokhwan@inha.ac.kr

Abstract. In this paper we present an effective pattern similarity match algorithm for multidimensional sequence data sets such as video streams and various analog or digital signals. To approximate a sequence of data points we introduce a *trend vector* that captures the moving trend of the sequence. Using the trend vector, our method is designed to filter out irrelevant sequences from a database and to find similar sequences with respect to a query. Experimental results show that it provides a lower reconstruction error and a higher precision rate compared to existing methods.

Keywords: Pattern match, Multidimensional data sequence, Trend vector.

1 Introduction

A multidimensional data sequence (MDS) is a sequence of data elements, each element being represented as a multidimensional vector. In [5], we have defined an MDS S with k points in d-dimensional space as a sequence of its component vectors, $S = \langle S_0, S_1, \ldots, S_{k-1} \rangle$, where each vector $S_i (0 \le i \le k-1)$ is composed of d scalar entries, $S_i = (S_i^1, S_i^2, \ldots, S_i^d)$. Time-series data, a sequence of *one-dimensional* real numbers, is obtained by replacing S_i with a single scalar value. An MDS is partitioned into segments of equal or varying lengths, each of which is represented as a *trend vector*. It holds the information on the moving trend of points within a segment. The similarity between segments is defined using the trend vector, and the similarity between sequences is defined in terms of the similarity between segments of the sequences. The pattern similarity search problem in this paper is defined as follows: Given a query Q, a set of data sequences SET, and a similarity threshold ζ ($0 \le \zeta \le 1$), the goal is to retrieve sequences S's from SET such that the similarity between S and Q is equal to or greater than ζ.

---

[*] This work was supported by Korea Research Foundation Grant (KRF-2004-041-D00665).
[**] Corresponding author.

Y. Shi et al. (Eds.): ICCS 2007, Part I, LNCS 4487, pp. 204–212, 2007.
© Springer-Verlag Berlin Heidelberg 2007

Recently, Keogh et al. [2] and Yi et al. [9] independently proposed a mean-based method approximating a time sequence by dividing it into equal-length segments and recording a mean value of data points that fall within the segment. In this method a sequence is represented as a feature vector that holds s means when the sequence is divided into s segments. By extending it Keogh et al. [3] introduced a new dimensionality reduction technique called an adaptive piecewise constant approximation (APCA) that approximates each time sequence by multiple segments of varying lengths such that their individual reconstruction errors are minimal. They showed that APCA could be indexed using a multidimensional index structure, and proposed two distance measures, D_{LB} and D_{AE}, to provide a lower bounding Euclidean distance approximation and a non-lower bounding tight approximation, respectively.

However, these methods basically address the similarity search for one-dimensional time-series data and thus do not handle multidimensional data sequences effectively. Furthermore, they represent each segment as a mean, causing the trend information of the segment to be lost. In mean-based approaches [2,3,9], the trend information may be maintained with a small segment size, but it increases the number of mean values, leading to a processing overhead.

2 Representation of a Sequence

The moving trend of a series of data points is usually represented as the slope of a straight line that approximates the points. Consider a segment *SEG* with a sequence of k points, $\langle P_0, P_1, ..., P_{k-1} \rangle$, as shown in Fig. 1. To represent the trend of *SEG* we usually use a *start-end vector SE*. It is however not desirable to choose *SE* since it causes a large reconstruction error with respect to an original sequence. We may compute an *optimal vector OP* that is parallel to *SE* and minimize a reconstruction error. But, to avoid a possible overhead to find *OP*, we use a vector that is parallel to *SE* and passes through the middle of a mean, called *trend vector TV*. Fig. 1 depicts *TV* for a one-dimensional time sequence. For convenience, we added a horizontal time-axis and thus the mean is drawn as a line. In multidimensional data space, the mean is represented as a point without the time axis. All data points in mean-based approach is approximated by a single value, *mean*, while in *TV*-representation points, $P_0, ..., P_{k-1}$ is approximated by points, $P_0', ..., P_{k-1}'$, on *TV* respectively.

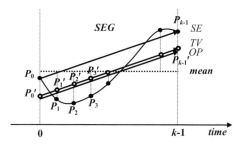

Fig. 1. Approximation of a sequence by mean, start-end vector (SE), optimal vector (OP), and trend vector (TV)

Definition 1: A trend vector **TV** in d-dimensional space is defined as $TV=\langle k, \boldsymbol{m}, \boldsymbol{\alpha}\rangle$, where k is the number of points that fall within a segment, $\boldsymbol{m}=(m_1, m_2, \ldots, m_d)$ is a mean point, each element of which is a mean of k points in each dimension, and $\boldsymbol{\alpha}=(\alpha_1, \alpha_2, \ldots, \alpha_d)$ is a slope of a start-end vector. ∎

By replacing $\boldsymbol{\alpha}$=NULL, this representation is reduced to APCA representation [3], while it is reduced to the segmented mean representation [2,9] by replacing k = NULL and $\boldsymbol{\alpha}$ = NULL. A trend-vector **TV** has various desirable properties: (1) It approximates well the moving trend of points in a segment, (2) It achieves a very low reconstruction error with respect to an original sequence, (3) Computing **TV** is straightforward and thus very fast, and (4) Using it for the similarity search prunes irrelevant sequences effectively. Fig. 2 shows sequences with a mean value approximation (dotted lines) and a trend-vector approximation (directional solid lines) for a one-dimensional sequence. We can observe that a trend-vector provides better approximation than the mean.

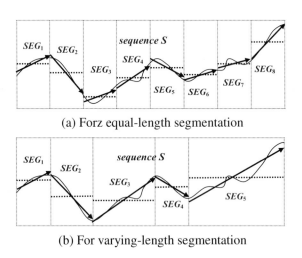

(a) Forz equal-length segmentation

(b) For varying-length segmentation

Fig. 2. Mean and trend-vector approximation of a sequence

3 Pattern Match Using a Trend Vector

3.1 Similarity Between Segments

We introduce two distance measures between segments. One is d_hr considering hyper-rectangles derived from **TV** to prune irrelevant segments from a database and the other is d_seg considering points on **TV** to refine candidate segments selected by d_hr. When segments are indexed and stored into a database for the post retrieval process, an MBR(minimum bounding rectangle) that bounds points in a segment is usually used for indexing since it provides the property of '*no-false dismissal*.' It however suffers from many '*false hits*,' and for our method, needs an extra storage to store low and high points of the rectangle. Thus we utilize the hyper-rectangle that bounds *TV* for indexing, called *TV-rectangle*. Even though it does not preserve the property of '*no-false dismissal*,' it reduces the search space, achieving a fast retrieval compared to MBR indexing. There exists a trade-off between the efficiency and the correctness. In reality many applications may not strongly insist on the correctness. The distance *d_hr* and *d_seg* are defined as follows:

Definition 2: Distance $d\_hr(SEG_1, SEG_2)$ between two segments, SEG_1 and SEG_2, is defined by the minimum Euclidean distance between two TV-rectangles that bounds TV_1 and TV_2 of SEG_1 and SEG_2, respectively. That is

$$d\_hr(SEG_1, SEG_2) = \sqrt{\sum_{j=1}^{d} u_j^2}, \tag{1}$$

$$\text{where } u_j = \begin{cases} \left| SEG_1.H_j - SEG_2.L_j \right| & \text{if } SEG_1.H_j < SEG_2.L_j \\ \left| SEG_1.L_j - SEG_2.H_j \right| & \text{if } SEG_2.H_j < SEG_1.L_j \\ 0 & \text{otherwise.} \end{cases}$$

where *SEG.L* and *SEG.H* are high and low points of a TV-rectangle of *SEG* respectively.   ∎

Observation 3: *d_hr* is shorter than the distance between any pair of points, one on TV_1 and the other on TV_2. Thus, when TR_1 and TR_2 is sets of points on TV_1 and TV_2 respectively, the following holds:

$$d\_hr(SEG_1, SEG_2) \leq \min_{P_1' \in TR_1, P_2' \in TR_2} d(P_1', P_2') \tag{2} \blacksquare$$

Definition 4: $d\_seg(SEG_1, SEG_2)$ between SEG_1 and SEG_2 of equal lengths, is defined as:

$$d\_seg(SEG_1, SEG_2) = \frac{1}{k} \cdot \sum_{i=0}^{k-1} d(P_{1,i}', P_{2,i}') \tag{3}$$

where $P_{1,i}'$ and $P_{2,i}'$ are *i*th points on TV_1 and TV_2, respectively.   ∎

Computing *d_seg* between different-length segments is straightforward. Since two segments cannot be compared using Equation 3, a shorter segment is compared to a longer one by sliding the former from the start to the end of the latter, in order to make a pair of equal-length comparison units. The shortest of produced distances is selected as the distance between the two segments.

Lemma 5 (*Lower Bounding Distance*: $d\_hr(SEG_q, SEG_s) \leq d\_seg(SEG_q, SEG_s)$):
$d\_hr(SEG_q, SEG_s)$ between two TV-rectangles of a query segment SEG_q and a database segment SEG_s is the lower bound of distance $d\_seg(SEG_q, SEG_s)$ of the two segments.

Proof: Let SEG_q, SEG_s have k data points, and P_q', P_s' be arbitrary points on trend vectors of SEG_q, SEG_s, respectively. Then:

$$d\_seg(SEG_q, SEG_s) = \frac{1}{k} \cdot \sum_{i=0}^{k-1} d(P_{q,i}', P_{s,i}') \geq \min_{P_q' \in TR_q, P_s' \in TR_s} d(P_q', P_s')$$

By Observation 3, the following holds:

$$d\_hr(SEG_q, SEG_s) \leq \min_{P_q' \in TR_q, P_s' \in TR_s} d(P_q', P_s')$$

Therefore, we conclude: $d\_hr(SEG_q, SEG_s) \leq d\_seg(SEG_q, SEG_s)$.   ∎

It is also straightforward to extend Lemma 5 for different-length segments. It implies we may use d_hr for index search *safely*, i.e., *without false-dismissal* with respect to d_seg, to prune irrelevant segments from a database. A set of segments obtained by d_hr is a superset of that obtained by d_seg. Since a data space is normalized in a $[0,1]^d$ hyper-cube, the maximum allowable distance is $\sqrt{d}$, a diagonal of the cube. Then the similarity $sim\_seg(SEG_1, SEG_2)$ between two segments SEG_1 and SEG_2 is obtained as follows:

$$sim\_seg(SEG_1, SEG_2) = 1 - \frac{1}{\sqrt{d}} \cdot d\_seg(SEG_1, SEG_2) \tag{4}$$

Accordingly a user-provided similarity threshold ζ can be transformed into a distance threshold ε for index search using a spatial access method and *vice versa*. That is, $\zeta = 1 - \varepsilon/\sqrt{d}$ and $\varepsilon = \sqrt{d}(1-\zeta)$.

3.2   Similarity Between Sequences and Pattern Similarity Match Algorithm

A sequence is partitioned into equal- or varying-length segments[1]. Consider two equal-length sequences S_1 and S_2, each of which is partitioned into varying-length segments. To compute $sim\_seq(S_1, S_2)$ between S_1 and S_2, we first divide a whole time interval into sub-intervals with respect to segment boundaries of both sequences. And next, we compute sim_seg's for the sub-intervals and merge them. A following example shows this.

Example 6: Consider two sequences S_1 and S_2 with 2 and 3 segments respectively as shown in Fig. 3: $S_1 = (SEG_{1,1}, SEG_{1,2})$ and $S_2 = (SEG_{2,1}, SEG_{2,2}, SEG_{2,3})$, where $SEG_{1,1} = \langle k_{1,1}, m_{1,1}, \alpha_{1,1} \rangle$, $SEG_{1,2} = \langle k_{1,2}, m_{1,2}, \alpha_{1,2} \rangle$, $SEG_{2,1} = \langle k_{2,1}, m_{2,1}, \alpha_{2,1} \rangle$, $SEG_{2,2} = \langle k_{2,2}, m_{2,2}, \alpha_{2,2} \rangle$, $SEG_{2,3} = \langle k_{2,3}, m_{2,3}, \alpha_{2,3} \rangle$. Assume that $k_{1,1} = 18$, $k_{1,2} = 14$, $k_{2,1} = 10$, $k_{2,2} = 14$, and $k_{2,3} = 8$. This segmentation yields 4 sub-intervals, I_1, I_2, I_3, and I_4. To compare

---

[1] For the varying-length segmentation, we use the technique that has been proposed in [6], in which the segmentation is based on geometric and semantic properties of a sequence. There are various optimal piecewise polynomial representations [1,7], but they need a considerable overhead. Meanwhile greedy approaches [4,6,8] are more efficient and thus realistic. Due to space limitation, we do not describe a segmentation algorithm in detail. Interested readers are referred to [6] for details.

segments we need to split $SEG_{1,1}$ to $SEG_{1,1,I_1}$ and $SEG_{1,1,I_2}$ in I_1 and I_2, $SEG_{1,2}$ to $SEG_{1,2,I_3}$ and $SEG_{1,2,I_4}$ in I_3 and I_4, and $SEG_{2,2}$ to $SEG_{2,2,I_2}$ and $SEG_{2,2,I_3}$ in I_2 and I_3, respectively. Then we get $k_{1,1,I_1} = k_{2,1} = 10$, $k_{1,1,I_2} = k_{2,2,I_2} = 8$, $k_{1,2,I_3} = k_{2,2,I_3} = 6$, and $k_{1,2,I_4} = k_{2,3} = 8$. By splitting segments we can get similarity values at each interval and compute $sim\_seq(S_1, S_2)$ as follows:

$sim\_seq(S_1, S_2)=[10 \cdot sim\_seg(SEG_{1,1,I_1}, SEG_{2,1}) + 8 \cdot sim\_seg(SEG_{1,1,I_2}, SEG_{2,2,I_2}) + 6 \cdot sim\_seg(SEG_{1,2,I_3}, SEG_{2,2,I_3}) + 8 \cdot sim\_seg(SEG_{1,2,I_4}, SEG_{2,3})] / 32$ ∎

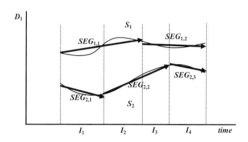

Fig. 3. Similarity between two sequences in each sub-interval

Consider $SEG = \langle k, m, \alpha \rangle$ to be split to r sub-segments, $SEG_1 = \langle k_1, m_1, \alpha_1 \rangle$, ..., $SEG_r = \langle k_r, m_r, \alpha_r \rangle$. We can approximate $SEG_l(1 \le l \le r)$ using a part of TV that falls in the sub-interval of SEG_l. Thus we obtain $k = \Sigma_{1 \le l \le r} k_l$, $\alpha = \alpha_1 = ... = \alpha_r$, and $m_l = (m_{l,1}, m_{l,2}, ..., m_{l,d})$ in d-dimensional space by a simple computation. When two sequences S_1, S_2 have k, l segments respectively, the number of sub-intervals T is: $T \le k+l-1$ where the inequality occurs when some segment boundaries of two sequences are identical. Thus the similarity $sim\_seq(S_1, S_2)$ is as follows:

$$sim\_seq(S_1, S_2) = \frac{1}{\sum_{t=1}^{T} k_{I_t}} \cdot \sum_{t=1}^{T} (k_{I_t} \cdot sim\_seg(SEG_{1,I_t}, SEG_{2,I_t}))$$ (5)

where $k_{I_t} = k_{1,I_t} = k_{2,I_t}$, and SEG_{1,I_t} and SEG_{2,I_t} are the segments of S_1 and S_2 in I_t, respectively. We can use a measure sim_seq safely to prune irrelevant sequences from a database with respect to a query, in the support of the following lemma:

Lemma 7: *If $sim\_seq(Q, S) \ge \zeta$, then there exists at least a pair of segments (SEG_q, SEG_s) such that $sim\_seg(SEG_q, SEG_s) \ge \zeta$, for $SEG_q \in Q$ and $SEG_s \in S$.*

Proof: *(By contradiction)* Let sequences Q, S have k, l segments respectively, i.e., $Q = (SEG_1, ..., SEG_k)$, $S = (SEG_1, ..., SEG_l)$. Without loss of generality, we assume $k \le l$, since we can switch Q and S if $k > l$. The number of intervals T generated from sequences Q and S is $T \le k+l-1$. Let us assume that when $sim\_seq(Q, S) \ge \zeta$ there does not exist any pair of segments (SEG_q, SEG_s) such that $sim\_seg(SEG_q, SEG_s) \ge \zeta$ for $1 \le q \le k$, $1 \le s \le l$. Then,

$$sim\_seq(Q,S) = \frac{1}{\sum_{t=1}^{T} k_{I_t}} \cdot \sum_{t=1}^{T} (k_{I_t} \cdot sim\_seg(SEG_{q,I_t}, SEG_{s,I_t}))$$

$$< \frac{1}{\sum_{t=1}^{T} k_{I_t}} \cdot (k_{I_1} \cdot \zeta + k_{I_2} \cdot \zeta + ... + k_{I_t} \cdot \zeta) = \zeta$$

The assumption causes $sim\_seq(Q, S) < \zeta$ that is a contradiction. Thus, Lemma 7 holds. ∎

Lemma 7 indicates that *sim_seg* is safe with respect to *sim_seq*, i.e., a set of candidate sequences obtained by *sim_seg* is a superset of that obtained by *sim_seq*. In this paper we present a similarity measure for equal-length sequences. The similarity for different-length sequences can be evaluated such that a shorter sequence is compared to a longer one by sliding the shorter one from the start to the end of the longer one.

Algorithm Pattern_Similarity_Match
$SET_{cand} \leftarrow \varnothing$ /* a set of candidate sequences   */
SET_{ans} $\leftarrow \varnothing$ /* a set of answer sequences   */
          /* i is for counting segments of a query sequence Q.   */
          /* j is for counting segments of a data sequence S.   */
          /* k is for counting data sequences in a database.   */
Step 0: /* Pre processing   */
          Partition data sequences S into one or more segments
          Extract a trend vector from each segment
          Index trend vectors and store them into a database
Step 1: /* Segmentation of a query sequence   */
          Partition a query sequence Q into one or more $SEG_{Q,i}$
          Extract a trend vector from each query segment $SEG_{Q,i}$
Step 2: /* 1st pruning by d_hr (index search) and *sim_seg*   */
          For each $SEG_{Q,i}$ of a query sequence Q
              Search an index based on the distance d_hr to find candidate segments $SEG_{k,j}$
              For each candidate segment $SEG_{k,j}$
                  if ($sim\_seg(SEG_{Q,i}, SEG_{k,j}) \geq \zeta$) then
                      $SET_{cand} \leftarrow SET_{cand} \cup \{S_k\}$
Step 3: /* 2nd pruning by the similarity measure *sim_seq* */
          For each selected sequence S_k in the set SET_{cand}
              if ($sim\_seq(Q, S_k) \geq \zeta$) then
                  $SET_{ans} \leftarrow SET_{ans} \cup \{S_k\}$
Step 4: Return SET_{ans}

Fig. 4. Pattern similarity match algorithm

Fig. 4 shows a pattern similarity match algorithm. First, a query sequence is partitioned into segments from which trend vectors are extracted. Next, for each segment of the query sequence, an index is searched to find candidate segments by d_hr, with respect to a threshold ε that is derived from a user-provided similarity threshold ζ. Those candidate segments are evaluated again by using *sim_seg*, to prune irrelevant segments further. Then, the sequences in which final candidate segments are contained will be candidate sequences. These sequences are evaluated with respect to the similarity measure between sequences, *sim_seq*, to determine a final set of answer sequences.

4 Experimental Evaluation

For experiments we generate 2500 MDS's from video streams by extracting colour features from pixels of each frame and averaging them. Each frame is mapped to a 3-dimensional point in the $[0, 1]^3$ unit cube and each sequence contains 64 to 1024 points. For each test, we have issued randomly selected queries from data sequences and taken the average of query results. To evaluate our method we compared the following: SS(sequential scanning method), MB-Diff(mean-based method for varying-length segments [3]), and TV-Diff(our method for varying-length segments).

To evaluate the effectiveness, we used the precision and recall that are well known in similarity search applications. We set the ground truth as a set of retrieved sequences by SS. That is, all sequences retrieved by it are regarded as *relevant* sequences. The experiment has been done with similarity threshold values, ranging from 0.60 to 0.95. This range is considered to be appropriate since two sequences with values below 0.60 are turned out to be 'different' in their pattern shapes in our experiment. From the experiment we observed that our method provides fairly better precisions (up to 1.9 times) than mean-based methods while they show almost the same recall rate (0.98-1.00), which implies that our method prunes more irrelevant sequences than other methods.

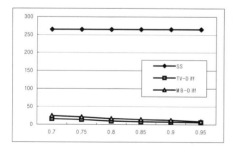 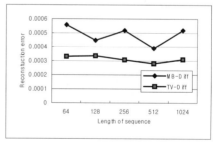

Fig. 5. Efficiency for SS, MB-Diff, and TV-Diff

Fig. 6. Reconstruction error for MB-Diff and TV-Diff

Fig. 5 and Fig. 6 show the response time in second for various threshold values and the reconstruction error. As we can see TV-Diff performs better than MB-Diff in all threshold ranges (1.34-1.69 times faster). The high precision causes more irrelevant sequences to be pruned, resulting in the less processing time. Meanwhile, the reconstruction error of TV-Diff is lower (59-74%) than that of MB-Diff, which indicates that a trend vector provides better approximation.

5 Conclusions

We addressed the problem of searching similar patterns for multidimensional sequences such as video streams. To solve the problem we represented each segment of a sequence as a trend vector that encapsulates the moving trend of points in the segment. Based on it we have defined similarity measures between segments and between sequences. Using the measures, our method prunes irrelevant sequences from a database with respect to a given query. The trend vector has the competitive strength since it provides better approximation and faster processing time than mean-based methods as shown in the experiment. Potential applications that are emphasized in this paper are the similarity search on video streams, but we believe other application areas can also benefit. As the future work, we plan to study on applying the proposed method to specialized application domains considering their own characteristics, such as voice signal matching and region-based image search.

References

1. C. Faloutsos, H.V. Jagadish, A. Mendelzon, and T. Milo. A signature technique for similarity-based queries. *SEQUENCES*, Italy, (1997).
2. E. Keogh, K. Chakrabarti, S. Mehrotra, and M. J. Pazzani. Dimensionality reduction for fast similarity search in large time series databases. *Journal of Knowledge and Information Systems*, (2000).
3. E. Keogh, K. Chakrabarti, S. Mehrotra, and M. J. Pazzani. Locally adaptive dimensionality reduction for indexing large time series databases. *Proc. of ACM SIGMOD*, pages 151-162, (2001).
4. E. Keogh and P. Smyth. A probabilistic approach to fast pattern matching in time series databases. *Proc. of Int'l Conference of Knowledge Discovery and Data Mining*, pages 20-24, (1997).
5. S. L. Lee, S. J. Chun, D. H. Kim, J. H. Lee, and C. W. Chung. Similarity search for multidimensional data sequences. *Proc. of IEEE ICDE*, pages 599-608, (2000).
6. S. L. Lee and C. W. Chung. Hyper-rectangle based segmentation and clustering of large video data sets. *Information Science*, Vol.141, No.1-2, pages 139-168, (2002).
7. T. Pavlidis. Waveform segmentation through functional approximation. *IEEE Transactions on Computers*, Vol. C-22, No. 7, (1976).
8. C. Wang and S. Wang. Supporting content-based searches on time series via approximation. *Int'l Conference on Scientific and Statistical Database Management*, (2000).
9. B. K. Yi and C. Faloutsos. Fast time sequence indexing for arbitrary Lp norms. *Proc. of Int'l Conference on VLDB*, pages 385-394, (2000).

GPU-Accelerated Montgomery Exponentiation

Sebastian Fleissner

Department of Computer Science and Engineering
The Chinese University of Hong Kong
seb@cse.cuhk.edu.hk

Abstract. The computing power and programmability of graphics processing units (GPUs) has been successfully exploited for calculations unrelated to graphics, such as data processing, numerical algorithms, and secret key cryptography. In this paper, a new variant of the Montgomery exponentiation algorithm that exploits the processing power and parallelism of GPUs is designed and implemented. Furthermore, performance tests are conducted and the suitability of the proposed algorithm for accelerating public key encryption is discussed.

Keywords: GPGPU, Montgomery Exponentiation, Encryption.

1 Introduction

Chips on consumer graphics cards have evolved into programmable graphics processing units (GPUs) capable of delivering real-time, photo-realistic effects in computer games and multimedia applications. According to an article by Manocha [1], the computational power of GPUs has a growth rate faster than Moore's law as it applies to other microprocessors. As a result, researchers attempt to exploit the processing power and programmability of GPUs for general purpose computing.

The purpose of this research is to investigate and realize the usage of GPUs for accelerating the Montgomery exponentiation algorithm, which is used by various public key encryption schemes, such as RSA and elliptic curves, for performing modular exponentiation of large integers. In particular, a new, GPU-accelerated Montgomery exponentiation algorithm is proposed and the performance of its implementation is evaluated. Furthermore, the suitability of the proposed GPU algorithm for accelerating public key encryption is discussed.

1.1 Paper Organization

The rest of the paper is organized as follows. Section 2 provides background information. Sections 3 and 4 specify and evaluate the proposed GPU-accelerated Montgomery exponentiation algorithm. Section 5 briefly discusses public key encryption schemes that are likely to benefit from the proposed algorithm.

Y. Shi et al. (Eds.): ICCS 2007, Part I, LNCS 4487, pp. 213–220, 2007.
© Springer-Verlag Berlin Heidelberg 2007

2   Background

2.1   General Purpose GPU Computations

General purpose GPU computing (GPGPU) [2] refers to the concept of exploiting the processing power of GPUs for performing general purpose calculations. Examples for GPGPU computing are data processing [3], evolutionary algorithms [4], and secret key cryptography [5].

A typical GPGPU application consists of two parts. One part, which is denoted as the *main program*, runs on the computer's main processor and is responsible for initializing the GPU, providing the GPU with input, and retrieving the results produced by the GPU. The other part, which is denoted as the *fragment program*, runs on the GPU and performs the desired general purpose calculation. The main program provides the operands for multiple calculations that are performed on the GPU in parallel. Since GPUs have only limited support for integer values, all data sent to and retrieved from GPUs is encoded using 32-bit floating point values.

The main program and fragment program exchange data via textures. Hence, the first step performed by the main program is to create several 2D textures inside the graphics memory, which serve as input and output buffers. After their creation, the textures serving as input buffers are filled with the operands for the general purpose calculation. After the textures are prepared, the main program invokes the fragment program on the GPU by drawing a rectangle with the same dimension as the textures. The GPU runs multiple instances of the fragment program in parallel, and each instance retrieves and processes the operands for a different calculation from the input textures. The results calculated by the fragment program instances are output as one or more 32 bit floating point values. These values are passed to the GPU processing pipeline and automatically stored in one or more designated output textures. The program on the main processor can then access the output textures to obtain the results.

2.2   The Montgomery Method

The Montgomery method [6] and its various improvements [7,8,9] are algorithms for efficient computation of modular multiplications $x = a \times b \bmod n$ and modular exponentiations $x = a^b \bmod n$, with a, b, n being k-bit large integers and n being odd.

As described in [10], the Montgomery algorithm consists of following steps:

1. **Perform pre-computation**
   Choose a large integer r as a power of 2, with $r > n$. Then n' is calculated, so that $r \times r^{-1} - n \times n' = 1$. Both r^{-1} and n' can be calculated using the extended Euclidian algorithm.
2. **Obtain Montgomery representation of a and b**
   This step uses the r generated during the pre-computation step to transform the operands a and b into Montgomery representation. Their Montgomery representations are obtained by calculating $\bar{a} := a \times r \bmod n$ and $\bar{b} := b \times r \bmod n$.

3. **Calculate Montgomery Product $\bar{x} = \bar{a} \times \bar{b}$**

   This step calculates the *Montgomery product* $\text{MonPro}(\bar{a}, \bar{b}) = \bar{x} := \bar{a} \times \bar{b} \times r^{-1}$ mod n. The Montgomery product is calculated as follows:

   (a) $t := \bar{a} \times \bar{b}$
   (b) $m := t \times n' \bmod r$
   (c) $x := (t + m \times n)/r$
   (d) if $x \geq n$ then set $x = x - n$

Montgomery Exponentiation. Because of the overhead caused by the pre-computation and representation transformation steps, the Montgomery method is used for modular exponentiation rather than a single modular multiplication. A common form of the Montgomery exponentiation algorithm, which is described in [10], uses the so-called binary *square-and-multiply* method to calculate $x = a^b$ mod n, where a, b, n are k-bit large integers and n is odd. With $|b|$ being the bit length of operand b, the algorithm consists of the following steps:

1. Use n to pre-compute n' and r.
2. Calculate $\bar{a} := a \times r \bmod n$.
3. Calculate $\bar{x} := 1 \times r \bmod n$.
4. For i := $|b|$ - 1 down to 0 do
   (a) Calculate $\bar{x} := \text{MonPro}(\bar{x}, \bar{x})$
   (b) If the i-th bit of b is set, then calculate $\bar{x} := \text{MonPro}(\bar{a}, \bar{x})$
5. Calculate $x = \text{MonPro}(\bar{x}, 1)$.

3   Proposed Algorithm

3.1   Overview

This section proposes a new, GPU-accelerated exponentiation algorithm based on the Montgomery method introduced in section 2.2. This algorithm, denoted as *GPU-MonExp*, exploits the parallelism of GPUs by calculating multiple modular exponentiations simultaneously. The exponentiation operands have a fixed bit size, which depends on the output capabilities of the GPU hardware.

Like the Montgomery exponentiation algorithm introduced in section 2.2, the GPU-MonExp algorithm depends on the Montgomery product. As a result, this section first describes a GPU-accelerated variant of the Montgomery product denoted as *GPU-MonPro*, which forms the basis of the GPU-MonExp algorithm.

As GPUs have limited support for integer values, the proposed GPU algorithms split large integers into k 24-bit chunks, and store each chunk in a 32-bit floating point value. The first chunk contains the least significant bits and the last chunk contains the most significant bits of the large integer. Hence, the representation of a large integer x is:

$$x = x[0], x[1], ..., x[k] \tag{1}$$

3.2   GPU Montgomery Product(GPU-MonPro)

The GPU-MonPro algorithm utilizes the GPU to calculate c Montgomery products in parallel:

$$\bar{x}_i := \bar{a}_i \times \bar{b}_i \times r^{-1} \bmod n_i \ , \ 1 \leq i \leq c \tag{2}$$

The GPU-MonPro algorithm uses large integers with a pre-defined, fixed bit size. Because of the fixed bit size, the maximum size of the n_i operands is known, and by considering that $r > n$ and $r = 2^z$ for some z, the value of the operand r can be pre-defined as well in order to simplify calculations. As graphics processing units do not provide any efficient instructions for performing bitwise operations, the GPU-MonPro algorithm uses an operand r that is a power of 256. Thus, multiplication and division operations by r can be implemented via byte shifting, which can be performed efficiently by GPUs. Since the large integers used by the algorithms in this chapter consist of k 24-bit chunks, the value of r is pre-defined as $r = 256^{3k}$.

As r is pre-defined, the input values for the GPU-MonPro algorithm are $\bar{a}_i$, $\bar{b}_i$, n_i, n'_i. These input values are supplied by the GPU-MonExp algorithm described in section 3.3.

The GPU-MonPro algorithm consists of two steps: Texture preparation and calculation of the Montgomery product.

Step 1: Texture preparation [Main Processor]. The GPU-MonPro algorithm uses multiple two-dimensional input and output textures in RGBA color format. An RGBA texel (texture element) consists of four 32-bit floating point values and can thus be used to encode four 24-bit chunks (96 bit) of a large integer. The algorithm uses four types of input textures corresponding to the four types of operands: tex-$\bar{a}$, tex-$\bar{b}$, tex-n, and tex-n'. Assuming input textures with a dimension of $w \times h$ to calculate $c = w \times h$ Montgomery products, the GPU-MonPro algorithm uses the following approach to store the operands $\bar{a}_i$, $\bar{b}_i$, n_i, n'_i in the input textures:

1. For each $0 \leq x < w$, $0 \leq y < h$ do:
2. $i := y \times w + x$
3. Store $\bar{a}_i$ in the tex-$\bar{a}$ texture(s)
   (a) tex-$\bar{a}_{[0]}$(x,y) $= \bar{a}_i[0,1,2,3]$ [1]
   (b) tex-$\bar{a}_{[1]}$(x,y) $= \bar{a}_i[4,5,6,7]$
   (c) ...
   (d) tex-$\bar{a}_{[k/4]}$(x,y) $= \bar{a}_i[k-4,k-3,k-2,k-1]$
4. Store $\bar{b}_i$ in the tex-$\bar{b}$ texture(s)
   (a) tex-$\bar{b}_{[0]}$(x,y) $= \bar{b}_i[0,1,2,3]$
   (b) ...
   (c) tex-$\bar{b}_{[k/4]}$(x,y) $= \bar{b}_i[k-4,k-3,k-2,k-1]$
5. Store n_i in the tex-n texture(s)
   (a) tex-$n_{[0]}$(x,y) $= n_i[0,1,2,3]$
   (b) ...
   (c) tex-$n_{[k/4]}$(x,y) $= n_i[k-4,k-3,k-2,k-1]$

---

[1] The term $\bar{a}_i[0,1,2,3]$ is an abbreviation for the four values $\bar{a}_i[0]$, $\bar{a}_i[1]$, $\bar{a}_i[2]$, $\bar{a}_i[3]$.

6. Store n'_i in the tex-n' texture(s)
   (a) tex-$n'_{[0]}$(x,y) = $n'_i[0, 1, 2, 3]$
   (b) ...
   (c) tex-$n'_{[k/4]}$(x,y) = $n'_i[k - 4, k - 3, k - 2, k - 1]$

Considering a large integer as k 24-bit chunks, the total number of textures required for storing the operands $\bar{a}_i$, $\bar{b}_i$, n_i, n'_i is $\frac{k}{4}4 = k$. (There are four operands and each texel can store four 24-bit chunks). The number of required output textures is $\frac{k}{4}$.

After the input and output textures are prepared, drawing commands are issued in order to invoke the fragment program instances on the GPU.

Step 2: Calculation of Montgomery product [GPU]. Each instance of the fragment program on the GPU calculates one modular product. Since the GPU hardware runs multiple instances of the fragment program in parallel, several modular products are calculated at the same time. Apart from the input textures containing the $\bar{a}_i$, $\bar{b}_i$, n_i, and n'_i values, each fragment program instance receives a (X, Y) coordinate pair that indicates which operands $\bar{a}$, $\bar{b}$, n, and n' should be retrieved from the input textures to calculate the Montgomery product.

The algorithm performed by the fragment program instances, which is essentially a standard Montgomery multiplication, is as follows:

1. Use the X and Y coordinates to obtain the four operands $\bar{a}_j$, $\bar{b}_j$, n_j, and n'_j for some specific j, with $1 \leq j \leq c$.
2. Calculate $t := \bar{a}_j \times \bar{b}_j$. Because the maximum bit size of the operands is pre-defined, multiplication can be implemented efficiently on GPUs using vector and matrix operations, which are capable of calculating multiple partial products in parallel.
3. Calculate $m := t \times n'_j \mod r$. Since r is a multiple of 256, the reduction by modulo r is achieved by byte shifting.
4. Calculate $\bar{x} := (t + m \times n_j)/r$. By using the vector and matrix operations of the GPU, addition can be implemented efficiently, since partial sums can be calculated in parallel. The division by r is achieved by byte shifting.
5. Output $\bar{x}$, which is automatically diverted and stored in the output textures.

3.3   GPU Montgomery Exponentiation (GPU-MonExp)

The GPU-MonExp algorithm calculates c modular exponentiations in parallel:

$$x_i = a_i^b \mod n_i , \ 1 \leq i \leq c \tag{3}$$

Each of the c exponentiations uses the same b, but different values for each a_i and n_i. With $|b|$ denoting the fixed bit size of operand b, the GPU-MonExp algorithm executes the following steps:

1. Execute the following loop on the main processor:
   For i := 1 to c do
   (a) Use n_i to pre-compute n_i'.
   (b) Calculate $\bar{x}_i := 1 \times r \bmod n_i$.
   (c) Calculate $\bar{a}_i := a_i \times r \bmod n_i$.
2. Execute the following loop on the main processor:
   For $l := |b| - 1$ down to 0 do
   (a) Invoke GPU-MonPro to calculate
       $\bar{x}_i := \bar{x}_i \times \bar{x}_i \times r^{-1} \bmod n_i$ on the GPU in parallel. $(1 \leq i \leq c)$
   (b) If the l-th bit of b is set, then invoke GPU-MonPro to calculate $\bar{x}_i :=$
       $\bar{a}_i \times \bar{x}_i \times r^{-1} \bmod n_i$ on the GPU in parallel. $(1 \leq i \leq c)$
3. Invoke GPU-MonPro to calculate the final results
   $x_i = \bar{x}_i \times 1 \times r^{-1} \bmod n_i$ on the GPU in parallel. $(1 \leq i \leq c)$

Step 1 Details. The pre-computation loop, which is executed on the main processor, calculates a corresponding n_i' for each n_i using the extended Euclidian algorithm. If all modular products to be calculated use the same n, then only one n' is computed, since n' only depends on n and not on the multiplicand and multiplier. Apart from the n_i' values, the loop determines the initial values for all $\bar{x}_i$ and the Montgomery representations of all a_i.

Step 2 Details. The main loop of the GPU-MonExp algorithm is run on the main processor and uses the square-and-multiply approach to calculate the modular exponentiations. The main loop first invokes GPU-MonPro with the $\bar{x}_i$ values as texture parameters in order to calculate the squares $\bar{x}_i \times \bar{x}_i \times r \bmod n_i$ on the GPU in parallel. Depending on whether the current bit of operand b is set, the algorithm invokes GPU-MonPro again with $\bar{x}_i$ and $\bar{a}_i$ as texture parameters to calculate the Montgomery products $\bar{x}_i \times \bar{a}_i \times r \bmod n_i$ on the GPU in parallel. After the main loop completes, the final results $x_i = a_i^b \bmod n_i$ are transferred back to the main processor.

4   Algorithm Evaluation

4.1   Performance Test Overview

This section introduces and analyzes the results of performance tests, which were conducted in order to evaluate the potential of the proposed GPU-MonExp algorithm. In order to obtain representative test data, three different hardware configurations were used to run the performance tests. The details of these three test systems, which are denoted as system A, B, and C, are shown in table 1.

The GPU-MonExp performance test works as follows. As a first step, the implementation of the GPU-MonExp algorithm is run with random 192-bit operands for 1 to 100000 exponentiations. The execution time T_{GPU} is measured and recorded. Then an implementation of the square-and-multiply Montgomery exponentiation algorithm described in section 2.2 is run with the same input

Table 1. Test Systems

System	Processor/Memory/Graphics Bus	GPU
A	Intel Pentium 4, 2.66 GHz 1 GB RAM, PCI-Express	NVIDIA GeForce 6500 256 MB RAM
B	Intel Celeron, 2.40 GHz 256 MB RAM, AGP	NVIDIA GeForce FX 5900 Ultra 256 MB RAM
C	Intel Pentium 4, 3.20GHz 1 GB RAM, PCI-Express	NVIDIA GeForce 7800 GTX 256 MB RAM

on the computer's main processor, and its execution time, denoted as T_{SQM}, is recorded as well. Using the execution times of both implementations, the following *speedup factor* of the GPU-MonExp algorithm is determined:

$$s = \frac{T_{SQM}}{T_{GPU}} \tag{4}$$

If the implementation of the GPU-MonExp algorithm runs faster than the square-and-multiply Montgomery exponentiation, then its execution time is shorter and $s > 1$.

The implementations of the GPU-MonExp and the underlying GPU-MonPro algorithm are based on OpenGL, C++, and GLSL (OpenGL Shading Language). The second step of GPU-MonPro described in section 3.2 is implemented as a GLSL fragment program. The parts of the GPU-MonExp algorithm running on the main processor are implemented in C++.

4.2 Test Results

Overall, the test results indicate that the GPU-MonExp algorithm is significantly faster than the square-and-multiply Montgomery exponentiation, if multiple results are calculated simultaneously. When a large amount of modular exponentiations is calculated simultaneously, the GPU-MonExp implementation is 136 - 168 times faster.

Table 2. GPU-MonExp Speedup Factors

Exponentiations	System A	System B	System C
1	0.2264	0.0783	0.1060
3	1.2462	0.4843	0.8271
10	4.0340	1.2878	2.9767
50	19.5056	6.4544	13.2813
100	33.2646	13.0150	25.6985
10000	138.0390	110.9095	138.8840
100000	168.9705	136.4484	167.1229

As shown in table 2, GPU-MonExp already achieves a performance gain when 3 to 10 exponentiations are calculated simultaneously. When calculating 100000 exponentiations simultaneously, the speedup factors are 168.9705 for system A, 136.4484 for system B, and 167.1229 for system C.

5   Conclusions

This paper introduces the concept of using graphics processing units for accelerating Montgomery exponentiation. In particular, a new GPU-accelerated Montgomery exponentiation algorithm, denoted as GPU-MonExp, is proposed, and performance tests show that its implementation runs 136 - 168 times faster than the standard Montgomery exponentiation algorithm.

Public key encryption algorithms that are based on elliptic curves defined over prime fields (prime curves) are likely to benefit from the proposed GPU-MonExp algorithm, which can serve as the basis for GPU-accelerated versions of the point doubling, point addition, and double-and-add algorithms. Signcryption schemes based on elliptic curves, such as [11,12], are a possible concrete application for the GPU-MonExp algorithm.

References

1. Manocha, D.: General-purpose computations using graphics processors. Computer **38**(8) (August 2005) 85–88
2. Pharr, M., Fernando, R.: GPU Gems 2 : Programming Techniques for High-Performance Graphics and General-Purpose Computation. Addison-Wesley (2005)
3. Govindaraju, N.K., Raghuvanshi, N., Manocha, D.: Fast and approximate stream mining of quantiles and frequencies using graphics processors. In: SIGMOD '05, New York, NY, USA, ACM Press (2005) 611–622
4. M. L. Wong, T.T.W., Foka, K.L.: Parallel evolutionary algorithms on graphics processing unit. In: IEEE Congress on Evolutionary Computation 2005. (2005) 2286–2293
5. Cook, D., Ioannidis, J., Keromytis, A., Luck, J.: Cryptographics: Secret key cryptography using graphics cards (2005)
6. Montgomery, P.L.: Modular multiplication without trial division. Mathematics of Computation **44**(170) (April 1985) 519–521
7. Gueron, S.: Enhanced montgomery multiplication. In: CHES '02, London, UK, Springer-Verlag (2003) 46–56
8. Walter, C.D.: Montgomery's multiplication technique: How to make it smaller and faster. Lecture Notes in Computer Science **1717** (1999) 80–93
9. WU, C.L., LOU, D.C., CHANG, T.J.: An efficient montgomery exponentiation algorithm for cryptographic applications. INFORMATICA **16**(3) (2005) 449–468
10. Koc, C.K.: High-speed RSA implementation. Technical report, RSA Laboratories (1994)
11. Zheng, Y., Imai, H.: Efficient signcryption schemes on elliptic curves. In: Proc. of IFIP SEC'98. (1998)
12. Han, Y., Yang, X.: Ecgsc: Elliptic curve based generalized signcryption scheme. Cryptology ePrint Archive, Report 2006/126 (2006)

Hierarchical-Matrix Preconditioners for Parabolic Optimal Control Problems

Suely Oliveira and Fang Yang

Department of Computer Science, The University of Iowa, Iowa City IA 52242, USA

Abstract. Hierarchical ($\mathcal{H}$)-matrices approximate full or sparse matrices using a hierarchical data sparse format. The corresponding $\mathcal{H}$-matrix arithmetic reduces the time complexity of the approximate $\mathcal{H}$-matrix operators to almost optimal while maintains certain accuracy. In this paper, we represent a scheme to solve the saddle point system arising from the control of parabolic partial differential equations by using $\mathcal{H}$-matrix LU-factors as preconditioners in iterative methods. The experiment shows that the $\mathcal{H}$-matrix preconditioners are effective and speed up the convergence of iterative methods.

Keywords: hierarchical matrices, multilevel methods, parabolic optimal control problems.

1 Introduction

Hierarchical-matrices ($\mathcal{H}$-matrices) [6], since their introduction [1,2,3,6], have been applied to various problems, such as integral equations and partial differential equations. The idea of $\mathcal{H}$-matrices is to partition a matrix into a hierarchy of rectangular subblocks and approximate the subblocks by low rank matrices (Rk-matrices). The $\mathcal{H}$-matrix arithmetic [1,3,4] defines operators over the $\mathcal{H}$-matrix format. The fixed-rank $\mathcal{H}$-matrix arithmetic keeps the rank of a Rk-matrix block below a fixed value, whereas the adaptive-rank $\mathcal{H}$-matrix arithmetic adjusts the rank of a Rk-matrix block to maintain certain accuracy in approximation. The operators defined in the $\mathcal{H}$-matrix arithmetic include $\mathcal{H}$-matrix addition, $\mathcal{H}$-matrix multiplication, $\mathcal{H}$-matrix inversion, $\mathcal{H}$-matrix LU factorization, etc. The computation complexity of these operators are almost optimal $O(n \log^a n)$.

The $\mathcal{H}$-matrix construction for matrices from discretization of partial differential equations depends on the geometric information underlying the problem [3]. The admission conditions, used to determine whether a subblock is approximated by a Rk-matrix, are typically based on Euclidean distances between the supports of the basis functions. For sparse matrices the algebraic approaches [4,11] can be used, which use matrix graphs to convert a sparse matrix to an $\mathcal{H}$-matrix by representing the off-diagonal zero blocks as Rk-matrices of rank 0.

Since the $\mathcal{H}$-matrix arithmetic provides cheap operators, it can be used with H-matrix construction approaches to construct preconditioners for iterative methods, such as Generalized Minimal Residual Method (GMRES), to solve systems of linear equations arising from finite element or meshfree discretizations of partial differential equations [4,8,9,10,11].

Y. Shi et al. (Eds.): ICCS 2007, Part I, LNCS 4487, pp. 221–228, 2007.
© Springer-Verlag Berlin Heidelberg 2007

In this paper we consider the finite time linear-quadratic optimal control problems governed by parabolic partial differential equations. To solve these problems, in [12] the parabolic partial differential equations are discretized by finite element methods in space and by θ-scheme in time; the cost function J to be minimized is discretized using midpoint rule for the state variable and using piecewise constant for the control variable in time; Lagrange multipliers are used to enforce the constraints, which result a system of saddle point type; then iterative methods with block preconditioners are used to solve the system.

We adapt the process in [12] and use $\mathcal{H}$-matrix preconditioners in iterative methods to solve the system. First we apply algebraic $\mathcal{H}$-matrix construction approaches to represent the system in the $\mathcal{H}$-matrix format; then $\mathcal{H}$-LU factorization in the $\mathcal{H}$-matrix arithmetic is adapted to the block structure of the saddle point system to compute the approximate $\mathcal{H}$-LU factors; at last, these factors are used as preconditioner in iterative methods to compute the approximate solutions. The numerical results show that the $\mathcal{H}$-matrix preconditioned approach is competitive and effective to solve the above optimal control problem.

This paper is organized as follows. In Sect. 2 we introduce the optimal control model problem and the discretization process; Section 3 is an introduction to $\mathcal{H}$-matrices; in Sect. 4, we review the algebraic approaches to $\mathcal{H}$-matrix construction; in Sect. 5 we present the scheme to build the $\mathcal{H}$-matrix preconditioners; finally in Sect. 6 we present the numerical results.

2   The Optimal Control Problem

The model problem [12] is to minimize the following quadratic cost function:

$$
\begin{aligned}
J(z(u), u) := &\frac{q}{2} \|z(v) - z_*\|^2_{L^2(t_0, t_f; L^2(\Omega))} + \frac{r}{2} \|v\|^2_{L^2(t_0, t_f; \Omega)} \\
&+ \frac{s}{2} \|z(v)(t_f, x) - z_*(t_f, x)\|^2_{L^2(\Omega)}
\end{aligned}
\tag{1}
$$

under the constraint of the state equation:

$$
\begin{cases}
\partial_t z + \mathcal{A} z = \mathcal{B} v, & t \in (t_0, t_f) \\
z(t, \partial \Omega) = 0 \\
z(t_0, \Omega) = 0
\end{cases}
,
\tag{2}
$$

where the state variable $z \in Y = H^1_0(\Omega)$ and the control variable $v \in U = L^2(t_0, t_f; \Omega)$. $\mathcal{B}$ is an operator in $\mathcal{L}(L^2(t_0, t_f; \Omega), L^2(t_0, t_f; Y'))$ and $\mathcal{A}$ is an uniformly elliptic linear operator from $L^2(t_0, t_f; Y)$ to $L^2(t_0, t_f; Y')$. The state variable z is dependent on v. z_* is a given target function.

2.1   Discretization in Space

The system is first discretized in space by fixing t. Considering the discrete subspace $Y_h \in Y$ and $U_h \in U$, then the discretized weak form of (2) is given as:

$$
(\dot{z}_h(t), \eta_h) + (\mathcal{A} z_h(t), \eta_h) = (\mathcal{B} u_h(t), \eta_h), \quad \forall \eta_h \in Y_h \text{ and } t \in (t_0, tf) .
\tag{3}
$$

Let $\{\phi_1, \phi_2, .., \phi_n\}$ be a basis of Y_h and $\{\psi_1, \psi_2, .., \psi_m\}$ be a basis of U_h, where $m \leq n$. Apply the finite element methods to (3), we obtain the following system of ordinary differential equations:

$$M\dot{y} + Ay = Bu, \quad t \in (t_0, t_f) . \tag{4}$$

Here $A_{i,j} = (\mathcal{A}\phi_j, \phi_i)$ is a stiffness matrix, $M_{i,j} = (\phi_j, \phi_i)$ and $R_{i,j} = (\psi_j, \psi_i)$ are mass matrices, and $B_{i,j} = (\mathcal{B}\psi_i, \phi_j)$. The semi-discrete solution is $z_h(t, x) = \sum_i y_i(t)\phi_i(x)$ with control function $u_h(t, x) = \sum_i u_i(t)\psi_i(x)$.

We can apply the analogous spatial discretization to the cost function (1), and obtain:

$$J(y, u) = \int_{t_f}^{t_0} e(t)^T Q(t)e(t) + u(t)^T R(t)u(t)\, dt + e(t_f)^T C(t)e(t_f), \tag{5}$$

where $e(\cdot) = y(\cdot) - y_*(\cdot)$ is the difference between the state variable and the given target function.

2.2   Discretization in Time

After spatial discretization, the original optimal problem is transferred into minimizing (5) under the constraint of n ordinary differential equations (4). θ-scheme is used to discretize the above problem.

First the time scale is subdivided into l intervals of length $\tau = (t_f - t_0)/l$. Let $F_0 = M + \tau(1 - \theta)A$ and $F_1 = M - \tau\theta A$. The discretization of equation (4) is given by:

$$Ey + Nu = f, \tag{6}$$

where

$$E = \begin{bmatrix} -F_1 & & \\ & \ddots & \ddots & \\ & & F_0 & -F_1 \end{bmatrix}, \quad N = \tau \begin{bmatrix} B & \\ & \ddots & \\ & & B \end{bmatrix}, \quad y \approx \begin{bmatrix} y(t_1) \\ \vdots \\ y(t_n) \end{bmatrix}, \quad \text{etc} . \tag{7}$$

Then discretize the cost function (5) by using piecewise linear functions to approximate the state variable and piecewise constant functions to approximate the control variable and obtain the following discrete form of (5):

$$J(\mathbf{y}, \mathbf{u}) = \mathbf{u}^T G\mathbf{u} + \mathbf{e}^T K\mathbf{e}, \tag{8}$$

where $\mathbf{e} = \mathbf{y} - \mathbf{y}_*$ and the target trajectory $z_*(t, x) \approx z_{*,h}(t, x) = \sum_i (y_*)_i(t)\phi_i(x)$. A Lagrange multiplier vector p is introduced to enforce the constraint of (6), and we have the Lagrangian

$$\mathcal{L}(\mathbf{y}, \mathbf{u}, \mathbf{p}) = \frac{1}{2}(\mathbf{u}^T G\mathbf{u} + \mathbf{e}^T K\mathbf{e}) + \mathbf{p}^T(E\mathbf{y} + N\mathbf{u} - \mathbf{f}) . \tag{9}$$

To find $\mathbf{y}$, $\mathbf{u}$ and $\mathbf{p}$ where $\nabla\mathcal{L}(\mathbf{y}, \mathbf{u}, \mathbf{p}) = 0$ in (9), we need to solve the following system:

$$\begin{bmatrix} K & 0 & E^T \\ 0 & G & N^T \\ E & N & 0 \end{bmatrix} \begin{bmatrix} \mathbf{y} \\ \mathbf{u} \\ \mathbf{p} \end{bmatrix} = \begin{bmatrix} M\mathbf{y}_* \\ 0 \\ \mathbf{f} \end{bmatrix} . \tag{10}$$

3   Hierarchical-Matrices

The concept and properties of $\mathcal{H}$-matrices are induced by the index cluster tree T_I and the block cluster tree $T_{I \times I}$ [6] . In the rest of this paper, $\#A$ denotes the number of elements of set A and $S(i)$ denotes the children of node i.

3.1   Index Cluster Tree and Block Cluster Tree

An index cluster tree T_I defines a hierarchical partition tree over an index set $I = (0, \ldots, n-1)$. Note that $(0, 1, 2) \neq (0, 2, 1)$.T_I has the following properties: its root is I; any node $i \in T_I$ either is a leaf or has children $S(i)$; the parent node $i = \bigcup_{j \in S(i)} j$ and its children are pairwise disjoint.

A block cluster tree $T_{I \times I}$ is a hierarchical partition tree over the product index set $I \times I$. Given tree T_I and an admissibility condition (see below), $T_{I \times I}$ can be constructed as follows: its root is $I \times I$; if $s \times t$ in $T_{I \times I}$ satisfies the admissibility condition, it is an Rk-matrix leaf; else if $\#s < N_s$ or $\#t < N_s$, it is a full-matrix leaf; otherwise it is partitioned into subblocks on the next level and its children (subblocks) are defined as $S(s \times t) = \{ i \times j \mid i, j \in T_I \text{ and } i \in S(s), \ j \in S(t) \}$. A constant $N_s \in [10, 100]$ is used to control the size of the smallest blocks.

An admissibility condition is used to determine whether a block to be approximated by an Rk-matrix. An example of an admissibility condition is:

$$s \times t \text{ is admissible if \& only if} : \ \min(\text{diam(s)}, \text{diam(t)}) \leq \mu \ \text{dist}(s, t), \quad (11)$$

where $\text{diam}(s)$ denotes the Euclidean diameter of the support set s, and $\text{dist}(s, t)$ denotes the Euclidean distance of the support set s and t. The papers [1,2,3] give further details on adapting the admissibility condition to the underlying problem or the cluster tree.

Now we can define an $\mathcal{H}$-matrix H induced by $T_{I \times I}$ as follows: H shares the same tree structure with $T_{I \times I}$; the data are stored in the leaves; for each leaf $s \times t \in T_{I \times I}$, its corresponding block $H_{s \times t}$ is a Rk-matrix, or a full matrix if $\#s < N_s$ or $\#t < N_s$.

Fig. 1 shows an example of T_I, $T_{I \times I}$ and the corresponding $\mathcal{H}$-matrix.

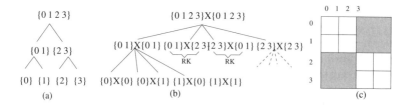

Fig. 1. (a) is T_I, (b) is $T_{I \times I}$ and (c) is the corresponding $\mathcal{H}$-matrix. The dark blocks in (c) are Rk-matrix blocks and the white blocks are full matrix blocks.

An $m \times n$ matrix M is called an Rk-matrix if $rank(M) \leq k$ and it is represented in the form of a matrix product $M = AB^T$, where M is $m \times n$, A is

$m \times k$ and B is $n \times k$. If M is not of rank k, then a rank k approximation can be computed in $O(k^2(n+m) + k^3)$ time by using a truncated Singular Value Decomposition (SVD) [1,3].

3.2 $\mathcal{H}$-Matrix Arithmetic

The following is a short summary of the $\mathcal{H}$-matrix arithmetic. A detailed introduction can be found in [1,2].

$\mathcal{H}$-matrix operations perform recursively; therefore it is important to define the corresponding operations on the leaf subblocks, which are either full or Rk-matrices. These operations are approximate as certain operations do not create Rk-matrices (such as adding two Rk-matrices). In such case, a truncation is performed using an SVD to compute a suitable Rk-matrix. For example, the sum of two rank k matrices can be computed by means of a $2k \times 2k$ SVD.

The computational complexity of the $\mathcal{H}$-matrix arithmetic strongly depends on the structure of $T_{I \times I}$. Under fairly general assumptions on the block cluster tree $T_{I \times I}$ the complexity of $\mathcal{H}$-matrix operators is $O(n \log^\alpha n)$ [3,6].

4   Algebraic Approaches for Hierarchical-Matrix Construction

Algebraic $\mathcal{H}$-matrix construction approaches can be applied to sparse matrices. These approaches take advantage that most entries of a sparse matrix are zeros. They build $\mathcal{H}$-matrix cluster tree by partitioning a matrix graph either bottom-up or top-down. The multilevel clustering based algorithm [11] constructs the cluster tree "bottom-up", i.e., starts with the leaves and successively clusters them together until the root is reached. Domain decomposition in [4] and bi-section are "top-down" algebraic approaches, which start with the root and successively subdivides a cluster into subsets.

4.1   Algebraic Approaches to Construct an Index Cluster Tree

In [11] we propose an $\mathcal{H}$-matrix construction approach based on multilevel clustering methods. To build clusters over the nodes in $G_i = (V(G_i), E(G_i))$, an algorithm based on Heavy Edge Matching (HEM) [7] is used. After building the clusters, a coarse graph G_{i+1} is constructed: such that for each cluster $C_k^{(i)} \subset V(G_i)$ there is a node $k \in V(G_{i+1})$; the edge weight w_{kt} of edge $e_{kt} \in E(G_{i+1})$ is the sum of the weights of all the edges, which connect the nodes in cluster $C_k^{(i)}$ to the nodes in cluster $C_t^{(i)}$ in graph G_i. Recursively applying the above coarsening process gives a sequence of coarse graphs $G_1, G_2, \ldots, G_h$. The index cluster tree T_I is constructed by making $k \in V(G_i)$ the parent of every $s \in C_k^{(i)}$. The root of T_I is the set $V(G_h)$, which is the parent to all nodes in G_h.

In [4], domain decomposition based clustering is used to build a cluster tree T_I. Starting from I, a cluster is divided into three sons, i.e. $S(c) = \{c_1, c_2, c_3\}$ and $c = c_1 \cup c_2 \cup c_3$, so that the domain-clusters c_1 and c_2 are disconnected and the interface-cluster c_3 is connected to both c_1 and c_2. Then the domain-clusters are successively divided into three subsets, and the interface-clusters are successively divided into two interface-clusters until the size of a cluster is small enough.

To build a cluster tree T_I based on bisection is straight forward. Starting from a root set I and a set is successive partitioned into two subsets with equal size. This construction approach is suitable for the sparse matrices where the none zero entries are around the diagonal blocks.

4.2   Block Cluster Tree Construction for Algebraic Approaches

The admissibility condition used to build $T_{I \times I}$ for the algebraic approaches is defined as follows: a block $s \times t \in T_{I \times I}$ is admissible if and only if s and t are not joined by an edge in the matrix graph; an admissible block corresponds to a zero block and is represented as a Rk-matrix of rank zero; an inadmissible block is partitioned further or represented by a full matrix.

5   Hierarchical-Matrix Preconditioners

The construction of $\mathcal{H}$-matrix preconditioners for a system of saddle point type is based on the block LU factorization.

To obtain a relative cheap yet good approximate LU factors, we replace the ordinary matrix operators by the corresponding $\mathcal{H}$-matrix operators[4,8,11].

First the matrix in (10) is converted to an $\mathcal{H}$-matrix. Since the nonzero entries of each subblock are centered around the diagonal blocks, we apply the bisection approach to the submatrix K, G, E and N respectively. Then we obtain the following $\mathcal{H}$-matrix, which is on the left side of the equation ($\mathcal{H}$ indicates a block in the $\mathcal{H}$-matrix format):

$$\begin{bmatrix} K_{\mathcal{H}} & 0 & E_{\mathcal{H}}^T \\ 0 & G_{\mathcal{H}} & N_{\mathcal{H}}^T \\ E_{\mathcal{H}} & N_{\mathcal{H}} & 0 \end{bmatrix} = \begin{bmatrix} L1_{\mathcal{H}} & 0 & 0 \\ 0 & L2_{\mathcal{H}} & 0 \\ M1_{\mathcal{H}} & M2_{\mathcal{H}} & L3_{\mathcal{H}} \end{bmatrix} \begin{bmatrix} U1_{\mathcal{H}} & 0 & M1_{\mathcal{H}}^T \\ 0 & U3_{\mathcal{H}} & M2_{\mathcal{H}}^T \\ 0 & 0 & U3_{\mathcal{H}} \end{bmatrix} . \qquad (12)$$

The block cluster tree $T_{I \times I}$ of $L1_{\mathcal{H}}$, $L2_{\mathcal{H}}$, $M1_{\mathcal{H}}$, and $M2_{\mathcal{H}}$ is same as the block cluster tree structure of $K_{\mathcal{H}}$, $G_{\mathcal{H}}$, $E_{\mathcal{H}}$, and $N_{\mathcal{H}}$ respectively. The block cluster tree structure of $L3_{\mathcal{H}}$ is based on the block tree structure of $E_{\mathcal{H}}$: the block tree of $L3_{\mathcal{H}}$ is symmetric; the tree structure of the lower-triangular of $L3_{\mathcal{H}}$ is same as the tree structure of the lower-triangular of $E_{\mathcal{H}}$; the tree structure of the upper-triangular of $L3_{\mathcal{H}}$ is the transpose of the tree structure of the lower triangular. $L1_{\mathcal{H}}$ and $L2_{\mathcal{H}}$ are obtained by apply $\mathcal{H}$-Cholesky factorization to $K_{\mathcal{H}}$ and $G_{\mathcal{H}}$: $K_{\mathcal{H}} = L1_{\mathcal{H}} *_{\mathcal{H}} U1_{\mathcal{H}}$ and $G_{\mathcal{H}} = L2_{\mathcal{H}} *_{\mathcal{H}} U2_{\mathcal{H}}$. Then use the $\mathcal{H}$-matrix upper triangular solve, we can get $M1_{\mathcal{H}}$ by solving $M1_{\mathcal{H}} U1_{\mathcal{H}} = E_{\mathcal{H}}$. $M1_{\mathcal{H}}$ have the same block tree as $E_{\mathcal{H}}$. In the same way we can compute $M2_{\mathcal{H}}$, which has the same block cluster tree structure as $N_{\mathcal{H}}$. At last we construct the block cluster tree for $L3_{\mathcal{H}}$ and then apply $\mathcal{H}$-LU factorization to get $L3_{\mathcal{H}}$: $L3_{\mathcal{H}} U3_{\mathcal{H}} = M1_{\mathcal{H}} *_{\mathcal{H}} M1_{\mathcal{H}}^T +_{\mathcal{H}} M2_{\mathcal{H}} *_{\mathcal{H}} M2_{\mathcal{H}}^T$.

6   Experimental Results

In this section, we present the numerical results of solving the optimal control problem (1) constrained by the following equation:

$$\begin{cases} \partial_t z - \partial_{xx} z = v, t \in (0,1), x \in (0,1) \\ \qquad z(t,0) = 0, z(t,1) = 0 \\ \qquad z(0,x) = 0, x \in [0,1] \end{cases} \tag{13}$$

with the target function $z_*(t,x) = x(1-x)e^{-x}$. The parameters in the control function J are $q = 1$, $r = 0.0001$ and $s = 0$.

Table 1. Time for computing $\mathcal{H}$-LU factors and GMRES iterations

n(n1/n2)	$L1_\mathcal{H}$	$L2_\mathcal{H}$	$M1_\mathcal{H}$	$M2_\mathcal{H}$	$L3_\mathcal{H}$	GMRES iteration time	number of GMRES iterations
592(240/112)	0	0	0.01	0	0.01	0	1
2464(992/480)	0.01	0	0.01	0	0.04	0	1
10048(4032/1984)	0.03	0.01	0.13	0.02	0.39	0.04	1
40576(16256/8064)	0.21	0.06	0.84	0.26	4.13	0.32	3
163072(65280/32512)	1.09	0.42	4.23	1.74	25.12	2.66	6

GMRES iteration stops where the original residuals are reduced by the factor of 10^{-12}. The convergence rate a ia defined as the average decreasing speed of residuals in each iteration. The fixed-rank $\mathcal{H}$-matrix arithmetic is used and we set the rank of each Rk-matrix block to be ≤ 2. The tests are performed on a Dell workstation with AMD64 X2 Dual Core Processors (2GHz) and 3GB memory.

Table 1 shows the time to compute the different parts of the $\mathcal{H}$-LU factors and the time of GMRES iterations (in seconds). n is the size of the problem, $n1$ and $n2$ is the number of rows of K and G respectively. Based on Table 1, the time to compute $L3_\mathcal{H}$ contributes the biggest part of the total time to set up the preconditioner.

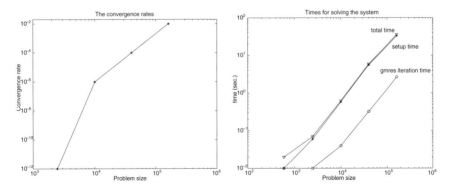

Fig. 2. (a) The convergence rates of GMRES (b) Total times for solving the system

Fig. 2-(a) shows the convergence rate of the $\mathcal{H}$-LU preconditioned GMRES and Fig. 2-(b), plotted on a log-log scale, shows the time for building the preconditioners and the time for the GMRES iterations.

Based the results, we can see that $\mathcal{H}$-LU speeds up the convergence of GMRES iteration significantly. The problem in our implementation is that the time to compute L_3 still consists a significant part of the LU-factorization time. In the future, more work needs to be done to reduce the complexity of computing L_3 further. More discussion about $\mathcal{H}$-matrix preconditioners and applications will come in [13].

References

1. Börm, S., Grasedyck, L., Hackbusch, W.: Introduction to hierarchical matrices with applications. Engineering Analysis with Boundary Elements. **27** (2003) 405–422
2. Börm, S., Grasedyck, L., Hackbush, W.: Hierarchical matrices. Lecture Notes No. 21. Max-Planck-Institute for Mathematics in the Sciences, Leipzig (2003)
3. Grasedyck, L., Hackbusch, W.: Construction and Arithmetics of H-matrices. Computing. **70** (2003) 295–334
4. Grasedyck, L. , Kriemann, R. , LeBorne, S.: Parallel Black Box Domain Decomposition Based H-LU Preconditioning. Mathematics of Computation, submitted. (2005)
5. Gravvanis, G.: Explicit approximate inverse preconditioning techniques Archives of Computational Methods in Engeneering. **9** (2002)
6. Hackbusch, W.: A sparse matrix arithmetic based on H-matrices. Part I: Introduction to H-matrices. Computing. **62** (1999) 89–108
7. Karypis, G., Kumar, V.: A fast and high quality multilevel scheme for partitioning irregular graphs. SIAM J. Sci. Comput. **20** (1999) 359–392
8. LeBorne, S.: Hierarchical matrix preconditioners for the Oseen equations. Comput. Vis. Sci. (2007)
9. LeBorne, S., Grasedyck, L.: H-matrix preconditioners in convection-dominated problems. SIAM J. Matrix Anal. Appl. **27** (2006) 1172–1183
10. LeBorne, S., Grasedyck, L., Kriemann, R.: Domain-decomposition based H-LU preconditioners. LNCSE. **55** (2006) 661–668
11. Oliveira, S., Yang, F.: An Algebraic Approach for H-matrix Preconditioners. Computing, submmitted. (2006)
12. Schaerer, C. and Mathew, T. and Sarkis, M.: Block Iterative Algorithms for the Solution of Parabolic Optimal Control Problems. VECPAR. (2006)
13. Yang, F.: H-matrix preconditioners and applications. PhD thesis. the University of Iowa

Searching and Updating Metric Space Databases Using the Parallel EGNAT

Mauricio Marin[1,2,3], Roberto Uribe[2], and Ricardo Barrientos[2]

[1] Yahoo! Research, Santiago, Chile
[2] DCC, University of Magallanes, Chile
[3] mmarin@yahoo-inc.com

Abstract. The Evolutionary Geometric Near-neighbor Access Tree (EGNAT) is a recently proposed data structure that is suitable for indexing large collections of complex objects. It allows searching for similar objects represented in metric spaces. The sequential EGNAT has been shown to achieve good performance in high-dimensional metric spaces with properties (not found in others of the same kind) of allowing update operations and efficient use of secondary memory. Thus, for example, it is suitable for indexing large multimedia databases. However, comparing two objects during a search can be a very expensive operation in terms of running time. This paper shows that parallel computing upon clusters of PCs can be a practical solution for reducing running time costs. We describe alternative distributions for the EGNAT index and their respective parallel search/update algorithms and concurrency control mechanism [1].

1 Introduction

Searching for all objects which are similar to a given query object is a problem that has been widely studied in recent years. For example, a typical query for these applications is the *range query* which consists on retrieving all objects within a certain distance from a given query object. That is, finding all *similar* objects to a given object. The solutions are based on the use of a data structure that acts as an index to speed up queries. Applications can be found in voice and image recognition, and data mining problems.

Similarity can be modeled as a metric space as stated by the following definitions.

Metric space. A metric space is a set X in which a distance function is defined $d : X^2 \rightarrow R$, such that $\forall\, x, y, z \in X$,

1. $d(x, y) \geq$ *and* $d(x, y) = 0$ iff $x = y$.
2. $d(x, y) = d(y, x)$.
3. $d(x, y) + d(y, z) \geq (d(x, z)$ (triangular inequality).

---

[1] This work has been partially funded by FONDECYT project 1060776.

Y. Shi et al. (Eds.): ICCS 2007, Part I, LNCS 4487, pp. 229–236, 2007.

Range query. Given a metric space *(X,d)*, a finite set $Y \subseteq X$, a query $x \in X$, and a range $r \in R$. The results for query x with range r is the set $y \in Y$, such that $d(x, y) \leq r$.

The k nearest neighbors: Given a metric space *(X,d)*, a finite set $Y \subseteq X$, a query $x \in X$ and $k > 0$. The k nearest neighbors of x is the set A in Y where $|A| = k$ and there is no object $y \in A$ such as *d(y,x)*.

The distance between two database objects in a high-dimensional space can be very expensive to compute and in many cases it is certainly the relevant performance metric to optimize; even over the cost secondary memory operations. For large and complex databases it then becomes crucial to reduce the number of distance calculations in order to achieve reasonable running times. This makes a case for the use of parallelism.

Search engines intended to be able to cope with the arrival of multiple query objects per unit time are compelled to using parallel computing techniques in order to reduce query processing running times. In addition, systems containing complex database objects may usually demand the use of metric spaces with high dimension and very large collections of objects may certainly require careful use of secondary memory.

The distance function encapsulates the particular features of the application objects which makes the different data structures for searching general purpose strategies. Well-known data structures for metric spaces are BKTree [3], MetricTree [8], GNAT [2], VpTree [10], FQTree [1], MTree [4], SAT [5], Slim-Tree [6]. Some of them are based on clustering and others on pivots. The EGNAT is based on clustering [7].

Most data structures and algorithms for searching in metric-space databases were not devised to be dynamic ones. However, some of them allow insertion operations in an efficient manner once the whole tree has been constructed from an initial set of objects. Deletion operations, however, are particularly complicated because in this strategies the invariant that supports the data structure can be easily broken with a sufficient number of deletions, which makes it necessary to re-construct from scratch the whole tree from the remaining objects.

When we consider the use of secondary memory we find in the literature just a few strategies which are able to cope efficiently with this requirement. A well-know strategy is the *M-Tree* [4] which has similar performance to the GNAT in terms of number of accesses to disk and overall size of the data structure. In [7] we show that the EGANT has better performance than the M-Tree and GNAT. The EGNAT is able to deliver efficient performance under high dimensional metric spaces and the use of secondary memory with a crucial advantage, namely it is able to handle update operations dynamically.

In this paper we propose the parallelization of the EGANT in the context of search engines for multimedia databases in which streams of read-only queries are constantly arriving from users together with update operations for objects in the database. We evaluate alternatives for distributing the EGANT data structure on a set of processors with local memory and propose algorithms for performing searches and updates with proper control of read-write conflicts.

2   The EGNAT Data Structure and Algorithms

The EGNAT is based on the concepts of Voronoi Diagrams and is an extension of the GNAT proposed in [2], which in turn is a generalization of the *Generalized Hyperplane Tree* (GHT) [8]. Basically the tree is constructed by taking k points selected randomly to divide the space $\{p_1, p_2, \ldots, p_k\}$, where every remaining point is assigned to the closet one among the k points. This is repeated recursively in each sub-tree D_{p_i}.

The EGNAT is a tree that contains two types of nodes, namely a node *bucket* and another *gnat*. All nodes are initially created as buckets maintaining only the distance to their fathers. This allows a significant reduction in space used in disk and allows good performance in terms a significant reduction of the number of distance evaluations. When a bucket becomes full it evolves from a bucket node to a gnat one by re-inserting all its objects into the newly created gnat node.

In the search algorithm described in the following we assume that one is interested in finding all objects at a distance $d \leq r$ to the query object q. During search it is necessary to determine whether it is a bucket node or a gnat node. If it is a bucket node, we can use the triangular inequality over the center associated with the bucket to avoid direct (and expensive) computation of the distances among the query object and the objects stored in the bucket. This is effected as follows,

- Let q be the query object, let p be the center associated with the bucket (i.e., p is a center that has a child that is a bucket node), let s_i be every object stored in the bucket, and let r be the range value for the search, then if holds

$$Dist(s_i, p) > Dist(q, p) + r$$

  or

$$Dist(s_i, p) < Dist(q, p) - r \, ,$$

  it is certain that the object s_i is not located within the range of the search. In other case it is necessary to compute the distance between q and s_i.

We have observed on different types of databases that this significantly reduces the total amount of distance calculations performed during searches.

For the case in which the node is of type gnat, the search is performed recursively with the standard GNAT method as follows,

1. Assume that we are interested in retrieving all objects with distance $d \leq r$ to the query object q (range query). Let P be the set of centers of the current node in the search tree.
2. Choose randomly a point p in P, calculate the distance $d(q, p)$. If $d(q, p) \leq r$, add p to the output set result.
3. $\forall\, x \in P$, if $[d(q, p) - r, d(q, p) + r] \cap \text{range}(p, D_x)$ is empty, the remove x from P.
4. Repeat steps 2 and 3 until processing all points (objects) in P.
5. For all points $p_i \in P$, repeat recursively the search in D_{p_i}.

3   Efficient Parallelization of the EGNAT

We basically propose two things in this section. Firstly, we look for a proper distribution of the tree nodes and based on that we describe how to perform searches in a situation in which many users submit queries to a parallel server by means of putting queries into a receiving broker machine. This broker routes the queries to the parallel server and receive from it the answers to pass on back the results to the users. This is the typical scheme for search engines. In addition, due to the EGNAT structure we employ to build a dynamic index in each processor, the parallel server is able to cope with update operations taking place concurrently with the search operations submitted by the users.

Secondly, we propose a very fast concurrency control algorithm which allows search and update operations to take place without producing the potential read/write conflicts arising in high traffic workloads for the server. We claim *very fast* based on the fact that the proposed concurrency control mechanism does not incur in the overheads introduced by the classical locks or rollback strategies employed by the typical asynchronous message passing model of parallel computation supported by the MPI or PVM libraries.

Our proposal is very simple indeed. The broker assigns a unique timestamp to each query and every processor maintains its arriving messages queue organized as a priority queue wherein higher priority means lower timestamp. Every processor performs the computations related to each query in strict priority order. The scheme works because during this process it is guaranteed that no messages are in transit and the processors are periodically barrier synchronized to send departing messages and receive new ones. In practical terms, the only overhead is the maintenance of the priority queue, a cost which should not be significant as we can profit from many really efficient designs proposed for this abstract data type so far.

The above described type of computation is the one supported by the bulk-synchronous model of parallel computing [9]. People could argue that the need to globally synchronize the processors could be detrimental and that there could be better ways of exploiting parallelism by means of tricks from asynchronous message passing methods. Not the case for the type of application we are dealing with in this paper. Our results show that even on very inefficient communication platforms such a group of PCs connected by a 100MB router switch, we are able to achieve good performance. This because what is really relevant to optimize is the load balance of distance calculations and balance of accesses to secondary memory in every processor. In all cases we have observed that the cost of barrier synchronizing the processors is below 1%.

Moreover, the particular way of performing parallel computing and communications allows processors further reduction of overheads by packing together into a large message all messages sent to a given processor. Another significant improvement in efficiency, which leads to super-linear speedups, is the relative increase of the size of disk-cache in every processor as a result of keeping a fraction N/P of the database in the respective secondary memory, where N is the total number of objects stored in the database and P the number of processors.

To show the suitability of the EGNAT data structure for supporting query processing in parallel, we evenly distributed the database among the P processors of a 10-processors cluster of PCs. Queries are processed in batches as we assume an environment in which a high traffic of queries is arriving to the broker machine. The broker routes the queries to the processors in a circular manner. We take batches as we use the BSP model of computing for performing the parallel computation and communication.

In the bulk-synchronous parallel (BSP) model of computing [9], any parallel computer (e.g., PC cluster, shared or distributed memory multiprocessors) is seen as composed of a set of P processor-local-memory components which communicate with each other through messages. The computation is organized as a sequence of *supersteps*. During a superstep, the processors may only perform sequential computations on local data and/or send messages to other processors. The messages are available for processing at their destinations by the next superstep, and each superstep is ended with the barrier synchronization of the processors.

The running time results shown below were obtained with three different metric space databases. (a) A 10-dimensional vector space with 100,000 points generated using a Gaussian distribution with average 1 and variance 0.1. The distance function among objects is the Euclidean distance. (b) Spanish dictionary with 86,061 words where the distance between two words is calculated by counting the minimum number of insertions, deletions and replacements of characters in order to make the two words identical. (c) Image collection represented by 100,000 vectors of dimension 15.

Searches were performed by selecting uniformly at random 10% of the database objects. For all cases the search of these objects is followed by the insertion of the same objects in a random way. After we searched 10 objects we randomly chose one of them and insert it into the database, and for every 10 objects inserted we delete one of them also selecting it at random. Notice that we repeated the same experiments shown below but without inserting/deleting objects and we observed no significant variation in the total running time. This means that the overheads introduced by the priority queue based approach for concurrency control we propose in this paper has no relevant effects in the total running time.

We used two approaches to the parallel processing of batches of queries. In the first case, we assume that a single EGNAT has been constructed considering the whole database. The first levels of the tree are kept duplicated in every processor. The size of this tree is large enough to fit in main memory. Downwards the tree branches or sub-trees are evenly distributed onto the secondary memory of the processors. A query starts at any processor and the sequential algorithm is used for the first levels of the tree. After this copies of the query "travel" to other processors to continue the search in the sub-trees stored in the remote secondary memories. That is queries can be divided in multiple copies to follow the tree paths that may contain valid results. These copies are processed in parallel when are sent to different processors. We call this strategy the *global index* approach.

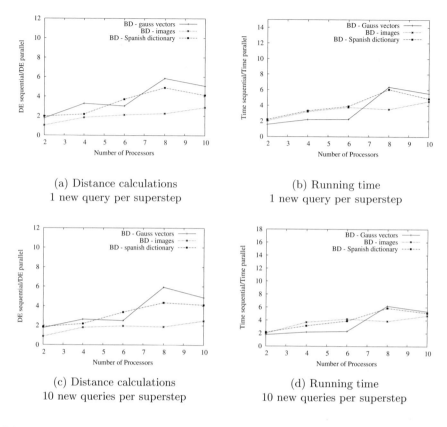

(a) Distance calculations
1 new query per superstep

(b) Running time
1 new query per superstep

(c) Distance calculations
10 new queries per superstep

(d) Running time
10 new queries per superstep

Fig. 1. Results for the global index approach. Figures (a) and (c) show the ratio number of sequential distance evaluations to parallel ones, and figures (b) and (d) show the respective effect in the running times.

Figure 1 shows running time and distance calculation measures for the global index approach against an optimized sequential implementation. The results show reasonable performance for small number of processors but not for large number of processors. This is because performance is mainly affected by the load imbalance observed in the distance calculation process. This cost is significantly more predominant over communication and synchronizations costs. The results for the ratio of distance calculations for the sequential algorithm to the parallel one show that there is a large imbalace in this process. In the following we describe the second case for parallelization which has better load balance. Notice that this case requires more communication because of the need for broadcasting each query to all processors.

In the second case, an independent EGNAT is constructed in the piece of database stored in each processor. Queries in this case start at any processor at the beginning of each superstep. The first step in processing any query is to send a copy of it to all processors including itself. At the next superstep the searching

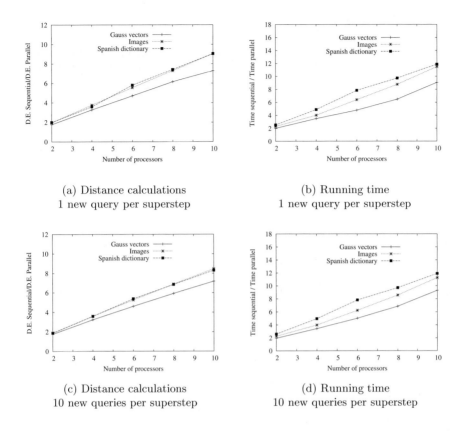

Fig. 2. Results for the local index approach. Figures (a) and (c) show the ratio number of sequential distance evaluations to parallel ones, and figures (b) and (d) show the respective effect in the running times.

algorithms is performed in the respective EGNAT and all solutions found are reported to the processor that originated the query. New objects are distributed circularly onto the processors and insertions are performed locally. We call this strategy the *local index* approach.

In the figures 2 we present results for running time and distance calculations for $Q=1$ and 10 new queries per superstep respectively. The results show that the local index approach has much better load balance and thereby it is able to achieve better speedups. In same cases, this speedup is superlinear because of the secondary memory effect. Notice that even processing batches of one query per processor is good enough to amortize the cost of communication and synchronization. Interestingly, the running times obtained in figure 2 are very similar to the case in which no write operations are performed in the index [7]. This means that the overhead introduced by the concurrency control method is indeed negligible.

4   Conclusions

We have described the efficient parallelization of the EGNAT data structure. As it allows insertions and deletions, we proposed a very efficient way of dealing with concurrent read/write operations upon an EGNAT evenly distributed on the processors. The local index approach is more suitable for this case as the dominant factor in performance is the proper balance of distance calculations taken place in parallel.

The results using different databases show that the EGNAT allows an efficient parallelization in practice. The results for running time show that it is feasible to significantly reduce the running time by the inclusion of more processors. This is because a number of distance calculations for a given query can take place in parallel during query processing. We emphasize that for the use of parallel computing to be justified we must put ourselves in a situation of a very high traffic of user queries. The results show that in practice just with a few queries per unit time it is possible to achieve good performance. That is, the combined effect of good load balance in both distance evaluations and accesses to secondary memory across the processors, is sufficient to achieve efficient performance.

References

1. R. Baeza-Yates and W. Cunto and U. Manber and S. Wu. Proximity matching using fixedqueries trees. 5th Combinatorial Pattern Matching (CPM'94), 1994.
2. S. Brin. Near neighbor search in large metric spaces. The 21st VLDB Conference, 1995.
3. W. Burkhard and R. Keller. Some approaches to best-match file searching. Communication of ACM, 1973.
4. P. Ciaccia and M. Patella and P. Zezula. M-tree: An efficient access method for similarity search in metric spaces. The 23st International Conference on VLDB, 1997.
5. G. Navarro and N. Reyes. Fully dynamic spatial approximation trees. In the 9th International Symposium on String Processing and Information Retrieval (SPIRE 2002), pages 254–270, Springer 2002.
6. C. Traina and A. Traina and B. Seeger and C. Faloutsos. Slim-trees: High performance metric trees minimizing overlap between nodes. VII International Conference on Extending Database Technology, 2000.
7. R. Uribe, G. Navarro, R. Barrientos, M. Marin, An index data structure for searching in metric space databases. International Conference on Computational Science (ICCS 2006), LNCS 3991 (part I) pp. 611-617, (Springer-Verlag), Reading, UK, May 2006.
8. J. Uhlmann. Satisfying general proximity/similarity queries with Metric Trees. Information Processing Letters, 1991.
9. L.G. Valiant. A bridging model for parallel computation. Comm. ACM, 1990.
10. P. Yianilos. Data structures and algoritms for nearest neighbor search in general metric spaces. 4th ACM-SIAM Symposium on Discrete Algorithms, 1993.

An Efficient Algorithm and Its Parallelization for Computing PageRank

Jonathan Qiao, Brittany Jones, and Stacy Thrall

Converse College
Spartanburg, SC 29302, USA
{Jonathan.Qiao,Brittany.Jones,Stacy.Thrall}@converse.edu

Abstract. In this paper, an efficient algorithm and its parallelization to compute PageRank are proposed. There are existing algorithms to perform such tasks. However, some algorithms exclude dangling nodes which are an important part and carry important information of the web graph. In this work, we consider dangling nodes as regular web pages without changing the web graph structure and therefore fully preserve the information carried by them. This differs from some other algorithms which include dangling nodes but treat them differently from regular pages for the purpose of efficiency. We then give an efficient algorithm with negligible overhead associated with dangling node treatment. Moreover, the treatment poses little difficulty in the parallelization of the algorithm.

Keywords: PageRank, power method, dangling nodes, algorithm.

1   Introduction

A significant amount of research effort has been devoted to hyperlink analysis of the web structure since Sergey Brin, Larry Page brought their innovative work [1] to the information science community in 1998. Brin and Page launched Google at the time when there were already a number of search engines. Google has succeeded mainly because it has a better way of ranking web pages, which is called PageRank by its founders.

The original PageRank model is solely based on the hyperlink structure. It considers the hyperlink structure of the web as a digraph. For each dangling node (a page without any out-link), edges are added so that it is connected to all nodes. Based on this, an adjacency matrix can be obtained. It is then to find the eigenvector of the adjacency matrix, whose elements are the ranking values of the corresponding web pages. Due to the size of the web graph, the dominant way in finding the eigenvector is the classic power method [2], which is known for its slow convergence. Because of this, a large amount of work has been done to speed up the PageRank computation since 1998.

One factor contributing substantially to the computational cost is the existence of dangling nodes. A single dangling node adds a row full of ones to the adjacency matrix. Including dangling nodes not only produces a significantly

Y. Shi et al. (Eds.): ICCS 2007, Part I, LNCS 4487, pp. 237–244, 2007.

larger matrix, but also makes it no longer sparse. Some existing algorithms exclude dangling nodes from consideration in order to speed up the computation. Those which consider them don't treat them as regular pages for purpose of efficiency. In either case, the original web structure is changed, and more importantly the information carried by dangling nodes is changed as well.

In this paper, we will aim at finding the PageRank vector using the power method with dangling nodes fully included and treated as regular pages. We will then propose an highly efficient algorithm which minimizes the overhead of our treatment of dangling nodes to be negligible in either centralized or distributed approach.

2   PageRank Review

As discussed in Sect. 1, the web structure is treated as a digraph, which gives a $N \times N$ column-wise adjacency matrix, where N is the number of total web pages. Let the matrix be A. Let D be a $N \times N$ diagonal matrix with each entry the reciprocal of the sum of the corresponding column of A. Then, AD is a column stochastic Markov matrix. Let $\mathbf{v} = AD\mathbf{e}$, where $\mathbf{e}$ is a vector of all ones. The ith entry of $\mathbf{v}$ tells how many times the ith page gets recommended by all web pages. Introducing the weight of a recommendation made by a page which is inversely proportional to the total number of recommendations made by that page gives the equation

$$\mathbf{v} = AD\mathbf{v} \ . \tag{1}$$

Equation (1) suggests that $\mathbf{v}$ is the ranking vector as well as an eigenvector of AD with the eigenvalue 1.

By nature of the web structure, the Markov matrix AD is not irreducible [2]. This implies that there are more than one independent eigenvectors associated with the eigenvalue 1. Google's solution to this problem is to replace AD by a convex combination of AD and another stochastic matrix $\mathbf{ee}^T/N$,

$$\mathbf{v} = \left(\alpha AD + \frac{1-\alpha}{N} \mathbf{ee}^T \right) \mathbf{v} \ , \tag{2}$$

where $0 < \alpha < 1$ (Google uses $\alpha = 0.85$). The underlying rational is, by probability 0.85, a random surfer chooses any out-link on a page arbitrarily if the page has out-links; by probability 0.15, or if a page is a dead end, another page is chosen at random from the entire web uniformly.

The convex combination in (2) makes the ranking vector $\mathbf{v}$ the unique eigenvector associated with the eigenvalue 1 up to scaling and the subdominant eigenvalue α, which is the rate of convergence when the power method is applied [3,4,2]. By normalizing the initial and subsequent vectors, a new power method formulation of (2) is

$$\mathbf{v}^{i+1} = \alpha AD\mathbf{v}^i + \frac{1-\alpha}{N} \mathbf{e} \ , \tag{3}$$

where $\mathbf{v}^0 = (1/N)\mathbf{e}$ because $\mathbf{e}^T\mathbf{v} = 1$.

3   Related Work

Most algorithms so far focus on the cost reduction of each power method iteration, although there have been efforts aiming at reducing the number of iterations, such as [5]. Because the major obstacle of speeding up each iteration lies in the size of data, it is natural to compress the data such that it fits into the main memory, as by [6,7,8]. However, as pointed out by [6], even the best compression scheme requires about .6 bytes per hyperlink, which still results in an exceedingly large space requirement. Others are to design I/O-efficient algorithms, such as [9,10], which can handle any size of data without any particular memory requirement. Also aiming at reducing the cost of each iteration, many other works, such as [11,12], combine some linear algebra techniques to reduce the computational cost of each iteration. Nevertheless, the I/O-efficiency must always be considered.

In [9], Haveliwala proposes two algorithms, *Naive Algorithm* and *Blocking Algorithm*, and demonstrates high I/O efficiency of the latter when the ranking vector does not fit in the main memory. Applying the computational model in (3), both algorithms use a binary link file and two vectors, the source vector holding the ranking values for the iteration i and the destination vector holding the ranking values for the iteration $i + 1$. Each diagonal entry of the matrix D is stored in the corresponding row of the link file as the number of out-links of each source page.

For purpose of the cost analysis, there are several more parameter families, M, $B(\cdot)$, K and $nnz(\cdot)$.

1. The total number of available memory pages will be denoted M.
2. The total number of pages of disk/memory resident will be denoted $B(\cdot)$.
3. The total number of dangling nodes will be denoted K.
4. The total number of links of a web digraph will be denoted $nnz(\cdot)$.

Unless specified, the cost is for a single iteration since we focus on reducing the cost of each iteration in this work.

3.1   Matrix-Vector Multiplication Treatment

Given the computation model in (3), computing a single entry of $\mathbf{v}^{i+1}$ requires reading a row of matrix A and the source vector. Since the link file is sorted by source node id's and all elements of a row of matrix A are spread out in the link file[1], a less careful implementation would result in one pass for calculating a single value. Preprocessing the link file so that it is sorted by destination node id's not only needs tremendous effort, but also adds N entries for each dangling node to the link file and thus significantly increases the storage requirement and the I/O cost. To get around of this difficulty, we could use the column version of the matrix-vector multiplication

$$A\mathbf{v} = v_0 A_{*0} + v_1 A_{*1} + \cdots + v_{n-1} A_{*(n-1)} \;, \tag{4}$$

---

[1] A column of A corresponds to a row in the adjacency list.

where each A_{*i} is the ith column of A. The computation model in (4) requires one pass of the link file and the source vector for finding the destination vector when the destination vector can be a memory resident.

3.2   The Blocking Algorithm

Let $\mathbf{v}'$ be the destination vector, $\mathbf{v}$ be the source vector and L be the adjacency list of the link file. To handle the memory bottleneck, *Blocking Algorithm* in [9] partitions $\mathbf{v}'$ evenly into b blocks so that each $\mathbf{v}'_i$ fits in $(M-2)$ pages. L is vertically partitioned into b blocks. Note, each L_i is then represented by a horizontal block A_i. The partition gives the equation

$$\mathbf{v}'_i = \alpha A_i D \mathbf{v} + \frac{1-\alpha}{N} \mathbf{e} \ , \tag{5}$$

where $i \in \{0, 1, \cdots, b-1\}$.

Based on (5), computing each $\mathbf{v}'_i$ requires one pass of $\mathbf{v}'_i$ L_i, $\mathbf{v}$. Updating the source vector at the end adds another pass of $\mathbf{v}$. Therefore, the cost (referred to as C_{block}) is

$$C_{block} = \sum_{i=0}^{b-1} B(L_i) + (b+1)B(\mathbf{v}) + \sum_{i=0}^{b-1} B(\mathbf{v}'_i) = (b+2)B(\mathbf{v}) + (1+\epsilon)B(L) \ , \tag{6}$$

where ϵ is a small positive number due to the partition overhead.

The cost in (6) does not scale up linearly when b is not a constant, which can happen given the remarkably fast growth of the web repository.

4   Dangling Nodes Treatment

The cost model in (6) only counts disk I/O's with an assumption that the in-memory computational cost is negligible compared to the I/O cost. The assumption is justifiable when A is sparse since the in-memory cost is approximately the number of 1's in A. This can be seen in [9], whose test data set which contains close to 80% of dangling nodes originally is preprocessed to exclude all dangling nodes.

Many web pages are by nature dangling nodes, such as, a PDF document, an image, a page of data, etc. In fact, dangling nodes are a increasingly large portion of the web repositories. For some subsets of the web, they are about 80% [2]. Some dangling nodes are highly recommended by many other important pages, simply throwing them away may result in a significant loss of information. This is why some existing works don't exclude dangling nodes, such as [12], which computes the ranking vector in two stages. In the first stage, dangling nodes are lumped into one; in the second stage, non-dangling nodes are combined into one. The global ranking vector is formed by concatenating two vectors.

One of goals of this work is to make improvements over [9] with dangling nodes included. Different from [12] and some other algorithms which include dangling nodes, our approach treats them as regular web pages. It can be easily seen that including dangling nodes does not add any storage overhead since a dangling node does not appear in the link file as a source *id*. Thus, our approach does not change the I/O cost model in (6).

To minimize the in-memory computational overhead imposed by inclusion of dangling nodes, we decompose A into $\hat{A} + \mathbf{e}\varDelta^T$, where $\hat{A}$ is an adjacency matrix of the original web graph (a row full of zeros for a dangling node), $\varDelta$ is a $N \times 1$ vector with the ith entry 1 if the ith node is a dangling node and 0 otherwise. Substituting the decomposition of A into (3), we have

$$\mathbf{v}^{i+1} = \alpha(\hat{A} + \mathbf{e}\varDelta^T)D\mathbf{v}^i + \mathbf{c} = \alpha\hat{A}D\mathbf{v}^i + \mathbf{c}^i + \mathbf{c} , \tag{7}$$

where $\mathbf{c} = \mathbf{e}(1-\alpha)/N$, a constant vector at all iterations, $\mathbf{c}^i$ is a constant vector at iteration i, whose constant entry is computed by adding all dangling node ranking values at the beginning of each iteration, which is

$$\sum_{out-degree(i)=0} (D\mathbf{v})_i . \tag{8}$$

This involves K (the number of dangling nodes) additions and one multiplication. In the implementation, we extract out-degree's from every L_i and save them in a separate file, which has N entries and can be used for computing (8). This also has an advantage of reducing the storage overhead caused by partition since each nonzero out-degree repeatedly appears in every L_i.

A substantial gain of the in-memory cost can be achieved. Let $\mathcal{G}$ be an original web digraph, then, the total number of floating number additions in (3) is

$$C_1 = nnz(\mathcal{G}) + KN + N = (r + K + 1)N , \tag{9}$$

where r is the average out-links, which varies from 5 to 15 [10]. Based on (7), this cost can be reduced to

$$C_2 = nnz(\mathcal{G}) + K = rN + K . \tag{10}$$

When K is large compared to N, which is often the case for the web data, C_1 is $\Theta(N^2)$ while C_2 is $\Theta(N)$.

5   The Parallelization of the Algorithm

The computation model in (7) can be readily parallelized without any extra parallel overhead associated with inclusion of dangling nodes.

When applying *Blocking Algorithm* directly, we may vertically partition the link file L into $L_0, L_1, \cdots, L_{b-1}$ and distribute them over b nodes in a cluster. Each node holds a partition of L, the source vector $\mathbf{v}$ and a partition of $\mathbf{v}'$. This gives the computation model at each node

$$\mathbf{v}'_i = \alpha A_i D\mathbf{v} + \mathbf{c}_i , \tag{11}$$

where $\mathbf{c_i}$ is a constant vector of the size N/b with $(1-\alpha)/N$ for all entries.

The above parallelization does not take the advantage given by the dangling nodes treatment. Combining (5) and (7), the new parallel formulation can be established as

$$\mathbf{v}'_i = \alpha A_i D \mathbf{v} + \hat{\mathbf{c}}_i + \mathbf{c}_i \ , \tag{12}$$

where $i \in \{0, 1, \cdots, b - 1\}$, $\hat{\mathbf{c}}_i$ is a constant vector at each iteration, and $\mathbf{c_i}$ is a constant vector at all iterations of the corresponding size as defined in (7). As discussed in Sect. 4, the vector representation of the matrix D is stored in a separated file, which is read b times at each node, the same as a partition of the link file.

Computing $\mathbf{v}'_i$ at the ith node is carried out in the same fashion as in the serial implementation. It needs to read L_i, $\mathbf{v}$ and to write $\mathbf{v}'_i$ to update the source vector. The I/O cost is $B(L_i) + B(\mathbf{v}) + B(\mathbf{v}'_i)$. One advantage of the distributed algorithm can be seen in normalizing the destination vector. A serial implementation needs to read the whole destination vector. In the distributed case, each partition of the destination vector is held in memory at its corresponding node and the normalization can be done concurrently. Therefore, the I/O cost at each node is

$$C_{I/O} = B(L_i) + (1 + \frac{1}{b})B(\mathbf{v}) \approx B(L)/b + B(\mathbf{v}) \ . \tag{13}$$

The in-memory computational cost does not have any parallel overhead and is therefore

$$C_{in-memory} \approx C_2/b \ . \tag{14}$$

The communication cost can be a bottleneck. To start a new iteration, every node in the cluster needs to collect one block of the updated source vector from every other node, which is $4N/b$ bytes. The total communication cost is

$$C'_{comm} = \sum_{i=0}^{b-1} \left(\sum_{j=0, j \neq i}^{b-1} 4N/b \right) = 4(b-1)N \ . \tag{15}$$

By making every pair of all nodes communicate concurrently, the communication cost is reduced to $\mathcal{O}(N)$ since

$$C_{comm} = C'_{comm} \times \frac{2}{b} \approx 8N \ . \tag{16}$$

The cost models in (13), (14) and (16) show we could achieve a near linear scale-up and a near linear speed-up provided that the data size, rN, is large compared to N since the communication cost is independent of r.

6   Experimental Evaluation

6.1   Experimental Setup

Experiments for the algorithm handling dangling nodes were conducted on Linux platform on a single dedicated machine with a 2.00 GHz Intel(R) Celeron(R)

Table 1. Data sets and the cost comparison

Name	Pages	K/N	r	Links	C_1	C_2	C_1/C_2
California	9.7K	48.00%	1.67	16K	26.3s	0.77s	34.2s
Stanford	281K	0.06%	8.45	2.4M	58.8s	30.7s	1.9s

CPU. Experiments for the parallel algorithm were conducted on Windows platform on dedicated machines each with a 2.8 GHz Intel Pentium(R)-4 CPU. The memory size in either case is ample in the sense that there is enough physical memory for the OS. The page size of disk access for all experiments is 512 KB. The implementations of both algorithms were in Java.

Table 1 shows the basic statistics for the two data sets. The first data set, *California*, is used solely to test the algorithm of handling dangling nodes. It was obtained from *http://www.cs.cornell.edu/Courses/cs685/2002fa/*. The second data set, *stanford*, is used for both algorithms. It was obtained from *http://www.stanford.edu/~sdkamvar/*.

6.2   Results for Handling Dangling Nodes

In Tab. 1, C_1 represents the cost based on the computation model in (3), C_2 represents the cost based on the computation model in (7), which handles dangling nodes using the proposed algorithm.

The data set *california* needs only one disk access due to its small size. Its I/O cost is then negligible. The dangling nodes in this data set are almost a half of the total pages. The speedup obtained by the proposed algorithm, which is the ratio of C_2 and C_1, is 34.2. This verifies the two cost models in (9) and (10). The data set *stanford* is about $11.2M$, which results in about 22 disk accesses. Even though the dangling nodes in the data set are only about 0.06%, the speedup obtained by the proposed algorithm is about 1.9.

6.3   Results for the Parallelization of the Algorithm

The parallel implementation uses the data set *stanford* only. The experiments were conducted on a cluster of different number of nodes. The experimental results in Tab. 2 show we have achieved a near linear speed-up. One reason for

Table 2. Parallel running times and the corresponding speed-up's

Number of Nodes	1	2	4	8
Elapsed Time	47.1s	23.7s	12.1s	6.2s
Speed-up	N/A	2.0	3.9	7.6

the nice results is that the average out-degree of the experimental data set is 8.45, which makes the data size much larger than the ranking vector size. Therefore, the I/O cost and the in-memory computational cost weigh significantly more than the communication cost.

7   Conclusions and Future Work

In this paper, we have derived an efficient algorithm and its parallelization for computing PageRank with dangling nodes fully preserved. Both the analysis and the experimental results demonstrate that our algorithm has little over-head associated with inclusion of dangling nodes. There are two areas for future work: conducting experiments on the larger data sets and at a larger parallel cluster; and exploring more fully dangling nodes' impact and their treatment to PagaRank computation.

References

1. Brin, S., Page, L., Motwami, R., Winograd, T.: The pagerank citation ranking: bringing order to the web. Technical report, Computer Science Department, Stanford University (1999)
2. Langville, A.N., Meyer, C.D.: Deeper inside pagerank. Internet Math **1** (2004) 335–380
3. Haveliwala, T.H., Kamvar, S.D.: The second eigenvalue of the google matrix. Technical report, Computer Science Department, Stanford University (2003)
4. Elden, L.: A note on the eigenvalues of the google matrix. Report LiTH-MAT-R-04-01 (2003)
5. Kamvar, S., Haveliwala, T., Manning, C., Golub, G.: Extrapolation methods for accelerating pagerank computations. Twelfth International World Wide Web Conference (2003)
6. Randall, K., Stata, R., Wickremesinghe, R., Wiener, J.: The link database: Fast access to graphs of the web. In: the IEEE Data Compression Conference. (March 2002)
7. Adler, M., Mitzenmacher, M.: Towards compressing web graphs. In: the IEEE Data Compression Conference. (March 2001)
8. Raghavan, S., Garcia-Molina, H.: Representing web graphs. In: the 19th IEEE Conference on Data Engineering, Bangalore, India (March 2003)
9. Haveliwala, T.H.: Efficient computation of pagerank. Technical report, Computer Science Department, Stanford University (Oct. 1999)
10. Chen, Y., Gan, Q., Suel, T.: I/o-efficient techniques for computing pagerank. In: Proc. of the 11th International Conference on Information and Knowledge Management. (2002)
11. Kamvar, S., Haveliwala, T., Golub, G.: Adaptive methods for the computation of pagerank. Technical report, Stanford University (2003)
12. Lee, C., Golub, G., Zenios, S.: A fast two-stage algorithm for computing pagerank and its extension. Technical report, Stanford University (2003)

A Query Index for Stream Data Using Interval Skip Lists Exploiting Locality

Jun-Ki Min

School of Internet-Media Engineering
Korea University of Technology and Education
Byeongcheon-myeon, Cheonan, Chungnam, Republic of Korea, 330-708
jkmin@kut.ac.kr

Abstract. To accelerate the query performance, diverse continuous query index schemes have been proposed for stream data processing systems. In general, a stream query contains the range condition. Thus, by using range conditions, the queries are indexed. In this paper, we propose an efficient range query index scheme QUISIS using a modified Interval Skip Lists to accelerate search time. QUISIS utilizes a locality where a value which will arrive in near future is similar to the current value.

Keywords: Stream Data, Query Index, Locality.

1 Introduction

Stream data management systems may receive huge number of data items from stream data source while large number of simultaneous long-running queries is registered and active[1,2]. In this case, if all registered queries are invoked whenever a stream data item arrives, the system performance degrades. Therefore, Query indexes are built on registered continuous queries [3]. Upon each stream data arrives, a CQ engine searches for matching queries using these indexes.

Existing query indexes simply maintain the all queries based on well known index structures such as the binary search tree and R-tree. However, some application of stream data processing such as stock market and temperature monitoring has a particular property, which is a locality. For example, the temperature in near future will be similar to the current temperature. Therefore, some or all queries which are currently invoked will be reused in the near future. Therefore, the locality of stream data should be considered in the query indexes.

In this paper, we present a range query index scheme, called *QUISIS* (QUery Index for Stream data using Interval Skip lists). Our work is inspired by BMQ-Index [4]. To the best of our knowledge, Interval Skip list [5] is the most efficient structure to search intervals containing a given point. Thus, QUISIS is based on the Interval Skip List in contrast to BMQ-Index. Using a temporal interesting list (TIL), QUISIS efficiently finds out the query set which can evaluate a newly arrived data item. The experimental results confirm that QUISIS is more efficient than the existing query index schemes.

Y. Shi et al. (Eds.): ICCS 2007, Part I, LNCS 4487, pp. 245–252, 2007.

2   Related Work

Some stream data management systems used balanced binary search trees for query indexes [6]. The query index allows to group query conditions combining all selections into a group-filter operator. As shown in Figure 1, a group filter consists of four data structure: a greater-than balanced binary tree, a less-than balanced binary tree, an equality hash-table, and inequality hash table.

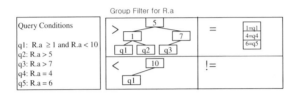

Fig. 1. An example for query indexes using binary search trees

When a data item arrives, balanced binary search trees and hash tables are probed with the value of the tuples. This approach is not appropriate to general range queries which have two bounded conditions. Each bounded condition is indexed in individual binary search tree. Thus, by search of each individual binary tree, unnecessary result may occur.

In addition, for query indexes, multi-dimensional data access methods such as R-Tree [7,8] and grid files can be used [9]. In general, the range conditions of queries are overlapped. These R-tree families are not appropriate for range query indexes since many nodes should be traversed due to a large amount of overlap of query conditions.

Recently, for the range query indexes, BMQ-Index has been proposed. BMQ-Index consists of two data structures: a DMR list, and a stream table. DMR list is a list $<DN_1, DN_2, \ldots, DN_n, DN_{n+1}>$ of DMR nodes. Let $Q = \{q_i\}$ be a set of queries. A DMR node DN_j is a tuple $<DR_j, +DQSet, -DQSet>$. DR_j is a matching region (b_{j-1}, b_j). $+DQSet$ is the set of queries q_k such that $l_k = b_{j-1}$ for each selection region (l_k, u_k) of Q_k. $-DQSet$ is the set of queries q_k such that $u_k = b_{j-1}$ for each selection region (l_k, u_k) of q_k. Figure 2 shows an example of BMQ-Index. A stream table keeps the recently accessed DMR node.

Let $QSet(t)$ be a set of queries for data v_t at time t and v_t be in the DN_j, and v_{t+1} is in the DN_h i.e., $b_{j-1} \leq v_t < b_j$ and $b_{h-1} \leq v_{t+1} < b_h$. $QSet(t+1)$ can be derived as follows:

$$
\begin{aligned}
&if \quad j < h, QSet(t+1) = QSet(t) \cup [\textstyle\bigcup_{i=j+1}^{h} +DQSet_i] - [\textstyle\bigcup_{i=j+1}^{h} -DQSet_i] \\
&if \quad j > h, QSet(t+1) = QSet(t) \cup [\textstyle\bigcup_{i=j}^{h+1} -DQSet_i] - [\textstyle\bigcup_{i=j}^{h+1} +DQSet_i] \\
&if \quad j = h, QSet(t+1) = QSet(t)
\end{aligned}
$$

$$(1)$$

The authors of BMQ-Index insist that only a small number of DRN nodes is retrieved, if the forthcoming data is not in the region due to the data locality.

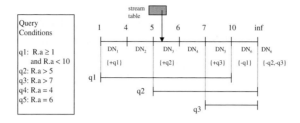

Fig. 2. An example of a BMQ-Index

However, BMQ-Index has some problem. First, if the forthcoming data is quite different from the current data, many DRN nodes should be retrieved like a linear search fashion. Second, BMQ-Index supports only (l, u) style conditions but does not support general condition such as [l,u] and (l, u]. Thus, as shown in Figure 2, q4 and q5 is not registered in BMQ-Index. In addition, BMQ-Index does not work correctly on boundary conditions. For example, if v_t is 5.5, the QSet(t) is {q1,q2}. Then, if v_{t+1} is 5, QSet(t+1) is also {q1,q2} by the above equation. However, the actual query set for v_{t+1} is q1.

3   QUISIS

In this section, we present the details of our proposed approach, QUISIS. As mentioned earlier, QUISIS is based on Interval Skip Lists[5]. Thus, we first introduce Interval Skip Lists, and then present our scheme.

3.1   Interval Skip Lists

Interval Skip Lists are similar to linked lists, except each node in the list has one or more forward pointers. The number of forward pointers of the node is called the level of the node. The level of a node is chosen at random. The probability a new node has k level is:

$$P(k) = \begin{cases} 0 & \text{for } k < 1 \\ (1 - p) \cdot p^{k-1} & \text{for } k \geq 1 \end{cases} \tag{2}$$

With p = 1/2, the distribution node levels will allocate approximately 1/2 of the nodes with one forward pointer, 1/4 with two forward pointers, and so on. A node's forward pointer at level l points to the next node with greater that or equal to l level.

In addition, nodes and forward pointers have markers in order to indicate the corresponding intervals. Consider I = (A,B) to be indexed. End points A and B are inserted in the list as nodes. Consider some forward edges from a node with value X to a node with value Y (i.e., X < Y). A marker containing the identifier of I will be placed on edge (X,Y) if and only if the following conditions hold:

– containment: I contains the interval (X,Y)
– maximality: There is no forward pointer in the list corresponding to an interval (X', Y') that lies within I and that contains (X,Y).

In addition, if a marker for I is placed on an edge, then the nodes of that edge and have a value contained in I will also have a marker (called eqMarker) placed on them for I.

The time complexity of Interval Skip Lists is known as O(log N) where N is the number of intervals. Since we present the extended version of the search algorithm in Section 3.2, we omit the formal description of the search algorithm for Interval Skip Lists.

3.2   Behavior of QUISIS

In Figure 3, QUISIS is shown when the current data item is 5.5. Given search key, the search procedure starts from Header in Interval Skip Lists. In stream data environment, a locality such that a data in the near future is similar to the current data occurs. By using this property, we devise the QUISIS based on Interval Skip Lists.

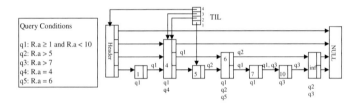

Fig. 3. An example of QUISIS

In order to keep the visited edges by the current data item, a temporal interesting list (TIL) is used. TIL records the nodes with level from MAX level to 1 whose forward pointer with level l represents an interval contains the current data item. As shown in Figure 3, the interval [5,6) represented by the node pointed by TIL with level 1 contains the current data item 5.5.

In Figure 3, we can interpret *Header* and *Null* such that *Header* represents the smallest value and *Null* represents the largest value. So *Header* is smaller than -∞ and *Null* is greater than ∞ in contrast to the conventional number system. Thus, the intervals represented by the nodes in TIL have the property such that:

Property 1. The interval by TIL with level i is contained in the interval by TIL with level $i + 1$.

For example, [5,6) by TIL with level 1 is contained in [4,6) by TIL with level 2 and also is contained in [4,*Null*) by TIL with level 3. By using this property, QUISIS reduces the search space efficiently compared with original Interval Skip Lists.

```
QSet //a query set for previous key
TIL // a list points to the nodes of QUISIS
Procedure findQuery(key )
begin
1.   if(TIL[1]->value = key) {
2.      for( i = TIL[1]->level; i ≥ 1; i−−) QSet := QSet - TIL[1]->markers[i]
3.      QSet := QSet ∪ TIL[1]->eqMarker
4.   } else if(TIL[1]->value < key and
              (TIL[1]->forward[1] = NULL or key < TIL[1]->forward[1]->key)) {
5.      QSet := QSet - TIL[1]->eqMarker
6.      for(i = TIL[1]->level; i ≥1; i-) QSet := QSet ∪ TIL[1]->markers[i]
7.   } else {
8.      QSet := QSet - TIL[1]->eqMarkers
9.      if(TIL[1]->forward[1] = NULL or key ≥ TIL[1]->forward[1]->value ) {
10.        for(i := 1; i ≤ maxLevel ; i++)
11.           if(TIL[i]->forward[i] = NULL or key < TIL[i]->forward[i]->value) break
12.           else QSet = QSet - TIL[i]->markers[i]
13.     } else {
14.        for(i = 1; i ≤ maxLevel; i++)
15.           if(TIL[i]= Header and key ≥ TIL[i]->value) break
16.           else QSet := QSet - TIL[i]->markers[i]
17.     }
18.     anode := TIL[−−i]
19.     while(i ≥ 1) {
20.        while(anode->forward[i] ≠ NULL and anode->forward[i]->value le key)
              anode = anode->forward[i]
21.        if(anode ≠ Header and anode->value ≠ key) QSet := QSet ∪ anode->markers[i]
22.        else if(anode ≠ Header) QSet := QSet ∪ anode->eqMarker[i]
23.        TIL[i] = anode;
24.        i:= i-1
25.     }
26. }
27. return QSet
end
```

Fig. 4. An algorithm of the event handler for `endElement`

In order to exploit TIL, we devised the procedure findQuery() using the Property 1. An outline of an implementation of findQuery() is shown in Figure 4.

Basically, the behavior of the procedure findQuery() is changed according to the condition of newly arrived data value (i.e., key) and the interval $[v_1, u_1)$ by TIL with level 1.

If key is equal to v_1 of the node n pointed by TIL with level 1 (Line 1-3 in Figure 4), all markers on the edges starting from n are removed (Line 2) since QSet may contain queries whose intervals are $(v_1, -)$. Instead, eqMarker of n is added (Line 3) since eqMarker of n contains the queries whose interval contains v_1 and the queries in eqMarker are on that edges which are ended or started at n. For example, when the data was 5.5, QSet was {q1,q2} and a new data item 5 arrives, the TIL[1] is the node 5. Thus, {q2} is removed from QSet by Line 2. And since eqMarker of the node 5 is ∅, the final query result is {q1}.

If key is in (v_1, u_1)(Line 4-6 in Figure 4), the queries in eqMarker of n are removed since the QSet may contain queries whose intervals are $(-, v_1]$. Instead, all markers on the edges starting from n are added.

If key is not in (v_1,u_1) (Line 7-27 in Figure 4), the procedure looks for the interval $[v_i, u_i)$ by TIL with level i which contains key (Line 8-17). This step is separated into two cases: key $\geq u_1$ (Line 9-12) and key $< v_1$ (Line (Line 13-17). Also, in this step, markers on the edges with level from 1 to i-1 are removed from QSet (Line 12 and 16). And then, the procedure gathers queries starting from the node (i.e., *anode*) whose value is v_i (Line 19-25). In this case, since the marker on the edge of *anode* with level i is already in QSet, level i decreases (Line 18).

If the interval represented by a forward pointer of *anode* with level i does not contain key, a search procedure traverses to the next node pointed by the forward pointer of a node with a level i (Line 21). If the value of *anode* is equal to key, eqMarker of *anode* is added (Line 22). Otherwise the marker on the forward pointer is added (Line 21). Then, the *anode* is set to TIL[i](Line 23) and i is dropped into $i-1$(Line 24). The search procedure continues until the level l is to be 1.

For example, when the data was 5.5, QSet was {q1,q2} and a new data item 13 arrives, [4, *Null*) represented by TIL[3] contains 13. Therefore, the procedure removes the search overhead starting from *Header*. {q2} and {q1} which are markers of TIL[1] and TIL[2], respectively, are removed from QSet (Line 9-12). Then, the procedure gathers queries starting from the node 4 with level 2 (Line 18). Since [4,6) does not contain 13, the procedure looks for next interval [6, *Null*) represented by node 6 with level 2. Since 13 is in [6, *Null*) but not equal to 6, a marker q2 is added. And TIL[2] points the node 6. Then, the procedure searches the list from the node 6 with level 1. Since [10, inf) contains 13, a marker {q3} is added. Finally, QSet is {q2, q3}.

In aspect of using the data locality, our proposed scheme QUISIS is similar to BMQ-Index. However, since QUISIS is based on Interval Skip Lists, QUISIS is much more efficient than BMQ-Index in general cases. Our experiments demonstrate the efficiency of QUISIS.

4   Experiments

In this section, we show the efficiency of QUISIS compared with the diverse query index techniques: BMQ-Index and Interval Skip Lists. The experiment performed on Windows-XP machine with a Pentium IV-3.0Ghz CPU and 1GB RAM. We evaluated the performance of QUISIS using the synthetic data over various parameters. We implemented a revised version of BMQ-Index which works correctly on the boundary conditions. The default experimental environment is summarized in Table 1. In Table 1, length of query range (W) denotes the average length of query condition normalized by the attribute domain and fluctuation level (FL) denotes the average distance of two consecutive data normalized by the attribute domain. Therefore, as FL decrease, the locality appears severely.

Table 1. Table of Symbols

Parameter	value
Attribute domain	$1 \sim 1,000,000$
# of Queries	100,000
Length of query range (W)	0.1% (= 1,000)
# of Data	$1,000 \sim 10,000$
Fluctuation level (FL)	0.01% (= 100)

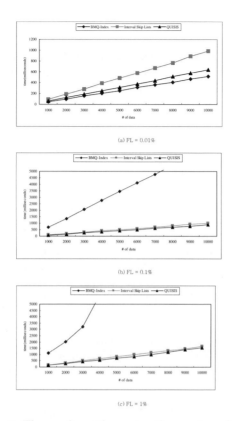

(a) FL = 0.01%

(b) FL = 0.1%

(c) FL = 1%

Fig. 5. The results with varying the number of data

We empirically performed experiments with varying parameters. However, due to the space limitation, we show only the experimental result when values of FL are 0.01%, 0.1% and 1%.

Our proposed index scheme QUISIS shows the best performance except the case when FL is 0.01%. Figure 5-(a), BMQ-Index shows the best performance when FL is 0.01% (i.e., high locality) due to its simple structure. In BMQ-Index, if the forthcoming data is different from the current data, many DMR nodes should be retrieved. Therefore, BMQ-Index shows the worst performance when FL is 0.1% (see Figure 5-(b)) and 1% (see Figure 5-(c)). In other words, BMQ-Index only fits on the high locality cases. In contrast to BMQ-Index, QUISIS shows good performance over all cases since QUISIS efficiently retrieves the query set using TIL and removes the overhead searching from *Header*. The performance of Interval Skip Lists does not affected by FL. As shown in Figure 5-(c), Interval Skip Lists shows the good performance when FL = 1.0%. Particulary, when FL are 0.1% and 1%, Interval Skip Lists is superior to BMQ-Index.

Consequently, QUISIS is shown to provide reasonable performance over diverse data locality.

5 Conclusion

In this paper, we present an efficient scheme for query indexing, called QUISIS which utilizes the data locality. QUISIS is based on Interval Skip Lists. In order to maintain the current locality, TIL (temporal interesting list) is equipped. To show the efficiency of QUISIS, we conducted an extensive experimental study with the synthetic data. The experimental results demonstrate that QUISIS is superior to existing query index schemes.

References

1. Arasu, A., Babcock, B., Babu, S., Datar, M., Ito, K., Motwani, R., Nishizawa, I., Srivastava, U., Thomas, D., Varma, R., Widom, J.: STREAM: The Stanford Stream Data Manager. IEEE Data Engineering Bulletin **26** (2003)
2. Chandrasekaran, S., Cooper, O., Deshpande, A., Franklin, M.J., Hellerstein, J.M., Hong, W., Krishnamurthy, S., Madden, S., Reiss, F., Shah, M.A.: TelegraphCQ: Continuous Dataflow Processing. In: ACM SIGMOD Conference. (2003)
3. Ross, K.A.: Conjunctive selection conditions in main memory. In: PODS Conference. (2002)
4. Lee, J., Lee, Y., Kang, S., Jin, H., Lee, S., Kim, B., Song, J.: BMQ-Index: Shared and Incremental Processing of Border Monitoring Queries over Data Streams. In: International Conference on Mobile Data Management (MDM'06). (2006)
5. Hanson, E.N., Johnson, T.: Selection Predicate Indexing for Active Databases Using Interval Skip Lists. Information Systems **21** (1996)
6. Madden, S., Shah, M.A., Hellerstein, J.M., Raman, V.: Continuously adaptive continuous queries over streams. In: ACM SIGMOD Conference. (2002)
7. Guttman, A.: R-Trees: A Dynamic Index Structure for Spatial Searching. In: ACM SIGMOD Conference. (1984)
8. Brinkhoff, T., Kriegel, H., Scheneider, R., Seeger, B.: The R*-tree: An Efficient and Robust Access Method for Points and Rectangles. In: ACM SIGMOD Conference. (1990)
9. Choi, S., Lee, J., Kim, S.M., Jo, S., Song, J., Lee, Y.J.: Accelerating Database Processing at e-Commerce Sites. In: International Conference on Electronic Commerce and Web Technologies. (2004)

Accelerating XML Structural Matching Using Suffix Bitmaps

Feng Shao, Gang Chen, and Jinxiang Dong

Dept. of Computer Science, Zhejiang University, Hangzhou, P.R. China
`microf_shao@msn.com,`
`cg@zju.edu.cn, djx@zju.edu.cn`

Abstract. With the rapidly increasing popularity of XML as a data format, there is a large demand for efficient techniques in structural matching of XML data. We propose a novel filtering technique to speed up the structural matching of XML data, which is based on an auxiliary data structure called *suffix bitmap*. The suffix bitmap captures in a packed format the suffix tag name list of the nodes in an XML document. By comparing the respective suffix bitmaps, most of the unmatched subtrees of a document can be skipped efficiently in the course of structural matching process. Using the suffix bitmap filtering, we extend two state-of-the-art structural matching algorithms: namely the *traversal matching algorithm* and the *structural join matching algorithm*. The experimental results show that the extended algorithms considerably outperform the original ones.

Keywords: XML, Suffix Bitmap, Structural Matching, Filtering.

1 Introduction

In the past decade, while XML has become the de facto standard of information representation and exchange over the Internet, efficient XML query processing techniques are still in great demand. The core problem of XML query processing, namely structural matching, still remains to be a great challenge. In this paper, we propose a novel acceleration technique for structural matching of XML documents. Our method utilizes an auxiliary data structure called suffix bitmap, which compresses the suffix tag names list of XML nodes in a packed format. In an XML document tree, each node corresponds to a sub-tree, which is rooted at the node itself. The suffix tag names list of an XML node contains all the distinct tag names in its corresponding sub-tree, which is described in [2]. For ease of implementation and store efficiency, we present a novel data structure called *suffix bitmap* to compress suffix tag names list. Suffix bitmap contains the non-structural information of XML sub-tree. As bitwise computation can be processed efficiently, we can skip most of the unmatched subtrees using bitwise suffix bitmap comparison. Therefore, the suffix bitmap can be deployed to filter the unmatched subtrees of XML documents.

In this paper, we will integrate suffix bitmap filtering into the traversal matching algorithm and the structural join matching algorithm. The experiments show that the extended matching algorithms considerably outperform original algorithms. Also, we present the construction procedure of suffix bitmap with linear time complexity. To

Y. Shi et al. (Eds.): ICCS 2007, Part I, LNCS 4487, pp. 253–260, 2007.

reduce the memory consumption further, we also present the variable-length suffix bitmap.

The rest of the paper is organized as follows. Section 2 describes the suffix bitmap and its construction algorithm. Section 3 integrates the suffix bitmap filtering into original matching algorithms. Section 4 compares the extended matching algorithms to the original algorithms. Section 5 lists some related work.

2 Suffix Bitmap

2.1 Global Order of Tag Names

Given a XML document, we define the global order of tag names. Each distinct tag name has a global sequence number, which is an incremental positive number and starts at 0. The assigned function is:

Definition 1

γ : *tagName* $\rightarrow$ *GSN*   *GSN is increasing numeric that starts from 0*

$GSN_{tagname1} < GSN_{tagname2}$ *iff tagname1 first appears before tagname2 on XML document import*
After global sequence numbers assigning, we get the global order of tag names. We represent the global order relation of tag names as an order set called tag names set.

Definition 2

$Set_{tagname} = \{ tagname_0, tagname_1, \dots , tagname_{n-1} \}$
The GSN of tagname$_i$ is i
For example, in figure1, tagname$_1$ is *bib*, tagname$_2$ is *book* and so on.

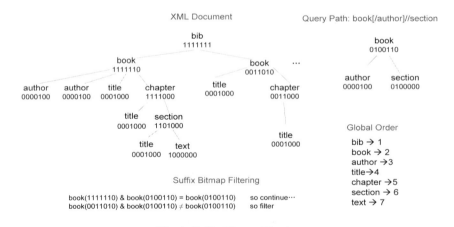

Fig. 1. Suffix Bitmap Filtering

2.2 Suffix Bitmap

We first give the description of suffix tag names list. In XML document tree, each node *n* corresponds to a sub-tree *SubTree$_n$*. The suffix tag names list of node *n* contains all distinct tag names appeared in *SubTree$_n$*. Due to the global order of tag names, we can represent suffix tag names as a bitmap, called suffix bitmap. We define the suffix bitmap as follows:

Definition3

Suffix Bitmap is the bitmap format of suffix tag names list. It has m bits.
Here m is the cardinality of $Set_{tagname}$
Given a bit Bit_i in suffix bitmap of node n

$$Bit_i = \begin{cases} & \text{if } tagname_i \text{ appears in } SubTree_n \\ 0 & \text{else} \end{cases}$$

For example, in figure1, *chapter(1111000)* has *title, chapter, section, text* in its subtree. We attach a suffix bitmap to each node in XML document tree. So the memory consumption of total suffix bitmaps is *node_count * m / 8* bytes, where *node_count* is the number of nodes in XML document.

2.3 Construction of the Suffix Bitmap

We give a preorder construction algorithm of the suffix bitmap, which runs on XML document import. The detail description is below:

Suffix Bitmap Construction Algorithm
1.   void StartDocument() {
2.   createStack()
3.   }
4.   void EndDocument() {
5.   destoryStack()
6.   }
7.   void StartElement(String name) {
8. $BSuffix_{curr} = 1 \ll getTagNameIdx(name)$
9.   pushStack($BSuffix_{curr}$)
10. }
11. void EndElement(String name) {
12. $BSuffix_{curr} = PopStack()$
13. $BSuffix_{parent} = StackTop()$
14.   IF $BSuffix_{parent} \neq NULL$
15. $BSuffix_{parent} = BSuffix_{parent} \mid BSuffix_{curr}$
16.   End IF
17. }

As shown in above algorithm, we use a stack to hold the in-coming suffix bitmap. When **StartElement** call, we create the corresponding suffix bitmap and push it into the stack. When **EndElement** call, we pop the corresponding suffix bitmap out and adapt parent node's suffix bitmap by *or* operation. Thus, the construction order is from bottom to top. The time complexity of construction is **O(n)**, where **n** is the number of nodes in XML document. The function **getTagNameIdx** inputs a tag name and outputs the corresponding global sequence number.

The readers maybe find a problem in our construction algorithm that is we can't determine the length of suffix bitmap during XML import. There are several methods to tackle this problem. One method is we implement the construction of suffix bitmap after XML import. The other method is we pre-estimate the length of suffix bitmap and if the real length of suffix bitmap is larger than the estimate length, we implement the construction again after XML import. Those methods are all inefficient. In the next subsection, we will propose variable-length suffix bitmap to solve this problem.

2.4   Variable-Length Suffix Bitmap

Since we can't determine the final cardinality of $Set_{tagname}$, the variable-length suffix bitmap is proposed. We first define the snapshot of $Set_{tagname}$, which is an order subset of $Set_{tagname}$, as follows:

Definition4

$Snapshot_k = \{tagname_i \mid i \leq k, tagname_i \in Set_{tagname}\}$

$Set_{tagname}$ has m Snapshots on xml document import, where m is the cardinality of $Set_{tagname}$

The variable-length suffix bitmap has the initial length that is equal to the cardinality of the $snapshot_k$ at the moment of creation. When the children suffix bitmap modify the parent suffix bitmap, the length of variable-length parent suffix bitmap is adaptive to the cardinality of current $snapshot_k$. Thus, we modify the length of suffix bitmap in line 8 and 15 of above procedure to corresponding $snapshot_k$ cardinality.

Because the global sequence number of each distinct tag name is invariable since creating, the variable-length suffix bitmap has the same effect as the fix-length suffix bitmap, except that we need to fill the variable-length suffix bitmap with 0s bits during the structural matching, which will be described in next section.

For some xml documents that contain many distinct tag names, our fixed-length suffix bitmap may be waste of large space. We can solve this problem using the variable-length suffix bitmap. To save space further, we present frequent query tag names to reduce the suffix bitmap, which employs some policies of data mining on xml queries.

3   Structural Matching Filtering

3.1   Traversal Filtering

The suffix bitmap contains the tag names information of each sub-tree, so it can judge the unmatched subtrees. Given a query path P and a node n_p on P, the suffix bitmap of query node is the same as the suffix bitmap of the document node. We have the filtering formula as follows:

Formula 1

Given the suffix bitmap BSuffix of element n, and the suffix bitmap QSuffix of query node n_p
IF BSuffix & QSuffix $\neq$ Qsuffix
Then there must be no matching occurrence in $Subtree_n$

In the formula 1, *QSuffix* must be the fix-length suffix bitmap, and *BSuffix* may be the variable-length suffix bitmap. However, it doesn't impact the correctness of the formula 1. For example, in figure 1, we filter xpath *book[/title]//section*. The QSuffix of path root *book* is 0100110. The element *book(0011010)* is filtered because it satisfies formula 1.

Based on the formula 1, we extend the classical traversal matching algorithm and integrate the suffix bitmap filtering into it. During matching traverse, we can skip most of the unmatched sub-tree by employing the formula 1. We give the modification detail of DFS algorithm, which is a state-of-the-art traversal algorithm.

Suffix Bitmap Filtering in DFS
1.   void DFS (Node QRoot, Element DocRoot) {
2.   primaryMatch(QRoot, DocRoot)
3.   }
4.   void PrimaryMatch(Node node, Element element) {
5.   IF (Filter(node, element))
6.   return
7.   // traversal matching process
8.   …
9.   // recursive PrimaryMatch call
10.   …
11. }
12. Boolean Filter(Node node, Element e) {
13.   IF $BSuffix_e$ & $BSuffix_{node}$!= $BSuffix_{node}$
14.   return true
15.   Else
16.   return false
17.   End IF
18. }

As shown in above algorithm, we call **Filter** function before each node matching. If the node satisfies the filtering condition, we skip the following matching process and recursive **PrimaryMatch** call again. Therefore, the bitmap suffix filtering is a highly efficient technique to skip most of the unmatched sub-trees. The efficiency of speedup is dependant on hit ratio of the structural matching. If hit ratio of the structural matching is fairly low, the filtering will be highly efficient. Since the cost of the function **Filter** is tiny, it adds very low overhead to matching process.

3.2 Cursor Stream Filtering

The structural join algorithms based on numbering schema are also the classical matching algorithms, which fall into the set-based category. The structural join algorithms utilize the structural numbering to judge the relationship of candidate nodes, which are obtained by the stream cursor. A disadvantage of structural join algorithms is that there often exist many useless intermediate results during processing because of uncertainty. Using the suffix bitmap filtering, we can eliminate most of useless intermediate results, therefore speed up the matching in structural join algorithms. The extra portion about integrating the suffix bitmap filtering into the structural join is described below:

Suffix Bitmap Filtering in Structural Join Algorithms
1.   Element getNextElement(Node node) {
2.   Element e
3.   do {
4.   e = getNextInStream(node)
5.   } while (e != null && Filter(node, e))
6.   return e
7.   }

As shown in pseudo-code, the function **getNextElement** is the basic function of cursor operation of structural join. We encapsulate it to a loop style, which skip the unmatched elements by the suffix bitmap filtering that has been described in above subsection. It can save costly time to filter the unmatched nodes. There has one difference between the filtering in traversal algorithms and the filtering in structural join algorithms. The former is to skip the unmatched sub-trees and the latter is to skip the unmatched nodes. However, there has the same effect that is the speedup.

4 Experiments

In this section, we present results and analyses of experiments on the comparison between extended algorithms and previous algorithms. All of our experiments were performed on a PC with Pentium4 2.4GHz CPU, 1GB memory and 120 GB IDE hard disk. The OS is Windows XP. We implemented all algorithms in JAVA. We implemented **DFS, FS-DFS, SJoin** and **FS-SJoin** algorithms. DFS is a depth-first traversal algorithm with inline predication checking. SJoin is a novel bottom-up structural join algorithm that enhances the holistic stack join algorithm. FS-DFS is our extended DFS algorithm with the suffix bitmap filtering. And FS-SJoin is our extended SJoin algorithm with the suffix bitmap filtering. The datasets include Nasa, XMark and Treebank [8]. There have three groups of queries: QN, QX and QT which correspond to Nasa, XMark and Treebank. The detail of datasets and queries is below:

Table 1. Datasets

Name	Type	Size(MB)	Elements	Level
Nasa	Small	24.5	476,646	8
XMark	Medium	113.8	1,666,315	12
Treebank	Large	84	2,437,666	36

We tested the performance of all algorithms by running each group of queries 1000 times. For the purpose of simplicity, we ran test codes all in memory and didn't refer to I/O. The figure 2 depicts the comparison of algorithm performance. As shown in figure 2, we find that our extended algorithms outperform previous algorithms 10%~300%. SJoin outperforms DFS when the query has many ancestor-descendant edges. DFS outperforms SJoin when the query has many parent-child edges.

Table 2. Queries

QID	Results	Query
QN1	2407	reference//source[/journal]//date
QN2	60663	tableHead[//tableLink/title]//name
QN3	23224	/datasets//fields[//definition]//para
QN4	9001	reference//journal[/title][/name]//author
QN5	813	dataset[/keywords]/reference[/related]/source/journal/name
QN6	13122	tableHead//footnote
QX1	1629	/site/people/person[/profile[age]/education]/phone
QX2	3274	/site/people/person[//age]//education
QX3	6409	people[//education]//age
QX4	6285	people//address[//city]//province
QX5	6000	europe/item[/name]/description
QX6	18579	closed_auction[//date]//annotation//text
QT1	3	VP[/DT]//PRP_DOLLAR_
QT2	151	S/VP/PP[/NP/VBN]/IN
QT3	16	S[//MD]//ADJ
QT4	5	S[/JJ]/NP
QT5	31	NP[//RBR_OR_JJR]//PP
QT6	21	NP/PP[//NNS_OR_NN]//NN

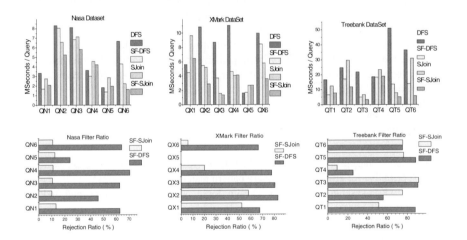

Fig. 2. Query Time and Filter Ratio

However, SF-DFS outperforms DFS at all queries. SF-SJoin outperforms SJoin as well. Also, figure 2 depicts the filter ratio about these three groups of queries. It shows that the efficiency of speedup depends on the filter ratio. When the filter ratio is higher, the efficiency of speedup is more desirable. We can find that the filtering adds very low overhead to matching process when the filter ratio is zero, which is shown in QX5.

5 Related Work

There has been much work in the area of xml structural matching. Structural join [1] and twig join [3] are efficient structural matching algorithms that have been extensively studied. These algorithms determine structural relationship of any two nodes by employing numbering schemes that have also been studied. DFS and BFS [7] are classical traversal algorithms for XML structural matching, which are extensively applied to xml summary graphs [5].

Also, there has been much previous work on speedup of xml structural matching. XR-Tree [4] and XB-Tree [3] are two index strategies to accelerate structural join. [6] proposed a look-ahead approach to skip useless intermediate results in holistic join. Recently, the Subtree Label is presented by [2] in native xml storage system. The Subtree Label is similar to our suffix bitmap, yet doesn't refer to the compression.

6 Conclusions and Future Work

In this paper, we present a novel filtering technology for structural matching, which we called suffix bitmap filtering. Based on the suffix bitmap filtering, we extend two state-of-the-art structural matching algorithms: the traversal matching algorithm and the structural join matching algorithm. For the traversal matching algorithm, we skip most of unmatched sub-trees in terms of the suffix bitmap comparison. For the structural join matching algorithm, we skip most of unmatched candidates as well. The experiments show that our extended algorithms perform significantly better than the previous algorithms, especially when the hit ratio of the matching is low.

In future, we will study the suffix bitmap of frequent tag names. Furthermore, the self-tuning technology along with several filtering policies will be considered to speed up the structural matching.

References

1. Al-Khalifa, S., Jagadish, H.V., Patel, J.M., Wu, Y., Koudas, N., Srivastava, D.: Structural joins: A primitive for efficient XML query pattern matching. ICDE (2002)
2. Boulos, J., Karakashian, S.: A New Design for a Native XML Storage and Indexing Manager. EDBT (2006)
3. Bruno, N., Koudas, N., Srivastava, D.: Holistic twig joins: Optimal XML pattern matching. SIGMOD Conference (2002)
4. Jiang, H., Lu, H., Wang, W., Ooi, B.: XR-Tree: Indexing XML Data for Efficient Structural Joins. ICDE (2003)
5. Kaushik, R., Shenoy, P., Bohannon, P., Gudes, E.: Exploiting Local Similarity for Indexing Paths in Graph-Structured Data. ICDE (2002)
6. Lu, J., Chen, T., Ling, T.: Efficient Processing of XML Twig Patterns with Parent Child Edges: A Look-ahead Approach. Proc of CIKM (2004)
7. Wang, W., Wang, H., Lu, H., Jiang, H., Lin, X., Li, J.: Efficient Processing of XML Path Queries Using the Disk-based F&B Index. VLDB (2005)
8. XML Data Repository In http://www.cs.washington.edu/research/xmldatasets/

Improving XML Querying with Maximal Frequent Query Patterns

Yijun Bei, Gang Chen, and Jinxiang Dong

College of Computer Science, Zhejiang University, Hangzhou, P.R. China 310027
alphabyj@yahoo.com.cn, cg@zju.edu.cn, djx@zju.edu.cn

Abstract. Querying on XML data is a computational-expensive process due to the complex nature of both the XML data and the query. In this paper, we propose an approach to expedite XML query processing by caching the results of a specific class of queries, namely the maximal frequent queries. We mine the maximal frequent query patterns from user-issued queries and cache the results of such queries. We propose a recursive algorithm for query processing using the cached query results. Query rewriting is employed to deal with four kinds of similar queries namely exact matching, exact containment, semantic matching and semantic containment. We perform experiments on the XMARK datasets and show that the proposed methods are both effective and efficient in improving the performance of XML queries.

Keywords: Caching, XML Query, Ming, Maximal Frequent Pattern.

1 Introduction

With the proliferation of XML as a standard for data representation and data exchanging, effective and efficient querying techniques of XML data becomes an important topic for the database community. For the sake of flexible query mechanisms, query languages like XPath [1] and XQuery [2] have been provided for XML data retrieval. XML queries represented by these query languages can be modeled as trees which are called query pattern trees.

Caching has played an important part in improving performance of XML query processing, especial for repeated or similar queries. Users can obtain answer right away if the query results have been computed and cached. Several caching approaches [3, 4, 5 and 6] have bee provided for XML queries. However these approaches only consider frequent query pattern trees rooted at the root of DTD. As a result, cache schemes in these approaches don't take effect on queries not rooted at the root of DTD. In our approach, we discover frequent query patterns rooted at any level of a DTD and can deal with arbitrary quires. One nice property of frequent trees is the apriori property, i.e., if a query pattern is frequent, then all of its query sub-patterns are also frequent. In case that we cache a query pattern, we will have to cache all its sub-patterns, which results in some unnecessary cache queries. In order to make full use of limited cache size, we will cache only maximal frequent XML queries.

In this paper we introduce the strategy of caching only maximal frequent XML queries. We present a recursive algorithm for obtaining query result by query

Y. Shi et al. (Eds.): ICCS 2007, Part I, LNCS 4487, pp. 261–269, 2007.
© Springer-Verlag Berlin Heidelberg 2007

rewriting and restructuring. To make full use of limited size of cache pool, we describe a cache replacement scheme that that utilizes recent accessing frequency as well as the query pattern support and query pattern discovered time.

The rest of the paper is organized as follows. Section 2 discusses previous work related to the XML caching and mining. In section 3, we present some concepts used in mining and caching process. In section 4 we describe an algorithm for query processing. Section 5 presents results of experiments and we conclude in section 6.

2   Related Work

Indexing strategy can accelerate XML query processing. However, it will cost much space to store the index in memory. Caching results of XML queries has been considered as a good strategy to improve performance of XML query processing. Semantic caching [3 and 4] has received attention in XML data retrieval. In [3] a compact structure called MIT is employed to represent the semantic regions of XML queries. XCache in [4] is a holistic XQuery-based semantic caching system. [5 and 6] incorporate mining approach to find frequent queries for caching. An efficient algorithm called FastXMiner in [5] is presented to discover frequent XML query patterns. Authors in [6] took into account the temporal features of user queries for frequent queries discovery and design an appropriate cache replacement strategy by finding both positive and negative association rules. However these two approaches only consider frequent query pattern tree rooted at the root of DTD.

For the problem of finding tree-like patterns, many efficient tree mining approaches have been developed. Basically, there are two steps for finding frequent trees in a database. Firstly, non-redundant candidate trees are generated. Secondly, the support of each candidate tree is computed. Asai presented a rooted ordered and a rooted unordered tree mining approach in [7] and [8] respectively. Zaki gave ordered and unordered embedded tree mining algorithms in [9, 10].

3   Background

3.1   Maximal Frequent Query Pattern Tree

Query Pattern Tree. A *query pattern tree* is a rooted tree $QPT = <N, E>$ where:
- N is the node set. Each node n has a label whose value is in {"*", "//"} $\cup$ labelSet where the labelSet is the label set of all elements, attributes and texts.
- E is the edge set. For each edge $e = (n_1, n_2)$, node n_1 is the parent of n_2.

Query Pattern Subtree. Given two query pattern trees T and S, we say that S is a *query pattern subtree* of T or T is a *query pattern supertree* of S, iff there exists a one-to-one mapping $\varphi: V_S \rightarrow V_T$, such that satisfies the following conditions:
- φ preserves the labels, i.e., $L(v) = L(\varphi(v))$, $\forall v \in V_S$.
- φ preserves the parent relation, i.e., $(u,v) \in E_S$ iff $(\varphi(u), \varphi(v)) \in E_T$.

Frequent Query Pattern Tree. Let Q denote all the query pattern trees of the issued queries and d_T be an indicator variable with $d_T(S) = 1$ if query pattern tree S is a query pattern subtree of T and $d_T(S) = 0$ if tree S is not a subtree of T. The support of query pattern tree S in Q can be defined as $\sigma(S) = \sum_{T \in Q} d_T(S) / \sum_{T \in Q}$, i.e., the percent of the number of trees in Q that contain tree S. A query pattern tree is frequent if its frequency is more than or equal to a user-specified *minimum support.*

Maximal Frequent Query Pattern Tree. A frequent query pattern tree is *maximal* if none of its proper supertrees is frequent.

We show a set of query pattern trees in Figure 1(a) and three query pattern subtrees in Figure 1(b). Given the minimum support 0.6, then the subtree S_1 is not a frequent one, while S_2 and S_3 are frequent query pattern trees. However, S_2 is not a maximal frequent query pattern tree, since one of its supertree S_3 is also a frequent one.

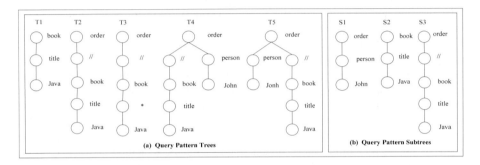

Fig. 1. Maximal Frequent Query Pattern Tree

3.2 Query Rewriting

Few users issue the exactly same queries when performing XML Querying. However, issued queries are often similar and have little difference with others. Therefore, when user issues a new similar query, a query rewriting should be performed to utilize cached queries. We introduce 4 relationships between two similar query pattern trees.

Exact Matching. If query pattern tree T_1 is a query pattern subtree T_2 and T_2 is also a query pattern subtree of T_1, then tree T_1 exactly matches tree T_2.

Exact Containment. If a query pattern tree T_2 is a query pattern subtree of T_1, then T_1 exactly contains tree T_2.

Semantic Matching. We employ a similar idea of *Extended Subtree Inclusion* [5] for the definition of *Semantic Matching.* Let T_1 and T_2 be two query pattern trees with root nodes t_1 and t_2. Let children(n) denotes the set of child nodes of n. We can recursive determine if tree T_1 semantically matches tree T_2, as follows:
$t_1 \leq t_2$ and satisfies:

- both t_1 and t_2 are leaf nodes; or
- t_1 is a leaf node and $t_2 = $ "//", then $\exists\ t_2' \in$ children(t_2) such that semanticmatch(T_1, T_2'); or
- both t_1 and t_2 are non-leaf nodes, and one of the following holds:

> 1) $\forall t_1' \in$ children(t_1), $\exists t_2' \in$ children(t_2) such that semanticmatch(T_1',T_2');
>
>   2) t_2 = "//" and $\forall t_1' \in$ children(t_1) we have semanticmatch(T_1',T_2');
>
>   3) t_2 = "//" and $\exists t_2' \in$ children(t_2) we have semanticmatch(T_1,T_2');

Semantic Containment. If any query pattern subtree of T_1 semantically matches query patterns T_2, then we have T_1 semantically contains tree T_2.

Figure 2 outlines 4 query rewriting cases. In Figure 2(a), result of the issued query can be easily acquired as the new query exactly matches the cached query. In Figure 2(b), the new query exactly contains the cached query, so result of left query subtree can be obtained from the cached one, and computation of the right one is enough. In Figure 2(c), since the new one semantically matches a cached one, result of the issued query can be retrieved through two steps. Firstly, results of semantic one is obtained. Secondly we compute the parent-child relation of nodes book and title, node title and java respectively instead of original grandparent-grandchild relation of node book and java. Using semantic matching, we reduce the search space and only need to compute relations on different labels between queries. Figure 2(d) presents a semantic containment case. Since the new query semantically contains the cached one, the result of left query subtree can be obtained with semantic matching method. Then we compute the right subtree and merge results of the two subtrees to achieve final result.

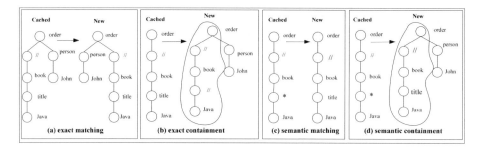

Fig. 2. Query Rewriting

4   Query Processing

4.1   Maximal Frequent Query Patterns Discovery

We use existent efficient mining algorithm for discovery of maximal frequent query patterns and perform the mining process automatically. We record all user queries and launch a mining process when the count of queries reaches our predefined value. Further more, the specified minimum support will be automatically tuned to be fit for cache pool. Suppose the mined maximal frequent queries are too many, which is larger than our cache pool size. We will automatically tune our minimum support threshold to be a larger one and discover less frequent queries next time. On the contrary, on condition that the frequent queries are less, we need decrease the minimum support to adapt to cache pool.

4.2 Querying Algorithm

Figure 3 outline an algorithm for XML querying using frequent patterns caching. The querying process comprises two parts: matched query discovery and general query performing. The input of the algorithm is a query pattern tree and the output is a set of matched root nodes. First of all, we try to find an exactly matched or semantically matched query from cache pool. If an exactly matched query is found, then result can

---

Algorithm Query(qpt)

Input: qpt (query pattern tree)

Output: nodeset (matched root node set)

matchedqpt = obtain query pattern exactly matching *qpt* from cache;

if (*matchedqpt* != null) // exactly matched query exist

    nodeset = result of *matchedqpt*;

    return;

end if

matchedqpt = obtain query pattern tree semantically matching *qpt* from cache;

if (*matchedqpt* != null)   // semantically matched qpt exist

    if (semantic match from ordinal label *l* to wildcard "*" or descendent path "//")

        compute child-parent relation between *l* and parent node of "*" (or "//");

        compute parent-child relation between *l* and child ndoe of "*" (or "//");

        merge join the two result set;

    else if (semantic match from wildcard "*" to descendent path "//")

        compute grandparent-grandchild relation between parent of "//" and child of "//";

    end if

    nodeset = result of *matchedqpt* satisfying previous relation computation;

    return;

end if

root = root node of *qpt*;

nodeset = get all nodes from XML document equals to label root;

for (each direct query pattern subtree *dqpst* of *root*)

    if (*nodeset* == Φ)

        return;

    end if

    subnodeset = query (*dqpst*);

    switch(type of root of dqpst)

        case ordinal_label:

            nodeset = merge join parent-child nodes(*nodeset*, *subnodeset*);

        case "*"

            nodeset = merge join grandparent-grandchild nodes(*nodeset*, *subnodeset*);

        case "//"

            nodeset = merge join ancestor-descendent nodes(*nodeset*, *subnodeset*);

    end switch

end for

return;

Fig. 3. Querying Algorithm

be acquired directly from cache. In case that a semantically matched query is found, result from cache is required a simple modification, as follows:

- Suppose semantic match is from an ordinal label l to wildcard "*" or descendent path "//". Two relations, i.e., child-parent relation between label l and parent of "*" (or "//"), parent-child relation between label l and child of "*" (or "//"), need computed.
- Suppose semantic match is from wildcard "*" to descendent path "//". In order to transform from ancestor-descendent relationship to grandparent-grandchild relationship, a grandparent-grandchild relationship should be computed between parent of "//" and child of "//".

From cache pool, we don't obtain exactly contained or semantic contained queries due to time-consuming computations of the two relationships. We adopt an approximate approach to simulate them. We perform a top to down search starting at the root of the query until we find exact matched or semantic matched query. In this way, we not only avoid complex computation, but also may enhance cache hit ratio in respect that a partial query is highly possible to be cached.

If there not exist a matched query in cache pool, a general querying process needs performed. We join the node set of root with result of each direct subtree of root. Assume the matched root node set denoted as *nodeset* and result of direct subtree is denoted as *subnodeset*, the merge join process is defined as follows:

- If the root of direct subtree has an ordinal label, parent-child relation computation is required to join the two node sets *nodeset* and *subnodeset*.
- If the root of direct subtree has a wildcard "*" label, grandparent-grandchild relation computation is required to join the node sets *nodeset* and *subnodeset*.
- If the root of direct subtree has a descendent path "//" label, ancestor-descendent relation computation is required to join the node sets *nodeset* and *subnodeset*.

4.3 Cache Replacement

We keep track of recent accessing frequency, support and discovered time of maximal frequent query patterns. Recent accessing frequency is considered as the most important factor for cache replacement. We divide accessing frequency of query patterns into several levels. Query patterns in lower levels will be selected as victims. Support and discovered time are also taken into account for cache replacement. For query patterns in the same level, we will need to consider the support and discovered time. When the cache is full, the replacement manager selects the query with least support and latest discovered one. Query pattern with a larger support should not be selected as a victim since more issued queries containing current query pattern. To avoid unnecessary computation, a lazy-result-retrieval scheme is employed. Result of frequent query is not retrieved until the query needs to be used.

5 Experimental Evaluation

In this section, we present experimental results of the prototype system of XML querying algorithm by caching maximal frequent query patterns. We implemented it in Java and carried out all experiments on an Intel Xeon 2GHz computer with 1GB of RAM running operating system RedHat Linux 9.0. We loaded XML dataset into memory and created indices for XML data before querying. Structural joins [11] approach is employed to decide node relationships when performing querying.

5.1 Dataset

We generate a XML document with size 116MB using tools provided by XMARK [12] for performance test. Queries are generated randomly using the following probabilities: site(0.5), regions(0.8), people(0.7), , africa(0.1), europe(0.5), asia(0.4), australia(0.2), item(0.5), item/name(0.3), incategory(0.4), quantity(0.2), person(0.6), person/name(0.5), emailaddress(0.6), *(0.1), //(0.1). The wildcard * and descendent path // are added to make the queries more complex. In the following experiments, we will set the cache size to 50 which means at most results of 50 queries will be cached.

5.2 Cache Performance

We evaluate the performance of caching scheme using maximal frequent query patterns (MFQP), and compare it with caching policy using frequent patterns (FQP), LRU caching policy (LRU) and MRU caching policy (MRU).

In Figure 4(a) we show the average response time for query processing with varying number of queries from 10,000 to 100,000, in which Q10K for total number of queries Q = 10,000, Q100K for Q = 100,000. The support of frequent queries mining is 0.1 for frequent queries cache policy. The average response time is defined as the ratio of total running time for answering a set of queries to the total number of queries in this set. MFQP is the most efficient compared to FQP, LRU and MRU. Caching policy with mining approach outperforms traditional LRU and MRU caching policy due to the discovery and store of useful queries. FQP policy will cache some small-size frequent patterns whose information may have been contained by large-size frequent patterns. In a sense, FQP policy wastes some cache space. Caching system employing MFQP policy also has a good scalability as the response time for querying doesn't vary much with varying number of queries from Q10K to Q100K.

In Figure 4(b), we show the average response time for query processing with varying support of queries from 0.01 to 0.1 for Q50K query set. We can see that the varying support dose not influence query response time much. However, when the

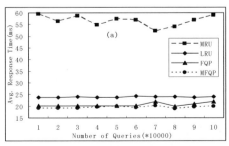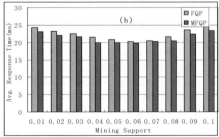

Fig. 4. Average Response Time for Querying

support is in the vicinity of 0.06, we have the minimum response time for XML querying. For a larger support less frequent queries will be mined and the cached results will be less. While for a less support the mining time will increase which will result in an additional cost for querying XML data.

Figure 5 (a) and (b) shows the hit ratio for our frequent queries caching system with varying number of queries and support respectively. From experiments we can see that MFQP policy has the highest hit ratio, which further demonstrates caching with MFQP policy outperform the other three policies.

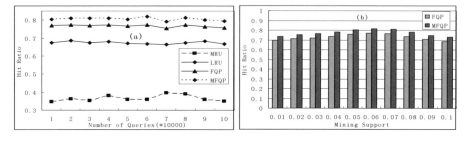

Fig. 5. Hit Ratio for Caching Queries

6   Conclusion

In this paper, we propose an approach for XML querying by caching maximal frequent query patterns. We discover the maximal frequent query patterns from user-issued queries and cache the results of such queries. We present a recursive algorithm for obtaining query results using discovered patterns. To deal with similar queries, we introduce four kinds of query relationships and employ a query rewriting process. Experiments show that our approach is efficient and outperforms the ordinal frequent patterns caching policy and traditional LRU, MRU caching policy.

References

1. Clark, J., DeRose, S.: XML path language (XPath) version 1.0 w3c recommendation. Technical Report REC-xpath-199991116, World Wide Web Consortium (1999)
2. Charmberlin, D., Florescu, D., Robie, J., Simon, J., Stefanescu, M.:XQuery: A query language for XML. W3C Working Draft. http://www.w3.org/TR/xquey (2001)
3. Hristidis, V., Petropoulos, M.: Semantic caching of xml databases. In Proc. Of the 5th WebDB (2002).
4. Chen, L., Rundensteiner, E.A., Wang S.,: Xcache-a semantic caching system for xml queries. In Demo in ACM SIGMOD (2002).
5. Yang, L. H., Lee, M.L., Hsu W.: Efficient mining of xml query patterns for caching. In Proc. of 29th VLDB (2003).
6. Chen, L., Bhowmick, S.S., Chia, L.T.: Mining Positive and Negative Association Rules from XML Query Patterns for Caching. In DASFAA (2005) 736-747.

7. Asai, T., Abe, K., Kawasoe, S., Arimura, H., Satamoto, H., Arikawa, S.: Efficient Substructure Discovery from Large Semi-structured Data, SDM (2002) 158-174.
8. Asai, T., Arimura, H., Uno, T., Nakano, S.: Discovering Frequent Substructures in Large Unordered Trees, 6th Int'l Conf. on Discovery Science (2003).
9. Zaki, M. J.: Efficiently Mining Frequent Trees in a Forest, 8th ACM SIGKDD Int'l Conf. Knowledge Discovery and Data Mining (2002).
10. Zaki, M. J.: Efficiently Mining Frequent Embedded Unordered Trees, Fundamenta Informaticae (2005)
11. S. Al-Khalifa, H. V. Jagadish, N. Koudas, J. M. Patel, D. Srivastava, and Y. Wu. Structural Joins: A Primitive for Efficient XML Query Pattern Matching. In Proceedings of the IEEE International Conference on Data Engineering (2002) 141-152
12. http://monetdb.cwi.nl/xml/

A Logic-Based Approach to Mining Inductive Databases

Hong-Cheu Liu[1], Jeffrey Xu Yu[2], John Zeleznikow[3], and Ying Guan[4]

[1] Department of Computer Science and Information Engineering,
Diwan University, Taiwan
hcliu@dwu.edu.tw
[2] Department of Systems Engineering and Engineering Management,
The Chinese University of Hong Kong, China
yu@se.cuhk.edu.hk
[3] School of Information Systems, Victoria University, Australia
John.Zeleznikow@vu.edu.au
[4] School of Information Technology and Computer Science,
University of Wollongong, Australia
yguan@uow.edu.au

Abstract. In this paper, we discuss the main problems of inductive query languages and optimisation issues. We present a logic-based inductive query language and illustrate the use of aggregates and exploit a new join operator to model specific data mining tasks. We show how a fixpoint operator works for association rule mining and a clustering method. A preliminary experimental result shows that fixpoint operator outperforms SQL and Apriori methods. The results of our framework could be useful for inductive query language design in the development of inductive database systems.

1 Introduction

Knowledge discovery from large databases has gained popularity and its importance is well recognised. Ever since the start of research in data mining, it has been realised that the knowledge discovery process should be supported by database technology. However, most efforts on data mining and knowledge discovery have focused on developing novel algorithms and data structures and these researches concentrated on examining the efficient implementation issues. While Data Base Management Systems (DBMS) and their enabling technology have evolved successfully to deal with most of the data-intensive application areas including decision support systems with OLAP queries, these techniques are still insufficient in a knowledge discovery environment. Therefore, databases today are still using primarily a cache-mining approach, where the data is first extracted from the database to a memory cache, which is then processed using procedural mining methods.

Research on inductive databases focuses on the integration of databases with data mining. Such integration has been formalised in the concept of inductive databases. The key ideas are that data and patterns (or models) are first class

Y. Shi et al. (Eds.): ICCS 2007, Part I, LNCS 4487, pp. 270–277, 2007.

citizen, i.e., they are handled in the same way. The emergence of Inductive DBMS research aims to improve the current state of cache mining approach and make it easy to mine databases by their query languages [1,2]. The one of crucial criteria for the promising success of inductive databases is reasoning about query evaluation and optimisation. In this paper, we will focus on inductive query languages and optimisation issues.

Several specialised inductive query languages have been proposed and implemented, such as MSQL [3], DMQL [4] and MINE RULE [5]. These projects have made a number of contributions including exploring and demonstrating some of the key features required in an Inductive DBMS. Currently, these researches have not led to significant commercial deployments due to performance concern and practical limitations.

The data mining query languages proposed in [3,4,5] require users to only provide high-level declarative queries specifying their mining goals. The underlying inductive database systems need sophisticated optimiser with aim to selecting suitable algorithms and query execution strategies in order to perform mining tasks. Another tight-coupling approach using SQL implementations gives unrealistic heavy-burden on the users to write complicated SQL queries [6]. So it is reasonable to explore alternative methods that make inductive databases realisable with current technology. Logic-based database languages provide a flexible model of representing, maintaining and utilising high-level knowledge. This motivates us to study a logic-based framework and develop relational operators (e.g., fixpoint) for intelligent data analysis.

In this paper, we consider the logic paradigm for association rule mining that use the idea of least fixpoint computation. We also demonstrate a logic query language for data mining tasks and discuss optimisation issues. Some preliminary experimental results show that our fixpoint algorithm outperforms SQL-based and the Apriori methods.

2   Inductive Query Languages

A desired feature of inductive database systems is the ability to support ad hoc and interactive data mining in order to facilitate flexible and effective knowledge discovery. Data mining query languages can be designed to support such a feature. In particular, declarative query language support acts an important role in the next generation of Web database systems with data mining functionality. Query systems should provide mechanism of obtaining, maintaining, representing and utilising high level knowledge in an unified framework. A knowledge discovery support environment should be an integrated mining and querying system capable of representing domain knowledge, extracting useful knowledge and organising in ontology [7].

Designing a comprehensive data mining language is challenging because data mining covers a wide spectrum of tasks and each task has different requirements. In this paper we provide some theoretical foundations for a logic-based data mining query language.

3   Relational Computation for Association Rules

In this section, we investigate relational computation methods and demonstrate a logic-based query paradigm for frequent itemset discovery that use the idea of least fixpoint computation.

3.1   Calculus+*Fixpoint*

We provide a non-inflationary extension of the complex value calculus with recursion and aggregate operation. We define a fixpoint operator which allows the iteration of calculus formulas up to a fixpoint. In effect, this allows us to define frequent itemsets inductively using calculus formulas.

Definition 1. *Let $\mathcal{S}^k(V)$ denote the set of all degree-k subset of V. For any two sets S and s, s is said to be a degree-k subset of S if $s \in \mathcal{P}(S)$ and $|s| = k$. $\mathcal{P}(S)$ denotes the powerset of S.*

The non-inflationary version of the fixpoint operator is presented as follows. Consider association rule mining from object-relational data. Suppose that raw data is first preprocessed to transform to an object-relational database. Let $D = (x, y)$ be a nested table in the mapped object-relational database. For example, $x = items$, $y = count$. *Items* is a set valued attribute. Let $S_x^k(D) = \{t \mid \exists u \in D, v = S^k(u[x]), t = (v, u[y])\}$. We develop a fixpoint operator for computing the frequent itemsets as follows. The relation J_n holding the frequent itemsets with support value greater than a threshold δ can be defined inductively using the following formulas

$$\varphi(T, k) = \sigma_{y \geq \delta}(\,_x\mathcal{G}_{sum(y)}\mathcal{S}_x^k(D)(x, y)) \longrightarrow T(x, y), \text{if } k = 1$$
$$\varphi(T, k) = T(x, y) \vee \sigma_{y \geq \delta}(\,_x\mathcal{G}_{sum(y)}(\exists u, v\{T(u, v)$$
$$\wedge(\mathcal{S}_x^k(D)(x, y)) \wedge u \subset x \longrightarrow T(x, y)\})), \text{if } k > 1$$

as follows: $J_0 = \emptyset$; $J_n = \varphi(J_{n-1}, n), n > 0$. Where $\mathcal{G}$ is the aggregation operator. Here $\varphi(J_{n-1}, n)$ denotes the result of evaluating $\varphi(T, k)$ when the value of T is J_{n-1} and the value of k is n. Note that, for each input database D, and the support threshold δ, the sequence $\{J_n\}_{n \geq 0}$ converges. That is, there exists some k for which $J_k = J_j$ for every $j > k$. Clearly, J_k holds the set of frequent itemsets of D. Thus the frequent itemsets can be defined as the limit of the forgoing sequence. Note that $J_k = \varphi(J_k, k+1)$, so J_k is also a fixpoint of $\varphi(T, k)$. The relation J_k thereby obtained is denoted by $\mu_T(\varphi(T, k))$. By definition, μ_T is an operator that produces a new nested relation (the fixpoint J_k) when applied to $\varphi(T, k)$.

3.2   Fixpoint Algorithm

We develop an algorithm for computing frequent itemset by using the above defined fixpoint operator. We define a new join operator called *sub-join*.

Definition 2. *Let us consider two relations with the same schemes $\{Item, Count\}$. The sub-join, $r \bowtie^{sub,k} s = \{t \mid \exists u \in r, v \in s \text{ such that } u[Item] \subseteq v[Item] \wedge \exists t' \text{ such that } (u[Item] \subseteq t' \subseteq v[Item] \wedge |t'| = k), t = <t', v[Count]>\}$*

Here, we treat the result of $r \bowtie^{sub,k} s$ as multiset meaning, as it may produce two tuples of t' with the same support value.

Example 1. *Given two relations r and s, the result of $r \bowtie^{sub,2} s$ is shown in Figure 1.*

r

Items	Support
$\{a\}$	0
$\{b,f\}$	0
$\{d,f\}$	0

s

Items	Support
$\{a,b,c\}$	3
$\{b,c,f\}$	4
$\{d,e\}$	2

$r \bowtie^{sub,2} s$

Items	Support
$\{a,b\}$	3
$\{a,c\}$	3
$\{b,f\}$	4

Fig. 1. An example of sub-join

Given a database $D = (Item, Support)$ and support threshold δ, the following fixpoint algorithm computes frequent itemset of D.

Algorithm. *fixpoint*
Input: An object-relational database D and support threshold δ.
Output: L, the frequent itemsets in D.
Method:

begin

$L_1 := \sigma_{Support \geq \delta}(\ _{Item}\mathcal{G}_{sum(Support)}\mathcal{S}^k_{Item}(D)))$
for $(k := 2; T \neq \emptyset; k++)$ {
$S := sub\_join(L_{k-1}, D)$
$T := \sigma_{Support \geq \delta}(\ _{Item}\mathcal{G}_{sum(Support)}(S)$
$L_k := L_{k-1} \cup T$
}
return L_k;

end

procedure *sub_join*
(T: frequent k-itemset; D: database)
for each itemset $l_1 \in T$,
    for each itemset $l_2 \in D$,
      $c = l_1 \bowtie^{sub,k} l_2$
      if *has_infrequent_subset* (c, T)
      then delete c else add c to S;
return S;

procedure *has_infrequent_subset*
(c: candidate k-itemset,
T: frequent $(k-1)$-itemsets);
for each $k-1$-subset s of c
    if s not $\in T$ then return TRUE;
return FALSE;

4 A Logic Query Language

In this section, we present a logic query language to model the association rule mining, naive Bayesian classification and partition-based clustering.

Association rule mining

We present an operational semantics for association rule mining queries expressed in $Datalog^{cv,\neg}$ program from fixpoint theory.

We present a Datalog program as shown below which can compute the frequent itemsets.

1. $cand(J, C)$ $\leftarrow freq(I, C), J \subset I, |J| = 1$
2. $large(J, C)$ $\leftarrow cand(J, C), C > \delta$
3. $T(x, C_2)$ $\leftarrow large(J, C_1), freq(I, C_2),$
                             $x \subset I, J \subset x, |x| = max(|J|) + 1,$
4. $T(genid(), x, C)$ $\leftarrow T(x, C), \neg has\_infrequent\_subset(x)$
5. $cand(x, sum < C >) \leftarrow T(id, x, C)$
6. $large(x, y)$ $\leftarrow cand(x, y), y > \delta$

The rule 1 generates the set of *1-itemset* from the input frequency table. The rule 2 selects the frequent *1-itemset* whose support is greater than the threshold. The program performs two kinds of actions, namely, join and prune. In the join component, the rule 3 performs the *sub-join* operation on the table *large* generated in the rule 2 and the input frequency table. The prune component (rule 4) employs the Apriori property to remove candidates that have a subset that is not frequent. The test for infrequent subsets is shown in procedure $has\_infrequent\_subset(x)$.

Datalog system is of set semantics. In the above program, we treat T facts as multisets, i.e., bag semantics, by using system generated *id* to simulate multiset operation. The rule 5 counts the sum total of all supports corresponding to each candidate item set generated in table T so far. Finally, rule 6 computes the frequent itemsets by selecting the itemsets in the candidate set whose support is greater than the threshold.

We now show the program that defines $has\_infrequent\_subset(x)$.

$$has\_infrequent\_subset(x) \leftarrow s \subset x, |s| = |x| - 1, \forall y[large(y, C), y \neq s]$$

Once the frequent itemset table has been generated, we can easily produce all association rules.

Naive Bayesian Classification

Let us consider a relation r with attributes $A_1, ..., A_n$ and a class label C. The Naive Bayesian classifier assigns an unknown sample X to the class C_i if and only if $P(C_i|X) > P(C_j|X)$, for $1 \leq j \leq m, j \neq i$.

We present a datalog program demonstrating that Naive Bayesian classification task can be performed in deductive environment. The program first evaluates the frequencies of the extension of r, each class and each pair of attribute A_i and class.

$freq\_r(A_1, ..., A_n, C, count(*)) \leftarrow r(A_1, ..., A_n, C)$
$freq\_class(C, count(*)) \qquad\quad \leftarrow r(A_1, ..., A_n, C)$
$freq\_A_i\_class(A_i, C, count(*)) \leftarrow r(A_1, ..., A_n, C)$

Then we obtain the probabilities of $P(A_i \mid C)$, as follows.

$$Pr\_class(C,p) \quad \leftarrow \quad freq\_r(A_1, ..., A_n, C, n_r), \; freq\_class(C, n_c), \; p = n_c/n_r$$
$$Pr\_A\_class(A,C,p) \leftarrow \quad freq\_A\_class(A, C, n_{A,C}), \; freq\_class(C, n_c), \; p = n_{A,C}/n_c$$

Finally, we get the answer predicate $Classifier(x_1, ..., x_k, class)$.

$$
\begin{aligned}
Pr(x_1, ..., x_k, class, p) \quad & \leftarrow \quad r(x_1, ..., x_k), \; Pr\_A\_class(A, class, p), \\
& \quad \exists t_i \in Pr\_A\_class, \; 1 \leq i \leq k, \\
& \quad x_1 = t_1.A, ..., x_k = t_k.A, \\
& \quad t_1.class = ... = t_k.class, \; p = \textstyle\prod t_i.p \\
P(x_1, ..., x_k, class, p) \quad & \leftarrow \quad Pr(x_1, ..., x_k, class, p_1), \\
& \quad Pr\_class(class, p_2), \; p = p_1 \times p_2 \\
Classifier(x_1, ..., x_k, class) & \leftarrow P(x_1, ..., x_k, class, p), \; p = max\{P.p\}
\end{aligned}
$$

Example 2. *We wish to predict the class label of an unknown sample using naive Bayesian classification, given a training data. The data samples are described by the attributes age, income, student, and credit-rating. The class label attribute, buys-computer, has two distinct values, namely, YES, NO. The unknown sample we wish to classify is* $X = (age = " <= 30", income = "medium", student = "yes", credit - rating = "fair")$.

The evaluation of the query $Classifier(x_1, ..., x_k, class)$ returns the answer predicate $Classifier(age = " <= 30", income = "medium", student = "yes", credit - rating = "fair", "YES")$.

Cluster analysis: partitioning method

Given a database of n objects and k, the number of clusters to form, a partitioning algorithm organises the objects into k partitions, where each partition represents a cluster. We present a deductive program performing the partitioning-based clustering task, as follows. $P(Y, C_i) \leftarrow r(X), \; 1 \leq i \leq k, \; Y_i = X$; $Cluster(Y, C_i, m_i) \leftarrow P(Y, C_i), \; m_i = mean\{Y\}$ Where *mean* is a function used to calculate the cluster mean value; *distance* is a similarity function. First, it randomly select k of the n objects, each of which initially represents a cluster mean. For each of the remaining objects, an object is assigned to the cluster to which it is the most similar, based on the distance between the object and the cluster mean. It then computes the new mean for each cluster. This process iterates until some criterion function converges, i.e., eventually, no redistribution of the objects in any cluster occurs and so the process terminates.

The following two rules show the clustering process. An operational semantics for the following datalog program P is fixpoint semantics. The immediate consequence operator, T_P is the mapping from instances of schema of P to instances of schema of P.

$$
\begin{aligned}
new\_cluster(X, C) \leftarrow \; & r(X), \; Cluster(Y, C, m), \; Cluster(Y, C', m'), \\
& c \neq c', \; distance(X, m) < distance(X, m'), \\
Cluster(X, C, m) \quad \leftarrow \; & new\_cluster(X, C), \; m = mean\{new\_cluster.X\}
\end{aligned}
$$

5   Performance and Optimisation Issues

Most performance experiments have shown that SQL-based data mining algorithms are inferior to cache-mining approach. The main-memory algorithms used in today's data mining application typically employ sophisticated data structures and try to scan the dataset fewer times compared to SQL-based algorithms which normally require many complex join operations between input tables. However, database support knowledge discovery is important when data mining applications need to analyse current data which is so large that the in-memory data structures grow beyond the size of main-memory.

An SQL3 expression mapping to a least fixpoint operator has been presented in [8]. The fixpoint operator of that article has significant strengths in term of iterative relational computation. But it also requires substantial optimisation and prune component to remove candidates that have a subset that is not frequent. The prune component can be implemented by a k-way join. The SQL3 implementation of a fixpoint operator discussed in [8] does not claim to have achieved performance level that is comparable to those SQL-based approaches, nor does it claim to have identified a query-optimisation strategy.

We argue that if SQL would allow expressing our sub-join, $\bowtie^{sub,k}$, in an intuitive manner and algorithms implementing this operator were available in a DBMS, this would greatly facilitate the processing of fixpoint queries for frequent itemset discovery.

A Datalog expression mapping to our fixpoint operator has more intuitive than SQL expressions. In our opinion, a fixpoint operator is more appropriate exploited in the deductive paradigm which is a promising approach for inductive database systems. The main disadvantage of the deductive paradigm for inductive query languages is the concern of its performance. However, optimisation techniques from deductive databases can be utilised and the most computationally intensive operations can be modularised.

There exist some opportunities for optimisation in the expressions and algorithms expressed in the deductive paradigm. Like in the algebraic paradigm, we may improve performance by exploiting relational optimisation techniques, for example, optimizing subset queries [9], index support, algebraic equivalences for nested relational operators [10]. Optimisation for data mining queries also requires novel techniques, not just extensions of object-relational optimisation technology.

We conducted several experiments on a PC with Pentium(R)4, 3.0G processor running Windows XP with 1GB memory, 1.2GB virtual memory and 40GB hard drive. These implementations were developed in C++. The preliminary experimental result shows that Calculus-Fixpoint alogrithm outperforms Apriori algorithm. We also chose SQL-92 implementation to compare with our algorithm. In SQL-92 implementation we used the sub-query approach. The experimental result shows that Calculus-Fixpoint program outperforms SQL sub-query approach.

6   Conclusion

Relational computation for association rule mining that uses the idea of least fixpoint computation has been demonstrated in this paper. We also present a logic-based inductive query language and illustrate the use of aggregates and exploit a new join operator to model specific data mining tasks. The results provide theoretical foundations for inductive database research and could be useful for data mining query language design in inductive database systems.

Acknowledgments

The work reported in this paper was supported by a grant (CUHK418205) from the Research Grants Council of the Hong Kong Special Administrative Region, China and by a Faculty of Commerce research grant from University of Wollongong. Part of the work performed while the first author was at University of Wollongong.

References

1. Raedt, L.D.: A perspective on inductive databases. SIGKDD Explorations **4** (2002) 69–77
2. Zaniolo, C.: Mining databases and data streams with query languages and rules. In: Proceedings of 4th Workshop on Knowledge Discovery in Inductive Databases, LNCS3933. (2005) 24–37
3. Imielinski, T., Virmani, A.: Msql: A query language for database mining. Data Mining and Knowledge Discovery **2** (1999) 373–408
4. Han, J., Fu, Y., Koperski, K., Wang, W., Zaiane, O.: Dmql: A data mining query language for relational databases. In: Proceedings of ACM SIGMOD Workshop on Research Issues on Data Mining and Knowledge Discovery. (1996)
5. Meo, R., Psaila, G., Ceri, S.: An extension to sql for mining association rules. Data mining and knowledge discovery **2** (1998) 195–224
6. Sarawagi, S., Thomas, S., Agrawal, R.: Integrating association rule mining with relational database systems: alternatives and implications. Data mining and knowledge discovery **4** (2000) 89–125
7. Giannotti, F., Manco, G., Turini, F.: Towards a logic query language for data mining. Database Support for Data Mining Applications, LNAI **2682** (2004) 76–94
8. Jamil, H.M.: Bottom-up association rule mining in relational databases. Journal of Intelligent Information Systems (2002) 1–17
9. Masson, C., Robardet, C., Boulicaut, J.F.: Optimizing subset queries: a step towards sql-based inductive databases for itemsets. In: Proceedings of the 2004 ACM symposium on applied computing. (2004) 535–539
10. Liu, H.C., Yu, J.: Algebraic equivalences of nested relational operators. Information Systems **30** (2005) 167–204

An Efficient Quantum-Behaved Particle Swarm Optimization for Multiprocessor Scheduling

Xiaohong Kong[1,2], Jun Sun[1], Bin Ye[1], and Wenbo Xu[1]

[1] School of Information Technology, Southern Yangtze University,
Wuxi 214122, China
nancykong@hist.edu.cn, sunjun_wx@hotmail.com, xwb@sytu.edu.cn
[2] Henan Institute Of Science and Technology,
Xinxiang, Henan 453003, China

Abstract. Quantum-behaved particle swarm optimization (QPSO) is employed to deal with multiprocessor scheduling problem (MSP), which speeds the convergence and has few parameters to control. We combine the QPSO search technique with list scheduling to improve the solution quality in short time. At the same time, we produce the solution based on the problem-space heuristic. Several benchmark instances are tested and the experiment results demonstrate much advantage of QPSO to some other heuristics in search ability and performance.

1 Introduction

Multiprocessor scheduling problem (MSP) is popularly modeled by a weighted directed acyclic graph (DAG) or micro-dataflow graph, and the objective of MSP is minimizing the parallel completion time or schedule length by properly assigning the nodes of the graph to the processors without violating the precedence constraints. Generally, the multiprocessor scheduling problem is NP-hard[1] except for some cases for which an optimal solution can be obtained in polynomial time. In the past decades, a myriad of heuristic algorithms have been investigated[2][3][4][5][6][7][8][9], but the results are constrained by efficiency and complexity. In this paper, the multiprocessor scheduling problem is tackled by establishing the priority list of tasks utilizing the Quantum-Behaved Particle Swarm Optimization (QPSO).

A directed acyclic weighted task graph (DAG) is defined by a collection $G = \{V, E, C, W\}$, where $V = (n_j; j = 1 : v)$ is the set of task nodes and E is the set of communication edges. The weight $c(n_i, n_j) \in C$ corresponds to the communication cost incurred while the task n_i and n_j are scheduled, which is zero if both nodes are assigned on the same processor. The weight $w(n_i) \in W$ is the execution cost of node $n_i \in V$. The edge $e(i, j) = (n_i, n_j) \in E$ represents the partial order between tasks n_i and n_j. The target system is commonly assumed to consist of p processors connected by an interconnection network based on a certain topology in which a message is transmitted through bidirectional links with the same speed.

Y. Shi et al. (Eds.): ICCS 2007, Part I, LNCS 4487, pp. 278–285, 2007.

2   Quantum-Behaved Particle Swarm Optimization

2.1   The Standard PSO

Particle Swarm Optimization (PSO)[10][11][12], originally proposed by Kennedy and Eberhart in 1995, is a novel evolutionary algorithm and a swarm intelligence computation technique. In PSO system, a particle or an individual depicted by its position vector X and its velocity vector V, is a candidate solution to the problem. To solve an optimal problem,a population of initialized solutions search through a multidimensional solution space, and each member continually adjusts its position and velocity learning its own experience and the experience of other members. Considering $D(d = 1 : D)$ dimensions as an example, the particles are manipulated according to the following formula [10][11]:

$$V_{i,d} = V_{id} + c_1 * r_1 * (P_{id} - X_{id}) + c_2 * r_2 * (P_{gd} - X_{id}) \tag{1}$$

$$X_{id} = X_{id} + V_{id} \tag{2}$$

Where c_1 and c_2 are learning factors, r_1,r_2 are random number uniformly distributed in the range $[0, 1]$. In equation(1),the vector P_i is the best position (the position giving the best evaluate value) of particle i , vector P_g is the position of the best particle among all the particles in the population.

2.2   Quantum-Behaved PSO

In 2004, J. Sun et al. proposed a Delta potential well model of PSO in quantum time-space framework[13][14]. With the quantum particles,the exact values of X and V cannot be determined simultaneously, and the quantum state of a particle is described by wavefunction $|\psi(X, t)|$. In Quantum-Behaved Particle Swarm Optimization (QPSO), the particle can only learn the probability of its appearance in position X from probability density function $|\psi(X, t)|^2$, the form of which depends on the potential field the particle lies in. The updates of the particles are accomplished according to the following iterative equations[13][14]:

$$X(t + 1) = P \pm \beta * |mbest - X(t)| * \ln(1/u) \tag{3}$$

$$mbest = \frac{1}{M} \sum_{i=1}^{M} P_i = (\frac{1}{M} \sum_{i=1}^{M} P_{i1}, \frac{1}{M} \sum_{i=1}^{M} P_{i2}, \cdots, \frac{1}{M} \sum_{i=1}^{M} P_{id}) \tag{4}$$

$$P = \varphi * P_{id} + (1 - \varphi) * P_{gd}, \varphi = rand() \tag{5}$$

$mbest$(Mean Best Position) is the mean value of all particles' best positions P_i, φ and u are a random number distributed uniformly on $[0,1]$ respectively; The only parameter in QPSO algorithm is β , called Contraction-Expansion Coefficient. As a new method, QPSO has its advantages such as simple concept, immediate accessibility for practical applications, simple structure, ease of use, speed to get the solutions, and robustness, and parallel direct search method. QPSO has successful application in optimization problem[13][14].

3   Multiprocessor Scheduling Based on QPSO

Since 1995, Particle Swarm Optimization has been reported in literature to solve a range of optimization problem[15][16][17], and scheduling is the new field of PSO algorithm in discrete space. In general, there are two forms of solution representations, namely permutation-based representation and priority-based representation[17][18].

3.1   Particle-Based Solution Representation

For the priority-based form, every dimension $d(d = 1 : D)$ represents the task index number, and the corresponding value of each element in the vector X means the priority a node is scheduled to start. We assign a higher priority to a task with a smaller element value, i.e. each task is scheduled in ascending order of each element value [18]. Fig.1 illustrates the priority-based solution representation of target individual with five dimensions: the 4th element is the smallest, so the task 4 is scheduled firstly, and so on.

Priorities vector	1.3	4.5	3.7	0.6	2.3
Task index	1	2	3	4	5
Scheduling list	4	1	5	3	2

Fig. 1. Priority-based solution representation

3.2   Permutation-Based Representation

For the permutation-based form, every dimension $d(d = 1 : D)$ in the vector means the order or sequence the node is scheduled, and the corresponding value of each element means a node index number.Used to stand for the index number of a task, the parameter of each element in the particle represented permutation should be an integer limited to $[1, D]$. For example, a permutation of 5 tasks for a scheduling can be represented as the particle in Fig. 2.

Priorities vector	4	1	5	3	2
scheduling sequence	1	2	3	4	5
Scheduling list	4	1	5	3	2

Fig. 2. Permutation-based solution representation

3.3   QPSO for Multiprocessor Scheduling

When designing the QPSO algorithm, we combine the priority-based QPSO with list scheduling. The proposed QPSO approach are adopted searching for optimal combination of the priority values. A ready list preserves the ready tasks whose parent has finished. The highest-priority ready task is removed from the ready list and selected for scheduling during the first step. Next, the task list is dispatched to processors using *the earliest start time* rule and satisfy precedence constraints. In case the earliest time of a task on two machines is the same, the algorithm breaks the tie by matching and scheduling the task to the first machine in the sequence. Finally, updating the ready task list, until all tasks are scheduled.

In addition, an initial population is a key factor for the solution quality so we produce an efficient initial set based on specific problem. *b_level* and *t_level* are two general attributes in DAG graph. *b_level* is total cost(including execution cost and communication cost)from the bottom node to specified node and *t_level* is total cost from top node to specified node[9]. As described above, the *ith* element of particle represents priority of a task n_i, and the X_{0i} of first particle X_0 is a set to the *b_level* of a task n_i. Every dimension of the rest of the initial population is generated by a random perturbation of the values of the first particle as follows:

$$X_{ji} = X_{0i} + uniform(t\_level/2, -t\_level/2) \tag{6}$$

Where X_{0i} is the priority of task n_i in the first particle,$i = (1; \cdots ; D); j = (1; \cdots ; Npop - 1)$; $uniform(t\_level/2, -t\_level/2)$ is a random number generated uniformly between $t\_level/2$ and $-t\_level/2$. This strategy has a good performance in the PSGA [19]. On the basis of above, the population evolves to search the optimal solution. After every iteration, we limit the position value range to interval $[b\_level + t\_level, b\_level]$. Here, each dimension has a different priority value for each task in different particles, so each particle guides the heuristic to generate a different schedule in this case. The Proposed QPSO algorithm for multiprocessor scheduling is depicted in Fig .3.

Initialize the population
Do {
    Ngen ++
    For i=1 to Npop(the particle population number)
       β linearly decreases
       Find out the mbest of the swarm
       Obtain the update priority according to euqation(3)(4)(5)
       Find possible permutation
    End for
    Re-evaluate the P_i and the P_g
While (the termination criterion is not met or generation ¡ Ngen(the generation number)
Output results

Fig. 3. QPSO algorithm for MSP

4   Experimental Results

Firstly, we execute the QPSO and some other list scheduling to a set of random graphs with different parameter, including CCR, density ρ, task number n and machine number p. Some classical algorithms including HLFET [2], ISH [3], MCP

[4], ETF [5] and DLS [6] are also implemented. CCR is defined to be the ration of the mean communication to the mean computation which represents the weight of the communication to the whole computation[9]. Edge density ρ is defined as the number of edges divided by the number of edges of a completely connected graph with the same number of nodes [20]. There, $CCR \in \{0.1, 0.5, 1.0, 5.0, 10\}$

$n \in \{20, 40, 60, 80, 100\}$

$p \in \{2, 4\}$

$\rho \in \{20, 40, 50, 60, 80\}$

For every combination (CCR, n, p, ρ), there are 5 graphs, amount to $5(CCR) \times 6(n) \times 2(p) \times 5(\rho) \times 5(graphs) = 1500(graphs) = 1500(graphs)$. In order to evaluate the performance, we normalize the actual scheduling length according to the following formula(7)[8][9]:

$$NSL = \frac{SL}{\sum\limits_{i \in CP} w(n_i)} \tag{7}$$

$\sum\limits_{i \in CP} w(n_i)$ is the sum of computation time of all the CP nodes.

First, we select the graphs ($n = 100$) with different density to investigate the QPSO and other algorithms. The performance of QPSO algorithm is computed after 10 runs ($5CCR \times 5graphs$), and the results are depicted in Fig.4. In our algorithm, the parameter β is set changing from 1.2 to 0.5[13][14], which decreased linearly. From the Fig.4, an apparent conclusion is that more improvement is achieved for sparse-task graphs, while little enhancement in solution quality is observed for dense ones. It appears that dense graphs spent more time waiting for predecessor tasks to complete their execution and transferring data between related tasks scheduled on different processors.

At the same time, we test the performance of graphs with different tasks. The results are depicted in Fig.5. With more tasks, the QPSO has always the comparable performance among these algorithms, which is due to global search ability

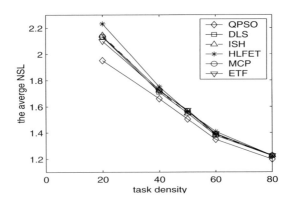

Fig. 4. The NSL of QPSO and a few algorithms to n=100 with different density ($p = 2$, $Npon = 20, Ngen = 30$)

of the QPSO to find near-optimal task list and the scalability. It appears that graphs with increasing tasks spent more time waiting for predecessor tasks to complete their execution and transferring data between related tasks scheduled on different processors.

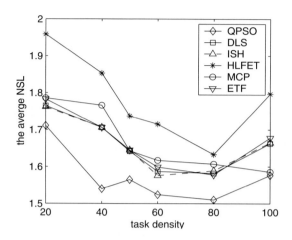

Fig. 5. The NSL of QPSO and a few algorithms different tasks number ($p = 4, Npon = 20, Ngen = 30$)

Finally, benchmark instances from website *(http://www.mi.sanu.ac.yu/ tanjad)*, and the results of a few algorithms, including algorithms CPES, LPTES, DC, LPTDC, LBMC, PPS are downloaded to test the algorithm. These algorithms are adjusted from other algorithms by T. Davidovic et al [20]. We select the $t200 - 80$ instances and investigate how the number of processors affects the scheduling results. The mean results of QPSO after 10 runs and other algorithms (download from *http://www.mi.sanu.ac.yu/ tanjad*) are listed as Table.1 and Table.2.

From the above results, in most cases QPSO attains better results than the other algorithms whether $p = 2$ or $p = 4$, but with the increasing of machines,

Table 1. The comparison of results between QPSO and other algorithms to t200-80 instances ($p = 2$)

instance	CPES	LPTES	DC	LPTDC	LBMC	PPS	QPSO
t200-80-1	3420	3433	3464	3454	4526	3687	3351
t200-80-2	5353	5535	5630	5600	7146	5977	5359
t200-80-3	3232	3244	3363	3365	4125	3441	3226
t200-80-4	3586	3618	3804	3864	4754	3990	3580
t200-80-5	1968	1979	2057	2069	2783	2151	1934
t200-80-6	2498	2522	2525	2549	3396	2807	2485

Table 2. The comparison of results between QPSO and other algorithms to t200-80 problem $(p = 4)$

instance	CPES	LPTES	DC	LPTDC	LBMC	PPS	QPSO
t200-80-1	3414	3427	3464	3453	5245	3875	3343
t200-80-2	5345	5475	5630	5581	7583	6225	5332
t200-80-3	3232	3244	3363	3365	4276	3413	3226
t200-80-4	3586	3618	3800	3684	4791	4040	3568
t200-80-5	1968	1979	2057	2069	3065	2705	1934
t200-80-6	2489	2522	2525	2549	3850	2995	2478

a little improvement is obtained. We analyze the reason behind the results: with more machines, more communications among machines are involved, the scheduling length is not proportional to machines.

5   Summary

As to be seen from the previous results of the performance-testing experiment, the performance of the Quantum Particle Swarm Optimization method in coping with scheduling problems is better than some classical algorithms in most cases. Our future work will focus on hybrid strategies of the QPSO algorithm with other optimization algorithms.

References

1. M.R. Garey, D.S. Johnson: Computers and Intractability: A Guide to the Theory of NP-Completeness. W. H. Freeman and Company (1979)
2. T. L. Adam, K. M. Chandy, J. Dickson: A comparison of list scheduling for parallel processing systems. Communication of the ACM **17** (1974) 685-690
3. B. Kruatrachue and T. G. Lewis: Duplication scheduling heuristic (DSH): A new precedence task scheduler for parallel processor systems. Technical Report, Oregon State University, Corvallis, OR 97331, (1987)
4. M.-Y. Wu, D. D. Gajski: Hypercool: a Programming Aid for Message-passing Systems. IEEE Trans. Parallel Distrib. Systems **1** (1990) 330-343
5. J. J. Hwang, Y. C. Chow, F. D. Anger, C. Y. Lee: Scheduling Precedence Graphs in Systems with Interprocessor Communication Times. SIAM Journal on Computing **18** (1989) 244-257
6. G. C. Sih, E. A. Lee: A Compile-time Scheduling Heuristic for Interconnection-constrained Heterogeneous Processor Architectures. IEEE Transactions Parallel and Distrib. Systems **4** (1993) 175-187
7. T. Yang, A. Gerasoulis: DSC: Scheduling Parallel Tasks on an Unbounded Number of processors. IEEE Transactions on Parallel and Distributed Systems **5** (1994) 951-967
8. Y.K. Kwok, I. Ahmad: A Static Scheduling Algorithm Using Dynamic Critical Path for Assigning Parallel Algorithms onto Multiprocessors. Proceedings of the 1994 International Conference on Parallel Processing (ICPP'94) **2** (1994) 155-159

9. Y.K Kwok, I.Ahmad: Benchmarking and Comparison of the Task Graph Schedul-
    ing Algorithms. Journal of Parallel and Distributed Computing **59** (1999) 381-422
10. J. Kennedy, R.c. Eberhart: Particle Swarm Optimization. Proc. IEEE Conf. On
    Neural Network **IV** (1995)1942-1948
11. R.C.Eberhart,J. Kennedy : A New Optimizer Using Particles Swarm Theory. Proc.
    Sixth Intemational Symposium on Micro Machine and Human Science (1995) 39-43
12. Y. Shi, R.C. Eberhart: Empirical Study of Particle Swarm Optimization. Proc.
    Congress on Evolutionary Computation (1999) 1945-1950.
13. J.Sun, B.Feng,W.B. Xu: Particle Swarm Optimization with Particles Having Quan-
    tum Behavior. Proc. 2004 Congress on Evolutionary Computation (2004) 325-331
14. J. Sun, W.B. Xu: A Global Search Strategy of Quantum-Behaved Particle Swarm
    Optimization. 2004 IEEE Conference on Cybernetics and Intelligent Systems
    (2004) 111-116
15. F. van den Bergh, A.P Engelbrecht: Cooperative Learning in Neural Networks using
    Particle Swarm Optimizers. South African Computer Journal **26** (2000) 84-90
16. S. C.Esquivel, C. A.Coello: On the Use of Particle Swarm Optimization with Mul-
    timodal Functions. Proceedings of IEEE Congress on Evolutionary Computation
    2003 (CEC 2003) 1130-1136
17. H. Zhang, X.D.Li, H.Li, F.L. Huang: Particle Swarm Optimization-based Schemes
    for Resource-constrained Project Scheduling. Automation in Construction **14**
    (2005) 393-404
18. M.F.Tasgetiren, Y.C. Liang, M.Sevkli, G. Gencyilmaz: Differential Evolution Al-
    gorithm for Permutation Flowshop Sequencing Problem with Makespan Criterion.
    4th Inter-national Symposium on Intelligent Manufacturing Systems (IMS 2004)
19. M. K. Dhodhi, Imtiaz Ahmad, A.Yatama,Ishfaq Ahmad: An Integrated Technique
    for Task Matching and Scheduling onto Distributed Heterogeneous Computing
    Systems. Journal of Parallel and Distributed Computing **62** (2002) 1338-1361
20. T. Davidovic, T. G. Crainic: Benchmark-problem Instances for Static Scheduling
    of Task Graphs with Communication Delays on Homogeneous Multiprocessor Sys-
    tems. Computers and Operations Research **33** (2006) 2155-2177

Toward Optimizing Particle-Simulation Systems

Hai Jiang[1], Hung-Chi Su[1], and Bin Zhang[2]

[1] Department of Computer Science, Arkansas State University,
Jonesboro, Arkansas 72401, USA
{hjiang,suh}@cs.astate.edu
[2] Department of Chemistry & Physics, Arkansas State University,
Jonesboro, Arkansas 72401, USA
bzhang@astate.edu

Abstract. Optimized event-driven particle collision simulation is on demand to study the behavior of systems consisted of moving objects. This
paper discusses the design and implementation issues of such simulation
systems with various optimizations such as discrete event handling, Lazy
Determination Strategy (LDS), and optimal cell number/size selection to
overcome the delay caused by dynamical dependencies among collisions.
Quantitative analyses and experiment results are provided to illustrate
the performance gains acquired from these optimizations.

Keywords: Particle collision; event-driven simulation; lazy determination.

1 Introduction

Computer simulation has been widely used to study interactions among moving
particles in atomic physics [1], molecular dynamics [2,3,4], and many other research fields [5,6]. These interactions change the microscopic states (positions,
momenta, etc) of individual particles and drive particle systems toward macroscopic equilibrium (that can be described by temperature, chemical potential,
etc.). However, realistic simulation process is computationally intensive. As long
as the simulation time is concerned, effective simulation programs are on demand.

In this paper, we will focus on the dynamical simulation of a system of partons
(quarks and gluons) in the Quark-Gluon Plasma produced in relativistic heavy
ion collisions. The algorithm used in the parton cascade model (developed by
Zhang [1]) will be employed for solution of Boltzmann equation involving large
number of particles. This paper makes the following contributions: First, the
breakdown of a simulation program is analyzed to identify candidates for optimization. Second, Lazy Determination Strategy (LDS) is proposed to speed up
simulation execution. Third, optimal cell number selection formula is provided
for better decision making. These generic optimizations improve both sequential
and parallel algorithms without introducing any error in the simulation results.

The remainder of this paper is organized as follows: Section 2 introduces
the background of many-particle simulation. Section 3 discusses how to manage

Y. Shi et al. (Eds.): ICCS 2007, Part I, LNCS 4487, pp. 286–293, 2007.
© Springer-Verlag Berlin Heidelberg 2007

local and global events. In Sections 4 and 5, we describe the Lazy Determination Strategy and optimal cell number/size selection, respectively. Some experimental results are given in Section 6. Our conclusions are presented in Section 7.

2   Background of Particle Simulation Systems

Time-driven simulation displaces all particles synchronously over a small discrete time period and checks the overlaps among particles whereas event-driven simulation displaces them over a serial of predicted collision events. The latter one's simulation time can advance to the next event time instead of crawling through all time periods in between [3,5,6,7]. This is more efficient, especially in sparse particle systems. Therefore,the event-driven approach is adopted by most simulation systems [1,8], including those parallelized ones which take the multi-tasking strategy for execution speedups [4].

A discrete-event simulation needs to find out all events and simulate them one-by-one. The particle simulation program consists of three steps:

1. *Collision Prediction* (CP) / *Collision Adjustment* (CA): determine the possible collision for each particle
2. *Collision Selection* (CS): search for the overall earliest next possible collisions
3. *Event Handling* (EH): simulate the actual collisions or other events

When a simulation starts, all particles participate in Step 1. For particle i, it can collide with any other ones, *i.e.,* there are $N-1$ possible collision events in an N-particle system. These events can be sorted according to their collision time and saved in a local event list: $e_{i1}e_{i2}e_{i3}...e_{i(N-1)}$. The process to derive and sort these events is called *Collision Prediction* (CP) whereas to update only one event, especially the first/earliest event e_{i1}, is called *Collision Adjustment* (CA). A local event list can contain as few as only the first/earliest collision event.

When a collision happens, the previously recorded events involving the colliding particles become invalid. Colliding particles need to go through CP. Other particles may have the colliding particles as recorded future collision partners. Whether these particles need to activate CP/CA depends on the determination strategy.

In Step 2 *Collision Selection* (CS), we need to find out the earliest collision. When each collision involves only two particles (note: three or more particles collisions can be more complicated), there are up to $N/2$ possibilities saved in a sorted global event list: $e_1e_2e_3...e_{N/2}$. Whenever an event takes place, it only filters out future ones with current colliding particles involved. If we maintain an event dependency graph, a subgraph will be reshaped and only affected events will be re-calculated. Obviously, graph maintenance is costly. But without it, all future collision events need to be checked for validation. Strong causal relationship among collisions imposes the major bottleneck in particle simulation systems.

In Step 3 *Event Handling* (EH), actual collisions or other events will be simulated. This step will update colliding particles' states, including formation time, position, and velocity, etc. Then, the system will loop back to Step 1.

3   Event Management

The *Collision Selection* (CS) in Step 2 is critical for simulation performance. The *Collision Prediction* (CP) in Step 1 normally employs linear search approach with complexity $O(N)$ for particle i's next collision time t_{c_i}. For cold systems, *i.e.*, newly started ones, all particles participate in CP and make complexity $O(N^2)$. If CS in Step 2 also uses linear search for the global minimal collision with $t_{min} = min(t_{c_1}, t_{c_2}, ...t_{c_N})$, the complexity is $O(N)$ too. If particles j and k collide, they will update their states in Step 3 and loop back to Step 1 to call CA for new t_{c_j} and t_{c_k}. Even if several other particles apply CP/CA, the complexity still remains as $O(N)$ [1] whereas EH in Step 3 keeps $O(1)$. The whole algorithm's complexity is $O(N) + O(N) + O(1) = O(N)$ [7].

Many algorithms have been proposed to speed up CS. The global event list can be mapped onto well known data structures such as *min heap*, *complete binary trees*, and other tree-like structures with complexity $O(\log N)$ [3,5,6]. The idea is to sort all possible collision events completely or partially according to their timestamps. It takes $O(1)$ time to remove the earliest event and $O(\log N)$ time to maintain the data structure state after event insertion/deletion. The assumption is that the N in CP can be reduced to N' so that for actual values, there exists $\alpha N' \leq \beta \log N$, where α and β are operational costs. Otherwise, the improvement in CS (Step 2) will be limited.

4   Lazy Determination Strategy (LDS)

The intuitive *Eager Determination Strategy* (EDS) in Step 1 intends to determine predictions as early as possible. However, because of the nature of dynamically causal relationship among collisions, this conservative scheme causes some unnecessary prediction updates. *Lazy Determination Strategy* (LDS) postpones prediction process to prevent from drawing a conclusion too early.

4.1   Cell Structure

Originally, each particle in CP needs to derive local event list $e_{i_1} e_{i_2} e_{i_3}...e_{i_{(N-1)}}$ with complexity $O(N)$. It is clear that particle systems hold strong collision locality characteristic, *i.e.*, particles are likely to collide with nearby ones first. If the local even list is created based on particles in neighborhood, its size will be much smaller, *i.e.*, $N' \ll N$. As mentioned before, when $\alpha N' \approx \beta \log N$, optimization in CS of Step 2 becomes much more effective.

A simulation space can be partitioned into a grid of 3D cells. If there are at least 3 cells along each dimension, a particle only needs to check collisions with particles in 27 neighboring cells (including the residing one). The cell number will be 9 in 2D case, as shown in Figure 1.

For two far-apart particles to collide, they need to enter each other's neighborhood first. Then cell structure has to introduce a new type of event, *Boundary Crossing* (BC), to detect a particle's entering and leaving actions. It moves particles from one cell to another and its cost is far less expensive than the one to

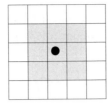

Fig. 1. The neighborhood of a particle

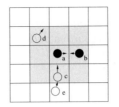

Fig. 2. Collision update scenario

detect whether two particles collide or not, called *Collision Determination* (CD), which requires several costly floating point operations. Thus, reducing frequency of CD can obtain performance gains. The less the BC cost is, the more cells one can afford to create without adversely affecting performance. Then, fewer particles will be involved in the updating of local event lists and less time is taken for CD/CP.

With naive EDS, a particle's position is updated after BC events. LDS has been widely used to postpone this since they can be derived easily according to the original position [5,6].

4.2 LDS for Collision Events

After a collision, colliding particles have to go through CP for next possible collision. Other particles in neighboring cells, called *third partners*, may or may not enter CP and CA processes.

Delayed *Collision Prediction*. If *third partners* intend to collide with current colliding particles in the future, such predictions are obviously out-of-date. With EDS, *third partners* need to issue CP immediately for new collisions. In Figure 2, particles a and b collide. *Third partner* c who planned to collide with old a will have to re-predict its next collision. Such recursive CP turns Step 1 into $O(mN')$, where m is determined by the dependency graph. For example, if the dependency graph is a shape of star, m will be $O(N')$ and the complexities of Steps 1 and 2 will be $O(N'^2)$ and $O(N' \log N)$, respectively. Because of dependency, each particle might call CP multiple times and results from early activations are overwritten by latter ones.

LDS keeps out-of-date collisions. Such *fake collisions* remain in the global event list in Step 2. Only when a *fake collision* becomes the next global event from CS, will CP be activated for the third partner. LDS with *fake collisions* can reduce some unnecessary CPs for performance gains.

Delayed *Collision Adjustment*. In colliding particles' CP process, if particles find that they might collide with a *third partner* earlier than its recorded next collision with others, EDS will update the *third partner*'s next collision by calling CA. For example, after the collision between a and b shown in Figure 2, particle a finds it can collide with *third partner* d at a time earlier than the one registered in d. LDS suggests not to call CA on d since particle a might find an even earlier collision with another particle. Only when particle a finishes its CP and it still intends to collide with particle d, will CA be called for particle d. Some immature decisions are avoided. This can achieve about 5% performance gain in our system.

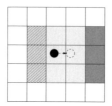   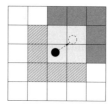

Fig. 3. Boundary crossing over one side   **Fig. 4.** Boundary crossing over a corner

4.3   LDS for Boundary-Crossing Events

If the next event from the global event list is a BC event, the moving particle will activate partial CP only for new territories such as the right column in Figure 3 and the upper-right area in Figure 4. Their opposite areas will be out of the moving particle's neighborhood, as the striped parts in Figures 3 and 4. EDS always checks these left-behind areas to clean up moving particle's effect whereas LDS prefers to leave them untouched. If an old particle still collides with the moving one, it will enter the latter one's neighborhood first. Otherwise, the collision will be a *fake collision* handled by calling CP when it is popped out of the CS process.

5   Optimal Cell Number/Size

Cell structure reduces the cost of CP/CA in Step 1 by narrowing down the scope of local event lists with minor BC cost. Assume with LDS, the real collision list is $e_1^c e_2^c e_3^c...$, and the overall event list is $e_1^c e_1^b...e_m^b e_2^c e_{m+1}^b...e_n^b e_3^c...$ with BC events e_i^b inserted between collisions. If no cell structure is applied, the time between two real collision events is:

$$t_{between} = t_{eh_{col}} + t_{cp_{col}} + t_{ca} + t_{cs}, \tag{1}$$

where $t_{eh_{col}}$ is the cost of state update, $t_{cp_{col}}$, t_{ca} and t_{cs} are the overheads for CP, CA, and CS. Since state update operation and event management algorithm are fixed, $t_{eh_{col}}$ and t_{cs} are almost constants. And the adjustment time t_{ca} is relatively smaller. So $t_{cp_{col}}$ dominates the whole system's behavior. With cell structure and extra costs for BC events, it will turn into

$$t'_{between} = t_{eh_{col}} + t'_{cp_{col}} + t_{ca} + t_{cs} + \lambda(t_{eh_{bc}} + t_{cp_{bc}} + t_{ca} + t_{cs}), \tag{2}$$

where λ is number of BC events, $t_{eh_{bc}}$ is time for next BC calculation, $t_{cp_{bc}}$ is the CP cost for BC, and for the same collision, CP cost $t'_{cp_{col}}$ is different from $t_{cp_{col}}$.

To speed up simulation, we need to ensure $t'_{between} < t_{between}$ and maximize their difference:

$$t_{diff} = t_{between} - t'_{between} = t_{cp_{col}} - t'_{cp_{col}} - \lambda(t_{eh_{bc}} + t_{cp_{bc}} + t_{ca} + t_{cs}). \quad (3)$$

Without cells, $t_{cp_{col}} = 2N\Delta$, where N is the particle number and Δ is the CD time. Since $t_{cp_{col}}$ is a constant, best simulation result comes from minimizing BC cost function $f = t'_{cp_{col}} + \lambda(t_{eh_{bc}} + t_{cp_{bc}} + t_{ca} + t_{cs})$. We assume that the box size is l^3, cell number n^3 (n along each dimension), particle's cross section is σ and its velocity is v ($v = 1$). The time between two collisions can be expressed as $\frac{l^3}{\sigma N^2}$. BC number λ within this period will be:

$$\lambda = \frac{l^3}{\sqrt{3}\sigma N^2} \cdot \frac{n}{l} = \frac{nl^2}{\sqrt{3}\sigma N^2}. \quad (4)$$

The cost function f changes according to cell number n, and can be refined as

$$f(n) = \frac{54N}{n^3}\Delta + \frac{nl^2}{\sqrt{3}\sigma N^2}\left(\frac{mN}{n^3}\Delta + \tau\right), \quad (5)$$

where $\tau = t_{eh_{bc}} + t_{ca} + t_{cs}$ which is almost a constant and m is the number of newly joined cells in BC events. Both Δ and τ can be determined in simulation programs. Two colliding particles need to run full CP with 27 neighboring cells and each BC particle runs partial CP on m (9~18) newly joined cells. The optimal cell number can be derived by equating the derivative (with respect to n) to zero:

$$\frac{d}{dn}f(n) = -\frac{162N\Delta}{n^4} - \frac{2ml^2\Delta}{\sqrt{3}\sigma Nn^3} + \frac{l^2\tau}{\sqrt{3}\sigma N^2} = 0, \quad (6)$$

or equivalently,

$$n^4 - \frac{2mN\Delta}{\tau}n - \frac{162\sqrt{3}\sigma N^3\Delta}{l^2\tau} = 0.$$

Since term $\frac{2mN\Delta}{\tau}n$ grows slower than the other one, the starting search point for the optimal cell number n_{opt} is:

$$n_{opt_0} = \sqrt[4]{\frac{162\sqrt{3}\sigma N^3\Delta}{l^2\tau}}. \quad (7)$$

From here, minor adjustment is applied to approach to n_{opt} quickly.

6   Experimental Results

We have conducted some experiments to illustrate various optimization results which indicate how much Lazy Determination Strategy (LDS) in Step 1 and event management algorithm in Step 2 will affect the simulation performance. Our test platform is a 3.8 GHz Pentium IV PC with 2 GB memory and 1MB-cache running SUSE 10. The simulation is conducted in a 10x10x10 $unit^3$ 3-D

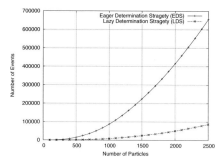

Fig. 5. Re-Prediction for third partners

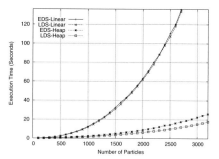

Fig. 6. Overall simulation time

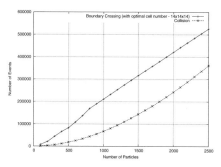

Fig. 7. Boundary Crossing vs. Collision

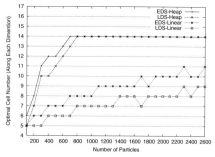

Fig. 8. Optimal cell number selection

box. Periodic Boundary Conditions (PBC) are used to reduce finite size effect. Particles' cross section is 0.5 $unit^2$. All tests are run to 10 time units.

LDS overcomes the conservativeness in EDS which intends to make the simulation system consistent all the time. Figure 5 shows the number of CP invocations called by third partners. LDS reduces such invocations dramatically and the only left-over ones are the *fake collisions*. Higher density cases will show the advantage of LDS even clearer. The lazy/relaxed consistency scheme reduces simulation time by avoiding unnecessary immature predictions.

EDS and LDS are the options for CP in Step 1 whereas linear search and heap are representative algorithms for CS in Step 2. Different combination results are shown in Figure 6. EDS and LDS adjust the m in $O(mN')$ whereas linear search and heap determine *Collision Selection* algorithms with complexities $O(N)$ and $O(\log N)$, respectively ($N' \ll N$). When linear search is adopted, Step 2 will dominate the performance as the top two curves in Figure 6 where LDS can only achieve minor improvement. However, with a heap, the effect from LDS becomes much clearer since the influence from Step 2 is reduced dramatically.

Cell structure determines the local event list length, *i.e.*, N'. But it causes BC event although its overhead is much smaller than the one of collision determination. The ratio is about 1:8 in our system and can be acquired at runtime. Figure 7 shows the numbers of these two kinds of events. Cheaper BC events are used to narrow down the search scope in CP. Cell structure is more important when heap is used (Step 2 is less dominant), as shown in Figure 8. This is clear in sparse cases before the top two curves hit the cell number limit (determined by the condition: cell size$>\sqrt{\sigma/\pi}$) on the top. EDS has a higher demand on cell number than LDS so that cell size will be smaller to reduce unnecessary prediction/update.

7   Conclusions

This paper analyzes and demonstrates different optimization schemes and their roles in overall simulation execution. Lazy Determination Strategy (LDS) contributes more performance gains whenever efficient event handling algorithms such as heaps are employed. Delayed prediction/updates avoid some immature decisions with no error introduced. Cell structure limits the search scope for next collision. The calculation of the initial optimal cell number is provided to help step toward the final optimal one quickly. These optimizations speed up executions without any negative impact on the accuracy and stability of simulation results. Their effectiveness is quantitatively analyzed. They can be utilized and even further exploited by any serial and parallelized particle simulations.

Acknowledgment

This research was partly supported by the U.S. National Science Foundation under grant No. PHY-0554930.

References

1. Zhang, B.: ZPC 1.0.1: a parton cascade for ultrarelativistic heavy ion collisions. Computer Physics Communications **109** (1998) 193–206
2. Donev, A., Torquato, S., Stillinger, F.: Neighbor list collision-driven molecular dynamics simulation for nonspherical hard particles. i. algorithmic details. Journal of Computational Physics **202**(2) (2005) 737–764
3. Isobe, M.: Simple and efficient algorithm for large scale modecular dynamics simulation in hard disk system. Intl. J. of Modern Physics **C**(10) (1999) 1281–1293
4. Miller, S., Luding, S.: Event-driven molecular dynamics in parallel. Journal of Computational Physics **193**(1) (2004) 306–316
5. Sigurgeirsson, H., Stuart, A., Wan, W.L.: Algorithms for particle-field simulations with collisions. Journal of Computational Physics **172**(2) (2002) 766–807
6. Lubachevsky, B.D.: How to simulate billiards and similar systems. Journal of Computational Physics **94**(2) (1991) 255–283
7. Su, H.C., Jiang, H., Zhang, B.: An empirical study on many-particle collision algorithms. In: Proc. of the Intl. Conf. on Comp. and Their App. (2007)
8. Kim, D.J., Guibas, L.J., Shin, S.Y.: Fast collision detection among multiple moving spheres. IEEE Trans. on Visualization and Comp. Graphics **4**(3) (1998) 230–242

A Modified Quantum-Behaved Particle Swarm Optimization

Jun Sun[1], C.-H. Lai[2], Wenbo Xu[1], Yanrui Ding[1], and Zhilei Chai[1]

[1]Center of Intelligent and High Performance Computing,
School of Information Technology, Southern Yangtze University,
No. 1800, Lihudadao Road, Wuxi,
214122 Jiangsu, China
{sunjun_wx, xwb_sytu}@hotmail.com, zlchai@gmail.com
2School of Computing and Mathematical Sciences,
University of Greenwich, Greenwich, London SE10 9LS, UK
C.H.Lai@gre.ac.uk

Abstract. Based on the previously introduced Quantum-behaved Particle Swarm Optimization (QPSO), a revised QPSO with Gaussian disturbance on the mean best position of the swarm is proposed. The reason for the introduction of this novel method is that the disturbance can effectively prevent the stagnation of the particles and therefore make them escape the local optima and sub-optima more easily. Before proposing the Revised QPSO (RQPSO), we introduce the origin and the development of the original PSO and QPSO. To evaluate the performance of the new method, the Revised QPSO, along with QPSO and Standard PSO, is tested on several well-known benchmark functions. The experimental results show that the Revised QPSO has better performance than QPSO and Standard PSO generally.

Keywords: Global Optimization, Swarm Intelligence, Particle Swarm, Quantum-behaved Particle Swarm and Gaussian distribution.

1 Introduction

The Particle Swarm Optimization (PSO), originally invented by J. Kennedy and R.C. Eberhart [7], is a member of a wider class of Swarm Intelligence methods used for solving Global Optimization (GO) problems. It was proposed as a population-based search technique simulating the knowledge evolvement of a social organism. In a PSO system, individuals (particles) representing the candidate solutions to the problem at hand, fly through a multidimensional search space to find out the optima or sub-optima. The particle evaluates its position to a goal (fitness) at every iteration, and particles in a local neighborhood share memories of their "best" positions. These memories are used to adjust particle velocities and their subsequent positions.

In the original PSO with M individuals, each individual is treated as an infinitesimal particle in D-dimensional space, with the position vector and velocity

Y. Shi et al. (Eds.): ICCS 2007, Part I, LNCS 4487, pp. 294–301, 2007.

vector of particle i, $X_i(t) = (X_{i1}(t), X_{i2}(t), \cdots, X_{iD}(t))$ and $V_i(t) = (V_{i1}(t), V_{i2}(t), \cdots, V_{iD}(t))$. The particle moves according to the following equations:

$$V_{ij}(t+1) = w \cdot V_{ij}(t) + c_1 \cdot r_1(P_{ij}(t) - X_{ij}(t)) + c_2 \cdot r_2 \cdot (P_{gj} - X_{ij}(t)) \tag{1}$$

$$X_{ij}(t+1) = X_{ij}(t) + V_{ij}(t+1) \tag{2}$$

for $i = 1,2,\cdots M; j = 1,2\cdots, D$. The parameters c_1 and c_2 are called the acceleration coefficients. Vector $P_i = (P_{i1}, P_{i2}, \cdots, P_{iD})$ known as the *personal best position*, is the best previous position (the position giving the best fitness value so far) of particle i; vector $P_g = (P_{g1}, P_{g2}, \cdots, P_{gD})$ is the position of the best particle among all the particles and is known as the *global best position*. The parameter w is the inertia weight and the parameters r_1 and r_2 are two random numbers distributed uniformly in $(0,1)$, that is $r_1, r_2 \sim U(0,1)$. Generally, the value of V_{ij} is restricted in the interval $[-V_{max}, V_{max}]$.

Many revised versions of PSO algorithm are proposed to improve the performance since its origin in 1995. For the detailed information about these variants of PSO, one may refer to the literature such as [1], [3], [8], [9], etc.

In our previous work, a Quantum-behaved Particle Swarm Optimization (QPSO), inspired by quantum mechanics was proposed in [10], [11] and [12]. In this paper, we introduce a method of improving QPSO by exerting a Gaussian disturbance on the mean best position of the swarm and therefore propose a revised version of QPSO.

The rest of the paper is structured as follows. In Section 2, the concept of QPSO is presented. The revised QPSO is proposed in Section 3. Section 4 gives the numerical results on some benchmark functions. Some concluding remarks and future work are presented in the last section.

2   Quantum-Behaved Particle Swarm Optimization

Trajectory analyses in [4] demonstrated the fact that convergence of PSO algorithm may be achieved if each particle converges to its local attractor $p_i = (p_{i1}, p_{i2}, \cdots p_{iD})$ with coordinates

$$p_{ij}(t) = (c_1 r_1 P_{ij}(t) + c_2 r_2 P_{gj}(t))/(c_1 r_1 + c_2 r_2), \text{ or } p_{ij}(t) = \varphi \cdot P_{ij}(t) + (1-\varphi) \cdot P_{gj}(t) \tag{3}$$

where $\varphi = c_1 r_1/(c_1 r_1 + c_2 r_2)$. It can be seen that the local attractor is a stochastic attractor of particle i that lies in a hyper-rectangle with P_i and P_g being two ends of its diagonal. We introduce the concepts of QPSO as follows.

Assume that each individual particle move in the search space with a δ potential on each dimension, of which the center is the point p_{ij}. For simplicity, we consider a particle in one-dimensional space, with point p the center of potential. Solving Schrödinger equation of one-dimensional δ potential well, we can get the probability distribution function D.

$$D(x) = e^{-2|p-x|/L} \tag{4}$$

Using Monte Carlo method, we obtain

$$x = p \pm \frac{L}{2} \ln(1/u) \, , \, u \sim U(0,1) \tag{5}$$

The above is the fundamental iterative equation of QPSO.

In [11], a global point called Mainstream Thought or Mean Best Position of the population is introduced into PSO. The global point, denoted as C, is defined as the mean of the personal best positions among all particles. That is

$$C(t) = (C_1(t), C_2(t), \cdots, C_D(t)) = \left(\frac{1}{M} \sum_{i=1}^{M} P_{i1}(t), \quad \frac{1}{M} \sum_{i=1}^{M} P_{i2}(t), \quad \cdots, \quad \frac{1}{M} \sum_{i=1}^{M} P_{iD}(t) \right) \tag{6}$$

where M is the population size and P_i is the personal best position of particle i. Then the value of L is evaluated by $L = 2\alpha \cdot \left| C_j(t) - X_{ij}(t) \right|$ and the position are updated by

$$X_{ij}(t+1) = p_{ij}(t) \pm \alpha \cdot \left| C_j(t) - X_{ij}(t) \right| \cdot \ln(1/u) \tag{7}$$

where parameter α is called Contraction-Expansion (CE) Coefficient, which can be tuned to control the convergence speed of the algorithms. Generally, we always call the PSO with equation (7) Quantum-behaved Particle Swarm Optimization (QPSO), where parameter α must be set as $\alpha < 1.782$ to guarantee convergence of the particle [12]. In most cases, α can be controlled to decrease linearly from α_0 to α_1 ($\alpha_0 < \alpha_1$).

3   The Proposed Method

Although QPSO possesses better global convergence behavior than PSO, it may encounter premature convergence, a major problem with PSO and other evolutionary algorithms in multi-modal optimization, which results in great performance loss and sub-optimal solutions. In a PSO system, with the fast information flow between particles due to its collectiveness, diversity of the particle swarm declines rapidly, leaving the PSO algorithm with great difficulties of escaping local optima. Therefore, the collectiveness of particles leads to low diversity with fitness stagnation as an overall result. In QPSO, although the search space of an individual particle at each iteration is the whole feasible solution space of the problem, diversity loss of the whole population is also inevitable due to the collectiveness.

From the update equations of PSO or QPSO, we can infer that all particles in PSO or QPSO will converge to a common point, leaving the diversity of the population extremely low and particles stagnated without further search before the iterations is over. To overcome the problem, we introduce a Revised QPSO by exerting a Gaussian disturbance on the mean best position when the swarm is evolving. That is,

$$m_j(t) = C_j(t) + \varepsilon \cdot Rn, \qquad j = 1, 2, \cdots, D \tag{8}$$

where ε is a pre-specified parameter and Rn is random number with Gaussian distribution with mean 0 and standard deviation 1. Thus the value of L is calculated as $L = 2\alpha \cdot \left| m_j(t) - X_{ij}(t) \right|$ and equation (6) becomes

$$X_{ij}(t+1) = p_{ij}(t) \pm \alpha \left| m_j(t) - X_{ij}(t) \right| \cdot \ln(1/u) \qquad (9)$$

The above iterative equation can effectively avoid the declination of the diversity and consequently the premature convergence. It is because that when the swarm is evolving, a consistent disturbance on mean best position can prevent the $L = 2\alpha \cdot \left| m_j(t) - X_{ij}(t) \right|$ decreasing to zero, maintaining the particle's vigor, which is particularly serviceable to make the particle escape local optima at the later stage of evolution, and able to result in a better performance of the algorithm overall. The revised QPSO is outlined as follows.

```
Revised QPSO Algorithm
Initialize particles with random position Xi=X[i][:];
Initialize personal best position by set Pi=Xi;
while the stop criterion is not met do
Compute the mean best position C[:] by equation (6);
Get the disturbed point m by equation (8)
    for i = 1 to swarm size M
        If f(Xi)<f(Pi) then Pi=Xi; Endif
        Find the Pg=arg min f(P[g][:]);
        for j=1 to D
            =rand(0,1); u=rand(0,1);
            p=*P[i][j]+(1- )*P[g][j];
            if (rand(0,1)>0.5)
                X[i][j]=p+α*abs(m[j]- X[i][j])*ln(1/u);
            Else
                X[i][j]=p-α*abs(m[j]- X[i][j])*ln(1/u);
            Endif
        Endfor
    Endfor
Endwhile
```

4 Numerical Experiments

In this section five benchmark functions listed in Table 1 are tested for performance comparisons of the Revised QPSO (RQPSO) with Standard PSO (SPSO) and QPSO algorithms. These functions are all minimization problems with minimum objective function values zeros. The initial range of the population is asymmetry, as used in [13], [14]. V_{max} for SPSO is set as the up-bound of the search domain.

Setting the fitness value as function value, we had 50 trial runs for every instance and recorded mean best fitness and standard deviation. In order to investigate the scalability of the algorithm, different population sizes M are used for each function with different dimensions. The population sizes are 20, 40 and 80. The maximum generation (iteration) is set as 1000, 1500 and 2000 corresponding to the dimensions 10, 20 and 30 for first four functions, respectively. The maximum generation is the last function is 2000. For SPSO, the acceleration coefficients are set to be $c_1=c_2=2$ and the inertia weight is decreasing linearly from 0.9 to 0.4 as in [13], [14].

Table 1. Expression of the five tested benchmark functions

	Function Expression	Search Domain	Initial Range
Sphere	$f_1(X) = \sum_{i=1}^{n} x_i^2$	(-100, 100)	(50, 100)
Rosenbrock	$f_2(X) = \sum_{i=1}^{n-1} (100 \cdot (x_{i+1} - x_i^2)^2 + (x_i - 1)^2)$	(-100, 100)	(15, 30)
Rastrigrin	$f_3(X) = \sum_{i=1}^{n} (x_i^2 - 10 \cdot \cos(2\pi x_i) - 10)$	(-10, 10)	(2.56, 5.12)
Greiwank	$f_4(X) = \frac{1}{4000} \sum_{i=1}^{n} x_i^2 - \prod_{i=1}^{n} \cos\left(\frac{x_i}{\sqrt{i}}\right) + 1$	(-600, 600)	(300, 600)
Shaffer's	$f_5(X) = 0.5 + \frac{(\sin(\sqrt{x_1^2 + x_2^2}))^2}{(1.0 + 0.001(x_1^2 + x_2^2))^2}$	(-100, 100)	(30, 100)

In experiments for QPSO and RQPSO, the value of CE Coefficient α varies from 1.0 to 0.5 linearly over the running of the algorithm as in [11], [12]. For Revised QPSO (RQPSO), the parameter ε is fixed at 0.001. The mean values and standard deviations of best fitness values for 50 runs of each function are recorded in Table 2 to Table 5.

The numerical results show that both QPSO and RQPSO are superior to SPSO except on most of the function. On Shpere Function the RQPSO has worse performance than QPSO and SPSO because maintaining the diversity by consistent disturbance on the mean best position leads to poor local convergence of the algorithm. On Rosenbrock function, the RQPSO outperforms the QPSO when the swarm size is 20, but it does not show the better performance than QPSO when the swarm size is 40 and 80. On Rastrigrin function, it is shown that the RQPSO generated better results than QPSO when the number of particles is 20, but its advantages over the QPSO are not remarkable if the sampling errors are considered. On Griewank function, RQPSO is superior to the QPSO in most cases, particularly when the dimension of the problem is high. On Shaffer's function, the RQPSO shows no improvements for QPSO. Generally speaking, the Revised QPSO has better global search ability than QPSO when the problem is complex.

The figure 1 shows the convergence process of the RQPSO and QPSO on the first four benchmark functions with dimension 30 and swarm size 20. It is shown that the RQPSO has comparable convergence speed with QPSO and both of them can converge more rapidly than SPSO.

Table 2. Numerical results on Sphere function

M	Dim.	Gmax	SPSO		QPSO		RQPSO	
			Mean Best	St. Dev.	Mean Best	St. Dev.	Mean Best	St. Dev.
20	10	1000	3.16E-20	6.23E-20	2.29E-41	1.49E-40	1.7412E-007	4.594E-008
	20	1500	5.29E-11	1.56E-10	1.68E-20	7.99E-20	1.8517E-006	3.387E-007
	30	2000	2.45E-06	7.72E-06	1.34E-13	3.32E-13	6.0118E-006	1.086E-006
40	10	1000	3.12E-23	8.01E-23	8.26E-72	5.83E-71	1.5035E-007	4.627E-008
	20	1500	4.16E-14	9.73E-14	1.53E-41	7.48E-41	1.3454E-006	2.572E-007
	30	2000	2.26E-10	5.10E-10	1.87E-28	6.73E-28	4.9908E-006	7.878E-007
80	10	1000	6.15E-28	2.63E-27	3.10E-100	2.10E-99	1.0779E-007	3.592E-008
	20	1500	2.68E-17	5.24E-17	1.56E-67	9.24E-67	1.3454E-006	2.572E-007
	30	2000	2.47E-12	7.16E-12	1.10E-48	2.67E-48	4.1723E-006	6.179E-007

Table 3. Numerical results on Rosenbrock function

M	Dim.	Gmax	SPSO		QPSO		RQPSO	
			Mean Best	St. Dev.	Mean Best	St. Dev.	Mean Best	St. Dev.
	10	1000	94.1276	194.3648	59.4764	153.0842	47.6904	99.3668
20	20	1500	204.337	293.4544	110.664	149.5483	70.7450	116.1039
	30	2000	313.734	547.2635	147.609	210.3262	103.7322	166.9141
	10	1000	71.0239	174.1108	16.2338	24.46731	20.1872	33.3523
40	20	1500	179.291	377.4305	46.5957	39.536	46.8270	51.7003
	30	2000	289.593	478.6273	59.0291	63.494	78.5551	71.2005
	10	1000	37.3747	57.4734	8.63638	16.6746	13.4288	14.5301
80	20	1500	83.6931	137.2637	35.8947	36.4702	29.8257	33.2984
	30	2000	202.672	289.9728	51.5479	40.849	54.1137	38.2911

Table 4. Numerical results on Rastrigrin function

M	Dim.	Gmax	SPSO		QPSO		RQPSO	
			Mean Best	St. Dev.	Mean Best	St. Dev.	Mean Best	St. Dev.
	10	1000	5.5382	3.0477	5.2543	2.8952	4.4489	2.2451
20	20	1500	23.1544	10.4739	16.2673	5.9771	15.9715	6.4180
	30	2000	47.4168	17.1595	31.4576	7.6882	27.6414	6.1376
	10	1000	3.5778	2.1384	3.5685	2.0678	3.2081	1.4512
40	20	1500	16.4337	5.4811	11.1351	3.6046	10.5817	3.6035
	30	2000	37.2796	14.2838	22.9594	7.2455	20.9748	5.2488
	10	1000	2.5646	1.5728	2.1245	2.2353	2.0922	1.5245
80	20	1500	13.3826	8.5137	10.2759	6.6244	8.4794	2.7922
	30	2000	28.6293	10.3431	16.7768	4.4858	16.4016	4.4312

Table 5. Numerical results on Griewank function

M	Dim.	Gmax	SPSO		QPSO		RQPSO	
			Mean Best	St. Dev.	Mean Best	St. Dev.	Mean Best	St. Dev.
	10	1000	0.09217	0.0833	0.08331	0.06805	0.0694	0.0641
20	20	1500	0.03002	0.03255	0.02033	0.02257	0.0201	0.0212
	30	2000	0.01811	0.02477	0.01119	0.01462	0.0091	0.0129
	10	1000	0.08496	0.0726	0.06912	0.05093	0.0531	0.0509
40	20	1500	0.02719	0.02517	0.01666	0.01755	0.0180	0.0165
	30	2000	0.01267	0.01479	0.01161	0.01246	0.0107	0.0137
	10	1000	0.07484	0.07107	0.03508	0.02086	0.0342	0.0423
80	20	1500	0.02854	0.0268	0.01463	0.01279	0.0141	0.0158
	30	2000	0.01258	0.01396	0.01136	0.01139	0.0066	0.0105

Table 6. Numerical results on Shaffer's f6 function

M	Dim.	Gmax	SPSO		QPSO		RQPSO	
			Mean Best	St. Dev.	Mean Best	St. Dev.	Mean Best	St. Dev.
20	2	2000	2.782E-04	0.001284	0.001361	0.003405	0.001433	0.0036
40	2	2000	4.744E-05	3.593E-05	3.891E-04	0.001923	2.1303E-008	6.726E-008
80	2	2000	2.568E-10	3.134E-10	1.723E-09	3.303E-09	3.5979E-009	1.251E-008

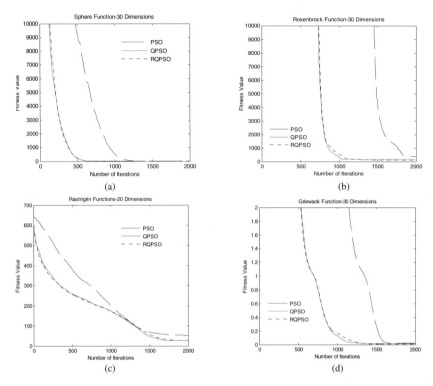

Fig. 1. Convergence process of the RQPSO and QPSO on the first four benchmark functions with dimension 30 and swarm size 20 averaged on 50 trail runs

5 Conclusions

In this paper, a Revised QPSO with the mean best position exerted by Gaussian disturbance is proposed. The reason for introduction of the disturbance is that it can avoid the declination of the diversity effectively, which is particularly serviceable for the particle's escaping local optima at later stage of the evolution. The iterative equation derives from the hybrid of exponential and normal distributions. The numerical results on benchmark functions show that the Revised QPSO enhance the global search ability of QPSO efficiently.

Our future work will focus on find out more efficient methods of improving QPSO. A promising method may be introducing another type of probability distribution into QPSO to replace exponential distribution. Moreover, we will also be devoted to applying the novel QPSO to many real world problems.

References

1. Angeline, P.J.: Using Selection to Improve Particle Swarm Optimization. Proc. 1998 IEEE International Conference on Evolutionary Computation. Piscataway, NJ (1998) 84-89
2. Van den Bergh, F.: An Analysis of Particle Swarm Optimizers. PhD Thesis. University of Pretoria, South Africa (2001)

3. Clerc, M.: The Swarm and Queen: Towards a Deterministic and Adaptive Particle Swarm Optimization. Proc. 1999 Congress on Evolutionary Computation. Piscataway, NJ (1999) 1951-1957
4. Clerc, M., Kennedy, J.: The Particle Swarm: Explosion, Stability, and Convergence in a Multi-dimensional Complex Space. IEEE Transactions on Evolutionary Computation, Vol. 6, No. 1. Piscataway, NJ (2002) 58-73
5. Eberhart, R.C., Shi, Y.: Comparison between Genetic Algorithm and Particle Swarm Optimization. Evolutionary Programming VII, Lecture Notes in Computer Science 1447, Springer-Verlag, Heidelberg (1998) 611-616
6. Holland, J.H.: Adaptation in Natural and Artificial Systems. The University of Michigan Press, Michigan (1975)
7. Kennedy, J., Eberhart, R.C.: Particle Swarm Optimization. Proc. IEEE 1995 International Conference on Neural Networks, IV. Piscataway, NJ (1995) 1942-1948
8. Kennedy, J.: Small worlds and Mega-minds: Effects of Neighborhood Topology on Particle Swarm Performance. Proc. 1999 Congress on Evolutionary Computation. Piscataway, NJ (1999) 1931-1938
9. Suganthan, P.N.: Particle Swarm Optimizer with Neighborhood Operator. Proc. 1999 Congress on Evolutionary Computation, Piscataway, NJ (1999) 1958-1962
10. Sun, J., Feng, B., Xu, W.-B.: Particle Swarm Optimization with Particles Having Quantum Behavior. Proc. 2004 Congress on Evolutionary Computation, Piscataway, NJ (2004) 325-331
11. Sun, J., Xu, W.-B., Feng, B.: A Global Search Strategy of Quantum-behaved Particle Swarm Optimization. Proc. 2004 IEEE Conference on Cybernetics and Intelligent Systems, Singapore (2004) 111-115
12. Sun, J., Xu, W.-B., Feng, B.: Adaptive Parameter Control for Quantum-behaved Particle Swarm Optimization on Individual Level. Proc. 2005 IEEE International Conference on Systems, Man and Cybernetics. Piscataway, NJ (2005) 3049-3054
13. Shi, Y., Eberhart, R.: Empirical Study of Particle Swarm Optimization. Proc. 1999 Congress on Evolutionary Computation. Piscataway, NJ (1999) 1945-1950
14. Shi, Y., Eberhart, R.C.: A Modified Particle Swarm. Proc. 1998 IEEE International Conference on Evolutionary Computation. Piscataway, NJ (1998) 69-73

Neural Networks for Predicting the Behavior of Preconditioned Iterative Solvers

America Holloway and Tzu-Yi Chen

Computer Science Department, Pomona College, Claremont CA 91711, USA
{america,tzuyi}@cs.pomona.edu

Abstract. We evaluate the effectiveness of neural networks as a tool for predicting whether a particular combination of preconditioner and iterative method will correctly solve a given sparse linear system $Ax = b$. We consider several scenarios corresponding to different assumptions about the relationship between the systems used to train the neural network and those for which the neural network is expected to predict behavior. Greater similarity between those two sets leads to better accuracy, but even when the two sets are very different prediction accuracy can be improved by using additional computation.

Keywords: iterative methods, preconditioners, neural networks.

1 Introduction

Preconditioned methods are generally used to solve large, sparse linear systems $Ax = b$ when the time or memory requirements of a direct solver are unacceptable. Ideally the preconditioner is inexpensive to compute and apply, and the subsequently applied iterative solver works more effectively on the preconditioned system. Unfortunately, because of the number and variety of existing preconditioners, as well as the lack of a good understanding of how well any preconditioned solver is likely to work for a given system, choosing a preconditioner is rarely straightforward. As a result, there is interest in helping users choose a preconditioner, with guidelines for setting parameter values, for their particular system. One option is to use the results of extensive experiments to suggest rules-of-thumb (eg, [1–3]) — this technique can suggest default settings but not when or how to adjust those defaults. Another, more recently explored option, uses machine learning techniques to extract meaningful features for predicting the behavior of preconditioned solvers (eg, [4,5]).

This paper contributes to the growing body of knowledge on using machine learning techniques for this problem by asking whether neural networks can be used to predict the behavior of preconditioned iterative solvers. In addition, this work explores a broad range of scenarios in which a user might wish to predict the behavior of a preconditioned solver. The first aspect considers the context for the problem and the amount of highly relevant data that is likely to be available. Is the system completely novel and therefore likely to be quite

Y. Shi et al. (Eds.): ICCS 2007, Part I, LNCS 4487, pp. 302–309, 2007.

different from other systems that have already been solved? Does the system closely resemble others that have already been solved and analyzed? Or has the user already tried to solve the exact system using other preconditioners in the past? The second aspect takes into consideration the relative cost of computing the preconditioner and performing the subsequent iterative solve. If the first is believed to be inexpensive, how much of an improvement can be gained by first computing the preconditioner and using some data from that computation in predicting the behavior of the subsequent solve?

After first describing how this work fits into the larger body of work on predicting the behavior of preconditioned iterative solvers, we describe the range of parameters and scenarios that was explored, and finally discuss the results. Overall, we find that neural networks are promising, even when the inputs consist of only easily computable matrix values. We explore the importance of training a neural network on matrices that are similar to those for which it is expected to predict results. In the case where the two sets of matrices are different and the accuracy suffers, we suggest a method for increasing the accuracy of the system using some additional computation.

2   Background

Predicting the behavior of preconditioned iterative solvers has been studied in papers including [3–5]. Throughout, assorted structural and numerical features are extracted from a matrix, and attempts are made to use those features to create a classifier that can take a matrix and recommend a particular preconditioned solver by determining whether it is likely to converge to a solution.

In [3] they solve each matrix using different versions of preconditioned GM-RES. Each matrix is labeled according to the behavior of the preconditioned solver (e.g. zero pivot encountered or solve successful). For each preconditioner, they describe the features of the matrices that correspond to a successful solve in order to give helpful guidelines to the user. In [5] a combination of support vector machines and clustering is used. After extracting assorted structural and numerical features from each matrix, they cluster the matrices based on these features and attempt to describe each cluster by its performance when different solvers are applied. Depending on the consistency of behavior within each cluster, they may also do further classification using support vector machines. In [4] boosting and alternating decision trees are used to create a classifier. In all cases, predictions and recommendations are generated for new matrices by first extracting the same set of features. Our work follows the same general framework, but we use a different machine learning technique, and we explore a broader range of scenarios that might be of interest to a user. Furthermore, our primary goal is not to describe a large software system, but simply to evaluate a single machine learning technique.

In particular, we consider neural networks, a supervised machine learning technique which has been used successfully in a variety of applications where the output is believed to be a complex function of possibly noisy input data.

Informally, a neural network is a web of interconnected nodes, each of which has a set of weighted input edges and output edges. Each node computes a linear function of the input signals and the edge weights, then passes the result to an activation function which determines the output value for the node. Training a neural network refers to learning the correct edge weights so that given an input signal, the correct output signal is achieved. For a more thorough introduction, see [6]. One of the disadvantages of neural networks is that, in general, little insight can be gained into why a network makes particular predictions. Thus in this initial study we focus primarily on making accurate predictions, rather than on also extracting rules for explaining the preconditioner behavior on a finer scale.

3   Methodology

Our primary goal is to train a neural network to distinguish between preconditioned systems that can be solved by a given iterative solver, and those that cannot. The network is first trained on examples, or instances, that each correspond to a specific preconditioned linear system. We then use the network to classify instances that were not a part of the training set.

In this section we describe the creation of the sample space, and the training of the network. We also describe the explicit construction of training and testing sets to model three scenarios users might encounter, as well as how information from the preconditioned system can be used to increase prediction accuracy.

3.1   Construction of the Sample Space

Our test suite consists of 260 matrices from the University of Florida Sparse Matrix Collection [7], on each of which we ran ILUTP_Mem [8] preconditioned GMRES(50) [9]. ILUTP_Mem allows the user to set the values of 3 parameters that together control the memory required, and the accuracy provided, by the preconditioner; we used a range of values for lfil (0,1,2,3,4,5), droptol (0,.001,.01,.1), and pivtol (0,.1,1). In addition, before computing the ILU factorization, matrices were first permuted and scaled using MC64 [10, 11] for stability, and then using COLAMD [12, 13] for sparsity. Note that by using value-based ILU preconditioners, instead of the level-based ILU preconditioners used in [5], we are working with a much larger number of parameter values. Each matrix was solved 72 different times, each time using a different combination of parameter values. This generated a sample space of over 18,000 instances. Of these instances 49.7% of them were successfully solved using preconditioned GMRES(50).

For input to the neural network we computed an assortment of statistics about each matrix. We used 35 input values including 32 structural and numerical statistics used in [5], as well as the values of the lfil, droptol, and pivtol input parameters to the ILUTP_Mem preconditioner. Thus an instance in the sample space is a 35-tuple: 32 extracted matrix features, and the values of the 3 parameters to the preconditioner. In some situations described below we used 3 additional input values that were computed by running the ILU preconditioner.

3.2   Neural Network Parameters

The neural network used consisted of an input layer, a hidden layer, and an output layer. The output layer contains a single node whose output was either a 0, indicating that the preconditioned solver should be successful, or a 1, indicating otherwise. The number of hidden nodes was initially chosen by the Baum-Haussler heuristic, although some experiments were done with modified values. The weights were initialized with random values from the range $[-.3, .3]$ and the learning rate was set at .3. (For more discussion of these terms, see [6].)

We use the Backpropagation algorithm to train the neural network because, despite its simplicity, it can yield powerful results for multilayer neural networks. After each training instance is propagated through the neural network, the actual output is compared to the desired output to obtain an error value. The error is propagated backward through the network and used to update the weight values maintained at each node. Once the entire training set has been propagated through the network, the network is tested on the validity set. If the error over the entire validity set has increased since the last iteration, the network patience is decreased. Training stops when the network patience reaches zero, or when the error on the validity set falls below some threshold value. Once the training stops, the weight values that correspond to the lowest error over the validity set are loaded into the neural network, and the accuracy of the neural network is evaluated on a set of test matrices.

Since a sample space size of 18,000 instances is smaller than the size recommended by the Baum-Haussler Rule, we used 17-fold cross validation. After choosing 1000 instances for the test set, we divide the remaining instances into 17 potential validity sets. We then train and test our neural networks 17 times, each time using the same test set, but a different validity and training set.

3.3   User Scenarios

When assigning instances to the testing, training, and validity sets, we considered three different scenarios that a user might be interested in:

Previously Solved: This corresponds to the situation where the system of interest has been studied before, but perhaps the user is curious about the likely effectiveness of a different preconditioner or of changes to user-set parameters. To simulate this scenario, we randomly choose 1000 instances from the entire sample space to create the test set. The remaining instances are used for the validity and training set. Thus for any single matrix, the instance generated from trying to solve it using a particular setting of the parameters might be in the testing set, while another instance with different parameter settings might end up in the validity or training sets.

Novel: This corresponds to the situation where the system comes from a novel application and is unlike any system seen before. In this scenario, we randomly choose 15 matrices to create the test set. If one of these matrices belongs to a family of matrices, we throw out these siblings so that they are not used in the training or validity sets.

Family: This corresponds to the situation where the system is similar to others seen before (e.g. resulting from a finer meshing of an existing problem). In this scenario, we randomly choose 15 matrices to create the test set. We allow other matrices in the same family to remain in the training and validity set. However for a given matrix in the testing set, any of the 72 instances generated by this matrix will only occur in the testing set and not in either the training or validity sets.

3.4   Using Information from the Preconditioner

Finally, we consider both the case where a user hopes to predict success based solely on information about the system, and the case where a user is willing to compute the preconditioner first and use information from this to predict the success of a subsequently applied iterative solver. In the latter case, we include information about the number of nonzeros in the computed preconditioner as well as a value indicating whether the preconditioner was successful. This creates an extended set of 38 input values to the neural network: 32 extracted matrix features, 3 parameters to the preconditioner, and 3 additional values from the preconditioned system. In both cases, whether information about the computed preconditioner is used or not, we consider all three scenarios described in the above section.

4   Results

In this section we first make some general observations, then present the results when only information from the system is used for prediction, and finally the results when additional information from computing the preconditioner is used. Since a user could be concerned about various forms of accuracy, we give the overall accuracy for all three user scenarios as well as the breakdown of incorrect responses into false positives (type 1 errors in which the network mistakenly predicts that a system is solvable) and false negatives (type 2 errors in which the network mistakenly predicts that a system is unsolvable).

4.1   General Observations

In each scenario we varied the number of hidden nodes, testing a range of values starting from the number recommended by the Baum-Haussler heuristic and increasing until the network performance stopped improving. We found that the number of hidden nodes affected the accuracy of the neural network, although more hidden nodes was not always better. Since the difference in accuracy was typically around 2%, and never more than 5%, in the rest of this section we always use the best result obtained over all the number of hidden nodes tried. In addition, since we used 17-fold cross validation we report on combined results across all 17 runs.

4.2   Using Information from Only the System

Initially we assume that the neural network is given only information about the matrix and the values of the `lfil`, `droptol`, and `pivtol` parameters given to the preconditioner. Table 1 summarizes the results under each of the three scenarios described in the previous section.

Table 1. Accuracy of classification with 35 input parameters

	Previously Solved	Novel	Family
Correctly classified	92.5%	67.9%	79.1%
False positives	3.9%	14.6%	11.4%
False negatives	3.7%	17.5%	9.6%

Looking at the first row of Table 1, the best performance is in the Previously Solved case, where 92.5% of instances in the test set were correctly classified. In this case, the number of hidden nodes had little effect. However in the other two situations, we found that the performance of the neural network increased to a certain point as the number of hidden nodes increased, and then began to decrease. For the Novel case the optimal number of hidden nodes was 36 and with the Family scenario it was 38.

The other rows of Table 1 shows the type of errors made in each scenario. Overall, the neural networks seem equally likely to mistakenly predict that the system is solvable versus unsolvable.

4.3   Using Information from the Preconditioner

We then repeated the experiments for all three scenarios above using three additional input parameters: whether the incomplete factorization was successful, and the number of nonzeros in the upper and lower triangular incomplete factors. Not surprisingly, as shown in the first row of Table 2, the accuracy increased in all three cases.

Table 2. Accuracy of classification with 38 input parameters

	Previously Solved	Novel	Family
Correctly classified	96.6%	73.5%	80.4%
False positives	1.1%	13.1%	11.2%
False negatives	2.3%	13.3%	8.3%

In the Previously Solved case, we correctly predicted the behavior of the preconditioned solver for 96.6% of the testing instances. This number was very consistent. Even though we report on the best performance (which was achieved using 38 hidden nodes), we note that the variance over all the different number of hidden nodes tried was only .006.

At the other extreme, in the Novel case we were only able to predict the behavior of the solver for 73.5% of the testing instances. The performance also varied more (with a variance of 2.39) as a function of the number of hidden nodes. The best performance here used 28 hidden nodes.

If we now look at the type of errors, we see that type 2 errors (false negatives) are more common than type 1 errors in the Previously Solved scenario. This means that the neural network is more likely to predict that a system cannot be solved when, in fact, it can — however, since examples in the first scenario are overwhelmingly correctly classified, this represents a very small number of incorrect predictions. In the Family case the neural network is slightly more likely to incorrectly predict that a system can be solved when it cannot, and in the Novel scenario there is little bias in either direction.

5   Discussion

In this paper we evaluate the effectiveness of neural networks as a tool for predicting the behavior of preconditioned iterative solvers under a range of scenarios, using matrix statistics that are relatively cheap to compute. The largest factor influencing prediction accuracy is the degree to which the training set is representative of the testing set. When the two sets are very similar, the neural network is highly accurate even when inexpensive matrix features, versus more expensive information about the preconditioner, are used as input. Not surprisingly, as the two sets become less similar, the accuracy drops. However, even when the two sets bear minimal resemblance to each other as in the Novel case discussed previously, the accuracy can be improved by adding information from the computed preconditioner. It is unfortunate that these neural networks do not seem strongly biased towards either type 1 or type 2 errors, since it means a user cannot make any assumptions about whether the prediction is likely to be conservative.

We are currently experimenting with using principal component analysis both to reduce the training time for the neural network, and to explain the behavior of the neural network. Preliminary results show that applying PCA to the sample space before training the neural network allows us to get the same accuracy (96% in the Previously Solved case), using only 15 inputs instead of 38. This points to a high level of redundancy and noise in the data. In addition, we plan to apply PCA to the weight values in the neural network. The hope is that this will determine which features are the most significant for prediction. In the situation where a user is interested in a completely novel system, this may enable us to create training sets of matrices that are similar to that system in a meaningful way.

Overall, we find these initial results encouraging, although the fact that we only considered ILU-preconditioned GMRES(50) limits the generalizability of these results. Clearly in the future it would also be desireable to perform these tests over a larger number of preconditioners, iterative solvers, and matrices.

Acknowledgements. This work was funded in part by the National Science Foundation under grant #CCF-0446604. Any opinions, findings, and conclusions or recommendations expressed in this material are those of the authors and do not necessarily reflect the views of the National Science Foundation.

References

1. T.-Y. Chen.: Preconditioning sparse matrices for computing eigenvalues and solving linear systems of equations. PhD thesis, University of California at Berkeley (December 2001)
2. E. Chow and Y. Saad.: Experimental study of ILU preconditioners for indefinite matrices. J. Comp. and Appl. Math. **16** (1997) 387–414
3. S. Xu, E. Lee, and J. Zhang.: An interim report on preconditioners and matrices. Technical Report Technical Report 388-03, Department of Computer Science, University of Kentucky (2003)
4. S. Bhowmick, V. Eijkhout, Y. Freund, E. Fuentes, and D. Keyes.: Application of machine learning to the selection of sparse linear solvers. Submitted (2006)
5. S. Xu and J. Zhang.: Solvability prediction of sparse matrices with matrix structure-based preconditioners. In Proceedings of Preconditioning 2005, Atlanta, Georgia (May 2005)
6. T. M. Mitchell.: Machine Learning. McGraw-Hill (1997)
7. T. Davis.: University of Florida sparse matrix collection. NA Digest, Vol .92,No. 42 (Oct. 16, 1994) and NA Digest, Vol. 96, No. 28 (Jul. 23, 1996) and NA Digest, Vol. 97, No. 23 (June 7, 1997) Available at: http://www.cise.ufl.edu/research/sparse/matrices/.
8. T.-Y. Chen.: ILUTP_Mem: A space-efficient incomplete LU preconditioner. In A.Laganà,M. L. Gavrilova, V. Kumar, Y. Mun, C. J. K. Tan, and O. Gervasi (eds.), Proceedings of the 2004 International Conference on Computational Science and its Applications. volume 3046 of LNCS. (2004) 31-39
9. Y. Saad and M. H. Schultz.: GMRES: A generalized minimal residual algorithm for solving nonsymmetric linear systems. SIAM J. Sci. Stat. Comput. **7(3)** (July 1986) 856–869
10. I. S. Duff and J. Koster.: The design and use of algorithms for permuting large entries to the diagonal of sparse matrices. SIAM J. Matrix Anal. Appl. **20(4)** (1999) 889–901
11. I. S. Duff and J. Koster.: On algorithms for permuting large entries to the diagonal of a sparse matrix. SIAM J. Matrix Anal. Appl. **22(4)**(2001) 973–996, .
12. T. Davis, J. Gilbert, S. Larimore, and E. Ng.: Algorithm 836: COLAMD, a column approximate minimum degree ordering algorithm. ACM Trans. on Math. Softw. **30(3)** (September 2004) 377–380
13. T. Davis, J. Gilbert, S. Larimore, and E. Ng. A column approximate minimum degree ordering algorithm. ACM Trans. on Math. Softw. **30(3)** (September 2004) 353–376

On the Normal Boundary Intersection Method for Generation of Efficient Front

Pradyumn Kumar Shukla

Institute of Numerical Mathematics
Department of Mathematics
Dresden University of Technology Dresden PIN 01069, Germany
`pradyumn.shukla@mailbox.tu-dresden.de`

Abstract. This paper is concerned with the problem of finding a representative sample of Pareto-optimal points in multi-objective optimization. The Normal Boundary Intersection algorithm is a scalarization scheme for generating a set of evenly spaced Efficient solutions. A drawback of this algorithm is that Pareto-optimality of solutions is not guaranteed. The contributions of this paper are two-fold. First, it presents alternate formulation of this algorithms, such that (weak) Pareto-optimality of solutions is guaranteed. This improvement makes these algorithm theoretically equivalent to other classical algorithms (like weighted-sum or ε-constraint methods), without losing its ability to generate a set of evenly spaced Efficient solutions. Second, an algorithm is presented so as to know beforehand about certain sub-problems whose solutions are not Pareto-optimal and thus not wasting computational effort to solve them. The relationship of the new algorithm with weighted-sum and goal programming method is also presented.

Keywords: Multi-objective optimization, Efficient front generation, Computationally efficient algorithm.

1 Introduction

There are usually multiple conflicting objectives in engineering problems (see for example [5,12,13,3,11]). Since the objectives are in conflict with each other there is a set of Pareto-optimal solutions. The main goal of multi-objective optimization is to seek Pareto-optimal solutions. Over the years there have been various approaches toward fulfillment of this goal. Usually it is not economical to generate the entire Pareto-surface, due to the high computational cost for function evaluations in engineering problems one aims for finding a representative sample of Pareto-optimal points. It have been observed that *convergence* and *diversity* are two conflicting criterion's which are desired in any algorithm which tries to generate the entire efficient front. There are many algorithms for generating the Pareto-surface for general non-convex multi-objective problems. See for example [4,10,3] and the references therein.

Most classical generating multi-objective optimization methods use an iterative scalarization scheme of standard procedures such as weighted-sum or

Y. Shi et al. (Eds.): ICCS 2007, Part I, LNCS 4487, pp. 310–317, 2007.

ε-constraint method as discussed in [4,10], the drawbacks of most of these approaches is that although there are results for convergence, diversity is hard to maintain. Thus we see that systematic variation of parameters in these scalarization techniques do not guarantee diversity in the solution sets as discussed in ([1]). The Normal Boundary Intersection (NBI) algorithm [2] gave a breakthrough by using scalarizing schemes that give a good diversity in the objective space. It start from equidistant points on the utopia plane (plane passing through individual function minimizers) and then go along a certain direction. The progress along the direction is measured by an auxiliary real variable. There are also some other algorithms also which try to generate evenly spaced Efficient solutions. Weck [7] developed adaptive weighted sum method for multi-objective optimization. Messac [9] developed the Normal Constraint method for getting even representation of the efficient frontier.

The developments in this paper are aimed at developing an improved algorithm which like direction-based algorithms produces evenly spaced points on the Efficient front and on the other hand like weighted-sum or ε-constraint method does not produce dominated points. This paper also proposes a way to jump over some of non-convex regions in which no Pareto-optimal solution lies. This amounts to solving a smaller number of sub-problems compared to existing methods.

The rest of the paper is structured as follows: in Section 2, the NBI algorithm is briefly described and an improved algorithm is also presented. A method to jump over dominated regions in non-convex cases is also presented. Section 3 discusses the relationship of the new algorithm with the weighted-sum and goal programming method. Finally conclusions are presented in the last section.

2   NBI Algorithm: Original and a New Formulation

The NBI method was developed by Das et. al. [2] for finding a uniformly spread Pareto-optimal solutions for a general nonlinear multi-objective optimization problem. The weighted-sum scalarization approach has a fundamental drawback of not being able to find a uniform spread of Pareto-optimal solutions, even if a uniform spread of weight vectors are used. The NBI approach uses a scalarization scheme with a property that a uniform spread in parameters will give rise to a near uniform spread in points on the efficient frontier. Also, the method is independent of the relative scales of different objective functions. The scalarization scheme is briefly described below.

Let us consider the following multi-objective problem (MP):

$$\begin{aligned} &\min_{\mathbf{x} \in S} F(\mathbf{x}), \\ &\text{where} \quad S = \{\mathbf{x} \mid h(\mathbf{x}) = 0; g(\mathbf{x}) \leq 0, a \leq \mathbf{x} \leq b\}. \end{aligned} \tag{1}$$

Let $F^* = (f_1^*, f_2^*, \ldots, f_m^*)^\top$ be the ideal point of the multi-objective optimization problem with m objective functions and n variables. Henceforth we shift the origin (in objective space) to F^* so that all the objective functions are non-negative. Let the individual minimum of the functions be attained at $\mathbf{x}_i^*$ for each $i = 1, 2, \ldots, m$. The convex hull of the individual minima is then obtained.

The simplex obtained by the convex hull of the individual minimum can be expressed as $\Phi\beta$, where $\Phi = (F(\mathbf{x}_1^*), F(\mathbf{x}_2^*), \ldots F(\mathbf{x}_m^*))$ is a $m \times m$ matrix and $\beta = \{(b_1, b_2, \ldots, b_m)^\top \mid \sum_{i=1}^m b_i = 1\}$. The original study suggested a systematic method of setting β vectors in order to find a uniformly distributed set of efficient points. The NBI scalarization scheme takes a point on the simplex and then searches for the maximum distance along the normal pointing toward the origin. The NBI subproblem (NBI$_\beta$) for a given vector β is as follows:

$$\begin{aligned} \max_{(\mathbf{x},t)} \quad & t, \\ \text{subject to } & \Phi\beta + t\hat{n} = F(\mathbf{x}), \\ & \mathbf{x} \in S, \end{aligned} \quad (2)$$

where $\hat{n}$ is the normal direction at the point $\Phi\beta$ pointing towards the origin. The solution of the above problem gives the maximum t and also the corresponding Pareto-optimal solution, $\mathbf{x}$. The method works even when the normal direction is not an exact one, but a quasi-normal direction. The following quasi-normal direction vector is suggested in Das et al. [2]: $\hat{n} = -\Phi e$, where $e = (1, 1, \ldots, 1)^T$ is a $m \times 1$ vector. The above quasi-normal direction has the property that NBI$_\beta$ is independent of the relative scales of the objective functions.

The main problem with NBI method is that the solutions of the sub-problems need not be Pareto-optimal (not even locally). This method aims at getting *boundary points* rather than Pareto-optimal points. Pareto-optimal points are a subset of boundary points. These obtained point may or may not be a Pareto-optimal point. Figure 1 shows the obtained solutions (i.e. a, b, c, d, e, f, g, h) corresponding to equidistant points (i.e. A, B, C, D, E, F, G, H) on the convex hull (i.e. line AH).

It can be seen that points d, e and f are not Pareto-optimal but are still found using the NBI method. It is because solutions of NBI sub-problem is only constrained to lie on a line, however for any point $\mathbf{x}^*$, to be Pareto-optimal, the existence of any point in the objective space which satisfies $F(x) \leq F(\mathbf{x}^*)$ need to be checked (since such a point dominates $F(\mathbf{x}^*)$).

Using the concept of the goal attaintment approach of Gembicki [6] we formulate a modified algorithm (called as mNBI) in which we solve the following sub-problem (mNBI$_\beta$) for a given vector β is as follows:

$$\begin{aligned} \max_{(\mathbf{x},t)} \quad & t, \\ \text{subject to } & \Phi\beta + t\hat{n} \geq F(\mathbf{x}), \\ & \mathbf{x} \in S, \end{aligned} \quad (3)$$

This change in formulation increases the feasible space so as to check for points which dominate the points on the line. Figure 2 shows the schematics. It can be seen that the points d, e and f which are not Pareto-optimal will not be found using the mNBI method. Points x, y and z dominate the corresponding points d, e and f (which results in larger value of t) and they are thus not optimal to $mNBI_\beta$. For the optimality of mNBI$_\beta$ subproblems we have the following result.

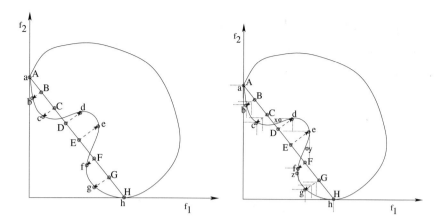

Fig. 1. Schematic showing obtained solutions using NBI method

Fig. 2. Schematic showing obtained solutions using a modified NBI method

Lemma 1. *Let (t^*, x^*) be the optimal solution of (3) subproblem then x^* is weakly efficient.*

Proof: The proof follows from [6] by using $\Phi\beta$ and $t\hat{n}$ as the goal vector and direction respectively. □

Another advantage of the mNBI algorithm is that since for non-convex problems all boundary points need not be Pareto-optimal, for some non-convex regions we can ignore (*jump* over the non-efficient regions in the objective space) solving certain sub-problems. Figure 3 shows the sketch of our proposed method to this. Suppose we start mNBI$_\beta$ from the individual minima A of f_1 and move on the line AB (convex hull of individual minima). When we reach point C the optimal solution of mNBI gives the (weakly) efficient point x. Note that the optimal solution of the NBI subproblem would have been y. From the constraints in mNBI subproblem we can calculate whether or not the constraints are active at optima. For example in this case the constraints $\Phi\beta + t\hat{n} \geq F(\mathbf{x})$ are not-active since the optima x does not lies on the line. Now since the obtained solution x will always weakly dominate the region $x + \mathbb{R}^m_+$ (Lemma 1). Since we know the direction $\hat{n}$ (constant for all sub-problems) we an easily find the point D on the line AB (convex hull of individual minima, in general a simplex), which corresponds to obtained solution x.

Lemma 2. *Let (t^*, x^*) be the optimal solution of (3). Let $w := \Phi\beta + t^*\hat{n}$. Let $\pi[S] \subseteq \mathbb{R}^{m-1}$ denote the projection (in the direction $\hat{n}$) of any surface $S \subseteq \mathbb{R}^m$ onto the simplex obtained by convex hull of individual minima. Then the solution of (3) for all $\beta \in \pi[(F(x) + \mathbb{R}^m_+) \bigcap (w - \mathbb{R}^m_+)]$ is not Pareto-optimal.*

Proof: The proof follows by noting that x weakly dominates the region $(F(x) + \mathbb{R}^m_+)$ and that there exist no feasible point in $(w - \mathbb{R}^m_+)$ that dominates w. □

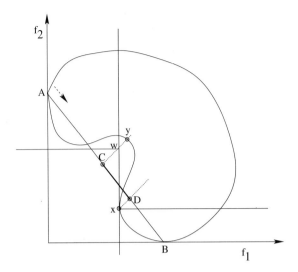

Fig. 3. Schematic showing how to avoid solving certain sub-problems whose solutions are not Pareto-optimal

For our example, all subproblems originating from region CD do not give Pareto-optimal solutions and thus we can ignore them. It is to be noted that this information is obtained before solving the sub-problems. This amounts to solving a smaller number of sub-problems compared to existing methods. This computational gain can be of significant interest in problems where the cost of function evaluation is high.

3   Relationship Between mNBI Subproblem and Weighted Sum Method

Given a Pareto-optimal solution $\mathbf{x}^*$, let $\overline{h}(\mathbf{x})$ denote the augmented vector of all constraints that are active at $\mathbf{x}^*$. We can think of the problem to be constrained only by $\overline{h}(\mathbf{x})$. Let $w \in \mathbb{R}^m$, $\sum_{i=1}^m w_i = 1$, denote positive weights for the objectives. We consider the weighted linear combination problem as follows

$$\begin{aligned}\max_{(\mathbf{x})} \quad & w^\top F(\mathbf{x}), \\ \text{subject to } & \overline{h}(\mathbf{x}) = 0.\end{aligned} \tag{4}$$

The problem (4)) will be denoted by LC_w. Part of first-order necessary KKT conditions for optimality of $(\mathbf{x}^*, \lambda^*)$ for LC_w is

$$\nabla_{\mathbf{x}} F(\mathbf{x}^*)w + \nabla_{\mathbf{x}}\overline{h}(\mathbf{x}^*)\lambda^* = 0. \tag{5}$$

In a similar way, the mNBI$_\beta$ subproblem can be written as

$$\begin{aligned}\min_{(\mathbf{x},t)} \quad & -t, \\ \text{subject to } & F(\mathbf{x}) - \Phi\beta - t\hat{n} \le 0, \\ & \overline{h}(\mathbf{x}) = 0,\end{aligned} \tag{6}$$

Part of first-order KKT condition for optimality of $(\mathbf{x}^*, t^*, \lambda^{(1)*}, \lambda^{(2)*})$ is

$$\nabla_{\mathbf{x}} F(\mathbf{x}^*) \lambda^{(1)*} + \nabla_{\mathbf{x}} \overline{h}(\mathbf{x}^*) \lambda^{(2)*} = 0,$$
$$-1 - \hat{n} \lambda^{(1)*} = 0, \tag{7}$$

where $\lambda^{(1)*} \in \mathbb{R}^m$, $\lambda^{(1)*} \geq 0$ represents the multipliers corresponds to the inequality constraints $F(\mathbf{x}) - \Phi\beta - t\hat{n} \leq 0$, and $\lambda^{(2)*}$ denote the multipliers of the equality constraints $\overline{h}(\mathbf{x}) = 0$.

Claim. Suppose $(\mathbf{x}^*, t^*, \lambda^{(1)*}, \lambda^{(2)*})$ is the solution of mNBI$_\beta$ and $\sum_{i=1}^m \lambda_i^{(1)*} \neq 0$. Then LC_w with the weight vector $w = \frac{1}{\sum_{i=1}^m \lambda_i^{(1)*}} \lambda^{(1)*}$ has the solution

$$\left[\mathbf{x}, \lambda^* = \frac{1}{\sum_{i=1}^m \lambda_i^{(1)*}} \lambda^{(2)*} \right]$$

Proof: Dividing both sides of first part of (7) by the positive scalar $\sum_{i=1}^m \lambda_i^{(1)*}$ and observing that $\overline{h}(\mathbf{x}) = 0$, the equivalence follows. □

Hence we can obtain the weighting of objectives that correspond to solution of mNBI$_\beta$ subproblem. Note that if one uses the original NBI method to get the equivalence, since the $F(\mathbf{x}) - \Phi\beta - t\hat{n} = 0$ are equality constraints the components of multiplier $\lambda^{(1)*} \in \mathbb{R}^m$ can be positive or negative. Thus one could obtain negative weights [2]. In such a case the solution of NBI$_\beta$ subproblem is non Pareto-optimal. This is because irrespective of whether the solution lies in a convex or a non-convex part of the boundary, it should satisfy the first-order necessary KKT conditions if it is Pareto-optimal. However, this does not occurs in mNBI$_\beta$ subproblems (the weights are always non-negative).

4   Relationship Between mNBI Subproblem and Goal Programming

A solution to an mNBI subproblem is also a solution to a goal programming problem.

Claim. Suppose $(\mathbf{x}^*, t^*, \lambda^{(1)*}, \lambda^{(2)*})$ is the solution of mNBI$_\beta$ and $\sum_{i=1}^m \lambda_i^{(1)*} \neq 0$. If $\lambda_k^{(1)*}$ is any strict positive component (note that all the components are non-negative), then $\mathbf{x}^*$ solves the following goal programming problem:

$$\begin{aligned} \min_{(\mathbf{x})} \quad & f_k(\mathbf{x}), \\ \text{subject to } \quad & f_i(\mathbf{x}) \leq \gamma_i, \forall i \neq k, \\ & \overline{h}(\mathbf{x}) = 0, \end{aligned} \tag{8}$$

with for all $i \neq k$ the goals given by

$$\gamma_i = \begin{cases} f_i(\mathbf{x}) & \text{if } \lambda_i^{(1)*} \neq 0; \\ \text{any finite number} \geq f_i(\mathbf{x}) & \text{if } \lambda_i^{(1)*} = 0. \end{cases}$$

Proof: Since $(\mathbf{x}^*, t^*, \lambda^{(1)*}, \lambda^{(2)*})$ is the solution of mNBI$_\beta$ it satisfies the first-order necessary KKT condition (7). Dividing both sides of (7) by $\lambda_k^{(1)*} > 0$ we obtain

$$\nabla_{\mathbf{x}} f_k(\mathbf{x}^*) w + \sum_{i=1, i \neq k}^{i=n} \nabla_{\mathbf{x}} f_i(\mathbf{x}^*) \frac{\lambda_i^{(1)*}}{\lambda_k^{(1)*}} + \nabla_{\mathbf{x}} \overline{h}(\mathbf{x}^*) \frac{\lambda^{(2)*}}{\lambda_k^{(1)*}} = 0. \tag{9}$$

Now with $\frac{\lambda_i^{(1)*}}{\lambda_k^{(1)*}} \geq 0$ as the multipliers of the $m-1$ inequality constraints in (8), the goals γ_i satisfy complementarity by definition since

$$\gamma_i = f_i(\mathbf{x}) \text{ whenever } \lambda_i^{(1)} \neq 0,$$
$$\frac{\lambda_i^{(1)*}}{\lambda_k^{(1)*}} (f_i(\mathbf{x}) - \gamma_i) = 0, \forall i \neq k. \tag{10}$$

Using the above together with feasibility of $\mathbf{x}$ for (8), we obtain that the point $\left(\mathbf{x}^*, \frac{\lambda_i^{(1)*}}{\lambda_k^{(1)*}}, \frac{\lambda^{(2)*}}{\lambda_k^{(1)*}} \right)$ satisfies the first order necessary optimality conditions of (8).

□

Note that as in the case of weighted-sum method if one uses the original NBI method to get the equivalence, due to the presence of equality constraints $F(\mathbf{x}) - \Phi\beta - t\hat{n} = 0$, the components of multiplier $\lambda^{(1)*} \in \mathbb{R}^m$ can be positive or negative. Thus one could obtain negative weights. In such a case the equivalence of NBI and goal programming problem requires additional assumption that the components of $\lambda^{(1)*}$ are of the same sign while no such assumptions are needed in the mNBI method.

5   Conclusions

In this paper, we presented an efficient formulation of the NBI algorithm for getting an even spread of efficient points. This method unlike NBI method does not produce dominated points and is theoretically equivalent to weighted-sum or ε-constraint method This paper also proposes a way to reduce the number of sub-problems to be solved for non-convex problems. We compare the mNBI method with other popular methods like weighted-sum method and goal programming methods using Lagrange multipliers. It turned out that mNBI method does not require any unusual assumption compared to relationship of NBI method with weighted-sum method and goal programming method. Lastly we would like to mention that since some other class of methods like the Normal-Constraint Method [9] or the Adaptive Weighted Sum Method [8] use similar line or inclined search based constraint in their sub-problems, the solutions of the sub-problems of these method are also in general not Pareto-optimal and hence the mNBI method presented in this paper is superior to them.

Acknowledgements

The author acknowledges the partial financial support by the Gottlieb-Daimler- and Karl Benz-Foundation.

References

1. I. Das and J.E. Dennis. A closer look at drawbacks of minimizing weighted sum of objecties for Pareto set generation in multicriteria optimization problems. *Structural Optimization*, 14(1):63–69, 1997.
2. I. Das and J.E. Dennis. Normal-boundary intersection: A new method for generating the Pareto surface in nonlinear multicriteria optimization problems. *SIAM Journal of Optimization*, 8(3):631–657, 1998.
3. K. Deb. *Multi-objective optimization using evolutionary algorithms*. Chichester, UK: Wiley, 2001.
4. M. Ehrgott. *Multicriteria Optimization*. Berlin: Springer, 2000.
5. H. Eschenauer, J. Koski, and A. Osyczka. *Multicriteria Design Optimization*. Berlin: Springer-Verlag, 1990.
6. F. W. Gembicki. *Performance and Sensitivity Optimization: A Vector Index Approach*. PhD thesis, Case Western Reserve University, Cleveland, OH, 1974.
7. I. Y. Kim and O.de Weck. Multiobjective optimization via the adaptive weighted sum method. In *10th AIAA/ISSMO Multidisciplinary Analysis and Optimization Conference*, 2004.
8. I. Y. Kim and O.de Weck. Adaptive weighted sum method for multiobjective optimization: a new method for Pareto front generation. *Structural and Multidisciplinary Optimization*, 31(2):105–116, 2005.
9. A. Messac and C. A. Mattson. Normal constraint method with guarantee of even representation of complete pareto frontier. *AIAA Journal*, 42(10):2101–2111, 2004.
10. K. Miettinen. *Nonlinear Multiobjective Optimization*. Kluwer, Boston, 1999.
11. Andrzej Osyczka. *Evolutionary Algorithms for Single and Multicriteria Design Optimization*. Physica Verlag, Germany, 2002. ISBN 3-7908-1418-0.
12. Eric Sandgren. Multicriteria design optimization by goal programming. In Hojjat Adeli, editor, *Advances in Design Optimization*, chapter 23, pages 225–265. Chapman & Hall, London, 1994.
13. R. B. Stanikov and J. B. Matusov. *Multicriteria Optimization and Engineering*. New York: Chapman and Hall, 1995.

An Improved Laplacian Smoothing Approach for Surface Meshes

Ligang Chen, Yao Zheng, Jianjun Chen, and Yi Liang

College of Computer Science, and Center for Engineering and Scientific Computation,
Zhejiang University, Hangzhou, Zhejiang, 310027, P.R. China
{ligangchen,yao.zheng,chenjj,yliang}@zju.edu.cn

Abstract. This paper presents an improved Laplacian smoothing approach (ILSA) to optimize surface meshes while maintaining the essential characteristics of the discrete surfaces. The approach first detects feature nodes of a mesh using a simple method, and then moves its *adjustable* or *free node* to a new position, which is found by first computing an optimal displacement of the node and then projecting it back to the original discrete surface. The optimal displacement is initially computed by the ILSA, and then adjusted iteratively by solving a constrained optimization problem with a quadratic penalty approach in order to avoid inverted elements. Several examples are presented to illustrate its capability of improving the quality of triangular surface meshes.

Keywords: Laplacian smoothing, surface mesh optimization, quadratic penalty approach.

1 Introduction

Surface triangulations are used in a wide range of applications (e.g. computer graphics, numerical simulations, etc.). For finite element methods, the quality of surface meshes is of paramount importance, because it influences greatly the ability of mesh generation algorithms for generating qualified solid meshes. Since surface meshes define external and internal boundaries of computational domains where boundary conditions are imposed, and thus they also influence the accuracy of numerical simulations.

Mesh modification and vertex repositioning are two main methods for optimizing surface meshes [1, 2]. While mesh modification methods change the topology of the mesh, the vertex repositioning, also termed as mesh smoothing, redistributes the vertices without changing its connectivity. This paper only focuses on smoothing techniques for surface mesh quality improvement.

Despite their popularity in optimizing 2D and 3D meshes [3, 4], smoothing methods for surface meshes present significant challenges due to additional geometric constraints, e.g. minimizing changes in the discrete surface characteristics such as discrete normals and curvature. When improving surface mesh quality by vertex repositioning, changes in the surface properties can usually maintained small by keeping the vertex movements small and by constraining the vertices to a smooth

Y. Shi et al. (Eds.): ICCS 2007, Part I, LNCS 4487, pp. 318–325, 2007.

surface underlying the mesh or to the original discrete surface. One approach commonly used to constrain nodes to the underlying smooth surface is to reposition each vertex in a locally derived tangent plane and then to pull the vertex back to the smooth surface [1, 5]. Another one is to reposition them in a 2D parameterization of the surface and then to map them back to the physical space [6, 7].

In this paper, an improved Laplacian smoothing approach (ILSA) is presented to enhance the quality of a surface mesh without sacrificing its essential surface characteristics. The enhancement is achieved through an iterative process in which each *adjustable* or *free node* of the mesh is moved to a new position that is on the adjacent elements of the node. This new position is found by first computing an optimal displacement of the node and then projecting it back to the original discrete surface. The optimal displacement is initially obtained by the ILSA, and then adjusted iteratively by solving a constrained optimization problem with a quadratic penalty approach in order to avoid inverted elements.

2 Outline of the Smoothing Procedure

The notations used in the paper are as follows. Let $T = (V, E, F)$ be a surface mesh, where V denotes the set of vertices of the mesh, E the set of edges and F the set of triangular faces. f_i, e_i and $\mathbf{v}_i$ represents the i'th face, edge and vertex of the mesh respectively. $A(b)$ denotes the set of all entities of type A connected to or contained in entity b, e.g., $V(f_i)$ is the set of vertices of face f_i and $F(\mathbf{v}_i)$ is the set of faces connected to vertex $\mathbf{v}_i$, which is also regarded as the local mesh at $\mathbf{v}_i$ determined by these faces. We also use $|S|$ to denote the number of elements of a set S.

The procedure begins with a simple method to classify the vertices of the mesh into four types: *boundary node*, *corner node*, *ridge node* and *smooth node*. The first two types of vertices are fixed during the smoothing process for feature preservation and the last two are referred to as *adjustable nodes*. More sophisticated algorithms for detecting salient features such as crest lines on discrete surfaces can be adopted [8]. Then in an iterative manner, a small optimal displacement is computed for each adjustable node using the ILSA, which accounts for some geometric factors. Moreover, for each smooth node its optimal displacement is adjusted by solving a constrained optimization problem so as to avoid inverted elements. Finally, all those redistributed vertices are projected back to the original discrete surface. The complete procedure is outlined as Algo. 1, of which the algorithmic parameters will be explained later.

3 Classifying the Vertices

The boundary nodes, if they exist, can be identified by examining the boundary edges that have only one adjacent element. For each interior node $\mathbf{v}_i$, let $m = |F(\mathbf{v}_i)|$, we first evaluate its discrete normal by solving the following linear equations

$$\mathbf{Ax} = \mathbf{1}. \tag{1}$$

where $\mathbf{A}$ is an $m \times 3$ matrix whose rows are the unit normals of $F(\mathbf{v}_i)$, and $\mathbf{1} = (1,1,...,1)^t$ is a vector of length m. Since $\mathbf{A}$ may be over- or under-determined, the solution is in least squares sense and we solve it by the singular value decomposition (SVD) method [9].

Algo. 1. The smoothing procedure.

Set the algorithmic parameters: *max_global_iter_num*, *max_smooth_iter_num*, *relax1*, *relax2*, μ, *wl*, *wa*;

Classify the vertices of the mesh into 4 types;

Initialize the smoothed mesh T_{new} as the original mesh T_{ori};

for *step* := 1 to *max_global_iter_num* **do**   //global iteration

Compute the normal of each vertex of T_{new};

Compute the optimal displacement of each ridge node of T_{new};

Compute the initial displacement of each smooth node of T_{new};

Adjust the displacement of each smooth node in order to avoid inverted elements;

Project the redistributed position of each adjustable node back to the original mesh T_{ori}, denote this new mesh as T'_{new};

Update T_{new} as T'_{new}

end for

Set the final optimized mesh as T_{new}.

The length of the resulting vertex normal has a geometric interpretation as an indicator of singularity of $\mathbf{v}_i$. Let $f_j \in F(\mathbf{v}_i)$, $1 \le j \le |F(\mathbf{v}_i)|$, $N(f_j)$ the unit normal of f_j and $N(\mathbf{v}_i) = \mathbf{x}$ the unknown vertex normal. The equation corresponding to f_j in Eq. (1) is

$$N(f_j)g\mathbf{x} = |\mathbf{x}| \cos \angle(N(f_j), \mathbf{x}) = 1. \tag{2}$$

Now it is obvious that, for some solution $\mathbf{x}$ of (1), the angles between $\mathbf{x}$ and $N(f_j)$, $1 \le j \le |F(\mathbf{v}_i)|$, would be approximately equal. Roughly speaking, if the local mesh $F(\mathbf{v}_i)$ is flat, the angles would be small, otherwise they would be large, consequently the length of the resulting vertex normal would be short and long, and the vertex will be regarded as a smooth node and a ridge node, respectively.

The ridge nodes will be further examined to determine whether they are corner nodes or not. Let e_i be an edge formed by two ridge nodes, if the bilateral angle between two faces attached to e_i is below a threshold angle ($8\pi/9$ in our algorithm), these two nodes are said to be *attached-sharp nodes* of each other. If the number of such nodes of a ridge node is not equal to two, this node is identified as a corner node. The geometric interpretation of the classification is self-evident.

4 Repositioning the Adjustable Vertices

4.1 Computing Displacements by the ILSA

In each global iteration of Algo. 1, the ILSA is employed to compute the initial optimal displacements for both ridge and smooth nodes. The procedure for treating these two types of nodes is similar. The major difference lies in that smooth nodes take all their adjacent nodes' effect into account, while ridge nodes consider only the effect of their two attached-sharp nodes. Algo. 2 illustrates the details.

Here if $\mathbf{v}_i$ is a ridge node, then $n = 2$ and $\{\mathbf{v}_j, j = 1, 2\}$ are two attached-sharp nodes of $\mathbf{v}_i$, otherwise $n = |V(\mathbf{v}_i)|$ and $\mathbf{v}_j \in V(\mathbf{v}_i)$. $\mathbf{d}(\mathbf{v}_j)$ is the current displacement of $\mathbf{v}_j$. Such treatment of ridge nodes tries to prevent the crest lines on surface meshes from disappearing. The adjusting vector $\mathbf{vec}$ takes the lengths of adjacent edges into consideration in order to obtain a smoother result.

4.2 Adjusting Displacements by a Quadratic Penalty Approach

Unfortunately, Laplacian smoothing for 2D mesh may produce invalid elements. When used for surface meshes, there are still possibilities of forming abnormal elements. To compensate for this, we adjust the displacement iteratively for each smooth node by solving a constrained optimization problem.

The idea originates in the minimal surface theory in differential geometry. Minimal surfaces are of zero mean curvature. Their physical interpretation is that surface tension tries to make the surface as "taut" as possible. That is, the surface should have the least surface area among all surfaces satisfying certain constraints like having fixed boundaries [10]. Every soap film is a physical model of a minimal surface. This motivates us, for the local mesh $F(\mathbf{v}_i)$ at a smooth node $\mathbf{v}_i$, to move $\mathbf{v}_i$ to minimize the overall area of the elements of $F(\mathbf{v}_i)$ in order to make this local mesh "taut" and thus to avoid invalid elements as much as possible. This new position $\mathbf{v}_i'$ is also softly constrained to be on a plane by a quadratic penalty approach.

Let $\mathbf{d}_{cur}(\mathbf{v}_i)$ and $\mathbf{N}(\mathbf{v}_i)$ be the current displacement and the discrete normal of $\mathbf{v}_i$ respectively. Initially $\mathbf{d}_{cur}(\mathbf{v}_i)$ is the result from Algo. 2. Let $\mathbf{x}$ be the new pending position of $\mathbf{v}_i$ and $\mathbf{d}_{new}(\mathbf{v}_i) = \mathbf{x} - \mathbf{v}_i$ the new adjusting displacement. Suppose $\mathbf{v}_j \in V(\mathbf{v}_i)$, $1 \le j \le n+1$, $n = |V(\mathbf{v}_i)|$ are the vertices surrounding $\mathbf{v}_i$ in circular sequence and $\mathbf{v}_{n+1} = \mathbf{v}_1$. $s(\mathbf{v}_1, \mathbf{v}_2, \mathbf{v}_3)$ represents the area of the triangle $\Delta \mathbf{v}_1 \mathbf{v}_2 \mathbf{v}_3$. Now the optimization problem can be formulated as follows

$$\min_{\mathbf{x}} g(\mathbf{x}) \quad \text{subject to} \quad c(\mathbf{x}) = 0 \tag{3}$$

where

$$g(\mathbf{x}) = wl \frac{1}{n} \sum_{j=1}^{n} |\mathbf{x} - \mathbf{v}_j|^2 + wa \sum_{j=1}^{n} \beta_{ij} s^2(\mathbf{x}, \mathbf{v}_j, \mathbf{v}_{j+1}) \tag{4}$$

Algo. 2. ILSA for adjustable nodes.

Initialize the vertex displacement $\mathbf{d}(\mathbf{v_i})$ of each adjustable node $\mathbf{v_i}$ of T_{new} as the Laplacian

coordinate of $\mathbf{v_i}$: $\mathbf{d}(\mathbf{v_i}) := \dfrac{1}{n}\sum_{j=1}^{n}\mathbf{v_j} - \mathbf{v_i}$;

for $k := 1$ to $max\_smooth\_iter\_num$ **do**

   for each adjustable node $\mathbf{v_i}$ **do**

      Compute a vector **vec** : $\mathbf{vec} := \dfrac{1}{S}\sum_{j=1}^{n}\alpha_{ij}\mathbf{d}(\mathbf{v_j})$,

      where $S = \sum_{j=1}^{n}\alpha_{ij}$ and $\dfrac{1}{\alpha_{ij}} = dist(\mathbf{v_i},\mathbf{v_j})$ is the distance

      between $\mathbf{v_i}$ and $\mathbf{v_j}$;

      Update $\mathbf{d}(\mathbf{v_i})$: $\mathbf{d}(\mathbf{v_i}) := (1 - relax1)\cdot\mathbf{d}(\mathbf{v_i}) + relax1\cdot\mathbf{vec}$;

   end for

   if <condition> **then**   // e.g. smooth enough

      break the iteration;

   end if

end for

and

$$c(\mathbf{x}) = \begin{cases} \mathbf{N}(\mathbf{v_i})\mathrm{gd}_{new}(\mathbf{v_i}) & if \ \mathbf{d}_{cur}(\mathbf{v_i}) = \mathbf{0} \\ \mathbf{d}_{cur}(\mathbf{v_i})\mathrm{gd}_{new}(\mathbf{v_i}) - |\mathbf{d}_{cur}(\mathbf{v_i})|^2 & if \ \mathbf{d}_{cur}(\mathbf{v_i}) \neq \mathbf{0} \end{cases} \tag{5}$$

Here wl and wa are two algorithmic parameters and $\beta_{ij} = 1/s(\mathbf{v_i},\mathbf{v_j},\mathbf{v_{j+1}})$. It can be observed that the constraint $c(\mathbf{x}) = 0$ is used to penalize the deviation of $\mathbf{x}$ from a plane. When $\mathbf{d}_{cur}(\mathbf{v_i}) = \mathbf{0}$, it is the tangent plane at $\mathbf{v_i}$, otherwise it is the plane vertical to $\mathbf{d}_{cur}(\mathbf{v_i})$ and passing through the node $\mathbf{v_i} + \mathbf{d}_{cur}(\mathbf{v_i})$. In other words, it tries to constrain $\mathbf{x}$ to be on the current smoothed discrete surface. It is also observed from Eq. (4) that we include another term related to the length $|\mathbf{x} - \mathbf{v_j}|$ and we use the square of area instead of area itself for simplicity. The area of a triangle can be calculated by a cross product $s(\mathbf{v_1},\mathbf{v_2},\mathbf{v_3}) = |(\mathbf{v_2} - \mathbf{v_1})\times(\mathbf{v_3} - \mathbf{v_1})|/2$. The quadratic penalty function $Q(\mathbf{x};\mu)$ for problem (3) is

$$Q(\mathbf{x};\mu) = g(\mathbf{x}) + \dfrac{1}{\mu}c^2(\mathbf{x}) \tag{6}$$

Algo. 3. Adjusting vertex displacement by a quadratic penalty approach.

for $k:=1$ to *max_smooth_iter_num* **do**

    for each smooth node $\mathbf{v}_i$ **do**

        Compute the adjusting displacement $\mathbf{d}_{new}(\mathbf{v}_i)$ by solving problem(3)

        Update the displacement: $\mathbf{d}_{cur}(\mathbf{v}_i):=relax2\cdot\mathbf{d}_{new}(\mathbf{v}_i)+(1-relax2)\cdot\mathbf{d}_{cur}(\mathbf{v}_i)$

        if <condition> //e.g. tiny change of vertex displacements

          break the iteration;

        end if

    end for

end for

where $\mu>0$ is the penalty parameter. Since $Q(\mathbf{x};\mu)$ is a quadratic function, its minimization can be obtained by solving a linear system, for which we again use the SVD method. This procedure of adjusting vertex displacement is given in Algo. 3.

4.3 Projecting the New Position Back to the Original Mesh

Once the final displacement of each adjustable node is available, the next step is to project the new position of the node back to the original discrete surface to form an optimized mesh. It is assumed that the displacement is so small that the new position of a node is near its original position. Thus, the projection can be confined to be on the two attached-ridge edges and the adjacent elements of the original node for ridge and smooth nodes, respectively.

5 Experimental Results

Two examples are presented to show the capability of our method. The aspect ratio is used to measure the quality of the elements.

The first example is a surface mesh defined on a single NURBS patch. The minimal and average aspect ratios of the original (resp. optimized) mesh are 0.09 and 0.81 (resp. 0.39 and 0.89).

The second example which is obtained from the Large Geometric Model Archives at Georgia Institute of Technology, is a famous scanned object named horse. The original mesh has 96966 triangles and 48485 nodes, and its average aspect ratio is 0.71, which has increased to 0.82 for the optimized counterpart. Note that the poor quality of the original mesh in several parts of the neck of the horse in Fig. 1(a) whose optimized result is given in Fig. 1(b). The details of horses' ears of both initial and optimized meshes have also shown that our surface smoothing procedure is capable of preserving sharp features.

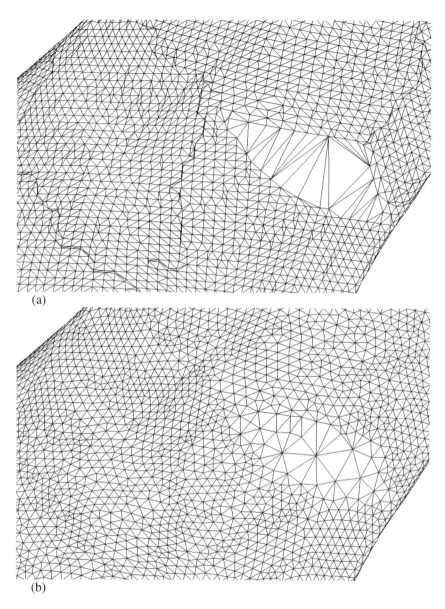

(a)

(b)

Fig. 1. Details of the neck of the horse for the initial (a) and optimized (b) meshes

6   Conclusion and Future Work

We have proposed an improved Laplacian smoothing approach for optimizing surface meshes. The nodes of an optimized mesh are kept on the initial mesh to avoid the shrinkage problem. A simple but effective procedure is also suggested to identify the feature points of a mesh in order to preserve its essential characteristics. Furthermore, to avoid the formation of inverted elements, we adjust the initial displacements by solving a constrained optimization problem with a quadratic penalty method.

In the future, this smoothing technique will be integrated into a surface remesher. A global and non-iterative Laplacian smoothing approach with feature preservation for surface meshes is also under investigation.

Acknowledgements. The authors would like to acknowledge the financial support received from the NSFC (National Natural Science Foundation of China) for the National Science Fund for Distinguished Young Scholars under grant Number 60225009, and the Major Research Plan under grant Number 90405003. The first author is very grateful to Mr. Bangti Jin of The Chinese University of Hang Kong for his valuable revision of this paper.

References

1. Frey, P.J., Borouchaki, H.: Geometric surface mesh optimization. Computing and Visualization in Science, 1(3) (1998) 113-121
2. Brewer, M., Freitag, L.A., Patrick M.K., Leurent, T., Melander, D.: The mesquite mesh quality improvement toolkit. In: Proc. of the 12th International Meshing Roundtable, Sandia National Laboratories, Albuquerque, NM, (2003) 239-250
3. Freitag, L.A., Knupp, P.M: Tetrahedral mesh improvement via optimization of the element condition number. International Journal of Numerical Methods in Engineering, 53 (2002) 1377-1391
4. Freitag, L.A., Plassmann, P.: Local optimization-based simplicial mesh untangling and improvement. International Journal of Numerical Methods in Engineering, 49 (2000) 109-125
5. Knupp, P. M.: Achieving finite element mesh quality via optimization of the jacobian matrix norm and associated quantities. Part 1 – a framework for surface mesh optimization. International Journal of Numerical Methods in Engineering, 48 (2000) 401-420
6. Garimella, R.V., Shashkov, M.J., Knupp, P.M.: Triangular and quadrilateral surface mesh quality optimization using local parametrization. Computer Methods in Applied Mechanics and Engineering, 193(9-11) (2004) 913-928
7. Escobar, J.M., Montero, G., Montenegro, R., Rodriguez, E.: An algebraic method for smoothing surface triangulations on a local parametric space. International Journal of Numerical Methods in Engineering, 66 (2006) 740-760.
8. Yoshizawa, S., Belyaev, A., Seidel, H.–P.: Fast and robust detection of crest lines on meshes. In: Proc. of the ACM symposium on Solid and physical modeling, MIT (2005) 227-232
9. William H.P., Saul A.T., William T.V., Brain P.F.: Numerical Recipes in C++. 2nd edn. Cambridge University Press, (2002)
10. Oprea, J.: Differential Geometry and Its Applications. 2nd edn. China Machine Press, (2005)

Red-Black Half-Sweep Iterative Method Using Triangle Finite Element Approximation for 2D Poisson Equations

J. Sulaiman[1], M. Othman[2], and M.K. Hasan[3]

[1] School of Science and Technology, Universiti Malaysia Sabah, Locked Bag 2073, 88999 Kota Kinabalu, Sabah, Malaysia
[2] Faculty of Computer Science and Info. Tech., Universiti Putra Malaysia, 43400 Serdang, Selangor D.E.
[3] Faculty of Information Science and Technology, Universiti Kebangsaan Malaysia, 43600 Bangi, Selangor D.E.
jumat@ums.edu.my

Abstract. This paper investigates the application of the Red-Black Half-Sweep Gauss-Seidel (HSGS-RB) method by using the half-sweep triangle finite element approximation equation based on the Galerkin scheme to solve two-dimensional Poisson equations. Formulations of the full-sweep and half-sweep triangle finite element approaches in using this scheme are also derived. Some numerical experiments are conducted to show that the HSGS-RB method is superior to the Full-Sweep method.

Keywords: Half-sweep Iteration, Red-Black Ordering, Galerkin Scheme, Triangle Element.

1 Introduction

By using the finite element method, many weighted residual schemes can be used by researchers to gain approximate solutions such as the subdomain, collocation, least-square, moments and Galerkin (Fletcher [4,5]). In this paper, by using the first order triangle finite element approximation equation based on the Galerkin scheme, we apply the Half-Sweep Gauss-Seidel (HSGS) method with the Red-Black ordering strategy for solving the two-dimensional Poisson equation.

To show the efficiency of the HSGS-RB method, let us consider the two-dimensional Poisson equation defined as

$$\frac{\partial^2 U}{\partial x^2} + \frac{\partial^2 U}{\partial y^2} = f(x,y), \quad (x,y) \in D = [a,b] \times [a,b] \tag{1}$$

subject to the Dirichlet boundary conditions

$$U(x,a) = g_1(x), \quad a \le x \le b$$
$$U(x,b) = g_2(x), \quad a \le x \le b$$
$$U(a,y) = g_3(y), \quad a \le y \le b$$
$$U(b,y) = g_4(y), \quad a \le y \le b$$

Y. Shi et al. (Eds.): ICCS 2007, Part I, LNCS 4487, pp. 326–333, 2007.

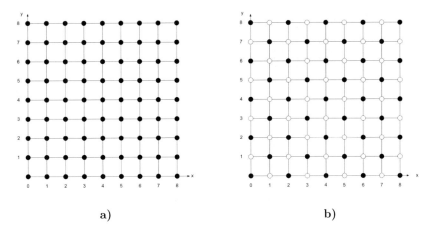

a)   b)

Fig. 1. a) and b) show the distribution of uniform node points for the full- and half-sweep cases respectively at $n = 7$

To facilitate in formulating the full-sweep and half-sweep linear finite element approximation equations for problem (1), we shall restrict our discussion onto uniform node points only as shown in Figure 1. Based on the figure, it has been shown that the solution domain, D is discretized uniformly in both x and y directions with a mesh size, h which is defined as

$$h = \frac{b-a}{m}, \quad m = n+1 \tag{2}$$

Based on Figure 1, we need to build the networks of triangle finite elements in order to facilitate us to derive triangle finite element approximation equations for problem (1). By using the same concept of the half-sweep iterative applied to the finite difference method (Abdullah [1], Sulaiman $et\ al.$ [13], Othman & Abdullah [8]), each triangle element will involves three node points only of type • as shown in Figure 2. Therefore, the implementation of the full-sweep and half-sweep iterative algorithms will be applied onto the node points of the same type until the iterative convergence test is met. Then other approximate solutions at remaining points (points of the different type) are computed directly (Abdullah [1], Abdullah & Ali [2], Ibrahim & Abdullah [6], Sulaiman $et\ al.$ [13,14], Yousif & Evans [17]).

2   Formulation of the Half-Sweep Finite Element Approximation

As mentioned in the previous section, we study the application of the HSGS-RB method by using the half-sweep linear finite element approximation equation based on the Galerkin scheme to solve two-dimensional Poisson equations. By considering three node points of type • only, the general approximation of the

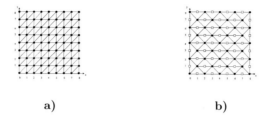

a)   b)

Fig. 2. a) and b) show the networks of triangle elements for the full- and half-sweep cases respectively at $n = 7$

function, $U(x, y)$ in the form of interpolation function for an arbitrary triangle element, e is given by (Fletcher [4], Lewis & Ward [7], Zienkiewicz [19])

$$\tilde{U}^{[e]}(x, y) = N_1(x, y)U_1 + N_2(x, y)U_2 + N_3(x, y)U_3 \qquad (3)$$

and the shape functions, $N_k(x, y)$, $k = 1, 2, 3$ can generally be stated as

$$N_k(x, y) = \frac{1}{\det A}(a_k + b_k x + c_k y), \quad k = 1, 2, 3 \qquad (4)$$

where,

$$\det A = x_1(y_2 - y_3) + x_2(y_3 - y_1) + x_3(y_1 - y_2),$$

$$\begin{bmatrix} a_1 \\ a_2 \\ a_3 \end{bmatrix} = \begin{bmatrix} x_2 y_3 - x_3 y_2 \\ x_3 y_1 - x_1 y_3 \\ x_1 y_2 - x_2 y_1 \end{bmatrix}, \quad \begin{bmatrix} a_1 \\ a_2 \\ a_3 \end{bmatrix} = \begin{bmatrix} a_1 \\ a_2 \\ a_3 \end{bmatrix}, \quad \begin{bmatrix} a_1 \\ a_2 \\ a_3 \end{bmatrix} = \begin{bmatrix} a_1 \\ a_2 \\ a_3 \end{bmatrix},$$

Beside this, the first order partial derivatives of the shape functions towards x and y are given respectively as

$$\left. \begin{array}{l} \frac{\partial}{\partial x}(N_k(x, y)) = \frac{b_k}{\det A} \\ \frac{\partial}{\partial y}(N_k(x, y)) = \frac{c_k}{\det A} \end{array} \right\} \quad k = 1, 2, 3 \qquad (5)$$

Again based on the distribution of the hat function, $R_{r,s}(x, y)$ in the solution domain, the approximation of the functions, $U(x, y)$ and $f(x, y)$ in case of the full-sweep and half-sweep cases for the entire domain will be defined respectively as (Vichnevetsky [16])

$$\tilde{U}(x, y) = \sum_{r=0}^{m} \sum_{s=0}^{m} R_{r,s}(x, y)U_{r,s} \qquad (6)$$

$$\tilde{f}(x, y) = \sum_{r=0}^{m} \sum_{s=0}^{m} R_{r,s}(x, y)f_{r,s} \qquad (7)$$

and

$$\tilde{U}(x, y) = \sum_{r=0,2,4}^{m} \sum_{s=0,2,4}^{m} R_{r,s}(x, y)U_{r,s} + \sum_{r=1,2,5}^{m-1} \sum_{s=1,3,5}^{m-1} R_{r,s}(x, y)U_{r,s} \qquad (8)$$

$$\tilde{f}(x,y) = \sum_{r=0,2,4}^{m} \sum_{s=0,2,4}^{m} R_{r,s}(x,y)f_{r,s} + \sum_{r=1,3,5}^{m-1} \sum_{s=1,3,5}^{m-1} R_{r,s}(x,y)f_{r,s} \qquad (9)$$

Thus, Eqs. (6) and (8) are approximate solutions for problem (1).

To construct the full-sweep and half-sweep linear finite element approximation equations for problem (1), this paper proposes the Galerkin finite element scheme. Thus, let consider the Galerkin residual method (Fletcher [4,5], Lewis & Ward [7]) be defined as

$$\int\int_D R_{i,j}(x,y)E(x,y)\ dxdy = 0, \quad i,j = 0,1,2,...,m \qquad (10)$$

where, $E(x,y) = \frac{\partial^2 U}{\partial x^2} + \frac{\partial^2 U}{\partial y^2} - f(x,y)$ is a residual function. By applying the Green theorem, Eq. 10 can be shown in the following form

$$\oint_\lambda \left(-R_{i,j}(x,y)\frac{\partial U}{\partial y}\ dx + R_{i,j}(x,y)\frac{\partial U}{\partial x}\ dy \right)$$
$$- \int_a^b \int_a^b \left(\frac{\partial R_{i,j}(x,y)}{\partial x}\frac{\partial U}{\partial x} + \frac{\partial R_{i,j}(x,y)}{\partial y}\frac{\partial U}{\partial y} \right)\ dxdy = F_{i,j} \qquad (11)$$

where,

$$F_{i,j} = \int_a^b \int_a^b R_{i,j}(x,y)f(x,y)\ dxdy$$

By applying Eq. (5) and substituting the boundary conditions into problem (1), it can be shown that Eq. (11) will generate a linear system for both cases. Generally both linear systems can be stated as

$$-\sum\sum K^*_{i,j,r,s}U_{r,s} = \sum\sum C^*_{i,j,r,s}f_{r,s} \qquad (12)$$

where,

$$K^*_{i,j,r,s} = \int_a^b \int_a^b \left(\frac{\partial R_{i,j}}{\partial x}\frac{\partial R_{r,s}}{\partial x} \right)\ dxdy + \int_a^b \int_a^b \left(\frac{\partial R_{i,j}}{\partial y}\frac{\partial R_{r,s}}{\partial y} \right)\ dxdy$$

$$C^*_{i,j,r,s} = \int_a^b \int_a^b (R_{i,j}(x,y)R_{r,s}(x,y))\ dxdy$$

Practically, the linear system in Eq. (12) for the full-sweep and half-sweep cases will be easily rewritten in the stencil form respectively as follows:

1. Full-sweep stencil (Zienkiewicz [19], Twizell [15], Fletcher [5])

$$\begin{bmatrix} 0 & 1 & 0 \\ 1 & -4 & 1 \\ 0 & 1 & 0 \end{bmatrix} U_{i,j} = \frac{h^2}{12}\begin{bmatrix} 0 & 1 & 1 \\ 1 & 6 & 1 \\ 1 & 1 & 0 \end{bmatrix} f_{i,j} \qquad (13)$$

2. Half-sweep stencil

$$
\begin{bmatrix} 1\,0 & & 1\,0 \\ 0 & -4\,0 & 0 \\ 1\,0 & & 1\,0 \end{bmatrix} U_{i,j} = \frac{h^2}{6} \begin{bmatrix} 1\,0\,1\,0 \\ 0\,5\,0\,1 \\ 1\,0\,1\,0 \end{bmatrix} f_{i,j}, \quad i = 1 \tag{14}
$$

$$
\begin{bmatrix} 0\,1\,0 & & 1\,0 \\ 0\,0 & -4\,0 & 0 \\ 0\,1\,0 & & 1\,0 \end{bmatrix} U_{i,j} = \frac{h^2}{6} \begin{bmatrix} 0\,1\,0\,1\,0 \\ 1\,0\,6\,0\,1 \\ 0\,1\,0\,1\,0 \end{bmatrix} f_{i,j}, \quad i \neq 1, n \tag{15}
$$

$$
\begin{bmatrix} 0\,1\,0 & & 1 \\ 0\,0 & -4\,0 & \\ 0\,1\,0 & & 1 \end{bmatrix} U_{i,j} = \frac{h^2}{6} \begin{bmatrix} 0\,1\,0\,1 \\ 1\,0\,5\,0 \\ 0\,1\,0\,1 \end{bmatrix} f_{i,j}, \quad i = n \tag{16}
$$

The stencil forms in Eqs. (13) till (16), which are based on the first order triangle finite element approximation equation, can be used to represent as the full-sweep and half-sweep computational molecules.

Actually, the computational molecules involve seven node points in formulating their approximation equations. However, two of its coefficients are zero. Apart of this, the form of the computational molecules for both triangle finite element schemes is the same compared to the existing five points finite difference scheme, see Abdullah [1], Abdullah and Ali [2], Yousif and Evans [17].

3 Implementation of the HSGS-RB

According to previous studies on the implementation of various orderings, it is obvious that combination of iterative schemes and ordering strategies which have been proven can accelerate the convergence rate, see Parter [12], Evans and Yousif [3], Zhang [18]. In this section, however, there are two ordering strategies considered in this paper such as the lexicography (NA) and red-black (RB) being applied to the HSGS iterative methods, called as HSGS-NA and HSGS-RB methods respectively. In comparison, the Full-Sweep Gauss-Seidel (FSGS) method with NA ordering, namely FSGS-NA, acts as the control of comparison of numerical results.

It can be seen from Figure 3 by using the half-sweep triangle finite element approximation equations in Eqs. (14) till (16), the position of numbers in the solution domain for $n = 7$ shows on how both HSGS-NA and HSGS-RB methods will be performed by starting at number 1 and ending at the last number.

4 Numerical Experiments

To study the efficiency of the HSGS-RB scheme by using the half-sweep linear finite element approximation equation in Eqs. [14] till [16] based on the Galerkin scheme, three items will be considered in comparison such as the number of

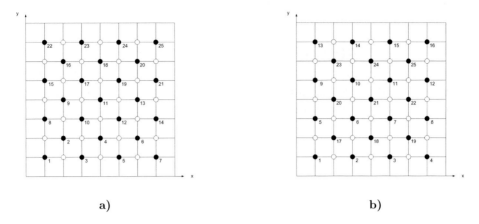

Fig. 3. a) and b) show the NA and RB ordering strategies for the half-sweep case at $n = 7$

Table 1. Comparison of number of iterations, execution time (in seconds) and maximum errors for the iterative methods

Number of iterations				
Methods	Mesh size			
	32	64	128	256
FSGS-NA	1986	7368	27164	99433
HSGS-NA	1031	3829	14159	52020
HSGS-RB	1027	3825	14152	52008
Execution time (seconds)				
Methods	Mesh size			
	32	64	128	256
FSGS-NA	0.14	2.08	30.51	498.89
HSGS-NA	0.03	0.63	9.08	218.74
HSGS-RB	0.03	0.56	8.19	215.70
Maximum absolute errors				
Methods	Mesh size			
	32	64	128	256
FSGS-NA	1.4770e-4	3.6970e-5	9.3750e-6	2.8971e-6
HSGS-NA	5.7443e-4	1.6312e-4	4.4746e-5	1.1932e-5
HSGS-RB	5.7443e-4	1.6312e-4	4.4746e-5	1.1932e-5

iterations, execution time and maximum absolute error. Some numerical experiments were conducted in solving the following 2D Poisson equation (Abdullah [1])

$$\frac{\partial^2 U}{\partial x^2} + \frac{\partial^2 U}{\partial y^2} = \left(x^2 + y^2\right)\exp(xy), \quad (x, y) \in D = [a, b] \times [a, b] \tag{17}$$

Then boundary conditions and the exact solution of the problem (17) are defined by

$$U(x, y) = exp(xy), \quad (x, y) = [a, b] \times [a, b] \tag{18}$$

All results of numerical experiments, obtained from implementation of the FSGS-NA, HSGS-NA and HSGS-RB methods, have been recorded in Table 1. In the implementation mentioned above, the convergence criteria considered the tolerance error, $\epsilon = 10^{-10}$.

5   Conclusion

In the previous section, it has shown that the full-sweep and half-sweep triangle finite element approximation equations based on the Galerkin scheme can be easily represented in Eqs. (13) till (16). Through numerical results collected in Table 1, the findings show that number of iterations have declined approximately $47.70 - 48.29\%$ and $47.68 - 48.09\%$ correspond to the HSGS-RB and HSGS-NA methods compared to FSGS-NA method. In fact, the execution time versus mesh size for both HSGS-RB and HSGS-NA methods are much faster approximately $56.76 - 78.57\%$ and $56.15 - 78.57\%$ respectively than the FSGS-NA method. Thus, we conclude that the HSGS-RB method is slightly better than the HSGS-NA method. In comparison between the FSGS and HSGS methods, it is very obvious that the HSGS method for both ordering strategies is far better than the FSGS-NA method in terms of number of iterations and the execution time. This is because the computational complexity of the HSGS method is nearly 50% of the FSGS-NA method. Again, approximate solutions for the HSGS method are in good agreement compared to the FSGS-NA method. For our future works, we shall investigate on the use of the HSGS-RB as a smoother for the halfsweep multigrid (Othman & Abdullah [8,9]) and the development and implementation of the Modified Explicit Group (MEG) (Othman & Abdullah [10], Othman et al. [11])and the Quarter-Sweep Iterative Alternating Decomposition Explicit (QSIADE) (Sulaiman et al. [14]) methods by using finite element approximation equations.

References

1. Abdullah, A.R.: The Four Point Explicit Decoupled Group (EDG) Method: A Fast Poisson Solver, Intern. Journal of Computer Mathematics, **38**(1991) 61-70.
2. Abdullah, A.R., Ali, N.H.M.: A comparative study of parallel strategies for the solution of elliptic pde's, Parallel Algorithms and Applications, **10**(1996) 93-103.
3. Evan, D.J., Yousif, W.F.: The Explicit Block Relaxation method as a grid smoother in the Multigrid V-cycle scheme, Intern. Journal of Computer Mathematics, **34**(1990) 71-78.
4. Fletcher, C.A.J.: The Galerkin method: An introduction. In. Noye, J. (pnyt.). Numerical Simulation of Fluid Motion, North-Holland Publishing Company,Amsterdam (1978) 113-170.

5. Fletcher, C.A.J.: Computational Galerkin method. Springer Series in Computational Physics. Springer-Verlag, New York (1984).
6. Ibrahim, A., Abdullah, A.R.: Solving the two-dimensional diffusion equation by the four point explicit decoupled group (EDG) iterative method. Intern. Journal of Computer Mathematics, **58**(1995) 253-256.
7. Lewis, P.E., Ward, J.P.: The Finite Element Method: Principles and Applications. Addison-Wesley Publishing Company, Wokingham (1991)
8. Othman, M., Abdullah, A.R.: The Halfsweeps Multigrid Method As A Fast Multigrid Poisson Solver. Intern. Journal of Computer Mathematics, **69**(1998) 219-229.
9. Othman, M., Abdullah, A.R.: An Effcient Multigrid Poisson Solver. Intern. Journal of Computer Mathematics, **71**(1999) 541-553.
10. Othman, M., Abdullah, A.R.: An Efficient Four Points Modified Explicit Group Poisson Solver, Intern. Journal of Computer Mathematics, **76**(2000) 203-217.
11. Othman, M., Abdullah, A.R., Evans, D.J.: A Parallel Four Point Modified Explicit Group Iterative Algorithm on Shared Memory Multiprocessors, Parallel Algorithms and Applications, **19(1)**(2004) 1-9 (On January 01, 2005 this publication was renamed International Journal of Parallel, Emergent and Distributed Systems).
12. Parter, S.V.: Estimates for Multigrid methods based on Red Black Gauss-Seidel smoothers, Numerical Mathematics, **52**(1998) 701-723.
13. Sulaiman. J., Hasan, M.K., Othman, M.: The Half-Sweep Iterative Alternating Decomposition Explicit (HSIADE) method for diffusion equations. LNCS 3314, Springer-Verlag, Berlin (2004)57-63.
14. Sulaiman, J., Othman, M., Hasan, M.K.: Quarter-Sweep Iterative Alternating Decomposition Explicit algorithm applied to diffusion equations. Intern. Journal of Computer Mathematics, **81**(2004) 1559-1565.
15. Twizell, E.H.: Computational methods for partial differential equations. Ellis Horwood Limited, Chichester (1984).
16. Vichnevetsky, R.: Computer Methods for Partial Differential Equations, Vol I. New Jersey: Prentice-Hall (1981)
17. Yousif, W.S., Evans, D.J.: Explicit De-coupled Group iterative methods and their implementations, Parallel Algorithms and Applications, **7**(1995) 53-71.
18. Zhang, J.: Acceleration of Five Points Red Black Gauss-Seidel in Multigrid for Poisson Equations, Applied Mathematics and Computation, **80(1)**(1996) 71-78.
19. Zienkiewicz, O.C.: Why finite elements?. In. Gallagher, R.H., Oden, J.T., Taylor, C., Zienkiewicz, O.C. (Eds). Finite Elements In Fluids-Volume, John Wiley & Sons,London **1**(1975) 1-23

Optimizing Surface Triangulation Via Near Isometry with Reference Meshes

Xiangmin Jiao, Narasimha R. Bayyana, and Hongyuan Zha

College of Computing
Georgia Institute of Technology, Atlanta, GA 30332, USA
{jiao,raob,zha}@cc.gatech.edu

Abstract. Optimization of the mesh quality of surface triangulation is critical for advanced numerical simulations and is challenging under the constraints of error minimization and density control. We derive a new method for optimizing surface triangulation by minimizing its discrepancy from a virtual reference mesh. Our method is as easy to implement as Laplacian smoothing, and owing to its variational formulation it delivers results as competitive as the optimization-based methods. In addition, our method minimizes geometric errors when redistributing the vertices using a principle component analysis without requiring a CAD model or an explicit high-order reconstruction of the surface. Experimental results demonstrate the effectiveness of our method.

Keywords: mesh optimization, numerical simulations, surface meshes, nearly isometric mapping.

1 Introduction

Improving surface mesh quality is important for many advanced 3-D numerical simulations. An example application is the moving boundary problems where the surfaces evolve over time and must be adapted for better numerical stability, accuracy, and efficiency while preserving the geometry. Frequently, the geometry of the evolving surface is unknown a priori but is part of the numerical solution, so the surface is only given by a triangulation without the availability of a CAD model. The quality of the mesh can be improved by mesh adaptation using edge flipping, edge collapsing, or edge splitting (see e.g. [1,2,3,4,5]), but it is often desirable to fix the connectivity and only redistribute the vertices, such as in the arbitrary Lagrangian-Eulerian methods [6]. In this paper, we focus on mesh optimization (a.k.a. mesh smoothing) with fixed connectivity.

Mesh smoothing has a vast amount of literature (for example, see [7,8,9,10,11]). Laplacian smoothing is often used in practice for its simplicity, although it is not very effective for irregular meshes. The more sophisticated methods are often optimization based. An example is the angle-based method of Zhou and Shimada [11]. Another notable example is the method of Garimella et al. [7], which minimizes the condition numbers of the Jacobian of the triangles against some reference Jacobian matrices (RJM). More recently, the finite-element-based method is used in [8], but

Y. Shi et al. (Eds.): ICCS 2007, Part I, LNCS 4487, pp. 334–341, 2007.

their method is relatively difficult to implement. Note that some mesh smoothing methods (such as the angle-based method) are designed for two-dimensional meshes, and the conventional wisdom is to first parameterize the surface locally or globally and then optimize the flattened mesh [2,12,13,14]. To preserve the geometry, these methods typically require a smooth or discrete CAD model and associated point location procedures to project the points onto the surface, which increase the implementation complexity.

The goal of this paper is to develop a mesh smoothing method that is as simple as Laplacian smoothing while being as effective as the sophisticated optimization-based methods without parameterizing the mesh. The novelty of our method is to formulate the problem as a near isometic mapping from an ideal reference mesh onto the surface and to derive a simple iterative procedure to solve the problem. Due to its variational nature, the method can balance angle optimization and density control. It also eliminates the needs of the preprocessing step of surface parameterization and the post-processing step of vertex projection, so it is much easier to implement and is well-suited for integration into numerical simulations.

The remainder of the paper is organized as follows. In Section 2, we formulate the mesh optimization problem and explain its relationship to isometric mappings. In Section 3, we describe a simple discretization of our method for triangulated surfaces with adaptive step-size control. In Section 4, we present some experimental results. Section 5 concludes the paper with discussions of future research directions.

2   Mesh Optimization and Isometric Mappings

Given a mesh with a set of vertices and triangles, the problem of mesh optimization is to redistribute the vertices so that the shapes of the triangles are improved in terms of angles and sizes. Surface parameterization is the procedure to map the points on one surface onto a parameter domain (such as a plane or sphere) under certain constraints such as preservation of areas or angles [15]. Although surface parameterization has been widely used in computer graphics and visualization [16,17,18,19,20,21], in this section we explore an interesting connection between it and mesh optimization.

More formally, given a 2-manifold surface $M \subset \mathbb{R}^3$ and a parameter domain $\Omega \subset \mathbb{R}^2$, the problem of *surface parameterization* is to find a mapping $f : \Omega \to M$ such that f is one-to-one and onto. Typically, the surface M is a triangulated surface, with a set of vertices $\boldsymbol{P}_i \in \mathbb{R}^3$ and triangles $T^{(j)} = (\boldsymbol{P}_{j1}, \boldsymbol{P}_{j2}, \boldsymbol{P}_{j3})$. The problem of *isometric parameterization* (or mapping) is to find the values of $\boldsymbol{p}_i$ such that $f(\boldsymbol{p}_i) = \boldsymbol{P}_i$, the triangles $t^{(j)} = (\boldsymbol{p}_{j1}, \boldsymbol{p}_{j2}, \boldsymbol{p}_{j3})$ do not overlap in Ω, and the angles and the areas of the triangles are preserved as much as possible.

We observe that the constraints of angle and area preservation in isometric parameterization are similar to angle optimization and density control in mesh optimization, except that the requirement of $\Omega \subset \mathbb{R}^2$ is overly restrictive.

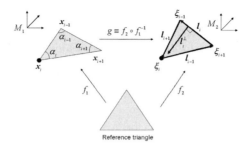

Fig. 1. Atomic mapping from triangle on M_1 to triangle on M_2

However, by replacing Ω with the input surface and replacing M by a "virtual" reference mesh with desired mesh quality and density (for example be composed of equilateral triangles of user-specified sizes, where the triangles need not fit together geometrically and hence we refer to it as a "virtual" mesh), we can then effectively use isometric parameterization as a computational framework for mesh optimization, which we describe as follows.

Let M_1 denote an ideal virtual reference mesh and M_2 the triangulated surface to be optimized. Let us first assume that M_2 is a planar surface with a global parameterization $\boldsymbol{\xi}$, and we will generalize the framework to curved surfaces in the next section. In this context, the problem can be considered as reparameterizing M_2 based on the triangles and the metrics of M_1. Consider an atomic linear mapping $g \equiv f_2 \circ f_1^{-1}$ from a triangle on M_1 onto a corresponding triangle on M_2, as shown in Figure 1, where f_1 and f_2 are both mappings from a reference triangle. Let $\boldsymbol{J}_1$ and $\boldsymbol{J}_2$ denote their Jacobian matrices with respect to the reference triangle. The Jacobian matrix of the mapping g is $\boldsymbol{A} = \boldsymbol{J}_2 \boldsymbol{J}_1^{-1}$. For g to be nearly isometric, $\boldsymbol{A}$ must be close to be orthogonal. We measure the deviation of $\boldsymbol{A}$ from orthogonality using two energies: angle distortion and area distortion. Angle distortion depends on the condition number of the matrix $\boldsymbol{A}$ in 2-norm, i.e., $\kappa_2(\boldsymbol{A}) = \|A\|_2 \|A^{-1}\|_2$. Area distortion can be measured by $\det(\boldsymbol{A}) = \det(\boldsymbol{J}_2)/\det(\boldsymbol{J}_1)$, which is equal to 1 if the mapping is area preserving. Let us define a function

$$\tau_p(s) \equiv \begin{cases} s^p + s^{-p} & \text{if } s > 0 \\ \infty & \text{otherwise,} \end{cases} \tag{1}$$

where $p > 0$, so τ_p is minimized if $s = 1$ and approaches ∞ if $s \leq 0$ or $s \to \infty$. To combine the angle- and area-distortion measures, we use the energy

$$E_I(T, \mu_A) \equiv (1 - \mu_A)\tau_1(\kappa_2(\boldsymbol{A})) + \mu_A \tau_{\frac{1}{2}}(\det(\boldsymbol{A})), \tag{2}$$

where μ_A is between 0 and 1 and indicates the relative importance of area preservations versus angle preservation. For feasible parameterizations, E_I is finite because $\det(\boldsymbol{A}) > 0$ and $\kappa_2(\boldsymbol{A}) \geq 1$. To obtain a nearly isometric mapping, we must find $\boldsymbol{\xi}_i = g(\boldsymbol{x}_i)$ for each vertex $\boldsymbol{x}_i \in M_1$ to minimize the sum of E_I over all triangles on M_1, i.e.,

$$E_I(M_1, \mu_A) \equiv \sum_{T \in M_1} E_I(T, \mu_A). \tag{3}$$

We refer to this minimization as *nearly isometric parameterization of surfaces* (*NIPS*), which balances the preservation of angles and areas. This formulation shares similarity with that of Degener et al. for surface parameterization [12], who also considered both area and angle distortions but used a different energy. Note that the special case with $\mu_A = 0$ would reduce to the most isometric parameterization (MIPS) of Hormann and Griener [18] and is closely related to the condition-number-based optimization in [7].

The direct minimization of E_I may seem difficult because of the presence of $\kappa_2(\boldsymbol{A})$. However, by using a result from [22] regarding the computation of Dirichlet energy $E_D(g) = \mathrm{trace}(\boldsymbol{A}^T \boldsymbol{A}) \det(J_1)/4$ as well as the fact that

$$\tau_1(\kappa_2(\boldsymbol{A})) = \kappa_F(\boldsymbol{A}) = \frac{\mathrm{trace}(\boldsymbol{A}^T \boldsymbol{A})}{\det(\boldsymbol{A})} = \frac{4E_D(g)}{\det(J_2)}, \tag{4}$$

one can show that

$$\tau_1(\kappa_2(\boldsymbol{A})) = \frac{1}{\det(J_2)} \sum_i \cot \alpha_i \|l_i\|_2^2, \tag{5}$$

where α_i and $l_i \equiv \boldsymbol{\xi}_{i-} - \boldsymbol{\xi}_{i+}$ are defined as in Figure 1, and $i-$ and $i+$ denote the predecessor and successor of i in the triangle.

To minimize E_I, we must evaluate its gradient with respect to $\boldsymbol{\xi}_i$. For a triangle T, let $l_i^{\perp} \equiv \hat{\boldsymbol{n}} \times l_i$ denote the 90° counter-clockwise rotation of the vector l_i on M_2. It can be verified that

$$\frac{\partial E_I(M, \mu_A)}{\partial \boldsymbol{\xi}_i} = \sum_{i \in T^{(j)}} \left(s_{i+}^{(j)} l_{i+}^{(j)} - s_{i-}^{(j)} l_{i-}^{(j)} + t_i^{(j)} l_i^{(j)\perp} \right), \tag{6}$$

where

$$s_{\pm} = (1-\mu_A)\frac{2 \cot \alpha_{\pm}}{\det(J_2)} \text{ and } t_i = (1-\mu_A)\frac{\kappa_F}{\det(J_2)} + \frac{1}{2}\mu_A\frac{\det(J_1) - \det(J_2)}{\det(J_1)^{\frac{1}{2}} \det(J_2)^{\frac{3}{2}}}. \tag{7}$$

Equation (6) is a simple weighted sum of its incident edges and 90° counter-clockwise rotation of opposite edges over all its incident triangles $T^{(j)}$.

3   Mesh Optimization for Curved Surfaces

From Eqs. (6) and (7), it is obvious that $\partial E_I/\partial \boldsymbol{\xi}_i$ does not depend on the underlying parameterization $\boldsymbol{\xi}_i$ of M_2, so it becomes straightforward to evaluate it directly on a curved surface. In particular, at each vertex i on M_1, let $\boldsymbol{V}_i \equiv (\hat{t}_1|\hat{t}_2)_{3\times 2}$ denote the matrix composed of the unit tangents of M_2 at the point. To reduce error, we constrain the point to move within the tangent space of

M_2 so that the displacements would be $\boldsymbol{V}_i\boldsymbol{u}_i$, where $\boldsymbol{u}_i \in \mathbb{R}^2$. Because of the linearity, from the chain rule we then have

$$\frac{\partial E_I(M_1, \mu_A)}{\partial \boldsymbol{u}_i} = \boldsymbol{V}_i^T \sum_{i\in T^{(j)}} \left(s_{i+}^{(j)} \boldsymbol{l}_{i+}^{(j)} - s_{i-}^{(j)} \boldsymbol{l}_{i-}^{(j)} + t_i^{(j)} \boldsymbol{l}_i^{(j)\perp} \right), \tag{8}$$

where $\boldsymbol{l}_i \in \mathbb{R}^3$ as defined earlier. This equation constrains the search direction within the local tangent space at vertex i without having to project its neighborhood onto a plane.

We estimate the tangent space as in [23]. In particular, at each vertex v, suppose v is the origin of a local coordinate frame, and m is the number of the faces incident on v. Let $\boldsymbol{N}$ be an $3 \times m$ matrix whose ith column vector is the unit outward normal to the ith incident face of v, and $\boldsymbol{W}$ be an $m \times m$ diagonal matrix with W_{ii} equal to the face area associated with the ith face. Let $\boldsymbol{A}$ denote $\boldsymbol{N}\boldsymbol{W}\boldsymbol{N}^T$, which we refer to as the *normal covariance matrix*. $\boldsymbol{A}$ is symmetric positive semi-definite with real eigenvalues. We use the vector space spanned by the eigenvectors corresponding to the two smaller eigenvalues of $\boldsymbol{A}$ as the tangent space. If the surface contains ridges or corners, we restrict the tangent space to contain only the eigenvector corresponding to the smallest eigenvalues of $\boldsymbol{A}$ at ridge vertices and make the tangent space empty at corners.

To solve the variational problem, one could use a Gauss-Seidel style iteration to move the vertices. This approach was taken in some parameterization and mesh optimization algorithms [3,7]. For simplicity, we use a simple nonlinear Jacobi iteration, which moves vertex i by a displacement of

$$\boldsymbol{d}_i \equiv -\boldsymbol{V}_i\boldsymbol{V}_i^T \frac{\sum_{i\in T^{(j)}} \left(s_{i+}^{(j)} \boldsymbol{l}_{i+}^{(j)} - s_{i-}^{(j)} \boldsymbol{l}_{i-}^{(j)} + t_i^{(j)} \boldsymbol{l}_i^{(j)\perp} \right)}{\sum_{i\in T^{(j)}} \left(s_{i+}^{(j)} + s_{i-}^{(j)} \right)}. \tag{9}$$

This Jacobi-style iteration may converge slower but it can be more efficient than the Gauss-Seidel style iteration, as it eliminates the need of reestimating the tangent spaces at the neighbor vertices of v after moving a vertex v.

The concurrent motion of the vertices in the Jacobi-style iterations may lead to mesh folding. To address this problem, we introduce an asynchronous step-size control. For each triangle $\boldsymbol{p}_i\boldsymbol{p}_j\boldsymbol{p}_k$, we solve for the maximum $\alpha \leq 1$ such that the triangle $\boldsymbol{p}_i^{(\alpha)}\boldsymbol{p}_j^{(\alpha)}\boldsymbol{p}_k^{(\alpha)}$ does not fold, where $\boldsymbol{p}_i^{(\alpha)} \equiv \boldsymbol{p}_i + \alpha\boldsymbol{d}_i$ [23]. We reduce $\boldsymbol{d}_i$ at vertex i by a factor equal to the minimum of the αs of its incident faces. After rescaling the displacement of all the vertices, we recompute α and repeat the rescaling process until $\alpha = 1$ for all vertices.

4   Experimental Results

In this section, we present some preliminary results using our method for static 2-D meshes and dynamic 3-D meshes.

Table 1. Comparative results of optimizing 2-D meshes. CN-based is our method with $\mu_A = 0$ and CNEA-based with $\mu_A = 1$. Symbol '-' indicates mesh folding.

	Minimum angle						Maximum angle					
	U1	U2	U3	R1	R2	R3	U1	U2	U3	R1	R2	R3
Original	12.0	8.8	8.5	0.11	1.1	0.043	129.5	156.9	147.5	179.7	176.9	179.9
Laplacian	33.1	31.4	30.8	5.8	4.9	4.5	99.4	105.9	109.2	168.3	170.8	171.0
Angle-based	32.9	30.9	29.5	3.9	-	-	96.0	103.7	105.9	170.3	-	-
CN-based	36.0	36.1	34.6	12.6	8.9	10.3	96.8	100.1	105.6	153.2	157.3	156.2
CNEA-based	35.7	35.3	34.2	12.7	7.8	8.9	97.0	101.3	105.8	153.9	163.4	160.6

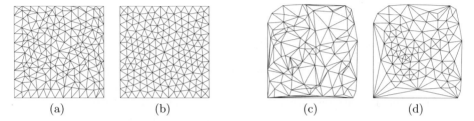

(a)   (b)   (c)   (d)

Fig. 2. Sample meshes (a,c) before and (b,d) after mesh optimization

4.1   Optimization of 2-D Meshes

We first compare the effectiveness of our method against the length-weighted Laplacian smoothing and the angle-based method of Zhou and Shimada [11]. Because these existing methods are better established for planar meshes, we perform our comparisons in 2-D. Table 1 shows the minimum and maximum angles of six different meshes before and after mesh optimization, including three relatively uniform meshes (denoted by U1, U2, and U3) and three meshes with random points (denoted by R1, R2, and R3). In our methods, we consider the virtual reference mesh to be composed of equilateral triangles with the average area of the triangles. Figure 2 shows the original and the optimized meshes using our method with $\mu_A = 1$ for U1 and R1. In nearly all cases, the condition-number based method (i.e., $\mu_A = 0$) performs substantially better in minimum angles for all cases, comparative or slightly better in maximum angles for uniform meshes, and significantly better in maximum angles for the random meshes. The area-equalizing optimization delivers slightly worse angles than condition-number based optimization, but the former still outperforms edge-weighted Laplacian smoothing and the angle-based methods while allowing better control of areas.

4.2   Optimization of Dynamic Meshes

In this test, we optimize a mesh that is deformed by a velocity field. Our test mesh discretizes a sphere with radius 0.15 centered at $(0.5, 0.75, 0.5)$ and contains 5832 vertices and 11660 triangles. The velocity field is given by

Fig. 3. Example of optimized dynamic surfaces. Colors indicate triangle areas.

$$u(x, y, z) = \cos(\pi t/T) \sin^2(\pi x)(\sin(2\pi z) - \sin(2\pi y)),$$
$$y(x, y, z) = \cos(\pi t/T) \sin^2(\pi y)(\sin(2\pi x) - \sin(2\pi z)),$$
$$z(x, y, z) = \cos(\pi t/T) \sin^2(\pi z)(\sin(2\pi y) - \sin(2\pi x)), \tag{10}$$

where $T = 3$, so that the shape is deformed the most at time $t = 1.5$ and should return to the original shape at time $t = 3$. We integrate the motion of the interface using the face-offsetting method in [23] while redistributing the vertices using the initial mesh as the reference mesh. In this test the angles and areas of the triangles were well preserved even after very large deformation. Furthermore, the vertex redistribution introduced very small errors and the surface was able to return to a nearly perfect sphere at time $t = 3$.

5   Conclusion

In this paper, we proposed a new method for optimizing surface meshes using a near isometry with reference meshes. We derived a simple discretization, which is easy to implement and is well suitable for integration into large-scale numerical simulations. We compared our method with some existing methods, showed substantial improvements in the maximum and minimum angles, and demonstrated its effective use for moving meshes. As a future direction, we plan to extend our method to optimize quadrilateral meshes and 3-D volume meshes.

Acknowledgments. This work was supported by a subcontract from the Center for Simulation of Advanced Rockets of the University of Illinois at Urbana-Champaign funded by the U.S. Department of Energy through the University of California under subcontract B523819 and in part by the National Science Foundation under award number CCF-0430349.

References

1. Alliez, P., Meyer, M., Desbrun, M.: Interactive geometry remeshing. ACM Trans. Graph. **21** (2002) 347–354 Proc. SIGGRAPH 2002.
2. Frey, P., Borouchaki, H.: Geometric surface mesh optimization. Comput. Visual. Sci. (1998) 113–121

3. Hormann, K., Labsik, U., Greiner, G.: Remeshing triangulated surfaces with optimal parametrizations. Comput. Aid. D's. **33** (2001) 779–788
4. Praun, E., Hoppe, H.: Spherical parametrization and remeshing. ACM Trans. Graph. **22** (2003) 340–349 Proc. SIGGRAPH 2003.
5. Surazhsky, V., Gotsman, C.: Explicit surface remeshing. In: Eurographics Symposium on Geometric Processing. (2003) 20–30
6. Donea, J., Huerta, A., Ponthot, J.P., Rodriguez-Ferran, A.: Arbitrary Lagrangian-Eulerian methods. In Stein, E., de Borst, R., Hughes, T.J., eds.: Encyclopedia of Computational Mechanics. Volume 1: Fundamentals. John Wiley (2004)
7. Garimella, R., Shashkov, M., Knupp, P.: Triangular and quadrilateral surface mesh quality optimization using local parametrization. Comput. Meth. Appl. Mech. Engrg. **193** (2004) 913–928
8. Hansen, G.A., Douglass, R.W., Zardecki, A.: Mesh Enhancement. Imperial College Press (2005)
9. Knupp, P., Margolin, L., Shashkov, M.: Reference Jacobian optimization based rezone strategies for arbitrary Lagrangian Eulerian methods. J. Comput. Phys. **176** (2002) 93–128
10. Shashkov, M.J., Knupp, P.M.: Optimization-based reference-matrix rezone strategies for arbitrary lagrangian-Eulerian methods on unstructured grids. In: Proc. the 10th International Meshing Roundtable. (2001) 167–176
11. Zhou, T., Shimada, K.: An angle-based approach to two-dimensional mesh smoothing. In: Proceedings of 9th International Meshing Roundtable. (2000) 373–384
12. Degener, P., Meseth, J., Klein, R.: An adaptable surface parameterization method. In: Proc. the 12th International Meshing Roundtable. (2003) 227–237
13. Knupp, P.: Achieving finite element mesh quality via optimization of the jacobian matrix norm and associated quantities. part i: a framework for surface mesh optimization. Int. J. Numer. Meth. Engrg. **48** (2000) 401–420
14. Sheffer, A., de Sturler, E.: Parameterization of faceted surfaces for meshing using angle-based flattening. Engrg. Comput. **17** (2001) 326–337
15. Floater, M.S., Hormann, K.: Surface parameterization: a tutorial and survey. In: Advances in Multiresolution for Geometric Modelling. Springer Verlag (2005) 157–186
16. Desbrun, M., Meyer, M., Alliez, P.: Intrinsic parameterizations of surface meshes. In: Proc. Eurographics 2002 Conference. (2002)
17. Gotsman, C., Gu, X., Sheffer, A.: Fundamentals of spherical parameterization for 3d meshes. ACM Trans. Graph. **22** (2003) 358–363 Proc. SIGGRAPH 2003.
18. Hormann, K., Greiner, G.: MIPS: An efficient global parametrization method. In Laurent, P.J., Sablonniere, P., Schumaker, L.L., eds.: Curve and Surface Design: Saint-Malo 1999. (2000) 153–162
19. Khodakovsky, A., Litke, N., Schröder, P.: Globally smooth parameterizations with low distortion. ACM Trans. Graph. **22** (2003) 350–357 Proc. SIGGRAPH 2003.
20. Kos, G., Varady, T.: Parameterizing complex triangular meshes. In Lyche, T., Mazure, M.L., Schumaker, L.L., eds.: Curve and Surface Design: Saint-Malo 2002, Modern Methods in Applied Mathematics. (2003) 265–274
21. Sheffer, A., Gotsman, C., Dyn, N.: Robust spherical parametrization of triangular meshes. In: Proc. the 4th Israel-Korea Bi-National Conference on Geometric Modeling and Computer Graphics. (2003) 94–99
22. Pinkall, U., Polthier, K.: Computing discrete minimal surfaces and their conjugates. Exper. Math. **2** (1993) 15–36
23. Jiao, X.: Face offsetting: a unified framework for explicit moving interfaces. J. Comput. Phys. **220** (2007) 612–625

Efficient Adaptive Strategy for Solving Inverse Problems

M. Paszyński[1], B. Barabasz[2], and R. Schaefer[1]

[1] Department of Computer Science
[2] Department of Modeling and Information Technology
AGH University of Science and Technology,
Al. Mickiewicza 30, 30-059 Cracow, Poland,
{paszynsk,schaefer}@agh.edu.pl,
barabasz@metal.agh.edu.pl
http://home.agh.edu.pl/~paszynsk

Abstract. The paper describes the strategy for efficient solving of difficult inverse problems, utilizing Finite Element Method (FEM) as a direct problem solver. The strategy consists of finding an optimal balance between the accuracy of global optimization method and the accuracy of an hp-adaptive FEM used for the multiple solving of the direct problem. The crucial relation among errors was found for the objective function being the energy of the system defining the direct problem. The strategy was applied for searching the thermal expansion coefficient (CTE) parameter in the Step-and-flash Imprint Lithography (SFIL) process.

Keywords: Inverse problems, Finite Element Method, hp adaptivity, Molecular Statics.

1 Introduction

Inverse parametric problems belong to the group of heaviest computational tasks. Their solution require a sequence of direct problem solutions, e.g. obtained by Finite Element Method (FEM), thus the accuracy of the inverse problem solution is limited by the accuracy of the direct problem solution. We utilize the fully automatic hp FEM codes [6,3] generating a sequence of computational meshes delivering exponential convergence of the numerical error with respect to the mesh size for solving direct problems. Using the maximum accuracy for the direct problem solve by each iteration of inverse solver leads to needles computational costs (see e.g. [4]). A better strategy is to balance dynamically the accuracy of both iterations.

However, to be able to execute such strategy we need to relate the error of optimization method defined as the uncorrectness of objective function value with the FEM solution error. We propose such relation and the detailed error balance strategy for the objective being the energy of the system that is described by the simple problem.

Y. Shi et al. (Eds.): ICCS 2007, Part I, LNCS 4487, pp. 342–349, 2007.

The strategy is tested on the Step-and-Flash Impring Lithography (SFIL) simulations. The objective of the inverse analysis is to find value of the thermal expansion coefficient enforcing shrinkage of the feature well comparable with experimental data. The energy used for the error estimation of the objective function was obtained from the experimental data and static molecular model calculations [5].

2   The Automatic *hp* Adaptive Finite Element Method

Sequential and parallel 3D *hp* adaptive FEM codes [6], [3] generate in fully automatic mode a sequence of *hp* FE meshes providing exponential convergence of the numerical error with respect to size of the mesh (number of degrees of freedom, CPU time). Given an initial mesh, called the coarse mesh, presented

Fig. 1. The coarse mesh with $p = 2$ and fine mesh with $p = 3$ on all elements edges, faces, and interiors. The optimal meshes after the first, second and third iterations. Various colors denote various polynomial orders of approximation.

on the first picture in Fig. 1, with polynomial orders of approximations $p = 2$ on elements edges, faces and interiors, we first perform global *hp* refinement to produce the fine mesh presented on the second picture in Fig. 1, by breaking each element into 8 son elements, and increasing the polynomial order of approximation by one. The direct problem is solved on the coarse and on the fine mesh. The energy norm (see e.g. [1]) difference between coarse and fine mesh solutions is then utilized to estimate relative errors over coarse mesh elements. The optimal refinements are then selected and performed for coarse mesh elements with high relative errors. The coarse mesh elements can be broken into smaller son elements (this procedure is called *h* refinement) or the polynomial order of approximation can be increased on element edges, faces or interiors (this procedure is called *p* refinement), or both (this is called *hp* refinement). For each finite element from the coarse mesh we consider locally several possible *h*, *p* or *hp* refinements. For each finite element the refinement strategy providing maximum error decrease rate is selected. The error decrease rate

$$rate = \frac{\|u_{h/2,p+1} - u_{hp}\| - \|u_{h/2,p+1} - w_{hp}\|}{nrdof_{added}} . \tag{1}$$

is defined as relative error estimation in energy norm divided by number of degrees of freedom added. Here u_{hp} stands for the coarse mesh solution, $u_{h/2,p+1}$

for the fine mesh solution and w_{hp} is the solution corresponding to proposed refinement strategy, obtained by the projection based interpolation technique [2]. The optimal mesh generated in such a way becomes a coarse mesh for the next iteration, and the entire procedure is repeated as long as the global relative error estimation is larger then the required accuracy of the solution (see [6] for more details). The sequence of optimal meshes generated by automatic hp-adaptive code from the coarse mesh is presented on third, fourth and fifth pictures in Figure 1. The relative error of the solution goes down from 15% to 5%.

3   The Relation Between the Objective Function Error and the Finite Element Method Error

We assume the direct problem is modeled by the abstract variational equation

$$\begin{cases} u \in u_0 + V \\ b(u,v) = l(v) \, \forall v \in V \end{cases} \tag{2}$$

where u_0 is the lift of the Dirichlet boundary conditions [2]. Functionals b and l depend on the inverse problem parameters $\mathbf{d}$. The variational problem (2) is equivalent with the minimization one (3), if b is symmetric and positive definite (see e.g. [2])

$$\begin{cases} u \in u_0 + V \\ E(u) = \frac{1}{2} b(u,u) - l(u) \longrightarrow min. \end{cases} \tag{3}$$

where $E(u) = \frac{1}{2} b(u,u) - l(u)$ is the functional of the total energy of the solution.

The Problem (2) may be approximated using the FEM on the finite dimensional subspace $V_{h,p} \subset V$

$$\begin{cases} u_{h,p} \in u_0 + V_{h,p} \\ b(u_{h,p}, v_{h,p}) = l(v_{h,p}) \, \forall v_{h,p} \in V_{h,p} \end{cases}. \tag{4}$$

For a sequence of meshes generated by the self-adaptive hp FEM code, for every coarse mesh, a coarse mesh space is a subset of the corresponding fine mesh space, $V_{h,p} \subset V_{h/2,p+1} \subset V$.

An absolute relative FEM error utilized by the self-adaptive hp FEM code is defined as the energy norm difference between the coarse and fine mesh solutions

$$err_{FEM} = \|u_{h,p} - u_{h/2,p+1}\|_E. \tag{5}$$

The inverse problem can be formulated as

$$Find \ \hat{\mathbf{d}} : \ |J_{h,p}(\hat{\mathbf{d}}) - J(\mathbf{d}^*)| = lim_{h \to 0, p \to \infty} min_{\mathbf{d}^k \in \Omega} |J_{h,p}(\mathbf{d}^k) - J(\mathbf{d}^*)| \tag{6}$$

where $\mathbf{d}^*$ denotes exact parameters of the inverse problem (exact solution of the variational formulation for these parameters is well comparable with experiment data), $\mathbf{d}^k$ denotes approximated parameters of the inverse problem, Ω is a set of all admissible parameters $\mathbf{d}^k$, $J(\mathbf{d}^*) = E(u(\mathbf{d}^*))$ is the energy of the exact solution $u(\mathbf{d}^*)$ of the variational problem (2) for exact parameters $\mathbf{d}^*$, $J_{h,p}(\mathbf{d^k}) = E(u_{h,p}(\mathbf{d^k}))$ is the energy of the solution $u_{h,p}(\mathbf{d^k})$ of the approximated problem (4) for approximated parameters $\mathbf{d^k}$.

Objective function error is defined as an energy difference between the solution of the approximated problem (4) for approximated parameter $\mathbf{d^k}$ and the exact solution of the problem (2) for exact parameter $\mathbf{d}^*$ (assumed to be equal to the energy of the experiment)

$$e_{h,p}(\mathbf{d^k}) = |J_{h,p}(\mathbf{d^k}) - J(\mathbf{d}^*)|. \tag{7}$$

In other words, the approximated parameter $\mathbf{d^k}$ is placed into the approximated formulation (4), the solution of the problem $u_{h,p}(\mathbf{d^k})$ (which depends on $\mathbf{d^k}$) is computed by FEM, and the energy of the solution $E(u_{h,p}(\mathbf{d^k}))$ is computed.

Lemma 1. $2\left[J_{h,p}(\mathbf{d^k}) - J_{h/2,p+1}(\mathbf{d^k})\right] = \|u_{h,p}(\mathbf{d^k}) - u_{h/2,p+1}(\mathbf{d^k})\|_E^2$

Proof: $2\left[J_{h,p}(\mathbf{d^k}) - J_{h/2,p+1}(\mathbf{d^k})\right] = 2\left[E(u_{h,p}(\mathbf{d^k})) - E(u_{h/2,p+1}(\mathbf{d^k}))\right] =$
$b(u_{h,p}(\mathbf{d^k}), u_{h,p}(\mathbf{d^k})) - 2l(u_{h,p}(\mathbf{d^k})) - b(u_{h/2,p+1}(\mathbf{d^k}), u_{h/2,p+1}(\mathbf{d^k})) +$
$2l(u_{h/2,p+1}(\mathbf{d^k})) = b(u_{h,p}(\mathbf{d^k}), u_{h,p}(\mathbf{d^k})) - b(u_{h/2,p+1}(\mathbf{d^k}), u_{h/2,p+1}(\mathbf{d^k})) +$
$2l(u_{h/2,p+1}(\mathbf{d^k}) - u_{h,p}(\mathbf{d^k})) = b(u_{h,p}(\mathbf{d^k}), u_{h,p}(\mathbf{d^k})) -$
$b(u_{h/2,p+1}(\mathbf{d^k}), u_{h/2,p+1}(\mathbf{d^k})) + 2b(u_{h/2,p+1}(\mathbf{d^k}), u_{h/2,p+1}(\mathbf{d^k}) - u_{h,p}(\mathbf{d^k})) =$

$b(u_{h/2,p+1}(\mathbf{d^k}) - u_{h,p}(\mathbf{d^k}), u_{h/2,p+1}(\mathbf{d^k}) - u_{h,p}(\mathbf{d^k})) =$
$\|u_{h,p}(\mathbf{d^k}) - u_{h/2,p+1}(\mathbf{d^k})\|_E^2$

where $V_{h,p} \subset V_{h/2,p+1} \subset V$ stand for the coarse and fine mesh subspaces. □

Lemma 2. $e_{h/2,p+1}(\mathbf{d^k}) \leq \frac{1}{2}\|u_{h/2,p+1}(\mathbf{d^k}) - u_{h,p}(\mathbf{d^k})\|_E^2 + |J_{h,p}(\mathbf{d^k}) - J(\mathbf{d}^*)|$, where $e_{h/2,p+1}(\mathbf{d^k}) := |J_{h/2,p+1}(\mathbf{d^k}) - J(\mathbf{d}^*)|$.

Proof: $e_{h/2,p+1}(\mathbf{d^k}) = |J_{h/2,p+1}(\mathbf{d^k}) - J(\mathbf{d}^*)| = |J_{h/2,p+1}(\mathbf{d^k}) - J_{h,p}(\mathbf{d^k}) +$
$J_{h,p}(\mathbf{d^k}) - J(\mathbf{d}^*)| \leq |J_{h/2,p+1}(\mathbf{d^k}) - J_{h,p}(\mathbf{d^k})| + |J_{h,p}(\mathbf{d^k}) - J(\mathbf{d}^*)| =$
$\frac{1}{2}\|u_{h/2,p+1}(\mathbf{d^k}) - u_{h,p}(\mathbf{d^k})\|_E^2 + |J_{h,p}(\mathbf{d^k}) - J(\mathbf{d}^*)|$. □

The objective function error over the fine mesh is limited by the relative error of the coarse mesh with respect to the fine mesh, plus the objective function error over the coarse mesh.

4   Algorithm

Lemma 2 motivates the following algorithm relating the inverse error with the objective function error. We start with random initial values of the inverse problem parameters

```
solve the problem on the coarse and fine FEM meshes
compute FEM error
inverse analysis loop
     Propose new values for inverse problem parameters
```

```
      solve the problem on the coarse mesh
      Compute objective function error
      if (objective function error < const * FEM error)
            execute one step of the hp adaptivity,
            solve the problem on the new coarse and fine FEM meshes
            compute FEM error
      if (inverse error < required accuracy) stop
end
```

Inverse error estimation proven in **Lemma 2** allows us to perform hp adaptation in the right moment. If the objective function error is much smaller than the FEM error, the minimization of the objective function error does not make sense on current FE mesh, and the mesh quality improvement is needed.

5   Step-and-Flash Imprint Lithography

The above algorithm will be tested on the SFIL process simulation. The SFIL is a modern patterning process utilizing photopolymerization to replicate the topography of a template into a substrate. It can be summarized in the following steps, compare Fig. 2: *Dispense* - the SFIL process employs a template / substrate alignment scheme to bring a rigid template and substrate into parallelism, trapping the etch barrier in the relief structure of the template; *Imprint* - the gap is closed until the force that ensures a thin base layer is reached; *Exposure* - the template is then illuminated through the backside to cure etch barrier; *Separate* - the template is withdrawn, leaving low-aspect ratio, high resolution features in the etch barrier; *Breakthrough Etch* - the residual etch barrier (base layer) is etched away with a short halogen plasma etch; *Transfer Edge* - the pattern is transferred into the transfer layer with an anisotropic oxygen reactive ion etch, creating high-aspect ratio, high resolution features in the organic transfer layer. The photopolymerization of the feature is often accompanied by the densification, see Fig. 2, which can be modeled by the linear elasticity with thermal expansion coefficient (CTE) [5]. We may define the problem: Find u - displacement vector field, such that

$$\begin{cases} u \in V \subset \left(H^1\left(\Omega\right)\right)^3 \\ b\left(u,v\right) = l\left(v\right) \forall v \in V \end{cases} . \tag{8}$$

where $V = \left\{v \in \left(H^1\left(\Omega\right)\right)^3 : tr(v) = 0 \, on \, \Gamma_D\right\}$, $\Omega \subset R^3$ stands for the cubic-shape domain, Γ_D is the bottom of the cube and $H^1(\Omega)$ is the Sobolev space.

$$b\left(u,v\right) = \int_\Omega \left(E_{ijkl} u_{k,l} v_{i,j}\right) dx; \quad l\left(v\right) = \alpha \int_\Omega v_{i,i} dx. \tag{9}$$

Here $E_{ijkl} = \mu\left(\delta_{ik}\delta_{jl} + \delta_{il}\delta_{jk}\right) + \lambda\delta_{ij}\delta_{kl}$ stands for the constitutive equation for the isotropic material, where μ and λ are Lame coefficients. The thermal expansion coefficient (CTE) $\alpha = \frac{\Delta V}{V \Delta T}$ is defined as a volumetric shrinkage of the edge barrier divided by 1 K.

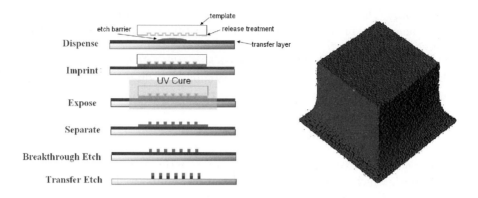

Fig. 2. Modeling of the Step-and-Flash Imprint Lithography process

6   Numerical Results

The proposed algorithm was executed for the problem of finding the proper value of the thermal expansion coefficient enforcing shrinkage of the feature comparable with experiments. The algorithm performed 43 iterations on the first optimal mesh (see the third picture in Fig. 1) providing 15% relative error of the direct problem solution. Then, the computational mesh was hp refined to increase the accuracy of the direct solver. The inverse algorithm continued by utilizing 8% relative error mesh (see the fourth picture in Fig.1) for the direct problem. After 39 iterations the mesh was again hp refined (see the fifth picture in Fig.1) to provide 5% relative error of the direct problem solution. After 35 iterations of the inverse algorithm on the most accurate mesh the inverse problem was solved. The history of the (CTE) parameter convergence on the first, second and third optimal meshes is presented in Fig. 3.

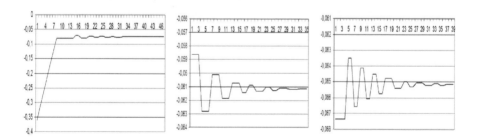

Fig. 3. History of convergence of CTE parameter on 3 meshes

We compared the total execution time equal to $0.1s + 43 \times 2 \times 0.1s + 1s + 39 \times 2 \times 1s + 10s + 35 \times 2 \times 10s = 8.7 + 79 + 710 = 797.7s$ with the classical algorithm, where the inverse problem was solved on the most accurate FEM mesh from the beginning. The classical algorithm required 91 iterations to obtain the same result. The execution time of the classical algorithm was $10s + 91 \times 2 \times 10s = 1830s$.

This difference will grow when the inverse algorithm will look for more inverse problem parameters at the same time, since number of direct problem solutions necessary to obtain the new propositions of the inverse parameters will grow.

7   The Molecular Static Model

The energy of the experimental data $J(d^*)$ was estimated from the molecular static model, which provides realistic simulation results, well comparable with experiments [5]. During the photopolymierization, the Van der Waals bound between particles forming polymer chain are converted into a stronger covalent bounds. The average distance between particles is decreasing and the volumetric contraction of the feature occurs. In the following, a general equations governing the equilibrium configurations of the molecular lattice structure after the densification and removing of the template are derived.

Let us consider an arbitrary pair of bonded molecules with indices α and β and given lattice position vector $p_\alpha = (\hat{x}_\alpha, \hat{y}_\alpha, \hat{z}_\alpha)$. The unknown equilibrium position vector of particle α, under the action of all their intermolecular bonds, is denoted $x_\alpha = (x_\alpha, y_\alpha, z_\alpha)$, the displacements from the initial position in the lattice to the equilibrium position is represented by the vector $u_\alpha = x_\alpha - p_\alpha$. Let $\| \cdot \|$ denote the vector norm or length in $\mathbb{R}^3$, let $r_{\alpha\beta} = \|\mathbf{x}_\beta - \mathbf{x}_\alpha\|$ be the distance between particles α and β in initial configuration. Then, the force $\mathbf{F}_{\alpha\beta}$, along the vector $x_\beta - x_\alpha$ is governed by the potential function $V(r_{\alpha\beta})$,

$$F_{\alpha\beta} = -\frac{\partial V(r_{\alpha\beta})}{\partial r_{\alpha\beta}} \frac{x_\beta - x_\alpha}{\| x_\beta - x_\alpha \|} . \tag{10}$$

where first term represents the magnitude and second term is the direction. If the indices of bonded neighboring particles of particle α are collected in the set $\mathcal{N}_\alpha$, then we obtain its force equilibrium by applying the following sum:

$$\sum_{\beta \in \mathcal{N}_\alpha} F_{\alpha\beta} = - \sum_{\beta \in \mathcal{N}_\alpha} \frac{\partial V(r_{\alpha\beta})}{\partial r_{\alpha\beta}} \frac{x_\beta - x_\alpha}{\| x_\beta - x_\alpha \|} = 0 . \tag{11}$$

The characteristics of the potential functions $\{V(r_{\alpha\beta})\}_{\beta \in \mathcal{N}_\alpha}$ are provided by the Monte Carlo simulation [5]. The covalent bounds are modeled by spring forces

$$F_{\alpha\beta} = C_1 r + C_2 . \tag{12}$$

Spring like potential $V(r)$ is quadratic. The Van der Waals bounds are model by non-linear forces and the Lennard-Jones potentials

$$V(r) = C_{\alpha\beta} \left[\left(\frac{\sigma_{\alpha\beta}}{r}\right)^{n_{\alpha\beta}} - \left(\frac{\sigma_{\alpha\beta}}{r}\right)^{m_{\alpha\beta}} \right] . \tag{13}$$

where $r = \| x_\beta - x_\alpha \|$.

The equilibrium equations are non-linear and the Newton-Raphson linearization procedure is applied to solve the system. The resulting shrinkage of the feature is presented in Figure 2.

8   Conclusions and Future Work

- The proper balance of errors of global optimization method and direct problem solvers allows for efficient speeding up the solution process of difficult inverse problems. The analytic relation among both errors is necessary.
- The relation between the objective function error and the relative error of the hp-adaptive FEM has been derived. The objective error was expressed as the energy difference between the numerical solution and experiment data.
- The strategy relating the convergence ratios of the inverse and direct problem solution has been proposed and successfully tested for searching value of the CTE parameter in the SFIL process. We obtained about 2.4 speedup in comparison to the solution without error balancing for the simple test example. The higher speedup may be expected for problems with larger dimension.
- The future work will include derivation of analytic relations between the hp-adaptive FEM error and objective function error defined in other ways. The possibilities of further speeding up of the solver will be tested by utilizing the parallel version of the hp-adaptive FEM codes [3].

Acknowledgments. The work reported in this paper was supported by Polish MNiSW grant no. 3 TO 8B 055 29.

References

1. Ciarlet P., The Finite Element Method for Elliptic Problems. North Holland, New York (1994)
2. Demkowicz L., Computing with hp-Adaptive Finite Elements, Vol. I. Chapman & Hall/Crc Applied Mathematics & Nonlinear Science, Taylor & Francis Group, Boca Raton London New York (2006)
3. Paszyński, M., Demkowicz, L., Parallel Fully Automatic hp-Adaptive 3D Finite Element Package. Engineering with Computers (2006) in press.
4. Paszyński, M., Szeliga, D., Barabasz, B. Macioł, P., Inverse analysis with 3D hp adaptive computations of the orthotropic heat transport and linear elasticity problems. VII World Congress on Computational Mechanics, Los Angeles, July 16-22 (2006)
5. Paszyński, M., Romkes, A., Collister, E., Meiring, J., Demkowicz, L., Willson, C. G., On the Modeling of Step-and-Flash Imprint Lithography using Molecular Statics Models. ICES Report 05-38 (2005) 1-26
6. Rachowicz, W., Pardo D., Demkowicz, L., Fully Automatic hp-Adaptivity in Three Dimensions. ICES Report 04-22 (2004) 1-52

Topology Preserving Tetrahedral Decomposition of Trilinear Cell

Bong-Soo Sohn

Department of Computer Engineering, Kyungpook National University
Daegu 702-701, South Korea
bongbong@knu.ac.kr
http://bh.knu.ac.kr/~bongbong

Abstract. We describe a method to decompose a cube with trilinear interpolation into a set of tetrahedra with linear interpolation, where isosurface topology is
preserved during decomposition for all isovalues. This method is useful for converting from a rectilinear grid into a tetrahedral grid in scalar data with topological correctness. We apply our method to topologically and geometrically accurate
isosurface extraction.

Keywords: volume visualization, isosurface, subdivision, topology.

1 Introduction

Scientific simulations and measurements often generate a real-valued volumetric data
in the form of function values sampled on a three dimensional (3D) rectilinear grid.
Trilinear interpolation is a common way to define a function inside each cell of the
grid. It is computationally simple and provides good estimation of a function between
sampled points in practice. *Isosurface extraction* is one of the most common techniques
for visualizing the volumetric data. An isosurface is a level set surface defined as $I(w) =$
$\{(x,y,z)|F(x,y,z) = w\}$ where F is a function defined from the data and w is an isovalue.
The isosurface I is often polygonized for modeling and rendering purposes. We call I a
trilinear isosurface to distinguish it from a polygonized isosurface when F is a trilinear
function.

Although the rectilinear volumetric data is the most common form, some techniques
[9,4,3,1,5] require a tetrahedral grid domain due to its simplicity. In order to apply such
techniques to rectilinear volumetric data, people usually decompose a cube into a set of
tetrahedra where a function is defined by linear interpolation. The decomposition may
significantly distort the function in terms of its level sets (e.g. isosurfaces) topology and
geometry. See [2] for examples. 2D/3D meshes with undesirable topology extracted
from the distorted function may cause a serious inaccuracy problem in various simulations such as Boundary Element and Finite Element Method, when the extracted meshes
are used as geometric domains for the simulation [10].

We describe a rule that decomposes a cube into a set of tetrahedra without changing
isosurface topology for all isovalues in the cube. The rule provides topological correctness to any visualization algorithms that run on tetrahedral data converted from
rectilinear data. The key idea is to add saddle points and connect them to the vertices

Y. Shi et al. (Eds.): ICCS 2007, Part I, LNCS 4487, pp. 350–357, 2007.

of a cube to generate tetrahedra. In case there is no saddle point, we perform a standard tetrahedral decomposition method [2] without inserting any points. The tetrahedra set converted from a cube involves a *minimal* set of points that can correctly capture the level set topology of trilinear function because level set topology changes only at critical points. Then, we apply our method to topologically and geometrically accurate isosurface triangulation for trilinear volumetric data.

The remainder of this paper is organized as follows. In section 2, we explain trilinear isosurface and its topology determination. In section 3, we describe topology preserving tetrahedral decomposition of a trilinear cell. Then, in section 4, we give applications and results. Finally, we conclude this paper in section 5.

2   Trilinear Isosurface Topology

The function inside a cube, F^c, is constructed by trilinear interpolation of values on eight vertices of the cube.

$$\begin{aligned}
F^c(x,y,z) = {} & F_{000}(1-x)(1-y)(1-z) + F_{001}(1-x)(1-y)z \\
& + F_{010}(1-x)y(1-z) + F_{011}(1-x)yz \\
& + F_{100}x(1-y)(1-z) + F_{101}x(1-y)z \\
& + F_{110}xy(1-z) + F_{111}xyz
\end{aligned}$$

This means the function on any face of a cell, F^f is a bilinear function computed from four vertices on the face.

$$F^f(x,y) = F_{00}(1-x)(1-y) + F_{01}(1-x)y + F_{10}x(1-y) + F_{11}xy$$

Saddle points, where their first partial derivative for each direction is zero, play important roles in determining correct topology of a trilinear isosurface in a cube. Computing the location of *face* and *body saddles* which satisfy $F_x^f = F_y^f = 0$ and $F_x^c = F_y^c = F_z^c = 0$ respectively is described in [6] and [8]. Saddle points outside the cube is ignored.

Marching Cubes (MC) [7] is the most popular method to triangulate a trilinear isosurface using sign configurations of eight vertices in a cube. It is well known that some of the sign configurations have ambiguities in determining contour connectivity. The papers [6,8] show that additional sign configurations of face and body saddle points can disambiguate the correct topology of a trilinear isosurface. Figure 1 shows every possible isosurface connectivity of trilinear function [6].

3   Topology Preserving Tetrahedral Decomposition

In this section, we describe a rule for decomposing a cube with trilinear interpolation into a set of tetrahedra with linear interpolation where isosurface topology is preserved for all isovalues during the decomposition. The tetrahedral decomposition is consistent for entire rectilinear volumetric data in the sense that the tetrahedra are seamlessly matched on a face between any two adjacent cubes. The rule is based on the analysis of

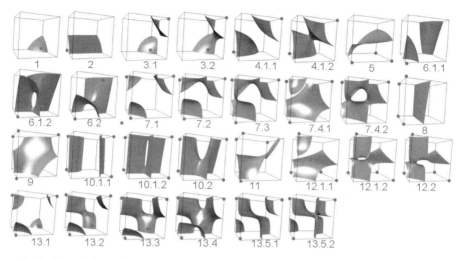

Fig. 1. Possible isosurface topology of trilinear interpolant. Numberings are taken from [6].

face and body saddles in a cell (e.g. cube). It is easy to understand the overall idea by looking at 2D case in Figure 2, which is much simpler than 3D case.

Let s_b and s_f be the number of body saddles and face saddles respectively. There are four cases based on the number of face saddles and body saddles, (i) $s_b = 0$ and $s_f = 0$, (ii) $s_b = 0$ and $1 \leq s_f \leq 4$, (iii) $s_b = 1$ and $0 \leq s_f \leq 4$, (iv) $s_b = 0$ and $s_f = 6$, and (v) $1 \leq s_b \leq 2$ and $s_f = 6$. The case where the number of face saddles is six is the most complicated case that requires careful treatment. Note that the number of face saddles cannot be five. The number of body saddles cannot be two unless the number of face saddles is six. The decomposition rule for each case is as follows :

- case (i) : Decompose a cube into 6 tetrahedra without inserting any point as in [2].
- case (ii) : Choose one face saddle and decompose a cube into five pyramids by connecting the face saddle into four corner vertices of each face except the face that contains the face saddle. If a face of a pyramid contains a face saddle, the face is decomposed into four triangles that share the face saddle and the pyramid is decomposed into four tetrahedra. If a face of a pyramid does not contain a face saddle, the pyramid is decomposed into two tetrahedra in a consistent manner. If the number of face saddles is three or four, we need to choose the second biggest face saddle.
- case (iii) : Decompose a cube into six pyramids by connecting a body saddle to four corner vertices of each face of a cube. Like in (ii), if a face of a pyramid contains a face saddle, the face is decomposed into four triangles and the pyramid is decomposed into four tetrahedra. Otherwise, the pyramid is decomposed into two tetrahedra.
- case (iv) : A diamond is created by connecting the six face saddles. The diamond is decomposed into four tetrahedra. Twelve tetrahedra are created by connecting two vertices of each twelve edge of a cube and two face saddles on the two faces which share the edge. Eight tetrahedra are created by connecting each eight face of the diamond and a corresponding vertex of the cube. This will decompose a cube into twenty four tetrahedra.

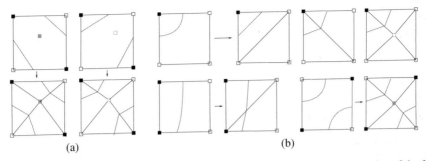

(a) (b)

Fig. 2. (a) Triangular decomposition of a face by connecting a face saddle to each edge of the face resolves an ambiguity in determining correct contour connectivity. (b) Triangular decomposition based on a saddle point preserves level sets topology of bilinear function for every case.

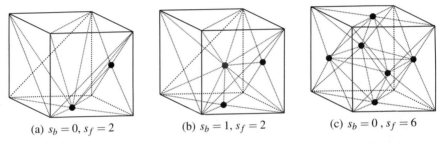

(a) $s_b = 0$, $s_f = 2$ (b) $s_b = 1$, $s_f = 2$ (c) $s_b = 0$, $s_f = 6$

Fig. 3. Example of rules for topology preserving tetrahedral decomposition

– case (v) : Figure 4 shows the cell decomposition when there are two body saddles and six face saddles. It generates two pyramids and four prisms where pyramids and prisms are further decomposed into tetrahedra. Choosing any two parallel faces that are connected to body saddles to form pyramids is fine. We classify saddle points as three *small face saddles*, a *small body saddle*, a *big body saddle*, and three *big face saddles* based on increasing order of the saddle values. Let *small/big corner vertex* be the vertex adjacent to three faces with small/big face saddles. The two parallel faces with small and big face saddles are connected to small and big body saddles respectively to form the two pyramids. The four prisms are decomposed into tetrahedra in a way that the small corner vertex should not be connected to the big body saddle and the big corner vertex should not be connected to the small body saddle. To satisfy this constraint, two types of decomposition of a prism are possible as shown in Figure 4 (c). In case $s_b = 1$, we consider a small or big body saddle moves to a face saddle of a pyramid that is connected to the body saddle and hence the pyramid is collapsed. In this case, the pair of parallel faces for forming the pyramids are chosen in a way that the face saddle of a collapsed pyramid should not be the smallest or the biggest face saddle.

Figure 3 shows several examples of applying the tetrahedral decomposition rule to a cube with different number of body and face saddles.

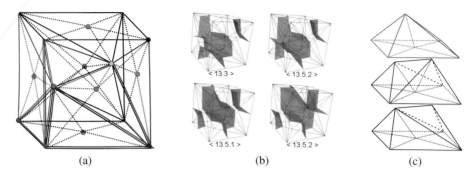

Fig. 4. (a) Cell decomposition when two body saddles exist. Dark blue, green, light blue, magenta, brown, and red circles represent small corner vertex, small face and body saddles, big body and face saddles, and big corner vertex. (b) Isosurface topology for different isovalues in case (a).

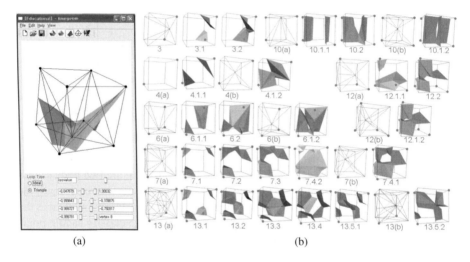

Fig. 5. (a) Demo program of topology preserving tetrahedral decomposition (b) Modified cell decomposition and triangulation for each case that has ambiguity

In appendix, we give a brief proof why topology of level sets of trilinear function in a cube is preserved during the decomposition for each case.

As shown in Figure 5(a), we implemented above method and produced a program that takes an isovalue and values of eight corner vertices of a cube, and displays tetrahedral decomposition with either a trilinear isosurface or a polygonized isosurface.

4 Trilinear Isosurface Triangulation

Readers may think that isosurface extraction for each tetrahedron, which is called Marching Tetrahedra (MT), after the tetrahedral decomposition would be simple and natural way. However, this direct application of the tetrahedral decomposition may cause extra cost in terms of the number of generated triangles and speed because even very simple case (e.g. type 1 in Figure 1) requires tetrahedral decomposition. We exploit the benefits of MC - mesh extraction with small number of elements and with high visual fidelity, and MT - ambiguity removal and simplicity by performing only necessary subdivisions.

We describe a modified cell decomposition method for resolving a triangulation ambiguity and reconstructing triangular approximation of an isosurface with correct topology. The main idea is to decompose a cube into a set of tetrahedral and pyramidal cells which do not cause a triangulation ambiguity while preserving contour topology in a cube. A facial ambiguity is resolved by decomposing a face cell into four triangles that share a face saddle point. Likewise, an internal ambiguity is resolved by decomposing a cube into six pyramids which share a body saddle point. If there is an internal ambiguity and the isosurface does not contain a tunnel shape (neck), a body saddle point is not required in the cell decomposition for reconstructing isosurface with correct topology.

There are four cases : (a) no face ambiguity with no tunnel, (b) face ambiguity (less than six) with no tunnel, (c) tunnel, and (d) six face ambiguities with no tunnel.

- case (a) : No decomposition is necessary. Just perform MC-type triangulation
- case (b) : Choose a face saddle on a face with ambiguity and decompose a cube into five pyramids by connecting the face saddle into four corner vertices of each face except the face that contains the face saddle. If a face of a pyramid contains face ambiguity, the face is decomposed into four triangles that share the face saddle and the pyramid is decomposed into four tetrahedra. If the number of face ambiguity is three, we need to choose second face saddle.
- case (c) : Decompose a cube into six pyramids by connecting a body saddle that is involved with a tunnel to four corner vertices of each face of a cube. Like in (b), if a face of a pyramid contains a face ambiguity, the face is decomposed into four triangles and the pyramid is decomposed into four tetrahedra.
- case (d) : We perform the same decomposition as the case (iv) in section 3, except that we decompose the diamond into two pyramids instead of four tetrahedra. (13(a) in Figure 5(b))

Isosurface configurations 1, 2, 5, 8, 9, and 11 in Figure 1 are the only possible cases which do not have an ambiguity in a cube. No decomposition is necessary for the cases. Triangulations for such cubes are listed in [7].

We implemented the modified decomposition method and applied it to each case in Figure 1, and extracted a triangular isosurface. Figure 5 shows the modified cell decomposition and its triangulation for every possible configuration of a trilinear isosurface that has an ambiguity. It confirms that the triangular isosurface generated from the cell decomposition is topologically equivalent to a trilinear isosurface from trilinear function by the comparison of Figure 1 and Figure 5(b).

5   Conclusion

We described a method for tetrahedral decomposition of a cube where level sets topology is preserved. We envision many visualization algorithms that can take only tetrahedral grid data can utilize our method for dealing with trilinear volumetric data instead of using standard tetrahedral decomposition methods that may significantly distort level sets topology and geometry.

Acknowledgments. The author is grateful to Prof. Chandrajit Bajaj who gave various inspirational ideas that are related to this work. This research was supported by Kyungpook National University Research Fund, 2006.

References

1. Bajaj, C. L., Pascucci V., Schikore D. R.: The contour spectrum. In IEEE Visualization Conference (1997) 167–1737
2. Carr, H., Möller, T., Snoeyink, J.: Simplicial subdivisions and sampling artifacts. In IEEE Visualization Conference (2001) 99–108
3. Carr, H., Snoeyink, J.: Path seeds and flexible isosurfaces using topology for exploratory visualization. In Proceedings of VisSym (2003) 49–58
4. Carr, H., Snoeyink, J., Axen, U.: Computing contour trees in all dimensions. Computational Geometry: Theory and Applications Vol. 24. (2003) 75–94
5. Edelsbrunner, H., Harer, J., Natarajan, V., Pascucci, V.: Morse complexes for piecewise linear 3-manifolds. In Proceeding of the 19-th ACM Symp. on Comp. Geom. (2003) 361–370
6. Lopes, A., Brodlie, K.: Improving the robustness and accuracy of the marching cubes algorithm for isosurfacing. IEEE Transactions on Visualization and Computer Graphics (2003) 19–26
7. Lorensen, W.E., Cline, H.E.: Marching cubes: A high resolution 3D surface construction algorithm. In ACM SIGGRAPH (1987) 163–169
8. Natarajan, B.K.: On generating topologically consistent isosurfaces from uniform samples. The Visual Computer, Vol. 11. (1994) 52–62
9. Nielson, G.M., Sung, J.: Interval volume tetrahedrization. In IEEE Visualization Conference (1997) 221–228
10. Zhang, J., Bajaj, C. L., Sohn, B.-S.: 3D finite element meshing from imaging data. Computer Methods in Applied Mechanics and Engineering (CMAME) Vol. 194. (2005) 5083–5106

Appendix: Proof of Topology Preservation During Decomposition

We call the isosurface extracted from a set of tetrahedra as PL isosurface. The proof is done by showing that, for any isovalue in a cube, the numbers of PL isosurface components and trilinear isosurface components are the same and PL isosurface components are topologically equivalent to trilinear isosurface components. Note that PL isosurface inside a cube should be always a manifold (except for a degenerate case). There is no isolated closed PL isosurface component in a cube.

First of all, it is easy to see that the decomposition preserves trilinear isosurface topology on each face. A face is decomposed into four triangles when there is a face saddle on the face and decomposed into two triangles when there is no face saddle. There are three possible contour connectivity except for the empty case where symmetric cases are ignored. Figure 2 shows that each isocontour connectivity of bilinear function on a face is preserved for any possible triangular mesh generated from our decomposition rule.

We classify corner vertices and saddles into *up-vertices* and *down-vertices* based on the check whether a function value of a vertex is bigger than or lower than an isovalue. We use a term, *component-vertices*, to indicate either up-vertices or down-vertices that have bigger or same number of connected components compared to the other one. Except for the configurations 13.5.1 and 13.5.2 in Figure 1, a connected component of component-vertices uniquely represents an isosurface component. Consider cases (i), (ii), and (iii). If there is no hole, connected components of component-vertices on faces are separated each other inside a cube to form a simple sheet (disk) for each connected component, which is consistent with actual trilinear isosurface topology. If there is a hole, the connected components of component-vertices on faces are connected through a saddle point inside a cube to form a tunnel isosurface. The reason why we choose the second biggest face saddle in the case (ii) with three or four face saddles and (v) with one body saddle is to avoid connecting components of component-vertices inside a cube that needs to be separated. For example, if we choose the smallest or the biggest face saddle in the configuration 7.2, two components of component-vertices on faces of a cube can be connected through an edge and hence two separate isosurface components would be connected with a tunnel.

In cases (iv) and (v) where the number of face saddles is six, the configurations except for 13.5.1 and 13.5.2 are proved in a similar way as the cases of (i), (ii), and (iii). The configurations 13.5.1 and 13.5.2 can be proved by taking out tetrahedra that contributes to the small isosurface component and apply the same proof of (i), (ii), and (iii) to the rest of isosurfaces for topological correctness.

FITTING: A Portal to Fit Potential Energy Functionals to *ab initio* Points

Leonardo Pacifici[2], Leonardo Arteconi[1], and Antonio Laganà[1]

[1] Department of Chemistry, University of Perugia,
via Elce di Sotto, 8 06123 Perugia, Italy
[2] Department of Mathematics and Computer Science, University of Perugia
via Vanvitelli, 1 06123 Perugia, Italy
`xleo,bodynet,lag@dyn.unipg.it`

Abstract. The design and the implementation in a Grid environment of an Internet portal devoted to best fitting potential energy functionals to *ab initio* data for few body systems is discussed. The case study of a generalized LEPS functional suited to fit reactive three body systems is discussed with an application to the NO_2 system.

Keywords: webportal, fitting, *ab initio* calculations, potential energy surfaces, multiscale simulations.

1 Introduction

Thanks to recent progress in computing technologies and network infrastructures it has become possible to assemble realistic accurate molecular simulators on the Grid. This has allowed us to develop a Grid Enabled Molecular Simulator (GEMS) [1,2,3,4] by exploiting the potentialities of the Grid infrastructure of EGEE [5]. The usual preliminary step of molecular approaches to chemical problems is the construction of a suitable potential energy surface (PES) out of the already available theoretical and experimental information on the electronic structure of the system considered. In the first prototype production implementation of GEMS, GEMS.0 [6], it is assumed that available information on the electronic structure of the system considered is formulated as a LEPS [7] PES. Unfortunately, extended use in dynamics studies have singled out the scarce flexibility of the LEPS functional in describing potential energy surfaces having bent (non collinear) minimum energy paths to reaction.

To progress beyond the limits of GEMS.0 an obvious choice was, therefore, not only to derive the value of the LEPS parameters from *ab initio* estimates of the electronic energies but also to add further flexibility to the LEPS functional form. For the former goal an Internet portal, SUPSIM, has been assembled as already discussed in the literature [8]. In the present paper we discuss the latter goal of making more general the LEPS functional. The paper deals in section 2 with a generalization of the functional representation of the LEPS and in section 3 with the assemblage of an Internet portal, called FITTING, devoted to the fitting of the LEPS to *ab initio* data. Finally, in section 4 the case study of the $N + O_2$ system is discussed.

Y. Shi et al. (Eds.): ICCS 2007, Part I, LNCS 4487, pp. 358–365, 2007.

2   A Generalization of the LEPS Potential

Most often molecular potentials are expressed as a sum of the various terms of a many body expansion [9,10]. In the case of three-atom systems such a sum is made by three one-body, three two-body and one three-body terms as follows:

$$
\begin{aligned}
V(r_{\mathrm{AB}}, r_{\mathrm{BC}}, r_{\mathrm{AC}}) = {}& V_{\mathrm{A}}^{(1)} + V_{\mathrm{B}}^{(1)} + V_{\mathrm{C}}^{(1)} + \\
& V_{\mathrm{AB}}^{(2)}(r_{\mathrm{AB}}) + V_{\mathrm{BC}}^{(2)}(r_{\mathrm{BC}}) + V_{\mathrm{AC}}^{(2)}(r_{\mathrm{AC}}) + \\
& V_{\mathrm{ABC}}^{(3)}(r_{\mathrm{AB}}, r_{\mathrm{BC}}, r_{\mathrm{AC}})
\end{aligned}
\tag{1}
$$

where the $V^{(1)}$ terms are the one-body ones (taken to be zero for atoms in ground state) while $V^{(2)}$ and $V^{(3)}$ terms are the two- and three-body ones and are usually expressed as polynomials in the related internuclear distances r_{AB}, r_{BC} and r_{AC}. These polynomials are damped by proper exponential-like functions of the related internuclear distances in order to vanish at infinity. More recently, use has been also made of Bond Order (BO) variables [11,12]. The n_{ij} BO variable is related to the internuclear distance r_{ij} of the ij diatom as follows:

$$
n_{ij} = \exp\left[-\beta_{ij}(r_{ij} - r_{eij})\right]
\tag{2}
$$

In Eq. 2 β_{ij} and r_{eij} are adjustable parameters (together with D_{ij}) of the best fit procedure trying to reproduce theoretical and experimental information of the ij diatomic molecule using the model potential

$$
V_{ij}^{(2)}(r_{ij}) = D_{ij} P(n_{ij})
\tag{3}
$$

where $P(n_{ij})$ is a polynomial in n_{ij}. The LEPS functional can be also written as a sum of two and three body BO terms. The usual LEPS can be written, in fact, as

$$
\begin{aligned}
V(r_{\mathrm{AB}}, r_{\mathrm{BC}}, r_{\mathrm{AC}}) = {}^{1}E_{\mathrm{AB}} + {}^{1}E_{\mathrm{BC}} + {}^{1}E_{\mathrm{AC}} - J_{\mathrm{AB}} - J_{\mathrm{BC}} - J_{\mathrm{AC}} \\
- \sqrt{J_{\mathrm{AB}}^{2} + J_{\mathrm{BC}}^{2} + J_{\mathrm{AC}}^{2} - J_{\mathrm{AB}} J_{\mathrm{BC}} - J_{\mathrm{AB}} J_{\mathrm{AC}} - J_{\mathrm{BC}} J_{\mathrm{AC}}}
\end{aligned}
\tag{4}
$$

where the J_{ij} terms are formulated as:

$$
J_{ij} = \frac{1}{2}({}^{1}E_{ij} - a_{ij}\,{}^{3}E_i)
\tag{5}
$$

with a_{ij} being an adjustable parameter (often expressed as $(1 - S_{ij})/(1 + S_{ij})$, where S_{ij} is the Sato parameter) and ^{1}E and ^{3}E being second order BO polynomials of the Morse

$$
{}^{1}E_{ij} = D_{ij} n_{ij}(n_{ij} - 2)
\tag{6}
$$

and antiMorse

$$
{}^{3}E_{ij} = \frac{D_{ij}}{2} n_{ij}(n_{ij} + 2)
\tag{7}
$$

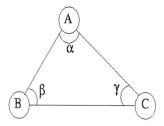

Fig. 1. A pictorial view of an atom-diatom system

type, respectively. Because of the choice of truncating the polynomial of Eq. 3 to the second order, β_{ij}, r_{eij} and D_{ij} correspond to the force constant, the equilibrium distance and the dissociation energy of the ij diatom, respectively. The two-body terms correspond therefore to the three $^1E_{ij}$ Morse potentials. The three body component $V^{(3)}$ of the potential is then worked out by subtracting the three diatomic terms to the *ab initio* data. The resulting values of the three body term are then fitted by optimizing the value of the a_{ij} parameters which are taken to be constant in the usual LEPS functional. In our generalization, as proposed some years ago by Takayanagi and Sato [13] and by Brown et al. [14], the Sato variables S_{ij} are made depend on the angle opposed to the bond considered (respectively γ, α and β as sketched in Fig. 1) to bear a kind of three body connotation. Accordingly, the a_{ij} coefficients of eq. 5 can be formulated as depending from the angle opposite to the ij diatom as follows:

$$a_{ab} = c_{\gamma 1} + c_{\gamma 2}cos\gamma + c_{\gamma 3}cos^2\gamma + c_{\gamma 4}cos^3\gamma + c_{\gamma 5}cos^4\gamma \qquad (8)$$

$$a_{bc} = c_{\alpha 1} + c_{\alpha 2}cos\alpha + c_{\alpha 3}cos^2\alpha + c_{\alpha 4}cos^3\alpha + c_{\alpha 5}cos^4\alpha \qquad (9)$$

$$a_{ac} = c_{\beta 1} + c_{\beta 2}cos\beta + c_{\beta 3}cos^2\beta + c_{\beta 4}cos^3\beta + c_{\beta 5}cos^4\beta \qquad (10)$$

3   The Internet Portal Structure

To handle the fitting procedure in an efficient way we developed a web interface (called FITTING) acting as an Internet portal and ensuring the advantages typical of a Grid based environment. This choice was motivated by the wish of being independent from the operating system available on the user side and therefore being able to modify and upgrade the software without the participation of the user. Other motivations were the user friendliness and the ubiquitous usability of the web graphical interfaces. For this purpose we created a cross-browser site using only server-side technologies. Accordingly, the end-user can utilize the FITTING web GUI (Graphical User Interface) by making use only of a W3-Compliant web browser [15]. The related Web Environment was implemented using the following elements:

1. A dynamic web server, based on the Apache Web [16] server containing the PHP4 module [17].
2. An RDBMS (MySQL [18] in our case) that handles the user data and supports the authentication phase.

The Portal was developed and tested using GPL Software and FreeSoftware (namely the Apache Web Server 1.3.32 and MySQL 4.1.3 powered by FreeBSD 5.4).

Because of the complexity of the workflow of FITTING, we produced a set of dynamically generated pages according to the following scheme:

1. registration of the user
2. selection of the atoms and the functional form
3. specification of the *ab initio* data
4. specification of additional *ab initio* data
5. generation of the best-fit parameters

These pages take care of managing the execution of the computational procedure by the Web server and help the user to define the input parameters of the fitting calculation through the GUI.

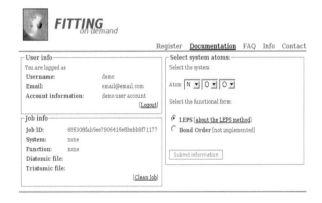

Fig. 2. Screenshot of a System configuration page of FITTING

As a first step the user registers through the GUI when first accessing the portal. After the verification of the identity, the user is assigned an account and the associated login and password. At this point the user can access the portal and run the fitting procedure. Because of the multiuser environment adopted, multiple requests to the web server are dealt using the Session support (enabled in PHP by default).

In the second step, the user selects, using the same GUI, the atoms composing the triatomic system considered and the fitting functional form to be used (see Fig. 2). In the third step, the server creates a dynamic web page which prompts the user to supply the name of the file of the *ab initio* data to be used during the

Fig. 3. Screenshot of a System configuration page of FITTING

calculation. In the fourth step, the same page allows the user to insert new *ab initio* data. The page asks the files of diatomic *ab initio* data (from one to three depending on the symmetry of the investigated system), as illustrated in Fig. 3. These files contain the *ab initio* values arranged in a two column format (the first column contains the value of the internuclear distance while the second column contains the corresponding value of the diatomic *ab initio* potential energy). The page prompts also the request for a file containing the *ab initio* triatomic data. This file contains in the first three columns the value of the three internuclear distances and in the fourth column the value of the corresponding triatomic *ab initio* potential energy. It is possible also to introduce other potential energy values to enforce some specific features of the potential or to constrain some input parameters. These choices depend on the functional form adopted for the fitting.

Finally, the best fit is carried out using the LMDER routine of MINPACK [20] which is based on an improved version of the Levemberg-Marquardt method [21] which solves non linear least squares problems. The calculated best-fit values are inserted, together with the already determined diatomic parameters, in the automatically generated source of the corresponding Fortran routine.

4   The N + O$_2$ Case Study

As already mentioned, in order to test the developed procedure, we considered the N + O$_2$ system for which a large set of accurate *ab initio* electronic energy values (calculated at both CASSCF and MR-SDCI level) are available from the literature [22]. CASSCF calculations were performed at various fixed values of the N$\hat{O}$O β attack angle (β= 135°, 110°, 90°, 70°, 45°). For each value of β, a matrix of geometries corresponding to regularly spaced values of $\rho_\beta = \sqrt{n_{NO}^2 + n_{OO}^2}$

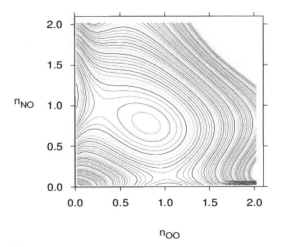

Fig. 4. Isoenergetic contours, plotted as a function of the n_{NO} (y axis) and n_{OO} (x axis) BO variables at $\beta = 135°$. Energy contours are drawn every 3 Kcal/mol.

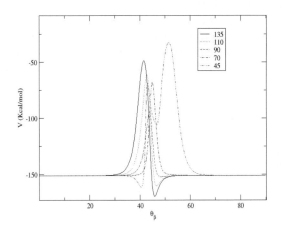

Fig. 5. Minimum energy paths of the generalized LEPS calculated at $\beta = 135°$, $110°$, $90°$, $70°$ and $45°$ plotted as a function of θ_β

(the radius of the polar version of the BO coordinates) and of the associated $\theta_\beta = sin^{-1}(n_{NO}/\rho_\beta)$ angle were considered for the *ab initio* calculations. Calculated CASSCF values were scaled to the MR-CI energies at the minimum of the fixed θ_β cut of the *ab initio* values.

Following the above mentioned procedure the asymptotic cuts of the *ab initio* points were fitted first to Morse diatomic potentials and the best-fit values of the parameters were used to compute the three body component of the potential. The computed three body component was then fitted using both three constant

Sato parameters (as in the usual extended LEPS functional) and the fifteen coefficients of our generalized angular dependent LEPS given in eqs. 8-10. Due also to the particular structure of the NO_2 PES we found the extended LEPS based on three constant Sato parameters to be scarcely flexible and to lead to a root mean square deviation of about 3.0 eV. Moreover, the isoenergetic contour plots in general poorly reproduce the *ab initio* values and have a wrong topology.

A much better reproduction of the *ab initio* data was obtained when using the generalized LEPS (the one which has angle dependent Sato parameters) which gave a root mean square deviation half that of the extended LEPS. This result, though still preliminary, can be considered highly satisfactory due to the fact that a non negligible fraction of the deviation is due to the already mentioned particular structure of the NO_2 PES whose two body component is not well reproduced by a Morse functional. The definitely better quality of the fitting carried out using the generalized LEPS functional can also be appreciated by inspecting the isoenergetic contours drawn at different fixed values of β and comparing them with the *ab initio* values. In particular, they not only always reproduce the topology of the fixed angle *ab initio* values (see for example the contours calculated at $\beta= 135°$ shown in Fig. 4) but they also reproduce in a quasi quantitative fashion the corresponding minimum energy paths (MEP). MEP plots (see Fig. 5) show, in fact, the large variability of the MEP and the peculiar double barrier structure of the MEP at some values of the approaching angle. Moreover, in agreement with the structure of the *ab initio* data we found also that when moving from large β values to 110° the barrier lowers to rise again in going from $\beta=110°$ to $\beta= 70°$.

5   Conclusions

In this paper the use of angle dependent LEPS functionals is proposed out and the development of an Internet portal called FITTING, aimed at inserting its fitting to *ab initio* data as part of the Grid Enabled Molecular Simulator (GEMS) implemented within EGEE, is illustrated. Using FITTING it is now possible to perform *ab initio* simulations starting from the generation of the potential energy values (for which the portal SUPSIM is already available) and continuing with their representation by a proper functional form to be used in dynamical calculations. This completes the workflow of GEMS for establishing a service of validation of LEPS potentials. Future work will be concerned with a further generalization of the angle dependent LEPS for its use as a three body component of force fields used in molecular dynamics.

References

1. Gervasi, O., Laganà, A., Cicoria, D., Baraglia, R.: Animazione e Calcolo Parallelo per lo Studio delle Reazioni Chimiche Elementari, Pixel **10** (1994) 19-26.
2. Bolloni, A.: Tesi di Laurea, Università di Perugia (1997).

3. Gervasi, O., Laganà, A., Lobbiani, M.: Lecture Notes in Computer Science, **2331** (2002) 956-964.
4. Gervasi, O., Dittamo, C., Laganà, A. A Grid Molecular Simulator for E-Science, Lecture Notes in Computer Science, **3470** (2005) 16-22
5. The Enabling Grids for E-sciencE (EGEE) project (http://public.eu-egee.org)
6. EGEE review Conference, Geneva, February 2005. http://indico.cern.ch/conferenceDisplay.py?confId=a043803
7. Polanyi, J.C., Schreiber, J.L.: The Dynamics of Bimolecular Reactions in Physical Chemistry an Advanced Treatise, Vol. VI, Kinetics of Gas Reactions, Eyring, H., Jost, W., Henderson, D. Ed. (Academic Press, New York, 1974) p. 383.
8. Storchi, L., Tarantelli, F., Laganà, A.: Computing Molecular Energy surfaces on a Grid, Lecture Notes in Computer Science, **3980**, 675-683, (2006).
9. Sorbie, K.S., Murrell, J.N., Mol. Phys. **29** (1975) 1387-1403.
10. Murrell, J.N., Carter, S., Farantos, S.C., Huxley, P., Varandas, A.J.C.: Molecular potential energy functions (Wiley, Chichester, 1984).
11. Laganà, A., García, E., Mol. Phys., **56** (1985) 621-628.
12. Laganà, A., García, E., Mol. Phys. **56** (1985) 629-641.
13. Takayanagi, T. and Sato, S., Chem. Phys. Lett., **144** (1988) 191-193
14. Brown, F.B., Steckler, R., Schwenke, D.W., Thrular, D.G. and Garrett, B.C., J. Chem. Phys., **82** (1985), 188.
15. World Wide Web Consortium (http://www.w3.org)
16. The Apache Software Foundation (http://www.apache.org)
17. PHP: Hypertext Preprocessor (http://www.php.net)
18. Popular Open Source Database (http://www.mysql.com)
19. Borgia, D.: Tesi di Laurea, Università di Perugia (2006)
20. Morè, J. J., Garbow, B. S., Hillstrom, K. E.: Argonne National Laboratory, (1980); MINPACK package can be obtained from http://www.netlib.org/minpack.
21. Morè, J. J., in Numerical Analysis, Lecture Notes in Mathematics, **630** (1977), 105.
22. G. Suzzi Valli, R. Orrù, E. Clementi, A. Laganà, S. Crocchianti, J. Chem. Phys., **102** (1995) 2825.

Impact of QoS on Replica Placement in Tree Networks

Anne Benoit, Veronika Rehn, and Yves Robert

Laboratoire LIP, ENS Lyon, France. UMR CNRS-INRIA-UCBL 5668
{Anne.Benoit|Veronika.Rehn|Yves.Robert}@ens-lyon.fr

Abstract. This paper discusses and compares several policies to place replicas in tree networks, subject to server capacity and QoS constraints. The client requests are known beforehand, while the number and location of the servers are to be determined. We study three strategies. The first two strategies assign each client to a unique server while the third allows requests of a client to be processed by multiple servers. The main contribution of this paper is to assess the impact of QoS constraints on the total replication cost. In this paper, we establish the NP-completeness of the problem on homogeneous networks when the requests of a given client can be processed by multiple servers. We provide several efficient polynomial heuristic algorithms for NP-complete instances of the problem.

Keywords: Replica placement, QoS constraints, access policies, heterogeneous platforms, complexity, placement heuristics.

1 Introduction

This paper deals with the problem of replica placement in tree networks with Quality of Service (QoS) guarantees. Informally, there are clients issuing several requests per time-unit, to be satisfied by servers with a given QoS. The clients are known (both their position in the tree and their number of requests), while the number and location of the servers are to be determined. A client is a leaf node of the tree, and its requests can be served by one or several internal nodes. Initially, there are no replicas; when a node is equipped with a replica, it can process a number of requests, up to its capacity limit (number of requests served by time-unit). Nodes equipped with a replica, also called servers, can only serve clients located in their subtree (so that the root, if equipped with a replica, can serve any client); this restriction is usually adopted to enforce the hierarchical nature of the target application platforms, where a node has knowledge only of its parent and children in the tree. Every client has some QoS constraints: its requests must be served within a limited time, and thus the servers handling these requests must not be too far from the client.

The rule of the game is to assign replicas to nodes so that some optimization function is minimized and QoS constraints are respected. Typically, this optimization function is the total utilization cost of the servers. In this paper we study this optimization problem, called REPLICA PLACEMENT, and we restrict

Y. Shi et al. (Eds.): ICCS 2007, Part I, LNCS 4487, pp. 366–373, 2007.

the QoS in terms of number of hops This means for instance that the requests of a client who has a QoS range of qos = 5 must be treated by one of the first five internal nodes on the path from the client up to the tree root.

We point out that the distribution tree (clients and nodes) is fixed in our approach. This key assumption is quite natural for a broad spectrum of applications, such as electronic, ISP, or VOD service delivery. The root server has the original copy of the database but cannot serve all clients directly, so a distribution tree is deployed to provide a hierarchical and distributed access to replicas of the original data. On the contrary, in other, more decentralized, applications (e.g. allocating Web mirrors in distributed networks), a two-step approach is used: first determine a "good" distribution tree in an arbitrary interconnection graph, and then determine a "good" placement of replicas among the tree nodes. Both steps are interdependent, and the problem is much more complex, due to the combinatorial solution space (the number of candidate distribution trees may well be exponential).

Many authors deal with the REPLICA PLACEMENT optimization problem. Most of the papers do not deal with QoS but instead consider average system performance such as total communication cost or total accessing cost. Please refer to [2] for a detailed description of related work with no QoS contraints.

Cidon et al [4] studied an instance of REPLICA PLACEMENT with multiple objects, where all requests of a client are served by the closest replica (*Closest* policy). In this work, the objective function integrates a communication cost, which can be seen as a substitute for QoS. Thus, they minimize the average communication cost for all the clients rather than ensuring a given QoS for each client. They target fully homogeneous platforms since there are no server capacity constraints in their approach. A similar instance of the problem has been studied by Liu et al [7], adding a QoS in terms of a range limit, and whose objective is to minimize the number of replicas. In this latter approach, the servers are homogeneous, and their capacity is bounded. Both [4,7] use a dynamic programming algorithm to find the optimal solution.

Some of the first authors to introduce actual QoS constraints in the problem were Tang and Xu [9]. In their approach, the QoS corresponds to the latency requirements of each client. Different access policies are considered. First, a replica-aware policy in a general graph with heterogeneous nodes is proven to be NP-complete. When the clients do not know where the replicas are (replica-blind policy), the graph is simplified to a tree (fixed routing scheme) with the *Closest* policy, and in this case again it is possible to find an optimal dynamic programming algorithm. In [10], Wang et al deal with the QoS aware replica placement problem on grid systems. In their general graph model, QoS is a parameter of communication cost. Their research includes heterogeneous nodes and communication links. A heuristic algorithm is proposed and compared to the results of Tang and Xu [9].

Another approach, this time for dynamic content distribution systems, is proposed by Chen et al [3]. They present a replica placement protocol to build a dissemination tree matching QoS and server capacity constraints. Their work focuses on Web content distribution built on top of peer-to-peer location services: QoS is defined by a latency within which the client has to be served, whereas server capacity is bounded by a fan-out-degree of direct children. Two placement algorithms (a native and a smart one) are proposed to build the dissemination tree over the physical structure.

In [2] we introduced two new access policies besides the *Closest* policy. In the first one, the restriction that all requests from a given client are processed by the same replica is kept, but client requests are allowed to "traverse" servers in order to be processed by other replicas located higher in the path (closer to the root). This approach is called the *Upwards* policy. In the second approach, access constraints are further relaxed and the processing of a given client's requests can be split among several servers located in the tree path from the client to the root. This policy with multiple servers is called *Multiple*.

In this paper we study the impact of QoS constraints on these three policies. On the theoretical side we prove the NP-completeness of *Multiple*/Homogeneous instance with QoS constraints, while the same problem was shown to be polynomial without QoS [2]. This result shows the additional combinatorial difficulties which we face when enforcing QoS constraints. On the practical side, we propose several heuristics for all policies. We compare them through simulations conducted for problem instances with different ranges of QoS constraints. We are also able to assess the absolute performance of the heuristics, by comparing them to the optimal solution of the problem provided by a formulation of the REPLICA PLACEMENT problem in terms of a mixed integer linear program. The solution of this program allows us to build an optimal solution [1] for reasonably large problem instances.

2   Framework and Access Policies

We consider a distribution tree $\mathcal{T}$ whose nodes are partitioned into a set of clients $\mathcal{C}$ and a set of nodes $\mathcal{N}$. The clients are leaf nodes of the tree, while $\mathcal{N}$ is the set of internal nodes. A client $i \in \mathcal{C}$ is making r_i requests per time unit to a database, with a QoS qos_i: the database must be placed not further than qos_i hops on the path from the client to the root.

A node $j \in \mathcal{N}$ may or may not have been provided with a replica of the database. A node j equipped with a replica (*i.e.* j is a server) can process up to W_j requests per time unit from clients in its subtree. In other words, there is a unique path from a client i to the root of the tree, and each node in this path is eligible to process some or all the requests issued by i when provided with a replica. We denote by $R \subseteq \mathcal{N}$ the set of replicas, and $\mathsf{Servers}(i) \subseteq R$ is the set of nodes which are processing requests from client i. The number of requests from client i satisfied by server s is $r_{i,s}$, and the number of hops between i and $j \in \mathcal{N}$ is denoted by $d(i,j)$. Two constraints must be satisfied:

- Server capacity: $\forall s \in R, \quad \sum_{i \in \mathcal{C} \mid s \in \mathsf{Servers}(i)} r_{i,s} \leq \mathsf{W}_s$
- QoS constraint: $\forall i \in \mathcal{C}, \forall s \in \mathsf{Servers}(i), \quad d(i,s) \leq \mathsf{qos}_i$

The objective function for the REPLICA PLACEMENT problem is defined as: Min $\sum_{s \in R} W_s$. When the servers are homogeneous, *i.e.* $\forall s \in \mathcal{N}, W_s = W$, the optimization problem reduces to finding a minimal number of replicas. This problem is called REPLICA COUNTING.

We consider three access policies in this paper. The first two are single server strategies, i.e. each client is assigned a single server responsible for processing all its requests. The *Closest* policy is the most restricted one: the server for client i is enforced to be the first server that can be found on the path from i upwards to the root. Relaxing this constraint leads to the *Upwards* policy. Clients are still assigned to a single server, but their requests are allowed to traverse one or several servers on the way up to the root, in order to be processed by another server closer to the root. The third policy is a multiple server strategy and hence a further relaxation: a client i may be assigned a set of several servers. Each server $s \in$ Servers(i) will handle a fraction $r_{i,s}$ of requests. Of course $\sum_{s \in \text{Servers}(i)} r_{i,s} = r_i$. This policy is referred to as the *Multiple* policy.

3   Complexity Results

Table 1 gives an overview of complexity results of the different instances of the REPLICA COUNTING problem (homogeneous servers). Liu et al [7] provided a polynomial algorithm for the *Closest* policy with QoS constraints. In [2] we proved the NP-completeness of the *Upwards* policy without QoS. This was a surprising result, to be contrasted with the fact that the *Multiple* policy is polynomial under the same conditions [2].

Table 1. Complexity results for the different instances of REPLICA COUNTING

	Homogeneous	**Homogeneous/QoS**
Closest	polynomial [4,7]	polynomial [7]
Upwards	NP-complete [2]	NP-complete [2]
Multiple	polynomial [2]	**NP-complete** (this paper)

An important contribution of this paper is the NP-completeness of the *Multiple* policy with QoS constraints. As stated above, the same problem was polynomial without QoS, which gives a clear insight on the additional complexity introduced by QoS constraints. The proof uses a reduction to 2-PARTITION-EQUAL [5]. Due to a lack of space we refer to the extended version of this paper [1] for the complete proof.

Theorem 1. *The instance of the* REPLICA COUNTING *problem with QoS constraints and the Multiple strategy is NP-complete.*

Finally, we point out that all three instances of the REPLICA PLACEMENT problem (heterogeneous servers with the *Closest*, *Upwards* and *Multiple* policies) are already NP-complete without QoS constraints [2].

4 Heuristics for the Replica Placement Problem

In this section several heuristics for the *Closest*, *Upwards* and *Multiple* policies are presented. As already pointed out, the quality of service is the number of hops that requests of a client are allowed to traverse until they have to reach their server. The code and some more heuristics can be found on the web [8]. All heuristics described below have polynomial, and even worst-case quadratic, complexity $O(s^2)$, where $s = |\mathcal{C}| + |\mathcal{N}|$ is the problem size.

In the following, we denote by $\mathsf{inreqQoS}_j$ the amount of requests that reach an inner node j within their QoS constraints, and by inreq_j the total amount of requests that reach j (including requests whose QoS constraints are violated).

Closest Big Subtree First - CBS. Here we traverse the tree in top-down manner. We place a replica on an inner node j if $\mathsf{inreqQoS}_j \leq \mathsf{W}_j$. When the condition holds, we do not process any other subtree of j. If this condition does not hold, we process the subtrees of j in non-increasing order of inreq_j. Once no further replica can be added, we repeat the procedure. We stop when no new replica is added during a pass.

Upwards Small QoS Started Servers First - USQoSS. Clients are sorted by non-decreasing order of qos_i (and non-increasing order of r_i in case of tie). For each client i in the list we search for an appropriate server: we take the next server on the way up to the root (i.e. an inner node that is already equipped with a replica) which has enough remaining capacity to treat all the client's requests. Of course the QoS-constraints of the client have to be respected. If there is no server, we take the first inner node j that satisfies $\mathsf{W}_j \geq r_i$ within the QoS-range and we place a replica in j. If we still find no appropriate node, this heuristic has no feasible solution.

Upwards Minimal Distance - UMD. This heuristic requires two steps. In the first step, so-called indispensable servers are chosen, i.e. inner nodes which have a client that must be treated by this very node. At the beginning, all servers that have a child client with $\mathsf{qos} = 1$ will be chosen. This step guarantees that in each loop of the algorithm, we do not forget any client. The criterion for indispensable servers is the following: for each client check the number of nodes eligible as servers; if there is only one, this node is indispensable and chosen. The second step of UMD chooses the inner node with minimal $(\mathsf{W}_j - \mathsf{inreqQoS}_j)$-value as server (if $\mathsf{inreqQoS}_j > 0$). Note that this value can be negative. Then clients are associated to this server in order of distance, i.e. clients that are close to the server are chosen first, until the server capacity W_j is reached or no further client can be found.

Multiple Small QoS Close Servers First - MSQoSC. The main idea of this heuristic is the same as for USQoSS, but with two differences. Searching for an appropriate server, we take the first inner node on the way up to the root which has some remaining capacity. Note that this makes the difference between *close* and *started* servers. If this capacity W_i is not sufficient (client c has more requests, $\mathsf{W}_i < r_c$), we choose other inner nodes going upwards to the root until all requests of the client can be processed (this is possible owing to the multiple-server relaxation). If we cannot find enough inner nodes for a client, this heuristic will not return a feasible solution.

Multiple Small QoS Minimal Requests - MSQoSM. In this heuristic clients are treated in non-decreasing order of qos_i, and the appropriate servers j are chosen by minimal $(W_j - inreqQoS_j)$-value until all requests of clients can be processed.

Multiple Minimal Requests - MMR. This heuristic is the counterpart of UMD for the *Multiple* policy and requires two steps. policy: servers Servers are added in the "indispensable" step, either when they are the only possible server for a client, or when the total capacity of all possible inner nodes for a client i is exactly r_i. The server chosen in the second step is also the inner node with minimal $(W_j - inreqQoS_j)$-value, but this time clients are associated in non-decreasing order of $\min(qos_i, d(i, r))$, where $d(i, r)$ is the number of hops between i and the root of the tree. Note that the last client that is associated to a server, might not be processed entirely by this server.

Mixed Best - MB. This heuristic unifies all previous ones, including those presented in [1]. For each tree, we select the best cost returned by the other heuristics. Since each solution for *Closest* is also a solution for *Upwards*, which in turn is a valid solution for *Multiple*, this heuristic provides a solution for the *Multiple* policy.

5   Experimental Plan

In this section we evaluate the performance of our heuristics on tree platforms with varying parameters. Through these experiments we want to assess the different access policies, and the impact of QoS constraints on the performance of the heuristics. We obtain an optimal solution for each tree platform with the help of a mixed integer linear program, see [1] for further details. We can compute the latter optimal solution for problem sizes up to 400 nodes and clients, using GLPK [6].

An important parameter in our tree networks is the load, i.e. the total number of requests compared to the total processing power: $\lambda = \frac{\sum_{i \in C} r_i}{\sum_{j \in N} W_j}$, where C is the set of clients in the tree and N the set of inner nodes. We tested our heuristics for $\lambda = 0.1, 0.2, ..., 0.9$, each on 30 randomly generated trees of two heights: in a first series, trees have a height between 4 and 7 (small trees). In the second series, tree heights vary between 16 and 21 (big trees). All trees have s nodes, where $15 \le s \le 400$. To assess the impact of QoS on the performance, we study the behavior (i) when QoS constraints are very tight ($qos \in \{1, 2\}$); (ii) when QoS constraints are more relaxed (the average value is set to half of the tree height $height$); and (iii) without any QoS constraint ($qos = height + 1$).

We have computed the number of solutions for each λ and each heuristic. The number of solutions obtained by the linear program indicates which problems are solvable. Of course we cannot expect a result with our heuristics for intractable problems. To assess the performance of our heuristics, we have studied the relative performance of each heuristic compared to the optimal solution. For each λ,

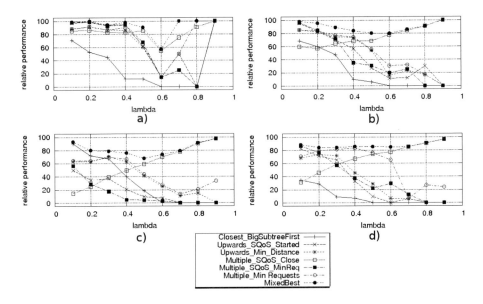

Fig. 1. Relative performance of small trees with (a) tight, (b) medium, (c) no QoS constraints. (d) Big trees with medium QoS constraints.

the cost is computed on those trees for which the linear program has a solution. Let T_λ be the subset of trees with a LP solution. Then, the relative performance for the heuristic h is obtained by $\frac{1}{|T_\lambda|} \sum_{t \in T_\lambda} \frac{cost_{LP}(t)}{cost_h(t)}$, where $cost_{LP(t)}$ is the optimal solution cost returned by the linear program on tree t, and $cost_h(t)$ is the cost involved by the solution proposed by heuristic h. In order to be fair versus heuristics that have a higher success rate, we set $cost_h(t) = +\infty$, if the heuristic did not find any solution.

Figure 1 gives an overview of our performance tests (see [1] for the complete set of results). The comparison between Fig. 1a and 1c shows the impact of QoS on the performance. The impact of the tree sizes can be seen by comparing Fig. 1b and 1d. Globally, all the results show that QoS constraints do not modify the relative performance of the three policies: with or without QoS, *Multiple* is better than *Upwards*, which in turn is better than *Closest*, and their difference in performance is not sensitive to QoS tightness or to tree sizes. This is an enjoyable result, that could not be predicted a priori. The MB heuristic returns very good results, being relatively close to the optimal in most cases. The best heuristic to use depends on the tightness of QoS constraints. Thus, for *Multiple*, MSQoSM is the best choice for tight QoS constraints and small λ (Fig. 1a). When QoS is less constrained, MMR is the best for λ up to 0.4. For big λ, MSQoSC is to prefer, since it never performs poorly in this case. Concerning the *Upwards* policy, USQoSS behaves the best for tight QoS, in the other cases UMD achieves better results. We kept only our best *Closest* heuristic on these curves, which outperforms the others [1].

6   Conclusion

In this paper we have dealt with the REPLICA PLACEMENT optimization problem with QoS constraints. We have proven the NP-completeness of *Multiple* with QoS constraints on homogeneous platforms, and we have proposed a set of efficient heuristics for the *Closest*, *Upwards* and *Multiple* access policies. To evaluate the absolute performance of our algorithms, we have compared the experimental results to the optimal solution of an integer linear program, and these results turned out quite satisfactory. In our experiments we have assessed the impact of QoS constraints on the different policies, and we have discussed which heuristic performed best depending upon problem instances, platform parameters and QoS tightness. We have also showed the impact of platform size on the performances. Although we studied the problem with a restricted QoS model, but we expect experimental results to be similar for more general QoS constraints.

As for future work, bandwidth and communication costs could be included in the experimental plan. Also the structure of the tree networks has to be studied more precisely. In this paper we have investigated different tree heights, but it would be interesting to study the impact of the average degree of the nodes onto the performance. In a longer term, the extension of the REPLICA PLACEMENT optimization problem to various object types should be considered, which would call for the design and evaluation of new efficient heuristics.

References

1. Benoit, A., Rehn, V., Robert, Y.: Impact of QoS on Replica Placement in Tree Networks. Research Report 2006-48, LIP, ENS Lyon, France, available at graal.ens-lyon.fr/~yrobert/ (2006).
2. Benoit, A., Rehn, V., Robert, Y.: Strategies for Replica Placement in Tree Networks. Research Report 2006-30, LIP, ENS Lyon, France, available at graal.ens-lyon.fr/~yrobert/ (2006)
3. Chen, Y., Katz, R.H, Kubiatowicz, J.D.: Dynamic Replica Placement for Scalable Content Delivery. In: Peer-to-Peer Systems: First International Workshop, IPTPS 2002, Cambridge, MA, USA (2002) 306–318
4. Cidon, I., Kutten, S., Soffer, R.: Optimal allocation of electronic content. Computer Networks **40** (2002) 205–218
5. Garey, M.R., Johnson, D.S.: Computers and Intractability, a Guide to the Theory of NP-Completeness. W.H. Freeman and Company (1979)
6. GLPK: GNU Linear Programming Kit. http://www.gnu.org/software/glpk/
7. Liu, P., Lin, Y.F., Wu, J.J.: Optimal placement of replicas in data grid environments with locality assurance. In: International Conference on Parallel and Distributed Systems (ICPADS). IEEE Computer Society Press, (2006)
8. Source Code for the Heuristics. http://graal.ens-lyon.fr/ vrehn/code/replicaQoS/
9. Tang, X., Xu, J.: QoS-Aware Replica Placement for Content Distribution. IEEE Trans. Parallel Distributed Systems **16(10)** (2005) 921–932
10. Wang, H., Liu, P., Wu, J.J.: A QoS-aware Heuristic Algorithm for Replica Placement. In: Proceedings of the 7th International Conference on Grid Computing, GRID2006. IEEE Computer Society, (2006) 96–103

Generating Traffic Time Series Based on Generalized Cauchy Process

Ming Li[1], S.C. Lim[2], and Huamin Feng[3]

[1] School of Information Science & Technology, East China Normal University,
Shanghai 200062, PR. China
mli@ee.ecnu.edu.cn, ming_lihk@yahoo.com
[2] Faculty of Engineering, Multimedia University, 63100 Cyberjaya, Selanger, Malaysia
sclim@mmu.edu.my
[3] Key Laboratory of Security and Secrecy of Information, Beijing Electronic Science and Tech
ology Institute, Beijing 100070, PR. China
fenghm@besti.edu.cn

Abstract. Generating traffic time series (traffic for short) is important in networking, e.g., simulating the Internet. In this aspect, it is desired to generate a
time series according to a given correlation structure that may well reflect the
statistics of real traffic. Recent research of traffic modeling exhibits that traffic is
well modeled by a type of Gaussian process called the generalized Cauchy (GC)
process indexed by two parameters that separately characterize the self-similarity
(SS), which is local property described by fractal dimension D, and long-range
dependence (LRD), which is a global feature that can be measured by the Hurst
parameter H, instead of using the linear relationship $D = 2 - H$ as that used in
traditional traffic model with a single parameter such as fractional Gaussian
noise (FGN). This paper presents a computational method to generate series
based on the correlation form of GC process indexed by 2 parameters. Hence, the
present model can be used to simulate realizations that flexibly capture the fractal phenomena of real traffic for both short-term lags and long-term lags.

Keywords: Random data generation, network traffic, self-similarity, long-range
dependence, generalized Cauchy process.

1 Introduction

Network traffic is a type of fractal series with both (local) self-similarity (SS) and
long-range dependence (LRD). Hence, it is a common case to exhibit fractal phenomena of time series, see e.g. [1], [2] and references therein. Its simulation is greatly
desired in the Internet communications [3]. Conventional method to generate random
series is either based on a given probability density function, see e.g. [4], or a given
power spectrum, see e.g. [5]. For traffic simulation, however, it is expected to accurately synthesize data series according to a predetermined correlation structure [3].
This is because not only the autocorrelation function (ACF) of traffic with LRD is an
ordinary function while the power spectrum of a series with LRD is a generalized
function but also ACF of arrival traffic greatly impacts the performances of queuing

Y. Shi et al. (Eds.): ICCS 2007, Part I, LNCS 4487, pp. 374–381, 2007.

systems [6]. In addition, performance analysis desires to accurately know how one packet arriving at time t statistically correlates to the other arriving at $t+\tau$ apart in future as remarked in [3, First sentence, Subparagraph 4, Paragraph 2, Section 6.1]. Therefore, this paper focuses on correlation based computational method.

As known, the statistics of synthesized traffic relies on traffic correlation model used in simulation. FGN with a single parameter is a widely used traditional traffic model, see e.g. [7], [8], [9]. Hence, traditional methods to synthesize traffic are based on FGN with a single parameter, see e.g. [10], [11], [12], [13], [14]. The realizations based on those methods, therefore, only have the statistical properties of FGN with a single parameter, which characterizes SS and LRD by the linear relationship $D = 2 - H$. Recall that SS and LRD are two different concepts, see e.g. [15], [16], [17], [18]. In fact, let $X(t)$ be traffic series. Then, $X(t)$ being of SS with SS index κ means

$$X(at) \triangleq a^{\kappa} X(t), a > 0, \tag{1}$$

where $\triangleq$ denotes equality in finite joint finite distribution. On the other hand, $X(t)$ is of LRD if its ACF, $r(\tau)$, is non-summable. That is, $r(\tau)$ follows power law given by

$$r(\tau) \sim c\tau^{-\beta} (\tau \rightarrow \infty), \ c > 0, \ 0 < \beta < 1. \tag{2}$$

In principle, D and H can be measured independently [15-18]. The former can be used to characterize SS while the later LRD. In the case of FGN introduced by [19], however, they are linearly related. ACF of FGN in the discrete case is given by

$$\frac{\sigma^2}{2} \left(\left\| \tau \right| + 1 \right|^{2H} - 2 \left| \tau \right|^{2H} + \left\| \tau \right| - 1 \right|^{2H} \right), \tag{3}$$

where $\sigma^2 = \dfrac{\Gamma(2 - H)\cos(\pi H)}{\pi H(2H - 1)}$ [1], [31]. The case $0.5 < H < 1$ corresponds to LRD, $0 < H < 0.5$ implies short-range dependence (SRD). FGN reduces to standard white noise when $H = 0.5$. FGN with a single parameter characterizes both SS and LRD of traffic by H, or equivalently D.

The limitation of single-parameter FGN in traffic modeling has been noticed as can be seen from the discussions in [9, Last sentence, Paragraph 4, §7.4], which stated that "it might be difficult to characterize the correlations over the entire traffic traces with a single Hurst parameter". As a matter of fact, [20] mentioned the shortcoming that the single parameter model, e.g., FGN, does not well fit series for short-term lags.

Recently, [16] introduced a class of stationary Gaussian processes indexed by two parameters, which separately characterizes SS and LRD of fractal time series. That model is in the Cauchy class since it can be regarded as extension of the generalized Cauchy (GC) process used in geostatistics [21]. In this regard, our recent papers [17], [18], [22] provided theoretical analysis with demonstrations and comparisons using real-traffic series to show that 2-parameter model is far agreement with real traffic in comparison with single parameter model such as FGN. In addition, our early work [14] gave a correlation-based method for the generation of LRD series with single parameter. Nevertheless, reports regarding the simulation of time series obeying the

GC model are rarely seen, to our best knowledge, letting along the simulation of traffic series. In this paper, we substantially extend our previous work [14] based on GC process model [17] to synthesize traffic series with 2 parameters so that the synthesized series has the property that SS and LRD are decoupled.

The remaining article is organized as follows. Section 2 briefs the concept of GC processes. Section 3 gives the computational model of traffic generation based on the Cauchy correlation model. Section 4 illustrates the demonstrations. Finally, Section 5 concludes the paper.

2 Concept of GC Process

$X(t)$ is called a GC process if it is a stationary Gaussian centred process with the following ACF:

$$r(\tau) = E[X(t+\tau)X(t)] = (1+\tau^{\alpha})^{-\beta/\alpha}, \tau > 0, \tag{4}$$

where $0 < \alpha \le 2$, and $\beta > 0$, [16], [17], [18].

Note that $r(\tau)$ is positive-definite for the above ranges of α and β, and it is a completely monotone for $0 < \alpha \le 1$, $\beta > 0$. When $\alpha = 2$ and $\beta = 1$, one gets the usual Cauchy process.

Clearly the GC process satisfies the LRD property for $\beta < 1$ since

$$\int_0^\infty r(\tau)d\tau = \int_0^\infty \left(1+|\tau|^{\alpha}\right)^{-\beta/\alpha} d\tau = \infty \text{ if } \beta < 1. \tag{5}$$

We note that the GC process is locally self-similar as can be seen from the following. As a matter of fact, it is a Gaussian stationary process and it is locally self-similar of order α since its ACF satisfies for $\tau \to 0$,

$$r(\tau) = 1 - \frac{\beta}{\alpha}|\tau|^{\alpha}\left\{1+O\left(|\tau|^{\gamma}\right)\right\}, \quad \gamma > 0. \tag{6}$$

The above expression is equivalent to the following more commonly used definition of locally self-similarity of a Gaussian process.

$$X(t) - X(a\tau) \triangleq a^{\kappa}[X(t) - X(\tau)], \quad \tau \to 0. \tag{7}$$

The equivalence can be shown by noting that for τ_1 and $\tau_2 \to 0$, (6) gives for $\kappa = \alpha/2$

$$E[(X(t+b\tau_1) - X(t))(X(t+b\tau_2) - X(t))] = \frac{\beta}{\alpha}\left[|b\tau_1|^{\alpha} + |b\tau_2|^{\alpha} - |b(\tau_1-\tau_2)|^{\alpha}\right]$$

$$= \frac{\beta b^{\alpha}}{\alpha}\left(|\tau_1|^{\alpha} + |\tau_2|^{\alpha} - |\tau_1-\tau_2|^{\alpha}\right) = E\left\{\left[b^{\alpha/2}(X(t+\tau_1) - X(t))\right]\left[b^{\alpha/2}(X(t+\tau_2) - X(t))\right]\right\}.$$

In order to determine the fractal dimension of the graph of $X(t)$, we consider the local property of traffic. The fractal dimension D of a locally self-similar traffic of order α is given by (see [17], [18], [23]).

$$D = 2 - \frac{\alpha}{2}. \tag{8}$$

From (2), therefore, one has

$$H = 1 - \frac{\beta}{2}. \tag{9}$$

From (8) and (9), we see that D and H may vary independently. $X(t)$ is of LRD for $\beta \in (0, 1)$ and SRD for $\beta > 1$. Thus, we have the fractal index α which determines the fractal dimension, and the index β that characterizes the LRD. In the end of this section, we discuss estimates of D and H of a GC process for the purpose of completing the description of GC processes though the focus of this paper does not relate to their estimators.

There are some techniques that are popularly used to estimate D, such as box-counting, spectral, and variogram-based methods, see e.g. [15], [24], [25]. Nevertheless, some of the more popular methods, e.g., box-counting, suffer from biases [26]. The method worth noting is called the variogram estimator that was explained in [27]. Denote $\gamma(d)$ the observed mean of the square of the difference between two values of the series at points that are distance d apart. Then, this estimator has the scaling law given by

$$\log \gamma(d) = \text{constant} + \alpha \log d + \text{error} \quad \text{for } d \to 0. \tag{10}$$

The above scaling law is suitable for stationary processes satisfying $1 - r(\tau) \sim |\tau|^{-\alpha}$ for $\tau \to 0$. The variogram estimator of D is expressed by $\hat{D} = 2 - \hat{\alpha}/2$, where $\hat{\alpha}$ is the slope in a log-log plot of $\gamma(d)$ versus d.

The reported estimators of H are rich, such as R/S analysis, maximum likelihood method, and so forth [2], [15], [28]. The method worth mentioning is called the detrended fluctuation analysis introduced in [29], [30]. With this technique, a series is partitioned into blocks of size m. Within each block, least-square fitting is used. Denote $v(m)$ the average of the sample variances. Then, the detrended fluctuation analysis is based on the following scaling law.

$$\log v(m) = \text{constant} + (2 - \beta) \log m + \text{error} \quad \text{for } m \to \infty. \tag{11}$$

The above is applicable for stationary processes satisfying (2), see [28] for details. The estimate $\hat{H}$ of H is half the slope in a log-log plot of $v(m)$ versus m.

3 Computational Model

Let $w(t)$, $W(\omega)$ and $S_w(\omega)$ be a white noise function, its spectrum and the power spectrum, respectively. Then, $W(\omega) = F[w(t)]$ and $S_w(\omega) = W\overline{W} = \text{Constant}$, where $\overline{W}$ is the complex conjugation of W and F is the operator of Fourier transform. Suppose $w(t)$ is the unit white noise. Then, $S_w(\omega) = 1$.

Let $h(t)$ and $H(\omega)$ respectively be the impulse function and system function of a linear filter, which we call simulator in this paper. Denote $y(t)$ the response under the excitation of $w(t)$. Then, $y = w * h$, where $*$ means the operation of convolution. Denote $S_y(\omega)$ the power spectrum of y. Then, under the excitation of w, one has $S_y(\omega) = |H(\omega)|^2$. Let y be the GC process X. Then, $|H(\omega)| = \sqrt{S_X(\omega)}$.

Denote $\psi(\omega)$ the phase function of $H(\omega)$. Then, $H(\omega) = \sqrt{S_X(\omega)} e^{-j\psi(\omega)}$, where

$$S_X(\omega) = F\left[\left(1 + |t|^\alpha\right)^{-\beta/\alpha} \right].$$

Without losing the generality, let $\psi(\omega) = 2n\pi (n = 0, 1, \cdots)$. Therefore, the impulse function of the simulator to generate traffic following GC processes under the excitation of white noise is

$$h(t) = F^{-1}\left\{ \left[F\left(1 + |t|^\alpha\right)^{-\beta/\alpha} \right]^{0.5} \right\}, \tag{12}$$

where F^{-1} is the inverse of F. Consequently, the output of simulator, i.e., the synthesized traffic obeying GC processes is given by

$$X(t) = w(t) * F^{-1}\left\{ \left[F\left(1 + |t|^\alpha\right)^{-\beta/\alpha} \right]^{0.5} \right\}. \tag{13}$$

Expressing α and β by D and H, we have

$$X(t) = w(t) * F^{-1}\left\{ \left[F\left(1 + |t|^{4-2D}\right)^{\frac{1-H}{2-D}} \right]^{0.5} \right\}. \tag{14}$$

In the above, $w(t) = F^{-1}[W(\omega)]$, where $W(\omega) = e^{j\vartheta(\omega)}$, where θ is a real random function with arbitrary distribution. In practice, traffic is band-limited. Thus, let

$$W(\omega) = \begin{cases} e^{j\phi(\omega)}, & |\omega| \leq \omega_c \\ 0, & \text{otherwise} \end{cases}, \tag{15}$$

where $\phi(\omega)$ is a real random function with arbitrary distribution and cutoff frequency ω_c is such that it completely covers the band of the traffic of interest.

In the discrete case, $w(n) = \text{IFFT}[W(\omega)]$, where IFFT represents the inverse of FFT (fast Fourier transform). Fig. 1 indicates the computation procedure.

4 Case Study

Simulated realizations are shown in Figs. 2-3. Due to the advantage of the separated characterizations of D and H by using the Cauchy correlation model, we can observe the distinct effects of D and H. In Fig. 2, the Hurst parameter H is constant 0.75 but the fractal dimension D decreases, ($D = 1.95$ and 1.20). Comparing Fig. 2 (a) and (b), we can see the realization in Fig. 2 (a) is rough while that in Fig. 2 (b) smooth.

In Fig. 3, fractal dimension D is constant 1 while H decreases ($H = 0.95$ and 0.55). From Fig. 3, we can evidently see the stronger persistence (i.e., stronger LRD) in (a) than that in (b).

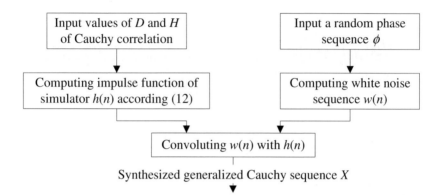

Fig. 1. Computation flow chart

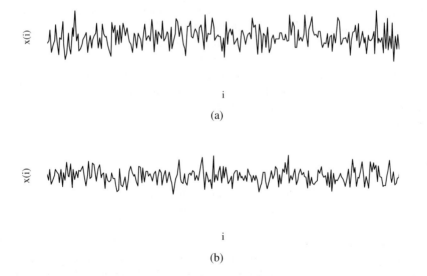

i

(a)

i

(b)

Fig. 2. Realizations. (a) $D = 1.95$, $H = 0.75$. (b) $D = 1.20$, $H = 0.75$.

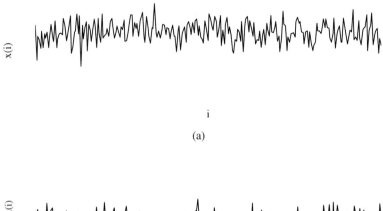

(a)

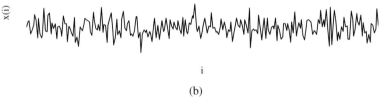

(b)

Fig. 3. Realizations. (a). $H = 0.95$, $D = 1$. (b). $H = 0.55$, $D = 1$.

5 Conclusions

We have given a computational model to generate traffic with separate parametrization of the self-similarity property and long-range dependence based on the correlation model of the generalized Cauchy processes. Since this correlation model can
separately characterize the fractal dimension and the Hurst parameter of a process, the
present method can be used to simulate realizations that have the same long-range
dependence with different fractal dimensions (i.e., different burstinesses from the
view point of networking). On the other hand, we can synthesize realizations that
have the same fractal dimension but with different long-range dependencies. Hence it
provides a flexible way to simulate realizations of traffic. These are key advantages of
the computational model presented.

Acknowledgements

This work was supported in part by the National Natural Science Foundation of China
under the project grant numbers 60573125 and 60672114, by the Key Laboratory of
Security and Secrecy of Information, Beijing Electronic Science and Technology
Institute under the project number KYKF 200606 of the open founding. SC Lim
would like to thank the Malaysia Ministry of Science, Technology and Innovation for
the IRPA Grant 09-99-01-0095 EA093, and Academy of Sciences of Malaysia for the
Scientific Advancement Fund Allocation (SAGA) P 96c.

References

1. Mandelbrot, B. B.: Gaussian Self-Affinity and Fractals. Springer (2001)
2. Beran, J.: Statistics for Long-Memory Processes. Chapman & Hall (1994)
3. Paxson, V., Floyd, S.: Why We Don't Know How to Simulate the Internet. Proc., Winter Simulation Conf. (1997) 1037-1044
4. Press, W. H., Teukolsky, S. A., Vetterling, W. T., Flannery, B. P.: Numerical Recipes in C: the Art of Scientific Computing. 2nd Edition, Cambridge University Press (1992)
5. Li, M.: Applied Mathematical Modelling 29 (2005) 55-63
6. Livny, M., Melamed, B., Tsiolis, A. K.: The Impact of Autocorrelation on Queuing Systems. Management Science 39 (1993) 322-339
7. Tsybakov, B., Georganas, N. D.: IEEE T. Information Theory 44 (1998) 1713-1725
8. Li, M., Zhao, W., Jia, W., Long, D. Y., Chi, C.-H.: Modeling Autocorrelation Functions of Self-Similar Teletraffic in Communication Networks based on Optimal Approximation in Hilbert Space. Applied Mathematical Modelling 27 (2003) 155-168
9. Paxson, V., Floyd, S.: IEEE/ACM T. Networking 3 (1995) 226-244
10. Paxson, V.: Fast Approximate Synthesis of Fractional Gaussian Noise for Generating Self-Similar Network Traffic. Computer Communication Review 27 (1997) 5-18
11. Jeong, H.-D. J., Lee, J.-S. R., McNickle, D., Pawlikowski, P.: Simulation. Modelling Practice and Theory 13 (2005) 233–256
12. Ledesma, S., Liu, D.: Computer Communication Review 30 (2000) 4-17
13. Garrett, M. W., Willinger, W.: Analysis, Modeling and Generation of Self-Similar VBR Traffic. Proc., ACM SigComm'94, London (1994) 269-280
14. Li, M., Chi, C.-H.: A Correlation-Based Computational Method for Simulating Long-Range Dependent Data. Journal of the Franklin Institute 340 (2003) 503-514
15. Mandelbrot, B. B.: The Fractal Geometry of Nature. W. H. Freeman (1982)
16. Gneiting, T., Schlather, M.: Stochastic Models That Separate Fractal Dimension and Hurst Effect. SIAM Review 46 (2004) 269-282
17. Li, M., Lim, S. C.: Modeling Network Traffic Using Cauchy Correlation Model with Long-Range Dependence. Modern Physics Letters B 19 (2005) 829-840
18. Lim, S. C., Li, M.: Generalized Cauchy Process and Its Application to Relaxation Phenomena. Journal of Physics A: Mathematical and General 39 (12) (2006) 2935-2951
19. Mandelbrot, B. B., van Ness, J. W.: Fractional Brownian Motions, Fractional Noises and Applications. SIAM Review 10 (1968) 422-437
20. Kaplan, L. M., Kuo C.-C. J.: IEEE T. Signal Processing 42 (1994) 3526-3530
21. Chiles, J-P., Delfiner, P.: Geostatistics, Modeling Spatial Uncertainty (Wiley) (1999)
22. Li, M.: Modeling Autocorrelation Functions of Long-Range Dependent Teletraffic Series based on Optimal Approximation in Hilbert Space-a Further Study. Applied Mathematical Modelling 31 (3) (2007) 625-631
23. Kent, J. T., Wood, A. T. A.: J. R. Statit. Soc. B 59 (1997) 579-599
24. Dubuc, B., Quiniou, J. F., Roques-Carmes, C., Tricot, C., Zucker, S. W.: Phys. Rev. A 39 (1989) 1500-1512
25. Hall P., Wood, A.: Biometrika 80 (1993) 246-252
26. Taylor, C. C., Taylor, S. J.: J. Roy. Statist. Soc. Ser. B 53 (1991) 353-364
27. Constantine, A. G., Hall, P.: J. Roy. Statist. Soc. Ser. B 56 (1994) 97-113
28. Taqqu, M. S., Teverovsky, V., Willinger, W.: Fractals 3 (1995) 785-798
29. Peng, C.-K., Buldyrev, S. V., Havlin, S., Simons, M., Stanley, H. E., Goldberger, A. L.: Mosaic Organization of DNA Nucleotides. Phys. Rev. E 49 (1994) 1685-1689
30. Kantelhardt, J. W., et al.: Phys. A 295 (2001) 441-454
31. Li, M., Lim, S. C.: A Rigorous Derivation of Power Spectrum of Fractional Gaussian Noise. Fluctuation and Noise Letters 6 (4) 2006, C33-36

Reliable and Scalable State Management Using Migration of State Information in Web Services[*]

Jongbae Moon, Hyungil Park, and Myungho Kim

#313, School of Computing, Soongsil University, Sangdo-5 Dong, Dongjak-Gu, Seoul,
156-743, Korea
{jbmoon, hgpark}@ss.ssu.ac.kr, kmh@ssu.ac.kr

Abstract. The WS-Resource framework (WSRF) was proposed as a reengineering and evolution of OGSI to be compatible with the current Web services conventions and specifications. Although WSRF separates state management from the stateless web service and provides a mechanism for state management, it still has some limitations. The tight-coupling between web services and their resource factories restricts the scalability. When the repository of stateful resources fails, stateful web services can not work. In this paper, we propose a new state management framework which is called State Management Web Service Framework (SMWSF) and implemented on Microsoft .NET framework. SMWSF provides reliability, flexibility, scalability, and security. SMWSF provides migration of state information and the service requestor can control the location of the repository of stateful resource. We also implemented a prototype system and conducted comparison experiments to evaluate performance.

Keywords: state management, web service, WSRF, migration.

1 Introduction

Web services are "software systems designed to support interoperable machine-to-machine interaction over a network" [1]. Though there has been some success of Web Service in the industry, Web service has been regarded as stateless and non-transient [2]. Recently, most of the web services, however, are stateful; state information is kept and used during the execution of applications.

To manage sate information within Web services framework, the Open Grid Services Infrastructure (OGSI) [3] and Web Service Resource Framework (WSRF) [4] were proposed. Both OGSI and WSRF are concerned with how to manipulate stateful resources. OGSI extends the power of Web services framework significantly by integrating support for transient, stateful service instances with existing Web services technologies. The Globus Toolkit 3 (GT3) [5] is an implementation of the OGSI specification and has become a de facto standard for Grid middleware. GT3 uses Grid service factory to create multiple instances of the Grid service, and the Grid service instances are stateful. However, because GT3 uses the same container for grid services, the

---

[*] This work was supported by the Soongsil University Research Fund.

Y. Shi et al. (Eds.): ICCS 2007, Part I, LNCS 4487, pp. 382–389, 2007.

service container, which is a Grid service factory, has to be restarted whenever a new service joins. This will affect all existing services in the same container.

WSRF was introduced as a refactoring and evolution of OGSI, and provides a generic, open framework for modeling and accessing stateful resources using Web services. WSRF uses different container to manage stateful resources and Web services; WSRF separates state management from the stateless web services. Therefore, there is no loss of state information and other service instances should continue to work although the execution of a service instance fails. WSRF, however, still has some limitations [2]. Each Web service accompanies a WS-Resource factory. The tight coupling between the web service and the resource factory leads to scalability problem. Moreover, WSRF does not provide the flexibility of choosing the location of the state repository. This may introduce security problems although the service itself is trusted by the requestor; the requestor and provider may have different security policies. Moreover, when the state repository fails, the stateful web service does not work.

In this paper, we propose a new state management framework, which is implemented on Microsoft .NET Web services, and implement a prototype system. The prototype system makes Web services and their state management loosely-coupled and Web services can use another state management service, which is in another service provider, to provide scalability. The state information can migrate to another repository to enhance reliability, security, and scalability while Web services are running. The migration of the state information also provides the flexibility of choosing the location of the state repository. To provide security for the state information, the state management stores the state information with an encryption key which is sent by the requestor. Moreover, whenever the state information migrates, it is transferred through a security channel.

The rest of this paper is organized as follows. Section 2 summarizes the existing researches regarding the state management in Web services. Section 3 proposes a system model. Section 4 describes how to implement a prototype system. Section 5 evaluates the performance of the proposed system by conducting comparison experiments, and Section 6 concludes the paper.

2   Related Works

Web services have a problem that it is difficult to maintain state because web services are built on top of the stateless HTTP protocol. While Web service implementations are typically stateless, their interfaces frequently provide a user with the ability to access and manipulate state. In [6], three different models to keep and manage state information are proposed. However, maintaining state information has restriction on scalability. Moreover, to provide security for the state, extra works are required.

OGSI enables access to stateful resources with the maintenance and retrieval of state information. OGSI introduces the resource factory to create multiple instances of the Grid service. Because OGSI uses the same container for Grid services, the service container has to be restarted whenever a new service joins.

WSRF provides standardized patterns and interfaces to manage state information. WSRF introduces the WS-Resource factory to create the instance of stateful resources. When a web service fails, WSRF can restore the state information by separating state management from a web service. WSRF, however, has some limitation to manage the state. The tight coupling between Web services and their factories leads to scalability problem.

In [2], a generic state management service, which separates state management from Web services, is introduced to overcome WSRF's limitations. Besides the reduction of work for service developers, scalability is enhanced by the flexible deployment of the state management service. However, once the state information is stored in a stateful resource, the requestor can not change the location of the repository. Therefore, although the Web service and state management service are trusted by the requestor, the repository may not guarantee security problems. Moreover, failure of the repository reduces the reliability and scalability.

3   Proposed State Management Model

In this section, we propose a new state management framework which overcomes the WSRF's limitation as mentioned above. The proposed framework is called the State Management Web Service Framework (SMWSF). SMWSF provides scalability, flexibility, security and reliability.

Fig. 1 shows the state management system model based on SMWSF. In this system model, state management, which is a Web service for creating and managing the instance of a stateful resource, is separated from web services. The service requestor can make web services use another state management service that exists in one of the other service providers. The state management service implements common interfaces to store the state information in some types of stateful resource. Therefore, service authors can easily develop web services regardless of the implementation of state management interfaces. Moreover, the state management service provides an interface that a stateful resource can migrate to another repository, as well as in another type. The state information is encrypted or decrypted before the state management service stores or restores it. Moreover, communications between the state management service and the stateful resource is established through a security channel.

A stateful resource is referred to as a WS-Resource in WSRF, and each WS-Resource is described by an XML document. In SMWSF, a stateful resource is implemented in some different types. The service requestor may want to change the location of the stateful resource because of security problems or failure of the repository. The state information can migrate to another repository chosen by the requestor, and the type of the stateful resource can be changed. In addition, migrating the stateful resource enhances reliability when the repository fails or does not work well. During the migration process, to guarantee security for the state information, communication is established through a security protocol, such as IPSec (IP Security).

Fig. 1 a) shows the process of using the Web service in SMWSF, and the details are described as follows. Fist of all, a service requestor sends a request including an encryption key to the service provider, choosing a type of the stateful resource and a location where the stateful resource is stored in. Then, the Web service generates an XML document including the state information, and sends a request with the generated XML document to the state management web service. After the state management service encrypts the state information with the requestor's encryption key, the encrypted state information is stored in a chosen place. Moreover as well as in a desired type. After that, the state management service returns the URI of the stateful resource to the Web service, and then the Web service returns the URI to the requestor. When the requestor requests the stateful resource to be moved, the Web service takes in the type and URI of the stateful resource from the requestor, and sends a migration request to the state management service. The state management service reads the corresponding resource, and then stores the state information in the specified location as well as specified type. In addition, Fig. 1 b) shows the details that the Web service changes the state management service with another one. The old state management service sends the URI of the stateful resource to the new state. After the new state management service gets in contact with the stateful resource, then the Web service communicate with the new state management service through a security channel.

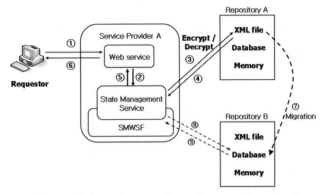

a) The state information can migrate from repository A to B.

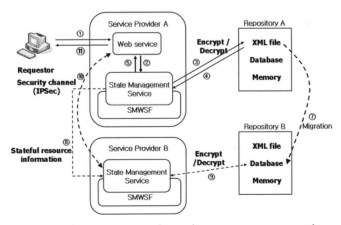

b) The service requestor can choose the state management service.

Fig. 1. Proposed State Management System Model

4 Implementation

We implemented a prototype system to test SMWSF. We use ASP.NET in C# to implement web service applications. Besides, we use Microsoft IIS (Internet Information Server) as a web application server. The prototype system is implemented on top of the Microsoft .NET framework. SMWSF implements basic WSRF specifications, such as WS-ResourceProperties [7] and WS-ResourceLifetime [8]. In the prototype system, a system is divided into three parts which are web services, a state management service including the proposed framework, and stateful resources. The system also provides libraries for implementing web services and lifetime management module for keeping track of the stateful resources created by the client requests.

In this system, every web service should be developed and implemented by importing libraries which is provided by SMWSF. Web services are ASP.NET web services. Service authors annotate their web service codes with metadata via .NET attributes. The port type attribute allows service authors to easily import the functionality that is defined by web service library into their service.

The web service libraries include state management interfaces, an XML generator, and a GUID generator. Web services use the state management interfaces, which are just defined but not implemented, to communicate with the state management service. The sate management web service is in charge of the implementation of the interfaces. Therefore, web services use another state management service which is provided by another service provider.

Web services generate service requestor's GUID (Global Unique ID) by using the GUID generator. The GUID, which is a 128bit integer number generated by using hash function, is used to grant access to a stateful resource. Web services generate an XML document to manipulate state information as a stateful resource by using the XML generator. In a web service, class-level data members are declared as part of the stateful resource via the [Resource] attribute to generate an XML document. The values of the class-level data members are saved into an XML document. The generated XML document includes a GUID and an encryption key in the <Header> element. The class-level data member's name and value that are described in the [Resource] attribute in the web service are set in the <Resource> element. This XML document is encrypted with the encryption key, and then stored in the repository.

The state management service manages web services' state information. This service implemented as a web service. Service authors can make web services communicate with another state management web service provided by another service provider. The state management web service implements interfaces for state management, such as store, restore, deletion, and migration of stateful resources. To do this, the port type attribute is also used to import functionality that is defined in SMWSF. In the case of the migration, the state management web service generates an XML document from the repository first by using one of the stateful resource repository management modules. The XML document is stored in a desired place, as well as in a desired type. When the service requestor wants web services to use another state management service, the XML document is sent to the selected state management service, and then stored in the repository.

A stateful resource must persist between service invocations until the end of its lifetime, meaning that the state of a stateful resource after one invocation should be the same as its state before the next. Persistence can be achieved by holding the resource in memory, writing it to disk, or storing it in a database. The memory model provides the best response-time performance but least fault-tolerant. The file system model provides slow performance than other models, but provides the ability to survive server failure at the cost of some performance. The database model is slower than memory model, but provides scalability, fault-tolerance, and access to powerful query/discover mechanisms that are not present in the file system model. Therefore, in the proposed system the state resource repository management implements the stateful resource in these three types.

5 Performance Evaluation

In this section, we conducted two experiments for comparison to evaluate performance of SMWSF; we compared the system based on SMWSF with other systems implementing WSRF specification: GT4 and WSRF.NET. First, we implemented a calculator web service, and estimated the response time for creating, deleting, storing, restoring, and migrating state information of a web service. We performed each operation 1000 times, and then measured the average of the response time. The calculator service is a web service providing computation functionalities. Second, we implemented an airline booking web service on each framework. Then, we compared the service time, which is measured by the time in seconds for the client to receive the full response, as the number of clients increase. The airline booking service is a web service that retrieves airline schedules and then books seats. To perform this experiment automatically, we made the airline booking service book the first retrieved flight and reserve a fixed seat on the plane.

We used four identically configured machines that have an Intel Pentium 4 3.0GHz CPU with 1GB Ram and 80GB E-IDE 7200 RPM HDD. Two machines for SMWSF and one for WSRF.NET ran Windows 2003 Server Standard Edition. One for GT4 ran Linux Fedora Core 4 (Linux kernel 2.6.11). In SMWSF, stateful resources were implemented in three types, which were XML file, database, and memory; MySQL was used as a database for this experiment. GT4 stores state information in system memory. WSRF.NET implements WS-Resources using SQL Server 2000.

Table 1 shows the average response time for basic operations of the calculator service. All the numbers are in milliseconds. The Remote SMWSF is the case when the calculator service uses a remote state management service provided by another service provider. In this case, the average response time was slower than SMWSF and other systems because an additional communication cost between a web services and its state management service is needed. As compared with GT4 implementing stateful resource in memory model, SMWSF was faster in every test because GT4 is implemented in Java. As compared with WSRF.NET implementing stateful resource in database model, SMWSF had similar performance although there were additional overheads during encryption and decryption. Therefore, we could see SMWSF has as good performance as WSRF.NET has.

Table 1. Average response time for basic operations

		Create	**Delete**	**Restore**	**Store**
GT4-Java		16.3 ms	23.6 ms	28.6 ms	24.9 ms
WSRF.NET		14.7 ms	21.4 ms	38.2 ms	24.4 ms
SMWSF	File System	15.3 ms	23.5 ms	32.8 ms	22.3 ms
	Memory	13.1 ms	20.1 ms	27.9 ms	19.5 ms
	Database	14.2 ms	21.8 ms	37.5 ms	24.0 ms
Remote SMWSF	File System	21.5 ms	34.4 ms	44.1 ms	35.2 ms
	Memory	19.4 ms	30.8 ms	37.4 ms	30.4 ms
	Database	20.8 ms	32.9 ms	47.4 ms	36.4 ms

Fig. 2 shows the service time of the airline booking web service as the number of clients increase from 50 to 400. In this experiment, the Remote SMWSF is considered only if memory model. As the number of clients increase, the service time of GT4 was the fastest, followed by SMWSF, Remote SMWSF, and WSRF.NET. As might have been expected, the systems using memory model was faster than the systems using database and file system model. Moreover, Remote SMWSF was slower than SMWSF and GT4 because of additional communication cost and encryption overhead. GT4 had stable performance even though the number of clients increased because GT4 implements HTTP connection caching which reuses HTTP connection that was previously created. In the first experiment, the response time of SMWSF and WSRF.NET were comparable. In this experiment, as compared with WSRF.NET, SMWSF was faster because of the overhead caused by web service extension; WSRF.NET uses Microsoft Web Services Enhancement to provide SOAP message exchange. In addition, there were more SOAP data than SMWSF.

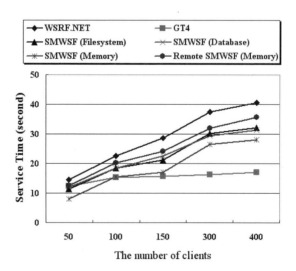

Fig. 2. Service time for an airline booking service according to the number of clients

6   Conclusions and Future Works

In this paper, we proposed a new state management framework which provides scalability, flexibility, security, and reliability. This framework is called State Management Web Service Framework (SMWSF). We also implemented a prototype system that is based on the Microsoft .NET framework. In the prototype system, the state management is separated from web services. The loosely-coupled between the web service and the state management provides scalability. The flexibility is provided by making service requestors choose another state management service among the other service providers. The state information can migrate to another repository, where the type of stateful resource can be changed from one type to another. The migration of the state information enhances reliability and security when the repository fails or does not work well.

Many issues still remain to be addressed. Because of the loosely-coupled between the web service and the state management, some communication overhead is occurred. Moreover, the communication between sate management service and the sateful resource is done through a security channel. We need to study for reducing this additional communication overhead. We need to implement many other components for the framework, especially WS-Notification specifications. In addition, more experiments must be conducted on fault-tolerance to evaluate performance the proposed system.

References

1. David, Booth, Hugo, Haas, Francis, McCabe, Eric, Newcomer, Michael, Champion, Chris, Ferris, David, Orchard: Web Services Architecture – W3C Working Draft 8 August 2003. http://www.w3.org/TR/2003/WD-ws-arch-20030808/
2. Y., Xie, Y.M., Teo: State Management Issues and Grid Services. International Conference on Grid and Cooperative Computing. LNCS, Vol. 3251 (2004) 17-25
3. S., Tuecke, K., Czajkowski, I., Foster, J., Frey, S., Graham, C., Kesselman, P., Vanderbilt: Open Grid Service Infrastructure (OGSI). (2002)
4. Czajkowski, K., Ferguson, D., Foster, I., Frey, J., Graham, S., Sedukhin, I., Snelling, D., Tuecke, S., Vambenepe, W.: The WS-Resource Framework. http://www.globus.org/wsrf/ (2004)
5. Globus Toolkit version 3. http://www.globus.org/
6. Xiang, Song, Namgeun, Jeong, Phillip, W., Hutto, Umakishore, Ramachandran, James, M., Rehg: State Management in Web Services. IEEE International Workshop on FTDCS'04 (2004)
7. Graham, S., Czajkwski, K., Ferguson, D., Foster, I., Frey, J., Leymann, F., Maguire, T., Nagaratnam, N., Nally, M., Storey, T., Sedukhin, I., Snelling, D., Tuecke, S., Vambenepe, W., Weerawarana, S.: WS-ResourceProperties. http://www-106.ibm.com/developerworks/library/ws-resource/ws-resourceproperties.pdf (2004)
8. Frey, J., Graham, S., Czajkowski, C., Ferguson, D., Foster, I., Leymann, F., Maguire, T., Nagaratnam, N., Nally, M., Storey, T., Sedukhin, I., Snelling, D., Tuecke, S., Vambenepe, W., Weerawarana, S.: WS-ResourceLifetime. http://www-106.ibm.com/developerworks/ library/ws-resource/ws-resourcelifetime.pdf (2004)

Efficient and Reliable Execution of Legacy Codes Exposed as Services

Bartosz Baliś[1], Marian Bubak[1,2], Kamil Sterna[1], and Adam Bemben[1]

[1] Institute of Computer Science, AGH, al. Mickiewicza 30, 30-059 Kraków, Poland
[2] Academic Computer Centre – CYFRONET, Nawojki 11, 30-950 Kraków, Poland
{bubak,balis}@agh.edu.pl
Phone: (+48 12) 617 39 64; Fax: (+48 12) 633 80 54

Abstract. In this paper, we propose a framework that enables fault tolerance and dynamic load balancing for legacy codes running as backends of services. The framework architecture is divided into two layers. The upper layer contains the service interfaces and additional management services, while the legacy backends run in the lower layer. The management layer can record the invocation history or save state of a legacy worker job that runs in the lower layer. Based on this, computing can be migrated to one of a pool of legacy worker jobs. Fault-tolerance in the upper layer is also handled by means of active replication. We argue that the combination of these two methods provides a comprehensive support for efficient and reliable execution of legacy codes. After presenting the architecture and basic scenarios for fault tolerance and load balancing, we conclude with performance evaluation of our framework.

Keywords: Legacy code, fault tolerance, load balancing, migration.

1 Introduction

Recently developed systems for conducting e-Science experiments evolve into Service-Oriented Architectures (SOA) that support a model of computation based on composition of services into workflows [4]. Legacy codes nevertheless, instead of being rewritten or reengineered, are often adapted to new architectures through exposing them as services or components, and remain the computational core of the application. Static and dynamic load balancing (LB) as well as fault tolerance (FT) are highly desired features of a modern execution environment, necessary to sustain the quality of service, high reliability and efficiency. The workflow model, in which the application logic is separated from the application itself, is for this very fact well-suited for LB and FT support, because the execution progress of a workflow is handled by generic services, the enactment engines. However, the presence of legacy jobs running in the backends of services complicates this support, as the legacy jobs are often not directly modeled in the workflow and thus not handled by the same generic enactment engines. While the subjects of FT and LB of parallel and distributed systems [2] are very well recognized for web services [10] [6] or components [8], similar problems for legacy codes running in service backends are still not well addressed.

Y. Shi et al. (Eds.): ICCS 2007, Part I, LNCS 4487, pp. 390–397, 2007.

This paper presents a solution to support FT and LB for legacy codes exposed as services. We propose a generic framework which enables seamless and transparent checkpointing, migration, and dynamic load balancing of legacy jobs running as backends of services. The proposed framework is based on our previous work, a system for virtualization of legacy codes as grid services, LGF (Legacy to Grid Framework) [1] [11]. Unlike our previous work which focused solely on exposing of legacy codes as services, this paper focuses on the aspects of reliable and efficient execution of legacy systems. We propose an architecture for the framework and justify our design choices. We present a prototype implementation of the framework and perform a feasibility study in order to verify whether our solution fulfills functional (FT and LB scenarios) and non-functional (performance overhead evaluation) requirements.

The remainder of this paper is as follows. Section 2 presents related work. Sections 3 and 4 describe the architecture and operation of our FT-LB framework, respectively. We conclude the paper in Section 5 which studies the impact of FT and LB mechanisms on the overall application performance.

2 State of the Art

Most existing approaches to adapting legacy codes to modern environments focus on command-line applications and follow a simple adapter pattern in which a web service or a component (e.g. CORBA [5]) wrapper is automatically generated for a legacy command line application, based on a specification of its input parameters [9]. Those approaches differ in terms of addressing other aspects of legacy system's adaptation such as security [7], automatic deployment or integration with a framework for workflow composition. In some cases even some brokering mechanisms are taken into account [3].

Of the available tools, relatively most comprehensive solution is presented by a tool CAWOM [12], wherein one can actually specify the format of legacy system's responses (using a formal language) which allows for more complex interactions with a legacy system, including synchronous and asynchronous calls.

Overall the mentioned tools, whether they offer simple wrapping, or more advanced frameworks with brokering, automatic deployment and workflow composition capabilities, neither take into account nor are designed to support fault tolerance and dynamic load balancing of legacy systems.

Our framework is designed to fill this gap. The separation into two layers, and operation of legacy jobs in a client instead of server fashion, solves many issues and enables flexible solutions of LB and FT problems. We describe those in the following sections of this paper.

3 LB-FT Framework Architecture

The architecture of our framework, presented in Fig. 1, is comprised of three main components, namely: Service Client, Service Manager which exposes interfaces, and Backend System on which the legacy code is actually executed.

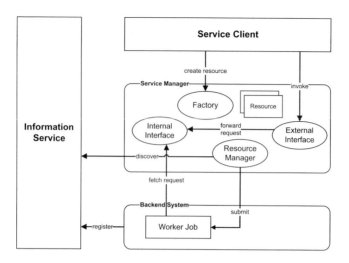

Fig. 1. Architecture of FT & LB Framework for Legacy Code

The heart of the system is a Service Manager which exposes the legacy code as a service (*External Interface*) to be invoked by a *Service Client*. The central concept that enables transparent migration of computation, being a prerequisite for FT and LB, is *decoupling* of service interface layer from actual legacy code, the latter being deployed as a job (*Worker Job*), on a (usually remote) machine (*Backend System*). An important design detail is that the legacy Worker Job acts as a client to the Service Manager and fetches Service Client requests from Internal Interface in the Service Manager. An alternative option would be to use notifications. However, in such a case Worker Jobs would have to act as servers which would make their deployment and migration much more difficult. The remaining component in the Service Manager is a *Resource Manager* whose main responsibility is to submit Worker Jobs to Backend Systems. In addition, we support stateful conversation with legacy software through WSRF-like services. To this end, a client can create a *Resource* using a Factory service. In this way, the framework enables stateful, session-based interaction with legacy services.

In the architecture, an external *Information System* is presented to which Backend Systems register while the Resource Manager acts as a resource broker deciding which Backend System to submit a new legacy Worker Job to, based on a list of available Backend Systems and corresponding monitoring information (such as current load). Alternatively, the Information Service could be replaced by an external Resource Broker to which all resource brokerage decisions would be delegated by the Resource Manager.

Thanks to decoupling of service interfaces and legacy back ends, multiple legacy worker jobs can be connected to a single service as a resource pool. The computation can be easily migrated from one back end to another in case of performance drop (load balancing) or failure (fault tolerance). The framework supports both *history-based* and *checkpoint-based* migration of stateful jobs. For the former case, the Service Manager records the invocation history of a given client and in case of migration to another worker job, the invocations can be repeated. The latter case is supported by allowing a legacy job to periodically send its state snapshot to the Service Manager; in the event of migration, the new worker job can restore the last state. The architecture enables both *low-level* and *high-level* checkpointing to create state snapshots, though the current implementation supports only the high-level one in which the worker jobs have to provide an implementation of code to create and restore snapshots.

In our framework, the service interface layer is thin and performs no processing, merely forwarding requests plus other management functions related to migration, state restoration, etc. However, this layer is also subject to failure, e.g. due to software aging of the underlying application containers. Consequently, we also take into account the fault-tolerance of this layer using the technique of *active replication*. Multiple Service Managers can be assigned to a single interaction between a client and a legacy back end. The client submits all requests to all Service Managers. Similarly, the backend worker job fetches requests from all Service Managers. In consequence, requests are received by the worker job multiple times, however, they are executed only once.

4   Fault Tolerance and Load Balancing Scenarios

Fault tolerance and load balancing scenarios for the backend side differ only in the way the migration is initiated. Both scenarios are shown in Fig. 2 (a) and (b) respectively.

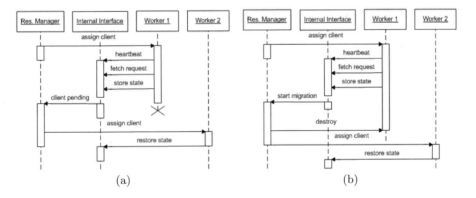

(a)   (b)

Fig. 2. Scenarios involving migration: (a) fault tolerance, (b) load balancing

The resource manager fetches a client request (not shown) and assigns a proper worker (*assign client*). The worker periodically signals the availability (*heartbeat*), retrieves the client request (*fetch request*) and stores checkpoints (*store state*). A migration is triggered either by a worker crash (fault tolerance) or a persistant node overload (load balancing). In the latter case, the worker is explicitly destroyed (*destroy*). In either case, the Service Manager assigns another worker (*assign client*), which restores the latest state (*restore state*) and the operation is carried on as it was.

For fault tolerance and software rejuvenation purposes, we employ the active replication mechanism at the upper layer. It is based on multiplying every operation on two (or more) service managers. An appropriate scheme is depicted in Fig. 3. The client calls two service managers at once (1,2). Similarly, the backend worker job fetches requests from both service managers (3, 5), however, only one request is actually executed (4). The result is returned to both service managers (6,7) and they forward it back to the client (8, 9). When one of service manager crashes (10), the operation is carried on with one service manager only (11-13).

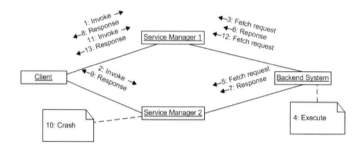

Fig. 3. Sequence of system operations for active replication model

5   Performance Evaluation

This section presents an evaluation of the impact of the framework on the overall application's performance. The framework prototype was developed in Java and it generates WSRF services based on Globus Toolkit 4. The following tests were conducted: (1) the impact of the interposition management layer on latency and throughput of a service as compared to direct invocation, (2) the latency of migration in a fault tolerance and load balancing scenario, and (3) the cost of active replication in terms of application's completion time.

For the evaluation, we used a simple algorithm computing a requested number in the Fibonacci sequence, exposed as a service. In total, four IA-32 machines with 1 GB RAM and 1.2 GHz CPU, running Linux Debian, Java 1.4 and Globus 3.2.1 were used.

Fig. 4 (a) shows the *overhead of the interposition layer*. We compared the performance of two functionally equivalent web services. One of them used a legacy library while the other was standalone. Both web services exposed a single method that calculated the length of a byte array passed as a parameter. Latency and bandwidth of both services were obtained based on the formula:

$$time(length) = length/bandwith + latency \tag{1}$$

Using least squares linear regression fitting, we have obtained figures for both services. As a result, we observed that while latency was increased 2.4 times, bandwidth was reduced only by 12%.

In the *fault tolerance scenario*, the Service Manager loses connection with one of the workers (at the moment of starting the service on a backend system). After the timeout, the backend system is considered to have undergone abnormal termination. Process migration is initiated, and the method invocations are delegated to another node. The result is shown in Fig. 4 (b).

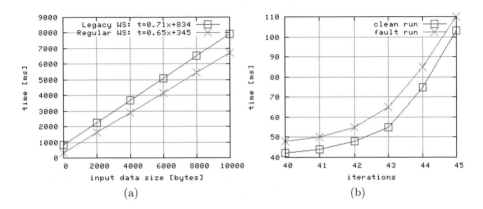

(a)   (b)

Fig. 4. (a) Interposition layer overhead (b) Migration overhead

We used history-based fault tolerance whose cost can be estimated at approximately 1-2 seconds. An additional cost is also connected with the number of lost heartbeats before a node is considered to be undergoing a failure.

The *load balancing scenario* is different only in terms of the cause which triggers the migration, which in this case is a node overload over a certain number of heartbeats. The Service Manager decides to migrate the resources to a node exposing better performance. The overhead proved to be quite similar and is not shown here separately.

Finally, we study the *active replication scenario*. The framework runs with two service managers present, running in separate containers and performing identical actions. One Service Manager crashes. The platform continues normal work; however the invocations are handled by one service manager only.

The overhead in this scenario is caused by additional processing: on the client side which performs more invocations and has to filter out redundant results, and the backend side where redundant requests have to be discarded and the results has to be returned to more than one Service Manager. Fig. 5 shows the processing time from a worker perspective, for a single, and for two service managers. The times for those two cases are practically the same which proves that the additional processing time – indeed, only limited to more invocations and discarding redundant operations – does not induce substantial overhead.

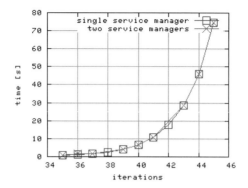

Fig. 5. Active replication overhead

At the same time we observe undisturbed operation of the system despite the crash of one Service Manager.

6   Conclusion

We have presented a framework for enabling fault tolerance and dynamic load balancing for legacy codes running in the backend of web services, typically as parts of e-Science workflows. We proposed a two layer architecture in which a management layer containing service interfaces is decoupled from a computational core which is deployed in separate worker jobs.

We use different, complementing strategies for FT & LB in the two system layers. In the backend layer, we use an efficient method based on a pool of worker jobs of a certain type among which the computing can be switched when there is a need to do so. Recovering state from snapshots or repeating of invocation history can be used in the case of migration of stateful services. In the front-end layer, we use the active replication of service interfaces. Though this method is expensive, it does not require further management such as state snapshots or heartbeat monitoring. Our investigation revealed that thanks to the very architecture of our framework in which the service layer is thin and contains no processing, the inherent overhead of active replication is compensated and is perfectly affordable.

Overall, the performed feasibility study shows the framework fulfills the functional and performance requirements and constitutes a comprehensive solution for reliable and efficient running of workflows that contain legacy code in the computational back ends, which is important for development of e-Science applications.

The future work encompasses, most importantly, the expansion of our system into data-centric workflows involving streaming between legacy jobs, and support for load balancing and fault tolerance scenarios for complex legacy systems, such as parallel MPI applications. Other aspects include integration with a security infrastructure.

Acknowledgements. This work is partly supported by EU-IST Project CoreGrid IST-2002-004265, and EU-IST Project Virolab IST-027446.

References

1. Balis, B., Bubak, M., Wegiel, M.: A Solution for Adapting Legacy Code as Web Services. In Proc. Workshop on Component Models and Systems for Grid Applications. 18th Annual ACM International Conference on Supercomputing, Saint-Malo, France, Kluwer (July 2004)
2. Cao, J., Spooner, D. P., Jarvis, S. A., Nudd, G. R.: Grid Load Balancing Using Intelligent Agents. Future Generation Computer Systems special issue on Intelligent Grid Environments: Principles and Applications, **21(1)** (2005) 135-149
3. Delaittre, T., Kiss, T., Goyeneche, A., Terstyanszky, G., Winter, S., Kacsuk, P.: GEMLCA: Running Legacy Code Applications as Grid Services. Journal of Grid Computing Vol. 3. No. 1-2. Springer Science + Business Media B.V. (June 2005) 75-90
4. E-Science 2006, Homepage: http://www.escience-meeting.org/eScience2006/
5. Gannod, G. C., Mudiam, S. V., Lindquist, T. E.: An Architecture Based Approach for Synthesizing and Integrating Adapters for Legacy Software. Proc. 7th Working Conference on Reverse Engineering. IEEE (November 2000) 128-137
6. Hwang, S., Kesselman, C.: A Flexible Framework for Fault Tolerance in the Grid. Journal of Grid Computing **1(3)**(2003) 251-272
7. Kandaswamy, G., Fang, L., Huang, Y., Shirasuna, S., Marru, S., Gannon, D.: Building Web Services for Scientific Grid Applications. IBM Journal of Research and Development, Vol. 50. No. 2/3 (March/May 2006) 249-260, .
8. Moser, L., Melliar-Smith, P., Narasimhan, P.: A Fault Tolerance Framework for CORBA. International Symposium on Fault Tolerant Computing (Madison, WI) (June 1999) 150-157
9. Pierce, M., Fox, G.: Making Scientific Applications as Web Services. Web Computing (January/Februray 2004)
10. Tartanoglu, F., Issarny, V., Romanovsky, A., Levy, N.: Dependability in the Web Services Architecture. Architecting Dependable Systems. LNCS 2677 (June 2003)
11. Wegiel, M., Bubak, M., Balis, B.: Fine-Grain Interoperability with Legacy Libraries Virtualized as Web Services. Proc. Grid-Enabling Legacy Applications and Supporting End Users Workshop within the framework 15th IEEE HPDC 15, Paris, France (June 2006)
12. Wohlstadter, E., Jackson, S., Devanbu, P.: Generating Wrappers for Command Line Programs: The cal-aggie wrap-o-matic Project. Proc. 23rd International Conference on Software Engineering (ICSE 2001). ACM (2001) 243-252

Provenance Provisioning in Mobile Agent-Based Distributed Job Workflow Execution

Yuhong Feng and Wentong Cai

School of Computer Engineering, Nanyang Technological University
Singapore 639798
{YHFeng, ASWTCai}@ntu.edu.sg

Abstract. Job workflow systems automate the execution of scientific applications, however they may hide how the results are achieved (i.e., the provenance information of the job workflow execution). This paper describes the development and evaluation of a decentralized recording and collection scheme for job workflow provenance in mobile agent-based distributed job workflow execution. A performance study was conducted to evaluate our approach against the one using a centralized provenance server. The results are discussed in the paper.

Keywords: Distributed Job Workflow Execution, Provenance Recording and Collection, Grid Computing.

1   Introduction

The provenance of some data is defined as the documentation of the process that led to the data [1]. The necessity of provenance for job workflow execution is apparent since provenance provides a traceable path on how a job workflow was executed and how the resulted data were derived. It is particularly important in Service Oriented Architecture (SOA) since shared services and data sets might be used in the course of the job workflow execution. Provenance information can be processed and used for various purposes, for example, for validation of e-Science experiments [2], credibility analysis of the results of workflow execution [3], fault-tolerance for service-based applications [4], and data sets regeneration for data intensive scientific applications [5].

The provenance information can be generated from the static information available in the original workflow specification (e.g., data dependencies) together with the runtime details obtained by tracing the execution of the workflow execution. The trace can be automatically generated by developing either a special "wrapping service" of the engine [6] or an "engine plugin" [7] to capture and record provenance related data directly from the workflow engine. The workflow trace can also be collected collectively by the services that execute the subjobs [8] or the services together with the workflow engine [9]. But, this puts the responsibility of provenance data recording to the service providers and may also require service modification.

Y. Shi et al. (Eds.): ICCS 2007, Part I, LNCS 4487, pp. 398–405, 2007.

No matter how the traces are collected, in general some special provenance services are used in the current systems to store the provenance data and to provide an interface for users to query the data. Thus, a protocol is needed for various service providers and the workflow engine to communicate with the provenance services during the provenance collection process [1]. A taxonomy of data provenance techniques can be found in [10], and a comprehensive documentation on provenance architecture can be found in [11].

Data intensive scientific applications often involve high volume, and distributed data sets. They can be generally expressed as a workflow of a number of analysis modules, each of which acts on specific sets of data and performs cross multidisplinary computations. To reduce the communication overhead caused by data movement and to provide decentralized control of execution during workflow enactment, the Mobile Code Collaboration Framework (MCCF) is developed to map the execution of subjobs to the distributed resources and to coordinate the subjobs' execution on runtime according to the abstract workflow provided by users [12]. LMA [13] is used in the MCCF for the purpose of separating functional description and executable code. The functional description of LMA is described using *Agent Core* (AC). An AC is essentially an XML file, containing the job workflow specification and other necessary information for agent creation and execution. An AC can be migrated from one resource to another. As for the executable code, to separate subjob specific code and common non-functional code (i.e., code for handling resource selection, subjob execution, agent communication, and AC migration), Code-on-Demand (CoD) [14] is used in the MCCF, that is, subjob specific code is downloaded to the computational resource and executed on demand. This enables an analysis module in the data intensive scientific applications to be executed at a computational resource close to where the required data set is. The execution of common non-functional code is carried out by a group of underlying *AC agents* (or agents in short).

The MCCF, which does not have a centralized engine, is different from the existing scientific workflow engines (e.g., Condor's DAGMan[1], and SCIRun[2]). Hence, the objective of this paper is to develop a provenance recording and collection algorithm so that mobile agents deployed in the execution of a job workflow can collectively collect a complete set of information about the job workflow execution.

2   Partner Identification in the MCCF

Job workflows in MCCF are modeled using *Directed Acyclic Graph* (DAG). A DAG can be denoted as $\mathcal{G} = (\mathcal{J}, \mathcal{E})$, where $\mathcal{J}$ is the set of vertices representing subjobs, i.e., $\mathcal{J}=\{J_0, J_1, ..., J_{n-1}\}$. $\mathcal{E}$ is the set of directed edges between subjobs. There is a directed edge from subjob J_i to J_j if J_j requires J_i's execution results as input. *Data dependency* (denoted as "$<$") between two subjobs exists if there

---

[1] http://www.cs.wisc.edu/condor/dagman/
[2] http://software.sci.utah.edu/scirun.html

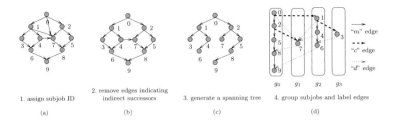

1. assign subjob ID

2. remove edges indicating indirect successors

3. generate a spanning tree

4. group subjobs and label edges

(a)   (b)   (c)   (d)

Fig. 1. Steps of the Preprocessing Algorithm

is a *path* between the subjobs. It is also assumed that a DAG representing a job workflow always has a unique staring node, J_0, and a unique end node, J_{n-1}.

When multiple data independent subjobs can be executed concurrently, replicas of an existing AC will be generated so that there is one AC for each subjob. MCCF uses a preprocessing algorithm to generate information for determining when and how to create AC replicas and generate information for the decision of which AC replica's agents will be responsible for the resource selection for a common successor [15]. The preprocessing algorithm performs the following steps: (i) **Assign subjob ids**: Topological order is used to assign subjob ids from 0 to $n-1$, as shown in Figure 1(a). (ii) **Remove indirect successors**. Let $\mathcal{S}(J_i)$ denote the successor set of subjob J_i. $\forall J_{j_1}, J_{j_2} \in \mathcal{S}(J_i)$, J_{j_2} is called J_i's *indirect successor* if $J_{j_1} < J_{j_2}$, otherwise, it is called J_i's *immediate successor*. If J_j is the indirect successor of J_i, the corresponding edge (J_i, J_j) will be removed from the DAG, as shown in Figure 1(b). (iii) **Generate spanning tree**: A spanning tree is generated based on a Depth-First Search (DFS)[3] on the DAG, as shown in Figure 1(c). (iv) **Group subjobs**: A pre-order traversal is applied to the spanning tree to generate a visit sequence of nodes, and the leaves in this sequence will partition it into a number of groups. For example, the visit sequence generated for the spanning tree shown in Figure 1(c), $< 0, 2, 5, 8, 9, 7, 1, 4, 6, 3 >$, is partitioned into groups: $\{0, 2, 5, 8, 9\}$, $\{7\}$, $\{1, 4, 6\}$ and $\{3\}$, as illustrated in Figure 1(d). (v) **Label edges**: The edges between two subjobs, J_i and J_j, in the same group are labeled "m" (denoted as $J_i \xrightarrow{m} J_j$), the edges between subjobs of different groups are labeled "c" (denoted as $J_i \xrightarrow{c} J_j$), and the edges in the original DAG but not in the spanning tree are labeled "d" (denoted as $J_i \xrightarrow{d} J_j$), as illustrated in Figure 1(d).

The preprocessed information is included in the AC and used during the dynamic job workflow execution. Suppose the current subjob is J_i, for one of its outgoing edges, (J_i, J_j): (i) If $J_i \xrightarrow{m} J_j$, the AC agents will select the resources for executing J_j. The AC replica will be migrated for J_j's execution after J_i completes its execution. (ii) If $J_i \xrightarrow{c} J_j$, similar to the last case, the AC agents will select the resources for executing J_j, but a new AC replica will be created

---

[3] We assume that the depth-first search and pre-order traversal algorithms visit child nodes in the order from right to left.

for J_j's execution. (iii) If $J_i \overset{d}{\to} J_j$, the AC agents need to communicate so that J_j can locate the location of J_i's execution result.

The MCCF uses a contact list based agent communication mechanism [16] for subjob result notification. Two agents communicating with each other are called *communication partners* (or partners in short). Each AC maintains a list of partners with their locations (that is, a *contact list*). Before an AC is migrated or discarded, its agents will notify its partners so that they can update their contact lists accordingly. The location of the subjob's execution result will also be notified to the partners. The partner identification is carried out dynamically during the job workflow execution. Assume $\mathcal{E}_d$ is a set of edges that are marked with "*d*" in $\mathcal{G} = (\mathcal{J}, \mathcal{E})$. Also assume that in the spanning tree generated by the preprocessing algorithm described above, a sub-tree rooted at J_i is denoted as $\mathcal{T}_{J_i}$. Suppose J_i and J_j are two subjobs that are currently under execution, if $\exists (J_{i'}, J_{j'}) \in \mathcal{E}_d$, and $J_{i'} \in \mathcal{T}_{J_i} \wedge J_{j'} \in \mathcal{T}_{J_j}$, then the AC agents executing the subjobs J_i and J_j are partners[4].

3   Provenance Recording and Collection Protocol

Let $G(J_i)$ be the group id of subjob J_i. It is easy to prove that the grouping algorithm described in the last section has the following property:

> In $\mathcal{G}$, $\forall J_i \in \mathcal{J}$ and $G(J_i) > 0$, there exists a path p from J_i to J_{n-1}, where for any two consecutive subjobs, e.g., J_{k_q} and $J_{k_{q+1}}$, on the path p, we have $G(J_{k_q}) \geq G(J_{k_{q+1}})$.

p is called a *propagation path*. It is obvious that the AC that is finally returned to the user (that is, the original AC created by the user) will contain a complete provenance information for the job workflow if partners with higher group id propagate provenance information to the partners with the lower group ids during the job workflow execution. This forms the basis for the development of the provenance recording and collection protocol.

Let $\mathcal{R}(J_i)$ denote a subset of J_i's successors, where for any subjob $J_j \in \mathcal{R}(J_i)$, we have either $(J_i \overset{m}{\to} J_j)$ or $(J_i \overset{c}{\to} J_j)$. Assuming that subjob J_i is under execution, the main steps of the protocol are:

- On receiving a communication message from its partner, J_i updates its AC to include the provenance information and updates its contact list accordingly.
- On completion of J_i's execution, J_i's corresponding AC agents will record the location of J_i's execution result into J_i's AC.
- If $\mathcal{R}(J_i) \neq \emptyset$, as stated in Section 2, for each subjob J_j, $J_j \in \mathcal{R}(J_i)$, AC agents corresponding to J_i will locate resources, that is, the computational resource, the input data sets from the distributed data repository, and the code from the code repository, for the execution of J_j. These information

---

[4] More detail about the preprocessing based dynamic partner identification algorithm can be found in [15].

will be recorded in J_i's AC. Then, if $|\mathcal{R}(J_i)| > 1$, $|\mathcal{R}(J_i)| - 1$ replicas of J_i's AC will be created, one for each subjob J_j, $J_j \in \mathcal{R}(J_i)$ and $(J_i \xrightarrow{c} J_j)$.

- Before J_i's corresponding AC is migrated (or discarded if $\mathcal{R}(J_i) = \emptyset$), J_i's AC agents will send a communication message to all its partners in the contact list for execution coordination and contact list updating. The message contains the location of J_i's execution result, and the scheduling information for each J_j, $J_j \in \mathcal{R}(J_i)$. The scheduling information for subjob J_j includes: subjob id, the id of the AC replica to be used to execute the subjob, locations of the selected computational resource, input data sets, and code for the execution of the subjob. In addition, if the partner has a smaller group id, provenance information received by J_i from its partners with a larger group id (recorded in J_i's AC replica) during J_i's execution is also piggyback on the message.

Using the above protocol, the provenance information will be recorded in the AC replicas and propagated along propagation paths during the distributed execution of job workflow. Eventually, the AC that is finally returned to the user will contain the complete provenance information about the job workflow execution.

4   Performance Evaluation

As explained in the last section, the provenance information is transmitted along with the messages for execution coordination and contact list updating. Although there is no additional message required, the size of message will be increased. There is no centralized server used during the provenance information recording and collection.

Execution provenance information can also be collected using a centralized provenance server which maintains a provenance repository. For each subjob executed, the AC agents need to notify the centralized server about the provenance information. After a job workflow completes its execution, users can then get the provenance information from the server. In this centralized approach, additional messages are required for AC agents to communicate with the provenance server. Assuming that there is no need to collect provenance information for the start and end nodes (since they are assumed to have zero computation cost), the traffic generated in the centralized model can be estimated by:

$$(n - 2) * \overline{msg} \tag{1}$$

where $\overline{msg}$ denotes the average size of a provenance message carrying a single provenance record.

To evaluate the performance of our distributed provenance collection algorithm, randomly generated Task Graphs (TG), that is, job workflows, were executed in the prototype MCCF system [12] on a cluster of computers. Six pseudo-random TGs were generated using TGFF[5]. As stated in Section 2, $\forall J_i \in$

---

[5] http://ziyang.ece.northwestern.edu/tgff/

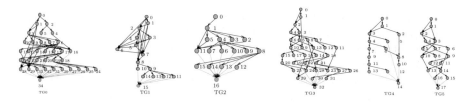

Fig. 2. Random Task Graphs

$\mathcal{J}, 0 < i < (n-1), J_0 < J_i < J_{n-1}$. Thus, when a TG has multiple subjobs that have no offspring, a hypothetical subjob is added, with no computation cost, to serve as the common immediate successor of such subjobs. The generated TGs are illustrated in Figure 2, where the dotted filled circles represent the added hypothetical subjobs, and the dotted edges represent the added edges.

The number of messages for execution coordination and contact list updating is illustrated in Table 1. These communication messages are required no matter whether decentralized or centralized method is used. During a job workflow execution, the messages that carry the propagated provenance records and the number of provenance records carried by such message were tracked. For a fair comparison with the centralized method, assuming that each propagated provenance record is carried by a separate provenance message, the total number of such messages in the decentralized method would be the summation of the number of propagated provenance records contained in all the tracked messages[6]. Each job workflow was executed 3 times, and the average numbers are shown in Table 1. Note that for the centralized method, formula (1) is used to calculate the number of messages generated for the provenance collection.

Table 1. Experiment Results

Task Graph	Msgs for Execution Coordination	Msgs for Provenance		% Improvement
		Decentralized	Centralized	
TG0	38	8	33	76%
TG1	5	0	14	100%
TG2	14	0	15	100%
TG3	44	11	31	65%
TG4	12	0	13	100%
TG5	21	7	16	56%

Table 1 also shows the percentage improvement of the decentralized method over the centralized one. From these results, we observe that the centralized model always generates higher number of messages for provenance information recording. Particularly, by recording the provenance information in the workflow

---

[6] Note that in general the bandwidth consumed by two or more messages sent separately is larger than that of sending them in a single bundle [17].

specification in the AC, provenance record does not always needs to be propagated during a job workflows' execution. For example, no additional provenance record is propagated during the execution of task graphs TG1, TG2, and TG4 (and thus there is no provenance message required). In the execution of these task graphs, the agents of the AC replicas created during the runtime are the partners of the agents of the original AC. Therefore, propagation of provenance information is not required. In this case, the percentage improvement of the decentralized model over the centralized model is 100%.

5   Conclusion and Future Work

Mobile agent-based distributed job workflow execution hides scientists from the Grid details, but also hides how the result is achieved (that is, the provenance of the job workflow execution). Since data processing in scientific computing may require some level of validation and verification, the information on the services and data sets used during the workflow execution is required. The provenance in many existing scientific workflow engines relies on a centralized provenance collection server for provenance recording and collection. However, in mobile agent-based distributed job workflow execution, there is no centralized workflow engine and thus naturally the provenance recording and collection should also be carried out in a distributed manner.

By studying the agent communication in the MCCF and the properties of the preprocessing algorithm for partner identification, a distributed provenance recording and collection protocol has been developed. The subjob provenance information is transmitted along the provenance propagation paths. Since provenance information is piggyback with the messages for execution coordination and contact list updating, there is no additional message required. To evaluate our approach, experimental study has been carried out on randomly generated job workflows. The results show that our approach has less communication overhead than the one using a centralized provenance server for provenance information recording and collection.

In the current algorithm, a subjob's provenance information might be propagated along multiple propagation paths. As a future work, the shortest and unique propagation path for a given subjob will be identified. This will further reduce the communication cost caused by propagation of provenance information. Although currently the execution coordination in the MCCF uses the contact list based mechanism, the provenance recording and collection protocol proposed in this paper should also work for other message-passing based execution coordination mechanisms (e.g., mailbox based mechanism [18]).

References

1. Groth, P., Luck, M., Moreau, L.: A protocol for recording provenance in service-oriented grids. In: 8th Intl Conf on Principle of Distributed Systems (PODIS2004), Grenoble, France (December 2004) 124–139

2. Wong, S.C., Miles, S., Fang, W.J., Groth, P., Moreau, L.: Provenance-based validation of e-science experiments. In: 4th Intl Semantic Web Conference. Volume 3729., Galway, Ireland (November 2005) 801–815

3. Rajbhandari, S., Wootten, I., Ali, A.S., Rana, O.F.: Evaluating provenance-based trust for scientific workflows. In: 6th IEEE Intl Symp on Cluster Computing and the Grid (CCGrid2006), Singapore (May 2006) 365–372

4. Townend, P., Groth, P., Xu, J.: A provenance-aware weighted fault tolerance scheme for service-based applications. In: 8th IEEE Intl Symp on Object-oriented Real-time Distributed Computing, USA (May 2005) 258–266

5. Foster, I., Voeckler, J., Wilde, M., Zhao, Y.: Chimera: A virtual data system for representing, querying, and automating data derivation. In: 14th Intl Conf on Scientific and Statistical Database Management. (July 2002) 37–46

6. Rajbhandari, S., Walker, D.W.: Support for provenance in a service-based computing Grid. In: UK e-Science All Hands Meeting 2004, UK (September 2004)

7. Zhao, J., Goble, C., Greenwood, M., Wroe, C., Stevens, R.: Annotating, linking and browsing provenance logs for e-Science. In: Wksp on Semantic Web Technologies for Searching and Retrieving Scientific Data (in conjunction with ISWC2003, CEUR Workshop Proceedings). Volume 83., Florida, USA (October 2003)

8. Bose, R., Frew, J.: Composing lineage metadata with XML for custom satellite-derived data products. In: 16th Intl Conf on Scientific and Statistical Database Management, Washington, DC, USA (2004) 275–284

9. Simmhan, Y.L., Plale, B., Gannon, D.: A framework for collecting provenance in data-centric scientific workflows. In: IEEE Intl Conf on Web Services 2006 (ICWS 2006), Chicago, USA (September 2006)

10. Simmhan, Y.L., Plale, B., Gannon, D.: A survey of data provenance in e-Science. SIGMOD Record **34**(3) (September 2005) 31–36

11. Groth, P., Jiang, S., Miles, S., Munrow, S., Tan, V., Tsasakou, S., Moreau, L.: An architecture for provenance systems. Technical report, Electronics and Computer Science, University of Southampton (October 2006)

12. Feng, Y.H., Cai, W.T.: MCCF: A distributed Grid job workflow execution framework. In: 2nd Intl Symposium on Parallel and Distributed Processing and Applications. Volume 3358., Hong Kong, China, LNCS (December 2004) 274–279

13. Brandt, R., Reiser, H.: Dynamic adaptation of mobile agents in heterogeneous environments. In: 5th Intl Conf on Mobile Agents (MA2001). Volume 2240., Atlanta, USA, LNCS (December 2001) 70–87

14. Fuggetta, A., Picco, G.P., Vigna, G.: Understanding code mobility. IEEE Trans on Software Engineering **24**(5) (1998) 342–361

15. Feng, Y.H., Cai, W.T., Cao, J.N.: Communication partner identification in distributed job workflow execution over the Grid. In: 3rd Intl Wksp on Mobile Distributed Computing (in conjunction with ICDCS05), Columbus, Ohio, USA (June 2005) 587–593

16. Cabri, G., Leonardi, L., Zambonelli, F.: Coordination infrastructures for mobile agents. Microprocessors and Microsystems **25**(2) (April 2001) 85–92

17. Berger, M.: Multipath packet switch using packet bundling. In: High Performance Switching and Routing (Workshop on Merging Optical and IP Technologies), Kobe, Japan (2002) 244–248

18. Cao, J.N., Zhang, L., Feng, X., Das, S.K.: Path pruning in mailbox-based mobile agent communications. J. of Info Sci and Eng **20**(3) (2004) 405–242

EPLAS: An Epistemic Programming Language for All Scientists

Isao Takahashi, Shinsuke Nara, Yuichi Goto, and Jingde Cheng

Department of Information and Computer Sciences, Saitama University,
255 Shimo-Okubo, Sakura-ku, Saitama-shi, Saitama, 338-8570, Japan
{isao, nara, gotoh, cheng}@aise.ics.saitama-u.ac.jp

Abstract. Epistemic Programming has been proposed as a new programming paradigm for scientists to program their epistemic processes in scientific discovery. As the first step to construct an epistemic programming environment, this paper proposes the first epistemic programming language, named 'EPLAS'. The paper analyzes the requirements of an epistemic programming language, presents the ideas to design EPLAS, shows the important features of EPLAS, and presents an interpreter implementation of EPLAS.

Keywords: Computer-aided scientific discovery, Epistemic process, Strong relevant logic, Scientific methodology.

1 Introduction

As a new programming paradigm, Cheng has proposed Epistemic Programming for scientists to program their epistemic processes in scientific discovery [3,4]. Conventional programming regards numeric values and/or character strings as the subject of computing, takes assignments as basic operations of computing, and regards algorithm as the subject of programming, but Epistemic Programming regards **beliefs** as the subject of computing, takes **primary epistemic operations** as basic operations of computing, and regards **epistemic processes** as the subject of programming [3,4].

Under the strong relevant logic model of epistemic processes proposed by Cheng, a belief is represented by a formula $A \in \mathrm{F}(\mathbf{EcQ})$ where $\mathbf{EcQ}$ is a predicate strong relevant logic [3,4] and $\mathrm{F}(\mathbf{EcQ})$ is the set of all well-formed formulas of $\mathbf{EcQ}$. The three primary epistemic operations are **epistemic deduction**, **epistemic expansion**, and **epistemic contraction**. Let $K \subseteq \mathrm{F}(\mathbf{EcQ})$ be a set of sentences to represent the explicitly known knowledge and current beliefs of an agent, and $\mathrm{T}_{\mathbf{EcQ}}(P)$ be a formal theory with premises P based on $\mathbf{EcQ}$. For any $A \in \mathrm{T}_{\mathbf{EcQ}}(K) - K$ where $\mathrm{T}_{\mathbf{EcQ}}(K) \neq K$, an epistemic deduction of A from K, denoted by $K^{\mathrm{d}+A}$, by the agent is defined as $K^{\mathrm{d}+A} =_{\mathrm{df}} K \cup \{A\}$. For any $A \notin \mathrm{T}_{\mathbf{EcQ}}(K)$, an epistemic expansion of K by A, denoted by $K^{\mathrm{e}+A}$, by the agent is defined as $K^{\mathrm{e}+A} =_{\mathrm{df}} K \cup \{A\}$. For any $A \in K$, an epistemic contraction of K by A, denoted by K^{-A}, by the agent is defined as $K^{-A} =_{\mathrm{df}} K - \{A\}$. An epistemic process of an agent is a sequence K_0, o_1, K_1, o_2, K_2, ..., K_{n-1},

Y. Shi et al. (Eds.): ICCS 2007, Part I, LNCS 4487, pp. 406–413, 2007.

o_n, K_n where K_i ($n \geq i \geq 0$) is an **epistemic state**, and $o_{i+1}(n > i \geq 0)$ is any of primary epistemic operations, and K_{i+1} is the result of applying o_{i+1} to K_i. In particular, K_0 is called the **primary epistemic state** of the epistemic process, K_n is called the **terminal epistemic state** of the epistemic process, respectively.

An **epistemic program** is a sequence of instructions such that for a primary epistemic state given as the initial input, an execution of the instructions produces an epistemic process where every primary epistemic operation corresponds to an instruction whose execution results in an epistemic state, in particular, the terminal epistemic state is also called the result of the execution of the program [3,4].

However, until now, there is no environment to perform Epistemic Programming and to run epistemic programs. We propose the first epistemic programming language, named 'EPLAS': an **E**pistemic **P**rogramming **L**anguage for **A**ll **S**cientists. In this paper, we analyze the requirements of an epistemic programming language at first, and then, present our design ideas for EPLAS and its important features. We also present an interpreter implementation of EPLAS.

2   Requirements

We define the requirements for an epistemic programming language and its implementation. We define R1 in order to write and execute epistemic programs.

R 1. *They should provide ways to represent beliefs and epistemic states as primary data, and epistemic operations as primary operations, and perform the operations.* Since, in Epistemic Programming, the subject of computing is a belief, basic operations of computing are epistemic operations, and the subject of the operations are epistemic states.

We also define R2, R3, and R4 in order to write and execute epistemic programs to help scientists with scientific discovery.

R 2. *They should represent and execute operations to perform deductive, inductive, and abductive reasoning, as primary operations.* Scientific reasoning is indispensable to any scientific discovery because any discovery must be previously unknown or unrecognized before the completion of discovery process and reasoning is the only way to draw new conclusions from some premises that are known facts or assumed hypothesis [3,4]. Therefore, reasoning is one of ways to get new belief when scientists perform epistemic expansion.

R 3. *They should represent and execute operations to help with dissolving contradictions as primary operations.* Scientists perform epistemic contraction in order to dissolve contradictions because beliefs may be inconsistent and incomplete.

R 4. *It should represent and execute operations to help with trial-and-error as primary operations.* Scientists do not always accurately know subjects of scientific discovery beforehand. Therefore, scientists must perform trial-and-error.

In a process of trial-and-error, scientists make many assumptions and test the assumptions by many different methods. It is a demanding work for scientists to make combinations of the assumptions and the methods. Furthermore, it is also demanding for scientists to test the combinations one at a time without omission.

3 EPLAS

EPLAS is designed as a typical procedural and strongly dynamic typed language. It has facilities to program control structures (they are *if-then statement, do-while statement,* and *foreach statement*) and procedures, and has a nested static-scope rule. With attribute grammar, we defined syntax and semantics of EPLAS and open EPLAS manual [1].

To satisfy R1, EPLAS should provide beliefs as a primary data type. For that purpose, EPLAS provides a way to represent beliefs as a primary data type. EPLAS also provides an operation to input a belief from standard input, which is denoted by '**input_belief**'. Conventional programming languages do not provide beliefs as a primary data type because their subjects of computing are lower-level data types. Then, to satisfy R1, EPLAS should provide epistemic states as a primary data type. Therefore, EPLAS provides sets of beliefs as set structured data type and assures that all epistemic states are numbered. EPLAS also provides an operation to get the i-th epistemic state, which is denoted by '**get_state**(i)'. Almost conventional programming languages do not provide a set structured data type and do not assure that all epistemic states are numbered. Furthermore, EPLAS should provide epistemic operations as primary operations to satisfy R1. Hence, EPLAS provides operations to perform epistemic deduction, epistemic expansion, and epistemic contraction by multiple beliefs as extensions as primary operations. An epistemic deduction is denoted by '**deduce**', and makes the current epistemic state K_i the next epistemic state $K_{i+1} = K_i \cup S$ for any $S \subseteq T_{\mathbf{EcQ}}(K_i) - K_i$ where $K_i \subseteq F(\mathbf{EcQ})$ and $T_{\mathbf{EcQ}} \neq K_i$. An epistemic expansion of multiple beliefs S is denoted by '**expand**(S)', and makes the current epistemic state K_i the next epistemic state $K_{i+1} = K_i \cup S$ for any $S \nsubseteq T_{\mathbf{EcQ}}(K_i)$ where $K_i \subseteq F(\mathbf{EcQ})$. An epistemic contraction by multiple beliefs S is denoted by '**contract**(S)', and makes the current epistemic state K_i the next epistemic state $K_{i+1} = K_i - S$ for any $S \subset K_i$ where $K_i \subseteq F(\mathbf{EcQ})$. Some conventional programming languages provide operations to perform epistemic expansion and epistemic contraction as set operations but any conventional programming languages do not provide an operation to perform epistemic deduction.

There are three forms of reasoning: deductive, inductive, and abductive reasoning. Therefore, an epistemic programming language should provide operations to perform reasoning by these three forms.

In order to satisfy R2, EPLAS should provide a way to represent various reasoning, at least, reasoning by the three forms. For that purpose, EPLAS provides inference rules as a primary data type. Inference rules are formulated with some schemata of well-formed formulas to reason by pattern matching, and consist of at least one schemata of well-formed formula as premises and at least one schemata of well-formed formulas as conclusions [7]. Let K be premises including '$\{P0(a0), P0(a1), P1(a0), \forall x0(P2(x0) \rightarrow P1(x0))\}$', ir_1 be an inference rule: '$P0(x0), P0(x1) \vdash \forall x2 P0(x2)$', ir_2 be an inference rule: '$P0(x0), P1(x0), P0(x1) \vdash P1(x1)$', and ir_3 be an inference rule: '$P2(x0) \rightarrow P1(x1), P1(x1) \vdash P2(x0)$'. ir_1 means an inductive generalization [5], ir_2 means an arguments from analogy [5], and ir_3 means an abductive reasoning [8]. '$\forall x2 P0(x2)$' is derived from K by ir_1, '$P1(a1)$' is derived from K by ir_2, and '$P2(a0)$' is derived from K by ir_3. EPLAS also provides an operation to input an inference rule from standard input, which is denoted by '**input_rule**'. Moreover, in order to satisfy R2, EPLAS should provide an operation to derive conclusions from beliefs in the current epistemic state by applying an inference rule to the beliefs and to perform epistemic expansion by the derived conclusions. A reasoning by an inference rule ir is denoted by '**reason**(ir)', and makes the current epistemic state K_i the next epistemic state $K_{i+1} = K_i \cup S$ where $K_i \subseteq F(\mathbf{EcQ})$, $S \subset R_{ir}(K_i)$, $S \neq \phi$, and $R_{ir}(K_i)$ is a set of beliefs derived from K_i by an inference rule ir. Any conventional programming languages do not provide inference rules as a primary data type and a reasoning operation as a primary operation.

As operations to help with dissolving contradictions, at least, an epistemic programming language should provide an operation to judge whether two beliefs are conflicting or not. Then, it should provide operations to output a derivation tree of a belief and to get all beliefs in a derivation tree of a belief in order for scientists to investigate causes of contradictions. It should also provide an operation to perform epistemic contraction of beliefs derived from a belief in order for scientists to reject beliefs derived from a conflicting belief.

In order to satisfy R3, EPLAS should provide an operation to judge whether two beliefs are conflicting or not. For that purpose, EPLAS provides the operation as a primary operation, which is denoted by '**\$\$**'. The binary operator '**\$\$**' is to judge whether two beliefs are conflicting or not, and returns true if and only if one belief A is negation of the other belief B, or false if not so. Then, in order to satisfy R3, EPLAS should provide operations to output a derivation tree of a belief. Hence, EPLAS provides an operation to output a derivation tree of a belief A from standard output, which is denoted by '**see_tree**(A)'. EPLAS should also provide an operation to get all beliefs in a derivation tree of a belief. Therefore, EPLAS provides an operation to get beliefs in the derivation tree of a belief A, which is denoted by '**get_ancestors**(A)'. Furthermore, in order to satisfy R3, EPLAS should provide an operation to perform epistemic contractions of all beliefs derived by the specific beliefs. For that purpose, EPLAS provide the operation to perform an epistemic contractions of beliefs derived by the specific beliefs S, which is denoted by '**contract_derivation**(S)', and which makes the current epistemic state K_i the next epistemic state $K_{i+1} = K_i - (T_{\mathbf{EcQ}}(K_i) - T_{\mathbf{EcQ}}(K_i - S))$ where $K_i \subseteq F(\mathbf{EcQ})$. Evidently, conventional programming languages do not provide these operations.

An epistemic programming language should provide operations to make combinations of beliefs and to test each combination in order to verify combinations of assumptions, at least, as operations to help with trial-and-error. It also should provide operations to make permutations of procedures and to test each permutation in order for scientists to test many methods by various turns. Furthermore, it should provide an operation to make the current epistemic state change into a past epistemic state in order for scientists to test assumptions by various methods in same epistemic state.

To satisfy R4, EPLAS should provide operations to make combinations of beliefs and to test each combination. For that purpose, EPLAS provides sets of sets as a set-set structured type and set operations, e.g., sum, difference, intersection, power, and direct product. Some conventional programming languages provide the structured data type and the set operations but almost conventional programming languages do not. Then, to satisfy R4, EPLAS should provide operations to make permutations of procedures and to test each permutation. Hence, EPLAS provide procedures as a primary data type and sequences as a seq structured type, and sequence operations, e.g., appending to the bottom, dropping from the bottom. A procedure is a name of a *procedure with arguments*, and is similar to a function pointer in C languages. In order to satisfy R4, furthermore, EPLAS should provide an operation to change the current epistemic state into a past one identified by a number. EPLAS provides the operation, which is denoted by '**return_to**(n)', and makes the current epistemic state K_i the next epistemic state $K_{i+1} = K_n$ where K_n is n–th epistemic state.

4   An Interpreter Implementation of EPLAS

We show an interpreter implementation of EPLAS. We implemented the interpreter with Java (J2SE 6.0) in order for the interpreter to be available on various computer environments. We, however, implemented the interpreter by naive methods because the interpreter is a prototype as the first step to construct an epistemic programming environment. The interpreter consists of the analyzer section and the attribute evaluation section. The analyzer section analyzes a program in an input source file and makes a parse tree. It has been implemented with SableCC [6]. Accordingly to semantic rules in the attributes grammar of EPLAS, the attribute evaluation section evaluates attributes on a parse tree made by the analyzer section. The attribute evaluation section has the symbol table, the beliefs manager, the epistemic states manager, and the reasoner to evaluate attributes.

The symbol table manages declaration of variables, functions, and procedures, data types and structured types of variables. It also assure that EPLAS is a strongly dynamic typed language. We implemented it with a hash table by a popular method.

The beliefs manager manages all input and/or derived beliefs and all their derivations trees, and provides functions to perform **input_belief**, **see_tree**, and **get_ancestors**. We implemented the beliefs manager as follows. The beliefs manager has a set of tuples where a tuple consists of a belief and a derivation tree of the belief. When performing **input_rule**, the beliefs manager analyzes

Table 1. Vocabulary of A Language Producing Beliefs

Vocabulary	Symbols
Constants	`a0, a1, ..., ai, ...`
Variables	`x0, x1, ..., xi, ...`
Functions	`f0, f1, ..., fi, ...`
Predicates	`P0, P1, ..., Pi, ...`
Connectives	`=>(entailment), &(and), !(negation)`
Quantifiers	`@(forall), #(exists)`
Punctuation	`(, ), ,`

an input string according to a belief form. The belief is formed by a language including vocabulary in Table 1 and the following Production Rules 1 and 2.

Production Rule 1. *Term*
(1) Any constant is a term and any variable is also a term.
(2) If `f` is a function and `t0, ..., tm` are terms then `f(t0, ..., tm)` is a term.
(3) Nothing else is a term.

Production Rule 2. *Formula*
(1) If `P` is a predicate and `t0, ..., tm` are terms then `P(t0, ..., tm)` is a formula.
(2) If `A` and `B` are formulas then `(A => B)`, `(A & B)`, and `(! A)` are formulas.
(3) If `A` is a formula and `x` is a variable then `(@xA)`, `(#xA)` are formulas.
(4) Nothing else is a formula.

The beliefs manager also adds new tuple of an input belief and a tree which has only root node denoting the input belief into the set. When performing **see_tree**(A), the beliefs manager outputs a derivation tree of A with JTree. When performing **get_ancestors**(A), the beliefs manager collects beliefs in a derivation tree of A by scanning the derivation tree and returns the beliefs.

The epistemic states manager manages all epistemic states from the primary epistemic state to the terminal epistemic state, and provides functions to perform **expand**, **contract**, **contract_derivation**, **return_to**, **get_state**, and **get_id**. We implemented the epistemic states manager as follows. The epistemic states has a sequence of sets of beliefs where a set of beliefs is an epistemic state, and the sequence is variable-length. When performing **expand**(S), the epistemic states manager appends a sum of a set of the bottom of the sequence and S to the sequence. When performing **contract**(S), the epistemic states manager appends a difference of a set of the bottom of the sequence and S. When performing **contract_derivation**(S), the epistemic states manager appends a difference of a set of the bottom of the sequence and all beliefs in derivation trees of S to the sequence. When performing **return_to**(i), the epistemic states manager appends

i-th epistemic state to the sequence. When performing **get_id**, the epistemic states manager returns a number of elements of the sequence. When performing **get_state**(i), the epistemic states manager returns a set of beliefs of the i-th element of the sequence.

The reasoner provides functions to perform **input_rule** and **reason**. We implemented the epistemic states manager as follows. There is an automated forward deduction system for general purpose entailment calculus EnCal [2,7]. EnCal automatically deduces new conclusions from given premises by applying inference rules to the premises and deduced results. Therefore, the reasoner has been implemented as an interface to EnCal. When performing **input_rule**, the reasoner analyzes an input string according to an inference rule form. The inference rule form is "$SLogicalSchema_1$, $\cdots$, $SLogicalSchema_n$, $SLogicalSchema_{n+1}$." "$SLogicalSchema$" is formed by a language including vocabulary in Table 2 and the following Production Rules 1 and 3.

Table 2. Vocabulary of A Language Producing Semi Logical Schema

Vocabulary	Symbols
Constants	a0, a1, ..., ti, ...
Variables	x0, x1, ..., xi, ...
Functions	f0, f1, ..., fi, ...
Predicates	P0, P1, ..., Pi, ...
Predicate Variable	X1, X1, ..., Xi, ...
Formula Variable	A0, A1, ..., Ai, ...
Connectives	=>(entailment), &(and), !(negation)
Quantifiers	@(forall), #(exists)
Punctuation	(,), ,

Production Rule 3. *Semi logical Schema*
(1) Any formula variable is a semi logical formula.
(2) If P is a predicate or a predicate variable and t0, ..., tm are terms then P(t0, ..., tm) is a semi logical formula.
(3) If A and B are semi logical formulas then (A => B), (A & B), and (! A) are semi logical formulas.
(4) If A is a semi logical formula and x is a variable then (@xA), (#xA) are semi logical formulas.
(5) Nothing else is a semi logical formula.

When performing **reason**(ir), the reasoner translates ir into an inference rule as an EnCal form and current beliefs which the epistemic states manager returns into formulas as an EnCal form, inputs these data to EnCal and executes EnCal, and then, gets the formulas derived by ir, translates the formulas into an EPLAS form, and makes the beliefs manager registers the formulas.

5 Concluding Remarks

As the first step to construct an epistemic programming environment, we proposed the first epistemic programming language, named 'EPLAS', and its interpreter implementation. EPLAS provides ways for scientists to write epistemic programs to help scientists with reasoning, dissolving contradictions, and trial-and-error. We also presented an interpreter implementation of EPLAS. We have provided the first environment to perform Epistemic Programming and run epistemic programs. In future works, we would like to establish Epistemic Programming methodology to make scientific discovery become a 'science' and/or an 'engineering'.

Acknowledgments

We would like to thank referees for their valuable comments for improving the quality of this paper. The work presented in this paper was supported in part by The Ministry of Education, Culture, Sports, Science and Technology of Japan under Grant-in-Aid for Exploratory Research No. 09878061, and Grant-in-Aid for Scientific Research (B) No. 11480079.

References

1. AISE Lab., Saitama University.: EPLAS Reference Manual. (2007)
2. Cheng, J.: Encal: An automated forward deduction system for general–purpose entailment calculus. In Terashima, N., Altman, E., eds.: Advanced IT Tools, IFIP World Conference on IT Tools, IFIP96 - 14th World Computer Congress. Chapman & Hall (1996) 507–517
3. Cheng, J.: Epistemic programming: What is it and why study it? Journal of Advanced Software Research **6**(2) (1999) 153–163
4. Cheng, J.: A strong relevant logic model of epistemic processes in scientific discovery. In Kawaguchi, E., Kangassalo, H., Jaakkola, H., Hamid, I.A., eds.: "Information Modelling and Knowledge Bases XI," Frontiers in Artificial Intelligence and Applications. Volume 61., IOS Press (2000) 136–159
5. Flach, P.A., Kakas, A.C.: Abductive and inductive reasoning: background and issues. In Flach, P.A., Kakas, A.C., eds.: Abduction and Induction: Essays on Their Relation and Integration. Kluwer Academic Publishers (2000)
6. Gagnon, E.M., Hendren, L.J.: Sablecc http://www.sablecc.org.
7. Nara, S., Omi, T., Goto, Y., Cheng, J.: A general-purpose forward deduction engine for modal logics. In Khosla, R., Howlett, R.J., Jain, L.C., eds.: Knowledge-Based Intelligent Information and Engineering Systems, 9th International Conference, KES2005, Melbourne, Australia, 14-16 September, 2005, Proceedings, Part II. Volume 3682., Springer-Verlag (2005) 739–745
8. Peirce, C.S.: Collected Papers of Charles Sanders Peirce. Harvard University Press (1958)

Translation of Common Information Model to Web Ontology Language

Marta Majewska[1], Bartosz Kryza[2], and Jacek Kitowski[1,2]

[1] Institute of Computer Science AGH-UST, Mickiewicza 30, 30-059 Krakow, Poland
[2] Academic Computer Centre Cyfronet-AGH, Nawojki 11, 30-950 Krakow, Poland
{mmajew, bkryza, kito}@agh.edu.pl

Abstract. This paper presents a brief overview of the work on translation of Common Information Model (CIM) to Web Ontology Language
(OWL) standard. The main motivation for the work is given, along with
discussion of major issues faced during this work. The paper contains
also comparison of existing approaches to conversion of CIM to OWL
and presents the CIM2OWL tool that performs the conversion of CIM
schema and allows convertion of CIM instances - representing for instance
configurations of particular systems - to OWL individuals.

Key words: Metadata, ontologies, Grid computing, Common Information Model, ontology translation.

1   Introduction

Several researchers have risen lately the issue of translation of existing and widely
recognized Distributed Management Task Force (DMTF) standard for resource
description called Common Information Model (CIM) [1] to Web Ontology Language (OWL) [2]. Especially in the Grid setting, where OWL could be used for
representation of semantic Grid metadata, the problem of the interoperability
appears. Among reference ontologies for modeling the hardware and software
computer resources the DMTF Common Information Model as well known, organizationally supported and regularly updated meta-model for the considered
area (e.g. popularly referred in software for management of systems, networks,
users and applications across multiple vendor environments) seemed promising.
CIM is a hybrid approach, inspired by the object oriented modeling and database information modeling. The CIM Schema consists in particular of Core and
Common Models as well as developed by users Extension Schemas. As it introduces the metadata for annotating model classes and instances, it is partially
not compliant with the UML methodology. OWL is W3C recommended ontology
language for the Semantic Web, which exploits many of the strengths of Description Logics, including well defined semantics and practical reasoning techniques.
OWL offers greater expressiveness of information content description then that
supported by XML, RDF, and RDF Schema, by providing additional vocabulary
along with some formal semantics.

Y. Shi et al. (Eds.): ICCS 2007, Part I, LNCS 4487, pp. 414–417, 2007.

2   CIM to OWL Mapping

In the beginning, the mapping between semantically equivalent constructs in MOF and OWL was established. That included class definition, inheritance (to some extent), data type attributes, cardinality constraints, comments, etc. The succeeding step in definition of the mapping was to extend the existing mapping with representations of MOF constructs, which do not have direct equivalents in OWL.

Table 1. The mapping definition from CIM to OWL

CIM Artifact	OWL Construct
Class	`<owl:Class>`
Generalization	`<rdfs:subClassOf>`

Association (Aggregation, Composition)	`<owl:Class rdf:ID="...">`
	`   <rdfs:subClassOf`
	`      rdf:resource="cim-meta:CIM_Association/>`
	`</owl:Class>`

Property	`<owl:DatatypeProperty>`
REF Property	`<owl:ObjectProperty>`
Method	`<cim-meta:hasMethod>`
Default Value	`<cim-meta:defaultValue>`
Override	`<rdfs:subPropertyOf>`
Key	`<owl:InverseFunctionalProperty>`
Min, Max	`<owl:minCardinality>`, `<owl:maxCardinality>`
ValueMap, Values	`<cim-meta:CIM_Value>` composed of `<cim-meta:value>` and `<cim-meta:valueMap>`
Deprecated	`<owl:deprecatedClass>` or `<owl:deprecatedProperty>`
Required	`<owl:minCardinality rdf:datatype="&xsd;int">1` `</owl:minCardinality>`
Experimental	`<cim-meta:Experimental>`
Alias	`<owl:equivalentClass>`, `<owl:equivalentProperty>` or `<owl:sameAs>`
ModelCorrespondence	`<rdfs:seeAlso>`
Read, Write	`<cim-meta:readable>`, `<cim-meta:writeable>`
Version	`<cim-meta:cimVersion>`
Abstract	`<cim-meta:Abstract>`
Units	`<rdfs:comment>`
Vectors	`<rdfs:comment>`

The lack of fully equivalent constructions or rules for expressing some MOF constructs in OWL implied that partial, simplified mappings had to be admitted. These unambiguous mappings concerned mainly data restrictions (e.g. qualifier Value, ValueMaps), distinctions (e.g. qualifier Key, Propagated, Weak), redefinitions (e.g. qualifier Override), access (e.g. qualifier Read, Write), versioning (qualifier Version), default values, abstracting (qualifier Abstract), and dynamics (e.g. procedures, qualifier IN, OUT). Unfortunately, many issues of the not fully semantically equivalent part schema mapping are out of the scope of this paper. To achieve optimal mappings a few approaches were used such as usage of various semantic constructs with an approximate and uncontradictory meaning (e.g. for qualifier Override), definition of an additional meta ontology for a provision of missing CIM vocabulary called cim-meta (e.g. for qualifiers Abstract, ValueMap, Values, default values) and usage of annotation properties and comments (e.g. for qualifier Units). An excerpt from mapping is given in Table 1 (see http://gom.kwfgrid.net for details). The resulting OWL ontology has OWL DL expressiveness except for Key qualifier which requires InverseFunctionalProperty to be applied to datatype values, which is not allowed in OWL DL. Thus using our tool, the user can choose to create OWL Full ontology with the Key qualifier, or stay in OWL DL dialect without the Key property. The instance mapping is provided based on the schema mapping. An occurrence of various qualifiers in class definitions usually does not influence the conversion of MOF instances to OWL and definition of new instances based on the OWL ontology.

3   Related Work

In K-Wf Grid project, a set of ontologies describing generic and domain specific aspects of the Grid were gathered and integrated for the thematic areas of workflows, Grid applications, services, data and resources [3]. Ontologies stored and managed by the component called Grid Organizational Memory (GOM) [4] are optimized for use in the Grid applications [5]. The authors of [6] propose to use XML-based ontology language to define resource management information for network management systems. In [7], authors present a generic and extensible ontology for describing several aspects of Grid environment. In [8] an idea of developing semantic descriptions of the web services based on the CIM specifications is presented. In [9] the mapping of CIM UML notation to RDF-S, and the extension to OWL, are presented. The author of [10] focuses on the troublesome CIM to OWL translation issues. Currently one translation tool is available called CIMTool [11]. Several generic knowledge representation translators also exist such as those described in [12] or [13].

4   Conclusions

As a result of our work we have created the mapping schema from CIM to OWL, the OWL resource ontology based on CIM and the CIM2OWL tool, whose advantage is the ability to translate not only CIM schema (i.e. CIM Core and Common Models) but also CIM instances. The large scope of the expressiveness and functionality was transformed from CIM to OWL.

Acknowledgments. This reasearch has been done in the framework of EU IST-2004-511385 K-Wf Grid project and EU-IST FP6-34363 Gredia project. AGH University of Science and Technology grant is also acknowledged.

References

1. DMTF, Common Information Model (CIM) Standards, http://www.dmtf.org/standards/cim/.
2. W3C, Web Ontology Language (OWL), http://www.w3.org/2004/OWL/.
3. Kryza, B., Pieczykolan, J., Babik, M., Majewska, M., Slota, R., Hluchy, L. and Kitowski, J. Managing Semantic Metadata in K-Wf Grid with Grid Organizational Memory. In: Bubak, M. et al. (eds), Proc. of the 5th Cracow Grid Workshop (CGW05), Krakow, Poland, ACC-CYFRONET AGH, 2006, pp. 66-73.
4. Kryza, B., Pieczykolan. J and Kitowski. J: Grid Organizational Memory - A Versatile Solution for Ontology Management in the Grid, In Proc. of 2nd Intl. Conf. on e-Science and Grid Computing, Dec. 4-6, 2006, Amsterdam, (C) IEEE Computer Society Press, ISBN 0-7695-2734-5.
5. Kryza, B., Slota, R., Majewska, M., Pieczykolan, J., Kitowski, J. Grid Organizational Memory - Provision of a High-Level Grid Abstraction Layer Supported by Ontology Alignment. The Intnl Journal of FGCS, Grid Computing: Theory, methods & Applications, vol. 23, issue 3, Mar 2007, Elsevier, 2007, pp. 348-358.
6. Lopez de Vergare, J.E., Villagra, V.A., Asensio, J.I., Berrocal, J. Application of the Web Ontology Language to define management information specifications. Proc. of the HP Openview University Association Workshop, France, 2004.
7. Xing, W., Dikaiakos, M. D., Sakellariou, R. A Core Grid Ontology for the Semantic Grid. In Proc. of 6th IEEE Intl. Sym. On Cluster Computing and the Grid (CCGRID'06) - Vol. 00, pp. 178-184. Washington, 2006
8. Keeney, J., Carey, K., Lewis, D., OSullivan, D., Wade, V. Ontology-based Semantics for Composable Autonomic Elements. Workshop of AI in Autonomic Communications at the 19th Intnl Joint Conf. on AI, Edinburgh, 2005.
9. Quirolgico, S., Assis, P., Westerinen, A., Baskey, M., Stokes, E. Toward a Formal Common Information Model Ontology. Proc. of Workshop on Intelligent Networked and Mobile Systems, WISE04, Australia, 2004, Springer, LNCS vol. 3307, pp. 11-21.
10. Heimbigner, D. DMTF CIM to OWL: A Case Study in Ontology Conversion. In: Maurer, F. and Ruhe, R. (eds), Proc. of 16th. Intl. Conf. on Software Engng. and Knowledge Engng. (SEKE2004), Canada, 2004, Knowledge Systems Institute, Skokie, pp. 470-474.
11. CIMTool project website, http://cimtool.org/.
12. Corcho, O., Gomez-Perez, A. ODEDialect: a set of declarative languages for implementing ontology translation systems. ECAI2004 Workshop on Semantic Intelligent Middleware for the Web and the Grid. Valencia, Spain, 2004.
13. Chalupsky, H. OntoMorph: a translation system for symbolic knowledge. In A.G. Cohn, et al. (eds), editors, Principles of Knowledge Representation and Reasoning: Proc. of the 7th Intnl Conf. (KR2000), San Francisco, CA, 2000. Morgan Kaufmann.

XML Based Semantic Data Grid Service

Hui Tan and Xinmeng Chen

Computer School, Wuhan University, Wuhan 430072, China
`journal@whu.edu.cn`

Abstract. This paper introduces a novel wrapper-mediator based semantic data grid service mechanism to solve the problem of Semantic heterogeneity and few compatible data sources. It uses ontology based semantic information to wrap the heterogeneous data source, and employs mediator structure to supply accessing interface for the data sources, and it extends semantic query, mapping and fusion languages to support semantic grid communication mechanism. The extension of XML algebra with semantic query enhanced is discussed to enable semantic querying on data gird environment.

1 Introduction

Data grid technology is the standard means of realizing the needs of integrating and querying distributed and heterogeneity information, especially semi-structured and non-structured information. However, the studies in data grid technology still have the shortcomings as follows: 1)The flexibility of the grid technology is limited. Taking OGSA-DAI[1] for example, it only supports the limited related database and native XML database. However, most information on Internet comes from web-based semi-structured data environment, such as company web application and XML-based e-commerce platform; furthermore, OGSA-DAI does not have the effective mechanism for other data sources to be integrated into the grid environment. 2) The individual node in the grid environment may exist in varied semantic environment; different data resource is constructed in accordance with different semantic standard. The present data grid does not take into consideration the semantic heterogeneity among different nodes. Many projects are focusing on these two topics. GridMiner[2] and OGSA-WEB[3] are two novel projects focusing on the first one, and DartGrid II[4] and SemreX[5] are excellent projects focusing on the second topic.

This paper focusses on these two topics too. It employs a mediator-wrapper framework to support different information sources and enable semantic information operation on different grid nodes. And it uses XML query style language to retrieve information from different grid nodes, because XML is rapidly becoming a language of choice to express, store and query information on the web. The remainder of this paper is structured as follows. Section 2 gives the general discussion about framework of the mediator-wrapper based semantic data grid. Section 3 discusses the knowledge communication mechanism to support

Y. Shi et al. (Eds.): ICCS 2007, Part I, LNCS 4487, pp. 418–425, 2007.

semantic querying and knowledge fusion. Section 4 discusses ontology enabled querying rewriting on XML based grid nodes. Section 5 summarizes the whole paper.

2   Mediator-Wrapper Based Semantic Data Grid Service

Semantic Data Grid (SDG) must satisfy the following requirements:

- The architecture must be opening and compatible with existing standard such as the framework of OGSA[6] or WSRF[7] considering compatible with OGSA-DAI;
- It must provide flexible method for integrating various data sources including relational databases, Native XML databases, or Web based application systems;
- It must support the global semantics to the users who access semantic data grid.

This paper uses a semantic grid adapter service to support semantic operation on the gird. This paper employs a mediator-wrapper method to construct the adapter service, which can be expressed by figure 1(a). The function of the wrapper of local grid nodes is to describe its semantics and its mapping relationship with other nodes, the information source of these nodes include both free and commercial databases, flat files services, web services or web based applications, HTML files and XML files, and the semantic information of every local gird node is described with the language based on its ontology. The mediator node constructs the global semantics of the local nodes, the semantic communication mechanism between the mediator and wrapper nodes is discussed in the following section.

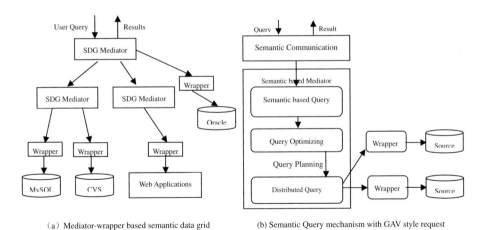

(a) Mediator-wrapper based semantic data grid   (b) Semantic Query mechanism with GAV style request

Fig. 1. Mediator-Wrapper based Semantic Data Grid

3 Communication Mechanism with Semantic Grid

It is very important to develop a knowledge communication and coordinating mechanism to support the ontology fusion and semantic query on different data grid nodes. This paper employs a Knowledge Communication and Manipulation Language for Semantic Grid, or KCML for short to support this mechanism, which is an extension of the KGOL[8] language. One function of KCML is to coordinate with each grid node to build the mediator-wrapper architecture dynamically. The other function is to build global knowledge on the mediator and enable semantic query. The communication language is build on SOAP, supporting SOAP over HTTP, HTTPS or other rock-bottom communication protocol. The language could describe as:

$$KCML ::= Ver|Operation|Sender|Receiver|Language|Content.$$

The field Ver is for keeping Expanding, showing which version language was used. The new version language has compatibility downwards, supporting the old communication mechanism; $Operation$ gives basic communication atom which will be described next; $Content$ describes what is communicated; $Sender$ defines sender's information, including user, address (such as IP ,e-mail,URL, port); $Receiver$ defines receiver's information (usually, receiver should be Web Service or Grid Service), ,including type (HOST, Web Service or Semantic Web Service), address(such as IP address, e-mail, URL, port, if receiver is Web Service, also including service address), identifier; $language$ defines which language is used this communication, including RDF/RDFs, DAML+OIL, OWL etc.

3.1 Basic Communication Atom

To illustrate the algorithm, we first define the ontology based knowledge on the mediators and wrappers.

Definition 1. *A knowledge schema is a structure $KB := (C_{KB}, R_{KB}, I, \iota_C, \iota_R)$ consisting of (1) two sets C_{KB} and R_{KB}, (2) a set I whose elements are called instance identifiers or instances, (3) a function $\iota_C : C_{KB} \to \Re(I)$ called concept instantiation, (4) a function $\iota_R : R_{KB} \to \Re(I^+)$ called relation instantiation.*

To simplify the content of this paper, we only discuss the atom of KCML language which support ontology fusion and semantic querying. The atom includes query operation, join operation and union operation etc. as following[9]:

- **Selection.** $\sigma_F(c) = \{x | x \in \iota_C(c) \land F(x) = true\}$ where F is composed of logic expression, supporting logic operation $\land, \lor, \neg, \forall, \exists, <, >, \leq, \geq, \neq, =$ and $\in$. c is concept element of knowledge instance;
- **Join.** $\bowtie (c_1, p, c_2) = \{x, y | x \in \iota_C(c_1) \land y \in \iota_C(c_2) \land p(x, y) = true\}$, where p is join condition, c_1 and c_2 is concept element;
- **Union.** $c_1 \cup c_2 = \{x | x \in \iota_C(c_1) \land x \in \iota_C(c_2)\}$, c_1 and c_2 is the same as above;
- **Minus.** $c_1 - c_2 = \{x | x \in \iota_C(c_1 \land \neg c_2)\}$, c_1 and c_2 is the same as above;
- **Projection.** $\pi_P(c) = \bigcup_{p_i \in P} \{y | \exists x, (x, y) \in \iota_R(p_i) \land x \in \iota_C(c)\}$, where c is concept element, P is a set of relationship and $P = \{p_1, p_2, \ldots, p_k\}$;

3.2   Semantic Fusion Atom

The mediator node constructs the global semantics of the local nodes based on ontology via ontology fusion mechanism[10] based on the ontology mapping patterns in gird environment, the patterns of ontology mapping can be categorized into four expressions: direct mapping, subsumption mapping, composition mapping and decomposition mapping[11], a mapping can be defined as:

Definition 2. *A **Ontology mapping** is a structure $\mathcal{M} = (\mathcal{S}, \mathcal{D}, \mathcal{R}, v)$, where $\mathcal{S}$ denotes the concepts of source ontology, $\mathcal{D}$ denotes the concepts of target ontology, $\mathcal{R}$ denotes the relation of the mapping and v denotes the confidence value of the mapping, $0 \leq v \leq 1$.*

The KCML language must support the mapping patterns between different semantic nodes on gird, we use **Match** atom to support it, it can be defined as $M(c, d, r) = \{(x, y) | x \in \iota_C(c) \wedge y \in \iota_C(d) \wedge (x, y) \in \iota_R(r)\}$, where c is different concept from d, r is relationship of mapping. The knowledge stored at mediator can be described as the ontology fusion connections list, which can be described as definition 3.

Definition 3.   *Fusion Connection* *is a structure $\mathcal{F}_c(O_1 : C_1, O_2 : C_2, \ldots, O_n : C_n, \mathcal{M})$, where C_1 denotes a concept or concept set of ontology O_1, C_2 denotes a concept or concept set of Ontology O_2, $\mathcal{M}$ denotes the mapping patterns between C_1 , C_2 , ... and C_n.*

4   Semantic XML Query Rewriting

The semantic query in a mediator-based SDG can be express as figure 1(b). The user's request is rewritten and modified accordingly based on the global semantics, and is due processed optimally. Corresponding operation plan is made and passed by the wrapper to each data source node for operation. From above description, we know that this paper employs the GAV(Global as View) method to process the user's query[12]. The query can be described as an XML query with semantic enhanced, which can be described as an extension of XML algebra, and it will be discussed in the following subsection. Because common XML query languages such as XQuery and XUpdate can be transferred into XML query algebra, so the extension is manageable.

This paper extended XML algebra TAX[13] to enable semantic querying on mediated gird nodes, TAX uses *Pattern Tree* to describe query language and *Witness Tree* to describe the result instances which satisfy the Pattern Tree. The extension of XML query algebra is discussed in paper [14]. The query planning is based on the semantic XML query rewriting technology. In order to simplify the discussion, this paper just pays attention to the query planning mechanism of the selection operation. Briefly, a selection operation can be expressed as $\sigma(X : S, Y) \{X \subseteq P_i \cup P_o, Y \subseteq PE\}$, where P_i is the input pattern tree, P_o is output pattern tree, PE is predication list, S denotes the site in which the query will be executed. We define two operators $\cup$ and $\bowtie$ to represent *Union* and *Join* operation separately, and define the operator $\Rightarrow$ to represent the query rewriting operation, and we use $\sigma(X : S_0, Y)$ or $\sigma(X, Y)$ to denote the user's query from the mediator site.

Firstly, we propose how to rewrite pattern tree (which is the X element of expression $\sigma(X, Y)$), there maybe several cases as follows:

1. X is one of the elements of input pattern tree or output pattern tree, and it is also a concept in the global ontology hierarchy. $X_i(1 \leq i \leq n)$ are the concepts for different local ontologies. X and X_i were combined into one concept in the integrated global ontology with strong direct mappings, which means that X and X_i can match each other, then we can rewrite X as $X \cup \bigcup\limits_{1 \leq i \leq n} X_i$. The responding selection rewriting can be expressed as:

$$\sigma(X, Y) \Rightarrow \sigma(X, Y) \cup \sigma(X_1 : S_1, Y) \cup \sigma(X_2 : S_2, Y) \ldots \cup \sigma(X_n : S_n, Y) \quad (1)$$

2. The concept of X is generated by the subsumption mapping or composition mapping of $X_i(1 \leq i \leq n)$, then we can rewrite X as $\bigcup\limits_{1 \leq i \leq n} X_i$. The responding selection rewriting can be expressed as:

$$\sigma(X, Y) \Rightarrow \sigma(X_1 : S_1, Y) \cup \sigma(X_2 : S_2, Y) \ldots \cup \sigma(X_n : S_n, Y) \quad (2)$$

And then, we propose how to rewrite the predication expressions (which is the Y element of the expression $\sigma(X, Y)$, there are also several cases, which can be described as follows:

1. If there are lots of concept $Y_i(1 \leq i \leq n)$ combined in the concept Y of global Ontology, we can rewrite Y as $Y \cup \bigcup\limits_{1 \leq i \leq n} Y_i$. The corresponding selection rewriting can be described as:

$$\sigma(X, Y) \Rightarrow \sigma(X, Y) \cup \sigma(X : S_1, Y_1) \cup \sigma(X : S_2, Y_2) \ldots \cup \sigma(X : S_n, Y_n) \quad (3)$$

2. If the concept Y is generated by the subsumption mapping of $Y_i(1 \leq i \leq n)$, we can rewrite Y as $\bigcup\limits_{1 \leq i \leq n} Y_i$. The corresponding selection rewriting can be described as:

$$\sigma(X, Y) \Rightarrow \sigma(X : S_1, Y_1) \cup \sigma(X : S_2, Y_2) \ldots \cup \sigma(X : S_n, Y_n) \quad (4)$$

3. If the concept Y is generated by the composition mapping of $Y_i(1 \leq i \leq n)$, suppose the composition condition is F, we can rewrite Y as $(Y_1 + Y_2 + \ldots Y_n) \cap F$. The corresponding selection rewriting can be described as:

$$\sigma(X, Y) \Rightarrow \sigma(X : S_1, Y_1 \wedge F) \bowtie \sigma(X : S_2, Y_2 \wedge F) \ldots \bowtie \sigma(X : S_n, Y_n \wedge F) \quad (5)$$

Algorithm 1. SEL_Rewrite_X(X)

Input: X is the pattern tree of selection query $\sigma(X, Y)$.

1 **foreach** $x \in X$ **do**
2   **switch** *Mappings of X node* **do**
3   **case** *funsion_node*
4              $x \leftarrow x \cup \bigcup_{1 \leq i \leq n} x_i$;
5              $\sigma(X, Y) \Rightarrow \sigma(X, Y) \cup \sigma(X_1, Y) \cup \sigma(X_2, Y) \ldots \cup \sigma(X_n, Y)$;
6   **foreach** x_i **do**
7   SEL_Rewrite_X(x_i);
8   **end**
9   **case** *subsumption or composition*
10             $x \leftarrow \bigcup_{1 \leq i \leq n} x_i$;
11             $\sigma(X, Y) \Rightarrow \sigma(X_1, Y) \cup \sigma(X_2, Y) \ldots \cup \sigma(X_n, Y)$;
12   **foreach** x_i **do**
13   SEL_Rewrite_X(x_i);
14   **end**
15   **end**
16   **end**
17 **end**

It is worth to point out that rewriting process may require a recursion in the transitivity property of semantic mapping. The process of rewriting pattern tree and predication expressions can be described as algorithm 1 and 2.

The query planning is a sequence, each node of the sequence can be denoted as $P_n = (Q_n, S_n, C_n, F_n)$, where Q_n is the query which is needed to rewrite, S_n is a set of sub query executed on different sites, C_n denotes the connection operator, in most time, it is $\cup$ or $\bowtie$ operator, F_n is the predication which denotes the connection conditions. P_n represents the query rewriting procedure of query Q_n. The query planning procedure of user's query $\sigma(X, Y)$ can be expressed in algorithm[14].

5   Discussion and Conclusion

Semantic data grid service mechanism we present in this paper wrapped various information source through ontology semantic, and used Mediator-Wrapper to support the heterogeneous data source, employed mediator structure to realize virtual data gird service which supports semi-structured information retrieving language. The extension of XML algebra with semantic query enhanced and semantic grid communication mechanism are also discussed to enable semantic accessing on data grid environment. However, query optimizing in distributed web sites and the capability of different nodes and network were not considered in the query planning mechanism discussed in this paper, future research will be focused on this topic.

Algorithm 2. SEL_Rewrite_Y(Y)

   Input: Y is the predication list of selection query $\sigma(X,Y)$.

1 **foreach** $y \in Y$ **do**

2   **switch** *Mappings of Y concept* **do**

3   **case** *funsion_node*

4            $y \leftarrow y \cup \bigcup_{1 \leq i \leq n} y_i$;

5            $\sigma(X,Y) \Rightarrow \sigma(X,Y) \cup \sigma(X,Y_1) \cup \sigma(X,Y_2) \ldots \cup \sigma(X,Y_n)$;

6   **foreach** y_i **do**

7   SEL_Rewrite_Y(y_i);

8   **end**

9   **case** *subsumption*

10            $y \leftarrow \bigcup_{1 \leq i \leq n} y_i$;

11            $\sigma(X,Y) \Rightarrow \sigma(X,Y_1) \cup \sigma(X,Y_2) \ldots \cup \sigma(X,Y_n)$;

12   **foreach** y_i **do**

13   SEL_Rewrite_Y(y_i);

14   **end**

15   **case** *decomposition*

16            $y \leftarrow (y_1 + y_2 + \ldots y_n) \cap F$;

17            $\sigma(X,Y) \Rightarrow \sigma(X,Y_1 \wedge F) \bowtie \sigma(X,Y_2 \wedge F) \ldots \bowtie \sigma(X,Y_n \wedge F)$;

18   **foreach** y_i **do**

19   SEL_Rewrite_Y(y_i);

20   **end**

21   **end**

22   **end**

23 **end**

Acknowledgment

This work was partially supported by a grant from the NSF (Natural Science Fundation) of Hubei Prov. of China under grant number 2005ABA235, and it was partially supported by China Postdoctoral Science Foundation under grant number 20060400275 and Jiangsu Postdoctoral Science Foundation under grant number 0601009B.

References

1. Antonioletti, M., Atkinson, M., Baxter, R., et al.: The design and implementation of Grid database services in OGSA-DAI. Concurrency and Computation: Practice and Experience **17** (2005) 357–376

2. Wöhrera, A., Brezanya, P., Tjoab, A.M.: Novel mediator architectures for Grid information systems. Future Generation Computer Systems **21** (2005) 107–114

3. Pahlevi, S.M., Kojima, I.: OGSA-WebDB: An OGSA-Based System for Bringing Web Databases into the Grid. In: Proceedings of International Conference on Information Technology: Coding and Computing (ITCC'04), IEEE Computer Society Press (2004) 105–110

4. Chen, H., Wu, Z., Mao, Y.: Q3: A Semantic Query Language for Dart Database Grid. In: Proceedings of the Third International Conference on Grid and Cooperative Computing (GCC 2004), Wuhan, China, LNCS 3251, Springer Verlag (2004) 372–380

5. Jin, H., Yu, Y.: SemreX: a Semantic Peer-to-Peer Scientific References Sharing System. In: Proceedings of the International Conference on Internet and Web Applications and Services (ICIW'06), IEEE Computer Society Press (2006)

6. Foster, I., Kesselman, C., Nick, J.M., Tuecke, S.: Grid Services for Distributed System Integration. IEEE Computer **35** (2002) 37–46

7. Czajkowski, K., Ferguson, D.F., Foster, I., et al.: The WS-Resource Framework. http://www.globus.org/wsrf/specs/ws-wsrf.pdf (2004)

8. Zhuge, H., Liu, J.: A Knowledge Grid Operation Language. ACM SIGPLAN Notices **38** (2003) 57–66

9. Sheng, Q.J., Shi, Z.Z.: A Knowledge-based Data Model and Query Algebra for the Next-Gereration Web. In: Proceedings of APWeb 2004, LNCS 3007 (2004) 489–499

10. Gu, J., Zhou, Y.: Ontology fusion with complex mapping patterns. In: Proceedings of 10th International Conference on Knowledge-Based, Intelligent Information and Engineering Systems, Bournemouth, United Kingdom, LNCS, Springer Verlag (2006) 738–745

11. KWON, J., JEONG, D., LEE, L.S., BAIK, D.K.: Intelligent semantic concept mapping for semantic query rewriting/optimization in ontology-based information integration system. International Journal of Software Engineering and Knowledge Engineering **14** (2004) 519–542

12. Levy, A.Y., Rajaraman, A., Ordille, J.J.: Query heterogeneous information sources using source descriptions. In: Proceedings of the 22nd VLDB Conference, Mumbai, India, Morgan Kaufmann Publishers Inc (1996) 251–262

13. H.V.Jagadish, L.V.S.Lakshmanan, D.Srivastava, et al: TAX: A Tree Algebra for XML. Lecture Notes In Computer Science **2379** (2001) 149–164

14. Gu, J., Hu, B., Zhou, Y.: Semantic Query Planning Mechanism on XML based Web Information Systems. In: WISE 2006 Workshops, LNCS 4256 (2006) 194–205

Communication-Aware Scheduling Algorithm Based on Heterogeneous Computing Systems

Youlin Ruan[1,2], Gan Liu[3], Jianjun Han[3], and Qinghua Li[3]

[1] School of Information Engineering, Wuhan University of Technology, 430070 Wuhan,
P.R. China
[2] State Key Laboratory for Novel Software Technology, Nanjing University, 210093 Nanjing,
P.R. China
[3] Department of Computer Science and Technology, Huazhong University of Science and
Technology, 430074 Wuhan, P.R. China
ruanyl@126.com

Abstract. This paper proposes a new scheduling model that integrates the communication awareness into task scheduling. A scheduling algorithm is proposed for the new model, which can produce optimal schedule by serializing the communications edges. Experiment results show the significantly improved accuracy and efficiency of the new model and algorithm.

Keywords: communication aware; concurrent communication; serializing.

1 Introduction

In order to exploit the heterogeneous computing power, a lot of works have focused on solving the task scheduling problem. This problem has been shown to be NP-complete[1]. There are several proposed algorithms dealing with the task scheduling problem in heterogeneous systems. They could be classified into a variety of categories, such as list-scheduling, clustering, duplication-based algorithms, and guided random search methods. The list-based algorithms provide good quality of schedules and their performance is comparable with other categories at a lower scheduling time. The Iso-Level Heterogeneous Allocation[2], the Heterogeneous Earliest-Finish-Time (HEFT) and Critical-Path-on-a-Processor (CROP) techniques[3] are all examples. However, most of the above-mentioned works have in common that they employ an idealized model of the target parallel system and don't deliberate on the impact of communication contentions. Intuition and experiments demonstrated that this model results in inaccurate and inefficient schedules[4]. Therefore, in order to exploit the potential of parallel processing, efficient scheduling algorithms under the consideration of the communication requirements must be developed. So, CATS was proposed to schedule parallel tasks by considering the communication requirements of the applications and the network bandwidth[5]. Sinnen investigated the incorporation of contention awareness into task scheduling by edge scheduling, which considered heterogeneous links and routing[6].

Y. Shi et al. (Eds.): ICCS 2007, Part I, LNCS 4487, pp. 426–429, 2007.

In this paper, we propose a communication-aware scheduling that extended the classic list scheduling. The rest of paper is organized as follows. In Section 2, we introduce some definitions and preliminaries. Section 3 describes the new contention model, and investigates the integration of the awareness for the contention in task scheduling. Experimental results are presented in section 4.

2   Preliminaries

In task scheduling, the program to be scheduled is represented by a directed acyclic graph(DAG) $G=(V, E, w, c)$, where V is the set of tasks and E is the set of edges. The size of task n_i is $w(n_i)$. An edge from n_i to task n_j is expressed as e_{ij} and size of the edge is expressed as c_{ij}. Given task n_i, the set of parent tasks is denoted as $pred(n_i)$; that is, $pred(n_i)=\{n_k \mid e_{ki} \in E \}$, and the set of children tasks is denoted as $succ(n_i)$; that is $succ(n_i)=\{ n_k \mid e_{ik} \in E \}$. A task n_i is called a join task if $\mid pred(n_i) \mid \geq 2$, an entry task if $\mid pred(n_i) \mid =0$, and an exit task if $\mid succ(n_i) \mid =0$.

To describe a schedule S of a DAG $G=(V, E, w, c)$, the following terms are defined: $est(n, P)$ and $ect(n, P)$ represent the earliest start time and completion time of n_i on processor P respectively, and $w(n, P)$ the execution time of node $n \in V$ on processor $P \in \textbf{P}$. Thus, the node's earliest completion time is given by $ect(n, P)=est(n, P)+w(n, P)$. Given an edge e_{ij}, $est(e_{ij})$ and $ect(e_{ij})$ represent the earliest start time and completion time of edge e_{ij} respectively. Thus, the edge's earliest completion time is given by $ect(e_{ij})= est(e_{ij})+ c_{ij}$ and $est(n_j, P) \geq ect(e_{ij})$, for $n_i, n_j \in V$, $e_{ij} \in E$.

3   The Proposed Algorithm

To make task scheduling contention aware, the communication edges must be serialized because communication cannot be concurrent. Thus, the awareness for communication is serialize achieved by scheduling edge.

Definition 1 (Contention Model): A target parallel system $M_{contention} =(\textbf{P}, w, comm)$ consists of a set of possibly heterogeneous processors $\textbf{P}$ connected by the communication network, where $comm(e_{ij}, P_{src}, P_{dst})$ denotes the communication cost of edge e_{ij} from processor P_{src} to P_{dst}.

When a communication, represented by the edge e_{ij}, is performed between two processors P_{src} and P_{dst}. Thus, the earliest completion time of e_{ij} is

$$ect(e_{ij}) = \begin{cases} ect(n_i) & if \ P_{src} = P_{dst} \\ ect(n_i) + comm(e_{ij}, P_{src}, P_{dst}) & otherwise \end{cases} \qquad (1)$$

where $P_{src}= proc(n_i)$ and $P_{dst}= proc(n_j)$. Thus, the edge earliest completion time $ect(e_{ij})$ is now the earliest completion time of n_i plus the communication of e_{ij} from P_{src} to P_{dst}, unless the communication is local. For n_a, we claim that a chain of communications can be found as $X_a: e_{1a} \rightarrow e_{2a} \rightarrow ... \rightarrow e_{ka}$, where $n_i \in pred(n_a)$ (i=1..k), such that $est(e_{1a}) \le est(e_{2a}) \le \ldots \le est(e_{ka})$, i.e., $ect(n_1) \le ect(n_2) \le \ldots \le ect(n_k)$. We define $\Omega(e_{ia})$ as the idle communication time before the communication edge e_{ia}, is issued as

$\Omega(e_{ia}) = est(e_{ia}) - ect(e_{(i-1)a})$, $(1 \le i \le k)$. Thus, $est(n_a) = ect(n_1) + \sum_{i=1}^{k} \{comm(c_{ia}, P_{src}, P_{dst}) + \Omega(e_{ia})\}$.

Given task n_a, $ect(n_a)$ is determined according to its parent tasks and communication edges. If task n_a has a

single parent task n_i, then $est(n_a) = ect(n_a)$ If n_a and n_i exist different processors, the execution of n_a is delayed by $ect(n_i) + comm(e_{ia}, proc(n_i), proc(n_a))$. If n_a and n_i exist same processor, the communication time c_{ia} is negligible and n_a can start at $ect(n_i)$. If task n_a has k parent task $n_1, n_2, \ldots, n_k$, where $ect(n_1) \le ect(n_2) \le ect(n_i) \le \ldots \le ect(n_k)$. At first, because of earliest completion of n_1, the start time of edge e_{1a} should be first scheduled to $proc(n_a)$, $ct_1(sc)$ is denoted as the completion time of scheduling the first communication. However, if the parent task n_2 cannot complete in advance, which means $ct_1(sc) = ect(n_1) + c_{1a} \le ect(n_2)$, there is an idle time Ω_1 between them. Thus, the edge e_{2a} must be delayed by $ect(n_2)$. Next, the procedure continue to schedules the edge e_{ia} $(2 \le i \le k)$ that satisfies the following condition: $t_s(e_{ia}) = \max(ct_{i-1}(sc), ect(n_i))$, where $ct_i(sc) = ct_{i-1}(sc) + \Omega_{i-1} + c_{ia}$ and

$$\Omega_i = \begin{cases} 0 & if \ ect(n_{i+1}) \le ct_i(sc) \\ ect(n_{i+1}) - ct_i(sc) & otherwise \end{cases} \tag{2}$$

If kth edge is scheduled, $ct_k(sc)$ and $est(n_a)$ can be determined. Therefore, $est(n_a) = ct_k(sc)$. We can easily prove that $est(n_a)$ is minimal.

The best known scheduling heuristic is list scheduling. A common and usually good priority is the node's bottom level bl, which is the length of the longest path leaving the node. Recursively defined, it is

$$bl(n_i) = w(n_i) + \max_{n_j \in succ(n_i)} \{c_{ij} + bl(n_j)\} \tag{3}$$

Based on above theory and analysis, we propose an extended list scheduling algorithm that considers the communication awareness by reasonable serializing the communication edges, which is listed as below.

```
Algorithm  List Scheduling With Communication Awareness
 Input: program G=(V, E, w, c) and model M =(P, w, comm);
 Output: A schedule with minimal completion time;
  begin
    Sort nodes n ∈E into list L, according to priority
    scheme bl and precedence constraints;
    for n_j∈L do
     { Find processor P∈P that allows earliest
        completion time of n_j;
       for each n_i∈pred(n_j) in ascend order ect(n_i) do
        if  proc(n_i)≠P then  schedule e_ij on P;
        schedule n_j on P;
     }
 end.
```

4   Experiment Results

This section presents experiments results to verify some aspects of the proposed contention aware scheduling. We adopt the same test methods with [6].In each experiment, the values of these parameters are assigned from the corresponding sets given below.

$SET_V = \{100, 200, 400, 800\}$, $SET_{in\,degree} = \{1, 2, 5, 10\}$ and $SET_{CCR} = \{0.1, 1, 5, 10\}$.

Fig. 1 and Fig.2 show the average accuracy and efficiency achieved under the two models over all DAGs on parallel systems. Experimental results demonstrated the significantly improved accuracy and efficiency than classic schedule schemes.

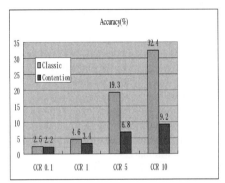   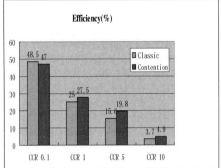

Fig. 1. Average accuracy   **Fig. 2.** Average efficiency

Acknowledgments. This work has been supported by the national natural science foundation of china under grant no. 60503048 and 60672059, the 863 program no. 2006AA01Z233.

References

1. Sarkar, V.: Partitionning and Scheduling Parallel Programs for Execution on Multiprocessors. MIT Press (1989)
2. Olivier, B., Vincent, B., Yves, R: The Iso-Level Scheduling Heuristic for Heterogeneous Processors, Proc. Of 10th Euromicro Workshop on Parallel, Distributed and Networkbased Processing (2002)
3. Topcuoglu, H., Hariri, S., Wu, M.: Performance-Effective and Low-Complexity Task Scheduling for Heterogeneous Computing. IEEE Trans. on Parallel and Distributed Systems, (2002) 260–274
4. Selvakumar, S. Murthy, C.S.R.: Scheduling Precedence Constrained Task Graphs with Non-Negligible Intertask Communication onto Multiprocessors, IEEE Trans. on Parallel and Distributed Systems, (1994) 328–336
5. Lai, G.L.: Scheduling Communication-Aware Tasks on Distributed Heterogeneous Computing Systems, Proceedings of 24th International Conference on Distributed Computing Systems Workshops (2004)
6. Sinnen, O., Sousa, L.A.: Communication Contention in Task Scheduling, IEEE Trans. Parallel and Distributed Systems (2005) 503–515

Macro Adjustment Based Task Scheduling in Hierarchical Grid Market

Peijie Huang[1], Hong Peng[1], and Xuezhen Li[2]

[1] College of Computer Science and Engineering, South China University of Technology,
Guangzhou 510640, P.R. China
[2] Department of Computer and Information Engineering, Guangdong Technical College of
Water Resources and Electric Engineering, Guangzhou 510635, P.R. China
scuthpj@yahoo.com.cn

Abstract. Hierarchical organization is suitable for computational Grid. Although a number of Grid systems adopt this organization, few of them have dealt with the task scheduling for the hierarchical architecture. In this paper, we introduce a hierarchical Grid market model, which maintains the autonomy of the Grid end users, but incorporates macro adjustment of Grid information center into hierarchical Grid task scheduling. Simulation experiments show that the proposed method can improve the inquiry efficiency for resource consumers and get better load balancing of the whole hierarchical Grid market.

Keywords: Grid computing, task scheduling, macro adjustment, hierarchical market.

1 Introduction

How to effectively match the Grid tasks with the available Grid resources is a challenge due to the dynamic, heterogeneous and autonomous nature of the Grid. One common solution is to design a hierarchical Grid market. Although a number of Grid models and mechanisms employ hierarchical organization [1], [2], few of them have dealt with the task scheduling for the hierarchical architecture. Many existing Grid literatures of task scheduling have been proposed and achieved better performance in a single Grid market [3], [4]. However, given the hierarchical market model, how to effectively schedule the Grid tasks remains an important issue.

Our solution is novel in the sense that we introduce a hierarchical Grid market model which maintains the autonomy of the Grid end users, but incorporates macro adjustment of Grid information center (GIC) into hierarchical Grid task scheduling, and thus achieve a better scheduling in hierarchical Grid market.

2 The Hierarchical Grid Market

The Grid system is a hierarchical architecture with several different levels of Grid markets. An example of two-level hierarchical Grid market is shown in Fig. 1.

Y. Shi et al. (Eds.): ICCS 2007, Part I, LNCS 4487, pp. 430–433, 2007.

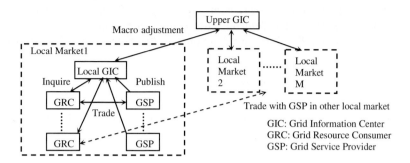

Fig. 1. The hierarchical Grid market

Each lowest level market covers certain local network area. In the local Grid markets, Grid service providers (GSPs) and Grid resource consumers (GRCs) trade with each other autonomously. Each local Grid market has a local GIC that allows GSPs to publish their services in order to attract GRCs to inquire for their tasks. Our market model extends the capability of the GICs, which also store and analyze the aggregate demands and supplies of their local Grid markets.

The network environment covering larger network area forms an upper Grid market, which consists of several lower-level Grid markets. Each upper Grid market has an upper GIC. The upper GIC stores the location of the lower-level GICs included and the aggregate demands and supplies of the lower-level Grid markets, and thus can provide inquiry across different lower-level markets within it and make macro adjustment to the lower-level GICs in this upper Grid market.

3   Macro Adjustment Based Grid Task Scheduling

Here, we present a method providing macro adjustment by off-line analysis.

Suppose that there are n different kinds of Grid resources, we can denote the supply and demand of each local market as an n-vector. During new statistical period, suppose that there are s GSPs in local market m, and p tasks submitted within local market m during peak demand period. We consider the following iterative procedure for GIC m to obtain new estimates of demand and supply, D_m and S_m:

$$s_{mi} = \alpha \cdot T \cdot \sum_{j=1}^{s} r_{ji} + (1-\alpha) \cdot (s_{mi})_{old} \qquad i = 1,2,\cdots,n \qquad (1)$$

$$d_{mi} = \beta \cdot \sum_{j=1}^{p} (u_{ji} \cdot t_{ji}) + (1-\beta) \cdot (d_{mi})_{old} \qquad i = 1,2,\cdots,n \qquad (2)$$

where s_{mi} and d_{mi} are respectively the component of the vectors S_m and D_m. r_{ji} is the published resource scale of resource i of GSP j, and T is the duration of the peak demand period. u_{ji} is the demand of task j for resource i, with occupy time t_{ji}. α and β are the factors between 0 and 1 to control the impact of the past estimates.

The excess supply of local market m is $Z_m = S_m - D_m$. For a certain resource combination c, we denote the demand ratio of this combination of the whole Grid market as an n-vector DR_c. Then, we can compute the excess supply capability for resource combination c of local market m, C_m^c as follows:

$$C_m^c = \max\{0, \min_{R_i \in c}(z_{mi}/dr_{ci})\} \qquad (3)$$

where z_{mi} and dr_{ci} are respectively the component of the vectors Z_m and DR_c. C_m^c is 0 indicates that local market m lacks excess supply of certain kinds of resources in combination c.

In the hierarchical Grid market, local GICs search within its local market with priority. When there is not any available resource for a certain submitted task with resource combination c, suppose that there are k local markets have not been searched, the probability that local market i is chosen to inquire is,

$$\rho_i^c = C_i^c \Big/ \sum_{j=1}^k C_j^c \qquad i = 1, 2, \cdots, k \qquad (4)$$

In next section, we compare the performance of our method, macro adjustment based task scheduling (*MA-Based* in short) with two contrastive methods:

Random Selection: In this method, without macro-adjusted guidance, when the local GICs cannot find an appropriate GSP in their own local markets, they choose another local market randomly with uniform probability to run a secondary inquiry.

Least Loaded: In the Least Loaded method, the local GICs choose the least loaded local market to run a secondary inquiry. Note, however, that in real environment the information is usually not up to date. We use a parameter p so that once a task is submitted to a local market, the load of that local market is updated only with probability p (for the results presented in this paper we use $p = 0.5$).

4   Simulation Experiments

We have developed a simulated Grid environment based on GridSim [3]. The simulated Grid system consists of 20 local markets. Each local market has 10 GSPs and random amount of GRCs, uniformly distributed between 200 and 1000. There are total 4 kinds of Grid resources in this simulated Grid system.

The experimental evaluations include the *inquiry efficiency* and the *load balancing* of the whole Grid market. The *inquiry efficiency* is measured by the average amount of local markets in which Grid tasks that need secondary inquiry inquire. Suppose that there are M local markets. The *load balancing* of the whole Grid market is measured by the balance degree during peak service period, which is denoted as:

$$BD = 1 - \sqrt{\frac{1}{M}\sum_{i=1}^M (O_i - \overline{O})^2} \Big/ \overline{O} \qquad (5)$$

where O_i is the resource occupy rate of local market i during the peak service period.

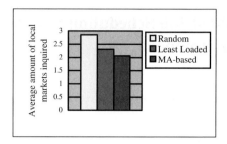   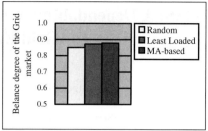

Fig. 2. The average amount of local markets in which Grid tasks that need secondary inquiry inquire

Fig. 3. The balance degree of the whole Grid market

First, we compare the inquiry efficiency. As we can see from Fig. 2, comparing to Least Loaded method, the performance gain of our method is about 15%. This is mainly due to probabilistic failure to update the load levels after each task submission while using Least Loaded method. While comparing to Random Selection, the performance of our method gain reaches as high as 27.9%.

We then observe the load balancing of the whole Grid market. As we can see from Fig. 3, our method guides most of the tasks to search in the local markets with big excess supply capability when they cannot be served in their own local markets, thus gets better load balancing than the contrastive ones. The Least Loaded method also outperforms the Random Selection. But it should pay for the large number of information exchanged between the upper GIC and the lower-level GICs.

5   Conclusions

This study is a first attempt to incorporate macro adjustment into Grid task scheduling. As we can see from the simulation results, our method outperforms the contrastive ones in improving the inquiry efficiency for resource consumers and getting better load balancing of the whole hierarchical Grid market.

Acknowledgements. This work is supported by the Provincial High-tech Program of Guangdong, China (A10202001, 2005B10101033).

References

1. Foster, I., Kesselman, C.: Globus: A Meta-computing Infrastructure Toolkit. International Journal of Supercomputer Applications, 1997
2. Chapin, S.J., Katramatos, D., Karpovich, J., et al: Resource Management in Legion. Future Generation Computer Systems, 1999, 15(5-6): 583-594
3. Buyya, R.: Economic-based Distributed Resource Management and Scheduling for Grid Computing. [Ph.D. Thesis]. Melbourne: Monash University, 2002
4. Li Chunlin, Li Layuan: QoS Based Resource Scheduling by Computational Economy in Computational Grid. Information Processing Letters. 2006, 98 (3): 119–126

DGSS: A Dependability Guided Job Scheduling System for Grid Environment[*]

Yongcai Tao, Hai Jin, and Xuanhua Shi

Services Computing Technology and System Lab
Cluster and Grid Computing Lab
School of Computer Science and Technology
Huazhong University of Science and Technology, Wuhan, 430074, China
hjin@hust.edu.cn

Abstract. Due to the diverse failures and error conditions in grid environments, node unavailability is increasingly becoming severe and poses great challenges to reliable job scheduling in grid environment. Current job management systems mainly exploit fault recovery mechanism to guarantee the completion of jobs, but sacrificing system efficiency. To address the challenges, in this paper, a node TTF (*Time To Failure*) prediction model and job completion prediction model are designed. Based on these models, the paper proposes a dependability guided job scheduling system, called DGSS, which provides failure avoidance job scheduling. The experimental results validate the improvement in the dependability of job execution and system resources utilization.

1 Introduction

In grid environment, resources can enter and depart without any prior notice. In addition, failures can occur at any moment by various reasons, e.g. the change of resource local policy and the breakdown of resources and network fabric [1]. Hence, jobs fail frequently and QoS can not be guaranteed. Node unavailability poses great challenges to grid job scheduling. To address these issues, existing job management systems mostly resort to fault recovery mechanism, such as checkpoint, retry and replication. Although relieving the challenges to some extent, this mechanism sacrifices system resources. For example, checkpoint usually requires extra disk space and network bandwidth to record the job running information, e.g. intermediate results and data for continuing job without starting from scratch. Replication policy runs the job at multiple available resources. The fault recovery mechanism belongs to compensating methodology and can not prevent the job failures in advance. To prevent the job failures proactively, the accurate information of temporal and spatial distribution of grid node availability in the future should be predicted. Thus, jobs can be scheduled onto nodes with long uptime instead of upcoming failure nodes.

So far, researches mainly give attention to model and predict the characteristics of cluster nodes' availability. As the scale is expanding rapidly, grid integrates not only

---

[*] This paper is supported by Nation Science Foundation of China under grant 90412010 and 60603058.

Y. Shi et al. (Eds.): ICCS 2007, Part I, LNCS 4487, pp. 434–441, 2007.
© Springer-Verlag Berlin Heidelberg 2007

cluster resources, but more wide-area personal resources which are mostly non-dedicated. Therefore, the characteristics of node unavailability in grid environment distinguish greatly from that in cluster environment. In the paper, the terms node unavailability and node failure are used interchangeably to refer to the event that nodes are inaccessible for nodes' leaving, hardware and software breakdown, and network connection malfunction, etc.

In this paper, a node TTF (*Time To Failure*) prediction model and job completion prediction model are designed. Based on these models, the paper proposes a dependability guided job scheduling system, called DGSS. Rational of DGSS is that if the completion time of job is less than the TTF of node, the job can expect to finish successfully; otherwise, the job is likely to fail and has to be restarted on the same node or somewhere else. Experiments are conducted and demonstrate that DGSS can enhance the dependability of job execution and resources utilization.

The rest of the paper is organized as follows. Section 2 reviews the related work. We introduce DGSS in section 3. The experimental evaluation is presented in section 4. Finally, we conclude this paper in section 5.

2 Related Work

There are a vast amount of related works on grid job scheduling. Existing strategies of grid job scheduling can be categorized into performance-driven, market-driven and trust-driven [2].

Papers [3, 4] adopt performance-driven strategy to submit jobs onto resources and achieve optimal performance for users and system (e.g. minimum job execution time and high throughput). The work in [5] utilizes market-driven strategy during job assignment. Recently, some research works have strived to address the scheduling problems with reliability optimization. In [6], trust-driven scheduling strategy is exploited to map jobs onto appropriate resources according to their trust levels. This strategy avoids selecting malicious nodes and non-reputable resources. However, it does not take the job completion prediction time into consideration. In [7], reliable scheduling adopts "Reliability Cost" as the indicator of how reliable a given system is when a group of tasks are assigned to it, but the failure rate of grid nodes and the communication links between them adopted in the reliability cost model is set experimentally.

To support reliable job scheduling, much attention is paid to model the characteristics of node failure to measure node reliability, which mainly focuses on homogeneous computing environments. Papers [8, 9] investigate the failure logs of cluster systems and conclude that the time between reboots of nodes is best modeled by a Weibull distribution with shape parameters of less than 1, implying that a node becomes more reliable the longer it has been operated. Based on the observed characteristic of node failure, researchers also design resources allocation strategy to improve service availability.

Due to the diverse failures and error conditions in grid environment, grid node failures are more stochastic than clusters. Hence, current models can not adapt well to grid environment.

3 Dependability Guided Job Scheduling System - DGSS

In this section, two critical components of DGSS, the node TTF prediction model and job completion time prediction model, are introduced. Then, the corresponding dependable scheduling algorithm is described in detail.

3.1 MC Based Node TTF Prediction Model

In many application fields, Markov model can be extended to model stochastic processes. *Discrete-time Markov chain* (DTMC) is defined at the basis of a set of finite state $M(S_1, S_2, ..., S_m)$ and $M \times M$ state transition matrix P. In matrix P, P_{ij} denotes the transition probability from state S_i to state S_j.

DTMC is mainly used to predict the state occurrence probability in the future. Suppose at time t_k, system state is S_i ($1 \leq i \leq M$) and the distribution of state S_i is $P_k(S_i)=e_i$, where e_i is $1 \times M$ row vector, the value at location i is 1, and others are 0. Thus, we can predict the distribution of S_i at next time: $P_{k+1}(S_i)=P_k(S_i)P=e_iP$. At time $k+2$, the distribution of S_i is: $P_{k+2}(S_i)=P_{k+1}(S_i)P=e_iP^2$. At time $k+n$, the distribution of S_i is: $P_{k+n}(S_i)=P_{k+n-1}(S_i)P=e_iP^n$. So, with DTMC, we can obtain the distribution of state S_i at this time and next time, and therefore we can get the occurrence probability of each state at each time.

In grid environment, TTF of nodes at each time is stochastic. Markov model can be used to model the node TTF stochastic process. In MC based node TTF prediction model, TTF can be seen as system state. In the Markov model described above, the M and P are invariable. The dynamic nature of grid requires large storage space for M and P, which make the model complex and unpractical. In order to address this issue, we present an adaptive MC based node TTF prediction model which can dynamically amend M and P.

When a node becomes inaccessible, new TTF is produced (called TTF_{new}). Then the M would be traversed. If there exists S_i whose absolute difference value and TTF_{new} is less than the specified value, S_i and TTF_{new} would be merged. New S_i is the average of S_i and TTF_{new} and the number of state transition would be added 1. Reversely if there does not exist this state, new state S_{m+1} would be created. At the same time, P would change correspondingly according to the following formula:

$$P_{ij} = n_{ij} \bigg/ \sum_K n_{ik} \qquad (1)$$

where n_{ij} is the transition number from state i to state j with K failures. $\sum_k n_{ik}$ is the all state transition number of K failures.

3.2 Job Completion Time Prediction Model

The job completion time prediction model is also a critical component and has great impact on the efficiency of scheduling system. To predict the completion time of jobs, we adopt the following model similar to the one proposed in [10].

$$CT_{ij} = OPT \times (WT_{ij} + ReT_i + PT_{ij} + WrT_i) \tag{2}$$

Some symbols used in the model are listed in Table 1.

Table 1. Parameters Definition

Symbol	Meaning
CT_{ij}	Completion time of job i at node j
BW_{ij}	Bandwidth between node i and j
$Parent(S_i)$	Parent set of node i
$Child(S_i)$	Child set of node i
$DVol(S_i)$	Data volume produced by job i
OPT	Correction parameter
Q_i	Number of jobs in waiting queue of node i
Q	Number of jobs submitted by user
P_i	Computation capability of node i
ReT_i	Time to read the input data for executing job i
WrT_i	Time to write the output data for executing job i
PT_{ij}	Processing time of job i at node j
WT_{ij}	Waiting time of job i at node j
S_j	The node on which the job is executed
TTF_i	Time to failure of node i
STT_i	Startup time of node i
$CURT$	Current time of system
$RTTF_i$	Remaining time to failure of node i
QSR	Array used to store qualified set of resources
$QDIFF$	Array used to keep the difference between job CT and qualified resource's $RTTF$
$UQSR$	Array used to store unqualified set of resources
$UQDIFF$	Array used to keep the difference between job CT and unqualified resource's $RTTF$
$SelectedR$	Array used to store selected resources

The model is composed of four parts. First, when the job is assigned to a specific resource, it must wait until all jobs in waiting queue are finished. WT_{ij} denotes the waiting time of job i at node j. Second, the node fetches the data needed during the job execution. ReT_i represents the time to read the input data. Third, PT_{ij} stands for the processing time of job i at node j. Finally, the node outputs the results of job execution and the relative data. WrT_i denotes the time to output the data. Therefore, the completion time of job can be obtained as following:

$$\text{CT}_{ij} = OPT \times (\sum_{p \in \text{Parent}(S_j)} \frac{\text{DVol(p)}}{\text{BW}_{pj}} + \sum_{p \in \text{Child}(S_j)} \frac{\text{DVol(p)}}{\text{BW}_{pj}} + \frac{Q_j}{P_j} + \frac{Q}{P_j}) \qquad (3)$$

where *OPT* is a parameter which can be adjusted to correct the predicted result, BW_{ij} can be obtained through the *Network Weather Service* (NWS).

3.3 Dependable Scheduling Algorithm

The 4-tuple <*R, M, J, Q*> is defined as the job scheduling including the resource performances metric and job QoS requirement description. Given a grid system with *m* resources: $R=\{R_1, R_2, ..., R_M\}$. $M=\{M_1, M_2, ..., M_M\}$ is the set of performance metric of corresponding resources such as reliability, charges. $J=\{J_1, J_2, ..., J_N\}$ is used to denote a set of job and $Q=\{Q_1, Q_2, ..., Q_N\}$ is the set of QoS requirements of each job, which can be time constraint, cost constraint and so on. The dependability guided scheduling algorithm is described in Algorithm 1.

```
Algorithm 1: Dependability Guided Scheduling
Input: a set of job requests
Output: a set of selected resources
  while ∃ job i not assigned do
    m = 1; n = 1;
    foreach available resource j do
        calculate CT_ij;   RTTF_j = (TTF_j─(CURT─STT_j));
        if CT_ij < RTTF_j then
          QSR[i][m++] = j; QDIFF[i][m++] = RTTF_i─CT_ij;
        else
          UQSR[i][n++] = j; UQDIFF[i][m++] = CT_ij─RTTF_i;
        endif
    endforeach
    if QSR is not null then
        SelectedR[i] = f_1(QSR[i][m++],QDIFF[i][m++]);
    else
        SelectedR[i] = f_1(UQSR[i][n++],UQDIFF[i][m++])
              or report fault;
    endif
  endwhile
```

In the above algorithm, the function f_1 is defined to adopt different heuristics. For example, for each job, it can select the node whose difference between *CT* and *RTTF* is largest to maximize the dependability of job execution, or the node whose difference is smallest to improve the overall system performance while guarantee the job execution. The function f_1 can also exploit users' QoS requirements as its secondary heuristics criteria such as the minimum completion time or the lowest cost.

For each job, at the beginning, the algorithm predicts the job completion time on each node and calculates the remained time to the failure of nodes. Then, by comparing the completion time of job with the *RTTF* of the node, system obtains the qualified resources and unqualified resources from the available resources pool. If *QSR* is not empty, the optimal resource will be chosen according to specific heuristic adopted by function f_l. If *QSR* is empty, the objective resource will be chosen from the UQSR or relative fault is reported. In this case, job may fail because it takes the risk of being executed on a node with its *RTTF* lower than the completion time of job. This process continues until all jobs are assigned appropriate resources.

4 Performance Evaluation

We compare DGSS with two common scheduling algorithms in terms of the amount of assigned jobs of nodes, success rate of jobs and system throughput. The two scheduling algorithms are as follows:

- RDS: Random Scheduling. In this scheme, jobs are submitted to specific resources which are randomly picked from the available resources.
- HCS: Heuristic Cost Scheduling. In this policy, jobs are assigned to the resources according to the ranking value of resources' cost while satisfying user's QoS requirements.

We perform the experiments in a real grid environment consisting of five grid nodes: three nodes at *Cluster and Grid Computing Lab* (CGCL), and two nodes at *National Hydro Electric Energy Simulation Laboratory* (NHEESL). The grid platforms are deployed with ChinaGrid Support Platform grid middleware (CGSP) [11]. We collect the running log of these five grid platforms for three months and use them to create MC based node availability prediction model to predict nodes' availability in the future. We test three types of applications. One is gene sequence matching application, FASTA, which deals with large size of database. The second is image processing application, which needs to process lots of images. The third is video conversion, which is both computing-intensive and data-intensive. We run 300 FASTA jobs, 300 image processing jobs and 300 video conversion jobs at different time interval.

Figure 1A, 1B, and 1C show the amount of three kinds of jobs assigned to different nodes. We can conclude that due to high uptime, nodes 2 and 3 are assigned more jobs than other nodes in DGSS. Meanwhile, nodes 4 and 5 are assigned more jobs because of lower charge in HCS mode, but the job failure ratio may be high for nodes' poor reliability. In addition, as FASTA requires longer execution time, the difference in amount of assigned jobs in Figure 1A is larger than that of Figure 1B and Figure 1C.

Figure 1D shows the change of system throughput of three different scheduling algorithms while the number of jobs increases. DGSS considers the node availability and job completion time while scheduling, so node loads are balanced and job execution is likely to success. Correspondingly the system throughput is high. Finally, we summarize the experimental data. The success rate of RDS is 51.65%, HCS is 68.57% and DGSS is 79.47%. Obviously, DGSS improves the success rate of job execution greatly. However, since there are still errors in the prediction models of job completion prediction and TTF of nodes, job failures may still occur. It can be seen that the accuracy of prediction models plays an important role on the efficiency of DGSS.

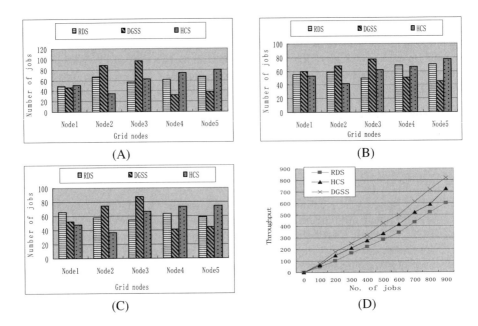

Fig. 1. (A)Amount of assigned FASTA jobs of nodes; (B) Amount of assigned image processing jobs of nodes; (C) Amount of assigned video conversion jobs of nodes; (D) Throughput comparison between three scheduling algorithms

5   Conclusions

In this paper, a dependability guided job scheduling system (DGSS) is proposed, which mainly consists of a node TTF prediction model, a job completion prediction model and corresponding dependable job scheduling algorithm. DGSS makes job scheduling decision by comparing the TTF of nodes with the job prediction completion time to avoid job failures. Experiment results evaluate DGSS against other scheduling algorithms and prove that DGSS can enhance the success ratio of job execution and resources utilization, and improve system performance.

References

1. Foster, I., Kesselman, C. eds.: The Grid: Blueprint for a New Computing Infrastructure. 2nd edition, Morgan Kaufmann, November (2003).
2. Krauter, K., Buyya, R., Maheswaran, M.: A Taxonomy and Survey of Grid Resource Management Systems for Distributed Computing. Software Practice and Experience, 32(2):135-164, February (2002).
3. Cooper, K., Dasgupata, A., Kennedy, K.: New Grid Scheduling and Rescheduling Methods in the GrADS Project. In Proceedings of NSF Next Generation Software Workshop at International Parallel and Distributed Processing Symposium, Santa Fe, IEEE CS Press, Los Alamitos, CA, USA, April (2004).

4. Cao, J., Jarvis, S. A., Saini, S., Nudd. G. R.: GridFlow: Workflow Management for Grid Computing. In Proceedings of 3rd International Symposium on Cluster Computing and the Grid (CCGrid'03), Tokyo, Japan, IEEE Computer Society Press, Los Alamitos, May 12-15, (2003).

5. Venugopal, S., Buyya, R., Winton, L.: A Grid Service Broker for Scheduling Distributed Data-Oriented Applications on Global Grids. In Proceedings of 2nd International Workshop on Middleware for Grid Computing (Middleware'04), Toronto, Ontario, Canada, ACM Press, New York, NY, USA, October 18, (2004).

6. Song, S. S., Kwok, Y. K., Hwang, K.: Trusted Job Scheduling in Open Computational Grids: Security-Driven Heuristics and A Fast Genetic Algorithm. In Proceedings of 19th IEEE International Parallel & Distributed Processing Symposium (IPDPS'05), Denver, CO, USA., IEEE Computer Society Press, Los Alamitos, CA, USA., April 4-8, (2005).

7. He, Y., Shao, Z., Xiao, B., Zhuge, Q., Sha, E.: Reliability Driven Task Scheduling for Heterogeneous Systems. In Proceedings of the Fifteenth IASTED International Conference on Parallel and Distributed Computing and Systems, pp.465-470, (2003).

8. Sahoo, R., Sivasubramaniam, A., Squillante, M., Zhang, Y.: Failure Data Analysis of a Large-scale Heterogeneous Server Environment. In Proceedings of the International Conference on Dependable Systems and Networks (DSN'04), Florence, Italy(2004).

9. Heath, T., Martin, R., Nguyen, T. D.: Improving Cluster Availability using Workstation Validation. In Proceedings of ACM SIGMETRICS 2002, Marina Del Rey, CA, (2002).

10. Jin, H., Shi, X., Qiang, W., Zou, D.: An Adaptive Meta-Scheduler for Data-Intensive Applications. International Journal of Grid and Utility Computing, Inderscience Enterprises Ltd., Vol.1, No.1, pp.32-37(2005),.

An Exact Algorithm for the Servers Allocation, Capacity and Flow Assignment Problem with Cost Criterion and Delay Constraint in Wide Area Networks

Marcin Markowski and Andrzej Kasprzak

Wroclaw University of Technology, Chair of Systems and Computer Networks
Wybrzeze Wyspianskiego 27, 50-370 Wroclaw, Poland
{marcin.markowski, andrzej.kasprzak}@pwr.wroc.pl

Abstract. The paper deals with the problem of simultaneously assignment of server's allocation, channel capacities and flow routes in the wide area network in order to minimize the criterion composed of the leasing capacity cost and the building cost of the network. An exact algorithm, based on the branch and bound method, is proposed to solve the problem. Some interesting features, observed during the computational experiments are reported.

Keywords: Server allocation, CFA problem, Wide Area Networks.

1 Introduction

The optimization problems, considered always when the wide area network must be built or modernized, consist in assignment of resource allocation (i.e. servers, replicas of servers), capacities of channels and flow routes. The optimal arrangement of resources and optimal choice of capacities and flow routes let us obtain the most efficient and economical solution. Designing of the wide area computer networks is always a compromise between the quality of service in the network [1] and the costs needed to build and to support the network. Then those two criterions are often used during the designing process, usually one of them is the optimizing criterion and the other is considered as constraint. In our previous papers [2, 3, 4] we have considered two-criteria and three-criteria problems based on designing the wide area networks, assuming that the maximal support cost of the network is bounded. Sometimes it is useful to formulate the optimizing problem in the other way – how to minimize the cost of the network, when the acceptable quality level is known. Then in the paper the problem of server's allocation, capacity and flow assignment with the cost criterion and delay constraint is considered. In our opinion it is well-founded to consider two kind of cost: the building cost of the network, borne once and the supporting cost, borne regularly. Then the criterion function is composed of two ingredients: the regular channel capacity leasing cost and the disposable server cost. We assume that maximal acceptable total average delay in the network is given as the constraint. Then the problem considered here may be formulated as follows:

Y. Shi et al. (Eds.): ICCS 2007, Part I, LNCS 4487, pp. 442–445, 2007.
© Springer-Verlag Berlin Heidelberg 2007

given:	user allocation at nodes, set of possible nodes for each server, maximal value of the total average delay in the network, traffic requirements user-user and user-server, set of capacities and their costs for each channel,
minimize:	linear combination of the capacity leasing cost and server cost,
over:	servers allocation, channel's capacity and multicommodity flow,
subject to:	multicommodity flow constraints, channel capacities constraints, server allocation constraints, total average delay constraint.

We consider the discrete cost-capacity function because it is the most important from practical point of view for the reason that the channels capacities can be chosen from the sequence defined by ITU-T recommendations. Such formulated problem is NP-complete as more general than the capacity and flow assignment problem (CFA) with discrete cost-capacity function which is NP-complete [5].

The literature focusing on the server's allocation, capacity and flow assignment problem is very limited. Some algorithms for this problem with different delay criterion may be found in [2, 3, 4, 6]. The formulated here problem uses cost criterion and take into account the maximal acceptable average delay in Wide Area Network as constraint. In the literature such formulated problem has not been considered yet.

2 Problem Formulation

Let X_r be the set of binary variables determining capacities of channels and Y_r be the set of binary variables determining servers allocation at nodes [3, 4]. (X_r, Y_r) is called a selection. Let $\mathfrak{R}$ be the family of all selections satisfying assumed constraint. Let $U(Y_r)$ be the server cost (the connecting cost of all servers at nodes) and $T(X_r, Y_r)$ be the minimal average delay per packet in WAN given by the Kleinrock's formula [7]. Let $T^{\max}$ be the maximal acceptable average delay per packet. Let $Q(X_r, Y_r) = \alpha D(X_r) + \beta U(Y_r)$, where α and β are positive coefficients $\alpha, \beta \in [0, 1]$, $\alpha + \beta = 1$. Then, the considered problem is formulated as follows:

$$\min_{(X_r, Y_r)} Q(X_r, Y_r) \tag{1}$$

subject to

$$(X_r, Y_r) \in \mathfrak{R} \tag{2}$$

$$T(X_r, Y_r) \leq T^{\max} \tag{3}$$

3 The Branch and Bound Algorithm

The problem (1-3) is NP-complete. Then the branch and bound method can be used to construct the exact algorithm for solving the considered problem. The detailed description of the calculation scheme of the branch and bound method may be found in the paper [8]. The branch and bound method involves constructing two important operations specific for the problem (1-3): lower bound and branching rules.

The lower bound of the criterion function for every possible successor (X_s, Y_s) generated from (X_r, Y_r) must be computed. Since traffic requirements in the network depend on server allocation, then obtaining the lower bound for the problem (1-3) is difficult. We propose, the lower bound may be obtained by relaxing some constraints and by approximating the discrete cost-capacity curves with the lower linear envelope [5].

The purposes of the branching rules is to find the variable from (X_r, Y_r) for complementing and generating a successor of the selection (X_r, Y_r) with the least possible value of criterion function Q. The choice criterions should be constructed in such a way that complementing reduces value of criterion Q and the increase of total average delay in the network is as minimal as possible [3, 4].

4 Computational Results

The presented algorithm was implemented in C++ code and extensive numerical experiments, for many different networks, have been performed. Experiments were conducted with two main purpose in mind: first, to examine the impact of various problem parameters on the solution (i.e. on the value of criterion Q) and second, to test the computational efficiency of the algorithm.

The typical dependence of the optimal value of Q on $T^{\max}$ for different values of parameters α and β is presented in the Fig. 1. It follows from computational experiments that the dependence of Q on $T^{\max}$ is decreasing function. The following conclusion follows from the computer experiments (Fig.1).

Conclusion 1. There exists such acceptable total average delay $T_*^{\max}$, that problem (1-3) has the same solution for each $T^{\max}$ greater or equal to $T_*^{\max}$.

Let $NT = ((T^{\max} - T_{\min})/(T_*^{\max} - T_{\min})) \cdot 100\%$ be the normalized maximal acceptable total average delay per packet in the network - problem (1-3) has no solution for $T^{\max} < T_{\min}$. Normalized value let us compare results obtained for different network topologies. Let $P(u,v)$, in percentage, be the arithmetic mean of the relative number of iterations for $NT \in [u,v]$ calculated for all considered networks. Fig. 2 shows the dependency of P on divisions [0%,10%), ..., [90%,100%] of NT. It follows from Fig. 2 that the exact algorithm is especially effective from computational point of view for $NT \in [50\%, 100\%]$.

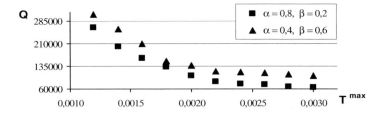

Fig. 1. Typical dependence of the optimal value of Q on $T^{\max}$

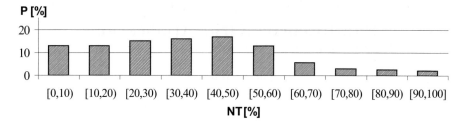

Fig. 2. The dependence of *P* on normalized maximal average delay per packet *NT*

5   Conclusion

In the paper the exact algorithm for solving the server allocation, channel capacities and flow assignments problem with cost criterion and delay constraint is proposed. Such formulated problem has not been considered in the literature yet. It follows from computational experiments that the presented algorithm is effective from computational point of view for greater values of acceptable average delay in the network (Fig. 2). Moreover, we are of the opinion that the Wide Area Network property formulated as conclusion 1 is very important from practical point of view. This property shows that the values of the acceptable average delay are limited.

This work was supported by a research project of The Polish State Committee for Scientific Research in 2005-2007.

References

1. Walkowiak, K.: QoS Dynamic Routing in Content Delivery Network, Lecture Notes in Computer Science, Vol. 3462 (2005), 1120-1132
2. Markowski, M., Kasprzak, A.: An Exact Algorithm for Host Allocation, Capacity and Flow Assignment Problem in WAN, Internet Technologies, Applications and Societal Impact, Kluwer Academic Publishers, Boston (2002), 73-82
3. Markowski, M., Kasprzak, A.: The Web Replica Allocation and Topology Assignment Problem in Wide Area Networks: Algorithms and Computational Results, Lecture Notes in Computer Science, Vol. 3483 (2005), 772-781
4. Markowski, M., Kasprzak, A., The Three-Criteria Servers Replication and Topology Assignment Problem in Wide Area Networks, Lecture Notes in Computer Science, Vol. 3982 (2006), 1119-1128
5. Pioro, M., Medhi, D.: Routing, Flow, and Capacity Design in Communication and Computer Networks, Elsevier, Morgan Kaufmann Publishers, San Francisco (2004)
6. Chari, K.: Resource Allocation and Capacity Assignment in Distributed Systems, Computers Ops Res., Vol. 23, No. 11 (1996), 1025-1041
7. Fratta, L., Gerla, M., Kleinrock, L.: The Flow Deviation Method: an Approach to Store-and-Forward Communication Network Design, Networks 3 (1973), 97-133
8. Wolsey, L.A.: Integer Programming. Wiley-Interscience, New York (1998)

Adaptive Divisible Load Model for Scheduling Data-Intensive Grid Applications

M. Othman*, M. Abdullah, H. Ibrahim, and S. Subramaniam

Department of Communication Technology and Network,
Faculty of Computer Science and Information Technology
University Putra Malaysia, 43400 UPM Serdang, Selangor D.E., Malaysia
mothman@fsktm.upm.edu.my, monabdullah@hotmail.com

Abstract. In many data grid applications, data can be decomposed into multiple independent sub datasets and schedule for parallel execution and analysis. Divisible Load Theory (DLT) is a powerful tool for modelling data-intensive grid problems where both communication and computation load is partitionable. This paper presents an Adaptive DLT (ADLT) model for scheduling data-intensive grid applications. This model reduces the expected processing time approximately 80% for communication intensive applications and 60% for computation intensive applications compared to the previous DLT model. Experimental results show that this model can balance the loads efficiently.

Keywords: Divisible Load Theory, Data grid, Scheduling, Load Balancing.

1 Introduction

Grid systems are interconnected collections of heterogeneous and geographically distributed resources. In data grid environments, many large-scale scientific experiments and simulations generate very large amounts of data in the distributed storages, spanning thousands of files and data sets [8].

Scheduling an application in such environments is significantly complicated and challenging because of the heterogeneous nature of a Grid system. Grid scheduling is defined as the process of making scheduling decisions involving allocating jobs to resources over multiple administrative domains [5].

Recently, DLT model has emerged as a powerful tool for modelling data-intensive grid problems [2]. DLT exploits the parallelism of a divisible application which is continuously divisible into parts of arbitrary size, by scheduling the loads in a single source onto multiple computing resources [3].

The load scheduling in Data Grid is addressed using DLT model with additional constraint that is each worker node receives the same load fraction from each data source [3]. However, this does not take into account the communication time and assumes that the communication time is faster than the computation time. Whereas, if we want to achieve high performance, we must consider both communication time and computation time [6,7].

---

* Institute of Mathematical Science Research (INSPEM), University Putra Malaysia.

Y. Shi et al. (Eds.): ICCS 2007, Part I, LNCS 4487, pp. 446–453, 2007.

In [4], the Constrained DLT (CDLT) is used for scheduling decomposable data-intensive applications. It is compared with results of Genetic Algorithm (GA). The same constraint was tested, which was suggested in [3] and each worker node receives the same load fraction from each data source. They considered the communication time but not in dividing the load. Firstly, they divided the load using DLT model then added the communication time to the expected processing time.

In this paper, an adaptive divisible load model which takes into account both computation time and communication time is proposed. The main objective of our model is to distribute loads over sites to achieve the desired performance level for large jobs - common in Data Grid applications such as Compact Muon Solenoid (CMS) experiment, see [1].

2   Scheduling Model

In [5], the target data intensive applications model can be decomposed into multiple independent subtasks and executed in parallel across multiple sites without any interaction among sub tasks. Lets consider job decomposition by decomposing input data objects into multiple smaller data objects of arbitrary size and processing them on multiple virtual sites. For example in theory, the High Energy Physic (HEP) jobs are arbitrarily divisible at event granularity and intermediate data product processing granularity [1]. Assume that a job requires a very large logical input data set (D) consists of N physical datasets and each physical dataset (of size L_k) resides at a data source (DS_k, for all $k = 1, 2, \ldots, N$) of a particular site. Fig 1 shows how the logical input data (D) is decomposed onto networks and their computing resources.

The scheduling problem is to decompose D into datasets (D_i for all $i = 1, 2, \ldots, M$) across M virtual sites in a Virtual Organization (VO) given its initial physical decomposition. Again, we assume that the decomposed data can be analyzed on any site.

2.1   Notations an Definitions

M	The total number of nodes in the system
N	The total number of data files in the system
L_i	The loads in data file i
L_{ij}	The loads that node j will receive from data file i
L	The sum of loads in the system, where $L = \sum_{i=1}^{N} L_i$
α_{ij}	The amount of load that node j will receive from data file i
α_j	The fraction of L that node j will receive from all data file
w_j	The inverse of the computing speed of node j
Z_{ij}	The link between node i and data source j
T_{cp}	The computing intensity constant.
$T(i)$	The processing time in node i.

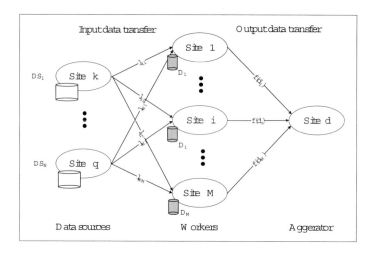

Fig. 1. Data decomposition and their processing

2.2   Cost Model

The execution time cost (T_i) of a subtask allocated to the site i and the turn around time $(T_{Turn\_Around\_Time})$ of a job J can be expressed as follows

$$T_i = T_{input\_cm}(i) + T_{cp}(i) + T_{output\_cm}(i,d)$$

and

$$T_{Turn\_Around\_Time} = \max_{i=1}^{M}\{T_i\},$$

respectively. The input data transfer $(T_{input\_cm}(i))$, computation $(T_{cp}(i))$, and output data transfer to the client at the destination site d $(T_{output\_cm}(i,d))$ are presented as a $\max_{k=1}^{N}\{l_{ki} \cdot \frac{1}{Z_{ki}}\}$, $d_i \cdot w_i \cdot ccRatio$ and $f(d_i) \cdot Z_{id}$, respectively. The function $f(d_i)$ is an output data size and $ccRatio$ is the non-zero ratio of computation and communication. The turn around time of an application is the maximum among all the execution times of the sub tasks.

The problem of scheduling a divisible job onto M sites can be stated as deciding the portion of original workload (D) to be allocated to each site, that is, finding a distribution of l_{ki} which minimizes the turn around time of a job. The proposed model uses this cost model when evaluating solutions at each generation.

3   Adaptive Scheduling Model

In all the literature related to the divisible load scheduling domain so far, an optimality criterion [7] is used to derive an optimal solution is as follows.

It states that in order to obtain an optimal processing time, it is necessary and sufficient that all the sites that participate in the computation must stop at the same time. Otherwise, load could be redistributed to improve the processing time. This optimality principle in the design of our load distribution strategy is used. The new proposed model is an improvement of DLT model, see [3]. The communication time fraction is added into the new model.

3.1   Computation Time Fraction

In [3], the authors have proposed the computation time fraction successfully. The optimal schedule is all nodes finish computation at the same moment of time. Based on that, we will have

$$\sum_{i=1}^{N} \alpha_{i,j} w_j T_{cp} = \sum_{i=1}^{N} \alpha_{i,j+1} w_{j+1} T_{cp}, \quad j = 1, 2, \dots, M-1. \tag{1}$$

As our objective is to determine the above optimal fractions $\alpha_{i,j}$, we impose the following condition in our strategy. Let $\alpha_{i,j} = \alpha_j L_i$, for all $i = 1, 2, \dots, N$ and $j = 1, 2, \dots, M$. This condition essentially assumes that each node receives a load that is proportional to the size of the load from the source. Moreover, each node receives the same load fraction (percentage of total load) from each source. Without this condition, the system of equation is under constrained, and additional constraints need to be added for a unique solution. With this condition, Eq. (1) can be derived as,

$$\sum_{i=1}^{N} \alpha_j L_i w_j = \sum_{i=1}^{N} \alpha_{j+1} L_i w_{j+1}, \quad j = 1, 2, \dots, M-1$$

$$\alpha_j w_j = \alpha_{j+1} w_{j+1}, \quad j = 1, 2, \dots, M-1 \tag{2}$$

$$\alpha_x w_x = \alpha_j w_j, \quad x = j+1, \quad j = 1, 2, \dots, M-1 \tag{3}$$

$$\sum_{x=1}^{M} \alpha_x = \sum_{x=1}^{M} \frac{\alpha_j w_j}{w_x}, \quad j = 1, 2, \dots, M \tag{4}$$

Using Eq. (4) together with $\sum_{i=1}^{M} \alpha_i = 1$, thus will leads to,

$$\sum_{x=1}^{M} \frac{\alpha_j w_j}{w_x} = 1, \quad j = 1, 2, \dots, M \tag{5}$$

$$\alpha_j = \frac{1}{w_j \left(\sum_{x=1}^{M} \frac{1}{w_x} \right)} \tag{6}$$

Hence, the fraction of load that should be given by data source i to node j is

$$\alpha_{i,j} = \frac{1}{w_j \left(\sum_{x=1}^{M} \frac{1}{w_x} \right)} L_i. \tag{7}$$

3.2 Communication Time Fraction

Scheduling has to consider the bandwidth availability of transfer between computational nodes to which a job is going to be submitted and the storage resource(s) from which the data required is to be retrieved.

In DLT, it is assumed that computation and communication loads can be partitioned arbitrarily among a number of processors and links, respectively. By considering communication time (Z) instead of computation time (w) in Eq. (3), we have

$$\alpha_x Z_{x,y} = \alpha_j Z_{i,j}, \quad x = j = 1, 2, \ldots, M. \tag{8}$$

As mentioned in section 3.1, all links must stop transmitting the fraction of load α_i at the same time. Eq. (8) shows that, it is true for all links and load fractions, in order to obtain the optimality. In order to derive the $\alpha_{i,j}$ from communication time, we have to compute the summation of all links at first as,

$$\sum_{i=1}^{N} \sum_{j=1}^{M} \frac{1}{Z_{i,j}} \tag{9}$$

From Eqs. (6) and (9), thus leads to

$$\alpha_j = \frac{1}{Z_{i,j} \sum_{x=1}^{N} \sum_{y=1}^{M} \frac{1}{Z_{x,y}}}. \tag{10}$$

3.3 The Final Form

After getting the computation time fraction and communication time fraction in (6) and (10) respectively, the final fraction will be given as

$$CM_{i,j} = \frac{1}{w_j \left(\sum_{x=1}^{M} \frac{1}{w_x} \right)} + \frac{1}{Z_{i,j} \sum_{x=1}^{N} \sum_{y=1}^{M} \frac{1}{Z_{x,y}}} \tag{11}$$

$$\alpha_j = \frac{CM_{i,j}}{\sum_{i=1}^{N} \sum_{j=1}^{M} CM_{i,j}} \tag{12}$$

$$\alpha_{i,j} = \frac{CM_{i,j}}{\sum_{i=1}^{N} \sum_{j=1}^{M} CM_{i,j}} L_i. \tag{13}$$

4 Numerical Experiments

To measure the performance of the proposed ADLT model against the previous model, randomly generated experimental configurations were used, see [4]. The network bandwidth between sites is uniformly distributed between 1Mbps and 10Mbps. The location of n data sources (DS_k) is randomly selected and each physical dataset size (L_k) is randomly selected with a uniform distribution in the range of 1GB to 1TB. We assume that the computing time spent in a site i to process a unit dataset of size 1MB is uniformly distributed in the range $1/r_{cb}$ to $10/r_{cb}$ seconds where r_{cb} is the ratio of computation speed to communication speed.

Example, consider a system with one data source and two worker nodes with parameters $w_0 = 0.1$ and $w_1 = 0.2$. Let the size of data source is 100MB. The bandwidth among links $Z_{0,0}$=3Mbps and $Z_{0,1}$=5Mbps. By using the model in [5] and proposed model (see Eq. (13)), we have the expected processing time as shown in Table 1.

Table 1. Expected Processing Time for the CDLT and ADLT Models

	CDLT	ADLT
$T(0)$	28.86	22.2
$T(1)$	13.32	19.2
$Max\{T(1), T(0)\}$	28.86	22.2

Based on CDLT model, the expected processing time $T(0)$(28.86 secs) is more than $T(1)$ (13.32 secs). In other words, the load is not divided equally. With the proposed ADLT model, it has better load balancing. Both node 0 and node 1 have nearly same load and the expected processing time is 22.2 secs. Thus, indicated that the proposed ADLT model performs better in load balancing.

Table 2. Expected Processing Time for the CDLT and ADLT Models

ccRatio	CDLT	ADLT
0.001	29013	5385
0.01	29014	5385
0.1	29018	5687
1	29059	5394
10	29468	5470
100	33563	6646
1000	74507	22091
10000	483954	198496

Table 2 shows the experimental results for the two models. We examined the overall performance of each model by running them under 100 randomly generated Grid configurations. We varied the parameters: ccRatio (0.001 to 10000), M (3 to 20), N (2 to 20) and r_{cb} (10 to 500). From the above values, the expected processing time of the proposed model seems to be the better one. It is clear that, our model balances the loads among the nodes more efficiently. It should be noted that as the *ccRatio* is less than 10, the differences of the expected processing time of the two models are near to be same due to the effect of *ccRatio* in cost model. This model reduced the expected processing time by 80% for communication intensive applications and by 60% for computation intensive applications on an average compared to previous DLT model, as shown in Fig 2.

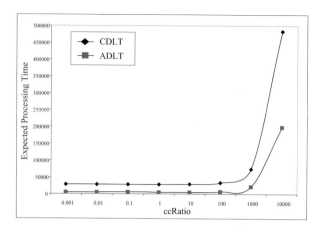

Fig. 2. Expected processing time for the CDLT and ADLT Models

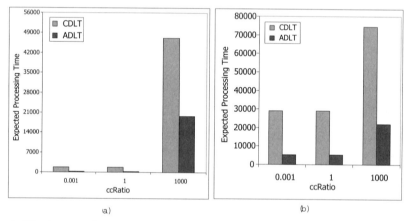

Fig. 3. The impact of output data size to input data size (a) $oiRatio = 0$: No output or small size of output (b) $oiRatio > 0.5$

The impact of the ratio of output data size to input data size is shown in Fig 3. ADLT model performs well for communication intensive applications that generate small output data compared to input data size (low $oiRatio$). For computation intensive applications, the ratio of output data size to input data size does not affect the performance of the algorithms much unless when $ccRatio$ is 10000.

5   Conclusion

The problem of scheduling data-intensive loads on grid platforms is addressed. We used the divisible load paradigm to derive closed-form solutions for processing time considering communication time. In the proposed model, the optimality principle was utilized to ensure an optimal solution. The experiment results of the proposed model ADLT show better performance as compared to the CDLT model in terms of expected processing time and load balancing.

References

1. Holtman K., *et al.*: CMS Requirements for the Grid, In proceeding of the International Conference on Computing in High Energy and Nuclear Physics, Science Press, Beijing China, (2001).
2. Robertazzi T.G.: Ten Reasons to Use Divisible Load Theory, IEEE Computer, **36(5)** (2003) 63-68.
3. Wong H. M., Veeravalli B., Dantong Y., and Robertazzi T. G.: Data Intensive Grid Scheduling: Multiple Sources with Capacity Constraints, In proceeding of the IASTED Conference on Parallel and Distributed Computing and Systems, Marina del Rey, USA, (2003).
4. Kim S., Weissman J. B.: A Genetic Algorithm Based Approach for Scheduling Decomposable Data Grid Applications, In proceeding of the International Conference on Parallel Processing, IEEE Computer Society Press, Washington DC USA, (2004).
5. Venugopal S., Buyya R., Ramamohanarao, K.: A Taxonomy of Data Grids for Distributed Data Sharing, Management and Processing, ACM Computing Surveys, **38(1)** (2006) 1-53.
6. Mequanint, M.: Modeling and Performance Analysis of Arbitrarily Divisible Loads for Sensor and Grid Networks, PhD Thesis, Dept. Electrical and Computer Engineering, Stony Brook University, NY 11794, USA, (2005).
7. Bharadwaj, V., Ghose D., Robertazzi T. G.: Divisible Load Theory: A New Paradigm for Load Scheduling in Distributed Systems, Cluster Computing, **6**, (2003) 7-17.
8. Jaechun N., Hyoungwoo, P.: GEDAS: A Data Management System for Data Grid Environments. In Sunderam *et al.* (Eds.): Computational Science. Lecture Notes in Computer Science, Vol. 3514. Springer-Verlag, Berlin Heidelberg New York (2005) 485-492.

Providing Fault-Tolerance in Unreliable Grid Systems Through Adaptive Checkpointing and Replication

Maria Chtepen[1], Filip H.A. Claeys[2], Bart Dhoedt[1], Filip De Turck[1], Peter A. Vanrolleghem[2], and Piet Demeester[1]

[1] Department of Information Technology (INTEC), Ghent University, Sint-Pietersnieuwstraat 41, Ghent, Belgium
{maria.chtepen, bart.dhoedt, filip.deturck}@intec.ugent.be
[2] Department of Applied Mathematics, Biometrics and Process Control (BIOMATH), Ghent University, Coupure Links 653, Ghent, Belgium
{filip.claeys, peter.vanrolleghem}@biomath.ugent.be

Abstract. As grids typically consist of autonomously managed subsystems with strongly varying resources, fault-tolerance forms an important aspect of the scheduling process of applications. Two well-known techniques for providing fault-tolerance in grids are periodic task checkpointing and replication. Both techniques mitigate the amount of work lost due to changing system availability but can introduce significant runtime overhead. The latter largely depends on the length of checkpointing interval and the chosen number of replicas, respectively. This paper presents a dynamic scheduling algorithm that switches between periodic checkpointing and replication to exploit the advantages of both techniques and to reduce the overhead. Furthermore, several novel heuristics are discussed that perform on-line adaptive tuning of the checkpointing period based on historical information on resource behavior. Simulation-based comparison of the proposed combined algorithm versus traditional strategies based on checkpointing and replication only, suggests significant reduction of average task makespan for systems with varying load.

Keywords: Grid computing, fault-tolerance, adaptive checkpointing, task replication.

1 Introduction

A typical grid system is an aggregation of (widespread) heterogeneous computational and storage resources managed by different organizations. The term "heterogeneous" addresses in this case not only hardware heterogeneity, but also differences in resources utilization. Resources connected into grids can be dedicated supercomputers, clusters, or merely PCs of individuals utilized inside the grid during idle periods (so-called desktop grids). As a result of this highly autonomous and heterogeneous nature of grid resources, failure becomes a commonplace feature that can have a significant impact on the system performance.

Y. Shi et al. (Eds.): ICCS 2007, Part I, LNCS 4487, pp. 454–461, 2007.

A failure can occur due to a resource or network corruption, temporary unavailability periods initiated by resource owners, or sudden increases in resource load. To reduce the amount of work lost in the presence of failure, two techniques are often applied: task checkpointing and replication. The checkpointing mechanism periodically saves the status of running tasks to a shared storage and uses this data for tasks restore in case of resource failure. Task replication is based on the assumption that the probability of a single resource failure is much higher than of a simultaneous failure of multiple resources. The technique avoids task recomputation by starting several copies of the same task on different resources. Since our previous work [1] has extensively studied the task replication issue, this paper is mainly dedicated to the checkpointing approach.

The purpose of checkpointing is to increase fault-tolerance and to speed up application execution on unreliable systems. However, as was shown by Oliner et al. [2], the efficiency of the mechanism is strongly dependent on the length of the checkpointing interval. Overzealous checkpointing can amplify the effects of failure, while infrequent checkpointing results in too much recomputation overhead. As may be presumed, the establishment of an optimal checkpointing frequency is far from a trivial task, which requires good knowledge of the application and the distributed system at hand. Therefore, this paper presents several heuristics that perform on-line adaptive tuning of statically provided checkpointing intervals for parallel applications with independent tasks. The designed heuristics are intended for incorporation in a dynamic scheduling algorithm that switches between job replication and periodic checkpointing to provide fault-tolerance and to reduce potential job delay resulting from the adoption of both techniques. An evaluation in a simulation environment (i.e. DSiDE [3]) has shown that the designed algorithm can significantly reduce the task delay in systems with varying load, compared to algorithms solely based on either checkpointing or replication. This paper is organized as follows: in Section 2 the related work is discussed; the assumed system model is presented in Section 3; Section 4 elaborates on the adaptive checkpointing heuristics and the proposed scheduling algorithm; simulation results are introduced in Section 5; Section 6 summarizes the paper and gives a short overview of future work.

2   Related Work

Much work has already been accomplished on checkpointing performance prediction and determination of the optimal checkpointing interval for uniprocessor and multi-processor systems. For uniprocessor systems, selection of such an interval is for the most part a solved problem [4]. The results for parallel systems are less straightforward [5] since the research is often based on particular assumptions, which reduce the general applicability of the proposed methods. In particular, it is generally presumed that failures are independent and identically distributed. Studies of real systems, however, show that failures are correlated temporally and spatially, are not identically distributed. Furthermore, the behavior of checkpointing schemes under these realistic failure distributions does not follow the behavior predicted by standard checkpointing models [2,6].

Since finding the overall optimal checkpointing frequency is a complicated task, other types of periodic checkpointing optimization were considered in literature. Quaglia [7] presents a checkpointing scheme for optimistic simulation, which is a mixed approach between periodic and probabilistic checkpointing. The algorithm estimates the probability of roll-back before the execution of each simulation event. Whenever the event execution is going to determine a large simulated time increment then a checkpoint is taken prior to this event, otherwise a checkpoint is omitted. To prevent excessively long suspension of checkpoints, a maximum number of allowed event executions between two successive checkpoint operations is fixed. Oliner [8] proposes so-called cooperative checkpointing approach that allows the application, compiler, and system to jointly decide when checkpoints should be performed. Specifically, the application requests checkpoints, which have been optimized for performance by the compiler, and the system grants or denies these requests. This approach has a disadvantage that applications have to be modified to trigger checkpointing periodically.

Job replication and determination of the optimal number of replicas are other rich fields of research [1,9]. However, to our knowledge, no methods dynamically altering between replication and checkpointing were introduced so far.

3   The System Model

A grid system running parallel applications with independent tasks is considered. The system is an aggregation of geographically dispersed sites, assembling collocated interchangeable computational (CR) and storage (SR) resources, and a number of services, such as a scheduler (GSched) and an information service (IS). It is assumed that all the grid components are stable except for CRs. The latter possess a varying failure and restore behavior, which is modelled to mimic reality as much as possible. As outlined by Zhang et al. [6], failures in large-scale distributed systems are mostly correlated and tend to occur in bursts. Besides, there are strong spatial correlations between failures and nodes, where a small fraction of the nodes incur most of the failures.

The checkpointing mechanism is either activated by the running application or by the GSched. In both cases, it takes W seconds before the checkpoint is completed and thus can be utilized for an eventual job restore. Furthermore, each checkpoint adds V seconds of overhead to the job run-time. Both parameters largely depend on size C of the saved job state. There is also a recovery overhead P, which is the time required for a job to restart from a checkpoint. Obviously, the overhead introduced by periodic checkpointing and restart may not exceed the overhead of the job restores without use of checkpointing data. To limit the overhead, a good choice of checkpointing frequency I is of crucial importance. Considering the assumed grid model, I_{opt}^j, which is the optimal checkpointing interval for a job j, is largely determined by the following function $I_{opt}^j = f(E_r^j, F_r, C^j)$, where E_r^j is the execution time of j on the resource r and F_r stands for the mean time between failures of r. Additionally, the value of

I_{opt} should be within the limits $V < I_{opt} < E_r$ to make sure that jobs make execution progress despite of periodic checkpointing.

The difficulty of finding I_{opt} is in fact that it is hard to determine the exact values of the application and system parameters. Furthermore, E_r and F_r can vary over time as a consequence of changing system loads and resource failure/restore patterns. This suggests that a statically determined checkpointing interval may be an inefficient solution when optimizing system throughput. In what follows, a number of novel heuristics for adaptive checkpointing are presented.

4   Adaptive Checkpointing Strategies

4.1   Last Failure Dependent Checkpointing (LFDC)

One of the main disadvantages of unconditional periodic task checkpointing (UTC) is that it performs identically whether the task is executed on a volatile or a stable resource. To deal with this shortcoming, LFDC adjusts the initial job checkpointing interval to the behavior of each individual resource r and the total execution time of the considered task j, which results in a customized checkpointing frequency I_r^j. For each resource a timestamp T_r^f of its last detected failure is kept. When no failure has occurred so far, T_r^f is initiated with the system "start" time. GSched evaluates all checkpointing requests and allows only these for which the comparison $T_c - T_r^f \leq E_r^j$ evaluates to true, where T_c is the current system time. Otherwise, the checkpoint is omitted to avoid unnecessary overhead as it is assumed that the resource is "stable". To prevent excessively long checkpoints suspension, a maximum number of checkpoint omissions can be defined, similar to the solution proposed in [7].

4.2   Mean Failure Dependent Checkpointing (MFDC)

Contrary to LFDC, MFDC adapts the initial checkpointing frequency in function of a resource mean failure interval (MF_r), which reduces the effect of an individual failure event. Furthermore, the considered job parameter is refined from the total job length to the estimation of the remaining execution time (RE_r^j). Each time the checkpointing is performed, MFDC saves the task state and modifies the initial interval I to better fit specific resource and job characteristics. The adapted interval I_r^j, is calculated as follows: if r appears to be sufficiently stable or the task is almost finished $(RE_r^j < MF_r)$ the frequency of checkpointing will be reduced by increasing the checkpointing interval $I_r^j = I_r^j + I$; in the other case it is desirable to decrease I_r^j and thus to perform checkpointing more frequently $I_r^j = I_r^j - I$. To keep I_r^j within a reasonable range, MFDC always checks the newly obtained values against predefined boundaries, in such a way that $I_{min} \leq I_r^j \leq I_{max}$. Both I_{min} and I_{max} can either be set by the application or initialized with default values $I_{min} = V + (E_r^j/100)$ and $I_{max} = V + (E_r^j/2)$. In both equations the V term ensures that time between consecutive checkpoints is never less than the time overhead added by each checkpoint, in which case more time is spent on checkpointing than performing useful computations. After the I_r^j interval expires, either the next checkpointing event is performed, or a flag is set indicating that the checkpointing can be accomplished as soon as the application is able to provide a consistent checkpoint.

In case of rather reliable systems, the calibration of checkpointing interval can be accelerated by replacing the default increment value I by a desirable percentage of total or remaining task execution time.

4.3 Adaptive Checkpoint and Replication-Based Scheduling (ACRS)

Checkpointing overhead can be avoided by providing another means to achieve system fault-tolerance. Replication is an efficient and almost costless solution, if the number of task copies is well chosen and there is a sufficient amount of idle computational resources [1]. On the other hand, when computational resources are scarce, replication is undesirable as it delays the start of new jobs. In this section, an adaptive scheme is proposed that dynamically switches between task checkpointing and replication, based on run-time information about system load. When the load is low, the algorithm is in "replication mode", where all tasks with less than R replicas are considered for submission to the available resources. Different strategies can be defined with regard to the assigment order. The strategy applied in this paper processes jobs according to ascending active replica numbers, which reduces the wait time of new jobs. The selected task replica is then submitted to a grid site s with minimal load $Load_s^{min}$ and the minimum number of identical replicas. The latter is important to reduce the chance of simultaneous replica failure. $Load_s^{min}$ is calculated as follows:

$$Load_s^{min} = min_{s \in S}((\sum_{r \in s} n_r)/(\sum_{r \in s} MIPS_r)), \tag{1}$$

where S is the collection of all sites, n_r is the number of tasks on the resource r; and $MIPS_r$ (Million Instructions Per Second) is the CPU speed of r. Inside the chosen site, the task will be assigned to the least loaded available resource with the smallest number of task replicas. The resource load $Load_r$ is determined as

$$Load_r = n_r/MIPS_r . \tag{2}$$

The algorithm switches to the "checkpointing mode" when idle resource availability IR drops to a certain limit L. In this mode, ACRS rolls back, if necessary, the earlier distributed active task replicas AR_j and starts task checkpointing. When processing the next task j the following situations can occur:

- **$AR_j > 0$:** start checkpointing of the most advanced active replica, cancel execution of other replicas
- **$AR_j = 0$** and **$IR > 0$:** start j on the least loaded available resource within the least loaded site, determined respectively by (2) and (1)
- **$AR_j = 0$** and **$IR = 0$:** select a random replicated job i if any, start checkpointing of its most advanced active replica, cancel execution of other replicas of i, submit j to the best available resource.

5   Simulation Results

Performance of the proposed methods was evaluated using the DSiDE simulator, on the bases of a grid model composed of 4 sites (3 CRs each) with varying availability. The availability parameter, which is defined to be the fraction of the total simulation time that the system spends performing useful work, is modelled as a variation of the CR's failure and restore events [1]. Distribution of these events is identical for each resource inside the same site and is depicted in Table 1. The table also shows the distributions with which the burst-based correlated nature of resource failures is approached. A failure event namely triggers the whole burst (see "Burst size") of connected resource malfunctions spread within a relatively short time interval (see "Burst distribution"). To simplify the algorithm comparisons, a workload composed of identical tasks with the following parameters was considered: $S = 30min, In(inputsize) = Out(outputsize) = 2MB, W = 9s, V = 2s, P = 14s$. Furthermore, each CR has 1 MIPS CPU speed and is limited to process at most 2 jobs simultaneously. Initially, the grid is heavily loaded since tasks are submitted in bursts of 6 up to 25 tasks followed by a short (5 to 20 min) idle period. It is also assumed that the application can start generating the next checkpoint, as soon as it is granted permission from GSched. The described grid model is observed during 24 hours of simulated time.

Table 1. Distributions of site failure and restore events together with distributions of the number and frequency of correlated failures

	Failure	Restore	Burst size	Burst distribution
Site 1	Uniform:1-300(s)	Uniform:1-300(s)	Uniform:1-3	Uniform:300-600(s)
Site 2	Uniform:1-900(s)	Uniform:1-300(s)	Uniform:1-3	Uniform:300-600(s)
Site 3	Uniform:1-2400(s)	Uniform:1-300(s)	Uniform:1-3	Uniform:300-600(s)
Site 4	No failure	-	-	-

The left part of Figure 1 shows the comparison, in terms of successfully executed tasks, between UPC, LFDC and MFDC checkpointing heuristics. The comparison is performed for a varying initial checkpointing frequency. For the MFDC algorithm I_{min} is set to the default value, while no limitation on the I_{max} is imposed. The results show that the performance of UPC strongly depends on the chosen checkpointing frequency. As can be seen on the UPC curve, excessively frequent checkpointing penalizes system performance to a greater extent than insufficient checkpointing. It can be explained by the fact that the considered grid model is relatively stable, with a total system availability of around 75%. In this case the checkpointing overhead exceeds the overhead of task recomputation. LFDC partially improves the situation by omitting some checkpoints. Since the algorithm doesn't consider checkpoint insertion, the performance for an excessively long checkpointing interval is the same as for UPC. Finally, the fully dynamic scheme of MFDC proves to be the most effective. Starting from a random checkpointing frequency it guarantees system performance close to the one provided by UPC with an optimal checkpointing interval.

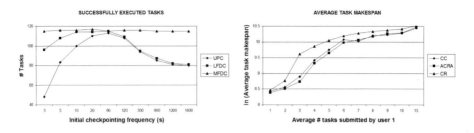

Fig. 1. (a) UPC, LFDC and MFDC checkpointing strategies performance; (b) CC, CR and ACRS scheduling performance

Finally, the ACRS ($R = 2, L = 7, I = 30$) is compared against common checkpointing (CC) and replication-based (CR) algorithms. The term "common checkpointing algorithm" refers to an algorithm monitoring resource downtimes and restarting failed jobs from their last saved checkpoint. The considered CC algorithm, as well as ACRS, makes use of the MFDC heuristic with $I = 30$ to determine the frequency of task checkpointing. The principle of the replication-based algorithm is quite straightforward: $R = 2$ replicas of each task are executed on preferably different resources, if a replica fails it is restarted from scratch [1]. It is clear that replication-based techniques can be efficient only when the system possesses some free resources. Fortunately, most of the observed real grids alternate between peak periods and periods with relatively low load. To simulate this behavior, the initial task submission pattern is modified to include 2 users: the first sends to the grid a limited number of tasks every 35-40 min; the second launches significant batch sizes of 15 up to 20 tasks every 3-5 hours. Figure 1 (right) shows the observed behavior of the three algorithms. During the simulations the number of tasks simultaneously submitted by the first user is modified as shown in the figure, which results in variations in system load among different simulations. When the system load is sufficiently low, ACRS and CC process an equal number of tasks, since each submitted task can be assigned to some resource. However, ACRS results in lower average task makespan. When the system load increases, ACRS switches to checkpointing mode after a short transitive phase. Therefore, the algorithm performs almost analogous to CC, except for a short delay due to the mode switch. Finally, CR considerably underperforms the other algorithms with respect to the number of executed tasks and average makespan. In the considered case ACRS provided for up to 15% reduction of the average task makespan compared to CC. The performance gain certainly depends on the overhead of checkpoints and the number of generated checkpointing events, which is close to optimal for the MFDC heuristic.

6   Conclusions and Future Work

This paper introduces a number of novel adaptive mechanisms, which optimize job checkpointing frequency as a function of task and system properties. The heuristics are able to modify the checkpointing interval at run-time reacting on dynamic system changes. Furthermore, a scheduling algorithm combining checkpointing and replication techniques for achieving fault-tolerance is introduced. The algorithm can significantly reduce task execution delay in systems with varying load by transparently switching, between both techniques.

In the following phase of the research the proposed adaptive checkpointing solutions will be further refined to consider different types of applications, with low and high checkpointing overhead. Also a more robust mechanism will be investigated dealing with temporary unavailability and low response rate of the grid scheduler (that is responsible for managing checkpointing events).

References

1. Chtepen, M., Claeys, F.H.A., Dhoedt, B., De Turck, F., Demeester, P., Vanrolleghem, P.A.: Evaluation of Replication and Rescheduling Heuristics for Grid Systems with Varying Availability. In Proc. of Parallel and Distributed Computing and Systems, Dallas (2006)
2. Oliner, A.J., Sahoo, R.K., Moreira, J.E., Gupta, M.: Performance Implications of Periodic Checkpointing on Large-Scale Cluster Systems. In Proc. of the 19th IEEE International Parallel and Distributed Processing Symposium (IPDPS'05), Washington (2005)
3. Chtepen, M., Claeys, F.H.A., Dhoedt, B., De Turck, F., Demeester, P., Vanrolleghem, P.A.: Dynamic Scheduling of Computationally Intensive Applications on Unreliable Infrastructures. In Proc. of the 2nd European Modeling and Simulation Symposium, Barcelona, Spain (2006)
4. Vaidya, N.H.: Impact of Checkpoint Latency on Overhead Ratio of a Checkpointing Scheme. IEEE Transactions on Computers, **46-8** (1997) 942–947
5. Wong, K.F., Franklin, M.: Checkpointing in Distributed Systems. Journal of Parallel and Distributed Systems, **35-1** (1996) 67–75
6. Zhang, Y., Squillante, M.S., Sivasubramaniam, A., Sahoo, R.K.: Performance Implications of Failures in Large-Scale Cluster Scheduling. In Proc. of the 10th Workshop on Job Scheduling Strategies for Parallel Processing, New York (2004)
7. Quaglia, F.: Combining Periodic and Probabilistic Checkpointing in Optimistic Simulation. In Proc. of the 13th workshop on Parallel and distributed simulation, Atlanta (1999)
8. Oliner, A.J.: Cooperative Checkpointing for Supercomputing Systems. Master Thesis. Massachusetts Institute of Technology (2005)
9. Li, Y., Mascagni, M.: Improving Performance via Computational Replication on a Large-Scale Computational Grid. In Proc. of the 3st International Symposium on Cluster Computing and the Grid, Washington (2003)

A Machine-Learning Based Load Prediction Approach for Distributed Service-Oriented Applications

Jun Wang, Yi Ren, Di Zheng, and Quan-Yuan Wu

School of Computer Science,
National University of Defence Technology,
Changsha, Hunan, China 410073
junwang@nudt.edu.cn

Abstract. By using middleware, we can satisfy the urgent demands of performance, scalability and availability in current distributed service-oriented applications. However to the complex applications, the load peak may make the system suffer extremely high load and the response time may be decreased for this kind of fluctuate. Therefore, to utilize the services effectively especially when the workloads fluctuate frequently, we should make the system react to the load fluctuate gradually and predictably. Many existing load balancing middleware use the *dampening* technology to make the load to be predicative. However, distributed systems are inherently difficult to manage and the dampening factor cannot be treated as static and fixed. The dampening factor should be adjusted dynamically according to different load fluctuate. So we have proposed a new technique based on machine learning for adaptive and flexible load prediction mechanism based on our load balancing middleware.

Keywords: Service-Oriented Applications, Load Prediction, Machine-Learning, Middleware.

1 Introduction

To service many increasing online clients those transmit a large, often busty, number of requests and provide dependable services with high quality constantly, we must make the distributed computing systems more scalable and dependable. Effective load balancing mechanisms must be made use of to distribute the client workload equitably among back-end servers to improve overall responsiveness. Load balancing mechanisms can be provided in any or all of the following layers:

- **Network-based load balancing:** This type of load balancing is provided by IP routers and domain name servers (DNS). Web sites often use network-based load balancing at the *network* layer (layer 3) and the *transport* layer (layer 4).
- **OS-based load balancing:** This type of load balancing is provided by distributed operating systems via *load sharing*, and *process migration* [1] mechanisms.
- **Middleware-based load balancing:** This type of load balancing is performed in middleware, often on a per-session or a per-request basis. The key enterprise

Y. Shi et al. (Eds.): ICCS 2007, Part I, LNCS 4487, pp. 462–465, 2007.

applications of the moment such as astronavigation, telecommunication, and finance all make use of the middleware based distributed software systems to handle complex distributed applications.

There are different realizations of load balancing middleware. For example, stateless distributed applications usually balance the workload with the help of naming service [2]. But this scheme of load balancing just support static non-adaptive load balancing and can't meet the need of complex distributed applications. For more complex applications, the adaptive load balancing schema [3, 4, 5] is needed to take into account the load condition dynamically and avoid override in some node. Many existing load balancing middleware use the *dampening* technology to make the load to be predicative. However, distributed systems are inherently difficult to manage and the dampening factor cannot be treated as static and fixed. So, in this paper we design and implement an efficient load prediction method for service-based applications. Clients can learn the best policy by employing reinforcement based learning technique. Reinforcement based learning is an un-supervised learning, in which agents learn through trial and error interactions with environment.

2 Load Prediction Method for Service-Oriented Applications

2.1 Model of the Load Balancing Middleware

Our middleware provides load balancing for the service-oriented applications, prevents bottlenecks at the application tier, balances the workload among the different services and enables replication of the service components in a scalable way. The core components are as follows:

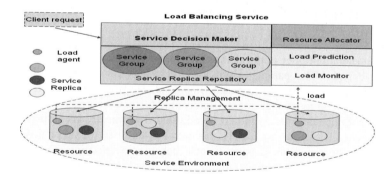

Fig. 1. Components of the Load Balancing Middleware

Service Replica Repository: Instances of services need to register with the Service Group. All the references of the groups are stored in the Service Replica Repository. A service group may include several replicas and we can add or remove replicas to the groups. The main purpose of the service group is to provide a view containing simple information about the references to the locations of all replicas registered with group. The uses need not to know where the replica is located.

Service Decision Maker: This component assigns the best replica in the group to service the request based on the algorithms configured in our load balancing policy [6]. The service decision maker acts as a proxy between the client and the dynamic services. It enables transparency between them without letting the client knowing about the multiple distributed service replicas.

Load Monitor: Load monitor collects load information from every load agent within certain time interval. The load information should be refreshed at a suitable interval so that the information provided is not expired.

Load Agent: The purpose of load agent is to provide load information of the hosts it resides when requested by load monitor. As different services might have replicas in the same host, it is hard to presume the percentage of resource is being used by which service at particular moment. Therefore, a general metric is needed to indicate the level of available resources at the machine during particular moment.

Resource Allocator: The purpose of this component is to dynamically adjust the resource to achieve a balance load distribution among different services. In fact, we control the resource allocation by managing the replicas of different services. For example, this component makes decisions on some mechanisms such as service replication, services coordination, dynamic adjustment and requests prediction.

2.2 Machine-Learning Based Load Prediction

As figure 2 shown, the host will suffer extreme loads at time T1 and T2, and if we computing the load basing on the numbers sampling from these two points we will make wrong load balancing decisions. A useful method is Using control theory technology called *dampening*, where the system minimizes unpredictable behavior by reacting slowly to changes and waiting for definite trends to minimize over-control decisions. The load will be computed in this way:

$$new\_load = multiplier * old\_load + (1 - multiplier) * new\_load \qquad (1)$$

The parameter multiplier is the dampening factor and the value of the multiplier is between 0 and 1. The parameter old_load represents for the former load and the new_load represents for the new sampling load. However, distributed systems are inherently difficult to manage and the dampening factor cannot be treated as static and fixed. The dampening factor should be adjusted dynamically according to different load fluctuate. Therefore, we use the machine-learning based load prediction method where the system minimizes unpredictable behavior by reacting slowly to changes and waiting for definite trends to minimize over-control decisions. The simple

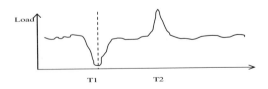

Fig. 2. Affection of the Peak Load

exponential smoothing method is based on a weighted average of current and past observations, with most weight to the current observation and declining weights to past observations. The formula for exponential moving average is given by equation (2): Where $0 \leq \theta \leq 1$ is know as dynamic dampening factor, L_n is the most recent load, φ_n stores the past history and φ_{n+1} is the predicted value of load.

$$\varphi_{n+1} = \theta L_n + (1 - \theta)\varphi_n \tag{2}$$

We maintained two exponentially weighted moving averages with different dynamic dampening factors. A slow moving average ($\theta \to 0$) is used to produce a smooth, stable estimate. A fast moving average ($\theta \to 1$) adapts quickly to changes in work load. The maximum of these two values are used as an account for current load on the service provider.

3   Conclusions

To utilize the services effectively especially when the workloads fluctuate frequently, we should make the system react to the load fluctuate gradually and predictably. Distributed systems are inherently difficult to manage and the dampening factor cannot be treated as static and fixed. The dampening factor should be adjusted dynamically according to different load fluctuate. So we have proposed and implemented a new technique based on machine learning for adaptive and flexible load prediction mechanism based on our load balancing middleware.

Acknowledgements

This work was funded by the National Grand Fundamental Research 973 Program of China under Grant No.2005cb321804 and the National Natural Science Foundation of China under Grant No.60603063.

References

1. Rajkumar, B.: High Performance Cluster Computing Architecture and Systems, ISBN7.5053-6770-6.2001
2. IONA Technologies, "Orbix 2000." www.iona-iportal.com/suite/orbix2000.htm.
3. Othman, O'Ryan, C., Schmidt, D. C.: The Design of an Adaptive CORBA Load Balancing Service. IEEE Distributed Systems Online(2001)
4. Othman, O., Schmidt, D. C.: Issues in the design of adaptive middleware load balancing. In: ACM SIGPLAN, ed. Proceedings of the ACM SIGPLAN workshop on Languages, Compilers and Tools for Embedded Systems. New York: ACM Press(2001)205-213
5. Othman, O., O'Ryan, C., Schmidt, D.C.: Strategies for CORBA middleware-based load balancing. IEEE Distributed Systems Online(2001) http://www.computer.org/dsonline
6. Gamma, E., Helm, R., Johnson, R., Vlissides, J.: Design Patterns: Elements of Reusable Object-Oriented Software. Reading: Addison-Wesley(2002)223-325

A Balanced Resource Allocation and Overload Control Infrastructure for the Service Grid Environment

Jun Wang, Yi Ren, Di Zheng, and Quan-Yuan Wu

School of Computer Science,
National University of Defence Technology,
Changsha, Hunan, China 410073
junwang@nudt.edu.cn

Abstract. For the service based Grid applications, the applications may be integrated by using the Grid services across Internet, thus we should balance the load for the applications to enhance the resource's utility and increase the throughput. To overcome the problem, one effective way is to make use of load balancing. Kinds of load balancing middleware have already been applied successfully in distributed computing. However, they don't take the services types into consideration and for different services requested by clients the workload would be different out of sight. Furthermore, traditional load balancing middleware uses the fixed and static replica management and uses the load migration to relieve overload. Therefore, we put forward an autonomic replica management infrastructure to support fast response, hot-spot control and balanced resource allocation among services. Corresponding simulation tests are implemented and results indicated that this model and its supplementary mechanisms are suitable to service based Grid applications.

Keywords: Web Service, Service based Grid Applications, Load Balancing, Adaptive Resource Allocation, Middleware.

1 Introduction

The Grid services [1, 2] conform to the specifications of the Web Service and it provides a new direction for constructing the Grid applications. The applications may be integrated across the Internet by using the Grid services and the distributed Grid services and resources must be scheduled automatically, transparently and efficiently. Therefore, we must balance the load of the diverse resources to improve the utilization of the resources and the throughput of the systems. Currently, load balancing mechanisms can be provided in any or all of the following layers in a distributed system:

- **Network-based load balancing:** This type of load balancing is provided by IP routers and domain name servers (DNS). However, it is somewhat limited by the fact that they do not take into account the content of the client requests.
- **OS-based load balancing:** At the lowest level for the hierarchy, OS-based load balancing is done by distributed operating system in the form of lowest system level scheduling among processors [3, 4].

Y. Shi et al. (Eds.): ICCS 2007, Part I, LNCS 4487, pp. 466–473, 2007.

- **Middleware-based load balancing:** This type of load balancing is performed in middleware, often on a per-session or a per-request basis. The key enterprise applications of the moment such as astronavigation, telecommunication, and finance all make use of the middleware based distributed software systems to handle complex distributed applications.

There are different realizations of load balancing middleware. For example, stateless distributed applications usually balance the workload with the help of naming service [6]. But this scheme of load balancing just support static non-adaptive load balancing and can't meet the need of complex distributed applications. For more complex applications, the adaptive load balancing schema [7, 8] is needed to take into account the load condition dynamically and avoid override in some node. However, many of the services are not dependable for loose coupling, high distribution in the Grid environment and traditional load balancing middleware pay no attentions to the resource allocation. Therefore, we put forward an autonomic replica management infrastructure based on middleware to support fast response, hot-spot control and balanced resource allocation among different services.

2 Architecture of the Load Balancing Middleware

Our load balancing service is a system-level service introduced to the application tier by using IDL interfaces. Figure 1 features the core components in our service as follows:

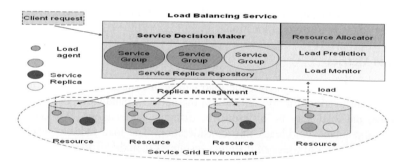

Fig. 1. Components of the Load Balancing Middleware

Service Replica Repository: Instances of services need to register with the Service Group. All the references of the groups are stored in the Repository. A service group may include several replicas and we can add or remove replicas to the groups.

Service Decision Maker: This component assigns the best replica in the group to service the request based on the algorithms configured in our load balancing policy [5]. The service decision maker acts as a proxy between the client and the dynamic services. It enables transparency between them without letting the client knowing about the multiple distributed service replicas.

Load Monitor: Load monitor collects load information (such as CPU utilization) from every load agent within certain time interval. The load information should be refreshed at a suitable interval so that the information provided is not expired.

Load Agent: The purpose of load agent is to provide load information of the Grid hosts it resides when requested by load monitor. As different services might have replicas in the same host, so a general metric is needed to indicate the level of available resources at the machine during particular moment.

Load Prediction: This module use the machine-learning based load prediction method where the system minimizes unpredictable behavior by reacting slowly to changes and waiting for definite trends to minimize over-control decisions.

Resource Allocator: The purpose of this component is to dynamically adjust the resource to achieve a balance load distribution among different services. In fact, we control the resource allocation by managing the replicas of different services.

3   Balanced Resource Allocation and Overload Control

In traditional load balancing middleware, a load balanced distributed application starts out with a given number of replicas and the requests just are distributed among these replicas to balance the workloads. However, different services may need different resources at all and in some occasions such as 911 or world cup the number of some kinds of requests will increase fast while it may be very small in most of the days. When demand increases dramatically for a certain service, the quality of service (QoS) deteriorates as the service provider fails to meet the demand. Therefore, depending on the availability of the resources, such as CPU load and network bandwidth, the number of replicas may need to grow or decrease over time. That is to say, the replicas should be created or destroyed on-demand. So we use an adaptive replica management approach to adjust the number of the replicas on demand to realize the adaptive resource allocation.

3.1   Load Metrics

In traditional distributed systems, the load can be measured by the CPU utilization, the memory utilization, the I/O utilization, the network bandwidth and so on. At the same time, the load may be different for different applications. We take multiple resources and mixed workloads into consideration. The load index of each node is composed of composite usage of different resources including CPU, memory, and I/O which can be calculated with:

$$L_j = \sqrt{\sum_{i=1}^{3} a_i\, k_i^{\,2}} \quad (0 \le a_i \le 1,\; \sum_{i=1}^{3} a_i = 1) \tag{1}$$

L_j denotes the load of host j and k_i denotes the percentage according resource has been exhausted. Furthermore, a_i denotes the weighted value of the certain load metric and the value can be configured differently for diverse applications.

3.2   On-Demand Creation and Destruction of the Replicas

At the beginning of the discussion let us give some definitions firstly. Let $H = \{h_1, h_2 ..., h_i\}$ where h_j represents the jth node of the system and let $S = \{s_1, s_2..., s_l\}$ where s_k represents the kth service of the system. Furthermore, let N_k represents the number of the replicas of the kth service of the system. So the set of the replicas of the kth service can be denoted by $R(S_k) = \{S_{k1},...S_{kN_k}\}$. At the same time, the host the mth replica of the lth service is residing in can be denoted by $H(S_{lm})$ whose load at time t is denoted by $L_{H(S_{lm})}(t)$.

The first problem is when to create the new replica. In normal conditions, new requests will be dispatched to the fittest replicas. But all the hosts the initial replicas residing in may be in high load and the new requests will cause them to be overload. So we should create new replicas to share the workload. We set the **replica_addition** threshold to help to make the decisions. For example, to the ith service of the system, if the equation (**2**) can be true, then new replica will be created.

$$\forall x(L_{H(S_{ix})}(t) \geq replica\_addition) \quad x \in R(S_i) \tag{2}$$

The second problem is where to create the new replica. As the load metrics we have discussed before we can compute the workloads of the hosts. Furthermore, the hosts may be heterogeneous and the workload of each host is different. In fact, we set the **replica_deployment** threshold for every host. According to the equation (**3**), if the workloads of some hosts don't exceed the threshold the new replica can be created on them. Otherwise no replica will be created because of the host will be overloaded soon and all the system will become unstable for the creation. Therefore the incoming requests should be rejected to prevent failures.

$$\exists x(L_{H_x}(t) < replica\_deployment) \quad x \in \{1,...i\} \tag{3}$$

The third problem is to create what kind of replicas. Because the applications may be composed of different services and the services may all need create new replicas. However, the services may have different importance and we should divide them with different priorities. Therefore, we classify the services as high priority services, medium priority services and low priority services according to the importance of them. The services having different priorities may have different maximum replicas. For example, supposing the number of the hosts is n, then the maximum number of the high priority can be n, the maximum number of the medium priority service can be $\lfloor \frac{n}{2} \rfloor$ and the maximum number of the low priority service can be $\lfloor \sqrt{n} \rfloor$. These configurations can be revised according to practical needs. Secondly, the Resource Allocator module maintains three different queues having different priorities. Each queue is managed by the FIFO strategy and the replicas of the services having higher priority will be created preferentially.

The last problem is the elimination of the replicas. For the coming of the requests may fluctuate. If the number of the requests is small, then monitoring multiple replicas and dispatching the requests among them is not necessary and wasteful. We should eliminate some replicas to make the system to be efficient and bring down the extra overhead. So we set the **replica_elimination** threshold to control the elimination of the replicas. For example, to the ith service of the system, if the equation (**4**) can be true, then some certain replica will be eliminated.

$$\exists x(L_{H(S_{ix})}(t) < replica\_elimination) \quad x \in R(S_i) \tag{4}$$

The elimination will be performed continuously until the equation (4) becomes false or the number of the replicas becomes one.

Furthermore, in some unusual occasions such as 911 and world cup some certain simple services will become hot spot. Therefore, we should adjust the priority of the services according to the number of the incoming requests to avoid overload. The low priority services may have higher priority with the increasing client request and more replicas will be created to response the requests. At the same time, once the number of the requests decreases, the priority of these services will be brought down and the exceeding replicas will be eliminated.

4 Performance Results

As depicted in figure 2, our middleware StarLB is running on RedHat Linux 9/Pentium4/256/1.7G. The clients are running on Windows2000/Pentium4/512M/2G. Furthermore, to compare the results easier the Grid hosts are as same as each other and all of the hosts are running on Windows2000/Pentium4/512M/2G.

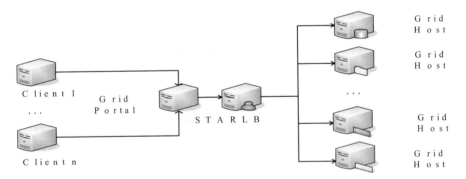

Fig. 2. Load Balancing Experiment Testbed

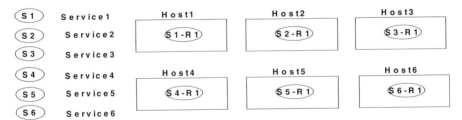

Fig. 3. Initial Replicas of different Services

At the beginning of this test, we used the services without the help of the autonomic replica management. Supposing there are six hosts and there are six different services. (This test just be used to analyze and present the mechanisms and more complex tests using many more hosts and services are ignored here.) Among these services there are two high priority services, two medium priority services and two low priority services. Among the six services the service1 and the service2 are high-priority services, the service3 and the service4 are medium-priority services and the remaining two are low-priority services. All the services have only one initial replica. Each replica resides in a host respectively and response to the client requests. The distribution of the replicas is as depicted in figure 3.

Furthermore, we set the low priority service can have at most two replicas, the medium priority service can have three replicas and the high priority service can have six replicas. As depicted in figure 4(a) and figure 4(b), according to our setting when the load index arrived at 85% new replica was created. From the two above broken lines in figure 4(a) we can see with the requests coming new replicas was created in the hosts the low priority services residing in. At the same time, because of the creation of new replicas the response time of the high priority services could be brought down and the throughput of these services was increased efficiently. Furthermore, the workload of all the hosts was balanced efficiently. All the creations are depicted in figure 5 and the services with higher priority may create new replicas more preferentially and have larger maximum replica number.

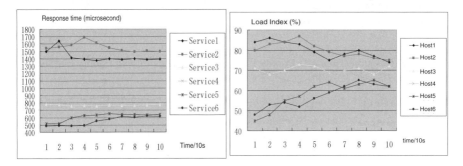

Fig. 4. (a) Response Time with Replica Management. (b)Load Index with Replica Management.

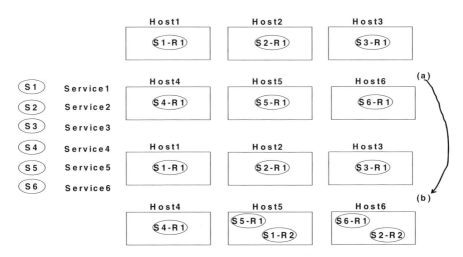

Fig. 5. Creation and Elimination of the Replicas

Then there is still a question to be discussed. That is the elimination of the extra replicas and the elevation of the priority. We deployed all the replicas as the initial state depicted in the figure 3 and made the service6 become hot spot. As the figure 6(a) and the figure 6(b) depicted, adding the number of the requests of the services6 gradually. Then the CPU utilization of the Host6 and the response time of the service6 increased too. According to our setting of the ***replica_addition*** threshold, when the Load index arrived at 85% new replica was created. For the load index of the Host5 was lowest, a new replica of the service6 was created in the Host5. By the creation of the new replica, the Load index of the Host6 decreased as well as the response time. At the same time, the response time of the service5 was just affected a little.

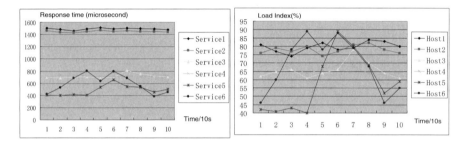

Fig. 6. (a) Response Time with Priority Elevation. (b)Load index with Priority Elevation.

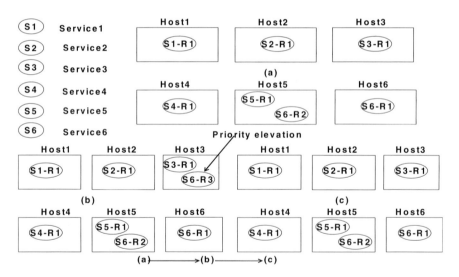

Fig. 7. Creation and Elimination of the Replicas with Priority Elevation

However, because the requests for the service 6 kept increasing and the Load index arrived at 85% again. As we have discussed before the service6 is low priority service and the maximum number of the replicas is two. So the priority of the service6 should be elevated to allow the creation of the new replicas. Then the new replica was created in the Host3 and the requests could be distributed with the help of the new replica. At last, when we decreased the requests of the service6 .The Load index decreased and too many replicas were not necessary. So the priority of the service6 should be decreased and unnecessary replicas should be eliminated. As depicted in the figure 6(b), when the Load index was below the ***replica_elimination*** threshold the replica in the Host3 was eliminated for the highest Load index among the three replicas. The remaining two replicas were keep dealing with the requests until the Load index shall become higher or lower. All the creation and elimination of the replicas with priority elevation are depicted in the figure 7.

5 Conclusions

Kinds of load balancing middleware have already been applied successfully in distributed computing. However, they don't take the services types into consideration and for different services requested by clients the workload would be different out of sight. Furthermore, traditional load balancing middleware uses the fixed and static replica management and uses the load migration to relieve overload. Therefore, we put forward an autonomic replica management infrastructure based on middleware to support fast response, hot-spot control and balanced resource allocation among different services. Corresponding simulation tests are implemented and their result s indicated that this model and its supplementary mechanisms are suitable to service based Grid applications.

Acknowledgments

This work was funded by the National Grand Fundamental Research 973 Program of China under Grant No.2005cb321804 and the National Natural Science Foundation of China under Grant No.60603063.

References

1. FOSTER, I., KESSELMAN, C., NICK, J.: Grid services for distributed system integration [J]. Computer(2002)37 -46
2. http://www.gridforum.org/ogsi-wg/drafts/GS_Spec_draft03_2002-07-17.pdf
3. Chow, R., Johnson, T.: Distributed Operating Systems and Algorithms, Addison Wesley Long, Inc.(1997)
4. Rajkumar, B.: High Performance Cluster Computing Architecture and Systems, ISBN7.5053-6770-6.2001.
5. Gamma, E., Helm, R., Johnson, R., Vlissides, J.: Design Patterns: Elements of Reusable Object-Oriented Software. Reading: Addison-Wesley(2002)223-325.
6. IONA Technologies, "Orbix 2000." www.iona-iportal.com/suite/orbix2000.htm.
7. Othman, C., O'Ryan, Schmidt, D. C.: The Design of an Adaptive CORBA Load Balancing Service. IEEE Distributed Systems Online, vol. 2, (2001)
8. Othman, O., Schmidt, D.C.: Issues in the design of adaptive middleware load balancing. In: ACM SIGPLAN, ed. Proceedings of the ACM SIGPLAN workshop on Languages, Compilers and Tools for Embedded Systems. New York: ACM Press(2001)205-213.

Recognition and Optimization of Loop-Carried Stream Reusing of Scientific Computing Applications on the Stream Processor

Ying Zhang, Gen Li, and Xuejun Yang

Institute of Computer, National University of Defense Technology, 410073 Changsha China
zhangying@nudt.edu.cn

Abstract. Compared with other stream applications, scientific stream programs are usually constrained by memory access. Loop-carried stream reusing means reusing streams across different iterations and it can improve the locality of SRF greatly. In the paper, we present algorisms to recognize loop-carried stream reusing and give the steps to utilize the optimization after analyzing characteristics of scientific computing applications. Then we perform several representative microbenchmarks and scientific stream programs with and without our optimization on Isim. Simulation results show that stream programs optimized by loop-carried stream reusing can improve the performance of memory-bound scientific stream programs greatly.

1 Introduction

Now conventional architecture has been not able to meet the demands of scientific computing[1][2]. In all state-of-the-art architectures, the stream processor[3](as shown in fig. 1) draws scientific researchers' attentions for its processing computation-intensive applications effectively[4-8]. Compared with other stream applications, scientific computing applications have much more data, more complex data access methods and more strong data dependence.

The stream processor has three level memory hierarchies[12] – local register files (LRF) near ALUs exploiting locality in kernels, global stream register files (SRF) exploiting producer-consumer locality between kernels, and streaming memory system exploiting global locality. The bandwidth ratios between three level memory hierarchies are large. In Imagine[9][10], the ratio is 1:13:218. As a result, how to enhance the locality of SRF and LRF and consequently how to reduce the chip-off memory traffics become key issues to improve the performance of scientific stream programs constrained by memory access. Fig. 2 shows a stream flows across three level memory hierarchies during the execution of a stream program. First, the stream is loaded from chip-off memory into SRF and distributed into corresponding buffer. Then it is loaded from SRF to LRF to supply operands to a kernel. During the execution of the kernel, all records participating in kernel and temporary results are saved in LRF. After the kernel is finished, the records are stored back to SRF. If there is producer-consumer locality between this kernel and its later kernel, the stream is saved in SRF. Otherwise, it is stored back to chip-off memory.

Y. Shi et al. (Eds.): ICCS 2007, Part I, LNCS 4487, pp. 474–481, 2007.

Loop-carried stream reusing means reusing streams across different iterations and it can improve the locality of SRF. In the paper, we present algorisms to recognize loop-carried stream reusing and give steps to use our method to optimize stream programs according to the analysis of typical scientific computing applications. We give the recognition algorism to decide what applications can be optimized and the steps how to utilize loop-carried stream reusing to optimize stream organization. Then we perform several representative microbenchmarks and scientific stream programs with and without our optimization on a cycle-accurate stream processor simulator. Simulation results show that the optimization method can improve scientific stream program performance efficiently.

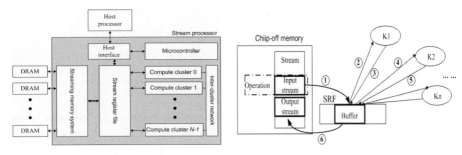

Fig. 1. Block diagram of a stream processor

Fig. 2. Stream flowing across the memory system

2 Loop-Carried Stream Reusing

Loop-carried stream reusing is defined as that between neighboring loop iterations of a stream-level program, input or output streams of kernels in the first iteration can be used as input streams of kernels in the second iteration. If input streams are reused, we call it loop-carried input stream reusing. Otherwise, we call it loop-carried output stream reusing. The essential of stream reusing optimization is to enhance the locality of SRF. Correspondingly, input stream reusing can enhance producer-producer locality of SRF while output stream reusing can enhance producer-consumer locality of SRF.

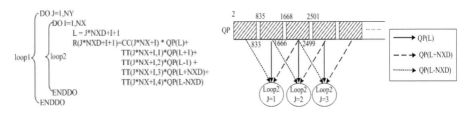

Fig. 3. Example code

Fig. 4. Data trace of array QP references

Then we take code in fig. 3 from stream program MVM as example to depict our methods, where NXD equals to NX+2. In the paper, we let NX and NY equal to 832.

Fig. 4 shows the data trace of QP(L), QP(L+NXD) and QP(L-NXD) participating in loop2 of fig. 3. QP(1668,2499) is QP(L+NXD) of loop2 when J=1, QP(L) of loop2 when J=2, and QP(L-NXD) of loop2 when J=3. So, stream QP can be reused between different iterations of loop1. If QP(1668,2499) is organized as a stream, it will be in SRF after loop2 with J=1 finishes. Consequently, when loop2 with J=2 or J=3 running, it doesn't get stream QP(1668,2499) from chip-off memory but SRF.

```
DO I₁ = L₁, U₁
DO I₂ = L₂, U₂
... ...
    DO I_D = L_D, U_D
I1  ... = A(F₁(I₁,...I_D),...,F_M(I₁,...I_D))
I2  ... = A(G₁(I₁,...I_D),...,G_M(I₁, ... I_D))
    ENDDO
    ... ...
ENDO
ENDDO
```

Fig. 5. A D-level nested loop

2.1 Recognizing Loop-Carried Stream Reusing

Fig. 5 shows a generalized perfect nest of D loops. The body of the loop nest reads elements of the m-dimensional array A twice. In the paper, we only consider linear subscription expressions, and the ith dimension subscription expression of the first array A reference is denoted as $F_i = \sum C_{i,j} * I_j + C_{i,0}$, where I_j is an index variable, $1 \leq j \leq D$, $C_{i,j}$ is the coefficient of I_j, $1 \leq j \leq D$, and $C_{i,0}$ is the remaining part of the subscript expression that does not contain any index variables. Correspondingly, the ith dimension subscription expression of the second array A reference is denoted as $G_i = \sum C'_{i,j} * I_j + C'_{i,0}$. If in the Pth level loop, the data trace covered by the leftmost Q dimensions[1] of one array A read references in the nest with $I_P=i$ is the same to that of the other array A read references in the nest with $I_P=i+d$, where d is a const, and they are different to each other in the same nest, such as the data trace of array QP in loop2 in fig. 3, the two array A read references can be optimized by input stream reusing in respect of loop P and dimension Q. Then we give the algorism of **R**ecognizing **L**oop-carried **I**nput **S**tream **R**eusing.

RLISR. Two array references in the same loop body can be optimized by input stream reusing in respect of loop P and dimension Q if:

(1) when $M = 1$, i.e. array A is a 1-D array, the subscript expressions of two array A references can be written as $F_1 = \sum C_{1,j} * I_j + C_{1,0}$ and $G_1 = \sum C'_{1,j} * I_j + C'_{1,0}$ respectively, and Q=1 now. The coefficients of F1 and G1 should satisfy following formulas:

$$\forall j : ((1 \leq j \leq D) \wedge C_{1,j} = C'_{1,j}) \tag{a}$$

$$C_{1,0} = C'_{1,0} \pm d * C'_{1,P} \tag{b}$$

$$\forall j : ((1 \leq j \leq D) \wedge (C_{1,j} \geq \sum_{j=1}^{D} C_{1,j} * (U_j - L_j)) \tag{c}$$

---

[1] In this paper, we assume the sequence of memory access is the leftmost dimension first just like as that in FORTRAN.

Formula (1a) and (1b) ensure when the loop indices of outmost P-1 loops are given and the loop body passes the space of innermost D-P loops, the data trace of one array A read reference in the nest with $I_P=i$ is the same to that of the other array A read references in the nest with $I_P=i+d$. d is a const and we specify d=1 or 2 according to[13]. When I_1, …, I_{P-1}, are given and I_P, …, I_D vary, formula(1c) restricts the data trace of one array A references in the nest with $I_P=i$ from overlapping that of the other in the nest with $I_P=i +d$. This formula ensures the characteristic of stream process, i.e. data as streams is loaded into the stream processor to process and reloaded into SRF in batches after process. For the stream processor, the cost of random access records of streams is very high.

(2) when $M \neq 1$, i.e. array A is a multi-dimensional array, the subscript expressions of two array A read references should satisfy following conditions:

(d) the Qth dimension subscript expression of one array is gotten by translating the index I_P in the dimension subscript expression of the other array by d, i.e. $G_Q = F_Q(I_P \pm d)$, and,

(e) all subscript expressions of one array A reference are the same with those of the other except the Qth dimension subscript expression, i.e. $\forall i((i \neq Q) \wedge (F_i = G_i))$, and,

(f) for the two array A reference, the innermost index variable in one subscript expression will not appear in any righter dimension subscript expressions, i.e.

$$\forall i(\ni j(C_{i,j} \neq 0 \wedge \forall j'(j'>j \wedge C_{i,j'} =0)) \rightarrow \forall i'(i'>i \wedge C_{i',j} =0)) \wedge \forall i(\ni j(C_{i,j} \neq 0 \wedge \forall j'(j'>j \wedge C_{i,j'} =0)) \rightarrow \forall i'(i'>i \wedge C_{i',j} =0)).$$

It can be proved that data access trace of two array references decided by condition (2) satisfies condition (1), and when U_j-I_j is large enough, they are equivalent.

The algorism of **R**ecognizing **L**oop-carried **O**utput **S**tream **R**eusing is similar to RLISR except that reusing stream mustn't change original data dependence. Then we give the RLOSR algorism without detailed specifications.

RLOSR. We denote the subscript expressions of read references as F_i and those of write references as G_i. Two array references in loop body can be optimized by output stream reusing in respect of loop P and dimension Q if:

(3) when $M = 1$, the coefficients of F1 and G1 should satisfy following formulas:

$$\forall j : ((1 \leq j \leq D) \wedge C_{1,j} = C'_{1,j}) \tag{g}$$

$$C_{1,0} = C'_{1,0} + d * C'_{1,P} \tag{h}$$

$$\forall j : ((1 \leq j \leq D) \wedge (C_{1,j} \geq \sum_{i=1}^{D} C_{1,j} * (U_j - L_j)) \tag{i}$$

(4) when $M \neq 1$, the subscript expressions of two array A read references should satisfy following formulas:

$$G_Q = F_Q(I_P + d) \tag{j}$$

$$\forall i((i \neq Q) \wedge (F_i = G_i)) \tag{k}$$

$$\forall i(\ni j(C_{i,j} \neq 0 \wedge \forall j'(j'>j \wedge C_{i,j} =0)) \rightarrow \forall i'(i'>i \wedge C_{i',j} =0)) \wedge \forall i(\ni j(C_{i,j} \neq 0 \wedge \forall j'(j'>j \wedge C_{i,j} =0)) \rightarrow \forall i'(i'>i \wedge C_{i',j} =0)) \tag{l}$$

2.2 Optimizing Loop-Carried Stream Reusing

Then we give the step of using input stream reusing method to optimize stream organization.

Step A. Organize different array A references in the innermost D-P loops as stream A1 and A2 according their data traces.

Step B. Organize all operations on array A in the innermost D-P loops as a kernel.

Step C. Organize all operations in the outmost P loops as stream-level program.

When the nest with $I_P=i$ of loop P in stream-level program operates on stream A1 and A2, one of them has been loaded into SRF by the former nest, which means that the kernel doesn't get it from chip-off memory but SRF. From the feature of the stream processor architecture, we can know the time to access chip-off memory is much larger than that to access SRF, so the method of stream reusing can improve stream program performance greatly.

The steps to use output stream reusing are analogous to steps above.

In stream program MVM unoptimized, we organize different array QP read in loop1 as three streams according their data trace ,and. organize operations in loop1 as a kernel The length of each stream is 832*832. When running, the stream program must load these three streams into SRF, the total length of which is 692224*3, nearly three times of that of array QP.

By the stream reusing method above, we organize different array QP read references in loop2 as three streams according their own data trace, organize operations in loop2 as a kernel, and organize operations in loop1 except loop2 as stream-level program. Thus there would be 832*3 streams in the stream program loop1, and the length of each is 832. So in stream program loop1, stream QP(L), QP(L+NXD) and QP(L-NXD) of neighboring iterations can be reused. As a result, the stream program only need load 832 streams with the length of 832 from chip-off memory to SRF, the total length of which is 692224, nearly 1/3 of that of unoptimized program.

3 Experiment

We compare the performance of microbenchmarks and several scientific applications optimized and unoptimized by stream reusing. All applications are run on a cycle-accurate simulator for a single-node Imagine stream processor, Isim[9][10].

Table 1 summarizes the test programs used for evaluation. Microbenchmarks listed in the upper half of the table stress particular aspects of loop-carried stream reusing, e.g., if there is an input stream reusing between adjacent loop nests in respect of loop 2 and dimension 2, the benchmark is named P2Q2d1. All microbenchmarks are stream programs of applications in fig. 6 in FORTRAN code. P2Q2d1, P3Q3d1, P3Q3d1O and P3Q3d2 are corresponding stream programs of 6(a), 6(b), 6(c) and

DO I1=1,N	DO I1=1,N	DO I1=1,N	DO I1=1,N
DO I2=1,N	DO I2=1,N	DO I2=1,N	DO I2=1,N
DO I3=1,N	DO I3=1,N	DO I3=1,N	DO I3=1,N
C(I3,I2,I1)=(A(I3,I2,I1)+B(I3,I2,I1)) *A(I3,I2+1,I1)	C(I3,I2,I1)=(A(I3,I2,I1)+B(I3,I2,I1)) *A(I3+1,I2,I1)	A(I3+1,I2,I1)=(A(I3,I2,I1)+B(I3,I2,I1)) *C(I3,I2,I1)	C(I3,I2,I1)=(A(I3,I2,I1)+B(I3,I2,I1)) *A(I3+2,I2,I1)
ENDDO	ENDDO	ENDDO	ENDDO
ENDDO	ENDDO	ENDDO	ENDDO
ENDDO	ENDDO	ENDDO	ENDDO
(a)	(b)	(c)	(d)

Fig. 6. FORTRAN code of applications to be optimized

Table 1. Benchmark programs

Name	Description
P2Q2d1	P=Q=2, d=1, and optimized by input stream reusing.
P2Q2d1l2	same application as P2Q2d1 except that we don't optimize it by stream reusing but organize array references of the innermost 2 loops as streams
P2Q2d1l3	same as P2Q2d1l2 except that array references of all 3 loops are organized as streams
P3Q3d1	P=Q=3, d=1, and optimized by input stream reusing
P3Q3d1O	same as P3Q3d1 except that it is optimized by output stream reusing
P3Q3d2	same as P3Q3d1 except that d=2
QMR.	ab. of QMRCGSTAB, a subspace method to solve large nonsymmetric sparse linear systems[14] whose coefficient array size is 800*800
MVM	a subroutine of a hydrodynamics application and computing band matrix multiplication with the size of 832*832
Laplace	calculating the central difference of two-dimension array whose size is 256*256

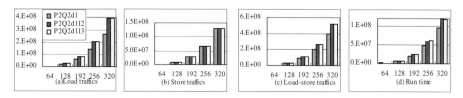

Fig. 7. With the increase of array size the performance of different stream implementations of the application in 6(a) in respect of memory traffics(bytes) and run time(cycles)

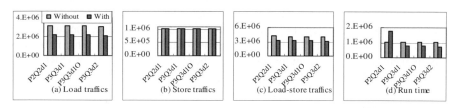

Fig. 8. Performance of P2Q2d1, P3Q3d1, P3Q3d1O and P3Q3d2 with array size of 64 in respect of memory traffics(bytes) and run time(cycles)

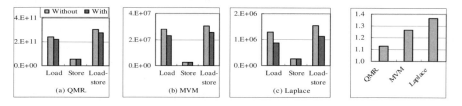

Fig. 9. Effects of stream reusing on the memory traffics of scientific programs

Fig. 10. Speedup of scientific programs with stream reusing

6(d), which is optimized by loop-carried stream reusing. P2Q2d1l2 and P2Q2d1l3 are corresponding stream programs of 6(a) without optimization. There are 2*N out of 4*N streams that can be reused as N stream in SRF in every microbenchmark except

P2Q2d1, in which there are $2*N^2$ out of $4*N^2$ streams that can be reused as N^2 stream in SRF . Scientific applications listed in the lower half of the table are all constrained by memory access. 14994 out of 87467 streams in QMR. can be reused as 4998 streams in SRF, 3 out of 8 streams in MVM can be reused as 1 stream in SRF, and 3 out of 5 streams in Laplace can be reused as 1 stream in SRF.

Fig. 7 shows the performance of different stream implementations of the application in 6(a) with the increase of array size. Fig. 7(a) shows chip-off memory load traffics, fig. 7(b) shows store traffics, fig. 7(c) shows the total chip-off memory traffics, and fig. 7(d) shows the run time of these implementations. In fig. 7(a), the load traffics of P2Q2d1 are nearly 2/3 of the other two implementations whatever the array size is. This is because input loop-carried stream reusing optimization finds the loop-carried stream reusing, improves the locality of SRF and consequently reduces the load memory traffics. In fig. 7(b) the store traffics of different implementations are the same because there is only input stream reusing, which has no effect on store traffics. From fig. 7(c), we can see that because loop-carried stream reusing reuses 2 input streams as one stream in SRF, it cut down the total memory traffics obviously. In fig. 7(d), when the array size is 64, the run time of P2Q2d1 is larger than the other two implementations. When the array size is 128, the run time of P2Q2d1 is a little larger than the other two implementations. The reason for above is that when the array size is small, the stream length of P2Q2d1 is much shorter than and the number of streams are larger than the other two implementations. As a result, the overheads to prepare to load streams from chip-off memory to SRF weigh so highly that they can't be hidden, including the time the host writes SDRs(Stream Descriptor Register) and MARs(Memory Access Register).With the increase of the array size, the run time of P2Q2d1 is smaller and smaller than the other two implementations. This is because with the increase of the stream length, the overheads to load streams into SRF weigh more and more highly and consequently the overheads to prepare loading streams can be hidden well. The memory traffics of P2Q2d1 are the least and consequently the P2Q2d1 program performance is the highest.

Fig. 8 shows the performance of P2Q2d1, P3Q3d1, P3Q3d1O and P3Q3d2 with array size of 64. Fig. 8(a) shows chip-off memory load traffics, fig. 8(b) shows store traffics, fig. 8(c) shows the total chip-off memory traffics, and fig. 8(d) shows the run time of them. These applications are representative examples of loop-carried stream reusing. In fig. 8(a), 8(b) and 8(c), chip-off memory load, store and total traffics have similar characteristics as those in fig. 7. In fig. 8(d), the performance of all applications except P2Q2d1 have been improved by stream reusing optimization. The reason for the reduction of P2Q2d1 performance has been given above. The results show that these representative applications optimized by loop-carried stream reusing all get similar performance increase as that in fig. 7.

Fig. 9 shows the effects of stream reusing on the memory traffics of scientific programs used in our experiments. Fig. 10 shows the speedup yielded by scientific applications with stream reusing over without. All these applications are optimized by input stream reusing. From results, we can see that because all these applications are constrained by memory access, the improvement of application performance brought by stream reusing is nearly in proportion to the amount of streams that can be reused.

4 Conclusion and Future Work

In this paper, we give a recognition algorism to decide what applications can be optimized and the steps how to utilize loop-carried stream reusing to optimize stream organization. Several representative microbenchmarks and scientific stream programs with and without our optimization are performed on Isim which is a cycle-accurate stream processor simulator. Simulation results show that the optimization method can improve the performance of scientific stream program constrained by memory access efficiently.

In the future, we are devoted to developing more programming optimizations to take advantage of architectural features of the stream processor for scientific computing applications.

References

1. W. A. Wulf, S. A. McKee. Hitting the memory wall: implications of the obvious. Computer Architecture News, 1995. 23(1): 20-24.
2. D. Burger, J. Goodman, A. Kagi. Memory bandwidth limitations of future microprocessors. In Proceedings of the 23rd International Symposium on Computer Architecture, Philadelphia, PA, 1996.78-89.
3. Saman Amarasinghe, William. Stream Architectures. In PACT 2003, September 27, 2003.
4. Merrimac – Stanford Streaming Supercomputer Project, Stanford University, http://merrimac.stanford.edu/
5. William J. Dally, Patrick Hanrahan, et al., "Merrimac: Supercomputing with Streams", SC2003, November 2003, Phoenix, Arizona.
6. Mattan Erez, Jung Ho Ahn, et al., "Merrimac - Supercomputing with Streams", Proceedings of the 2004 SIGGRAPH GP^2 Workshop on General Purpose Computing on Graphics Processors, June 2004, Los Angeles, California.
7. Wang Guibin, Tang Yuhua, et al., "Application and Study of Scientific Computing on Stream Processor", Advances on Computer Architecture (ACA'06), August 2006, Chengdu, China.
8. Du Jing, Yang Xuejun, et al., "Implementation and Evaluation of Scientific Computing Programs on Imagine", Advances on Computer Architecture (ACA'06), August 2006, Chengdu, China.
9. MScott Rixner, Stream Processor Architecture. Kluwer Academic Publishers. Boston, MA, 2001.
10. Peter Mattson. A Programming System for the Imagine Media Processor. Dept. of Electrical Engineering. Ph.D. thesis, Stanford University, 2002
11. 11.Ola Johnsson, Magnus Stenemo, Zain ul-Abdin. Programming & Implementation of Streaming Applications. Master's thesis, Computer and Electrical Engineering Halmstad University, 2005.
12. Ujval J. Kapasi, Scott Rixner, et al., Programmable Stream Processor, IEEE Computer, August 2003
13. Goff, G., Kennedy, K., and Tseng, C.-W. 1991. Practical dependence testing. In Proceedings of the SIGPLAN '91 Conference on Programming Language Design and Implementation. ACM, New York.
14. A Quasi-Minimal Residual Variant Of The Bi-Cgstab Algorithm For Nonsymmetric Systems (1994) T. F. Chan, E. Gallopoulos, V. Simoncini, T. Szeto, C. H. TongSIAM Journal on Scientific Computing.

A Scalable Parallel Software Volume Rendering Algorithm for Large-Scale Unstructured Data

Kangjian Wangc and Yao Zheng*

College of Computer Science, and Center for Engineering and Scientific Computation,
Zhejiang University, Hangzhou, 310027, P.R. China
kangjian.wang@hotmail.com, yao.zheng@zju.edu.cn

Abstract. In this paper, we develop a highly accurate parallel software scanned cell projection algorithm (PSSCPA) which is applicable of any classification system. This algorithm could handle both convex and non-convex meshes, and provide maximum flexibilities in applicable types of cells. Compared with previous algorithms using 3D commodity graphics hardware, it introduces no the volume decomposition and rendering artifacts in the resulting images. Finally, high resolution images generated by the algorithm are provided, and the scalability of the algorithm is demonstrated on a PC Cluster with modest parallel resources.

Keywords: Parallel volume rendering, cell projection, software volume rendering.

1 Introduction

Traditionally, parallel volume rendering algorithms were designed to run on expensive parallel machines like SGI Power Challenge, IBM SP2, or SGI Origin 2000 [1, 2, 3, 4]. Recently, however, the decreasing cost and high availability of commodity PCs and network technologies have enabled researchers to build powerful PC clusters for large-scale computations. Scientists can now afford to use clusters for visualization calculations either for runtime visual monitoring of simulations or post-processing visualization. Therefore, parallel software volume rendering on clusters is becoming a viable solution for visualizing large-scale data sets.

We develop a highly accurate Parallel Software Scanned Cell Projection Algorithm (PSSCPA) in this paper. The algorithm employs a standard scan-line algorithm and a partial pre-integration method proposed by Moreland and Angel [5], and thus supports the rendering of data with any classification system. It could handle meshes composed of tetrahedra, bricks, prisms, wedges, and pyramids, or the complex of these types of cells. The PSSCPA runs on the distributed-memory parallel architectures, and uses asynchronous *send/receive* operations and a multi-buffer method to reduce communication overheads and overlap rendering computations. Moreover, we use a hierarchical spatial data structure and an A-Buffer [6] technique to allow early ray-merging to take place within a local neighborhood.

---

* Corresponding author.

Y. Shi et al. (Eds.): ICCS 2007, Part I, LNCS 4487, pp. 482–489, 2007.

The remainder of this paper is organized as follows. In the next section we relate our work to previous research. Section 3 describes the PSSCPA. In section 4 we present the experiment results of the PSSCPA. The paper is concluded in section 5, where some proposals for future work are also presented.

2 Related Work

Cell projection is a well-known volume rendering technique for unstructured meshes. A scalar field is formed by specifying scalar values at all vertices of a mesh, and then visualized by mapping it to colors and opacities with feasible transfer functions.

To efficiently visualize large-scale unstructured grid volume data, parallel processing is one of the best options. Ma [3] presented a parallel ray-casting volume rendering algorithm on distributed memory architectures. This algorithm needs explicit connectivity information for each ray to march from one element to the next, which incurs considerable memory usage and computational overheads. Nieh and Levoy [7] developed a parallel volume rendering system. Their algorithm, however, was tested on a distributed-shared memory architecture, so did the PZSweep algorithm proposed by Farias et al. [8]. Farias et al. [9] soon enhanced the PZSweep routine [8] to configure it on a PC cluster, and strived to find a feasible load balancing strategy applicable for the new architecture. Chen et al. [10] presented a hybrid parallel rendering algorithm on SMP clusters, which make it easier to achieve efficient load balancing. Ma et al. [4] presented a variant algorithm without requirement for connectivity information. Since each tetrahedral cell is rendered independently of other cells, data can be distributed in a more flexible manner. The PSSCPA investigated in this paper is just originated from this algorithm. However, several improvements have been made, as introduced in Section 3.

3 The Parallel Software Scanned Cell Projection Algorithm

Fig. 1 shows the parallel volume rendering pipeline performed by the PSSCPA. The exterior faces of the volume and the volume data are distributed in a round robin fashion among processors. The image screen is divided using a simple scan-line interleaving scheme. Then a parallel k-d tree and an A-Buffer are constructed. They are used in the rendering step to optimize the compositing process and to reduce runtime memory consumption. Each processor scan converts its local cells to produce many ray segments, and send them to their final destinations in image space for merging. A multi-buffer scheme is used in conjunction with asynchronous communication operations to reduce overheads and overlap communications of ray segments with rendering computations. When scan conversion and ray-segment mergence are finished, the master node receives competed sub-images from all slavers and then assembles them for display. Our algorithm can handle convex meshes, non-convex meshes, and meshes with disconnected components.

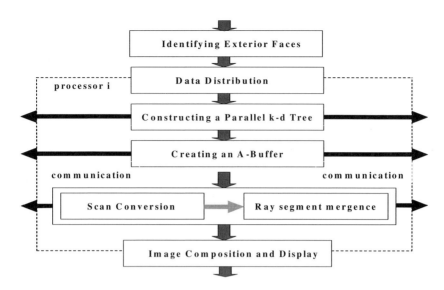

Fig. 1. The parallel volume rendering pipeline

3.1 Data Distribution

The ideal goal of a data distribution scheme is that each processor incurs the same computational load and the same amount of memory usage. However, it is prevented being obtained by several factors. First, there are some computational costs to scan convert a cell. Variations in the number of cells assigned to each processor will produce variations in workloads. Second, cells come in different sizes and shapes. The difference in size can be as large as several orders of magnitude due to the adaptive nature of the mesh. As a result, the projected image area of a cell can vary dramatically, which produces similar variations in scan conversion costs. Finally, the projected area of a cell also depends on the viewing direction.

Generally, nearby cells in object space are often similar in size, so that grouping them together exacerbates load imbalances, making it very difficult to obtain satisfactory result. We have therefore chosen to take the round-robin scheme, dispersing connected cells as widely as possible among the processors. Thus with sufficient enough cells, the computational requirements for each processor tend to average out, producing an approximate load balance. The approach also satisfies our requirement for flexibility, since the data distribution can be computed trivially for any number of processors, without the need for an expensive preprocessing time.

We also need to evenly distribute the pixel-oriented ray-merging operations. Local variations in cell sizes within the mesh lead directly to variations in depth complexity in image space. Therefore we need an image partitioning strategy to disperse the ray-merging operations as well. Scan-line interleaving, which assigns successive image scan-lines to processors in the round-robin fashion, generally works well as long as the image's vertical resolution is several times larger than the number of processors. In our current implementation, we use this strategy.

3.2 Parallel k-d Tree

Our round-robin data distribution scheme completely destroys the spatial coherence among neighboring mesh cells, making an unstructured dataset even more irregular. We would like to restore some ordering so that the rendering step may be performed more efficiently. We are to have all processors render the cells in the same neighborhood at about the same time. Ray segments generated for a particular region will consequently arrive at their image-space destinations within a relatively short window of time, allowing them to be merged early. This early merging reduces the length of the ray-segment list maintained by each processor, which benefits the rendering process in two ways: first, a shorter list reduces the cost of inserting a ray segment in its proper position within the list; and second, the memory needed to store unmerged ray segments is reduced.

To provide the desired ordering, a parallel k-d tree should be constructed cooperatively so that the resulting spatial partitioning is exactly the same on each processor. After the data cells are initially distributed, all processors participate in a synchronized parallel partitioning process. A detailed description of the algorithm has been given by Ma et al. [4].

3.3 Creating an A-Buffer

For a convex mesh, a parallel k-d tree should be constructed for ray-segments to be merged early. However, for a non-convex mesh or a mesh with disconnected components, only the k-d tree is not sufficient. The ray-gaps, i.e. a segment of a ray between a point on an exterior face of a mesh where the ray leaves the mesh, and another such point where the ray reenters a mesh, will clag ray-segments to be merged early. Our approach to this problem is to add an assistant A-Buffer.

We can identify exterior faces of a mesh and evenly distribute them to each processor. Then the exterior faces will scan-converted, and ray-face intersections are sent to their image-space destinations. Each processor saves the ray-face intersections received from other processor in an A-Buffer type of data structure along each ray.

An A-Buffer is created, implemented as an array of pixel (PIX) lists, one per pixel in image space. Fig. 2 shows an A-Buffer on a slice across a scan line through an unstructured mesh with disconnected components.

A PIX list consists of a series of PIX list entry records (PIX entries), as described below. As each pixel p of an exterior face f is enumerated by the scan conversion, a new PIX entry is created containing the distance z from the screen to p, a pointer to f, and a next pointer for the PIX list. The PIX entry is then inserted into the appropriate PIX list in the A-Buffer, in the order of increasing z. At each pixel location, we maintain a linked list of ray segments, which are merged to form the final pixel value. The ray segments in each linked list assisted by a PIX list are allowed to be merged early in a non-convex mesh or a mesh with disconnected components.

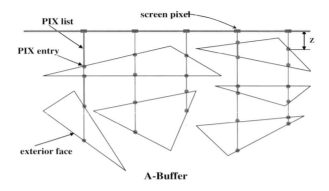

Fig. 2. Diagram of an A-Buffer on a 2D slice

3.4   Task Management with Asynchronous Communication

To reduce computational overheads and improve parallel efficiencies, we adopt an asynchronous communication strategy first suggested by Ma et al. [4].

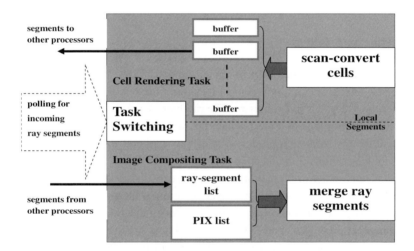

Fig. 3. Task management with asynchronous communications

This scheme allows us to overlap computation and communication, which hides data transfer overheads and spreads the communication load over time. During the course of rendering, there are two main tasks to be performed: scan conversion and image composition. High efficiency is obtained if we can keep all processors busy within either of these two tasks. We employ a polling strategy to interleave the two tasks, and thus achieve a good performance. Fig. 3 illustrates at a high level the management of the two tasks and the accompanying communications. Each processor starts by scan converting one more data cells. Periodically the processor checks to see if incoming ray segments are available; if so, it switches to the merging task, sorting and merging incoming rays until no more input is pending.

Because of the large number of ray segments generated, the overheads for communicating each of them individually would be prohibitive in most parallel machines. Instead, it is better to buffer them locally and send many ray segments together in one operation. This strategy is even more effective when multi-buffers are provided for each destination. When a send operation is pending for a full buffer, the scan conversion process can be placing ray segments in other buffers.

3.5 Image Composition

Since we divide the image space using a simple scan-line interleaving scheme, the image composition step is very simple. Tiles do not overlap, and pixels are generated independently. So the tiles rendered by each processor correspond directly to sub-images, and there is no need for depth composition. When the master processor receives from other processors the sub-images rendered, it simply pastes it on the final image.

4 Experimental Results

We implemented our PSSCPA in the C++ language using MPI message passing for inter-processor communication. All the experiments presented subsequently are conducted on a Dawning PC Cluster (Intel CPUs: 48*2.4GHz; Memory: 48GB; Network: 1-Gbit Fast Ethernet) at the Center for Engineering and Scientific Computation (CESC), Zhejiang University (ZJU).

We have used two three different datasets in our experiments: G1 and FD. G1 represents the stress distribution on a three-dimensional gear. FD represents the evolution of the structure of interface in three-dimensional Rayleigh-Taylor instability. Both G1 and FD are unstructured grids composed of tetrahedral cells. G1 is a

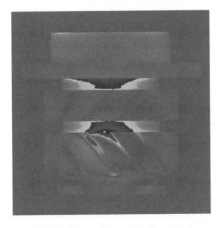
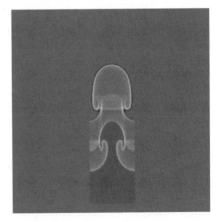

Fig. 4. Volume rendering of G1   **Fig. 5.** Volume rendering of FD

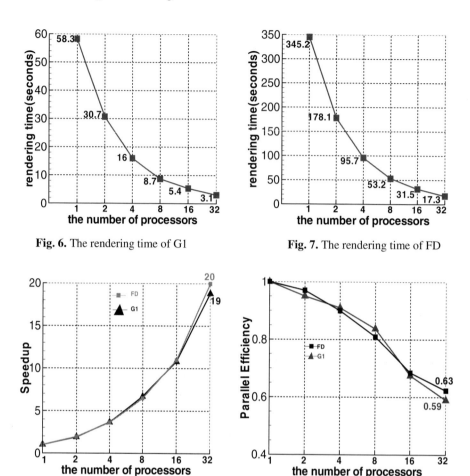

Fig. 6. The rendering time of G1

Fig. 7. The rendering time of FD

Fig. 8. Speedup

Fig. 9. Parallel efficiency

non-convex mesh with 0.5M cells; FD is a convex mesh with 3M cells. The image size in our experiments is 512*512. Figs. 4 and 5 show two volume-rendered views.

Figs.6 and 7 plot the rendering time versus the number of processors. With 32 processors involved we can render 0.5M tetrahedral cells in 3.1 seconds per frame.

Figs.8 and 9 show the speedups and parallel efficiencies obtained as the number of processors varies from 1 to 32, respectively While 32 processors are involved, we achieve a speedup of 20, and a parallel efficiency of 63%, for FD.

5   Conclusions and Future work

In this paper, by combining a k-d tree and an A-Buffer with an asynchronous communication strategy, we have developed a volume renderer for unstructured meshes which employs inexpensive static load balancing to achieve good performance. Because the partial pre-integration method is adopted in the volume renderer, our system supports the rendering of data with any classification system. By employing the A-Buffer technique, our system also can handle convex meshes, non-convex meshes, or meshes with disconnected components.

In the future, we will extend the PSSCPA to the problem of rendering time-varying data.

Acknowledgements. The authors would like to thank the National Natural Science Foundation of China, for the National Science Fund for Distinguished Young Scholars under grant No.60225009 and the Major Program of the National Natural Science Foundation of China under Grant No.90405003. The first author is grateful to the simulation data provided by Jianfeng Zou, the constructive discussions with Dibin Zhou, and the valuable suggestions from Jianjun Chen, Lijun Xie, Jian Deng.

References

1. C. Hofsetz and K.-L. Ma.: Multi-threaded rendering unstructured-grid volume data on the SGI origin 2000. In Third Eurographics Workshop on Parallel Graphics and Visualization, (2000)
2. L. Hong and A. Kaufman.: Accelerated ray-casting for curvilinear volumes IEEE Visualization'98, October (1998) 247–254.
3. K.-L. Ma.: Parallel volume ray-casting for unstructured-grid data on distributed-memory architectures. IEEE Parallel Rendering Symposium, October (1995) 23–30
4. K.-L. Ma and T. Crockett.: A scalable parallel cell-projection volume rendering algorithm for three-dimensional unstructured data. IEEE Parallel Rendering Symposium, November (1997) 95–104
5. Moreland, K. Angel, E.: A fast high accuracy volume renderer for unstructured data. In Proceedings of IEEE Symposium on Volume Visualization and Graphics 2004, October (2004)9-16
6. L. Carpenter.: The A-buffer, an antialiased hidden surface method. In Computer Graphics Proc., SIGGRAPH'84, July (1984) 103-108
7. J. Nieh and M. Levoy.: Volume rendering on scalable shared-memory mimd architectures. In 1992 Workshop on Volume Visualization Proceedings, October (1992) 17–24
8. R. Farias, and C. Silva.: Parallelizing the ZSWEEP Algorithm for Distributed-Shared Memory Architectures. In International Workshop on Volume Graphics, October (2001) 91–99
9. R. Farias, C. Bentes, A. Coelho, S. Guedes, L. Goncalves.: Work distribution for parallel ZSweep algorithm. In XVI Brazilian Symposium on Computer Graphics and Image Processing (SIBGRAPI'03), (2003) 107- 114
10. L. Chen, I. Fujishiro, and K. Nakajima.: Parallel performance optimization of large-scale unstructured data visualization for the earth simulator. In Proceedings of the Fourth Eurographics Workshop on Parallel Graphics and Visualization, (2002) 133-140

Geometry-Driven Nonlinear Equation with an Accelerating Coupled Scheme for Image Enhancement[*]

Shujun Fu[1,2,**], Qiuqi Ruan[2], Chengpo Mu[3], and Wenqia Wang[1]

[1] School of Mathematics and System Sciences, Shandong University, Jinan, 250100, China
[2] Institute of Information Science, Beijing Jiaotong University, Beijing, 100044, China
[3] School of Aerospace Science and Technology, Beijing Institute of Technology, Beijing, 100081, China
[**] shujunfu@163.com

Abstract. In this paper, a geometry-driven nonlinear shock-diffusion equation is presented for image denoising and edge sharpening. An image is divided into three-type different regions according to image features: edges, textures and details, and flat areas. For edges, a shock-type backward diffusion is performed in the gradient direction to the isophote line (edge), incorporating a forward diffusion in the isophote line direction; while for textures and details, a soft backward diffusion is done to enhance image features preserving a natural transition. Moreover, an isotropic diffusion is used to smooth flat areas simultaneously. Finally, a shock capturing scheme with a special limiter function is developed to speed the process with numerical stability. Experiments on real images show that this method produces better visual results of the enhanced images than some related equations.

1 Introduction

Main information of an image resides in such features as its edges, local details and textures. Image features are not only very important to the visual quality of the image, but also are significant to image post processing tasks, for example, image segmentation, image recognition and image comprehension, etc. Among image features, edges are the most general and important, which partition different objectives in an image. Because of some limitations of imaging process, however, edges may not be sharp in images. In addition to noise, both small intensity difference across edge and big edge width would result in a weak and blurry edge [1].

In the past decades there has been a growing amount of research concerning partial differential equations (PDEs) in image enhancement, such as anisotropic diffusion filters [2-5] for edge preserving noise removal, and shock filters [6-9] for edge sharpening. A great deal of successful applications of nonlinear evolving PDEs in "low level" image

[*] This work was supported by the natural science fund of Shandong province, P.R. China (No. Y2006G08); the researcher fund for the special project of Beijing Jiaotong University, P.R. China (No. 48109); the open project of the National Laboratory of Pattern Recognition at the Institute of Automation of the Chinese Academy of Sciences, P.R. China; the general program project of School of Mathematics and System Sciences of Shandong University, P.R. China (No. 306002).

Y. Shi et al. (Eds.): ICCS 2007, Part I, LNCS 4487, pp. 490–496, 2007.

processing can mainly be attributed to their two basic characteristics: "to be local" and "to be iterative". The word "differential" means that an algorithm is of local processing, while the word "evolving" means that it is of iterative one when numerically implemented.

One of most influential work in using partial differential equations (PDEs) in image processing is the anisotropic diffusion (AD) filter, which was proposed by P. Perona and J. Malik [3] for image denoising, enhancement, sharpening, etc. The scalar diffusivity is chosen as a non-increasing function to govern the behaviour of the diffusion process. Different from the nonlinear parabolic diffusion process, L. Alvarez and L. Mazorra [7] proposed an anisotropic diffusion with shock filter (ADSF) equation by adding a hyperbolic equation, called shock Filter which was introduced by S.J. Osher and L.I. Rudin [6], for noise elimination and edge sharpening.

In image enhancement and sharpening, it is crucial to preserve and even enhance image features when one removes image noise and sharpens edges at the same time. Therefore, image enhancement is composed of two steps: features detection and the processings by corresponding tactic according to different features.

In this paper, incorporating anisotropic diffusion with shock filter, we present a geometry-driven nonlinear shock-diffusion equation to remove image noise, and to sharpen edges by reducing their width simultaneously.

An image comprises regions with different features. Utilizing the techniques of differential geometry, we partition local structures and features of image into flat areas, edges, details such as corners, junctions and fine lines, and textures. These structures should be treated differently to obtain a better result in an image processing task. In our algorithm, for edges between different objects, a shock-type backward diffusion is performed in the gradient direction to the isophote line (edge), incorporating a forward diffusion in the isophote line direction. For textures and details, shock filters with the sign function enhance image features in a binary decision process, which produce unfortunately a false piecewise constant result. To overcome this drawback, we use a hyperbolic tangent function to control softly changes of gray levels of the image. As a result, a soft shock-type backward diffusion is introduced to enhance these features while preserving a natural transition in these areas. Finally, an isotropic diffusion is used to smooth flat areas simultaneously.

In order to solve effectively the nonlinear equation to obtain discontinuous solution with numerical stability, after we have discussed the difficulty of the numerical implementation to this type equation, a shock capturing scheme is developed with a special limiter function to speed the process.

This paper is organized as follows. In section 2, some related equations are introduced for enhancing images: anisotropic diffusions and shock filters. Then, we propose a geometry-driven shock-diffusion equation. In section 3, we implement the proposed method and test it on real images. Conclusions are presented in section 4

2 Geometry-Driven Shock-Diffusion Equation

2.1 Differentials of a Typical Ramp Edge and Edge Sharpening

We first analyze differential properties of a typical ramp edge. In Fig.1 (a), a denotes the profile of a ramp edge, whose center point is o, and, b and c denote its first and second differential curves respectively. It is evident that b increases in value from 0 gradually, reaches its maximum at o, and then decreases to 0; while c changes its sign at o, from positive to negative in value. Here we control changes of gray level beside the edge center o. More precisely, we reduce gray levels of pixels on the left of o (whose second derivatives are positive), while increase those on the right of o (whose second derivatives are negative), by which the edge can be sharpened by reducing its width (see Fig.1 (b)). Shock-type diffusions later are just based on above analysis.

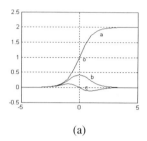
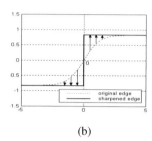

(a)   (b)

Fig. 1. Differentials of a typical ramp edge and edge Sharpening. (a) Differentials of 1D typical ramp edge a, with center o, and the first and second differentials b, c respectively; (b) Edge sharpening process (the solid line), compared with original edge (the broken line).

2.2 Local Differential Structure of Image

Consider image as a real function $u(x, y)$ on a 2D rectangular domain Ω, and image edge is isophote line (level set line), along which the image intensity is constant. Image gradient is a vector, $u_N = \nabla u = (u_x, u_y)$. If $\nabla u \neq 0$, then a local coordinates system can be defined at a point $\mathbf{o}$:

$$\bar{N} = \nabla u / |\nabla u| \ , \ \bar{T} = \nabla u^\perp / |\nabla u^\perp|$$

where $\nabla u^\perp = (-u_y, u_x)$. The Hessian matrix of image function $u(x, y)$ is :

$$Hu = \begin{pmatrix} u_{xx} & u_{xy} \\ u_{xy} & u_{yy} \end{pmatrix}$$

For two vectors X and Y , we define $Hu(X, Y) = X^T Hu Y$. Thus, we have the second directional derivatives in the directions $\bar{N}$ and $\bar{T}$:

$$u_{NN} = Hu(\bar{N}, \bar{N}) = (u_x^2 u_{xx} + u_y^2 u_{yy} + 2u_x u_y u_{xy}) / |\nabla u|^2$$

$$u_{TT} = Hu(\bar{T}, \bar{T}) = (u_x^2 u_{yy} + u_y^2 u_{xx} - 2u_x u_y u_{xy}) / |\nabla u|^2$$

By calculating, the curvature of isophote line at a point o is :

$$k = div(\bar{N}) = u_{TT} / |\nabla u| = (u_x^2 u_{yy} + u_y^2 u_{xx} - 2u_x u_y u_{xy}) / |\nabla u|^{\frac{3}{2}}$$

where *div* is a divergence operator.

2.3 The Geometry-Driven Shock-Diffusion Equation

An image comprises regions with different features, such as edges, textures and details, and flat areas, which should be treated differently to obtain a better result in an image processing task. We divide an image into three-type regions by its smoothed gradient magnitude.

For edges between different objects, a shock-type backward diffusion is performed in the gradient direction, incorporating a forward diffusion in the isophote line. For textures and details, in equations (3) and (4), to enhance an image using the sign function $\text{sign}(\mathbf{x})$ is a binary decision process, which is a hard partition without middle transition. Unfortunately, the obtained result is a false piecewise constant image in some areas producing bad visual quality. We notice that the change of texture and detail is gradual in these areas. In order to approach this change, we use a hyperbolic tangent membership function $\text{th}(\mathbf{x})$ to guarantee a natural smooth transition in these areas, by controlling softly changes of gray levels of the image. As a result, a soft shock-type backward diffusion is introduced to enhance these features. Finally, an isotropic diffusion is used to smooth flat areas simultaneously.

Thus, incorporating shock filter with anisotropic diffusion, we develop a nonlinear geometry-driven shock-diffusion equation (GSE) process to reduce noise, and to sharpen edges while preserving image features simultaneously: Let $(x, y) \in \Omega \subset R^2$, and $t \in [0, +\infty)$, a multi-scale image $u(x, y, t): \Omega \times [0, +\infty) \to R$,

$$\begin{cases} u_G = G_\sigma * u \\ \dfrac{\partial u}{\partial t} = c_N u_{NN} + c_T u_{TT} - w(u_{NN})\text{sign}((u_G)_{NN})|u_N| \end{cases} \tag{1}$$

with Neumann boundary condition, where the parameters are chosen as follows according to different image regions:

	c_N	c_T	$w(u_{NN})$
$\left\|(u_G)_N\right\| > T_1$	0	$1/(1+l_1 u_{TT}^2)$	1
$T_2 < \left\|(u_G)_N\right\| \le T_1$	0	$1/(1+l_1 u_{TT}^2)$	$\left\|\text{th}(l_2 u_{NN})\right\|$
else	1	1	0

where G_σ is defined in previous section, c_N and c_T are the normal and tangent flow control coefficients respectively. The tangent flow control coefficient is used to prevent excess smoothness to smaller details; l_2 is a parameter to control the gradient of the membership function $\text{th}(x)$; T_1 and T_2 are two thresholds; l_1 and l_2 are constants. 3. Numerical Implementation and Experimental Results

3   Numerical Implementation and Experimental Results

3.1   A Shock Capturing Scheme

Nonlinear convection-diffusion evolution equation is a very important model in the fluid dynamics, which can be used to depict transmission processes of momentum, energy and mass of fluid. Because of its hyperbolic characteristic, the solution to the convection-diffusion equation often has discontinuity even if its initial condition is very smooth. Mathematically only weak solution can be obtained here. If a weak solution satisfies the entropic increase principle for an adiabatic irreversible system, then it is called a shock wave.

When one solves numerically a convection-diffusion equation using a difference scheme, he may find some annoying problems in numerical simulation, such as instability, over smoothing, spurious oscillation or wave shift of a scheme. The reason for above is that, despite the original equation are deduced according to some physical conversation laws, its discrete equation may deviate from these laws, which can bring about numerical dissipation, numerical dispersion and group velocity of wave packets effects in numerical solutions specially for the hyperbolic term. Therefore, the hyperbolic term must be discretized carefully so that the flow of small scale and shock waves can be captured accurately.

Besides of satisfying consistence and stability, a good numerical scheme also need to capture shock waves. One method to capture shock waves is to add artificial viscosity term to the difference scheme for controlling and limiting numerical fluctuations near shock waves. But by this method it is inconvenient to adjust free parameters for different tasks, and the resolution of shocks can also be affected. Another method is to try to stop from numerical fluctuations before them appear, which is based on the TVD (Total Variation Diminishing) and nonlinear limiters. Their main idea is to use a limiter function to control the change of the numerical solution by a nonlinear way, and the corresponding schemes satisfy the TVD condition and eliminate above disadvantage effects, which guarantee of capturing shock waves with a high resolution.

In a word, when solving numerically a nonlinear convection-diffusion equation like (1) using a difference scheme, the hyperbolic term must be discretized carefully because discontinuity solutions, numerical instability and spurious oscillation may appear. Shock capturing methods with high resolution are effective tools. For more details, we refer the reader to the book [10]. Here, we develop a speeding scheme by using a proper limiter function.

An explicit Euler method with central difference scheme is used to approximate equation (1) except the gradient term $|u_N|$. Below we detail a numerical approach to it. On the image grid, the approximate solution is to satisfy:

$$u_{ij}^n \approx u(ih, jh, n\Delta t), \ i, j, n \in Z^+ \tag{2}$$

where h and $\triangle t$ are the spatial and temporal step respectively. Let $h = 1$, $\delta^+ u_{ij}^n$ and $\delta^- u_{ij}^n$ are forward and backward difference schemes of u_{ij}^n respectively. A limiter function MS is used to approximate the gradient term:

$$|u_N| = \sqrt{(MS(\delta_x^+ u_{ij}^n, \delta_x^- u_{ij}^n))^2 + (MS(\delta_y^+ u_{ij}^n, \delta_y^- u_{ij}^n))^2} \tag{3}$$

where

$$MS(x, y) = \begin{cases} x, & |x| < |y| \\ y, & |x| > |y| \\ x, & |x| = |y| \ \text{and} \ xy > 0 \\ 0, & |x| = |y| \ \text{and} \ xy \leq 0 \end{cases} \tag{4}$$

The *MS* function bears fewer 0 in value than the *minmod* function does in the *x-y* plane, which also make the scheme satisfy the numerical stability. Because the gradient term represents the transport speed of the scheme, the *MS* function makes our scheme evolve faster with a bigger transport speed than those with the *minmod* function.

In [8], other than above flux limitation technique, a fidelity term $(u - u_0)$ is used to carry out the stabilization task, and they also displayed that the SNRs of results tend towards 0 if $a_f = 0$. However, this is not enough to eliminate overshoots, and this term also affect its performance.

3.2 The Coupled Iteration

Based on preceding discussion, when implementing iteratively equation (1), we find that the shock and diffusion forces will cancel mutually in a single formula. We split equation (1) into two formulas and propose the following coupled scheme by iterating with time steps:

$$\begin{cases} v^0 = u^0, \ u_G = G_\sigma * u \\ v^{n+1} = u^n + \Delta t(-w(u_{NN}^n)\text{sign}((u_G^n)_{NN})|u_N|) \\ u^{n+1} = v^{n+1} + \Delta t(c_N v_{NN}^{n+1} + c_T v_{TT}^{n+1}) \end{cases} \quad (5)$$

where Δt is the time step, u^0 is an original image. By computing iteratively in the order of $u^0 \to v^0 \to v^1 \to u^1 \to v^2 \to u^2 \to \cdots$, we obtain the enhanced image after some steps.

3.3 Experiments

We present results obtained by using our scheme (5), and compare its performance with those of above related methods, where the parameters are selected such that best results are obtained for all methods.

We compare performances of related methods on the blurred Cameraman image (Gaussian blur, $\sigma = 2.5$) with added high level noise (SNR=14dB). In this case, weaker features are smeared by big noise in the image, which are difficult to be restored completely. As it can be seen, although the AD method denoises the image well specially in the smoother segments, it produces the blurry image with unsharp edges, whose ability to sharpen edges is limited, because of its poor sharpening process with the improper diffusion coefficient along the gradient direction. Moreover, with the diffusion coefficient in inverse proportion to the image gradient magnitude along the tangent direction, it does not diffuse fully in this direction and presents rough contours.

For the ADSF method, though it sharpens edges very well, in a binary decision process they yield the false piecewise constant images, which look unnatural with a discontinueous transition in the homogenous areas. Further, it cannot reduce noise well only by a single directional diffusion in the smoother regions.

The best visual quality is obtained by enhancing the image using GSE, which enhances most features of the image with a natural transition in the homogenous areas, and produces pleasing sharp edges and smooth contours while denoising the image effectively.

Finally, we discuss the performances of these methods in smoothing image contours on bigger gradients in the tangent direction of edges. As we explain above, image contours obtained by AD are not smooth with blurry edges in the gradient direction. The results obtained using ADSF and GSE respectively all present smooth contours in the tangent direction.

4 Conclusions

This paper deals with image enhancement for noisy blurry images. By reducing the width of edges, a geometry-driven nonlinear shock-diffusion equation is proposed to remove noise and to sharpen edges.

Our model performs a powerful process to noise blurry images, by which we not only can remove noise and sharpen edges effectively, but also can smooth image contours even in the presence of high level noise. Enhancing image features such as edges, textures and details with a natural transition in interior areas, this method produces better visual quality than some relative equations.

References

1. Castleman, K.R.: Digital Image Processing, Prentice Hall (1995).
2. Aubert, G., Kornprobst, P.: Mathematical Problems in Image Processing: Partial Differential Equations and the Calculus of Variations, vol.147 of Applied Mathematical Sciences, Springer-Verlag (2001).
3. Perona, P., Malik, J.: Scale-space and edge detection using anisotropic diffusion. IEEE Trans. Pattern Anal. Machine Intell., 12(7)(1990) 629-639.
4. Nitzberg, M., Shiota, T.: Nonlinear image filtering with edge and corner enhancement. IEEE Transactions on PAMI, 14(8)(1992) 826-833.
5. You, Y.L., Xu, W., Tannenbaum, A., Kaveh, M.: Behavioral analysis of anisotropic diffusion in image processing. IEEE Trans. on Image Processing, 5(11)(1996) 1539-1553.
6. Osher, S.J., Rudin, L.I.: Feature-oriented image enhancement using shock filters. SIAM J. Numer. Anal., 27(1990) 919-940.
7. Alvarez, L., Mazorra, L.: Signal and image restoration using shock filters and anisotropic diffusion. SIAM J. Numer. Anal., 31(2)(1994) 590-605.
8. Kornprobst, P., Deriche, R., Aubert, G.: Image coupling, restoration and enhancement via PDE's. IEEE ICIP, 2(1997) 458-461.
9. Gilboa, G., Sochen, N., Zeevi, Y.Y.: Image Enhancement and denoising by complex diffusion processes. IEEE Transactions on PAMI, 26(8)(2004) 1020-1036.
10. Liu, R.X., Shu, Q.W.: Some new methods in Computing Fluid Dynamics, Science Press of China, Beijing (2004).

A Graph Clustering Algorithm Based on Minimum and Normalized Cut

Jiabing Wang[1], Hong Peng[1], Jingsong Hu[1], and Chuangxin Yang[1,2]

[1] School of Computer Science and Engineering, South China University of Technology
Guangzhou 510641, China
[2] Guangdong University of Commerce, Guangzhou 510320, China
{jbwang, mahpeng, cshjs}@scut.edu.cn

Abstract. Clustering is the unsupervised classification of patterns into groups. In this paper, a clustering algorithm for weighted similarity graph is proposed based on minimum and normalized cut. The minimum cut is used as the stopping condition of the recursive algorithm, and the normalized cut is used to partition a graph into two subgraphs. The algorithm has the advantage of many existing algorithms: nonparametric clustering method, low polynomial complexity, and the provable properties. The algorithm is applied to image segmentation; the provable properties together with experimental results demonstrate that the algorithm performs well.

Keywords: graph clustering, minimum cut, normalized cut, image segmentation.

1 Introduction

Clustering is the unsupervised classification of patterns (observations, data items, or feature vectors) into groups (clusters) [1-2]. It groups a set of data in a way that maximizes the similarity within clusters and minimizes the similarity between two different clusters. Due to its wide applicability, the clustering problem has been addressed in many contexts; this reflects its broad appeal and usefulness as one of the steps in exploratory data analysis.

In the past decade, one of the most active research areas of data clustering methods has been spectral graph partition, e.g. [3-7], because of the following advantages: does not need to give the number of clusters beforehand; low polynomial computational complexity, etc. In spectral graph partition, the original clustering problem is first transformed to a graph model; then, the graph is partitioned into subgraphs using a linear algebraic approach.

Minimum cut in similarity graphs were used by Wu and Leahy [8], Hartuv and Shamir [9]. The minimum cut often causes an unbalanced partition; it may cut a portion of a graph with a small number of vertices [3-4]. In the context of graph clustering, this is, in general, not desirable. To avoid partitioning out a small part of a graph by using edge-cut alone, many graph partitioning criteria were proposed, such as ratio cut [10], normalized cut [4], min-max cut [11], etc. Soundararajan and Sarkar

Y. Shi et al. (Eds.): ICCS 2007, Part I, LNCS 4487, pp. 497–504, 2007.

[12] have made an in-depth research to evaluate the following partitioning criteria: minimum cut, average cut, and normalized cut.

Normalized cut proposed by Shi and Malik [4] is a spectral clustering algorithm and has been successfully applied to many domains [13-18], especially in image segmentation. However, when applying normalized cut to recursively partitioning data into clusters, a real number parameter—the stopping condition—must be given beforehand [4]. To our knowledge, there are no theoretic results about how to select the parameter. If the parameter is inappropriate, the clustering result is very bad (see an example as shown in Fig. 1 and Fig. 2 in section 2).

In this paper, we propose a clustering algorithm based on minimum and normalized cut. By a novel definition of a cluster for weighted similarity graph, the algorithm does not need to give the stopping condition beforehand and holds many good properties: low polynomial complexity, the provable properties, and automatically determining the number of clusters in the process of clustering.

The rest of the paper is organized as follows. In section 2, we give some basic definitions, a brief review of normalized cut, and the description of our algorithm. In section 3, we prove some properties of the algorithm. In section 4, we apply the algorithm to image segmentation and give some preliminary results. This paper concludes with some comments.

2 The MAN-C Algorithm

A weighted, undirected graph $G = (V, E, W)$ consists of a set V of vertexes, a set E of edges, and a weight matrix W. The positive weight w_{ij} on an edge connecting two nodes i and j denotes the similarity between i and j. For a weighted, undirected graph G, we also use n to denote the number of vertexes of G, m the number of edges of G, T the sum of weights of all edges of G.

The *distance* $d(u,v)$ between vertices u and v in G is the minimum length of a path joining them, if such path exists; otherwise $d(u,v) = \infty$ (the *length* of a path is the number of edges in it). The *degree* of vertex v in a graph, denoted $deg(v)$, is the number of edges incident on it.

We say that A and B partition the set V if $A \cup B = V$ and $A \cap B = \varnothing$. We denote the partition by the unordered pair (A, B). The *cost* of a partition for a graph is the sum of the weights of the edges connecting the two parts cut by the partition, i.e.,

$$Cut(A,B) = \sum_{i \in A, j \in B} w_{ij} \ .$$

The graph *minimum cut* (abbreviated *min-cut*) problem is to find a partition (A, B) such that the cost of the partition is minimal, i.e.,

$$minCut(A,B) = min \sum_{i \in A, j \in B} w_{ij} \ .$$

Shi and Malik [4] proposed a normalized similarity criterion to evaluate a partition. They call this criterion normalized cut:

$$Ncut = \frac{Cut(A,B)}{assoc(A,V)} + \frac{Cut(B,A)}{assoc(B,V)} \ .$$

where $assoc(A,V) = \sum_{i \in A, j \in V} w_{ij}$ is the total connection from nodes in A to all the nodes in the graph, and $assoc(B, V)$ is similarly defined. It is clear that the optimal partition can be achieved by minimizing *Ncut*.

The theoretical attraction of the normalized cut lies in its analytical solution. The near-optimal solution of the normalized cut can be obtained by solving the relaxed generalized eigensystem. One key advantage of using normalized cut is that a good approximation to the optimal partition can be computed very efficiently. Let D represent a diagonal matrix such that $D_{ii} = \sum_{j \in V} w_{ij}$, i.e., D_{ii} is the sum of the weights of all the connections to node i. Then, the problem of minimizing *Ncut* can be written as the expression (1), which can be reduced to a generalized eigenvalue system (2) [4].

$$MinNcut = \min_{x} \frac{x^T (D-W)x}{x^T Dx} . \tag{1}$$

$$(D-W)x = \lambda Dx . \tag{2}$$

where x represents the eigenvectors, in the real domain, which contain the necessary segmentation information. The eigenvector with the second smallest eigenvalue, called Fiedler vector, is often used to indicate the membership of data points in a subset. Fiedler vector provides the linear search order for the splitting point that can minimize the *Ncut* objective.

Just as statements in introduction, when applying normalized cut to recursive partitioning data into clusters, a parameter—the stopping condition—must be given beforehand. If the parameter is inappropriate, the clustering result is very bad. An example is given in Fig. 1 and Fig. 2 that show clustering results when given different parameters.

In Fig. 1, the leftmost is the original image, the second one is the clustering result when given the stopping condition *Ncut* < 0.25, and the other two images is the

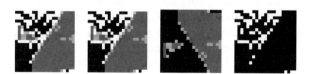

Fig. 1. The leftmost is the original image. The second one is the clustering result when *Ncut* < 0.25. The other two images is the clustering result when *Ncut* < 0.5.

Fig. 2. The clustering result when *Ncut* < 1

clustering result when given the stopping condition *Ncut* < 0.5. Fig. 2 shows the clustering result when given the stopping condition *Ncut* < 1. We can see that different parameters result in different clustering results. Especially, when *Ncut* < 0.25, the normalized cut cannot segment the original image. So, selection of the stopping condition is very important for applying normalized cut to data clustering. However, to our knowledge, there are no theoretic results on how to select the parameter.

In order to avoid the issue of parameter selection, we give the following definition:

Definition 1. For a graph $G = (V, E, W)$, G is a *cluster* if and only if $min\text{-}cut(G) \geq (nT)/(2m)$, where n is the number of vertexes of G, m is the number of edges of G, and T is the sum of weights of all edges of G.

We will see that such a definition of a cluster results in good properties as described in section 3.

According to the definition 1, we have the clustering algorithm MAN-C (Minimum And Normalized Cut) as shown in Fig. 3.

> **Algorithm** MAN-C(G)
> begin
> $C \leftarrow$ Min-Cut(G);
>   if($C < (nT_G)/(2m_G)$)
> $(H, H') \leftarrow$ Ncut(G);
>   MAN-C(H);
>   MAN-C(H');
>   else
>   return G;
>   end if
> end

Fig. 3. The MAN-C algorithm

In Fig. 3, n_G is the number of vertexes of G, m_G is the number of edges of G, and T_G the sum of weights of all edges of G. The procedure *Min-Cut(G)* returns the minimum cut value C, and *Ncut(G)* returns two subgraphs H and H' by implementation of normalized cut. Procedure MAN-C returns a graph in case it identifies it as a cluster, and subgraphs identified as clusters are returned by lower levels of the recursion. Single vertices are not considered clusters and are grouped into a *singletons* set S. The collection of subgraphs returned when applying MAN-C on the original graph constitutes the overall solution.

The running time of the MAN-C algorithm is bounded by $N \times(2f_1(n, m)+f_2(n))$, where N is the number of clusters found by MAN-C, $f_1(n, m)$ is the time complexity of computing a minimum cut, and $f_2(n, m)$ is the time complexity of computing a normalized cut in a graph with n vertices and m edges .

The usual approach to solve the *min-cut* problem is to use its close relationship to the maximum flow problem. Nagamochi and Ibaraki [19] published the first deterministic *min-cut* algorithm that is not based on a flow algorithm, has the fastest running time of $O(nm)$, but is rather complicated. Stoer and Wagner [20] published a *min-cut* algorithm with the same running time as Nagamochi and Ibaraki's, but is very simple.

The normalized cut can be efficiently computed using Lanczos method using the running time $O(kn)+O(kM(n))$ [4], where k is the maximum number of matrix-vector computations required and $M(n)$ is the cost of a matrix-vector computation of Ax, where $A= D^{-1/2}(D-W)D^{-1/2}$ (see formula (1)).

3 Properties of MAN-C Algorithm

In this section we prove some properties of the clusters produced by the MAN-C algorithm. These demonstrate the homogeneity of the solution.

Definition 2. For a graph $G = (V, E, W)$, a vertex $x \in V$, we define the *average edge weight* (AEW) as formula (3). That is, the AEW of a vertex x is the average weight of all edges incident on x.

$$\varpi_x = \frac{\sum_{v \in V} w_{xv}}{\deg(x)} \tag{3}$$

Theorem 1. For a *cluster* $G = (V, E, W)$, the following properties hold:
1. For each pair vertices v_1 and v_2, if $\varpi_{v1} \le T/m$ and $\varpi_{v2} \le T/m$, then the distance between v_1 and v_2 is at most two.
2. For each vertex $x \in V$, $\varpi_x > T/(2m)$.
3. There are $O(n^2)$ edges in G.

Proof. Assertion (1) When all edges incident on a vertex are removed, a disconnected graph results. Therefore the following inequality holds according to the definition 1:

$$\sum_{v \in V} w_{xv} \ge \frac{n}{2} \frac{T}{m}, \quad \text{for each} \quad x \in V .$$

$$\text{equivalently,} \quad \deg(x)\varpi_x \ge \frac{n}{2} \frac{T}{m}, \quad \text{for each} \quad x \in V ,$$

i.e.,

$$\deg(x) \ge \frac{n}{2} \frac{T}{m} \frac{1}{\varpi_x}, \quad \text{for each} \quad x \in V \tag{4}$$

So, if $\varpi_{v1} \le T/m$ and $\varpi_{v2} \le T/m$, then $\deg(v_1) \ge n/2$, and $\deg(v_2) \ge n/2$. Since $\deg(v_1) + \deg(v_2) \ge n$, therefore, the distance between v_1 and v_2 is at most two, as they have a common neighbor.

Assertion (2) By formula (4), we have:

$$\varpi_x \ge \frac{n}{2\deg(x)} \frac{T}{m} .$$

Since $\deg(x) \le n - 1$ for each $x \in V$, we have

$$\varpi_x \ge \frac{n}{2(n-1)} \frac{T}{m} > \frac{T}{2m} . \tag{5}$$

Assertion (3) By formula (4), summing over all vertexes in V we get:

$$\sum_{x \in V} deg(x) \bar{\omega}_x \geq n \frac{T}{2} \frac{T}{m} = \frac{n^2 T}{2m} \ .$$

Equivalently,

$$2T \geq \frac{n^2 T}{2m} \ .$$

That is,

$$m \geq \frac{n^2}{4} \ . \tag{6}$$

That is, there are $O(n^2)$ edges in G.   □

The provable properties of MAN-C as shown in Theorem 1 are strong indication of homogeneity. By Theorem 1, the average edge weight of each vertex in a cluster G must be no less than half of average edge weight of G, and if the average edge weight of some vertex is small, then it must have more neighbors. Moreover, Theorem 1 shows that each cluster is at least half as dense as a clique, which is another strong indication of homogeneity.

4 Experimental Results

Image segmentation is a hot topic in image processing. We have applied MAN-C algorithm to image segmentation. In order to apply the MAN-C to image segmentation, a similarity graph must be constructed. In our experiments, the similarity graph is constructed as follows:

1. Construct a weighted graph G by taking each pixel as a node and connecting each pair of pixels by an edge;
2. For gray image, using just the brightness value of the pixels, we can define the edge weight connecting the two nodes i and j as formula (7);
3. For color image, using HSV values of the pixels, we can define the edge weight connecting the two nodes i and j as formula (8).

$$w_{ij} = e^{\frac{-(I(i)-I(j))^2}{\sigma_I}} \ . \tag{7}$$

where $I(i)$ is the intensity value for a pixel i, and σ_I is a parameter set to 0.1 [4].

$$w_{ij} = e^{\frac{-\|F(i)-F(j)\|_2^2}{\sigma_I}} \tag{8}$$

where $F(i) = [v, v{\cdot}s{\cdot}sin(h), v{\cdot}s{\cdot}cos(h)](i)$, and h, s, v are the HSV values [4].

Considering the space limitation, here we give three examples as shown in Fig. 4, Fig. 5, and Fig. 6. Note that the cluster is drawn using black as background, so the cluster with background is omitted.

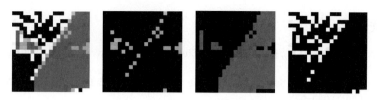

Fig. 4. The leftmost is original image and other images are clusters produced by MAN-C

Fig. 5. The leftmost is original image and other images are clusters produced by MAN-C

Fig. 6. The leftmost is original image and other images are clusters produced by MAN-C

From the results, we can see that the MAN-C algorithm performs well on all images.

5 Conclusions

By a novel definition of a cluster, a graph-theoretic clustering algorithm MAN-C is proposed based on minimum and normalized cut. The minimum cut is used as the stopping condition of the recursive algorithm, and the normalized cut is used to partition a graph into two subgraphs. The MAN-C algorithm has the advantage of many existing algorithm: nonparametric clustering method, low polynomial complexity, and the provable properties. The provable properties of MAN-C together with experimental results demonstrated that the MAN-C algorithm performs well.

Acknowledgements

This work was supported by Natural Science Foundation of China under Grant No 60574078, Natural Science Foundation of Guangdong Province under Grant No 06300170 and Natural Science Youth Foundation of South China University of Technology under Grant No B07-E5050920.

References

1. Jain A.K., Murty M.N., Flynn P.J.: Data Clustering: A Review. ACM Computing Surveys, 31 (1999) 264–323.
2. Xu R., Wunsch II D.: Survey of Clustering Algorithms. IEEE Trans. on Neural Networks, 16(3) (2005) 645–678.
3. Kannan R., Vempala S., Vetta A.: On Clusterings: Good, Bad and Spectral. J. ACM, 51(3) (2004) 497–515.
4. Shi J., Malik J.: Normalized Cuts and Image Segmentation. IEEE Trans. on Pattern Analysis and Machine Intelligence, 22(8) (2000) 888–905.
5. Qiu H., Hancock E. R. Graph Matching and Clustering Using Spectral Partitions. Pattern Recognition, 39 (2006) 22–34.
6. Tritchler D., Fallah S., Beyene J.: A Spectral Clustering Method for Microarray Data. Computational Statistics & Data Analysis, 49 (2005) 63–76.
7. Van Vaerenbergh S., Santamaría I.: A Spectral Clustering Approach to Underdetermined Postnonlinear Blind Source Separation of Sparse Sources. IEEE Trans. on Neural Networks, 17(3) (2006) 811–814.
8. Wu Z., Leahy R.: An Optimal Graph Theoretic Approach to Data Clustering: Theory and Its Application to Image Segmentation, IEEE Trans. on Pattern Analysis and Machine Intelligence, 15 (11) (1993) 1101–1113.
9. Hartuv E., Shamir R.: A Clustering Algorithm Based on Graph Connectivity. Information Processing Letters, 76 (2000) 175–181.
10. Hagen L., Kahng, A. B.: New Spectral Methods for Ratio Cut Partitioning and Clustering. IEEE Trans. on Computer-Aided Design, 11(9) (1992) 1074–1085.
11. Ding H., He X., Zha H., et al: A Min-Max Cut Algorithm for Graph Partitioning and Data Clustering. In: Proceedings of IEEE 2001 International Conference on Data Mining, IEEE Computer Society Press, Los Almitos, CA, (2001) 107-114.
12. Soundararajan P., Sarkar S.: An In-Depth Study of Graph Partitioning Measures for Perceptual Organization. IEEE Trans. on Pattern Analysis and Machine Intelligence, 25(6) (2003) 642–660.
13. Carballido-Gamio J., Belongie S., Majumdar J. S.: Normalized Cuts in 3-D for Spinal MRI Segmentation. IEEE Trans. on Medical Imaging, 23(1) (2004) 36–44.
14. Duarte A., Sánchez Á., Fernández F., et al: Improving Image Segmentation Quality through Effective Region Merging Using a Hierarchical Social Metaheuristic. Pattern Recognition Letters, 27 (2006) 1239–1251.
15. He X., Zha H., Ding C. H.Q., et al: Web Document Clustering Using Hyperlink Structures. Computational Statistics & Data Analysis, 41 (2002) 19–45.
16. Li H., Chen W., Shen I-F.: Segmentation of Discrete Vector Fields. IEEE Trans. on Visualization and Computer Graphics, 12(3) (2006) 289–300.
17. Ngo C., Ma Y., Zhang H.: Video Summarization and Scene Detection by Graph Modeling. IEEE Trans. on Circuits and Systems for Video Technology, 15(2) (2005) 296–305.
18. Yu Stella X., Shi J.: Segmentation Given Partial Grouping Constraints. IEEE Trans. on Pattern Analysis and Machine Intelligence, 26(2) (2004) 173–183.
19. Nagamochi H., Ibaraki T.: Computing Edge-Connectivity in Multigraphs and Capacitated Graphs. SIAM J. Discrete Mathematics, 5 (1992) 54–66.
20. Stoer M., Wagner F.: A Simple Min-Cut Algorithm. J. ACM, 44(4) (1997) 585–591.

A-PARM: Adaptive Division of Sub-cells in the PARM for Efficient Volume Ray Casting

Sukhyun Lim and Byeong-Seok Shin

Inha University, Dept. Computer Science and Information Engineering,
253 Yonghyun-dong, Nam-gu, Inchon, 402-751, Rep. of Korea
slim@inhaian.net, bsshin@inha.ac.kr

Abstract. The PARM is a data structure to ensure interactive frame rates on a PC platform for CPU-based volume ray casting. After determining candidate cells that contribute to the final images, it partitions each candidate cell into several sub-cells. Then, it stores trilinearly interpolated scalar value and an index of encoded gradient vector for each sub-cell. Since the information that requires time-consuming computations is already stored in the structure, the rendering time is reduced. However, it requires huge memory space because most precomputed values are loaded in the system memory. We solve it by adaptively dividing candidate cells into different sub-cells. That is, we divide a candidate cell in which the gradient is strictly changed into a large number of sub-cells, and vice versa. By this approach, we acquire moderate images while reducing the memory size.

1 Introduction

Volume visualization is a research area that deals with various techniques to extract meaningful and visual information from volume data [1], [2]. Volume datasets can be represented by *voxels*, and adjacent eight voxels form a cube called a *cell*. One of the most frequently applied techniques is direct volume rendering, producing high-quality 3D rendered images directly from volume data without intermediate representation. Volume ray casting is a well-known direct volume rendering method [1]. In general, it is composed of two steps [3]: after a ray advances through transparent region (*leaping step*), the ray integrates colors and opacities as it penetrates an object boundary (*color computation step*). Although volume ray casting produces high-quality images, the rendering speed is too slow [1], [2], [3].

Several software-based acceleration techniques for direct volume rendering have been proposed to reduce rendering time. Yagel and Kaufman exploited the use of a ray template to accelerate ray traversal, using spatial coherence of the ray trajectory [4]. However, it is difficult to apply their method to applications involving perspective projection. The shear-warp method introduced by Lacroute and Levoy rearranges the voxels in memory to allow optimized ray traversal (shearing the volume dataset) [5]. Although this method can produce images from reasonable volumes at quasi-interactive frame rates, image quality is sacrificed because bilinear interpolation is used as a convolution kernel. Parker et al. demonstrated a full-featured ray tracer on a workstation with large shared memory [6]. Unfortunately, this method requires 128 CPUs for interactive

Y. Shi et al. (Eds.): ICCS 2007, Part I, LNCS 4487, pp. 505–512, 2007.
© Springer-Verlag Berlin Heidelberg 2007

rendering and is specialized for isosurface rendering. Although Knittel presented the interleaving of voxel addresses and deep optimizations (by using the MMX™ instruction) to improve the cache hit ratio [7], it generates only 256×256 pixel images. If we want a high-resolution image, the 256×256 image can be magnified by using graphics hardware. Mora et al. proposed a high performance projection method that exploits a spread-memory layout called object-order ray casting [8]. However, this also does not support perspective projection. Grimm et al. introduced acceleration structures [9]. They used a gradient cache, and a memory-efficient hybrid removal and skipping technique for transparent regions. This method achieves fast rendering on a commodity notebook PC, but it is not applicable to perspective projection.

We call a cell of which the enclosing eight voxels are all transparent as a *transparent cell*, and a cell with eight opaque voxels is an *opaque cell*. The *candidate cell* means a cell that contributes to the final images. After dividing each candidate cell into N_{sc}^3 cells, where N_{sc} is a representative on one axis to divide one candidate cell into several cells, we call the resulting cell as a *sub-cell*. That is, a single candidate cell contains N_{sc}^3 sub-cells. Fig. 1 shows the structure of a candidate cell.

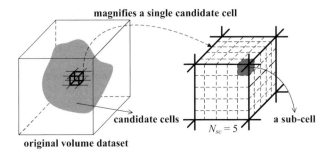

Fig. 1. Structure of a single candidate cell. A candidate cell is composed of several sub-cells. In this example, one candidate cell is composed of 125 sub-cells (that is, N_{sc}=5).

Recently, Lim and Shin proposed the PARM (Precomputed scAlar and gRadient Map) structure to accelerate the color computation step of the volume ray casting [10]. This structure stores the information required to compute color values of the CPU-based volume ray casting. At preprocessing time, after determining candidate cells that contribute to the final images, it divides each candidate cell into several sub-cells. Then, it stores trilinearly interpolated scalar value and an index of encoded gradient vector for each sub-cell. During the rendering stage, when a ray lies on an arbitrary sample point in a candidate cell after skipping over transparent region, a color value can be determined without time-consuming computations because most values to compute a color value are already stored in the PARM structure. However, because it stores the values in the system memory, it is a burden to the memory when the number of candidate cells is increased.

If a sampling rate is reduced in the candidate cell in which the gradient is strictly changed, we acquire aliased or deteriorated images. Although the previous PARM structure assigns fixed numbers of sub-cells for entire candidate cells, our method adaptively divide the candidate cells into different numbers of sub-cells according to the gradient variations. So, we call it *A-PARM* (the initial '*A*' means that 'adaptive'). We explain our data structure in detail in Section 2, and experimental results are presented in Section 3. Finally, we conclude our work.

2   A-PARM (Adaptive-Precomputed scAlar and gRadient Map)

Our method focuses on reducing the memory size for the PARM structure while preserving moderate image quality. Through the next section, we can recognize the method in detail.

2.1   PARM

The obvious way to accelerate the color computation step of the CPU-based ray casting is that we precompute required values in preprocessing step and refer to the values in rendering step. For that, we proposed the PARM structure in our previous work [10]. The first generation step of the PARM is to determine candidate cells from the entire volume [10]. Then, we compute required values. However, because computing the positions of sample points is not feasible because the points can lie in an arbitrary position in a candidate cell, we compute the required values per sub-cell after dividing each candidate cell into N_{sc}^3 sub-cells. The first requirement value is trilinearly interpolated scalar value, and this value is used to determine the opacity value.

The next value is a trilinearly interpolated gradient vector. However, since a vector is composed of three components (x-, y- and z-element), we require three bytes per a sub-cell. So, we apply the *gradient encoding method* proposed by Rusinkiewicz and Levoy [11]. This method maps the gradient vectors on a grid in each of the six faces of a cube. In conventional method [11], they exploited 52x52x6=16,224 different gradient vectors. However, we increase the grid size as 104x104x6 (approximately 2 bytes) to reduce visual artifacts. It leads to a mean error of below 0.01 radian. As a result, after computing trilinear interpolated gradient vector, we store an index.

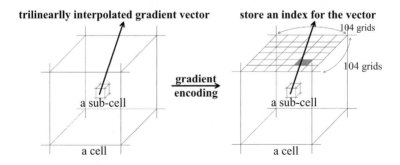

Fig. 2. By the gradient encoding method, we store an index of the gradient vector instead of three components of interpolated vector

The ray-traversal procedure when we exploit the PARM is as follows: first, a ray is fired from each pixel on an image plane. After traversing transparent cells using the conventional space-leaping methods, it reaches a candidate cell. Second, after select-ing the nearest sub-cell, it refers to the precomputed scalar value from the PARM. Because the scalar value on a sample point is already interpolated, it does not require time-consuming trilinear interpolation. If the value is within the transparent range, the ray advances to the next sample point. Otherwise (that is, if the scalar value is re-garded as nontransparent), the ray also refers to the encoded gradient vector index. To decode representable vector from the index quickly, we use a lookup table [11]. Lastly, a color value is determined from the referred gradient vector. Those steps continue for all rays in the image plane.

2.2 A-PARM

When we exploit the PARM structure, we accelerate the color computations of the CPU-based volume ray casting. However, because most required values in the PARM are calculated in preprocessing stage, it requires long preprocessing time and huge memory size. One of the challenging issues of the volume rendering is to acquire high-quality images. When we exploit the PARM structure, the simple way to satisfy it is that we can increase the number of sub-cells. However, unfortunately, the pre-processing time is increased according to increase the number of them, and we can also require large memory space.

   The obvious way to solve it while increasing the image quality is to adaptively di-vide candidate cells into different numbers of sub-cells. In our method, therefore, we increase the number of them in which the gradient is strictly changed. That is, if the gradient is over the user-defined threshold (τ), we increase the number of them, and if the gradient is below τ, we reduce the number of them. The steps are as follows; at first, we estimate three components of the gradient vector $\nabla f(x) = \nabla f(x,y,z)$ for all candidate cells. To determine the candidate cells, we used original method [10]. We assume that a volume dataset is composed of N voxels, and each voxel $\mathbf{v}$ is indexed by $\mathbf{v}(x, y, z)$ where $x,y,z = 0,\ldots,N\text{-}1$. Eq. (1) shows our approach when we use the central difference method [3]. It is the most common approach for gradient estimation in volume graphics [3]. If one of the eight voxels in a candidate cell is satisfied with the Eq. (1), we increase the number of sub-cells.

$$\nabla f(x, y, z) \approx \frac{1}{2}\begin{pmatrix} f(x+1, y, z) - f(x-1, y, z) \\ f(x, y+1, z) - f(x, y-1, z) \\ f(x, y, z+1) - f(x, y, z-1) \end{pmatrix} > \tau \ . \tag{1}$$

   In implement aspect, the PARM is composed of two data structures; *INDEX BUFFER* and *DATA BUFFER*. The *DATA BUFFER* stores precomputed values for each sub-cell. And, to refer the values from it, a data structure stored on the indices of the candidate cells is required, and this is called *INDEX BUFFER*. Because the N_{sc} value is already determined (once again, N_{sc} is a representative on one axis to divide one candidate cell into several cells), we refer expected values from the *DATA BUFFER* quickly. That is, if a ray lies on a sub-cell S_j ($0 \leq S_j < N_{sc}$) in a candidate cell, we refer an index C_i of the candidate cell from the *INDEX BUFFER*. Then, we acquire a precomputed scalar value and an index of the encoded gradient vector from the *DATA BUFFER* by referring the index of $N_{sc}\mathbf{x}(C_i\text{-}1)+S_j$.

Those two buffers are optimal for the original PARM structure. However, since we divide candidate cells into different numbers of sub-cells according to the gradient variations, we cannot exploit them. One of the simple methods is to generate several *INDEX BUFFERs*. However, it is a burden to the system memory because the size of one *INDEX BUFFER* is identical to the volume dataset. Therefore, we require other smart approach.

To solve it, we generate a modified *INDEX BUFFER*, named *M-INDEX BUFFER*. Firstly, from the previous *INDEX BUFFER*, we rearrange indices according to gradient variations, by using the sorting algorithm. We assume that a candidate cell is divided into N_{sc1}^3, N_{sc2}^3 or N_{sc3}^3 sub-cells according to gradients. And we divide a candidate cell into N_{sc1}^3 sub-cells if the gradient is steep, and when it is moderate we divide it into N_{sc2}^3 sub-cells. Otherwise (that is, if it is gentle), we partition it into N_{sc3}^3 sub-cells. Secondly, after searching all candidate cells from the previous *INDEX BUFFER*, we rearrange indices as the sequence from N_{sc1}^3 to N_{sc3}^3 sub-cells. By this approach, only one size of the *INDEX BUFFER* is required. Of course, it requires additional generation time compared with the conventional method. However, its time is marginal against the increment of memory size because only the $\mathbf{O}(N)$ time is necessary. Fig. 3 shows our *M-INDEX BUFFER* generation method.

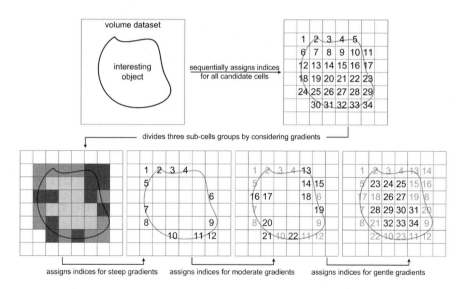

Fig. 3. (Upper) when we use conventional data structure, we require three *INDEX BUFFERs*. Therefore, it is not adequate for our method. (Lower) although we divide candidate cells into three numbers of sub-cells according to the gradient variations, we can require only one *INDEX BUFFER* size by using the *M-INDEX BUFFER*. The red, blue, and black squares mean candidate cells as the gradients are steep, moderate, and gentle, respectively.

Besides the *M-INDEX BUFFER*, we store total numbers of candidate cells related to the N_{sc1}, N_{sc2} or N_{sc3}, and we call them as #N_{sc1}, #N_{sc2} or #N_{sc3} (for example, in Fig. 3, they are 12, 10, and 12, respectively). In ray traversal, if a ray reaches a candidate cell and its index of the *M-INDEX BUFFER* is below #N_{sc1} (that is, $0 \leq C_i <$ #N_{sc1}), we acquire expected precomputed structure by referring Eq. (2). When the index of it is over $\#N_{sc1}$ and below $\#N_{sc2}$ (that is, $\#N_{sc1} \leq C_i < (\#N_{sc1}+\#N_{sc2})$), we reference the index by Eq. (3). Otherwise (that is, $\#N_{sc1}+\#N_{sc2} \leq C_i < (\#N_{sc1}+\#N_{sc2}+\#N_{sc3})$), we refer the Eq. (4). Other remaining steps are identical with the previous PARM (see Sect. 2.1).

$$N_{sc1} \times (C_i - 1) + S_j \; . \tag{2}$$

$$\left(N_{sc1} \times \# N_{sc1}\right) + \left(N_{sc2} \times (C_i - \# N_{sc1} - 1) + S_j\right) \; . \tag{3}$$

$$\left(N_{sc1} \times \# N_{sc1} + N_{sc2} \times \# N_{sc2}\right) + \left(N_{sc3} \times \left(C_i - (\# N_{sc1} + \# N_{sc2}) - 1\right) + S_j\right) \; . \tag{4}$$

3   Experimental Results

All the methods were implemented on a PC equipped with an AMD Athlon64x2™ 4200+ CPU, 2 GB main memory, and GeForce™ 6600 graphics card. The graphics card capabilities are only used to display the final images. The first and second dataset were an engine block and a bonsai with resolutions of 512^3, respectively. The third dataset was an aneurysm of a human brain vessel with a resolution of 768^3. Fig. 4 shows OTFs (Opacity Transfer Functions) for the datasets, and we divide candidate cells into three different numbers of sub-cells (τ) according to the gradient variations. If one scalar value of the three elements in $\nabla f(x,y,z)$ for a candidate cell is over 150, we divide the candidate cell into 64 sub-cells (that is, $N_{sc1}^3=4^3=64$). When the gradient is from 50 to 149, we partition the cell into 27 sub-cells ($N_{sc2}^3=3^3=27$). Otherwise, we divide into $N_{sc3}^3=2^3=8$ sub-cells.

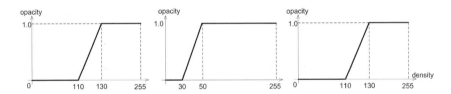

Fig. 4. The OTFs for the engine block, bonsai, and aneurysm dataset, respectively

The rendering times to produce the perspective view are shown in Table 1. Our method is at least 50% faster than the method using the previous hierarchical min-max octree, and this time is almost same with the previous PARM structure (the tolerance is below 5%). The hierarchical min-max octree is widely used data structure for empty-space skipping [1], [7]. [10]. Table 2 shows the preprocessing time and required memory for each dataset. Compared with the previous PARM, we can reduce the preprocessing time and memory storage as the amount of 26%.

Table 1. Rendering improvements for each dataset before/after our method (unit: %). The time for space leaping is included in our method. All results are rendered to a perspective projection.

dataset	engine	bonsai	aneurysm	**average**
improvements	55	49	46	**50**

Table 2. Preprocessing time and required memory for each dataset. The previous PARM divides all candidate cells into 64 sub-cells.

dataset	engine	bonsai	aneurysm
the number of candidate cells (voxels)	5,520,128	4,865,792	2,579,232
candidate cell determination time (A) (secs)	7.24	6.01	3.30
generation time of the PARM (B) (secs)	168.02	147.84	76.97
generation time of the A-PARM (C) (secs)	114.09	107.7	58.89
total preprocessing time of the PARM (D=A+B) (secs)	175.26	153.85	80.27
total preprocessing time of the A-PARM (E=A+C) (secs)	121.33	113.71	62.19
efficiency of the preprocessing (D/E) (%)	**30.8**	**26.1**	**22.5**
required memory of the PARM (F) (MB)	1328	1203	764
required memory of the A-PARM (G) (MB)	907	896	602
efficiency of the required memory (F/G) (%)	**31.7**	**25.5**	**21.2**

Fig. 5 shows the image quality. As you can see, the image quality using our structure is almost same with that using the previous PARM. We compute the *Root Mean Square Error* (RMSE) [12] for accurate comparisons. Compared with the results of the previous PARM, the mean value of RMSE results is about 2.04. This is marginal against reducing the preprocessing time and memory storage.

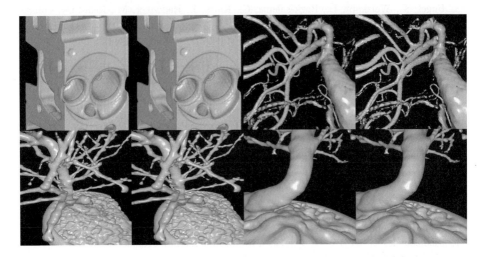

Fig. 5. Comparisons of image quality: the previous PARM (left) and A-PARM (right) for each dataset. The RMSE results are 1.15, 2.89, and 2.09. The engine is rendered to parallel projection and the bonsai and aneurysm are projected to perspective viewing. And, in case of bonsai, we magnify the soil region for accurate comparisons.

4 Conclusion

The most important issue in volume visualization is to produce high-quality images in real time. To achieve interactive frame rates on a PC platform, the PARM structure is proposed in our previous work. Although it reduces the rendering time by precomputing a trilinearly interpolated scalar value and an index of encoded gradient vector for each sub-cell, the preprocessing time is increased. Moreover the memory storage is also increased since most precomputed values are loaded in the system memory. To solve them, we adaptively divide candidate cells into different numbers of sub-cells according to the gradient variations. The experimental results show that our method reduces rendering time and produces high-quality images while reducing the preprocessing time and memory storage.

Acknowledgment

This work was supported by IITA through IT Leading R&D Support Project.

References

1. Levoy, M.: Display of Surface from Volume Data. IEEE Computer Graphics and Applications, Vol. 8, No. 3 (1988) 29-37
2. Kaufman, A.: Volume Visualization. 1st ed., Ed. IEEE Computer Society Press (1991)
3. Engel, K., Weiskopf, D., Rezk-salama, C., Kniss, J., Hadwiger, M.: Real-time Volume Graphics, AK Peters (2006)
4. Yagel, R., Kaufmann, A.: Template-based Volume Viewing. Computer Graphics Forum, Vol. 11 (1992) 153-167
5. Lacroute, P., Levoy, M.: Fast Volume Rendering Using a Shear-Warp Factorization of the Viewing Transformation. Proc. SIGGRAPH 1994 (1994) 451-457
6. Parker, S., Shirley, P., Livnat, Y., Hansen, C., Sloan, P.: Interactive Ray Tracing for Isosurface Rendering. Proc. IEEE Visualization 1998 (1998) 233-238
7. Knittel, G.: The UltraVis System. Proc. IEEE Volume Visualization 2000 (2000) 71-79
8. Mora, B. Jessel, J.P., Caubet, R.: A New Object Order Ray-Casting Algorithm. Proc. IEEE Volume Visualization 2002 (2002) 203-210
9. Grimm, S., Bruckner, S., Kanitsar, A., Gröller, E.: Memory Efficient Acceleration Structures and Techniques for CPU-based Volume Raycasting of Large Data. Proc. IEEE Volume Visualization 2004 (2004) 1-8
10. Lim, S., Shin, B-S: PARM: Data Structure for Efficient Volume Ray Casting. Lecture Notes in Computer Science, Vol. 4263 (2006) 296-305
11. Rusinkiewicz, S., Levoy, M.: QSplat: A Multiresolution Point Rendering System for Large Meshes. Proc. SIGGRAPH 2000 (2000) 343-352
12. Kim, K., Wittenbrink, C.M., Pang, A.: Extended Specifications and Test Data Sets for Data Level Comparisons of Direct Volume Rendering Algorithms. IEEE Trans. on Visualization and Computer Graphics, Vol. 7 (2001) 299-317

Inaccuracies of Shape Averaging Method Using Dynamic Time Warping for Time Series Data

Vit Niennattrakul and Chotirat Ann Ratanamahatana

Department of Computer Engineering, Chulalongkorn University
Phayathai Rd., Pathumwan, Bangkok 10330 Thailand
{g49vnn, ann}@cp.eng.chula.ac.th

Abstract. Shape averaging or signal averaging of time series data is one of the prevalent subroutines in data mining tasks, where Dynamic Time Warping distance measure (DTW) is known to work exceptionally well with these time series data, and has long been demonstrated in various data mining tasks involving shape similarity among various domains. Therefore, DTW has been used to find the *average* shape of two time series according to the optimal mapping between them. Several methods have been proposed, some of which require the number of time series being averaged to be a power of two. In this work, we will demonstrate that these proposed methods cannot produce the *real* average of the time series. We conclude with a suggestion of a method to potentially find the shape-based time series average.

Keywords: Time Series, Shape Averaging, Dynamic Time Warping.

1 Introduction

The need to find the template or the data representative from a group of time series is prevalent in major data mining tasks' subroutines [2][6][7][9][10][14][16][19]. These include query refinement in Relevance Feedback [16], finding the cluster centers in k-means clustering algorithm, and template calculation in speech processing or pattern recognition. Various algorithms have been applied to calculate these data representations, often times we simply call it a data average. A simple averaging technique uses Euclidean distance metric. However, its one-to-one mapping nature is unable to capture the actual average shape of the two time series. In this case, shape averaging algorithm, Dynamic Time Warping, is much more appropriate [8].

In shape-based time series data, shape averaging method should be considered. However, most work involving time series averaging appear to avoid using DTW in spite of its dire need in the shape-similarity-based calculation [2][5][6][7][9][10][13][14][19] without providing sufficient reasons other than simplicity. For those who use k-means clustering, Euclidean distance metric is often used for time series average. This is also true in other domains such as speech recognition and pattern recognition [1][6][9][14], which perhaps is a good indicator flagging problems in DTW averaging method.

Y. Shi et al. (Eds.): ICCS 2007, Part I, LNCS 4487, pp. 513–520, 2007.
© Springer-Verlag Berlin Heidelberg 2007

Despite many proposed shape averaging algorithms, most of them provide method for specific domains [3][11][12], such as evoked potential in medical domains. In particular, after surveying related publications in the past decade, there appears to be only one proposed by Gupta et al. [8], who introduced the shape averaging using Dynamic Time Warping, and has been the basis for all subsequent work involving shape averaging. As shown in Figure 1 (a), the average is done in pairs, and the averaged time series in each level are hierarchically combined until the final average is achieved. Otherwise, another method – sequential hierarchical averaging – has been suggested, as shown in Figure 1 (b). Many subsequent publications inherit this method under the restriction of having the power of two time series data. In this paper, we will show that the proposed method in [8] does not have associative property as claimed.

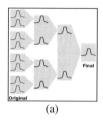   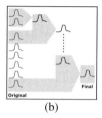

(a)   (b)

Fig. 1. Two averaging method – (a) balanced hierarchical averaging and (b) sequential hierarchical averaging

The rest of the paper is organized as follows. Section 2 explains some of important background involving shape averaging. Section 3 reveals the problems with current shape averaging method by extensive set of experiments. Finally, in section 4, we conclude with some discussion of potential causes of these inaccuracies, and suggest possible treatment to shape-based time series averaging problem.

2 Background

In this section, we provide brief details of Dynamic Time Warping (DTW) distance measure, its properties, time series averaging using DTW.

2.1 Distance Measurement

Distance measure is extensively used in finding the similarity/dissimilarity between time series. The two well known measures are Euclidean distance metric and DTW distance measure. As a distance metric, it must satisfy the four properties – symmetry, self-identity, non-negativity, and triangular inequality.

A distance measure, however, does not need to satisfy all the properties above. Specifically, the triangular inequality does not hold for the DTW distance measure, which is an important key to the explanation why we have such a hard time in shape averaging using Dynamic Time Warping.

2.2 Dynamic Time Warping Distance

DTW [15] is a well-known similarity measure based on shape. It uses dynamic programming technique to find all possible paths, and selects the one with the minimum distance between two time series. To calculate the distance, it creates a distance matrix, where each element in the matrix is cumulative distance of the minimum of three surrounding neighbors. Suppose we have two time series, a sequence Q of length n ($Q = q_1, q_2, \ldots, q_i, \ldots, q_n$) and a sequence C of length m ($C = c_1, c_2, \ldots, c_j, \ldots, c_m$). First, we create an n-by-m matrix where every (i, j) element of the matrix is the cumulative distance of the distance at (i, j) and the minimum of three neighbor elements, where $0 < i \leq n$ and $0 < j \leq m$. We can define the (i, j) element as:

$$e_{ij} = d_{ij} + \min\{e_{(i-1)(j-1)}, e_{(i-1)j}, e_{i(j-1)}\} \tag{1}$$

where $d_{ij} = (c_i - q_j)^2$ and e_{ij} is (i, j) element of the matrix which is the summation between the squared distance of q_i and c_j, and the minimum cumulative distance of three elements surrounding the (i, j) element. Then, to find an optimal path, we choose the path that has minimum cumulative distance at (n, m), which is defined as:

$$D_{DTW}(Q, C) = \min_{\forall w \in P}\left\{\sqrt{\sum_{k=1}^{K} d_{w_k}}\right. \tag{2}$$

where P is a set of all possible warping paths, and w_k is (i, j) at k^{th} element of a warping path and K is the length of the warping path.

2.3 Dynamic Time Warping Averaging

Shape averaging exploits DTW distance [8] to find the appropriate mapping for an average. More specifically, the algorithm needs to create a DTW distance matrix and find an optimal warping path. After the path is found, an averaged time series is calculated along this path by using the index (i, j) of each data point w_k on the warping path, which corresponds to the data points q_i and c_j on the time series $Q = q_1, q_2, \ldots, q_i, \ldots, q_n$ and $C = c_1, c_2, \ldots, c_j, \ldots, c_m$, respectively. An optimal warping path W with length K is defined as

$$W = w_1, w_2, \ldots, w_k, \ldots, w_K \tag{3}$$

In addition, w_k, which is mapped with index (i, j), is calculated by the mean value between time series whose indices are i and j. Note that in query refinement, where two time series may have different weights, weight α_Q for a sequence Q and weight α_C for a sequence C, the equation above may then be simply generalized according to the desired weight below

$$w_k = \frac{(\alpha_Q \cdot q_i + \alpha_C \cdot c_j)}{\alpha_Q + \alpha_C} \tag{4}$$

3   Experiment Evaluation

To validate our hypotheses, we set up 4 experiments to disprove the claims of the current shape averaging method. The first experiment tests whether reordering of the sequences will have any effect on the averaged result. The second experiment tests whether DTW averaging of two time series will give the real average. The third experiment tests whether the averaged result is in fact at the center of all original time series. Finally, in the fourth experiment, we test our overall hypotheses by running k-means clustering and demonstrate its failure in returning meaningful results.

3.1   Does Reordering Make Any Differences?

This experiment tests whether reordering of the sequences of balanced hierarchical averaging will affect the final averaged time series. According to [8], they claim the associative property under 2^n data constraint, and explicitly state that no matter how we rearrange the data, it will not make any difference in the final averaged outcome.

To show the associative property, we use the Cylinder-Bell-Funnel (CBF) [17], Leaf, Face, Gun, and ECG dataset, from the UCR time series data mining archive [http://www.cs.ucr.edu/~eamonn/time_series_data/]. The well-known 3-class CBF dataset contains 64 instances with the length of 128 data points. The last 3 datasets are multimedia data transformed into time series [16]. The Face dataset contains 112 total normalized instances of 350 data points. The Leaf dataset contains 442 instances of rescaled lengths of 150 data points. Gun dataset has two classes, with 100 instances each, and each instance has the same length of 150 data points. ECG dataset consists of two different heart-pulse classes; each class contains 28 instances with normalized length of 205 data points. Examples of CBF, Leaf, Face, Gun, and ECG data show in Figures 2, 3, 4, 5, and 6.

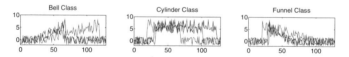

Fig. 2. Examples of CBF data with variations in time axis, i.e. the onset and ending positions

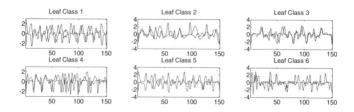

Fig. 3. Examples of six species of Leaf data using time series representation

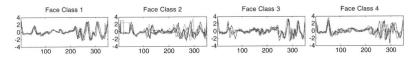

Fig. 4. Examples of six different Face classes

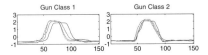

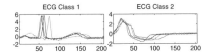

Fig. 5. Examples of Gun data   **Fig. 6.** Examples of ECG data

Even when $n = 1$, we cannot guarantee the commutative property of the averaging method, i.e., $DTW\_Avg(Q,C)$ may or may not give the same result as $DTW\_Avg(C,Q)$, though its symmetric property will give the same DTW distance. When n is larger than one, we want to test whether shuffling the sequences would affect the (balanced hierarchical) averaged result. We first test by averaging only instances within their own class. We compute the distance of *every possible pairings*, then reshuffle the data and repeat the computation (100 runs). We then compare whether the distances among all averaged results from each variation using DTW distance are in fact equal. It is very surprising to see that the averaged time series from each run do not have the same shape, giving the discrepancy among each of the averaging results from different runs much larger than zero. The result are shown in Table 1.

Table 1. Mean and standard deviation of discrepancy distance

	CBF	Leaf	Gun	ECG	Face
Discrepancy distance	227.95±17.23	99.32±10.66	142.50±13.31	6.17±0.90	24.58±2.49

3.2 Correctness of DTW Averaging Between Two Time Series

In this experiment, we demonstrate that the averaged time series, when comparing back to the two original time series, does not have the same distance. Our general intuitive hypothesis is that if we average two time series, the averaged result should *equally* contain characteristics from both original time series. To examine this, we compute the DTW distance from the averaged result back to *both* original time series, and we *should* get the same distance. If this property does not hold, the large number of data mining algorithms that have used this averaging method would probably have worked incorrectly, especially in [8] itself (balanced hierarchical averaging by pairing of 2^n time series). For evaluation, mean percentage errors in all possible pairs in each dataset are computed. Suppose we have two original time series, Q and C, and their resulting averaged time series X. The percentage error is defined as

$$PercentageError(Q,C,X) = \frac{\left|D_{DTW}(Q,X) - D_{DTW}(C,X)\right|}{\max\{D_{DTW}(Q,X), D_{DTW}(C,X)\}} \tag{5}$$

Table 2 shows the experiment results with mean and the standard deviation of percentage errors in each dataset. Note that the percentage error should be 0%.

Table 2. The percentage error from the average results in each dataset

	CBF	Leaf	Gun	ECG	Face
Discrepancy distance	227.95±17.23	99.32±10.66	142.50±13.31	6.17±0.90	24.58±2.49

3.3   Can Cluster Center Drift Out of the Cluster?

In this experiment, we demonstrate an undesirable phenomenon that could happen when we average more than two time series using DTW. We first test on the simplest case where there are only 3 objects to average. We combine hierarchical averaging and sequential averaging methods proposed by [8], to make sure that all three time series are used in the averaging process. We then determine whether the averaged result is in the middle of the group. The example is shown in Figure 7 (a). We first average data, A-B-C from all three data points – A, B, and C. In Figure 7 (b), we calculate average results between each pair of the data points – A-B, A-C, and B-C. If DTW distances between: A-B and C is less than that between A-B-C and C, A-C and B is less than that between A-B-C and B, or B-C and A is less than that between A-B-C and A, then that means the averaged result, A-B-C, is not in the center of the data points. Figure 7 (c) shows the averaged result satisfying the above assumption, but Figure 7 (d) shows the averaged result violating the assumption.

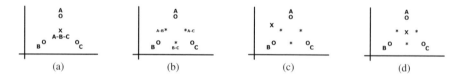

Fig. 7. The cluster center from DTW average, that may drift from the actual cluster center

Table 3 shows the percentage of occurrences that averaged results from all possible three time series that are outside the data group (unsatisfied averaging). Note that the unsatisfied averaging percentage must be 0% to verify correctness of this averaging method.

Table 3. Percentages of average results that are outside the group

	CBF	Leaf	Gun	ECG	Face
Unsatisfied averaged result (%)	0.008%	0.031%	8.936%	23.521%	0.006%

3.4   Failure in *K*-Means Clustering with DTW

This last experiment shows the use of *k*-means clustering using DTW distance to find the cluster center (shape averaging), comparing with an unproblematic *k*-medoids clustering with DTW distance. In this experiment, we will show that if *k*-means method is used in clustering, there is a high probability of failure (and that is probably why we do not see much of k-means clustering with DTW averaging in the literature. We will show this by reporting the average number of iteration up to the point where k-means fails, which is when the averaged cluster center happens to drift outside of the cluster, as we discussed in Experiment 3. We run k-means clustering with the same datasets. In each dataset, we choose k to be its actual numbers of class. For clearer evaluation, we compare the results with the number of iteration obtained using the k-medoids methods (which always succeed). We run the experiment 1,000 times for each dataset. Table 4 shows the mean and standard deviation of the number of iteration when k-means fails to give meaningful clustering results for each dataset.

Table 4. The mean and standard deviation of number of iteration when k-means fails compares to k-medoids successes for each dataset

	CBF	Leaf	Gun	ECG	Face
Failure: iteration (k-means)	2.16±1.13	1.32±0.34	5.16±1.71	1.76±0.76	1.72±0.83
Success: iteration (k-medoids)	3.87±0.94	4.19±0.90	4.06±0.93	2.50±0.51	3.61±0.72

4 Discussion, Conclusion, and Future Works

In search of the remedies, we can categorize the problems into three parts, i.e., a distance measure, an averaging method, and dataset properties. First, since DTW is the distance measure that has no triangular inequality property, the averaged time series may not be the actual mean because DTW cannot guarantee the position of averaged result in Euclidean space. Second, in finding a new the averaging method, we suggest that a new averaging method should satisfy various criteria in our proposed experiments. And third, to satisfy triangular inequalities, it also depends on the properties of the data at hands (generally, only a handful of data within a dataset would violate the triangular inequalities). It is possible to first split the data into groups that triangular inequalities hold within. We can simply find the DTW average for each group, and then finally merge those averages together.

In conclusion, we have empirically demonstrated various counterexamples to current shape averaging method using Dynamic Time Warping distance. From these experiments' findings, we have confirmed that the current DTW averaging is inaccurate and should not be used as a subroutine where shape averaging is needed due to lacks of several properties discussed earlier. We conjecture that the reason to this undesirable phenomenon is the triangular inequality that DTW also lacks of. Therefore, DTW averaging cannot guarantee the correctness of the averaging result.

In this paper, we intend to make a first attempt in pointing out some misunderstanding and misuse of current DTW averaging method. As our future work, from these findings, we will investigate how these problems can be resolved and come up with a remedy in accurately averaging shape-based time series data.

References

1. Abdulla, W.H., Chow, D., and Sin, G. Cross-words reference template for DTW-based speech recognition systems. In Proc. of TENCON 2003 (2003)
2. Bagnall, A. and Janacek, G. 2005. Clustering Time Series with Clipped Data. Mach. Learn. 58 (2005) 151-178
3. Boudaoud, S., Rix, H., and Meste, O. Integral shape averaging and structural average estimation. IEEE Trans. on Signal Processing, vol.53, no10 (2005) 3644-3650
4. Bradley, P. S., and Fayyad, U.M. Refining Initial Points for K--Means Clustering. In Proceedings of the 15th Int'l Conference on Machine Learning (1998) 91-99
5. Caiani, E.G., Porta, A., Baselli, G., Turiel, M., Muzzupappa, S., Pieruzzi, F., Crema, C., Malliani, A. and Cerutti, S. Warped-average template technique to track on a cycle-bycycle basis the cardiac filling phases on left ventricular volume. IEEE Computers in Cardiology (1998)
6. Corradini, A. Dynamic Time Warping for Off-Line Recognition of a Small Gesture Vocabulary. In Proc. of the IEEE ICCV Workshop on Ratfg-Rts. Washington, DC (2001)
7. Chu, S., Keogh, E., Hart, and D., Pazzani, M. Iterative deepening dynamic time warping for time series. In Proceedings of SIAM International Conference on Data Mining (2002)
8. Gupta, L., Molfese, D.L., Tammana, R., and Simos, P.G. Nonlinear alignment and averaging for estimating the evoked potential. IEEE Trans. on Biomed. Eng. vol.43, no.4 (1996) 348-356
9. Hu, J. and Ray, B. An Interleaved HMM/DTW Approach to Robust Time Series Clustering. IBM T.J. Watson Research Center (2006)
10. Keogh, E. and Smyth, P. An enhanced representation of time series which allows fast classification, clustering and relevance feedback. KDD (1997) 24-30
11. Lange, D.H., Pratt, H., and Inbar, G.F. (1997). Modeling and estimation of single evoked brain potential components. IEEE Trans. on Biomed. Eng. Vol.44 (1997) 791-799
12. Mor-Avi, V., Gillesberg, I.E., Korcarz, C., Sandelski, J., Lang, R.M. Signal averaging helps reliable noninvasive monitoring of left ventricular dimensions based on acoustic quantification. Computers in Cardiology (1994) 21-24
13. Oates, T., Firoiu, L., and Cohen, P.R. Using Dynamic Time Warping to Bootstrap HMM-Based Clustering of Time Series. In Sequence Learning Paradigms, Algorithms, and Applications. Volume 1828 of Lecture Notes in Computer Science (2001) 35-52
14. Rabiner, L. R., Levinson, S. E., Rosenberg, A. E., and Wilpon, J. G. Speaker-independent recognition of isolated words using clustering techniques. In Readings in Speech Recognition, CA (1990) 166-179
15. Ratanamahatana, C.A. and Keogh, E. Everything you know about Dynamic Time Warping is Wrong. In Proc. of KDD Workshop on Mining Temporal and Sequential Data (2004)
16. Ratanamahatana., C.A. and Keogh, E. Multimedia Retrieval Using Time Series Representation and Relevance Feedback. In Proc. of 8th ICADL (2005)
17. Saito, N. Local feature extraction and its application using a library of bases. PhD thesis, Yale University (1994)
18. Salvador, S. and Chan, P. FastDTW: Toward Accurate Dynamic Time Warping in Linear Time and Space. In Proc. of KDD Workshop on Mining Temporal and Sequential Data (2004)
19. Wilpon, J. and Rabiner, L. A modified K-means clustering algorithm for use in isolated work recognition. IEEE Trans. on Signal Processing. vol.33 (1985) 587-594

An Algebraic Substructuring Method for High-Frequency Response Analysis of Micro-systems

Jin Hwan Ko[1] and Zhaojun Bai[2]

[1]Dept. of Aerospace Information Engineering, Konkuk University, Seoul, Korea
jhko@invest.konkuk.ac.kr, jhko@cs.ucdavis.edu
[2]Dept. of Computer Science, University of California, Davis CA 95616, USA
bai@cs.ucdavis.edu

Abstract. High-frequency response analysis (Hi-FRA) is required to predict the resonant behavior of modern microsystems operated over a high frequency range. Algebraic substructuring (AS) method is a powerful numerical technique for FRA. However, the existing AS method is developed for low-FRA, say over the range 1Hz–2KHz. In this work, we extend the AS method for FRA over a given frequency range $[\omega_{min}, \omega_{max}]$. Therefore, it can be efficiently applied to systems operated at high frequency, say over the range 1MHz–2MHz. The success of the proposed method is demonstrated by Hi-FRA of a microgyroscope.

Keywords: High-Frequency Response Analysis, Algebraic Substructuring, Micro-Systems, Frequency Sweep Algorithm.

1 Introduction

Frequency Response Analysis (FRA) studies structural responses to steady-state oscillatory excitation to predict the resonant behavior in an operation (excitation) range of frequencies. Resonant sensors in microelectromechanical systems (MEMS) and other microscale structures are designed to catch the resonant behavior over a higher frequency range. Therefore, the Hi-FRA is typically required for the microscale structures.

The discretized model of a structure we consider in this paper is a continuous single-input single-output second-order system of the form

$$\begin{cases} M\ddot{x}(t) + D\dot{x}(t) + Kx(t) = bu(t) \\ y(t) = l^T x(t) \end{cases} \tag{1}$$

with the initial conditions $x(0) = x_0$ and $\dot{x}(0) = v_0$. Here t is the time variable, $x(t) \in \mathcal{R}^N$ is a state vector, N is the degree of freedom (DOF). $u(t)$ is the input excitation force and $y(t)$ is the output measurement function. $b \in \mathcal{R}^N$ and $l \in \mathcal{R}^N$ are the input and output distribution vectors. $M, K, D \in \mathcal{R}^{N \times N}$ are system mass, stiffness and damping matrices. It is assumed that the M and K

Y. Shi et al. (Eds.): ICCS 2007, Part I, LNCS 4487, pp. 521–528, 2007.

are symmetric positive definite. The input-output behavior of the model (1) is characterized by the transfer function

$$H(\omega) = l^T(-\omega^2 M + \mathrm{i}\omega D + K)^{-1}b, \tag{2}$$

where ω is the frequency and $\mathrm{i} = \sqrt{-1}$. Mathematically, the low FRA is on the computation of the transfer function $H(\omega)$ for ω over the range $[1, \omega_{\max}]$, where $\omega_{\max}$ is small, say at KHz. The Hi-FRA is about the computation of $H(\omega)$ for ω over the range $[\omega_{\min}, \omega_{\max}]$, where $\omega_{\min}$ and $\omega_{\max}$ are large, say at MHz.

Due to the large DOF of the model (1), it is prohibitive to directly compute $H(\omega)$ over a large number of frequency points ω_k over the range of interest. A popular approach of the FRA is based on an eigensystem analysis, called the mode superposition (MS) method. One first extracts n eigenpairs (λ_k, q_k) of the matrix pair (K, M):

$$Kq_k = \lambda_k M q_k, \tag{3}$$

where $q_k^T K q_k = \lambda_k$ and $q_k^T M q_k = 1$. Then by projecting the transfer function $H(\omega)$ onto the subspace $\mathrm{span}\{Q_n\} = \mathrm{span}\{[q_1, q_2, \ldots, q_n]\}$, it yields

$$H_n(\omega) = l_n^T(-\omega^2 I_n + \mathrm{i}\omega D_n + \Lambda_n)^{-1}b_n, \tag{4}$$

where $\Lambda_n = \mathrm{diag}(\lambda_1, ..., \lambda_n)$, $D_n = Q_n^T D Q_n$, $l_n = Q_n^T l$ and $b_n = Q_n^T b$. The shift-and-invert Lanczos (SIL) method as an eigensolver has been the method of choice for decades. However, the continual and compelling need for the FRA of very large model (1) challenges the computational efficiency of the method. Substructuring approaches, initially developed in early 1960s, are being studied in recent years. The automated multi-level substructuring (AMLS) method [1,2] is one of substructuring approaches, in which the structure is recursively divided of many of subdomains, and these subdomains can be handled efficiently and in parallel. An algebraic analysis of the AMLS method, referred to as the algebraic structure (AS) method, is studied in [10,5]. The AMLS has been successfully used for low FRA in which the smallest eigenmodes are required [2]. However, the direct application of the AMLS to the Hi-FRA would require a large number of eigenmodes starting from the smallest to the large ones to match the high frequencies. It is computationally inefficient. In this paper, we propose an extension of the AMLS method for Hi-FRA application. Since the implementation of the AMLS is a proprietary software, we will use the AS method presented in [10,5] as an eigensolver, and then present the FRA method that is the an extension of the AMLS frequency sweeping algorithm [2].

2   Algebraic Substructuring

For Hi-FRA, the eigenmodes corresponding to the natural frequencies closest to the operation range are most important. Hence, we begin with a shifted eigenproblem of (3):

$$K^\sigma q = \lambda^\sigma M q, \tag{5}$$

where $K^\sigma = K - \sigma M$ and $\lambda^\sigma = \lambda - \sigma$. σ is a prescribed shift related to the frequency range $[\omega_{min}, \omega_{max}]$. The choice of the shift σ is to be discussed later. We assume that the matrix pair (K^σ, M) is of the partition

$$
K^\sigma = \begin{array}{c} N_1 \\ N_2 \\ N_3 \end{array} \begin{array}{ccc} N_1 & N_2 & N_3 \\ \left[\begin{matrix} K_{11}^\sigma & & K_{13}^\sigma \\ & K_{22}^\sigma & K_{23}^\sigma \\ K_{31}^\sigma & K_{32}^\sigma & K_{33}^\sigma \end{matrix} \right] \end{array}, \quad
M = \begin{array}{c} N_1 \\ N_2 \\ N_3 \end{array} \begin{array}{ccc} N_1 & N_2 & N_3 \\ \left[\begin{matrix} M_{11} & & M_{13} \\ & M_{22} & M_{23} \\ M_{31} & M_{32} & M_{33} \end{matrix} \right] \end{array}, \quad (6)
$$

where (K_{11}^σ, M_{11}) and (K_{22}^σ, M_{22}) are two substructures that are connected by the interface (K_{33}^σ, M_{33}). For simplicity, we only show in single-level substructuring. A multi-level extension is performed for the shifted matrices through the same process which is described in [5].

By performing a block LDL^T factorization of the matrix K^σ, i.e., $K^\sigma = L\widehat{K}^\sigma L^T$, the shifted eigenproblem (5) is transformed to the eigenproblem

$$\widehat{K}^\sigma \widehat{q} = \lambda^\sigma \widehat{M} \widehat{q}, \quad (7)$$

where $\widehat{K}^\sigma$ and $\widehat{M}$ are in the Craig-Bampton form [3]:

$$\widehat{K}^\sigma = L^{-T} K^\sigma L^{-1} \quad \text{and} \quad \widehat{M} = L^{-T} M L^{-1}.$$

The next step of the AS is to extract the eigenmodes (called *local modes*) of the interior substructures and interface specified by the *local cutoff values* μ_{min}^σ and μ_{max}^σ. The subspace spanned by the column of the matrix

$$
S = \begin{array}{c} N_1 \\ N_2 \\ N_3 \end{array} \begin{array}{ccc} m_1 & m_2 & m_3 \\ \left[\begin{matrix} S_1 & & \\ & S_2 & \\ & & S_3 \end{matrix} \right] \end{array} \quad (8)
$$

where S_1, S_2 and S_3 consist of extracted eigenvectors of substructures and the interface, respectively.

By projecting the eigenproblem (7) onto the subspace span S, then we have a reduced eigensystem of order $m = m_1 + m_2 + m_3$:

$$K_m^\sigma \phi = \theta^\sigma M_m \phi, \quad (9)$$

where $K_m^\sigma = S^T \widehat{K}^\sigma S$ and $M_m = S^T \widehat{M} S$. The eigenmodes ϕ are referred to as the *global modes*. These global modes are grouped into *retained modes* and *truncated modes* determined by *(left and right) global cutoff values* λ_{min}^σ and λ_{max}^σ. If we write $\Phi = [\phi] = [\Phi_l \ \Phi_n \ \Phi_r]$, then Φ_n are the retained modes, Φ_l and Φ_r are the truncated modes corresponding to the eigenvalues smaller and larger than the cutoff values λ_{min}^σ and λ_{max}^σ, respectively. Φ_t are all truncated modes, $\Phi_t = [\Phi_l \ \Phi_r]$. The subspace spanned by the columns of the matrix $L^{-1}S$ is referred to as an *AS subspace*.

3   Frequency Response Analysis

With the assumption of Rayleigh damping $D = \alpha M + \beta K$ and the introduction of the shift σ, the transfer function $H(\omega)$ can be written as

$$H(\omega) = l^T[\gamma_1 K^\sigma + \gamma_2 M]^{-1}b, \tag{10}$$

where $\gamma_1 = \gamma_1(\omega) = 1 + \mathrm{i}\omega\beta$ and $\gamma_2 = \gamma_2(\omega,\sigma) = -\omega^2 + \sigma + \mathrm{i}\omega(\alpha + \sigma\beta)$. Projecting $H(\omega)$ onto the AS subspace, we have

$$H_m(\omega) = l_m^T[\gamma_1 K_m^\sigma + \gamma_2 M_m]^{-1}b_m = l_m^T p_m(\omega), \tag{11}$$

where $l_m = (L^{-1}S)^T l$ and $b_m = (L^{-1}S)^T b$, and $p_m(\omega)$ is the solution of the parameterized linear system of the order m:

$$G_m(\omega)p_m(\omega) = b_m. \tag{12}$$

where $G_m(\omega) = \gamma_1 K_m^\sigma + \gamma_2 M_m$. It is typical that by the AS method, the order m is still too large to apply for computing frequency responses. In AMLS, a so-called frequency sweep (FS) algorithm is introduced [2]. The FS algorithm retains only the low frequency modes and truncate all high frequency modes. However, for efficient Hi-FRA, it is important to be able to retain those modes corresponding to the frequency range $[\omega_{\min}, \omega_{\max}]$ of interest. To do so, let the vector $p_m(\omega)$ be written as $p_m(\omega) = p_n(\omega) + p_t(\omega)$ where $p_n(\omega)$ are in the subspace spanned by the retained global modes Φ_n and $p_t(\omega)$ in the subspace spanned by the truncated modes Φ_l and Φ_r. Write $p_n(\omega) = \Phi_n\eta_n(\omega)$ for some coefficient vector $\eta_n(\omega)$, then the equation (12) becomes

$$G_m(\omega)(\Phi_n\eta_n(\omega) + p_t(\omega)) = b_m. \tag{13}$$

Pre-multiplying the equation by Φ_n^T, then by the orthogonality of the global modes, the vector $p_n(\omega)$ is immediately given by the n uncoupled equations:

$$p_n(\omega) = \Phi_n(\Phi_n^T G_m(\omega)\Phi_n)^{-1}\Phi_n^T b_m = \Phi_n(\gamma_1\Theta_n^\sigma + \gamma_2 I)^{-1}\Phi_n^T b_m. \tag{14}$$

Subsequently, the equation (13) can be written as a parameterized linear system for $p_t(\omega)$:

$$G_m(\omega)p_t(\omega) = b_m - G_m(\omega)p_n(\omega). \tag{15}$$

Since it is anticipated the effect of the truncated modes for the accuracy of FRA is marginal, we employ a simple iterative refinement scheme for computing $p_t^{\ell-1}(\omega)$: $p_t^\ell(\omega) = p_t^{\ell-1}(\omega) + \Delta p_t^\ell(\omega)$ where the correction term $\Delta p_t^\ell(\omega)$ is the solution of the refinement equation

$$G_m(\omega)\Delta p_t^\ell(\omega) = r_m^{\ell-1}(\omega), \tag{16}$$

and $r_m^{\ell-1}(\omega) = b_m - G_m(\omega)(p_n(\omega) + p_t^{\ell-1}(\omega))$, the $(\ell - 1)$-th residual vector.

To solve the refinement equation (16), we use a Galerkin subspace projection technique, namely, seek $\Delta p_t^\ell(\omega)$ such that

$$\Delta p_t^\ell(\omega) \in \mathrm{span}\{\Phi_t\} \quad \text{and} \quad G_m(\omega)\Delta p_t^\ell(\omega) - r_m^{\ell-1}(\omega) \perp \mathrm{span}\{\Phi_t\}.$$

By some algebraic manipulation, we have

$$\Delta p_t^\ell(\omega) = \Phi_t(\gamma_1 \Theta_t^\sigma + \gamma_2 I)^{-1} \Phi_t^T r_m^{\ell-1}(\omega)$$
$$= \left[(\gamma_1 K_m^\sigma + \gamma_2 M_m)^{-1} - \Phi_n(\gamma_1 \Theta_n^\sigma + \gamma_2 I)^{-1} \Phi_n^T \right] r_m^{\ell-1}(\omega).$$

For computational efficiency, noting that K_m^σ is diagonal, we simply use the following approximation for computing the correction term $\Delta p_t^\ell(\omega)$:

$$\Delta p_t^\ell(\omega) \approx \Phi_t(\gamma_1 \Theta_t^\sigma)^{-1} \Phi_t^T r_m^{\ell-1}(\omega) \qquad (17)$$

Subsequently, we derive the following iterative refinement iteration for computing the vector $p_t(\omega)$:

$$p_t^\ell(\omega) = p_t^{\ell-1}(\omega) + \frac{1}{\gamma_1} \left[(K_m^\sigma)^{-1} - \Phi_n(\Theta_n^\sigma)^{-1} \Phi_n^T \right] r_m^{\ell-1}(\omega) \qquad (18)$$

for $\ell = 1, 2, \ldots$, with the initial guess $p_t^0(\omega)$. A practical stopping criterion is to test the relative residual error $\|\Delta p_t^\ell(\omega)\|_2 / \|(\gamma_1 K_m^\sigma)^{-1} b_m\|_2 \le \epsilon$ for a given tolerance ϵ. The convergent solution is denoted as $p_t^*(\omega)$.

Assume that it is required to calculate the n_f frequency points: $\omega_{\min} \le \omega_1 < \omega_2 < \cdots < \omega_{n_f} \le \omega_{\max}$. Then we can determine the initial guess $p_t^0(\omega_k)$ at the frequency ω_k by a linear extrapolation for $k = 3, 4, \ldots, n_f$ with $p_t^0(\omega_1) = 0$ and $p_t^0(\omega_2) = p_t^*(\omega_1)$. Then all initial guess $p_t^0(\omega_k) \in \text{span}\{\Phi_t\}$. $p_t^\ell(\omega)$ by the iteration (18) is guaranteed to be orthogonal to the vector $p_n(\omega)$.

Now we turn to investigate the relationship between the frequency range $[\omega_{\min}, \omega_{\max}]$ of interest and the interval $[\lambda_{\min}^\sigma, \lambda_{\max}^\sigma]$ for the global modes to be retained to guarantee the convergence of the iteration (18). By the equation (17), we have

$$\|\Delta p_t^\ell(\omega)\|_2 \approx \|\Phi_t(\gamma_1 \Theta_t^\sigma)^{-1} \Phi_t^T r_m^{\ell-1}(\omega)\|_2 \le \|\Phi_t(\gamma_1 \Theta_t^\sigma)^{-1}\|_2 \|\Phi_t^T r_m^{\ell-1}(\omega)\|_2.$$

The term $\Phi_t^T r_m(\omega)$ is referred to as a *truncated modal residual*. By some algebraic manipulation, we see that two consecutive truncated modal residuals satisfy the relation

$$\Phi_t^T r_m^\ell(\omega) = -\frac{\gamma_2}{\gamma_1} \begin{bmatrix} \Theta_l^\sigma & \\ & \Theta_r^\sigma \end{bmatrix}^{-1} \Phi_t^T r_m^{\ell-1}(\omega)$$

Therefore, if we introduce a positive constant ξ, referred to as *contraction ratio*, such that

$$\left| \frac{\phi_k^T r_m^\ell(\omega)}{\phi_k^T r_m^{\ell-1}(\omega)} \right| = \frac{d(\omega, \sigma)}{|\theta_k^\sigma|} \le \frac{d_{\max}}{|\theta_k^\sigma|} \le \xi < 1 \qquad (19)$$

where $\phi_k \in \Phi_t$, $d(\omega, \sigma) = |-\gamma_2/\gamma_1|$, and $d_{\max} = \max\{d(\omega_k, \sigma), 1 \le k \le n_f\}$. Then the components of the truncated modal residual are contracted, i.e., the norm of the correction term $\Delta p_t(\omega)$ decreases and the iteration (18) converges.

By (19), it derives that the global modes outside the interval $[-d_{\max}/\xi, d_{\max}/\xi]$ can be "cut off", i.e., the global cutoff values are determined by

$$\lambda_{\min}^{\sigma} = -d_{\max}/\xi \quad \text{and} \quad \lambda_{\max}^{\sigma} = d_{\max}/\xi \tag{20}$$

When there is no shift, i.e., $\sigma = 0$, the low global cutoff value is less than zero. It means that all low frequency modes smaller than $\lambda_{\max}^{\sigma}$ are retained.

4   ASFRA Algorithm

By combining the AS method for extracting eigenpairs and the frequency sweep iteration, we derive an algorithm for computing the frequency responses $H(\omega_k)$. The algorithm is referred to as the ASFRA algorithm. In this section, we briefly discuss the choice of parameters in ASFRA. Detail will be presented in a full paper elsewhere.

It is necessary that the shift $\sigma \in [\omega_{\min}^2, \omega_{\max}^2]$. In order to minimize the range of the global modes to be retained/extracted, the center of the frequency range is used, $\sigma = \frac{1}{2}(\omega_{\max}^2 + \omega_{\min}^2)$.

By (20), the global cutoff values $\lambda_{\min}^{\sigma}$ and $\lambda_{\max}^{\sigma}$ are essentially determined the contraction ratio ξ. To improve the convergence of the FS iteration (18), the contraction ratio ξ should be small. However, it makes the number of retained global modes large. From our numerical experiments, we found that a good choice is $\xi = 0.5$.

How to retain the local modes for a desired number and accuracy of global modes has been an important issue in the study of the AS algorithm [10]. To achieve a desired level of accuracy of the global modes, a large number of local modes are required. the local cutoff values $\mu_{\min}^{\sigma}$ and $\mu_{\max}^{\sigma}$ are typically chosen proportionally to the global cutoff values $\lambda_{\min}^{\sigma}$ and $\lambda_{\max}^{\sigma}$, namely $\mu_{\min}^{\sigma} = c_l \lambda_{\min}^{\sigma}$ and $\mu_{\max}^{\sigma} = c_u \lambda_{\max}^{\sigma}$, where c_l and c_u are relaxation factors. As c_l and c_u increase, the accuracy of the global modes is typically improved. We use $c_l = c_u = 10$ as default by referring to the previous research in [1].

We implemented ASFRA based on the ASEIG [5]. The multilevel partition is done by METIS [6]. The global modes and the local modes of the substructure blocks are computed by ARPACK [7] with SuperLU [4] and the local modes of the interface are solved by LAPACK.

We will compare the performance of ASFRA with three other methods in the next section. Let us briefly review these methods. The first method is the so-called *direct method*. It computes the frequency responses $H(\omega_k)$ by solving the underlying linear system (2) by a direct sparse method. Specifically, we use the SuperLU method. The second method is to use the shift-and-invert Lanczos (SIL) method from ARPACK to extract n eigenmodes and then approximate $H(\omega_k)$ by $H_n(\omega_k)$ as defined in (4). The shift is $\sigma = 0$ and the eigenmodes are determined by upper cutoff value $\lambda_{\max}$ and the residual flexibility vectors are supplemented [9]. The upper cutoff value $\lambda_{\max}$ is determined by $\lambda_{\max} = (\chi\omega_{\max})^2$, where χ is a multiplication factor. Typically, $\chi = 2$ or 3, when there are no residual flexibility vectors. Otherwise, χ can be smaller, say 1.11. The

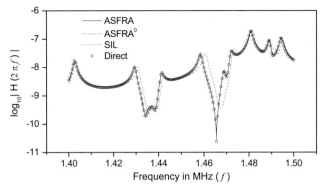

Fig. 1. The frequency responses of a butterfly gyro

third method is a special case of ASFRA with the zero shift $\sigma = 0$, and the lower cutoff value $\lambda^\sigma_{\min} = 0$. The frequency response $H(\omega_k)$ is approximated by $H_m(\omega_k)$ in eq. (11). It is denoted as $ASFRA^0$. $ASFRA^0$ is essentially the AMLS with frequency sweep iteration as presented in [2].

5   Numerical Experiment

MEMS resonators under an electrostatic actuation are utilized in various MEMS devices such as angular rate sensors and bandpass filters. Specifically, we consider a FE model with solid elements of a butterfly gyro which is an angular rate sensor using the MEMS resonator [8]. The order of the system K and M matrices is $N = 17631$. The Rayleigh damping parameters are set by $\alpha = 0.0$, $\beta = 10^{-10}$. Frequency responses changes rapidly near the resonances in the range $[f_{\min}, f_{\max}] = [\omega_{\min}/(2\pi), \omega_{\max}/(2\pi)] = [1.4, 1.5]$MHz. Numerical experiment is conducted on an Intel Itanium 2 Server with Linux OS.

The substructuring level of the AS is 3. The tolerance of the frequency sweep iteration is $\epsilon = 10^{-5}$. By Figure 1, ASFRA shows better accuracy than $ASFRA^0$ with the given parameters. The detailed results are listed in Table 1. All methods calculate the responses at $n_f = 201$ frequencies ω_k in an equal space on the interval. ASFRA is 2.3 times faster than $ASFRA^0$. SIL is more expensive than ASFRA because it needs to compute eigenmodes of the full-size eigensystem. Finally, we note that the performance of ASFRA and $ASFRA^0$ does not change significantly when the parameters ξ and c_l, c_u are changed slight from the present choice.

Table 1. The dimension of AS subspace, numbers of retained modes, total FS iteration, and the elapsed time

	ASFRA	$ASFRA^0$	SIL	Direct
m (AS subspace dim.)	213	651	-	–
n (retained modes)	20	175	156	–
Total FS iteration	238	51	-	–
Elapsed Time(sec.)	26.77	62.94	80.42	754.6

6   Conclusion

In this paper, we presented an algebraic substructuring based frequency response analysis (ASFRA) algorithm to calculate the frequency response of a large dynamic system between two specified frequency $\omega_{\min}$ and $\omega_{\max}$. ASFRA can be efficiently applied to Hi-FRA, as demonstrated by a microelectomechanical sensor operated at 1MHz–2MHz. Future work includes the optimal choice of parameters and parallelization techniques.

Acknowledgments. JHK was supported by Korea Research Foundation Grant KRF-2005-214-D00015. ZB was supported in part by the NSF grant DMS-0611548. Most of this work was done while JHK visiting the University of California, Davis.

References

1. Bennighof, J. K., Kim, C. K. (eds.): An addaptive multi-level substururing method for efficient modeling of complex strucutures. Proceedings of the AIAA 33rd SDM Conference, Dallas, Texas, (1992) 1631–1639
2. Bennighof, J. K., Kaplan, M. F. (eds.): Frequency sweep analysis using multi-level substructuring, global modes and iteration. Proceedings of 39th AIAA/ASME/ASCE/AHS Structures, Structural Dynamics and Materials Conference. (1998)
3. Craig, Jr. R. R., Bampton, M.C.C. (eds.): Coupling of substructures for dynamic analysis. AIAA Journal, Vol. 6(7). (1968) 1313–1319
4. Demmel, J.W., Eisenstat, S. C., Gilbert, J. R. , Li, X. S., Liu, J. W. H. (eds.): A supernodal approach to sparse partial pivoting. SIAM J. Matrix Anal. Appl., Vol. 20(3). (1999) 720–755
5. Gao, W., Li, X.S., Yang, C., Bai, Z. (eds.): An implementation and evaluation of the AMLS method for sparse eigenvalue problems. Technical Report LBNL-57438, Lawrence Berkeley National Laboratory. (2006)
6. Karypis, G.: METIS. Department of Computer Science and Engineering at the University of Minnesota, http://www-users.cs.umn.edu/ karypis/metis/metis/index.html. (2006)
7. Lehoucq, R., Sorensen, D. C., Yang, C.: ARPACK User's Guide: Solution of Large-Scale Eignevalue Problems with Implicitly Restarted Arnoldi Methods. SIAM. Philadelphia (1998)
8. Lienemann, J., Billger, D., Rudnyi, E. B., Greiner, A., Korvink, J. G.: MEMS compact modeling meets model order reduction: Examples of the application of arnoldi methods to microsystem devices. The Technical Proceedings of the 2004 Nanotechnology Conference and Trade Show, Nanotech 04. (2004)
9. Thomas, B., Gu, R. J.: Structural-acoustic mode synthesis for vehicle interior using finite-boundary elements with residual flexibility. Int. J. of Vehicle Design, Vol. 23. (2000) 191–202
10. Yang, C., Gao, W., Bai, Z., Li, X., Lee, L., Husbands, P., Ng, E.: An algebraic substructuring method for large-scale eigenvalue calculations. SIAM J. Sci. Comput., Vol. 27(3). (2005) 873–892

Multilevel Task Partition Algorithm for Parallel Simulation of Power System Dynamics

Wei Xue and Shanxiang Qi

Department of Computer Science and Technology, Tsinghua University,
Beijing, China, 100084
xuewei@tsinghua.edu.cn, qishanxiang@gmail.com

Abstract. Nowadays task partition for parallel computing is becoming more and more important. Particular in power system dynamic simulation, it is critical to design an efficient partition algorithm to reduce the communication and balance the computation load [1]. This paper presents a novel multilevel partition scheme based on the graph partition algorithm. By introducing regional characteristic into the partition, improving the weights of nodes and edges, proposing an objective function to evaluate the partition results and some other schemes, we can efficiently improve the defects in the traditional partition methods. With 12 CPUs for a large scale power system with 10188 nodes, the parallel efficiency with our new algorithm was 63% higher than that with METIS, a well-known program used for partitioning graphs. The proposed algorithm will satisfy the requirement for large scale power grid dynamic simulation.

Keywords: power system, parallel simulation, task partition, graph partition.

1 Introduction

Power system dynamic simulation is an important tool in power system research. With the development of power system scale, the computation tasks are becoming heavier and more complex. Up to now, there have been many research topics related to parallel algorithms and parallel simulation of power system dynamics [2-10]. To accomplish the task of power system dynamic simulation, a set of differential algebraic equations (DAEs) have to be solved.

$$\begin{cases} \dot{\mathbf{X}} = f(\mathbf{X}, \mathbf{V}) = \mathbf{A}\mathbf{X} + \mathbf{B}u(\mathbf{X}, \mathbf{V}) \\ 0 = I - Y(\mathbf{X}) * \mathbf{V} \end{cases} \qquad (1)$$

In (1), the first nonlinear differential equation group describes the dynamic characteristics of the power devices, and the second nonlinear equation group represents the restriction of the power network, where X is the state vector of individual dynamic devices, I is the vector of current injected from the devices into the network, V is the node voltage vector, $Y(X)$ is the complex sparse matrix, which is not constant with time, and is the function of X and V.

The most commonly used sequential algorithm for parallel simulation of power system dynamics is the interlaced alternating implicit approach (IAI algorithm). The

Y. Shi et al. (Eds.): ICCS 2007, Part I, LNCS 4487, pp. 529–537, 2007.

IAI algorithm uses a trapezoidal rule in the difference process and solves differential equations and algebraic equations alternately and iteratively. It not only maintains the advantages of the implicit integration approach, but also has modeling and computing flexibility.

Now the time consumed for simulation increases super-linearly as the power system's size increases. The performance of sequential simulation is not adequate for real-time simulation of large scale power grids, which is critical for power system planning and dispatching. Its limitations become increasingly serious as the integrated power grid continues to develop. Therefore, it is very important to study practical parallel algorithms and software.

The key problem of solving the equations is to solve a sparse linear system, which can be briefly formulated as follows:

$$\begin{bmatrix} Y & M' \\ M & Z_{CF} \end{bmatrix} \begin{bmatrix} U \\ I \end{bmatrix} = \begin{bmatrix} I_p \\ 0 \end{bmatrix} \tag{2}$$

In which,

$$M = \begin{bmatrix} \mathbf{M}_{CF-1p} & \mathbf{M}_{CF-2p} & & & & \\ & & \mathbf{M}_{CF-1n} & \mathbf{M}_{CF-2n} & & \\ & & & & \mathbf{M}_{CF-1z} & \mathbf{M}_{CF-2z} \end{bmatrix}$$

$$Y = \begin{bmatrix} Y_{1p} & & & & & \\ & Y_{2p} & & & & \\ & & Y_{1n} & & & \\ & & & Y_{2n} & & \\ & & & & Y_{1z} & \\ & & & & & Y_{2z} \end{bmatrix} \quad U = \begin{bmatrix} \mathbf{U}_{1p} \\ \mathbf{U}_{2p} \\ \mathbf{U}_{1n} \\ \mathbf{U}_{2n} \\ \mathbf{U}_{1z} \\ \mathbf{U}_{2z} \end{bmatrix}$$

$$I_p = \begin{bmatrix} I_{1p} \\ I_{2p} \\ 0 \\ 0 \\ 0 \\ 0 \end{bmatrix} \quad M' = \begin{bmatrix} M_{1p-CF} & & & \\ M_{2p-CF} & & & \\ & M_{1n-CF} & & \\ & M_{2n-CF} & & \\ & & M_{1z-CF} \\ & & M_{2z-CF} \end{bmatrix}$$

$$\mathbf{Z} = \mathbf{Z}_{CF}, \quad \mathbf{I} = \mathbf{I}_{CF}$$

Where the subscripts 1 and 2 represent the sub-area number, and the subscripts p, n, z represent the positive, negative, and zero sequence networks respectively. Y_{1p}, Y_{2p}, Y_{1n}, Y_{2n}, Y_{1z} and Y_{2z} are the admittance matrices of three (positive, negative and zero) sequence networks respectively. Z_{CF} is the impedance matrix for cutting branches (the branches between different sub-areas) and fault branches, and also the coefficient matrix of boundary equations in the BBDF computation. $\mathbf{M}$ and $\mathbf{M'}$ are the associated matrices between $\mathbf{Y}$ and $\mathbf{Z}$ respectively.

As shown in (2), the equations have Block Bordered Diagonal Form and can match the concept of parallelism. Each sub-area calculates its part of the solution. But obviously, each sub-area should calculate in serial with the boundary system represented by $\mathbf{Z}$. Global communication like gather/scatter is used between the processes of sub-areas and boundary system. Meanwhile, the boundary system is formed due to the connections between sub-areas. So the more connections between sub-areas, the larger boundary system as well as the more communication between processes in parallel simulation.

It is an essential and critical step to appropriately partition the power system so as to map the computational load onto the processors. Generally, there are two main objectives for the partition:

A. Minimization of the connections between different sub-areas;

B. Maintain an equal computational load for each sub-area.

Power systems are large scale networks with connections related to geographical positions. The problem of network decomposition can be changed into the graph partition problem, which is well developed in HPC field [12]. However, if a graph partition algorithm is simply used; more communication time will be introduced into the parallel simulation because of more cutting edges during graph partitioning [13]. On the other hand, because power systems are developed with connections of regional networks, the connections in the regional network are much tighter than those between regions. So partitioning based on this regional characteristic of power networks can reduce the communication cost in parallel simulation, but it may introduce more load imbalance.

Based on the observation above, this paper presents a novel multilevel partition scheme, which integrates the advantages of graph partition algorithms and the partition method based on power network regional characteristics. This scheme efficiently reduces the boundary system computation as well as the communication in power system dynamic simulation.

2 Multilevel Partition Algorithm

There are four main steps in our new scheme as shown by Fig 1:

1) Establish the layered regional model for the power network. This model indicates the weak connections in a power network and gives the guide for successive partition. Thank to the development of power system, we can get the model easily by its obvious regional nature.

2) Network coarsing. During the coarsing phase, aggregating the nodes based on the reasonable level (such as the highest level) of regional model successively decreases the size of the power network graph. The partition problem of power networks is formulated into a weighted graph partition problem, in which the weights of the vertex and edge represent the computation load of the sub-network in the vertex and the amount of communication between sub-networks. Following network coarsing, the multilevel partition algorithm can identify the weak connections in a power network easily and effectively. The time consumed in the partition process is also reduced sharply.

3) Multi-level partition. In the multilevel partitioning phase, the derived small weighted graph is broken down into a specific number of sub-graphs with a graph partition algorithm (bisection method is used here). As we know, the small derived graph may limit the freedom in partitioning and may bring more load

imbalance into the final solution. The multilevel scheme is, therefore, proposed to evaluate the quality of partition results, decompose the sub-area with the maximum computation load and partition the new graph again. This process continues to run until the optimal solution is found.

4) Refine results. After the weighted graph partition, the results have to be applied to the original network. Further refinement and adjustment are performed to improve the quality of final partition results.

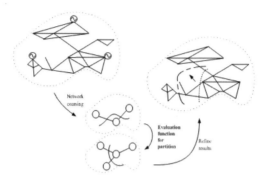

Fig. 1. The process of multilevel partition algorithm

The following key points in the new method can be described as follows:

1) How to construct graph in order to describe the amount of computation and communication.
2) How to evaluate the partition results and then control the partitioning process.
3) How to refine results for partition if possible.

2.1 Establish the Layered Model

Due to the nature of the power system, we can establish the layered model according to the regionalism.

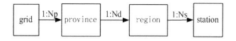

Fig. 2. Layered Power Network Model

2.2 Weights for Graph Vertex and Edge

In this paper, we introduce the weights of the vertex and edge to represent the computation load of the sub-network in the vertex and the amount of computation in boundary system.

Simulation for power system dynamics is floating point computation dominated. So the weight of the vertex describes the amount of floating point computation of the devices and network. So the vertex V_i's weight WV_{V_i} is shown below:

$$WV_{V_i} = N_n + 2N_{bC} + \sum_{D_i \in VD_i} Path(D_i) \tag{3}$$

In which, N_n is the number of the nodes in the coarsing vertex. N_{bC} is the number of the edges between nodes in vertex. $N_n + N_{bC}$ represents the floating point computation of network in vertex. VDi is the set of the devices related to V_i. $Path(D_i)$ is the estimation of computation for device D_i, which can be regarded as the number of edges in relation graph between variables for specific device mathematical model.

The weight of the edge describes the contribution of the cutting branches to the boundary system. With our parallel algorithm, the nodes related to the cutting edges will be included in the boundary system. But the size of boundary system is not equal to twice of the number of cutting edges. So the edge E_i's weight WE_{E_i} is shown below:

$$WE_{E_i} = \frac{1}{DE(N_p)} + \frac{1}{DE(N_q)} \tag{4}$$

In which, $DE(N_p)$ and $DE(N_q)$ are the degree of the nodes. This weight can give an accurate estimation of size of boundary system, which is critical for BBDF parallel algorithm.

2.3   Evaluation Function for Partition

Our evaluation function is based on the implementation of BBDF parallel algorithm.

$$F(p) = \underset{i=1,\dots,p}{Max}(CompCost_i) + CompCost_{Border} \tag{5}$$

In which $CompCost_i$ is the computation of the sub-area, and $CompCost_{Border}$ is the computation of the boundary system described below. In the evaluation function, the sum of $CompCost_i$ and $CompCost_{Border}$ represents the total floating point computation in the critical path and almost the overall time consumed in the parallel simulation. It is noted that the more communication occurs between subtasks, the more computation is introduced into the boundary system. So the influence of communication has been taken into account by $CompCost_{Border}$.

According to this problem, we suggest that:

$$CompCost_i = VW_i + N_{Ti}^2 + 2*\sum_{j \in Ti} Degree(j) \tag{6}$$

$$CompCost_{Border} = \sum_{i=1}^{p} N_{Ti}^2 + N_C \tag{7}$$

In which VW_i is the floating point computation of the internal network (computation related to Y in formula 2), which can be evaluated by formula 3. N_{Ti} is the number of nodes included in boundary system. An interface matrix has to be computed by sub-area i and be communicated to boundary system. Due to the interface matrix is almost dense, the computation can be evaluated by the square of the matrix dimension N_{Ti}. And the computation related to M and M^T in formula 2 in sub-area i is about the degree of nodes, which related to boundary system. With the same idea, we can get the computation evaluation of boundary system using formula 7.

In fact, it is difficult to give the accurate computation overhead in sparse matrix operation before factoring. Based on our analysis and testing, for comparing the different partition results, our method gives a trade-off between accuracy and speed to evaluate the overhead of parallel computation effectively.

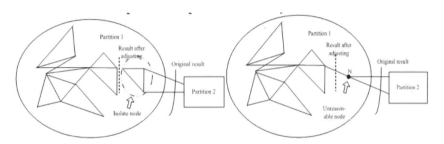

Fig. 3. Partition improvement for isolated node set   **Fig. 4.** Partition improvement for Unreasonable node

2.4 Refine Results

After the graph partitioning phase, it may have two similar problems: isolate node sets and unreasonable node related to the cutting branches. These two problems are shown above.

We identify the isolated node sets by topological analysis and analyze the connections of the nodes in the boundary system to improve the quality of final partition results.

3 Test Results

In this study, two power systems shown in Table 1 have been tested on a SMP Cluster. In the cluster, each node is a SMP computer and has 4 Intel Xeon PIII700MHz CPUs and 1 gigabyte of memory. The communication device between SMP nodes is Myrinet with a bandwidth of 2.56Gb/s. The software environments are Redhat Linux 7.2 (kernel version 2.4.7-10smp), MPICH-1.2.1..7 and gm-1.5pre4.

The detail computation models including 5 order generator model, typical exciter and governor model, and induction motor model, are concerned [13] for all test cases. An A-phase fault on single 220kV branch is assumed for Case 1, the fault occurs at 0s, and the branch trips at 0.16s. It is a typical non-symmetric fault case. A three-phase fault is assumed for Case 2, the fault occurs at 0s, and the branch trips at 0.08s.

It is a serious symmetric fault case with numerical difficult. In the simulations, the fixed time step 0.01s is used, and the simulation time is 10s. The convergence tolerance is 10^{-4} pu. For all the test cases, the same code for parallel algorithm and complier flags are used as well as the code tuning schemes.

Table 1. Network information for two power systems

	System	Case 1	Case 2
	number of nodes	706	10188
Scale	number of branches	1069	13499
	number of generators	88	1072
	number of loads	459	3003

In the figures below, Sp stands for speedup, which is the ratio of the time required for parallel simulations with partitions to the time required for sequential simulations without partitions. And the efficiency is expressed as E=Sp/P, where P is the number of CPUs.

For Case 1, experiment result shows the maximum deviation of node voltage with the parallel program presented in this paper and the software named PSASP. PSASP is the standard sequential software package developed by EPRI China, which is widely used for power system simulation in China [14, 18]. According to Fig.5, the maximum deviation of node voltage between our algorithm and PSASP was less than 10^{-5} pu. This proves that our algorithm is accurate.

Table 2 shows that the multilevel partition scheme results in much fewer cutting edges than the recursive bisection algorithm in METIS [15] but leads to a little more imbalance between sub-areas. In the table, the "Max/Min" represents the ratio of the number of nodes in the maximal partition to the number in the minimal partition, and "CutBrn" represents the number of branches between sub-areas.

Table 2. Comparison of the multi-level scheme and algorithm in METIS (Case 1)

Partition number	Multilevel scheme		METIS algorithm	
	Max/Min	CutBrn	Max/Min	CutBrn
2	422/284	4	379/327	12
4	301/121	7	207/156	28
8	119/68	34	145/74	54

Following experiments suggest that the multilevel scheme achieves higher efficiency in power system dynamic simulations on cluster systems, and its performance does not noticeably suffer from the imbalance of sub-areas. Furthermore, higher efficiency is achieved with the increase of partition number. With eight CPUs, the efficiency of our algorithm was about 70% higher than that of METIS.

Table 3. Comparison of the multi-level scheme and algorithm in METIS (Case 2)

Partition number	Multilevel scheme		METIS algorithm	
	Max/Min	CutBrn	Max/Min	CutBrn
2	5258/4930	2	5102/5086	2
4	3205/762	6	2552/2535	18
8	2004/628	9	1284/1266	64
10	1374/762	19	1035/1008	78
12	1143/694	27	861/835	79
16	1116/305	42	644/621	121
20	1088/156	40	531/496	140

Table 3 also shows that the multilevel partition scheme gets much less cutting edges than the recursive bisection algorithm in METIS, but leads to more imbalance between sub-areas.

With 12 CPUs, the efficiency of our algorithm is about 63% higher than that of METIS. Besides, we can also find that if the partition number is small, our method is not obviously more efficient than that of METIS. The reason is that the METIS considers the graph containing 10188 nodes in Case 2 for partition, while our method uses the evaluation function to decompose the vertex in a simplified graph. Therefore, in the cases with the small partition number, METIS has more freedom than our method. But in contrast, in the cases with large partition number, our method is better than METIS because of considering the accurate behaviour in the parallel simulation.

In conclusion, the multilevel partition scheme fits cluster-based BBDF algorithm well, especially when more CPUs are used. The multilevel partition scheme and the BBDF algorithm give an integrated solution to parallel dynamic simulation and can get very satisfying results.

Acknowledgements. This work was supported in part by the National Natural Science Foundation of China (Grant No.90612018).

Reference

1. P. Zhang, J. R. Martí, et al, Network Partitioning for Real-time Power System Simulation. International Conference on Power Systems Transients, No.177
2. W. Xue, J. W. Shu, et al, Advance of parallel algorithm for power system transient stability simulation, Journal of system simulation, 2002, 14(2), 177-182.
3. IEEE Committee Report. Parallel processing in power systems computation. IEEE Trans. on Power Systems, 1992, 7(2): 629-638.
4. I.C.Decker, D.M.Falcao, et al. Conjugate gradient methods for power system dynamic simulation on parallel computers. IEEE Trans. on PWRS, 1996,11(3), 1218-1227
5. M. La Scala, G. Sblendorio, A. Bose, J. Q. Wu. Comparison of algorithms for transient stability simulations on shared and distributed memory multiprocessors. IEEE Trans. on Power Systems, 1996, 11(4), 2045-2050.

6. G. Aloisio, M. A. Bochicchio, et al. A distributed computing approach for realtime transient stability analysis. IEEE Trans. on Power Systems, 1997, 12(2), 981-987.

7. K.W.Chan, R.C.Dai, C.H.Cheung. A coarse grain parallel solution method for solving large set of power systems network equations. 2002 international conference on power system technology, 2002.Volume: 4: 2640 -2644

8. A. H. Jorge, R. M. Jose Real time network simulation with PC-Cluster. IEEE Trans. on power systems, 2003, 18(2), 563-569.

9. Y. L. Li, X. X. Zhou, Z. X. Wu, Parallel algorithms for transient stability simulation on PC cluster. PowerCon 2002, vol.3, 1592-1596.

10. Y. L. Li, et al, A Parallel Complex Fault Computation Algorithm for Large Scale Power System Digital Simulation. Proceedings of CSEE, 2003, 23(12), 1-5

11. H.Simon, S. Teng. How good is recursive bisection? SIAM J. Scientific Computing, 1997, 18(5):1445-1463

12. Kirk Schloegel, George Karypis, Vipim Kumar. Graph Partitioning for High Performance Scientific Simulations. CRPC Parallel Computing Handbook. Morgan Kaufmann, 2000

13. Jiwu Shu, Wei Xue, Weimin Zheng, An Optimal Partition Scheme of Transient Stable Parallel Computing in Power System, Automation of Electric Power Systems, 2003, 27(19), 6-10.

14. EPRI China, Power System Analysis Software Package, Fundamental database user manual, 2001.

15. K. George, V. P. Kumar, METIS-a software package for partitioning unstructured graphs, partitioning meshes, and computing fill-reducing orderings of sparse matrices, version 4.0 [EB/OL], http:// www.cs.umn.edu/?karypis, 1998,9.

16. M.H.M.Vale, D.M.Falcao, E.Kaszkurewicz, Electrical power network decomposition for parallel computations. Proceedings of the IEEE Symposium on Circuits and Systems. San Diego, CA, May 1992. 2761-2764

17. Banerjee.P., Jones.M.H., Sargent J.S.. Parallel simulated annealing algorithms for cell placement on hypercube multiprocessors. IEEE Trans.,1990,PDS-1:91-106

18. Cheng Hua, Xu Zheng, Comparison of mathematical models for transient stability calculation in PSASP and PSS/E and corresponding calculation results, Power system teconology, 2004, 28(5), 1-4.

An Extended Implementation of the Great Deluge Algorithm for Course Timetabling

Paul McMullan

School of Electronics, Electrical Engineering and Computer Science,
Queen's University of Belfast, Northern Ireland
p.p.mcmullan@qub.ac.uk

Abstract. Course Scheduling consists of assigning lecture events to a limited set of specific timeslots and rooms. The objective is to satisfy as many soft constraints as possible, while maintaining a feasible solution timetable. The most successful techniques to date require a compute-intensive examination of the solution neighbourhood to direct searches to an optimum solution. Although they may require fewer neighbourhood moves than more exhaustive techniques to gain comparable results, they can take considerably longer to achieve success. This paper introduces an extended version of the Great Deluge Algorithm for the Course Timetabling problem which, while avoiding the problem of getting trapped in local optima, uses simple Neighbourhood search heuristics to obtain solutions in a relatively short amount of time. The paper presents results based on a standard set of benchmark datasets, beating over half of the currently published best results with in some cases up to 60% of an improvement.

Keywords: Course Scheduling, Great Deluge, Timetabling, Neighbourhood Search.

1 Introduction

The Course Timetabling problem deals with the assignment of course (or lecture events) to a limited set of specific timeslots and rooms, subject to a variety of hard and soft constraints. All hard constraints must be satisfied, obtaining a feasible solution, while as many soft constraints as possible must be satisfied. The feasible solution which satisfies all of the soft constraints as is possible can be considered optimal. The quality of any course timetable solution is based on how close to this goal it approaches. The set of all feasible solutions is vast; given the number of permutations possible for even a modestly sized problem, an exhaustive search for an optimum solution will take an impractical amount of time. Directed searches which attempt to improve the quality of one or more initial solutions can cut down the amount of time needed to reach a solution of acceptable quality. However, although compute-intensive Neighbourhood search heuristics can provide acceptable results, the time taken to obtain them may be impractical for real-world applications [1]. This paper proposes an extension of an improvement algorithm based on the Great Deluge

Y. Shi et al. (Eds.): ICCS 2007, Part I, LNCS 4487, pp. 538–545, 2007.

[2] which will employ simple (non-compute-intensive) heuristics while providing better or comparable results to successful techniques outlined in the current literature.

2 The University Course Timetabling Problem

The problem involves the satisfaction of a variety of hard and soft constraints [3]. The following Hard Constraints, which must be satisfied in order to achieve a feasible timetable solution, are presented:

- *No student can be assigned to more than one course at the same time.*
- *The room should satisfy the features required by the course.*
- *The number of students attending the course should be les than or equal to the capacity of the room.*
- *No more than one course is allowed to a timeslot in each room.*

The following soft constraints are used as measure of the overall penalty for a solution, calculated in terms of the total number of students who:

- *Have a course scheduled in the last timeslot of the day.*
- *Have more than 2 consecutive courses.*
- *Have a single course on a day.*

In general, the problem has the following characteristics:

- *A set of N courses, e = {e1,...,eN}*
- *45 timeslots*
- *A set of R rooms*
- *A set of F room features*
- *A set of M students.*

An improvement in the quality of any solution for the course timetabling problem is concerned with the minimisation of soft constraint violations, which directly includes the number of students affected by the violations. The objective cost function is calculated as the sum of the number of violations of the soft constraints.

A full description of the problem has been listed by the EU Metaheuristics Network [4]. A set of eleven specification data sets were subsequently released [3] for the purposes of Benchmarking, and in 2002 an international timetabling competition was run using another twenty sets of data [5] in order to encourage the production of innovative techniques in the automation of course timetable construction.

Recent literature concentrated in automating the process of course timetable creation has described the application of several techniques and heuristics, with varying measures of success. [6] provides an introduction to recent work in the area. Hill-Climbing [7] is the simplest local search algorithm, which is involved in improving a current solution by accepting a candidate solution from the neighbourhood only if it of a better or equivalent fitness to the current. This has the disadvantage of getting trapped quite quickly in local optima, most often with a poor quality solution.

Extensions to the simple Hill-climbing algorithm involve allowing the acceptance of worse solutions in order to at some point get better ones, or directing Neighbourhood moves according to some criteria dictated by initial or iteration-dependent parameters. Simulated annealing [8] accepts worse solutions with a probability: $P = e^{-\Delta P/T}$, where $\Delta P = f(s^*)-f(s)$ and the parameter T represents the temperature, analogous to the temperature in the process of annealing. The temperature is reduced according to a cooling rate which allows a wider exploration of the solution space at the beginning, avoiding getting trapped in a local optimum. This has been applied to a number of application areas [9]. Tabu-Search [10] modifies the neighbourhood structure of each solution as searching progresses, using a tabu list to exclude solutions or attributes of solutions in order to escape local optima.

Variable Neighbourhood Search (VNS), introduced by Mladevonic and Hansen [11] uses more than one neighbourhood structure which change systematically during the search. This increases the probability of escaping local optima, in that neighbourhoods which are different from the current solution can be considered. The use of composite neighbourhood structures is described in [12], and involves choosing solutions based on a Monte Carlo acceptance criteria [13] from the most effective of a choice of multiple neighbourhood structures.

Evolutionary techniques such as Genetic Algorithms have been employed in this problem area [14] to improve multiple "populations" of course timetable solutions based on selection of the fittest, using selection, crossover and mutation operators. Further experimentation involves the use of hybrid evolutionary approaches and memetic algorithms [14], [15]. These have also been combined with Meta- and Hyper-heuristic techniques [16], [17] which employ high-level heuristics to dictate the choice of heuristics when searching the solution space. Socha et al investigate the use of ant colony optimisation methodologies to the course timetabling problem [18], [19], using the idea of pheromone trails to reinforce successful orderings or moves over those less successful in improving solutions. Other techniques involve the use of Multi-Objective or Multi-Criteria approaches to reaching an optimal solution [15] whereby the quality of a solution is considered as a vector of individual objectives rather than as a single composite sum, weighted according to the importance of each. Fuzzy methodologies have been successfully applied to examination scheduling [20] and more recently to course timetabling problems [21].

The Great Deluge (also known as Degraded Ceiling) was introduced by Dueck [2] as an alternative to Simulated Annealing. This uses a Boundary condition rather than a probability measure with which to accept worse solutions. The Boundary is initially set slightly higher than the initial solution cost, and reduced gradually through the improvement process. New solutions are only accepted if they improve on the current cost evaluation or are within the boundary value. This has been applied successfully to construction and improvement techniques in timetabling [22], [23]. [15] describes a modification of the Great Deluge algorithm for Multi-criteria decision making.

3   The Extended Great Deluge Algorithm

The standard Great Deluge algorithm has been extended to allow a reheat, similar to that employed with simulated annealing in timetabling [15]. The aim of this approach is to both improve the speed at which an optimal solution can be found and at the same time utilise the benefits of this technique in avoiding the trap of local optima. In order to reduce the amount of time taken, relatively simple neighbourhood moves are employed.

Generally, the Great Deluge or Simulated Annealing processes will terminate when a lack of improvement has been observed for a specified amount of time, as the most optimal solution will have been reached. Rather than terminating, the Extended GD will employ the reheat to widen the boundary condition to allow worse moves to be applied to the current solution. Cooling will continue and the boundary will be reduced at a rate according to the remaining length of the run. The algorithm for the Extended Great Deluge is presented in Figure 1.

```
Set the initial solution s using a construction heuristic;
Calculate initial cost function f(s)
Set Initial Boundary Level B₀ = f(s)
Set initial decay Rate ΔB based on Cooling Parameter
While stopping criteria not met do
Apply neighbourhood Heuristic S* on S
      Calculate f(s*)
         If f(s*) <= f(s) or (f(s*) <= B Then
           Accept s = s*
         Lower Boundary B = B - ΔB
         If no improvement in given time T Then
           Reset Boundary Level B₀ = f(s)
             Set new decay rate ΔB based on Secondary
                Cooling Parameter
```

Fig. 1. Extended Great Deluge Algorithm

The initial solution construction is handled with an Adaptive (Squeaky-Wheel) ordering heuristic [22] technique. This utilises a weighted order list of the events to be scheduled based on the individual penalties incurred during each iteration of construction. The adaptive heuristic does not attempt to improve the solution itself, but simply continues until a feasible solution is found.

The first parameter used within the Extended Great Deluge is the initial decay rate, which will dictate how fast the Boundary is reduced and ultimately the condition for accepting worse moves is narrowed. The approach outlined in this paper uses a Decay Rate proportional to 50% of the entire run. This will force the algorithm to attempt to reach the optimal solution by, at the very most, half-way through the process. Generally, a continuous lack of improvement will occur before this is reached, at which time the re-heat mechanism is activated.

The 'wait' parameter which dictates when to activate the re-heat mechanism due to lack of improvement can be specified in terms of percentage or number of total moves in the process. Through experimentation with a number of data set instances a general value for this parameter can be established. After reheat the Boundary ceiling is once again set to be greater than the current best evaluation by a similar percentage to that applied in the initial boundary setting. The subsequent decay is set to a 'quicker' rate than with the initial decay, in order to increase the speed of the exploration of neighbouring solutions for improvement. The general setting chosen for the algorithm outlined is set to 25% of the remaining time, with the improvement wait time remaining unchanged. The neighbourhood structures employed in the process are deliberately kept simple, to ensure the process is not protracted with time-consuming explorations of the neighbouring solution space. The two heuristics used are Move (random event is moved to a random timeslot) and Swap (two random events swap timeslots), while ensuring that a feasible solution is maintained. There are approximately 2 Moves made for each Swap.

4 Experiments and Results

The algorithm was tested on a range of standard benchmark data sets [3]. These are divided into a number of categories accordingly on the amount of data required to be taken into account as part of the scheduling problem. This gives some indication of the comparative complexity of each, and hence the amount of hard and soft constraints which must be considered. For instance, Small data sets require 100 Courses to be scheduled, whereas Medium and Large have 400 Courses.

Table 1. Results for Multiple-run tests of Algorithm

Data Set	EGD		CNS	VNS with Tabu	Local Search	Ant	Tabu HH	Grph HH
	Best	Avg						
Small1	0 (x 4)	0.8	0	0	8	1	1	6
Small2	0 (x 2)	2	0	0	11	3	2	7
Small3	0 (x 4)	1.3	0	0	8	1	0	3
Small4	0 (x 5)	1	0	0	7	1	1	3
Small5	0 (x 8)	0.2	0	0	5	0	0	4
Medium1	80	101.4	242	317	199	195	146	372
Medium2	105	116.9	161	313	202.5	184	173	419
Medium3	139	162.1	265	357	77.5% Inf	248	267	359
Medium4	88	108.8	181	247	177.5	164.5	169	348
Medium5	88	119.7	151	292	100% Inf	219.5	303	171
Large	730	834.1	100 % Inf	100 % Inf	100% Inf	851.5	80% Inf 1166	1068

The algorithm was implemented on a Pentium PC Machine using Visual Studio .NET on a Windows XP Operating System. For each benchmark Data Set the algorithm was run for 200,000 evaluations with 10 test-runs to obtain an average value. Table 1 shows the comparison of our approach with others available in the literature. They include a local search method and ant algorithm [18], [19], Tabu-search hyper-heuristic and Graph hyper-heuristic [10], [20], Variable Neighbourhood Search with Tabu list and Composite Neighbourhood Search [12].

The Extended Great Deluge has beaten all current results for the *Medium* data sets, with approximately 60% improvement given the average cost evaluation obtained over the 10 test runs. The *Large* data set has a marginal improvement on the previous best. Results for the *Small* data sets do not improve on the previously published results, but do compare favourably. It should be noted that in all cases for *Small*, evaluations of zero were obtained, with some achieved more frequently than others.

Between each of the *Small* data sets, if we compare the frequency of achieving zero evaluation, it can be noted that our algorithm achieves zero more often on those which could be considered less difficult (based on the results obtained by the other techniques included in the table) and vice versa. The time taken to achieve the best solution for the data sets ranged between 15 and 60 seconds.

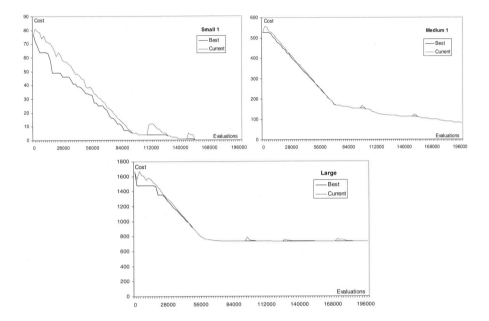

Fig. 2a, b and c. Performance of boundary for Small, Medium and Large datasets

Figures 2a, b and c illustrate the effectiveness of the reheat and rapid cooling mechanism. The initial state at which further improvement is no longer achieved is reached before the process is 50% complete. The reheat clearly indicates that further improvement is achieved. In the case of *Small*, in most cases this allows the technique to explore a number of severe non-improvement moves and achieve a zero evaluation. The technique is also very effective for the *Medium* data sets, with the graphs clearly indicating that as much as a further reduction of 50% in the cost evaluation can be achieved. This is less effective for the *Large* data set, although some improvement as a result of the reheat can be observed. Some initial investigation was also carried out on the Competition data instances [5]. Adhering to the time restrictions dictated by the rules of the competition, it was observed that the Extended Great Deluge performed comparably to results obtained by the top 3 entrants.

The nature of how the Boundary conditions are set within the Great Deluge algorithm can explain to some extent the problems in achieving a zero evaluation consistently for the *Small* data sets. The Boundary is set as a percentage of the current or best evaluation initially or when a reheat takes place. A common percentage value for setting the boundary is 110% of the current best evaluation. When the evaluation tends towards zero, any new boundary value will be quite close or equal to the best evaluation (in terms of rounding to whole numbers). Therefore the boundary becomes ineffective in allowing worse moves to be accepted towards escaping local optima.

5 Conclusions and Future Work

This paper has introduced a variant of the Great Deluge (Degraded Ceiling) algorithm which attempts to maximise the inherent advantages with this technique in escaping from local optima while also maintaining a relatively simple set of neighbourhood moves, and hence reducing the time required for an iterative search process. As has been shown, this approach has been successful in achieving a general improvement in solution evaluation as compared to currently published results for the course timetabling benchmark data sets, given a common termination criterion. This represents a significant contribution to the area of Course timetabling.

A much wider analysis of the technique is proposed in order to determine whether further improvement can be achieved by modification of the associated parameters and variables used in the process, while retaining generality for all data sets. In order to help ensure generality is maintained, it is intended to include the Competition data sets along with the standard benchmarks for comparison. The variables which may affect the results include Construction Time and Initial Ordering heuristics, Initial and post-reheat decay rates, Non-Improvement 'wait' time, Neighbourhood search heuristics (initially will involve modification of the Move / Swap ratio and pattern), Initial Boundary evaluation factor and boundary / multiple non-improvement heuristics.

It is also clear that the technique will need to resolve the issues identified when processing smaller data sets or indeed larger problems for which solutions approaching an evaluation of zero are found. Current experimentation is involved with activating a more directed set of neighbourhood structures in these circumstances, at the expense of some computation time, towards the goal of achieving zero evaluation. Further results from all experimentation outlined will be published in due course.

References

1. B. McCollum, "University Timetabling: Bridging the Gap between Research and Practice", Proceedings of the 6th international conference on the Practice and Theory of Automated Timetabling, Brno, (2006), 15-35.
2. G. Dueck, "Threshold Accepting: A General Purpose Optimization Algorithm Appearing Superior to Simulated Annealing", J. Computational Physics, Vol. 90, (1990), 161-175.
3. B. Paechter, Website: http://www.dcs.napier.ac.uk/~benp/
4. Metaheuristic Network Website: http://www.metaheuristics.org/
5. International Timetabling Competition Website: http://www.idsia.ch/Files/ttcomp2002/
6. E. K. Burke, , M. Trick, (eds), The Practice and Theory of Automated Timetabling V: Revised Selected Papers from the 5th International conference, Pittsburgh 2004, Springer Lecture Notes in Computer Science, Vol. 3616. (2005).

7. J. S. Appleby, D. V. Blake, E. A. Newman, "Techniques for Producing School Timetables on a Computer and their Application to other Scheduling Problems", The Computer Journal, Vol. 3, (1960), 237-245.
8. S. Kirkpatrick, J. C. D. Gellat, M. P. Vecci, "Optimization by Simulated Annealing", Science, Vol. 220, (1983), 671-680.
9. C. Koulmas, S. R. Antony, R. Jaen, "A Survey of Simulated Annealing Applications to Operational Research Problems", Omega International Journal of Management Science, Vol. 22, (1994), 41-56.
10. E.K. Burke, G. Kendall, E. Soubeiga, "A Tabu-Search Hyperheuristic for Timetabling and Rostering", Journal of Heuristics, 9(6): (2003), 451-470.
11. N. Mladenovic, P. Hansen, "Variable Neighbourhood Search", Computers and Operations Research, 24(11), (1997), 1097-1100.
12. S. Adbullah, E. K. Burke, B. McCollum, "Using a Randomised Iterative Improvement Algorithm with Composite Neighbourhood Structures for University Course Timetabling", Comp. Science Interfaces Book Series (Eds. Doerner, K.F., Gendreau, M., Greistorfer, P., Gutjahr, W.J., Hartl, R.F. & Reimann, M.), Springer Operations Research (2006).
13. M. Ayob, G. Kendall, "A Monte Carlo Hyper-Heuristic to Optimise Component Placement Sequencing for Multi-head Placement Machine", Proceedings of the International Conference on Intelligent Technologies, InTech '03, (2003), 132-141.
14. E. K. Burke, J. P. Newall, "A Multi-Stage Evolutionary Algorithm for the Timetable Problem", IEEE Transactions on Evolutionary Computation, Vol 3.1, (1999), 63-74.
15. S. Petrovic, E. K. Burke, "University Timetabling", Handbook of Scheduling: Algorithms, Models and Performance Analysis, CRC Press, Chapter 45, (2004)
16. O. Rossi-Doria, B. Paechter, "An Hyperheuristic Approach to Course Timetabling Problem using an Evolutionary Algorithm", Technical Report, Napier University, Edinburgh, Scotland, (2003)
17. C. Head, S. Shaban, "A heuristic approach to simultaneous course/student timetabling", Computers & Operations Research, 34(4): (2007), 919-933.
18. K. Socha, J. Knowles, M. Sampels, "A Max-Min Ant System for the University Course Timetabling Problem", Proceedings of the 3rd International Workshop on Ant Algorithms, ANTS 2002, Springer Lecture Notes in Computer Science, Vol. 2463, Springer-Verlag, (2002), 1-13.
19. K. Socha, M. Sampels, M. Manfrin, "Ant Algorithms for the University Course Timetabling Problem with regard to the State-of-the-Art", Proceedings of the 3rd International Workshop on Evolutionary Computation in Combinatorial Optimization (EVOCOP 2003), United Kingdom, (2003), 335-345.
20. E. K. Burke, B. McCollum, A. Meisels, S. Petrovic, R. Qu, "A Graph-based Hyper Heuristic for Timetabling Problems", European Journal of Operational Research, 176: (2007), 177-192.
21. H. Asmuni, E. K. Burke, J. M. Garibaldi, "Fuzzy Multiple Heuristic Ordering for Course Timetabling", Proceedings of 5th UK Workshop on Computational Intelligence (UKCI '05), (2005), 302-309.
22. E. K. Burke, Y. Bykov, J. P. Newall, S. Petrovic, "A Time-Predefined Approach to Course Timetabling", Yugoslav J. of Operational Research (YUJOR), 13(2): (2003) 139-151.
23. E. K. Burke, Y. Bykov, J. P. Newall, S. Petrovic, "A Time-Predefined Local Search Approach to Exam Timetabling Problems", IIE Transactions, 36(6), (2004), 509-528.

Cubage-Weight Balance Algorithm for the Scattered Goods Loading with Two Aims

Liu Xiao-qun[1], Ma Shi-hua[2], and Li Qi[3]

[1] National Earthquake Infrastructure Service, Beijing 100036, China
[2] School of Management, Huazhong University of Sci.&Tec., Wuhan 430074, China
[3] Graduate School of Chinese Academy of Sciences, Beijing, 100049, China
{hustlxq,stevenmai,casliqi}@126.com

Abstract. To utilize the volume and carrying capacity to solve the low efficiency of truck loading, a loading planning was made by using optimization methods based on combinatorics. The mathematic model of goods loading was set up and a polynomial approximation algorithm was proposed according to the characteristics of the goods transportation. With the volume and load of the truck taken into consideration, this algorithm can converge at a satisfactory solution and has a good effect in application.

Keywords: loading, optimization, combinatorics, polynomial approximation.

1 Introduction

Loading is a typical 2-Dimensional Packing Problem, which needs to allocate a set of rectangular items to larger rectangularly standardized stock (such as train, steamship and truck) units by minimizing the waste. The truck loading is important because it is reported that the road transportation is of particularly higher transportation costs and lower efficiency when comparing with other transportations. It is known that the utilization of trucks is only about 60% in China. Most algorithms for 2-DPP are heuristic. General reviews on packing and loading problems can be found in Dowsland and *et.al* [1]. Results on the worst-case performance and the probabilistic analysis of packing algorithms can be found in Coffman et al [2].

There are two common approaches for the solution of loading problem. The first approach is based on lower bounds and heuristics based on combinatorial considerations. Chen and Srivastava [3] studied a generalization of the Martello-Toth lower bounding procedure based on a famous lower bound algorithm, and Stephen [4] described a column-generation based on branch-and-bound algorithms. These algorithms are very fast in some cases but not effective enough to provide optimal solutions when embedded within a branch-and-bound scheme.

The second approach is based on an integer programming formulation with a huge number of variables, whose linear programming relaxation can be solved by column generation, typically requiring a considerable time, but obtaining extensive information about the optimal solution of the problem. An integer program was implemented to solve a kind of rectangular packing problem, and Berghammer

Y. Shi et al. (Eds.): ICCS 2007, Part I, LNCS 4487, pp. 546–553, 2007.

developed a linear approximation algorithm for bin packing with absolute approximation factor [5]. One kind of container and multi-container loading problem was solved by Scheithauer [6] based on improved LP. While for bipartite vertex packing problem, strongly polynomial simplex algorithm was introduced by Armstrong [7]. But just as shown in the study of Woeginger [8], we are not aware of any exact algorithm for 2-DPP in the literature even if 2-DPP does not have an asymptotic polynomial time approximation scheme [9] unless $P=NP$.

This article introduces a C-W (short for Cubage-Weight) balance method characteristic by little counters and quick constringency to optimize the solutions mainly based on A_k [1] and First-Fit [2] algorithm [10]. It is proved by numerical simulation that C-W balance method can converge to satisfactory solution effectively and has a good effect in practice because it takes into account the truck's volume and carrying capacities at the same time compared with the methods which consider one of two elements preferentially.

2 Model Description, Assumptions and Notation

The problem can be described as follows: there are some trucks in the distribution center. Now scattered goods need to be delivered. How does one make the loading plan rationally in order to utilize the most of truck's cubage and carrying capacity? In this problem, it is assumed that: a) the flow direction of all these scattered goods is the same; b) no goods need to be loaded specially; c) in the process of loading, only the cubage and the carry capacity restrictions are considered. The following additional notation will be used below in the algorithm designing:

$N(i)$: Set of all scattered goods, i is the number of goods i ($i=1,2,\cdots,n$),

$N(v_i)$: Volume set of scattered goods, v_i is the volume of goods i ($i \in N(i)$),

$N(g_i)$: Weight set of scattered goods, g_i is the weight of goods i ($i \in N(i)$),

$M(j)$: Set of trucks, j is the number of truck j ($j=1,2,\cdots,m$),

$M(V_j)$: Cubage set of trucks, V_j is the cubage of truck j ($j \in M(j)$),

$M(G_j)$: Carry capacity set of trucks, G_j is the carry capacity of truck j ($j \in M(j)$),

x_{ij} : 0-1 Variable, if goods i is loaded into truck j , then $x_{ij}=1$, else then $x_{ij}=0$,

y_j : 0-1 Variable, if truck j is selected to loading scattered goods, then $y_j=1$, else then $y_j=0$,

$s_j^{(k)}(i)$: Loaded goods set of truck j in the kth iteration of the algorithm, $j \in M(j)$, $i \in N(i)$,

---

[1] A_k algortihm: Slect goods set s (no more than k elemtens) from $N(i)$, if $\sum_{i \in s} g_i \leq G$ and $\sum_{i \in s} v_i \leq V$, loaded the goods increasingly from s until the truck is full loaded.

[2] First-Fit algortihm: Seek the most suitable goods for the left truck with the most loading capacity until all the trucks are full loaded. See more in reference [10].

$s_j^*(i)$: Satisfying or optimization solution of truck j, $j \in M(j)$, $i \in N(i)$,

$S^{(k)}$: Loaded goods set of truck set $M(j)$ in the kth iteration of the algorithm, $S^{(k)} = (s_1^{(k)}(i), s_2^{(k)}(i), ..., s_m^{(k)}(i))$,

S^*: Satisfying or optimization solution of truck set $M(j)$.

The model of scattered goods loading problem is:

$$MinZ = \sum_{j=1}^{m} y_j, \quad MaxZ_g = \sum_{y_j=1}^{n}\sum_{i=1}^{n} g_i x_{ij}, \quad MaxZ_v = \sum_{y_j=1}^{n}\sum_{i=1}^{n} v_i x_{ij}$$

Subject to:

$$\sum_{j=1}^{m} x_{ij} \leq 1 \qquad i = 1, 2, \cdots, n$$

$$\sum_{i=1}^{n} g_i x_{ij} \leq G_j y_j \qquad j = 1, 2, \cdots, m$$

$$\sum_{i=1}^{n} v_i x_{ij} \leq V_j y_j \qquad j = 1, 2, \cdots, m$$

$$x_{ij} = 0 \text{ or } 1 \qquad i = 1, 2, \cdots, n, \ j = 1, 2, \cdots, m$$

$$y_j = 0 \text{ or } 1 \qquad j = 1, 2, \cdots, m$$

The solution is to find goods sets $s_1^{(k)}(i)$, $s_2^{(k)}(i)$, ..., $s_m^{(k)}(i)$ in $N(i)$ (Here, $s_1^{(k)}(i)$, $s_2^{(k)}(i)$, ..., $s_m^{(k)}(i)$ should satisfy these conditions: $\forall \ i$, j, $s_i^{(k)}(i) \cap s_j^{(k)}(i)$ =empty set, $\bigcup s_i^{(k)}(i) = \bigcup s_j^{(k)}(i) \subseteq N(i)$ ($i \neq j$) in order to realization $MinZ$, $MaxZ_g$ and $MaxZ_v$ under the condition of $\forall j \in M$, $\sum_{i=1}^{n} g_i x_{ij} \leq G_j y_j$ and $\sum_{i=1}^{n} v_i x_{ij} \leq V_j y_j$.

3 Method and Algorithm

3.1 One Truck Loading

Defining the notation as follows:

c_i: The ratio of goods volume v_i and weight g_i, $c_i = v_i / g_i$;

C: The ratio of truck's cubage V and weight G, $C = V / G$;

V_{sum}: The total loaded goods volume;

G_{sum}: The total loaded goods weight;

c': The ratio of V_{sum} and weight G_{sum}, $c' = V_{sum} / G_{sum}$;

r: The truck's utilization rate, $r = c' / C$.

The one truck loading algorithm is as follows: a) compare c_i $(i \in N(i))$ with C, where $N(i)$ is sorted increasingly according to numerical value of $|c_i - C|$ and $(c_i - C) \times (c_{i+1} - C) \leq 0$, $(c_i - C) \times (c_{i+2} - C) \geq 0$, $|c_i - C| \leq |c_{i+1} - C|$, $(i = 1, 2, ..., n-2)$; b) mark the sorted set of $N(i)$ as P; c) choose the goods from P orderly. If goods can be loaded in the truck, then loads the goods, otherwise the loading process is over. In order not to lost generality, assuming $\max\{g_1, g_2, ..., g_n\} \leq G$ and $\max\{v_1, v_2, ..., v_n\} \leq V$ is tenable.

Algorithm of one truck loading

```
Step 1: Input  N(g_i) = {g_1, g_2, ···, g_n} , N(v_i) = {v_1, v_2, ···, v_n} , G , V .
Step 2: Calculate  c(i) = {c_1, c_2, ···, c_n} (c_i = v_i / g_i (i = 1, 2, ···, n)) , C = V / G .
Step 3: Set  P = empty set, V_sum = 0 , G_sum = 0 , c' = 0 , r = 0 .
Step 4: Compare  C with  c_i (i ∈ N(i)) , sorted  N(i)
        increasingly according to numerical value of |c_i - C| and
        (c_i - C) × (c_{i+1} - C) ≤ 0 , (c_i - C) × (c_{i+2} - C) ≥ 0 , |c_i - C| ≤ |c_{i+1} - C| ,
        (i = 1, 2, ..., n - 2) and mark the sorted set of  N(i) is  P .
Step 5: Supposing P[1] = i ,  if  g_i ≤ (G - G_sum) and v_i ≤ (V - V_sum) ,
        then run step 6; else  P\P[1] → P ,  run step 7.
Step 6: Let  V_sum + v_i → V_sum , G_sum + g_i → G_sum , c' = V_sum / G_sum , r = c'/C .
Step 7: If P == empty set, then run step 8, else run step 5.
Step 8: Output P , G_sum , V_sum , c' , r .
```

3.2 M $(M > 1)$ **Truck Loading**

To M truck loading problem, the heuristic algorithm is designed as follows.

Stage 1: Truck Selection
In this stage, the suited truck is selected according to the most patchable C-W ratio between the total scattered goods and all trucks. Basic function is as follows:

```
Step 1: Input  N(g_i) , N(v_i) , M(G_j) , M(V_j) .
Step 2: Calculate  M(C) = {C_1, C_2, ···, C_m} (C_j = V_j / G_j (j = 1, 2, ···, m)) .
Step 3: Calculate  c(c = Σ_{i∈N(i)} v_i / Σ_{i∈N(i)} g_i ).
Step 4: Compare  C with  C_j (j ∈ M(j)) , mark the element
        satisfying the function  Min|c - C_j| as  j , then the truck
        j  is selected to load scattered goods.
```

Stage 2: One Truck Loading
To the truck j selected in the stage 1, loads the scattered goods using the one truck loading algorithm and marks the final loaded goods is $s_j^{(1)}(i)$.

Stage 3: Solution Optimization
The solution obtained in stage 2 is optimized in this stage. Namely, the calculation methods of c in stage 1 is adjusted from the $c = \sum_{i \in N(i)} v_i / \sum_{i \in N(i)} g_i$ to the

$c^{(2)} = \sum_{i \in Us_j} v_i / \sum_{i \in Us_j} g_i$. Marking the element satisfying the function $Min \, | \, c^{(2)} - C_j \, |$ as

$j^{(2)}$, the truck $j^{(2)}$ is selected from trucks set $M(j)$ to load scattered goods, and the loaded goods set in the No. 2 iterative algorithm is $s_j^{(2)}(i)$.

The set of all loaded goods in the kth iteration of the algorithm is $S^{(k)}(i) = Us_j^{(k)}(i)$. If the truck's utilization ratio in the two consecutive iterations get very close

$\left(\left| \sum_{i \in S^{(k+1)}} v_i \middle/ \sum_{j \in M^{(k+1)}(j)} V_j - \sum_{i \in S^{(k)}} v_i \middle/ \sum_{j \in M^{(k)}(j)} V_j \right| \leq \varepsilon \right.$, ε is the solution difference level), the algorithm calculation is finished.

4   Numerical Investigation and Discussion

4.1   One Truck Loading

Comparing the C-W balance algorithm of one truck loading with other optimization methods, the concrete data of reference [10] is selected. The result is shown as follows.

Through simulating the computational algorithm to a large number of numerical investigations, some conclusions are drawn:

1) Comparing with the $A_k \, (k = 0)$ algorithm, the C-W balance algorithm can obtain better solutions under the same computational complexity. Generally speaking, the truck's loading efficiency of the C-W balance algorithm, comparing with the $A_k \, (k = 0)$ algorithm, can be improved by 5% to 20%.

2) Comparing with the $A_k \, (k = 1)$ algorithm, the C-W balance algorithm can obtain satisfactory solutions in equal efficiency. But the computational complexity of $A_k \, (k = 1)$ algorithm is $O(kn^{k+1})$, while the computational complexity of C-W balance algorithm is just $O(n \log n)$. With the increasing of scattered goods, the C-W balance algorithm runs fast.

4.2 $M \, (M > 1)$ Truck Loading

Through numerical simulation, the C-W balance algorithm is compared with the First-Fit algorithm. The relevant data of simulation are as follows: a) trucks: the cubage and

Table 1. One truck loading algorithm

Algorithm	$A_k\,(k=0)$	$A_k\,(k\geq1)$	C-W balance algorithm
Input	Goods set: $N(i)=\{1,2,3,4,5,6,7,8\}$		
	Goods volume set: $N(v_i)=\{110,108,96,80,49,50,40,7\}$		
	Goods weight set: $N(g_i)=\{64,52,50,41,22,20,14,2\}$		
	Truck: cubage $V=250$, carry capacity $G=110$		
Loaded goods set	$p=\{1,6,7,8\}$	$P=\{2,6,7,5\}$	$P=\{2,6,7,5\}$
Total loaded cubage	207	247	247
Cubage utilization rate	90%	98%	98%
Total loaded weight	100	108	108
Weight utilization rate	82%	98%	98%
Computational complexity	$O(n\log n)$	$O(kn^{k+1})\ (k\geq1)$	$O(n\log n)$

the weight are generated at random from the array (200-400), (100-200) respectively, and the quantity of trucks is 100; b) scattered goods: the volume and the weight are generated at random from the array (20-60), (10-40) respectively, and the quantity of all scattered goods is 5000. The simulation results are shown in Fig.1 and Fig.2.

The following observation can be seen in Fig.1 and Fig.2.

(1) For truck No. 75, 52, 88, 57, 62, 48, 16, 84, 39, 30, 96, the cubage utilization ratio (CUR) of First-Fit algorithm is slightly higher than that of the C-W balanced algorithm, and the First-Fit utilization is up to 8% higher than C-W balanced algorithm. For other trucks, the CUR of C-W balanced algorithm is higher. Especially in the later stage of problem solving, the disparity is more pronounced. Among these trucks, the CUR of truck No. 61, 13, 81, 77, 32, 46, and 87 is more than 20% higher. And to truck No. 61, the utilization is 28% higher using C-W balanced algorithm.

(2) For truck No. 11, 39, 30 the weight utilization ratio (WUR) of First-Fit algorithm is slightly higher than that of the C-W balanced algorithm, and the largest difference (truck No.11) is nearly 1.3%. For other trucks, the WUR of C-W balanced algorithm is higher, and this disparity is more obvious in the later stages of problem solving. For truck No. 99, 54, 13, 77, 32, 91, 46, 1, 70, 78, 24, 33, 92, 56, the WUR of C-W balanced algorithm is 15%. For truck No. 87, the disparity is as much as 21%.

(3) The numerical simulation results show the stability of C-W balance algorithm is better than that of the First-Fit algorithm. As shown in the trend-lines, the CUR curve and the WUR curve of C-W balance algorithm are relatively steady, and the fluctuation of the First-Fit algorithm is relatively large. Especially to the later stage of problem solving, the efficiency of the First-Fit algorithm drops significantly below that of C-W balance algorithm.

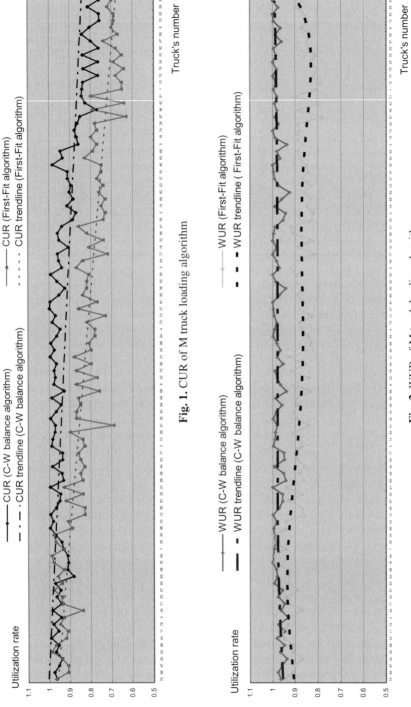

Fig. 1. CUR of M truck loading algorithm

Fig. 2. WUR of M truck loading algorithm

5   Conclusion and Application

In practice, because there are a great deal of loaded goods and loading trucks and calculating by hand is impossible, we can take some measures as follows: 1) using computer. Using the data of loaded goods and loading trucks as input, we run the computer program and output the loading plan according to the loaded goods. 2) When the condition is poor and no computer can be available, we can assemble the goods with maximum C-W rate and the goods with minimum C-W rate together among all loaded goods until no goods can be selected, then make a loading plan through calculating by hand. Because the optimization plan cannot be gained in each loading problem, so seeking a satisfying solution can simplify the problem and save the calculation time. Sometimes, it is necessary to simplify the loading restriction to speed up the loading and also can improve the integrative effect. The simplest way of the loading problem maybe as following: loading the big C-W rate goods and the small one together, laying the big C-W rate goods at the bottom of the truck and pile up with those small C-W rate goods, and loading the rest goods directly until the truck is filled. But at this time, these unloaded earlier should be loaded at the truck's border.

Acknowledgments

This work was supported by research grants from the National Natural Science Foundation of China (No.70332001, 70672040), National High Technology R&D Project of China (No. 2006AA04Z153) and HUST Doctor Foundation (No. D0540).

References

1. Dowsland, K.A., Dowsland, W.B.: Packing Problems. European Journal of Operational Research, 56(1992) 2-14
2. Coffman Jr., E.G., Garey, M.R., Johnson, D.S.: Approximation algorithms for bin packing-an updated survey, in: Hochbaum, D.S. (Ed.), Approximation Algorithms for NP-Hard Problems, PWS Publishing Company, Boston (1997)
3. Chen, B., Srivastava, B.: An improved lower bound for the bin-packing problem. Discrete Applied Mathematics, 66 (1996) 81-94
4. Stephen, M.H., Liu, Y.-H.: The application of integer linear programming to the implementation of a graphical user interface: a new rectangular packing problem. Applied Mathematical Modeling, 4(1995) 244-254
5. Berghammer, R., Reuter, F.: A linear approximation algorithm for bin packing with absolute approximation factor. Science of Computer Programming, 1(2003) 67-80
6. Scheithauer, G.: LP-based bounds for the container and multi-container loading problem. International Transactions in Operational Research, 6(1999) 199-213
7. Armstrong, R.D., Jin, Z.-Y.: Strongly polynomial simplex algorithm for bipartite vertex packing. Discrete Applied Mathematics, 2(1996) 97-103
8. Woeginger, G.J.: There is no asymptotic PTAS for two-dimensional vector packing. Information Process Letter, 64 (1997) 293-297
9. Chan, T.M.: Polynomial-time approximation schemes for packing and piercing fat objects. Journal of Algorithms, 2(2003) 178-189
10. Sun, Y., Li, Z.-Z.: The Polynomial Algorithm for the Allocation Problem with Two Aims (In Chinese). Changsha Railway University Transaction, 2(1997) 33-39

Modeling VaR in Crude Oil Market:
A Multi Scale Nonlinear Ensemble Approach
Incorporating Wavelet Analysis and ANN

Kin Keung Lai[1,2], Kaijian He[1,2], and Jerome Yen[3]

[1] Department of Management Sciences, City University of Hong Kong, Tat Chee Avenue,
Hong Kong
mskklai@cityu.edu.hk
[2] School of Business Administration, Hunan University, Changsha, Hunan, 410082, China
paulhekj@gmail.com
[3] Department of Finance, Hong Kong University of Science and Technology, Hong Kong
risksolution@gmail.com

Abstract. Price fluctuations in the crude oil markets worldwide have attracted significant attentions from both, industries and academics, due to their profound impact on businesses and governments. Proper measurement and management of risks due to unexpected price movements in the markets has been crucial from both, operational and strategic perspectives. However, risk measurements from current approaches offer insufficient explanatory power and performance due to the complicated non-linear nature of risk evolutions. This paper adopts a VaR approach to measure risks and proposes multi-scale non-linear ensemble approaches to model the risk evolutions in WTI crude oil market. The proposed WDNEVaR follows a semi-parametric paradigm, incorporating both, wavelet analysis and artificial neural network techniques. Experiment results from empirical studies suggest that the proposed WDNEVaR is superior to traditional approaches. It provides VaR estimates of higher reliability and accuracy. It also brings significantly more flexibility during the modeling attempts.

Keywords: Value at Risk, Wavelet Analysis, Nonlinear Ensemble Algorithm, Artificial Neural Network.

1 Introduction

Crude oil is a globally important industrial input. On one hand, crude oil price fluctuations affect every sector in regional and global economies. On the other hand, crude oil price forecasting and the allied risk measurement serve as important inputs and form the basis for macro-econometric models, financial markets' risk assessments and derivatives' pricing, etc. Therefore, demand for risk management in the crude oil market is on the rise from both, academics and industries.

Among numerous risk measures, this paper adopts Value at Risk to measure market risks in the crude oil markets. Due to its simplicity, VaR is widely accepted as a risk measure by both, academics and industries [1]. Although different approaches

Y. Shi et al. (Eds.): ICCS 2007, Part I, LNCS 4487, pp. 554–561, 2007.

have been developed to estimate VaRs, the essential multi-scale heterogeneous data structures remain unexplored. Therefore, this paper takes a semi-parametric approach and proposes the novel wavelet based non-linear ensemble approach. Empirical studies are conducted in the representative US West Intermediate Taxes (WTI) crude oil market. The experiment results suggest that the proposed WDNEVaR model outperforms the more traditional ARMA-GARCH based model.

The rest of the paper develops as follows: the second section reviews the relevant theories. The third section lays down the algorithm for wavelet based non-linear ensemble approach for VaR estimates. The fourth section presents and interprets results of empirical studies in WTI crude oil market. The fifth section concludes.

2 Background and Theories

VaR is a single number describing the maximal potential loss of a portfolio of financial instruments during a given period of time for a given probability. Estimation of appropriate VaR is a difficult challenge that has been tackled mainly from three approaches over the years: parametric, non-parametric, and semi-parametric approach [1]. The semi-parametric approach is the emerging methodology that strikes the balance between the parametric and non parametric approaches. These include many inter-disciplinary techniques such as wavelet analysis, extreme value theory, etc [2].

To verify the modeling adequacy, the model's past performance are often taken as the indication of the future performance [1]. Kupiec frequency based backtesting procedure is the most dominant hypothesis based approach, which takes the consistency level of observed frequency of losses as indication of the level of model adequacy. When the sample size is sufficiently large, the Kupiec backtesting procedures would offer an acceptable level of discriminatory power between good and bad models. Therefore, it is adopted in this paper to backtest and evaluate different models.

As the emerging new tool for estimating VaRs, wavelet analysis is the latest development in the harmonic analysis field. It possesses the capability to project data into time-scale domain for analysis [2]. Mathematically, wavelets are continuous functions that satisfy admissibility conditions as in $C_\psi = \int_0^\infty \frac{|\Psi(f)|}{f} df < \infty \Leftrightarrow \int_{-\infty}^\infty \psi(t)dt = 0$ and unit energy condition as in $\int_{-\infty}^\infty |\psi(t)|^2 dt = 1$, where φ is the Fourier transform of ψ.

There are different families of wavelets designed, each with their own special characteristics [2]. Typical wavelet families include the Haar wavelet, the Daubechies wavelets, and the coiflets wavelets, etc[2].

The wavelet transform is conducted as $W(u,s) = \int_{-\infty}^\infty x(t)\psi_{u,s}(t)dt$ while the inverse (synthesis) operation is performed as in $x(t) = \frac{1}{C_\psi} \int_0^\infty \int_{-\infty}^\infty W(u,s)\psi_{u,s}(t)du \frac{ds}{s^2}$. Thus together they form the basis for multi-resolution analysis as in $f(t) \approx S_J(t) + D_J(t) + D_{J-1}(t) + ... + D_1(t)$, where $S_J(t)$ refers to the smooth signals and equals $\sum_k s_{j,k}\phi_{j,k}(t)$. $D_J(t)$ refers to the detail signals and equals $\sum_k d_{j,k}\psi_{j,k}(t)$.

Since Ramsey (1999) introduced wavelet analysis into the mainstream literature, wavelet analysis has been applied in three different areas: firstly, wavelet analysis has been applied for economic and financial relationship identification to reveal the non-linear multi-scale data structure[3]. Secondly, wavelet analysis has been used extensively for denoising financial data, which extracts the maximal level of useful information [4]. Thirdly wavelet analysis has been used extensively in the forecasting model building process to decompose the original data series for further modeling attempts[5]. However, wavelet analysis has received little attention in the risk management field. Fernandez (2006a, 2006b) has used wavelet analysis to investigate the risk distribution across time scales in the Chilean stock markets and found concentration of risk in high frequency parts of data[6, 7]. But, these approaches mainly focus on investigating the distribution of potential market losses embedded in VaR numbers, across the time horizon. Their approach is based on the assumption that wavelet decomposed variances at different scales represent investors' preferences. However, they seem to have ignored the impact of different wavelet families chosen during their analysis, which leaves their findings largely inconclusive.

Another methodology proposed for estimating VaR is the ensemble approach. The Ensemble approach pools the partial information sets employed by different individual forecasters into a unified one as in $\hat{y}_t = \sum_{i=1}^{n} w_t \hat{x}_{i,t}$, where $\hat{y}_t$ is the output at time t. w_t is the weight assigned to different forecasts. $\hat{x}_{i,t}$ is the forecast at time t. n is the number of forecasts. Among different techniques introduced to optimize weights, artificial neural network is the most popular one. It is an important connectionist model for estimating non-linear data patterns and has distinguished itself as a universal function approximator for empirical studies. Artificial neural network consists of nodes and connections between the nodes. Neurons are grouped in three different types of layers: input layers, hidden layers and output layers. Depending on the direction of information flow, the neural network structure can be grouped into feed-forward and recurrent types. The network structure takes the matrix form as in (1):

$$y_t = a_0 + \sum_{j=1}^{q} w_j f(a_j + \sum_{i=1}^{p} w_{i,j} x_i) + \varepsilon_t \tag{1}$$

where a_j refers to the bias on the jth unit, w refers to the connection weight between different layers, f(.) refers to the transfer function of the hidden layer, p is the total number of input nodes while q is the total number of hidden nodes.

3   Wavelet Decomposed Nonlinear Ensemble Value at Risk (WDNEVaR)

When the sample is of sufficient length, VaR can be estimated by the parametric approach as in (2):

$$VaR_t = \mu_t + \sigma_t G^{-1}(\alpha) \tag{2}$$

where $G^{-1}(\alpha)$ refers to the inverse of the cumulative normal distribution.

The estimation process boils down to estimation of conditional mean and conditional standard deviation, which involves three steps:

Firstly, the original data series are decomposed using wavelet analysis based on the chosen wavelet families as in (3):

$$r_t = r_{A^J,t} + \sum_{j=1}^{J} r_{D^j,t}$$

(3)

Secondly, the conditional mean at each scale is assumed to follow ARMA processes and it is forecasted at each individual scales as in (4).

$$\hat{\mu}_{t,s} = a_{0,s} + \sum_{i=1}^{r} a_{i,s} r_{t-i,s} + \sum_{j=1}^{m} b_{j,s} \varepsilon_{t-j,s}$$

(4)

where $\hat{\mu}_{t,s}$ is the conditional mean at time t, s is the wavelet scales. $r_{t-i,s} (i = 1...r)$ is the lag r returns with parameter $a_{i,s}$, and $\varepsilon_{t-j,s} (j = 1...m)$ is the lag m residuals in the previous period with parameter $b_{j,s}$.

The conditional standard deviation at each scale is assumed to follow GARCH processes and it is estimated at each individual scales as in (5):

$$\hat{\sigma}_{t,s}^2 = \omega_s + \sum_{i=1}^{p} a_{i,s} \varepsilon_{t-i,s}^2 + \sum_{j=1}^{q} \beta_{j,s} \sigma_{t-j,s}^2$$

(5)

where $\hat{\sigma}_{t,s}^2$ is the conditional variance at time t, $\sigma_{t-j,s}^2 (j = 1...q)$ is the lag q variance with parameter $a_{i,s}$, and $\varepsilon_{t-i,s}^2 (i = 1...p)$ is the lag p squared errors with parameter $\beta_{j,s}$ in the previous period, during which the mean return is set to be zero.

Thirdly, as wavelet transform decorrelates the decomposed data, the conditional mean can be aggregated from the conditional means estimated at each individual scales as in (6).

$$\hat{\mu}_t = \hat{\mu}_{A^J,t} + \sum_{j=1}^{J} \hat{\mu}_{D^j,t}$$

(6)

The conditional standard deviations at each scale are synthesized, using wavelet analysis, to produce the aggregated estimates as in (7). The synthesis process pools the partial information set at each scale together and retains its authenticity based on the "preservation of energy" property of wavelet analysis.

$$\hat{\sigma}_t^2 = \hat{\sigma}_{A^J,t}^2 + \sum_{j=1}^{J} \hat{\sigma}_{D^j,t}^2$$

(7)

In the next step, since previous VaRs are estimated based on certain wavelet families, to reduce the biases introduced by arbitrary choice of wavelet families, the non-linear ensemble algorithm is proposed, to combine VaRs estimated based on different wavelet families into the optimal one utilizing artificial neural network, as in (8).

$$\sigma_{WDNEVaR} = \sum_{i \in F} Weight_i * \sigma_i \tag{8}$$

where F refers to the set of wavelet families attempted. Weight refers to the weight of the standard deviation estimated using different wavelet families.

When training artificial neural network to ensemble different volatility forecasts, a historical volatility measure is needed to provide an objective during the training process. There are mainly three dominant volatility measures identified in the literature: squared returns as $Volatility_{squaredReturns} = \sqrt{r_t^2}$, realized volatilities as $RV = \sqrt{\sum_{t=1}^{n} r_i^2}$ and historical volatilities $HV = \sqrt{\frac{1}{N} \sum_{t=1}^{N} (r_i - \bar{r})^2}$.

4 Empirical Studies

Data for this study were obtained from Global Financial Data (GFD) and consists of 5840 daily closing price for US West Intermediate Taxes (WTI) crude oil in dollars/barrel, which transforms to 1212 weekly observations. The date range for the data is from 4th April, 1983 to 30th June, 2006. As for ARMA-GARCH and WDVaR model fitting, 60% of the data, i.e. 727 observations, are reserved as the training set, while the remaining serves as the test set. As for WDNEVaR estimates, 40% of the test set, i.e. 194 observations, is reserved for training the neural network structure. The rolling window method is used to reflect the arrival of new information. Three different ex-post measures of volatilities, i.e. squared returns, historical volatility and realized volatilities are calculated using daily closing prices in the markets. VaRs are estimated using the average weekly prices as calculated in $P_w = \frac{1}{l} \sum_{i=1}^{l} P_i$, where P_w is the weekly price. P_i is the daily price within the week. l is the length of the week. Performance of different models is evaluated by both Mean Square Error (MSE) for its accuracy and Backtesting procedures for reliability. The returns are log differenced as $r_t = \ln(P_t / P_{t-1})$. For each experiment, a portfolio of one asset position worth $1 is assumed.

Several stylized facts about the data series can be summarized from table 1 as follows: price series exhibit significant fluctuations. The standard deviation is significantly positive, which indicates unignorable level of risks in the markets. There are considerable gains, as well as losses, and the market return is asymmetric and loss intending, as suggested by negative skewness levels. There is also high probability of extreme events occurrence as suggested by significant excess kurtosis levels.

Table 1. Descriptive Statistics for crude oil returns series in US WTI crude oil markets

Crude Oil market	Mean	Maximum	Minimum	Medium	Standard Deviation	Skewness	Kurtosis	Jarque-Bera Test (p value)	BDS Test (p value)
WTI	0.0007	0.2440	-0.1967	0.0019	0.0412	-0.2107	6.7580	0	0

Rejection of both, Jarque-Bera test of normality and BDS test of independence, suggests that the data deviate significantly from the normal distribution and are non-linear dependent in nature.Therefore, the market is characterized by the leptokurtic (fat-tail and heteroscedasticity) phenomenon, where non-linear dynamics dominate.

Table 2. Experiment results for ARMA-GARCH VaR in US WTI crude oil market

Crude Oil market	Confidence Level	Exceedances	MSE	Kupiec Test Statistics	P-Value	Model Acceptance
US WTI Crude Oil	99%	8	0.0131	1.7281	0.1187	√
	97.5%	14	0.0099	0.2835	0.5944	√
	95%	21	0.0076	0.4793	0.4887	√

Experiment results in table 2 show that the ARMA-GARCH approach offers acceptable levels of reliability in the tested market. The ARMA-GARCH VaR is accepted across all confidence levels. However, under fierce competition and increasingly tight profit margins, the ARMA-GARCH approach no longer suffices due to the following reasons. Firstly, as indicated by the generally low level of failure rates, the ARMA-GARCH provides conservative VaR estimates that achieve reliability at the cost of accuracy. This may result in "more than enough" idle capital and increase operational costs. Secondly, when higher levels of reliability are demanded by market participants due to increasing competition level, this linear combination approach offers little room for further performance.

Table 3. Experiment results for WDVAR(x, 1) in WTI crude oil market

Crude Oil market	Confidence level	WDVaR(Haar, 1)			WDVaR(Db2, 1)			WDVaR(Coif2, 1)		
		Exceedances	MSE	P-Value	Exceedances	MSE	P-Value	Exceedances	MSE	P-Value
US WTI Crude Oil	99.0%	14	0.0123	0.0007	14	0.0148	0.0007	18	0.0137	0.0000
	97.5%	16	0.0093	0.2857	19	0.0111	0.0656	24	0.0104	0.0023
	95.0%	28	0.0072	0.4516	32	0.0085	0.1256	33	0.0079	0.0852

The notion WDVaR(x, i) in table 3 refers to the VaR estimated based on x wavelet families and the decomposition level (scale) i. Experiment results above show that by switching from Coif2 to Db2 and Haar wavelet family, both accuracy and reliability level can be improved. E.g. when switching from Coif2 to Haar wavelet families, both reliability and accuracy level of VaR estimates increase significantly. As experiment results suggest, the performance difference is closely related to the characteristics of the wavelets used as different wavelet families lead to different trade-offs between accuracy and reliability levels in VaR estimates. Clearly the advantage of WDVaR lies in the additional flexibility introduced into the modeling process. By tuning the two parameters, WDVaR provide different perspectives into the underlying risk evolution and is capable of striking a balance between reliability and accuracy for VaR estimates.

Since VaR estimated based on different wavelet families tracks the underlying risk evolution based on partial information set extraction, to estimate VaR at the highest level of reliability and accuracy, it would be necessary to combine partial information sets during forecasting. The artificial neural network based non-linear ensemble algorithm is used to derive the optimal weights for combining individual forecasts. The back propagation feedforward neural network with structure (3, 6, 1) is used. Through trial and error method, the historical volatility is selected to be the volatility proxy since WDNEVaR using it provides the best performance. Experiment results are listed in table 4.

Table 4. Experiment results for WDNEVAR in WTI crude oil market

Crude Oil Markets	Volatility Proxy	Confidence Level	Exceedances	MSE	Kupiec Test Statistics	P-value	Model Acceptance
US WTI Crude Oil	Historical Volatility	99.0%	3	0.0120	0.0022	0.9626	√
		97.5%	6	0.0091	0.2525	0.6153	√
		95.0%	12	0.0069	0.5175	0.4719	√

Experiment results in table 4 suggest that WDNEVaR achieve higher levels of reliability and accuracy than the ARMA-GARCH based approach. Reliability level of VaR estimated improves as the p value improves in most of the situations, except the slight inferior performance in WTI market at 95% confidence level. More importantly, the accuracy level increases drastically, when measured by significantly lower MSE value uniformly. The improved accuracy in risk tracking implies less idle funds and better allocation of capital resources for financial institutions.

5   Conclusions

The contribution of this paper is two fold. Firstly, this paper incorporates the multi-scale heterogeneous data structure into the modeling process. Separation and integration of different layers in the data are conducted using wavelet analysis. When data are decomposed into the underlying influencing factors in the time scale domain, models are fitted to more stationary data with fewer violations of assumptions and improved goodness of fit. Wavelet analysis has demonstrated its capability to explicitly reveal and track the multi-scale time varying patterns of the data in these experiments. Secondly, in light of the estimation biases resulting from arbitrary selection of different sets of parameters, this paper proposes Artificial Neural Network (ANN) based non-linear ensemble algorithm to reduce estimation bias introduced by different wavelet families. Given the lack of consensus on the suitability and correspondence of certain wavelet families to particular data features in the current literature, non-linear ensemble approaches have proven to be a promising approach to minimize biases introduced by different wavelet families during the forecasting process. It should be noted that the performance of the proposed

WDNEVaR is sensitive to volatility proxies and parameter settings, as suggested by experiment results. More refined volatility proxies could further improve the model performance.

Acknowledgement. The work described in this paper was supported by a grant from City University of Hong Kong (No. 9610058).

References

1. Dowd, K.: Measuring Market Risk. John Wiley, Chichester (2005)
2. Gençay, R., Selçuk, F., Whitcher, B.: An introduction to wavelets and other filtering methods in finance and economics. Academic Press, San Diego, CA (2002)
3. Ramsey, J.B.: The contribution of wavelets to the analysis of economic and financial data. Philosophical Transactions of the Royal Society of London Series a-Mathematical Physical and Engineering Sciences 357 (1999) 2593-2606
4. Aminghafari, M., Cheze, N., Poggi, J.-M.: Multivariate denoising using wavelets and principal component analysis. Computational Statistics & Data Analysis 50 (2006) 2381-2398
5. Yamada, H.: Wavelet-based beta estimation and Japanese industrial stock prices. Appl. Econ. Lett. 12 (2005) 85-88
6. Fernandez, V.P.: The international CAPM and a wavelet-based decomposition of value at risk. Stud. Nonlinear Dyn. Econom. 9 (2005) -
7. Fernandez, V.: The CAPM and value at risk at different time-scales. International Review of Financial Analysis 15 (2006) 203-219

On the Assessment of Petroleum Corporation's Sustainability Based on Linguistic Fuzzy Method

Li-fan Zhang

College of Economics and Management, Nanjing University of Posts and Telecommunications,
Nanjing 210003
{Li-fan ZHANG,ZLF9977}@163.COM

Abstract. By use of the relationship between uncertain fuzzy number and linguistic number, a linguistic fuzzy method to assess the sustainability of Petroleum Corporation as one kind of hybrid multi-attribute decision making problems is put forward. Simultaneously, the attribute weight of decision making problem is derived from the maximum entropy model. The OWA operator is used to aggregate the attribute information. Finally, it is illustrated by a numerical example that this method is feasible and effective.

Keywords: sustainability, hybrid multiple attribute decision making, linguistic fuzzy number, petroleum corporation.

1 Introduction

The assessment of Petroleum Corporation's sustainability involves economy, society and environment. The assessment attributes may be quantitative and qualitative. Other researchers usually assess it with crisp numbers. However, because of the complex geological structure, it would be practical to employ uncertain number to express some of the attributes. For the qualitative attributes, decision-makers sometimes use fuzzy linguistic terms [1] as the assessment information. Consequently, the assessment of Petroleum Corporation's sustainability is the hybrid attributes decision making issue. For solving the problem, a linguistic fuzzy method is proposed. Firstly, using the relationship between the uncertainty fuzzy number and the fuzzy linguistic terms, we transform the hybrid attributes into a uniform fuzzy linguistic attribute in order to make the assessment results being in good according with our understanding. Secondly, when selecting the weight of decision attributes, the attitude value is used for us to fully consider the risk attitude of decision makers. Finally, attribute values are aggregated by OWA (Ordered weighted averaging) operator.

2 The Maximum Entropy Model with the Attitude of Decision

2.1 OWA Operator

Let $X = \{x_1, x_2, \cdots, x_m\}$ be a set of alternatives, $\{p_1, p_2, \cdots, p_n\}$ be a set of attributes.
Definition 1 [2]: Let $OWA: L^n \mapsto L$, if

Y. Shi et al. (Eds.): ICCS 2007, Part I, LNCS 4487, pp. 562–566, 2007.

$$OWA_\omega(a_1, a_2, \cdots a_n) = \sum_{j=1}^{n} \omega_j b_j \tag{1}$$

Where $\omega = (\omega_1, \omega_2, \cdots, \omega_n)^T$ is a weighting vector associated with OWA operator, satisfied $\omega_i \in [0,1], \sum_{i=1}^{n} \omega_i = 1$; b_j is the jth largest element of $\{a_1, a_2, \cdots, a_n\}$; L is a number set; OWA is called an order weighted averaging operator, that is OWA operator.

If the weight vector of attributes is taken as a kind of attitudinal-character probability, under a certain attitude, the attribute weight should be selected which can maximize the entropy as the following model (*):

$$Max\ E(v) = -\sum_{i=1}^{n} v_i \ln v_i \tag{2}$$

$$\alpha = \frac{1}{n-1} \sum_{i=1}^{n} (n-i) v_i \tag{3}$$

Where $\alpha \in [0,1]$ is the degree of optimism and $\sum_{i=1}^{n} v_i = 1$, $v_i \in [0,1]$, $i = 1, \cdots, n$.

2.2 The 2-Tuple Linguistic Term and Uncertainty Fuzzy Number

Let $S = \{S_0, \cdots, S_g\}$ be a linguistic term set, where $2 \le g \le 14$. The semantics of the terms is given by fuzzy numbers defined in $[0,1]$ interval. A way to characterize a fuzzy number is to use a representation based on parameter of its membership function. Sometimes the linguistic terms can be representing by triangle fuzzy number, such as $S_i = (a_i, b_i, c_i)$, where $a_i = \frac{i-1}{g} (1 \le i \le g)$, $b_i = \frac{i}{g} (0 \le i \le g)$, $c_i = \frac{i+1}{g} (0 \le i \le g-1), a_0 = 0, c_g = 1$.

The linguistic information is represented by a 2-tuple linguistic model [3] (S_i, α_i). The 2-tuple linguistic operator can be defined as follow:

Definition 2 [3]: Let $S = \{S_0, S_1, \cdots, S_g\}$ be a linguistic term set and $\beta \in [0, g]$ is a value supporting the result of a symbolic aggregation operation, then the 2-tuple that expresses the equivalent information to β is obtained with the following function Δ:

$$\Delta : [0, g] \rightarrow S \times [-0.5, 0.5)$$

$$\Delta(\beta) = (S_i, \alpha) \tag{4}$$

Where $i = round(\beta), \alpha = \beta - i, \alpha \in [-0.5, 0.5)$ and "round" is the rounding operation.

Definition 3 [3]: Let I be an uncertain fuzzy number. Then I can be transforming into linguistic 2-tuple set by the function $\tau_{IS_T} : [0,1] \rightarrow F(S_T)$.

$$\tau_{IS_T}(I) = \{(S_k, \omega_k)/k \in \{0, \cdots, g\}\} \tag{5}$$

$$\omega_k = \max_y min\{\mu_I(y), \mu_{S_k}(y)\} \tag{6}$$

Where $\mu_I(.), \mu_{S_k}(.)$ are the membership functions associated with the terms I and S_k.

Definition 4[4]: Let $\tau(I) = \{(S_0, \omega_0), (S_1, \omega_1), \cdots, (S_g, \omega_g)\}$ be a fuzzy set that represents a numerical value I over the linguistic term set S. we obtain a numerical value that represents the information from the fuzzy set by means of the function $\chi : F(S_T) \rightarrow [0, g]$.

$$\chi\tau(I) = \chi(F(S_T)) = \chi\{(S_j, \omega_j), j = 0, 1, \cdots, g\} = \sum_{j=0}^{g} j\omega_j (\sum_{j=0}^{g} \omega_j)^{-1} = \beta \qquad (7)$$

The value β is easily transformed into a linguistic 2-tuple using the definition 2.

3 The Aggregation Approach

Consider a multi-attribute decision making problem with m alternatives $x_1, x_2, \cdots; x_m$, and n decision attributes $p_1, p_2, \cdots, p_n$. Each alternative is assessed with respect to each attribute. The assessment scores assigned to the attributes are the components of decision matrix denoted by $R = (\tilde{r}_{ij})_{m \times n}$, where $\tilde{r}_{ij}$ is a numerical value or uncertainty fuzzy number. The decision making steps are given as follows:

Step 1. Construct the decision matrix.

Step 2. Normalize the decision matrix $R = (\tilde{r}_{ij})_{m \times n}$ to matrix $R' = (r_{ij})_{m \times n}$. For instance, the elements $\tilde{r}_{ij} = (a_{ij}, b_{ij}, c_{ij})$ of the decision matrix $R = (\tilde{r}_{ij})_{m \times n}$ are triangular fuzzy numbers, then $r_{ij} = (\frac{a_{ij}}{c_j^+}, \frac{b_{ij}}{c_j^+}, \frac{c_{ij}}{c_j^+}), j \in B$, and $c_j^+ = \max_i c_{ij}, j \in B$. Any decision matrix with entries being numerical values can be normalized using the same method mentioned above.

Step 3. Using equations (4-7), transform the attribute values into 2-tuple linguistic terms. We get a decision matrix $P = (p_{ij})_{m \times n}$ with entries being 2-tuple linguistic values.

Step 4. Calculate the comprehensive attribute values. The weight vector $\omega^* = (\omega_1^*, \cdots, \omega_k^*, \cdots, \omega_n^*)^T, k = 1, \cdots, n$ can be got by model $(*)$. Then the comprehensive attributes value $t_i, i = 1, 2, \ldots, m$ of each alternative can be aggregated by the OWA operator:

$$t_i = \phi_Q(p_{i1}, \cdots, p_{in}) = \Delta(\sum_{k=1}^{n} \omega_k^* \Delta^{-1}(p'_{ik})) \qquad (8)$$

Where p'_{ik} is the kth largest value in the set $\{p_{i1}, \cdots, p_{in}\}$.

Step 5. Rank the alternatives $x_1, x_2, \cdots, x_m$.

4 Numerical Example

Let $X = \{x_1, \cdots, x_9\}$ denotes nine Petroleum Corporations set, which are BP、Exxon-Mobil、Shell、Anni、CNPC、SINOPEC、CNOOC、Statoil、Petrobras. The attributes of sustainability of Petroleum Corporation are given in table 1.

The values of sub-attributes are given according to the following principle: The numerical values are given by 2004 annually report directly or through simple calculated. Especially, the data of oil and gas surplus workable reserves attributes take the form of the triangle fuzzy numbers. These two triangles fuzzy numbers sets are constructed according to the reserves data in 2003 annals and in 2004 annals and the reserves in 2003 annals go through 3%'s fluctuation. The society benefit and environment harmonious attributes gain from investigation.

Table 1. The sustainability attributes of Petroleum Corporation

Attributes	Scale of resource	Acquire the resource	Capital increment	Society benefit	Environment harmonious
Sub-attributes	Total assets	Oil output	Net profit	Satisfied product need	Provide new energy ability
	Oil surplus workable reserves	Gas output	Return on asset	Provide jobs ability	Avoid venture ability
	Gas surplus workable reserves	Refine ability	Return on equity	Guarantee safe ability	Protect the environment

Suppose that the decision makers hold the normal risk attitude, then attributes weight can be derived from the model $(*)$, which is $\omega^* = (.1620, .1791, .1980, .2190, .2421)^T$. By (8), the ranking order of nine petroleum corporations is $x_2 \succ x_1 \succ x_3 \succ x_5 \succ x_9 \succ x_4 \succ x_6 \succ x_7 \succ x_8$.

The appraisal result is reasonable. We know the sustainability of Petroleum Corporation is close connecting with its comprehensive ability. This ranking order is the same as the order of their comprehensive ability that US "petroleum intelligence weekly" magazine given in 2003. It can be conclude that the assessment method given is objective and valid.

5   Conclusions

This paper develops an assessment approach of Petroleum Corporation's sustainability, and it is illustrated by a numerical example that this method is feasible and effective. Using the relationship between the uncertainty fuzzy number and the linguistic fuzzy term, we transform the hybrid attribute values into the 2-tuple linguistic values which makes the assessment results being more intuitionist and more understandable. In the process of selecting the weight of decision attributes, we consider the attitudes of decision-makers, and in the meantime, we aggregated the attribute values by utilizing the *OWA* operator which makes the assessment results being more objective and scientific.

References

1. Yongqi, Xia, Qizong, Wu: A Technique of Order Preference by Similarity to Ideal Solution for Hybrid Multiple Attribute Decision Making Problems. Journal of Systems Engineering, Vol 19. (2004) 630-634.

2. Yager, R. R.: Applications and Extensions of OWA Aggregations. International Journal of Man-Machine Studies, Vol 37. (1992) 103-132.
3. Herrera, F, Martinez, L.: Sanchez Managing Non-Homogeneous Information in Group Decision Making. European Journal of Operational Research, Vol 166. (2005) 115-132.
4. Herrera, F, Martinez, L.: An Approach for Combing Linguistic and Numerical Information Based on the 2-tuple Fuzzy Linguistic Representation Model in Decision Making. International Journal of Uncertainty, Fuzziness and Knowledge-Based Systems, Vol 8. (2000) 539-562.

A Multiagent Model for Supporting Tourism Policy-Making by Market Simulations

Arnaldo Cecchini and Giuseppe A. Trunfio

Laboratory of Analysis and Modelling for Planning
Department of Architecture and Planning - University of Sassari
Palazzo del Pou Salit, Piazza Duomo, 6 I07041 Alghero (SS), Italy
cecchini@uniss.it, trunfio@uniss.it

Abstract. Sustainable tourism development at a destination requires an understanding of the many interrelated factors that influence risks and returns. For this reason, operators in tourism management and public policy makers are increasingly interested in effective tools for exploring the consequences of their decisions. With this purpose, this paper describes a multiagent system which enables evaluation of the impact of some possible interventions and policies at a destination by computer simulations and provides an insight into the functioning of the market.

1 Introduction

As tourism is a primary industry for many geographical areas, strategic management of tourist destinations is becoming more and more important. On the other hand, the success of this economic activity strongly depends on the decisions taken by the actors in the field [1]. For example, tourism managers have to take decisions on segmentation and market positioning, product design, pricing, retail strategy and advertising, while public administrators have to implement policies regarding the accessibility of the destination and of the single attractions, the organization of events or the economic incentive of specific market segments. Given the difficulty of predicting the consequences of decisions regarding a system which involves many variables interrelated in a non-linear manner, it is recognized that models for the simulation of market dynamics can provide valuable support. Some models have already been developed for such purpose (e.g. see [2,3]), exploiting various techniques. In this paper we propose a multiagent model [4] which, differently from most of the existing ones, has together the following characteristics: (*i*) it is based on spatially-situated agents; (*ii*) the agents are heterogeneous and with memory of their past experiences; (*iii*) the agents constitute a social network. As in [5], the basic assumption of the model is that tourists attempt to maximize utility from their holiday. The system is built in such a way that a tourist who is satisfied with his tourist destination is more likely to return to that destination than one who obtained lower utility. Moreover, in the model as in the real world, a satisfied tourist can influence other individuals with similar characteristics to choose the same destination, whereas an unsatisfied tourist can influence other potential tourists to avoid that destination.

Y. Shi et al. (Eds.): ICCS 2007, Part I, LNCS 4487, pp. 567–574, 2007.

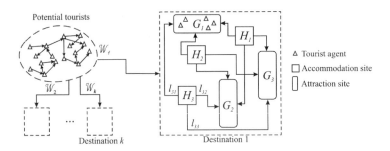

Fig. 1. The multiagent model overview

2   Tourist Destination

As shown in Figure 1, the main elements of the environment where agents operate are a set of *tourist destinations*, that is cities, towns or small geographical areas, the economy of which is to some extent dependent on revenues coming from the tourist industry. In the model a tourist destination is represented as:

$$Destination = \langle \mathcal{W},\ \mathcal{H},\ \mathcal{G},\ \mathcal{L} \rangle \tag{1}$$

where $\mathcal{W} = \{W_1, \dots, W_p\}$ represents accessibility and each element:

$$W_i = \langle price,\ time,\ capacity \rangle \tag{2}$$

corresponds to the i-th means of transport available for reaching the resort (e.g. train, bus, ship, car or a combination of these) and groups together the price per person, the duration of the journey and the daily capacity, respectively.

Set $\mathcal{H} = \{H_1, \dots, H_n\}$ of Equation 1 collects the accommodation sites:

$$H_i = \langle \mathbf{x},\ type,\ quality,\ price,\ capacity \rangle \tag{3}$$

where $\mathbf{x}$ is the position in a bi-dimensional space, *type* is the kind of accommodation, *quality* is a real number belonging to the interval [0, 1] which summarizes the degree of excellence of the significant features of H_i, *price(t)* provides the price per day and per person as a function of the tourist season t_s and *capacity* is the number of tourists that can be simultaneously accommodated;

Set $G = \{G_1, \dots, G_m\}$ of Equation 1 collects the *attractions* (i.e. places of interest that tourists visit) present at the destination, which are defined as:

$$G_i = \langle \mathbf{x},\ type,\ quality,\ price,\ capacity \rangle \tag{4}$$

where $\mathbf{x}$ is the position in a bi-dimensional space, *type* is the kind of attraction, *quality* $\in [0, 1]$ is a synthetic indicator of the quality of the significant features of G_i (e.g. view, water and sand quality for an uncontaminated beach or the quality of the services and facilities for an equipped beach), *price(t)* provides the entrance price per person given season t_s and *capacity* is the number of people who can simultaneously enjoy G_i.

Set $\mathcal{L}$ of Equation 1 contains links:

$$l_{ij} = \langle H_i, \ G_j, \ type, \ price, \ time \rangle \tag{5}$$

specifying the money and time costs which are required for reaching the attraction site G_j from the accommodation site H_i, using the means of transport specified by the attribute *type*. Thus, the elements of $\mathcal{H}$ and $\mathcal{G}$ are the edges of a graph representing the environment in which the model agents operate.

For each destination, an *accessibility agent* answers the tourist agent's requests, that is, he provides the first available departure date, given the period, the access mode and the duration of the stay. Besides, each accommodation site H_i and attraction site G_i holds a *manager agent* that has three main tasks: (i) to keep a schedule of admittance in order to accept or deny booking requests on the basis of the site's capacity, by interacting with tourist agents (ii), to answer questions about the current state of the site (e.g. the number of people currently hosted) (iii), to adapt his prices to the demand coming from tourists.

3 Tourist Agent

Each tourist, who it is assumed takes one vacation per year, has the main aim of maximizing his utility and is capable of remembering the experiences had at a destination. In addition, he can receive information from other tourists or from advertising messages. More formally, the tourist agent is defined as:

$$A = \langle \mathcal{P}, \ \mathcal{S}, \ \mathcal{M}, \ \mathcal{E}, \ \mathcal{V}, \ \mathcal{B} \rangle \tag{6}$$

In the latter Equation $\mathcal{P}$ is a set of properties that influences the tourist's behaviour (e.g. *Age*, *Gender*, *Family size*, *Income*, *Education*).

Set $\mathcal{S} = \{t_0, \ T^{(h)}, \ Acc, \ Att\}$ collects some properties that concur in defining the agent's state. In particular, t_0 is the instant of the start of the journey, $T^{(h)}$ is the holiday duration, while Acc and Att are two links which, during the temporal interval $[t_0, \ t_0 + T^{(h)}]$, indicate the accommodation site and the attraction site currently holding the agent, respectively.

Set $\mathcal{V}$ in Equation 6 constitutes the information concerning potential tourist destinations possessed by the agent. Each piece of information is represented as:

$$V = \langle Dest, \ St, \ p_G, \ \mathcal{G}_V \rangle \tag{7}$$

where *Dest* refers to the destination, *St* is the *information strength*, p_G is an estimate of the cost per day of a stay excluding accommodation and $\mathcal{G}_V$ id the set of attractions known by the agent in *Dest*. In each element $g_i = \langle G_k, \ n_i, \ \alpha_i \rangle \in \mathcal{G}_V$, the value of n_i is the number of visits of the attraction G_k in a previous stay in *Dest* while α_i is an evaluation parameter assigned to the same attraction. The quantity $St \in [0, \ 1]$ represents a measure of information reliability perceived by the agent (e.g. $St = 1$ in the case of information originating from direct knowledge). During the simulation, information strength St varies according to two mechanisms: at each time-step k it decreases according to the relation $St^{(k+1)} = \sigma \, St^{(k)}$, with $\sigma \in [0, 1]$ and it changes in the case the agent receives new elements of information about $Dest$.

Set $\mathcal{E} = \{E_1, \ldots, E_q\}$ in Equation 6 groups together elements E_i defined as:

$$E_i = \langle Att_i, \; n_i^{(s)}, \; n_i^{(f)}, \; p_i, \; r_i, \; \alpha_i \rangle \tag{8}$$

representing the *agent's memory* of attractions experienced during a vacation. In particular, Att_i is the link to the attraction, $n_i^{(s)}$ is the number of accesses to Att_i, $n_i^{(f)}$ is the number of failures in accessing (e.g. an access request may be refused when Att_i is full), p_i is the total expenditure made for the attraction, r_i is the sum of all ratings attributed to each of the $n_i = n_i^{(s)} + n_i^{(f)}$ requests, α_i represents the overall evaluation which is attributed by the agent to the n_i experiences on the whole. For $t = t_0$ (i.e. the holiday start) the set $\mathcal{E}$ is empty. During the holiday, the agent fills the set $\mathcal{E}$ of experiences with evaluations α_i. At the holiday end, values $n_i^{(s)}$ and α_i are transferred in the corresponding variables of the set V related to the destination (see Equation 7).

Set $\mathcal{M}$ in Equation 6 collects links to other potential tourists. As explained below, the network of tourist agents determines the exchange of information related to potential tourist destinations.

The agent's behaviour is defined by the actions collected in set $\mathcal{B}$ of Equation 6. During the simulation, agents have to make decisions about actions to undertake (e.g. which destination to visit). We assume that such decisions are made probabilistically, according to a logistic model [6]. In particular, given a choice set $\mathcal{C}$, the agent associates a utility value $U(c_i)$ to each alternative $c_i \in \mathcal{C}$ [7]. The utility values are then transformed into probabilities $Pr(c_i|\mathcal{C}) = e^{U(c_i)}/\sum_{j=0}^{\sharp\mathcal{C}} e^{U(c_j)}$, where $\sharp\mathcal{C}$ is the number of alternatives. The actual choice is made comparing a drawn random number in $[0, 1]$ with the cumulative probabilities $\sum_{j=0}^{i} P(c_j|\mathcal{C})$ associated to the alternatives c_i. In some cases, the choice account for a *satisfier* threshold U_{th}, that is, in the choice set $\mathcal{C}$, the alternatives c_k with $U(c_k) < U_{th}$ are not considered by the agent.

If the current time t is greater than $t_0 + T^{(h)}$, at each time-step of the simulation and with probability p_b, the tourist agent tries to book a new holiday. The algorithm proceeds by choosing first the preferred season and the holiday duration $T^{(h)}$, then the trip destination together with the means of transport and the accommodation site. At the end the agent tries to book for a specific departure date t_0. In the follows only the most relevant agent's actions are illustrated.

Choice of destination, accommodation and means of transport. When the agent decided to take a vacation, that is, the period of the year and the duration $T^{(h)}$ have been chosen, for each piece of information regarding potential destinations possessed by the agent (i.e. contained in $\mathcal{V}$), and for each combination $\langle W_k, H_j \rangle$, the agent computes two utility functions:

$$U_D = \frac{T^{(n)} \left[\sum_{i=1}^{m} (1 + St\,\alpha_i)\beta_i(n_i)q_i + \beta_{acc}q_{acc} \right]}{T\,(2\,m + 1)} \tag{9}$$

$$U_P = 1 - \frac{p_{jour} + T^{(n)} (p_{acc} + p_G)}{p_{\max}(T)} \tag{10}$$

which take values in the interval $[0, 1]$. In Equations (9) and (10):

- for each of the $m = \sharp \mathcal{G}_V$ different attractions known by the agent in the potential destination (i.e. in the set $\mathcal{G}$) there are: a function $\beta_i(n_i) \in [0, 1]$ accounting for the degree of attractive influence on the agent and depending on the number n_i of previous visits to the attraction, the coefficient $q_i = quality(G)$ (see Equation 4) and the coefficient α_i (see Equation 7);
- St is the strength of the information on the basis of which the agent is considering the destination (see Equation 7);
- $\beta_{acc} \in [0, 1]$ is a parameter expressing the degree of liking of the agent for the accommodation considered, having the quality $q_{acc} = quality(H_j)$;
- $T^{(n)} = T^{(h)} - time(W_k)$ is the net time available for enjoying the stay (i.e. the total holiday duration minus the journey time);
- $p_{jour} = price(W_k)$ is the total cost of the journey, according to the means of transport W_k chosen, $p_{acc} = price(H_j)$ is the cost per day of accommodation H_j and p_G is the estimated cost per day of a stay at the destination considered; $p_{\max}$ is the maximum cost that the agent is willing to spend.

Agents compute their utility as the product of U_D and U_P. The values of utility are then transformed into probabilities according to the logit model and a satisfier threshold $U_{th}^{(D)}$ is used. for the choice.

Given the destination $Dest$, the access mode W_k and the accommodation H_j, the trip start date t_0 is determined by interacting with the *accessibility agent* and the *manager agent* (see Section 2).

The holiday experience. During the stay the tourist's agent attempts to maximize utility from his holiday. In particular, at $t = t_0$ the set $\mathcal{E}$, which represents the history of the holiday, is empty. The agent's accommodation site being H_i, every time-step of the simulation the agent chooses an attraction $G_j \in \mathcal{G}$ by using the utility function:

$$U_G = \frac{1}{2} \left(1 - \frac{\pi(G_j, l_{ij})}{\pi_{\max}(G_j)} \right) (1 + \alpha) \beta q \tag{11}$$

$$\pi(G_j, l_{ij}) = price(G_j) + price(l_{ij}) + \mu \ time(l_{ij}) \tag{12}$$

where β and q have the same meaning so that in Equation 9, while the parameter α, which is kept in memory by the agent as stated in Equation 8, expresses an overall evaluation relative to the attraction site on the basis of previous experiences of the same stay, μ is the cost of the time which depends on the agent's characteristics [8] and $\pi_{\max}(G_j)$ is the maximum cost that the agent is willing to spend for enjoying the attraction. The choice is made transforming values (11) into probabilities and using a satisfier threshold $U_{th}^{(G)}$. Let Att be the attraction chosen, the corresponding entry $E \in \mathcal{E}$ (see Equation 8) is updated in order to reflect the new experience. In particular:

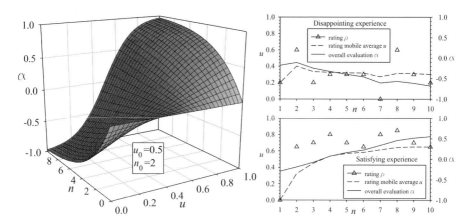

Fig. 2. The evaluation function used by the tourist agent

1. $n^{(s)}$ (or $n^{(f)}$ in the case the request of access is refused) is incremented by one, the total expenditure p is updated adding the new value $\pi(G_j, l_{ij})$ and the total rating r is updated with the new rating $\rho \in [0, 1]$;
2. the evaluation $\alpha(u, n)$ of the attraction site is computed as a function of the average rating $u = r/n$ and of the number of requests $n = n^{(s)} + n^{(f)}$.

The evaluation function $\alpha(u, n)$ is defined as:

$$\alpha(u, n) = \frac{e^{\frac{u}{u_0} \ln b(n)} - b(n)}{e^{\frac{u}{u_0} \ln b(n)} + b(n)}, \quad \text{with} \quad b(n) = 2e^{\frac{n}{n_0} \ln 2} - 1 \qquad (13)$$

where the parameters n_0 and u_0 have the following meaning: (i) if the average rating u is still null after n_0 access requests, this corresponds to an overall evaluation of $\alpha = -0.5$, that is, to the halving of the next estimation of the utility component U_E for that attraction site; (ii) u_0 is the average rating corresponding to the neutral evaluation $\alpha = 0.0$. Figure 2 reports two examples of agent's experience for an attraction site using Equations 13. It can be seen that ratings above u_0 contribute, throughout the history of the experience, to increase the overall evaluation α, making the experience satisfying. On the contrary, a sequence of ratings below u_0 produces, in agent convincement, a bad overall evaluation of the attraction site. The rating ρ should reflect the actual agent's experience at the attraction site. Thus, its determination depends on both the agent's interactions and the interaction between agents and the environment. At the end of the holiday, total expenditures p, number of accesses $n_i^{(s)}$ and evaluations α_i contained in the agent's memory $\mathcal{E}$, are transformed into updated pieces of information for the set $\mathcal{V}$.

Influencing other individuals. As shown in Figure 1, potential tourists constitute a social network in the sense that each agent can influence the decisions of other agents. The latter are the ones pointed by the links contained in the set $\mathcal{M}$ of each agent. The network is randomly initialized at the simulation start by

Table 1. The main characteristics of the focused destination

Id	type	quality	price	capacity
Accommodation sites				
H_{1-5}	four-star hotel	0.8	130	200
H_{6-15}	three-star hotel	0.8	80	240
H_{16-35}	bed & breakfast	0.6	35	10
Attraction sites				
G_{1-4}	beach	0.6	2	500
G_{5-7}	equipped beach	0.6	10	600
G_8	historical center	0.5	5	5000
G_{10}	museum	0.8	4	600

Accessibility			
Id	price	time	capacity
W_1	300	3	360
W_3	100	12	1000
W_3	60	4	180

assigning for each agent the number of elements in $\mathcal{M}$ according to the characteristics in $\mathcal{P}$ and by assuring that the links are predominantly directed towards agents having similar characteristics. In particular, an agent A_i which is not in his holiday interval, selects with probability p_e a group of neighbouring agents. Then, to each of the selected agents A_j, agent A_i communicates one of his pieces of information $V^{(i)} \in \mathcal{V}$ which is probabilistically selected. When the agent A_j receives the element of information $V^{(i)}$, he updates his corresponding element $V^{(j)}$ which refers to the same destination as:

$$\hat{St}^{(j)} = \max(St^{(j)}, \lambda_{ij}\, St^{(i)}) \quad \text{and} \quad \hat{a}_k^{(j)} = (a_k^{(j)} + \lambda_{ij}\, St^{(j)}\, a_k^{(i)})/2 \quad (14)$$

where $\hat{V}^{(j)}$ is the updated information and the coefficient $\lambda_{ij} \in [0, 1]$ is the degree of trust of the agent A_i in the eyes of A_j.

4   A Preliminary Test

The model, implemented in C++ language, was tested with a hypothetical but plausible scenario. A set of potential tourists was initialized with 100.000 agents having random characteristics and relationships. Five destinations, each having various access modes, accommodation and attraction sites, were included in the artificial marketplace. The time-step size was set at 12 hours and it was assumed each tourist would visit one attraction per step during vacations. The destination highlighted was composed in particulary of elements H_i and G_i having the characteristics shown in Table 1. For each couple H_i and G_j three links l_{ij}, were generated with random properties appropriate to represent walking, moving by bus and by taxi, respectively. The attractions present at the destinations were constantly populated by local visitors (i.e. visitors not accommodated at any of the sites in $\mathcal{H}$) for one half of their capacity.

The test presented here was aimed at assessing the influence of a modification in accessibility $\mathcal{W}$. To this end, two groups of four runs were executed averaging the probabilistic results. Each run was composed of 2920 steps, corresponding to four years. In the first group the set $\mathcal{W}$ for the destination under study was composed of the two elements W_1 and W_2 shown in Table 1. In the second group, the element W_3 shown in Table 1, which represents a daily flight of a low-cost company,

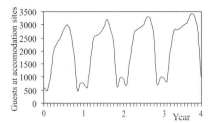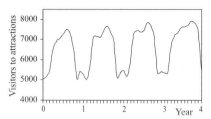

Fig. 3. Effects of the introduction of a low-cost daily flight

was added to the set $\mathcal{W}$ at the end of the first year. The results are illustrated in Figure 3, where both the effects in terms of guests at accommodation sites and visitors to attraction sites are reported. According to the model, the effects in terms of number of visiting tourists affect both the peaks in high season and, to a lower extent, the visitors in low season. The increment of tourists grows in the second year by the introduction of E_3 and tends to become stable after two years.

5   Conclusions and Future Work

The model allows evaluation of the effect of advertising messages, the effect of pricing strategies, market segmentation approaches, introduction or enhancement of new attractions or the interrelated consequences of many simultaneous actions. On the other hand, a drawback is represented by the difficulties which may arise in the initial setup of the model for an actual destination. Indeed, this phase requires a time-consuming activity involving the definitions of parameter values and the validation by comparison with real historical trends for similar situations. Future work will focus on contextualizing the model in a real geographical area, for which sufficient data are available.

References

1. Sainaghi, R.: From contents to processes: Versus a dynamic destination management model (DDMM). Tourism Management **27** (2006) 1053–1063
2. Velthuijsen, J.W., Verhagen, M.: A simulation model of the dutch tourism market. Annals of Tourism Research **21**(4) (1994) 812–827
3. Kandelaars, P.: A dynamic simulation model of tourism and environment in the yucatan peninsula. Working papers, Int. Inst. for Applied Systems Analysis (1997)
4. Wooldridge, M.: An Introduction to MultiAgent Systems. John Wiley & Sons (2002)
5. Hernández-López, M., Cáceres-Hernández, J.J.: Forecasting tourists characteristics by a genetic algorithm with a transition matrix. Tourism Management **28** (2007) 290–297
6. Sirakayaa, E., Woodsideb, A.G.: Building and testing theories of decision making by travellers. Tourism Management **26** (2005) 815–832
7. Russel, S., Norvig, P.: Artificial Intelligence: A Modern Approach. Prentice-Hall (1995)
8. Meignan, D., Simonin, O., Koukam, A.: Multiagent approach for simulation and evaluation of urban bus networks. In: 5rd AAMAS Conference

An Improved Chaos-Based Image Encryption Scheme

Chong Fu[1], Zhen-chuan Zhang[1], Ying Chen[2], and Xing-wei Wang[1]

[1] School of Information Science and Engineering, Northeastern University,
Shenyang 110004, China
[2] School of Economy Management, Shenyang Institute of Chemical Technology,
Shenyang 110142, China
fu_chong@sohu.com

Abstract. Chaos theory provides a new approach to image encryption technology. The key stream generator is the key design issue of an image encryption system, it directly determines the security and efficiency of the system. This paper proposes an improved chaos-based key stream generator to enlarge the key space, extend the period and improve the linear complexity of the key stream under precision restricted condition so as to enhance the security of a chaos-based image encryption system. The generator is constructed by three Logistic maps and a nonlinear transform. The balance and correlation properties of the generated sequence are analyzed. The sequence is proved to be a binary Bernoulli stochastic sequence and the distribution of the differences between the amounts of 0 and 1 is analyzed. The side lobes of auto and cross correlation are proved to obey normal distribution $N(0, 1/N)$. The experimental results indicate that the scheme has advantages of long period and strong anti various attack ability over conventional chaos-based encryption system.

1 Introduction

Image is one of the most important information representation styles and above 80% information we obtained is from vision apperceiving. With the fast development of computer network technology, more and more sensitive images such as in medical, military, financial etc. fields need effective protection in open network environment and image encryption technology has become an important branch of cryptography. Conventional symmetric encryption algorithm such as DES, IDEA, AES, etc. is not suitable for image encryption due to the special storage characteristics of an image. Most of the classic image encryption algorithms are position permutation based, such as Arnold transform, magic square transform, IFS transform and scan pattern, etc. These methods have advantages of fast encryption speeds but the security completely depends on the secrecy of the algorithm that we use, which do not satisfy the requirement of a modern cryptography system [1-2]. Besides, the encrypted images are only position permutated and the grayscale of each pixel still remains its original value, which is insecure against known plaintext attack.

A chaos system has the properties of extreme initial value sensitive, nonperiodic, unpredictable and Gaussian like correlation properties, while it is deterministic, so it is very suitable to be used as key stream generator for image encryption system [3-5]. The basic idea of chaotic image encryption algorithm is producing two chaos-based

Y. Shi et al. (Eds.): ICCS 2007, Part I, LNCS 4487, pp. 575–582, 2007.

pseudorandom sequences, one is for position permutation and the other is for gray-scale substitution. The security of a chaos based image encryption system depends on the unpredictability of the key stream generated by the chaos system [6-9].

Unfortunately, the concrete implementation of single chaos map based encryption system is by far ideal as the abstract model with infinite and not even countable cardinality. The iteration of a chaos map will work with a finite set of rational numbers because no processor is precision unrestricted, which makes the key space limited. The maximum calculation precision of common PC processor is 16, so the key space is $10^{16} \approx 2^{53}$, which is a little smaller than DES(2^{56}) and by far smaller than AES(2^{128}). At the same time, the raw chaotic sequence will be periodic due to limited calculation precision, which will cause the quantified sequence also being periodic and making the correlation properties worse. Besides, single chaos map based key stream is easy to be attacked by using adaptive parameter chaos synchronization method [10-11]. In this paper, an improved key stream generator based on nonlinear transform of three basic Logistic maps is proposed which enlarges the key space to 2^{158}, extends the period and improves the linear complexity of the key stream under precision restricted condition, thus enhances the security of a chaos-based image encryption system.

2   The Statistical Properties of Logistic Map

Logistic map is defined as:

$$x_{i+1} = \mu x_i (1 - x_i), \tag{1}$$

μ is usually set to 4 for full map, in which $x_n \in [0, 1]$. The probability density of chaotic orbit generated by Eq.1 is [12]

$$\rho(x) = \begin{cases} \dfrac{1}{\pi \sqrt{x(1-x)}} & 0 \le x \le 1, \\ 0 & otherwise. \end{cases} \tag{2}$$

Let $\{x_i\}$ be chaotic sequence generated by Eq.1, we can get the following three properties based on Eq.2.

Property 1. The mean of $\{x_i\}$ is:

$$\bar{x} = \lim_{N \to \infty} \frac{1}{N} \sum_{i=0}^{N-1} x_i = \int_0^1 x \rho(x) dx = 0.5. \tag{3}$$

Property 2. The normalized auto correlation function of $\{x_i\}$ is:

$$AC(m) = \lim_{N \to \infty} \frac{1}{N} \sum_{i=0}^{N-1} (x_i - \bar{x})(x_{i+m} - \bar{x}) = \int_0^1 x f^m(x) \rho(x) dx - \bar{x}^2 = \begin{cases} 0.125 & m = 0, \\ 0 & m \ne 0. \end{cases} \tag{4}$$

Property 3. For any two different initial values $\{x_{i1}\}$ and $\{x_{i2}\}$, the normalized cross correlation function of the two generated sequences is:

$$CC_{12}(m) = \lim_{N \to \infty} \frac{1}{N} \sum_{i=0}^{N-1} (x_{i1} - \bar{x})(x_{(i+m)2} - \bar{x})$$

$$= \int_0^1 \int_0^1 x_1 f^m(x_2) \rho(x_1) \rho(x_2) dx_1 dx_2 - \bar{x}^2 = 0. \tag{5}$$

Property 2 and 3 indicate that Logistic map is excellent to be used as key generator for image encryption system. However, above 3 properties are under ideal conditions and the sequence length is usually finite in practical use. Let $\{x_i\}$, $\{x_{i1}\}$ and $\{x_{i2}\}$ be chaotic sequences with finite length N.

The normalized auto correlation function of $\{x_i\}$ is defined as:

$$AC(m) = \begin{cases} \dfrac{1}{N-|m|} \sum_{i=0}^{N-1-|m|}(x_i - \bar{x})(x_{i+m} - \bar{x}) & 1-N \leq m \leq N-1, \\ 0 & N \leq |m|. \end{cases} \tag{6}$$

The normalized cross correlation function of $\{x_{i1}\}$ and $\{x_{i2}\}$ is defined as:

$$CC_{12}(m) = \begin{cases} \dfrac{1}{N-|m|} \sum_{i=0}^{N-1-m}(x_{i1} - \bar{x})(x_{(i+m)2} - \bar{x}) & 1-N \leq m \leq N-1, \\ 0 & N \leq |m|. \end{cases} \tag{7}$$

3 Key Stream Generator Construction and Its Performance Analysis

3.1 Key Stream Generator Construction

The nonlinear transform based key stream generator is shown in Fig. 1.

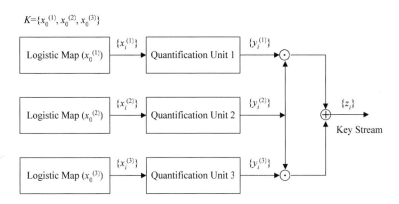

Fig. 1. Nonlinear transform based key stream generator

Three Logistic map with different initial values $x_0^{(1)}$, $x_0^{(2)}$ and $x_0^{(3)}$ produce three independent pseudorandom analog sequences, and three quantification units quantify each input analog sequence $\{x_i^{(k)}\}(k=1,2,3)$ to binary form $\{y_i^{(k)}\}(k=1,2,3)$ by using Eq.8.

$$y_i = \begin{cases} 0 & x_i < 0.5 \\ 1 & x_i \geq 0.5 \end{cases} \tag{8}$$

Quantification unit2 is used as control unit, when unit2 output 1, the unit2 is connected to unit1; otherwise it is connected to unit3, so the output key stream $\{z_i\}$ is

$$z_i = y_i^{(1)} y_i^{(2)} + y_i^{(3)} \overline{y_i^{(2)}} = y_i^{(1)} y_i^{(2)} + y_i^{(3)} y_i^{(2)} + y_i^{(3)}. \tag{9}$$

The key is constructed by three decimal numbers $k = \{x_0^{(1)}, x_0^{(2)}, x_0^{(3)}\}$, so the key space size is $10^{48} \approx 2^{158}$, which is larger than the acknowledged most security AES standard. Let the period of chaotic binary sequences $\{y_i^{(k)}\}(k = 1,2,3)$ under precision restricted condition be N, from the Eq.9 we can see that the period of $\{z_i\}$ is $2N^2+N$. The period of the output sequence is greatly extended and the linear complexity is improved, so the algorithm is more secure against various attacks such as adaptive parameter chaos synchronization and reverse iteration reconstruction. The balance and correlation properties are the two most important factors to evaluate the performance of a key stream, which will be analyzed in the following sections.

3.2 Balance Performance Analysis

In order to analyze the balance and correlation performance of generated key stream, lemma 1 is proposed.

Lemma 1. The generated key stream $\{z_i\}$ is a binary Bernoulli distributed pseudorandom sequence. For sequence with length N, the distribution of the differences between the amounts of 0 and 1 is shown in Table 1.

Table 1. The distribution of the differences between the amounts of 0 and 1

Differences	N	$N-2$	$N-4$	...	$-N+4$	$-N+2$	$-N$
Probability	$(\frac{1}{2})^N$	$C_N^1(\frac{1}{2})^N$	$C_N^2(\frac{1}{2})^N$	...	$C_N^{N-2}(\frac{1}{2})^N$	$C_N^{N-1}(\frac{1}{2})^N$	$(\frac{1}{2})^N$

The simulation result is shown in Fig. 2. 1000 sequences with length $N=8192$ are generated and the initial values are selected independently. The mean of differences is -0.9480 and the standard variance is 91.8419, while the theoretical values are 0 and 90.5097, which indicate that the binary chaotic key stream have excellent balance performance.

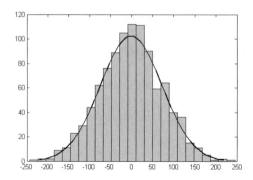

Fig. 2. Distribution of the differences between amounts of 0 and 1

3.3 Correlativity Performance Analysis

Lemma 2. For any two different key stream $\{z_{k1}\}$ and $\{z_{k2}\}$, they satisfy: When N is large enough and m is relatively small, the side lobes of auto correlation and the values of cross correlation of $\{z_{k1}\}$ and $\{z_{k2}\}$ obey normal distribution $N(0, 1/N)$.

The simulation results of the auto/cross correlation functions and the distribution of auto/cross correlation side lobes are shown in Fig. 3 and 4. The sequence length is 8192 and the initial values are selected as 0.60000 and 0.60001 to identify its initial parameter sensitive property.

3.4 Security Performance Comparison

Security performance comparison of different image encryption schemes is shown in Table 2. Attack (I) represents the neural network based adaptive parameter synchronous attack. Attack (II) represents the reverse iteration based chaotic system reconstruction attack. Attack (III) represents the known plaintext attack.

Table 2. Security comparison of different image encryption schemes

Schemes	DES	AES	Pure position permutation	Single chaos based	This paper Proposed
Type	block cipher	block cipher	—	stream cipher	stream cipher
Suitable for image	N	N	Y	Y	Y
Key space	2^{64}	2^{128}	—	2^{53}	2^{158}
Period	—	—	—	M	$2M^2 + M$
Anti attack(I)	—	—	—	N	Y
Anti attack(II)	—	—	—	N	Y
Anti attack(III)	Y	Y	N	Y	Y

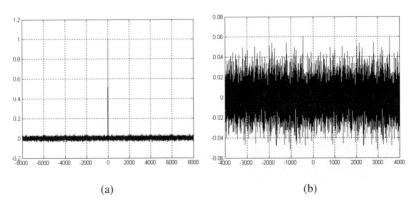

(a)   (b)

Fig. 3. (a) The auto correlation function, (b) The cross correlation function

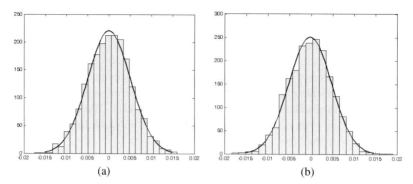

Fig. 4. (a) Distribution of auto correlation side lobes, (b) Distribution of cross correlation

From Table 2 we can see, compared with the single chaos map based ones, the nonlinear transform based key stream generator proposed in this paper greatly enlarges the key space and extends the period of the key stream under precision restricted condition, which make the complete key search impossible and improves the correlation performance of the key stream. The nonlinear transform of three logistic maps also enhances the complexity of the key stream, which makes it more secure against neural network based adaptive parameter synchronous attack and reverse iteration based chaotic system reconstruction attack.

4   Experimental Results

We take 256×256 size 8 bits Lenna image as example, the two pseudorandom key streams used for position permutation and grayscale substitution are generated by scheme proposed in Section 3.1. Original image and its histogram are shown in Fig.5, and the encrypted image and its histogram is shown in Fig. 6. The initial parameters (key pair) are selected as $k = \{0.6, 0.7, 0.8\}$, note k must be far more complex in practical use. From Fig.6 we can see, the grayscale distribution of the encrypted image has good balance property, which is secure against known plaintext attack.

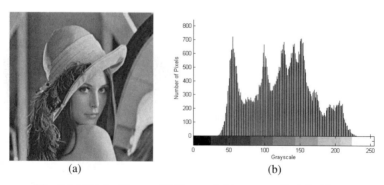

Fig. 5. (a) Original image, (b) Grayscale histogram of original image

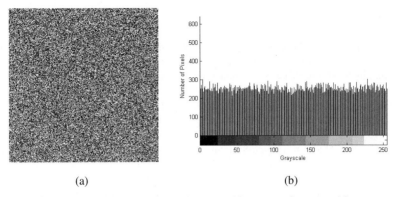

(a)   (b)

Fig. 6. (a) Encrypted image, (b) Grayscale histogram of encrypted image

Fig. 7(a) is correct decrypted image and Fig. 7(b) is incorrect decrypted image with key $k' = \{0.60001, 0.7, 0.8\}$, from which we can see that the original image can not be restored even if with a minus difference due to the extreme initial parameter sensitive property of a chaos system.

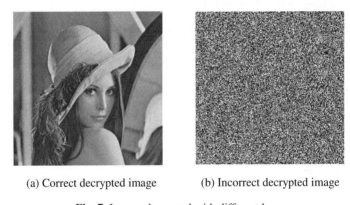

(a) Correct decrypted image   (b) Incorrect decrypted image

Fig. 7. Images decrypted with different keys

5   Conclusion

The improved chaos-based key stream generator for image encryption system proposed in this paper greatly extends the key space and improves the linear complexity of the key stream under precision restricted condition over conventional single chaos based schemes. The generator has strong anti attack ability to the commonly used chaos system attack methods such as adaptive parameter chaos synchronization and reverse iteration reconstruction, which enhances the security of a chaos-based encryption system. The nonlinear part of the generator is general and other one dimension chaotic maps such as Chebyshev, Tent, etc. can also be used as analog pseudorandom sequence source.

References

1. Alexopoulos, C. N., Bourbakis G., Ioannou N.: Image Encryption Method Using a Class of Fractals. Journal of Electronic Imaging, Vol. 4 (1995), 251–259
2. Maniccam S. S., Bourbakis N. G.: Image and Video encryption using SCAN patterns. Pattern Recognition, Vol. 37 (2004), 725–737
3. Baptista M. S.: Cryptography with Chaos. Physics Letters A, Vol. 240 (1998), 50–54
4. Alvarez E., Fernandez A.: New Approach to Chaotic Encryption. Physics Letters A, Vol. 263 (1999), 373–375
5. Masuda, N., Aihara K.: CryptoSystems with Discretized Chaotic Maps. IEEE Transactions on Circuits and Systems, Vol. 49 (2002), 28–40
6. Fu C., Wang P. R., MA X.M.: A Fast Pseudo Stochastic Sequence Quantification Algorithm Based on Chebyshev Map and Its Application in Data Encryption. Lecture Notes in Computer Science, Vol. 3991 (2006), 826–829
7. Fridrich J.: Image Encryption Based on Chaotic Maps. IEEE International Conference on Systems, Man, and Cybernetics (1997), 1105–1110
8. Guan Z. H., Huang F. J., Guan W. J.: Chaos-based Image Encryption Algorithm. Physics Letters A, Vol. 346 (2005), 153–157
9. Zhang L. H., Liao X. F., Wang X. B.: An Image Encryption Approach Based on Chaotic Maps. Chaos, Solitons and Fractals, Vol. 24 (2005), 759–765
10. Chen G., Mao Y.: A Symmetric Image Encryption Scheme Based on 3D Chaotic Cat Maps. Chaos, Solitons and Fractals, Vol. 21 (2004), 749–761
11. Jakimoski G., Kocarev L.: Analysis of Some Recently Proposed Chaos-based Encryption Algorithms. Physics Letters A, Vol. 291 (2001), 381–384
12. Li S. J., Mou X. Q., Cai Y. L.: On the Security of a Chaotic Encryption Scheme: Problems with Computerized Chaos in Finite Computing Precision Computer. Physics Communications, Vol. 153 (2003), 52–58
13. Lian S. H., Sun J. S., Wang Z. Q.: Security Analysis of a Chaos-based Image Encryption Algorithm. Physica A, Vol. 351 (2005), 645–661
14. Kohda T., Tsuneda A.: Statistics of Chaotic Binary Sequences. IEEE Transactions on Information Theory, Vol. 43 (1997), 104–112

A Factory Pattern in Fortran 95

Viktor K. Decyk[1] and Henry J. Gardner[2]

[1] Department of Physics and Astronomy, University of California
Los Angeles, CA, 90095-1547, USA
decyk@physics.ucla.edu
[2] Computer Science, FEIT, CECS, Australian National University
Canberra, ACT 0200, Australia
Henry.Gardner@anu.edu.au

Abstract. This paper discusses the concept and application of software
design patterns in Fortran 95-based scientific programming. An example
of a factory pattern is given together with a context in Particle in Cell
plasma simulation.

Keywords: Fortran, design patterns, object oriented, Particle in Cell.

1   Introduction

Object-oriented (OO) design patterns are an aspect of modern software
engineering which is widely accepted as best practice in commercial software
development. The seminal work on this subject is the book by Gamma, Helm,
Johnson, and Vlissides[1], which identifies 23 recurring design patterns together
with example contexts and a discussion of their strengths and disadvantages.
Although many other design patterns texts have appeared subsequently most
of these have had a focus on commercial applications and graphical user interfaces. In the scientific computing and parallel programming communities, there
has been some discussion of design patterns for scientific applications in C++
and Java [1] as well as the recent book by Mattson et al[2]. But, until now, the
treatment of Fortran seems to have been relegated to how legacy code might
be integrated into frameworks written in other languages rather than how patterns might be used in Fortran itself. A discussion of Fortran 95 as a serious
implementation language for design patterns has only just begun to appear[3,4].

Fortran 95 programs can be written in an "object-based" fashion using the
`module` construct to define the fundamental object-oriented concept of a *class*
[5,6,7,8,9,10]. But inheritance is not a feature of Fortran 95, so a fully OO
approach is not possible. Instead module "composition", where one module has
an association, or link, to another with the keyword `uses`, is used to emulate
inheritance. According to Gamma et al.[1], one of the fundamental principles of
OO design patterns is to "favor composition over inheritance" so this motivates

---

[1] For example, in conference series such as the International Symposium on Object-
Oriented Computing in Parallel Environments (ISCOPE).

Y. Shi et al. (Eds.): ICCS 2007, Part I, LNCS 4487, pp. 583–590, 2007.

an exploration of how the essential nature of design patterns might be captured in programs written in Fortran 95.

In this paper, we describe one important pattern together with its computational science context: It has been used to build part of a large software framework for Particle in Cell (PIC) plasma simulation written in Fortran. (For comparison, the design and implementation of an object-oriented Particle in Cell simulation in Java is given in [11].)

2   An Object-Based Electrostatic Particle Simulation

Particle in Cell plasma codes[12] integrate the self-consistent equations of motion of a large number (up to billions!) of charged particles in electromagnetic fields. Their basic structure is to calculate the density of charge, and possibly current, on a fixed grid. Maxwell's equations, or a subset thereof, are solved on this grid and the forces on all particles are calculated using Newton's Law and the Lorentz force. Particle motion is advanced and new densities are calculated at the next time step.

It is a common practice for scientists to build a set of PIC models to study plasma phenomena at differing levels of complexity. At the basic level, an *electrostatic* code models particles that respond to Coulomb forces only. This is then extended to treat *electromagnetic* particles which correspond to both electric and magnetic fields. As the details of the physics are refined, the models can incorporate relativistic effects, differing boundary conditions, differing field solutions, multispecies plasmas and so on. A framework for building PIC models would allow all of these submodels to be generated and for common code to be maintained and reused between them.

We start by creating a Fortran 95 class for electrostatic particles (which respond to Coulomb forces only). This class uses the Fortran 95 `module` to wrap up and reuse legacy subroutines written in Fortran 77. It has the following structure: a type, followed by functions which operate on that type, and, perhaps, shared data[6]. The type declaration describes properties of particles, but it does not actually contain the particle position and velocity data which are stored elsewhere in normal Fortran arrays and are passed to the class in the subroutine argument "`part`". The type stores a particle's charge, `qm`, charge to mass ratio, `qbm`, and the number of particles of that type, `npp`:

```
module es_particles_class
type particles
    integer :: npp
    real :: qm, qbm
end type
contains
subroutine new_es_particles(this, qm, qbm)
! 'this' is of type 'particles'
! set this%npp, this%qm, this%qbm
    ...
subroutine initialize_es_particles(this, part, idimp, npp)
```

```
! initialize particle positions and velocities
      ...
    subroutine charge_deposit (this , part , q )
! deposit particle charge onto mesh
      ...
    subroutine es_push (this , part , fxyz , dt )
! advance particles in time from forces
      ...
    subroutine particle_manager (this , part )
! handle boundary conditions
      ...
    end module es_particles_class
```

Most of the subroutines provide a simple interface to some legacy code. For example, the initialization subroutine assigns initial positions and velocities to the particle array, `part`:

```
    subroutine initialize_es_particles (this , part , idimp , npp )
! initialize positions and velocities
    implicit none
    type ( particles ) :: this
    real , dimension ( : , : ) , pointer :: part
    integer :: idimp , npp
    allocate ( part ( idimp , npp ))
    this%npp = npp
! call legacy initialization subroutine for part
      ...
```

The iteration loop in the main program consists of a charge deposit, a field solver, a particle push subroutine, and a boundary condition check:

```
    program es_main
! main program for electrostatic particles
    use es_particles_class
    implicit none
    integer :: i, idimp = 6, npp = 32768, nx = 32, ny = 32, nz = 32
    integer :: nloop = 1
    real :: qm = 1.0, qbm = 1.0, dt = 0.2
    type ( particles ) :: electrons
    real , dimension ( : , : ) , pointer :: part
    real , dimension ( : , : , : ) , pointer :: charge_density
    real , dimension ( : , : , : , : ) , pointer :: efield
! initialization
    call new_es_particles (electrons , qm, qbm )
    call initialize_es_particles (electrons , part , idimp , npp )
    allocate ( charge_density (nx , ny , nz ) , efield (3 , nx , ny , nz ))
! main loop over number of time steps
    do i = 1, nloop
        call charge_deposit (electrons , part , charge_density )
! omitted : solve for electrostatic fields
        call es_push (electrons , part , efield , dt )
```

```
        call particle_manager(electrons , part)
    enddo
!
    end program es_main
```

3 Extension to Electromagnetic Particles

Now let us consider particles which respond to both electric and magnetic forces. The push is different, and there is a current deposit in addition to the charge deposit. But the initialization, charge deposit, and particle manager are the same as in the electrostatic class and they can be reused. An electromagnetic particle class can be created by "using" the electrostatic class and adding the new subroutines as follows. (This "using" is an example of using object composition in a place where inheritance might be employed in an OO language.)

```
    module em_particles_class
    use es_particles_class
    contains
    subroutine em_current_deposit(this , part , cu , dt)
! deposit particle current onto mesh
    ...
    subroutine em_push(this , part , fxyz , bxyz , dt)
! advance particles in time from electromagnetic forces
    ...
    end module em_particles_class
```

A program where one could select which type of particles to use might first read a flag, emforce, and then use this flag to choose the appropriate routine to execute:

```
    program generic_main
    use es_particles_class
    use em_particles_class
    integer , parameter :: ELECTROSTATIC = 0 , ELECTROMAGNETIC = 1
    ...
    call new_es_particles(electrons ,qm,qbm)
    call initialize_es_particles(electrons , part , idimp ,npp)
    allocate(charge_density(nx , ny , nz) , efield(3,nx , ny , nz))
    if (emforce==ELECTROMAGNETIC) then
        allocate(current(3,nx , ny , nz) , bfield(3,nx , ny , nz))
    endif
    do i = 1, nloop !loop over number of time steps
        if (emforce==ELECTROMAGNETIC) then
            call em_current_deposit(electrons , part , current , dt)
        endif
        call charge_deposit(electrons , part , charge_density)
! omitted: solve for electrostatic or electromagnetic fields
        select case (emforce)
        case (ELECTROSTATIC)
            call es_push(electrons , part , efield , dt)
        case (ELECTROMAGNETIC)
```

```
      call em_push(electrons ,part ,efield ,bfield ,dt)
    end select
    call particle_manager(electrons ,part)
enddo
end program generic_main
```

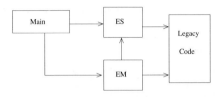

Fig. 1. An electrostatic/electromagnetic particle simulation which reuses es_particles
_class (denoted by "ES")

The design of this program is shown schematically in Fig. 1 where, with
exception of the legacy code, the boxes represent modules and the open arrows
represent "uses" associations between modules. This design is disciplined and
it reuses much of the previous, electrostatic, code, but the widespread use of
select case or if statements can make the main program difficult to read and
also necessitates keeping track of all the different choices if the code should be
extended further.

4   A Factory Pattern

The essential idea of a factory pattern is to encapsulate "creational" logic inside
a dedicated class, or collection of classes. We propose that a factory pattern can
be incorporated into the above example by creating a "generic particle" class
which will create storage for particles of the relevant type and will then ensure
that the correct type of push and current deposit subroutines are chosen for a
given particle type. This can be done by reusing almost all of the earlier version
of the software without modification - save for the addition of the flag, emforce,
into the particles type. The first part of this new class would read:

```
module particles_class
use es_particles_class
use em_particles_class
contains
subroutine new_particles(this ,emforce ,qm,qbm)
implicit none
type (particles) :: this
integer :: emforce
real :: qm, qbm
call new_es_particles(this ,qm,qbm)
this%emforce = emforce
end subroutine new_particles
  ...
```

Within `particles_class`, the particle push routine looks like:

```
subroutine push_particles(this,part,fxyz,bxyz,dt)
! advance particles in time
  implicit none
  type (particles) :: this
  real, dimension(:,:), pointer :: part
  real, dimension(:,:,:,:), pointer :: fxyz, bxyz
  real :: dt
  select case(this%emforce)
  case (ELECTROSTATIC)
     call es_push(this,part,fxyz,dt)
  case (ELECTROMAGNETIC)
     call em_push(this,part,fxyz,bxyz,dt)
  end select
  write (*,*) 'done push_particles'
end subroutine push_particles
```

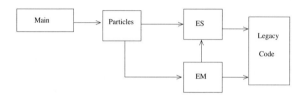

Fig. 2. Representation of the Fortran 95 factory pattern described in the text

The main loop of our refactored program now has the `if` and `select case` statements omitted and the decision making has been delegated to the `particles_class` module. The listing follows and the block diagram is shown in Fig. 2.

```
program main
! main program for various kinds of particles
  use particles_class
  ...
  call new_particles(electrons, emforce, qm, qbm)
  ...
! loop over number of time steps
  do i = 1, nloop
     call current_deposit(electrons, part, current, dt)
     call charge_deposit(electrons, part, charge_density)
! omitted: solve for electrostatic or electromagnetic fields
     call push_particles(electrons, part, efield, bfield, dt)
     call particle_manager(electrons, part)
  enddo
end program
```

How much work would it be to add a third type of particle? For example, suppose we wanted to create relativistic, electromagnetic particles. Relativistic particles need a new component in the particles type, the speed of light, as well as a different push and current deposit subroutine. These two subroutines, as well as a new emforce value, RELATIVISTIC, would be incorporated into a new, *rel_particles_class* class. Two new lines would be added in the generic particles class, to the push and current-deposit subroutines, to allow the selection of relativistic particles. In addition, the constructor would have a new optional argument, the speed of light. Except for the additional argument in the constructor, the main loop would not change at all.

5   Discussion

Figure 3 shows a conventional, object-oriented factory pattern. A *client* has an association with an instance of a *factory* class which is responsible for creating an object from an inheritance hierarchy of target objects. The factory returns a handle on the desired object which, thereafter, calls methods directly on that object. The target object often implements an *interface* which is defined by the top of an inheritance hierarchy. The pattern shown in Fig. 2 differs from the conventional OO factory pattern because of the lack of inheritance, and the lack of conventional, OO interfaces, in Fortran 95. In our Fortran 95 pattern, the particles class is responsible for creating the object of the desired type and also for funneling calls to the correct subroutines for that particular type after it has been created. Still, the present pattern can be recommended to Fortran 95 programmers because it encapsulates and reuses significant portions of code and it manages these away from the main program logic.

In general, the rule for design patterns is to encapsulate what varies. We did this first by writing a general main program which had the if and select case statements explicit. We then encapsulated these statements inside a special class, particles. We have thus demonstrated a simple programming pattern together how it might be used in a believable process of iterative software development. In the complete framework, this process has been extended to model relativistic, multispecies plasmas with varying boundary conditions as well as with varying parallel-programming models.

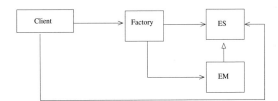

Fig. 3. Representation of a more conventional, object-oriented factory pattern

Acknowledgments

Viktor Decyk acknowledges the support of the US Department of Energy under the SCIDAC program. Henry Gardner acknowledges the support of the Australian Partnership for Advanced Computing (APAC) National Facility and the APAC Education, Outreach and Training program. Further details of this pattern and other patterns in Fortran 95, as well as an introduction to object-based programming in Fortran 95, will be available on the APAC-EOT website (www.apac.edu.au).

References

1. Gamma, E., Helm, R., Johnson, R., Vlissides, J.: Design Patterns: Elements of Reusable Object Oriented Software. Addison-Wesley (1995) ISBN 0201633612.
2. Mattson, T.G., Sanders, B.A., Massingill, B.L.: Patterns for Parallel Programming. Addison-Wesley (2005) ISBN 0321228111.
3. Markus, A.: Design patterns and fortran 90/95. SIGPLAN Fortran Forum **25** (2006) 13–29
4. Gardner, H.J., Decyk, V.K.: Comments on the arjen markus article: Design patterns and fortran. SIGPLAN Fortran Forum **25** (2006) 8–11
5. Gray, M.G., Roberts, R.M.: Object-based programming in fortran 90. Computers in Physics **11** (1997) 355–361
6. Decyk, V.K., Norton, C.D., Szymanski, B.K.: How to express c++ concepts in ₁ortran 90. Scientific Programming **6** (1997) 363–390
7. Decyk, V.K., Norton, C.D., Szymanski, B.K.: Expressing object-oriented concepts in Fortran 90. ACM Fortran Forum **16** (1997) 13–18
8. Cary, J.R., Shasharina, S.G., Cummings, J.C., Reynders, J.V., Hinkler, P.J.: Comparison of c++ and fortran 90 for object-oriented scientific programming. Computer Physics Communications **105** (1997) 20–36
9. Machiels, L., Deville, M.O.: Fortran 90: An entry to object-oriented programming for the solution of partial differential equations. ACM Transactions on Mathematical Software **23** (1997) 32–49
10. Decyk, V.K., Norton, C.D., Szymanski, B.K.: How to support inheritance and runtime polymorphism in fortran 90. Computer Physics Communications **115** (1998) 9–17
11. Markidis, S., Lapenta, G., VanderHeyden, W.: Parsek: Object oriented particle-in-cell implementation and performance issues. In: Proceedings of joint ACM-ISCOPE conference on Java Grande, Seattle, Washington, USA, 3-5 November, 2002, ACM, New York (2002) 141–147
12. Birdsall, C.K., Langdon, A.B.: Plasma Physics via Computer Simulation. Institute of Physics Publishing (1991) ISBN 0750301171.

Mapping Pipeline Skeletons onto Heterogeneous Platforms

Anne Benoit and Yves Robert

LIP, UMR CNRS-INRIA-UCBL 5668, ENS Lyon, France
{Anne.Benoit|Yves.Robert}@ens-lyon.fr

Abstract. Mapping applications onto parallel platforms is a challenging problem, that becomes even more difficult when platforms are heterogeneous –nowadays a standard assumption. A high-level approach to parallel programming not only eases the application developer's task, but it also provides additional information which can help realize an efficient mapping of the application. In this paper, we discuss the mapping of pipeline skeletons onto different types of platforms (from fully homogeneous to heterogeneous). We assume that a pipeline stage must be mapped on a single processor, and we establish new theoretical complexity results for two different mapping policies: a mapping can be either one-to-one (at most one stage per processor), or interval-based (interval of consecutive stages per processor). We provide several efficient polynomial heuristics for the most important policy/platform combination, namely interval-based mappings on platforms with identical communication links but different speed processors.

Keywords: Pipeline skeleton, scheduling algorithms, throughput optimization, heterogeneous platforms, complexity results.

1 Introduction

Mapping applications onto parallel platforms is a difficult challenge. Several scheduling and load-balancing techniques have been developed for homogeneous architectures (see [8] for a survey) but the advent of heterogeneous clusters has rendered the mapping problem even more difficult. Typically, such clusters are composed of different-speed processors interconnected either by plain Ethernet (the low-end version) or by a high-speed switch (the high-end counterpart), and they constitute the experimental platform of choice in most academic or industry research departments. In this context, a structured programming approach rules out many of the problems which the low-level parallel application developer is usually confronted to, such as deadlocks or process starvation. Moreover, many real applications draw from a range of well-known solution paradigms, such as pipelined or farmed computations. High-level approaches based on algorithmic skeletons [4,6] identify such patterns and seeks to make it easy for an application developer to tailor such a paradigm to a specific problem.

In this paper, we focus on the pipeline skeleton, which is one of the most widely used. Consecutive tasks are input to the first stage and progress from

Y. Shi et al. (Eds.): ICCS 2007, Part I, LNCS 4487, pp. 591–598, 2007.

stage to stage until the final result is computed. Each stage has its own communication and computation requirements: it reads an input file from the previous stage, processes the data and outputs a result to the next stage. The pipeline operates in synchronous mode: after some latency due to the initialization delay, a new task is completed every period. The period is defined as the longest cycle-time to operate a stage. The problem of mapping pipeline skeletons onto parallel platforms has received some attention. In particular, Subhlok and Vondran [9,10] have dealt with this problem on homogeneous platforms. In this paper, we extend their work and target heterogeneous clusters. Our main goal is to assess the additional complexity induced by the heterogeneity of processors, and/or of communication links. As in [9], we aim at deriving optimal mappings, i.e. mappings which minimize the period of the system. Each pipeline stage can be seen as a sequential procedure which may perform disc accesses or write data in the memory for each task. This data may be reused from one task to another, hence tasks must be processed in sequential order within a stage. Moreover, due to the possible local memory accesses, a given stage must be mapped onto a single processor: we cannot process half of the tasks on a processor and the remaining tasks on another without exchanging intra-stage information, which might be costly and difficult to implement. Therefore a processor that is assigned a stage will execute the operations required by this stage for all the tasks fed into the pipeline. The optimization problem can be stated informally as follows: which stage to assign to which processor? We consider two main variants, in which we require the mapping to be one-to-one (a processor is assigned at most one stage), or interval-based (a processor is assigned an interval of consecutive stages).

In addition to these two mapping strategies, we target three different platform types: (i) *Fully Homogeneous* platforms have identical processors and interconnection links. (ii) *Communication Homogeneous* platforms, with identical links but different speed processors; and (iii) *Fully Heterogeneous* platforms, with different speed processors and different capacity links. The main objective of the paper is to assess the complexity of each mapping variant onto each platform type. We establish several new complexity results for this important optimization problem, and we derive efficient polynomial heuristics for the most important policy/platform combination, namely interval-based mappings on *Communication Homogeneous* platforms.

The rest of the paper is organized as follows. Section 2 presents target optimization problems. Next in Section 3 we proceed to the complexity results. In Section 4 we introduce several heuristics to solve the mapping problem. These heuristics are compared through simulations, whose results are analyzed in Section 5. Due to lack of space, related work is not discussed in this paper: please refer to the extended version [1] for a full overview of relevant literature.

2   Framework

Applicative Framework. We consider a pipeline of n stages $\mathcal{S}_k$, $1 \leq k \leq n$, as illustrated on Figure 1. The k-th stage $\mathcal{S}_k$ receives an input from the previous

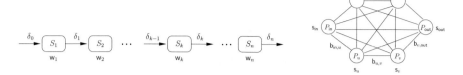

Fig. 1. The application pipeline

Fig. 2. The target platform

stage, of size δ_{k-1}, performs a number of w_k computations, and outputs data of size δ_k to the next stage.

Target Platform. We target a heterogeneous platform (see Figure 2), with p processors P_u, $1 \leq u \leq p$, fully interconnected as a (virtual) clique. There is a bidirectional link $\text{link}_{u,v} : P_u \to P_v$ between any processor pair P_u and P_v, of bandwidth $b_{u,v}$. Communications contention is taken care of by enforcing the *one-port* model [2]. In this model, a given processor can be involved in a single communication at any time-step, either a send or a receive. However, independent communications between distinct processor pairs can take place simultaneously. The one-port model seems to fit the performance of some current MPI implementations, which serialize asynchronous MPI sends as soon as message sizes exceed a few megabytes [7]. The speed of processor P_u is denoted as s_u, and it takes X/s_u time-units for P_u to execute X floating point operations. We also enforce a linear cost model for communications, hence it takes $X/b_{u,v}$ time-units to send (resp. receive) a message of size X to (resp. from) P_v. We assume that additional processors P_{in} and P_{out} are devoted to input/output data.

We classify particular platform categories: (i) *Fully Homogeneous*– Platforms with identical processors ($s_u = s$) and links ($b_{u,v} = b$). They represent typical parallel machines. (ii) *Communication Homogeneous*– Platforms with different-speed processors ($s_u \neq s_v$) interconnected by links of same capacities ($b_{u,v} = b$). They correspond to networks of workstations with plain TCP/IP interconnects or other LANs. (iii) *Fully Heterogeneous*– Platforms with $s_u \neq s_v$ and $b_{u,v} \neq b_{u',v'}$. Hierarchical platforms made up with several clusters interconnected by slower backbone links can be modeled this way.

Mapping Problem. The general mapping problem consists in assigning application stages to platform processors. Two different mapping strategies are discussed below.

ONE-TO-ONE MAPPING. Assume that each stage $\mathcal{S}_k$ of the application pipeline is mapped onto a distinct processor $P_{\text{alloc}(k)}$ (which is possible only if $n \leq p$). What is the period of $P_{\text{alloc}(k)}$, i.e. the minimum delay between the processing of two consecutive tasks? To answer this question, we need to know which processors the previous and next stages are assigned to. Let $t = \text{alloc}(k-1)$, $u = \text{alloc}(k)$ and $v = \text{alloc}(k+1)$. P_u needs $\delta_{k-1}/b_{t,u}$ to receive the input data from P_t, w_k/s_u to process it, and $\delta_k/b_{u,v}$ to send the result to P_v, hence a cycle-time of

$\delta_{k-1}/b_{t,u} + w_k/s_u + \delta_k/b_{u,v}$ for P_u. The *period* achieved with the mapping is the maximum of the cycle-times of the processors.

The optimization problem can be stated as follows: determine a one-to-one allocation function alloc : $[1,n] \rightarrow [1,p]$ (augmented with alloc(0) = in and alloc($n+1$) = out) such that the period T_{period} is minimized, where:

$$T_{\text{period}} = \max_{1 \leq k \leq n} \left\{ \frac{\delta_{k-1}}{b_{\text{alloc}(k-1),\text{alloc}(k)}} + \frac{w_k}{s_{\text{alloc}(k)}} + \frac{\delta_k}{b_{\text{alloc}(k),\text{alloc}(k+1)}} \right\} \quad (1)$$

INTERVAL MAPPING. One-to-one mappings may be unduly restrictive. A natural extension is to search for interval mappings, i.e. allocation functions where each participating processor is assigned an interval of consecutive stages. Intuitively, assigning several consecutive tasks to the same processors will increase their computational load, but may well dramatically decrease communication requirements. In fact, the best interval mapping may turn out to be a one-to-one mapping, or instead may enroll only a very small number of fast computing processors interconnected by high-speed links. Interval mappings constitute a natural and useful generalization of one-to-one mappings. To express this optimization problem, we need to enforce that the intervals achieve a partition of the original set of stages $\mathcal{S}_1$ to $\mathcal{S}_n$. We search for a partition of $[1..n]$ into m intervals $I_j = [d_j, e_j]$ such that $d_j \leq e_j$ for $1 \leq j \leq m$, $d_1 = 1$, $d_{j+1} = e_j + 1$ for $1 \leq j \leq m-1$ and $e_m = n$. Interval I_j is mapped onto processor $P_{\text{alloc}(j)}$, and the period is expressed as

$$T_{\text{period}} = \max_{1 \leq j \leq m} \left\{ \frac{\delta_{d_j - 1}}{b_{\text{alloc}(j-1),\text{alloc}(j)}} + \frac{\sum_{i=d_j}^{e_j} w_i}{s_{\text{alloc}(j)}} + \frac{\delta_{e_j}}{b_{\text{alloc}(j),\text{alloc}(j+1)}} \right\} \quad (2)$$

Here, we assume that alloc(0) = in and alloc($m+1$) = out. The search is over all possible partitions into intervals, and over all processor assignments.

3   Complexity Results

To the best of our knowledge, this work is the first to study the complexity of the ONE-TO-ONE MAPPING and INTERVAL MAPPING strategies, for each of the different platform categories (*Fully Homogeneous, Communication Homogeneous* and *Fully Heterogeneous*). Table 1 summarizes all our new results.

For *Fully Homogeneous* or *Communication Homogeneous* platforms, determining the optimal ONE-TO-ONE MAPPING can be achieved through a binary search over possible periods, invoking a greedy algorithm at each step. The problem surprisingly turns out to be NP-hard for *Fully Heterogeneous* platforms. The INTERVAL MAPPING problem is more complex to deal with. For *Fully Homogeneous* platforms we simply recall the optimal dynamic programming algorithm of Subhlok and Vondran [9]. For *Communication Homogeneous* platforms the problem turns out to be NP-hard. Quite interestingly, this result is a consequence of the fact that the natural extension of the chains-to-chains problem [5] to different-speed processors is NP-hard[1]. Finally, both optimization problems are NP-hard for *Fully Heterogeneous* platforms.

The proof of all these results, as well as the description of the binary search and dynamic programming algorithms, are provided in the extended version [1].

Table 1. Complexity results for the different instances of the mapping problem

	Fully Homogeneous	*Comm. Homogeneous*	*Fully Heterogeneous*
One-to-one	polyn. (bin. search)	polyn. (bin. search)	NP-complete
Interval	polyn. (dyn. prog. [9])	NP-complete[1]	NP-complete

4   Heuristics

In this section several heuristics for *Communication Homogeneous* platforms are presented. We restrict to such platforms because, as already pointed out in Section 1, clusters made of different-speed processors interconnected by either plain Ethernet or a high-speed switch constitute the typical experimental platforms in most academic or industry research departments.

BS121: Binary Search for One-to-One Mapping– This heuristic implements the optimal polynomial algorithm for the ONE-TO-ONE MAPPING case [1]. When $p < n$, we cut the application in fixed intervals of size $L = \lceil n/p \rceil$.

SPL: Splitting Intervals – This heuristic sorts the processors by decreasing speed, and starts by assigning all the stages to the first processor in the list. This processor becomes used. Then, at each step, we select the used processor j with the largest period and we try to split its interval of stages, giving some stages to the next fastest processor j' in the list (not yet used). This can be done by splitting the interval at any place, and either placing the first part of the interval on j and the remainder on j', or the other way round. The solution which minimizes $max(period(j), period(j'))$ is chosen if it is better than the original solution. Splitting is performed as long as we improve the period of the solution.

BSL and BSC: Binary Search (Longest/Closest) – The last two heuristics perform a binary search on the period of the solution. For a given period P, we study if there is a feasible solution, starting with the first stage ($s = 1$) and constructing intervals (s, s') to fit on processors. For each processor u, and each $s' \geq s$ we compute the period (s, s', u) of stages $s..s'$ running on processor u and check whether it is smaller than P (then it is a possible assignment). The first variant **BSL** choose the longest possible interval (maximizing s') fitting on a processor for a given period, and in case of equality, the interval and processor

---

[1] For the sake of completeness, this result is mentioned here, but it was not available at the date of the original submission. Please refer to the extended version [1] for further information on the chains-to-chains problem [3,5], and the NP-completeness of its generalization to heterogeneous platforms.

with the closest period to the solution period. The second variant **BSC** does not take into account the length of the interval, but only finds out the closest period.

The code for these heuristics and more heuristics described in [1] can be found on the Web at: `http://graal.ens-lyon.fr/~abenoit/code/sh.c`

5   Experiments

Several experiments have been conducted in order to assess the performance of the previous heuristics. We have generated a set of random applications with $n = 1$ to 50 stages, and two sets of random platforms, one set with $p = 10$ processors and the other with $p = 100$ processors. The first case corresponds to a situation in which several stages are likely to be mapped on the same processor because there are much fewer processors than stages. However, in the second case, we expect the mapping to be a ONE-TO-ONE MAPPING, except when communications are really costly. The heuristics have been designed for *Communication Homogeneous* platforms, so we restrict to such platforms in these experiments. Although there are four categories of parameters to play with, i.e. the values of δ, w, s and b, we can see from equation (2) that only the relative ratios $\frac{\delta}{b}$ and $\frac{w}{s}$ have an impact on the performance.

Each experimental value reported in the following has been calculated as an average over 100 randomly chosen application/platforms pairs. We report four main sets of experiments. For each of them, we vary some key application/platform parameter to assess the impact of this parameter on the performance of the heuristics. The first two experiments deal with applications where communications and computations have the same order of magnitude, and we study the impact of the degree of heterogeneity of the communications: in the first experiment the communication are homogeneous, while in the second one they are heterogeneous. The last two experiments deal with imbalanced applications: the third experiment assumes large computations (large value of the w to δ ratio), while the fourth one reports results for small computations.

Please refer to the extended version [1] for a detailed description of each set of experiments and the full results description. We first point out that our heuristics are always much more efficient than random or greedy mappings described in [1], and which do not appear in these curves for readability reasons. Each of our heuristics may turn out to be the most efficient, depending upon the application and platform characteristics. When there are more processors than pipeline stages, we can expect that a ONE-TO-ONE MAPPING is a good choice. If communications are homogeneous or less important than the computational part, the optimal binary search BS121 should be used (Experiments 1 and 3, with $p = 100$, see for instance Figure 3). With similar communications but fewer processors (Experiments 1 and 3, $p = 10$, see for instance Figure 4), it is necessary to share the computation load between processors, and the decision where to split intervals can be really relevant. Since BS121 is using intervals of fixed length, it cannot make any clever choices. In such cases, BSC is the most efficient heuristic. The third case is when communications are costly or with a

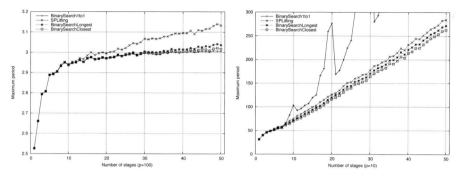

Fig. 3. Experiment 1, $p = 100$ **Fig. 4.** Experiment 3, $p = 10$

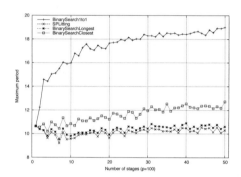

Fig. 5. Experiment 2, $p = 100$

high degree of heterogeneity, as in Experiments 2 and 4 (see for instance Figure 5). Then BS121 do not return satisfying results, because it may cut intervals on costly links between stages (independently of the number of processors). In such cases, the splitting heuristic SPL is the best choice.

6   Conclusion

In this paper, we have thoroughly studied the difficult problem of mapping applications which have a pipeline structure onto heterogeneous platforms. To the best of our knowledge, it is the first time that pipeline mapping is studied from a theoretical perspective, while it is quite a standard and widely used pattern in many real-life applications. The optimal algorithms and complexity results provided in this paper for the different mapping strategies and platform types represent a significant contribution to the theoretical analysis of this important optimization problem. In addition to the optimal algorithms for *Fully Homogeneous* platforms, we designed a set of efficient polynomial heuristics for the ONE-TO-ONE MAPPING and INTERVAL MAPPING strategies on *Communication Homogeneous* platforms. Finally, we point out that the absolute performance of the heuristics is quite good, since their result is close to the optimal solution returned by an integer linear program. Due to a lack of space, this last important result is discussed in [1].

There remains much work to extend the results of this paper. On the practical side, we still need to design efficient heuristics for *Fully Heterogeneous* platforms, which is a challenging problem. Also, we would like to further assess the interest of fully general mappings, where a processor may be assigned several nonconsecutive stages (hence which do not form an interval). Such mappings, which are discussed in the extended version [1], require a multi-port communication model instead of the one-port model used in this paper. In the longer term, we plan to perform real experiments on heterogeneous platforms, using an already-implemented skeleton library, in order to compare the effective performance of the application for a given mapping (obtained with our heuristics) against the theoretical performance of this mapping. A natural extension of this work would be to consider other widely used skeletons: when there is a bottleneck in the pipeline operation due to a stage which is both computationally-demanding and not constrained by internal dependencies, we can split the workload of this stage among several processors. Extending our mapping strategies to automatically identify opportunities for such *deal* skeletons [4], and implement these, is a difficult but very interesting perspective.

References

1. Benoit, A., Robert, Y.: Mapping pipeline skeletons onto heterogeneous platforms. Research Report 2007-05, LIP, ENS Lyon, France, available at graal.ens-lyon.fr/~yrobert/ (2007)
2. Bhat, P., Raghavendra, C., Prasanna, V.: Efficient collective communication in distributed heterogeneous systems. Journal of Parallel and Distributed Computing **63** (2003) 251–263
3. Bokhari, S.H.: Partitioning problems in parallel, pipeline, and distributed computing. IEEE Trans. Computers **37(1)** (1988) 48–57
4. Cole, M.: Bringing Skeletons out of the Closet: A Pragmatic Manifesto for Skeletal Parallel Programming. Parallel Computing **30(3)** (2004) 389–406
5. Pinar, A., Aykanat, C.: Fast optimal load balancing algorithms for 1D partitioning. J. Parallel Distributed Computing, **64(8)** (2004) 974–996
6. Rabhi, F., Gorlatch, S.: Patterns and Skeletons for Parallel and Distributed Computing. Springer Verlag (2002)
7. Saif, T., Parashar, M.: Understanding the behavior and performance of non-blocking communications in MPI. In: Proceedings of Euro-Par 2004: Parallel Processing, LNCS 3149, Springer (2004) 173–182
8. Shirazi, B.A, Hurson, A.R, Kavi, K.M.: Scheduling and load balancing in parallel and distributed systems. IEEE Computer Science Press (1995)
9. Subhlok, J., Vondran, G.: Optimal mapping of sequences of data parallel tasks. In: Proc. 5th ACM SIGPLAN Symposium on Principles and Practice of Parallel Programming, PPoPP'95, ACM Press (1995) 134–143
10. Subhlok, J., Vondran, G.: Optimal latency-throughput tradeoffs for data parallel pipelines. In: ACM Symposium on Parallel Algorithms and Architectures SPAA'96, ACM Press (1996) 62–71

On the Optimal Object-Oriented
Program Re-modularization

Saeed Parsa and Omid Bushehrian

Faculty of Computer Engineering, Iran University of Science and Technology
{parsa,bushehrian}@iust.ac.ir

Abstract. In this paper a new criterion for automatic re-modularization of object-oriented programs is presented. The aim of re-modularization here is to determine a distributed execution of a program over a dedicated network of computers with the shortest execution time. To achieve this, a criterion to quantitatively evaluate performance of a re-modularized program is presented as a function. This function is automatically constructed while traversing the program call flow graph once before the search for the optimal re-modularization of the program and considers both synchronous and asynchronous types for each call within the call flow graph.

1 Introduction

With the increasing popularity of using clusters and network of low cost computers in solving computationally intensive problems, there is a great demand for system and application software that can provide transparent and efficient utilization of the multiple machines in a distributed system [2][3][5]. There are a number of such application softwares including middle-wares and utility libraries which support parallel and distributed programming over a network of machines. A distributed program written using these middle-wares comprises a number of modules or distributed parts communicating by means of message passing or asynchronous method calls.

Our aim has been to develop automatic techniques to obtain maximum execution concurrency among distributed parts or modules of a program. To reach this end, the main difficulty is to determine theses distributed parts, or equivalently, the architecture of a distributed program code. The architecture of a program can be reconstructed using software reverse engineering and re-modularization techniques [1][4].

2 The Optimal Re-modularization of a Program

Each clustering of a program call graph, which is a modularization of that program, represents a subset of program method calls, named *remote-call* set, to be converted to remote asynchronous calls. For instance consider the modularized call graph of four classes in Figure 1. This modularization corresponds to *remote-call* set {c1, c2, c4}.

Y. Shi et al. (Eds.): ICCS 2007, Part I, LNCS 4487, pp. 599–602, 2007.
© Springer-Verlag Berlin Heidelberg 2007

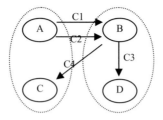

Fig. 1. Re-modularization of a program call graph

3   Performance Estimation of a Re-modularized Program

In order to evaluate a re-modularized program performance, the *remote-call* set corresponding to that re-modularization is obtained and evaluated by applying a function called *Estimated Execution Time* (*EET*). For a given *remote-call* set r, $EET_I(r)$ calculates a value which is an estimation of the amount of execution time of method call I with respect to r. Each *EET* formula is generated from the program call flow graph (CFG). CFG shows the flow of method calls among program classes. Each node in this graph represents a method body in an abstract way by means of a sequence of symbols. Each symbol in this sequence indicates one of these concepts: a method invocation, a synchronization point between caller and callee methods or an ordinary program instruction which are denoted by I_i, S_i and W_i respectively. Symbol S_i indicates the first program location which is data dependent to a method invocation I_i in the CFG node sequence. Symbol W_i represents any collection of ordinary program statements with estimated execution time i. Below in Figure 2 is a sample Java code and its corresponding CFG. EET function for a program is generated automatically by traversing the program CFG. Since each method invocation I_i in the program CFG may be executed either synchronously or asynchronously, depending on the specified modularization of the program classes, the EET function includes time estimation for both synchronous and asynchronous execution types for each invocation I_i. For instance the EET function for CFG in Figure 2 is generated as follows:

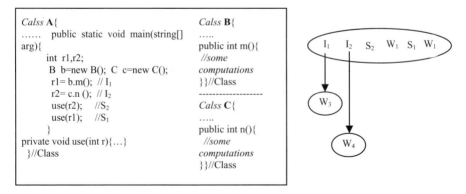

Fig. 2. A sample program including three classes and its CFG

$$EET_{main}(r) = a_1 * EET_{I1}(r) + a_2 * EET_{I2}(r) + (1-a_2) * T(S_2) + W_1 + (1-a_1) * T(S_1)$$
$$+ W_1 \quad EET_{I1}(r) = W_3 , EET_{I2}(r) = W_4 , \tag{1}$$

In this relation, depending on the execution type of invocations I_1 and I_2, asynchronous or synchronous, coefficients a_1 and a_2 are set to 0 or 1 respectively. S_1 and S_2 are synchronization points of calls I_1 and I_2 respectively and $T(S_i)$ indicates the amount of time that should be elapsed at synchronization point S_i until invocation I_i is completed.

The general form of an EET relation for a program is as follows:

$$EET_m(r) = \sum w_i + \sum a_i * EET_{Ii}(r) + \sum (1-a_i) * T(S_i) \tag{2}$$

In the above formula coefficients a_i are determined by *remote-call* set r as follows:

$$a_i = \begin{cases} 1 & : I_i \notin r \\ 0 & : I_i \in r \end{cases}$$

As described above, $T(S_i)$ is the amount of time that should be elapsed at synchronization point S_i until invocation I_i is completed. $T(S_i)$ is calculated by the following relation:

$$T(S_i) = \max((EET_{Ii}(r) + O_i) - t_i , 0) \tag{3}$$

Where, t_i is estimated execution time of the program fragment between symbols I_i and S_i. Since each asynchronous method invocation I_i imposes a communication overhead on the overall program execution time, this overhead which is denoted by O_i, is added to the estimated execution time of I_i. Since it is assumed that CFG is cycle free, $EET_m(r)$ can be solved by recursively replacing EET terms until $EET_m(r)$ contains only a_i coefficients, W_i terms, O_i terms and *max* operators.

4 Conclusions

The main difficulty in obtaining a distributed execution of a program with minimum execution time is to find the smallest set of program invocations to be converted to remote asynchronous invocations. Program re-modularization can be applied as an approach to reach this end. Program re-modularization is used to reconstruct program architecture with respect to one ore more quality constraints such as performance or maintainability. In this paper a new criterion for performance driven re-modularization of a program has been proposed. This criterion is used to quantitatively estimate the performance of a re-modularized program with a function which is generated automatically from the program call flow graph (CFG). This function includes time estimations for both asynchronous and synchronous execution types of each method in the program call flow graph.

References

1. Berndt Bellay, Harald Gall, "Reverse Engineering to Recover and Describe a Systems Architecture", Development and Evolution of Software Architectures for Product Families, Lecture Notes in Computer Science(1998), Volume 1429 .
2. Bushehrian Omid, Parsa Saeed, "Formal Description of a Runtime Infrastructure for Automatic Distribution of Programs", The 21th International Symposium on Computer and Information Sciences, Lecture Notes in Computer Science(2006), Vol. 4263.
3. Jameela Al-Jaroodi, Nader Mahamad, Hong Jiang, David Swanson, "JOPI: a Java Object Passing Interface", Concurrency Computat. : Pract. Exper. (2005); 17:775–795
4. Parsa S. , Bushehrian O., "The Design and Implementation of a Tool for Automatic Software Modularization", Journal of Supercomputing, Volume 32, Issue 1, April (2005).
5. Parsa Saeed, Khalilpour Vahid, "Automatic Distribution of Sequential Code Using JavaSymphony Middleware", SOFSEM06, Lecture Notes in Computer Science(2006), Vol. 3831,.

A Buffered-Mode MPI Implementation for the Cell BE™ Processor

Arun Kumar[1], Ganapathy Senthilkumar[1], Murali Krishna[1], Naresh Jayam[1],
Pallav K. Baruah[1], Raghunath Sharma[1], Ashok Srinivasan[2], and Shakti Kapoor[3]

[1] Dept. of Mathematics and Computer Science, Sri Sathya Sai University,
Prashanthi Nilayam, India.
[2] Dept. of Computer Science, Florida State University.
[3] IBM, Austin

Abstract. The Cell Broadband Engine™ is a heterogeneous multi-core architecture developed by IBM, Sony and Toshiba. It has eight computation intensive cores (SPEs) with a small local memory, and a single PowerPC core. The SPEs have a total peak single precision performance of 204.8 Gflops/s, and 14.64 Gflops/s in double precision. Therefore, the Cell has a good potential for high performance computing. But the unconventional architecture makes it difficult to program. We propose an implementation of the core features of MPI as a solution to this problem. This can enable a large class of existing applications to be ported to the Cell. Our MPI implementation attains bandwidth up to 6.01 GB/s, and latency as small as 0.41 µs. The significance of our work is in demonstrating the effectiveness of intra-Cell MPI, consequently enabling the porting of MPI applications to the Cell with minimal effort.

Keywords: MPI, Cell processor, heterogeneous multi-core processors.

1 Introduction

The Cell is a heterogeneous multi-core processor targeted at the gaming industry. There is also much interest in using it for high performance computing. Some compute intensive math kernels have shown very good performance on the Cell [1], demonstrating its potential for scientific computing. However, due to its unconventional programming model, applications need to be significantly changed in order to exploit the full potential of the Cell. As a solution to the programming problem, we provide an implementation of core features of MPI 1, which uses each SPE as if it were a node for an MPI process. This will enable the running of the large code base of existing MPI applications.

Each SPE has a small (256 KB) local store that it can directly operate on, and access to a larger common main memory from/to which it can DMA data. If one attempts to directly port an application to the SPE, then the small size of the SPE local store poses a significant problem. For applications with data larger than the local store size, a software-controlled cache approach is feasible, wherein the data is actually in the main memory and moved to the local store as needed. Some features to automate this process are expected in the next releases of the compiler and operating system [2,3]. These features would allow the porting of applications in a more generic

Y. Shi et al. (Eds.): ICCS 2007, Part I, LNCS 4487, pp. 603–610, 2007.

fashion, except for the parallel use of all the SPEs. Our MPI implementation handles the parallelization aspect too. We have hand-coded some large-data applications in order to analyze the performance of our implementation. For small memory applications, we have modified our MPI implementation so that such applications can be ported right now, without needing the compiler and operating system support expected in the near future.

Existing MPI implementations for the shared memory architectures cannot be directly ported to the Cell, because the SPEs have somewhat limited functionality. For example, they cannot directly operate on the main memory – they need to explicitly DMA the required data to local store and then use it. In fact, they cannot even dynamically allocate space in main memory.

Heterogeneous multi-core processors show much promise in the future. In that perspective, we expect the impact of this work to be much broader than just for the Cell processor, by demonstrating the feasibility of MPI on heterogeneous multi-core processors.

The rest of the paper is organized as follows. In §2, the architectural features of the Cell processor that are relevant to the MPI implementation are described. Our MPI implementation is described in §3. The performance results are presented in §4. We discuss related work in §5. We then describe limitations of the current work, and future plans to overcome these limitations, in §6, followed by conclusions in §7.

2   Cell Architecture

The Cell processor consists of a cache coherent PowerPC core and eight SPEs running at 3.2 GHz. All of them execute instructions in-order. It has a 512 MB to 2 GB external main memory, and an XDR memory controller provides access to it at a rate of 25.6 GB/s. The PPE, SPEs, DRAM controller, and I/O controllers are all connected via four data rings, collectively known as the EIB. Multiple data transfers can be in progress concurrently on each ring, including more than 100 outstanding DMA memory requests between main storage and the SPEs. Simultaneous transfers on the same ring are also possible. The EIB's maximum intra-chip bandwidth is 204.8 GB/s.

Each SPE has its own 256 KB local memory from which it fetches code and reads and writes data. Access latency to and from local store is 6 cycles [4] (page 75, table 3.3). All loads and stores issued from the SPE can only access the SPE's local memory. Any data needed by the SPE that is present in the main memory must be moved into the local store explicitly, in software, through a DMA operation. DMA commands may be executed out-of-order.

In order to use the SPEs, a process running on the PPE can spawn a thread that runs on the SPE. The SPE's local store and registers are mapped onto the effective address space of the process that spawned the SPE thread. Data can be transferred from the local store of one SPE to the local store or special registers of another SPE by obtaining these memory mapped addresses.

3   MPI Design

In this section, we describe our basic design for the blocking point to point communication. We also describe the application start-up process. We have not described the handling of errors, in order to present a clearer high-level view of our implementation.

3.1 MPI Initialization

A user can run an MPI application, provided it uses only features that we have currently implemented, by compiling the application for the SPE and executing the following command on the PPE:

mpirun –n <N> executable arguments

where $<N>$ is the number of SPEs on which the code is to be run. The mpirun process spawns the desired number of threads on the SPEs. Note that only one thread can be spawned on an SPE, and so $<N>$ cannot exceed eight on a single processor or sixteen for a blade. We have not considered latencies related to the NUMA aspects of the architecture in the latter case.

Note that the data for each SPE thread is distinct, and not shared, unlike in conventional threads. The MPI operations need some common shared space through which they can communicate, as explained later. This space is allocated by mpirun. This information, along with other information, such as the rank in MPI_COMM_WORLD, and the command line arguments, are passed to the SPE threads by storing them in a structure and sending a mailbox message[1] with the address of this structure. The SPE threads receive this information during their call to MPI_Init. The PPE process is not further involved in the application until the threads terminate, when it cleans up allocated memory and then terminates. It is important to keep the PPE as free as possible for good performance, because it can otherwise become a bottleneck.

3.2 Point-to-Point Communication

We describe the implementation of point-to-point communication. Collective operations were implemented on top of the point-to-point operations.

Communication Architecture. Let N be the number of MPI nodes. The mpirun process allocates N message buffers in main memory in its address space. Since the SPE threads are part of the mpirun process, they can access the buffers allocated by mpirun. Buffer P_i is used by SPE i to copy its data, when SPE i sends a message. Thus, even though SPEs cannot dynamically allocate space in main memory, they can use space allocated by mpirun. SPE i manages the piece of memory P_i, to allocate space within P_i for its send calls.

The sender and receiver processes communicate information about messages through meta-data queues maintained in SPE local stores. There are $N*(N-1)$ queues in total. Queue Q_{ij} is used by SPE i to send information about a message to SPE j. Q_{ij} is present in the local store of SPE j (the receiver). Thus each SPE maintains $N-1$ queues. These queues are organized as circular arrays. Each entry in Q_{ij} contains information about the location of the message within P_i, the message tag, the data type, message size, communicator identifier and flag bits. The total size of an entry is 16 bytes. The flag field is either *valid*, indicating that the entry contains information on a sent message that has not been received, or is *invalid*, indicating that it is free to be written by the sending SPE. All entries are initialized to *invalid*.

---

[1] A mailbox message is a fast communication mechanism for 32-bit messages.

Send Protocol. The send operation from SPE i to SPE j, shown in fig. 1 (a), proceeds as follows: The send operation first finds the offset of a free block in buffer P_i, managed by it. The message data is copied into this location in P_i. The copying is done through DMA operations. Since the SPE's data and P_i are in main memory, the data is first DMA-ed into the local store from main memory in pieces, and then DMA-ed out to P_i. The send operation does a single DMA or a series of DMAs, depending on the size of the message (a single DMA transfer can be of 16KB maximum). SPE i then finds the next entry in the meta-data queue Q_{ij} and waits until it is marked *invalid*. This entry is updated by DMA-ing the relevant information, with flag set to *valid*. Send returns after this DMA completes. Note that we use blocking DMAs in copying data, to ensure that P_i contains the entire data before the corresponding entry has flag set to *valid*.

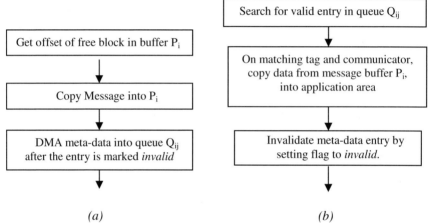

(a)   *(b)*

Fig. 1. Execution of (a) send and (b) receive for a message from SPE i to SPE j

Receive Protocol. The receive operation has four flavors. (i) It can receive a message with a specific tag from a specific source, (ii) it can receive a message with any tag (MPI_ANY_TAG) from a specific source, (iii) it can receive a message with a specific tag from any source (MPI_ANY_SOURCE), or (iv) it can receive a message with any tag from any source. Case (i) is shown in fig. 1 (b).

The receive operation on SPE j for a message from SPE i proceeds as follows, in case (i). The meta-data queue Q_{ij}, is searched in order to find a valid entry with the specific tag and communicator value. The searching is done from the logical front of the circular array to the logical end, in a linear order, to avoid overtaking of an older entry by a newer one with the same tag value. The search is repeated until a matching entry is found. Once a matching entry is found, the location of the message in main memory is obtained from the location field of the meta-data entry and the data is copied from P_i into the location for the application variable, in a similar manner as in the send operation. Finally, the meta-data entry is marked invalid.

In case of MPI_ANY_TAG by a receiver j from a specific source i, the first valid entry in Q_{ij} with the same communicator is matched. MPI_ANY_SOURCE has two cases similar to the above, except that queues Q_{ij}, for each i, are searched.

Lock Free Data Structure. The meta-data queues are handled in a lock free approach. Each queue is an array of entries which is filled by the sender in a circular fashion. The receiver maintains the range of the entries to be searched, in its local store. The need for locks has been avoided by using the fact that the local store is single ported. That is, at a given clock cycle, either a DMA operation can access the local store or the load store unit of the SPE can access it. DMA writes are in units of 128 bytes. The meta-data entry is less than 128 bytes and will therefore be seen in full, or not seen at all, by the receive operation when the receiver's search for the entry and the sender's DMA of it are taking place simultaneously. (The size of each entry is 16 bytes, which divides 128, and so all entries are properly aligned too.) Therefore the send and receive can operate in a lock free fashion.

Communication modes. MPI_Send may be implemented in either buffered mode, as in the description above, or in synchronous mode. In the latter, the send can complete only after the matching receive has been posted. The rendezvous protocol is typically used, where the receive copies the data directly, without an intermediate buffer, and then both send and receive complete. A safe application should not make any assumption on the choice made by the implementation [5]. Implementations typically use the buffered mode for small messages, and switch to synchronous mode for large messages [6]. We too switch to synchronous mode for large messages. The send then writes the address of the data in main memory into the meta-data entry, and blocks until the receive operation copies this data.

4   Performance Evaluation

We evaluated the performance of our MPI implementation, in order to determine the bandwidth and latency as a function of the message size. We also evaluated the performance of a parallel matrix-vector multiplication kernel using our MPI implementation. We performed our experiments on a 3.2 GHz Rev 2 Cell blade with 1 GB main memory running Linux 2.6.16 at IBM Rochester. We had dedicated access to the machine while running our tests.

Figures 2 and 3 show the latency and bandwidth results respectively, using the pingpong test from mpptest [7]. We switch from buffered mode to synchronous mode for messages larger than 2 KB. The pingpong test was modified to place its data in main memory, instead of in local store. We timed the operations by using the decrementer register available in the SPE, which is decremented at a frequency of 14.318 MHz, or, equivalently, around every 70 ns. The latency is comparable to that on good shared memory implementations, such as around 1.1 μs for MPICH with Nemesis on Xeon [8] and around 0.3 μs for the same on an Opteron [6]. The peak bandwidth is 6.01 GB/s, compared with around 0.65 GB/s and around 1.5 GB/s in the latter two respectively. Thus, the performance is comparable to good shared memory implementations on full-fledged cores, even though the SPEs have limited functionality.

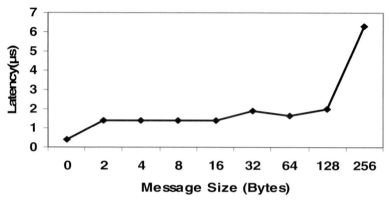

Fig. 2. Latency results

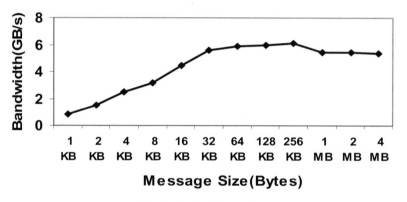

Fig. 3. Bandwidth results

We also studied the performance of a parallel double precision matrix-vector kernel using a simple 1-dimensional decomposition. We transformed the application by placing the data in the main memory and inserting DMA instructions wherever necessary. We did not use SPE intrinsics to optimize this application, nor was the algorithm chosen optimal, because our focus was on the parallelization. The only MPI communication in this application is an *MPI_Allgather* call. We got a throughput of 6.08 Gflops for a matrix of dimension 512 and throughput of 7.84 Gflops for a square matrix of dimension 1024. The same implementation yields 0.292 Gflops on a single core 2 GHz Opteron, and 0.553 Gflops on 8 Opteron processors/4 nodes connected with Gigabit Ethernet. An optimized BLAS implementation for this kernel on a single 3.2 GHz 3GB RAM Xeon processor at NCSA yields 3 Gflops.

5   Related Work

Conventional shared memory MPI implementations run a separate process on each processor. These processes have distinct address spaces. However, operating systems provide some mechanisms for processes to be allocated shared memory regions, which can be used for fast communication between processes. There are a variety of techniques that can be used, based on this general idea. They differ in scalability, effects on cache, and latency overhead. A detailed comparison of popular techniques is presented in [9]. The TMPI implementation takes a slightly different approach, by spawning threads instead of processes [10]. Since the global variables of these threads are shared (unlike that of the SPE threads in our implementation), some program transformation is required to deal with these. They too use $O(N^2)$ lock-free queues, but the implementation differs from ours. Note that some implementations on conventional processors need memory barriers to deal with out of order execution, which is common on modern processors[11]. In-order execution on the Cell avoids such problems.

Much work is being performed to make the programming of the Cell processor easier, such as developing frameworks that will enable the programming of the Cell at an abstract level [12,13,14]. Work has also been done to port a number of computational kernels like DGEMM, SGEMM, 1D FFT and 2D FFT to the Cell processor [1], and close to peak performance is obtained on DGEMM. Results on other kernels too show much superior performance to those on conventional processors.

6   Limitations and Future Work

We have implemented some core features of MPI 1, but not the complete set. We plan to implement more features. If the code size is too large, then we intend to provide overlaying facilities in the library itself, which will bring in code to the local store as needed. Also, we intend to implement a customized software cache and study the impact of the latencies introduced due to the use of a software cache and NUMA aspects of the blade. We intend to implement non-blocking calls. Also, we plan to optimize the collective communication calls using Cell-specific features. The non-MPI portion of the application still needs some compiler and OS support in order to be ported without changes to the code for large memory applications. We expect this to be accomplished by other groups.

7   Conclusions

We have shown the feasibility of an efficient MPI implementation on the Cell processor, using the SPEs as MPI nodes. Applications using only the core features of MPI can use our implementation right now, without any changes to the code if the application fits into the local store memory. Large applications can either make some hand-coded changes, or wait for compiler and OS support that is expected in the near future. Our approach, therefore, reduces the programming burden, which is considered a significant obstacle to the use of the Cell processor. Furthermore, our implementation demonstrates that simple cores for future generation heterogeneous multicore processors may run MPI applications efficiently.

Acknowledgments. We thank several employees at IBM Bangalore for running the performance tests on the Cell hardware. Most of all, we express our gratitude to Bhagawan Sri Sathya Sai Baba, Chancellor of Sri Sathya Sai University, for bringing us all together to perform this work, and for inspiring and helping us toward our goal.

References

1. Williams, S., Shalf, J., Oliker, L., Kamil, S., Husbands, P., Yelick, K.: The Potential of the Cell Processor for Scientific Computing, Proceedings of the ACM International Conference on Computing Frontiers, (2006)
2. An Introduction to Compiling for The Cell Broadband Engine Architecture, Part 4: Partitioning Large Tasks, (2006) http://www-128.ibm.com/developerworks/edu/pa-dw-pacbecompile4-i.html
3. An Introduction to Compiling for The Cell Broadband Engine Architecture, Part 5: Managing Memory, (2006) http://www-128.ibm.com/developerworks/edu/pa-dw-pacbecompile5-i.html
4. Cell Broadband Engine Programming Handbook, Version 1.0, April (2006) http://www-306.ibm.com/chips/techlib/techlib.nsf/techdocs/9F820A5FFA3ECE8C8725716A0062585 F/$file/BE_Handbook_v1.0_10May2006.pdf
5. Snir, M., Otto, S., Huss-Lederman, S., Walker, D., Dongarra, J.: MPI – The Complete Reference, Volume 1, The MPI Core, second edition, MIT Press (1998)
6. Buntinas, D., Mercier, G., Gropp, W.: Implementation and Shared-Memory Evaluation of MPICH2 over the Nemesis Communication Subsystem. Proceedings of the Euro PVM/MPI Conference, (2006)
7. Gropp, W., Lusk, E.,: Reproducible Measurements of MPI Performance Characteristics, Argonne National Lab Technical Report ANL/MCS/CP-99345, (1999)
8. Buntinas, D., Mercier, G., Gropp, W.: Design and Evaluation of Nemesis, a Scalable, Low-Latency, Message-Passing Communication Subsystem, Proceedings of the International Symposium on Cluster Computing and the Grid, (2006)
9. Buntinas, D., Mercier, G., Gropp, W.: Data Transfers Between Processes in an SMP System: Performance Study and Application to MPI, Proceedings of the International Conference on Parallel Processing, (2006) 487-496
10. Tang, H., Shen, K., Yang, T.: Program Transformation and Runtime Support for Threaded MPI Execution on Shared-Memory Machines, ACM Transactions on Programming Languages and Systems, 22 (2000) 673-700
11. Gropp, W., Lusk, E.,: A High-Performance MPI Implementation on a Shared-Memory Vector Supercomputer, Parallel Computing, 22 (1997) 1513-1526
12. Fatahalian, K., Knight, T.J., Houston, M., Erez, M.,: Sequoia: Programming the Memory Hierarchy, Proceedings of SC2006, (2006)
13. MultiCore Framework, Harnessing the Performance of the Cell BE™ Processor, Mercury Computer Systems, Inc., (2006) http://www.mc.com/literature/literature_files/MCF-ds.pdf
14. Ohara, M., Inoue, H., Sohda, Y., Komatsu, H., Nakatani, T.: MPI Microtask for Programming the Cell Broadband Engine™ Processor, IBM Systems Journal, 45 (2006) 85-102

Implementation of the Parallel Superposition in Bulk-Synchronous Parallel ML

Frédéric Gava

Laboratory of Algorithms, Complexity and Logic, University of Paris XII
gava@univ-paris12.fr

Abstract. Bulk-Synchronous Parallel ML (BSML) is a functional data-parallel language to code Bulk-Synchronous Parallel (BSP) algorithms. It allows an estimation of execution time, avoids deadlocks and nondeterminism. This paper presents the implementation of a new primitive for BSML which can express divide-and-conquer algorithms.

Keywords: BSP Functional Programming, divide-and-conquer.

1   Introduction

Bulk-Synchronous Parallel ML (*BSML*) is an extension of ML to code *Bulk-Synchronous Parallel* (BSP) algorithms as functional programs in direct mode. BSP is a parallel programming model (we refer to [2,9] for an gentle introduction to BSP) which offers a high degree of abstraction and allows scalable and predictable performance on a wide variety of architectures with a realistic cost model based on a structured parallelism. Deadlocks and non-determinism are avoided. BSP programs are portable across many parallel architectures and BSML expresses them with a small set of primitives. These primitives are implemented as a parallel library (http://bsmllib.free.fr) for the functional programming language Objective Caml (OCaml). Using a safe high-level language as ML to programming BSP algorithms allows performance, scalability and expressivity.

The BSP model does not allow to synchronize a subset of the processors. This is often considered as an obstacle to express divide-and-conquer algorithms in the BSP model. Nevertheless it is showed in [9] that for typical applications, exploiting the loss of efficiency due to the lack of computation-communication overlapping is outweighed by the advantages of bulk data transfer, while making programming and debugging much more difficult.

Nevertheless, [10] argues that the divide-and-conquer paradigm fits naturally into the BSP model, without any need of subset synchronization. It proposes a method which is fully compliant with the BSP model. This method is based on sequentially interleaved threads of BSP computation, called *superthreads* (more explanation what superthreads are and how they support the divide-and-conquer scheme can be found in [10]). [7] presents a new primitive for BSML called *parallel superposition* and its associated cost model. This primitive is based on a notion similar to Tiskin's superthreads and was only informally described as equivalent to pairing. Adding this primitive in BSML allows to programming

Y. Shi et al. (Eds.): ICCS 2007, Part I, LNCS 4487, pp. 611–619, 2007.

bsp_p: unit→int   **bsp_l**: unit→float   **bsp_g**: unit→float
mkpar: (int→α)→α **par**   **apply**: (α →β)**par**→α **par**→β **par**
put: (int→α option)**par**→(int→α option)**par** **proj**: α option **par**→int→α option
type α option = None | Some **of** α **super**: (unit →α) →(unit →β) →$\alpha * \beta$

Fig. 1. The core BSMLlib library

more easily divide-and-conquer BSP algorithms. In the current paper we present
the behavior of the parallel implementation.

2   Functional Bulk-Synchronous Parallel ML

BSML does not rely on SPMD (Single Program Multiple Data) programming.
Programs are identical to usual sequential OCaml programs but work on a parallel
data structure. Some of the advantages are better readability.

The BSMLlib library is based on the elements given in Figure 1. It gives access
to the BSP parameters of the underling architecture. For example, **bsp_p**() is
p, the *static* number of processes. There is an abstract polymorphic type α **par**
which represents the type of p-wide parallel vectors of objects of type α one per
processor. Those parallel vectors are created by **mkpar** so that (**mkpar** f) stores
(f i) on process i for i between 0 and $p-1$: **mkpar** f = $\boxed{(f\ 0)}\ \cdots\ \boxed{(f\ i)}\ \cdots\ \boxed{(f\ (p-1))}$

Asynchronous phases are programmed with **mkpar** and with **apply** such that
(**apply** (**mkpar** f) (**mkpar** e)) stores ((f i)(e i)) on process i:

$$\textbf{apply}\ \boxed{\cdots\ \boxed{f_i}\ \cdots}\ \boxed{\cdots\ \boxed{v_i}\ \cdots} = \boxed{\cdots\ \boxed{(f_i\ v_i)}\ \cdots}$$

The **put** primitive expresses communication and synchronization phases. Consider the expression: **put(mkpar(fun** i→fs$_i$)) ($*$). To send a value v (resp. no
value) from process j to process i, the function fs$_j$ at process j must be such as
(fs$_j$ i) evaluates to Some v (resp. None). The expression ($*$) evaluates to a parallel
vector containing functions fd$_i$ of delivered messages. At process i, (fd$_i$ j) evaluates to Some v (resp. None) if process j sent the value v (resp. no value) to the
process i. The **proj** primitive also expresses communication and synchronization
phases. (**proj** vec) returns a function f such that (f n) returns the nth value of the
parallel vector vec. If this value is the empty value None then process n sends no
message to the other processes. Otherwise this value is broadcast. Without this
primitive, the global control cannot take into account data computed locally.

The primitive **super** effects *parallel superposition*, which allows the evaluation
of two BSML expressions as interleaved threads of BSP computationsin. From
the programmer's point of view, the functional semantics of the superposition is
the same as pairing but of course the evaluation of **super** E_1 E_2 is different from
the evaluation of (E_1, E_2) [4]). The phases of asynchronous computation of E_1
and E_2 are run. Then the communication phase of E_1 is merged with that of of
E_2. The messages are obtained by concatenation of the messages and only one
barrier occurs. If the evaluation of E_1 needs more supersteps than that of E_2
then the evaluation of E_1 continues (and *vice versa*).

3   Implementation

The implementation of the superthreads needed for the superposition uses the thread feature of OCaml. Each superthread is defined as a thread associated with an identifier and a channel of communication (à la Concurrent ML) to sleep or wake up a superthread. A specific scheduler is thus needed. We also need an environment of communication defined as a hash table (where keys are the identifier of the superthreads).

3.1   New Implementation of the Primitive of BSML

In this new implementation of the BSMLlib library, the core module which contains the primitives presented in section 2, is implemented in SPMD style using a lower level communication library, a module for the scheduling of the superthreads called Scheduler, a module for the environment of communication called EnvComm and a module of generic operators for the parallel superposition called SuperThread. The implementation of all the other modules of the BSMLlib library is independent of the actual implementation of these modules. The module of communication called Comm is based on the following elements [8]:

pid: unit→int   nprocs: unit→int   send: α option array→α option array

There are several implementations of Comm based on MPI (which used the MPI MPI_Alltoall C operator), PUB [3] and TCP/IP (only using the TCP/IP features of OCaml). The meaning of pid and nprocs is obvious. The function Comm.send takes on each process an array of size nprocs() of optional values. If at process j the value contained at index i is (Some v) then the value v will be sent from process j to process i. If the value is the None value, nothing will be sent. The result is an array of sending values. A global synchronization occurs inside this communication function.

The implementation of the abstract parallel vectors, **mkpar** and **apply** is as follows (rules 7.2 and 7.3, page 121 of the small-steps semantics in [4]):

type α **par** = α let mkpar f = f (Comm.pid())   let apply f v = (f v)

The communication primitives of the BSMLlib, i.e, the **put** and **proj** primitives are also implemented as in the small-steps semantics (rules 7.4 and 7.5):

```
let send v = let id=(Scheduler.pid_superthread_run SuperThread.our_schedule) in
    EnvComm.add SuperThread.envComm id v; (SuperThread.rcv())
let mkfuns = (fun res i →if ((0<=i)&&(i<(!nprocs))) then res.(i) else None)
let put f = mkfuns (send (Array.init (!nprocs) f))
let proj v = put (fun _ →v)
let super f1 f2 = let t=(SuperThread.create_child f2) and v=f1 ()
                in (v, SuperThread.wait t)
```

The send operator (rule 7.8) takes the identifier (Scheduler.pid_superthread_run) of the current active superthread (from the scheduler SuperThread.our_schedule) and puts the values to send in the environment of communication (EnvComm). Then it returns the received values after the communication phase.

The primitive **super** is also implemented as in the semantics (rule 7.10 page 122), i.e., we build a pair where in the first component, we compute f1 and in the

module SuperThread:**functor**(HowComm:**sig val** make_comm:(unit→unit) **ref end**)→**sig**
val envComm : unit EnvComm.t   **val** our_schedule : Scheduler.t
val rcv : unit →α **type** β data_of_thread
val create_child: (unit→β)→β data_of_thread **val** wait: β data_of_thread →β
end

Fig. 2. The Super Module

second component we run a child as another superthread to compute f2 and we
wait for its result. SuperThread.create_child run another superthread and return
the identifier of the new superthread which is the argument of the **wait** operator.

In [4], we describe how we implemented the module Scheduler which helps to
schedule the superthreads (with the strategy of the small-steps semantics) and
the module EnvComm of the environment of communication.

3.2   Functions for the Implementation of the Superposition

The module for implementing the parallel superposition called SuperThread is
a functor based on the elements given in Figure 2. We have an environment of
communication called envComm, a scheduler called our_schedule, the abstract
type β data_of_thread of a superthread and the functions used above.

This functor is parameterized by a module which contains a reference to a
function that makes the communication. This reference would be affected at
the initialization with a function that would manipulate the environment of
communication (iterating the hash-table) and perform the communications using
the function send of the module Comm. Now, we will describe the implementation
of the principal elements of this module.

The first one is the rcv operator which works as in rule 7.9 of the small-
steps semantics i.e., this operator returns and deletes from the environment of
communication the values received by the superthread at the current superstep.
rcv is implemented using the functions of the scheduler which works as follow.
We first test if the current superthread is the last superthread of the superstep.
If it is the case, communications are done because we are at the end of the
superstep. Then, we test if the current superthread is not the only one to run. If
it is the case, those superthreads need to be run (strategy of the semantics: after
the blocked communications, superthreads need to continue their works) and
thus, we sleep the current superthread and wake up the first of them. Note that
in this case the scheduler would give the hand to the current active superthread
in the future. If not i.e., the current superthread is the only one, we finish by
returning the value read from the environment. of communication.

The second one is the **wait** operator which returns the final result of the
superthread child. It works as follow. First we test if the child has finished its
computation. If not the father has to wait this result and thus we remove the
superthread from the scheduler. We also wake up the next superthread and make
sleep the current superthread. Its child would wake up it in the future. To finish,
the result of the child is returning.

The last one, is the creation of a child (a new superthread) with the create
_child operator which works as follow. First, a new superthread is created as
a new OCaml's thread which first sleeps (strategy of the semantics), computes
the final result and to finish. The end of the child works as follow. First, we
test if the father has ended its computation or not. If it is the case, then the
child wakes up its father. Else, the superthread is removing from the scheduler
and the next superthread is waked up. We also test if the superthread child is
the last superthread of the superstep. If it is the case, communications are done
because we are at the end of the superstep.

4   Example and Benchmarks

4.1   Calculus of the Prefix Using the Parallel Superposition

The example presented below is a divide-and-conquer version of the scan program using the parallel superposition. In this version of the calculus of the prefix,
the processors are divided into two parts and the scan is recursively applied to
those parts (Figure 3). The value held by the last processor of the first part
is broadcast to all the processors of the second part, then this value and the
values held locally are combined together by the operator op on the second
part.

In our benchmarks, we will make a performance comparison of the divide-
and-conquer version of the computation of the prefix with two other versions.
The first one, is the direct version and the second one is the binary computation
using $\log(p)$ supersteps [2] those coded in Figure 4.

```
let inbounds first last n = (n>=first)&&(n<=last) (* inbounds: α →α →α →bool *)
let within_bounds = inbound 0 (bsp_p()−1) (* mix: int→α par * α par→α par *)
let mix m (v1,v2)=let f pid v1 v2=if pid<=m then v1 else v2
                    in apply (apply (mkpar f) v1) v2
let replicate e = mkpar (fun _ →e) (* replicate: α →α par *)
let parfun f v = apply (replicate f) (* parfun:(α →β )→α par→β par *)
let parfun2 f v1 v2 = apply (parfun f v1) v2

(* scan: (α →α →α )→α →α par→α par *)
let scan_super op vec =
 let rec scan' fst lst op vec = if fst>=lst then vec else
 let mid=(fst+lst)/2 in
 let vec'=mix mid (super(fun()→scan' fst mid op vec)(fun()→scan'(mid+1) lst op vec))
    in let msg vec = apply (mkpar(fun i v→
           if i=mid then fun dst→if inbounds (mid+1) lst dst then Some v else None
             else fun dst→ None)) vec
    and parop = parfun2(fun x y→match x with None→y|Some v→op v y) in
    parop (apply(put(msg vec')) (replicate mid)) vec' in
 scan' 0 (bsp_p()−1) op vec
```

Fig. 3. Code of the divide-and-conquer algorithm of the parallel prefix

```
let scan_direct op e vv =
 let mkmsg pid v dst=if dst<pid then None else Some v in
 let procs_lists=mkpar(fun pid→from_to 0 pid) in
 let rcv_msgs=put(apply(mkpar mkmsg) vv) in
 let values_lists= parfun2 List.map (parfun (compose noSome) rcv_msgs) procs_lists in
  applyat 0 (fun _ →e) (List.fold_left op e) values_lists

let scan_logp op vec =
 let rec scan_aux n vec =
  if n >= (bsp_p()) then vec else
   let msg = mkpar(fun pid v dst→
    if ((dst=pid+n)or(pid mod (2*n)=0))&&(within_bounds (dst−n))
     then Some v else None)
   and senders = mkpar(fun pid→natmod (pid−n) (bsp_p()))
   and op' = fun x y→match y with Some y'→op y' x | None →x in
    let vec' = apply (put(apply msg vec)) senders in
    let vec''= parfun2 op' vec vec' in
     scan_aux (n*2) vec'' in
 scan_aux 1 vec
```

Fig. 4. Code of the direct and binary algorithm of the parallel prefix

4.2 Benchmarks

We did some preliminary experiments on a cluster with 10 Pentium IV nodes (with 1 Go of main memory per node) interconnected with a Gigabit Ethernet network. Both need $\log p$ supersteps except the direct version which need one superstep. The values were arrays of floats representing polynomials. The binary operation is the sum of two polynomials.

The MPI and TCP/IP implementations of the Comm module were used. These programs ran 100 consecutively times with initial randomized polynomials and this 5 times. The native code compiler of OCaml was used. In Figures 5, diagrams show the average of the results with increasing size of polynomials. In (a) and (c) (resp. (b) and (d)), we give the performances using the MPI (resp. TCP/IP) implementation of BSML.

The direct version is the faster for small polynomials. However, the version using the superposition and TCP/IP seems to be the faster one for big polynomials. For all the versions, the scalability of the BSP model is well-preserved.

We have also perform some performance comparisons of the direct and binary versions of the scan with a BSMLlib which does not contain the superposition and with our new one which supports superposition. We have show that the overhead, for programs which do not use superposition, is negligible.

5   Related works

The superthread way to divide-and-conquer in the framework of an object-oriented language was presented in [10]. There is no formal semantics and no

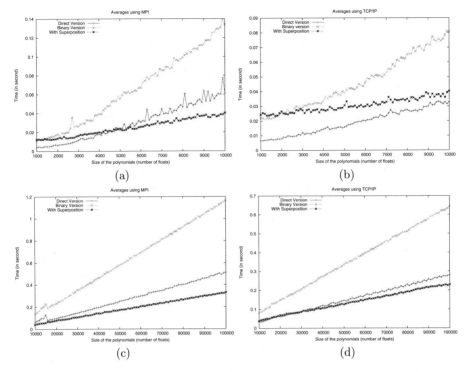

Fig. 5. Experiments of parallel prefix of polynomial using MPI or TCP/IP

implementation from now on. An algorithmic skeletons language which offers divide-and-conquer skeletons was designed in [12]. Nevertheless, the cost model is not really the BSP model but the D-BSP model which allows subset synchronization. We follow [9] to reject such a possibility.

A general data-parallel formulation for a class of divide-and-conquer problems was evaluated in [1]. A combination of techniques are used to reorganize the algorithmic data-flow, providing great flexibility to efficiently exploit data locality and to reduce communications. But those techniques are only define for a low-level parallel language, High Performance Fortran.

In [5], the proposed approach distinguished three levels of abstraction and their instantiations. (1), a small language, as an extension of ML, defines the static parallel parts of the programs. The language comes with a partial evaluator which acts as a code transformer using MetaOCaml. (2), an implementation of a divide-and-conquer skeleton demonstrates how meta-programming can generate the appropriate set of communications for a particular process from an abstract specification. (3), the application programmer composes the program using skeletons, without then need to consider details of parallelism. However, cost prediction nor native (efficient) code generation are possible.

6   Conclusion

The parallel superposition is a new primitive of BSML and it allows divide-and-conquer algorithms to be expressed easily, without breaking the BSP execution model. Compared to the parallel juxtaposition [6], this new primitive has not the drawbacks of its predecessor: the cost model is a compositional one and it can be seen as a purely functional primitive. We have presented in this paper how implements this new primitive with the help of the low-level semantics and makes some benchmarks of a classical BSP algorithms.

The ease of use of the superposition will be experimented by implementing BSP algorithms described as divide-and-conquer algorithms in the literature. An implementation of the superthreads using fault tolerant threads of MPI and static cost analysis as in [11] are also another directions of research.

Acknowledgments. This work is supported by the ACI Jeunes Chercheurs program from the French Ministry of Research, under the project PROPAC wwwpropac.free.fr. The author wishes to thanks Anne Benoît and Julien Signoles for their comments.

References

1. M. Aumor, F. Arguello, J. Lopez, O. Plata, and L. Zapata. A data-parallel formulation for divide-and-conquer algorithms. *The Computer Journal*, 44(4):303–320, 2001.
2. R. H. Bisseling. *Parallel Scientific Computation. A structured approach using BSP and MPI*. Oxford University Press, 2004.
3. O. Bonorden, B. Juurlink, I. Von Otte, and O. Rieping. The Paderborn University BSP (PUB) library. *Parallel Computing*, 29(2):187–207, 2003.
4. F. Gava. *Approches fonctionnelles de la programmation parallèle et des méta-ordinateurs ; Sémantiques, implantation et certification*. PhD thesis, University of Paris XII, 2005.
5. C. A. Herrmann. Functional meta-programming in the construction of parallel programs. *Parallel Processing Letters*, 2006. to appear.
6. F. Loulergue. Parallel Juxtaposition for Bulk Synchronous Parallel ML. In H. Kosch, L. Boszorményi, and H. Hellwagner, editors, *Euro-Par 2003*, number 2790 in LNCS, pages 781–788. Springer Verlag, 2003.
7. F. Loulergue. Parallel Superposition for Bulk Synchronous Parallel ML. In Peter M. A. Sloot and al., editors, *International Conference on Computational Science (ICCS 2003), Part III*, number 2659 in LNCS, pages 223–232. Springer Verlag, june 2003.
8. F. Loulergue, F. Gava, and D. Billiet. Bulk-Synchronous Parallel ML: Modular Implementation and Performance Prediction. In V. S. Sunderem, G. D. van Albada, P. M. A. Sloot, and J. Dongarra, editors, *International Conference on Computational Science (ICCS), Part II*, number 3515 in LNCS, pages 1046–1054. Springer, 2005.
9. D. B. Skillicorn, J. M. D. Hill, and W. F. McColl. Questions and Answers about BSP. *Scientific Programming*, 6(3):249–274, 1997.

10. A. Tiskin. A New Way to Divide and Conquer. *Parallel Processing Letters*, 11(4):409–422, 2001.
11. P. B. Vasconcelos and K. Hammond. Inferring Cost Equations for Recursive, Polymorphic and Higher-Order Functional Programs. In *IFL'02*, LNCS, pages 110–125. Springer Verlag, 2003.
12. A. Zavanella. *Skeletons and BSP : Performance Portability for Parallel Programming*. PhD thesis, Universita degli studi di Pisa, 1999.

Parallelization of Generic Libraries Based on Type Properties

Prabhanjan Kambadur, Douglas Gregor, and Andrew Lumsdaine

Open Systems Laboratory
Indiana University
Bloomington, IN 47405
{pkambadu, dgregor, lums}@osl.iu.edu

Abstract. This paper describes a new approach for parallelizing generic software libraries. Generic algorithms are expressed in terms of *type properties,* which allows them to work with entire families of types rather than specific types. Despite this generality, generic algorithms can be made as efficient as their hand-coded variants through the use of *specialization,* which provides algorithm variants tuned for types with certain properties. Our approach leverages the specialization mechanism of generic programming to effect parallelism. We illustrate the process of specializing generic algorithms for parallelism using common algorithms in the C++ Standard Template Library. When the resulting algorithms are invoked with types that have the required properties for parallelization, the parallel variants of these algorithms are executed. We illustrate that our parallelization strategy is simple, practical, and efficient, using tools and techniques available in most commercial compilers.

Keywords: Parallel Algorithms, Specialization, Type Properties.

1   Introduction

The advent of many-core and multi-core processors will significantly broaden the applicability of parallel computing. While this new era of ubiquitous parallelism will surely bring new languages and tools built from the ground up for parallelism, a vast majority of tomorrow's potential parallel applications are today's sequential programs. The primary challenge to ubiquitous parallelism is the need to ease the transition from sequential to parallel, allowing existing applications to be parallelized without requiring them to be rewritten or refactored.

The ultimate goal for enabling parallelization of sequential applications is the fully automatic parallelizing compiler. Although there has been research on such compilers for many years, practical solutions have remained elusive. An alternate solution to parallelization is to provide user-level libraries to aid in writing explicitly parallel applications. Various user-level libraries such as pthreads and MPI have provided services that enable parallelization of applications. However, the programming interface that these libraries offer is very low-level. Thus, they are both difficult to use and often lack support for the abstractions of relatively high-level languages like C++.

Y. Shi et al. (Eds.): ICCS 2007, Part I, LNCS 4487, pp. 620–627, 2007.

A first step towards parallelizing existing sequential applications is to parallelize those components (libraries) that are most commonly used by these applications. When parallel versions of commonly used libraries become available, previously sequential applications are "automatically" parallelized. This approach has been successfully followed in many scientific libraries (for example, BLAS [1] [2]). In this paper, we describe the methodology by which we are able to accurately parallelize applications that use generic libraries. For example, we are able to parallelize the following code segment:

```
std::vector<int> vec(100);
// Initialize vec...
std::accumulate (vec.begin(), vec.end(), 1, std::multiplies<int>());
```

This code segment calls accumulate(), a *generic* algorithm implemented in the C++ Standard Template Library (STL) [3]. The STL is an excellent example of *Generic Programming*, an important programming paradigm that is used for the development of highly efficient and maximally reusable software [4]. Moreover, the STL is a widely-used portion of the C++ Standard Library, so parallelizing it can have an immediate impact on many existing C++ applications.

In this paper, we illustrate the parallelization of the generic algorithms in the STL through the use of type properties, and demonstrate how our approach automatically parallelizes calls to STL algorithms. Our parallelization is based on OpenMP, a set of compiler directives and library routines that enable users to express shared-memory parallelism. We choose OpenMP because it is simple, intuitive to use and is available in most modern C++ compilers. This last point is particularly important: OpenMP, C++, and the STL are all existing, widely-deployed and widely-used technologies. By building on these technologies, our OpenMP-parallel implementation of the STL can be used to parallelize existing applications without requiring developers or end-users to install new tools or run-time systems; merely recompiling with our OpenMP-parallel STL is sufficient.

2   Related Work

ROSE [5] is a C++ infrastructure for source-to-source translation that provides an interface for writing customized translators that optimize user-defined high-level abstractions. One such translator automatically inserts OpenMP statements into user code. In principle, ROSE is akin to what we intend to achieve. However, the methodologies are vastly different. ROSE is an external tool that all developers must install whereas we advocate a minimal approach based only on C++ and OpenMP, both of which are widely available.

STAPL [6] [7] is designed in much the same spirit as the STL with containers, iterators and algorithms. STAPL also allows users to specify their own scheduling policies, data composition and data dependence environment. STAPL is designed to operate on both shared and distributed memory systems and, therefore, it includes an advanced runtime system. Existing applications need to be rewritten to benefit from STAPL, and end-users must install the STAPL runtime system to use these applications.

HPC++ [8] is a C++ class and template library that portably supports shared-memory and distributed-memory parallel applications. HPC++ supports both task and data parallel applications. Unlike our approach, HPC++ delivers parallelism through a Java-like class library and consequently suffers from many of the same drawbacks as STAPL. HPC++ is no longer in active development.

Threading Building Blocks (TBB) [?] is a library-based solution to shared-memory parallelism that is tailored to the high-level of C++ abstractions. TBB offers a wide variety of customizable parallel algorithms and concurrent data-structures that facilitate development of parallel applications. However, parallelizing applications with TBB encumbers some amount of rewriting.

3   Generic Programming

Most modern software libraries consist of collections of related functions and data types, often as archives of object files in conjunction with header files that define interfaces to components in the library. Although the goal of these libraries is to promote reuse, the requirements for invoking functions in these libraries are often over-specified. This leads to a less than optimal reuse of the components.

With the explicit goal of improving the reuse of software libraries, generic programming has emerged as an important paradigm for the development of highly efficient and maximally reusable software [4]. Generic programming involves the implementation of generic (abstract) algorithms that operate correctly and efficiently on a wide range of data types—including those not known at algorithm design time—so long as the data types meet certain requirements expressed as *type properties*. Related type properties are grouped together into *concepts*, which identify domain-specific abstractions such as a Matrix, Vector, Graph, or Iterator on which generic algorithms will operate.

C++ does not provide an inherent way to specify concepts and, therefore, there exists no direct means of checking if a type models a particular concept. However, we can approximate this behavior through the use of *traits* classes. Traits are a C++ template technique that allow one to provide additional information about a set of types without modifying the types themselves [9]. A key advantage of using traits classes is that type properties can be attached retroactively. This enables attachment of properties to types whose definitions are not available.

In many cases, we can improve the performance of a generic algorithm by exploiting additional type properties. For example, consider a generic algorithm that computes the circumference of an arbitrary polygon: it will require linear time to sum the edge lengths. However, if we restrict the algorithm to equilateral polygons, we can compute the circumference in constant time. A generic library typically provides the most generic algorithm (e.g., for arbitrary polygons) along with several *specializations* of the algorithm (e.g., for equilateral polygons). When a user invokes the generic algorithm, the library will automatically select the most specialized algorithm at compile time, ensuring that the library provides the best possible performance. In practice "selecting" the right

```
// Specialized Version
template <typename InIter, typename Func>
typename enable_if<is_parallel_safe<Func>::value
                    && is_random_access_iterator<InIter>::value, Func>::type
for_each(InIter first, InIter last, Func f) {
    typedef typename std::iterator_traits<InIter>::difference_type difference_type;
    const register difference_type range = last−first;

#pragma omp parallel for schedule(static)
    for (register difference_type i=0; i<range; i++) { f(first[i]);}
    return f;
}
// Fallback version
typename disable_if<is_parallel_safe<Func>::value
                    && is_random_access_iterator<InIter>::value, Func>::type
for_each(InIter first, InIter last, Func f) {
    for (; first!=last; ++first) f(*first);
    return f;
}
```

Fig. 1. Parallel implementation of the STL for_each() algorithm

algorithm often involves accessing traits classes to ensure that the input types to the *specialized* algorithm meet all the type requirements.

4   Parallelizing Libraries

Specializations based on type properties can be used to extract optimal performance out of generic algorithms. Type properties serve as gate keepers that ensure that input types meet the requirements stipulated by specialized versions of algorithms. In this section, we elaborate through examples, the process of parallelizing generic algorithms based on type properties.

First, consider the STL algorithm for_each(), which applies a unary function to all the values in the range [first, last).

```
template <typename InIter, typename Func>
Func for_each (InIter first, InIter last, Func f);
```

This algorithm stipulates that InIter model the InputIterator concept, Func model the UnaryFunction concept (i.e., have operator ( ) defined that accepts one argument and returns a result) and that InIter's value type be convertible to the argument type of Func. Although these type requirements are necessary and sufficient for sequential correctness, to parallelize for_each() using OpenMP, input types must possess additional properties. As OpenMP can only parallelize for loops that are expressed in a simple single induction variable form, InIter must model the RandomAccessIterator concept. Similarly, Func has to be reentrant and give the same results regardless of the order in which the values are supplied to it, i.e., Func must model the ParallelSafe concept. This particular requirement ensures that it is safe to execute Func in parallel.

To check for these two additional type properties (i.e., meeting the ParallelSafe and RandomAccessIterator concept requirements), we make use of the metafunctions is_parallel_safe and is_random_access_iterator. is_parallel_safe is a simple templated structure that by default specifies that no function object is safe to be executed in parallel. To specify that the type plugged into Func models the ParallelSafe concept, users need to *fully specialize* the is_parallel_safe structure. This is nothing but a means of specifying *traits* for function objects. For convenience, is_parallel_safe has already been specialized for all the STL function objects. Similarly, is_random_access_iterator is used to determine whether a particular iterator models the RandomAccessIterator concept. Like is_parallel_safe, is_random_access_iterator is pre-defined for all the STL iterators. The metafunctions enable_if and disable_if [10] are used to ensure that only one of the versions of for_each() is "turned on" when instantiated with concrete types. The code fragment that implements this specialization is given in Figure 1. To demonstrate when the parallel for_each is invoked, consider the following example.

```
struct square_f { void operator()(int& value) {value *= value;}};

template <> struct is_parallel_safe <square_f> {static const bool value = true; };

std::vector<int> vec(100);
std::list<int> lst(100);
for_each(vec.begin(), vec.end(), square_f());// PARALLEL
for_each(lst.begin(), lst.end(), square_f());// SEQUENTIAL
```

The first call to for_each() is supplied iterators to std::vector and a function object of type square_f. Since std::vector iterators support random access and square_f is safely invokable in parallel (as specified through the specialization of the is_parallel_safe structure), this call is parallelized. The second call to for_each() uses the same function object as the first, but is instantiated with iterators to std::list, which do not support random access. Consequently, this call is not parallelized and is executed sequentially instead.

Consider accumulate(), an STL algorithm that reduces based on some reduction operator, all the values in the iterator range [first,last). For example, the code fragment given below computes the sum of all the elements in the array.

```
int array[] = {1,2,3,4};
std::cout << accumulate (a, a+4, 10, std::plus<int>());
```

This fragment outputs 20, which is the sum of all the elements of the array plus the initial value of 10. The prototype of accumulate() is given below.

```
template <typename InIter, typename T, typename Func>
T accumulate (InIter first, InIter last, T init, Func binary_op);
```

The type requirements for accumulate() stipulate that InIter model the InputIterator concept, T be Assignable, T be convertible to value type of InIter and Func model the BinaryFunction concept (i.e, have operator () defined that accepts two arguments and returns a result). As in the case of for_each(), to parallelize this algorithm using

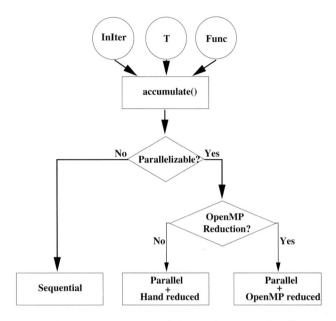

Fig. 2. Decision Logic followed for accumulate()

OpenMP, InIter is required to model the RandomAccessIterator concept and Func to model the ParallelSafe concept. However, if we also want to use OpenMP's reduction clause, we require T to be a fundamental type such as an int and Func to be one of $\{+,-,* \&,\ |,\ \hat{}\ \&\&,\ ||\}$ operators.

Therefore, we can parallelize accumulate() at two levels. First, we decide on whether the input types meet the basic criteria for parallelization. If they do meet these requirements, we then decide if we can use OpenMP's reduction clause. If it cannot be used, we manually reduce the results from the various threads at the very end. Thus, we provide two different parallel specializations of the accumulate() algorithm, the most suitable of which will be chosen at compile time. Had we provided only one of these, we would have either limited the parallelism (e.g., by only providing the more limited OpenMP reduction) or limited the potential for better performance (e.g., by using a manually-coded reduction where an architecture-optimized OpenMP reduction would be better).

The decision process for accumulate() is illustrated by means of a flowchart in Figure 2. To illustrate the decision logic, we give some of code examples and explain which final version of accumulate() is used. First, consider the following code that uses some STL iterators and STL function objects.

```
std::vector<int> array(100); ...
int sum = accumulate (array.begin(), array.end(), 1, std::multiplies<int>());
```

Here, iterator to std::vector is plugged in for InIter, int for T and std::multiplies<> for Func. As iterators to std::vector model RandomAccessIterator concept, int is a fundamental type and std::multiplies<int> can be mapped to the primitive

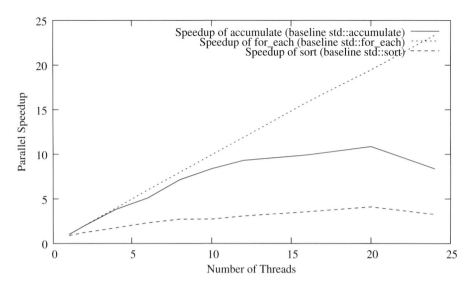

Fig. 3. Parallel speedup achieved for several STL algorithms on a 32-way SMP

∗ operator, the parallel version with OpenMP reduction is used. Consider the following example that uses fundamental types and user-defined function objects.

```
std::vector<int> array(100); ...
int sum = accumulate (array.begin(), array.end(), 0, my_bin_op());
```

Although fundamental types are used, the iterators support random access, and calls to my_bin_op can be parallelized (assuming is_parallel_safe is specialized), OpenMP reductions cannot be applied because there is no primitive operator to which we can map my_bin_op. Instead, we parallelize this invocation using a hand-coded reduction. The same logic applies if T were a non-fundamental type and a "known" function object such as std::plus<> were used. Furthermore, if, in this example, array were to be a std::list instead of a std::vector, no parallelization would be possible since iterators to std::list do not support random access.

5   Performance Evaluation

We have developed parallel specializations for several algorithms in the STL. In this section, we give performance results for three of the algorithms: for_each(), sort() and accumulate(). Tests were run on a 32-way IBM AIX p690 SMP using 1.3 GHz Power4 processors with 192 GB of memory. IBM's LoadLeveler was used for batch queuing. XLC compiler version 7.0 was used with the −qsmp and −O3 flags. The performance results are given in Figure 3. All results are for 1 million randomly generated floating point numbers that were stored in a std::vector. sort() showed a maximum speedup of around 4x at 20 threads, accumulate() showed a speedup of 11x for 20 threads. for_each(), which is an embarassingly parallel algorithm, scales linearly with the number of threads for this particular problem size.

6   Conclusions

In this paper, we have presented a new methodology for parallelizing generic libraries based on type properties. Using modern C++ idioms such as specialization and traits, we can isolate candidate invocations of generic algorithms for automatic parallelization at the library level. Unlike most approaches to automatic parallelization or library-level parallelization, which require new compilers, source-to-source translation tools, or new run-time systems, our solution works with existing, widely-deployed technologies (OpenMP and C++).

Preliminary results show that this approach gives excellent parallel speedup for commonly used algorithms such as accumulate(), sort() and for_each() on large shared-memory machines; we have seen similar results on commodity desktops and servers as well.

Acknowledgements. This work was supported by NSF grants EIA-0202048, EIA-0131354, a grant by the Lilly Endowment and in part by Shared University Research grants from IBM, Inc. to Indiana University.

References

1. Alpatov, P., Baker, G., Edwards, C., Gunnels, J., Morrow, G., Overfelt, J., van de Geijn, R.: PLAPACK: Parallel linear algebra package. In: SIAM Parallel Processing Conference. (1997)
2. Blackford, L.S., Choi, J., Cleary, A., D'Azevedo, E., Demmel, J., Dhillon, I., Dongarra, J., Hammarling, S., Henry, G., Petitet, A., Stanley, K., Walker, D., Whaley, R.C.: ScaLAPACK: a linear algebra library for message-passing computers. In: Proceedings of the Eighth SIAM Conference on Parallel Processing for Scientific Computing (Minneapolis, MN, 1997). (1997)
3. Stepanov, A.A., Lee, M.: The Standard Template Library. Technical Report X3J16/94-0095, WG21/N0482, ISO Programming Language C++ Project (1994)
4. Musser, D.R., Stepanov, A.A.: Generic programming. In Gianni, P.P., ed.: Symbolic and algebraic computation: ISSAC '88, Rome, Italy, July 4–8, 1988: Proceedings. Volume 358 of Lecture Notes in Computer Science., Berlin, Springer Verlag (1989) 13–25
5. Quinlan, D.: ROSE: Compiler support for object-oriented frameworks. Parallel Processing Letters **10**(2,3) (2000) 215–226
6. An, P., Jula, A., Rus, S., Saunders, S., Smith, T., Tanase, G., Thomas, N., Amato, N., Rauchwerger, L.: STAPL: A standard template adaptive parallel C++ library. In: Int. Wkshp on Adv. Compiler Technology for High Perf. and Embedded Processors. (2001) 10
7. An, P., Jula, A., Rus, S., Saunders, S., Smith, T., Tanase, G., Thomas, N., Amato, N., Rauchwerger, L.: STAPL: An adaptive, generic parallel programming library for C++. In: Wkshp. on Lang. and Comp. for Par. Comp. (LCPC). (2001) 193–208
8. Gannon, D., Beckman, P., Johnson, E., Green, T., Levine, M.: HPC++ and the HPC++Lib Toolkit. (The High Performance C++ consortium)
9. Myers, N.C.: Traits: a new and useful template technique. C++ Report (1995)
10. Järvi, J., Willcock, J., Hinnant, H., Lumsdaine, A.: Function overloading based on arbitrary properties of types. C/C++ Users Journal **21**(6) (2003) 25–32

Traffic Routing Through Off-Line LSP Creation

Srecko Krile and Djuro Kuzumilovic

University of Dubrovnik, Department of Electrical Engineering
and Computing, Cira Carica 4, 20000 Dubrovnik, Croatia
srecko.krile@unidu.hr

Abstract. In the context of dynamic bandwidth allocation the QoS path provisioning for coexisted and aggregated traffic could be very important element of resource management. As we know all traffic in DiffServ/MPLS network is distributed among LSPs (Label Switching Path). Packets are classified in FEC (Forwarding Equivalence Class) and can be routed in relation to CoS (Class of Service). In the process of resource management we are looking for optimal LSP, taking care of concurrent flows traversing the network simultaneously. For better load-balancing purposes and congestion avoidance the LSP creation can be done off-line, possible during negotiation process. The main difference from well known routing techniques is that optimal LSP need not to be necessarily the shortest path solution as it is in the case of typical on-line routing (e.g. with OSPF protocol).

Keywords: on-demand resource allocation, dynamical bandwidth management, end-to-end QoS routing, SLA in DiffServ/MPLS networks.

1 Introduction

Using dynamic service negotiation approach for SLA (Service Level Agreement) the problem of QoS path provisioning has to be in firm correlation with bandwidth management; see [1]. Aggregated flow consisted of numerous LSPs (Label Switching Path) is coming to LSR (Label Switched Router) and has to be routed to egress router. All traffic, traversing simultaneously a DiffServ/MPLS domain, is distributed among LSPs as the resultant of the routing protocol. Some of LSPs generally traverse through the same path across the network, so they coexist on the same link with congestion probability; see [2]. In that sense the network operator has to find the optimal LSP for each aggregated flow without any possible congestion in the network; see [6]. In the context of on-demand resource allocation the load balancing technique is necessary. The main condition is: the sufficient network resources must be available for the priority traffic at any moment. In this paper such off-line routing technique is proposed. Proposed heuristic algorithm helps in bottleneck detection on the path and maintains high network resource utilization efficiency.

A brief explanation: During SLA negotiation the RM (Resource Manager) module can apply any shortest path-based routing algorithm (e.g. OSPF - Open Shortest Path First) to generate the initial LSP. At the next step all simultaneous flows caused by former contracted SLAs have to be taken in calculation, too (correlation with other LSPs); see [4]. The BB (Bandwidth Broker) will therefore check if there are enough resources to satisfy the requested CoS in the specified path. With such congestion control/expansion algorithm the RM can predict sufficient link resources to satisfy all

Y. Shi et al. (Eds.): ICCS 2007, Part I, LNCS 4487, pp. 628–631, 2007.

traffic demands. If the optimal routing sequence has any congested link, that link has to be eliminated from initial LSP and procedure starts again. Alternatively, adding capacity arrangement (short-term) can produce extra cost [6]. If the calculation finds the path without congestions on the path, new SLA can be accepted and related LSP is stored in database of BB. In the moment of service invocation stored LSP can be easily distributed from BB to the MPLS network to support explicit connection [5]. In opposite the SLA cannot be accepted or it must be re-negotiated.

The proposed routing technique can be seen as the capacity expansion problem (CEP) without shortages. The mathematical model explanation is given in the section 2. Heuristic algorithm development and testing results are discussed in the section 3.

2   CEP for Load Balancing Purpose

The routing problem explained above can be seen as the capacity expansion problem (CEP) without shortages and it can be formulated as Minimum Cost Multi-Commodity Flow Problem (MCMCF). Such problem (NP-complete) can be easily represented by multi-commodity the single (common) source multiple destination network. The diagram on fig. 1. shows the network flow representation of such problem with multiple QoS levels. Links connect M core routers (LSR) on the path.

Transmission link capacities on the path are capable to serve traffic demands for N different QoS levels (service classes). The N levels are ranked from $i = 1, 2,..., N$, from higher to lover quality level. Order number of the link is denoted with m, $m = 1, ..., M+1$. $r_{i,m}$ denotes traffic demand increment for additional capacity for each link on the path. $I_{i,m}$ represents the relative amount of idle capacity on the link; $I_{i1} = 0$, $I_{i,M+1} = 0$. $x_{i,m}$ represents the amount of used capacity and $y_{i,j,m}$ the amount of capacity for quality level i on the link m redirected to satisfy the lower quality level j.

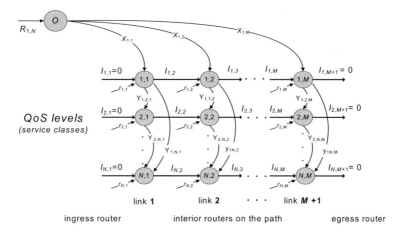

Fig. 1. The network flow representation of the CEP with application for congestion control

Let $G(A, E)$ denote a network topology, where E is the set of nodes, representing link capacity states and A, the set of arcs representing traffic flows between routers. In this paper we talk about one-dimensional link weight vectors for $M+1$ links on the path $\{w_{i,m}, m \in A, i = 1, \ldots, N\}$ with only one constraint, the capacity bounds, denoted with $L_{i,m}$ $(L_{1,m}$ $L_{2,m}, \ldots L_{N,m})$. For a non-additive measure (e.g. bandwidth) definition of the single-constrained problem is to find a path from ingress to egress node with minimum link weight along the path. The flow situation on the link depends of expansion and conversion values $(x_{i,m}, y_{i,j,m})$. It means that the link weight (cost) is the function of used capacity: lower amount of used capacity gives lower weight. If the link expansion cost corresponds to the amount of used capacity, the objective is to find optimal routing policy that minimizes the total cost on the path. Generalizing the concept of the capacity states we can define *a capacity point*:

$$\alpha_m = (I_{1,m}, I_{2,m}, \ldots, I_{N,m}) \tag{1}$$

$$\alpha_1 = \alpha_{M+1} = (0, 0, \ldots, 0) \tag{2}$$

Each column on the flow diagram (fig. 1.) represents a capacity point, consisting of N capacity state values for each. i-th QoS level. Each link capacity is capable to serve different service classes. The capacity class labeled with i is primarily used to serve demands of that service class but it can be used to satisfy traffic of lower QoS level j $(j > i)$. Formulation (2) implies that idle capacities or capacity shortages are not allowed on the link between edge and interior router.

The network optimization can be divided in two steps. In the first step we are calculating the minimal expansion weights between all pairs of capacity points. The calculation of each weight value we call it: capacity expansion sub-problem (CES).

Let C_m be the number of capacity point values at router position m. The total number of all sub-problem is:

$$N_d = \sum_{i=1}^{M} C_i \cdot \left[\sum_{j=i+1}^{M+1} C_j \right] \tag{3}$$

In the CEP we have to find many cost values $d_{u,v}(\alpha_u, \alpha_{v+1})$ that emanate two capacity points, from each node (u, α_u) to node $(v+1, \alpha_{v+1})$ for $v \geq u$. It is very important to reduce that number; see [3]. In the second step of network optimization we are looking for the shortest path in acyclic network with former calculated weights between node pairs (capacity points). It has to be noted that the optimal routing sequence for traffic (caused by new SLA) need not to be the shortest path solution.

3 Heuristic Approach and Results

The optimal flow theory enables separation of such extreme flows that cannot be a part of an optimal expansion solution from those which can be. So, many of expansion solutions are not acceptable. It can be shown that a feasible flow in the network (given in fig. 1.) corresponds to an extreme point solution only if it is not the part of any cycle (loop) with positive flows. This result holds for all single source networks. With such heuristic approach (denoted with BasicH) the near-optimal result is possible but significant computational savings appear, on average over than 50 %.

In real situation we can introduce some limitations on capacity state values, talking about different heuristic options: a) the total capacity of a capacity point can be negative but in acceptable limits (adding capacity from network provider). That option is denoted with AH. b) The capacity state value inside of a capacity point can be negative, but the total sum of the link capacity is positive (denoted with RH). c) Only no-negative capacity values are allowed (denoted with PH). d) Only null capacity values are allowed (TH - Trivial Heuristic option).

The heuristic algorithm with above mentioned options was tested in relation to efficiency of algorithm with exact approach. The basic heuristic approach (BasicH) requires effort of $O(M^3 N^4 R_i^{2(N-1)})$. For other heuristic options the complexity is significantly reduced (see fig. 2) but in all cases it is much over than $O(M^3 N^2)$, that is the complexity for a trivial heuristic solution (TH). For only few test-examples all algorithm options found the best expansion sequence, providing the minimal cost no matter of which heuristic approach is used. Only in the trivial heuristic option the significant deterioration of the best result is present.

The proposed heuristic algorithm with different options can be efficiently applied in off-line routing and on-demand resource allocation, helping us in congestion avoidance and load-balancing purpose. After the starting LSP is defined (e.g. with shortest path-based algorithm) we can check congestion probabilities on each link on the path. We can start with the algorithm of low complexity (TH or PH option). If the congestion appears we can apply more complex algorithm (RH, AH or BasicH). If the congestion still exists the adding capacity is necessary or the SLA negotiation has to be continued.

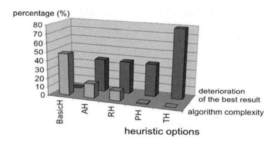

Fig. 2. Trend of the complexity changes for each heuristic option and the average value of the result deterioration

References

1. Sarangan, V., J.-C. Chen: Comparative Study of Protocols for Dynamic Service Negotiation in the Next-Generation Internet, IEEE Comm. Mag.,Vol.44, No.3 (2006) 151-156
2. Haddadou and all: Designing Scalable On-Demand Policy-Based Resource Allocation in IP Networks, IEEE Comm. Mag., Vol.44, No.3, (2006) 142-149
3. Krile, S., Kuzumilovic, D.: The Application of Bandwidth Optimization Technique in SLA Negotiation Process, Proc. of 11th CAMAD, Trento, (2006) 115-121
4. Bhatnagar, S.; Ganguly, S.; Nath, B.: Creating Multipoint-to-Point LSPs for Traffic Engineering, IEEE Comm. Mag., Vol.43., No.1, (2005) 95-100
5. Yu Cheng and all: Virtual Network Approach to Scalable IP Service Deployment and Efficient Resource Management, IEEE Comm. Mag., Vol.43, No. 10, (2005) 76-84
6. Anjali, T., Scoglio, C., de Oliveira, J.C.: New MPLS Network Management TechniquesBased on Adaptive Learning, Trans. on Neural Networks, Vol.16, No.5, (2005) 1242 - 1255

Simulating Trust Overlay in P2P Networks

Yan Zhang[1], Wei Wang[2], and Shunying Lü[1]

[1] Faculty Mathematics and Computer Science, Hubei University, 430062, Wuhan, China
[2] Department of computer Science and Technology, Tongji University, 201804, Shanghai, China
`willtongji@gmail.com`

Abstract. Though research on the overlay network has progressed at a steady pace, its promise has yet to be realized. One major difficulty is that, by its very nature, the overlay networks is a large, uncensored system to which anyone may contribute. This raises the question of how much dependability to give each information source. Traditional overlay network simulators provide accurate low-level models of the network hardware and protocols, but none of them deal with the issue of trust and reliability in the large scale overlay networks. We tackle this problem by employing a trust overlay simulator, which offer a viable solution to simulate trustworthy behavior in overlay networks. With this simulator, we can examine varies kinds of trust and reputation mechanisms in a large scale overlay environment.

Keywords: overlay networks, trust, simulator, security, peer-to-peer.

1 Introduction

Over the past two decades, researchers have proposed lots of solutions to extend the Internet's functionality, and to improve its resilience and security [1]. After sustained efforts to add new functions such as mobility and multicast to IP, researchers have recently turned their attention to developing new network architectures [2, 3] and using overlays to address the Internet's limitations.

Overlay networks tend to be large, heterogeneous systems with complex interactions between the physical machines, underlying network, application and user. Hence, developing and testing an overlay-oriented algorithm or protocol in a realistic environment is often not feasible. So, it is possible to use a simulation of an overlay network to evaluate the proposed applications and protocols in controlled environment.

However, though research on these works has progressed at a steady pace, its promise has yet to be realized. One major difficulty is that, by its very nature, the overlay networks is a large, uncensored system to which anyone may contribute, for example using a P2P infrastructure for information sharing. This raises the question of how much dependability to give each information source. Traditional network simulators only provide low-level models of the network hardware and protocols and do not consider trustworthy behavior of the overlays. So, simulating trustworthy behavior in overlay networks is a challenge task in recent simulation research. The aim of this paper is to motivate the need for overlay network simulators for developing and testing trust and reputation-based schemas.

Y. Shi et al. (Eds.): ICCS 2007, Part I, LNCS 4487, pp. 632–639, 2007.

In this paper, we propose a novel *trust overlay simulator* (TOSim). TOSim is developed based on PeerSim which is wildly used in simulating overlay networks [8, 9]. We extend this simulator by introducing varies kinds of components into it, which is important when simulating trust and reputation mechanisms. The advantage of this proposed simulator is that it can easily simulate varies kinds of dynamic behavior in overlay networks which can be used to evaluate trust and reputation mechanism proposed by other researches. With this simulator, research can develop and evaluate their proposed trust-related protocols easily in an overlay enrolment.

The rest of the paper is organized as follows. We review some related work in Section 2. Section 3 addresses the detail of the trust overlay simulator design. Section 4 describes the threat model in the proposed simulator. Section 5 makes simulation experiments. Then the paper concludes in Section 6.

2 Related Work

Traditional network simulators provide accurate low-level models of the network hardware and protocols but are too detailed to be effective in analysis of large scale overlay networks. For example, the *NS2 simulator* [5] is perhaps the most widely used networking simulator. In addition, there is a number of existing P2P simulators. But none of them can simulate the need of trust and reputation mechanisms, or trustworthy behavior, in overlay networks.

Packet-level P2P [6] models on an otherwise packet simulator. It creates wrappers that translate P2P level events into commands to the underlying packet simulator. *SimP2* [7] is a graph-based simulator for analysis of ad-hoc P2P networks. The analysis is based on a non-uniform random graph model, and is limited to studying basic properties such as reach ability and nodal degree. *Peersim* [8] is a Java based search framework that allows modeling of P2P overlay search algorithms, which is under the GPL open source license. *DHTSim* [17] is a discrete event simulator for structured overlays, specifically DHTs. It is intended as a basis for teaching the implementation of DHT protocols, and as such it does not include much functionality for extracting statistics. *P2PSim* [18] is a discrete event packet level simulator that can simulate structured overlays only. It contains implementations of six candidate protocols: Chord, Accordion, Koorde, Kelips, Tapestry and Kademlia. *PlanetSim* [19] is a discrete-event overlay network simulator, written in Java. It supports both structured and unstructured overlays, and is packaged with Chord- SIGCOMM and Symphony implementations.

On the other hand, varies kinds of trust and reputation models are proposed to deal with the problem of lacking trust in overlay networks. EigenTrust algorithm [13] is proposed to decrease the number if downloads of insecurity files in P2P networks that assigns each peer a unique global trust value, based on the peer's history of uploads. The framework for trust reasoning in distributed systems (FTRDS) in [14] is based on sociological studies. A reputation information exchange amongst members of the community assists on trust decisions. Elements of the work have been incorporated into Sun's JXTA framework and Ericsson Research's prototype model. The mathematical framework for modeling trust and reputation (MFMTR) in [15] is rooted in finding from the social science. The framework makes explicit the importance of social information (i.e. indirect channels of inference) in aiding members of a social network choose whom they want to partner with or to avoid. We also proposed our own Bayesian trust model in networked computing environment [4, 10].

In this paper, we propose a novel trust overlay simulator (TOSim) which can simulate the need of trust and reputation mechanisms in overlay networks.

3   Trust Overlay Simulator Design

The architecture of overlay networks is illustrated in Figure 1. According to it, TOSim has been designed to be highly modular and configurable, without incurring in excessive overhead both in terms of memory and time. The simulated network is composed of a collection of nodes, and each of them may host one or more protocols.

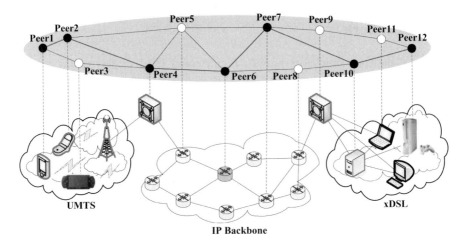

Fig. 1. Architecture of the overlay networks

A trust overlay simulator for P2P systems may have very different objectives from general-purpose networks simulators:

- **Extreme scalability.** Simulated networks may be composed of millions of nodes. This may be obtained only if a careful design of the memory layout of the simulator is performed. Being able to store data for a large number of nodes, however, is not the only requirement for large-scale simulations; the simulation engine must be optimized as well, trying to reduce, whenever possible, any form of overhead.
- **Support for dynamicity.** The simulator must be capable to deal with nodes that join and leave the network, either definitively or temporarily. This has some implications on memory management in the simulator, requiring mechanisms for removing nodes that are not useful any more.
- **Simulation all kinds of threads model.** As for developing and testing trust and reputation-based applications, the proposed simulator should simulate all kinds of behavior in the overlay network. We perform this by developing some kinds of thread models which will be described in section 4.

In addition to these requirements, the modular approach we are pursuing in this paper must be reflected in the architecture of the simulation environment as well. The idea is to provide a composition mechanism that enables the construction of simulations as collections of components. Every component of the simulation (for example, proto-cols or the environment) must be easily replaceable through simple configuration files. The flexibility offered by this mechanism should enable developers to re-implement, when needed, every component of the system, with the freedom of re-using existing components for fast prototyping.

Network model: We consider a typical P2P network (other overlay scheme will be developed in the future) : Interconnected, file-sharing peers are able to issue queries for files, peers can respond to queries, and files can be transferred between two peers to conclude a search process. When a query is issued by a peer, it is propagated by broadcast with hop-count horizon throughout the network; peers which receive the query forward it and check if they are able to respond to it.

Node model: Our simulator consists of good nodes (normal nodes, participating in the network to download and upload files) and malicious nodes (adversarial nodes, participating in the network to undermine its performance). We consider different threat models, where a threat model describes the behavior of a malicious peer in the network. Threat models will be described in more detail later on. Note also that some good nodes in the network are appointed as highly trusted nodes.

Simulation engine: The simulation engine is the module that will actually perform the simulation; its task is to orchestrate the execution of the different components loaded in the system. The engine adopts a time-stepped simulation model instead of more complex and expensive event-based architecture. At each time step, all nodes in the system are selected in a random order, and a callback method is invoked on each of the protocols included in that node.

Content distribution model: Interactions between peers – i.e., which queries are issued and which queries are answered by given peers – are computed based on a probabilistic content distribution model. Briefly, peers are assumed to be interested in a subset of the total available content in the network, i.e., each peer initially picks a number of content categories and shares files only in these categories. Reference [11] has shown that files shared in a P2P network are often clustered by content categories. Also, we assume that within one content category files with different popularities exist, governed by a Zipf distribution. The number of files shared by peers and other distributions used in the model are taken from measurements in real-world P2P net-works [12].

Simulation execution: The simulation of a network proceeds in simulation cycles: Each simulation cycle is subdivided into a number of query cycles. In each query cycle, a peer in the network may be actively issuing a query, inactive, or even down and not responding to queries passing by. Upon issuing a query, a peer waits for in-coming responses, selects a source among those nodes that responded and starts downloading the file.

Some of these goals may appear contradictory. A careful design is needed trying to obtain the best equilibrium.

4   Threat Model

In order to simulate behaviors related to trust, we consider several strategies of malicious peers to cause insecurity upload [13]. In short, malicious peers operating under *threat model A* simply try to upload insecurity files and assign high trust values to any other malicious peer they get to interact with while participating in the network. In *threat model B*, malicious peers know each other upfront and deterministically give high local trust values to each other. In *threat model C*, malicious peers try to get some high local trust values from good peers by providing authentic files in some cases when selected as download sources. Under *threat model D*, one group of malicious peers in the network provides only authentic files and uses the reputation they gain to boost the trust values of another group of malicious peers that only provides insecurity files.

Threat Model A: *Individual Malicious Peers.* Malicious peers always provide an insecurity file when selected as download source. For example, you can set Malicious peers local trust values to be *insecurity* file downloads instead of authentic file downloads.

Threat Model B: *Malicious Collectives.* Malicious peers always provide an insecurity file when selected as download source. Malicious peers form a malicious collective by assigning a single trust value of 1 to another malicious peer in the network. In terms of the probabilistic interpretation of our scheme, malicious peers form a collective out of which a random surfer or agent, once it has entered the collective, will not be able to escape, thus boosting the trust values of all peers in the collective.

Threat Model C: *Malicious Collectives with Camouflage.* Malicious peers provide an insecurity file in all cases when selected as download source. Malicious peers form a malicious collective as described above.

Threat Model D: *Malicious Spies.* Malicious peers answer a small fraction of the most popular queries and provide a good file when selected as download source. Malicious peers of type D assign trust constant values to all malicious nodes of type B in the network.

Threat Model E. *Sybil Attack.* An adversary initiates thousands of peers on the network. Each time one of the peers is selected for download, it sends an insecurity file, after which it disconnected and replaced with a new peer identity.

Threat Model F: *Virus-Disseminators.* (Variant of threat model C) A malicious peer sends one virus-laden (insecurity) copy of a particular file every 100th request. At all other times, the authentic file is sent.

   All of these threat models can be easily simulated in TOSim by setting the appropriate parameters. And researchers can carry their experiments on the simulator under different kinds of threat situations.

5   Simulation Experiments and Analysis

The presented TOSim in the previous section was conceived in such a way that existing trust models could be easily simulated in the simulator. In the following, we present the mapping of the local trust models to the introduced model and suggest some algorithmic adaptations. The investigated algorithms discussed here are proposed by Abdul-Rahman et al. [14], Lik Mui et al. [15] and Sepandar D. Kamvar et al. [13] separately, which have been introduced in related work section.

To assess the quality of the trust algorithms presented above, a series of test scenarios was developed. Each scenario simulates a different behavior pattern of trustor and trustee as a list of ratings. This pattern is then reflected by each single trust algorithm as trust dynamics. A test scenario can bring forward a specific feature or a malfunction of a trust update algorithm.

We want to stress the fact that due to the subjectiveness of trust in general, also the quality estimation of the behavior reflected in the trust dynamics is subjective. Therefore, we do not offer a ranking of trust algorithms, but instead point out the distinctive features of the algorithms, so that each user can simulator the algorithm that most closely reflects his expected behavior.

Based on the previous research [16], we choose the following two scenarios for our simulator:

MinMaxRatings: First, a series of minimal ratings is given, which is followed by a series of maximal ratings. We would expect the trust dynamics to decrease and eventually approach the minimal trust value. After switching to maximal ratings, trust should rise again.

MaxMinRatings: First, a series of maximal ratings is given, followed by a series of minimal ratings. We expect trust to rise at first. When the series of minimal ratings starts, trust should decrease again.

In MinMaxRatings (Figure 2) when the maximal ratings start, trust starts rising again in all analyzed algorithms but FTRDS. In this latter case, trust remains at the lowest level until as much maximal ratings as minimal ratings have been received.

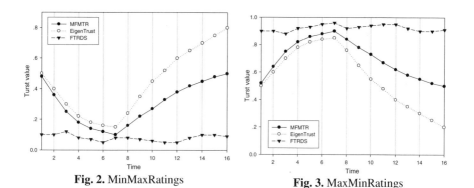

Fig. 2. MinMaxRatings   **Fig. 3.** MaxMinRatings

The MaxMinRatings test scenario (Figure 3) shows similar results as the previous scenario. When the minimal ratings start, trust drops with all but FTRDS's algorithm. Here, trust suddenly drops from maximum to the minimal value at the end of the scenario which is the point when more minimal than maximal ratings are recorded in the history. In both scenarios we notice that EigenTrust shows a quick reaction to the pattern change in the ratings.

6 Conclusions

In this paper, aiming at the characteristic of trust overlay networks, we presented a trust mechanism oriented overlay simulator. The presented simulator can simulate varies kinds of peers' behavior with different kinds of thread models. Besides, we simulate some experiments with three different kinds of trust mechanisms, and analyze the results. The ever increasing demand of new applications has led researchers to propose new network architectures that address limitations of the current Internet. Given the rigidity of the Internet today, overlay networks are used to implement such architectures, in the hope of gaining a large user base.

References

1. Joseph, D., Kannan, J., et al.: OCALA: An Architecture for Supporting Legacy Applications over Overlays, (2005)
2. Andersen, D., Balakrishnan, H., Kaashoek, F., et al.: Resilient Overlay Networks. In Proceedings of SOSP'01, (2001)
3. Stoica, I., Adkins, D., Zhuang, S., et al.: Internet Indirection Infrastructure. In Proceedings of SIGCOMM'02, (2002)
4. Wang, W., Zeng, G. S., Yuan, L. L.: A Semantic Reputation Mechanism in P2P Semantic Web, In: Proceedings of the 1st Asian Semantic Web Conference (ASWC), LNCS 4185 (2006) 682-688
5. Berkeley/LNBL/ISI. The NS-2 Network Simulator. http://www.isi.edu/nsnam/ns/
6. He, Q., Ammar, M., Riley, G., et al.: Mapping Peer Behavior to Packet-level Details: A Framework for Packet-level Simulation of Peer-to-Peer Systems. In Proceedings of MASCOTS 2003, Orlando, FL, (2003)
7. Kant, K. Iyer, R.: Modeling and Simulation of Adhoc/P2P Resource Sharing Networks. In Proceedings of TOOLS'03, (2003)
8. Jelasity, M., Montresor, Jesi, A. G. P.: Peersim Peer-to- Peer Simulator, (2004). http://peersim.sourceforge.net/
9. Mark J., Alberto M., Ozalp B. Gossip-based aggregation in large dynamic networks. ACM Transactions on Computer Systems, 23(3) (2005) 219–252
10. Wang, W., Zeng, G. S., Liu, T.: An Autonomous Trust Construction System Based on Bayesian Method, In: Proceedings of the IEEE/WIC/ACM International Conference on Intelligent Agent Technology (IAT'06), Hong Kong, China: IEEE Computer Society Press, December 18-22 (2006) 357–362
11. Arumugam, M., Sheth, A., Arpinar, I. B.: Towards Peer-to-Peer Semantic Web: A Distributed Environment for Sharing Semantic Knowledge on the Web. Technical report, Large Scale Distributed Information Systems Lab, University of Georgia, (2001)
12. Saroiu, S., Gummadi, P. K., Gribble, S. D.: A Measurement Study of Peer-to-Peer File Sharing Systems. In Proceedings of Multimedia Computing and Networking 2002 (MMCN '02), San Jose, CA, USA, January (2002)
13. Kamvar, S. D., Schlosser, M. T., Garcia-Molina, H.: The Eigentrust algorithm for reputation management in p2p networks, In Proceedings of the 12th International Conference on World Wide Web, ACM Press, (2003) 640–651
14. Abdul-Rahman, A.: A Framework for Decentralized Trust Reasoning. PhD thesis, Department of Computer Science, University College London, (2004)

15. Mui, L.: Computational Models of Trust and Reputation: Agents, Evolutionary Games, and Social Networks. PhD thesis, Massachusetts Institute of Technology, (2003)

16. Kinateder, M., Baschny, E., Rothermel, K.: Towards a Generic Trust Model – Comparison of Various Trust Update Algorithms. iTrust 2005, LNCS 3477 (2005) 177–192

17. 17."DHTSim"   accessed   01-May-2006.   [Online].   Available: http://www.informatics.sussex.ac.uk/users/ianw/teach/dist-sys/

18. "P2Psim: A Simulator for Peer-to-Peer (P2P) Protocols," 2006. [Online]. Available: http://pdos.csail.mit.edu/p2psim/

19. Virgili, U. R.: "PlanetSim: An Overlay Network Simulation Framework," (2006). [Online]. Available: http://planet.urv.es/planetsim/

Detecting Shrew HTTP Flood Attacks for Flash Crowds

Yi Xie and Shun-Zheng Yu

Department of Electrical and Communication Engineering Sun Yat-Sen University,
Guangzhou 510275, P.R. China
xieyicn@163.com

Abstract. Countering network attacks is becoming ever more challenging. Web-based vulnerabilities represent a substantial portion of the security exposures of computer networks. In order to detect a new Web-based assault named shrew Distributed Denial of Service attacks based on HTTP flood, Principle Component Analysis and Independent Component Analysis are applied to abstract the multivariate observation vector. A novel anomaly detector based on hidden semi-Markov model is proposed. Experiment results based on real traffic trace and emulated attacks show, the scheme can be used effectively to implement the detection of the shrew HTTP flood attacks embedded in the normal flash crowd of large-scale Website; and the detection is not dependent on the intensity of attack traffic.

1 Introduction

In the past ten years, Web-based applications have become a popular way to provide information and services dynamically. Electronic commerce, ranging from internet shopping malls to Web-based banking and trading, has become an integral component of modern society. At the same time, Web servers and Web-based applications are popular attack targets. CVE [1] reported Web-based vulnerabilities accounted for more than 25% of the total number of security flaws during 1999 to 2005.

For these reasons, studies have been done to detect Web-based attacks. However, it isn't enough for most current works to detect the special new Web-based attacks known as "Application layer Distribute Denial of Service" (App-DDoS) attacks. The main difference between the App-DDoS attacks and the traditional DDoS attacks is that they are protocol-compliant, non-intrusive, and utilize legitimate Web requests to overwhelm system resources. App-DDoS attacks utilize the weakness enabled by the standard practice of opening services such as HTTP (TCP port 80) and HTTPS (TCP port 443) through most firewalls to launch the flood. Many applications and protocols, both legitimate and illegitimate, can use these openings to tunnel through firewalls by connecting over standard TCP port 80 (e.g., Code Red virus) or encapsulating in SSL tunnels (HTTPS). Attack packets aiming at these services pass through the firewall without being identified. Thus, most current intrusion detections designed for Web-based attacks and DDoS attacks become invalid when they are used to block this type of sophisticated attacks. This is especially the case when the attackers secrete the offensive traffic in the flash crowds of special heavily-accessed Web server environments (e.g., Olympic Games Website, e-commerce Website).

Y. Shi et al. (Eds.): ICCS 2007, Part I, LNCS 4487, pp. 640–647, 2007.

This paper focuses on the detection of shrew HTTP flood attacks under unstable and bursty background traffic (e.g., flash crowd). Shrew HTTP flood is a special type of App-DDoS attacks which is a compound of shrew attack and HTTP flood. The difficulties of its detection include: (i) in stead of constantly injecting traffic flows with huge rates into the network, shrew attackers launch assaults by sending burst pulses periodically, which has high peak rate while maintaining a low average rate to exhibit "stealthy" behavior; and (ii) burst traffic and high volume are the common characteristics of HTTP flood and flash crowd, which make the detection much more difficult. This paper meets this challenge by a novel detection scheme.

We carry out the anomaly detection by file popularity which has been proved to have significant temporal stability in [7]. The main contributions of this paper are (i) development of a new efficient mechanism to detect the shrew HTTP flood attacks which mimic (or mix with) the normal flash crowds and utilizes legitimate requests to overwhelm system resources, (ii) applying principal component analysis (PCA) [8] and independent component analysis (ICA) [9] to abstract multivariate observation vector, (iii) using hidden semi-Markov model (HsMM) [10] to construct the anomaly recognizer, and (iv) experimentation of a real traffic of heavily-accessed Web server and an emulated shrew HTTP flood attack to evaluate the model.

The rest of the paper is organized as follows. In section 2, we describe related work in Web-based attacks detection. In section 3, we present the rationale of this work. In section 4, we conduct experiments to validate our detection algorithms. Finally, in section 5, we conclude our work.

2   Related Work

The detection of Web-based attacks has recently received considerable attention because of the increasingly critical role that Web-based services are playing, e.g., [3][4]. However, current detection systems are mostly misuse-based and therefore suffer from the problem of not being able to detect new attacks. Recently, Juan [5] accomplished the detection by analyzing each incoming HTTP payloads, which may lead to high computational complexity. In [6], Tombini proposed a serial architecture where Web-related events are first passed through an anomaly detection component. Then, the events that are identified as neither normal nor intrusive are passed on to a misuse detection component, but it requires extensive manual analysis to evaluate the characteristics of the events being analyzed.

A valuable recent work on App-DDoS can be found in [2][11]. In [2], the authors developed a counter-mechanism which assigns a suspicion measure to a session in proportion to its deviation from legitimate behavior. However, this work impliedly assumes the incoming traffic is stable, which is not always true in large-scale heavily-accessed Websites. In [11], Kandula et al. designed a system to protect a Web cluster from DDoS attacks by designing a probabilistic authentication mechanism using "puzzles". Unfortunately, requiring all users to solve graph puzzles has the possibility of annoying users and introducing additional service delays for legitimate users. This also has the effect of denying Web crawlers access to the site and as a result search engines may not be able to index the content. Finally, new techniques may render the graphical puzzles solvable using automated methods [12]. Other researchers have explored the defense of shrew attacks, e.g., [13]. However, authors of these literatures merely aimed at the TCP attacks in stead of HTTP flooding attacks. Furthermore, in contrast to our burst and unstable background traffic of detection scenario, their common implied hypothesis is that the background traffic is stable, which may not help in this paper.

3   Detection Algorithms

In contrast to the prior work, the detection scheme proposed in this paper uses passive detection which applies statistical methods to detect shrew HTTP flood attacks. Our observation data is file popularity which can be derived from the Web server log easily. In order to perform efficient, unsupervised, learning-based anomaly detection, this paper proposes a novel detection scheme based on multivariate statistical analysis and outlier detection. PCA and ICA are used to abstract the observed data, and then the HsMM is applied to construct the anomaly detector.

3.1   File Popularity Matrix

We define the $N \times T$ dimensional file popularity matrix as:

$$A_{N \times T} = \begin{bmatrix} a_{11} & a_{12} & \cdots & a_{1T} \\ a_{21} & a_{22} & \cdots & a_{2T} \\ \cdots & \cdots & \cdots & \cdots \\ a_{N1} & a_{N2} & \cdots & a_{NT} \end{bmatrix} = \begin{bmatrix} \mathbf{a}_1 \\ \mathbf{a}_2 \\ \cdots \\ \mathbf{a}_N \end{bmatrix} = \begin{bmatrix} \vec{a}_1 \\ \vec{a}_2 \\ \cdots \\ \vec{a}_T \end{bmatrix}^{\mathrm{T}} \tag{1}$$

where a_{it} denotes the popularity of the ith document at t^{th} time unit, $\mathbf{a}_i = \begin{bmatrix} a_{i1} & \cdots & a_{iT} \end{bmatrix}$ is the time series of ith document's popularity, $\vec{a}_t = [a_{1t} \quad \cdots \quad a_{Nt}]^{\mathrm{T}}$ denotes the popularities of N documents at t^{th} time unit, which forms the N-dimensional observation vector. Especially, if we consider the documents as a subspace and time as another subspace, the file popularity matrix reflects the spatial-temporal patterns of all users' access behaviors. Distribution of $\vec{a}_t$ presents the spatial distribution of popularity at t^{th} time unit while the distribution of a_i presents the temporal distribution of the ith document's popularity. For brevity of notation, we call them spatial distribution and temporal distribution, respectively.

   In [7], the authors have found, although the domains from which clients access the popular content are unstable, there is significant temporal stability in file popularity of busy Website. Thus, file popularity is used as detection signal in this paper.

3.2   HsMM

Because of the high dimensional document space (N), we can not monitor the shrew HTTP flood attacks merely by checking every document's popularity distribution. Furthermore, detection through checking popularity distribution of each document solely may give wrong warning signals because changes of individual document's popularity distribution don't imply attacks. Hence, we consider the joint distribution of N-dimensional document space to implement numerical and effective detection in this paper. We extend the HsMM [10] to capture the dynamics of file popularity matrix and monitor the shrew HTTP flood attacks during flash crowd event.

HsMM is a Hidden Markov Model (HMM) with variable state duration. It has been widely used in recognition and detection because it is very suitable to describe most practical physical signals (e.g., speech recognition, character recognition and DNA sequences clustering). The HsMM is a stochastic finite state machine, specified by (S, π, A, P) where S is a discrete set of hidden states with cardinality M; π is the probability distribution for the initial state $\pi_i \equiv \Pr[s_1 = i]$ and satisfy $\sum_i \pi_i = 1$, s_t denotes the state that the system takes at time t and $i \in S$; A is the state transition matrix with probabilities $a_{ij} \equiv \Pr[s_t = j | s_{t-1} = i]$ and satisfy $\sum_j a_{ij} = 1$, $i, j \in S$; P is the state duration matrix with probabilities $p_i(d) \equiv \Pr[\tau_i = d | s_t = i]$ and satisfy $\sum_d p_i(d) = 1$, τ_i denotes the remaining (or residual) time of the current state s_t, $i \in S$, $d \in \{1, ..., D\}$, D is the maximum interval between any two consecutive state transitions. For brevity of notation, we denote the complete set of model parameters: $\lambda = (\{\pi_i\}, \{a_{ij}\}, \{b_i(k)\}, \{p_i(d)\})$. The entropy ($En$) of observation fitting to the HsMM and the average logarithmic entropy (ALE) per observation is defined as the measure criteria as following, respectively:

$$En = \Pr[\bar{o}_1^T | \lambda] = \sum_{i,d} \Pr[\bar{o}_1^T, (s_T, \tau_T) = (i, d) | \lambda] , \qquad (2)$$

$$ALE = \ln(\Pr[\bar{o}_1^T | \lambda]) / T , \qquad (3)$$

In our previous work [10], we have developed efficient parameter reestimate algorithm and dynamic update algorithm for HsMM. Based on the work, HsMM is applied to construct the anomaly detector for shrew HTTP flood attacks. We consider the file popularity matrix as an N-dimensional stochastic process which is controlled by an underlying semi-Markov process. For an HsMM whose parameters are given, hidden state s_t of the HsMM can be used to describe a representative spatial distribution of popularity of N documents at t^{th} time unit. Transitions of the hidden states (i.e., from s_{t-1} to s_t) can be considered as the characteristics of access behaviors from one spatial distribution (s_{t-1}) to another spatial distribution (s_t). Duration (τ_t) of a hidden state (s_t) can be considered as the time unit amount that the current spatial distribution (s_t) will persists.

Considering the HsMM will be very complex when the observation is a multidimensional vector with dependent elements, we use the Principal component analysis [8] to reduce the dimensionality of file popularity matrix and apply the Independent component analysis [9] to obtain independent vector.

3.3 PCA and ICA

PCA is a well-established technique for dimensionality reduction and multivariate analysis. Examples of its many applications include data compression, image processing, visualization, exploratory data analysis, pattern recognition, time series prediction and distributed sensor network.

PCA aims to find a linear orthogonal transformation:

$$\bar{y}_t = E\bar{a}_t , \qquad (4)$$

where the rows of E are the eigenvectors of the covariance matrix of the original N-dimensional data $\bar{x}$ (assumed to have 0 mean, condition that can always be easily achieved by subtracting the mean estimated over the training set). The jth principal component (PC, the jth element of $\bar{y}_t$) of a vector $\bar{a}_t$ corresponds to its projection along the direction of the jth eigenvector of the data covariance matrix, and the principal components are non-correlated. The eigenvectors are ordered so that, if σ_i^2 is the eigenvalue related to the ith eigenvector, then $\sigma_1^2 \geq \sigma_2^2 \geq \cdots \geq \sigma_m^2$. The eigenvalue accounts for the data variance along the direction of the corresponding eigenvector. The first eigenvector indicates the direction of highest variance in the data, the second one the direction orthogonal to the first one with the highest variance and so on. Often the last eigenvectors account for very small variance and the corresponding principal components can be eliminated. Thus, we choose the first K eigenvectors having the largest eigenvalues. This implies that K is the inherent dimensionality of the subspace governing the "signal" while the remaining $(N-K)$ dimensions generally contain noise. The dimensionality of the subspace K can be determined by:

$$\sum_{i=1}^{K} \lambda_i / \sum_{i=1}^{N} \lambda_i \geq \alpha , \tag{5}$$

where α is the ratio of variation in the subspace to the total variation in the original space. This allows a reduction of the dimensionality making the following anomaly detector based on HsMM faster and better trainable (the number of parameters is reduced as it depends on the size of the observation vector). A complete discussion of PCA can be found in [8].

ICA is a new statistical signal processing technique which is widely used in blind source separation (BSS). In contrast to the PCA which is sensitive to high-order relationships, the basic idea of ICA is to represent a set of random variables using basis function, where the components are statistically independent and non-gaussianity as possible. The main aim of ICA is to find the linear transformation:

$$\bar{y}_t = W\bar{a}_t , \tag{6}$$

where W is called unmixing matrix which satisfies $p(y_i,y_j)=p(y_i)p(y_j)$ for $i{\neq}j$, where y_i is the ith element of $\bar{y}_t$. Many algorithms can be used to solve this non-Gaussian problem of ICA. This paper uses the FastICA algorithm [9] which has good performance in simulations and fast convergence during the parameter reestimate.

3.4 Detection Architecture

The overall procedure of our detection architecture is illustrated in Fig. 1. The scheme includes training phase and detection phase. The training phase includes the following five steps: (i) compute the file popularities of each document; (ii) input the file popularity matrix into PCA module and compute the orthogonal transformation E and PCs; (iii) remain the first K PCs by (5), α=80% in this paper; (iv) input the first K PCs into the ICA module and compute the unmixing matrix W and ICs; and (v) use the ICs to estimate the parameters of HsMM.

The detection is implemented as the follows: (i) compute the file popularities of each document; (ii) compute the first K PCs of the observed data based on E; (iii) compute the ICs of the first K PCs based on W; and (iv) ompute the entropy of observed data fitting to the HsMM by (2) and (3);

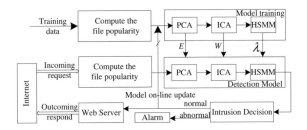

Fig. 1. Anomaly detector

In the practical implementation, the model is first trained by the stable and low volume Web workload whose normality can be ensured by most existing anomaly detection systems, and then it is used to monitor the following Web workload for a period of T minutes. When the period is past, model will be updated by the new collected Web workload whose normality is ensured by its entropy fitting to the model. After the update, the model enters the next cycle for App-DDoS attacks detection. Web traffic is reported as anomalous if the difference between its entropy and normal level is larger than the predefined threshold

4   Experiments

We use the trace of FIFA 1998 WorldCup (http://ita.ee.lbl.gov/html/contrib /WorldCup.html) and emulate shrew HTTP flood attacks to validate the model.

In contrast to the traditional shrew attacks with constant attack parameters, we emulate a more stealthy type of shrew attacks of HTTP flood whose attack parameters are all chosen randomly by attack nodes. As shown in Fig. 2, a single source stochastic pulsing attack is modeled as a square waveform HTTP request stream with an attack period of T, length of the burst l, and the burst rate H. Among the variables, T, l, H and Δt are stochastic variables reset by each attack node randomly before it is going to fire the next pulsing attack, which makes the different attack nodes have different attack parameters and produces different attack parameters at different attack period. Thus, the attacks exhibit a fluctuating rate oscillating between different H and zero and dynamics of a pulsing attack appear as a stochastic ON/OFF pattern.

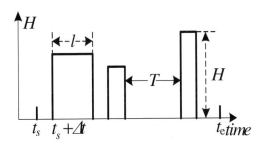

Fig. 2. Shrew HTTP flood attack

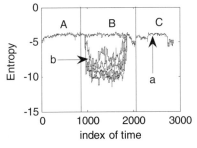

Fig. 3. Pulsing rate attack

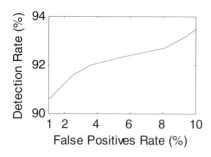

Fig. 4. ROC curve

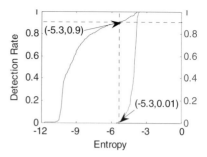

Fig. 5. Entropy, DR and FPR

Fig. 3 shows the entropy varies with the time, where curve a represents the entropy of normal flash crowd and curve cluster b represents the entropy of normal flash crowd embedded with shrew HTTP flood attacks in zone C. As shown in the figure, although the flash crowd is burst and high volume, the document popularity is stable because of the steady access behavior of individual Web surfer, thus, the entropy series is stable. But, when the vicious requests are added into the flash crowd, the original popularity distribution of documents is destroyed, which cause the entropy series deviates from the normal level. Thus, the potential shrew HTTP flood attacks can be detected by the entropy of document popularity fitting to the model.

Fig. 4 is the receiver operating characteristics (ROC) curve showing the performance of our detection model. Fig. 5 shows, if we take -5.3 for threshold value of normal Web traffic's average entropy, the False Negative Ratio (FNR) is about 1%, and the Detection rate is about 90%.

5   Conclusions and Future Work

Based on PCA, ICA and HsMM, this paper proposes a novel proposal to detect the shrew HTTP flood attacks occurring during the flash crowd by monitoring the file popularity of Web server. Experiment results based on a real traffic and emulated attacks show the model could capture shift of Web traffic caused by the attacks occurring during flash crowd and might be both practical and helpful for triggering more focused detection and filtering in victim network.

One disadvantage of our scheme is the technique cannot readily distinguish and filter the malicious sources from the normal ones, which make it merely serve as an alert function to trigger more detailed monitoring mechanisms. We will improve it in our further work.

Acknowledgments

This work was supported by National Natural Science Foundation of China under grant no. 90304011, Guangdong Natural Science Foundation under grant no. 04009747 and Higher Education Foundation for Ph.D Program under grant no. 20040558043.

References

1. Common Vulnerabilities and Exposures. http://www.cve.mitre.org/, 2005
2. S. Ranjan, R. Swaminathan, M. Uysal, and E. Knightly. "DDoS-Resilient Scheduling to Counter Application Layer Attacks under Imperfect Detection" in Proceedings of IEEE INFOCOM. April 2006. online: http://www-ece.rice.edu/ networks/papers/dos- sched.pdf
3. M. Almgren, U. Lindqvist, "Application-integrated data collection for security monitoring" in Proceedings of Recent Advances in Intrusion Detection, Davis, CA, October 2001, LNCS, Springer, 2001, pp.22–36.
4. G. Vigna, W. Robertson, V. Kher, R.A. Kemmerer, "A stateful intrusion detection system for world-wide Web servers" in the Proceedings of the Annual Computer Security Applications Conference, Las Vegas, NV, December 2003, pp. 34–43.
5. Juan M. Estévez-Tapiador, Pedro García-Teodoro, Jesús E. Díaz-Verdejo. "Detection of Web-based Attacks through Markovian Protocol Parsing" in the Proceedings of the 10th IEEE Symposium on Computers and Communications.
6. E. Tombini, H. Debar, L. Me, and M. Ducasse. "A Serial Combination of Anomaly and Misuse IDSes Applied to HTTP Traffic" In Proceedings of the Twentieth Annual Computer Security Applications Conference, Tucson, Arizona, December 2004.
7. Venkata N. Padmanabhan and Lili Qiu. "The content and access dynamics of a busy Web site: Findings and implications" In the proceeding of ACM SIGCOMM 2000, Stockholm, Sweden, Aug. 2000.
8. Lindsay I Smith. "A tutorial on Principal Components Analysis" [EB/OL]. http://www.snl.salk.edu/~shlens/pub/ notes/pca.pdf, 2003-03.
9. A. Hyvärinen. "Survey on Independent Component Analysis" Neural Computing Surveys, 1999,2. pp.94-128.
10. S.-Z. Yu, and H. Kobayashi, "An Efficient Forward-Backward Algorithm for an Explicit Duration Hidden Markov Model" IEEE Signal Processing Letters, Vol. 10, No. 1, January 2003, pp. 11-14.
11. S. Kandula, D. Katabi, M. Jacob, and A. W. Berger. "Botz-4-sale: Surviving organized DDoS attacks that mimic flash crowds" Technical Report TR-969, MIT., October 2004. Online: http://www.usenix.org/events/nsdi05/tech /kandula/kandula.pdf
12. G.Mori and J.Malik. "Recognizing objects in adversarial clutter: Breaking a visual captcha" IEEE Computer Vision and Pattern Recognition, 2003.
13. X. Luo, and R. K. C. Chang, "On a New Class of Pulsing Denial-of-Service Attacks and the defense," Network and Distributed System Security Symposium (NDSS'05), San Diego, CA., Feb. 2-5, 2005.

A New Fault-Tolerant Routing Algorithm for m-ary n-cube Multi-computers and Its Performance Analysis

Liu Hongmei

College of Science, China Three Gorges University
Hubei Yichang 443002, PRC
liuhm@ctgu.edu.cn

Abstract. A new algorithm for fault-tolerant routing based on detour and backtracking techniques is developed for m-ary n-cube multi-computer networks. We analyzed its performance under the condition that when an arbitrary number of components have been damaged and derived some exact expressions for the probability of routing messages via optimal paths from the source node to obstructed node. The probability of routing messages via an optimal path between any two nodes is a special case of our results, and can be obtained by replacing the obstructed node with the destination node.

1 Introduction

m-ary n-cube usually denoted by $Q_n(m)$ is a kind of generalized hypercube and has drawn many attention. [1] showed that the restricted connectivity is $2n(m-1)-m$ and the restricted fault diameter can be controlled to be less than $n+3$. [2] and [3] concluded that there are $n(m-1)$ disjoint paths linking any two nodes and the $n(m-1)-$wide diameter is $n+1$. Two deterministic shortest and fastest routing algorithm have been proposed in [3]. [4]and [5]designed the fault-tolerant routing schemes for $Q_n(m)$. [6] discussed its fault tolerance and transmitting delay. Other parameters can be found in [7].

We shall develop a routing scheme for $Q_n(m)$, in which each message is accompanied with a stack which keeps track of the history of the path travelled as it is routed toward its destination, and tries to avoid visiting a node more than once unless a backtracking is enforced, each node is required to know only the condition (faulty or not) of its adjacent components. This routing algorithm is analyzed rigorously. Similar discussion has been made for hypercube ([8]).

The first node in the message's route that is aware of the nonexistence of an optimal path from itself to the destination is called obstructed node. At the obstructed node, the message has to take a detour. In this paper, we derive exact expressions for the probabilities of optimal path routing from the source node to a given obstructed node in the presence of components failures. Note that determination of the probability for optimal path routing between any two nodes can be viewed as an obstructed node that is 0 hop away from the destination node.

Y. Shi et al. (Eds.): ICCS 2007, Part I, LNCS 4487, pp. 648–651, 2007.

$Q_n(m)$ has vertex set $V(Q_n(m)) = \{x_1x_2\ldots x_n,\ x_i \in \{0,1\ldots,m-1\}\}$, x and y are adjacent if and only if they differ by exactly one bit. For $x = x_1x_2\ldots x_n$, the leftmost coordinate of the address will be referred to as 1-dimension, and the second to the leftmost coordinate as 2-dimension, and so on.

Suppose $x = x_1x_2x_3\ldots x_n$ and $y = y_1y_2y_3\ldots y_n$ be two nodes of $Q_n(m)$, xy is an edge of $i-$dimension if $x_j = y_j$ for $j \neq i$ and $x_i \neq y_i$. From the definition, $Q_n(m)$ contains m^n vertices and $m^n n(m-1)/2$ edges and $Q_n(m)$ is $n(m-1)$-regular with diameter n.

An optimal path is a path whose length is equal to the Hamming distance between the source and destination. We call the routing via an optimal path the optimal path routing. An incident link of node x is said to be toward another node y if the link belongs to one of the optimal path from x to y and call y the forward node of x.

A given path of length k between x and y in $Q_n(m)$ can be described by a coordinate sequence $C = [c_1, c_2, \ldots, c_k]$ where $1 \leq c_i \leq n$, the coordinate sequence is a sequence of ordered pairs. A coordinate sequence is said to be simple if any dimension does not occur more than once in that sequence. It is easy to see that a path is optimal if and only if its coordinate sequence is simple. For example, $[0002, 0000, 0010, 2010]$ is an optimal path from 0002 to 2010, and can be represented by a coordinate sequence $[4, 3, 1]$.

The number of inversions of a simple coordinate sequence $C = [c_1, c_2, \ldots, c_k]$ denoted by $V(C)$, is the number of pairs (c_i, c_j) such that $1 \leq i < j \leq k$ but $c_i > c_j$. For example $V([4,3,1]) = 3$.

2   Routing Algorithm

Algorithm A: Fault-tolerant Routing Algorithm

Step 1. If $u = d$, the message is reached destination, Stop.

Step 2. If the forward adjacent node v of u is normal and the link uv is normal and $v \notin TD$, select such a vertex v satisfied $i = \min\{i : uv$ is an edge of $i-$ dimension$\}$, then

send (message TD) to v, $TD = TD \cup \{u\}$, $u = v$.

Step 3. If v is a adjacent node of u and satisfies the following condition:

1. $v \notin TD$
2. v is not a forward node node of u to d
3. v and uv are normal components

select such a vertex v satisfied that $j = \min\{j : uv$ is an edge of $j-$ dimension$\}$, then send (message TD) to v, $TD = TD \cup \{v\}$, $u = v$. Go to Step 1.

Step 4. If the algorithm is not terminated yet, then Backtracking is taken, the message must be returned to the node from which this message was originally received. Go to Step 3.

3   Performance Analysis of Routing Algorithm

Theorem 1. *Suppose x and y are respectively the source and destination in $Q_n(m)$, $H(x,y) = n$. Then the number of fault components required for the simple coordinate sequence $C = [c_1, c_2, \ldots, c_t]$ to be the path chosen by algorithm A to an obstructed node located j hops away from y is $V(C) + W(c_1, c_2, \ldots, c_t) - \sum_{i=1}^{t} i + j$, where $t = n - j$, $W(c_1, c_2, \ldots, c_t) = \sum_{i=1}^{t} c_i$.*

Let $S(n, r)$ be the set of combinations of r different numbers out of $\{1, \ldots, n\}$ and $I_n(r)$ denote the number of permutations of n numbers with exactly r inversions.

Theorem 2. *Suppose there are f fault links in a $m-ary$ $n-cube$ computer network, and a message is routed by A from node x to node y where $H(x,y) = n$. Let h_L be the Hamming distance between obstructed node and the destination node. Then*

$$P(h_L = j) = \frac{1}{C_L^f} \sum_{\sigma \in S(n,t)} \sum_{k=0}^{\min\{\frac{n(n-1)}{2}, f-j\}} I_t(\alpha) C_{L-n-k}^{f-j-k}$$

where $\alpha = k - W(\sigma) + \frac{t(t+1)}{2}$ and $P(A)$ is the probability of event A, $L = n(m-1)m^n/2$ and $t + j = n$.

The probability of an optimal path routing can be viewed as a special case of Theorem 2 by setting the obstructed node to the destination node, namely, $P(h_L = 0)$.

Corollary 1. *The probability for a message to be routed in an $Q_n(m)$ with f fault links via an optimal path to a destination node which is n hops away can be expressed as*

$$P(h_L = 0) = \frac{1}{C_{\frac{n(m-1)m^n}{2}}^f} \sum_{k=0}^{\min\{\frac{n(n-1)}{2}, f\}} I_n(k) C_{\frac{n(m-1)m^n}{2} - n - k}^{f-k}$$

Theorem 3. *Suppose there exist h faulty nodes in a $Q_n(m)$, and a message is routed by A from x to y where $H(x,y) = n$. Let h_N be the Hamming distance between obstructed node and the destination node. Then for $2 \leq j \leq \min\{h, n\}$, we have,*

$$P(h_N = j) = \frac{1}{C_{m^n-2}^h} \sum_{\sigma \in S(n,t)} \sum_{k=0}^{\min\{\frac{n(n-1)}{2}, h-j\}} I_t(k - W(\sigma) + \frac{t(t+1)}{2}) C_{m^n-2-n-k}^{h-j-k}$$

where $t + j = n$.

Corollary 2. *Under algorithm A, the probability for a message to be routed in a $Q_n(m)$ with h faulty nodes via an optimal path to a destination located n hops away is*

$$P(h_N = 0) = \frac{1}{C_{m^n-2}^h} \sum_{k=0}^{\min\{\frac{n(n-1)}{2}, h\}} I_n(k) C_{L=m^n-1-n-k}^{h-k}.$$

4   Conclusion

This paper proposed a new fault-tolerant routing algorithm for m-ary n-cube. This algorithm is based on the detour and backtracking technique. The knowledge on the number os inversions of a given permutation is used to analyze the performance of this routing. The number of faulty components required for a coordinate sequence to become the coordinate sequence of a path toward a given obstructed node is determined.Probability for routing messages via optimal path to given obstructed node location are determined.

Acknowledgment

This work is supported by NSFC (10671081) and The Science Foundation of Hubei Province(2006AA412C27).

References

1. Hongmei Liu: The restricted connectivity and restricted fault diameter in m-ary n-cube systems. The 3rd international conference on impulsive dynamic systems and applications. July 21-23. **4** (2006) 1368–1371
2. Liu Hongmei: Topological properties for m-ary n-cube. Journal of Wuhan University of Technology (Transportation Science and Engineering) **30** (2006) 340–343
3. Liu Hongmei: The routing algorithm for generalized hypercube. Mathematics in Practice and Theory **36** (2006) 258–261
4. Wu J., Gao G. H.: Fault tolerant measures for m-ary n-dimensional hypercubes based on forbidden faulty sets. IEEE Transaction Comput. **47** (1988) 888–893
5. Dhabaleswar K., Panda, Sanjay Singal, and Ram Kesavan: Multidestination message passing in wormhole k-ary n-cube networks with based routing conformed paths. IEEE Transaction on parallel and distributed systems **(10)** (1999) 76–96
6. Xu Junming: Fault tolerance and transmission delay of generalized hypercube networks. Journal of China University of Science and Technology **31** (2001) 16–20
7. Xu Junming: Topological Structure and Analysis of Interconnecting Networks. Dordrecht/Boston/London:Kluwer Academic Publishers (2001)
8. Ming-Syan Chen, Kang G. Shin: Depth-first search approach for fault-tolerant routing in hypercube multicomputers. IEEE Transactions on parallel and distributed systems **1** (1990) 152–129

CARP: Context-Aware Resource Provisioning for Multimedia over *4G* Wireless Networks

Navrati Saxena[1], Abhishek Roy[2], and Jitae Shin[1]

[1] School of Info. & Comm. Eng. Sungkyunkwan University, Suwon, Korea
[2] WiBro System Lab, Samsung Electronics, Suwon, South Korea
{navrati,jtshin}@ece.skku.ac.kr, abhishek.roy@samsung.com

Abstract. In this paper a context-aware resource provisioning framework is developed for multi-system, heterogeneous *4G* wireless networks. The framework envisions that each individual sub-network is fairly independent, and uses Bayesian learning scheme for capturing user's mobility profiles. The concept of Asymptotic Equipartition Property (AEP) helps to predict the most likely path-segments, that the mobile user (MU) is going to follow in near future, with very good accuracy. This helps in proactive resource management along the MU's future paths and locations providing the optimal resource (e.g., bandwidth) reservation. Simulation results on a synthetic wireless traces corroborate this high prediction success and demonstrate sufficient improvement in delay, blocking and throughput of ongoing wireless multimedia sessions.

Keywords: 4G Heterogeneous wireless networks, Bayesian learning, resource management, Information Theory, mobile multimedia, AEP.

1 Introduction

The concept of packet-based, next-generation *4G* wireless networks [2] lies in the co-existence of multiple cellular access networks. These individual cellular networks might possess significant differences in topology, coverage area and transmission range. However, by using multiple radio interfaces, future mobile nodes can seamlessly switch between these technologies, thus achieving the ultimate goal of global roaming. The *4G* network can thus be conceptually visualized as a collection of multiple independent access *sub-networks*, each of which has its own cellular layout and separate location tracking technique. "Context-awareness" is perhaps the key characteristic of these next-generation wireless networks and associated applications. *Mobility and location information* is the most important "context" in these networks, because the information needed by the vast majority of MUs depend strongly on their current or near future location.

In this paper we have developed a new Context-Aware Resource Provisioning scheme *CARP* across mobile-user's most frequent routes in heterogeneous *4G* cellular networks. Operating in *symbolic space*, *CARP* uses Bayesian algorithm to capture the movement profiles of individual MU across such networks. Although, in such systems, there exists a wide number of possible routes from a

Y. Shi et al. (Eds.): ICCS 2007, Part I, LNCS 4487, pp. 652–659, 2007.

particular source to a destination, a MU usually follows his/her most likely paths. A similar analogy, dealing with the *asymptotic equipartition property* (AEP) [11] states that among all the long-range sequences consisting of random variables, there exists a fairly small *typical set* [11] which contains most of the probability mass and controls the average behavior of all such sequences. Using this concept, CARP captures the MU's *typical path segments* and reserves bandwidth along the typical paths. This helps in seamless transmission of ongoing real-time wireless communication with specific Quality of Service (QoS) guarantee. Using a priority-based service classification scheme CARP is also capable of guaranteeing significant improvement in blocking of real-time multimedia sessions.

Section 2 reviews the major existing works in location and resource management in 4G networks. A symbolic domain description of such networks is shown in Section 3. The new concept of capturing MU's frequently used routes is developed in Section 4. An efficient resource reservation and admission control scheme based on this route formation is discussed in Section 5. Simulation results in Section 6 demonstrates the improvement of QoS in ongoing wireless sessions. Finally Section 7 concludes the paper.

2   Related Works: Mobility in 4G Wireless Systems

Given the large body of work on resource management in a single cellular access system (e.g., PCS), we survey only the major contributions related to our context. The guard channel policy [8] and fractional guard channel policy [9] determine the number of guard channels for hand-offs. The shadow cluster scheme [6] estimates future resource requirements in a collection of cells in which a mobile is likely to visit in the future. An efficient scheme for mobility-based predictive admission control is also proposed in [12] for single-system wireless networks.

Some recent work has reported on the problem of location and resource management in a multi-system environment. It is shown in [1] that an integrated location management strategy can significantly outperform an independent operation of each sub-system's location management algorithm. Similarly recent researches have pointed out a significant trend towards seamless inter-networking of wireless LANs and cellular networks [3]. The issues of network selection [10] in heterogeneous wireless networks are also under investigation. The problem of optimal location management [7] in heterogeneous wireless networks is also recently pointed out. However, the problem of resource reservation and management for guaranteed QoS in a generic multi-system environment is yet to be effectively addressed. This motivates the development of *CARP*.

3   Symbolic Representation of 4G Wireless Systems

Figure 1 shows an example of an *integrated 4G wireless network*, comprising a collection of satellite, PCS and campus area (IEEE 802.11 WLAN and Bluetooth) based independent sub-networks. Clearly, the coverage area of each individual sub-system can be discontinuous. Accordingly, the set of sub-networks that can

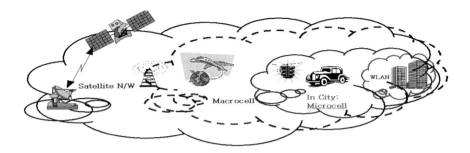

Fig. 1. A Multi-System Heterogeneous Wireless Network

be accessed concurrently by a mobile node is not constant but a function of its current location. Let, the $4G$ network consists of N sub-networks, where each sub-network is a collection of (either partitioned or overlapping) cells, such that C_i^j represents the jth cell in the ith sub-network. If the mobile node is currently out of range of sub-network S_i, its location vector includes the cell C_i^ϕ.

The movement pattern of the mobile node (MN) can then be represented as a N-dimension random vector $\bar{X}$, where the ith element of the vector corresponds to the current cell of sub-network S_i. For instance, if $\bar{X}(2) = 4$, the MN is currently located in the 4th cell of sub-network S_2. Characterizing the mobility as a probabilistic sequence suggests that it can be defined as a stochastic, vector-valued process $\chi = \{\bar{X}_n\}$. where the "..." imply that the corresponding random variable can take any possible value within its range.

4   Context-Awareness in Heterogenous Wireless Networks

The MU's location uncertainty in heterogeneous wireless networks leads to uncertainty in the resources allocation and reservation policies. Once the location is managed properly, only then the network will be smart enough to reserve suitable wireless bandwidth. While the existing shadow cluster [6] approach provides the idea to preserve resources along the mobile's surrounding cells, it fails to give any generalized way to estimate these shadow clusters, and only assumes either some definite knowledge of user mobility or totally random users. The novelty of *CARP* lies in the fact that it not only captures the MU's locations, but also accumulates and predicts the user's *typical* routes. Thus, effectively *CARP* extends the concept of shadow cluster into heterogenous wireless networks to form a more smart resource management scheme.

4.1   Probabilistic Estimation of User-Mobility

CARP uses an online *Bayesian Aggregation* technique [4] to make the system knowledgeable enough about the MU's mobility profiles. Figure 2 provides a pseudo-code of our mobility prediction scheme. Each cell-vector $\bar{X}_i$ in the heterogenous wireless system is assigned to a weight w_i. Initially, all the weights are

---

Probabilistic Mobility Prediction
1. initialize $m := 1$; assign weight $w_i := 1/n$ to each instant $\bar{X}_i$;
3. repeat steps below until $(X_p - X_a \leq \varepsilon)$;
4.   predict the next cell-vector (location) $\bar{X}_p := \sum_{i=1}^{n} \frac{w_i \times \bar{X}_i}{w_i}$;
5.   continue it's own movement; 6. Capture the actual location $\bar{X}_a$;
7.   compute the individual (l_i) and overall (l_o) prediction inaccuracy
      $l_i = -|\bar{X}_a| \ln |\bar{X}_i| - (1 - |\bar{X}_a|) \ln(1 - |\bar{X}_i|)$ and $l_o = \sum_{i=1}^{n} l_i$;
8.   update the weights as: $w_i^{m+1} = w_i^m e^{-l_i}$;

Fig. 2. Bayesian Algorithm for Capturing User-Mobility

same, i.e., $w_i = 1/n$. The mobile-node now estimates it's own future-location by computing a *weighted measure* of these individual instances of cell-vectors (locations). Formally, the predicted location $\bar{X}_p$ is:

$$\bar{X}_p = \frac{\sum_{i=1}^{n} w_i \times \bar{X}_i}{\sum_{i=1}^{n} w_i} \tag{1}$$

The objective of *CARP* is to predict the future location correctly, so that *the system becomes cognizant of user's mobility*. If, $\bar{X}_a$ represents the actual location, the objective is to make the system cognizant enough so that $\bar{X}_p = \bar{X}_a$. The most fair measure of the prediction-inaccuracy (error) associated with this process is *entropic loss* [4]. This *entropic loss* l_i for each location and its cumulative estimate in m iterations l_i^c is given by

$$l_i = -|\bar{X}_a| \ln |\bar{X}_i| - (1 - |\bar{X}_a|) \ln(1 - |\bar{X}_i|), \ and \ l_i^c = \sum_{i=1}^{m} l_i \tag{2}$$

At every iteration m, the weights associated with every time-instant T_i are now updated as

$$w_i^{m+1} = w_i^m e^{-l_i}. \tag{3}$$

This procedure is iterated until $\bar{X}_i - \bar{X}_a \leq \varepsilon$, where ε is a predefined precision. The effects of locations far from accuracy are reduced by exponentially decreasing their associated weights. In time the predicted location $\bar{X}_i$ approaches $\bar{X}_a$. The overall expected deviation form the optimality (l_o) is given by:

$$l_o \leq \min_i l_i^c + \ln(m) \tag{4}$$

4.2   Context-Aware Route Collection

A close look into the life-style of a particular MU reveals that he/she typically follows only a small subset of all the paths, which essentially is guided by his life-style in the long run. The concept of *typical set* and *asymptotic equipartition property* (AEP) from information theory helps in obtaining this *small set* of *highly probable* routes maintained by a particular MU.

Definition 1. *The type [4] of a vector-sequence* $\bar{\mathbf{x}} = \{\bar{x}_1, \bar{x}_2, \ldots, \bar{x}_n\}$ *is the relative proportion of occurrences of each vector-symbol. Also, the set of all vector-sequences of a particular length and type is referred as the* type class.

The essential power of the method of types arises from the fact that the number of types is *always at-most polynomial* [11]. Hence, the crucial point to note here is that, there are polynomial number of types, but exponential number of sequences.

Result 1. *If* $\bar{\mathcal{V}}_1, \bar{\mathcal{V}}_2, \ldots, \bar{\mathcal{V}}_n$ *denotes the set of vector-sequences drawn according to a specific distribution, then the probability of the sequence depends on* relative difference *between the type and the original distribution.*

Hence, the type classes and the corresponding sequences that are far from the original distribution have exponentially smaller probabilities. This leads to the concept of *typical set*:

Definition 2. *For a given* $\epsilon > 0$, *a typical set* T_Q^ϵ *of sequences for the distribution* Q^n *is defined as:* $T_{\bar{\mathbf{v}}}^\epsilon = \{\bar{\mathbf{v}} : |Type - Distribution| < \epsilon\}$.

All we now need is the probabilistic estimation of these typical sequences. Fortunately, this is exactly what the *Shannon-McMillan-Brieman* theorem [4,11] provides. According to this theorem, if $H(\chi)$ is the entropy of finite-valued stochastic ergodic process χ, then

$$-\frac{1}{n} \lg \left[p \left(\bar{V}_0, \bar{V}_1, \ldots, \bar{V}_{n-1} \right) \right] \to H(\chi) \qquad (5)$$

Equation 5 provides the basis of *asymptotic equipartition property* (AEP) [11] for any stationary, ergodic, stochastic process.

Result 2. *AEP states that for a fixed* $\epsilon > 0$, *as* $n \to \infty$, $Pr\{\chi \in T_{\bar{\mathbf{v}}}^\epsilon\} \to 1$.

This is basically similar to the *weak law of large numbers* [4] for ergodic, stochastic sequences, assuring that asymptotically almost all the probability mass of $\mathcal{V}$ is concentrated in the *typical set*. It encompasses the mobile user's most likely paths and determines the *average nature* of the large route-sequences.

5   Advanced Resource Provisioning

Resource reservation and admission control for multimedia services is a challenging research topic in the field of current and future generation wireless networks. The novelty of our approach lies in designing the bandwidth reservation scheme along the MU's typical (most probable) set of routes to ensure strict guarantee of QoS parameters like delay, throughput and blocking of real-time wireless applications; which had been missing from literature. The wireless audio-video applications are now capable of following the MU along his/her typical routes. Thus, the MU will be able to listen the news and watch a streaming video as he/she moves from one location to another. An estimation of reserved and free bandwidth in any heterogeneous cell $\bar{z}$ is determined by:

$$B_{res}(\bar{z}, m) = Pr[\bar{z}] \times B_m, = \sum_{\bar{X}^n \in T_e^Q} \left[Pr[\bar{X}^n] \times \frac{f(\bar{z})}{|\bar{X}^n|} \right] \times B_m, \qquad (6)$$

$$B_{free}(\bar{z}, m) = B_{tot} - B_{res}(\bar{z}, m) \qquad (7)$$

where, $f(\bar{z})$ represents frequency of occurrence of $\bar{z}$, B_{tot}, B_m, B_{resv} and B_{free} are respectively the total bandwidth requirement of the mobile m, bandwidth requirement at $\bar{z}$, bandwidth reserved at $\bar{z}$ and free bandwidth at $\bar{z}$.

On arrival of a particular media packet in any $\bar{z}$, the bandwidth availability in that zone is checked for admission of the real-time video and voice stream. The unavailability of bandwidth leads to blocking of the media stream. On the other hand, if the bandwidth is available, the availability is checked only along the typical routes of the MU. The admission control of the multimedia packets is determined by the following condition:

$$\sum_{\bar{z} \in \mathcal{V}^n} Pr[\bar{z}] \times \beta(\bar{z}, B_m) \geq \tau \sum_{\bar{z} \in \mathcal{V}^n} Pr[\bar{z}], \qquad (8)$$

$$\beta(\bar{z}, B_m) = \begin{cases} 1, & B_{free}(\bar{z}) \geq B_m \\ \frac{B_{free}(\bar{z}, m)}{B_m}, & otherwise \end{cases} \qquad (9)$$

where τ is the admission threshold. Clearly, τ varies for voice, video and data packets. The data packets are given lower priority, and are admitted only if sufficient bandwidth is left (according to equations (6)–(9)) after admitting the real-time voice and video traffic.

6   Simulation Results

We have developed an object-oriented, discrete-event simulation environment for supporting MU's movements, associated prediction of typical routes and resource management for supporting location-aware multimedia services. The wireless voice services is modelled with exponentially distributed 'on' and 'off' periods (mean lengths 0.6 sec and 0.8 sec respectively). We have taken fluid queues having Weibullian inter-arrivals with heavy tails. The video sequences, are assumed to be encoded with ITU's (International Telecommunication Unit) H.264 [5] specifications with 176×144 pixel resolution and 7–30 frames/sec.

CARP reserves the highest bandwidth for ongoing voice (40%–45%-high priority) and the lowest for data traffic (10%–15%-low priority) along the MU's typical routes. Figure 3 demonstrates that the such a mobility-aware, proactive bandwidth reservation results in almost 50%, 65% and 30% improvements respectively, in blocking of wireless voice, video and data streams, with increasing session arrival rates. In order to investigate into the mutual effect of the real-time voice and video streams over the other, we have looked into the throughput of voice and video traffic. Figure 4 shows that the with increasing session arrival rate the throughput of voice and video traffic lies around 65%–85% of the actual transmission data rate. Thus, we can conclude that such a location-aware

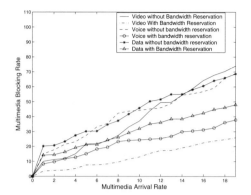

Fig. 3. Gain in Blocking of Mobile Multimedia

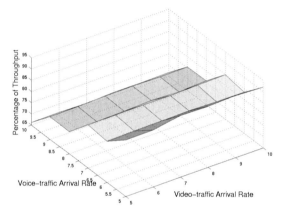

Fig. 4. Throughput of Voice and Video Traffic

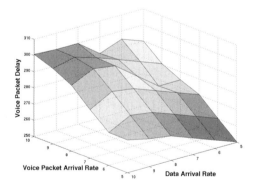

Fig. 5. Mutual Effects of Voice and Data Traffic

bandwidth reservation scheme has the potential for providing steady throughput of multimedia streams. Another point of concern is the mutual effects of traffic over the delay experienced by the MU in receiving the multimedia streams, while moving from one location to the another along the typical routes. Figure 5 shows that using this delay can be kept quite low for voice ($\approx$ 600 msec) services.

7 Conclusions

In this paper we have looked into the advanced resource reservation scheme for location-aware multimedia in heterogenous wireless networks. The developed system is quite cognizant about the MU's daily movement patterns and is capable of performing advanced reservation of wireless bandwidth along the MUs typical routes. This proposed framework aids in development of location-oriented multimedia with specific QoS (delay and throughput) guarantee, reduced blocking and seamless transmission along the MU's typical routes.

Acknowledgments. This work was supported by the Korea Research Foundation Grant funded by the Korean Government (MOEHRD)(KRF-2006-331-D00358).

References

1. I. F. Akyildiz and W. Wang, "A Dynamic Location Management Scheme for Next-Generation Multitier PCS Systems", *IEEE Transactions on Wireless Communications*, vol. 1, no. 1, pp. 178-189, Jan 2002.
2. R. Berezdivin, R. Breinig and R. Topp, "Next-Generation Wireless Communications Concepts and Technologies", *IEEE Comm. Mag.*, vol. 40, no. 3, March 2002.
3. D. Cavalcanti, D. Agarwal, C. Cordeiro, B. Xie and A. Kumar, "Issues in Integrating Cellular Networks, WLANS, and MANETS: A Futuristic Heterogeneous Wireless Network", *IEEE Wireless Communications*, pp. 30–41, 2005.
4. T. M. Cover and J. A. Thomas, *Elements of Information Theory*, John Wiley, '91.
5. ITU-T H.264; "Advanced video coding for generic audiovisual services", *ITUT Specifications for Video Coding*
6. D. A. Levine, I. F. Akyildiz and M. Nagshineh, "A resource estimation and call admission algorithm for wireless multimedia networks using the shadow cluster concept," *IEEE Transactions on Networking*, vol. 5, no. 1, pp. 1-12, February '97.
7. A. Misra, A. Roy and S. K. Das, "An Information Theoretic Approach for Optimal Location Tracking in Multi-System 4G Wireless Networks", *IEEE InfoCom*, '04.
8. E. C. Posner and R. Guerin, Traffic Policies in Cellular Radio that Minimize Blocking of Handoff Calls, *11th Intl. Teletraffic Conference*, Kyoto, Japan, Sept. 1985.
9. R. Ramjee, R. Nagarajan and Don Towsley, On Optimal Call Admission Control in Cellular Networks, *Proceedings of IEEE INFOCOM*, San Francisco, CA, 1996.
10. Q. Song and A. Jamalipour, "Network Selection in an Integrated Wireless LAN and UMTS Environment Using Mathematical Modeling and Computing Techniques", *IEEE Wireless Communications*, pp. 42 - 48, 2005.
11. A. Wyner and Z. Ziv, "Some Asymptotic Properties of the Entropy of a Stationary Ergodic Data Source with Applications to Data Compression", *IEEE Transactions on Information Theory*, vol. 35, no. 6, pp. 1250-1258, Nov. 1989.
12. F. Yu and V. Leung, "Mobility-based Predictive Call Admission Control and Bandwidth Reservation in Wireless Cellular Networks", *Computer Networks*, vol. 38.

Improved Fast Handovers for Mobile IPv6 over IEEE 802.16e Network

Sukyoung Ahn and Youngsong Mun

School of Computing Soongsil University, Seoul, Korea
ahnsukyoung@sunny.ssu.ac.kr,
mun@computing.ssu.ac.kr

Abstract. The purpose of Fast Handovers for Mobile IPv6(FMIPv6) is to reduce the overall handover latency resulting from standard Mobile IPv6(MIPv6) procedures. The standard FMIPv6 procedure comprises layer 2 and layer 3 handover procedures, and these are performed serially. This causes a long latency problem which is not negligible for real-time service such as Voice over IP(VoIP) over IEEE 802.16e network. We revise FMIPv6 messages to reduce the overall latency on handover over IEEE 802.16e networks. A Layer 2 handover message gives layer 3 handover messages a piggyback, and so the mobile node(MN) which is operated by our scheme, is able to finish overall handover earlier than standard MIPv6 and FMIPv6. We evaluate the performance of the scheme by analyzing the overall cost. Compared to FMIPv6 handover scheme, the proposed scheme gains about 27% of improvement.

Keywords: FMIPv6, MIPv6, IEEE 802.16e, Performance analysis.

1 Introduction

The advent of WiBro in Korea let people aware of Broadband Wireless Access(BWA) service. And that have people think it is going to be able to access the Internet at anywhere no matter what vehicle they are riding on. It is certain that the concept of services of the Internet in the future will be totally different from today's one. As the features of the Internet such as data rates and error rates are improved, the services on the Internet are changed as well. When we talk about the applications over the Internet now, we cannot talk about those without real-time services. It is extremely important to reduce handover latency for the real-time service applications.

In order to support mobility for the Internet services, the Internet Engineering Task Force(IETF) proposed Mobile IPv6[2]. The IETF again proposed FMIPv6[3] and Hierarchical Mobile IPv6(HMIPv6)[4] to reduce the latency for handovers. It is however not enough to satisfy real-time traffic yet. We propose an improved fast handover scheme (IFH) to lower the handover latency, by modifying the FMIPv6 messages and messages for network re-entry procedure on IEEE 802.16e network[1]. The main idea of FMIPv6 is tunneling from Previous Access Router(PAR) to New Care-of Address(NCoA), so the packets which may be

Y. Shi et al. (Eds.): ICCS 2007, Part I, LNCS 4487, pp. 660–667, 2007.

lost in standard MIPv6 operation can be preserved. At the same time, the MN becomes able to receive packets sent by Correspondent Node(CN), as soon as the MN sends Fast Neighbor Advertisement message(FNA). When the time for tunneling gets longer and longer the packets coming through the tunnel become a burden to the New Access Router(NAR), since the NAR must proxy those packets until it receives Fast Neighbor Advertisement(FNA) message from MN.

We focus on those two points (i.e., shorten overall latency and time for tunneling) and propose the IFH scheme to solve those two problems at the same time.

This document is organized as follows. In section 2, FMIPv6 procedure over the IEEE 802.16e network and few efforts to reduce the handover latency on IEEE 802.16e and IEEE 802.11 are presented. In section 3, we describe the improved handover scheme(IFH). We set up a simplified model to analyze the costs and benefits of using IFH, in section 4. Finally, in section 5, we conclude discussion with future study.

2   Fast Handovers for MIPv6 over IEEE 802.16e

MN cannot stay at only one Base Station(BS), and hence obviously it need to perform handover to neighbor BS. Fig.1 shows an example of handover in IEEE 802.16e network.

For FMIPv6 procedure over the IEEE 802.16e, Serving BS broadcasts MOB _NBR-ADV message periodically which has information about its neighbor BS[5]. If the MN found a new BS(NBS) in MOB_NBR-ADV message, it may perform scanning to measure the signal power of the NBS. Once the MN decides handover to the new neighbor BS, the MN and the Previous BS(PBS) exchange MOB_MNHO-REQ/MOB_BSHO-REP messages to initiate handover. On reception of MOB_BSHO-REP from PBS, the MN exchanges Fast Binding Update(FBU) and Fast Binding Acknowledgement(FBAck) messages with PAR. Receiving FBU, the PAR establishes a tunnel to NCoA with the NAR by exchanging Handover Initiate(HI) and Handover Acknowledgement(HAck) messages. Before send HAck, the NAR verifies availability of NCoA. Layer 2 of the MN sends MOB_HO-IND to PBS as a final indication of handovers, when the MN receives FBAck from PAR. Then, the MN closes the connection with PBS and performs network re-entry procedure(Hard handover). When the network

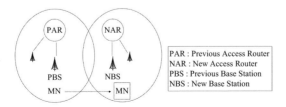

Fig. 1. 802.16e Deployment Architecture

re-entry procedure is completed, MN issues FNA message to NAR via NBS. As soon as NAR receives FNA message, NAR delivers MN the buffered packets sent via tunnel from PAR. The network re-entry procedure comprises three phases roughly. In the first phase, MN starts synchronization with new downlink, ranging and synchronization with uplink. The first phase is all about adjusting uplink/downlink parameters and timing offsets. The second phase is authorization phase to protect the MN from malicious node spoofing. As the last phase, MN and NBS perform re-registration process.

3   The Proposed Scheme

We propose a method, called Improved Fast Handovers for MIPv6(IFH), to reduce the overall handover latency and the time for tunneling between the PAR and the NCoA. We perceive that BS has an IP address, which means the BS can manipulate Layer 3 messages. We consider that MN sends few Layer 3 messages, i.e. Binding Update message to Home Agent, Home Test Init(HoTI) message, and Care-of Test Init(CoTI) message in encapsulated form in one of the Layer 2 handover messages.

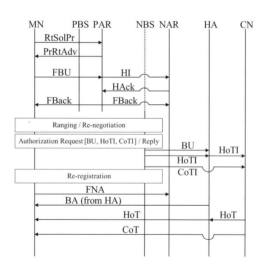

Fig. 2. The improved fast handover scheme(IFH)

For security consideration, it is regarded that it should be proceeded right after Layer 2 authorization phase finished. With the IFH, the MN can register its NCoA to Home Agent(HA) and CN earlier than the FMIPv6. In doing the IFH, the time for maintaining tunneling between PAR to NCoA would also decrease, which makes unburden NAR more quickly.

4   Performance Evaluation

4.1   System Modeling

We set up the network model for our paper as Fig.3. In our system model, we consider a,b,c and d as one hop distance. Any connection to the outside of PAR subnet is considerd to have five hop distance except e. We assume that NAR and PAR are connected to each other in Hierarchical manner, so we consider connection e has two hop distance. We assume that the CN generates and sends data packets to the MN at a mean rate λ. The cost for transmitting a data packet is l times bigger than that for signaling packet, where $l = l_{data}/l_{signal}$. l_{data} is the average length of data packets, and l_{signal} is the average length of signaling packets. The delay to process packets in any node is very short comparing to the transport delay, and IFH and FMIPv6 have almost same amount of packet processing. So we ignore the packet processing delay for this performance evaluation.

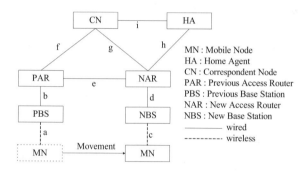

Fig. 3. System model

4.2   Cost Analysis

In this paper, we compare the IFH with the FMIPv6 to show the improved performance of the IFH. We define overall handover latency as

$$C_{overall} = C_{signal} + C_{data} , \tag{1}$$

where C_{signal} is the sum of costs to transport signaling messages, and C_{data} is the sum of costs to transport data packets. The FMIPv6 procedure comprises six phases[6] as shown in Fig.4. T_{fast} is the delay for MN to acquire and verify the NCoA, and during this time the tunnel between the PAR and NCoA is established. T_{L2} is the delay for Layer 2 handover. T_{IP} is the delay to send FNA message. T_{HABU} is the delay for the Binding Update(BU) and the Binding Acknowledgement(BA) messages to HA. T_{RR} is the delay during the Return Routability Procedure[3], and T_{CNBU} is the delay to send the BU and the BA messages to CN.

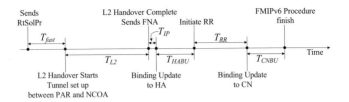

Fig. 4. Timing diagram in the fast handover

The signaling cost of the FMIPv6 is defined as

$$C_{signal\_FH} = C_{sig\_fast} + C_{sig\_L2} + C_{sig\_IP} + C_{sig\_HABU} + C_{sig\_RR} + C_{sig\_CNBU}. \quad (2)$$

The cost to transport data packets is defined as

$$C_{data\_FH} = P_{success} \cdot \lambda \cdot C_{tunnel\_FH} \cdot T_{tunnel\_FH}$$
$$+ P_{fail} \cdot \lambda \cdot \{(C_{loss} \cdot T_{loss})$$
$$+ (C_{reactive} \cdot T_{reactive})\}. \quad (3)$$

The $P_{success}$ denote the probability of MN to perform FMIPv6 in predictive mode successfully, and the P_{fail} denote the probability of failure as well[11]. According to [3], there is no certain guarantee that the packets sent by CN would not be lost, in case of failure of the predictive mode. Hence we assume those packets would be lost and are needed to be sent again. Those C_{loss} and T_{loss} are about the lost packets because of the failure of the predictive mode. $C_{reactive}$ and $T_{reactive}$ are about the retransmission and the reactive mode. The C_{tunnel_FH} is the cost for the CN to send data packets through the tunnel between PAR and NCoA, and it can be calculated by l times of the signaling cost between the CN and the NCoA. And the T_{tunnel_FH} can be described as

$$T_{tunnel\_FH} = T_{L2} + T_{IP} + T_{HABU} + T_{RR} + T_{CNBU}. \quad (4)$$

In case of failure, the C_{loss} can be calculated by l times of the signaling cost between the CN and the PCoA, and the T_{loss} can be described as

$$T_{loss} = T_{L2} + T_{IP} + T_{FBU\_reactive} + T_{Notice}, \quad (5)$$

where $T_{FBU\_reactive}$ is the time for the FBU from the NAR to the PAR, which was encapsulated in FNA message. T_{Notice} is the time to let the PAR aware of the Handover of the MN.

$C_{reactive}$ is same as C_{tunnel_FH}, because the path of the data packets is same as the path of the predictive mode. $T_{reactive}$ is given by

$$T_{reactive} = T_{L2} + T_{Handover\_Init\_reactive} + T_{HABU} + T_{RR} + T_{CNBU}, \quad (6)$$

where $T_{Handover\_Init\_reactive}$ is the time for FNA, FBU, HI, HAck and FBAck messages of the reactive mode of FMIPv6.

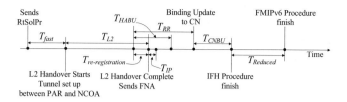

Fig. 5. Timing diagram in the Improved Fast Handover(IFH)

Fig.5 depicts the IFH in time line manner, and it is similar to the time line of the FMIPv6. The two phases, T_{HABU} and T_{RR} however, start at the same time. Those two phases begin right after the authentication phase of Layer 2. The overall handover latency of the fast handover is reduced by the IFH as much as $T_{Reduced}$. The signaling cost and the cost to transport data packets are calculated in the same manner as we do for the FMIPv6. The T_{tunnel_FH} however is replaced by

$$T_{tunnel\_IFH} = T_{L2} + \max\left(T_{IP}, T_{RR} - T_{re-registration}\right) + T_{CNBU} \qquad (7)$$

The rest of the cost to transport data packets is same as FMIPv6.

4.3   Numerical Results and Analysis

In our paper, we leverage formulas derived from experiential communication delay model. In the scenario used for modeling, a CN, a HA and ARs are connected to a wired network. The network consists of switched 10 Mbps Ethernet LAN segments, IP routers and the Bellcore network backbone. Regression analysis of the collected data for wired network produce

$$T_{wired-RT}(h, k) = 3.63k + 3.21(h - 1) , \qquad (8)$$

where k is the length of the packet in KB, h is the number of hops[9]. We assume $T_{wireless-RT}$ as $10ms$, because a unit of time frame is $5ms$ in case of Wibro[8],[10].

The variation of a signaling and a data packet with variation of λ. The signaling cost is a constant value, so it is not affected by the value of λ. The IFH needs less signaling cost than the standard FMIPv6, since IFH embeds three Layer 3 handover messages, i.e. BU for HA, HoTI and CoTI messages, inside of the Layer 2 handover signaling message. The data cost in two different schemes increases in proportion to λ.

Cost ratio with variation of λ. Both overall cost for the FMIPv6 ($C_{overall\_FMIPv6}$) and overall cost for the IFH($C_{overall\_IFH}$) are calculated by (1). The relationship of overall cost for the FMIPv6($C_{overall\_FMIPv6}$) and overall cost for the IFH($C_{overall\_IFH}$) are illustrated as Fig.6.

Table 1. System parameters

Parameter	Value	Parameter	Value
l_{signal}	200 bytes	l_{data}	1024 bytes
T_{fast}	25.751 $msec$	T_{L2}	55 $msec$
T_{IP}	5.363 $msec$	T_{HABU}	24.292 $msec$
T_{RR}	37.858 $msec$	T_{CNBU}	24.292 $msec$
$T_{re-registration}$	10 $msec$	$T_{Handover\_Init\_reactive}$	13.235 $msec$
$T_{FBU\_reactive}$	1.968 $msec$	T_{Notice}	7.425 $msec$

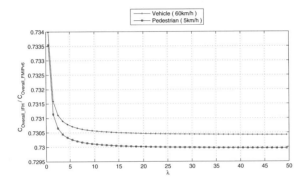

Fig. 6. Cost ratio of IFH against FMIPv6

In case of pedestrian speed($5km/h$),

$$\lim_{\lambda \to \infty} \frac{C_{overall\_IFH}}{C_{overall\_FMIPv6}} = \lim_{\lambda \to \infty} \frac{C_{signal\_IFH} + C_{data\_IFH}}{C_{signal\_FMIPv6} + C_{data\_FMIPv6}} \approx 0.7299, \quad (9)$$

where $\lambda > 50$.

In case of vehicle speed($60km/h$),

$$\lim_{\lambda \to \infty} \frac{C_{overall\_IFH}}{C_{overall\_FMIPv6}} = \lim_{\lambda \to \infty} \frac{C_{signal\_IFH} + C_{data\_IFH}}{C_{signal\_FMIPv6} + C_{data\_FMIPv6}} \approx 0.7304, \quad (10)$$

where $\lambda > 50$.

In Fig.6 describes the variation of cost ratio against the FMIPv6, respectively when the radius of a cell is $1km$. These two numerical expressions (9) and (10) show a small difference of the cost ratio between pedestrian speed and vehicle speed. This shows that IEEE 802.16e network is designed to support especially the mobility of the MN. So the probability of predictive mode of FMIPv6 is not affected by the speed of the MN. The mean of cost ratio of (9) and (10) is about 0.73, and hence we claim that our scheme (IFH) improves the performance of FMIPv6 by 27%.

5 Conclusions and Discussion

It is very obvious that we need to solve technical problems about delays of MIPv6 and FMIPv6, prior to deploy those two techniques over IEEE 802.16e network. The service over IEEE 802.16e network would require the MN to support seamless real-time connectivity. Using only FMIPv6 however, is not enough for the seamless connectivity yet. We propose an Improved Fast Handovers(IFH) to reduce the overall latency on the handover and to unburden the PAR earlier.

We proved the IFH achieves improved performance by comparison against FMIPv6. Compared to the standard FMIPv6, the IFH acquired 27% improvement. For more realistic evaluation of our scheme, we plan to investigate with the Wibro equipments about the effect of the IFH through various applications.

Acknowledgement

This work was supported by the Korea Research Foundation Grant funded by the Korean Government(MOEHRD) (KRF-2006-005-J03802).

References

1. IEEE Standard 802.16,IEEE Standard for Local and metropolitan area networks - Part 16: Air Interface for Fixed Broadband Wireless Access Systems (2004)
2. Johnson, D., Perkins, C., Arkko, J.: Mobility Support in IPv6 , RFC 3775 (2004)
3. Koodli, R.: Fast Handovers for Mobile IPv6, RFC 4068 (2005)
4. Soliman, H., Malki, K. El: Hierarchical Mobility management, RFC 4140 (2004)
5. Jang, H., Jee, J., Han, Y., Park, D. S., Cha, J.: Mobile IPv6 Fast Handovers over IEEE 802.16e Networks, Internet-Draft, draft-jang-mipshop-fh80216e-01.txt (2005)
6. Ryu, S., Lim, Y., Ahn, S., Mun, Y.: Enhanced Fast Handover for Mobile IPv6 based on IEEE 802.11 Network (2004)
7. Choi, S., Hwang, G., Kwon, T., Lim, A., Cho, D.: Fast Handover Scheme for Real-Time Downlink Services in IEEE 802.16e BWA System (2005)
8. Http://www.tta.or.kr
9. Jain, R., Raleigh, T., Graff, C., Bereschinsky, M.: Mobile Internet Access and QoS Guarantees using Mobile IP and RSVP with Location Registers. ICC'98 Conf. (1998)
10. Http://www.etri.re.kr
11. Mun, Y., Park, J.: Layer 2 Handoff for Mobile-IPv4 with 802.11 (2003)

Advanced Bounded Shortest Multicast Algorithm for Delay Constrained Minimum Cost

Moonseong Kim[1], Gunu Jho[2], and Hyunseung Choo[1,*]

[1] School of Information and Communication Engineering
Sungkyunkwan University, Korea
{moonseong, choo}@ece.skku.ac.kr
[2] Telecommunication Network Business
Samsung Electronics, Korea
jhogunu@daum.net

Abstract. The Bounded Shortest Multicast Algorithm (BSMA) is a very well-known one of delay-constrained minimum-cost multicast routing algorithms. Although the algorithm shows excellent performance in terms of generated tree cost, it suffers from high time complexity. For this reason, there is much literature relating to the BSMA. In this paper, the BSMA is analyzed. The algorithms and shortcomings are corrected, and an improved scheme is proposed without changing the main idea of the BSMA.

Keywords: Multicast Routing Algorithm, Delay-Bounded Minimum Steiner Tree (DBMST) Problem, and Bounded Shortest Multicast Algorithm (BSMA).

1 Introduction

Depending on the optimization goals, which include cost, delay, bandwidth, delay-variation, reliability, and so on, the multicast routing problem exists at varying levels of complexity. A Delay-Bounded Minimum Steiner Tree (DBMST) problem deals with the minimum-cost multicast tree, satisfying the delay-bounds from source to destinations. The Bounded Shortest Multicast Algorithm (BSMA) is a routing algorithm which solves the DBMST problem for networks with asymmetric link characteristics [1,2]. The BSMA starts by obtaining a minimum delay tree, calculated by using Dijkstra's shortest path algorithm. It then iteratively improves the cost, by performing, the delay-bounded path switching. The evaluation performed by the Salama *et al.* [3] demonstrates that the BSMA is one of the most efficient algorithms for the DBMST problem, in terms of the generated tree cost. However, the high time complexity presents a major drawback of BSMA, because a k-shortest path algorithm is used iteratively for path switching. There are also several approaches to improve the time complexity of BSMA [4,5]. However, among them, none can peer with the BSMA in terms of cost.

---

* Corresponding author.

Y. Shi et al. (Eds.): ICCS 2007, Part I, LNCS 4487, pp. 668–675, 2007.
© Springer-Verlag Berlin Heidelberg 2007

The subsequent sections of this paper are organized as follows. In Section 2, the BSMA is described. In Section 3, then the problems with the BSMA are described and the fact that the BSMA can perform inefficient patch switching in terms of delays from source to destinations without reducing the tree cost, are presented. A new algorithm is proposed, furthermore, to substitute for the k-shortest path algorithm, considering the properties of the paths used for the path switching. Finally, this paper is concluded in Section 4.

2   Bounded Shortest Multicast Algorithm

BSMA constructs a DBMST by performing the following steps:

1) Initial step: Construct an initial tree with minimum delays from the source to all destinations.
2) Improvement step: Iteratively minimize the cost of the tree while always satisfying the delay bounds.

The initial tree is minimum-delay tree, which is constructed using Dijkstra's shortest path algorithm. If the initial tree could not satisfy the given delay bounds, some negotiation would be required to relax the delay bounds of DDF (Destination Delay-bound Function). Otherwise, tree construction cannot succeed in satisfying the DDF.

BSMA's improvement step iteratively transforms the tree topology to decrease its cost monotonically, while satisfying the delay bounds. The transformation performed by BSMA at each iteration of the improvement step consists of a delay-bounded path switching. The path switching replaces a path in tree T_j by a new path with smaller cost, resulting in a new tree topology T_{j+1}. It involves the following:

1) Choosing the path to be taken out of T_j and obtaining two disjoint subtrees T_j^1 and T_j^2
2) Finding a new path to connect T_j^1 and T_j^2, resulting in the new tree topology T_{j+1} with smaller cost, while the delay bounds are satisfied.

A candidate paths in T_j for path switching is called a *superedge*. Removing a superedge from a multicast tree corresponds to removing all of the the tree edges and internal nodes in the superedge. From the definition of a superedge [1,2], a destination node or a source node cannot be an internal node of a superedge. This prevents the removal of destination nodes or the source node from the tree as a result of a path switching.

At the beginning of the improvement step, BSMA sets all superedges unmarked and selects the superedge p_h with the highest cost among all unmarked superedges. Removing the p_h in T_j breaks T_j into two disjoint subtrees T_j^1 and T_j^2, where $T_j = p_h \cup T_j^1 \cup T_j^2$. A delay-bounded minimum-cost path p_s between T_j^1 and T_j^2 is used to reconnect T_j^1 and T_j^2 to obtain the new tree topology T_{j+1}, where $T_{j+1} = p_s \cup T_j^1 \cup T_j^2$. The cost of p_s is not higher than that of p_h.

The search for the p_s starts with the minimum-cost path between the two trees. If the minimum-cost path results in a violation of delay bounds, BSMA uses an incremental k-shortest path algorithm [6] to find p_s. The k-shortest path problem consists of finding kth shortest simple path connecting a given source-destination pair in a graph. k-shortest path in BSMA is k-minimum-cost path between two trees and is equivalent to finding the k-shortest path between the two nodes. Because BSMA uses a k-shortest path algorithm for the path switching, its high time complexity is the major drawback. For this reason, the improvement algorithms are proposed in [4,5]. While these reduce the execution time, the performance loss in terms of tree cost is also happened.

3   Difficulties in BSMA

3.1   Undirected Graph Model for BSMA

It is not mentioned clearly in [1] whether the network model for BSMA is a directed or an undirected graph. Although the figures in [1] implicate that it is undirected, later version describing the BSMA [2] and other literatures [3,4,5] related to the BSMA simulation state that the network is modeled as a directed graph. But, we argue that it should be an undirected graph.

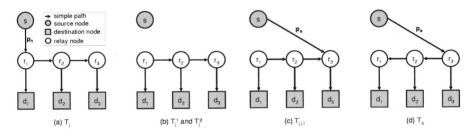

Fig. 1. A case that is able to happen at the step for delay-bounded path switching

The following case in Fig. 1 can be happened during the delay-bounded path switching of BSMA and shows that there could be a problem. Fig. 1(a) shows a tree T_j before the path switching, of which the highest cost superedge p_h is a path from s to r_1, as shown. There are two disjoint subtree T_j^1 and T_j^2 in Fig. 1(b) which are calculated by removing the highest cost superedge p_h in T_j. At this step of the path switching, a delay-bounded shortest path p_s is searched. By reconnecting the p_s (Fig. 1(c)), T_{j+1} is obtained.

As shown in the Fig. 1(c), T_{j+1} is a wrong tree, because the source s cannot send anything to destinations d_1 and d_2 using T_{j+1}. To convert T_{j+1} in Fig. 1(c) to T_c in Fig. 1(d), that is what the algorithm wants, it must be guaranteed that both path-delays and path-costs of the paths from r_1 to r_2 and from r_2 to r_3 are the same as those of paths from r_2 to r_1 and from r_3 to r_2, respectively. If the algorithm must guarantee this, it becomes overhead to check all the links in a subtree without source s that is T_j^1 or T_j^2 at every step it performs the path switching. Simultaneously it can severely reduce the possible cases that can make the path switching as many as there are asymmetric links in the tree. Of course, there is no routine to handle this case in BSMA. To avoid this and to contribute to the main idea of the BSMA, the network model should be assumed as the undirected graph. From now on, we use the undirected links, so that $(u, v) = (v, u) \in E$ with the same link-delay and link-cost values.

3.2   Meaning of 'Unmark' the Superedge

The issue of this subsection is about the superedge. There are five superedges in Fig. 1(a), those are $p(s, r_1)$, $p(r_1, d_1)$, $p(r_1, r_2)$, $p(r_2, d_2)$, and $p(r_2, d_3)$. After the path switching, there are different superedges in Fig. 1(d), those are $p(s, r_3)$, $p(r_3, d_3)$, $p(r_3, r_2)$, $p(r_2, d_2)$, and $p(r_2, d_1)$. You can see that the superedges change, as the tree changes. The simple paths in Fig. 2(a), which redraws the Fig. 1(d), are all superedges. The tree in Fig. 2(b) is the result by another path switching. And the tree in Fig. 2(c) shows the same tree where the simple paths are all superedges.

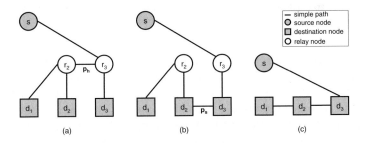

Fig. 2. Changes of superedges as the tree changes

BSMA marks the highest-cost unmarked superedge p_h when the superedge is on path switching. If the p_h is switched to a delay-bounded minimum-cost path p_s, BSMA *unmarks* all marked superedges [1,2]. If the flag value of a superedge is marked, it means there is no path to substitute for the superedge to reduce the current tree T_j. When p_h is switched to p_s, the tree is changed from T_j to T_{j+1}. Of course, BSMA must recalculate the superedges in the new tree T_{j+1} with initializing them as unmarked.

3.3   The Delay-Bounded Minimum-Cost Path p_s and the Function to Get the p_s Between T_j^1 and T_j^2

According to [1,2], one of two cases must happen when the delay-bounded minimum-cost path p_s is obtained:

1. path p_s is the same as the path p_h; or
2. path p_s is different from the path p_h.

If the first case occurs, the p_h has been examined without improvement of the tree cost. If the second case occurs, the tree T_{j+1} would be more cost-effective tree than T_j. But it is possible in the second case to generate the tree T_{j+1} whose cost is the same as that of T_j and end-to-end delay between a source and each destination is worse than that of T_j. The former cost-effective tree T_{j+1} is what the algorithm wants. But the latter tree is not. This is because a path with the same cost as that of p_h could be searched as p_s.

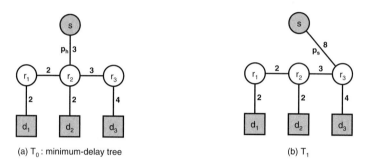

(a) T_0 : minimum-delay tree   (b) T_1

Fig. 3. Ineffective path switching in terms of end-to-end delay without any improvement of tree cost

Fig. 3 shows the example for that. The values on the simple paths in the Fig. 3 are path-delays. The tree T_0 in Fig. 3(a) is a minimum-delay tree generated by using Dijkstra's algorithm. So, it is sure that the end-to-end delays between the source s and each destination d_1, d_2, and d_3 are the minimum values. The tree T_1 in Fig. 3 (b) is a tree after the path switching. The p_s is a simple path whose path-delay is 8 and path-cost is the same as that of p_h. As a result of the path switching, BSMA generates the ineffective tree T_1 in terms of end-to-end delay without any improvement of tree cost.

If BSMA considers only whether the p_s is equal to p_h or not, this ineffective path switching is always possible as long as there could be a path whose cost is the same as that of p_h while satisfying the delay-bounds. So BSMA must select a path with smaller cost than that of p_h, as the p_s. (We must note here that the ineffective path switching does not happen in *BSMA based on greedy heuristic*, since it performs the path switching when the *gain* is larger than zero.)

Additionally, we need to think about the procedure to determine whether a path is delay-bounded or not. Whenever BSMA finds the p_s, it has to determine whether a path is delay-bounded or not. That is to say, BSMA has to perform one of the followings:

1. construction of a tree from T_j^1, T_j^2 and a candidate path for p_s, for every candidate until finding the p_s; or
2. pre-calculation of end-to-end delays between a source and each destination for all cases that the p_s would connect T_j^1 with T_j^2.

The total cost of a tree can be calculated without considering how the nodes in the tree are connected by the links. Because it is the sum of link-costs in the tree, we only need information about which links are in the tree and how

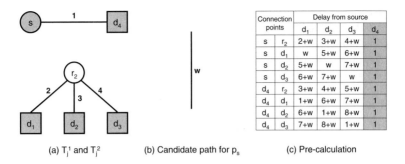

Connection points		Delay from source			
		d_1	d_2	d_3	d_4
s	r_2	2+w	3+w	4+w	1
s	d_1	w	5+w	6+w	1
s	d_2	5+w	w	7+w	1
s	d_3	6+w	7+w	w	1
d_4	r_2	3+w	4+w	5+w	1
d_4	d_1	1+w	6+w	7+w	1
d_4	d_2	6+w	1+w	8+w	1
d_4	d_3	7+w	8+w	1+w	1

(a) T_j^1 and T_j^2 (b) Candidate path for p_s (c) Pre-calculation

Fig. 4. Calculation of end-to-end delays in tree

much their costs are. But end-to-end delay between two tree nodes, such as a source and a destination, is only able to be calculated after considering the tree structure.

Fig. 4(a) shows subtrees on BSMA's path switching and (b) is a candidate path for p_s. If the values on the simple paths in Fig. 4(a) and (b) are the path-costs, the total cost of a tree T_{j+1} $(= T_j^1 \cup T_j^2 \cup p_s)$ would be determined when the path-cost w of the candidate is determined (that is $1+2+3+4+w$), without considering how the candidate connects T_j^1 with T_j^2. But if the values indicate the path-delays, it is impossible to determine the end-to-end delays between the source s and each destination d_1, d_2, or d_3 without considering the tree structures of T_j^1 and T_j^2.

Obviously there are two ways to determine whether the candidate is delay-bounded or not; one is to construct the tree by connecting T_j^1 and T_j^2 with the candidate; the other is to perform a pre-calculation such like Fig. 4(c). Of course, both must consider the structures of two subtrees.

3.4   Inefficient Use of k-Shortest Path Algorithm

As the literatures [4,5] mentioned, the k-shortest path algorithm [6] used for finding the p_s is the major drawback of BSMA. The time complexity of the k-shortest path algorithm is $O(kn^3)$. The k value can be set to a fixed value to reduce execution time of BSMA. However this also reduces the performance of BSMA in terms of generated tree cost. In this subsection we propose another algorithm to substitute the k-shortest path algorithm. The proposed algorithm finds candidate paths for p_s within some path-cost range and does not deteriorate the performance of BSMA.

According to what we described in subsection 3.3, BSMA do not need the paths whose costs are equal to or larger than that of p_h while finding p_s. And obviously we cannot find any path with smaller cost than that of minimum-cost path calculated by Dijkstra's algorithm. Consequently, candidate paths for p_s are the paths with the cost range that is equal to or larger than that of the minimum-cost path and smaller than that of p_h. The following is the pseudo code of the proposed algorithm.

Description for internal variables and functions

$p[0..(|V|-1)]$, $q[0..(|V|-1)]$: an array containing the node sequence for a path

$index_p$: the index for p

$cost_p$: the path-cost of p

P: the set of searched paths

Q: queue containing paths on searching

PUSH(Q, p): insert p to Q

POP(Q): take a path out of Q

PROCEDURE PathSearch_SameCostRange$(T_j^1, T_j^2, minCost, maxCost, G)$

1. $P \leftarrow \emptyset$, $Q \leftarrow \emptyset$;
2. **for** each node $i \in T_j^1$ {
3. $p[0] \leftarrow i$, $index_p \leftarrow 0$, $cost_p \leftarrow 0$;
4.   **PUSH**(Q, p);
5. }
6. **while** $Q \neq \emptyset$ {
7. $p \leftarrow$ **POP**(Q);
8.   **for** each neighbor node n of $p[index_p]$ {
9.   **if** (n is in the array p) **then continue**;
10.   \\ *Because we are looking for simple paths*
11.   **if** ($n \in T_j^1$) **then continue**;
12.   \\ *Because we are looking for paths between T_j^1 and T_j^2*
13.   **if** ($cost_p+$ link-cost of $(p[index_p], n) \geq maxCost$) **then continue**;
14. $q \leftarrow p$, $cost_q \leftarrow cost_p$, $index_q \leftarrow index_p$; \\ *Copy p to q*
15. $q[index_q + 1] \leftarrow n$;
16. $cost_q \leftarrow cost_q+$ link-cost of $(q[index_q], n)$;
17. $index_q \leftarrow index_q + 1$;
18.   **if** ($n \in T_j^2$ **AND** $cost_q \geq minCost$) **then** $P \leftarrow P \cup \{q\}$;
19.   **if** ($n \notin T_j^2$) **then PUSH**(Q, q);
20.   }
21. }
22. **return** P;

Although a number of candidates for p_s are searched, BSMA use only the one satisfying the delay-bounds with smallest cost, that is p_s. So BSMA do not need to find all the candidates at once. Therefore the arguments $minCost$ and $maxCost$ are not minimum path-cost between T_j^1 and T_j^2, and path-cost of p_h, respectively. The half closed interval [*minimum path-cost between T_j^1 and T_j^2, path-cost of p_h*) is divide into several intervals to be the $minCost$ and $maxCost$. BSMA iteratively increases the $minCost$ and $maxCost$ until either finding p_s or recognizing there is no path to substitute p_h.

When the network size is large with many links, the memory required to calculate the candidates is also heavy as well as the high time complexity. So, this is quite practical and dose not provide any limitation to BSMA's performance. The difference between $minCost$ and $maxCost$ can be adjusted according to characteristic of modeled link-cost. (*i.e.* If the link-costs is modeled as integer values, *minCost* and *maxCost* can be the integer value x and $x+1$, where x is some starting point of divided interval and 1 stands for the characteristic.) In the next section, the characteristic is notated as the *sys*.

4 Conclusion

BSMA is very well-known one of delay-constrained minimum-cost multicast routing algorithms. Although, its performance is excellent in terms of generated tree cost, the time complexity is very high. There are many literatures related to BSMA for this reason. We have shown that BSMA has fallacies and ambiguities then, modified it. We start on the describing BSMA [2]. Then, we show that the BSMA has fallacies, and that the BSMA can perform inefficient patch switching in terms of delays from source to destinations without reducing the tree cost. Hence, we propose an algorithm to substitute for the k-shortest path algorithm considering the properties of the paths which are used for the path switching.

Acknowledgment

This research was supported by the MIC(Ministry of Information and Communication), Korea, under the ITRC(Information Technology Research Center) support program supervised by the IITA(Institute of Information Technology Assessment), IITA-2006-(C1090-0603-0046).

References

1. Zhu, Q., Parsa, M., Garcia-Luna-Aceves, J. J.: A Source-Based Algorithm for Delay-Constrained Minimal-Cost Multicasting. Proceeding of INFOCOM. IEEE (1995) 377-385
2. Parsa, M., Zhu, Q., Garcia-Luna-Aceves, J. J.: An Iterative Algorithm for Delay-Constrained Minimum-Cost Multicasting. IEEE/ACM Transactions Networking, Vol. 6, Issue 4. IEEE/ACM (1998) 461-474
3. Salama, H. F., Reeves, D. S., Viniotis, Y.: Evaluation of Multicast Routing Algorithms for Real-Time Communication on High-Speed Networks. Journal of Selected Areas in Communications, Vol. 15, No. 3. IEEE (1997) 332-345
4. Gang, F., Kia, M., Pissinoul, N.: Efficient Implementations of Bounded Shortest Multicast Algorithm. Proceeding of ICCCN. IEEE (2002) 312-317
5. Gang, F.: An Efficient Delay Sensitive Multicast Routing Algorithm. Proceeding of the International Conference on Communications in Computing. CSREA (2004) 340-348
6. Lawler, E.: Combinatorial Optimization: Networks and Matroids. Holt, Rinehart and Winston (1976)

Efficient Deadlock Detection in Parallel Computer Systems with Wormhole Routing

Soojung Lee

GyeongIn National University of Education
6-8 Seoksu-dong, Anyang, Korea 430-739
sjlee@gin.ac.kr

Abstract. Wormhole routing has been popular in massively parallel computing systems due to its low packet latency. However, it is subject to deadlock, where packets are waiting for resources in a cyclic form indefinitely. Current deadlock detection techniques are basically dependent on the time-out strategy, thus yielding unignorable number of false deadlock detections especially in heavy network loads or with long packets. Moreover, several packets in a deadlock may be marked as deadlocked, which would saturate the resources allocated for recovery. This paper proposes a simple but more accurate deadlock detection scheme which is less dependent on the time-out value. The proposed scheme presumes deadlock only when a cyclic dependency among blocked packets exists. Consequently, the suggested scheme considerably reduces the probability of detecting false deadlocks over previous schemes, thus enabling more efficient deadlock recovery and higher network throughput. Simulation results are provided to demonstrate the efficiency of the proposed scheme.

1 Introduction

Wormhole routing has been quite popular in parallel computing systems with interconnection networks, because it can significantly reduce packet latency and the requirement of packet buffers is obviated [1]. In wormhole routing, a packet is split into several *flits* for transmission. A header flit leads the route and the remaining flits follow in a pipelined fashion. However, it is susceptible to deadlock, where a set of packets may become blocked forever. This situation occurs when each packet in the set requests a channel resource held by another packet in the set in a circular way.

Deadlock avoidance has been a traditional approach in handling deadlock problem [2]. In this approach, routing is restricted in a way that no cyclic dependency exists between channels. For example, the turn model [3,6] prohibits turns that may form a cycle. However, such design of routing algorithm results in low adaptivity and increased latency. A routing algorithm is said to be adaptive if a routing path is selected based on dynamic network conditions.

A way to have higher throughput while avoiding deadlock is using the virtual channel. A number of virtual channels share a physical channel, thereby composing virtual networks and facilitating adaptive routing algorithms. In [4], virtual

Y. Shi et al. (Eds.): ICCS 2007, Part I, LNCS 4487, pp. 676–683, 2007.

channels are divided into two classes; one for dimension-order routing with no cyclic dependency and the other for fully adaptive minimal routing. Although this scheme can provide more flexibility, it is only partially adaptive.

The frequency of deadlock occurrence is reported to be very low with a fully adaptive routing algorithm [11]. Hence, it is wasteful to limit routing adaptivity for rarely occurring deadlocks. This motivated a new approach to handling deadlocks, deadlock detection and recovery. The criteria for determining deadlock is basically time-out. That is, a packet is presumed as deadlocked if it has been waiting for longer than a given threshold [7,10] or if all of its requested channels are inactive for longer than the threshold [9]. Although these schemes can detect all deadlocks, they may misinterpret simply-congested packets as deadlocked. A more sophisticated method to determine deadlock was proposed in [8], which, to our knowledge, performs best in detecting deadlocks accurately. It notices a sequence of blocked packets as a tree whose root is a packet that is advancing. When the root becomes blocked later, only the packet blocked due to the root is eligible to recover. However, the accuracy of the mechanism in [8] relies on the dependency configuration of blocked packets as well as the threshold value. In general, deadlock is recovered by ejecting deadlocked packets from the network [8] or by forwarding them through a dedicated deadlock-free path [10].

Deadlock frequency determines the performance of deadlock detection and recovery schemes. In heavily loaded networks, those packets presumed as deadlocked will saturate the recovery resources, thus degrading performance considerably. Therefore, it is required that only real-deadlocked packets use the resources, as their occurrence frequency is low [11]. However, previous schemes [7,8,9] cannot distinguish between real deadlocked and blocked packets waiting longer than the given threshold. Also, they force all the packets in deadlock to recover, although it is sufficient to choose only one packet to break the deadlock.

The performance of a fully adaptive routing algorithm relies on the effectiveness of the deadlock detection mechanism associated with it. We propose simple but effective deadlock detection mechanisms which employ a special control packet named *probe* to detect deadlock. A blocked packet initiates a probe when all of its requested channels are inactive for the threshold and propagates it along the path of inactive channels. The presence of deadlock is presumed, if a cyclic dependency among blocked packets is suspected through probe propagation, thereby reducing the number of packets detected as deadlocked considerably over previous schemes. The performance of our schemes is simulated and compared with that of a previous scheme [8], known to be most efficient in reducing the number of false deadlock detections.

2   The Proposed Mechanism

We first describe our scheme for mesh networks. To depict resource dependencies at a point of time, the *channel wait-for graph* (CWFG) can be used, where vertices represent the resources (either virtual channels or physical channels for networks with no virtual channel) and edges represent either 'wait-for' or 'owned-after' relations [5,11,12]. A wait-for edge (c_i, c_j) represents that there exists a message occupying channel c_i and waiting for channel c_j. An owned-after edge (c_i, c_j) implies the temporal order in which the channels were occupied, i.e., c_j is owned after c_i. In a network with virtual channels, a packet header may proceed if any of the virtual channels associated with its requested channel is available. Hence, in such network, there are multiple wait-for edges outgoing from a vertex, while there always exists only one owned-after edge, as data flits in wormhole networks simply follow their previous flits in a pipelined fashion. For the description of the proposed mechanism, we introduce the following notation.

Notation 1. Assume that blocked packet m holds c and an edge (c, c') exists in the CWFG. We refer to c' as a *predecessor* of c with respect to m and denote the set of predecessors of c with respect to m as $pred(c)|_m$. Also, let $dim(c)$ and $dir(c)$ denote the dimension and direction of a physical or virtual channel c, respectively. $\qquad\qquad\square$

As a cycle is a necessary condition to form a deadlock, our scheme is motivated by a simple observation that a cycle involves at least four blocked packets in a minimal routing. From this observation, one may think of an idea that a cycle is detected by counting the number of blocked packets in sequence. That is, if the number counts up to at least four, one concludes that a potential deadlock exists. This idea obviously reduces the number of false deadlock detections over those schemes which simply measure the channel inactivity time for time-out for deadlock detection; these schemes would yield deadlock detections as much as the number of blocked packets.

Obviously, our idea may detect deadlock falsely. For instance, consider a sequence of blocked packets residing within one dimension only, without turning to other dimensions; note that such sequence of blocked packets cannot form a cycle in meshes. Our idea would declare deadlock in such case, although there is none. Therefore, in order to further reduce the number of false deadlock detections, we take a different view of identifying a cycle. Namely, we focus on the number of corners, rather than on the number of blocked packets in sequence. If the number of corners formed by a sequence of blocked packets counts up to four or more, the presence of a deadlock is presumed.

The above criteria for detecting deadlock based on the number of turns are likely to detect deadlock falsely. However, the frequency of deadlock occurrence is reported to be very low with a fully adaptive routing algorithm [11]. Hence, it is believed that a complex cycle would rarely occur except in a heavy network condition. Moreover, it is more important to quickly dissipate congestion by resolving simple cycles before they develop into complex ones.

To implement the above idea, we employ a special control packet named *probe* to traverse along inactive channels for deadlock detection. Basically a probe is initiated upon time-out. However, in order not to initiate probes repetitively along the same channel, a bit, named PIB (Probe Initiation Bit), is allocated for each physical channel to indicate that a probe is initiated and transmitted through the channel. The bit is reset when the physical channel becomes active. Specifically, probes are initiated and propagated according to the following rules.

Rule 1. A router initiates a probe if
(i) there is a blocked packet,
(ii) all the channels requested by the blocked packet are inactive for threshold TO due to other blocked packets holding the channels, and
(iii) PIB of any one of the channels requested by the blocked packet is zero.

Let c be a channel with zero PIB, requested by the blocked packet. Also, let m be one of the packets holding c. Upon initiation, the router transmits probe(m) through c and sets the PIB of c to one.   □

Rule 2. When a packet is delivered along a channel with PIB of one, set the PIB to zero.   □

Rule 3. Let c be the input channel through which probe(m) is received. Let e be (c, c'), where $c' \in pred(c)|_m$. If e is an owned-after edge, simply forward probe(m) through c'. Otherwise if e is a wait-for edge, check if all the channels requested by m are inactive for threshold TO_F due to other blocked packets holding the channels. If yes, transmit probe(m') through c', where m' is one of the packets holding c'. If no, discard the received probe(m).   □

By Rule 3, a probe follows the path through which a blocked packet is routed until the header of the packet is met. At that moment, all the channels requested by the header are checked for their inactivity time. When the time exceeds TO_F threshold for each of the channels, the probe is forwarded through one of the channels. Unlike TO, one may set TO_F to a small value, in order not to delay the probe transmission.

Let us call the process of initiation and transmission of a probe *probing*. The probe carries the information on the number of turns made by blocked packets which hold the channels on the probing path; a packet is said to make *turn* if it changes its routing dimension. The number of turns is represented by *count*. When a router receives a probe, it examines *count* carried by the probe If *count* is at least four, the router presumes the presence of deadlock. As *count* is carried by probes, we name this mechanism COUNTING scheme. Specifically, the mechanism manipulates *count* as follows.

Rule C1. Upon initiation of a probe for blocked packet m waiting on input channel c, if the probe is to be transmitted along channel c', then
(i) if $dim(c) = dim(c')$, then transmit the probe carrying *count* of zero along c'.
(ii) otherwise transmit the probe carrying *count* of one along c'.   □

Rule C2. Upon receiving a probe through channel c, if the probe is to be transmitted along channel c', then
(i) if $dim(c) \neq dim(c')$, then increase the received *count* by one.
(ii) if $count \geq 4$ and (c, c') is a wait-for edge, then declare deadlock; otherwise, transmit the probe carrying *count* along c'.   □

In Rules C1 and C2, whether to send the probe or not and which channel to send the probe through are all determined by Rules 1 and 3. Rule C2 allows deadlock declaration when the probe encounters a packet header. This is to recover the potential deadlock by selecting that packet as a victim.

Note that COUNTING scheme does not consider directions of turns. It may misinterpret a non-cyclic sequence of blocked packets involving turns of the same direction as deadlock. In order to better distinguish deadlock, we suggest a slight modification to COUNTING scheme that reflects the direction of turns. A bit is used for each direction in each dimension. For example, four bits are used for 2D networks; two bits for positive and negative directions in dimension zero and another two bits in dimension one. In general, $2n$ bits are used for nD networks. These bits are carried by probes as *count* is carried in COUNTING scheme. The basic idea of the modified scheme is set the bits corresponding to the turn direction and declare deadlock when at least four bits corresponding to any two dimensions are set. We call this modified scheme BITSET scheme. The basic operations for probe and PIB management are the same as described in Rules 1 to 3. Hence, we present only bit operations below.

Notation 2. Probes in BITSET scheme carry $2n$ TBs (Turn Bits) in nD networks. Specifically, $TB_{d,\,+}$ and $TB_{d,\,-}$ represent bits for positive and negative directions in dimension d, respectively. □

Rule B1. Upon initiation of a probe for blocked packet m waiting on input channel c, if the probe is to be transmitted along channel c', then
(i) if $dim(c) = dim(c')$, then transmit the probe along c' carrying zero TBs.
(ii) otherwise transmit the probe along c' carrying TBs with $TB_{dim(c),\,dir(c)}$ and $TB_{dim(c'),\,dir(c')}$ set, where $dir(c) = +$ if packet m was sent along positive direction of c. Otherwise, $dir(c) = -$. $dir(c')$ is set similarly. □

Rule B2. Upon receiving a probe through channel c, if the probe is to be transmitted along channel c', then
(i) if $dim(c) \neq dim(c')$, then set $TB_{dim(c),\,dir(c)}$ and $TB_{dim(c'),\,dir(c')}$.
(ii) if $TB_{d1,\,+}$, $TB_{d1,\,-}$, $TB_{d2,\,+}$, and $TB_{d2,\,-}$, for any two dimensions $d1$ and $d2$, are set and (c, c') is a wait-for edge, then declare deadlock; otherwise, transmit the probe carrying TBs along c'. □

In k-ary n-cube networks, deadlock can be formed involving wraparound channels. This type of deadlock may not be detected by the rules above if it does not include sufficient number of turns. We take a simple approach for detecting such deadlock by regarding wraparound channel usage along the same dimension as a 180-degree turn, thus increasing *count* by two for COUNTING scheme. For BITSET scheme, it is treated as if the packet is making a 180-degree turn through one higher dimension, thus setting TBs corresponding to those two dimensions. The detailed description is omitted due to the space constraint.

3   Performance

This section presents simulation results of the proposed schemes and the scheme in [8] which is considered, to our knowledge, as most efficient in reducing the

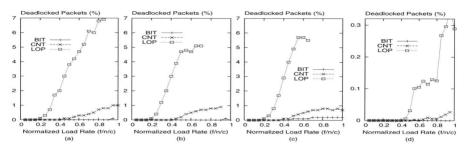

Fig. 1. Percentage of packets detected as deadlocked (a) TO=16 cycles, 16x16 meshes. (b) TO=16 cycles, 8x8x8 meshes. (c) TO=16 cycles, 16x16 tori. (d) TO=128 cycles, 16x16 meshes.

number of false deadlock detections. The simulations are run on 16x16 and 8x8x8 mesh and torus networks. Channels are with three virtual channels of buffer depth of two flits. The routing algorithm is minimal and fully adaptive. Packets are 32 flit-sized and their destinations are assumed uniformly distributed. We assume one clock cycle each for transmission of a flit over a channel, decoding a control flit, and crossing a switch. The statistics have been gathered after executing the program for 50000 clock cycles. The result of the first 10000 cycle is discarded for the initial transient period. Packets are generated exponentially with varying injection rate where the same rate is applied to all nodes. A packet presumed as deadlocked is ejected from the network and re-injected later when any of its requested channel resources is available.

We measured the percentage of packets detected as deadlocked by each strategy for varying normalized load rate of flits per node per cycle (f/n/c). Figure 1 shows the results for two TO thresholds of 16 and 128 clock cycles. The results of [8] are indicated with the legend 'LOP' and those of COUNTING and BITSET schemes with 'CNT' and 'BIT', respectively. TO_F threshold for forwarding probes is set to two cycles for all experiments. The four figures show similar behaviors approximately. As expected, the percentage increases with the load rate. BIT detects almost no deadlock except for torus networks. Overall, CNT performs better than LOP, its percentage being approximately as much as eight times lower for meshes and eleven times lower for tori. It is shown that for large TO such as 128 cycles, LOP yields less than 0.3 percentage of packets presumed as deadlocked, even at high loads for meshes. The results for CNT for the three network types are almost comparable, although packets would turn more often in 3D networks.

Normalized accepted traffic measured in flits per node per cycle is presented in Figure 2 for 2D meshes. All three schemes perform comparably in most cases except for high loads and TO of 16 cycles, at which the throughput of LOP drops drastically. This is because LOP detects too many packets as deadlocked for that network condition, as shown in Figure 1(a). Note that for other network conditions, the difference in the number of deadlocked packets has no significant effect on throughput.

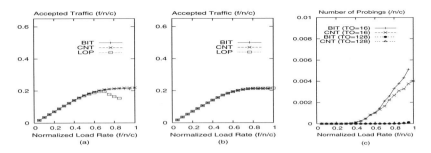

Fig. 2. 16x16 meshes (a) Normalized accepted traffic when TO=16 cycles. (b) Normalized accepted traffic when TO=128 cycles. (c) Mean number of probe initiations per node per clock cycle.

Table 1. Mean number of probe transmissions per probing for COUNTING scheme

Network	Normalized Load Rate								
Configurations	0.1	0.2	0.3	0.4	0.5	0.6	0.7	0.8	0.9
16x16 meshes and TO=16 cycles	N/A	5.6	6.6	6.2	6.3	6.4	6.6	6.6	6.6
16x16 meshes and TO=128 cycles	N/A	N/A	N/A	N/A	6.3	5.5	6.7	7.3	8.5
8x8x8 meshes and TO=16 cycles	N/A	4.5	3.9	3.8	3.9	4.1	4.2	4.3	4.4
8x8x8 meshes and TO=128 cycles	N/A	N/A	N/A	2.8	3.4	4.0	4.3	4.4	4.5

As the proposed schemes utilize a control packet, we measured its load on the router through the number of probings and the number of probe transmissions per probing. The number of probings initiated by a node per cycle is shown in Figure 2(c) for 2D meshes. Obviously, a node tends to initiate probings more often with higher loads but less frequently with higher thresholds. In particular, there is a significant difference between the results for the two thresholds. It is observed for TO of 128 cycles that there is virtually no probing activity regardless of the network load. For TO of 16 cycles, it is noted for both schemes that a node initiates approximately two probings per 1000 clock cycles at the saturated load and no more than five probings at extremely high loads. The reason for the slight difference between the results of the two schemes at high loads for TO of 16 cycles is due to the fact that CNT detects more deadlocks than BIT as shown in Figure 1(a). That is, it facilitates blocked packets to proceed, since more packets are ejected from the highly loaded network. This reduces the need for probe initiations. BIT and CNT schemes showed similar results for the other network configurations.

Table 1 shows the number of probe transmissions for a probing. In general, more probes are transmitted for 2D meshes than for 3D meshes. This is simply because packets have more routing adaptivity and are less blocked in 3D meshes. It is noted that the number of probes tends to increase with the load rate for both networks, especially for TO of 128 cycles. For TO of 16 cycles, blocked packets are dissipated promptly by more frequent probings than for 128-cycle TO, which leads to lower possibility of forwarding probes.

4 Conclusions

This paper proposed enhanced mechanisms for deadlock detection in wormhole-routed direct networks. Different from the previous schemes, the proposed schemes do not solely rely on the threshold value. A control packet propagates to find out the presence of deadlock. As the control packets traverse only along inactive channels, they virtually do not disturb normal packet progression. Simulation studies are conducted to compare the performance of the proposed schemes with that of the scheme which, to our knowledge, is most efficient in reducing the number of false deadlock detections. The simulation results demonstrate that the suggested schemes yield a substantial decrease in the number of deadlock detections in various network conditions. Consequently, our schemes outperform the previous scheme in terms of the network throughput irrespective of the time-out threshold.

References

1. Al-Tawil, K.M., Abd-El-Barr, M., Ashraf, F.: A survey and comparison of wormhole routing techniques in a mesh network. IEEE Network **11**(2) (1997) 38–45
2. Park, H., Agrawal, D.P.: A generic design methodology for deadlock-free routing in multicomputer networks. Journal of Parallel and Distributed Computing **61**(9) (2001) 1225–1248
3. Chiu, G.M.: The odd-even turn model for adaptive routing. IEEE Trans. Parallel and Distributed Systems **11**(7) (2000)
4. Duato, J.: A general theory for deadlock-free adaptive routing using a mixed set of resources. IEEE Trans. Parallel and Distributed Systems **12**(12) (2001) 1219–1235
5. Duato, J.: A necessary and sufficient condition for deadlock-free adaptive routing in wormhole networks. IEEE Trans. Parallel and Distributed Systems **6**(10) (1995) 1055–1067
6. Glass, C.J., Ni, L.M.: The turn model for adaptive routing. Journal of the ACM **41**(5) (1994) 874–902
7. Kim, J., Liu, Z., Chien, A.: Compressionless routing: a framework for adaptive and fault-tolerant routing. IEEE Trans. Parallel and Distributed Systems **8**(3) (1997) 229–244
8. Martinez, J.M., Lopez, P., Duato, J.: FC3D: flow control-based distributed deadlock detection mechanism for true fully adaptive routing in wormhole networks. IEEE Trans. Parallel and Distributed Systems **14**(8) (2003) 765–779
9. Martinez, J.M., Lopez, P., Duato, J.: A cost-effective approach to deadlock handling in wormhole networks. IEEE Trans. Parallel and Distributed Systems **12**(7) (2001) 716–729
10. Pinkston, T.M.: Flexible and efficient routing based on progressive deadlock recovery. IEEE Trans. Computers **48**(7) (1999) 649–669
11. Pinkston, T.M., Warnakulasuriya, S.: Characterization of deadlocks in k-ary n-cube networks. IEEE Trans. Parallel and Distributed Systems **10**(9) (1999) 904–921
12. Schwiebert, L., Jayasimha, D.N.: A necessary and sufficient condition for deadlock-free wormhole routing. Journal of Parallel and Distributed Computing **32** (1996) 103–117

Type-Based Query Expansion for Sentence Retrieval

Keke Cai, Chun Chen, Jiajun Bu, and Guang Qiu

College of Computer Science, Zhejiang University
Hangzhou, 310027, China
{caikeke, chenc, bjj, qiuguang}@zju.edu.cn

Abstract. In this paper, a novel sentence retrieval model with type-based expansion is proposed. In this retrieval model, sentences expected to be relevant should meet with the requirements both in query terms and query types. To obtain the information about query types, this paper proposes a solution based on classification, which utilizes the potential associations between terms and information types to obtain the optimized classification results. Inspired by the idea that relevant sentences always tend to occur nearby, this paper further re-ranks each sentence by considering the relevance of its adjacent sentences. The proposed retrieval model has been compared with other traditional retrieval models and experiment results indicate its significant improvements in retrieval effectiveness.

Keywords: Sentence retrieval, query type identification, query expansion.

1 Introduction

Sentence retrieval is to retrieve query-relevant sentences in response to users' queries. It has been widely applied in many traditional applications, such as passage retrieval [1], document summarization [2], question answering [3], novelty detection [4] and content-based retrieval presentation [5]. A lot of different approaches have been proposed for sentence retrieval. Most of them, however, have not been proven efficient enough. The main reason is due to the limited information expressed in sentences. To improve the performance of sentence retrieval, besides the key words in queries and sentences, additional features that are helpful for indicating sentences' relevance should be explored.

Query type, which expresses relevant information satisfying users' information need, has been effectively used in some applications involving sentence retrieval, such as question-answering that looks for sentences containing the expected type of answer, and novelty detection where sentences involving information of special type will be considered more relevant. However, little efforts have been made to incorporate such information into the process of keyword-based information retrieval, where the most difficulty is the identification of query-relevant information types.

This paper proposes a new sentence retrieval model, in which the information about query types is explored and incorporated into the retrieval process. The idea is similar to that of query expansion. The difference is that this model expands each query with the relevant information types instead of the concreted terms. In this paper,

Y. Shi et al. (Eds.): ICCS 2007, Part I, LNCS 4487, pp. 684–691, 2007.

query types are defined as the types of the expected information entities, such as persons, locations, numbers, dates, times and etc, which are considered necessary for satisfying user's request or information need. Therefore, in the retrieval process, sentences expected to be relevant should meet with the requirements both in query terms and query types. To achieve such a retrieval model, the most important factor is the identification of query types. This paper proposes a solution based on classification. This classification model makes a full use of the theory of information association, with the purpose to utilize the potential associations between terms and information types to obtain the optimized classification results. In addition to term and type information described above, another type of information is also considered in the evaluation of sentence relevance, that is, the proximity between sentences. The idea underneath is that relevant sentences always tend to occur nearby. Then, each sentence is further re-ranked by considering the relevance of its adjacent sentences.

The remainder of the paper is structured as follows: Section 2 introduces the related studies in sentence retrieval. Section 3 describes the proposed sentence retrieval model and the classification approach for query type identification. In Section 4, the experimental results are presented. Section 5 concludes the paper.

2 Related Works

Most exiting approaches for sentence retrieval are based on term matching between query and sentence. They are essentially the applications of algorithms designed for document retrieval [6] [7] [8]. However, compared with documents, sentences are much smaller. Thus, the performance of typical document retrieval systems on the retrieval of sentences is significantly worse. Some systems try to utilize linguistic or other features of sentences to facilitate the detection of relevant sentences. In the study of [5], factors used for ranking sentences include the position of sentence in the source document, the words contained in sentence and the number of query terms contained in sentence. In [9], semantic and lexical features are extracted from the initial retrieved sentences to filter out possible non-relevant sentences.

In addition to the mining of features in sentences, some systems concentrate on the studies of features in queries. One of the most significant features about queries is the query type. In most cases, query type is defined as the entity types of the expected answer. For example, in [4], queries are described as patterns that include both query words and the required answer types. Then, these patterns are used to retrieve sentences. Sentences without the expected type of named entities will be considered irrelevant. In the domain of question answering, query type is also an important factor for sentence relevance evaluation. Given a question, the question is analyzed for their expected answer type and then submitted to retrieve sentences that contain the key words from the question or the tokens or phrase that are consistent with the expected answer type of the question. Studies have shown the positive effects of query type on sentence retrieval, which however, in the context of keyword-based retrieval, has not been fully utilized. The most difficulty is the identification of query types, which becomes one of the focuses of our studies in this paper.

3 Sentence Retrieval Model Involved with Query Type

The proposed sentence retrieval model with type-based query expansion measures sentence relevance from two perspectives: lexical similarity and type similarity. Lexical similarity is an indicator of the degree to which query and sentence discuss the same topic. It is always evaluated according to term matching between query and sentence. Given two terms, their similarity can be viewed from different point of views, such as, synonymy, variant, co-occurrence or others. Since this paper expects to pay more attention to the application of query type, we adopt the most basic definition for lexical similarity. If two terms are exactly the same, they are lexical similarity. Type similarity is actually to evaluate the coincidence between the query types and information types related to sentence. From sentence perspective, the related information types are defined as the types of entities contained in sentence and can be identified by using the existing named entity recognition technique. However, from query perspective, the identification of query types is a little more difficult. In this paper, this problem is solved by a solution based on classification.

3.1 Information Association Based Classification

Inspired by the theory of Hyperspace Analogue to Language (HAL) [10], a novel classification approach is proposed to solve the problem of query type identification. In this approach, a special type of information association is explored, with the purpose of reflecting the dependencies between terms and information types. When such kind of associations is incorporated into query classification, information types that have the most probabilities to be associated with a query can be identified. The implementation of this classification is based on the information model reflecting the Associations between Terms and Information Types (ATIT).

Construction of the ATIT Model. The construction of the ATIT model consists of two steps.

The first step is to construct the HAL model given the large document corpus. The HAL model is constructed in the traditional way (See [10] for more details). Let $T = \{t_1, t_2 \ldots t_n\}$ be the term set of the document corpus, the constructed HAL is finally represented as a $n*n$ matrix M_{HAL}, in which each term of T is described as a n-dimension term vector $V_i^{HAL} = \{M_{HAL}(i,1), \ldots, M_{HAL}(i,n)\}$, where $M_{HAL}(i, j)$ describes the association strength between the terms t_i and t_j.

The second step realizes the construction of ATIT model based on HAL model. In the ATIT model, each term t_i is expected to be described as a m-dimension vector $V_i^{ATIT} = \{M_{ATIT}(i,1), \ldots, M_{ATIT}(i,m)\}$, where $M_{ATIT}(i,j)$ represents the association strength between the term t_i and the information type c_j. The construction of ATIT can be further divided into three sub-steps: Firstly, entity type related to each term t_i is discovered. In this paper, it is realized by named entity recognition (NER) [11]. However, it is noted that not all entities of all types can be well identified. To solve this problem, the manual NER approach is adopted, which is to use human generated entity annotation results to realize a part of these entity recognitions. Secondly, based on the association information provided by HAL, the association strengths between terms and information types are calculated. Let terms of information type c_j be $T_{c_j} = \{ t_{c_j 1}, \ldots, t_{c_j k} \}$, the association strength between term t_i and c_j is calculated as:

$$M_{ATIT}(i, j) = \sum_{p=c_j 1}^{c_j k} M_{HAL}(i, p) \;. \tag{1}$$

Implementation of Classification. The ATIT information model makes it possible to identify the relationships between query terms and information types. Then, the probability of query Q being relevant to information type c_j can be evaluated by:

$$P(c_j | Q) = \sum_{t_i \in Q} P(c_j | t_i) P(t_i | Q) \;. \tag{2}$$

where $P(t_i | Q)$ means the probabilities of t_i in Q. Since that queries are normally short, each $P(t_i | Q)$ can be approximately assigned an equal value, i.e., $P(t_i | Q) = 1/|Q|$, where $|Q|$ is the number of terms in the query. The probability $P(c_j | t_i)$ represents the association strength of the information type c_j with respect to the term t_i. According to Bayesian formula, it can be transformed into:

$$P(c_j | t_i) = \frac{P(t_i | c_j) * P(c_j)}{P(t_i)} \;. \tag{3}$$

where $P(c_j)$ and $P(q_i)$ are respectively the prior probability of category c_j and query term q_i. Here, we set them to be constants. $P(t_i | c_j)$ represents the conditional probability of t_i. Based on the previous constructed ATIT model, it is defined as:

$$P(q_i | c_j) = \frac{M_{ATIT}(i, j)}{\sum_{i=1}^{n} M_{ATIT}(i, j)} \;. \tag{4}$$

Probability of query Q being relevant to information type c_j can be evaluated by:

$$P(c_j | Q) \stackrel{rank}{=} \sum_{t_i \in Q} \frac{M_{ATIT}(i, j)}{\sum_{i=1}^{n} M_{ATIT}(i, j)} \;. \tag{5}$$

3.2 Relevance Ranking of Sentence

Experiments in [12] show that although there is no significant difference in the performance of traditional retrieval models when implemented in sentence retrieval, TFIDF technique performs consistently better than the others across different query sets. Thus, in this paper, we decided to use the vector space model (VSM) with tf.idf weighting as the retrieval framework. The retrieval model is formulated as:

$$sim(S,Q) = \lambda \sum_{l \in L_S \wedge L_Q} W_{S,l} * W_{Q,l} + (1-\lambda) \sum_{t \in T_S \wedge T_Q} W_{S,t} * W_{Q,t} \ . \tag{6}$$

where, the parameter λ is used to control the influence of each component to sentence retrieval. Let A be the sentence S or the query Q, then L_A and T_A respectively represent the term vector and the type vector with respect to A. $W_{A,l}$ denotes the weight of term l in A and can be defined as $\log(Ltf_l+1)*\log(N/Ldf_l+1)$, where Ltf_l is the frequency of term l in A, N is the total number of sentences and Ldf_l is the number of sentences containing the term l. $W_{S,t}$ denotes the weight of the information type t in sentence S and can be defined as $\log(Ttf_t+1)*\log(N/Tdf_t+1)$, where Ttf_t is the frequency of entities in S with type t and Tdf_t is the number of sentences containing entities of type t. $W_{Q,t}$ denotes the weight of the information type t in query Q. It is the normalized probability of each query type and defined as:

$$W_{Q,t} = \frac{P(t \mid Q)}{\sum_{t \in T_Q} p(t \mid Q)} \ . \tag{7}$$

In our relevant sentence retrieval model, a sentence that contains not only query terms but also the entities of the expected information types will be considered more relevant. This is the main difference between our type-based expansion retrieval model and other word-based approaches. Since this method makes an estimation of the most possible query context by query relevant information types, the more accurate relevance judgment is expected. However, it is noted that in most cases this approach is more effective for long sentences than short sentences. To solve this problem, further optimization is considered. Conclusions in [9] show that query relevant information always appear in several continuous sentences. Inspired by this idea, this paper proposes to re-rank the sentence according to the following rules.

For sentence S_i that has information of the expected types, but contain few query terms, if the sentence S_j that is before or after S_i has many terms in common with the query, but do not contain the expected information types, the ranks of sentences S_i and S_j will be improved by: sim'(S_i, Q) = sim(S_i, Q)+α*sim(S_j, Q), sim'(S_j, Q) = sim(S_j, Q)+β*sim(S_i, Q), where $0 \leq \alpha$, $\beta \leq 1$ and sim(S_i, Q) and sim(S_j, Q) are the initial relevance values of S_i and S_j evaluated by formula 6.

4   Experiments

We use the collections of the TREC Novelty Track 2003 and 2004 for the evaluation of our proposed retrieval method. To validate its applicability to keyword-based retrieval, we select the TREC topic titles to formulate the queries.

The first part of experiments is to verify the potential associations between query and information of certain types. Table 1 gives the statistical results. In this table, the second and third row shows the distribution of entity information in relevant sentences. Signification information about information entities has been discovered in most relevant sentences. It implies its large potentialities in sentence retrieval. In Table 1, P(N) represents the probability of queries concerned with N types of entities. It is discovered that most queries involve one or two different types of information entities. This further approves the assumption of the underlying associations between query and information types, which, when considered carefully, will improve the effectiveness of sentence retrieval.

Table 1. Statistical information about entities and entity types in relevant sentences

	TREC 2003				TREC 2004			
# of relevant sentences per query	311.14				166.86			
# of entities per relevant sentence	1.703				1.935			
P(N)	N=0	N=1	N=2	N>2	N=0	N=1	N=2	N>2
	18%	38%	30%	14%	32%	30%	26%	12%

Another part of experiments is to compare the classification performance of our proposed information association based approach (IA) with word pattern based approach (WP) applied in [4] and the machine learning classification approach (ML), which in this paper is support vector machine approach. We select the data obtained at the website (http://l2r.cs.uiuc.edu/~cogcomp/Data/QA/QC/) as the training and test data. Since these data are in the form of question, to make them applicable to our classification case, we have transformed them to be the form of keyword query. These formed queries contain the kernel terms in questions excepting the interrogative terms. Table 2 respectively shows the experimental results measured by macro precision, macro recall and macro F1-measure. As shown in Table 2, word pattern based approach has a higher classification precision, but a lower recall. It on one hand shows the relative accuracy of the defined word pattern and on the other hand its limited coverage. Compared with the machine learning, our proposed approach has a similar classification precision, but achieves a higher recall. In our experiments, we discovered that some valued relationships between term and information type have not been identified as expected. It is for that reason that a large amount of needed entities information cannot be properly recognized by the applied entity recognition technique. The study of entity recognition technique is out of the scope of this paper. However, it is believed that with the perfection of entity recognition technique, information association based classification approach can achieve better performance.

The purpose of the results shown in Tables 3 is to compare the performance of the proposed sentence retrieval model with type-based query expansion (TPQE) to other three retrieval approaches, including TFIDF model (TFIDF), BIR model (BIR), KL-divergence model (KLD) and the traditional term-based expansion sentence retrieval model (TQE). Statistical analysis of the retrieval results shows that in the context of sentence retrieval the application of traditional document retrieval algorithm is actually to detect the existence of query terms in sentence or not. As shown in Table 3 there is no significant difference in the performances of these traditional retrieval models. However, since most sentences have fewer words, TQE approach always add noise to the sentence retrieval process. Sentences containing expansion terms may not be as relevant to the original query as sentences containing the original query terms. Table 3 shows that performance of this approach. Comparatively speaking, our proposed method considers query expansion from another perspective. It is helpful to identify the most relevant information of query and therefore avoids the introduction of large noise. As shown in Table 3, our proposed approach does do better than all other approaches.

Table 2. Performances of different classification approaches

	Macro precision	Macro recall	Macro F1
WP	0.7230	0.4418	0.4990
ML	0.6980	0.5181	0.5778
IA	0.6895	0.5903	0.6081

Table 3. Performance comparison in finding relevant sentences for 50 queries in TREC 2003 & TREC 2004

Database	50 queries in TREC 2003	50 queries in TREC 2004
TFIDF	0.349	0.228
BIR	0.292	0.178
KLD	0.330	0.2833
TQE	0.304	0.212
TPQE	0.385	0.249

Table 4. Retrieval performance with or without consideration of proximity

Database	50 queries in TREC 2003		50 queries in TREC 2004	
Methods	TPQE	TPQE_PX	TPQE	TPQE_PX
P@5	0.6324	0.6989	0.3920	0.4380
P@10	0.6559	0.6860	0.3875	0.4200
P@20	0.5326	0.6639	0.3704	0.4020
P@30	0.5919	0.6400	0.3680	0.4160
P@50	0.5534	0.5896	0.3542	0.3976
P@100	0.5026	0.5438	0.3185	0.3413

With the hypothesis that relevant sentences always exist in close proximity to each other, we further propose the proximity-based optimization method for re-ranking the sentences. This optimization scheme focuses on the distributions of query terms and query types involved in the proximity sentences, with the hope to reveal some relevant sentences, which when work together, can provide the integrated information satisfying user's information need. Table 4 illustrates the experimental results, where P@n means the precision at the top n ranked documents. As shown in this table 4, retrieval with consideration of sentence proximity achieves clear improvement in retrieval effectiveness. It further validates the relevancies among adjacent sentences.

5 Conclusion

Compared with the traditional sentence retrieval model, the features of this proposed retrieval model include: it views the information of query type as a factor for identifying sentences' relevance; it re-ranks each sentence by considering the relationships between adjacent sentences. The proposed retrieval model has been compared with other traditional retrieval models. Experiment results indicate that it produces significant improvements in retrieval effectiveness.

References

1. Salton G., Allan J., Buckley C.: Automatic structuring and retrieval of large text files. Communication of the ACM, Vol. 37(2). (1994) 97-108
2. Daumé III H., Marcu D.: Bayesian query-focused summarization. In Proceedings of the 21st International Conference on Computational Linguistics and 44th Annual Meeting of the Association for Computational Linguistics. Sydney, Australia (2006) 305–312
3. Li X.: Syntactic Features in Question Answering. In Proceedings of 26th Annual International ACM SIGIR Conference on Research and Development in Information Retrieval. Toronto, Canada. (2003) 383-38
4. Li X., Croft W.: Novelty detection based on sentence level patterns. In Proceedings of 2005 ACM CIKM International Conference on Information and Knowledge Management. Bremen, Germany. (2005) 744-751
5. White R., Jose J., Ruthven I.: Using top-ranking sentences to facilitate effective information access. Journal of the American Society for Information Science and Technology. Vol. 56(10). (2005) 1113-1125
6. Larkey L., Allan J., Connell M., Bolivar A., Wade, C.: UMass at TREC 2002: Cross Language and Novelty Tracks. In Proceedings of 11th Text REtrieval Conference. Gaithersburg, Maryland. (2002) 721–732
7. Schiffman B.: Experiments in Novelty Detection at Columbia University. In Proceedings of 11th Text REtrieval Conference. Gaithersburg, Maryland. (2002) 188-196
8. Zhang M., Lin C., Liu Y., Zhao L., Ma S.: THUIR at TREC 2003: Novelty, Robust and Web. In Proceedings of 12th Text REtrieval Conference. Gaithersburg, Maryland. (2003) 556-567
9. Collins-Thompson K., Ogilvie P., Zhang Y., Callan J.: Information filtering, Novelty Detection, and Named-Page Finding. In Proceedings of 11th Text REtrieval Conference. Gaithersburg, Maryland. 107-118
10. Lund K., Burgess C. Producing High dimensional Semantic Spaces from Lexical Co-occurrence. Behavior Research Methods, Instruments, & Computers, Vol. 28, (1996) 203-208
11. Andrew B. A Maximum Entropy Approach to Named Entity Recognition. Ph.D. thesis, New York University, (1999)
12. Allan J., Wade C., Bolivar A.: Retrieval and Novelty Detection at the Sentence Level. In Proceedings of 26th Annual International ACM SIGIR Conference on Research and Development in Information Retrieval. Toronto, Canada. (2003) 314-321

An Extended R-Tree Indexing Method
Using Selective Prefetching in Main Memory

Hong-Koo Kang, Joung-Joon Kim, Dong-Oh Kim, and Ki-Joon Han

School of Computer Science & Engineering, Konkuk University,
1, Hwayang-Dong, Gwangjin-Gu, Seoul 143-701, Korea
{hkkang,jjkim9,dokim,kjhan}@db.konkuk.ac.kr

Abstract. Recently, researches have been performed on a general method that can improve the cache performance of the R-Tree in the main memory to reduce the size of an entry so that a node can store more entries. However, this method generally requires additional processes to reduce information of entries. In addition, the cache miss always occurs on moving between a parent node and a child node. To solve these problems, this paper proposes the SPR-Tree (Selective Prefetching R-Tree), which is an extended R-Tree indexing method using selective prefetching according to node size in the main memory. The SPR-Tree can produce wider nodes to optimize prefetching without additional modifications on the R-Tree. Moreover, the SPR-Tree can reduce the cache miss that can occur in the R-Tree. In our simulation, the search, insert, and delete performance of the SPR-Tree improved up to 40%, 10%, 30% respectively, compared with the R-Tree.

Keywords: SPR-Tree, Extended R-Tree, Cache Performance, Cache Miss, Main Memory.

1 Introduction

Recently, with the speed gap being broader between the processor and the main memory, how effectively to use the cache memory in the main memory-based index is making a critical impact on the performance of the entire system[1,5]. The R-Tree is similar to the B-Tree, but is used for spatial access methods for indexing multi-dimensional data[2]. Since the R-Tree is originally designed to reduce disk I/O effectively for the disk-based index, the node size is optimized for disk block.

However, the R-Tree is not suitable for the cache memory with a small block. Delay time caused by cache miss accounts for a significant part of the entire performance time[10]. Especially, when the R-Tree, as in a main memory DBMS, resides in the main memory, disk I/O does not affects the entire performance seriously. Consequently, studies on the index structure and algorithms with the improved cache performance are being carried out by numerous researchers in many ways[3,5-8].

Rao and Ross pointed out the importance of the cache performance in designing a main memory index and proposed the CSS-Tree(Cache-Sensitive Search Tree) which has a faster search performance than the Binary Search Tree or the T-Tree in the

Y. Shi et al. (Eds.): ICCS 2007, Part I, LNCS 4487, pp. 692–699, 2007.
© Springer-Verlag Berlin Heidelberg 2007

read-only OLAP environment[6]. They also proposed the CSB+-Tree which is an extension of the CSS-Tree and can improve the cache performance of the B+-Tree[7].

Sitzmann and Stuckey proposed the pR-Tree(partial R-Tree), which adjusts the size of the R-Tree node to that of cache block and deletes unnecessary information within MBR(Minimum Bounding Rectangle) to store more information in a node[8]. Kim and Cha proposed the CR-Tree(Cache-conscious R-Tree) which compresses MBR of an entry to include more entries in a node[3]. The typical approach for cache performance improvement is to minimize cache misses by reducing the size of the entry to increase the fanout and storing more entries in a node. But, in this approach, the update performance is generally lowered due to additional operations to recover the compressed entry information and cache miss occurring when moving between nodes still results in the lowered performance of the entire system.

In order to solve the above problems, this paper proposes the SPR-Tree(Selective Prefetching R-Tree), an extended R-Tree indexing method, which applies the selective prefetching to the R-Tree in the main memory. The SPR-Tree loads the child node onto the cache memory in advance to extend the size of the node to be optimized for prefetching without transforming the R-Tree radically and reduce cache misses occurring when moving between nodes. The performance improvement of the SPR-Tree using selective prefetching is in proportion to the size and the number of the nodes to access. Therefore, it is more effective in the range query than in the point query.

The rest of this paper is organized as follows. Chapter 2 introduces selective prefetching and then analyzes the existing cache conscious index methods. Chapter 3 explains the structure of the SPR-Tree and algorithms for the SPR-Tree. In Chapter 4, the performance of the SPR-Tree is analyzed and the results of the SPR-Tree evaluation are presented. Finally, the conclusion is provided in Chapter 5.

2 Related Works

This chapter will introduce selective prefetching and analyze the various existing cache conscious index methods.

2.1 Selective Prefetching

The cache memory is used to provide data to the processor in a fast way. Located between the main memory and the processor, the cache memory generally consists of 2 layers; L1 cache and L2 cache. L1 cache is located between the register and L2 cache, while L2 cache is located between L1 cache and the main memory[4]. When the processor is accessing data, if the data is present in the cache memory, it is called "cache hit" and if the data is not present, it is called "cache miss".

The cache block is the basic transfer unit between the cache memory and the main memory. The current systems tend to have bigger size of the cache block and larger-capacity of the cache memory. Typical cache block size ranges from 32 bytes to 128 bytes. Generally, the data cache follows the basic principle of the data locality. A tree structure has the low data locality as data to refer to is accessed through the pointer. Therefore, in order to improve the cache performance in the tree structure, the amount of data to access should be reduced or selective prefetching should be executed.

The selective prefetching is a technique to selectively load data into the cache memory in advance to accelerate the program execution. Especially, the selective prefetching can reduce cache misses by loading data which does not exist in the cache memory before the processor requests it. In order to reduce cache misses in the R-Tree, the selective prefetching should be used to reduce memory delay occurring when accessing nodes overall. The selective prefetching is controlled in two ways; the hardware-based prefetching where the prefetching is automatically carried out by the processor and the software-based prefetching where a prefetching command is inserted into the program source code[9].

2.2   Cache Conscious Index Methods

The CSB+-Tree is a variant of the B+-Tree, removing all child node pointers except the first child node pointer to store child nodes consecutively in order to reduce cache misses in the B+-Tree[7]. But, this method of eliminating pointers is not so effective in the R-Tree where pointers account for a relatively small part. And since child nodes are consecutively stored in the CSB+-Tree, every update operation requires reorganization of consecutively arranged child nodes.

The pR-Tree is a variant of the R-Tree, removing child MBR's coordinate values overlapped with those of parent MBR to reduce cache misses in the R-Tree[8]. This method also eliminates the pointers, like in the CSB+-Tree, and shows better performance when the number of entries is small. However, this method has worse performance as the number of entries increases, since the number of child MBR's coordinate values overlapped with those of parent MBR is decreased. In addition, due to the elimination of overlapped child MBR's coordinate values, additional operations are needed for reorganization of the eliminated coordinate values, which lowers the update performance.

The CR-Tree is a kind of the R-Tree that compresses MBR, which accounts for most of indexes, and uses the compressed MBR as a key[3]. In the CR-Tree, MBR is compressed according to the following procedure; MBR of the child node is represented in relative coordinates to MBR of the parent node and it is quantized so that it can be represented in definite bits. However while compressing MBR in the CR-Tree, a small error can occur and this may produce a wrong result(i.e., false hit). Moreover, additional operations for reorganization of the compressed MBR in the update operation can lower the update performance.

3   SPR-Tree

This chapter will describe the SPR-Tree, a main memory-based R-Tree using selective prefetching. First, the structure and characteristics of the SPR-Tree will be given and then the algorithms used in the SPR-Tree also will be suggested.

3.1 Structure

The SPR-Tree, similar to the R-Tree, has the root node, the intermediate node, and the leaf node. All operations on the SPR-Tree start from the root node and the references to real data objects exist only in the leaf node. Figure 1 illustrates the node structure of the SPR-Tree. The SPR-Tree uses a *rectangle*, which is a rectilinear shape that can completely contain other rectangles or data objects.

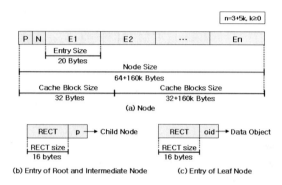

Fig. 1. Node Structure of the SPR-Tree

In Figure 1(a), *P* and *N* represent the node level and the number of entries in a node, respectively. Each of *E1*, *E2*, ... , *En* ($n=3+5k$, $k \geq 0$) represents an entry which has two types, that is, an entry for the root node or the intermediate node and an entry for the leaf node. Figure 1(b) shows the entry for the root node or the intermediate node, where *RECT* is a rectangle which completely contains all rectangles in the child node's entries and *p* is an address of a child node. Figure 1(c) represents the entry for the leaf node, where *RECT* is a rectangle which completely contains a data object and oid refers to the data object.

Since the SPR-Tree nodes adjust the number of entries suited to the cache block; the SPR-Tree decides the node size in proportion to the cache block size. Generally, the cache block size can be 32 or 64 bytes. If the cache block size is 32 bytes, the node size becomes $64+160k$ ($k \geq 0$) bytes and if it is 64 bytes, the node size becomes $64+320k$ ($k \geq 0$).

Figure 2 shows an example of the SPR-Tree. As the Figure 2 shows, rectangles can enclose a single data object or one or more rectangles. For example, rectangle R8, which is at the leaf level of the SPR-Tree, contains data object O. Rectangle R3, which is at the intermediate level of the SPR-Tree, contains rectangles R8, R9, and R13. Rectangle R1, which is at the root level, contains rectangles R3 and R4. In Figure 2, a prefetching node group enclosed by a dotted line is determined according to the node size.

3.2 Algorithms

This section will describe the search, node split insert, and delete algorithms of the SPR-Tree in detail.

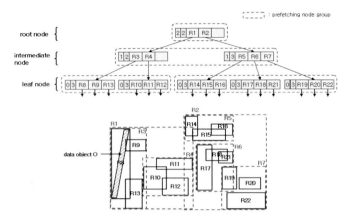

Fig. 2. Example of the SPR-Tree

3.2.1 Insert Algorithm

The insert operation repeats, from the root node down to the leaf node, a process of using lower node's rectangle information contained in entries of each node to determine whether the size expansion of the node can be minimized or not when an object is inserted into the leaf node. At this time, if the leaf node becomes full, then a node split occurs. In the insert algorithm of the SPR-Tree, prefetching is carried out while looking for the leaf node to insert an entry. Figure 3 shows the insert algorithm of the SPR-Tree.

3.2.2 Delete Algorithm

The delete operation repeats, from the root node down to the leaf node, a process of using lower node's rectangle information contained in entries of each node to determine whether a query region is contained or overlapped in the lower nodes. At this time, if an entry is deleted and the number of remaining entries is below the minimum number of entries in the leaf node, then the leaf node is deleted and its remaining entries are reinserted into the SPR-Tree. The delete algorithm of the SPR-Tree uses a prefetching command based on the node size. The child node to be accessed is prefetched after the current node according to the node size. Figure 4 shows the delete algorithm of the SPR-Tree.

3.2.3 Search Algorithm

The search operation descends the SPR-Tree from the root node to the leaf node. And, it repeats a process of using lower node's rectangle information contained in entries of each node to determine whether the lower node contains or overlaps a query region or not. If the lower node is contained or overlapped with the query region, the search operation follows the lower node as the root node until it reaches the leaf node. The search algorithm of the SPR-Tree uses a prefetch command to prefetch a child node to be accessed after the current node. If the node has few entries, the SPR-Tree makes a prefetching node group using some nodes at the same level and prefetches it. While the node has many entries, it prefetches only a child node to be accessed into the cache memory. Figure 5 shows the search algorithm of the SPR-Tree.

```
Insert Algorithm ─────────────────
  Input)    R: SPR-Tree Node
            IR: Rectangle to be Inserted
  Output) The New SPR-Tree

  begin
    if (R is not a leaf node) then
      for each entry (p, RECT) of R do
        PREFETCH(p, num_entry);
        if (RECT's expansion is minimum) then
          CHILD = node pointed to by p;
        end if
      end for
      INSERT(IR, CHILD);
      if (CHILD overflows) then
        SPLIT(CHILD);
      end if
    else
      INSERT(IR, R);
      if (R overflows) then
        SPLIT(R);
      end if
    end if
  end
```

```
Delete Algorithm ─────────────────
  Input)    R: SPR-Tree Node
            RR: Rectangle to be removed
  Output) The New SPR-Tree

  begin
    if (R is not a leaf node) then
      for each entry (p, RECT) of R do
        PREFETCH(p, num_entry);
        if (RECT overlaps RR) then
          CHILD = node pointed to by p;
          DELETE(CHILD, RR);
          if (CHILD underflows) then
            REINSERT(CHILD's entries);
          end if
        end if
      end for
    else
      DELETE(RR, R);
      if (CHILD underflows) then
        REINSERT(CHILD's entries);
      end if
    end if
  end
```

Fig. 3. Insert Algorithm   **Fig. 4.** Delete Algorithm

3.2.4 Node Split Algorithm

When a leaf node is full during the execution of an insert operation in the SPR-Tree, the node split operation must be executed. First, the entries in the node are divided into two nodes with minimum rectangle expansion. If the number of entries in the parent node exceeds the maximum number of entries in the parent node due to the node split, the parent node also must be split. The node split algorithm prefetches the current node before split and creates two new nodes to distribute the entries of the current node. Figure 6 shows the node split algorithm of the SPR-Tree.

```
Search Algorithm ─────────────────
  Input)    R: SPR-Tree Node, W: Qeury rectangle
  Output) Results: All Objects overlapping W

  begin
    Results = Φ
    if (R is not a leaf node) then
      if (entry's child pointer is not NULL) then
        for each entry (p, RECT) of R do
          PREFETCH(p, num_entry);
          if (RECT overlaps W) then
            CHILD = node pointed to by p;
            SEARCH(CHILD, W ∩ RECT);
          end if
        end for
      end if
    else
      if (entry's child pointer is not NULL) then
        for each entry (p, RECT) of R do
          if (RECT overlaps W) then
            OBJECT = object pointed to by p;
            Results += OBJECT;
          end if
        end for
      end if
    end if
    return Results;
  end
```

```
Node Split Algorithm ──────────────
  Input)    R: SPR-Tree Node
  Output) The New SPR-Tree

  begin
    PREFETCH(p, num_entry);
    (S1, S2) = PARTITION(R);
    RECT = MBROF(R);  P = POINTEROF(R);
    RECT1 = MBROF(S1);  RECT2 = MBROF(S2);
    R1 = CREATENODE(RECT1);
    R2 = CREATENODE(RECT2);
    for each nodes Rk(Pk, RECTk) of R do
      if (RECT1 contains RECTk) then ADD(Rk, R1);
      else if (RECT2 contains RECTk) then ADD(Rk, R2);
      end if
    end for
    if (R is root node) then
      NR = CREATENODE(RECT1 ∪ RECT2);
      ADD(R1, NR);  ADD(R2, NR);  ROOT = NR;
    else
      PR = parent node of R;
      ADD(R1, PR);  ADD(R2, PR);
      if (PR overflows) then
        SPLIT(PR);
      end if
    end if
  end
```

Fig. 5. Search Algorithm   **Fig. 6.** Node Split Algorithm

4 Performance Evaluation

The system used in the performance evaluation was equipped with Intel Pentium III 1GHz, 1GB main memory, and L1 and L2 caches whose block size is 32 bytes. As a test data, we created 10,000 objects, a square with side length of 0.0001 on the average, uniformly distributed in a square with side length of 1 as the whole area.

Figure 7 shows the performance results of the search operation. The query region was supposed to occupy 30%~70% of the whole area. In Figure 7, the SPR-Tree has better search performance than the R-Tree and improvement through prefetching appears more consistent, as memory delay is reduced while accessing nodes. The search performance of the SPR-Tree was improved up to 35% over the R-Tree.

Figure 8 shows the performance results of the search operation in a skewed data set. As shown in Figure 8, the larger the node size is, the better search performance it has. This is because there is more reduced memory delay time due to prefetching, as the spatial objects are skewed, which increases overlapping between nodes and the number of nodes to access. The search performance of the SPR-Tree was improved up to 40% over the R-Tree for skewed data set.

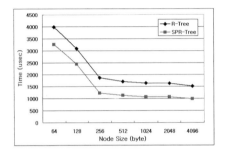

Fig. 7. Performance of Search Operations

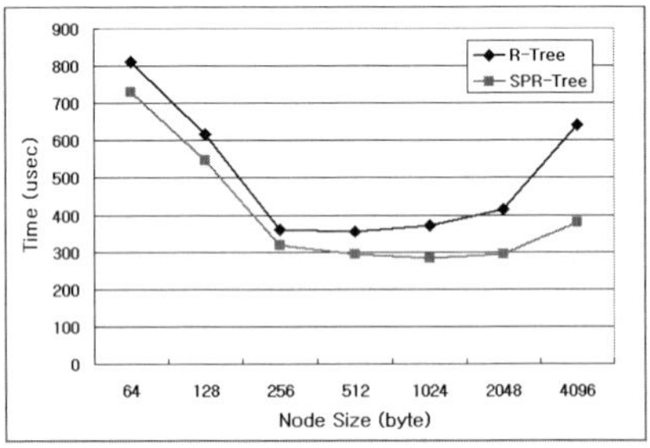

Fig. 8. Performance of Search Operations in Skewed Data Set

Figure 9 shows the performance results of the insert operation. The spatial objects were inserted and the side length of the objects was 0.0001 on the average. As shown in Figure 9, when the node size is larger, the insert time increases, but we can see that the performance improvement rate increases due to prefetching. This is because when prefetching is used, larger node size brings higher performance. The insert performance of the SPR-Tree showed up to 10% improvement over the R-Tree.

Figure 10 shows the performance results of the delete operation. We deleted objects involved in the region whose side length was 0.001 on the average. In Figure 10, the larger node size generally leads to the better performance of the delete operation and the performance improvement through prefetching is consistent, as memory delay

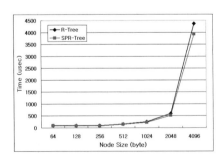

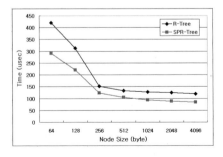

Fig. 9. Performance of Insert Operations

Fig. 10. Performance of Delete Operations

time reduced by prefetching is consistent while accessing nodes. In the evaluation, the delete performance of the SPR-Tree was improved up to 30% over the R-Tree.

5 Conclusion

Recently an approach that can improve the main memory-based R-Tree index structure by reducing the node size was proposed. However, in this approach, the update performance is lowered due to additional operations to recover the compressed entry information and, still, cache misses occurring when moving between nodes contributes to the lowered performance of the entire system.

In order to solve the above problems, this paper proposes the SPR-Tree which applies the selective prefetching to the R-Tree to reduce cache misses as well as eliminate additional cost in the update operation. The SPR-Tree optimizes the node size for prefetching and minimizes cache misses by prefetching child nodes when moving between nodes. In the performance evaluation, the SPR-Tree was improved up to 40% in the search operation, up to 10% in the insert operation, and up to 30% in the delete operation over the R-Tree.

Acknowledgements

This research was supported by the MIC(Ministry of Information and Communication), Korea, under the ITRC(Information Technology Research Center) support program supervised by the IITA(Institute of Information Technology Assessment).

References

1. Chen, S., Gibbons, P. B., Mowry, T. C., Valentin, G.: Fractal Prefetching B+-Trees : Optimizing Both Cache and Disk Performances. Proceedings of ACM SIGMOD Conference (2002) 157-168.
2. Guttman, A.: R-Trees: a Dynamic Index Structure for Spatial Searching. Proceedings of ACM SIGMOD Conference (1984) 47-54.
3. Kim, K. H., Cha, S. K., Kwon, K. J.: Optimizing Multidimensional Index Tree for Main Memory Access. Proceedings of ACM SIGMOD Conference (2001) 139-150.
4. Mowry, T. C., Lam, M. S., Gupta, A.: Design and Evaluation of a Compiler Algorithm for Prefetching. Proceedings of International Conference on Architectural Support for Programming Languages and Operating Systems (1992) 62-73.
5. Park, M. S., Lee, S. H.: A Cache Optimized Multidimensional Index in Disk-Based Environments. IEICE Transactions on Information and Systems, Vol.E88-D (2005) 1940-1947.
6. Rao, J., Ross, K. A.: Cache Conscious Indexing for Decision-Support in Main Memory. Proceedings of International Conference on VLDB (1999) 78-89.
7. Rao, J., Ross, K. A.: Making B+-Trees Cache Conscious in Main Memory. Proceedings of ACM SIGMOD Conference (2000) 475-486.
8. Sitzmann, I., Stuckey, P. J.: Compacting Discriminator Information for Spatial Trees. Proceedings of Australasian Database Conference (2002) 167-176.
9. VanderWiel, S. P., Lilja, D. J.: Data Prefetch Mechanisms. ACM Computing Surveys, Vol.32 (2000) 174-199.
10. Zhou, J., Ross, K. A.: Buffering Accesses of Memory-Resident Index Structures. Proceedings of International Conference on VLDB (2003) 405-416.

Single Data Copying for MPI Communication Optimization on Shared Memory System

Qiankun Miao[1], Guangzhong Sun[1,*], Jiulong Shan[2], and Guoliang Chen[1]

[1] Anhui Province-MOST Key Co-Lab of High Performance Computing
and Its Applications, Department of Computer Science,
University of Science and Technology of China, Hefei, 230027, P.R. China
[2] Microprocessor Technology Lab, Intel China Research Center, Beijing, China
miao@mail.ustc.edu.cn, gzsun@ustc.edu.cn, jiulong.shan@intel.com,
glchen@ustc.edu.cn

Abstract. Shared memory system is an important platform for high
performance computing. In traditional parallel programming, message
passing interface (MPI) is widely used. But current implementation of
MPI doesn't take full advantage of shared memory for communication.
A double data copying method is used to copy data to and from system
buffer for message passing. In this paper, we propose a novel method to
design and implement the communication protocol for MPI on shared
memory system. The double data copying method is replaced by a single data copying method, thus, message is transferred without the system buffer. We compare the new communication protocol with that in
MPICH an implementation of MPI. Our performance measurements indicate that the new communication protocol outperforms MPICH with
lower latency. For Point-to-Point communication, the new protocol performs up to about 15 times faster than MPICH, and it performs up to
about 300 times faster than MPICH for collective communication.

1 Introduction

Message Passing Interface (MPI) [1] is a standard interface for high performance
parallel computing. MPI can be used on both distributed memory multiprocessor
and shared memory multiprocessor. It has been successfully used to develop
many parallel applications on distributed memory system.

Shared memory system is an important architecture for high performance
computing. With the advent of multi-core computing [4], shared memory system will draw more and more attention. There are several programming models
for shared memory system, such as OpenMP [2] and Pthread [3]. However, we
still need to program on shared memory system with MPI. First, it is required
to reuse the existing MPI code on distributed memory system for quickly developing applications on shared memory system. Second, applications developed
using MPI are compatible across a wide range of architectures and need only a
few modifications for performance tune-up. Third, the future computation will

---

* Corresponding author.

Y. Shi et al. (Eds.): ICCS 2007, Part I, LNCS 4487, pp. 700–707, 2007.

shift to multi-core computation. Consequently, exploiting the parallelism among these cores becomes especially important. We must figure out how to program on them to get high performance. MPI can easily do such works due to its great success on parallel computing today.

It is difficult to develop an efficient MPI application on shared memory system, since the MPI programming model does few considerations of the underlying architecture. In the default communication protocol of MPI, receiver copies data from system buffer after sender copied the data to the system buffer. Usually a locking mechanism is used for synchronization on shared memory system, which has a high overhead. Therefore, MPI programs suffer severe performance degradation on shared memory system. In this paper, we propose techniques to eliminate these performance bottlenecks. We make use of the shared memory for direct communication bypass buffer copying instead of double data copying. Meanwhile, A simple busy-waiting polling method is used to reduce expense of locking for synchronization. To evaluate these techniques, some experiments are carried out on basic MPI primitives. We make comparisons of the performance between the new implementation and the original implementation in MPICH [5]. Results indicate that the new implementation is able to achieve much higher performance than MPICH do.

There are other works studying on the optimized implementation of the communication in MPI programs on shared memory system. A lock-free scheme is studied on NEC SX-4 machine [6]. They use a layered communication protocol for a portable consideration. A shared memory communication device in MPICH is conducted for Sun's UltraEnterprise 6000 servers [7]. They use a static memory management similar as we describe in Section 2. TMPI [8] use a thread-level approach for communication optimization. TMPI can achieve significant performance advantages in a multiprogrammed environment. We consider the condition only a process per processor. This condition is usually happened in real computation. Our implementation is not restricted to a certain machine. We could achieve lower latency than the native MPI implementation.

The rest of the paper is organized as follows. Next section introduces the default communication implementation of MPI and discusses its drawbacks on shared memory system. Section 3 describes the design and implementation of the new approach used in this paper, which makes efficient communication on shared memory system. Section 4 describes our evaluation methodology and presents the evaluation results. Section 5 concludes the paper and presents the future work.

2   Motivation

In traditional implementation of MPI, a general shared memory device is needed, which can be used on many differently configured systems. There is a limitation that only a small memory space shared by all processes can be allocated on some systems. As a result, the communication between two processes is through a shared memory system buffer [9]. The system buffer can be accessed by all the processes in a parallel computer. We indicate this data transmission procedure in Fig. 1.

Shared System Buffer

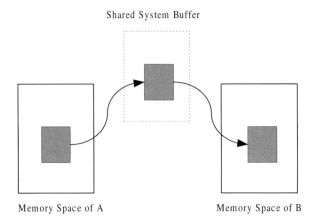

Memory Space of A   Memory Space of B

Fig. 1. Communication in original MPI, the message packet is transmitted by copying them into and out of the shared system buffer

A typical implementation of communication between two processes could be as follows. Suppose process A intends to send a message to process B. We use A to denote process A and B to denote process B following. There is a free shared memory segment pre-allocated and a synchronization flag can be accessed by A and B. The synchronization flag is used to indicate whether the shared memory segment is ready for A to write to or for B to read from. At the beginning, A checks the synchronization flag. If the shared memory segment is ready for writing, A copies data to be transmitted from its user buffer into the shared memory segment. Then A sets the synchronization flag after finishing the data copying. This indicates that the data in the shared memory segment is ready to be copied to B's user buffer. A can either wait for an acknowledgement after copying or it can continue execution. If the shared memory segment is not ready for A, it chooses a back-off strategy, and tries again later. Meanwhile, B can check for a forthcoming data packet by probing the synchronization flag. When it finds the flag is set, B copies the data in shared memory segment to its own user buffer. After the data transmitted completely, B clears the synchronization flag. At this time new communication can start. MPICH implements the above mechanism as an abstract device interface *ch_shmem* [7].

The communication strategy above is inefficient for a few reasons.

1. A double data copying protocol is required for each data exchange. But this is unnecessary and will increase the burden of memory/cache system. It will result in high latency of the communication.
2. A shared system buffer is required where the data is copied to and from. This is an extreme waste of capacity of memory. On shared memory system, local memory and cache capacity as well as aggregate bandwidth will limit the scalability of MPI implementation.

3. The cost of back-off strategy for synchronization is extremely large. The synchronization cost can adversely affect communication performance for short message.

4. Furthermore, for those collective communications (e.g. MPI_Bcast), the extra copy and complex synchronization will aggravate these problems.

In a word, the critical problem for the communication on shared memory system is how to reduce the memory/cache capacity and the waiting time for synchronization. In the next section, we will propose techniques to solve these problems.

3   Design and Implementation

To solve the problems existing in MPI for shared memory system, we design new communication protocols. We use primitives (IPC/shm) to create process level shared memory segment for message to be transmitted. Communication among processes uses a single data copying protocol. Considering usually only one process per processor we choose a simple busy-waiting polling strategy for synchronization.

3.1   Optimized Communication Protocol

In Section 2, we have described the default implementation of communication in MPI. In order to transmit data among processes, a double data copying protocol is required. One is used to copy the data into shared memory, and the other is used to copy the data from the shared memory. However, the double data copying can be reduced to only one data copying when the sender process need retain an independent copy of the data. Even no data copying is required when the sender process needn't retain a copy of the data or all the processes share the same data structure and do operation on disjoint parts of the data structure. We can allocate a shared memory segment for the data to be transmitted. Thus, every process can access this shared segment. Data in the shared memory segment can be copied to the receiver process's user buffer either in the shared memory or in its private space. If all the processes deal with disjoint parts of the same data structure, we need no data transmission because after one process updated the data all other processes can see the update immediately and they can use IPC/shm to directly read the data. This technique is able to obtain lower latency and less memory consumption, because it needs only one or no data copying and no extra system buffer. Fig. 2 illustrates this technique.

According to the above discussion, we can model the required time for sending a message packet with size of n bytes by the following formulas. We use $T_{original}$ to indicate the communication time of the original version and $T_{optimized}$ to indicate that of our optimized implementation.

Original MPI device using two data copying:

$$T_{original} = 2T_{datacpy} + T_{syn} \tag{1}$$

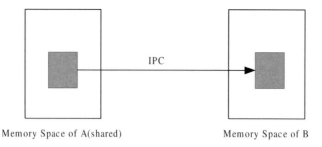

Memory Space of A(shared)   Memory Space of B

Fig. 2. Optimized communication with only a single buffer copying by employing IPC

Optimized implementation using single data copying:

$$T_{optimiezed} = T_{datacpy} + T_{alloc} + T_{free} + T_{bwsyn} \tag{2}$$

Optimized implementation using no data copying:

$$T_{optimiezed} = T_{alloc} + T_{free} + T_{bwsyn} \tag{3}$$

Where $T_{datacpy}$ represents the time for one data copying, T_{syn} represents the time for communication synchronization in original version, T_{alloc} and T_{free} represent the time for shared memory allocating and deleting respectively, T_{bwsyn} represents the time for communication synchronization using busy-waiting strategy in our implementation.

In general case, the cost of the shared memory segment allocation and free is very little compared with the cost of one data copying. So, $T_{original}$ is larger than $T_{optimized}$. This indicates that we could achieve higher performance using the optimized communication protocol.

3.2   Busy-Waiting Polling Strategy

A synchronization mechanism is used to prevent the processes from interfering with each other during the communication procedure. It defines the time when sender process makes data ready and the time when receiver process completes the data copying. Usually the mechanism is provided by locks and semaphores on shared memory system. Unfortunately, typical implementations of lock tend to cause a system call, which is often very time-consuming. So, the cost of synchronization by using a lock is too expensive to provide high performance. When the lock is not free, a back-off strategy is used. This may delay the starting time of communication. Though the lock is already free, it may still need to wait for the process to be active again.

With the assumption that there is usually only one application process per processor and processor resources are not needed for other tasks, we use a simple busy-waiting polling approach to replace exponential back-off strategy. An application process repeatedly tests the synchronization flag as frequently as possible to determine when it may access the shared memory. Once the process has found the synchronization flag switched, it would immediately detect the change. Consequently, the data transmission can start without delay. This polling strategy would reduce the time for synchronization on shared memory system when there is only one process per processor.

4   Performance Evaluation

We conduct experiments to study the performance of our implementation. Latency time is collected for multiple repetitions of each operation over various message sizes between 0 byte and 1 megabytes. We use latency time here to denote the total transmission time of a message among processes. For 0 byte message, latency time is only the overhead of communication start-up and end. All the results given are averaged over multiple runs.

4.1   Platform Configuration

The target system is a 16-way shared memory multiprocessor system running Suse Linux 9.0 platform. It has 16 x86 processors running at 3.0 GHz, 4 levels of cache with each 32MB L4 cache shared amongst 4 CPUs. As for the interconnection, the system uses two 4x4 crossbars. We use MPICH-1.2.5.2 (affiliated with device ch_shmem and gcc 3.3.3) library form Argonne National Laboratory to generate the executables. Each process is bound to a processor to prevent operating system from migrating processes among processors.

4.2   Performance Evaluation Results

Point-to-Point communication transmits data between a pair of processes. We choose the standard MPI_Send/Recv for our test. Collective communication is often used in real applications. Usually it involves a large number of processes. This means that there is a large amount of data to be transmitted among the involved processes. So, its performance heavily depends on the implementation. We investigate MPI_Bcast, MPI_Gather, MPI_Scatter, MPI_Alltoall, MPI_Reduce to evaluate the collective communication with new communication approach. We only present the results of MPI_Bcast due to the limited paper size. In MPI_Bcast a root process broadcasts a message to all the other processors.

We present the experimental results in Fig. 3. We compare two implementations of Send/Recv in the left of Fig. 3, one is the default version in MPICH which is labeled Original Send/Recv, the other is the optimized version which is labeled New Send/Recv. In the right of Fig. 3, we compare two implementations of Bcast which are labeled Original Bcast and New Bcast respectively.

From Fig. 3, we can see that our new implementation outperforms the default implementation in MPICH. Our implementation has lower latency on various size message for both point-to-point communication and collective communication. Our implementation is almost six times faster than MPICH for short

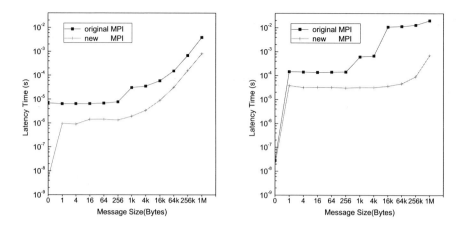

Fig. 3. Left: Latency time of original MPICH Send/Receive (upper line) and optimized implementation (lower line). Right: Latency time of original MPICH Bcast (upper line) and optimized implementation (lower line).

message and up to fifteen times faster for long message. Our implementation gains great success for collective communication. The optimized implementation is about five times faster than MPICH for short message. It is two orders of magnitude faster than MPICH when the message size is larger than 1kBytes. For message size from 1 bytes to 4kBytes, the transmission time of broadcast is almost no increase in the new implementation. That's because all other processes can directly copy data from the sender process's user buffer simultaneously, and the transmission time is determined by the cost of *memcpy()* which is almost same for various short data sizes. As the message size is large, memory bandwidth contention will limit the communication performance, the transmission time increases with message size. When the message is so large that the system buffer pre-allocated for communication of the original MPI implementation can't hold the total message at one time, it is required to split the message to several small pieces and transmits one piece at a time. We can see a step-up increase of the transmission time in MPICH from Fig. 3. However, all receivers can still directly copy from the sender's buffer in the optimized implementation, since the optimized implementation isn't limited on message size.

5   Conclusion

In this paper, we present an optimized implementation of the communication significantly improves the performance of MPI communication on shared memory system. The optimized methods include a single data copying protocol and a busy-waiting polling strategy for synchronization. Some experiments are conducted to evaluate the performance of a few basic MPI primitives with the optimized communication. The experimental results indicate that the primitives with optimized communication achieved lower latency than the native version in MPICH.

For future work, we intend to use the methods to develop more real-world applications on shared memory system and make a source to source translator to translate the MPI programs written for distributed system into an efficient shared memory version. We will also compare the overall performance of an application written using optimized message passing with that of an application written using shared memory programming model such as OpenMP.

Acknowledgements

This work is supported by the National Natural Science Foundation of China No.60533020. This work is partially finished at Intel China Research Center. We would like to thank the anonymous referees for their useful suggestions to improve the presentation of this paper.

References

1. The MPI Forum. *The MPI Message-Passing Interface Standard.* http://www.mpi-forum.org/. (1995).
2. OpenMP Standards Board. *OpenMP: A Proposed Industry Standard API for Shared Memory Programming.* http://www.openmp.org/openmp/mp-documents/paper/paper.html. (1997).
3. Pthread interface. *ANSI/IEEE Standard 1003.1.* (1996)
4. M. Greeger. *Multicore CPUs for the Masses.* ACM Queue, **3**(7) (2005):63–64
5. W. Gropp, E. Lusk: *Skjellum A, Doss N. MPICH: A high-performance, portable implementation for the MPI message-passing interface.* Parallel Computing 1996; **22**, (1996) pp. 789–928.
6. W. Gropp, E. Lusk: *A high-performance MPI implementation on a shared-memory vector supercomputer.* Parallel Computing , **22**, (1997) pp. 1513–1526.
7. B. V. Protopopov, A. Skjellum: *Shared-memory communication approaches for an MPI message-passing library.* Concurrency: Practice and Experience, **12**(9), (2000) pp. 799–820.
8. H. Tang, K. Shen, and T. Yang: *Compile/Run-time Support for Threaded MPI Execution on Multiprogrammed Shared Memory Machines.* In Proceedings of ACM SIGPLAN Symposium on Principles and Practice of Parallel Programming, (1999) pp. 107–118.
9. D. Buntinas, M. Guillaume, W. Gropp: *Data Transfers between Processes in an SMP System: Performance Study and Application to MPI.* International Conference on Parallel Processing (ICPP), Columbus, Ohio, USA, (2006), pp. 487–496.

Adaptive Sparse Grid Classification Using Grid Environments

Dirk Pflüger, Ioan Lucian Muntean, and Hans-Joachim Bungartz

Technische Universität München, Department of Informatics,
Boltzmannstr. 3, 85748 Garching, Germany
{pflueged, muntean, bungartz}@in.tum.de

Abstract. Common techniques tackling the task of classification in data mining employ ansatz functions associated to training data points to fit the data as well as possible. Instead, the feature space can be discretized and ansatz functions centered on grid points can be used. This allows for classification algorithms scaling only linearly in the number of training data points, enabling to learn from data sets with millions of data points. As the curse of dimensionality prohibits the use of standard grids, sparse grids have to be used.

Adaptive sparse grids allow to get a trade-off between both worlds by refining in rough regions of the target function rather than in smooth ones. We present new results for some typical classification tasks and show first observations of dimension adaptivity. As the study of the critical parameters during development involves many computations for different parameter values, we used a grid environment which we present.

Keywords: data mining, classification, adaptive sparse grids, grid environment.

1 Introduction

Today, an ever increasing amount of data is available in various fields such as medicine, e-commerce, or geology. Classification is a common task making use of previously known data and making predictions for new, yet unknown data. Efficient algorithms that can process vast datasets are sought for. The basics of sparse grid classification have already been described in [1,2], for example. Therefore, in this section, we summarize the main ideas only very briefly and refer to the references cited above for further information.

We focus on binary classification. Given is a preclassified set of M data points for training, $S = \{(\boldsymbol{x}_i, y_i) \in [0,1]^d \times \{-1,1\}\}_{i=1}^{M}$, normalized to the d-dimensional unit hypercube. The aim is to compute a classifier $f : [0,1]^d \rightarrow \{-1,1\}$ to obtain a prediction of the class -1 or $+1$ for previously unseen data points. To compute f, we follow the Regularization Network approach and minimize the functional

$$H[f] = \frac{1}{M} \sum_{i=1}^{M} (y_i - f(\boldsymbol{x}_i))^2 + \lambda \|\nabla f\|_{L_2}^2,$$

Y. Shi et al. (Eds.): ICCS 2007, Part I, LNCS 4487, pp. 708–715, 2007.

with the cost function $(y_i - f(\boldsymbol{x}_i))^2$ ensuring good approximation of the training data by f and the regularization operator $\|\nabla f\|_{L_2}^2$ guaranteeing that f is somehow smooth, which is necessary as the classifier should generalize from S. The regularization parameter λ stirs the trade-off between accuracy and smoothness.

Rather than common algorithms which employ mostly global ansatz functions associated to data points, scaling typically quadratically or worse in M, we follow a somehow data-independent approach and discretize the feature space to obtain a classification algorithm that scales linearly in M: We restrict the problem to a finite dimensional space V_N spanned by N basis functions ϕ_j to obtain our classifier $f_N(\boldsymbol{x}) = \sum_{j=1}^{N} \alpha_j \phi_j(\boldsymbol{x})$, in our case the space of d-linear functions. Minimization of $H[f]$ leads to a linear system with N unknowns,

$$\left(\lambda M C + B \cdot B^T\right) \boldsymbol{\alpha} = B\boldsymbol{y}, \tag{1}$$

with $C_{ij} = (\nabla\phi_i(\boldsymbol{x}), \nabla\phi_j(\boldsymbol{x}))_{L_2}$ and $B_{ij} = \phi_i(\boldsymbol{x}_j)$.

To counter the curse of dimensionality and to avoid N^d unknowns in d dimensions, we use sparse grids, described for example in [3]. Regular sparse grids $V_n^{(1)}$ up to level n in each direction base on a hierarchical formulation of basis functions and an a priori selection of grid points – needing only $\mathcal{O}(N \log(N)^{d-1})$ grid points with just slightly deteriorated accuracy. Sparse grids have been used for classification via the combination technique [1], where the sparse grid solution is approximated by a combination of solutions for multiple, but smaller and regular grids. Sparse grids have the nice property that they are inherently adaptive, which is what we will make use of in the following sections.

Using sparse grids, the system of linear equations can be solved iteratively, each iteration scaling only linearly in the number of training data points and grid points, respectively. The underlying so-called UpDown algorithm which was shown in [2] bases on traversals of the tree of basis functions.

2   Grid-Based Development

For classification using regular sparse grids there are two important parameters determining the accuracy of the classifier to be learned. First, we have n, the maximum level of the grid. For low values of n there are usually not enough degrees of freedom for the classifier to be able to learn the underlying structure, whereas large n lead to overfitting, as each noisy point in the training data set can be learnt. Second and closely related, there is the regularization parameter λ, steering the trade-off between smoothness and approximation accuracy. Given enough basis functions, λ determines the degree of generalization.

To find good values for these parameters, heuristics or experience from other problems can be used. For a fixed n the accuracy is maximized for a certain value of λ and decreases for higher or lower values, oscillating only little, for example. Especially during the development stage, heuristics are not sufficient; an optimal combination of both parameters for a certain parameter resolution is of interest. To make things even harder, systems under development are usually not optimized for efficiency, but rather designed to be able to be flexible so that different approaches can be tested.

Such parameter studies typically demand a significant computational effort, are not very time critical, but should be performed within a reasonable amount of time. This requires that sufficient computational power and/or storage space is available to the developer at the moment when needed. Grid environments grant users access to such resources. Additionally, they provide an easy access and therefore simplify the interaction of the users – in our setting the algorithm developers – with various resources [4]. In grid middleware, for example the Globus Toolkit [5], this is achieved through mechanisms such as single sign-on, delegation, or by providing extensive support for job processing. Parameter studies are currently among the most widespreaded applications of grid computing.

For the current work, we used the grid research framework GridSFEA (Grid-based Simulation Framework for Engineering Applications) [6]. This research environment aims at providing various engineering simulation tasks, such as fluid dynamics, in a grid environment. GridSFEA uses the Globus Toolkit V4. It comprises both server-side components (grid tools and services) and client-side tools (such as an OGCE2 grid portal, using the portlet technology as a way to customize the access to the grid environment).

The grid portal was extended by a portlet adapted to the requirements of sparse grid classification. Thus, support was introduced for sampling the parameter space, for automatic generation and submission of corresponding jobs, and for collecting and processing of results. Figure 1 shows a snapshot of our current portlet, running a sparse grid classification for various values of λ.

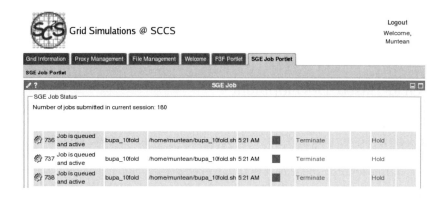

Fig. 1. Performing sparse grid parameter studies using a portlet of the GridSFEA portal

The usage of this grid portal enables the algorithm developer to focus on the design of the algorithm and the interpretation of the results, and to leave the management of the computations to the framework. Furthermore, the portal of GridSFEA allows the user to choose the grid environment to be used. This way, the computations – here parameter studies – can be performed at different computing sites, such as supercomputing centres.

3   Adaptivity and Results

The need for a classification algorithm that scales only linearly in the number of training data points forces us in the first place to trade the possibility to make use of the structure of the training data for a somehow data-independent discretization of the feature space. Therefore we have to deal with a higher number of basis functions than common classification algorithms which try to fit the data with as few basis functions as possible. Even though one iteration of a sparse grid solver scales only linearly in the number of unknowns, this still imposes restrictions on the maximum depth of the underlying grid.

The idea of adaptive sparse grid classification is to obtain a trade-off between common, data-dependent algorithms and the data independence of sparse grid classification. The aim is to reduce the order of magnitude of unknowns while keeping the linear time training complexity: Especially those grid points should be invested that contribute most to the classification boundary, as the zero-crossing of the classifier determines the class prediction on new data. For the classification task, the exact magnitude of the approximation error is not important in regions that are purely positive or negative. It is reasonable to spend more effort in rough regions of the target function than in smooth ones.

As we already showed in [2], special consideration has to be put on treating the boundary of the feature space. Normally, in sparse grid applications, grid points have to be invested on the boundary to allow for non-zero function values there. Employing adaptivity in classification allows us to neglect those grid points as long as it is guaranteed that no data points are located exactly on the boundary. This corresponds to the assumption that the data belonging to a certain class is somehow clustered together – and therefore can be separated from other data – which means that regions towards the border of the feature space usually belong to the same class. In regions where the classification boundary is close to the border of the feature space, adaptivity will take care of this by refining the grid where necessary and by creating basis functions that take the place of common boundary functions.

Thus we normalize our data to fit in a domain which is slightly smaller than the d-dimensional unit hypercube, the domain of f_N. This way we can start the adaptive classification with the d-dimensional sparse grid for level 2, $V_2^{(1)}$, without the boundary and therefore only $2d + 1$ grid points, rather than using grid points on the boundary, which would result in $\sum_{j=0}^{d} \binom{d}{j} 2^{d-j}(1 + 2j) = 3^d + 2d \cdot 3^{d-1} \in \mathcal{O}(d \cdot 3^d)$ unknowns.

For the remaining part of this section we proceed as follows: Starting with $V_2^{(1)}$, we use a preconditioned Conjugated Gradient method for a few iterations to train our classifier. Out of all of the refinement candidates we choose the grid point with the highest surplus, add all the children to our current grid and repeat this until satisfied. One has to take care not to violate one of the basic properties of sparse grids: For each basis function all parents in the hierarchical tree of basis functions have to exist. All missing ancestors have to be created, which can be done recursively.

3.1   Classification Examples

The first example, taken from Ripley [7], is a two-dimensional artificial dataset which was constructed to contain 8% of noise. It consists of 250 data points for training and 1000 points to test on. Being neither linear separable nor very complicated, it shows typical characteristics of real world data sets. Looking for a good value of λ we evaluated our classifier after six refinement steps for $\lambda = 10^{-i}$, $i = 0 \ldots 6$, took the ones with the best and the two neighbouring accuracies, namely 0.01, 0.001 and 0.0001, and looked once more for a better value of lambda in between those by evaluating for $\lambda = 0.01 - 0.001 \cdot i$, $i = 1 \ldots 9$ and $\lambda = 0.001 - 0.0001 \cdot i$, $i = 1 \ldots 9$. The best λ we got this way was 0.004.

Figure 2 shows the training data (left) as well as the classification for $\lambda = 0.004$ and the underlying adaptive sparse grid after only 7 refinement steps (right). Even though there are only a few steps of refinement, it can clearly be observed that more grid points are spent in the region with the most noise and that regions with training data belonging to the same class are neglected. Note that one refinement step in x_1-direction (along $x_2 = 0.5$) causes the creation of two children nodes in x_2-direction (at $x_2 = 0.25$ and $x_2 = 0.75$).

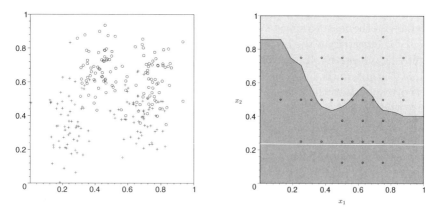

Fig. 2. Ripley dataset: training data and classification for $\lambda = 0.004$

Table 1 shows some results comparing different sparse grid techniques. For the combination technique and sparse grids with and without grid points on the boundary, we calculated the accuracy for level one to four and $\lambda := 10^{-j} - i \cdot 10^{-j-1}$, $i = 0 \ldots 8, j = 0 \ldots 5$, each. Increasing the level further does not improve the results as this quickly leads to overfitting. We show the number of unknowns for each grid, the best value of λ, and the best accuracy achieved on the test data. A general property of sparse grid classification can be observed: The higher the number of unknowns, the more important the smoothness functional gets and therefore the value for the regularization parameter λ increases. Using a coarser grid induces a higher degree of smoothness, which is a well-known phenomenon.

Table 1. Ripley dataset: accuracies [%] obtained for the combination technique, regular sparse grids with and without boundary functions, and the adaptive technique

	comb. techn.			sg boundary			sg			adapt. sg $\lambda = 0.004$	
n	|grids|	λ	acc.	|grid|	λ	acc.	|grid|	λ	acc.	|grid|	acc.
1	9	$6 \cdot 10^{-5}$	90.3	9	$6 \cdot 10^{-5}$	90.3	1	*	50.0	5	89.9
2	39	0.0005	90.8	21	0.0004	90.7	5	0.0005	90.3	9	90.2
3	109	0.005	91.1	49	0.006	91.2	17	0.005	91.2	13	89.8
4	271	0.007	91.1	113	0.005	91.2	49	0.007	91.2	19	91.1
										21	91.2
										24	91.2
										28	91.3
										32	91.4
										35	91.5

Another observation is that the assumption that grid points on the boundary can be neglected, which holds here even for regular grids. Of course, the sparse grid on level 1 can do nothing but guess one class value for all data points and five unknowns are not enough to achieve an accuracy of 90.7%, but already for level 3 the same accuracy as for the grid with boundary values is reached.

For the adaptive sparse grid classification we show the results only for $\lambda = 0.004$ for eight times of refinement. With less grid points than the sparse grid with boundary on level 3, we achieved an excellent accuracy of 91.5% – 0.3% higher than what we were able to get using regular sparse grids. This is due to the fact that increasing a full level results in a better classification boundary in some parts, but leads to overfitting in other parts at the same time. Here, adaptivity helps to overcome this problem.

We used our grid environment to compute and gather the results for these 180 runs for each level, even though the problem sizes are quite low: Our implementation of the combination technique has just been done for comparing it with the direct sparse grid technique, for example, and it is therefore far from being efficient. We neither use an efficient solver as a multigrid technique for example, nor care about the amount of memory we consume. We are just interested in getting eventually some results, a problem suited for grid environments. As the portlet mentioned in Sec. 2 allows us to easily specify a set of parameters, the adaptation for the use in GridSFEA was just a matter of minutes.

A second, 6-dimensional dataset we used is a real-world dataset taken from [8] which contains medical data of 345 patients, mainly blood test values. The class value indicates whether a liver disease occurred or not. Because of the limited data available we did a stratified 10-fold cross-validation. To find a good value for λ we calculated the accuracy for different values of the regularization parameter as we did above for the regular sparse grids. This time we calculated the accuracy of the prediction on the 10-fold test data. Again we used GridSFEA to compute and gather the results. Theoretically one could even submit a job for each fold of the 10-fold cross-validation and gather the different accuracies afterwards, but as the coefficients computed for the first fold can be used as a very good starting vector for the PCG in the other 9 folds, this was neglected.

Table 2 shows some results, comparing the results of the adaptive technique (zero boundary conditions) with the best ones taken from [9], the combination technique (anisotropic grid, linear basis functions), and from [10], both linear and non-linear support vector machines (SVM). The adaptive sparse grid technique

Table 2. Accuracies [%] (10-fold testing) for the Bupa liver dataset

adapt. sg $\lambda = 0.3$			comb. techn. lin. anisotrop.	SVM linear	SVM non-linear
# refinements	|grid|	acc.	acc.	acc.	acc.
7	485	70.71	73.9	70.0	73.7
8	863	75.64			
9	1302	76.25			

was able to outperform the other techniques. For a relatively large value of λ and after only 9 refinement steps we got a 10-fold testing accuracy which is more than 2.3% higher than our reference results – with only 1302 unknowns. Again it proved to be useful to refine in the most critical regions while neglecting parts of the domain where the target function is smooth.

3.2   Dimension Adaptivity

Similar observations can be made for a third dataset, the Pima diabetes dataset, taken again from [8]. 769 women of Pima Indian heritage living near Phoenix were tested for diabetes mellitus. 8 features describe the results of different examinations, such as blood pressure measurements. For reasons of shortness we will focus only on some observations of the dimension adaptivity of the adaptive sparse grid technique.

As assumed, the adaptive refinement neglects dimensions containing no information but only noise quite well. Extending the diabetes dataset for example by one additional feature with all values set to 0.5 leads to two more grid points for $V_2^{(1)}$. As the two additional surpluses are close to zero, the effects on BB^T and By are just minor, see (1). But extending the dimensionality modifies the smoothness functional. There are stronger impacts on C and therefore the trade-off between smoothness and approximation error changes. One could expect that the same effects could be produced by changing the value of λ suitably. And in fact, for training on the first 567 instances and testing on the remaining 192 data points, almost identical accuracies (about 74.5%) can be achieved when changing λ to 0.0002 compared to extending the dimensionality by one for $\lambda = 0.001$ – at least for the first few refinements.

Further improvement could be achieved by omitting the "weakest" dimension, the one with the lowest surpluses, during refinement. When some attributes are expected to be less important than others, this could lead to further improvements considering the number of unknowns needed.

4   Summary

We showed that adaptive sparse grid classification is not only possible, but useful. It makes use of both worlds: the data-independent and linear runtime world and the data-dependent, non-linear one that reduces the number of unknowns.

If increasing the level of a regular sparse grid leads to overfitting in some regions but improves the quality of the classifier in others, adaptivity can take care of this by refining just in rough regions of the target function. In comparison to regular sparse grids, the adaptive ones need far less unknowns. This can reduce the computational costs significantly.

Considering dimension adaptivity we pointed out first observations. Further research has to be invested here.

Finally we demonstrated the use of GridFSEA, our grid framework, for parameter studies, especially during algorithm development stage.

Theoretically, even high dimensional classification problems could be tackled by adaptive sparse grid classification as we showed that we can start with a number of grid points which is linear in the dimensionality. Practically, our current smoothness functional permits high dimensionalities by introducing a factor of 2^d in the number of operations. Therefore we intend to investigate the use of alternative smoothness functionals in the near future to avoid this.

References

1. Garcke, J., Griebel, M., Thess, M.: Data mining with sparse grids. Computing **67**(3) (2001) 225–253
2. Bungartz, H.J., Pflüger, D., Zimmer, S.: Adaptive Sparse Grid Techniques for Data Mining. In: Modelling, Simulation and Optimization of Complex Processes, Proc. Int. Conf. HPSC, Hanoi, Vietnam, Springer (2007) to appear.
3. Bungartz, H.J., Griebel, M.: Sparse grids. Acta Numerica **13** (2004) 147–269
4. Foster, I., Kesselman, C.: The Grid: Blueprint for a New Computing Infrastructure. Morgan Kaufmann Publishers (2005)
5. Foster, I.: Globus toolkit version 4: Software for service-oriented systems. In: IFIP International Conference on Network and Parallel Computing. Volume 3779 of LNCS., Springer-Verlag (2005) 2–13
6. Muntean, I.L., Mundani, R.P.: GridSFEA - Grid-based Simulation Framework for Engineering Applications. http://www5.in.tum.de/forschung/grid/gridsfea (2007)
7. Ripley, B.D., Hjort, N.L.: Pattern Recognition and Neural Networks. Cambridge University Press, New York, NY, USA (1995)
8. Newman, D., Hettich, S., Blake, C., Merz, C.: UCI repository of machine learning databases. http://www.ics.uci.edu/~mlearn/MLRepository.html (1998)
9. Garcke, J., Griebel, M.: Classification with sparse grids using simplicial basis functions. Intelligent Data Analysis **6**(6) (2002) 483–502
10. Fung, G., Mangasarian, O.L.: Proximal support vector machine classifiers. In: KDD '01: Proceedings of the seventh ACM SIGKDD international conference on Knowledge discovery and data mining, New York, NY, USA, ACM Press (2001) 77–86

Latency-Optimized Parallelization of the FMM Near-Field Computations

Ivo Kabadshow[1] and Bruno Lang[2]

[1] John von Neumann Institute for Computing,
Central Institute for Applied Mathematics, Research Centre Jülich, Germany
i.kabadshow@fz-juelich.de
[2] Applied Computer Science and Scientific Computing Group,
Department of Mathematics, University of Wuppertal, Germany

Abstract. In this paper we present a new parallelization scheme for the FMM near-field. The parallelization is based on the Global Arrays Toolkit and uses one-sided communication with overlapping. It employs a purely static load-balancing approach to minimize the number of communication steps and benefits from a maximum utilization of data locality. In contrast to other implementations the communication is initiated by the process owning the data via a `put` call, not the process receiving the data (via a `get` call).

1 Introduction

The simulation of particle systems is a central problem in computational physics. If the interaction between these particles is described using an electrostatic or gravitational potential $\sim 1/r$, the accurate solution poses several problems. A straightforward computation of all pairwise interactions has the complexity $\mathcal{O}(N^2)$. The Fast Multipole Method (FMM) developed by Greengard and Rokhlin [1] reduces the complexity to $\mathcal{O}(N)$. A detailed depiction of the FMM would be beyond the scope of this paper and can be found elsewhere [2]. We will only outline the most important details for the parallelization.

The FMM proceeds in five passes.

- Sort all particles into boxes.
- Pass 1: Calculation and shifting of multipole moments.
- Pass 2: Transforming multipole moments.
- Pass 3: Shifting Taylor-like coefficients.
- Pass 4: Calculation of the far-field energy.
- Pass 5: Calculation of the near-field energy.

The most time-consuming parts of the algorithm are Pass 2 and Pass 5, each contributing approximately 45% to the overall computing time. Since we have different calculation schemes for Pass 2 and 5, different parallelization schemes are necessary. This paper deals with Pass 5 only, the direct (near-field) pairwise interaction. The sequential version of the near-field computation involves the following steps.

Y. Shi et al. (Eds.): ICCS 2007, Part I, LNCS 4487, pp. 716–722, 2007.

- Create the particle–box relation `ibox` in skip-vector form (see Sec. 3.1).
- Calculate all interactions of particles contained in the same box i (routine `pass5inbox`).
- Calculate all interactions of particles contained in different boxes i and j (routine `pass5bibj`).

2   Constraints for the Parallelization

In order to match the parallelization of the remaining FMM passes and to achieve reasonable efficiency, the following constraints had to be met.

1. Use the Global Arrays Toolkit (GA) for storing global data to preserve the global view of the data.

   GA provides a parallel programming model with a global view of data structures. GA uses the ARMCI communication library offering one-sided and non-blocking communication. A detailed description of all GA features can be found in Ref. [4]. ARMCI is available on a wide range of architectures. Additionally GA comes with its own memory allocator (MA). To use the available memory efficiently, our implementation will allocate all large local scratch arrays dynamically by MA routines instead of direct Fortran allocation.

2. Minimize the number of communication steps.

   All tests were performed on the Jülich Multi Processor (JUMP); a SMP cluster with 41 nodes with 32 IBM Power 4+ processors and 128 GB RAM per node. The nodes are connected through a high performance switch with peak bandwidth of 1.4GB/s and measured average latencies of $\approx 30\mu s$. Shared memory can be accessed inside a node at the cost of $\approx 3\mu s$.

   Ignoring the latency issue on such machines can have a dramatic impact on the efficiency; see Fig. 1

3. To reduce the communication costs further, try to overlap communication and calculation and use one-sided communication to receive (`ga_get`) and send (`ga_put`) data.

3   Implementation

In this section we describe some details of the parallel implementation. Related work and other implementation schemes can be found elsewhere [7,8,9].

3.1   Initial Data Distribution

The most memory consuming data arrays are the Cartesian coordinates (`xyz`), the charges (`q`), the particle–box relation stored in the `ibox` vector, and two output arrays storing the data for the potential (`fmmpot`) and the gradient (`fmmgrad`). These arrays have to be distributed to prevent redundancy.

The `ibox` vector is a special auxiliary data structure mapping all charges to the corresponding boxes. To enable fast access in a box-wise search this structure is stored in a skip-vector form (see Fig.2).

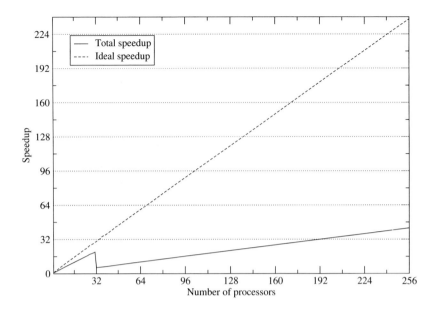

Fig. 1. This figure highlights the latency bottleneck. The speedup for a system with $8^7 \approx 2 \cdot 10^6$ homogeneously distributed particles is shown. Communication is implemented using blocking communication. Each box is fetched separately. The latency switching from $3\mu s$ to $30\mu s$ can be clearly seen at 32 processors.

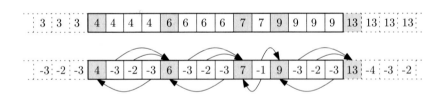

Fig. 2. The diagram illustrates the box management. Empty boxes (5,8,10,11,12) are not stored. The `ibox` vector associates the box number with each particle. To allow fast access to neighboring boxes the `ibox` vector is modified into a skipped form. Only the first particle in a box holds the box number, all subsequent charges link forward to the first particle of the next box. Finally, the last particle in a box links back to the first particle of the same box.

3.2   Design Steps

The parallel algorithm can be divided into 5 steps:

- Align boxes on processors
- Prepare and initiate communication
- Compute local contributions

- Compute remote contributions
- Further communication and computation steps, if small communication buffers make them necessary.

Step 1 - Box Alignment. In Passes 1 to 4 of the FMM scheme the particles of some boxes may be stored in more than one processor. These boxes have to be aligned to only one processor. While this guarantees that the Pass 5 subroutine `pass5inbox` can operate locally, it may introduce load imbalance. However, assuming a homogeneous particle distribution, the load-imbalance will be very small, since the workload for one single box is very small.

The alignment can be done as follows:

- Compute size and indices of the "leftmost" and "rightmost" boxes (w.r.t. Morton ordering; cf. Fig. 3) and store this information in a global array of size $\mathcal{O}(\texttt{nprocs})$.
- Gather this array into a local array `boxinfo` containing information from each processor.
- Assign every box such that the processor owning the larger part gets the whole box.
- Reshape the irregular global arrays `ibox`, `xyz`, `q`, `fmmgrad`, `fmmpot`.
- Update `boxinfo`.

After alignment the `pass5inbox` subroutine can be called locally.

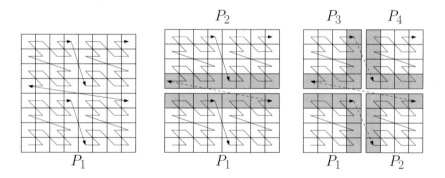

Fig. 3. The data is stored along a Morton-ordered three-dimensional space filling curve (SFC)[6]. The diagram shows a 2D Morton-ordered SFC for the sequential version and a two and four processor parallel version. Since the data is distributed homogeneously the particle/box split is exactly as shown here.

Step 2 - Prepare and Initiate Communication. Like in the sequential program, each processor determines the neighbors of the boxes it holds, i.e. the boxes which lie geometrically above, to the right or in front of the respective box. All other boxes do not need to be considered, since these interactions were already taken into account earlier. This is a special feature of the Morton-ordered data structure. Boxes not available locally will be sent by the owner of the box to assure a minimal amount of communication steps. The remote boxes are stored in a local buffer. If the buffer is too small to hold all neighboring boxes, all considered interactions will be postponed until the buffer can be reused.

In order to achieve the $\mathcal{O}(\texttt{nprocs})$ latency, foreign boxes are put by the processor that holds them into the local memory of the processor requiring them. Each processor has to perform the following steps:

- Determine local boxes that are needed by another processor.
- Determine the processors they reside on.
- Compute the total number of 'items' that every processor will receive.
- Check local buffer space against total size.
- If possible, create a sufficiently large send buffer.
- Issue a `ga_put()` command to initiate the data transfer.

Step 3 - Compute Local Contributions. Now, that the communication is started, the local portion of the data can be used for computation. All in-box and box–box interactions that are available locally are computed. All computations involving remote/buffered boxes are deferred to Steps 4/5. Step 3 comprises the the calculations of the total Coulomb energy E_c, the Coulomb forces $\boldsymbol{F}_c(\boldsymbol{r}_k)$ and the Coulomb potential $\phi_c(\boldsymbol{r}_k)$. Let n_i denote the number of particles in box i, let (i, k) be the kth particle in box i, and let $q_{i,k}$ and $\boldsymbol{r}_{i,k}$ be its charge and position, respectively. Then the total Coulomb energy between all particles in boxes i and j, the total Coulomb forces of all particles in box j to a certain particle (i, k) in box i, and the corresponding Coulomb potential are given by

$$E_c = \frac{1}{2} \sum_{k=1}^{n_i} \sum_{l=1}^{n_j} \frac{q_{i,k} q_{j,l}}{|\boldsymbol{r}_{i,k} - \boldsymbol{r}_{j,l}|} \tag{1}$$

$$\boldsymbol{F}_c(\boldsymbol{r}_{i,k}) = q_{i,k} \sum_{l=1}^{n_j} \frac{q_{j,l}}{(\boldsymbol{r}_{i,k} - \boldsymbol{r}_{j,l})^3} (\boldsymbol{r}_{i,k} - \boldsymbol{r}_{j,l}) \tag{2}$$

$$\phi_c(\boldsymbol{r}_{i,k}) = \sum_{l=1}^{n_j} \frac{q_{j,l}}{|\boldsymbol{r}_{i,k} - \boldsymbol{r}_{j,l}|} \tag{3}$$

respectively. Note that for the in-box case, $i = j$, the terms corresponding to the interaction of a particle with itself must be dropped.

Step 4/5 - Compute Remote Contributions. Since the communication request to fill the buffer was issued before the start of step 3, all communication should be finished by now. After a global synchronization call `ga_sync`, the buffer is finally used to calculate interactions with the remote/buffered boxes.

4 Results

Three different test cases were studied. All test cases contain equally distributed particles (each box contains 8 particles). This guarantees that all boxes are occupied with particles and therefore all possible communication at the processors'

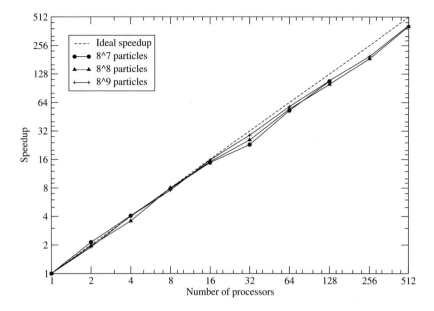

Fig. 4. Three different test cases were computed. The speedup is almost independent from the number of particles even for larger processor numbers. Thus, especially the smallest test case with 8^7 particles benefits from the reduced communication steps.

border will indeed take place. This represents a worst case communication pattern for any number of processors since empty boxes would not be communicated over the network.

The key features can be summarized as follows: All expensive operations (box search, calculation) are done locally. Especially gathering all needed data for each processor is done by the processor owning the data (`ga_put`), not by the processor receiving the data (`ga_get`). This saves unnecessary communication time and thus latency. Even for small test cases with few particles per processor the communication can be hidden behind the calculation. The static load-balancing approach outperforms the dynamic approach especially for clusters with long latency times. Compared to the model described in Sec. 1, this approach scales well beyond 32 processors. There is no visible break in the speedup at 32 processors.

5   Outlook

The presented scheme assumes an approximately homogeneous distribution of the particles, but can be extended to handle inhomogeneous distributions by introducing "splitting boxes", i.e., dividing large boxes further until a certain box granularity is reached; hence, the workload can again be distributed equally over all processors.

Table 1. Computation times in seconds/speedups for three different particle systems. For 8^7 particles calculation was done with processor numbers up to 128, since the total computation time was already below 0.25 seconds and hence results with more processors would suffer from the low measurement precision. The upper limit of 512 processors was due to limitations in our GA implementation.

procs	8^7 particles		8^8 particles		8^9 particles	
1	26.78	—	211.74	—	1654.13	—
2	12.51	2.14	107.92	1.96	869.53	1.90
4	6.43	4.16	59.19	3.58	409.70	4.04
8	3.33	8.04	26.44	8.01	218.22	7.58
16	1.76	15.22	14.03	15.09	105.01	15.75
32	1.11	24.13	8.25	25.67	57.26	28.89
64	0.45	59.51	3.89	54.43	28.59	57.86
128	0.21	127.52	2.13	99.41	15.23	108.61
256	—	—	1.14	185.74	8.41	196.69
512	—	—	0.52	407.19	3.96	417.71

Acknowledgements

The authors acknowledge the support by H. Dachsel for his sequential FMM code, as well as B. Kuehnel for his work on the parallel code as a guest student at research centre Jülich.

References

1. L. Greengard and V. Rokhlin: A fast algorithm for particle simulations. J. Comput. Phys. **73**, No.2 (1987) 325–348
2. C. A. White and M. Head-Gordon: Derivation and efficient implementation of the fast multipole method. J. Chem. Phys. **101** (1994) 6593–6605
3. H. Dachsel: An error-controlled Fast Multipole Method. (in preparation)
4. J. Nieplocha, B. Palmer, V. Tipparaju, M. Krishnan, H. Trease and E. Apra: Advances, Applications and Performance of the Global Arrays Shared Memory Programming Toolkit. IJHPCA, **20**, No. 2, (2006) 203–231
5. J. Nieplocha, V. Tipparaju, M. Krishnan, and D. Panda: High Performance Remote Memory Access Communications: The ARMCI Approach. IJHPCA, **20**, No. 2, (2006) 233–253
6. M.F. Mokbel, W.G. Aref and I. Kamel: Analysis of multi-dimensional space-filling curves. Geoinformatica **7**, No.3 (2003) 179–209
7. L. Greengard and W.D. Gopp: A parallel version of the multipole method. Computers Math. Applic. **20**, No. 7 (1990) 63–71
8. J. Kurzak and B.M. Pettitt: Communications overlapping in fast multipole particle dynamics methods. J. Comput. Phys. **203** (2005) 731–743
9. J. Kurzak and B.M. Pettitt: Massively parallel implementation of a fast multipole method for distributed memory machines. J. Par. Dist. Comp. **65** (2005) 870–881

Efficient Generation of Parallel Quasirandom Faure Sequences Via Scrambling

Hongmei Chi[1,2] and Michael Mascagni[1,3]

[1] School of Computational Science, Florida State University,
Tallahassee, FL 32306-4120, USA
chi@scs.fsu.edu
[2] Department of Computer and Information Sciences, Florida A& M University,
Tallahassee, FL 32307-5100, USA
[3] Department of Computer Science, Florida State University,
Tallahassee, FL 32306-4530, USA

Abstract. Much of the recent work on parallelizing quasi-Monte Carlo methods has been aimed at splitting a quasirandom sequence into many subsequences which are then used independently on the various parallel processes. This method works well for the parallelization of pseudorandom numbers, but due to the nature of quality in quasirandom numbers, this technique has many drawbacks. In contrast, this paper proposes an alternative approach for generating parallel quasirandom sequences via scrambling. The exact meaning of the digit scrambling we use depends on the mathematical details of the quasirandom number sequence's method of generation. The Faure sequence is scramble by modifying the generator matrices in the definition. Thus, we not only obtain the expected near-perfect speedup of the naturally parallel Monte Carlo methods, but the errors in the parallel computation is even smaller than if the computation were done with the same quantity of quasirandom numbers using the original Faure sequence.

Keywords: Parallel computing, Faure sequence, quasi-Monte Carlo, scrambling, optimal sequences.

1 Introduction

One major advantage of Monte Carlo methods is that they are usually very easy to parallelize; this leads us to all Monte Carlo methods "naturally parallel" algorithms. This is, in principal, also true of quasi-Monte Carlo (QMC) methods. As with ordinary Monte Carlo, QMC applications have high degrees of parallelism, can tolerate large latencies, and usually require considerable computational effort, making them extremely well suited for parallel, distributed, and even Grid-based computational environments. Parallel computations using QMC require a source of quasirandom sequences, which are distributed among the individual parallel processes. In these environments, a large QMC problem can be broken up into many small subproblems. These subproblems are then scheduled

Y. Shi et al. (Eds.): ICCS 2007, Part I, LNCS 4487, pp. 723–730, 2007.

on the parallel, distributed, or Grid-based environment. In a more traditional instantiation, these environments are usually a workstation cluster connected by a local-area network, where the computational workload is carefully distributed. Recently, peer-to-peer and Grid-based computing, the cooperative use of geographically distributed resources unified to act as a single powerful computer, has been investigated as an appropriate computational environment for Monte Carlo applications [9,10]. QMC methods can significantly increase the accuracy of the likelihood-estimated over regular MC [7]. In addition, QMC can improve the convergence rate of traditional Markov chain Monte Carlo [16]. This paper explores an approach to generating parallel quasirandom sequences for parallel, distributed, and Grid-based computing.

Like pseudorandom numbers, quasirandom sequences are deterministically generated, but, in contrast, are constructed to be highly uniformly distributed. The high level of uniformity is a global property of these sequences, and something as innocent sounding as the deletion of a single quasirandom point from a sequence can harm its uniformity quite drastically. The measure of uniformity used traditionally in quasirandom number is the star-discrepancy. The reason for this is the Koksma-Hlawka inequality. This is the fundamental result in QMC theory, and motivates the search and study of low-discrepancy sequences. This result states that for any sequence $X = \{x_0, \ldots, x_{N-1}\}$ and any function, f, with bounded variation defined on the s-dimensional unit cube, I^s, the integration error over I^d is bounded as,

$$\left| \int_{I^s} f(x)\, dx - \frac{1}{N} \sum_{i=1}^{N} f(x_i) \right| \leq V[f] D_N^*, \tag{1}$$

where $V[f]$ is the total variation of f, in the sense of Hardy-Krause, D_N^*[1] is the star discrepancy of sequence X [20]. As we are normally given a problem to solve, it is only D_N^* that we can control. This is done by constructing low-discrepancy sequences for use in QMC.

However, the successful parallel implementation of a quasi-Monte Carlo application depends crucially on various quality aspects of the parallel quasirandom sequences used. Randomness can be brought to bear on otherwise deterministic quasirandom sequence by using various scrambling techniques. These methods randomize quasirandom sequences by using pseudorandom numbers to scramble the order the of quasirandom numbers generated or to permute their digits. Thus by the term "scrambling" we are referring more generally to the randomization of quasirandom numbers. Scrambling provides a natural way to parallelize quasirandom sequences, because scrambled quasirandom sequences form a stochastic family which can be assigned to different processes in a parallel computation. In

---

[1] For a sequence of N points $X = \{x_0, \ldots, x_{N-1}\}$ in the d-dimensional unit cube I^s, and for any box, J, with one corner at the origin in I^s, the star discrepancy, D_N^*, is defined as $D_N^* = \sup_{J \in I^s} |\mu_x(J) - \mu(J)|$, where $\mu_X(J) = \frac{\#\text{of points in } J}{N}$ is the discrete measure of J, i. e., the fraction of points of X in J, and $\mu(J)$ is the Lebesgue measure of J, i. e., the volume of J.

addition, there are many scrambling schemes that produce scrambled sequences with the same D_N^* as the parent, which means that we expect no degradation of results with these sorts of scrambling. These scrambling schemes are different from other proposed QMC parallelization schemes such as the leap-frog scheme [2] and the blocking scheme [18], which split up a single quasirandom sequence into subsequences.

The Faure sequence is one of the most widely used quasirandom sequences. The original construction of quasirandom sequences was related to the van der Corput sequence, which itself is a one-dimension quasirandom sequence based on digital inversion. This digital inversion method is a central idea behind the construction of many current quasirandom sequences in arbitrary bases and dimensions. Following the construction of the van der Corput sequence, a significant generalization of this method was proposed by Faure [4] to the sequences that now bear his name. In addition, an efficient implementation of the Faure sequence was published shortly afterwards [6]. Later, Tezuka [20] proposed a generalized Faure sequence, GFaure, which forms a family of randomized Faure sequences. We will discuss the scrambling of Faure sequences in this paper.

The organization of this paper is as follows. In §2, an overview of scrambling methods and a brief introduction to the theory of constructing quasirandom sequences is given. Parallelizations and implementations of quasirandom Faure sequences are presented in §3. The consequences of choosing a generator matrix, the resulting computational issues, and a some numerical results are illustrated in §4, and conclusions are presented in §5.

2   Scrambling

Before we begin our discussion of the scrambled Faure sequence, it behooves us to describe, in detail, the standard and widely accepted methods of Faure sequence generation. The reason for this is that construction of Faure sequence follows from the generation of the Van der Corput and Soboĺ sequences. Often QMC scrambling methods are combined with the original quasirandom number generation algorithms.

The construction of quasirandom sequences is based on finite-field arithmetic. For example, the Soboĺ sequences [19] are constructed using linear recurring relations over $\mathcal{F}_2$. Faure used the Pascal matrix in his construction, and Niederreiter used the formal Laurent series over finite fields to construct low-discrepancy sequences that now bear his name.

We now briefly describe the construction of the above "classical" quasirandom sequences.

- **Van der Corput sequences:** Let $b \geq 2$ be an integer, and n a non-negative integer with $n = n_0 + n_1 b + ... + n_m b^m$ it b-adic representation. Then the nth term of the Van der Corput sequence is $\phi_b(\mathbf{n}) = \frac{n_0}{b} + \frac{n_1}{b^2} + ... + \frac{n_m}{b^m}$. Here $\phi_b(\mathbf{n})$ is the radical inverse function in base b and and $\mathbf{n} = (n_0, n_1, ..., n_m)^T$

is the digit vector of the b-adic representation of n. $\phi_b(\cdot)$ simply reverses the digit expansion of n, and places it to the right of the "decimal" point. The Van der Corput sequence in s-dimensions, more commonly called the Halton sequence, is one of the most basic one-dimension quasirandom sequences, and can be rewritten in the following form:

$$(\phi_{b_1}(\mathbf{Cn}), \phi_{b_2}(\mathbf{Cn})..., \phi_{b_s}(\mathbf{Cn}))$$

Here the "generator matrix" $\mathbf{C}$ is the identity matrix and the nth Halton sequence is defined as $(\phi_{b_1}(\mathbf{n}), \phi_{b_2}(\mathbf{n})..., \phi_{b_s}(\mathbf{n}))$ where the bases, $b_1, b_2, ..., b_s$, are pairwise coprime. The other commonly uses quasirandom sequences can be similarly defined by specifying different generator matrices.

- **Faure and GFaure sequences:** The nth element of the Faure sequence is expressed as
  $(\phi_b(P^0\mathbf{n}), \phi_b(P^1\mathbf{n}), ..., \phi_b(P^{s-1}\mathbf{n}))$,
  where b is prime and $b \geq s$ and $P$is Pascal matrix whose (i,j) element
  is equal to $\binom{i-1}{j-1}$ Tezuka [22] proposed the generalized Faure sequence,
  GFaure, with the jth dimension generator matrix $C^{(j)} = A^{(j)}P^{j-1}$, where
  $A^{(j)}$ are a random nonsingular lower triangular matrices. Faure [5] extended
  this idea to bigger set. Also a subset of GFaure is called GFaure with the
  i-binomial property [21], where $A^{(j)}$ is defined as:

$$A^{(j)} = \begin{pmatrix} h_1 & 0 & 0 & 0 & ... \\ g_2 & h_1 & 0 & 0 & ... \\ g_3 & g_2 & h_1 & 0 & ... \\ g_4 & g_3 & g_2 & h_1 & ... \\ & \cdot & \cdot & \cdot & ... \\ & \cdot & \cdot & \cdot & ... \\ & \cdot & \cdot & \cdot & ... \end{pmatrix},$$

where h_1 is uniformly distributed on $\mathcal{F}_b - \{0\}$ and g_i is uniformly distributed on $\mathcal{F}_b$. For each $A^{(j)}$, there will be a different random matrix in the above form.

We will generate parallel Faure sequences by randomizing the matrix $A^{(j)}$. However, arbitrary choice of matrix could lead to high correlation between different quasirandom streams. We will address this concern in next section and show how to choose the matrix $A^{(j)}$ so that correlation between streams is minimized.

3   Parallelization and Implementations

Much of the recent work on parallelizing quasi-Monte Carlo methods has been aimed at splitting a quasirandom sequence into many subsequences which are then used independently on the various parallel processes. This method works well for the parallelization of pseudorandom numbers, but due to the nature of quality in quasirandom numbers, this technique has many drawbacks. In contrast, this paper proposes an alternative approach for generating parallel quasirandom sequences. Here we take a single quasirandom sequence and provide different random digit scramblings of the given sequence. If the scrambling preserves certain equidistribution properties of the parent sequence, then the result will be high-quality quasirandom numbers for each parallel process, and an overall successful parallel quasi-Monte Carlo computation as expected from the Koksma-Hlawka inequality [11].

Parallelization via splitting [18] uses a single quasirandom sequence and assigns subsequences of this quasirandom sequence to different processes. The idea behind splitting is to assume that any subsequence of a given quasirandom sequence has the same uniformity as the parent quasirandom sequence. This is an assumption that is often false [8]. In comparison to splitting, each scrambled sequence can be thought of as an independent sequence and thus assigned to a different process, and under certain circumstances it can be proven that the scrambled sequences are as uniform as the parent [15]. Since the quality (small discrepancy) of quasirandom sequences is a collective property of the entire sequence, forming new sequences from parts is potentially troublesome. Therefore, scrambled quasirandom sequences provide a very appealing alternative for parallel quasirandom sequences, especially where a single quasirandom sequence is scrambled to provide all the parallel streams. Such a scheme would also be very useful for providing QMC support to the computational Grid [9,10].

We will consider a variety of generator matrices for parallelizing and implementing parallel Faure sequences. The generation of the original Faure sequences is fast and easy to implement. Here, we applied the same methods as was used in Halton sequences because the original Halton sequence suffers from correlations [3] between radical inverse functions with different bases used for different dimensions. These correlations result in poorly distributed two-dimensional projections. A standard solution to this phenomenon is to use a randomized (scrambled) version of the Halton sequence. We proposed a similar algorithm to improve quality of the Faure sequences. We will use the the same approach here to randomize the generator matrix and thus generate parallel Faure sequences. The strategy used in [3] is to find an optimal h_i and g_i in base b to improve the quality of the Faure sequence. Most importantly, the two-most significant digits of each Faure point are only scrambled by h_1 and g_2. Thus the choice of these two elements is crucial for producing good scrambled Faure sequences. After these choices, the rest of elements in matrix $A^{(j)}$ could be chosen randomly.

Whenever scrambled methods are applied, pseudorandom numbers are the "scramblers". Therefore, it is important to find a good pseudorandom number generator (PRNG) to act as a scrambler so that we can obtain well scrambled quasirandom sequences. A good parallel pseudorandom number generator such as SPRNG [13,12] is chosen as our scrambler.[2]

---

[2] The SPRNG library can be downloaded at http://sprng.fsu.edu.

4 Applications

FINDER [17], a commercial software system which uses quasirandom sequences to solve problems in finance, is an example of the successful use of GFaure, as a modified GFaure sequence is included in FINDER, where matrices $A^{(j)}$ are empirically chosen to provide high-quality sequences.

Assume that an integrand, f, is defined over the s-dimensional unit cube, $[0,1)^s$, and that $I(f)$ is defined as: $I(f) = \int_{I^s} f(x)\,dx$. Then the s-dimensional integral, $I(f)$, in Equation (2) may be approximated by $Q_N(f)$ [14]: $Q_N(f) = \sum_{i=1}^{N} \omega_i f(x_i)$, where x_i is in $[0,1)^s$, and the $\omega_i's$ are weights. If $\{x_1, \ldots, x_N\}$ is chosen randomly, and $\omega_i = \frac{1}{N}$, then $Q_N(f)$ becomes the standard Monte Carlo integral estimate, whose statistical error can be estimated using the Central Limit Theorem. If $\{x_1, \ldots, x_N\}$ are a set of quasirandom numbers, then $Q_N(f)$ is a quasi-Monte Carlo estimate and the above mentioned Koksma-Hlawka inequality can be appealed to for error bounds.

To empirically test the quality of our proposed parallel Faure sequence, we evaluate the test integral discussed in [6] with $a_i = 1$

$$\int_0^1 \cdots \int_0^1 \Pi_{i=1}^s \frac{|4x_i - 2| + a_i}{1 + a_i}\,dx_1 \ldots dx_s = 1. \tag{2}$$

Table 1. Estimation of The Integral $\int_0^1 \cdots \int_0^1 \Pi_{i=1}^s \frac{|4x_i - 2| + 1}{2}\,dx_1 \ldots dx_s = 1$

r	N	$s = 13$	$s = 20$	$s = 25$	$s = 40$
10	1000	0.125	0.399	0.388	0.699
10	3000	0.908	0.869	0.769	0.515
10	5000	0.912	0.985	1.979	0.419
10	7000	0.943	1.014	1.342	0.489
10	30000	0.988	1.097	1.171	1.206
10	40000	1.014	1.118	1.181	1.118
10	50000	1.006	1.016	1.089	1.034

In Table 1, we presented the average of ten ($r = 10$) Faure parallel streams for computing the numerical values of integral. The accuracy of quasi-Monte Carlo integration depends not simply on the dimension of the integrands, but on their effective dimension [23]. It is instructive to publish the results of these numerical integrations when we instead use the original Faure sequence to provide the same number of points as were used in the above computations. Table 2 shows these results. The astonishing fact is that the quality numerical integration using the original is lower than the parallel scrambled Faure sequence. We do not report speedup results, as this is a naturally parallel algorithm, but we feel it important to stress that using scrambling in QMC for parallelization has a very interesting consequence: the quality of the parallel sequences is actually better than the original sequences. The authors are not aware of similar situations,

Table 2. Estimation of The Integral $\int_0^1 \ldots \int_0^1 \Pi_{i=1}^s \frac{|4x_i - 2| + 1}{2} dx_1 \ldots dx_s = 1$

r	N	$s = 13$	$s = 20$	$s = 25$	$s = 40$
10	1000	0.412	0.456	1.300	0.677
10	3000	1.208	0.679	0.955	0.775
10	5000	0.919	0.976	1.066	0.871
10	7000	1.043	1.214	0.958	0.916
10	30000	0.987	1.097	0.989	1.026
10	40000	1.010	1.103	1.201	0.886
10	50000	0.996	1.006	1.178	0.791

where parallelization actually increases the quality of a computation, but are interested to learn about more examples in the literature.

5   Conclusions

A new scheme for parallel QMC streams using different `GFaure` sequences is proposed. The advantage of this scheme is that we can provide unlimited independent streams for QMC in heterogeneous computing environment. This scheme is an alternative for generating parallel quasirandom number streams. The obtained are very interesting as the scrambled versions used in individual processes are of higher quality than the original Faure sequence. We need to carry out a big, yet feasible, computations that will provide the data required for a parallel generator based on these ideas. More Numerical experiments, such as application in bioinformatics [1], need to done to further validate this parallel methods. However, this parallelization has a very interesting property. The parallel version of the quasirandom numbers are of better quality than the original sequence. This is due to the fact that the scrambling done for parallelization also increases the quality of the sequences. Thus, not only do we have a faster parallel computation, but the parallel computation is simultaneous more accurate without any extra computational effort.

References

1. P. Beerli and H. Chi. Quasi-markov chain Monte Carlo to improve inference of population genetic parameters. *Mathematics and Computers in Simulation*, page In press, 2007.
2. B.C. Bromley. Quasirandom number generators for parallel Monte Carlo algorithms. *Journal of Parallel and Distributed Computing*, **38**(1):101–104, 1996.
3. H. Chi, M. Mascagni, and T. Warnock. On the optimal Halton sequences. *Mathematics and Computers in Simulation*, **70/1**:9–21, 2005.
4. H. Faure. Discrepancy of sequences associated with a number system(in dimension s). *Acta. Arith*, **41**(4):337–351, 1982[French].
5. H. Faure. Variations on(0,s)-sequences. *Journal of Complexity*, **17**(4):741–753, 2001.

6. B. Fox. Implementation and relative efficiency of quasirandom sequence generators. *ACM Trans. on Mathematical Software*, **12**:362–376, 1986.

7. W. Jank. Efficient simulated maximum likelihood with an application to online retailing. *Statistics and Computing*, 16:111–124, 2006.

8. L. Kocis and W. Whiten. Computational investigations of low discrepancy sequences. *ACM Trans. Mathematical software*, **23**:266–294, 1997.

9. Y. Li and M. Mascagni. Analysis of large-scale grid-based Monte Carlo applications. *International Journal of High Performance Computing Applications*, **17**(4):369–382, 2003.

10. K. Liu and F. J. Hickernell. A scalable low discrepancy point generator for parallel computing. In *Lecture Notes in Computer Science 3358*, volume **3358**, pages 257–262, 2004.

11. W. L. Loh. On the asymptotic distribution of scrambled net quadrature. *Annals of Statistics*, 31:1282–1324, 2003.

12. M. Mascagni and H. Chi. Parallel linear congruential generators with Sophie-Germain moduli. *Parallel Computing*, **30**:1217–1231, 2004.

13. M. Mascagni and A. Srinivasan. Algorithm 806: SPRNG: A scalable library for pseudorandom number generation. *ACM Transactions on Mathematical Software*, **26**:436–461, 2000.

14. H. Niederreiter. *Random number generation and quasi-Monte Carlo methods.* SIAM, Philadephia, 1992.

15. A. B. Owen. Randomly permuted (t, m, s)-nets and (t, s)-sequences. *Monte Carlo and Quasi-Monte Carlo Methods in Scientific Computing*, 106 in Lecture Notes in Statistics:299–317, 1995.

16. A. B. Owen and S. Tribble. A quasi-Monte Carlo metroplis algorithm. *Proceedings of the National Academy of Sciences of the United States of America*, 102:8844–8849, 2005.

17. A. Papageorgiou and J. Traub. Beating Monte Carlo. *RISK*, **9**:63–65, 1997.

18. W. Schmid and A. Uhl. Techniques for parallel quasi-Monte Carlo integration with digital sequences and associated problems. *Math. Comput. Simulat*, **55**(1-3):249–257, 2001.

19. I. M. Soboĺ. Uniformly distributed sequences with additional uniformity properties. *USSR Comput. Math. and Math. Phy.*, **16**:236–242, 1976.

20. S. Tezuka. *Uniform Random Numbers, Theory and Practice.* Kluwer Academic Publishers, IBM Japan, 1995.

21. S. Tezuka and H. Faure. I-binomial scrambling of digital nets and sequences. *Journal of Complexity*, **19**(6):744–757, 2003.

22. S. Tezuka and T. Tokuyama. A note on polynomial arithmetic analogue of Halton sequences. *ACM Trans. on Modelling and Computer Simulation*, **4**:279–284, 1994.

23. X. Wang and K.T. Fang. The effective dimension and quasi-monte carlo. *Journal of Complexity*, 19(2):101–124, 2003.

Complexity of Monte Carlo Algorithms for a Class of Integral Equations*

Ivan Dimov[1,2] and Rayna Georgieva[2]

[1] Centre for Advanced Computing and Emerging Technologies
School of Systems Engineering, The University of Reading
Whiteknights, PO Box 225, Reading, RG6 6AY, UK
i.t.dimov@reading.ac.uk
[2] Institute for Parallel Processing, Bulgarian Academy of Sciences
Acad. G. Bonchev 25 A, 1113 Sofia, Bulgaria
ivdimov@bas.bg, rayna@parallel.bas.bg

Abstract. In this work we study the computational complexity of a class of *grid* Monte Carlo algorithms for integral equations. The idea of the algorithms consists in an approximation of the integral equation by a system of algebraic equations. Then the Markov chain iterative Monte Carlo is used to solve the system. The assumption here is that the corresponding Neumann series for the iterative matrix does not necessarily converge or converges slowly. We use a special technique to accelerate the convergence. An estimate of the computational complexity of Monte Carlo algorithm using the considered approach is obtained. The estimate of the complexity is compared with the corresponding quantity for the complexity of the *grid-free* Monte Carlo algorithm. The conditions under which the class of *grid* Monte Carlo algorithms is more efficient are given.

1 Introduction

Monte Carlo method (MCM) is established as a powerful numerical approach for investigation of various problems (evaluation of integrals, solving integral equations, boundary value problems) with the progress in modern computational systems. In this paper, a special class of integral equations obtained from boundary value problems for elliptic partial differential equations is considered. Many problems in the area of environmental modeling, radiation transport, semiconductor modeling, and remote geological sensing are described in terms of integral equations that appear as integral representation of elliptic boundary value problems. Especially, the approach presented in this paper is of great importance for studying environmental security. There are different Monte Carlo algorithms (MCAs) for solving integral equations. A class of grid Monte Carlo algorithms (GMCAs) falls into the range of the present research. The question: *Which Monte Carlo algorithm is preferable to solve a given problem?* is of great importance in computational mathematics. That is why the purpose of this paper is to study

* Partially supported by NATO grant "Monte Carlo Sensitivity Studies of Environmental Security" (PDD(TC)-(ESP.EAP.CLG 982641), BIS-21++ project funded by the European Commission (INCO-CT-2005-016639) as well as by the Ministry of Education and Science of Bulgaria, under grant I-1405/2004.

Y. Shi et al. (Eds.): ICCS 2007, Part I, LNCS 4487, pp. 731–738, 2007.

the conditions under which the class of algorithms under consideration solves a given problem more efficiently with the same accuracy than other MCAs or is the only applicable. Here we compare the efficiency of *grid* MCAs with known **grid-free M**onte **C**arlo **a**lgorithms (GFMCAs), called spherical process (see [5]). A measure of the efficiency of an algorithm is its complexity (computational cost), which is defined as the mean number of operations (arithmetic and logical) necessary for computing the value of the random variable for a transition in a Markov chain.

2 Formulation of the Problem

We consider a special class of Fredholm integral equations that normally appears as an integral representation of some boundary-value problems for differential equations. As an example which has many interesting applications we consider an elliptic boundary value problem:

$$\left|\begin{array}{ll} \mathcal{M}u = -\phi(\mathbf{x}), \ \mathbf{x} \in \Omega \subset \mathbb{R}^d, \\ u = \omega(\mathbf{x}) \qquad \mathbf{x} \in \partial\Omega, \end{array}\right. \tag{1}$$

where $\mathcal{M} = \displaystyle\sum_{i=1}^{d} \left(\dfrac{\partial^2}{\partial x^{(i)2}} + v_i(\mathbf{x}) \dfrac{\partial}{\partial x^{(i)}} \right) + w(\mathbf{x}), \quad \mathbf{x} = (x^{(1)}, x^{(2)}, \ldots, x^{(d)}).$

Definition 1. *The domain Ω belongs to the class $\mathbf{A}^{(n,\nu)}$ if it is possible to associate a hypersphere $\Gamma(\mathbf{x})$ with each point $\mathbf{x} \in \partial\Omega$, so that the boundary $\partial\Omega$ can be presented as a function $z^{(d)} = \zeta(z^{(1)}, \ldots, z^{(d-1)})$ in the neighborhood of $\mathbf{x}$ for which $\zeta^{(n)}(z^{(1)}, z^{(2)}, \ldots, z^{(d-1)}) \in \mathbf{C}^{(0,\nu)},$ i.e. $|\zeta^{(n)}(\mathbf{z_1}) - \zeta^{(n)}(\mathbf{z_2})| \leq const\,|\mathbf{z_1} - \mathbf{z_2}|^\nu,$ where the vectors $\mathbf{z_1} = (z_1^{(1)}, z_1^{(2)}, \ldots, z_1^{(d-1)})$ and $\mathbf{z_2} = (z_2^{(1)}, z_2^{(2)}, \ldots, z_2^{(d-1)})$ are $(d-1)$-dimensional vectors and $\nu \in (0,1]$.*

If in the bounded domain $\bar{\Omega} \in \mathbf{A}^{(1,\nu)}$ the coefficients of the operator $\mathcal{M}$ satisfy the conditions $v_j, w(\mathbf{x}) \in \mathbf{C}^{(0,\nu)}(\bar{\Omega}), w(\mathbf{x}) \leq 0$ and $\phi \in \mathbf{C}^{(0,\nu)}(\Omega) \cap \mathbf{C}(\bar{\Omega}), \omega \in \mathbf{C}(\partial\Omega),$ the problem (1) has an unique solution $u(\mathbf{x}) \in \mathbf{C}^2(\Omega) \cap \mathbf{C}(\bar{\Omega}).$ The conditions for uniqueness of the solution can be found in [9].

An integral representation of the solution is obtained using the Green's function for standard domains $B(\mathbf{x}), \mathbf{x} \in \Omega$ (for example - sphere, ball, ellipsoid), lying inside the domain Ω taking into account that $B(\mathbf{x})$ satisfies required conditions (see [9]). Therefore, the initial problem for solving an elliptic differential task (1) is transformed into the following Fredholm integral equation of the second kind with a spectral parameter λ ($\mathcal{K}$ is an integral operator, $\mathcal{K} : L_p \longmapsto L_p$):

$$u(\mathbf{x}) = \lambda \int_{B(\mathbf{x})} k(\mathbf{x}, \mathbf{t})\, u(\mathbf{t})\, \mathrm{d}\mathbf{t} + f(\mathbf{x}), \ \mathbf{x} \in \Omega \ (\text{or} \ u = \lambda \mathcal{K}u + f), \tag{2}$$

where $k(\mathbf{x}, \mathbf{t})$ and $f(\mathbf{x})$ are obtained using Levy's function and satisfy:

$$k(\mathbf{x}, \mathbf{t}) \in L_p^{\mathbf{x}}(\Omega) \bigcap L_q^{\mathbf{t}}(B(\mathbf{x})), \quad f(\mathbf{x}) \in L_p(\Omega), \quad p, q \in \mathbb{Z}, p, q \geq 0, \quad \frac{1}{p} + \frac{1}{q} = 1.$$

The unknown function is denoted by $u(\mathbf{x}) \in L_p(\Omega), \mathbf{x} \in \Omega, \mathbf{t} \in B(\mathbf{x}).$

We are interested in Monte Carlo method for evaluation with a *priori* given error ε of linear functionals of the solution of the integral equation (2) of the following type:

$$J(u) = \int_\Omega \varphi(\mathbf{x})\, u(\mathbf{x})\, d\mathbf{x} = (\varphi, u) \qquad \text{for} \quad \lambda = \lambda_*. \tag{3}$$

It is assumed that $\varphi(\mathbf{x}) \in L_q(\Omega)$, $q \geq 0$, $q \in \mathbb{Z}$.

3   A Class of Grid Monte Carlo Algorithms for Integral Equations

The investigated *grid* Monte Carlo approach for approximate evaluating of the linear functional (3) is based on the approximation of the given integral equation (2) by a system of **l**inear **a**lgebraic **e**quations (SLAE). This transformation represents the initial step of the considered class of *grid* MCAs. It is obtained using some approximate cubature rule (cubature method, Nystrom method, [1,7]). The next step is to apply the resolvent MCA [2,3] for solving linear systems of equations.

3.1   Cubature Method

Let the set $\{A_j\}_{j=1}^m$ be the weights and the points $\{\mathbf{x}_j\}_{j=1}^m \in \Omega$ be the nodes of the chosen cubature formula. Thus, the initial problem for evaluating of (φ, u) is transformed into the problem for evaluating of the bilinear form (h, y) of the solution y of the obtained SLAE:

$$y = \lambda L y + b, \quad L = \{l_{ij}\} \in \mathbb{R}^{m \times m}, \quad y = \{y_i\},\ b = \{b_i\},\ h = \{h_i\} \in \mathbb{R}^{m \times 1} \tag{4}$$

with the vector $h \in \mathbb{R}^{m \times 1}$. The following notation is used:

$$l_{ij} = A_j\, k(\mathbf{x}_i, \mathbf{x}_j), \quad y_i = u(\mathbf{x}_i), \quad b_i = f(\mathbf{x}_i), \quad h_i = A_i\, \varphi(\mathbf{x}_i), \quad i, j = 1, \ldots, m.$$

The error in the approximation on the first step is equal to:

$$\lambda \sum_{i=1}^m h_i\, \rho_1(\mathbf{x}_i; m, k, u) + \rho_2(m, \varphi, u),$$

where $\rho_1(\mathbf{x}_j; m, k, u)$ and $\rho_2(m, \varphi, u)$ are the approximation errors for the integral in equation (2) at the node $\mathbf{x}_i$ and linear functional (3), respectively.

Some estimations for the obtained errors ρ_1, ρ_2 from the approximation with some quadrature formula in the case when Ω is an interval $[a, b] \subset \mathbb{R}$ are given below. The errors depend on derivatives of some order of the functions $k(x, t)\, u(t)$ and $\varphi(x)\, u(x)$. Estimations for these quantities obtained after differentiation of the integral equation and using Leibnitz's rule are given in the works of Kantorovich and Krylov [7]. Analogous estimations are given below:

$$\left| \frac{\partial^j}{\partial t^j} [\, k(x_i, t)\, u(t)] \right| \leq \sum_{l=0}^j \binom{j}{l} F^{(l)}\, K_t^{(j-l)} + |\lambda|(b - a) \sum_{l=0}^j \binom{j}{l} K_x^{(l)}\, K_t^{(j-l)}\, U^{(0)},$$

$$\left| (\varphi u)^{(j)} \right| \leq \sum_{l=0}^j \binom{j}{l} F^{(l)}\, \Phi^{(j-l)} + |\lambda|(b - a) \sum_{l=0}^j \binom{j}{l} K_x^{(l)}\, \Phi^{(j-l)}\, U^{(0)},$$

where $$K_x^{(j)} = \max_{t \in B(x)} \left| \frac{\partial^j k(x,t)}{\partial x^j} \right|_{x=x_i}, \qquad K_t^{(j)} = \max_{t \in B(x)} \left| \frac{\partial^j k(x_i,t)}{\partial t^j} \right|,$$

$$F^{(j)} = \max_{x \in \Omega} |f^{(j)}(x)|, \quad U^{(j)} = \max_{x \in \Omega} |u^{(j)}(x)|, \quad \Phi^{(j)} = \max_{x \in \Omega} |\varphi^{(j)}(x)|.$$

The quantity $U^{(0)}$, which represents the maximum of the solution u in the interval $\Omega = [a,b]$, is unknown. We estimate it using the original integral equation, where the maximum of the solution in the right-hand side is estimated by the maximum of the initial approximation: $U^{(0)} \le (1 + |\lambda|(b-a)K^{(0)}) F^{(0)}$.

3.2   Resolvent Monte Carlo Method for SLAE

Iterative Monte Carlo algorithm is used for evaluating a bilinear form (h, y) of the solution of the SLAE (4), obtained after the discretization of the given integral equation (2).

Consider the discrete Markov chain $T : k_0 \longrightarrow k_1 \longrightarrow \ldots \longrightarrow k_i$ with m states $1, 2, \ldots, m$. The chain is constructed according to initial probability $\pi = \{\pi_i\}_{i=1}^m$ and transition probability $P = \{p_{ij}\}_{i,j=1}^m$. The mentioned probabilities have to be normalized and tolerant to the vector h and the matrix L respectively.

It is known (see, for example, [5,11]) that the mathematical expectation of the random variable, defined by the formula

$$\theta[h] = \frac{h_{k_0}}{\pi_{k_0}} \sum_{j=0}^{\infty} W_j \, b_{k_j}, \qquad \text{where } W_0 = 1, \quad W_j = W_{j-1} \frac{l_{k_{j-1}k_j}}{p_{k_{j-1}k_j}},$$

is equal to the unknown bilinear form, i.e. $\mathbf{E}\theta[h] = (h, y)$.

Iterative MCM is characterized by two types of errors:

– *systematic* error r_i, $i \ge 1$ (obtained from truncation of Markov chain) which depends on the number of iterations i of the used iterative process:

$$|r_i| \le \alpha^{i+1} \|b\|_2 / (1 - \alpha), \qquad \alpha = |\lambda| \|L\|_2, \quad b = \{b_j\}_{j=1}^m, \quad b_j = f(x_j)$$

– *statistical* error r_N, which depends on the number of samples N of Markov chain:
$$r_N = c_\beta \, \sigma^2(\theta[h]) N^{-1/2}, \qquad 0 < \beta < 1, \ \beta \in \mathbb{R}.$$

The constant c_β (and therefore also the complexity estimates of algorithms) depends on the confidence level β. Probable error is often used, which corresponds to a 50% confidence level.

The problem to achieve a good balance between the systematic and statistical error has a great practical importance.

4   Estimate of the Computational Complexity

In this section, computational complexity of two approaches for solving integral equations is analysed. These approaches are related to iterative Monte Carlo methods and they have similar order of computational cost. That is why, our main goal is to compare the coefficients of leading terms in the expressions for complexity of algorithms under consideration. The values of these coefficients (depending on the number of operations necessary for every move in Markov chain) allow to determine the conditions when the considered *grid* MCA has higher computational efficiency than the mentioned *grid-free* MCA.

4.1   A Grid Monte Carlo Algorithm

To estimate the performance of MCAs, one has to consider the mathematical expectation $\mathbf{E}T(\mathcal{A})$ of the time required for solving the problem using an algorithm $\mathcal{A}$ (see [4]). Let l_A and l_L be the number of suboperations of the arithmetic and logical operations, respectively. The time required to complete a suboperation is denoted by τ (for real computers this is usually the clock period).

Cubature Algorithm. The computational complexity is estimated for a given cubature rule:

$$T(\mathcal{CA}) > \tau \left[c^s (p_k + 1)\varepsilon^{-s} + c^{-s/2}(p_f + p_\varphi + p_{node})\varepsilon^{-s/2} + p_{coef} \right] l_A,$$

where the constant c depends on the following quantities

$$c = c\left(\lambda, K_x^{(r)}, K_t^{(r)}, F^{(r)}, \Phi^{(r)}\right), \quad r = 1, \ldots, ADA + 1, \quad s = s(ADA) \quad (5)$$

(ADA is the algebraic degree of accuracy of the chosen cubature formula). The number of arithmetic operations required to compute one value of the functions $k(\mathbf{x}, \mathbf{t})$, $f(\mathbf{x})$ and $\varphi(\mathbf{x})$ and one node (coefficient) is denoted by p_k, p_f and p_φ, respectively and by $p_{node}(p_{coef})$.

The degree s and the constants p_{node} and p_{coef} depend on the applied formula. For instance:

$$s = \begin{cases} 1 & \text{for rectangular and Trapezoidal rule;} \\ 1/2 & \text{for Simpson's rule.} \end{cases} \quad (6)$$

Resolvent Monte Carlo Algorithm. Firstly, the case when the corresponding Neumann series converges (the supposition for slow convergence is allowed) is considered. The following number of operations is necessary for one random walk:

 – generation of one random number : k_A arithmetic and k_L logical operations;
 – modeling the initial probability π to determine the initial **or** next point in the Markov chain: μ_A arithmetic and μ_L logical operations ($\mathbf{E}\,\mu_A + 1 = \mathbf{E}\,\mu_L = \mu$, $1 \le \mu \le m - 1$);
 – computing one value of the random variable: 4 arithmetic operations.

To calculate in advance the initial π and transition P probabilities (a vector and a square matrix, respectively), it is necessary a number of arithmetic operations, proportional to the matrix dimension m: $2m(1 + m)$.

To ensure a statistical error ε, it is necessary to perform i transitions in the Markov process, where i is chosen from the inequality

$$i > \ln^{-1} \alpha \left(\ln \varepsilon + \ln (1 - \alpha) - \ln \|b\|_2\right) - 1 \quad (\text{assuming } \|b\|_2 > \varepsilon (1 - \alpha)),$$

where $\alpha = |\lambda| \|L\|_2$ and the initial approximation is chosen to be the right-hand side b.

To achieve a probable error ε, it is necessary to do N samples depending on the inequality $N > c_{0.5}\, \sigma^2(\theta)\, \varepsilon^{-2}$, $c_{0.5} \approx 0.6745$, where θ is the random variable, whose mathematical expectation coincides with the desired linear functional (3).

Therefore, the following estimate holds for the mathematical expectation of the time required to obtain an approximation with accuracy ε using the considered *grid* MCA:

$$\mathbf{E}\,T(\mathcal{RMCA}) > \tau\,[(k_A + \mu + 3)\,l_A + (k_L + \mu)\,l_L]\,\frac{[c_\beta\,\sigma(\xi_j^{\mathcal{R}}[h])]^2}{\ln^3 \alpha}\,\frac{(\ln^3 \varepsilon + a)}{\varepsilon^2}$$
$$+\ 2\tau m(m+1)l_A,$$

where $a = \ln(1 - \alpha) - \ln \|b\|_2$, $m > \sqrt{c^s}\varepsilon^{-s/2}$ (the constants are given by (5) and (6)), and $\xi_j^{\mathcal{R}}$ is the unbiased estimate of the j-th iteration of the matrix L, obtained using the resolvent MCA.

Consider the case when the corresponding Neumann series does not converge. The convergence of the Monte Carlo algorithm for solving the SLAE (4) can be ensured (or accelerated) by application of an analytical continuation of the Neumann series by substituting of the spectral parameter λ (mapping) (see [2,6,7,8,10]). The main advantage of this approach for acceleration of convergence of an iterative process is its inessential influence over the computational complexity of the algorithm. The computational complexity on every walk is increased only with one arithmetic operation required for multiplication by the coefficients $g_j, j \geq 0$, that ensures convergence (on the supposition that these coefficients are calculated with a high precision in advance). To obtain the computational complexity of the modified algorithm, it is necessary to estimate the variation of the new random variable: $\theta[h] = \dfrac{h_{k_0}}{\pi_{k_0}} \displaystyle\sum_{j=0}^{\infty} \mathbf{g_j}\,W_j\,b_{k_j}$.

We will use the following statement for a class of mappings proved in [10]: *The conformal mapping $\lambda = \psi(\eta) = a_1\eta + a_2\eta + \ldots$ has only simple poles on its boundary of convergence $|\eta| = 1$. If $Var\,\xi_k^{\mathcal{R}} \leq \sigma^2$ and $q = \bar{a}\,|\eta_*|/(1 - |\eta_*|) < 1$, then the complexity estimate of the algorithm has an order $O(|\ln \varepsilon|^4/\varepsilon^2)$, where $\bar{a}$ is such a constant that $|a_i| \leq \bar{a}, i = 1, 2, \ldots$, λ_* is the value of the spectral parameter in the integral equation (2) (respectively SLAE (4)) and $\eta_* = \psi^{-1}(\lambda_*)$.*

In general, a computational estimate of this class of *grid* MCAs can be obtained if the behavior of g_j and $Var\,\xi_j^{\mathcal{R}}$ is known.

4.2   A Grid-Free Monte Carlo Algorithm

The computational complexity of the *grid* MCA under consideration is compared with the computational complexity of a *grid-free* Monte Carlo approach. This approach is based on the use of a local integral representation (assuming that such a representation exists, [9,12]) of the solution of an elliptic boundary value problem. Existence of this representation allows to construct a Monte Carlo algorithm, called spherical process (in the simplest case) for computing of the corresponding linear functional. As a first step of this algorithm an ϵ-strip $\partial\Omega_\epsilon$ of the boundary $\partial\Omega$ is chosen (on the supposition that the solution is known on the boundary) to ensure the convergence of the constructed iterative process. The following number of operation is necessary for one random walk:

- generation of n (this number depends on initial probability π) random numbers to determine the initial point in the Markov chain: $n(k_A + k_L)$ operations (k_A and k_L are the arithmetic and logical operations necessary for the generation of one random

number) **or** modeling of an isotropic vector that needs of the order of $R*n(k_A+k_L)$ operations (the constant R depends on the efficiency of the modeling method and transition probability);
- calculating the coordinates of the initial **or** next point: p_{next} (depends on the modeling method and the dimension d of the domain $B(x)$);
- calculating one value of functions: p_f; p_π, p_φ **or** p_k, p_P;
- calculating one sample of the random variable: 4 arithmetic operations;
- calculating the distance from the current point to the boundary $\partial\Omega$: γ_A arithmetic and γ_L logical operations (depends on the dimension d of the domain Ω);
- verification if the current point belongs to the chosen δ-strip $\partial\Omega_\delta$.

The following logarithmic estimate for the average number $\mathbf{E}i$ of spheres on a single trajectory holds for a wide class of boundaries [5]:

$$\mathbf{E}\,i \le const\,|\ln\delta|, \qquad const > 0, \tag{7}$$

where $const$ depends on the boundary $\partial\Omega$.

Calculating the linear functional with a preliminary given accuracy ε and attainment of a good balance between the *statistical* and the *systematic* error is a problem of interest to us.

Let us to restrict our investigation of the statistical error to the domain $\Omega \equiv [a, b]$. To ensure a statistical error ε, it is necessary to do i transitions in the Markov process, where i is chosen from the inequality:

$$i > \ln^{-1}\alpha\,(\ln\varepsilon + \ln(1-\alpha) - \ln F^{(0)}) - 1 \quad (\text{assuming } F^{(0)} > \varepsilon\,(1-\alpha)),$$

where $\alpha = |\lambda|\,V_{B(x)}\,K$, $\quad K = \max\limits_{x,t}|k(x,t)|$ and the initial approximation is chosen to be the right-hand side $f(x)$. On the other hand, the estimate (7) depending on the chosen ϵ-strip of the boundary is done. Then, an expression for δ according to the number of transition i is obtained from these two estimates: $\delta \approx e^{-i/const}$. Therefore, the following estimate holds for the mathematical expectation of the time required to obtain an approximation with accuracy ε using the considered *grid-free* MCA:

$$\mathbf{E}\,T(\mathcal{GFMCA}) > \tau\,[(n\,k_A + p_{next} + p_f + p_\pi + p_\varphi + \gamma_A + 4)\,l_A$$
$$+(n\,k_L + \gamma_L + 1)\,l_L + ((R\,n\,k_A + p_{next} + p_f + p_k + p_P + 4 + \gamma_A)\,l_A$$
$$+(R\,n\,k_L + \gamma_L + 1)\,l_L) \times \frac{(\ln\varepsilon + \ln(1-\alpha) - \ln^3 F^{(0)}}{\ln^3\alpha}]\,\frac{[c_\beta\,\sigma(\xi_j^S[h])]^2}{\varepsilon^2}.$$

Obtained expressions for coefficients in computational complexity for MCAs under consideration allow us to define some conditions when the GMCA is preferable to the GFMCA:

- the functions that define the integral equation (2) $(k(x,t), f(x), \varphi(x))$ have comparatively small maximum norm in the corresponding domain and their values can be calculated with a low complexity

- the initial and transition probability are complicated for modeling (acceptance-rejection method)
- large dimension of the integration domain

It has to be noted the fact that the *grid* MCAs under consideration are admissible only for integral equations with smooth functions, but some techniques of avoiding singularities of this kind exist (see [1]).

5   Concluding Discussion

In this paper we deal with performance analysis of a class of Markov chain *grid* Monte Carlo algorithms for solving Fredholm integral equations of second kind. We compare this class of algorithms with a class of *grid-free* algorithms. *Grid-free* Monte Carlo uses the so-called spherical process for computing of the corresponding linear functional. Obviously, the *grid* approach assumes higher regularity of the input data since it includes an approximation procedure described in Section 4.1. The (*grid-free*) approach does not need additional approximation procedure and directly produces a bias approximation to the solution. But the *grid-free* algorithm is more complicated and its implementation needs more routine operations (like checking the distance from a given point to the boundary) that decrease the efficiency of the algorithm. Analyzing the regularity of the problem one may chose either *grid* or *grid-free* algorithm is preferable. Especially, if the input data has higher regularity ($k(x,t)$, $f(x)$, $\varphi(x)$ have comparatively small maximum norm) than the *grid* algorithm should be preferred.

References

1. Bahvalov, N.S., Zhidkov, N.P., Kobelkov, G.M.: Numerical Methods. Nauka, Moscow (1987)
2. Dimov, I.T., Alexandrov, V.N., Karaivanova, A.N.: Parallel Resolvent Monte Carlo Algorithms for Linear Algebra Problems. Mathematics and Computers in Simulation **55** (2001) 25–35
3. Dimov, I.T., Karaivanova, A.N.: Iterative Monte Carlo Algorithms for Linear Algebra Problems. In: Vulkov, L., Wasniewski, J., Yalamov, P. (eds.): Lecture Notes in Computer Science, Vol. 1196. Springer-Verlag, Berlin (1996) 150–160
4. Dimov, I.T., Tonev, O.I.: Monte Carlo Algorithms: Performance Analysis for Some Computer Architectures. J. Comput. Appl. Math. **48** (1993) 253–277
5. Ermakov, S.M., Mikhailov, G.A.: Statistical Modeling. Nauka, Moscow (1982)
6. Kantorovich, L.V., Akilov, G.P.: Functional Analysis in Normed Spaces. Pergamon Press, Oxford (1964)
7. Kantorovich, L.V., Krylov, V.I.: Approximate Methods of Higher Analysis. Physical and Mathematical State Publishing House, Leningrad (1962)
8. Kublanovskaya, V.N.: Application of Analytical Continuation by Substitution of Variables in Numerical Analysis. In: Proceedings of the Steklov Institute of Mathematics (1959)
9. Miranda, C.: Partial Differential Equations of Elliptic Type. Springer-Verlag, Berlin Heidelberg New York (1970)
10. Sabelfeld, K.K.: Monte Carlo Methods in Boundary Value Problems. Springer-Verlag, Berlin Heidelberg New York London (1991)
11. Sobol̀, I.M.: The Monte Carlo Method. The University of Chicago Press, Chicago (1974)
12. Vladimirov, S.V.: Equations of Mathematical Physics. Nauka, Moscow (1976)

Modeling of Carrier Transport in Nanowires[*]

T. Gurov, E. Atanassov, M. Nedjalkov, and I. Dimov

IPP, Bulgarian Academy of Sciences, Sofia, Bulgaria
{gurov,emanouil}@parallel.bas.bg, ivdimov@bas.bg
Institute for Microelectronics, TU Vienna, Austria
mixi@iue.tuwien.ac.at
Centre for Advanced Computing and Emerging Technologies
School of Systems Engineering, The University of Reading
Whiteknights, PO Box 225, Reading, RG6 6AY, UK
i.t.dimov@reading.ac.uk

Abstract. We consider a physical model of ultrafast evolution of an initial electron distribution in a quantum wire. The electron evolution is described by a
quantum-kinetic equation accounting for the interaction with phonons. A Monte
Carlo approach has been developed for solving the equation. The corresponding
Monte Carlo algorithm is NP-hard problem concerning the evolution time. To
obtain solutions for long evolution times with small stochastic error we combine
both variance reduction techniques and distributed computations. Grid technologies are implemented due to the large computational efforts imposed by the quantum character of the model.

1 Introduction

The Monte Carlo (MC) methods provide approximate solutions by performing statistical sampling experiments. These methods are based on simulation of random variables
whose mathematical expectations are equal to a functional of the problem solution.

Many problems in the transport theory and related areas can be described mathematically by a second kind Fredholm integral equation.

$$f = I\!K(f) + \phi. \tag{1}$$

In general the physical quantities of interest are determined by functionals of the type:

$$J(f) \equiv (g, f) = \int_G g(x) f(x) dx, \tag{2}$$

where the domain $G \subset I\!R^d$ and a point $x \in G$ is a point in the Euclidean space $I\!R^d$.
The functions $f(x)$ and $g(x)$ belong to a Banach space X and to the adjoint space X^*,
respectively, and $f(x)$ is the solution of (1).

The mathematical concept of the MC approach is based on the iterative expansion of
the solution of (1): $f_s = I\!K(f_{s-1}) + \phi$, $s = 1, 2, \ldots$, where s is the number of iterations.

[*] Supported by the Bulgarian Ministry of Education and Science, under grant I-1405/2004.

Thus we define a Neumann series $f_s = \phi + I\!K(\phi) + \ldots + I\!K^{s-1}(\phi) + I\!K^s(f_0)$, $s > 1$, where $I\!K^s$ means the s-th iteration of $I\!K$. In the case when the corresponding infinite series converges then the sum is an element f from the space X which satisfies the equation (1). The Neumann series, replaced in (2), gives rise to a sum of consecutive terms which are evaluated by the MC method with the help of random estimators. A random variable ξ is said to be a MC estimator for the functional (2) if the mathematical expectation of ξ is equal to $J(f)$: $E\xi = J(f)$. Therefore we can define a MC method $\bar{\xi} = \frac{1}{N}\sum_{i=1}^{N}\xi^{(i)} \xrightarrow{P} J(f)$, where $\xi^{(1)}, \ldots, \xi^{(N)}$ are independent values of ξ and $\xrightarrow{P}$ means stochastic convergence as $N \longrightarrow \infty$. The rate of convergence is evaluated by the *law of the three sigmas*: convergence (see [1]): $P\left(|\bar{\xi} - J(f)| < 3\frac{\sqrt{Var\xi}}{\sqrt{N}}\right) \approx$ 0.997, where $Var(\xi) = E\xi^2 - E^2\xi$ is the variance. It is seen that, when using a random estimator, the result is obtained with a statistical error [1,2]. As N increases, the statistical error decreases as $N^{-1/2}$.

Thus, there are two types of errors - systematic (a truncation error) and stochastic (a probability error) [2]. The systematic error depends on the number of iterations of the used iterative method, while the stochastic error is related to the the probabilistic nature of the MC method. The MC method still does not determine the computation algorithm: we must specify the modeling function (sampling rule) $\Theta = F(\beta_1, \beta_2, \ldots,)$, where $\beta_1, \beta_2, \ldots$, are uniformly distributed random numbers in the interval $(0, 1)$. Now both relations the MC method and the sampling rule define a MC algorithm for estimating $J(f)$. The case when $g = \delta(x - x_0)$ is of special interest, because we are interested in calculating the value of f at x_0, where $x_0 \in G$ is a fixed point.

Every iterative algorithm uses a finite number of iterations s. In practice we define a MC estimator ξ_s for computing the functional $J(f_s)$ with a statistical error. On the other hand ξ_s is a biased estimator for the functional $J(f)$ with statistical and truncation errors. The number of iterations can be a random variable when an ε-criterion is used to truncate the Neumann series or the corresponding Markov chain in the MC algorithm. The stochastic convergence rate is approximated by $O(N^{-1/2})$. In order to accelerate the convergence rate of the MC methods several techniques have been developed. Variance reductions techniques, like antithetic varieties, stratification and importance sampling [1], reduce the variance which is a quantity to measure the probabilistic uncertainly. Parallelism is an another way to accelerate the convergence of a MC computation. If n processors execute n independence of the MC computation using nonoverlaping random sequences, the accumulated result has a variance n times smaller than that of a single copy.

2   Physical Model

The early time dynamics of highly non-equilibrium semiconductor electrons is rich of quantum phenomena which are basic components of the physics of the modern micro- and nano-electronics [4]. Confined systems are characterized by small spatial scales where another classical assumption - this for electron-phonon scattering occuring at a well defined position - loses its validity. These scales become relevant for the process of evolution of an initial electron distribution, locally excited or injected into a semiconductor nanowire. Beyond-Boltzmann transport models for wire electrons have been recently derived [3] in terms of the Wigner function. They appear as inhomogeneous conterparts of Levinson's and the Barker-Ferry's equations. The later is a physically more refined model, which accounts for the finite lifetime of the electrons due to the interaction with the phonons.

The corresponding physical process can be summarized as follows. An initial non-equilibrium electron distribution is created locally in a nanowire by e.g. an optical excitation. The electrons begin to propagate along the wire, where they are characterized by two variables: the position z and the component of the wave vector is k_z. We note that these are Wigner coordinates so that a classical interpretation as "position and momentum of the electron" is not relevant. A general, time-dependent electric field $E(t)$ can be applied along the nanowire. The field changes the wave vector with the time: $k_z(t') = k_z - \int_{t'}^t eE(\tau)/\hbar d\tau$ where k_z is the wave vector value at the initialization time t. Another source of modifications of the wave vector are dissipative processes of electron-phonon interaction, responsible for the energy relaxation of the electrons.

The following considerations simplify the problem thus allowing to focus on the numerical aspects: *(i)* We regard the Levinson type of equation which bears the basic numerical properties of the physical model:

$$\left(\frac{\partial}{\partial t} + \frac{\hbar k_z}{m}\nabla_z\right) f_w(z, k_z, t) = \int d\mathbf{k}' \int_0^t dt' \qquad (3)$$

$$\left\{ S(k_z, k_z', \mathbf{q}_\perp', t, t') f_w(z(t'), k_z'(t'), t') - S(k_z', k_z, \mathbf{q}_\perp', t, t') f_w(z(t'), k_z(t'), t') \right\},$$

where $\int_G d^3 k' = \int d\mathbf{q}_\perp' \int_{-Q_2}^{Q_2} dk_z'$ and the domain G is specified in the next section. The spatial part $z(t')$ of the trajectory is initialized by the value z at time t: $z(t') = z - \frac{\hbar}{m} \int_{t'}^t \left(k_z(\tau) - \frac{q_z'}{2} \right) d\tau; q_z' = k_z - k_z'$. The scattering function S is

$$S(k_z, k_z', \mathbf{q}_\perp', t, t') = \frac{2V}{(2\pi)^3}|\Gamma(\mathbf{q}_\perp')\mathcal{F}(\mathbf{q}_\perp', k_z - k_z')|^2$$

$$\times \left[\left(n(\mathbf{q}') + \frac{1}{2} \pm \frac{1}{2} \right) \cos \left(\int_{t'}^t \frac{(\epsilon(k_z(\tau)) - \epsilon(k_z'(\tau))) \pm \hbar\omega_{\mathbf{q}'}) d\tau}{\hbar} \right) \right].$$

The electron-phonon coupling $\mathcal{F}$ is chosen for the case of Fröhlich polar optical interaction:

$$\mathcal{F}(\mathbf{q}_\perp', k_z - k_z')$$

$$= - \left[\frac{2\pi e^2 \omega_{\mathbf{q}'}}{\hbar V} \left(\frac{1}{\varepsilon_\infty} - \frac{1}{\varepsilon_s} \right) \frac{1}{(\mathbf{q}')^2} \right]^{\frac{1}{2}}; \Gamma(\mathbf{q}_\perp') = \int d\mathbf{r}_\perp |\Psi(\mathbf{r}_\perp)|^2 e^{i\mathbf{q}_\perp' \cdot \mathbf{r}_\perp},$$

where (ε_∞) and (ε_s) are the optical and static dielectric constants. $\Gamma(\mathbf{q}_\perp')$ is the Fourier transform of the square of the ground state wave function. Thus we come to the second simplifying assumption: *(ii)* Very low temperatures are assumed, so that the electrons reside in the ground state $\Psi(\mathbf{r}_\perp)$ in the plane normal to the wire. The phonon distribution is described by the Bose function, $n(\mathbf{q}') = (e^{\frac{\hbar\omega_{\mathbf{q}'}}{KT}} - 1)^{-1}$ with K the Boltzmann constant and T is the temperature of the crystal. $\hbar\omega_{\mathbf{q}'}$ is the phonon energy which generally depends on $\mathbf{q}' = (\mathbf{q}'_\perp, q'_z)$, and $\varepsilon(k_z) = (\hbar^2 k_z^2)/2m$ is the electron energy. *(iii)* Finally we consider the case of no electric field applied along the wire. Next we need to transform the transport equation into the corresponding integral form (1):

$$f_w(z, k_z, t) = f_{w0}(z - \frac{\hbar k_z}{m}t, k_z) \tag{4}$$

$$+ \int_0^t dt'' \int_{t''}^t dt' \int_G d^3\mathbf{k}' \{K_1(k_z, k'_z, \mathbf{q}'_\perp, t', t'') f_w (z + z_0(k_z, q'_z, t, t', t''), k'_z, t'')\}$$

$$+ \int_0^t dt'' \int_{t''}^t dt' \int_G d^3\mathbf{k}' \{K_2(k_z, k'_z, \mathbf{q}'_\perp, t', t'') f_w (z + z_0(k_z, q'_z, t, t', t''), k_z, t'')\},$$

where $z_0(k_z, q'_z, t, t', t'') = -\frac{\hbar k_z}{m}(t - t'') + \frac{\hbar q'_z}{2m}(t' - t'')$,

$$K_1(k_z, k'_z, \mathbf{q}'_\perp, t', t'') = S(k'_z, k_z, \mathbf{q}'_\perp, t', t'') = -K_2(k'_z.k_z, \mathbf{q}'_\perp, t', t'').$$

We note that the evolution problem becomes inhomogeneous due to the spatial dependence of the initial condition f_{w0}. The shape of the wire modifies the electron-phonon coupling via the ground state in the normal plane. If the wire cross section is chosen to be a square with side a, the corresponding factor Γ becomes: $|\Gamma(q'_x)\Gamma(q'_y)|^2 =$

$$|\Gamma(\mathbf{q}'_\perp)|^2 = \left(\frac{4\pi^2}{q'_x a((q'_x a)^2 - 4\pi^2)}\right)^2 4\sin^2(aq'_x/2) \left(\frac{4\pi^2}{q'_y a((q'_y a)^2 - 4\pi^2)}\right)^2 4\sin^2(aq'_y/2).$$

We note that the Neumann series of such type integral equations as (4) converges [5] and it can be evaluated by a MC estimator [2].

3   The Monte Carlo Method

The values of the physical quantities are expressed by the following general functional of the solution of (4):

$$J_g(f) \equiv (g, f) = \int_0^T \int_D g(z, k_z, t) f_w(z, k_z, t) dz dk_z dt, \tag{5}$$

by a MC method. Here we specify that the phase space point (z, k_z) belongs to a rectangular domain $D = (-Q_1, Q_1) \times (-Q_2, Q_2)$, and $t \in (0, T)$. The function $g(z, k_z, t)$ depends on the quantity of interest. In particular the Wigner function, the wave vector and density distributions, and the energy density are given by:

$$(i)\quad g_W(z, k_z, t) = \delta(z - z_0)\delta(k_z - k_{z,0})\delta(t - t_0),$$
$$(ii)\quad g_n(z, k_z, t) = \delta(k_z - k_{z,0})\delta(t - t_0),$$
$$(iii)\quad g_k(z, k_z, t) = \delta(z - z_0)\delta(t - t_0),$$
$$(iv)\quad g(z, k_z, t) = \epsilon(k_z)\delta(z - z_0)\delta(t - t_0)/g_n(z, k_z, t),$$

We construct a biased MC estimator for evaluating the functional (5) using backward time evolution of the numerical trajectories in the following way:

$$\xi_s[J_g(f)] = \frac{g(z, k_z, t)}{p_{in}(z, k_z, t)} W_0 f_{w,0}(., k_z, 0) + \frac{g(z, k_z, t)}{p_{in}(z, k_z, t)} \sum_{j=1}^{s} W_j^{\alpha} f_{w,0}\left(., k_{z,j}^{\alpha}, t_j\right),$$

where

$$
f_{w,0}\left(., k_{z,j}^{\alpha}, t_j\right)
$$
$$
= \begin{cases} f_{w,0}\left(z + z_0(k_{z,j-1}, k_{z,j-1} - k_{z,j}, t_{j-1}, t'_j, t_j), k_{z,j}, t_j\right), & \text{if } \alpha = 1, \\ f_{w,0}\left(z + z_0(k_{z,j-1}, k_{z,j} - k_{z,j-1}, t_{j-1}, t'_j, t_j), k_{z,j-1}, t_j\right), & \text{if } \alpha = 2, \end{cases}
$$

$$W_j^{\alpha} = W_{j-1}^{\alpha} \frac{K_{\alpha}(k_{zj-1}, \mathbf{k}_j, t'_j, t_j)}{p_{\alpha} p_{tr}(\mathbf{k}_{j-1}, \mathbf{k}_j, t'_j, t_j)}, \quad W_0^{\alpha} = W_0 = 1, \quad \alpha = 1, 2, \quad j = 1, \dots, s.$$

The probabilities $p_{\alpha}, (\alpha = 1, 2)$ are chosen to be proportional to the absolute value of the kernels in (4). $p_{in}(z, k_z, t)$ and $p_{tr}(\mathbf{k}, \mathbf{k}', t', t'')$, which are an initial density and a transition density, are chosen to be tolerant[1] to the function $g(z, k_z, t)$ and the kernels, respectively. The first point (z, k_{z0}, t_0) in the Markov chain is chosen using the initial density, where k_{z0} is the third coordinate of the wave vector $\mathbf{k}_0$. Next points $(k_{zj}, t'_j, t_j) \in (-Q_2, Q_2) \times (t_j, t_{j-1}) \times (0, t_{j-1})$ of the Markov chain:

$$(k_{z0}, t_0) \to (k_{z1}, t'_1, t_1) \to \dots \to (k_{zj}, t'_j, t_j) \to \dots \to (k_{zs}, t'_s, t_s), \quad j = 1, 2, \dots, s$$

do not depend on the position z of the electrons. They are sampled using the transition density $p_{tr}(\mathbf{k}, \mathbf{k}', t', t'')$. The z coordinate of the generated wave vectors is taken for the Markov chain, while the normal coordinates give values for $\mathbf{q}_{\perp}$. As the integral on $\mathbf{q}_{\perp}$ can be assigned to the kernel these values do not form a Markov chain but are independent on the consecutive steps. The time t'_j conditionally depends on the selected time t_j. The Markov chain terminates in time $t_s < \varepsilon_1$, where ε_1 is related to the truncation error introduced in first section. Now the functional (5) can be evaluated by N independent samples of the obtained estimator, namely, $\frac{1}{N} \sum_{i=1}^{N} (\xi_s[J_g(f)])_i \xrightarrow{P} J_g(f_s) \approx J_g(f)$, where $\xrightarrow{P}$ means stochastic convergence as $N \to \infty$; f_s is the iterative solution obtained by the Neumann series of (4), and s is the number of iterations. To define a MC algorithm we have to specify the initial and transition densities, as well the modeling function (or sampling rule). The modeling function describes the rule needed to calculate the states of the Markov chain by using uniformly distributed random numbers in the interval $(0, 1)$. The transition density is chosen: $p_{tr}(\mathbf{k}, \mathbf{k}', t', t'') = p(\mathbf{k}'/\mathbf{k}) p(t, t', t'')$, where $p(t, t', t'') = p(t, t'') p(t'/t'') = \frac{1}{t} \frac{1}{(t - t'')} \, p(\mathbf{k}'/\mathbf{k}) = c_1/(\mathbf{k}' - \mathbf{k})^2$], and c_1 is the normalized constant. Thus, by simulating the Markov chain under consideration, the desired physical quantities (values of the Wigner function, the wave vector relaxation, the electron and energy density) can be evaluated simultaneously.

4   Grid Implementation and Numerical Results

The computational complexity of an MC algorithm can be measured by the quantity $CC = N \times \tau \times M(s_{\varepsilon_1})$. The number of the random walks, N, and the average number

---

[1] $r(x)$ is tolerant of $g(x)$ if $r(x) > 0$ when $g(x) \neq 0$ and $r(x) \geq 0$ when $g(x) = 0$.

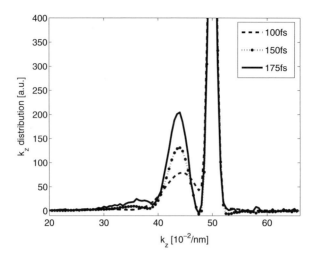

Fig. 1. Wave vector relaxation of the highly non-equilibrium initial condition. The quantum solution shows broadening of the replicas. Electrons appear in the classically forbidden region above the initial condition.

of transitions in the Markov chain, $M(s_{\varepsilon_1})$, are related to the stochastic and systematic errors [5]. The mean time for modeling one transition, τ, depends on the complexity of the transition density functions and on the sampling rule, as well as on the choice of the random number generator (rng). It is proved [5,6] that the stochastic error has order $O(\exp(c_2 t)/N^{1/2})$, where t is the evolution time and c_2 is a constant depending on the kernels of the quantum kinetic equation under consideration. This estimate shows that when t is fixed and $N \to \infty$ the error decreases, but for large t the factor $\exp(c_2 t)$ looks ominous. Therefore, the algorithm solves an NP-hard problem concerning the evolution time. To solve this problem for long evolution times with small stochastic error we have to combine both MC variance reduction techniques and distributed or parallel computations. By using the Grid environment provided by the EGEE-II project middleware[2] [8] we were able to reduce the computing time of the MC algorithm under consideration. The simulations are parallelized on the existing Grid infrastructure by splitting the underlying random number sequences. The numerical results discussed in Fig. 1 are obtained for zero temperature and GaAs material parameters: the electron effective mass is 0.063, the optimal phonon energy is $36meV$, the static and optical dielectric constants are $\varepsilon_s = 10.92$ and $\varepsilon_\infty = 12.9$. The initial condition is a product of two Gaussian distributions of the energy and space. The k_z distribution corresponds to a generating laser pulse with an excess energy of about $150meV$. This distribution was estimated for 130 points in the interval $(0, 66)$, where $Q_2 = 66 \times 10^7 m^{-1}$. The z distribution is centered around zero (see Figures 2-3) and it is estimated for 400 points in the interval $(-Q_1, Q_1)$, where $Q_1 = 400 \times 10^9 m^{-1}$. The side a of the wire is chosen to be 10 nanometers. The SPRNG library has been used to produce independent and

[2] The Enabling Grids for E-sciencE-II (EGEE-II) project is funded by the EC under grand INFSO-RI-031688. For more information see http://www.eu-egee.org/.

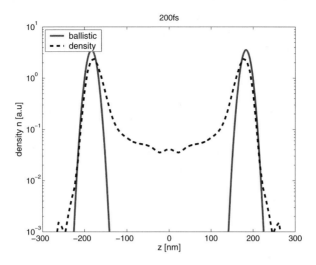

Fig. 2. Electron density along the wire after 200 fs. The ballistic curve outlines the largest distance which can be reached by classical electrons. The quantum solution reaches larger distances due to the electrons scattered in the classically forbidden energy region.

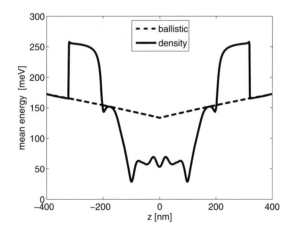

Fig. 3. Energy density after 175fs evolution time. A comparison with the ballistic curve shows that the mean kinetic energy per particle is lower in the central region. On contrary hot electrons reside in the regions apart from the center. These electrons are faster than the ballistic ones and thus cover larger distances during the evolution.

non-overlapping random sequences [9]. Successful tests of the algorithm were performed at the Bulgarian SEE-GRID[3] sites. The MPI implementation was MPICH 1.2.6, and the execution is controlled from the Computing Element via the Torque batch

---

[3] South Eastern European GRid-enabled eInfrastructure Development-2 (SEE-GRID-2) project is funded by the EC under grand FP6-RI-031775. For more information see http://www.see-grid.eu/.

Table 1. The CPU time (seconds) for all points (in which the physical quantities are estimated), the speed-up, and the parallel efficiency. The number of random walks is $N = 100000$. The evolution time is $100\,fs$.

Number of CPUs	CPU Time (s)	Speed-up	Parallel Efficiency
2	9790	-	-
4	4896	1.9996	0.9998
6	3265	2.9985	0.9995

system. The timing results for evolution time t=100 femtoseconds are shown in Table 1. The parallel efficiency is close to 100%.

5   Conclusion

A quantum-kinetic model for the evolution of an initial electron distribution in a quantum wire has been introduced in terms of the electron Wigner function. The physical quantities, expressed as functionals of the Wigner function are evaluated within a stochastic approach. The developed MC method is characterized by the typical for quantum algorithms computational demands. The stochastic variance grows exponentially with the evolution time. The importance sampling technique is used to reduce the variance of the MC method. The suggested MC algorithm evaluates simultaneously the desired physical quantities using simulation of Markov chain under consideration. Grid technologies are implemented to reduce computational efforts.

References

1. M.A. Kalos, P.A. Whitlock, *Monte Carlo methods*, Wiley Interscience, New York (1986).
2. G.A. Mikhailov, New MC Methods with Estimating Derivatives, Utrecht, The Netherlands (1995).
3. M. Nedjalkov et al., "Wigner transport models of the electron-phonon kinetics in quantum wires", *Physical Review B*, vol. 74, pp. 035311-1–035311-18, (2006).
4. T.C. Schmidt and K. Moehring, "Stochastic Path-Integral Simulation of Quantum Scattering", *Physical Review A*, vol. 48, no. 5, pp. R3418–R3420, (1993).
5. T.V. Gurov, P.A. Whitlock, "An Efficient Backward Monte Carlo Estimator for Solving of a Quantum Kinetic Equation with Memory Kernel", *Mathematics and Computers in Simulation* (60), pp. 85-105, (2002).
6. T.V. Gurov et al., "Femtosecond Relaxation of Hot Electrons by Phonon Emission in Presence of Electric Field", *Physica B* (314), pp. 301-304, (2002).
7. T.V. Gurov, I.T. Dimov, "A Parallel Monte Carlo Method For Electron Quantum Kinetic Equation", *Lect. Notes in Comp. Sci.* (2907), Springer-Verlang, pp.151-161, (2004).
8. EGEE Grid Middleware, *http://lcg.web.cern.ch/LCG/Sites/releases.html*.
9. Scalable Parallel Random Number Generators Library for Parallel Monte Carlo Computations, *SPRNG 1.0 and SPRNG 2.0 – http://sprng.cs.fsu.edu* .

Monte Carlo Numerical Treatment of Large Linear Algebra Problems*

Ivan Dimov, Vassil Alexandrov, Rumyana Papancheva,
and Christian Weihrauch

Centre for Advanced Computing and Emerging Technologies
School of Systems Engineering, The University of Reading
Whiteknights, PO Box 225, Reading, RG6 6AY, UK
{i.t.dimov, v.n.alexandrov, c.weihrauch}@reading.ac.uk
Institute for Parallel Processing, Bulgarian Academy of Sciences
Acad. G. Bonchev 25 A, 1113 Sofia, Bulgaria
ivdimov@bas.bg, rumi@parallel.bas.bg

Abstract. In this paper we deal with performance analysis of Monte Carlo algorithm for large linear algebra problems. We consider applicability and efficiency of the Markov chain Monte Carlo for large problems, i.e., problems involving matrices with a number of non-zero elements ranging between one million and one billion. We are concentrating on analysis of the almost Optimal Monte Carlo (MAO) algorithm for evaluating bilinear forms of matrix powers since they form the so-called Krylov subspaces.

Results are presented comparing the performance of the Robust and Non-robust Monte Carlo algorithms. The algorithms are tested on large dense matrices as well as on large unstructured sparse matrices.

Keywords: Monte Carlo algorithms, large-scale problems, matrix computations, performance analysis, iterative process.

1 Introduction

Under *large* we consider problems involving dense or general sparse matrices with a number of non-zero elements ranging between one million and one billion.

It is known that Monte Carlo methods give statistical estimates for bilinear forms of the solution of systems of linear algebraic equations (SLAE) by performing random sampling of a certain random variable, whose mathematical expectation is the desired solution [8]. The problem of variance estimation, in the optimal case, has been considered for extremal eigenvalues [7,9]. In [5,6] we analyse the errors of iterative Monte Carlo for computing bilinear forms of matrix powers. If one is interested to apply Markov chain Monte Carlo for large

---

* Partially supported by the Bulgarian Ministry of Education and Science, under grant I-1405/2004. The authors would like to acknowledge the support of the European Commission's Research Infrastructures activity of the Structuring the European Research Area programme, contract number RII3-CT-2003-506079 (HPC-Europa).

Y. Shi et al. (Eds.): ICCS 2007, Part I, LNCS 4487, pp. 747–754, 2007.

problems, then the applicability and robustness of the algorithm should be studied. The run of large-scale linear algebra problems on parallel computational systems introduce many additional difficulties connected with data parallelization, distribution of parallel subtasks and parallel random number generators. But at the same time one may expect that the influence of unbalancing of matrices is a lot bigger for smaller matrices (of size 100, or so) (see [5]) and with the increasing the matrix size robustness is also increasing. It is reasonable to consider large matrices, and particularly large unstructured sparse matrices since such matrices appear in many important real-live computational problems.

We are interested in the bilinear form of matrix powers since it is a basic subtask for many linear algebra problems:

$$(v, A^k h). \tag{1}$$

If x is the solution of a SLAE $Bx = b$, then

$$(v, x) = \left(v, \sum_{i=0}^{k} A^i h \right),$$

where the Jacobi Over-relaxation Iterative Method has been used to transform the SLAE into the problem $x = Ax + h$.

For an arbitrary large natural number k the Rayleigh quotient can be used to obtain an approximation for λ_1, the dominant eigenvalue, of a matrix A:

$$\lambda_1 \approx \frac{(v, A^k h)}{(v, A^{k-1} h)}.$$

In the latter case we should restrict our consideration to real symmetric matrices in order to deal with real eigenvalues. Thus it is clear that having an efficient way of calculating (1) is important. This is especially important in cases where we are dealing with large matrices.

2   Markov Chain Monte Carlo

The algorithm we use in our runs is the so-called Almost Optimal Monte Carlo (MAO) algorithm studied in [1,3,4]. We consider a Markov chain $T = \alpha_0 \rightarrow \alpha_1 \rightarrow \alpha_2 \rightarrow \ldots \rightarrow \alpha_k \rightarrow \ldots$ with n states. The random trajectory (chain) T_k of length k starting in the state α_0 is defined as follows: $T_k = \alpha_0 \rightarrow \alpha_1 \rightarrow \ldots \rightarrow \alpha_j \rightarrow \ldots \rightarrow \alpha_k$, where α_j means the number of the state chosen, for $j = 1, \ldots, k$. Assume that $P(\alpha_0 = \alpha) = p_\alpha$, $P(\alpha_j = \beta | \alpha_{j-1} = \alpha) = p_{\alpha\beta}$, where p_α is the probability that the chain starts in state α and $p_{\alpha\beta}$ is the transition probability to state β after being in state α. Probabilities $p_{\alpha\beta}$ define a transition matrix P.

In all algorithms used in this study we will consider a special choice of density distributions p_i and p_{ij} defined as follows:

$$p_i = \frac{|v_i|}{\| v \|}, \quad \| v \| = \sum_{i=1}^{n} |v_i| \quad \text{and} \quad p_{ij} = \frac{|a_{ij}|}{\| a_i \|}, \quad \| a_i \| = \sum_{j=1}^{n} |a_{ij}|. \tag{2}$$

The specially defined Markov chain induces the following product of matrix/vector entrances and norms:

$$A_v^k = v_{\alpha_0} \prod_{s=1}^{k} a_{\alpha_{s-1}\alpha_s}; \quad \| A_v^k \| = \| v \| \times \prod_{s=1}^{k} \| a_{\alpha_{s-1}} \| .$$

We have shown in [5] that the value

$$\bar{\theta}^{(k)} = \frac{1}{N} \sum_{i=1}^{N} \theta_i^{(k)} = sign\{A_v^k\} \| A_v^k \| \frac{1}{N} \sum_{i=1}^{N} \{h_{\alpha_k}\}_i \tag{3}$$

can be considered as a MC approximation of the form $(v, A^k h)$. For the probability error of this approximation one can have:

$$R_N^{(k)} = \left| (v, A^k h) - \bar{\theta}^{(k)} \right| = c_p \sigma \{\theta^{(k)}\} N^{-\frac{1}{2}},$$

where c_p is a constant.

In fact, (3) together with the sampling rules using probabilities (2) defines the MC algorithm used in our runs. Naturally, the quality of the MC algorithm depends on the behaviour of the standard deviation $\sigma\{\theta^{(k)}\}$. So, there is a reason to consider a special class of *robust MC algorithms*. Following [5] under *robust MC algorithms* we assume algorithms for which the standard deviation does not increase with the increasing of the matrix power k. So, robustness in our consideration is not only a characteristic of the quality of the algorithm. It also depends on the input data, i.e., on the matrix under consideration. As better balanced is the iterative matrix, and as smaller norm it has, as bigger a chances to get a robust MC algorithm.

3   Numerical Experiments

In this section we present results on experimental study of quality of the Markov chain Monte Carlo for large matrices. We run algorithms for evaluating mainly bilinear forms of matrix powers as a basic Monte Carlo iterative algorithms as well as the algorithm for evaluating the dominant eigenvalue. In our experiments we use dense and unstructured sparse matrices of sizes

- $n = 1000$, $n = 5000$, $n = 10000$, $n = 15000$, $n = 20000$, $n = 40000$.

We can control some properties of random matrices. Some of the matrices are well balanced (in some sense matrices are *close* to stochastic matrices), some of them are not balanced, and some are completely unbalanced. Some of the iterative matrices are with small norms which makes the Markov chain algorithm robust (as we showed in Section 2), and some of the matrices have large norms. Since the balancing is responsible for the variance of the random variable dealing with unbalanced matrices we may expect higher values for the stochastic error.

In such a way we can study how the stochastic error propagates with the number of Monte Carlo iterations for different matrices. Dealing with matrices of small norms we may expect high robustness of the Markov chain Monte Carlo and a high rate of convergence. To be able to compare the accuracy of various runs of the algorithm for computing bilinear forms of matrix powers we also compute them by a direct deterministic method using double precision. These runs are more time consuming since the computational complexity is higher than the complexity for Markov chain Monte Carlo, but we accept the results as "exact results" and use them to analyse the accuracy of the results produced by our Monte Carlo code.

For sparse matrices we use the Yale sparse matrix format [2]. We exploit the sparsity in sense that the used almost optimal Monte Carlo algorithm only deals with non-zero matrix entrances. The Yale sparse matrix format is very suitable since is allows to present large unstructured sparse matrices in a compact form in the processor's memory [2]. The latter fact allows to perform *jumps* of the Markov chain from one to another non-zero elements of a given matrix very fast.

We also study the scalability of the algorithms under consideration. We run our algorithms on different computer systems. The used systems are given below:

- IBM p690+ Regatta system cluster of IBM SMP nodes, containing a total of 1536 IBM POWER5 processors;
- Sun Fire 15K server with 52x0,9GHz UltraSPARC III processors;
- SGI Prism equipped with 8 x 1.5 GHz Itanium II processors and 16 GByte of main memory.

On Figure 1 we present results for the Monte Carlo solution of bilinear form of a dense matrix of size $n = 15000$ from the matrix power k (the matrix power corresponds to the number of moves in every Markov chain used in computations). The Monte Carlo algorithm used in calculations is robust. For comparison we present *exact* results obtained by deterministic algorithm with double precision. One can not see any differences on this graph. As one can expect the error of the robust Monte Carlo is very small and it decreases with increasing the matrix power k. In fact the stochastic error exists but it increases rapidly with increasing of k. This fact is shown more precisely on Figure 2.

The Monte Carlo probability error $R_N^{(k)}$ and the Relative Monte Carlo probability error $Rel_N^{(k)}$ was computed in the following way:

$$R_N^{(k)} = \left| \frac{1}{N} \sum_{i=1}^{N} \theta_i^{(k)} - \frac{(v, A^k h)}{(v, h)} \right|, \quad Rel_N^{(k)} = \frac{(v, h)}{(v, A^k h)} R_N^{(k)}.$$

In the robust case the relative MC error decreases to values smaller than 10^{-22} when the number of MC iterations is 20, while for in the non-robust case the corresponding values slightly increase with increasing of matrix power (for $k = 20$ the values are between 10^{-3} and 10^{-2} (see Figure 2).

If we apply both the robust and the non-robust Markov chain Monte Carlo to compute the dominant eigenvalue of the same dense matrices of size $n = 15000$ then we get the result presented on Figure 3.

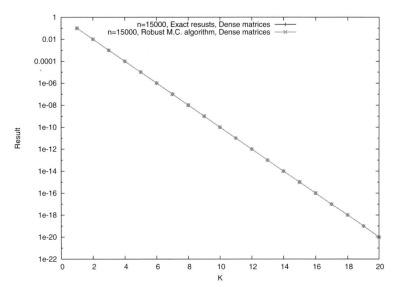

Fig. 1. Comparison of Robust Monte Carlo algorithm results with the *exact* solution for the bilinear form of a dense matrix of size $n = 15000$

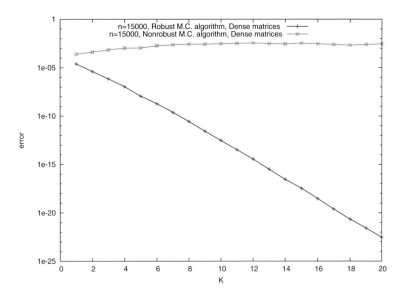

Fig. 2. Comparison of the Relative MC error for the robust and non-robust algorithms. Matrices of size $n = 15000$ are used.

We should stress on the fact that the random matrices are very similar. Only the difference is that in the robust case the matrix is well balanced. From Figure 3 one can see that the oscillations of the solution are much smaller when the matrix

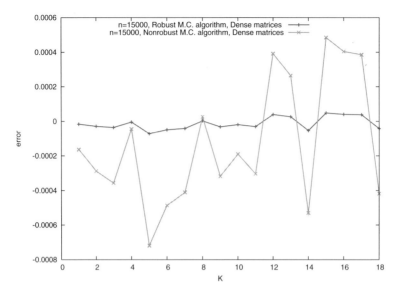

Fig. 3. The relative MC error for the robust and non-robust algorithms. The matrix size is $n = 15000$.

is well balanced. The reason for that is that the variance for the well balanced matrix is much smaller than for non-balanced matrix.

Very similar results are obtained for matrices of size $1000, 5000, 10000, 20000$, and 40000. Some results for sparse matrices are plotted on Figures 4, and 5.

Results of Monte Carlo computation of the bilinear form of an unstructured sparse matrix of size 10000 are plotted on Figure 4. The MC results are compared with the *exact* results. On this graph one can not find any differences between MC results and the *exact* solution. One can see that if the robust algorithm is applied for solving systems of linear algebraic equations or for computing the dominant eigenvalue of real symmetric matrices (in order to get real eigenvalues), then just 5 or 6 Monte Carlo iterations are enough to get fairly accurate solution (with 4 right digits). If we present the same results for the same sparse matrix in a logarithmic scale, then one can see that after 20 iterations the relative MC error is smaller than 10^{-20} since the algorithm is robust and with increasing the number of iterations the stochastic error decreases dramatically. Similar results for a random sparse matrix of size 40000 are shown on Figure 5.

Since the algorithm is robust and the matrix is well balanced the results of MC computations are very closed to the results of deterministic algorithm performed with a double precision.

Our observation from the numerical experiments performed are that the error increases linearly if k is increasing. The larger the matrix is, the smaller the influence of non-balancing is. This is also an expected result since the stochastic error is proportional to the standard deviation of the random variable computed as a weight of the Markov chain. When a random matrix is very large it is becoming

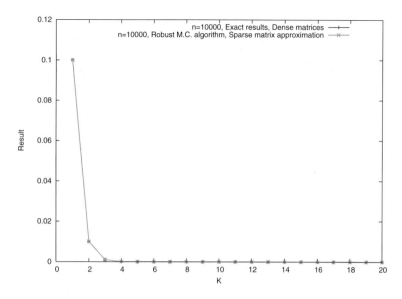

Fig. 4. Comparison of the MC results for bilinear form of matrix powers for a sparse matrix of size $n = 10000$ with the exact solution. 5 or 6 Monte Carlo iterations are enough for solving the system of linear algebraic equations or for computing the dominant eigenvalue for the robust Monte Carlo.

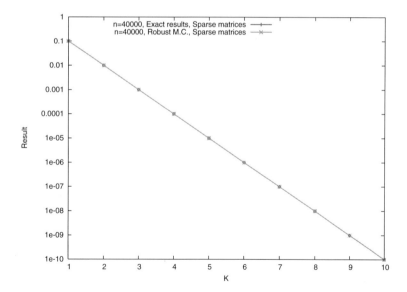

Fig. 5. Comparison of the MC results for bilinear form of matrix powers for a sparse matrix of size $n = 40000$ with the exact solution

more close (in some sense) to the stochastic matrix and the standard deviation of the random variable statistically decreases which statistically increases the accuracy.

4 Conclusion

In this paper we have analysed the performance of the proposed MC algorithm for linear algebra problems. We are focused on the computing bilinear form of matrix powers $(v, A^k h)$ as a basic subtask of MC algorithms for solving a class of Linear Algebra problems. We study the applicability and robustness of Markov chain Monte Carlo. The robustness of the Monte Carlo algorithm with large dense and unstructured sparse matrices has been demonstrated. It's an important observation that the balancing of the input matrix is very important for MC computations since it decreases the stochastic error and improved the robustness.

References

1. V. Alexandrov, E. Atanassov, I. Dimov: *Parallel Quasi-Monte Carlo Methods for Linear Algebra Problems*, Monte Carlo Methods and Applications, Vol. 10, No. 3-4 (2004), pp. 213-219.
2. R. E. Bank and C. C. Douglas: *Sparse matrix multiplication package (SMMP)*, Advances in Computational Mathematics, Vol. 1, Number 1 / February (1993), pp. 127-137.
3. I. Dimov: *Minimization of the Probable Error for Some Monte Carlo methods*. Proc. Int. Conf. on Mathematical Modeling and Scientific Computation, Albena, Bulgaria, Sofia, Publ. House of the Bulgarian Academy of Sciences, 1991, pp. 159-170.
4. I. Dimov: *Monte Carlo Algorithms for Linear Problems*, Pliska (Studia Mathematica Bulgarica), Vol. 13 (2000), pp. 57-77.
5. I. Dimov, V. Alexandrov, S. Branford, and C. Weihrauch: *Error Analysis of a Monte Carlo Algorithm for Computing Bilinear Forms of Matrix Powers*, Computational Science (V.N. Alexandrov et al. Eds.), Lecture Notes in Computing Sciences, Springer-Verlag Berlin Heidelberg, Vol. 3993, (2006), 632-639.
6. C. Weihrauch, I. Dimov, S. Branford, and V. Alexandrov: *Comparison of the Computational Cost of a Monte Carlo and Deterministic Algorithm for Computing Bilinear Forms of Matrix Powers*, Computational Science (V.N. Alexandrov et al. Eds.), Lecture Notes in Computing Sciences, Springer-Verlag Berlin Heidelberg, Vol. 3993, (2006), 640-647.
7. I. Dimov, A. Karaivanova: *Parallel computations of eigenvalues based on a Monte Carlo approach*, Journal of Monte Carlo Method and Applications, Vol. 4, Nu. 1, (1998), pp. 33-52.
8. J.R. Westlake: *A Handbook of Numerical matrix Inversion and Solution of Linear Equations*, John Wiley & Sons, inc., New York, London, Sydney, 1968.
9. M. Mascagni, A. Karaivanova: *A Parallel Quasi-Monte Carlo Method for Computing Extremal Eigenvalues*, Monte Carlo and Quasi-Monte Carlo Methods (2000), Springer, pp. 369-380.

Simulation of Multiphysics Multiscale Systems: Introduction to the ICCS'2007 Workshop

Valeria V. Krzhizhanovskaya[1] and Shuyu Sun[2]

[1] Section Computational Science, Faculty of Science, University of Amsterdam,
Kruislaan 403, 1098 SJ Amsterdam, The Netherlands
valeria@science.uva.nl
[2] Department of Mathematical Sciences, Clemson University
O-221 Martin Hall, Clemson, SC 29634-0975, USA
shuyu@clemson.edu

Abstract. Modeling and simulation of multiphysics multiscale systems poses a grand challenge to computational science. To adequately simulate numerous intertwined processes characterized by different spatial and temporal scales (often spanning many orders of magnitude), sophisticated models and advanced computational techniques are required. The aim of the workshop on Simulation of Multiphysics Multiscale Systems (SMMS) is to facilitate the progress in this multidisciplinary research field. We provide an overview of the recent trends and latest developments, with special emphasis on the research projects selected for presentation at the SMMS'2007 workshop.

Keywords: Multiphysics, Multiscale, Complex systems, Modeling, Simulation, ICCS, Workshop.

1 Introduction

Real-life processes are inherently multiphysics and multiscale. From atoms to galaxies, from amino-acids to living organisms, nature builds systems that involve interactions amongst a wide range of physical phenomena operating at different spatial and temporal scales. Complex flows, fluid-structure interactions, plasma and chemical processes, thermo-mechanical and electromagnetic systems are just a few examples essential for fundamental and applied sciences. Numerical simulation of such complex multiscale phenomena is vital for better understanding Nature and for advancing modern technologies. Due to the tremendous complexity of multiphysics multiscale systems, adequate simulation requires development of sophisticated models and smart methods for coupling different scales and levels of description (nano-micro-meso-macro). Until recently, such coupled modeling has been computationally prohibitive. But spectacular advances in computer performance and emerging technologies of parallel distributed grid computing have provided the community of computational physicists with the tools to break the barriers and bring simulation to a higher level of detail and accuracy. On the other hand, this progress calls for new efficient numerical algorithms and advanced computational techniques specific to the field where coupling different models or scales within one simulation is essential.

Y. Shi et al. (Eds.): ICCS 2007, Part I, LNCS 4487, pp. 755–761, 2007.

In the last decade, modeling and simulation showed a clear trend away from simpli-fied models that treat the processes on a single scale toward advanced self-adapting multiscale and multi-model simulations. The importance of such advanced computer simulations is recognized by various research groups and supported by national and international projects, e.g. the Dutch Computational eScience Initiative [1], the SCaLeS initiative in the USA [2] and the Virtual Physiological Human EU project [3]. Many significant developments were accomplished as a result of joint efforts in the multidisci-plinary research society of physicists, biologists, computational scientists and computer experts. To boost scientific cross-fertilization and promote collaboration of these diverse groups of specialists, we have launched a series of mini-symposia on Simulation of Multiphysics Multiscale Systems (SMMS) in conjunction with the International Confer-ence on Computational Sciences (ICCS) [4].

The fourth workshop in this series, organized as part of ICCS'2007, expands the scope of the meeting from physics and engineering to biological and biomedical ap-plications. This includes computational models of tissue- and organo-genesis, tumor growth, blood vessel formation and interaction with the hosting tissue, biochemical transport and signaling, biomedical simulations for surgical planning, etc. The topics traditionally addressed by the symposium include modeling of multiphysics and/or multiscale systems on different levels of description, novel approaches to combine different models and scales in one problem solution, advanced numerical methods for solving multiphysics multiscale problems, new algorithms for parallel distributed computing specific to the field, and challenging multiphysics multiscale applications from industry and academia.

A large collection of rigorously reviewed papers selected for the workshops high-lights modern trends and recent achievements [5]. It shows in particular the progress made in coupling different models (such as continuous and discrete models; quantum and classical approaches; deterministic and stochastic techniques; nano, micro, meso and macro descriptions) and suggests various coupling approaches (e.g. homogeniza-tion techniques, multigrid and nested grids methods, variational multiscale methods; embedded, concurrent, integrated or hand-shaking multiscale methods, domain bridg-ing methods, etc.). A selected number of papers have been published in the special issues of International Journal for Multiscale Computational Engineering [6-7], col-lecting state-of-the-art methods for multiscale multiphysics applications covering a large spectrum of topics such as multiphase flows, discharge plasmas, turbulent com-bustion, chemical vapor deposition, fluid-structure interaction, thermo-mechanical and magnetostrictive systems, and astrophysics simulation. In this paper we overview the latest developments in modeling and simulation of multiphysics multiscale sys-tems exemplified by the research presented at the SMMS'2007 workshop.

2   Overview of Work Presented in This Workshop

The papers presented in this workshop cover state-of-the-art simulations of mul-tiphysics multiscale problems; they represent ongoing research projects on various important topics relevant to the modeling and computation of these complex systems. Numerical simulations of these problems require two essential components. The first one is the development of sophisticated models for each physical process, characterized by its own specific scales and its own mechanisms, and integration of these models into one seamless simulation. Coupling or extension of atomistic and continuum models studied in [8-14] shows that sophisticated modeling is essential to accurately represent the physical world. Similarly, works in [15-18] demonstrate that biological or biomedical systems have intrinsically multiscale nature and require multiscale modeling. The second essential component for numerical simulation of multiphysics and multiscale problems includes efficient numerical algorithms and advanced computational techniques. Computational methodologies and programming tools [19-21] and advanced mathematical and numerical algorithms [22-25] are indispensable for efficient implementation of multiscale multiphysics models, which are computationally very intensive and often intractable using ordinary methods. Cellular automata [11,26,27] and the lattice Boltzmann method [28-31], which can be considered both as modeling tools and as numerical techniques, prove to be very powerful and promising in modeling complex flows and other multiscale complex systems.

The projects in [8-10] investigate computationally efficient yet physically meaningful ways of coupling discrete and continuum models across multiple scales. Another way of treating multiscale problems is to develop single-scale approximation models. Papers [12,13] present development and analysis of models at an atomic or molecular scale, while project [11] couples multiple continuum models at a macroscopic scale. In [8], an adaptively coupled approach is presented for compressible viscous flows, based on the overlapped Schwarz coupling procedure. The continuum domain is described by Navier-Stokes equations solved using a finite volume formulation in compressible form to capture the shock, and the molecular domain is solved by the Direct Simulation Monte Carlo method. Work conducted in [9] leads to development and application of two computational tools linking atomistic and continuum models of gaseous systems: the first tool, a Unified Flow Solver for rarefied and continuum flows, is based on a direct Boltzmann solver and kinetic CFD schemes, and the second tool is a multiscale computational environment integrating CFD tools with Kinetic Monte Carlo and Molecular Dynamics tools. Paper [10] describes an application of the Unified Flow Solver (UFS) for complex gas flows with rarefied and continuum regions. The UFS methodology is based on the direct numerical solution of the Boltzmann equation for rarefied flow domains and the kinetic schemes of gas dynamics (for the Euler or Navier-Stokes equations) for continuum flow domains.

In [13], molecular dynamics simulations are extended to slow dynamics that could arise in materials science, chemistry, physics and biology. In particular, the hyperdynamics method developed for low-dimension energy-dominated systems is extended to simulate slow dynamics in atomistic general systems. In [12], a new isothermal quantum Euler model is derived and the asymptotic behavior of the quantum Euler system is formally analyzed in the semiclassical and zero-temperature limits. To simulate the process of biomass conversion [14], a submodel is developed for reverse combustion process in a solid fuel layer on the grate. It gives good predictions for the velocity of combustion front and spatial profiles of porosity, oxygen fraction and temperature, which are essential inputs for NO_x calculations.

Multiscale approaches proved to be very useful for modeling and simulation of biological and biomedical systems [15-18]. In [15], a multiscale cell-based model is presented that addresses three stages of cancer development: avascular tumor growth, tumor-induced angiogenesis, and vascular tumor growth. The model includes the following three levels that are integrated through a hybrid MPI parallel scheme: the intracellular regulations that are described by Boolean networks, the cellular level growth and dynamics that are described by a lattice Monte Carlo model, and the extracellular dynamics of the signaling molecules and metabolites that are described by a system of reaction-diffusion equations. The work [17] is related to the analysis of dynamics of semi-flexible polymers, such as DNA molecules. A new efficient approximate technique predicts material properties of the polymeric fluids accounting for internal viscosity. The results explain the phenomenon of shear thinning and provide better predictions compared to the traditional techniques. In [16], coupled auto-regulated oscillators in a single- and multi-cellular environment are modeled, taking into consideration intrinsic noise effects in genetic regulation, characterized by delays due to the slow biochemical processes. Diverse disciplines including physiology, biomechanics, fluid mechanics and simulation are brought together in [18] to develop a predictive model of the behavior of a prosthetic heart valve by applying simulation techniques for the study of cardiovascular problems, such as blood clotting. A commercial finite volume computational fluid dynamics code ANSYS/CFX is used for the 3D components of the model.

Advanced mathematical and numerical algorithms are required for effective coupling of various models across multiple scales and for efficient reduction of the computations needed for fine scale simulations without loss of accuracy [22-25]. As a significant extension to the classical multiscale finite element methods, paper [24] is devoted to the development of a theoretical framework for multiscale Discontinuous Galerkin (DG) methods and their application to efficient solutions of flow and transport problems in porous media with interesting numerical examples. In this work, local DG basis functions at the coarse scale are first constructed to capture the local properties of the differential operator at the fine scale, and then the DG formulations using the newly constructed local basis functions instead of conventional polynomial functions are solved on the coarse scale elements. In [23], an efficient characteristic finite element method is proposed for solving the magnetic induction equation in magnetohydrodynamics, with numerical results exhibiting how the topological structure and energy of the magnetic field evolve for different resistivity scales. Paper [22] includes a fast Fourier spectral technique to simulate the Navier-Stokes equations with no-slip boundary conditions, enforced by an immersed boundary technique called volume-penalization. In [25], a deflation technique is proposed to accelerate the iterative processing of the linear system built from discretization of the pressure Poisson equation with bubbly flow problems.

A number of computational methodologies and programming tools have been developed for simulations of multiscale multiphysics systems [19-21]. In the mesh generation technique presented in [21], surface reconstruction in applications involving complex irregular domains is considered for modeling biological systems, and an efficient and relatively simple approach is proposed to automatically recover a high quality surface mesh from low-quality non-consistent inputs that are often obtained via 3-D acquisition systems like magnetic resonance imaging, microscopy or laser scanning. In [20], a new methodology for two-way connection of microscopic model and macroscopic model, called Macro-Micro Interlocked simulation, is presented for multiscale simulations, together with demonstration of the applicability of the methodology for the various phenomena, such as cloud formation in atmosphere, gas detonation, aurora, solid friction, and onset of solar flares. Paper [19] addresses the challenge arising from the intercomponent data exchanges among components of multiscale models and the language interoperability between their various constituent codes. This work leads to the creation of a set of interlanguage bindings for a successful parallel coupling library, the Model Coupling Toolkit.

Automaton, a mathematical model for a finite state machine, has been studied as a paradigm for modeling multiscale complex systems [11,26,27]. Systems considered in [26] arise from the modeling of weed dispersal. In this work, the systems are approximated by pathways through a network of cells, and the results of simulations provide evidence that the method is suitable for modeling weed propagation mechanisms using multiple scales of observation. In [27], complex automata are formalized with identification of five classes of scale separation and further investigation of the scale separation map in relation with its capability to specify its components. Efforts are spent in [11] on the application of macroscopic modeling with cellular automata to simulation of lava flows, which consist of unconfined multiphase streams, the characteristics of which vary in space and time as a consequence of many interrelated physical and chemical phenomena.

The lattice Boltzmann method, being a discrete computational method based upon the Boltzmann equation, is a powerful mesoscopic technique for modeling a wide variety of complex fluid flow problems. In addition to its capability to accommodate a variety of boundary conditions, this approach is able to bridge microscopic phenomena with the continuum macroscopic equations [28-31]. In [28], the problem of mixed convection in a driven cavity packed with porous medium is studied. A lattice Boltzmann model for incompressible flow in porous media and another thermal lattice Boltzmann model for solving energy equation are proposed based on the generalized volume-averaged flow model. Project [31] presents a model for molecular transport effects on double diffusive convection; in particular, this model is intended to access the impact of variable molecular transport effects on the heat and mass transfer in a horizontal shallow cavity due to natural convection of a binary fluid. In [29], a multiscale approach is applied to model the polymer dynamics in the presence of a fluid solvent, combining Langevin molecular dynamics techniques with a mesoscopic lattice Boltzmann method for the solvent dynamics. This work is applied in the interesting context of DNA translocation through a nanopore. In [30], the lattice Boltzmann method for convection-diffusion equation with source term is applied directly to solve some important nonlinear complex equations by using complex-valued distribution function and relaxation time.

3   Conclusions

The progress in understanding physical, chemical, biological, sociological and even economical processes is strongly dependent on adequacy and accuracy of numerical simulation. All the systems important for scientific and industrial applications are essentially multiphysics and multiscale: they are characterized by the interaction of a great number of intertwined processes that operate at different spatial and temporal scales. Modern simulation technologies make efforts to bridge the gaps between different levels of description, and to seamlessly combine the scales spanning many orders of magnitude in one simulation. The progress in developing multiphysics multiscale models and specific numerical methodologies is exemplified by the projects presented in the SMMS workshops [4].

Acknowledgments. We would like to thank the participants of our workshop for their inspiring contributions, and the members of the Program Committee for their diligent work, which led to the very high quality of the conference. Special thanks go to Alfons Hoekstra for his efficient and energetic work on preparing the SMMS'2007 workshop. The organization of this event was partly supported by the NWO/RFBR projects # 047.016.007 and 047.016.018, and the Virtual Laboratory for e-Science Bsik project.

References[*]

1. P.M.A. Sloot et al. White paper on Computational e-Science: Studying complex systems in silico. A National Research Initiative. December 2006 http://www.science.uva.nl/research/pscs/papers/archive/Sloot2006d.pdf
2. SCaLeS: A Science Case for Large Scale Simulation: http://www.pnl.gov/scales/
3. N. Ayache et al. Towards Virtual Physiological Human: Multilevel modelling and simulation of the human anatomy and physiology. White paper. 2005. http://ec.europa.eu/information_society/activities/health/docs/events/barcelona2005/ec-vph-white-paper2005nov.pdf
4. http://www.science.uva.nl/~valeria/SMMS
5. LNCS V. 3992/2006 DOI 10.1007/11758525, pp. 1-138;
   LNCS V. 3516/2005 DOI 10.1007/b136575, pp. 1-146;
   LNCS V. 3039/2004 DOI 10.1007/b98005, pp. 540-678
6. V.V. Krzhizhanovskaya, B. Chopard, Y.E. Gorbachev (Eds.) Simulation of Multiphysics Multiscale Systems. Special Issue of International Journal for Multiscale Computational Engineering. V. 4, Issue 2, 2006. DOI: 10.1615/IntJMultCompEng.v4.i2
7. V.V. Krzhizhanovskaya, B. Chopard, Y.E. Gorbachev (Eds.) Simulation of Multiphysics Multiscale Systems. Special Issue of International Journal for Multiscale Computational Engineering. V. 4, Issue 3, 2006. DOI: 10.1615/IntJMultCompEng.v4.i3
8. G. Abbate, B.J. Thijsse, and C.R.K. Kleijn. Coupled Navier-Stokes/DSMC for transient and steady-state flows. Proceedings ICCS'2007.
9. V.I. Kolobov, R.R. Arslanbekov, and A.V. Vasenkov. Coupling atomistic and continuum models for multi-scale simulations. Proceedings ICCS'2007.
10. V.V. Aristov, A.A. Frolova, S.A. Zabelok, V.I. Kolobov, R.R. Arslanbekov. Simulations of multiscale gas flows on the basis of the Unified Flow Solver. Proceedings ICCS'2007.
11. M.V. Avolio, D. D'Ambrosio, S. Di Gregorio, W. Spataro, and R. Rongo. Modelling macroscopic phenomena with cellular automata and parallel genetic algorithms: an application to lava flows. Proceedings ICCS'2007.
12. S. Gallego, P. Degond, and F. Mhats. On a new isothermal quantum Euler model: Derivation, asymptotic analysis and simulation. Proceedings ICCS'2007.

---

[*] All papers in Proceedings of ICCS'2007 are presented in the same LNCS volume, following this paper.

13. X. Zhou and Y. Jiang. A general long-time molecular dynamics scheme in atomistic systems. Proceedings ICCS'2007.

14. R.J.M. Bastiaans, J.A. van Oijen, and L.P.H. de Goey. A model for the conversion of a porous fuel bed of biomass. Proceedings ICCS'2007.

15. Y. Jiang. Multiscale, cell-based model for cancer development. Proceedings ICCS'2007.

16. A. Leier and P. Burrage. Stochastic modelling and simulation of coupled autoregulated oscillators in a multicellular environment: the her1/her7 genes. Proceedings ICCS'2007.

17. J. Yang and R.V.N. Melnik. A new model for the analysis of semi-flexible polymers with internal viscosity and applications. Proceedings ICCS'2007.

18. V. Diaz Zuccarini. Multi-physics and multiscale modelling in cardiovascular physiology: Advanced users methods of biological systems with CFX. Proceedings ICCS'2007.

19. E.T. Ong et al. Multilingual interfaces for parallel coupling in multiphysics and multiscale systems. Proceedings ICCS'2007.

20. K. Kusano, A. Kawano, and H. Hasegawa. Macro-micro interlocked simulations for multiscale phenomena. Proceedings ICCS'2007.

21. D. Szczerba, R. McGregor, and G. Szekely. High quality surface mesh generation for multi-physics bio-medical simulations. Proceedings ICCS'2007.

22. G.H. Keetels, H.J.H. Clercx, and G.J.F. van Heijst. A Fourier spectral solver for wall bounded 2D flow using volume-penalization. Proceedings ICCS'2007.

23. J. Liu. An efficient characteristic method for the magnetic induction equation with various resistivity scales. Proceedings ICCS'2007.

24. S. Sun. Multiscale discontinuous Galerkin methods for modeling flow and transport in porous media. Proceedings ICCS'2007.

25. J.M. Tang and C. Vuik. Acceleration of preconditioned Krylov solvers for bubbly flow problems. Proceedings ICCS'2007.

26. A.G. Dunn and J.D. Majer. Simulating weed propagation via hierarchical, patch-based cellular automata. Proceedings ICCS'2007.

27. A.G. Hoekstra. Towards a complex automata framework for multi-scale modelling formalism and the scale separation map. Proceedings ICCS'2007.

28. Z. Chai, B. Shi, and Z. Guo. Lattice Boltzmann simulation of mixed convection in a driven cavity packed with porous medium. Proceedings ICCS'2007.

29. S. Melchionna, E. Kaxiras, and S. Succi. Multiscale modeling of biopolymer translocation through a nanopore. Proceedings ICCS'2007.

30. B.C. Shi. Lattice Boltzmann simulation of some nonlinear complex equations. Proceedings ICCS'2007.

31. X.M. Yu, B.C. Shi, and Z.L. Guo. Numerical study of molecular transport effects on double diffusive convection with lattice Boltzmann method. Proceedings ICCS'2007.

Simulating Weed Propagation Via Hierarchical, Patch-Based Cellular Automata

Adam G. Dunn and Jonathan D. Majer

Alcoa Research Centre for Stronger Communities and
Department of Environmental Biology, Curtin University of Technology,
Western Australia
{A.Dunn, J.Majer}@curtin.edu.au

Abstract. Ecological systems are complex systems that feature heterogeneity at a number of spatial scales. Modelling weed propagation is
difficult because local interactions are unpredictable, yet responsible for
global patterns. A patch-based and hierarchical cellular automaton using
probabilistic connections suits the nature of environmental weed dispersal mechanisms. In the presented model, weed dispersal mechanisms,
including human disturbance and dispersal by fauna, are approximated
by pathways through a network of cells. The results of simulations provide evidence that the method is suitable for modelling weed dispersal
mechanisms using multiple scales of observation.

Keywords: Environmental weeds, hierarchical patch dynamics, cellular
automata, multiscale heterogeneity

1 Introduction and Context

Weed establishment and spread is a significant issue in Western Australia (WA),
especially in the region that is known as an international biodiversity hotspot [1].
Biodiversity describes the state of an ecosystem in terms of the natural complexity through which it evolves over time. *Environmental weeds* are plants that
degrade an ecosystem via simplification; these weeds compete more effectively
than their native counterparts in an environment that is foreign to them [2].
Along the south coast of WA, local and state governments, the Gondwana Link
initiative [3] and other community groups benefit from a predictive analysis of
the landscape-scale effects of their decisions on the propagation of weeds.

An ecological system is a complex system [4] and the intricate web of interactions between species of flora and fauna form a resilient system. A system's
degradation by weeds makes it more susceptible to further degradation.

Weeds propagate by dispersal and are constrained by competition with other
plants and by predation [5]. Significant dispersal vectors include seed-eating birds
and grazing animals, human transport networks, watercourses, wind and agricultural disturbances. Each of these dispersal vectors acts in a different manner
and often at different spatial and temporal scales.

Y. Shi et al. (Eds.): ICCS 2007, Part I, LNCS 4487, pp. 762–769, 2007.

The approach to modelling the landscape described here is novel because it extends the cellular automata formalism to create a cellular space that is hierarchical and irregular. As a result, the structure captures the patchy nature of the landscape realistically, as well as being capable of representing GIS data with a variety of levels granularity in a single structure.

2   Background

Ecosystems are complex systems in which the spatial dynamics may be described by a hierarchy of patches. Existing models of spatial ecosystem dynamics vary considerably in structure; two-dimensional raster grids that discretise mechanisms of competition or propagation are typical. Hierarchical patch dynamics is a modelling hypothesis for ecosystems that matches the physical spatial structure of ecosystems and other landscape patterns, such as land use and urbanisation.

Ecological systems are heterogeneous, complex systems that are neither completely regular nor completely random [6]. They exhibit a natural *patchiness* [7] that is defined by the processes of dispersal and competition, and are influenced by soil structure and natural resources, fire and forestry regimes, other human disturbances and climate. Distinct and abrupt changes between vegetation type are typical (from woodland to grassland, for example) and a mosaic of vegetation types over a landscape is a natural feature of an ecological system, besides being a feature of human-modified landscapes.

Ecological systems display *multiscale heterogeneity* — when viewing a system at a single level of spatial granularity, the landscape may be segmented into a set of *internally homogeneous* patches. To be internally homogeneous, a patch need only have an aggregable set of properties that are considered to be consistent throughout the patch at that level. For example, a coarse granularity view of a landscape may be segmented into regions of forest, woodland and grassland, whereas an individual patch of forest may be segmented by the distribution of species or even by individual plants.

In the case of weed propagation, the data quality issue is one of uncertainty at the microscale. In broad terms, weed propagation is the combination of the processes of germination, competition for resources and seed dispersal, as well as the uncertain long distance dispersal (LDD) events [8]. The typical approach for modelling the dispersal of any plant is to use empirical information to build a function that aggregates the several modes of seed dispersal by likelihood for a homogeneous landscape. The 'observational reality' of weed dispersal phenomena is one of uncertainty and multiple scales. At the scale of individual plants, the effects of individual birds, other animals, water flow, or human disturbances are essentially 'random events' in the colloquial sense of the phrasing. At a slightly coarser scale, the dispersal mechanisms for weeds become more stable in the sense that a model recovers some notional sense of predictability; mechanisms are described using a radius of spread or a likely direction of spread, and this is the *operational scale* of weed propagation [9].

Given the need to capture multiscale heterogeneity and the patchiness of ecological systems, the *hierarchical patch dynamics formalism* [10,11] may be used to model weed propagation. Hierarchical patch dynamics is a modelling method that captures the patchiness and multiscale heterogeneity of ecological systems as a hierarchy of interacting, heterogeneous (and internally homogeneous) patches.

The hierarchical patch dynamics method may be implemented as a *cellular automaton* [12], modified to be irregular, probabilistic and hierarchical. Coarse patches comprise several finer patches in a hierarchy. Regular-grid cellular automata introduce a bias in propagation [13], in the absence of any other method to avoid it, such as stochastic mechanisms in timing or update. Introducing stochasticity implicitly via irregularity or asynchronicity, or explicitly via a probabilistic update can restore isotropy in a cellular automata model of propagation [14]. The model described in the following section includes stochasticity to match the spatial nature of an ecological landscape by following the hierarchical patch dynamics concept.

3   Method

The model is constructed as a network of both interlevel connections forming a hierarchy and intralevel connections forming a graph. The structure of the model captures multiple scales of observation from GIS data, relating the levels to each other via the process of abstraction. Simulations are run by traversing the structure to determine the updated state of the network, which involves discovering the likelihood of dispersal through the heterogeneous and multiscale landscapes. The approach is analogous to existing models that use dispersal curves [15] but they additionally capture the dispersal modes explicitly rather than aggregating them into a single function, and they effectively manage the multiscale heterogeneity.

A *cell* is defined by its state, a set of connections to its *neighbourhood* and its Cartesian coordinates $i, j \in \mathbb{R}$. The *state* of a cell $s \in \mathcal{S}$, is defined by the static physical properties taken from GIS data of a specific level of abstraction associated with its level of abstraction; $\mathcal{S}$ represents the union of the set of possible states from each level (for example, $|\mathcal{S}| = 13$ in Fig. 2). The extent of a cell is defined by the Voronoi decomposition [16] of the landscape at the level to which the cell belongs and is therefore dependent on the locations of the nodes in the cell's neighbourhood. The intralevel neighbourhood of a cell is defined by the Delaunay triangulation of the subset of nodes in the graph and the interlevel neighbourhood is defined by the process of abstraction in which the hierarchy is created by physical state information.

An *abstraction* necessarily involves three homogeneous cells belonging to a single triangle in the Delaunay triangulation. Using this definition, the cells composed to create a new cell are guaranteed to be neighbours, they will represent an internally homogeneous environment (for the given level of abstraction) and will create a new level of abstraction in which fewer cells are used to represent the landscape and the average size of a cell increases. A single abstraction is illustrated in Fig. 1. The extents of the cells are given by the Voronoi decomposition and the connectivity is given by the Delaunay triangulation. In the figure, the cell d is now related to cells a, b and c with a 'composed of' relationship and three interlevel connections are formed, contributing to the structure of the hierarchy. Note that the choice of ternary trees over other forms is the simplest choice given the triangulation method; other forms produce the same shapes using a centroid method for defining the new cell's node location.

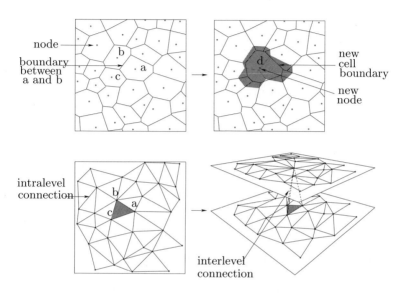

Fig. 1. A single abstraction is shown on a detail view of one landscape. The top two illustrations are the Voronoi decomposition of a set of nodes. The bottom two illustrations are a subset of the Delaunay triangulation for a set of nodes. The cells a, b and c in the left illustrations are abstracted to form cell d as given on both of the illustrations on the right. Cell d has a larger size than cells a, b and c (taken individually) and the distance to the nearest cells increases.

The result of multiple abstractions creates a hierarchy that builds a set of complete ternary trees. A set of points is initially distributed as a set of Halton points and each abstraction is made for a group of three internally homogeneous cells using the GIS data as a guide. The abstraction process continues until there are no more groups of homogeneous cells that may be abstracted such that the newly created distances between nodes are above a critical distance (representative of the operational scale). The above abstraction process is repeated for the several levels of GIS data until the coarsest GIS data is captured within the hierarchical structure. In Fig. 2 an example landscape is presented for for three levels of fictitious GIS data captured as raster information (see Fig. 3 for two intermediate levels of abstraction within the hierarchical structure that is built for this data).

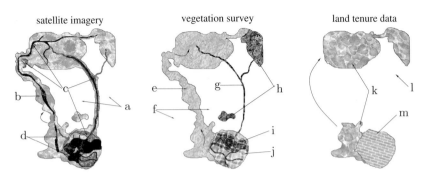

Fig. 2. Fictitious satellite imagery, vegetation survey data and land tenure data of the same landscape. Three levels of an example landscape are illustrated with the following categorisations: *a* pasture and crops, *b* natural vegetation, *c* bare or disturbed, *d* impermeable or built up, *e* type 1 vegetation, *f* crop or pasture, *g* road, *h* type 2 vegetation *i* urban, *j* peri-urban, *k* nature reserve, *l* private agricultural, *m* private residential.

The structure of the landscape model provides a useful multiscale description of the environment for multiple scales of observation (based on GIS data) but it does not, alone, provide a method for simulating weed propagation for the variety of dispersal modes, each with their own operational scale. The simulation method chosen for the implementation described here is to associate a probability with each of the connections in the *network* — the interlevel hierarchy and intralevel graphs combined. At each time step (a typical synchronous timing mechanism is adopted) and for each cell, a breadth-first search through the network is used to discover the avenues of possible dispersal and the likelihood value. This path mimics the seed dispersal curves and offers a unique approach to modelling LDD via its inherent uncertainty.

Probabilities for each of the connections are determined by the state information of the connected cells and the distance between their nodes. For example, given the higher density of seed rain (dispersal by seed-eating birds perching on trees) near the periphery of remnant vegetation [17] the likelihood of dispersal between remnant vegetation and most other types of landscape is given a higher value than between two remnant vegetation cells at the vegetation survey scale. The operational scale here has a finer granularity than human dispersal modes, which occur over both longer distances and with higher uncertainty. This is analogous to dispersal curve models of short and LDD.

There are three significant assumptions made about the landscape and the process of dispersal for the implementation described here. Firstly, it is assumed that several processes including dispersal, seed predation, germination, seasonal variance and seed bank dynamics may be aggregated into a single dispersal mechanism. Secondly, it is assumed that the operational scale of the variety of dispersal modes exist within a reasonable expanse of orders of magnitude. Computer simulation would be too computation-intensive if the operational scales included millimetres and hundreds of kilometres (and their associated time scales). Lastly, it is assumed for these experiments that the landscape is static except for the weed propagation — *ceteris paribus* has been invoked in a heavy fashion for this experiment.

4   Results and Discussion

In testing this implementation of hierarchical patch dynamics, the results of simulations demonstrate that the method is sensitive to the choice of granularity, but that the methodology is capable of representing the multiple dispersal mechanisms and the dynamics of a weed population. Importantly, the simulations demonstrate that there is advantage in the rigorous mapping between the operational scales of the dispersal phenomena and the observational scales captured by the GIS data. The approach is useful specifically because it captures dispersal modes and multiple scales explicitly, combining them in a framework that provides the aforementioned rigorous mapping.

The results of one simulation are presented in Fig. 3 — the simulation uses the fictitious GIS data depicted in Fig. 2. In this example, an environmental weed is given properties that are typical of a fruit-bearing weed whose dispersal is dependent on birds, small mammals and human disturbance routines. A weed of this type is likely to spread along the periphery of remnant vegetation because of the desirability of the perching trees for birds in these regions, and be found more commonly in and near regions with higher populations and greater frequencies of disturbance. The probabilities associated with remnant vegetation ↔ crop/pasture are relatively high, crop/pasture ↔ crop/pasture are zero and any connections within or between frequently disturbed patches are associated with relatively high probabilities. Since the weed and the landscape are both fictitious, the results of the experiment are not easily verified; instead this experiment represents a proof of concept for the modelling formalism and represents a typical set of operational scales that a model may be expected to represent.

Experiments in both homogeneous and heterogeneous landscapes suggest two specific issues with the specific implementation of aggregation/abstraction. The method does not exhibit the same bias caused by a regular lattice [13] but there is an apparent degradation in the approximation to isotropy in homogeneous experiments where the highest density of points is too coarse. In simpler terms, a minimum density of points is required to maintain isotropic propagation. An issue apparent in the heterogeneous experiments is that thin corridors through which weeds may spread are sensitive to maximum densities; a model structure must represent the thinnest corridor through which a weed may propagate. It is therefore important to accurately capture the grain and extent of the operational scale — for example, some fauna involved in the dispersal may ignore thin corridors, whilst other fauna may actively use thin corridors.

By combining different forms of data (such as vegetation surveys and land tenure information), the model captures the effects of a range of dispersal mechanisms in the same simulation, predicting the combined effects of human and

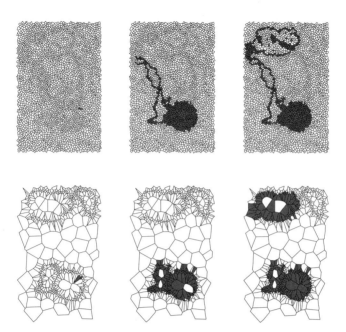

Fig. 3. The extent of cells depicted for three intermediate time steps and two levels of abstraction in the hierarchy. Elapsed time increases from left to right, the three sub-figures above are for the satellite imagery level of abstraction and the bottom three sub-figures are for the land tenure data level of abstraction. The increased rate of spread along the periphery is apparent in the top sub-figures and the dispersal over a disconnection is apparent in the bottom sub-figures.

natural dispersal mechanisms. A final assessment suggests that a more rigorous method for mapping from the dispersal mechanisms to the likelihood values is required. Specifically, the behaviour of the system is sensitive to the choice of both aggregation values between levels of abstraction and intralevel connection probabilities.

By creating a spatial model of the ecosystem in the south-west of WA using this implementation of hierarchical patch dynamics, predictive analysis of risk may be undertaken using spatial information about government policy and community group choices. Hierarchical patch dynamics is used as a model of multiscale ecological heterogeneity and the simulation presented here demonstrates a rigorous implementation of hierarchical patch dynamics. The fundamental approach to combining a variety of levels of abstraction (for human observation as well as organism perception) is the key advantage of using this approach over existing approaches.

The results of simulations suggest that the methodology is useful in creating a physically realistic (and therefore useful) model of the uncertain phenomena of weed propagation, but it is also suggestive of the need for introducing stronger links to *a priori* information about individual dispersal mechanisms in practical solutions to the issue of environmental weeds, rather than adopting a purely empirical approach.

Acknowledgments. The authors acknowledge the Alcoa Foundation's Sustainability & Conservation Fellowship Program for funding the postdoctoral fellowship of the senior author (http://strongercommunities.curtin.edu.au) and two anonymous reviewers for helpful comments.

References

1. Myers, N., Mittermeier, R.A., Mittermeier, C.G., Fonseca, G.A.B.d., Kent, J.: Biodiversity hotspots for conservation priorities. Nature **403** (2000) 853–858
2. Ellis, A., Sutton, D., Knight, J., eds.: State of the Environment Report Western Australia draft 2006. Environmental Protection Authority (2006)
3. Anon.: Gondwana Link Website (2006) [online] http://www.gondwanalink.org, last accessed 05/10/2006.
4. Bradbury, R.H., Green, D.G., Snoad, N.: Are ecosystems complex systems? In Bossomaier, T.R.G., Green, D.G., eds.: Complex Systems. Cambridge University Press, Cambridge (2000) 339–365
5. van Groenendael, J.M.: Patchy distribution of weeds and some implications for modelling population dynamics: a short literature review. Weed Research **28** (1988) 437–441
6. Green, D., Klomp, N., Rimmington, G., Sadedin, S.: Complexity in Landscape Ecology, Landscape Series. Volume 4. Springer (2006)
7. Greig-Smith, P.: Pattern in vegetation. Journal of Ecology **67** (1979) 755–779
8. Nathan, R., Perry, G., Cronin, J.T., Strand, A.E., Cain, M.L.: Methods for estimating long-distance dispersal. Oikos **103** (2003) 261–273
9. Wu, J.: Effects of changing scale on landscape pattern analysis: scaling relations. Landscape Ecology **19** (2004) 125 – 138
10. Wu, J.: From balance-of-nature to hierarchical patch dynamics: a paradigm shift in ecology. Q. Rev. Biol. **70** (1995) 439–466
11. Wu, J., David, J.L.: A spatially explicit hierarchical approach to modeling complex ecological systems: theory and applications. Ecological Modelling **153** (2002) 7–26
12. Chopard, B., Droz, M.: Cellular Automata Modeling of Physical Systems. Monographs and Texts in Statistical Physics. Cambridge University Press (1998)
13. O'Regan, W., Kourtz, P., Nozaki, S.: Bias in the contagion analog to fire spread. Forest Science **22** (1976) 61–68
14. Schönfisch, B.: Anisotropy in cellular automata. BioSystems **41** (1997) 29–41
15. Higgins, S.I., Richardson, D.M.: Predicting plant migration rates in a changing world: The role of long-distance dispersal. The American Naturalist **153**(5) (1999) 464–475
16. Okabe, A., Boots, B., Sugihara, K.: Spatial Tessellations — Concepts and Applications of Voronoi Diagrams. John Wiley & Sons (1992)
17. Buckley, Y.M., Anderson, S., Catterall, C.P., Corlett, R.T., Engel, T., Gosper, C.R., Nathan, R., Richardson, D.M., Setter, M., Spiegel, O., Vivian-Smith, G., Voigt, F.A., Weir, J.E.S., Westcott, D.A.: Management of plant invasions mediated by frugivore interactions. Journal of Applied Ecology **43** (2006) 848–857

A Multiscale, Cell-Based Framework for Modeling Cancer Development

Yi Jiang

Theoretical Division, Los Alamos National Laboratory, Los Alamos, NM 87545, USA
jiang@lanl.gov

Abstract. We use a systems approach to develop a predictive model that medical researchers can use to study and treat cancerous tumors. Our multiscale, cell-based model includes intracellular regulations, cellular level dynamics and intercellular interactions, and extracellular chemical dynamics. The intracellular level protein regulations and signaling pathways are described by Boolean networks. The cellular level growth and division dynamics, cellular adhesion, and interaction with the extracellular matrix are described by a lattice Monte Carlo model. The extracellular dynamics of the chemicals follow a system of reaction-diffusion equations. All three levels of the model are integrated into a high-performance simulation tool. Our simulation results reproduce experimental data in both avascular tumors and tumor angiogenesis. This model will enable medical researchers to gain a deeper understanding of the cellular and molecular interactions associated with cancer progression and treatment.

1 Introduction

Since 2002, cancer has become the leading cause of death for Americans between the ages of 40 and 74 [1]. But the overall effectiveness of cancer therapeutic treatments is only 50%. Understanding the tumor biology and developing a prognostic tool could therefore have immediate impact on the lives of millions of people diagnosed with cancer. There is growing recognition that achieving an integrative understanding of molecules, cells, tissues and organs is the next major frontier of biomedical science. Because of the inherent complexity of real biological systems, the development and analysis of computational models based directly on experimental data is necessary to achieve this understanding. Our model aims to capture knowledge through the explicit representation of dynamic biochemical and biophysical processes of tumor development in a multiscale framework.

Tumor development is very complex and dynamic. Primary malignant tumors arise from small nodes of cells that have lost, or ceased to respond to, normal growth regulatory mechanisms, through mutations and/or altered gene expression [2]. This genetic instability causes continued malignant alterations, resulting in a biologically complex tumor. However, all tumors start from a relatively simpler, avascular stage of growth, with nutrient supply by diffusion from the surrounding tissue. The restricted supply of critical nutrients, such as oxygen and glucose, results in marked gradients within the cell mass. The tumor

Y. Shi et al. (Eds.): ICCS 2007, Part I, LNCS 4487, pp. 770–777, 2007.

cells respond both through induced alterations in physiology and metabolism, and through altered gene and protein expression [3], leading to the secretion of a wide variety of angiogenic factors.

Angiogenesis, formation of new blood vessels from existing blood vessels, is necessary for subsequent tumor expansion. Angiogenic growth factors generated by tumor cells diffuse into the nearby tissue and bind to specific receptors on the endothelial cells of nearby pre-existing blood vessels. The endothelial cells become activated; they proliferate and migrate towards the tumor, generating blood vessel tubes that connect to form blood vessel loops that can circulate blood. With the new supply system, the tumor will renew growth at a much faster rate. Cells can invade the surrounding tissue and use their new blood supply as highways to travel to other parts of the body. Members of the vascular endothelial growth factor (VEGF) family are known to have a predominant role in angiogenesis.

The desire to understand tumor biology has given rise to mathematical models to describe the tumor development. But no mathematical models of tumor growth yet can start from a single cell and undergo the whole process of tumor development. The state of the art in this effort employs hybrid approaches: Cristini et al. simulated the transition from avasular tumor growth to angiogenesis to vascular and invasive growth using an adaptive finite-element method coupled to a level-set method [4]; and Alarcon et al. used a hybrid of cellular automata and continuous equations for vascular tumor growth [5,6].

We have developed a multiscale, cell-based model of tumor growth and angiogenesis [7,8]. This paper aims to review and promote this model framework. The Model section describes our model at three levels. The Parallelization section describes the hybrid scheme that makes the model a high-performance simulation tool. The Results section shows that the model reproduces quantitatively experimental measurements in tumor spheroids, and qualitatively experimental observations in tumor angiogenesis. We conclude by commenting on the broad applicability of this cell-based, multiscale modeling framework.

2   Model

Our model consists of three levels. At the intracellular level, a simple protein regulatory network controls cell cycle. At the cellular level, a Monte Carlo model describes cell growth, death, cell cycle arrest, and cell-cell adhesion. At the extracellular, a set of reaction-diffusion equations describes for chemical dynamics. The three levels are closely integrated. The details of the avascular tumor model has been described in [7].

The passage of a cell through its cell cycle is controlled by a series of proteins. Since experiments suggest that more than 85% of the quiescent cells are arrested in the G1 phase [11], in our model, the cells in their G1 phase have the highest probability of becoming quiescent. We model this cell-cycle control through a simplified protein regulatory network [12], which controls the transition between G1 and S phases [7]. We model these proteins as on or off. By default this

Boolean network allows the cell transition to S phase. However, concentrations of growth and inhibitory factors directly influence the protein expression. If the outcome of this Boolean regulatory network is zero, the cell undergoes cell-cycle arrest, or turns quiescent. When a cell turns quiescent, it reduces its metabolism and stops its growth.

The cellular model is based on the Cellular Potts Model (CPM) [9,10]. The CPM adopts phenomenological simplification of cell representation and cell interactions. CPM cells are spatially extended with complex shapes and without internal structure, as domains on the lattice with specific cell ID numbers. Most cell behaviors and interactions are in terms of effective energies and elastic constraints. Cells evolve continuously to minimize the effective energy; the CPM employs a modified Metropolis Monte-Carlo algorithm, which chooses update sites randomly and accepts them with a Boltzmann probability. The total energy of the tumor cell system includes an interfacial energy between the cells that describes the cell-type dependent adhesion energy, a volume constraint that keeps the cell volume to a value that it "intends" to maintain; and an effective chemical energy that describes the cell's ability to migrate up or down the gradient of the chemical concentration. When the cell's clock reaches the cell cycle duration and the cell volume reaches the target volume, the cell will divide. The daughter cells inherit all properties of their parent with a probability for mutation.

Cells also interact with their environment, which is characterized by local concentrations of biochemicals. We consider two types of chemicals: the metabolites and the chemoattractants. The former includes nutrients (oxygen and glucose), metabolic waste (lactate), and growth factors and inhibitors that cells secret and uptake. The latter corresponds to the chemotactic signaling molecules. The chemicals follow the reaction-diffusion dynamics:

$$\frac{\partial M_i}{\partial t} = d_i \nabla^2 M_i - a_i. \tag{1}$$

$$\frac{\partial C_i}{\partial t} = D_i \nabla^2 C_i - \gamma C_i + B. \tag{2}$$

Here the metabolite (concentration M) diffuses with the diffusion constant d and is consumed (or produced) at a constant rate a. The chemoattractant, i.e. VEGF secreted by tumor cells to activate endothelial cells, diffuses with diffusion constant D and decays at a rate γ. The local uptake function B describes that an endothelial cell could bind as much as VEGF is available until its surface receptors saturate. Both the metabolic rates and the uptake function B are functions of time and space. By assuming that (1) inside the tumor the diffusion coefficients are constant; and (2) each cell is chemically homogeneous, while different cells might have different chemical concentrations, we can solve the equations on a much coarser lattice than the lattice for CPM.

We use parameters derived from multicellular tumor spheroid experiments, the primary in vitro experimental model of avascular tumors [2]. Like tumors in vivo, a typical spheroid consists of a necrotic core surrounded by layers of quiescent and proliferating tumor cells, as shown in Fig. 2(c). It is critical to emphasize that this multicellular tumor spheroid experimental model recapitulates all major characteristics of the growth, composition, microenvironment, and therapeutic response of solid tumors in humans [2].

To model tumor angiogenesis, the simulation domain now corresponds to the stroma between the existing blood vessel and an avascular tumor. The tumor is a constant source for VEGF, whose dynamics follow Eqn. (3) and establish a concentration gradient across the stroma. Each endothelial cell becomes activated when the local VEGF concentration exceeds a threshold value. The activated vascular endothelial cells not only increase proliferation, decrease apoptosis rate, but also migrate towards the higher concentration of the VEGF signal.

The stroma consists of normal tissue cells and extracellular matrix (ECM). We model the ECM explicitly as a matrix of fibers. They are a special "cell" with a semi-rigid volume. The interstitial fluid that fills the space amongst the normal tissue cells and the fibers is more deformable than the fibers. When compressed by the growing endothelial sprout, the normal cells can undergo apoptosis and become part of the interstitial fluid. The endothelial cells can modify the local fibronectin concentration and re-organizing the fiber structure, as well as migrate on the matrix through haptotaxis, or follow the gradient of adhesion from the gradient of fibronectin density.

3   Parallelization

The underlying structure for the cell system is a 3D lattice. As all operations in the CPM are local, domain decomposition method is a natural scheme for parallelizing the CPM. We divide the physical domain into subdomains, one for each processor. Each processor then model the subsystem, storing only the data associated with the subdomain, and exchanging knowledge zone information with processors handling its neighboring subdomains. We adopt 1D domain decomposition based on two considerations. First, it will decrease the communication overhead when the knowledge zones are more continuous. Second, this simple decomposition allows us to store the cell information in two layers: the lattice layer and the cell information layer. As each cell occupies many lattice sites (over 100), this two-layer information structure far more efficient than storing all cell information on the lattice.

In the Monte Carlo update, to avoid expensive remote memory access, we use openMP, a parallel approach suited for shared memory processors. Before the Monte Carlo update of the lattice, the Master Node gathers subdomain lattice information and cell information from slave nodes. The Master Node performs Monte Carlo operations parallelly using openMP, and distributes subdomain data to corresponding slave nodes for next operation (Fig. 1) [13].

We solve the reaction-diffusion equations that govern the chemical dynamics on a chemical lattice that is coarser than the cell lattice. This lattice is similarly decomposed and the equations solved in parallel within each subdomain. To accelerate long time simulations, we use implicit PDE sovling schemes based on BoxMG, an MPI-based multigrid PDE solver [14].

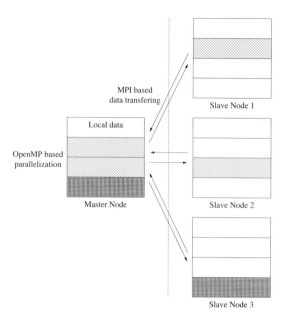

Fig. 1. OpenMP based parallelization in Monte Carlo operation

4   Result

In our simulations, a single tumor cell evolves into a layered structure consisting of concentric spheres of proliferating and quiescent cells at the surface and the intermediate layer respectively, and a necrotic core at the center of the spheroid, reproducing the experimental structure (Fig. 2).

Fig. 3 shows the comparison between the growth curves of a simulated solid tumor and two sets of spheroid experimental data. With 0.08 mM oxygen and 5.5 mM glucose kept constant in the medium, the number of cells (Fig. 3a) and the tumor volume (Fig. 3b) first grow exponentially in time for about 5–7 days.

Fig. 2. Snapshots of a simulated solid tumor at 10 days (a), and 18 days (b) of growth from a single cell. Blue, cyan, yellow and crimson encode cell medium, proliferating, quiescent, and necrotic cells. (c) A histological crosssection of a spheroid of mouse mammary cell line, EMT6/R0.

The growth then slows down, coinciding with the appearance of quiescent cells. In both the experiments [15,16] and simulations, spheroid growth saturates after around 28–30 days. We fit both the experimental and the simulation data to a Gompertz function, in order to objectively estimate the initial doubling times and the spheroid saturation sizes [3]. The doubling times for cell volume in experiments and simulations differ by a factor of two, over almost 5 orders of magnitude. The agreement was very good.

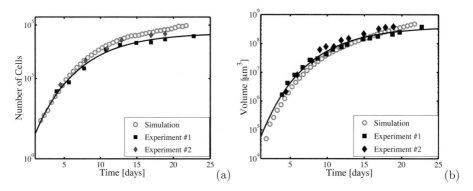

Fig. 3. The growth curves of a spheroid with 0.08 mM O2 and 5.5 mM glucose in the medium: (a) the number of cells and (b) the volume of spheroid in time. The solid symbols are experimental data for EMT6/Ro[15,16]; the circles are simulation data. The solid lines are the best fit with a Gompertz function for experimental data.

In order to test the robustness of our model, we kept all the parameters in the model fixed at the values determined to produce the best fit to the growth of spheroids in 0.08 mM oxygen and 5.5 mM glucose. We then varied only the nutrient concentrations in the medium, as was done in spheroid experiments. Our simulations still showed good agreement between simulation and experimental growth curves when the external conditions were changed to 0.28 mM O2 and 16.5 mM glucose in the medium [7].

In fitting our model to the experimental data, we predicted a set of conditions for the cell to undergo necrosis and the diffusion coefficients for the growth promoters and inhibitors to be in the order of 10^{-7} and 10^{-6} cm^2/hr, respectively. These predictions can be tested experimentally.

In tumor angiogenesis, Fig. 4 shows that our model is able to capture realistic cell dynamics and capillary morphologies, such as preferential sprout migration along matrix fibers and cell elongation, and more complex events, such as sprout branching and fusion, or anastomosis [8], that occur during angiogenesis. Our model constitutes the first cell-based model of tumor-induced angiogenesis with the realistic cell-cell and cell-matrix intereactions. This model can be employed as a simulation tool for investigating mechanisms, testing existing and formulating new research hypotheses. For example, we showed that freely diffusing VEGF would result in broad and swollen sprouts, while matrix bound VEGF typically generates thin sprouts [8], supporting the recent experimental interpretations.

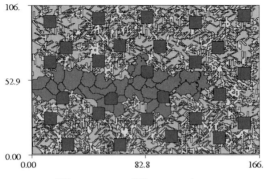

Distance toward Tumor - microns

Fig. 4. Tumor angiogenesis: a typical simulated vessel sprout. The left edge of the simulation domain is the blood vessel wall, while the right edge is the source of VEGF. Endothelial (red) cells grow and migrate in the stroma consists of normal cells (blue squares), matrix fibers (yellow) and interstitial fluids (green).

5   Discussion and Outlook

This multiscale approach treats cells as the fundamental unit of cancer development. We will further develop the model and investigate the growth of vessels into and inside tumor, as well as tumor growth and invasion. With this framework, we will model the development of cancer from beginning to full metastasis. We will also be able to test the effects of drugs and therapeutic strategies. Combined with the extant data (e.g. in vitro spheroid data and in vivo angiogenesis data), this type of model will help construct anatomically accurate models of a tumor and its vascular system. If successfully implemented, the model can guide experimental design and interpretation. Continuously revised by new information, the final model could potentially enable us to assess tumor susceptibility to multiple therapeutic interventions, improve understanding of tumor biology, better predict and prevent tumor metastasis, and ultimately increase patient survival. Furthermore, most biomedical problems involves systems level interactions. Genome, or molecular, or single cell studies cannot possibly provide systems level behaviors. This cell-based, multiscale modeling framework is applicable to a number of problems, e.g. biofilm formation, organogenesis, where cell-cell and cell-environment interactions dictate the collective behavior.

Acknowledgments

This work was supported by the National Nuclear Security Administration of the U.S. Department of Energy at Los Alamos National Laboratory under Contract No. DE-AC52-06NA25396.

References

1. Jemal, A.:The Journal of the American Medical Association, **294** (2005) 1255–1259.
2. Sutherland, R.M. Cell and environment interactions in tumor microregions: the multicell spheroid model. Science **240** (1988) 177–184.
3. Marusic, M., Bajzer Z, Freyer J.P., and Vuk-Pavlovic S: Analysis of growth of multicellular tumour spheroid by mathematical models. Cell Prolif. **27** (1994) 73.
4. Zheng, X., Wise S.M., and Cristini V.: Nonlinear simulation of tumor necrosis, neovascularization and tissue invasion via an adaptive finite-element/level-set method, Bull. Math. Biol. **67** (2005) 211-259
5. Alarcon, T., Byrne H.M., and Maini P.K.: A Multiple Scale Model for Tumor Growth, Multiscale Modeling and Simulation. **3** (2004) 440–475.
6. Alarcon, T., Byrne H.M., and Maini P.K.: Towards whole-organ modelling of tumour growth, Progress in Biophysics & Molecular Biology. **85** (2004) 451 – 472.
7. Jiang, Y., Pjesivac J., Cantrell C, and Freyer J.P.: A multiscale model for avascular tumor growth, Biophys. J. **89** (2005) 3873–3883 .
8. Bauer, A.L., Jackson T.L., and Jiang Y.: A cell-based model exhibiting branching and anastomosis during tumor-induced angiogenesis. Biophys. J. **92** (2007) in press.
9. Glazier, J.A. and Garner F: Simulation of the differential adhesion driven rearrangement of biological cells. Phys. Rev. E **47** (1993): 2128-2154.
10. Jiang, Y., Levine H. and Glazier J.A.: Differential adhesion and chemotaxis in mound formation of Dictyostelium, Biophys. J. **75**(1998) 2615 –2625.
11. LaRue, K.E., Kahlil M., and Freyer J.P.: Microenvironmental regulation of proliferation in EMT6 multicellular spheroids is mediated through differential expression of cycline-dependent kinase inhibitors. Cancer Res. **64**(2004) 1621–1631.
12. Data from the Kyoto Encyclopedia of Genes and Genomes (kegg.com).
13. He, K., Dong S., and Jiang Y.: Parallel Cellular Potts Model (2007) in preparation.
14. Austin T., Berndt M., et al. Parallel, Scalable, and Robust Multigrid on Structured Grids. Los Alamos Research Report (2003): LA-UR-03-9167.
15. Freyer, J.P. and Sutherland R.M.: Regulation of growth saturation and development of necrosis in EMT6/R0 multicellular spheroids by the glucose and oxygen supply. Cancer Res. **46**(1986) 3504-3512.
16. Freyer, J.P. and R.M. Sutherland. Proliferative and clonogenic heterogeneity of cells from EMT6/Ro multicellular spheroids induced by the glucose and oxygen supply. Cancer Res. **46** (1986) 3513-3520.

Stochastic Modelling and Simulation of Coupled Autoregulated Oscillators in a Multicellular Environment: The her1/her7 Genes

André Leier, Kevin Burrage, and Pamela Burrage

Advanced Computational Modelling Centre, University of Queensland, Brisbane,
QLD 4072 Australia
{leier,kb,pmb}@acmc.uq.edu.au

Abstract. Delays are an important feature in temporal models of genetic regulation due to slow biochemical processes such as transcription and translation. In this paper we show how to model intrinsic noise effects in a delayed setting. As a particular application we apply these ideas to modelling somite segmentation in zebrafish across a number of cells in which two linked oscillatory genes her1 and her7 are synchronized via Notch signalling between the cells.

Keywords: delay stochastic simulation algorithm, coupled regulatory systems, multicellular environment, multiscale modelling.

1 Introduction

Temporal models of genetic regulatory networks have to take account of time delays that are associated with transcription, translation and nuclear and cytoplasmic translocations in order to allow for more reliable predictions [1]. An important aspect of modelling biochemical reaction systems is intrinsic noise that is due to the uncertainty of knowing when a reaction occurs and which reaction it is. When modelling intrinsic noise we can identify three modelling regimes. The first regime corresponds to the case where there are small numbers of molecules in the system so that intrinsic noise effects dominate. In this regime the Stochastic Simulation Algorithm (SSA) [2] is the method of choice and it describes the evolution of a discrete nonlinear Markov process representing the number of molecules in the system. The intermediate regime is called the Langevin regime and here the framework for modelling chemical kinetics is that of a system of Itô stochastic differential equations. In this regime the numbers of molecules are such that we can talk about concentrations rather than individual numbers of molecules but the intrinsic noise effects are still significant. The final regime is the deterministic regime where there are large numbers of molecules for each species. This regime is given by the standard chemical kinetic rate equations that are described by ordinary differential equations. In some sense this third regime represents the mean behaviour of the kinetics in the other two regimes. It is vital to model the chemical kinetics of a system in the most appropriate regime otherwise the dynamics may be poorly represented.

Y. Shi et al. (Eds.): ICCS 2007, Part I, LNCS 4487, pp. 778–785, 2007.

In order to take proper account of both intrinsic randomness and time delays, we have developed the delay stochastic simulation algorithm (DSSA) [3]. This algorithm very naturally generalises the SSA in a delayed setting.

Transcriptional and translational time delays are known to drive genetic oscillators. There are many types of molecular clocks that regulate biological processes but apart from circadian clocks [4] these clocks are still relatively poorly characterised. Oscillatory dynamics are also observed for Notch signalling molecules such as Hes1 and Her1/Her7. The hes1 gene and the two linked genes her1 and her7 are known to play key roles as molecular clocks during somite segmentation in mouse and zebrafish, respectively.

In zebrafish the genes her1 and her7 are autorepressed by their own gene products and positively regulated by Notch signalling that leads to oscillatory gene expression with a period of about 30 min, generating regular patterns of somites (future segments of the vertebrate) [5]. In both cases the transcriptional and translational delays are responsible for the oscillatory behaviour.

In a recent set of experiments Hirata et al. [6] measured the production of hes1 mRNA and Hes1 protein in mouse. They measured a regular two hour cycle with a phase lag of approximately 15 minutes between the oscillatory profiles of mRNA and protein. The oscillations are not dependent on the stimulus but can be induced by exposure to cells expressing delta. This work led to a number of modelling approaches using the framework of Delay Differential Equations (DDEs) [1,7]. However, in a more recent work Barrio et al. used a discrete delay simulation algorithm that took into account intrinsic noise and transcriptional and translational delays to show that the Hes1 system was robust to intrinsic noise but that the oscillatory dynamics crucially depended on the size of the transcriptional and translational delays.

In a similar setting Lewis [5] and Giudicelli and Lewis [8] have studied the nature of somite segmentation in zebrafish. In zebrafish it is well known that two linked oscillating genes her1/her7 code for inhibitory gene regulatory proteins that are implicated in the pattern of somites at the tail end of the zebrafish embryo. The genes her1 and her7 code for autoinhibitory transcription factors Her1 and Her7 (see Fig. 1). The direct autoinhibition causes oscillations in mRNA and protein concentrations with a period determined by the transcriptional and translational delays.

Horikawa et al. [9] have performed a series of experiments in which they investigate the system-level properties of the segmentation clock in zebrafish. Their main conclusion is that the segmentation clock behaves as a coupled oscillator. The key element is the Notch-dependent intercellular communication which itself is regulated by the internal hairy oscillator and whose coupling of neighbouring cells synchronises the oscillations. In one particular experiment they replaced some coupled cells by cells that were out of phase with the remaining cells but showed that at a later stage they still became fully synchronised. Clearly the intercellular coupling plays a crucial role in minimising the effects of noise to maintain coherent oscillations.

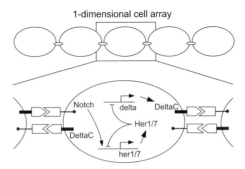

Fig. 1. Diagram showing the inter- and intracellular Delta-Notch signalling pathway and the autoinhibition of her1 and her7 genes. DeltaC proteins in the neighboring cells activate the Notch signal within the cell.

Both Lewis and Horikawa have used a stochastic model to understand the above effects. But this model is very different from our approach. The Lewis model for a single cell and two coupled cells is generalised by Horikawa et al. to a one-dimensional array of cells. In both approaches they essentially couple a delay differential equation with noise associated with the uncertainty of proteins binding to the operator sites on the DNA. In our case we are rigorously applying the effects of intrinsic noise, in a delayed setting, at all stages of the chemical kinetics. We also note that this is the first stage in developing a truly multi-scaled approach to understanding the effects of delays in a multi-celled environment. Such a multi-scaled model will require us to couple together delay models in the discrete, stochastic and deterministic regimes - see, for example, attempts to do this in Burrage et al. [10].

Section 2 gives a brief description of our DSSA implementation along with a mathematical description of the coupled Her1/Her7 Delta-Notch system for a linear chain of cells. Section 3 presents the numerical results and the paper concludes with discussion on the significance of our approach.

2   Methods

The SSA describes the evolution of a discrete stochastic chemical kinetic process in a well stirred mixture. Thus assume that there are m reactions between N chemical species, and let $X(t) = (X_1(t), \cdots, X_N(t))^\top$ be the vector of chemical species where $X_i(t)$ is the number of species i at time t. The chemical kinetics is uniquely characterised by the m stoichiometric vectors $\nu_1, \cdots, \nu_m$ and the propensity functions $a_1(X), \cdots, a_m(X)$ that represent the unscaled probabilities of the reactions to occur. The underlying idea behind the SSA is that at each time step t a step size θ is determined from an exponential waiting time distribution such that at most one reaction can occur in the time interval $(t, t+\theta)$. If the most likely reaction, as determined from the relative sizes of the propensity functions, is reaction j say, then the state vector is updated as $X(t + \theta) = X(t) + \nu_j$.

Algorithm 1. DSSA

Data: reactions defined by reactant and product vectors, consuming delayed
reactions are marked, stoichiometry, reaction rates, initial state $X(0)$,
simulation time T, delays

Result: state dynamics

begin

    while $t < T$ **do**

        generate U_1 and U_2 as $U(0,1)$ random variables

        $a_0(X(t)) = \sum_{j=1}^{m} a_j(X(t))$

        $\theta = \frac{1}{a_0(X(t))} \ln(1/U_1)$

        select j such that

        $\sum_{k=1}^{j-1} a_k(X(t)) < U_2 a_0(X(t)) \le \sum_{k=1}^{j} a_k(X(t))$

        if *delayed reactions are scheduled within* $(t, t + \theta]$ **then**

            let k *be the delayed reaction scheduled next at time* $t + \tau$

            if k *is a consuming delayed reaction* **then**

                $X(t + \tau) = X(t) + \nu_k^p$ (update products only)

            else

                $X(t + \tau) = X(t) + \nu_k$

            $t = t + \tau$

        else

            if j *is not a delayed reaction* **then**

                $X(t + \theta) = X(t) + \nu_j$

            else

                record time $t + \theta + \tau_j$ for delayed reaction j with delay τ_j

                if j *is a consuming delayed reaction* **then**

                      $X(t + \theta) = X(t) + \nu_j^s$ (update reactants)

            $t = t + \theta$

end

In a delayed setting, the SSA loses its Markov property and concurrent events
become an issue as non-delayed instantaneous reactions occur while delayed reactions wait to be updated. In our implementation [3] (see Algorithm 1), the
DSSA proceeds as the SSA as long as there are no delayed reactions scheduled
in the next time step. Otherwise, it ignores the waiting time and the reaction
that should be updated beyond the current update point and moves to the scheduled delayed reaction. Furthermore, in order to avoid the possibility of obtaining
negative molecular numbers, reactants and products of delayed consuming reactions must be updated separately, namely when the delayed reaction is selected
and when it is completed, respectively.

Our model is based on the chemical reaction models of Lewis and Horikawa
et al. but our implementation is entirely different as intrinsic noise is represented

Table 1. Model parameters used for DDE and DSSA. Parameter values [5].

Parameter	Description	Rate constant
b_{h1}, b_{h7}	Her1/Her7 protein degradation rate	0.23 min^{-1}
b_d	DeltaC protein degradation rate	0.23 min^{-1}
c_{h1}, c_{h7}	her1/her7 mRNA degradation rate	0.23 min^{-1}
c_d	deltaC mRNA degradation rate	0.23 min^{-1}
a_{h1}, a_{h7}	Her1/Her7 protein synthesis rate (max.)	4.5 min^{-1}
a_d	DeltaC protein synthesis rate (max.)	4.5 min^{-1}
k_{h1}, k_{h7}	her1/her7 mRNA synthesis rate (max.)	33 min^{-1}
k_d	deltaC mRNA synthesis rate (max.)	33 min^{-1}
P_0	critical no. of Her1 + Her7 proteins/cell	40
D_0	critical no. of Delta proteins/cell	1000
τ_{h1m}, τ_{h7m}	time to produce a single her1/her7 mRNA molecule	12.0, 7.1 min
τ_{h1p}, τ_{h7p}	time to produce a single Her1/Her7 protein	2.8, 1.7 min
τ_{dm}	time to produce a single deltaC mRNA molecule	16.0 min
τ_{dp}	time to produce a single DeltaC protein	20.5 min

correctly for each reaction. In the initial state the number of molecules for each species is set to zero.

For the 5-cell model we get 30 different species and a set of 60 reactions. The corresponding rate constants are listed in Table 1. Denote by M_{h1}, M_{h7}, M_d, P_{h1}, P_{h7} and P_d the six species her1 mRNA, her7 mRNA, deltaC mRNA, Her1 protein, Her7 protein and deltaC protein in a particular cell i. For each cell we have 6 (non-delayed) degradations

$$\{M_{h1}, M_{h7}, M_d, P_{h1}, P_{h7}, P_d\} \longrightarrow 0$$

with reaction rate constants c_{h1}, c_{h7}, c_d , b_{h1}, b_{h7}, and b_d, respectively, and propensities $a_{R_1} = c_{h1}M_{h1}$, $a_{R_2} = c_{h7}M_{h7}$, $a_{R_3} = c_dM_d$, $a_{R_4} = b_{h1}P_{h1}$, $a_{R_5} = b_{h7}P_{h7}$, and $a_{R_6} = b_dP_d$. The three translation reactions with delays τ_{h1p}, τ_{h7p}, and τ_{dp} are

$$\{M_{h1}, M_{h7}, M_d\} \longrightarrow \{M_{h1} + P_{h1}, M_{h7} + P_{h7}, M_d + P_d\}$$

with reaction rate constants a_{h1}, a_{h7} and a_d and propensities $a_{R_7} = a_{h1}M_{h1}$, $a_{R_8} = a_{h7}M_{h7}$, and $a_{R_9} = a_dM_d$. The three regulated transcription reactions with delays τ_{h1m}, τ_{h7m}, and τ_{dm} are

$$\{P_{h1}, P_{h7}, P_d\} \longrightarrow \{M_{h1} + P_{h1}, M_{h7} + P_{h7}, M_d + P_d\}$$

with reaction rate constants k_{h1}, k_{h7}, and k_d and corresponding propensities $a_{R_{10}} = k_{h1}f(P_{h1}, P_{h7}, \tilde{P}_D)$, $a_{R_{11}} = k_{h7}f(P_{h1}, P_{h7}, \tilde{P}_D)$, and $a_{R_{12}} = k_dg(P_{h1}, P_{h7})$. For cells 2 to 4 the Hill function f is defined by

$$f(P_{h1}, P_{h7}, \tilde{P}_D) = r_h\frac{1}{1 + (P_{h1}P_{h7})/P_0^2} + r_{hd}\frac{1}{1 + (P_{h1}P_{h7})/P_0^2}\frac{\tilde{P}_D/D_0}{1 + \tilde{P}_D/D_0}$$

with $\tilde{P}_D = (P_D^{n1} + P_D^{n2})/2$ (the average number of P_D for the two neighboring cells $n1$ and $n2$). The parameters r_h and r_{hd} are weight parameters that determine the balance of internal and external contribution of oscillating molecules. With $r_h + r_{hd} = 1$ the coupling strength r_{hd}/r_h can be defined. In our experiments we set $r_{hd} = 1$, that is, the coupling is 100% combinatorial. In accordance with the Horikawa model we used the Hill functions

$$f(P_{h1}, P_{h7}, P_D) = \frac{1}{1 + (P_{h1}P_{h7})/P_0^2} \frac{P_D/D_0}{1 + P_D/D_0},$$

$$f(P_{h1}, P_{h7}, P_D) = \frac{1}{1 + (P_{h1}P_{h7})/P_0^2} \frac{500/D_0}{1 + 500/D_0}$$

for cell 1 and 5, respectively. The Hill function g is given by $g(P_{h1}, P_{h7}) = \frac{1}{1+(P_{h1}P_{h7})/P_0^2}$.

The single cell, single-gene model consists only of 2 species (her1 mRNA and Her1 protein) and 4 reactions. The two degradation and the single translation reactions correspond to those in the 5-cell model. For the inhibitory regulation of transcription we assume a Hill function with Hill coefficient 2 (P_{h1} acts as a dimer). The Hill function takes the form $f(P_{h1}) = 1/(1 + (P_{h1}/P_0)^2)$.

3   Results and Discussion

In this section we present individual simulations of a system of 5 coupled cells, so that the dimension of the system is 30, in both the DSSA and DDE cases.

Figure 2 (a,b) shows the dynamics for a single cell. In the DDE case after an initial overshoot, the amplitudes are completely regular and the oscillatory period is approximately 40 minutes. In the intrinsic noise case there are still sustained oscillations but there is some irregularity in the profiles and the oscillatory period is closer to 50 minutes. The time lag (5-7 min) between protein and mRNA is about the same in both cases. In Fig. 2 (c,d) we present DSSA simulations of the 5 coupled cells and give the profiles for mRNA and protein at deltaC and her1 for cell 3. Now the period of oscillation is closer to 45 minutes and the lag between protein and mRNA is about 25 minutes for deltaC and about 7 minutes for her1. Thus we see that the coupling has some effect on the period of oscillation. In Fig. 3 we mimic an experiment by Horikawa et al. In both the DDE and the DSSA setting we disturb cell 3 after a certain time period (500 minutes in the DSSA case and 260 minutes in the DDE case). This is done by resetting all the values for cell 3 to zero at this point. This is meant to represent the experiment of Horikawa et al. in which some of the cells are replaced by oscillating cells that are out of phase. They then observed that nearly all the cells become resynchronized after three oscillations (90 min.). Interestingly, in the DDE setting it only takes about 60 minutes for the onset of resynchronization while in the DSSA setting it takes about 180 minutes. The difference can be partly due to the larger number of cells that are experimentally transplanted.

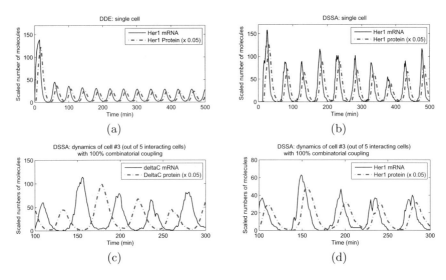

Fig. 2. (a) DDE solution and (b) single DSSA run for the Her1/Her7 single cell model. (c,d) DSSA simulation of five Delta-Notch coupled cells, showing the dynamics of deltaC mRNA and protein and her1 mRNA and protein in cell three.

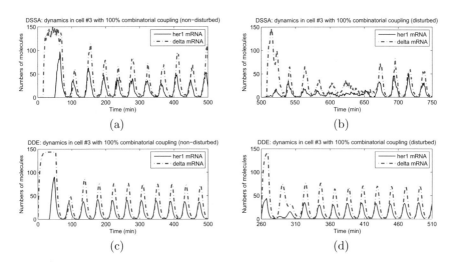

Fig. 3. DSSA simulation result and DDE solution for the 5-cell array in the non-disturbed and disturbed setting. The graphs show the dynamics of deltaC and her1 mRNA in cell three. (a,c) DSSA and DDE results in the non-disturbed setting, respectively. (b,d) DSSA and DDE results in the disturbed setting. Initial conditions for cell 3 are set to zero. All other initial molecular numbers stem from the non-disturbed DSSA and DDE results in (a,c) after 500 and 260 minutes, respectively.

4   Conclusions

In this paper we have simulated Delta-Notch coupled her1/her7 oscillators for 5 cells in both the deterministic (DDE) and delayed, intrinsic noise setting (DSSA). We have shown that there are some similarities between the dynamics of both but the intrinsic noise simulations do make some predictions that are different to the deterministic model (see Fig. 3) that can be verified experimentally. Thus it is important that both intrinsic noise delayed models and continuous deterministic delay models are simulated whenever insights into genetic regulation are being gleaned. However, since the time steps in the DSSA setting can be very small, there are considerable computational overheads in modelling even a chain of 5 cells. In fact, one simulation takes about 90 minutes on a Pentium 4 PC (3.06 GHz) using MatLab 7.2. If we wish to extend these ideas to large cellular systems then we need new multiscale algorithms which will still model intrinsic noise in a delayed setting but will overcome the issues of small stepsizes. This has been considered in the non-delay case by for example Tian and Burrage [11] through their use of τ-leap methods, and similar ideas are needed in the delay setting. This is the subject of further work, along with considerations on how to combine spatial and temporal aspects when dealing with the lack of homogeneity within a cell.

References

1. Monk, N.A.M.: Oscillatory expression of hes1, p53, and nf-κb driven by transcriptional time delays. Curr Biol **13** (2003) 1409–1413
2. Gillespie, D.T.: Exact stochastic simulation of coupled chemical reactions. J Phys Chem **81** (1977) 2340–2361
3. Barrio, M., Burrage, K., Leier, A., Tian, T.: Oscillatory regulation of hes1: discrete stochastic delay modelling and simulation. PLoS Comput Biol **2** (2006) e117
4. Reppert, S.M., Weaver, D.R.: Molecular analysis of mammalian circadian rhythms. Annu Rev Physiol **63** (2001) 647–676
5. Lewis, J.: Autoinhibition with transcriptional delay: a simple mechanism for the zebrafish somitogenesis oscillator. Curr Biol **13** (2003) 1398–1408
6. Hirata, H., Yoshiura, S., Ohtsuka, T., Bessho, Y., Harada, T., Yoshikawa, K., Kageyama, R.: Oscillatory expression of the bhlh factor hes1 regulated by a negative feedback loop. Science **298** (2002) 840–843
7. Jensen, M.H., Sneppen, K., Tiana, G.: Sustained oscillations and time delays in gene expression of protein hes1. FEBS Lett **541** (2003) 176–177
8. Giudicelli, F., Lewis, J.: The vertebrate segmentation clock. Curr Opin Genet Dev **14** (2004) 407–414
9. Horikawa, K., Ishimatsu, K., Yoshimoto, E., Kondo, S., Takeda, H.: Noise-resistant and synchronized oscillation of the segmentation clock. Nature **441** (2006) 719–723
10. Burrage, K., Tian, T., Burrage, P.: A multi-scaled approach for simulating chemical reaction systems. Prog Biophys Mol Biol **85** (2004) 217–234
11. Tian, T., Burrage, K.: Binomial leap methods for simulating stochastic chemical kinetics. J Chem Phys **121** (2004) 10356–10364

Multiscale Modeling of Biopolymer Translocation Through a Nanopore

Maria Fyta[1], Simone Melchionna[2], Efthimios Kaxiras[1], and Sauro Succi[4]

[1] Department of Physics and Division of Engineering and Applied Sciences
Harvard University, Cambridge MA 02138, USA
mfyta@physics.harvard.edu, kaxiras@physics.harvard.edu
[2] INFM-SOFT, Department of Physics, Università di Roma *La Sapienza*
P.le A. Moro 2, 00185 Rome, Italy
Simone.Melchionna@Roma1.infn.it
[3] Istituto Applicazioni Calcolo, CNR, Viale del Policlinico 137, 00161, Rome, Italy
succi@iac.rm.cnr.it

Abstract. We employ a multiscale approach to model the translocation of biopolymers through nanometer size pores. Our computational
scheme combines microscopic Langevin molecular dynamics (MD) with
a mesoscopic lattice Boltzmann (LB) method for the solvent dynamics,
explicitly taking into account the interactions of the molecule with the
surrounding fluid. Both dynamical and statistical aspects of the translocation process were investigated, by simulating polymers of various initial
configurations and lengths. For a representative molecule size, we explore
the effects of important parameters that enter in the simulation, paying
particular attention to the strength of the molecule-solvent coupling and
of the external electric field which drives the translocation process. Finally, we explore the connection between the generic polymers modeled
in the simulation and DNA, for which interesting recent experimental
results are available.

1 Introduction

Biological systems exhibit a complexity and diversity far richer than the simple
solid or fluid systems traditionally studied in physics or chemistry. The powerful
quantitative methods developed in the latter two disciplines to analyze the behavior of prototypical simple systems are often difficult to extend to the domain
of biological systems. Advances in computer technology and breakthroughs in
simulational methods have been constantly reducing the gap between quantitative models and actual biological behavior. The main challenge remains the wide
and disparate range of spatio-temporal scales involved in the dynamical evolution
of complex biological systems. In response to this challenge, various strategies
have been developed recently, which are in general referred to as "multiscale
modeling". These methods are based on composite computational schemes in
which information is exchanged between the scales.

We have recently developed a multiscale framework which is well suited to
address a class of biologically related problems. This method involves different

Y. Shi et al. (Eds.): ICCS 2007, Part I, LNCS 4487, pp. 786–793, 2007.
© Springer-Verlag Berlin Heidelberg 2007

levels of the statistical description of matter (continuum and atomistic) and is able to handle different scales through the spatial and temporal coupling of a *mesoscopic* fluid solvent, using the lattice Boltzmann method [1] (LB), with the atomistic level, which employs explicit molecular dynamics (MD). The solvent dynamics does not require any form of statistical ensemble averaging as it is represented through a discrete set of pre-averaged probability distribution functions, which are propagated along straight particle trajectories. This dual field/particle nature greatly facilitates the coupling between the mesoscopic fluid and the atomistic level, which proceeds seamlessy in time and only requires standard interpolation/extrapolation for information-transfer in physical space. Full details on this scheme are reported in Ref. [2]. We must note that to the best of our knowledge, although LB and MD with Langevin dynamics have been coupled before [3], this is the first time that such a coupling is put in place for long molecules of biological interest.

Motivated by recent experimental studies, we apply this multiscale approach to the translocation of a biopolymer through a narrow pore. These kind of biophysical processes are important in phenomena like viral infection by phages, inter-bacterial DNA transduction or gene therapy [4]. In addition, they are believed to open a way for ultrafast DNA-sequencing by reading the base sequence as the biopolymer passes through a nanopore. Experimentally, translocation is observed *in vitro* by pulling DNA molecules through micro-fabricated solid state or membrane channels under the effect of a localized electric field [5]. From a theoretical point of view, simplified schemes [6] and non-hydrodynamic coarse-grained or microscopic models [7,8] are able to analyze universal features of the translocation process. This, though, is a complex phenomenon involving the competition between many-body interactions at the atomic or molecular scale, fluid-atom hydrodynamic coupling, as well as the interaction of the biopolymer with wall molecules in the region of the pore. A quantitative description of this complex phenomenon calls for state-of-the art modeling, towards which the results presented here are directed.

2   Numerical Set-Up

In our simulations we use a three-dimensional box of size $N_x \times N_x/2 \times N_x/2$ in units of the lattice spacing Δx. The box contains both the polymer and the fluid solvent. The former is initialized via a standard self-avoiding random walk algorithm and further relaxed to equilibrium by Molecular Dynamics. The solvent is initialized with the equilibrium distribution corresponding to a constant density and zero macroscopic speed. Periodicity is imposed for both the fluid and the polymer in all directions. A separating wall is located in the mid-section of the x direction, at $x/\Delta x = N_x/2$, with a square hole of side $h = 3\Delta x$ at the center, through which the polymer can translocate from one chamber to the other. For polymers with up to $N = 400$ beads we use $N_x = 80$; for larger polymers $N_x = 100$. At $t = 0$ the polymer resides entirely in the right chamber at

$x/\Delta x > N_x/2$. The polymer is advanced in time according to the following set of Molecular Dynamics-Langevin equations for the bead positions $\boldsymbol{r}_p$ and velocities $\boldsymbol{v}_p$ (index p runs over all beads):

$$M_p \frac{d\boldsymbol{v}_p}{dt} = -\sum_q \partial_{\boldsymbol{r}_p} V_{LJ}(\boldsymbol{r}_p - \boldsymbol{r}_q) + \gamma(\boldsymbol{u}_p - \boldsymbol{v}_p) + M_p \boldsymbol{\xi}_p - \lambda_p \partial_{\boldsymbol{r}_p} \kappa_p \quad (1)$$

These interact among themselves through a Lennard-Jones potential with $\sigma = 1.8$ and $\varepsilon = 10^{-4}$:

$$V_{LJ}(r) = 4\varepsilon \left[\left(\frac{\sigma}{r}\right)^{12} - \left(\frac{\sigma}{r}\right)^6 \right] \quad (2)$$

This potential is augmented by an angular harmonic term to account for distortions of the angle between consecutive bonds. The second term in Eq.(1) represents the mechanical friction between a bead and the surrounding fluid, $\boldsymbol{u}_p$ is the fluid velocity evaluated at the bead position and γ the friction coefficient. In addition to mechanical drag, the polymer feels the effects of stochastic fluctuations of the fluid environment, through the random term, $\boldsymbol{\xi}_p$. This is related to the third term in Eq.(1), which is an incorrelated random term with zero mean. Finally, the last term in Eq.(1) is the reaction force resulting from $N-1$ holonomic constraints for molecules modelled with rigid covalent bonds. The bond length is set at $b = 1.2$ and M_p is the bead mass equal to 1.

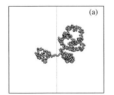 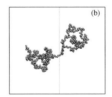 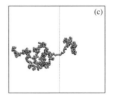

Fig. 1. Snapshots of a typical event: a polymer ($N = 300$) translocating from the right to the left is depicted at a time equal to (a) 0.11, (b) 0.47, and (c) 0.81 of the total time for this translocation. The vertical line in the middle of each panel shows the wall.

Translocation is induced by a constant electric force (F_{drive}) which acts along the x direction and is confined in a rectangular channel of size $3\Delta x \times \Delta x \times \Delta x$ along the streamwise (x direction) and cross-flow (y, z directions). The solvent density and kinematic viscosity are 1 and 0.1, respectively, and the temperature is $k_B T = 10^{-4}$. All parameters are in units of the LB timestep Δt and lattice spacing Δx, which we set equal to 1. Additional details have been presented in Ref. [2]. In our simulations we use $F_{drive} = 0.02$ and a friction coefficient $\gamma = 0.1$. It should be kept in mind that γ is a parameter governing both the structural relation of the polymer towards equilibrium and the strength of the coupling with the surrounding fluid. The MD timestep is a fraction of the timestep for the LB part $\Delta t = m\Delta t_{MD}$, where m is a constant typically set at $m = 5$. With this parametrization, the process falls in the fast translocation regime, where the total translocation time is much smaller than the Zimm relaxation time. We refer to this set of parameters as our "reference"; we explore the effect of the most important parameters for certain representative cases.

3   Translocation Dynamics

Extensive simulations of a large number of translocation events over $100 - 1000$ initial polymer configurations for each length confirm that most of the time during the translocation process the polymer assumes the form of two almost compact blobs on either side of the wall: one of them (the untranslocated part, denoted by U) is contracting and the other (the translocated part, denoted by T) is expanding. Snapshots of a typical translocation event shown in Fig. 1 strongly support this picture. A radius of gyration $R_I(t)$ (with $I = U, T$) is assigned to each of these blobs, following a static scaling law with the number of beads N_I: $R_I(t) \sim N_I^\nu(t)$ with $\nu \simeq 0.6$ being the Flory exponent for a three-dimensional self-avoiding random walk. Based on the conservation of polymer length, $N_U + N_T = N_{tot}$, an effective translocation radius can be defined as $R_E(t) \equiv (R_T(t)^{1/\nu} + R_U(t)^{1/\nu})^\nu$. We have shown that $R_E(t)$ is approximately constant for all times when the static scaling applies, which is the case throughout the process except near the end points (initiation and completion of the event) [2]. At these end points, deviations from the mean field picture, where the polymer is represented as two uncorrelated compact blobs, occur. The volume of the polymer also changes after its passage through the pore. At the end, the radius of gyration is considerably smaller than it was initially: $R_T(t_X) < R_U(0)$, where t_X is the total translocation time for an individual event. For our reference simulation an average over a few hundreds of events for $N = 200$ beads showed that $\lambda_R = R_T(t_X)/R_U(0) \sim 0.7$. This reveals the fact that as the polymer passes through the pore it is more compact than it was at the initial stage of the event, due to incomplete relaxation.

The variety of different initial polymer realizations produce a scaling law dependence of the translocation times on length [8]. By accumulating all events for each length, duration histograms were constructed. The resulting distributions deviate from simple gaussians and are skewed towards longer times (see Fig. 2(a) inset). Hence, the translocation time for each length is not assigned to the mean, but to the most probable time (t_{max}), which is the position of the maximum in the histogram (noted by the arrow in the inset of Fig. 2(a) for the case $N = 200$). By calculating the most probable times for each length, a superlinear relation between the translocation time τ and the number of beads N is obtained and is reported in Fig. 2(a). The exponent in the scaling law $\tau(N) \sim N^\alpha$ is calculated as $\alpha \sim 1.28 \pm 0.01$, for lengths up to $N = 500$ beads. The observed exponent is in very good agreement with a recent experiment on double-stranded DNA translocation, that reported $\alpha \simeq 1.27 \pm 0.03$ [9]. This agreement makes it plausible that the generic polymers modeled in our simulations can be thought of as DNA molecules; we return to this issue in section 5.

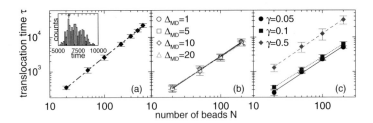

Fig. 2. (a) Scaling of τ with the number of beads N. Inset: distribution of translocation times over 300 events for $N = 200$. Time is given in units of the LB timestep. The arrow shows the most probable translocation time for this length. Effect of the various parameters on the scaling law: (b) changing the value of the MD timestep (Δt_{MD}); (c) changing the value of the solvent-molecule coupling coefficient γ.

4   Effects of Parameter Values

We next investigate the effect that the various parameters have on the simulations, using as standard of comparison the parameter set that we called the "reference" case. For all lengths and parameters about 100 different initial configurations were generated to assess the statistical and dynamical features of the translocation process. As a first step we simulate polymers of different lengths ($N = 20 - 200$). Following a procedure similar to the previous section we extract the scaling laws for the translocation time and their vatiation with the friction coefficient γ and the MD timestep Δt_{MD}. The results are shown in Fig. 2(b) and (c). In these calculations the error bars were also taken into account. The scaling exponent for our reference simulation ($\gamma = 0.1$) presented in Fig. 2(a) is $\alpha \simeq 1.27 \pm 0.01$ when only the lengths up to $N = 200$ are included. The exponent for smaller damping ($\gamma = 0.05$) is $\alpha \simeq 1.32 \pm 0.06$, and for larger ($\gamma = 0.5$) $\alpha \simeq 1.38 \pm 0.04$. By increasing γ by one order of magnitude the time scale rises by approximately one order of magnitude, showing an almost linear dependence of the translocation time with hydrodynamic friction; we discuss this further in the next section. However, for larger γ, thus overdamped dynamics and smaller influence of the driving force, the deviation from the $\alpha = 1.28$ exponent suggests a systematic departure from the fast translocation regime. Similar analysis for various values of Δt_{MD} shows that the exponent becomes $\alpha \simeq 1.34 \pm 0.04$ when Δt_{MD} is equal to the LB timestep ($m = 1$); for $m = 10$ the exponent is $\alpha \simeq 1.32 \pm 0.04$, while for $m = 20$, $\alpha \simeq 1.28 \pm 0.01$ with similar prefactors.

We next consider what happens when we fix the length to $N = 200$ and vary γ and the pulling force F_{drive}. For all forces used, the process falls in the fast translocation regime. The most probable time (t_{max}) for each case was calculated and the results are shown in Fig. 3. The dependence of t_{max} on γ is linear related to the linear dependence of τ on γ, mentioned in the previous section. The variation of t_{max} with F_{drive} follows an inverse power law: $t_{max} \sim 1/F_{drive}^{\mu}$,

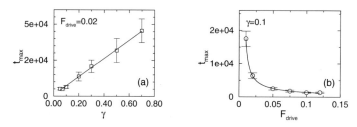

Fig. 3. Variation of t_{max} with (a) γ, and (b) F_{drive} for $N = 200$ beads

with μ of the order 1. The effect of γ is further explored in relation to the effective radii of gyration R_E, and is presented in Fig. 4. The latter must be constant when the static scaling $R \sim N^{0.6}$ holds. This is confirmed for small γ up to about 0.2. As γ increases, R_E is no more constant with time, and shows interesting behavior: it increases continuously up to a point where a large fraction of the chain has passed through the pore and subsequently drops to a value smaller than the initial $R_U(0)$. Hence, as γ increases large deviations from the static scaling occur and the translocating polymer can no longer be represented as two distinct blobs. In all cases, the translocated blob becomes more compact. For all values of γ considered, λ_R is always less than unity ranging from 0.7 (γ=0.1) to 0.9 (γ=0.5) following no specific trend with γ.

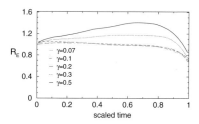

Fig. 4. The dependence of the effective radii of gyration $R_E(t)$ on γ ($N = 200$). Time and R_E are scaled with respect to the total translocation time and $R_U(0)$ for each case.

5   Mapping to Real Biopolymers

As a final step towards connecting our computer simulations to real experiments and after having established the agreement in terms of the scaling behavior, we investigate the mapping issue of the polymer beads to double-stranded DNA. In order to interpret our results in terms of physical units, we turn to the persistence length (l_p) of the semiflexible polymers used in our simulations. Accordingly, we use the formula for the fixed-bond-angle model of a worm-like chain [10]:

$$l_p = \frac{b}{1 - \cos\langle\theta\rangle} \qquad (3)$$

where $\langle \theta \rangle$ is complementary to the average bond angle between adjacent bonds. In lattice units (Δx) an average persistence length for the polymers considered, was found to be approximately 12. For λ-phage DNA $l_p \sim 50$ nm [11] which is set equal to l_p for our polymers. Thereby, the lattice spacing is $\Delta x \sim 4$ nm, which is also the size of one bead. Given that the base-pair spacing is ~ 0.34 nm, one bead maps approximately to 12 base pairs. With this mapping, the pore size is about ~ 12 nm, close to the experimental pores which are of the order of 10 nm. The polymers presented here correspond to DNA lengths in the range $0.2 - 6$ kbp. The DNA lengths used in the experiments are larger (up to ~ 100kbp); the current multiscale approach can be extended to handle these lengths, assuming that appropriate computational resources are available.

Choosing polymer lengths that match experimental data we compare the corresponding experimental duration histograms (see Fig. 1c of Ref. [9]) to the theoretical ones. This comparison sets the LB timestep to $\Delta t \sim 8$ nsec. In Fig. 5 the time distributions for representative DNA lengths simulated here are shown. In this figure, physical units are used according to the mapping described above and promote comparison with similar experimental data [9]. The MD timestep for $m = 5$ will then be $t_{MD} \sim 40$ nsec indicating that the MD timescale related to the coarse-grained model that handles the DNA molecules is significantly stretched over the physical process. Exact match to all the experimental parameters is of course not feasible with coarse-grained simulations. However, essential features of DNA translocation are captured, allowing the use of the current approach to model similar biophysical processes that involve biopolymers in solution. This can become more efficient by exploiting the freedom of further fine-tuning the parameters used in this multiscale model.

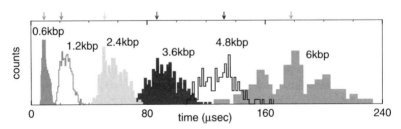

Fig. 5. Histograms of calculated translocation times for a large number of events and different DNA lengths. The arrows link to the most probable time (t_{max}) for each case.

6   Conclusions

In summary, we applied a multiscale methodology to model the translocation of a biopolymer through a nanopore. Hydrodynamic correlations between the polymer and the surrounding fluid have explicitly been included. The polymer obeys a static scaling except near the end points for each event (initiation and completion of the process) and the translocation times vary exponentially with the polymer length. A preliminary exploration of the effects of the most important parameters used in our simulations was also presented, specifically the values of the friction coefficient and the pulling force describing the effect of the external electric field that drives the translocation. These were found to significantly affect the dynamic features of the process. Finally, our generic polymer models were directly mapped to double-stranded DNA and a comparison to experimental results was discussed.

Acknowledgments. MF acknowledges support by Harvard's Nanoscale Science and Engineering Center, funded by NSF (Award No. PHY-0117795).

References

1. Wolf-Gladrow, D. A.: Lattice gas cellular automata and lattice Boltzmann models. Springer Verlag, New York 2000; Succi, S.: The lattice Boltzmann equation. Oxford University Press, Oxford 2001; Benzi, R. Succi, S., and Vergassola, M.:, The lattice Boltzmann-equation - Theory and applications. Phys. Rep. **222** (1992) 145–197.
2. Fyta, M. G., Melchionna, S., Kaxiras, E., and Succi, S.: Multiscale coupling of molecular dynamics and hydrodynamics: application to DNA translocation through a nanopore. Multiscale Model. Simul. **5** (2006) 1156–1173.
3. Ahlrichs, P. and Duenweg, B.: Lattice-Boltzmann simulation of polymer-solvent systems. Int. J. Mod. Phys. C **9** (1999) 1429–1438; Simulation of a single polymer chain in solution by combining lattice Boltzmann and molecular dynamics. J. Chem. Phys. **111** (1999) 8225–8239.
4. Lodish, H., Baltimore, D., Berk, A., Zipursky, S., Matsudaira, P., and Darnell, J.: Molecular Cell Biology, W.H. Freeman and Company, New York (1996).
5. Kasianowicz, J. J., et al: Characterization of individual polynucleotide molecules using a membrane channel. Proc. Nat. Acad. Sci. USA **93** (1996) 13770–13773; Meller, A., et al: Rapid nanopore discrimination between single polynucleotide molecules. **97** (2000) 1079–1084; Li, J., et al: DNA molecules and configurations in a solid-state nanopore microscope. Nature Mater. **2** (2003) 611–615.
6. Sung, W. and Park, P. J.: Polymer translocation through a pore in a membrane. Phys. Rev. Lett. **77** (1996) 783–786.
7. Matysiak, S., et al: Dynamics of polymer translocation through nanopores: Theory meets experiment. Phys. Rev. Lett. **96** (2006) 118103.
8. Lubensky, D. K. and Nelson, D. R.: Driven polymer translocation through a narrow pore. Biophys. J. **77** (1999) 1824–1838.
9. Storm, A. J. et al: Fast DNA translocation through a solid-state nanopore. Nanolett. **5** (2005) 1193–1197.
10. Yamakawa, H.: Modern Theory of Polymer Solutions, Harper & Row, NY 1971.
11. Hagerman, P. J.: Flexibility of DNA. Annu. Rev. Biophys. Biophys. Chem. **17** (1988) 265–286; Smith, S., Finzi, L., and Bustamante, C.: Direct mechanical measurement of the elasticity of single DNA molecules by using magnetic beads. Science **258** (1992), 1122–1126.

Multi-physics and Multi-scale Modelling in Cardiovascular Physiology: Advanced User Methods for Simulation of Biological Systems with ANSYS/CFX

V. Díaz-Zuccarini[1], D. Rafirou[2], D.R. Hose[1], P.V. Lawford[1], and A.J. Narracott[1]

[1] University of Sheffield, Academic Unit of Medical Physics, Royal Hallamshire Hospital, Glossop Road, S10 2JF, Sheffield, UK
[2] Electrical Engineering Department / Biomedical Engineering Center, Technical University of Cluj-Napoca, 15, C. Daicoviciu Street, 400020 Cluj-Napoca, Romania
v.diaz@sheffield.ac.uk

Abstract. This work encompasses together a number of diverse disciplines (physiology, biomechanics, fluid mechanics and simulation) in order to develop a predictive model of the behaviour of a prosthetic heart valve in vivo. The application of simulation, for the study of other cardiovascular problems, such as blood clotting is also discussed. A commercial, finite volume, computational fluid dynamics (CFD) code (ANSYS/CFX) is used for the 3D component of the model. This software provides technical options for advanced users which allow user-specific variables to be defined that will interact with the flow solver. User-defined functions and junction boxes offer appropriate options to facilitate multi-physics and multi-scale complex applications. Our main goal is to present a 3D model using the internal features available in ANSYS/CFX coupled to a multiscale model of the left ventricle to address complex cardiovascular problems.

Keywords: Multi-scale models, cardiovascular modelling & simulation, mechanical heart valves, coupled 3D-lumped parameter models, simulation software.

1 Introduction

Modelling and simulation can be used to explore complex interactions which occur in the human body. In order to model these biological processes, a multi-scale approach is needed. One of the "success stories" in bioengineering is the study of the fluid dynamics of blood (haemodynamics) within the cardiovascular system and the relationship between haemodynamics and the development of cardiovascular disease [1].

Cardiovascular models present a particular challenge in that they require both a multi-scale and a multi-physics approach. Using finite elements for the whole system is computationally prohibitive thus, a compromise is needed. The most sophisticated fluid-solid interaction structures [2] provide exquisite detail in the fluid domain, but are limited in that the boundary conditions are prescribed. An alternative multiscale solution is to couple lumped parameter models of the boundary conditions with a finite element model of the part where detail and accuracy are needed [3]. Significant improvement can be made in terms of the understanding of the underlying physics if the lumped parameter approach includes more physiologically representative

Y. Shi et al. (Eds.): ICCS 2007, Part I, LNCS 4487, pp. 794–801, 2007.

mechanisms rather than traditional "black-box" models. Usually, boundary conditions are expressed in terms of pressure and flow. These are the macroscopic expression of physiological or pathological conditions. These macroscopic variables can be related to the microscopic level of the physiology/pathology through molecular/cellular aspects of the process with the definition of a new range of variables. These new variables will not be available to the user as normal boundary conditions, as they are totally dependant on the level of modelling chosen for the problem under study.

2 ANSYS/CFX Special Features: Crossing the Flow Interface Towards Biological Modelling

ANSYS/CFX is highly specialized software, built in such a way that interaction with specific features of the solver is kept to a minimum. Whilst this is appropriate for many applications, in order to run multi-scale applications, the user must be able to interact with the solver at a lower level. The user needs to be able to define where, when and how the equations and variables imposed as boundaries (i.e. mathematical models) will participate in the solution. In this paper we describe two options which ANSYS/CFX provides for advanced users: Command Expression Language (CEL) functions/subroutines and Junction Boxes. We will use CEL functions/subroutines at the flow interface to solve the equations of the boundary conditions at different scales and Junction Boxes to provide a structure to update the variables.

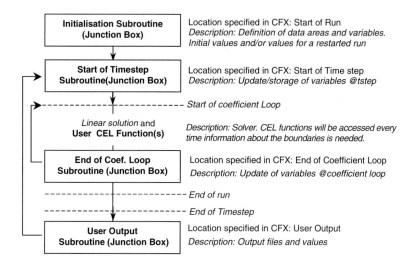

Fig. 1. Structure of an ANSYS/CFX model using Junction Boxes and CEL Subroutines

Definitions [4]:
- *CEL Subroutines*: These are used in conjunction with User CEL Functions to define quantities in ANSYS CFX-Pre based on a FORTRAN subroutine. The User CEL Routine is used to set the calling name, the location of the Subroutine and the location of the Shared Libraries associated with the Subroutine.

- *CEL Functions*: User CEL Functions are used in conjunction with User CEL Routines. User Functions set the name of the User CEL Routine associated with the function, and the input and return arguments.
- *Junction Box Routines*: Are used to call user-defined FORTRAN subroutines during execution of the ANSYS CFX-Solver. A Junction Box Routine object must be created so that the ANSYS CFX-Solver knows the location of the Subroutine, the location of its Shared Libraries and when to call the Subroutine.

A general form of a transient and structured model in ANSYS/CFX is shown in Fig. 1. When solving differential or partial differential equation models as boundary conditions for an ANSYS/CFX model the equations describing the boundary condition must be discretized in such a way that variables are solved and passed to ANSYS/CFX at each time-step and updated within coefficient loops. Using ANSYS/CFX functions, the value of physical quantities at specific locations of the 3D modelare passed as arguments to the boundary condition model and a specific value is returned to ANSYS/CFX to update the boundary condition for that region. Using this approach, the boundary conditions of the 3D model are updated at each coefficient loop of the flow solver, providing close coupling of the 3D ANSYS/CFX model and the boundary condition model.

3   Case Study: Coupling a Model of the Left Ventricle to an Idealized Heart Valve

This section describes the development of more complex boundary conditions to represent the behaviour of the left ventricle (LV). We also describe the coupling of the LV model to a fluid-structure interaction (FSI) model of an idealized heart valve.

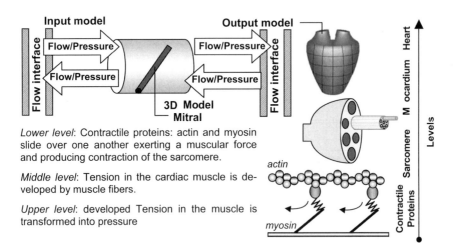

Fig. 2. Representation of the LV model and its different physical scales. Cardiac contraction starts at protein level in the sarcomere. Protein contraction produces tension in the cardiac wall that then is translated into pressure.

Fig. 2 shows the sub-levels of organisation of the ventricle that are taken into account to give a "more physiologically realistic" model of the LV. Following the procedure described in *ex vivo* isolated LV preparations, a constant pressure source is connected to the ventricle via the mitral valve (input model). The blood fills the LV (output model) and ejects a volume of blood into the arterial network via the aortic valve. Contraction in the cardiac muscle is described starting from the microscopic level of contractile proteins (actin and myosin), up to the tissue (muscle level) and to the LV level to finally reach the haemodynamic part of the LV and its arterial load.

3.1 A Multi-scale Model of the Left Ventricle

The contractile proteins (actin and myosin) acting at the level of the sarcomere produce cardiac contraction. These slide over each other [5], attaching and detaching by means of cross-bridges, to produce shortening of the sarcomere and hence contraction of the ventricular wall (Fig. 2). To represent the chemical reaction from the detached state (X_D) to the attached one (X_A) and vice-versa we use the simplest model available:

$$X_D \underset{k_d}{\overset{k_a}{\rightleftarrows}} X_A \tag{1}$$

The periodically varying kinetic constants of attachment and detachment are k_a and k_d. The chemical potential of a thermodynamic system is the amount by which the energy of the system would change if an additional particle were introduced, with the entropy and volume held fixed. The modified Nernst formula provides a first approximation:

$$\mu_a = \left(A_a + B_a(T)\ln(X_A)\right)\beta(l_m) \tag{2}$$

$$\mu_d = A_d + B_d(T)\ln(X_D) \tag{3}$$

The only non-classical feature is the β factor of equation (2). For a more detailed explanation, the reader is referred to [6],[7]. Using the usual expression of the reaction rate in chemical thermodynamics, the reaction flow may be expressed as:

$$\frac{dX_A}{dt} = k_A(t)\cdot e^{(\mu_d - A_d)/B_d} - k_D(t)\cdot e^{(\mu_a - \beta A_a)/B_a} \tag{4}$$

This can be reduced to the following expression:

$$\frac{dX_A}{dt} = k_A(t)(1 - X_A(t)) - k_D(t)X_D(t)^\beta \tag{5}$$

We can now relate the chemical and mechanical aspects of the model, to describe the *chemo-mechanic* transduction in the cardiac muscle. A single muscle equivalent to all the muscles of the heart is considered. Its length is l_m and X_A is its number of attached cross bridges. Neglecting the energy in free cross bridges, the energy of the muscle is given by:

$$E_m = E_m(l_m, X_A) \tag{6}$$

The mechanical power is obtained by derivation:

$$P_m = \frac{\partial E_m}{\partial l_m}\frac{dl_m}{dt} + \frac{\partial E_m}{\partial X_a}\frac{dX_A}{dt} = f_m v_m + \mu_a \dot{X}_A \tag{7}$$

As shown in equation (7), the mechanical force is obtained by derivation of the energy equation of the muscle with respect to the attached cross-bridges (E_X) and the length of muscle fibers (l_m).

$$f_m = E_X(X_A)E'_a(l_m) + (1 - E_X(X_A))E'_r(l_m) \tag{8}$$

E'_a and E'_r are respectively, the derivative of the energy of the muscle in active (a) and resting (r) state respect to l_m and they are expressed as functions of the muscular force f_m. Full details of the formulation and assumptions may be found in [6],[7]. In our model, the cardiac chamber transforms muscular force into pressure and volumetric flow into velocity of the muscle. We assume an empiric relationship φ_G between volume Q_{LV} and equivalent fibre length l_m

$$Q_{LV}(t) = \varphi_G(l_m(t)) \leftrightarrow l_m(t) = \varphi_G^{-1}(Q_{LV}(t)) \tag{9}$$

Assuming φ_G is known and by differentiation of (9), a relationship between the velocity of the muscle dl_m/dt and the ventricular volumetric flow dQ_{LV}/dt is obtained:

$$\frac{dQ_{lv}(t)}{dt} = \psi_G \frac{dl_m}{dt} \tag{10}$$

Where $\psi_G = \psi_G(Q_{LV})$ is the derivative of φ_G with respect to l_m. If the transformation is power-conservative, mechanical power (depending on muscular force f_m) =hydraulic power (depending on pressure P).

$$P \cdot \frac{dQ_{lv}}{dt} = N \cdot f_m \cdot \frac{dl_m}{dt} \tag{11}$$

Where we have supposed that pressure P is created from the forces of N identical fibres. Equation (12) is then obtained:

$$P = \frac{1}{\psi_G(l_m)} N.f_m \tag{12}$$

A simple multi-scale model of the left ventricle has been presented, starting from the mechanisms of cardiac contractions at the protein scale. In our case, the ventricular pressure P defined (12) will be the input to the 3D model shown in the next section.

3.2 Computational Model of the Mitral Valve and Coupling with the LV Model

The computational approach presented here is based on the use of a fluid-structure interaction model, built within ANSYS/CFX v10. It describes the interaction of a blood-like fluid flowing under a pressure gradient imposed by ventricular contraction, and the structure of a single leaflet valve. A generic CAD model, consisting of a flat leaflet (the occluder) which moves inside a mounting ring (the housing), was built to represent the geometry of the valve (Fig. 3). The thickness of the disc-shaped occluder is h=2 mm and its diameter is d=21 mm. In the fully closed position, the gap between the mounting ring and the occluder is g=0.5 mm. The movement of the occluder is considered to be purely rotational, acting around an eccentric axis situated at distance   from its centroid (OX axis). The occluder travels 80 degrees from the fully open to the fully closed position,. To complete the CAD model, two cylindrical chambers were added to represent the atrium and ventricle. The atrial chamber, positioned in the positive OZ direction, is 66 mm in length whilst the ventricular chamber, positioned in the negative OZ direction, is 22 mm in length. Only half of the valve was considered in the 3D model. Blood is considered an incompressible Newtonian fluid with density ρ =1100 kg/m3 and dynamic viscosity μ=0.004 kg/ms.

The unsteady flow field inside the valve is described by the 3D equations of continuity and momentum (13) with the boundary conditions suggested by Fig. 3 (left).

$$\rho\frac{\partial \vec{u}}{\partial t}+\left(\vec{u}\cdot\nabla\right)\vec{u}\rho=-\nabla p+\mu\Delta\vec{u}\ ;\ \nabla\cdot\vec{u}=0 \tag{13}$$

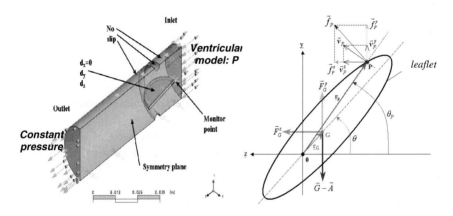

Fig. 3. Left: CAD model and boundary conditions: Snapshot of CFXPre implementation. At the inlet the ventricular pressure (P) is coming from the ventricular model. A constant pressure is applied at the outlet. Right: General representation of a rigid body demonstrating the principles of fluid-structure interaction and rotation of a leaflet. The variables are defined in the text.

The principle used to model the fluid-structure interaction is outlined in Fig 3 (right). The occluder rotates under the combined effects of hydrodynamic, buoyancy and gravitational forces acting on it. At every time step, the drag, $\vec{f}_P^z$, and lift, $\vec{f}_P^y$, forces exerted by the flowing fluid on an arbitrary point P of the leaflet's surface, are reduced to the centroid, G. The total contribution of drag, $\vec{F}_G^z$, and lift $\vec{F}_G^y$ are added to the difference between the gravitational and buoyancy forces. Conservation of the kinetic moment is imposed, resulting in the dynamic equation that describes the motion.

$$d\theta=\varpi_{old}dt+\frac{M}{J}dt^2 \tag{14}$$

In equation 14, $d\theta$ represents the leaflet's angular step, dt is the time step, ϖ_{old} is the angular velocity at the beginning of the current time step and M is the total momentum acted on the leaflet from the external forces:

$$M = r_G \left[F_G^z \sin\theta + (F_G^y - G + A)\cos\theta \right] \qquad (15)$$

The total moment of inertia is J=6.10^{-6} kg.m^2.

Some simulation results of the model are shown in Figure 4:

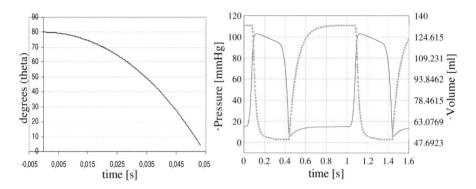

Fig. 4. Left: Closure degree vs. time. Right: Ventricular pressureand Volume vs. time.

It is interesting to notice that from the implementation point of view, closure is continuous and smooth, although when compared to experimental data, closure time is too long [8] and closure is reported too early during systole. This could be due to several reasons, including an insufficient pressure rise and/or an overestimation of the mass of the occluder. This issue is being addressed and will be the object of a future publication. As automatic remeshing tools are not available in the software, remeshing was carried out by hand and the simulation was stopped several times before full closure. Re-start and interpolation tools are available within the software and are appropriate to tackle this particular problem. A restriction in the use of the commercial software chain is that efficient implementation of an automatic remeshing capability is very difficult without close collaboration with the software developers.

4 Towards Other Biological Applications

Another feature of the ANSYS/CFX solver enables the user to define additional variables which can be used to represent other properties of the flow field. Previous work described the use of such additional variables to model the change in viscosity of blood during the formation of blood clots [9]. This is based on the work of Friedrich and Reininger [10] who proposed a model of the clotting process based on a variable blood viscosity determined by the residence time of the blood t, the viscosity at time 0, μ_0, and the rate constants k_1 and k_2, which are dependent on thrombin concentration. Fluid residence time can be modelled within ANSYS/CFX by the introduction of an additional variable which convects throughout the fluid domain. A source term of 1 is defined for the additional variable, resulting in the value of the additional variable for each element of the fluid increasing by a unit for every unit of time that it remains within the domain. A model of blood clotting was implemented within CFX 5.5.1 using an additional variable, labelled AGE, to represent the residence time. Due to the method of implementation in CFX, the residence time is expressed as a density and has units of kgm^{-3}. Additional user functions allow the viscosity of the fluid to be defined as a function of the additional variable. Whilst this is a very simple model, it demonstrates the power of custom routines that can be utilized in such modeling. This technique has possible applications for other convective-diffusive processes.

5 Conclusions and Perspectives

In this paper, we presented a multi-scale and fully coupled model of the LV with a 3D model of the mitral valve in ANSYS/CFX. The preliminary results are encouraging. Use of advanced features of the software allows describing biological applications. A second application demonstrates the versatility of the software.

Acknowledgement

The authors acknowledge financial support under the Marie Curie project C-CAReS and thank Dr I. Jones and Dr J. Penrose, ANSYS-CFX, for constructive advice and training.

References

1. Taylor, C.A., Hughes, T.J., Zarins, C.K. *Finite element modeling of three-dimensional pulsatile flow in the abdominal aorta: relevance to atherosclerosis.* Annals of Biomedical Engineering (1998) 975–987.
2. De Hart J., Peters G.., Schreurs P., Baaijens F. *A three-dimensional computational analysis of fluid–structure interaction in the aortic valve.* J. of Biomech, (2003)103-112.
3. Laganà K, Balossino R, Migliavacca F, Pennati G, Bove E.L, de Leval M, Dubini G. *Multiscale modeling of the cardiovascular system: application to the study of pulmonary and coronary perfusions in the univentricular circulation.* J. of Biomec.(2005)1129-1141.
4. ANSYS/CFX 10.0 manual, © 2005 ANSYS Europe Ltd (http://www.ansys.com).
5. Huxley AF. *Muscle structures and theories of contraction.* Prog Biophys. Chem, (1957) 279-305.
6. Díaz-Zuccarini V, ,LeFèvre J. *An energetically coherent lumped parameter model of the left ventricle specially developed for educational purposes.* Comp in Biol and Med. (In press).
7. Díaz-Zuccarini, V. *Etude des Conditions d'efficacite du ventricule gauche par optimisation teleonomique d'un modele de son fonctionnement.* PhD Thesis, EC-Lille, (2003)
8. Chandran K. B, Dexter, E. U, Aluri S, Richenbacher W. *Negative Pressure Transients with Mechanical Heart-Valve Closure: Correlation between In Vitro and In Vivo Results.* Annals of Biomedical Engineering. Volume 26, Number 4 / July, (1998)
9. Narracott A, Smith S, Lawford P, Liu H, Himeno R, Wilkinson I, Griffiths P, Hose R. *Development and validation of models for the investigation of blood clotting in idealized stenoses and cerebral aneurysms.* J Artif Organs. (2005) 56-62.
10. Friedrich P, Reininger AJ. Occlusive thrombus formation on indwelling catheters: In vitro investigation and computational analysis. Thrombosis and Haemostasis (1995) 66-72.

Lattice Boltzmann Simulation of Mixed Convection in a Driven Cavity Packed with Porous Medium

Zhenhua Chai[1], Zhaoli Guo[1], and Baochang Shi[2,*]

[1] State Key Laboratory of Coal Combustion,
Huazhong University of Science and Technology, 430074 Wuhan P.R. China
hustczh@126.com,
zlguo@mail.hust.edu.cn
[2] Department of Mathematics,
Huazhong University of Science and Technology, 430074 Wuhan P.R. China
sbchust@126.com

Abstract. The problem of mixed convection in a driven cavity packed with porous medium is studied with lattice Boltzmann method. A lattice Boltzmann model for incompressible flow in porous media and another thermal lattice Boltzmann model for solving the energy equation are proposed based on the generalized volume-averaged flow model. The present models have been validated by simulating mixed convection in a driven cavity (without porous medium) and it is found that the numerical results predicted by present models are in good agreement with available data reported in previous studies. Extensive parametric studies on mixed convection in a driven cavity filled with porous medium are carried out for various values of Reynolds number, Richardson number and Darcy number. It is found that the flow and temperature patterns change greatly with variations of these parameters.

Keywords: Lattice Boltzmann method; Mixed convection; Porous medium.

1 Introduction

Fluid flow and heat transfer in a driven cavity have recently received increasing attention because of its wide applications in engineering and science [1,2]. Some of these applications include oil extraction, cooling of electronic devices and heat transfer improvement in heat exchanger devices [3]. From a practical point of view, the research on mixed convection in a driven cavity packed with porous medium is motivated by its wide applications in engineering such as petroleum reservoirs, building thermal insulation, chemical catalytic reactors, heat exchangers, solar power collectors, packed-bed catalytic reactors, nuclear energy

---

* Corresponding author.

Y. Shi et al. (Eds.): ICCS 2007, Part I, LNCS 4487, pp. 802–809, 2007.
© Springer-Verlag Berlin Heidelberg 2007

systems and so on [3,4]. These important applications have led to extensive investigations in this area [3,5,6,7].

In this paper, the problem of mixed convection in a driven cavity packed with porous medium is studied with the lattice Boltzmann method (LBM). The aim of the present study is to examine the effects of Reynolds number (Re), Richardson number (Ri) and Darcy number (Da) on characteristics of the flow and temperature fields. The numerical results in present work indicate that the flow and temperature patterns change greatly with variations of the parameters mentioned above. Furthermore, as these parameters are varied in a wide range, some new phenomena is observed.

2   Numerical Method: The Lattice Boltzmann Method

In the past two decades, the LBM achieved great success in simulating complex fluid flows and transport phenomena since its emergence [8]. In the present work, the LBM is extended to study mixed convection in a driven cavity filled with porous medium.

The dimensionless generalized volume-averaged Navier-Stokes equations and energy equation are written as [9,10]

$$\nabla \cdot \mathbf{u} = 0, \tag{1}$$

$$\frac{\partial \mathbf{u}}{\partial t} + \mathbf{u} \cdot \nabla(\frac{\mathbf{u}}{\epsilon}) = -\nabla(\epsilon p) + \frac{1}{Re_e}\nabla^2 \mathbf{u} + \mathbf{F}, \tag{2}$$

$$\sigma\frac{\partial T}{\partial t} + \mathbf{u} \cdot \nabla T = \frac{1}{PrRe}\nabla^2 T, \tag{3}$$

where $\mathbf{u}$ and p are the volume-averaged velocity and pressure, respectively; ϵ is the porosity of the medium, Re_e is the effective Reynolds number, Pr is the Prandtl number, $\sigma = \epsilon + (1 - \epsilon)\rho_s c_{ps}/\rho_f c_{pf}$ represents the ratio between the heat capacities of the solid and fluid phases, with $\rho_s(\rho_f)$ and $c_{ps}(c_{pf})$ being the density and capacity of the solid (fluid) phase, respectively. $\mathbf{F} = -\frac{\epsilon}{DaRe}\mathbf{u} - \frac{\epsilon F_\epsilon}{\sqrt{Da}}|\mathbf{u}|\mathbf{u} + \frac{\epsilon Gr}{Re_e^2}\mathbf{k}T$, where Da is the Darcy number, Re is the Reynolds number, which is assumed to equal Re_e in the present work, we would like to point out that this assumption is widely used in engineering; Gr is the Grashof number, for simplicity, the Richardson number (Ri) is introduced, and defined as $\frac{Gr}{Re_e^2}$, $\mathbf{k}$ is unit vector in the y-direction, F_ϵ is geometric function, and defined as [11]

$$F_\epsilon = \frac{1.75}{\sqrt{150 \times \epsilon^3}}.$$

The evolution equations of the single particle density distribution and temperature distribution can be written as [10,12]

$$f_i(\mathbf{x} + c_i\delta t, t + \delta t) - f_i(\mathbf{x}, t) = -\frac{1}{\tau_f}[f_i(\mathbf{x}, t) - f_i^{(eq)}(\mathbf{x}, t)] + \delta t F_i, \tag{4}$$

$$g_i(\mathbf{x} + c_i\delta t, t + \delta t) - g_i(\mathbf{x}, t) = -\frac{1}{\tau_g}[g_i(\mathbf{x}, t) - g_i^{(eq)}(\mathbf{x}, t)], \tag{5}$$

where δt is time step, $f_i(\mathbf{x}, t)$ and $g_i(\mathbf{x}, t)$ are density distribution function and temperature distribution function, respectively, τ_f and τ_g are the dimensionless relaxation times. $f_i^{(eq)}(\mathbf{x}, t)$, $g_i^{(eq)}(\mathbf{x}, t)$ are the equilibrium distribution functions corresponding to $f_i(\mathbf{x}, t)$ and $g_i(\mathbf{x}, t)$, which are given as

$$f_i^{(eq)}(\mathbf{x}, t) = \omega_i \rho [1 + \frac{c_i \cdot \mathbf{u}}{c_s^2} + \frac{(c_i \cdot \mathbf{u})^2}{2\epsilon c_s^4} - \frac{|\mathbf{u}|^2}{2\epsilon c_s^2}], \qquad (6)$$

$$g_i^{(eq)}(\mathbf{x}, t) = \omega_i T [\sigma + \frac{c_i \cdot \mathbf{u}}{c_s^2}], \qquad (7)$$

where ω_i is weight coefficient and c_s is the sound speed. Unless otherwise stated, σ in Eqs. (3) and (7) is assumed to equal 1, and the same treatment can be found in Ref. [9]. In the present work, we choose two-dimensional nine-bit model where the discrete velocities are given as $c_0 = (0, 0)$, $c_i = (\cos[(i-1)\pi/2], \sin[(i-1)\pi/2])c$ $(i = 1 - 4)$, $c_i = (\cos[(2i-9)\pi/4], \sin[(2i-9)\pi/4])\sqrt{2}c$ $(i = 5 - 8)$, where $c = \delta x / \delta t$, δx is lattice spacing. The sound speed in this D2Q9 model is given as $c_s = c/\sqrt{3}$, and the weights are $\omega_0 = 4/9$, $\omega_i = 1/9$ $(i = 1 - 4)$, $\omega_i = 1/36$ $(i = 5 - 8)$.

The forcing term F_i in Eq. (4) is given as [12]

$$F_i = \omega_i \rho (1 - \frac{1}{2\tau}) [\frac{c_i \cdot F}{c_s^2} + \frac{\mathbf{u} F : (c_i c_i - c_s^2 I)}{\epsilon c_s^4}]. \qquad (8)$$

The volume-averaged density and velocity are defined by

$$\rho = \sum_{i=0}^{8} f_i(\mathbf{x}, t), \quad \mathbf{u} = \frac{\mathbf{v}}{d_0 + \sqrt{d_0^2 + d_1 \mathbf{v}}},$$

where d_0, d_1 and $\mathbf{v}$ are defined as $d_0 = \frac{1}{2}(1 + \frac{\delta t}{2} \frac{\epsilon}{DaRe})$, $d_1 = \frac{\delta t}{2} \frac{\epsilon F_\epsilon}{\sqrt{Da}}$, $\rho \mathbf{v} = \sum_{i=0}^{8} c_i f_i + \frac{\delta t}{2} \rho \frac{\epsilon Gr}{Re_e} \mathbf{k} T$.

Through the Chapman-Enskog expansion, and in the incompressible limit, we can derive the macroscopic equations (1)-(3). The further detailed analysis on this procedure can be found in Refs. [10,12]. In addition, the boundary conditions should be treated carefully, here the non-equilibrium extrapolation scheme proposed in Ref. [13] is used, this is because this scheme has exhibited better numerical stability in numerical simulations.

3   Numerical Results and Discussion

The configuration described in present study is shown in Fig. 1. The geometry is a square cavity with a length $L = 1$. The cavity is packed with a porous material that is homogeneous and isotropic.

The present models were validated by simulating mixed convection in a square cavity with a driving lid for $Re = 100$, 400, 1000 and the corresponding $Ri = 1 \times 10^{-2}$, 6.25×10^{-4}, 1×10^{-4}. The results were compared with available data

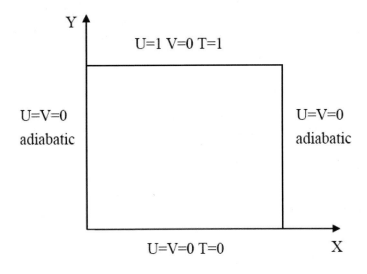

Fig. 1. The configuration of the problem under consideration

reported in previous studies in the absence of porous medium and heat generation for $Pr = 0.71$. It should be noted that the solution of mixed convection in the cavity without porous medium can be derived if $Da \to \infty$, $\epsilon \to 1$ and $\sigma = 1$. As shown in Tables 1 and 2, the present numerical results are in good agreement with those reported in previous studies. In the following parts, we will focus on

Table 1. Comparisons of the average Nusselt number $(\overline{Nu})$ at the top wall between the present work and that reported in previous studies

Re	Present	Ref.[2]	Ref.[6]	FIDAP[6]	Ref.[5]
100	1.97	1.94	2.01	1.99	2.01
400	4.03	3.84	3.91	4.02	3.91
1000	6.56	6.33	6.33	6.47	6.33

Table 2. Comparisons of the maximum and minimum values of the horizontal and vertical velocities at the center lines of the cavity between the present work and those reported in previous studies

	$Ri = 1.0 \times 10^{-2}$				$Ri = 6.25 \times 10^{-4}$			
	Present	Ref.[2]	Ref.[6]	Ref.[5]	Present	Ref.[2]	Ref.[6]	Ref.[5]
u_{min}	-0.2079	-0.2037	-0.2122	-0.2122	-0.3201	-0.3197	-0.3099	-0.3099
u_{max}	1.0000	1.0000	1.0000	1.0000	1.0000	1.0000	1.0000	1.0000
v_{min}	-0.2451	-0.2448	-0.2506	-0.2506	-0.4422	-0.4459	-0.4363	-0.4363
v_{max}	0.1729	0.1699	0.1765	0.1765	0.2948	0.2955	0.2866	0.2866

investigating the effect of variations of parameters, including Re, Ri and Da, on flow and temperature fields. The porosity of the medium is set to be 0.5, and the Prandtl number is set to equal 0.71. Numerical simulations are carried out on a 257×257 lattice.

3.1   Effect of the Reynolds Number (Re)

The range of Re is tested from 400 to 5000 under the condition of $Ri = 0.0001$ and $Da = 0.01$. As shown in Fig. 2, the variation of Re has an important impact on flow and temperature fields. It is found that the qualitative character of flow is similar to the convectional lid-driven cavity flow of non-stratified fluid, a primary vortex is formed in center region of the cavity, and small vortexes are visible near the bottom corners with increasing Re. For temperature field, it is observed that there is a steep temperature gradient in the vertical direction near the bottom, and a weak temperature gradient in the center region. It is important that the convective region is enlarged with Re increases.

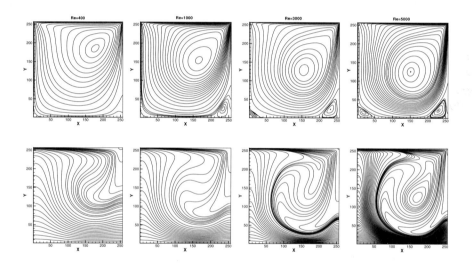

Fig. 2. The streamlines (top) and isothermals (bottom) for Re=400, 1000, 3000, 5000

3.2   Effect of the Richardson Number (Ri)

The Richardson number is defined as the ratio of Gr/Re^2, which provides a measure of the importance of buoyancy-driven natural convection relative to lid-driven cavity force convection. It reflects a dominant conduction mode as $Ri \geq O(1)$, while it resembles similar driven cavity flow behavior for a non-stratified fluid if $Ri \leq O(1)$ [2,6]. In the present work, Ri is varied in the range

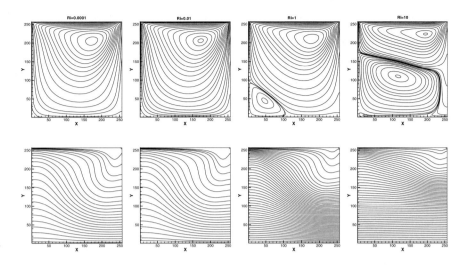

Fig. 3. The streamlines (top) and isothermals (bottom) for Ri=0.0001, 0.01, 1, 10

of $0.0001 - 10$ for $Re = 100$, $Da = 0.01$. As shown in Fig. 3, the flow and temperature fields change with the variations of Ri. As $Ri \leq 0.01$, the buoyancy effect is overwhelmed by the mechanical effect of the sliding lid, and only a primary vortex close to the top boundary is observed. As Ri increases to 1.0, beside the primary vortex mentioned above, another vortex is formed near the bottom corners. However, with increasing Ri, the vortex near the bottom corner moves toward the geometric center, and finally comes to the center of the cavity as Ri is up to 10. It is also shown that, as $Ri \geq O(1)$, i.e., the effect of buoyancy is more prominent than mechanical effect of the sliding lid, the phenomena in driven cavity is more complex. The isothermals in Fig. 3 show the fact that heat transfer is mostly conductive in the middle and bottom parts of the cavity. The relative uniform temperature is only formed in a small region in the top portion of the cavity, where the mechanically induced convective activities are appreciable. However, it should be noted that this convective region is decreasing with increasing Ri.

3.3   Effect of the Darcy Number (Da)

The Da in the present work is varied in the range of $0.0001 - 0.1$ for $Ri = 0.0001$, $Re = 100$. As shown in Fig. 4, the variation of Da significantly affects the flow and temperature fields. It is obvious that the increase of Da induces flow activity deeper into the cavity, which leads to more energy to be carried away from the sliding top wall toward bottom, and consequently, the convective region in the top portion of the cavity is enlarged. However, as Da is decreased to 0.0001, the primary vortex in the cavity is compelled to move toward left wall and a new

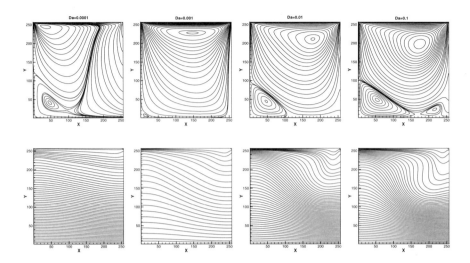

Fig. 4. The streamlines (top) and isothermals (bottom) for Da=0.0001, 0.001, 0.01, 0.1

vortex near the bottom corner is formed. In fact, as Da is small enough, the effect of the nonlinear term in Eq. (2) is prominent, which may induces some new phenomena as observed in present work. Finally, we would like to point out that the numerical results derived in this paper qualitatively agree well with those reported in Ref. [5,6].

4   Conclusion

In the present work, the problem of mixed convection in a driven cavity filled with porous medium is studied with LBM. The influence of the Reynolds number, Richardson number and Darcy number on the flow and temperature fields are investigated in detail. As these parameters are varied in a wide range, some new phenomena is observed. Through comparisons with the existing literature, it is found that the LBM can be used as an alternative approach to study this problem. Compared with traditional numerical methods, LBM offers flexibility, efficiency and outstanding amenability to parallelism when modelling complex flows, and thus it is more suitable for computation on parallel computers; but to derive the same accurate results, larger number of grid may be needed. A recent comparison between LBM and finite different method for simulating natural convection in porous media can be found in Ref. [14].

Acknowledgments. This work is supported by the National Basic Research Program of China (Grant No. 2006CB705804) and the National Science Foundation of China (Grant No. 50606012).

References

1. Shankar, P.N., Deshpande, M.D.: Fluid mechanics in the driven cavity. Annu. Rev. Fluid Mech. **32** (2000) 93–136
2. Iwatsu, R., Hyun, J.M., Kuwahara, K.: Mixed convection in a driven cavity with a stable vertical temperature gradient. Int. J. Heat Mass Transfer **36** (1993) 1601–1608
3. Oztop, H.F.: Combined convection heat transfer in a porous lid-driven enclosure due to heater with finite length. Int. Commun. Heat Mass Transf. **33** (2006) 772–779
4. Vafai, K.: Convective flow and heat transfer in variable-porosity media. J. Fluid Mech. **147** (1984) 233–259
5. Khanafer, K.M., Chamkha, A.J.: Mixed convection flow in a lid-driven enclosure filled with a fluid-saturated porous medium. Int. J. Heat Mass Transfer **42** (1999) 2465–2481
6. Al-Amiri, A.M.: Analysis of momentum and energy transfer in a lid-driven cavity filled with a porous medium. Int. J. Heat Mass Transfer 43 (2000) 3513–3527
7. Jue, T.C.: Analysis of flows driven by a torsionally-oscillatory lid in a fluid-saturated porous enclosure with thermal stable stratification. Int. J. Therm. Sci. **41** (2002) 795–804
8. Chen, S., Doolen, G.: Lattice Boltzmann method for fluid flow. Annu. Rev. Fluid Mech. **30** (1998) 329–364
9. Nithiarasu, P., Seetharamu, K.N., Sundararajan, T.: Natural convective heat transfer in a fluid saturated variable porosity medium. Int. J. Heat Mass Transfer **40** (1997) 3955–3967
10. Guo, Z., Zhao T.S.: A lattice Boltzmann model for convection heat transfer in porous media. Numerical Heat Transfer, Part B **47** (2005) 157–177
11. Ergun, S.: Fluid flow through packed columns. Chem. Eng. Prog. **48** (1952) 89–94
12. Guo, Z., Zhao, T.S.: Lattice Boltzmann model for incompressible flows through porous media. Phys. Rev. E **66** (2002) 036304
13. Guo, Z., Zheng, C., Shi, B.: Non-equilibrium extrapolation method for velocity and pressure boundary conditions in the lattice Boltzmann method. Chin. Phys. **11** (2002) 366–374
14. Seta, T., Takegoshi, E., Okui, K.: Lattice Boltzmann simulation of natural convection in porous media. Math. Comput. Simulat. **72** (2006) 195–200

Numerical Study of Cross Diffusion Effects on Double Diffusive Convection with Lattice Boltzmann Method

Xiaomei Yu[1], Zhaoli Guo[1], and Baochang Shi[2],*

[1] State Key Laboratory of Coal Combustion,
Huazhong University of Science and Technology,
Wuhan 430074 P.R. China
yuxiaomei_hust@126.com,
zlguo@mail.hust.edu.cn
[2] Department of Mathematics,
Huazhong University of Science and Technology,
Wuhan 430074 P.R. China
sbchust@126.com

Abstract. A lattice Boltzmann model is proposed to asses the impact of variable molecular transport effects on the heat and mass transfer in a horizontal shallow cavity due to natural convection. The formulation includes a generalized form of the Soret and Dufour mass and heat diffusion (cross diffusion) vectors derived from non-equilibrium thermodynamics and fluctuation theory. Both the individual cross diffusion effect and combined effects on transport phenomena are considered. Results from numerical simulations indicate that Soret mass flux and Dufour energy flux have appreciable effect and sometimes are significant. At the same time, the lattice Boltzmann model has been proved to be adequate to describe higher order effects on energy and mass transfer.

Keywords: lattice Boltzmann model; Soret effect; Dufour effect; Natural convection.

1 Introduction

Transport phenomena and thermo-physical property in fluid convection submitted to the influence of thermal and concentration horizontal gradients arise in many fields of science and engineering. The conservation equations which describe the transport of energy and mass in these fluid systems are well developed [1-3]. The energy flux includes contributions due to a temperature gradient (Fourier heat conduction), concentration gradient(Dufour diffusion) and a term which accounts for the energy transport as a results of each species having different enthalpies (species interdiffusion). The mass flux consists of terms due

---

* Corresponding author.

Y. Shi et al. (Eds.): ICCS 2007, Part I, LNCS 4487, pp. 810–817, 2007.

to a concentration gradient (Fickian diffusion), temperature gradient (Soret diffusion), Pressure gradient (pressure diffusion) and a term which accounts for external forces affecting each species by a different magnitude. However, most studies concerned on transport phenomena only considered the contributions due to Fourier heat conduction and Fickian diffusion.

The Soret mass flux and Dufour energy flux become significant when the thermal diffusion factor and the temperature and concentration gradients are large. Actually, Rosner [4] has stressed that Soret diffusion is significant in several important engineering applications. Similarly, Atimtay and Gill [5] have shown Soret and Dufour diffusion to be appreciable for convection on a rotating disc. An error as high as 30% for the wall mass flux is introduced when the Soret effect is not accounted for. Of particular interest, crystal growth from the vapor is sometimes carried out under conditions conductive to Soret and Dufour effects. As greater demands are made for tighter control of industrial processes, such as in microelctromechanical systems (MEMS) [6], second-order effects such as Soret and Dufour diffusion may have to be included.

In regard to actual energy and mass transport in double diffusive convection fluid systems which include Soret and Dufour cross diffusion effect, little, if any, complete and detailed work has been done. However, a few studies have considered the convection, within a vertical cavity, induced by Soret effects. The first study on this topic is due to Bergman and Srinivasan [7]. Their numerical results indicate that the Soret-induced buoyancy effects are more important when convection is relatively weak. The particular case of a square cavity under the influence of thermal and solutal buoyancy forces, which are opposing and of equal intensity, has been investigated by Traore and Mojtabi [8]. The Soret effect on the flow structures was investigated numerically. The same problem was considered by Krishnan [9] and Gobin and Bennacer [10] for the case of an infinite vertical fluid layer. The critical Rayleigh number for the onset of motion was determined by these authors. More recently, Ouriemi et al. [11] considered the case of a shallow layer of a binary fluid submitted to the influence of thermal and concentration horizontal gradients with Soret effects. Moreover, few studies have been concerned with the Soret and Dufour effects at the same time. Weaver and Viskanta numerically simulates these effects in a cavity [12]. Malashetty et al.[13] have performed an stability analysis of this problem. Because of the limited number of studies available, the knowledge concerning the influence of these effects on the heat and mass transfer and fluid flow is incomplete.

The lattice Boltzmann method (LBM) is a new method for simulating fluid flow and modeling physics in fluids. It has shown its power and advantages in wide range of situations for its mesoscopic nature and the distinctive computational features [14,15,16]. Contrary to the conventional numerical methods, the LBM has many special advantages in simulating physical problems. The objective of the present study is to develop an effective lattice Boltzmann model to examine the influences and the contributions of the Soret and Dufour effects on the natural convection with simultaneous heat and mass transfer across a horizontal shallow cavity. Based on the idea of the double distribution function models for a fluid flow involving heat transfer, a generalized lattice Boltzmann model is proposed to solve the control equations with Soret and Dufour effects.

2   The Lattice Boltzmann Method for the Formulation of the Problem

The configuration considered in this study is a horizontal shallow cavity, of width H and length L. The top and bottom end walls were assumed to be adiabatic and impermeable to heat and mass transfer while the boundary conditions along the right and left side walls were subjected by Dirichlet conditions. Considering the Soret effects and Dufour effects, the dimensionless incompressible fluid equations for the conservation of mass, momentum, solutal concentration, and temperature, with the inclusion of the Boussinesq approximation for the density variation, are written as.

$$\nabla \cdot \mathbf{u} = 0, \tag{1}$$

$$\frac{\partial \mathbf{u}}{\partial t} + \mathbf{u} \cdot \nabla(\mathbf{u}) = -\nabla p + (\frac{Pr}{Ra})^{1/2}\nabla^2\mathbf{u} + Pr(T + \varphi C), \tag{2}$$

$$\frac{\partial T}{\partial t} + \mathbf{u} \cdot \nabla T = \frac{1}{\sqrt{Rapr}}(\nabla^2 T + D_{CT}\nabla^2 C), \tag{3}$$

$$\frac{\partial C}{\partial t} + \mathbf{u} \cdot \nabla C = \frac{1}{Le\sqrt{Rapr}}(\nabla^2 C + S_{TC}\nabla^2 T) \tag{4}$$

along with the boundary conditions where ρ and $\mathbf{u}$ are the fluid density and velocity, respectively. T is the temperature of the fluid.

The dimensionless variables are defined as

$$\mathbf{X} = \frac{\mathbf{X}}{H}, \ t = \frac{t\sqrt{Ra}}{H^2/\alpha}, \ \mathbf{u} = \frac{\mathbf{u}}{(\alpha/H)\sqrt{Ra}}, \ p = \frac{p}{(\alpha^2/H^2)Ra}, \ T = \frac{T - T_0}{\Delta T},$$

$$C = \frac{C - C_0}{\Delta C}, \ Da = \frac{K}{H^2}, \ Pr = \frac{\nu}{\alpha}, \ Le = \frac{\alpha}{D}, \ Ra = \frac{g\beta_T\Delta T H^3}{\nu\alpha}, \ \varphi = \frac{\beta_C\Delta C}{\beta_T\Delta T}$$

Here, ν, α, D are the kinematic viscosity, thermal diffusivity and diffusion coefficient, respectively. The remaining notation is conventional. From the above equations it is observed that the present problem is governed by the thermal Rayleigh number Ra, buoyancy ratio φ, Lewis number Le, Prandtl number Pr, Dofour factor D_{CT} and Soret factor S_{TC}. In this paper, we concerned on the effect of Soret and Dufour factor. So other dimensionless variables will be kept at certain values.

Based on the idea of TLBGK model in [17], we take the temperature and concentration as passive scalar. Then the control equations of velocity, temperature and concentration can be treated individually. The evolution equation for the velocity field is similar to Ref. [18]

$$f_i(\mathbf{x} + \mathbf{c}_i\Delta t, t + \Delta t) - f_i(\mathbf{x}, t) = -\frac{1}{\tau}(f_i(\mathbf{x}, t) - f_i^{eq}(\mathbf{x}, t)) \\ + \Delta t F_i(\mathbf{x}, t) \quad i = 0, \ldots, b - 1, \tag{5}$$

where $\mathbf{c}_i$ is the discrete particle velocity, F_i represents the force term, $f_i(\mathbf{x}, t)$ is the distribution function (DF) for the particle with velocity $\mathbf{c}_i$ at position $\mathbf{x}$ and time t, Δt is the time increment. τ is the nondimensional relaxation time and $f_i^{(eq)}$ is the equilibrium distribution function (EDF). The EDF must be defined appropriately such that the mass and momentum are conserved and some symmetry requirements are satisfied in order to describe the correct hydrodynamics of the fluid. In the $DnQb$[19] models, the EDF is now defined as

$$f_i^{(eq)} = \alpha_i p + \omega_i \Big[\frac{\mathbf{c}_i \cdot \mathbf{u}}{c_s^2} + \frac{\mathbf{uu} : (\mathbf{c}_i \mathbf{c}_i - c_s^2 \mathbf{I})}{2c_s^4}\Big]. \tag{6}$$

$$F_i = \frac{(\delta_{i2} + \delta_{i4})}{2c^2} \mathbf{c}_i Pr(T + \varphi C) \tag{7}$$

where ω_i is the weight and c_s is the sound speed. Both ω_i and c_s depend on the underlying lattice. Take the $D2Q9$ model for example, the discrete velocities are given by $\mathbf{c}_0 = 0$, and $\mathbf{c}_i = \lambda_i(\cos\theta_i, \sin\theta_i)c$ with $\lambda_i = 1$, $\theta_i = (i-1)\pi/2$ for $i = 1-4$, and $\lambda_i = \sqrt{2}$, $\theta_i = (i-5)\pi/2 + \pi/4$ for $i = 5-8$, and $c_s = c/\sqrt{3}$. $\omega_0 = 4/9$ and $\alpha_0 = -4\sigma/c^2$, $\omega_i = 1/9$ and $\alpha_i = \lambda/c^2$ for $(i = 1, 2, 3, 4)$, $\omega_i = 1/36$ and $\alpha_i = \gamma/c^2$ for $(i = 5, 6, 7, 8)$, respectively. Here, $\sigma = \frac{5}{12}, \lambda = \frac{1}{3}, \gamma = \frac{1}{12}$ is set [20] with best computation effects.

The fluid velocity $\mathbf{u}$ and pressure p are defined by the DF f_i

$$\mathbf{u} = \sum_i \mathbf{c}_i f_i = \sum_i \mathbf{c}_i f_i^{(eq)}, \quad p = \frac{c^2}{4\sigma}\Big[\sum_{i \neq 0} f_i - \frac{2}{3}\frac{|\mathbf{u}|^2}{c^2}\Big]. \tag{8}$$

Through the Chapman-Enskog expansion, the Eqs.(1),(2) can be obtained. The kinetic viscosity are given by $(\frac{Pr}{Ra})^{1/2} = \frac{(2\tau-1)}{6}c^2\Delta t$.

Similarly, the LBGK equations for temperature and concentration fields with $D2Q5$ lattice are written as follows,

$$T_i(\mathbf{x} + \mathbf{c}_i\Delta t, t + \Delta t) - T_i(\mathbf{x}, t) = -\frac{1}{\tau_T}[T_i(\mathbf{x}, t) - T_i^{(0)}(\mathbf{x}, t)] \tag{9}$$

$$S_i(\mathbf{x} + \mathbf{c}_i\Delta t, t + \Delta t) - S_i(\mathbf{x}, t) = -\frac{1}{\tau_S}[S_i(\mathbf{x}, t) - S_i^{(0)}(\mathbf{x}, t)] \tag{10}$$

where T_i and S_i are the DF for the temperature field and concentration field, respectively. τ_T and τ_S are relaxation time corresponding to temperature DF and concentration DF trend to equilibrium. Just like the thoughts of constructing generalized LBGK model for Burgers equation [21], the EDF of temperature and concentration are defined as

$$\begin{aligned} T_i^{(0)} &= \frac{T}{4}[1 - d_0 + 2\frac{\mathbf{c}_i \cdot \mathbf{u}}{c^2}] + \frac{D_{ST}S(1-d_0)}{4}, \quad i \neq 0, \\ T_0^{(0)} &= Td_0 - D_{ST}S(1 - d_0), \quad i = 0. \end{aligned} \tag{11}$$

$$\begin{aligned} S_i^{(0)} &= \frac{S}{4}[(1 - l_0)\phi + 2\frac{\mathbf{c}_i \cdot \mathbf{u}}{c^2}] + \frac{S_{TS}T(1-l_0)\phi}{4}, \quad i \neq 0, \\ S_0^{(0)} &= Sl_0\phi - S_{TS}T(1 - l_0)\phi, \quad i = 0. \end{aligned} \tag{12}$$

The fluid temperature and concentration can be expressed by DFs,

$$T = \sum_i T_i = \sum_i T_i^{(eq)}, \ S = \sum_i S_i = \sum_i S_i^{(eq)}. \tag{13}$$

It can be proved that the macroscopic temperature and concentration diffusion equations (3) and (4) can be recovered from LBE(9),(10). The corresponding thermal conduction coefficient, diffusion factor are

$$\frac{1}{\sqrt{RaPr}} = \frac{c^2}{2}(\tau_T - \frac{1}{2})\Delta t(1 - d_0), \tag{14}$$

$$\frac{1}{Le\sqrt{RaPr}} = \frac{c^2}{2}(\tau_T - \frac{1}{2})\Delta t(1 - d_0). \tag{15}$$

Here, d_0 and l_0 are adjustable parameters. Obviously, the lattice Boltzmann model can simulate the double diffusive not only including Fourier heat conduction and Fickian diffusion but also second-order cross diffusion by choosing adequate parameters.

3   Numerical Results and Discussion

Using the lattice Boltzmann model, we simulate the Soret and Dufour effects in double diffusive due to temperature and concentration gradient. Since the Soret and Dufour effects are diffusive processes, the Rayleigh number should not too high in order to decrease the advective flux relative to the diffusive flux. Furthermore, the buoyancy ratio is set to be $\varphi = 1$, which preserves the buoyancy forces induced by the thermal and solutal effects are of equal intensity. Simulation tests are conducted with varied Soret and Dufour factor and fixed Le and Pr number equal to 1 and 0.71, respectively. The lattice size is 128×256 with Ra equal to 10^3.

Streamlines, isothermals and iso-concentration profiles varied with Dufour and Soret coefficients are shown in Fig. 1. The effect on mass, momentum and concentration transports made by varied cross diffusion coefficients can be observed obviously. The center velocity, local (averaged) Nusselt and Sherwood number on the left wall are showed in Fig. 2, Fig. 3 and Table 1 for detailed discussion, respectively. A positive (negative) value of Dufour and Soret coefficients decreases (increases) the value of the maximum velocity, and extends (reduces) the range of the velocity field. Moreover, Positive values for the Dufour and Soret coefficients yield solute concentrations below that of the free stream and so is the temperature, the opposite being also true. But when the Dufour coefficient is certain, the local Nusselt numbers decrease with Soret coefficients but local Sherwood number increase. While the Soret coefficient is certain, the local Nusselt numbers increse and the Sherwood number decrease instead. All curves show the expected behavior.

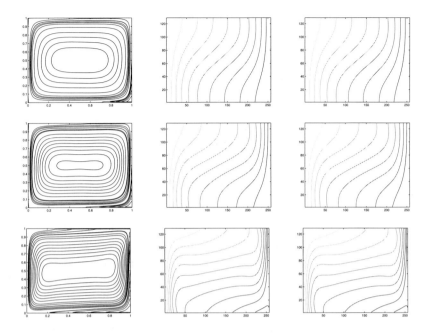

Fig. 1. Streamline, isothermals and iso-concentrations from left to right. From top to bottom, the Dufour and Soret coefficients are (0, 0),(0,0.5),(-0.5,-0.5),(-0.9,-0.9), respectively.

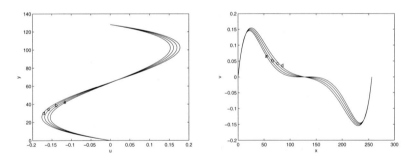

Fig. 2. The velocity components, u and v, along the vertical and horizontal lines through the center at different cross diffusion numbers (D_{CT}, S_{TC}): a=$(0, -0.5)$; b=$(0, 0.0)$; c=$(0, 0.5)$; d=$(0, 0.9)$

Table 1. Averaged Nusselt and Sherwood number on the left wall veried with Soret and Dufour factors

(D, S)	(0,-0.5)	(0,0)	(0,0.5)	(0,0.9)	-0.5,0)	(0.5,0)	(0.9,0)	(-0.5,-0.5)	(0.5,0.5)	(0.9,0.9)
Nu	0.9570	0.7698	0.5608	0.3785	0.7465	0.7965	0.8200	1.0838	0.6532	0.6053
Sh	0.7465	0.7698	0.7965	0.8200	0.9570	0.5608	0.3785	1.0838	0.6532	0.6053

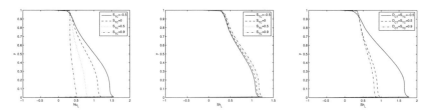

Fig. 3. Local Nusselt and Sherwood number on the left wall varied with cross diffusion numbers (The D_{CT} is equal 0 at the left two figures)

4   Summary

In this paper, a lattice Boltzmann model is proposed and used to asses the Soret and Dufour effects on the heat and mass transfer in a horizontal shallow cavity due to natural convection. The proposed model is constructed in the double distribution function framework. Cross diffusion effects are studied under the buoyancy counteracting flows and augmenting flows in shallow cavity with the lattice Boltzmann model. The numerical results show that Soret and Dufour effects have contributions to mass and energy flux. Sometimes, it may appears to significant. Simulation results agree well with previous work. Moreover, it indicates that the lattice Boltzmann model is adequate to simulate molecular transport which even includes second-order effects. It seems that the lattice Boltzmann method is effective in describing higher order physical effects, which are important as the industry develops.

Acknowledgments. This work is supported by the National Basic Research Program of China (Grant No. 2006CB705804) and the National Science Foundation of China (Grant No. 50606012).

References

1. Hirshfelder J. O., Curtiss C. F., Bird R. B.: Molecular Theory of Gases and Liquids. Wiley, New York (1960)
2. Groot S. R., Mazur P.: Thermodynamics of Trreversible Processes. Dover, New York (1984)
3. Chapman S., Cowling T. G.: The Mathematical Theory of Non-Unifrom Gases. 3rd edn. Cambridge University, Cambridge (1970)
4. Rosner D. E.: Thermal (Soret) Diffusion Effects on Interfacial Mass Transport Rates. PhysicoChem. Hydordyn. **1** (1980) 159–185
5. Atimtay A. T., Gill W. N.: The Effect of Free Stream Concentration on Heat and Binary Mass Transfer with Thermodynamic Coupling in Convection on a Rotating Disc. Chem. Engng Commun. **34** (1985) 161–185
6. Karniadakis G. E., Beskok A.: Micro flows, Fundamentals and Simulation. Springer, Nw York (2001)
7. Bergman T. L., Srinivasan R.: Numerical Simulation of Soret-induced Double Diffusion in an Initiallly Uniform Concentration Binary Fluid. Int. J. Heat Mass Transfer **32** (1989) 679–687

8. Traore Ph., Mojtabi A.: Analyse de l'effect soret en convection thermosolutale, Entropie **184/185** (1989) 32–37

9. Krishnan R.: A Numerical Study of the Instability of Double-Diffusive Convection in a Square Enclosure with Horizontal Temperature and Concentration Gradients, Heat transfer in convective flows. HTD ASME National Heat Transfer conference, vol 107. Philadelphia (1989)

10. Gobin D., Bennacer R.: Double-diffusion Convection in a Vertical Fluid Layer: Onset of the Convection Regime. Phys. Fluids **6** (1994) 59–67

11. Ouriemi M., Vasseur P., Bahloul A., Robillard L.: Natural Convection in a Horizontal Layer of a Binary Mixture. Int. J. Thermal Sciences **45** (2006) 752–759

12. Weaver J. A., Viskanta R.: Natural Convection Due to Horizontal Temperature and Concentration Gradients -2. Species Interdiffusion, Soret and Dufour Effects. Int. J. Heat Mass Transfer **34** (1991) 3121–3133

13. Malashetty M. S., Gaikward S. N., Effects of Cross Diffusion on Double Diffusive Convection in the Presence of Horizontal Gradients. Int. J. Engineering Science **40** (2002) 773–787

14. Benzi R., Succi S., Vergassola M.: The Lattice Boltzmann Equation: Theory and Applications. Phys. Report. **222** (1992) 145–197

15. Qian Y. H., Succi S., Orszag S.: Recent Advances in Lattice Boltzmann computing. Annu. Rev. Comp. Phys. **3** (1995) 195–242

16. Chen S. Y., Doolen G.: Lattice Boltzmann Method for Fluid Flows. Annu. Rev. Fluid Mech. **30** (1998) 329–364

17. Guo Z. L., Shi B. C., Zheng C. G.: A Coupled Lattive BGK Model for the Bouessinesq Equation. Int. J. Num. Meth. Fluids **39** (2002) 325–342

18. Deng B., Shi B. C., Wang G. C: A New Lattice-Bhatnagar-Gross-Krook Model for the Convection-Diffusion Equation with a Source Term. Chinese Phys. Lett. **22** (2005) 267–270

19. Qian Y. H., D'Humières D., Lallemand P.: Lattice BGK Models for Navier-Stokes Equation. Europhys. Lett. **17** (1992) 479–484

20. Guo Z. L., Shi B. C., Wang N. C., Lattice BGK model for incompressible Navier-Stokes Equation. J. Comput. Phys. **165** (2000) 288–306

21. Yu X. M., Shi B. C.: A Lattice Bhatnagar-Gross-Krook model for a class of the generalized Burgers equations. Chin. Phys. Soc. **15** (2006) 1441–1449

Lattice Boltzmann Simulation of Some Nonlinear Complex Equations

Baochang Shi

Department of Mathematics, Huazhong University of Science and Technology,
Wuhan 430074, PR China
sbchust@126.com

Abstract. In this paper, the lattice Boltzmann method for convection-diffusion equation with source term is applied directly to solve some important nonlinear complex equations, including nonlinear Schrödinger (NLS) equation, coupled NLS equations, Klein-Gordon equation and coupled Klein-Gordon-Schrödinger equations, by using complex-valued distribution function and relaxation time. Detailed simulations of these equations are carried out. Numerical results agree well with the analytical solutions, which show that the lattice Boltzmann method is an effective numerical solver for complex nonlinear systems.

Keywords: Lattice Boltzmann method, nonlinear Schrödinger equation, Klein-Gordon equation, Klein-Gordon-Schrödinger equations.

1 Introduction

The lattice Boltzmann method (LBM) is an innovative computational fluid dynamics (CFD) approach for simulating fluid flows and modeling complex physics in fluids [1]. Compared with the conventional CFD approach, the LBM is easy for programming, intrinsically parallel, and it is also easy to incorporate complicated boundary conditions such as those in porous media. The LBM also shows potentials to simulate the nonlinear systems, including reaction-diffusion equation [2,3,4], convection-diffusion equation [5,6], Burgers equation [7] and wave equation [3,8], etc. Recently, a generic LB model for advection and anisotropic dispersion equation was proposed [9]. However, almost all of the existing LB models are used for real nonlinear systems. Beginning in the mid 1990s, based on quantum-computing ideas, several types of quantum lattice gases have been studied to model some real/complex mathematical-physical equations, such as Dirac equation, Schrödinger equation, Burgers equation and KdV equation [10,11,12,13,14,15,16,17]. Although these work are not in the classical LBM framework, they bring us an interesting problem: how does the classical LBM work when used to model complex equations? Very recently, Linhao Zhong, Shide Feng, Ping Dong, et al. [18] applied the LBM to solve one-dimensional nonlinear Schrödinger (NLS) equation using the idea of quantum lattice-gas model [13,14] for treating the reaction term. Detailed simulation results in Ref.[18] have shown that the LB schemes proposed have accuracy that is better than or at least comparable to the Crank-Nicolson finite difference scheme. Therefore, it is necessary to study the LBM for nonlinear complex equations further.

Y. Shi et al. (Eds.): ICCS 2007, Part I, LNCS 4487, pp. 818–825, 2007.

In this paper, using the idea of adopting complex-valued distribution function and relaxation time [18], the LBM for n-dimensional (nD) convection-diffusion equation (CDE) with source term is applied directly to solve some important nonlinear complex equations, including nonlinear Schrödinger (NLS) equation, coupled NLS equations, Klein-Gordon (KG) equation and coupled Klein-Gordon-Schrödinger (CKGS) equations. Detailed simulations of these equations are carried out for accuracy test. Numerical results agree well with the analytical solutions, which show that the LBM is also an effective numerical solver for complex nonlinear systems.

2   Lattice Boltzmann Model

The nD CDE with source term considered in this paper can be written as

$$\partial_t \phi + \nabla \cdot (\phi \mathbf{u}) = \alpha \nabla^2 \phi + F(\mathbf{x}, t), \tag{1}$$

where ∇ is the gradient operator with respect to the spatial coordinate $\mathbf{x}$ in n dimensions. ϕ is a scalar function of time t and position $\mathbf{x}$. $\mathbf{u}$ is a constant velocity vector. $F(\mathbf{x},t)$ is the source term. When $\mathbf{u}=0$, Eq.(1) becomes the diffusion equation (DE) with source term, and several of such equations form a reaction-diffusion system (RDS).

2.1   LB Model for CDE

The LB model for Eq.(1) is based on the DnQb lattice [1] with b velocity directions in nD space.

The evolution equation of the distribution function in the model reads

$$f_j(\mathbf{x}+\mathbf{c}_j \Delta t, t+\Delta t) - f_j(\mathbf{x}, t) = -\frac{1}{\tau}(f_j(\mathbf{x}, t) - f_j^{eq}(\mathbf{x}, t)) + \Delta t F_j(\mathbf{x}, t), j=0, \dots, b-1, \tag{2}$$

where $\{\mathbf{c}_j, j = 0, \dots, b-1\}$ is the set of discrete velocity directions, Δx and Δt are the lattice spacing and the time step, respectively, $c = \Delta x/\Delta t$ is the particle speed, τ is the dimensionless relaxation time, and $f_i^{eq}(\mathbf{x}, t)$ is the equilibrium distribution function which is determined by

$$f_j^{eq}(\mathbf{x}, t) = \omega_j \phi (1 + \frac{\mathbf{c}_j \cdot \mathbf{u}}{c_s^2} + \frac{(\mathbf{uu}):(\mathbf{c}_j\mathbf{c}_j - c_s^2 \mathbf{I})}{2c_s^4}) \tag{3}$$

such that

$$\sum_j f_j = \sum_j f_j^{eq} = \phi, \sum_j \mathbf{c}_j f_j^{eq} = \phi\mathbf{u}, \sum_j \mathbf{c}_j\mathbf{c}_j f_j^{eq} = c_s^2 \phi\mathbf{I} + \phi\mathbf{uu} \tag{4}$$

with $\alpha = c_s^2(\tau - \frac{1}{2})\Delta t$, where $\mathbf{I}$ is the unit tensor, ω_j are weights and c_s, so called sound speed in the LBM for fluids, is related to c and ω_j. They depend on the lattice model used.

For the D1Q3 model, $\omega_0 = 2/3$, $\omega_{1\sim2} = 1/6$, for the D2Q9 one, $\omega_0 = 4/9$, $\omega_{1\sim4} = 1/9$, $\omega_{5\sim8} = 1/36$, then $c_s^2 = c^2/3$ for both of them.

F_j in Eq.(2), corresponding to the source term in Eq.(1), is taken as

$$F_j = \omega_j F (1 + \lambda \frac{\mathbf{c}_j \cdot \mathbf{u}}{c_s^2}) \tag{5}$$

such that $\sum_j F_j = F, \sum_j \mathbf{c}_j F_j = \lambda F \mathbf{u}$, where λ is a parameter which is set to be $\frac{\tau - \frac{1}{2}}{\tau}$ in the paper. It is found that the LB model using this setting has better numerical accuracy and stability when $\mathbf{u} \neq \mathbf{0}$.

The macroscopic equation (1) can be derived though the Chapman-Enskog expansion (See Appendix for details). It should be noted that the equilibrium distribution function for above LBM was often used in linear form of $\mathbf{u}$, which is different from that in Eq.(3). However, from the appendix we can see that some additional terms in the recovered macroscopic equation can be eliminated if Eq.(3) is used. Moreover, if $\mathbf{u}$ is not constant, the appropriate assumption on it is needed in order to remove the additional term(s) when recovering Eq.(1).

2.2   Version of LB Model for Complex CDE

Almost all of the existing LB models simulate the real evolutionary equations. However, from the Chapman-Enskog analysis, we can find that the functions in CDE and related distribution function can be both real and complex without affecting the results. In general, for the complex evolutionary equations, let us decompose the related complex functions and relaxation time into their real and imaginary parts by writing

$$f_j = g_j + \mathrm{i} h_j, f_j^{eq} = g_j^{eq} + \mathrm{i} h_j^{eq}, F_j = G_j + \mathrm{i} H_j, w = \frac{1}{\tau} = w_1 + \mathrm{i} w_2, \tag{6}$$

where $\mathrm{i}^2 = -1$.

Now we can rewrite Eq.(2) as

$$\begin{aligned}
g_j(\mathbf{x} + \mathbf{c}_j \Delta t, t + \Delta t) - g_j(\mathbf{x}, t) = &- w_1(g_j(\mathbf{x}, t) - g_j^{eq}(\mathbf{x}, t)) \\
+ w_2(h_j(\mathbf{x}, t) - h_j^{eq}(\mathbf{x}, t)) + \Delta t G_j(\mathbf{x}, t),& \\
h_j(\mathbf{x} + \mathbf{c}_j \Delta t, t + \Delta t) - h_j(\mathbf{x}, t) = &- w_2(g_j(\mathbf{x}, t) - g_j^{eq}(\mathbf{x}, t)) \\
- w_1(h_j(\mathbf{x}, t) - h_j^{eq}(\mathbf{x}, t)) + \Delta t H_j(\mathbf{x}, t),& \\
j = 0, \dots, b - 1.&
\end{aligned} \tag{7}$$

Eq.(7) is the implemental version of the LB model proposed for complex CDE. It should be noted that Eq.(7) reflects the coupling effect of real and imaginary parts of unknown function in complex CDE through the complex-valued relaxation time in a natural way.

3   Simulation Results

To test the LB model proposed above, numerical simulations of some CDEs with source term are carried out. Here we select three types of complex nonlinear evolutionary equations with analysis solutions to test mainly the numerical order of accuracy of the LBM. In simulations, the initial value of distribution function is taken as that of its equilibrium part at time $t = 0$, which is a commonly used strategy, and $\Delta x^2/\Delta t(= \Delta x \times c)$ is set to be constant for different grids due to some claims that LB schemes are second-order accurate in space and first-order accurate in time.

Example 3.1. We show an accuracy test for the 2D NLS equation

$$iu_t + u_{xx} + u_{yy} + \beta|u|^2 u = 0, \tag{8}$$

which admits a progressive plane wave solution [19]

$$u(x, y, t) = A \exp(i(c_1 x + c_2 y - \omega t)), \tag{9}$$

where $\omega = c_1^2 + c_2^2 - \beta|A|^2$; A, c_1 and c_2 are constants.

In simulations, a D2Q9 LB model is used. We set $A = c_1 = c_2 = 1, \beta = 2$ and use the periodic boundary condition is used in $[0, 2\pi] \times [0, 2\pi]$ as in Ref.[19], while the initial condition for u is determined by the analytical solution (9) at $t = 0$. The global relative L^1 error and the numerical order of accuracy are contained in Table 1 at time $t = 1$. From the table we can see that the LBM for the NLS equation (8) has about second-order of accuracy, but the order of accuracy decreases as the resolution increases.

Table 1. Accuracy test for the NLS equation (8)

Grid $N \times N$	c	Real part of u L^1 err	Order	Imaginary part of u L^1 err	Order
62×62	50	2.56e-2	-	2.57e-2	-
124×124	100	5.71e-3	2.1646	5.71e-3	2.1702
248×248	200	1.34e-3	2.0913	1.34e-3	2.0913
496×496	400	4.15e-4	1.6910	4.15e-4	1.6910

Example 3.2. We show an accuracy test for the 1D coupled NLS equation

$$\begin{aligned} iu_t + i\alpha u_x + \tfrac{1}{2}u_{xx} + (|u|^2 + \beta|v|^2)u = 0, \\ iv_t - i\alpha v_x + \tfrac{1}{2}v_{xx} + (\beta|u|^2 + |v|^2)v = 0, \end{aligned} \tag{10}$$

with the soliton solutions [19]

$$\begin{aligned} u(x,t) = A\mathrm{sech}(\sqrt{2a}(x - \gamma t)) \exp(i((\gamma - \alpha)x - \omega t)), \\ v(x,t) = A\mathrm{sech}(\sqrt{2a}(x - \gamma t)) \exp(i((\gamma + \alpha)x - \omega t)), \end{aligned} \tag{11}$$

where $A = \sqrt{\frac{a}{1+\beta}}, \omega = \frac{\gamma^2 - \alpha^2}{2} - a$; a, γ, α and β are real constants.

We can use two LB evolutionary equations based on D1Q3 lattice to simulate Eq.(10). We set $a = 1, \gamma = 1, \alpha = 0.5, \beta = 2/3$ and use the periodic boundary condition in $[-25, 25]$ as in Ref.[19], while the initial condition is determined by the analytical solution (11) at $t = 0$. The global relative L^1 errors are plotted in Fig. 1 (left) at time $t = 1$ for $\Delta x = 1/10$ to $1/320$, and $c = 20$ to 640. It is found that the LBM for the coupled NLS equations (10) has the second-order of accuracy, and the order of accuracy is nearly fixed to 2.0 for different grid resolution. To test the LBM further the error evolution with time is also plotted in Fig.1 (right) for $\Delta x = 1/100$, and $\Delta t = 10^{-4}$ and 10^{-5}, respectively. The errors for $\Delta t = 10^{-4}$ are about 7.0 to 9.0 times those for $\Delta t = 10^{-5}$.

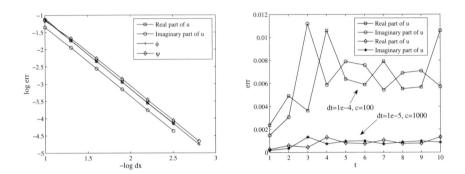

Fig. 1. Left: Global relative errors vs. space steps at $t = 1.0$; Right: Error evolves with time

Example 3.3. Consider the 1D coupled KGS equations

$$i\psi_t + \tfrac{1}{2}\psi_{xx} + \psi\phi = 0,$$
$$\phi_{tt} - \phi_{xx} + \phi - |\psi|^2 = 0, \tag{12}$$

with the soliton solutions [20]

$$\psi(x,t) = \frac{3\sqrt{2}}{4\sqrt{1-v^2}}\operatorname{sech}^2\frac{1}{2\sqrt{1-v^2}}(x - vt - x_0)\exp(i(vx + \tfrac{1-v^2+v^4}{2(1-v^2)}t)),$$
$$\phi(x,t) = \frac{3}{4(1-v^2)}\operatorname{sech}^2\frac{1}{2\sqrt{1-v^2}}(x - vt - x_0), \tag{13}$$

where v is the propagating velocity of wave and x_0 is the initial phase.

Note that the KG equation in Eq.(12) is different from RD equation, the special case of Eq.(1), due to the second time differential of ϕ. In order to solve this equation by LBM, we must modify the LB model in section 2.

Consider the nD KG equation

$$\phi_{tt} - \alpha\nabla^2\phi + V(\phi) = 0, \tag{14}$$

in the spacial region Ω, where $V(\phi)$ is some nonlinear function of ϕ. The the initial conditions associated with Eq.(14) are given by

$$\phi(\mathbf{x}, 0) = f(\mathbf{x}), \phi_t(\mathbf{x}, 0) = g(\mathbf{x}), \tag{15}$$

where the boundary conditions can be found according to the given problem.

Using the idea in Ref.[8], we modify Eq.(3) as follows

$$f_j^{eq}(\mathbf{x}, t) = \omega_j(\phi_t + \frac{c_s^2(\phi-\phi_t)\mathbf{I}:(\mathbf{C}_j\mathbf{C}_j - c_s^2\mathbf{I})}{2c_s^4})$$
$$= \omega_j(\phi_t + \frac{(\phi-\phi_t)(\mathbf{C}_j^2 - nc_s^2)}{2c_s^2}), \tag{16}$$

such that

$$\sum_j f_j = \sum_j f_j^{eq} = \phi_t, \sum_j \mathbf{c}_j f_j^{eq} = \mathbf{0}, \sum_j \mathbf{c}_j \mathbf{c}_j f_j^{eq} = c_s^2 \phi \mathbf{I} \qquad (17)$$

with $\alpha = c_s^2(\tau - \frac{1}{2})\Delta t$. We use the difference scheme for computing ϕ_t to obtain ϕ. For instance, $\phi(\mathbf{x}, t + \Delta t) = \Delta t \sum_j f_j(\mathbf{x}, t + \Delta t) + \phi(\mathbf{x}, t)$.

Now we can use two LB evolutionary equations based on D1Q3 lattice to simulate Eq.(12), one for NLS equation and the other for KG equation. In simulation, the initial and boundary conditions are determined by the analytical solution (13) in $[-10, 10]$. The non-equilibrium extrapolation scheme [21] is used for treating the boundary conditions. The global relative L^1 errors are also plotted in Fig. 1 (left) for $v = 0.8, x_0 = 0$ at time $t = 1$ for $\Delta x = 1/10$ to $1/640$, and $c = 10$ to 640. It is found that the LBM for the coupled KGS equations (12) has also the second-order of accuracy, which is comparable to the multisymplectic scheme in Ref.[20]. We also find that the order of accuracy for ϕ increases from 1.7308 to 2.0033 as the grid resolution increases. This may be because that we use the first-order difference scheme to compute ϕ_t.

4   Conclusion

In this paper the LBM for nD CDE with source term has been applied directly to solve some important nonlinear complex equations by using complex-valued distribution function and relaxation time, and the LB model for nD KG equation is derived by modifying the method for CDE. Simulations of 2D NLS equation, 1D coupled NLS equations and 1D coupled KGS equations are carried out for accuracy test. Numerical results agree well with the analytical solutions and the second-order accuracy of the LBM is confirmed. We found that to attain better accuracy the LBM for the test problems requires a relatively small time step Δt and $\Delta t = 10^{-4}$ is a proper choice. Since the Chapman-Enskog analysis shows that this kind of complex-valued LBM is only the direct *translation* of the classical LBM in complex-valued function, the LBM can be applied directly to other complex evolutionary equations or real ones with complex-valued solutions.

Although the preliminary work in this paper shows that the classical LBM has also potentials to simulate complex-valued nonlinear systems, some problems still need to be solved, such as how to improve the accuracy and efficiency of complex LBM and that how does the complex LBM for CDE work when $\mathbf{u}$ is not constant.

References

1. Qian, Y. H., Succi, S., Orszag, S.: Recent advances in lattice Boltzmann computing. Annu. Rev. Comput. Phys. **3** (1995) 195–242
2. Dawson, S. P., Chen S. Y., Doolen,G. D.: Lattice Boltzmann computations for reaction-diffusion equations. J. Chem. Phys. **98** (1993) 1514–1523
3. Chopard, B., Droz, M.: Cellular automata modeling of physical systems. Cambridge University Press, Cambridge (1998)

4. Blaak, R., Sloot, P. M.: Lattice dependence of reaction-diffusion in lattice Boltz-mann modeling. Comput. Phys. Comm. **129** (2000) 256–266

5. Van der Sman, R. G. M., Ernst, M. H.: Advection-diffusion lattice Boltzmann scheme for irregular lattices. J. Comput. Phys. **160**(2) (2000) 766–782

6. Deng, B., Shi, B. C., Wang, G. C.: A new lattice Bhatnagar-Gross-Krook model for convection-diffusion equation with a source term. Chin. Phys. Lett. **22** (2005) 267–270

7. Yu, X. M., Shi, B. C.: A lattice Bhatnagar-Gross-Krook model for a class of the generalized Burgers equations. Chin. Phys. **25**(7) (2006) 1441–1449

8. Yan, G. W.: A lattice Boltzmann equation for waves. J. Comput. Phys. (2000) **161**(9) (2000) 61–69

9. Ginzburg, I.: Equilibrium-type and link-type lattice Boltzmann models for generic advection and anisotropic-dispersion equation. Advances in Water Resources. **28**(11) (2005) 1171–1195

10. Meyer, D. A.: From quantum cellular automata to quantum lattice gas. J. Stat. Phys. **85** (1996) 551–574

11. Succi, S., Benzi, R.: The lattice Boltzmann equation for quantum mechanics. Physica D **69** (1993) 327–332

12. Succi, S.: Numerical solution of the Schrödinger equation using discrete kinetic theory. Phys. Rev. E **53** (1996) 1969–1975

13. Boghosian, B. M., Taylor IV, W.: Quantum lattice gas models for the many-body Schrödinger equation. Int. J. Mod. Phys. C **8** (1997) 705–716

14. Yepez, J., Boghosian, B.: An efficient and accurate quantum lattice-gas model for the many-body Schrödinger wave equation. Comput. Phys. Commun. **146** (2002) 280–294

15. Yepez, J.: Quantum lattice-gas model for the Burgers equation. J. Stat. Phys. **107** (2002) 203–224

16. Vahala, G., Yepez, J., Vahala, L.: Quantum lattice gas representation of some classical solitons. Phys. Lett. A **310** (2003) 187–196

17. Vahala, G., Vahala, L., Yepez, J: Quantum lattice representations for vector solitons in external potentials. Physica A **362** (2006) 215–221

18. Zhong, L. H., Feng, S. D., Dong, P., et al.: Lattice Boltzmann schemes for the nonlinear Schrödinger equation. Phys. Rev. E. **74** (2006) 036704-1–9

19. Xu, Y., Shu, C.-W.: Local discontinuous Galerkin methods for nonlinear Schrödinger equations. J. Comput. Phys. **205** (2005) 72–97

20. Kong, L. H., Liu, R. X., Xu, Z. L.: Numerical solution of interaction between Schrödinger field and Klein-Gordon field by multisymplectic method. Appl. Math. Comput. **181** (2006) 242–350

21. Guo, Z. L., Zheng, C. G., Shi, B. C.: Non-equilibrium extrapolation method for velocity and pressure boundary conditions in the lattice Boltzmann method. Chin. Phys. **11** (2002) 366–374

Appendix: Derivation of Macroscopic Equation

To derive the macroscopic equation (1), the Chapman-Enskog expansion in time and space is applied:

$$f_j = f_j^{eq} + \epsilon f_j^{(1)} + \epsilon^2 f_j^{(2)}, F = \epsilon F^{(1)}, \partial_t = \epsilon \partial_{t_1} + \epsilon^2 \partial_{t_2}, \nabla = \epsilon \nabla_1, \qquad (18)$$

where ϵ is the Knudsen number, a small number.

From Eqs(18), (4) and (5), it follows that

$$\sum_j f_j^{(k)} = 0 (k \geq 1), \quad \sum_j F_j^{(1)} = F^{(1)}, \quad \sum_j \mathbf{c}_j F_j^{(1)} = \lambda F^{(1)} \mathbf{u}, \tag{19}$$

where $F_j^{(1)} = \omega_j F^{(1)}(1 + \lambda \frac{\mathbf{c}_j \cdot \mathbf{u}}{c_s^2})$ and λ is a parameter specified later.

Applying the Taylor expansion and Eq.(18) to Eq.(2), we have

$$O(\epsilon) : D_{1j} f_j^{eq} = -\frac{1}{\tau \Delta t} f_j^{(1)} + F_j^{(1)}, \tag{20}$$

$$O(\epsilon^2) : \partial_{t_2} f_j^{eq} + D_{1j} f_j^{(1)} + \frac{\Delta t}{2} D_{1j}^2 f_j^{eq} = -\frac{1}{\tau \Delta t} f_j^{(2)}. \tag{21}$$

where $D_{1j} = \partial_{t_1} + \mathbf{c}_j \cdot \nabla_1$.

Applying Eq.(20) to the left side of Eq.(21), we can rewrite Eq.(21) as

$$\partial_{t_2} f_j^{eq} + (1 - \frac{1}{2\tau}) D_{1j} f_j^{(1)} + \frac{\Delta t}{2} D_{1j} F_j^{(1)} = -\frac{1}{\tau \Delta t} f_j^{(2)}. \tag{22}$$

Summing Eq.(20) and Eq.(22) over j and using Eq.(4) and Eq.(19), we have

$$\partial_{t_1} \phi + \nabla_1 \cdot (\phi \mathbf{u}) = F^{(1)}, \tag{23}$$

$$\partial_{t_2} \phi + (1 - \frac{1}{2\tau}) \nabla_1 \cdot \sum_j \mathbf{c}_j f_j^{(1)} + \frac{\Delta t}{2} (\partial_{t_1} F^{(1)} + \nabla_1 \cdot (\lambda F^{(1)} \mathbf{u})) = 0. \tag{24}$$

Since $\mathbf{u}$ is a constant vector, using Eqs (20), (4), (19) and (23), we have

$$\begin{aligned}
\sum_j \mathbf{c}_j f_j^{(1)} &= -\tau \Delta t \sum_j \mathbf{c}_j (D_{1j} f_j^{eq} - F_j^{(1)}) \\
&= -\tau \Delta t (\partial_{t_1} (\phi \mathbf{u}) + \nabla_1 \cdot (\phi \mathbf{u} \mathbf{u} + c_s^2 \phi \mathbf{I}) - \lambda F^{(1)} \mathbf{u}) \\
&= -\tau \Delta t (\mathbf{u} (\partial_{t_1} \phi + \nabla_1 \cdot (\phi \mathbf{u})) + c_s^2 \nabla_1 \phi - \lambda F^{(1)} \mathbf{u}) \\
&= -\tau \Delta t (c_s^2 \nabla_1 \phi + (1 - \lambda) F^{(1)} \mathbf{u}).
\end{aligned} \tag{25}$$

Then substituting Eq.(25) into Eq.(24), we obtain

$$\partial_{t_2} \phi = \alpha \nabla_1^2 \phi + \Delta t (\tau - \frac{1}{2} - \lambda \tau) \nabla_1 \cdot (F^{(1)} \mathbf{u}) - \frac{\Delta t}{2} \partial_{t_1} F^{(1)}. \tag{26}$$

where $\alpha = c_s^2 (\tau - \frac{1}{2}) \Delta t$. Therefore, combining Eq.(26) with Eq.(23) and taking $\lambda = \frac{\tau - \frac{1}{2}}{\tau}$, we have

$$\partial_t \phi + \nabla \cdot (\phi \mathbf{u}) = \alpha \nabla^2 \phi + F - \frac{\Delta t}{2} \epsilon \partial_{t_1} F. \tag{27}$$

Neglecting the last term in the right side of Eq.(27), the CDE (1) is recovered. If we use the LB scheme proposed in Ref.[6], this term can be fully eliminated.

A General Long-Time Molecular Dynamics Scheme in Atomistic Systems: Hyperdynamics in Entropy Dominated Systems

Xin Zhou and Yi Jiang

Los Alamos National Laboratory, Los Alamos, NM 87545, USA
xzhou@lanl.gov

Abstract. We extend the hyperdynamics method developed for low-dimensional energy-dominated systems, to simulate slow dynamics in more general atomistic systems. We show that a few functionals of the pair distribution function forms a low-dimensional collective space, which is a good approximation to distinguish stable and transitional conformations. A bias potential that raises the energy in stable regions, where the system is at local equilibrium, is constructed in the pair-correlation space on the fly. Thus a new MD scheme is present to study any time-scale dynamics with atomic details. We examine the slow gas-liquid transition of Lennard-Jones systems and show that this method can generate correct long-time dynamics and focus on the transition conformations without prior knowledge of the systems. We also discuss the application and possible improvement of the method for more complex systems.

1 Introduction

The atomistic molecular dynamics (MD) simulations are typically limited to a time scale of less than a microsecond, so many interesting slow processes in chemistry, physics, biology and materials science cannot be simulated directly. Many new methods, such as kinetic monte carlo [1], transition path ensemble methods [2], minimal action/time methods [3]have been developed to study the slow processes (for a review see [4]). They all require prior knowledge of the system, which is often hard to obtain, and they can only deal with a few special processes inside a small part of configurational space of the system.

In many systems, the interesting slow dynamics are governed by the infrequent, fast transitions between (meta-) stable regions; yet the systems spend most of their time in the stable regions, whose dynamics can be well described by some time-average properties. Hence an alternative approach to describing the long-time dynamic trajectory would be some suitable time propagator entering in/out the stable regions as well as the real short transition trajectory among the regions. In other words, we would coarse-grain the stable regions while keeping the needed details outside the stable regions. This natural coarse-graining technique is different from the usual method that average some degrees of freedom. The averaging process in the latter method usually changes the dynamics, although the static properties of systems might remain correct.

Y. Shi et al. (Eds.): ICCS 2007, Part I, LNCS 4487, pp. 826–833, 2007.

Hyperdynamics, originally developed by Voter [5],is an example of such a coarse-graining method. The hyperdynamics method treats the potential basins as the stable configurational regions that are separated by saddle surfaces. A bias potential is designed to lift the potential energy of the system in the basins, while keeping the saddle surfaces intact. Dynamics on the biased system lead to accelerated evolution from one stable region to another. Based on transition state theory (TST), the realistic waiting time t_{real} before a transition from the basins can be re-produced statistically:

$$t_{real} = \Delta t \sum_i \exp(\beta \Delta V(r_i)), \qquad (1)$$

where Δt is MD time step, and $\Delta V(r_i)$ is the bias potential at the conformation r_i, $\beta = 1/k_B T$, k_B is the Boltzmann constant and T is the temperature of the system; r refers to the $3N$-dimensional conformation vector throughout this paper. The method has been applied successfully to systems in which the relevant states correspond to deep wells in the potential energy surface (PES), with dividing surfaces at the energy ridge tops separating these states. This is typical of solid-state diffusion systems [6].However, in systems where entropy is not negligible, the basins of PES do not completely correspond to the long-time stable regions. Hyperdynamics cannot readily apply. An extreme example is hard sphere systems: all physical permitted conformations have the same zero potential energy, but some conformations, which correspond to transition regions among stable regions, are rarely visited. A complication occurs even when applying hyperdynamics in solids with fairly clearly defined stable regions: after applying the bias potential, the energy landscape becomes much flatter and the system can start to have entropic-like characteristics. These effects limit the improvement in the simulation rate that can be achieved by the hyperdynamics method over the direct MD approach [5]. Thus, although there have been some attempts [7,8] to apply hyperdynamics to enhance conformational sampling in biological systems, generally, accurate slow dynamics or kinetics can only be expected for relatively simple or low-dimensional systems.

Here, we derive a more general hyperdynamics method that can be used to access very long (possibly all) time scale in fluids where both entropy and potential energy are important. A description of part of this method has appeared in [9]. We first present explicit conditions for applying this method without using TST. Then we give some possible collective variables for identifying the stable and transition conformations to design suitable bias potential. We then examine the performance of this method in simple fluids. Finally we discuss the further development and application of this method in more general (and complex) systems.

2   Theory

We begin this process by introducing a time-compressing transformation,

$$d\tau = a(r)dt, \qquad (2)$$

where $d\tau$ is a pseudo-time step, dt is the real time step, and the local dimensionless compression factor is given by a conformational function $a(r)$, which is ≤ 1. Thus the trajectory $r(t)$ can be rewritten as $r(\tau) = r(\tau(t))$ in a shorter pseudo-time interval τ, and we have,

$$\tau = \int_0^t dt' a(r(t')) = t \int dr D(r; r(t); t) a(r), \tag{3}$$

where $D(r; r(t); t)$ is the density probability of $r(t)$ in the conformational space in interval $[0, t]$,

$$D(r; r(t); t) = \frac{1}{t} \int_0^t dt' \delta(r - r(t')) \tag{4}$$

The compressed trajectory $r(\tau)$ will satisfy a new equation of motion [9]. If we use the usual Langevin equation to simulate the evolution of system in NVT ensemble, the new equation is,

$$\frac{d}{d\tau} R_i = \frac{\mathcal{P}_i}{M_i}$$
$$\frac{d}{d\tau} \mathcal{P}_i = -\frac{\partial U(r)}{\partial R_i} - \zeta'(r)\mathcal{P}_i + f_i(\tau) - \sum_j \mathcal{L}_{ij} \frac{\partial \Delta V(r)}{\partial R_j}, \tag{5}$$

where R_i is the ith component of conformational vector r, $M_i = m_i a^2(r)$ is the pseudo-mass of the particle with the real mass m_i. It is an equation of motion of particles with smaller conformation-dependent mass $M(r)$ under new potential $U(r) = V(r) + \Delta V(r)$ where $\Delta V(r) = -k_B T \ln a(r)$, as well as new friction coefficient, if we neglect the zero ensemble-average term

$$\mathcal{L}_{ij} = \frac{1}{k_B T} \frac{\mathcal{P}_i \mathcal{P}_j}{M_j} - \delta_{ij},$$

where δ_{ij} is the kroneck δ-symbol. Similarly, the zero-mean white noise friction force $f(\tau)$ satisfies the fluctuation-dissipation theorem of the new system,

$$< f_i(\tau) f_j(\tau') > = 2k_B T M_i \zeta' \delta_{ij} \delta(\tau - \tau'). \tag{6}$$

where $\zeta'(r) = \zeta/a(r)$ and ζ is the friction coefficient of the original system.

At the first glance, it would not appear to be advantageous over directly generate $r(\tau)$ from (5), as very short time steps are necessary due to the small mass M. However, if we only focus on the long-time dynamics, we can use a smoother pseudo-trajectory $\mathcal{R}(\tau)$ to replace $r(\tau)$, provided that the reproduced time from $\mathcal{R}(\tau)$ is the same as that from $r(\tau)$. Thus, a sufficient condition to replace $r(\tau)$ with $\mathcal{R}(\tau)$ is that their density probability are the same. Actually, if the compressed factor is defined as function of some collective variables, notated as X, rather than that of r, a simpler condition is,

$$D(X; \mathcal{R}(\tau); \tau) = D(X; r(\tau); \tau), \tag{7}$$

where $D(X; \mathcal{R}(\tau); \tau)$ and $D(X; r(\tau); \tau)$ are the density probability of the trajectory $\mathcal{R}(\tau)$ and $r(\tau)$ in the X space, respectively.

In the time-consuming regions (stable regions), similar conformations would be visited repeatedly many times even during a finite simulation time. Thus we can assume the distribution can be approximated by $D(r; r(t); t) \propto \exp(-\beta V(r))$. Many methods might be used to generate $\mathcal{R}(\tau)$ with the required distribution. One of them is to use a realistic trajectory corresponding to the local equilibrium of a new potential $U(r) = V(r) - k_B T \ln a(r)$. Actually, if we select $a(r) < 1$ only in the potential wells, we effectively have the hyperdynamics presented by Voter [5].

In many cases, some collective variables X can be used to identify transition conformations. The transition conformations related to slow dynamics always locate in some regions of X space, so we can select $a(r) = a(X(r)) < 1$ outside these regions that involve transition conformations, thus the new potential is $U(r) = V(r) - k_B T \ln a(X(r))$. For example, in polymers, the transition regions of slow conformational transitions can be identified in the soft torsion angles space. We require that the density probability of pseudo-trajectory in the collective-variable space equals to that of the compressed trajectory. Thus, if we use a bias potential to generate the pseudo-trajectory, a simple design of the bias potential is

$$\Delta V(r) = k_B T f_+(\ln(D(X(r))/D_c)), \tag{8}$$

where $f_+(y) = y\Theta(y)$, $\Theta(y)$ is the step function. $D(X)$ is the density probability of a segment of trajectory, and D_c is a pre-selected threshold value. The design means we compress the trajectory to make the density probability reach D_c if the original density is larger than the value. We can repeat the biasing process: simulating a segment of trajectory to get a bias potential, then simulating another segment of trajectory under the designed bias potentials to get another bias potential. The sum of all bias potentials forms the total bias potential to reach (almost) any time scale. In practice, we also add a correction the definition of $f_+(y)$ near $y = 0$ to get continuous bias force.

The key to apply hyperdynamics successfully is to design suitable bias potentials $\Delta V(r)$. Obviously, $\Delta V(r)$ should have the same symmetry as $V(r)$. For a simple case of N identical particles, the conformational vector $\{\mathbf{R}_j\}$ $(j = 1, \cdots, N)$ can be rewritten as a density field, $\hat{\rho}(\mathbf{x}) = \sum_j \delta(\mathbf{x} - \mathbf{R}_j)$. Here both $\mathbf{x}$ and $\mathbf{R}$ are the normal 3-dimension spatial vectors. Since the neighboring conformations are equivalent in the viewpoint of slow dynamics, $\hat{\rho}(\mathbf{x})$ can be averaged to get a smooth function, $\bar{\rho}(\mathbf{x})$, for example, by using a Gaussian function to replace the Dirac-δ function. If the width of the Gaussian function is small, $\bar{\rho}(\mathbf{x})$ can be used to identify different conformations. Another similar description is the $k-$space density field, $\hat{\rho}(\mathbf{k}) = \sum_i \exp(i\mathbf{k} \cdot \mathbf{R}_i)$. If neglecting the effects of multi-body correlations and directional correlations, we can approximate the density field $\bar{\rho}(\mathbf{x})$ to the radial pair distribution function $g(z)$, which is defined as,

$$g(z) = \frac{1}{4\pi \rho z^2 N} \sum_i \sum_{j \neq i} \delta(r_{ij} - z), \tag{9}$$

where $r_{ij} = |\mathbf{R}_i - \mathbf{R}_j|$, or more exactly, some bin-average values of $g(z)$ along z,

$$g_p = 2\pi\rho \int g(z)h_p(z)z^2dz, \tag{10}$$

Where $h_p(z)$ is a two-step function, unity in a special z range (a_p, b_p) and zero otherwise. Thus, each conformation corresponds to a group of g_p (g vector), the spatial neighborhood of the conformation and their symmetric companions also correspond to the same g vector. If we select enough g_p, all conformations with the same $\{g_p\}$ are thought to be identical in the viewpoint of slow dynamics. Therefore, we can define the bias potential in the low-dimension g space, $\Delta V(\mathbf{R}^N) = \Delta V(\{g_p(\mathbf{R}^N)\})$. To better identify conformations with not-too-small bin size, we can add some important dynamics-related physical variables, such as, potential $V(\mathbf{R}^N)$, to the collective variable group. Another important variable is the two-body entropy s_2, presented first by H. S. Green [10], defined as,

$$s_2 = -2\pi\rho \int [g(z)\ln g(z) - g(z) + 1]z^2dz. \tag{11}$$

The two-body entropy, which forms the main part ($85\% - 95\%$) of macroscopic excess entropy, has been studied widely [11,12]. Actually, both g_p and s_2 are functional of $g(z)$, similarly, it is also possible to use some another functional of $g(z)$ to form the collective variables. In some special systems, it may be also useful to add some special order parameters O_q to take into account possible multi-body correlations. Finally, we have a group (of order 10) of general collective variables $X = \{X^p\}$, which might involve V, s_2, some $\{g_p\}$ and some possible $\{O_q\}$, to identify conformations and form an appropriate bias potential $\Delta V(X(\mathbf{R}^N)) = k_BTf_+(\ln D(X)/D_c)$. The corresponded bias force on each particle can be calculated by the chain rule of differentiation,

$$\Delta\mathbf{f}_i = -\sum_p \frac{\partial\Delta V}{\partial X^p}\frac{\partial X^p}{\partial \mathbf{R}_i}. \tag{12}$$

For example, from the (9), we have,

$$\frac{\partial g(z)}{\partial \mathbf{R}_k} = \frac{1}{2\pi\rho z^2 N}\sum_{j\neq k}\hat{r}_{kj}\frac{\partial}{\partial z}\delta(r_{kj} - z) \tag{13}$$

where $\hat{r}_{kj}$ is the corresponding unit vector of the distance vector $\mathbf{r}_{kj} = \mathbf{R}_j - \mathbf{R}_k$, and $r_{kj} = r_{jk}$ is the length of $\mathbf{r}_{kj}$. Thus, for any functional of $g(z)$, for example, s_2 and g_p, we can easily calculate its derivative, and hence the bias force.

In practice, in order to get continuous bias forces, we replace the Dirac-δ function by some smooth functions, for example, Epanechnikov kernel function,

$$K_e(z) = \frac{3}{4\sqrt{5}\epsilon}(1 - \frac{z^2}{5\epsilon^2}), \tag{14}$$

if $-\sqrt{5}\epsilon \leq z \leq \sqrt{5}\epsilon$, otherwise, $K_e(z) = 0$. While $\epsilon \to 0$, the $K_e(z) \to \delta(z)$. Under the replacement, we redefine

$$g_p = \frac{1}{4\pi\rho\delta Nz_p^2}\sum_i\sum_{j\neq i}\int_{z_p-\delta/2}^{z_p+\delta/2} K_e(z - r_{ij})dz, \tag{15}$$

where δ and z_p is the size and the center of the pth bin, respectively. It is easy to know, while $\epsilon \to 0$, eq.(15) is same as the normal formula of pair correlation function in bins. The s_2 is redefined as,

$$s_2 = -2\pi\rho\delta \sum_{p=1}^{p_m} z_p^2(g_p \ln g_p - g_p + 1),$$ (16)

where p_m is the maximal index of the bins and we have already set a cutoff along z, g_p is the bin-average value of $g(z)$ at the pth bin, defined by the (15). Therefore, the derivative of the s_2 and g_p is continuous. For example,

$$\frac{\partial s_2}{\partial \mathbf{R}_k} = -\frac{1}{N} \sum_{j\neq k} \hat{r}_{kj} c(r_{kj}),$$ (17)

$$c(z) = \sum_p \ln g_p[K_e(z_p + \delta/2 - z) - K_e(z_p - \delta/2 - z].$$

Here $c(z) \to -\frac{d}{dz} \ln g(z)$ under the limit $\epsilon \to 0$.

Besides choosing some functionals of pair distribution function $g(z)$ as the collective variables, another selection is possible, and may be better in some special systems. For more complex systems, for example, multi-component mixtures or macromolecular systems, etc., we should identify different kinds of atoms and calculate different pair distribution function $g_{AB}(z)$, where A and B are the types of atoms. An alternative method is to use energy distribution function. For any pair-interactive potential $E = u(z)$, such as Lennard-Jones interaction, coulomb interaction, the distance r_{ij} of atom pairs can be replace by the interactive energy $u(r_{ij})$, thus we can define the pair distribution function in energy space,

$$G(E) \propto \sum_i \sum_{j\neq i} \delta(E - u(r_{ij})).$$ (18)

Actually, $G(E)$ is a transformation of $g(z)$. Since the interactive energy is more directly related to the dynamics, using $G(E)$ replace $g(z)$ might be a good idea in identifying conformations. In addition, while existing many kinds of interaction between atoms, we can define the $G(E)$ for different kinds of interactive energy, for example $G(E_{bond})$, $G(E_{angle})$, $G(E_{torsion})$, $G(E_{lj})$, $G(E_{coul})$ etc., thus it is possible to take into account higher-order correlation. In the $G(E)$ cases, the bias force can be calculated similarly as that in the $g(z)$ cases. We are testing this idea in water condensation, details will appear elsewhere.

We have examined the general hyperdynamics method in a simple system of N identical Lennard-Jones (LJ) particles and studied the slow gas/liquid transition in the NVT ensemble. The main results and simulation details are published in ref. ([9]). In such a system, the potential energy of transitional conformations (liquid drops with critical size) is lower than that of the stable gas phase. Thus, simple bias potential based on potential energy [13] cannot work at all. In general, for entropy-important systems, using the potential alone is not sufficient to identify the transitional conformations and then form the bias potential. We also found that with only two functionals of $g(z)$, the potential energy V and the two-body entropy s_2, as the collective variables, we were able to correctly produce the slow dynamics. By designing suitable bias potentials, we could reach almost any slow timescale gas-liquid transition in reasonable time frame: we used a 10^6 time boost factor to find the transition in the system with very small saturation shown in the Fig. (1) [9]).

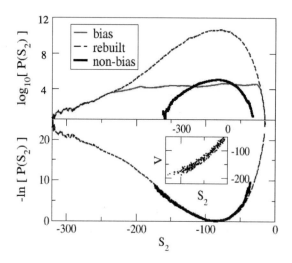

Fig. 1. Top: the distributions of two body entropy S_2 from non-biased and biased simulations, and the rebuilt distribution of the bias simulation. Here, the simulated system starts from gas phase with very small saturation. Bottom: the free energy profiles from the non-biased and biased simulations are compared. The inset shows the simulated samples in the (S_2, V) space. The observed liquid phase does not show here.

To summarize, we have extended the hyperdynamics method to more general cases by inhomogeneously compressing time. Our approach directly generates an explicit method to design the bias potential. In simple systems, two-body entropy s_2 as a functional of the pair distribution function provide a good approximation of the density field in identifying the important conformations and for constructing the bias potential without prior knowledge of the conformational space. The method can be applied in complex fluids, such as glass transition and liquid/solid transition of single or multi component Lennard-Jones fluids, water etc., and s_2 should be still the leading collective variable in the complex systems. For more complex cases, for example, polymers, biological systems, in whole conformational space, it is possible that we will need too many collective variables in identifying transitions, so that it is very difficult to estimate the density probability in the higher-dimension collective variables. A possible improvement is to divide the whole conformational space into some small parts, and use a few collective variables locally in each part.

Acknowledgments

This work was supported by the US DOE under contract No. DE-AC52-06NA25396. We are grateful to K. Kremer, H. Ziock, S. Rasmussen and A. F. Voter for stimulating discussions, comments and suggestions.

References

1. Bortz, A. B. and Kalos, M. H. and Lebowitz, J. L.: A new algorithm for Monte Carlo simulation of Ising spin systems. J. Comp. Phys. **17**, (1975) 10-18.
2. Bolhuis, P. G. and Chandler, D. and Dellago, C. and Geissler, P. L.: Transition Path Sampling: Throwing Ropes over Mountain Passes in the Dark. Ann. Rev. Phys. Chem. **53**, (2002) 291-318.
3. Olender, R. and Elber, R.: Calculation of classical trajectories with a very large time step: Formalism and numberical examples. J. Chem. Phys. **105**, (1996) 9299-9315.
4. Elber, R.: Long-timescale simulation methods. Curr. Opin. Struct. Bio. **15**, (2005) 151-156 .
5. Voter, A. F.: A method for accelerating the molecular dynamics simulation of infrequent events. J. Chem. Phys. **106**, (1997) 4665-4677.
6. Miron, R. A. and Fichthorn, K. A.: Accelerated molecular dynamics with the bond-boost method. J. Chem. Phys. **119**, (2003) 6210-6216.
7. Rahman, J. A. and Tully, J. C.: Puddle-skimming: An efficient sampling of multi-dimensional configuration space. J. Chem. Phys. **116**, (2002) 8750-8760.
8. Hamelberg, D. and Shen, T.-Y. and McCammon, J. A.: Phosphorylation effects on cis/trans isomerization and the backbone conformation of serine-proline Motifs: accelerated molecular dynamics analysis. J. Am. Chem. Soc. **127**, (2005) 1969-1974.
9. Zhou, X. and Jiang, Y. and Kremer, K. and Ziock, H. and Rasmussen, S.: Hyperdynamics for entropic systems: Time-space compression and pair correlation function approximation. Phys. Rev. E **74**, (2006) R035701.
10. Green, H. S.: The Molecular Theory of Fluids. (North-Holland, Amsterdam, 1952);
11. Baranyai, A. and Evans, D. J.: Direct entropy calculation from computer simulation of liquids. Phys. Rev. A **40**, (1989) 3817-3822.
12. Giaquinta, P. V. and Giunta, G. and PrestipinoGiarritta, S.: Entropy and the freezing of simple liquids., Phys. Rev. A **45** (1992) R6966 - R6968.
13. Steiner, M. M. and Genilloud, P.-A and Wilkins, J. W.: Simple bias potential for boosting molecular dynamics with the hyperdynamics scheme. Phys. Rev. B **57**, (1998) 10236-10239.

A New Constitutive Model for the Analysis of Semi-flexible Polymers with Internal Viscosity

Jack Xiao-Dong Yang[1,2] and Roderick V.N. Melnik[2]

[1] Department of Engineering Mechanics,
Shenyang Institute of Aeronautical Engineering, Shenyang 110136, China
[2] Mathematical Modelling & Computational Sciences, Wilfrid Laurier University,
Waterloo, Ontario, Canada N2L 3C5
{jyang,rmelnik}@wlu.ca

Abstract. The analysis of dynamics of semi-flexible polymers, such as
DNA molecules, is an important multiscale problem with a wide range
of applications in science and bioengineering. In this contribution, we
show how accounting for internal viscosity in dumbbell-type models may
render physically plausible results with minimal computational cost. We
focus our attention on the cases of steady shear and extensional flows
of polymeric solutions. First, the tensors with moments other than the
second order moment are approximated. Then, the nonlinear algebraic
equation for the second moment conformation tensor is solved. Finally,
substituting the resulting conformation tensor into the Kramers equation of Hookean spring force, the constitutive equations for the model
are obtained. The shear material properties are discussed in the context of different internal viscosities and our computational results are
compared with the results of other methods applicable for high shear or
extensional rates.

Keywords: Polymeric fluid; Dumbbell model; Internal viscosity.

1 Introduction

The dynamics of polymeric fluids is an important multiple time scale problem, involving the quick speed of the small solvent molecules, fast movement of
atomic particles that constitute the polymers and slow orientation of the polymer deformation [1]. Brownian dynamic simulation is the technique taking into
account all the motions in the polymeric fluids. Coarse-grained models are often
used in the analysis of the rheological properties of polymeric fluids. From the
mesoscopic view, the fast movements of the atomic particles of the polymers are
neglected in order to relieve the complexity. In the investigation of the mathematical problems for the coarse-grained models, multiscale method is also an
efficient technique to obtain the approximate solutions [2].

The simplest, albeit extremely useful in applications, model for polymer solutions is the Hookean dumbbell model proposed by Kuhn [3], where a polymer
molecule in dilute solution is represented by two beads connected by a spring

Y. Shi et al. (Eds.): ICCS 2007, Part I, LNCS 4487, pp. 834–841, 2007.

force. This mathematical simplification has contributed a lot to the development of constitutive models for investigating properties of polymeric fluids with dumbbell-type models [1,4,5]. In order to obtain a better agreement with experimental results, a few additions have been incorporated into the standard dumbbell model. Of particular importance for our further discussion are two of them: finitely extensible nonlinear elastic (FENE) property and internal viscosity of the spring. Taking into account these properties leads to a situation where the governing equations of conformation tensors become nonlinear and the resulting problem has no closed form solution unless an appropriate approximation is made.

Brownian dynamic simulations have been used widely in applications of the FENE dumbbell model to the analysis of fluids and semi-flexible polymers. The results obtained on the basis of such simulations are often more accurate compared to approximate theoretical methodologies discussed in [6,7,8]. Nevertheless, Brownian dynamic simulations are time consuming and do not render a straightforward technique for the explanation of the underlying physical properties of fluids and polymers. A consequence of this is intensive research efforts in developing semi-analytical methodologies which should be combined with computational experiments in order to shed further light on such properties. From a mathematical point of view, these efforts are reducible to the construction of some closure forms for the equation written with respect to the conformation tensor in such a way that the governing constitutive equations for the system (after certain approximations) become analytically solvable. The main idea is to apply an ensemble averaging methodology to non-second moment terms so that closed form solutions could be obtained [2,9,10]. However, for the dumbbell model with internal viscosity there is a forth moment tensor in the governing equation which makes it difficult to find the closed form solutions. Therefore, a numerical methodology should be developed to simulate the dynamics of the polymeric system. In the context of rheological properties of complex fluids, this issue was previously discussed by a number of authors (e.g., [11,15]). Booij and Wiechen [12] used a perturbation solution approach to calculate the first order approximation of the internal viscosity parameter. More recently, based on the Gaussian closure method, the second moment tensor and higher moment tensors have been calculated by integration over the conformation distribution space [13,14,15].

In this contribution, a new approximation scheme is proposed to solve the governing equations analytically. The forth moment tensor is approximated by an expression involving the second moment tensor in order to obtain a set of nonlinear algebraic equations. Based on the analytical solutions of such equations, the material properties of the polymeric fluids in steady-state shear flows and extensional viscosities in extensional flows are discussed. The phenomena of shear thinning and attenuation of pressure drop have been found and the results have been compared to the results obtained by Brownian dynamics simulations and the Gaussian closure method in the context of high shear or extensional rates. Our results can explain the phenomenon of shear thinning by introducing the internal viscosity and obtaining better predictions compared to the traditional technique.

2   The Governing Equations

For the polymers in a Newtonian solvent with viscosity η_s described with the bead-spring-bead dumbbell model it is assumed that there is no interaction between the beads. Let us denote the viscous drag coefficient due to the resistance of the flow by ζ. For the dumbbell model with internal viscosity (IV), the spring force is a function of the configuration vector, $\mathbf{Q}$, and configuration velocity $\dot{\mathbf{Q}}$. The force law in this case can be expressed as:

$$\mathbf{F}\left(\mathbf{Q}, \dot{\mathbf{Q}}\right) = H\mathbf{Q} + K\left(\frac{\mathbf{Q} \otimes \mathbf{Q}}{Q^2}\right)\dot{\mathbf{Q}}, \tag{1}$$

where Q is the length of vector $\mathbf{Q}$, H is the spring coefficient of the dumbbell model and K is a constant denoting the measurement of the IV. The dot indicates differentiation with respect to time t, so that $\dot{\mathbf{Q}}$ represents the velocity vector of the dumbbells. By substituting equation (1) and the equation of motion of one bead into the continuity equation [1]

$$\frac{\partial \psi}{\partial t} = -\frac{\partial}{\partial \mathbf{Q}} \cdot \dot{\mathbf{Q}}\psi, \tag{2}$$

we can derive the diffusion equation

$$\frac{\partial \psi}{\partial t} = -\frac{\partial}{\partial \mathbf{Q}} \cdot \left\{\left(\boldsymbol{\delta} - g\left\langle\frac{\mathbf{Q} \otimes \mathbf{Q}}{Q^2}\right\rangle\right) \cdot \left([\boldsymbol{\kappa} \cdot \mathbf{Q}]\,\psi - \frac{2kT}{\zeta}\frac{\partial \psi}{\partial \mathbf{Q}} - \frac{2H}{\zeta}\mathbf{Q}\right)\right\}, \tag{3}$$

where $\boldsymbol{\delta}$ is a unit matrix, $g = 2\varepsilon/(1+2\varepsilon)$, and ε is the relative internal viscosity. Note that $\varepsilon = K/2\zeta$ and it can formally range from zero to infinity. Furthermore, for $g = 0$, equation (3) recovers the form of the diffusion equation for Hookean dumbbells without IV.

The second moment conformation tensor $\langle\mathbf{Q} \otimes \mathbf{Q}\rangle$ is of the most interest when calculating the stress tensor. The governing equation for the conformation tensor can be developed by multiplying the diffusion equation by the dyadic product $\mathbf{Q} \otimes \mathbf{Q}$ and integrating over the entire configuration space:

$$\langle\mathbf{Q} \otimes \mathbf{Q}\rangle_{(1)} = \frac{4kT}{\zeta}\left(\boldsymbol{\delta} - 3g\left\langle\frac{\mathbf{Q} \otimes \mathbf{Q}}{Q^2}\right\rangle\right) - \frac{4H}{\zeta}(1-g)\langle\mathbf{Q} \otimes \mathbf{Q}\rangle$$
$$-2g\boldsymbol{\kappa} : \left\langle\frac{\mathbf{Q} \otimes \mathbf{Q} \otimes \mathbf{Q} \otimes \mathbf{Q}}{Q^2}\right\rangle. \tag{4}$$

The subscript "(1)" denotes convected derivatives. In homogeneous flows the convected derivative is defined as

$$\mathbf{A}_{(1)} = \frac{\partial}{\partial t}\mathbf{A} - \left\{\boldsymbol{\kappa} \cdot \mathbf{A} + \mathbf{A} \cdot \boldsymbol{\kappa}^T\right\}. \tag{5}$$

Unfortunately, it is not possible to calculate the second moment tensor $\langle \mathbf{Q} \otimes \mathbf{Q} \rangle$ directly because there are other moment terms, e.g., $\left\langle \frac{\mathbf{Q} \otimes \mathbf{Q}}{Q^2} \right\rangle$ and $\langle \mathbf{Q} \otimes \mathbf{Q} \otimes \mathbf{Q} \otimes \mathbf{Q} \rangle$. Hence, in order to cast the governing equation into a form amenable to the analytical solution, the higher order terms should be approximated. This is done as follows:

$$\left\langle \frac{\mathbf{Q} \otimes \mathbf{Q}}{Q^2} \right\rangle \approx \frac{\langle \mathbf{Q} \otimes \mathbf{Q} \rangle}{\langle Q^2 \rangle_{eq}}, \tag{6}$$

$$\left\langle \frac{\mathbf{Q} \otimes \mathbf{Q} \otimes \mathbf{Q} \otimes \mathbf{Q}}{Q^2} \right\rangle \approx \frac{\langle \mathbf{Q} \otimes \mathbf{Q} \rangle \otimes \langle \mathbf{Q} \otimes \mathbf{Q} \rangle}{\langle Q^2 \rangle_{eq}}. \tag{7}$$

The equations (6), (7) are key approximations allowing us to make the governing equation analytically solvable. Note that equation (6) is similar to the Perterlin approximation used in FENE dumbbell model. By using equations (6) and (7), the governing equation (2) can be cast in the following form:

$$\langle \mathbf{Q} \otimes \mathbf{Q} \rangle_{(1)} = \frac{4kT}{\zeta} \left(\delta - 3g \frac{\langle \mathbf{Q} \otimes \mathbf{Q} \rangle}{\langle Q^2 \rangle_{eq}} \right) - \frac{4H}{\zeta} (1 - g) \langle \mathbf{Q} \otimes \mathbf{Q} \rangle$$
$$-2g\kappa : \frac{\langle \mathbf{Q} \otimes \mathbf{Q} \rangle \otimes \langle \mathbf{Q} \otimes \mathbf{Q} \rangle}{\langle Q^2 \rangle_{eq}}. \tag{8}$$

In the steady state flow case, when all the time-dependent terms can be neglected, equation (8) becomes a nonlinear algebraic equation with respect to $\langle \mathbf{Q} \otimes \mathbf{Q} \rangle$. In the next section, we will seek a closed form solution to this governing equation, followed by the material properties discussion in the case of steady state shear flow.

3   Results and Examples

3.1   The Material Coefficients in Steady Shear Flows

First, we consider the steady state shear flow with the velocity vector given by

$$\mathbf{v} = (v_x, v_y, v_z) = (\dot{\gamma}y, 0, 0), \tag{9}$$

where $\dot{\gamma}$ is the shear rate. The transpose of velocity vector gradient is

$$\kappa = (\nabla \mathbf{v})^T = \begin{pmatrix} 0 & \dot{\gamma} & 0 \\ 0 & 0 & 0 \\ 0 & 0 & 0 \end{pmatrix}. \tag{10}$$

The average value of the square of the end-to-end distance in equilibrium in shear flow with shear rate $\dot{\gamma}$ can be represented as [1]

$$\langle Q^2 \rangle_{eq} = \frac{3kT}{H} + \frac{2kT}{H} (\lambda_H \dot{\gamma})^2, \tag{11}$$

where the time constant λ_H is defined by $\lambda_H = \zeta/4H$. For convenience, we use the following notation for the conformation tensor

$$\mathbf{A} = \begin{pmatrix} A_{xx} & A_{xy} & A_{xz} \\ A_{yx} & A_{yy} & A_{yz} \\ A_{zx} & A_{zy} & A_{zz} \end{pmatrix} = \frac{H}{kT} \langle \mathbf{Q} \otimes \mathbf{Q} \rangle, \tag{12}$$

so that its convected differentiation in steady state situations gives

$$\mathbf{A}_{(1)} = -\left\{ \boldsymbol{\kappa} \cdot \mathbf{A} + \mathbf{A} \cdot \boldsymbol{\kappa}^T \right\}. \tag{13}$$

Substituting equations (11)-(13) into (8) and equating corresponding tensor elements in the result, we conclude that the nonzero elements of $\mathbf{A}$ can be calculated as follows

$$A_{xx} = \frac{2\lambda_H \dot{\gamma} A_{xy} + 1}{1 - \frac{2(\lambda_H \dot{\gamma})^2 - 2\lambda_H \dot{\gamma} A_{xy}}{3 + 2(\lambda_H \dot{\gamma})^2} g}, \quad A_{yy} = A_{zz} = \frac{1}{1 - \frac{2(\lambda_H \dot{\gamma})^2 - 2\lambda_H \dot{\gamma} A_{xy}}{3 + 2(\lambda_H \dot{\gamma})^2} g}, \tag{14}$$

where A_{xy} is the real root of

$$\frac{4 (\lambda_H \dot{\gamma})^2}{\left[3 + 2 (\lambda_H \dot{\gamma})^2\right]^2} g^2 A_{xy}^3 + \frac{4\lambda_H \dot{\gamma}}{3 + 2 (\lambda_H \dot{\gamma})^2} \left[1 - \frac{2 (\lambda_H \dot{\gamma})^2}{3 + 2 (\lambda_H \dot{\gamma})^2} g\right] A_{xy}^2$$

$$+ \left[1 - \frac{2 (\lambda_H \dot{\gamma})^2}{3 + 2 (\lambda_H \dot{\gamma})^2} g\right]^2 A_{xy} - \lambda_H \dot{\gamma} = 0. \tag{15}$$

For convenience, we use the Kramers equation for the stress tensor in the spring model:

$$\frac{\tau_p}{nkT} = -\frac{H}{kT} \langle \mathbf{Q} \otimes \mathbf{Q} \rangle + \boldsymbol{\delta} = -\mathbf{A} + \boldsymbol{\delta}. \tag{16}$$

The three material functions of interest, namely, the viscosity $\eta(\dot{\gamma})$, the first-normal stress coefficient $\Psi_1(\dot{\gamma})$, and the second-normal stress coefficient $\Psi_2(\dot{\gamma})$ are connected with the stress components by the following relationships:

$$\begin{aligned} \tau_{xy} &= -\eta(\dot{\gamma}) \dot{\gamma}, \\ \tau_{xx} - \tau_{yy} &= -\Psi_1(\dot{\gamma}) \dot{\gamma}^2, \\ \tau_{yy} - \tau_{zz} &= -\Psi_2(\dot{\gamma}) \dot{\gamma}^2. \end{aligned} \tag{17}$$

Substituting equation (16) into (14), we obtain the material coefficients via the following representations:

$$\begin{aligned} \frac{\eta(\dot{\gamma})}{nkT\lambda_H} &= \frac{A_{xy}}{\lambda_H \dot{\gamma}}, \\ \frac{\Psi_1(\dot{\gamma})}{nkT\lambda_H^2} &= \frac{2A_{xy}}{\left[1 - \frac{2(\lambda_H \dot{\gamma})^2 - 2\lambda_H \dot{\gamma} A_{xy}}{3 + 2(\lambda_H \dot{\gamma})^2} g\right] \lambda_H \dot{\gamma}} \\ \Psi_2 &= 0. \end{aligned} \tag{18}$$

Based on equations (15) and (18), we calculate the material properties for different internal viscosities. The plots presented in Figure 1 demonstrate comparison results between our approximate solutions (AS), the results obtained by Brownian dynamics simulations (BD), and the results obtained with the Gaussian closure technique (GC) [15]. The internal viscosity was chosen as $\varepsilon = 1$. In Figure 1, the solid lines represent our algebraic solutions to the governing equations for the material coefficients, the dots represent the data obtained by Brownian dynamics simulations, and the dot-dash lines represent the results obtained by the Gaussian closure method. For both viscosity and first-normal stress coefficients, our algebraic solutions exhibit the plateau values appearing also in the case of Brownian dynamics simulations. Observe that compared to the Gaussian closure method, our methodology has a wider range of applicability.

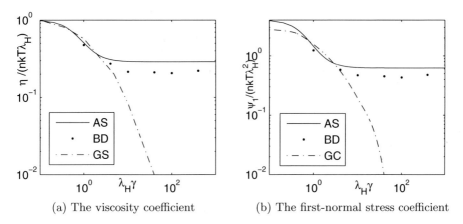

(a) The viscosity coefficient   (b) The first-normal stress coefficient

Fig. 1. Comparison of analytical results with Brownian dynamics simulations and the Gaussian closure methodology ($\varepsilon = 1$)

3.2   The Extensional Viscosity in Steady Extensional Flows

Next, we consider the steady extensional flow with the velocity vector given by

$$\mathbf{v} = (v_x, v_y, v_z) = \left(-\frac{1}{2}, -\frac{1}{2}, 1\right)\dot{\varepsilon}, \tag{19}$$

where $\dot{\varepsilon}$ is the extensional rate in the z direction. The transpose of velocity vector gradient is

$$\boldsymbol{\kappa} = (\nabla \mathbf{v})^T = \begin{pmatrix} -\frac{1}{2} & 0 & 0 \\ 0 & -\frac{1}{2} & 0 \\ 0 & 0 & 1 \end{pmatrix}\dot{\varepsilon}. \tag{20}$$

In the steady uniaxial extensional flow with strain rate $\dot{\varepsilon}$, the extensional viscosity is defined as [16]

$$\mu_e = \frac{2\tau_{zz} - \tau_{xx} - \tau_{yy}}{6\dot{\varepsilon}}. \tag{21}$$

By using the equation (8), we get the solutions of the equation with respect to the conformation tensor by the same procedure used in the steady state shear flow case. Then the extensional viscosity is calculated as

$$\frac{\mu_e}{nkT\lambda_H} = \frac{2A_{zz} - A_{xx} - A_{yy}}{6\dot{\varepsilon}}, \tag{22}$$

where A_{xx}, A_{yy} and A_{zz} are determined by the following set of algebraic equations:

$$
\begin{aligned}
(\lambda_H\dot{\varepsilon} + 1) A_{xx} + \tfrac{4}{3}g\lambda_H\dot{\varepsilon}(A_{zz} - A_{11}) &= 1, \\
(-2\lambda_H\dot{\varepsilon} + 1) A_{xx} + \tfrac{4}{3}g\lambda_H\dot{\varepsilon}(A_{zz} - A_{11}) &= 1, \\
A_{yy} &= A_{xx}.
\end{aligned}
\tag{23}
$$

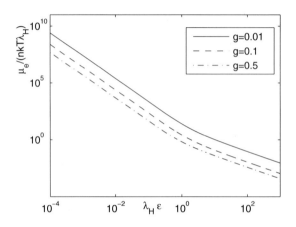

Fig. 2. The viscosity coefficient in the steady extensional flow

Figure 2 demonstrates decrease in extensional viscosity with higher rates of extensional flows. This explains the attenuation of the pressure drop in strong extensional flows.

4   Conclusions

In this contribution, we developed a set of approximate semi-analytical solutions for the dumbbell model with IV without integration of the Gaussian distribution. Our concise equations can predict the material coefficients of polymeric fluid well, qualitatively and also quantitatively. The shear thinning phenomena are described well with the new developed model deduced from the dumbbell model with internal viscosity. The effect of internal viscosity in the extensional flow case has also been demonstrated. By comparing our computational results with Brownian dynamic simulations and the Gaussian closure methodology, we demonstrated the efficiency of the proposed approximate technique for a wider range of high shear or extensional flow rates.

Acknowledgment. This work was supported by NSERC.

References

1. Bird R.B., Curtiss C.F., Armstrong R.C., Hassager O.: Dynamics of Polymer Liquids Vol. 2 Kinetic Theory. John Wiley & Sons (1987)
2. Nitsche L.C., Zhang W., Wedgewood L.E.: Asymptotic basis of the L closure for finitely extensible dumbbells in suddenly started uniaxial extension. J. Non-Newtonian Fluid Mech. **133** (2005) 14-27
3. Kuhn W. :Über die Gestalt fadenförmiger Moleküle in Lösungen. Kolloid Z. **68** (1934) 2-11
4. Bird R.B., Wiest J.M.: Constitutive equations for polymeric liquids. Annu. Rev. Fluid Mech. **27** (1995) 169-193
5. Öttinger H.C.: Stochastic Processes in Polymeric Fluids: Tools and Examples for Developing Simulation Algorithms. Springer-Verlag, Berlin Heidlberg New York (1996)
6. Hur J.S., Shaqfeh E.S.G.: Brownian dynamics simulations of single DNA molecules in shear flow. J. Rheol. **44** (2000) 713-742
7. Hu X., Ding Z., Lee L.J.: Simulation of 2D transient viscoelastic flow using the CONNFFESSIT approach. J. Non-Newtonian Fluid Mech. **127** (2005) 107-122
8. Lozinski A., Chauviere C.: A fast solver for Fokker-Planck equation applied to viscoelastic flows calculations: 2D FENE model. J. Computational Physics **189** (2003) 607-625
9. Herrchen M., Öttinger H.C.: A detailed comparison of various FENE dumbbell models. J. Non-Newtonian Fluid Mech. **68** (1997) 17-42
10. Keunings R.: On the Peterlin approximation for finitely extensible dumbbells. J. Non-Newtonian Fluid Mech. **68** (1997) 85-100
11. Hua C.C., Schieber J.D.: Nonequilibrium Brownian dynamics simulations of Hookean and FENE dumbbells with internal viscosity. J. Non-Newtonian Fluid Mech. **56** (1995) 307-332
12. Booij H.C., Wiechen, P.H.V.: Effect of internal viscosity on the deformation of a linear macromolecule in a sheared solution. J. Chem. Phys. **52** (1970) 5056-5068
13. Manke C.W., Williams M.C.: The internal-viscosit dumbbell in the high-IV limit: Implications for rheological modeling. J. Rheol. **30** (1986) 019-028
14. Schieber J.D.: Internal viscosity dumbbell model with a Gaussian approximation. J. Rheol. **37** (1993) 1003-1027
15. Wedgewood L.E.: Internal viscosity in polymer kinetic theory: shear flows. Rheologica Acta **32** (1993) 405-417
16. Tirtaatmadja V., Sridhar T.: A filament stretching device for measurement of extensional viscosity. J. Rheol. **37** (1993) 1081-1102

Coupled Navier-Stokes/DSMC Method for Transient and Steady-State Gas Flows

Giannandrea Abbate[1], Barend J. Thijsse[2], and Chris R. Kleijn[1]

[1] Dept. of Multi-Scale Physics & J.M.Burgers Centre for Fluid Mechanics, Delft
University of Technology,
Prins Bernhardlaan 6, Delft, The Netherlands
G.Abbate@klft.tn.tudelft.nl, C.R.kleijn@tudelft.nl
http://www.msp.tudelft.nl
[2] Dept. of Material Science and Engineering, Delft University of Technology,
Mekelweg 2, Delft, The Netherlands
B.J.Thijsse@tnw.tudelft.nl
http://www.3me.tudelft.nl

Abstract. An adaptatively coupled continuum-DSMC approach for compressible, viscous gas flows has been developed. The continuum domain is described by the unsteady Navier-Stokes equations, solved using a finite volume formulation in compressible form to capture the shock. The molecular domain is solved by the Direct Simulation Monte Carlo method (DSMC). The coupling procedure is an overlapped Schwarz method with Dirichlet-Dirichlet boundary conditions. The domains are determined automatically by computing the Kn number with respect to the local gradients length scale. The method has been applied to simulate a 1-D shock tube problem and a 2-D expanding jet in a low pressure chamber.

Keywords: Direct Simulation Monte Carlo; Coupled Method; Hybrid Method; Rarefied Gas Flow; Navier-Stokes solver.

1 Introduction

In several applications we are faced with the challenge to model a gas flow transition from continuum to rarefied regime. Examples include: flow around vehicles at high altitudes [1], flow through microfluidic gas devices [2], small cold thruster nozzle and plume flows [3], and low pressure thin film deposition processes or gas jets [4].

It is always very complicated to describe this kind of flows; in the continuum regime ($Kn \ll 1$), Navier-Stokes equations can be used to model the flow, whereas free molecular flow ($Kn \gg 1$) can be modelled using Molecular Dynamics models. For the intermediate Knudsen number ranges ($Kn = 0.01 - 10$), neither of the approaches is suitable. In this regime the best method to use is DSMC (Direct Simulation Monte Carlo). The computational demands of DSMC, however, scale with Kn^{-4} and when the Knudsen number is less than ~ 0.05, its time and memory expenses become inadmissible.

Y. Shi et al. (Eds.): ICCS 2007, Part I, LNCS 4487, pp. 842–849, 2007.
© Springer-Verlag Berlin Heidelberg 2007

Different solutions have been proposed to compute such flows. The most standard uses a continuum solver with analytical slip boundary conditions [5]. This method is suitable only in conditions where $Kn < 0.1$ and the precise formulation of the slip boundary conditions is strongly geometry dependent. For this reason, several hybrid continuum/molecular models have been proposed, for instance: Molecular Dynamics (MD) and Navier-Stokes (N-S) equations [6], Boltzmann and N-S equations [7], Direct Simulation Monte Carlo (DSMC) and Stokes equations [2], DSMC and incompressible N-S equations [8], and DSMC and N-S equations [9,10,11,12,13].

In particular, Garcia et al.[9] constructed a hybrid particle/continuum method with an adaptive mesh and algorithm refinement. It was a flux-based coupling method with no overlapping between the continuum and the DSMC regions. On the contrary, Wu and al. [10] and Schwartzentruber and al. [11,12] proposed an 'iterative' coupled CFD-DSMC method where the coupling is achieved through an overlapped Schwarz method with Dirichlet-Dirichlet type boundary conditions. However, both Wu and al. [10] and Schwartzentruber and al. [11,12] methods are only suitable for gas flow simulations under steady-state conditions, while the method that we propose has been applied both to transient and steady-state gas flow simulations.

We consider, in the continuum regime, the compressible N-S equations and, in the transitional regime, DSMC because it is several order of magnitude more efficient than MD and the Boltzmann equations solvers. The coupling of the two models is reached through an overlapped Schwarz method [10] with Dirichlet-Dirichlet boundary conditions. It is an adaptive method in which, during the computations, the Kn number with respect to the local gradients is computed to determine and divide the CFD (Computational Fluid Dynamics) domain from the DSMC one.

2   The Coupling Method

2.1   The CFD Solver

The CFD code used is a 2-D, unsteady code based on a finite volume formulation in compressible form to capture the shock. It uses an explicit, second-order, flux-splitting, MUSCL scheme for the Navier-Stokes equations.

Because a high temperature flow has to be modelled, a power-law temperature dependence was used for the viscosity μ and a model coming from kinetic gas theory for the thermal conductivity κ. The density was computed from the ideal gas law.

2.2   The Molecular Algorithm: DSMC

The 2-D DSMC code developed is based on the algorithm described in [14].

A "particle reservoirs" approach was used to implement the inlet (outlet) boundary conditions. A Maxwell-Boltzmann or a Chapman-Enskog velocity distributions can be used to generate molecules in those reservoirs.

2.3   Schwarz Coupling

We describe in this section two different strategies developed and implemented to couple the Navier-Stokes based CFD code to the DSMC code: One for steady state flow simulation, the other for unsteady flow simulations.

Steady Formulation: We propose a hybrid coupling method based on the Schwarz method [10] and consisting of two stages.

In the first stage the unsteady N-S equations are integrated in time on the entire domain Ω until a steady state is reached. From this solution, local Kn numbers with respect to the local gradients length scales [15] are computed according to

$$Kn_Q = \frac{\lambda}{Q} | \bigtriangledown Q | \tag{1}$$

where λ is the mean free path length and Q is a flow property (density, temperature etc.); The values of Kn_Q are used to split Ω in the subdomains Ω_{DSMC} $(Kn > Kn_{split} - \Delta Kn)$, where the flow field will be evaluated using the DSMC technique, and Ω_{CFD} $(Kn < Kn_{split})$, where N-S equation will be solved. For Kn_{split} a value of 0.05 was used. Between the DSMC and CFD regions an overlap region is considered, where the flow is computed with both the DSMC and the CFD solver; the value of ΔKn can be chosen in "ad hoc" manner in order to vary the overlap region size.

In the second stage, DSMC and CFD are run in their respective subdomains with their own time steps (Δt_{DSMC} and Δt_{CFD}, respectively), until a steady state is reached. First DSMC is applied; molecules are allocated in the DSMC subdomain according to the density, velocity and temperature obtained from the initial CFD solution. A Maxwell-Boltzmann or a Chapmann-Enskog distributions can be chosen to create molecules. It is important to say that the grid is automatically refined in the DSMC region in order to respect the DSMC requirements. The boundary conditions to the DSMC region come from the solution in the CFD region. As described in the previous section for the inlet (outlet) boundary, outside the overlapping region some "particle reservoirs" are considered. In these cells molecules are created according to density, velocity, temperature and their gradients of the solution in the CFD region, with a Maxwell-Boltzmann or a Chapmann-Enskog distributions. After running the DSMC, the N-S equations are solved in the CFD region. The boundary conditions comes from the solution in the DSMC region averaged over the CFD cells.

Once a steady state solution has been obtained in both the DSMC and N-S region, the local Kn_Q numbers are re-evaluated and a new boundary between the two regions is computed. This second stage is iterated until in the overlapping region DSMC and CFD solutions differ less than a prescribed value.

We made an extensive study of the influence of various coupling parameters, such as the size of the overlap region ($4 - 59$ mean free path lengths) and the amount of averaging applied to the reduce DSMC statistical scatter (averaging over 5, 30 and 50 repeated runs). The influence of these parameters on the final solution was found to be small.

Unsteady Formulation: In the unsteady formulation, the described coupling method is re-iterated every coupling time step $\Delta t_{coupling} >> \Delta t_{DSMC}, \Delta t_{CFD}$, starting on the solution at the previous time step. As expected, in order to avoid instabilities, it was necessary to keep the Courant number (based on the coupling time step, the molecules most probable velocity, and the CFD grid cell size) below one.

In the second stage, after every coupling step, the program compares the predicted DSMC region with the one of the previous step. In the cells that still belong to the DSMC region, we consider the same molecules of the previous time step whose properties were recorded. Molecules that are in the cells that no longer belong to the DSMC region are deleted. In cells that have changed from CFD into a DSMC cell, new molecules are created with a Maxwell-Boltzmann or a Chapmann-Enskog distribution, according to the density, velocity and temperature of the CFD solution at the previous time step.

At the end of the every coupling step molecule properties are recorded to set the initial conditions in the DSMC region for the next coupling step.

3   Results

3.1   1-D Shock-Tube Problem

The unsteady coupling method was applied to the unsteady shock tube test case (fig.1).

Fig. 1. Shock tube test case

The code models a flow of Argon inside an $0.5m$ long tube between two tanks in different thermo-fluid-dynamic conditions. In the left tank there are a pressure of $2000Pa$ and a temperature of $12000K$, while in the tube and in the right tank there are a pressure of $100Pa$ and a temperature of $2000K$. When the membrane that separates the two regions breaks a shock travels in the tube from left to right. Upstream from the shock, the gas has high temperature and pressure, but gradient length scales are very small. Downstream of it both temperature and pressure are lower, but gradient length scales are large. As a result, the local Kn number Kn_Q is high upstream of the shock and low downstream. In the hybrid DSMC-CFD approach, DSMC is therefore applied upstream, and CFD downstream. The continuum grid is composed of 100 cells in x direction and 1 cell in y direction, while the code automatically refines the mesh in the DSMC region to fulfill its requirements. In the DSMC region molecules were created with the Chapman-Enskog distribution. It was demonstrated, in fact, that in a hybrid DSMC/CFD method a Chapman-Enskog distribution is required when the viscous fluxes are taken into account, while a simple Maxwellian distribution is adequate when the continuum region is well approximated by the Euler equations [9]. The particle cross section was evaluated using the Variable Soft Sphere (VSS) model because it is more accurate than the Variable Hard Sphere (VHS) one to model viscous effect. The coupling time step is $\Delta t_{coupling} = 2.0 \times 10^{-6} sec.$ and the ensemble averages of the DSMC solution to reduce the scattering were made on 30 repeated runs. In addition to the hybrid approach, the problem was also solved using CFD only and DSMC only (which was feasible because of the 1-D nature of the problem). The latter is considered to be the most accurate. In fig.2 the pressure inside the tube after $3.0 \times 10^{-5} sec.$, evaluated with the hybrid (Schwarz coupling) method is compared with the results of the full DSMC simulation and the full CFD simulation.

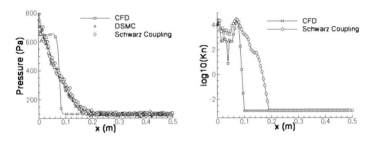

Fig. 2. Pressure and Kn number in the tube after $3.0 \times 10^{-5} sec$

In the same picture also local Knudsen number Kn_Q, computed using the hybrid method, is compared with the full CFD simulation.

From the results shown in fig.2, it is clear that the full CFD approach fails due to the high values of the local Kn number caused by the shock presence. The full CFD approach predicts a shock thickness less than 1 cm, which is unrealistic considering the fact that the mean free path near the shock is of the order of several centimeters. In the full DSMC approach, therefore, the shock is smeared over almost 20 cm. The results obtained with the hybrid approach are virtually identical to those obtained with the full DSMC solver, but they were obtained in less than one fifth of the CPU time.

3.2   2-D Expanding Jet in a Low Pressure Chamber

The steady-state coupling method was applied to a steady state expanding neutral gas jet in a low pressure chamber (fig.3).

The code models an Argon jet, at a temperature of $6000K$ and Mach number 1, injected from the top in a 2-D chamber of dimensions $0.32m \times 0.8m$, through a slot of $0.032m$. The pressure inside the chamber is kept at a value of $10Pa$ through two slots of $0.04m$ wide disposed on its lateral sides at a distance $0.6m$ from the top. Walls are cooled at a temperature of $400K$. The continuum grid

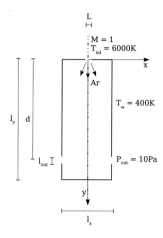

Fig. 3. Expanding jet in a low pressure deposition chamber test case

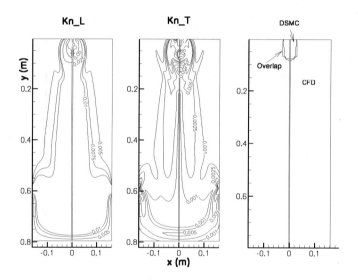

Fig. 4. Kn number and CFD/DSMC domains splitting

is composed of 50 cells in x direction and 160 in y direction while the code automatically refines the mesh in the DSMC region to fullfill its requirements. In the DSMC region, molecules were created with the Chapman-Enskog distribution and the particle cross section was evaluated using the VSS model. Fig.4 shows the Knudsen number in the chamber, respectively evaluated with reference to the inlet dimension (Kn_L) and to the local temperature gradient length scale (Kn_T). The pressure does not change a lot in the domain. Around the inlet the temperature is high and, because of the presence of a shock, gradient length scales are small. In the rest of the chamber the temperature is lower and

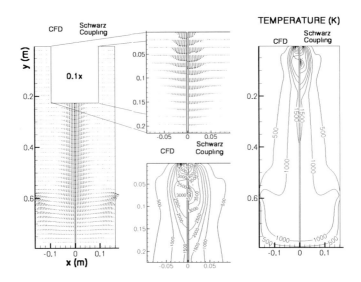

Fig. 5. Velocity and temperature fields in the deposition chamber

gradient length scales are large. As a result the Kn number is high around the inlet and low in the rest of the domain. In the right-hand side of Fig.4, the resulting division between the DSMC, CFD and overlapping regions is shown.

In fig.5 the velocity and temperature fields, evaluated with the hybrid (Schwarz coupling) method, are compared with the results of a full CFD simulation. It is evident that the DSMC region influences the flow field and its effects are present in a region wider then the DSMC and overlapping regions alone. Far away from the DSMC region, however, the full CFD and the hybrid method give the very similar results.

4 Conclusions

A hybrid continuum-rarefied flow simulation method was developed to couple a Navier-Stokes description of a continuum flow field with a DSMC description of a rarefied one. The coupling is achieved by an overlapped Schwarz method implemented both for steady state and transient flows. Continuum subdomain boundary conditions are imposed on the molecular subdomain via particle reservoirs. The molecular subdomain boundary conditions are imposed on the continuum subdomain using simple averaging. The subdomains are determined automatically by computing the Kn number with respect to the local gradients length scale on a preliminary Navier-Stokes solution.

The method has been applied to a shock tube and to a 2-D expanding jet in a low pressure chamber problems showing its capability of predicting the flow field even where a CFD solver fails.

Acknowledgments. We thank Profs. D.C.Schram and M.C.M.Van de Sanden for usefull discussions and the DCSE (Delft Centre for Computational Science and Engineering) for financial support.

References

1. F.Sharipov, Hypersonic flow of rarefied gas near the brazilian satellite during its re-entry into atmosphere, Brazilian Journal of Physics, vol.33, no.2, June 2003
2. O.Aktas, N.R.Aluru, A combined Continuum/DSMC technique for multiscale analysis of microfluidic filters, Journal of Computational Physics 178, 342–372 (2002)
3. C.Cai, I.D.Boyd, 3D simulation of Plume flows from a cluster of plasma thrusters, 36th AIAA Plasmadynamics and Laser Conference, 6-9 June 2005, Toronto, Ontario, Canada, AIAA-2005-4662
4. M.C.M.van de Sanden, R.J.Severens RJ, J.W.A.M.Gielen et al. (1996), Deposition of a-Si:H and a-C:H using an expanding thermal arc plasma, Plasma sources Science and Technology 5 (2): 268–274
5. B.Alder, Highly discretized dynamics, Physica A 240 (1997) 193-195
6. N.G.Hadjiconstantinou, Hybrid Atomistic-Continuum formulations and moving contact-line problem, Journal Computational Physics 154, 245–265 (1999)
7. P.Le Tallec, F.Mallinger, Coupling Boltzmann and Navier-Stokes Equations by half fluxes, Journal Computational Physics 136, 51–67 (1997)
8. H.S.Wijesinghe, N.G. Hadijconstantinou, Discussion of hybrid Atomistic-Continuum methods for multiscale hydrodynamics, International Journal for multiscale Computational Engineering, 2(2)189-202 (2004)
9. A.L.Garcia, J.B.Bell, W.Y.Crutchfield, B.J.Alder, Adaptive mesh and algorithm refinement using Direct Simulation Monte Carlo, Journal of Computational Physics 154, 134-155 (1999)
10. J.S.Wu, Y.Y.Lian, G.Cheng. R.P.Koomullil, K.C.Tseng, Development and verification of a coupled DSMC-NS scheme using unstructured mesh, Journal of Computational Physics 219, 579-607 (2006)
11. T.E.Schwartzentruber, L.C.Scalabrin, I.D. Boyd, Hybrid Particle-Continuum Simulations of Non-Equilibrium Hypersonic Blunt Body Flows, AIAA-2006-3602, June 2006, San Francisco, CA.
12. T.E.Schwartzentruber, I.D. Boyd, A hybrid particle-continuum method applied to shock waves, Journal of Computational Physics, 215, No. 2, 402-416 (2006).
13. A.J.Lofthouse, I.D.Boyd, M.J.Wright, Effects of Continuum Breakdown on Hypersonic Aerothermodynamics, AIAA-2006-0993, January 2006, Reno, NV.
14. G.A.Bird, Molecular gas dynamics and Direct Simulation Monte Carlo, Claredon Press Oxford Science, 1998
15. Wen-Lan Wang, I.D.Boyd, Continuum Breakdown in Hypersonic Viscous Flows, 40th AIAA Aerospace Sciences Meeting and Exhibit, January 14-17, 2002, Reno, NV

Multi-scale Simulations of Gas Flows with Unified Flow Solver

V.V. Aristov[1], A.A. Frolova[1], S.A. Zabelok[1],
V.I. Kolobov[2], and R.R. Arslanbekov[2]

[1] Dorodnicyn Computing Center of the Russian Academy of Sciences
Vavilova str., 40, 119991, Moscow, Russia
{aristov,afrol,serge}@ccas.ru
[2] CFD Research Corporation, 215 Wynn Drive,
Huntsville, AL , 35803, USA
{vik,rra}@cfdrc.com

Abstract. The Boltzmann kinetic equation links micro- and macroscale descriptions of gases. This paper describes multi-scale simulations of gas flows using a Unified Flow Solver (UFS) based on the Boltzmann equation. A direct Boltzmann solver is used for microscopic simulations of the rarefied parts of the flows, whereas kinetic CFD solvers are used for the continuum parts of the flows. The UFS employs an adaptive mesh and algorithm refinement procedure for automatic decomposition of computational domain into kinetic and continuum parts. The paper presents examples of flow problems for different Knudsen and Mach numbers and describes future directions for the development of UFS technology.

Keywords: Boltzmann equation, Rarefied Gas Dynamics, direct Boltzmann solver, kinetic CFD scheme, multiscale flows, adaptive mesh.

1 Introduction

The Boltzmann kinetic equation is a fundamental physical tool, which links two levels of description of gases. The first one is the microscopic level at atomistic scale and the second one is the macroscopic level at continuum scale. The kinetic equation takes into account interactions among gas particles under the molecular chaos assumption to describe phenomena on the scale of the mean free path. In principle, the Boltzmann equation can also describe macroscopic phenomena, but its usage at the macroscopic scale becomes prohibitory expensive and not necessary. The continuum equations (Euler or Navier-Stokes) derived from the Boltzmann equations can be used at this scale. The present paper describes mathematical methods, which allow one to fully realize this property of the Boltzmann equation. The Unified Flow Solver (UFS) is a variant of hybrid methods for simulation of complex gas (or maybe liquid) flows in the presence of multiscale phenomena including nonequilibrium and nonlinear processes. The UFS methodology combines direct numerical solution of the Boltzmann equation with the kinetic schemes (asymptotic case of the direct methods of solving the Boltzmann equation), which approximate the Euler or Navier-Stokes (NS) equations.

Y. Shi et al. (Eds.): ICCS 2007, Part I, LNCS 4487, pp. 850–857, 2007.

The development of hybrid solvers combining kinetic and continuum models has been an important area of research over the last decade (see Ref. [1] for review). Most researchers applied traditional DSMC methods for rarefied domains and identified statistical noise inherent to the particle methods as an obstacle for coupling kinetic and continuum solvers [2]. The UFS combines direct Boltzmann solver [3,4] with kinetic CFD schemes to facilitate coupling kinetic and continuum models based on continuity of half-fluxes at the boundaries [5]. The UFS uses Cartesian grid in physical space, which is generated automatically around objects embedded in computational domain. A continuum solver is running first and the computational grid in physical space is dynamically adapted to the solution. Kinetic domains are identified using continuum breakdown criteria, and the Boltzmann solver replaces the continuum solver where necessary. The Boltzmann solver uses Cartesian mesh in velocity space and employs efficient conservative methods of calculating the collision integral developed by Tcheremissine [6]. The parallel version of the UFS enforces dynamic load balance among processors. Domain decomposition among processors is performed using space-filling curves (SFC) with different weights assigned to kinetic and continuum cells depending on CPU time required for performing computations in the cell [7].

This paper presents illustrative examples of UFS applications for supersonic and subsonic problems and outlines directions for future development of UFS methodology for simulating complex multi-scale flows.

2 Examples of Multi-scale Simulations with UFS

The UFS has already been demonstrated for a variety of steady-state laminar flows with different Knudsen and Mach numbers [5]. In this Section, we present some examples to illustrate important features of the UFS.

2.1 High Speed Flows

A comparison of UFS results with available experimental data for supersonic flows of single component atomic gases was performed for a 1D shock wave structure and 2D flow around a circular cylinder [5]. Here we show some new results for hypersonic flows and extensions for gas mixtures. Figure 1 shows an example of 2D solutions by

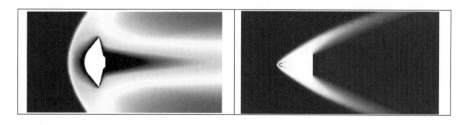

Fig. 1. Axial velocity for OREX at M=27, Kn=0.1 (on the left). Gas pressure for a prism at M=18, Kn=0.25 (on the right).

the kinetic Navier-Stokes solver for the Orbital Reentry Experiment [9] (left) and flow around a prism (right). Kinetic scheme for the NS equation is derived from the Boltzmann equation as described in [8, 5]

Figure 2 illustrates interaction of supersonic jets of two gases with different masses. The kinetic Euler solver for gas mixtures is based on the model of Ref. [10]. Two free jets at Mach number $M=3$ are considered with the ratio of inlet pressure to bulk pressure of 10, and the mass ratio of 2. The distributions of gas density and longitudinal velocity are shown in Fig. 2 for a steady state. An asymmetry with respect to y=0 can be explained by the different masses of molecules in the top and below jets.

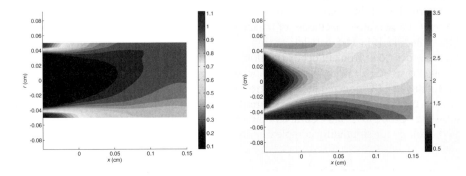

Fig. 2. Density (left) and longitudinal velocity (right) for a steady regime

The UFS is currently being enhanced with non-equilibrium chemistry (based on the model kinetic equation by Spiga & Groppi [11]) coupled to radiation transport and plasma capabilities. This will bring the fidelity of modeling high speed flows of molecular gases to a next level and enable accurate prediction of aerothermal environment around trans-atmospheric vehicles.

2.2 Low Speed Flows

There is a class of low-speed continuum flows, which are not described by the traditional NS equations [12]. The Ghost and non-NS effects present themselves in well-known classical fluid problems such as Benard and Taylor-Couette problems. These effects can have important practical implications for low-pressure industrial processing reactors [13]. The statistical DSMC methods are not well suited for simulation of low speed problems, especially for studies of instabilities and transient effects due to large statistical noise.

Figure 3 illustrates a low-speed flow induced by temperature gradients between two non-uniformly heated plates. This flow is absent according to the traditional NS equations with slip boundary conditions. Both the direct Boltzmann solver and the kinetic NS solver produce correct physical picture of the flow shown in Fig. 3. The temperature T of surfaces goes from 1.7 to 1 (hot bottom and cold top), bottom is symmetry, top and left boundaries have T=1.

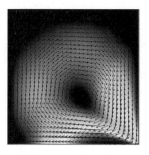 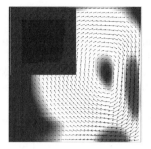 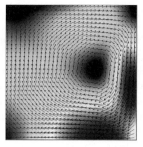

Fig. 3. Temperature driven vortex: temperature and velocity fields for three values of Knudsen numbers (Kn=0.01, 0.07, 0.3 from left to right). Kinetic and continuum zones are shown in the middle Figure corner: dark – continuum, grey – kinetic zones.

The UFS has been used for simulations of gas flows in micro-channels and nozzles. We have recently confirmed the Knudsen minimum in the dependence of the mass flow rate on Knudsen number for a 2D isothermal flow in a channel using the Bolzmann solver with hard sphere molecules.

3 Future Directions

The Boltzmann kinetic equation provides an appropriate physical and mathematical apparatus for description of gas flows in different regimes. The UFS approach can serve the instrument for practical simulations of complex multi-scale flows including unstable and turbulent regimes. For small Kn numbers (large Re numbers) the kinetic Euler solver can be used for both laminar flows and unstable coherent large-scale structures. For smaller scales, we can select NS solver or kinetic solver according to specific criteria. In any case we assume that the kinetic solver can adequately describe micro scales for unstable and turbulent flows (but can be very expensive computationally).

From the kinetic point of view, the transition to unstable turbulent regimes can be treated as appearance of virtual structures in phase space with different scales in physical and velocity spaces. Strong deviations from equilibrium in velocity space can explain rapid growth of dissipation in turbulent flows. The applicability of NS equations in these regimes and the necessity to replace the reological linear model by more complex nonlinear models remain discussion topics [14]. The kinetic description is expected to describe correctly all flow properties at microscopic level since the smallest vortices of turbulent flows remain of the order of 100 mean free paths. The unstable flows already analyzed with the Boltzmann equation [15-17] required sufficiently fine velocity grid, but the spatial and temporal scales remained macroscopic, although smaller than for the laminar flows. For some parameters (frequencies of chaotic pulsations and the density amplitude) there was agreement with the experimental data for a free supersonic jet flows and computations from [15-17, 3]. The capabilities of the kinetic approach have been confirmed in independent investigations of different unstable and turbulent gas flows by the DSMC schemes [18, 19] and by the BGK equation [20].

3.1 Instabilities

3D solutions of the Boltzmann equation for two Kn numbers for stable and unstable regimes are shown in Fig. 4 for a free underexpanded supersonic jet flow. The gas flows from left to right through a circular orifice, the Mach number at the orifice is M=1. The pressure ratio of flowing gas (to the left from the orifice) to the gas at rest (to the right from the orifice p0/p1=10. One can see in Fig. 4 that for the large Reynolds number (small Knudsen number) the gas flow is unstable. The mechanism of this instability is connected with the appearance of so-called Taylor-Gertler vortices. The Boltzmann equation has been solved by the direct conservative method with the spatial steps larger than the mean free path. The calculated system of vortices has been observed in experiments [21], and the kinetic simulations reproduced the experimentally observed macroscopic structures.

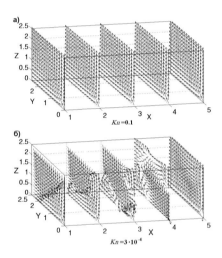

Fig. 4. Velocity fields in cross-sections for 5 values of x-coordinates. There is no vorticity in the plot for the larger Knudsen number, - a). Pairs of the longitudinal vortices are shown - б). A quarter for each cross section is depicted.

3.2 Multi-scale Simulation of Turbulent Flows

Modern approach to simulation of turbulent flows uses Euler, NS and Boltzmann models for different parts of the flows [22]. The Euler models are used for large-scale structures, whereas kinetic models are used for small-scale stochastic background. The Boltzmann equation for the velocity distribution of gas particles is used for compressive gas flows. For liquids, there is no universal kinetic equation for the probability density of instant flow velocity. Often, kinetic equation for turbulent background is used in the form resembling Boltzmann transport equation for gases.

The UFS can serve a useful tool for the first principle simulations of turbulent gas flows. The macro-scale eddies can be described by the Euler or NS equations and the Boltzmann solver used for micro-scale phenomena. For liquids, semi-empirical kinetic equations of the Onufriev-Lundgren type could be used at the micro-scale. Molecular dynamic simulations can help justify and improve these equations for liquids. Additional research is needed to understand how to properly expand the UFS methodology for complex turbulent flows. The macro-scale coherent structures of turbulent flows can be well described by the Euler or NS equations. Fig. 5 shows an example of 2D UFS simulations (here the kinetic NS solver is used) of unstable phenomena appeared in the wake behind a prism at M=3, Kn=10^{-5} for the angle of attack of 3 degrees.

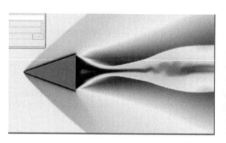 

Fig. 5. Instantaneous Mach number (left) in a range 9 10^{-3} <M <5 and temperature (right) in a range 0.4 <T< 4.14 for a gas flow around a prism

3.3 Multiscale Flow Structures in Open Systems

The UFS can be extended for multi-scale analysis of chemical and biological structures. An important kinetic model of such problems is the nonuniform relaxation problem [23,3]. Very interesting physical effects have been observed in such systems, e.g. the heat flux has the same sign as the temperature gradient. This model could describe properties of complex natural objects [24]. Biological structures are open flux systems, which often possess two key processes: convection and chemical reactions. The scales of these processes are different: reactions are usually slower than advection. The analysis of 1D nonuniform relaxation problem for monatomic one-component gas has shown that nonequilibrium in velocity space is "projected" onto physical space and results in spatial structures. A generalization for mixtures showed [25] that it is possible to change the character of the spatial structures by changing the nonequilibrium distribution functions for gas components at the boundaries.

4   Conclusion

We have described a Unified Flow Solver for simulations of complex multi-scale gas (and possibly liquid) flows in different regimes. We considered some examples of supersonic and subsonic flows including unstable complex regimes for which the UFS and its components show reliable results. Perspectives of future development of the UFS methodology have been outlined. New approach to some important problems, such as non-uniform nonequilibrium structures for complex flows in open systems has been discussed.

Acknowledgments. This work is partially supported by the Program N15 of the Presidium of the Russian Academy of Sciences and by the US Air Force SBIR Project F33615-03-M-3326.

References

1. Kolobov, V.I, Bayyuk, S.A., Arslanbekov, R.R., Aristov, V.V., Frolova, A.A., Zabelok, S.A.: Construction of a Unified Continuum/Kinetic Solver for Aerodynamic Problems. AIAA Journal of Spacecraft and Rockets **42**, (2005) 598-605
2. Carlson, H.A., Roveda, R., Boyd, I.D., Candler, G.A.: A Hybrid CFD-DSMC Method of Modeling Continuum-Rarefied Flows. AIAA Paper 2004-1180 (2004)
3. Aristov, V.V.: Direct Methods for Solving the Boltzmann Equation and Study of Non-Equilibrium Flows. Kluwer Academic Publishers, Dordrecht (2001)
4. Tcheremissine, F.G.: Direct Numerical Solution of the Boltzmann Equation In: M.Capitelli (ed.): Rarefied Gas Dynamics. AIP Conference Proceedings, Vol. 762. Melville. New York (2005) 677-685
5. Kolobov, V.I., Arslanbekov, R.R., Aristov, V.V., Frolova, A.A., Zabelok, S.A.: Unified Solver for Rarefied and Continuum Flows with Adaptive Mesh and Algorithm Refinement, J. Comput. Phys. (2006), doi:10.1016/j.jcp.2006.09.021
6. Tcheremissine, F.G.: Solution of the Boltzmann Kinetic Equation for High-Speed Flows. Comp. Math. Math. Phys. **46** (2006) 315-327
7. Zabelok, S.A. et al.: Parallel Implementation of the Unified Flow Solver, 25th Intern Symp. on Rarefied Gas Dynamics. Book of Abstracts. St. Petersburg (2006) 62
8. Li, O., Fu, S., Xu, K.: A Compressible Navier-Stokes Flow Solver with Scalar Transport. J. Comput.Phys. **204** (2005) 692-709
9. Moss, J.N., Gupta, R.N., Price, J.M.: DSMC Simulations of OREX Entry Conditions. In: Shen, C. (ed.): Rarefied Gas Dynamics. Peking University Press, Peking (1997) 459-464
10. Andries, P., Aoki, K., Perthame, B.: J. Stat. Phys. **106** (2002) 993-1018.
11. Groppi, M., Spiga, G. A BGK Model for a Mixture of Chemically Reacting Gases. In: M.Capitelli (ed.): Rarefied Gas Dynamics. AIP Conference Proceedings, Vol. 762. Melville. New York (2005) 842-847
12. Sone, Y.: Molecular Gas Dynamics. Birkhauser, Boston (2007)
13. Arslanbekov, R.R., Kolobov, V.I.: Simulation of Low Pressure Plasma Processing Reactors: Kinetics of Electrons and Neutrals, 25th Intern. Symp. Rarefied Gas Dynamics. Book of Abstracts. St. Petersburg, (2006) 63
14. Astarita, J., Maruchhi, L.: Basis of Hydromechanics of Non-newtonian Fluids. Mir, Moscow (1978)
15. Aristov, V.V.: Numerical Analysis of Free Jets at Small Knudsen Numbers.: In: Harvey, J., Lord, G. (eds.): Rarefied Gas Dynamics. Oxford University Press, Oxford, Vol.2. (1995) 1293-1299
16. Aristov, V.V.: Stability and Instability Analysis of Free Jets Based on the Boltzmann Equation. Fluid. Dynam. **33** (1998) 280-283
17. Aristov, V.V.: Solving the Boltzmann Equation at Small Knudsen Numbers.: Comp. Math. Math. Phys. **44** (2004) 1069-1081
18. Cercignani, C., Stefanov, S.: Vorticities and Turbulence in Hypersonic Rarefied Flows. In: Harvey, J., Lord, G. (eds.): Rarefied Gas Dynamics. Oxford University Press, Oxford, Vol.2. (1995) 1147-1153

19. Bird, G.A. The Initiation of Centrifugal Instabilities in an Axially Symmetric Flow. In: Shen, C. (ed.): Rarefied Gas Dynamics. Peking University Press, Peking (1997) 149-154

20. Sakurai, A., Takayama, S.: Vorticity and Shock Wave in a Compressible Turbulent Flow of Complanar Kinetic Gas Model. In: Shen, C. (ed.): Rarefied Gas Dynamics. Peking University Press, Peking (1997) 291-296

21. Arnette, S.A., Samimy, M., Elliot, G.S.: On Streamwise Vortices in High Reynolds Number Supersonic Axisymmetric Jets. Phys. Fluids. A 5 (1993) 187-202

22. Belotserkovskii, O.M., Oparin, A.M., Chechetkin, V.M.: Turbulence: New Approaches. Cambridge International Science Publishing, Boston (2005)

23. Aristov, V.V.: A Steady State, Supersonic Flow Solution of the Boltzmann Equation. Phys. Letters A. 250 (1998) 354-359

24. Aristov, V.V.: Dissipative Structures Describing by the Boltzmann Equation and Relaxation Model Kinetic Equation. In: Topics in Biomathematics. World Scientific, Singapore (1993) 109-112

25. Aristov, V.V.: Nonuniform Relaxation Problem for Mixtures and Effects of Anomalous Heat Transfer. In: M.Capitelli (ed.): Rarefied Gas Dynamics. AIP Conference Proceedings, Vol. 762. Melville. New York (2005) 288-293

Coupling Atomistic and Continuum Models for Multi-scale Simulations of Gas Flows

Vladimir Kolobov, Robert Arslanbekov, and Alex Vasenkov

CFD Research Corporation, 215 Wynn Drive,
Huntsville, AL , 35803, USA
{vik, rra, avv}@cfdrc.com

Abstract. This paper describes two computational tools linking atomistic and continuum models of gaseous systems. The first one, a Unified Flow Solver (UFS), is based on a direct Boltzmann solver and kinetic CFD schemes. The UFS uses an adaptive mesh and algorithm refinement procedure for automatic decomposition of computational domain into kinetic and continuum parts. The UFS has been used for a variety of flow problems in a wide range of Knudsen and Mach numbers. The second tool is a Multi-Scale Computational Environment (MSCE) integrating CFD tools with Kinetic Monte Carlo (KMC) and Molecular Dynamics (MD). The MSCE was applied for analysis of catalytic growth of vertically aligned carbon nanotubes (CNT) in a C_2H_2/H_2 inductively coupled plasma. The MSCE is capable of predicting paths for delivering carbon onto catalyst/CNT interface, formation of single wall or multi-wall CNTs depending on the shape of catalyst, and transition from nucleation to the steady growth of CNTs.

Keywords: Rarefied Gas Dynamics, Boltzmann solver, Kinetic Monte Carlo, Molecular Dynamics, carbon nanotubes, Unified Flow Solver.

1 Introduction

The need for computational analysis at different scales appears in different fields of science and engineering ranging from aerospace applications to material processing. Recent advances of nanotechnology demand understanding of manufacturing processes at atomistic scales for design smart materials and structures. The development of hybrid solvers combining kinetic and fluid models has been an important area of research over the last decade (see Refs. [1,2] for reviews). The challenges of developing multi-scale methods are physical and numerical. Refining computational mesh in continuum simulations down to the atomic scale results in the time step governed by the smallest element of the mesh. Using such small time steps in large cells result in an enormous waste of computing time since continuum dynamics usually evolves on a much larger time scale. Efficient multiscale methods should use different time steps in atomistic and continuum regions. Another major problem in coupling atomistic and continuum models appears in pathological wave reflections at an interface between kinetic and continuum regions, induced by high frequency processes in the kinetic regions.

Y. Shi et al. (Eds.): ICCS 2007, Part I, LNCS 4487, pp. 858–865, 2007.

This paper describes two computational tools linking atomistic and continuum models for gaseous systems. The first one is a Unified Flow Solver (UFS) for simulations of rarefied, transitional and continuum flows. The UFS solves the Boltzmann and continuum equations in appropriate parts of computational domain. Domain decomposition and coupling algorithm are based on the adaptive mesh and algorithm refinement methods. The computational mesh in physical space is dynamically adapted to the solution and geometry using a tree-based data structure. The UFS separates non-equilibrium and near-equilibrium domains using continuum breakdown criteria and automatically selects appropriate continuum and kinetic solvers. The UFS was used for a variety of flow problems in a wide range of Knudsen and Mach numbers and has recently been extended to multicomponent mixtures of molecular gases with internal degrees of freedom.

The second tool is a Multi-Scale Computational Environment (MSCE) integrating CFD tools with Kinetic Monte Carlo (KMC) and Molecular Dynamics (MD) tools [3]. The atomistic KMC and MD solvers are used in regions where atoms self-assemble into molecular structures. For coupling atomistic and continuum models, macroscopic fluxes from the CFD solver are transferred to KMC, which calculates velocity distributions of gas species near surfaces for nanomaterial fabrication. The MD simulations are used for calculation of surface reaction rates used in KMC. The MSCE was applied for analysis of catalytic growth of vertically aligned carbon nanotubes (CNT) in a C_2H_2/H_2 inductively coupled plasma. We have demonstrated that MSCE is capable of predicting paths for delivering carbon onto catalyst/CNT interface, formation of single wall or multi-wall CNTs depending on the shape of catalyst, and transition from nucleation to the steady growth of CNTs [4].

2 Unified Flow Solver

The development of hybrid solvers combining kinetic and continuum models for gas flows has been an important area of research over the last decade. The key parameter defining the choice of the appropriate physical model is the local Knudsen number, *Kn*, defined as the ratio of the mean free path to the characteristic size of the system. Particle methods such as DSMC or Molecular Dynamics are used in regions with strong deviations from equilibrium, and continuum (Euler or NS) solvers are used elsewhere. Statistical noise inherent to DSMC has been identified as an obstacle for the development of hybrid solvers [5]. Direct methods of solving Boltzmann equation have recently emerged as a viable alternative to the particle methods [6-8]. They are preferable for coupling kinetic and continuum models because similar numerical techniques are used for solving both the Boltzmann and continuum equations. The Unified Flow Solver (UFS) uses the direct Boltzmann solver and kinetic schemes for the continuum solvers to facilitate coupling of the solutions based on continuity of half-fluxes in the finite volume approach.

2.1 UFS Solution Procedure

The UFS architecture is described in details in [9]. Here, we only briefly outline some important features. Geometry of the problem is defined using standard files and a

Cartesian mesh is generated automatically around objects embedded in computational domain. A continuum solver is running first and the computational grid is dynamically adapted to the solution. Kinetic domains are identified using continuum breakdown criteria, and the Boltzmann solver replaces the continuum solver in the kinetic domains. The coupling of kinetic and continuum solvers at the interface occurs seamlessly based on the continuity of half-fluxes.

For solving the Boltzmann equation, a Cartesian mesh in velocity space is introduced with a cell size $\Delta\xi$ and nodes ξ_β. Using this velocity mesh, the Boltzmann equation for each gas component is reduced to linear hyperbolic system of transport equations in physical space with a nonlinear source term

$$\frac{\partial f_\beta}{\partial t} + \nabla_{\mathbf{r}} \cdot (\xi_\beta f_\beta) = I(f_\beta, f_\beta) \tag{1}$$

The solution of the system (1) can be found by standard CFD methods for arbitrary mesh in physical space. We use efficient conservative methods of calculating the Boltzmann collision term $I(f,f)$ developed by Tcheremissine [10] using Korobov's sequences [11] for the evaluation of multi-dimensional integrals.

The parallel version of the UFS enforces a dynamic load balance among processors. The procedure of domain decomposition is performed using space-filling curves. Different weights are assigned to kinetic and continuum cells depending on CPU time required for performing computations in the cells [12]. Figure 1 illustrates an example of 3D simulations of the Inflatable Reentry Vehicle Experiment (IRVE) [13] for $Kn = 0.01$ and $M = 3.94$ performed on an 8 node cluster.

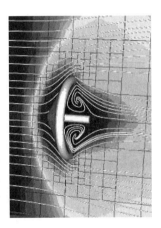 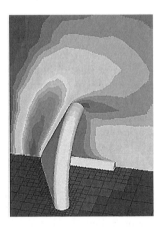

Fig. 1. Streamlines, Mach number, and computational mesh (on the left). Gas temperature in the vertical plane, kinetic (light) and continuum (dark) domains in the horizontal plane (on the right).

2.2 Molecular Gases and Gas Mixtures

The UFS has been extended to molecular gases and gas mixtures. Figure 2 shows an example of supersonic flow of a binary mixture over a cylinder at $M=2$, for 3

Knudsen numbers. Two species with the mass ratio of 2 with no chemical reactions are considered. The Hard Sphere model of molecular interactions is used for the Boltzmann solver. The domain decomposition is performed using continuum breakdown criterion based on the gradient of the total gas density. The temperatures of species become different in the kinetic domains while they are equal in the continuum domains. For these simulations, the typical number of velocity nodes (in each direction) is 15-20 depending on the Mach number and mass ratio of gas species.

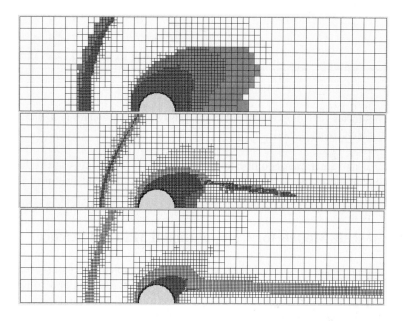

Fig. 2. The computational mesh and kinetic/continuum domains for binary mixture of monatomic gases at $M=2$, for Kn numbers 0.125, 0.025, and 0.0125

The kinetic solver has been extended to molecular gases with internal degrees of freedom following the work [14]. Figure 3 shows the shock wave (SW) structure in Nitrogen at $M=12.9$ with the center of SW located at $X=0$. On the left are distributions of gas density, translational and rotational temperatures, on the right are rotational spectra for 25 levels at different points along the wave front. On the x-axis is the rotational number; on the y-axis is the population of the rotational levels. It is clearly seen that the rotational equilibrium inside the SW does not exist for such a high Mach number. The typical CPU time for these SW simulations is about 30 hours on an AMD64 3000 processor. Although accurate calculations require considerable CPU time that can be reduced by using crude discretization of the velocity space and reduced number of levels of the rotational energy spectrum [15].

We have also extended the kinetic solver for vibrationally excited molecules. For future development of the UFS, we plan to take into consideration that vibrational equilibration in VT and VV collisions occur considerably slower compared to

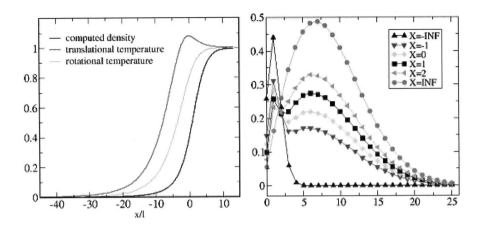

Fig. 3. SW structure in Nitrogen at $M=12.9$: gas density and temperatures (left) and rotational spectra at different points (right)

translational and rotational equilibration. As a result, in most cases, it is adequate to model computational domains where vibrational relaxation and chemical reactions take place using continuum models.

3   Multi-scale Models for Computational Material Science

Computational material science deals with strongly coupled processes occurring at disparate length and time scales at an interface between gaseous and solid phases. The gas-phase processes (chemical reactions, heat transfer, etc.) are typically several orders of magnitude slower than the surface processes. Continuum models or atomistic models are used for simulations of reactive gas flows at the reactor scale. These models are not appropriate for modeling fast surface processes such as surface diffusion of adsorbed molecules, chemical reactions or atomic relaxation occurring with characteristic times of 0.1 ns and spatial scales of 0.1 nm. Such processes are modeled using atomic-scale models such as Kinetic Monte Carlo (KMC) or Molecular Dynamics (MD). Over the last several years, there has been a rapidly growing interest in the development of computational tools with different levels of detail. The major challenge here is the development of algorithms capable of bridging atomistic and continuum models working on disparate time and length scales.

3.1   Bridging Disparate Length and Time Scales

For bridging disparate spatial scales, Kevrekidis et al. [16] introduced the "Gap-tooth" method involving atomistic modeling in tiny teeth, defined as regions where continuum models are not available or are invalid, and a less detailed continuum simulation within large gaps referred to as regions where nanoscale phenomena are not important. This method is designed to approximate a so-called "time-stepper" for an unavailable macroscopic equation in a macroscopic domain. The authors have demonstrated the ability of this model to solve selected multiscale hydrodynamic problems such as boundary layer behavior and shock formation. It was also shown that it is possible to obtain a convergent scheme without constraining the microscopic code, by introducing buffer regions around the computational boxes. The Gap-tooth approach was successfully adapted for integrating KMC modeling with CFD reactor-scale simulations in the MSCF described in the next section.

Bridging the time scales is an equally difficult and fundamentally different problem than bridging the length scales. The major issue here is that the dynamics of complex systems is often driven by rare but important events [17]. The simulations of these events often require long run times, and rapidly become prohibitively expensive due to the exponential scaling with inverse temperature. Several groups reported significant progress in the development of new techniques for accelerated modeling of rate events [18,19]. Kevrekidis et al. introduced a coarse model for mapping a mesoscopic description onto atomistic description [20]. They assumed that a coarse model (e.g. continuum model) exists in closed form for the fine scale simulator (e.g. KMC model). Based on this assumption, a Coarse Controller was constructed for KMC simulator, and the developed framework was applied for a simple surface catalyst. In subsequent studies, this group further developed the Coarse Controller and used it to accelerate MD modeling. The projection operator formalism was introduced to integrate the dynamics of rare events with fastest molecular processes of bond vibration and atomic collisions [21]. This was achieved by estimating both the thermodynamic driving forces for slow motions and their dynamic properties. Information about the slow coarse dynamics was extracted from the projected motions of many, appropriately initialized, but otherwise independent and unbiased replica simulations. The advantage of the coarse MD is that it can be used for any type of intermolecular potentials and for a wide range of temperatures.

3.2 Multi-scale Computational Framework

The Multi-Scale Computational Framework (MSCF) has been developed for material processing applications [3]. The MSCF integrates continuum models with atomistic KMC and MD models, using the coarse timestepper and gap-tooth modules as bridging algorithms. The reactor-scale continuum simulator is used in large gaps where details of atomic motion are not important, while atomistic KMC and MD solvers are applied to tiny regions, teeth, defined as areas where atoms self-assemble into molecular structures. Macroscopic fluxes from the continuum solver are transferred to the gap-tooth module for estimating microscopic fluxes towards the substrate. These microscopic fluxes are used in the KMC solver for prescribing velocity distributions of ions and neutrals at the source plane. The gap-tooth module also accounts for the influence of KMC modeling in one tooth on those in the neighboring teeth by establishing a relationship between incoming microscopic fluxes to each tooth and outgoing fluxes from the neighboring teeth. The velocity distributions of neutral species in KMC are chosen as isotropic Maxwellian distributions; the ions are launched with mono-directional and mono-energetic distributions. The Coarse Timestepper module projects, in time, the rates and probabilities of chemical reactions computed by the MD solver. Figures 4 and 5 illustrate the MSCF application to analysis of catalytic growth of vertically aligned carbon nanotubes (CNT) in a C_2H_2/H_2 inductively coupled plasma [4].

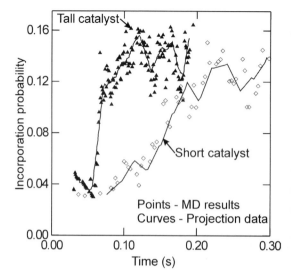

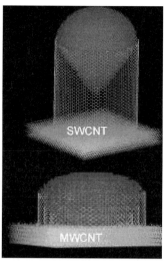

Fig. 4. Transition from nucleation to steady growth of CNTs is illustrated by time-dependent incorporation probabilities for the tall and short catalyst

Fig. 5. The growth of single wall and multi-wall CNTs on the catalysts of different shape

4 Conclusion

We have described two computational tools linking atomistic and continuum models of gaseous systems. The UFS was originally developed for aerospace applications. We believe that with additional physical models and further improvements, the UFS will be valuable for a wide range of applications from trans-atmospheric flights to low-pressure material processing and semiconductor manufacturing. The MSCF, originally developed for CNT applications, will be further extended into processing of biocompatible polymers and other applications.

Acknowledgments. The UFS was developed in collaboration with Drs. V.V. Aristov, A.A. Frolova, S.A. Zabelok, and F.G. Tcheremissine from the Russian Academy of Sciences. The work is partially supported by USAF and NSF SBIR Projects

References

1. Wijesinghe, H.S., Hadjiconstantinou, N.G.: A discussion of hybrid atomistic-continuum methods for multiscale hydrodynamics. Int. J. Multiscale Comput. Eng. **2** (2004), 189
2. Kolobov, V.I., Bayyuk, S.A., Arslanbekov R.R., Aristov, V.V., Frolova, A.A., Zabelok, S.A.: Construction of a Unified Continuum/Kinetic Solver for Aerodynamic Problems, AIAA Journal of Spacecraft and Rockets **42** (2005), 598-605
3. Vasenkov, A.V., Fedoseyev, A. I., Kolobov, V. I., Choi, H.S., Hong, K.-H., Kim, K., Kim, J., Lee, H. S., Shin, J. K.: Computational Framework for Modeling of Multi-Scale Processes, J. Comput. Theor. Nanoscience **3** (2006), 453-458

4. Vasenkov, A.V., Kolobov, V. I.: Modeling of multi-scale processes during the growth of carbon nanotubes, in: Technical Proceedings of the 2006 NSTI Nanotechnology Conference and Trade Show. Vol. 1 (2006), 617-621

5. Carlson, H.A., Roveda, R., Boyd, I.D., Candler, G.V.: A Hybrid CFD-DSMC Method of Modeling Continuum-Rarefied Flows, AIAA Paper 2004-1180 (2004)

6. Rogier, F., Schneider, J.A.: A direct method for solving the Boltzmann equation. Transport Theory Stat. Phys. 23 (1994), 313-338

7. Aristov,V.V.: Direct Methods for Solving the Boltzmann Equation and Study of Non-Equilibrium Flows, Kluwer Academic Publishers, Dordrecht, (2001)

8. Tcheremissine, F.G.: Direct Numerical Solution of the Boltzmann Equation, in: M.Capitelli (ed.) Rarefied Gas Dynamics, AIP Conference Proceedings, N.Y. vol. 762 (2005), 677-685

9. Kolobov, V.I., Arslanbekov, R.R., Aristov, V.V., Frolova, A.A., Zabelok, S.A.: Unified solver for rarefied and continuum flows with adaptive mesh and algorithm refinement, J. Comput. Phys. (2006), doi:10.1016/j.jcp.2006.09.021

10. Tcheremissine, F.G.: Solution to the Boltzmann kinetic equation for high-speed flows. Comp. Math. Math. Phys. 46 (2006) 315

11. Korobov, N.M.: Exponential Sums and Their Applications, Springer, (2001), 232p

12. Zabelok, S.A, Aristov, V.V., Frolova, A.A., Kolobov, V.I., Arslanbekov, R.R.: Parallel implementation of the Unified Flow Solver. in: Rarefied Gas Dynamics, AIP Conference Proceedings (2007)

13. Moss, J.N. et al.: Low-Density Aerodynamics of the Inflatable Reentry Vehicle Experiment (IRVE), AIAA Paper 2006-1189 (2006)

14. Tcheremissine, F.G.: Solution of the Wang-Chang-Uhlenbeck master equation. Doklady Physics 47 (2002), 872-875

15. Tcheremissine, F.G., Kolobov, V.I., Arslanbekov, R.R.: Simulation of Shock Wave Structure in Nitrogen with Realistic Rotational Spectrum and Molecular Interaction Potential. in Rarefied Gas Dynamics, AIP Conference Proceedings (2007)

16. Samaey, G., Kevrekidis, I.G., Roose, D.: in *Multiscale Modeling and Simulation*, Springer, Verlag, (2004), 93

17. Hanggi, P., Talkner, P., and Borkovec, M.: Reaction-rate theory: fifty years after Kramers. Rev. Mod. Phys. **62** (1990), 251-341

18. Sorensen, M. R., Voter, A.F.: Temperature-accelerated dynamics for simulation of infrequent events. J. Chem. Phys. **112** (2000), 9599-9606

19. Dellago, C., Bolhuis, P. G., Chandler, D.: Efficient transition path sampling: Application to Lennard-Jones cluster rearrangements. J. Chem. Phys. **108** (1998), 9236-9245

20. Siettos, C. I., Armaou, A., Makeev, A. G., Kevrekidis, I. G.: Microscopic/stochastic timesteppers and "coarse" control: A KMC example. AIChE Journal **49** (2003), 1922-1926

21. Sriraman, S., Kevrekidis, I. G., Hummer, G.: Coarse Master Equation from Bayesian Analysis of Replica Molecular Dynamics Simulations. J. Phys. Chem. B **109** (2005), 6479-6484

Modelling Macroscopic Phenomena with Cellular Automata and Parallel Genetic Algorithms: An Application to Lava Flows

Maria Vittoria Avolio[1], Donato D'Ambrosio[2], Salvatore Di Gregorio[2], Rocco Rongo[1], William Spataro[2], and Giuseppe A. Trunfio[3]

[1] Department of Earth Sciences, University of Calabria, 87036 Rende (CS), Italy
[2] Department of Mathematics, University of Calabria, 87036 Rende (CS), Italy
[3] DAP, University of Sassari, 07041 Alghero (SS), Italy
{avoliomv, d.dambrosio, dig, spataro}@unical.it, trunfio@uniss.it

Abstract. Forecasting through simulations the shape of lava invasions in a real topography represents a challenging problem, especially considering that the phenomenon usually evolves for a long time (e.g. from a few to hundreds of days) and on very large areas. In the latest years, Cellular Automata (CA) have been well recognized as a valid computational approach in lava flow modelling. In this paper we present some significant developments of SCIARA, a family of deterministic CA models of lava flows which are optimized for a specific scenario through the use of a parallel genetic algorithm. Following a calibration-validation approach, the model outcomes are compared with three real events of lava effusion.

Keywords: Modelling and Simulation, Cellular Automata, Lava flows.

1 Introduction

Lava flows consist of unconfined multiphase streams, the characteristics of which (e.g. temperature, rheologic parameters, velocity, phase state and shape) vary in space and time as a consequence of many interrelated physical and chemical phenomena (e.g. the flow dynamic itself, the loose of thermal energy through radiation, convecting air and conduction, the change in viscosity and the solidification processes). Thus, forecasting the shape of lava invasions in a real topography, given the lava effusion rate (i.e. the volumetric flux of lava from the vent), requires the ability to perform accurate simulations of a typical multiphysics system. Besides, considering that the phenomenon usually evolves for long time and on very large areas, standard approaches, such as those based on differential equations and related approximate techniques, result of difficult use.

On the other hand, it is well recognized that Cellular Automata (CA) can represent a valid alternative in lava flow modelling. A family of deterministic CA models specifically developed for simulating lava flows is SCIARA [1,2,3]. The model, which is optimized for a specific scenario through a parallel Genetic Algorithm (GA), accounts for the relevant physical processes involved in the macroscopic phenomenon and enables for the fast production of accurate forecasting of lava invasions. In this paper some characteristics of the latest version,

Y. Shi et al. (Eds.): ICCS 2007, Part I, LNCS 4487, pp. 866–873, 2007.

named SCIARA-fv, which introduces many model improvements and computational optimizations, is illustrated. Given the availability of historical data
regarding various eruptive episodes of Mount Etna (Italy), which is the most
active volcano in Europe, a cross-validation methodology is exploited for both
the parameter calibration and assessment of the model reliability in that specific
scenario. To this end, a new measure which takes into account simultaneously
the spatial and temporal dimensions has been employed as fitness function for
the evaluation of the simulation results with respect to the real events. A parallel
Master-Slave GA is used for the model calibration in a training-case while the
validation is carried out on a test set constituted by different real episodes.

2   Model Specification

SCIARA-fv is the latest release in a family of bi-dimensional CA models for
lava flows simulation, based on hexagonal cells. The model follows the Macroscopic Cellular Automata (MCA) approach [4], where: (i) the state of the cell
is decomposed in *substates*, each one representing a particular feature of the
phenomenon to be modelled; (ii) the transition function is split in *elementary
processes*, each one describing a particular aspect of the considered phenomenon.
Moreover, some *parameters* are generally considered, which allow to "tune" the
model for reproducing different dynamical behaviours. Finally, several *external
influences* can be considered in order to model features which are not easy to be
described in terms of local interactions. Even though principally derived from
the SCIARA-hex1 version [3], SCIARA-fv embeds a better management of several aspects with respect to the original one, some of which will be described
later. Formally, the model is defined as:

$$\text{SCIARA-fv} = \langle R, L, X, Q, P, \lambda, \gamma \rangle \tag{1}$$

where:

- R is the set of hexagonal cells covering the finite region where the phenomenon evolves;
- $L \subset R$ specifies the lava source cells (i.e. vents);
- $X = \{Center, NW, NE, E, SE, SW, W\}$ identifies the hexagonal pattern
  of cells that influence the cell state change (i.e. the cell itself, *Center*, and the
  North-West, North-East, East, South-East, South-West and *West* neighbors);
- $Q = Q_a \times Q_t \times Q_T \times Q_f^6$ is the finite set of states, considered as Cartesian
  product of "substates". Their meanings are: cell altitude, cell lava thickness,
  cell lava temperature, and outflows lava thickness (from the central cell toward the six adjacent cells), respectively. It is worth to observe that, in the
  model, the third dimension, the height, is included as a property of the state
  of the cell (i.e. a substate);
- $P = \{p_s, p_{Tv}, p_{Tsol}, p_{adv}, p_{adsol}, p_{cool}, p_a\}$ is the finite set of parameters
  (invariant in time and space), which affect the transition function (see Tab.
  1 for the meanings of parameters in P);

- $\lambda : Q^7 \rightarrow Q$ is the cell deterministic transition function;
- $\gamma : Q_t \times \mathbb{N} \rightarrow Q_t$ specifies the emitted lava thickness from the source cells at each step $k \in \mathbb{N}$.

The main elementary processes which compose the transition function are outlined, in order of application, in the following subsections.

Lava Flows Computation. Lava rheological resistance is modelled in terms of an adherence effect, measured by ν, which represents the amount of lava (i.e. thickness) that cannot flow out of a cell towards any neighboring ones. In particular, since lava rheological resistance increases as temperature decreases, the model is expressed by means of the inverse exponential function [5]:

$$\nu = k_1 \, e^{k_2 \, T} \tag{2}$$

where $T \in Q_T$ is the current lava temperature, while k_1 and k_2 are parameters depending on lava rheological properties [6]. The values of k_1 and k_2 are simply obtained by solving the system of equations:

$$\begin{cases} p_{adv} = k_1 e^{-k_2 \, p_{Tv}} \\ p_{adsol} = k_1 e^{-k_2 \, p_{Tsol}} \end{cases} \tag{3}$$

Since the third spatial dimension, i.e. the height, is included in the model as a substate, cell outflows can be computed by the algorithm for the minimization of differences described in [4]. The minimization algorithm, which is based on the iterative fulfilment of the hydrostatic equilibrium principle through the minimization of the differences in height, guarantees the mass conservation.

Let $s[k]$ indicate the current value of the substate s for the k-th cell of the neighborhood, with $k = 0$ for the cell under consideration. Let also $a \in Q_a$ and $t \in Q_t$ be the cell altitude and cell lava thickness, respectively. In order to compute lava outflows from the central cell towards its neighboring ones, the minimisation algorithm is applied to the following quantities: the *unmovable part* $u[0] = a[0] + \nu$ and the *mobile part* $m[0] = t[0] - \nu$ (i.e. the only part which can be distributed to the adjacent cells). Let $\phi[x, y]$ denote the flow from cell x to cell y, thus $m[0]$ can be written as:

$$m[0] = \sum_{i=0}^{\sharp X} \phi[0, i] \tag{4}$$

Table 1. The parameters used in SCIARA-fv model

Parameter	Definition
p_s	time corresponding to a CA step
p_{Tv}	lava temperature at the vent
p_{Tsol}	lava temperature at solidification
p_{adv}	lava adherence at the vent
p_{adsol}	lava adherence at solidification
p_{cool}	the cooling parameter
p_a	cell apothem

where $\phi[0,0]$ is the part which is not distributed, and $\sharp X$ is the number of cells belonging to the neighborhood X. During the minimization, the quantities in the adjacent cells, $u[i] = a[i] + t[i]$ ($i = 1, 2, ..., \sharp X$) are considered unmovable. Let $c[i] = u[i] + \phi[0,i]$ $\forall i$ be the new quantity content in the i-th neighboring cell after the distribution and let c_{min} be the minimum value of $c[i]$ $\forall i$. The outflows are computed in order to minimise the expression $\sum_{i=0}^{\sharp X} (c[i] - c_{min})$. In [4] it is shown that the simultaneous application of the minimization principle to each cell gives rise to the global equilibrium of the system.

Thermal Exchanges and Lava Solidification. It is assumed that a lava flow can be treated as if thermally well mixed, so that each cell has a temperature field constant in space. Besides, only thermal energy loss due to lava surface irradiation is considered. Although such a cooling model may not always be realistic, it does yield a characteristic cooling time scale comparable to representative emplacement times of some important typology of lava flows.

Under this hypotheses, the new cell temperature $T_{t+\Delta t}$ is determined in two phases. In the first phase, which accounts for the energy exchange directly related to the mass exchange, the cell temperature T_{avg} is obtained as weighted average of residual lava inside the cell and lava inflows from neighboring ones:

$$T_{avg} = \frac{t_r\, T_t[0] + \sum_{i=1}^{6} \phi[i,0]\, T_t[i]}{t_r + \sum_{i=1}^{6} \phi[i,0]} \tag{5}$$

where $t_r \in Q_t$ is the residual lava thickness inside the central cell after the outflows distribution, $T_t[k] \in Q_T$ is the lava temperature at the time step t for the k-th cell of the neighborhood (i.e. $k = 0$ for the cell under consideration) and $\phi[i,0]$ the lava inflow from the i-th neighboring cell. Note that $\phi[i,0]$ corresponds to the lava outflow from the i-th neighboring cell towards the central one, computed by means of the minimisation algorithm.

The final step updates the temperature given by (5) considering thermal energy loss due to lava surface irradiation, which can be expressed by the Stefan–Boltzmann equation. In particular, starting from a temperature T_t a good approximation of the temperature $T_{t+\Delta t}$ is given by [6]:

$$T_{t+\Delta t} = T_t \left(1 + \frac{3\, T_t^3\, \varepsilon\, \sigma\, A\, \Delta t}{\rho\, c\, V}\right)^{-\frac{1}{3}} \tag{6}$$

where ρ is the lava density, c the specific heat, V the volume, σ the Stephan-Boltzmann constant, T_t the absolute temperature at time t, A is the air exposed area of the cell, ε is the surface emissivity and Δt is the step duration of the CA. Thus, according to (6), the temperature T_{avg} computed in (5) is updated through the following relation:

$$T_{t+\Delta t}[0] = T_{avg} \left(1 + \frac{p_{cool}\, T_{avg}^3\, A\, p_s}{V}\right)^{-\frac{1}{3}} \tag{7}$$

where p_{cool} accounts for the relevant lava's physical properties and $p_s = \Delta t$ (see Tab. 1). It is worth to note that the parameter p_{cool}, the measurement of which would be quite difficult, is directly determined during the calibration process exploiting the availability of real spatio-temporal data regarding past events of the effusion of a specific kind of lava. Finally, if the temperature drops below the threshold p_{Tsol}, the lava solidifies, that is, its thickness is added to the geographical altitude of the cell and the lava thickness is reset to zero.

3   Model Calibration

Once that a MCA model has been defined, two stages are needed to assess its reliability: the calibration and validation phases. The former searches a set of parameters able to adequately reproduce a specific real case; the latter tests the model on a sufficient number of cases (which must be different of those considered in the calibration phase, though similar in terms of physical and geological properties), permitting to give a final response on the model goodness.

The calibration phase is performed by means of a GA similar to that applied in [7], where parameters to be optimised were encoded as bit strings. Moreover, the GA adopts a steady-state and elitist model, so that at each step only the worst individuals are replaced. The remaining ones, required to form the new population, are copied from the old one, choosing the best. In order to select the individuals to be reproduced, the "binary-tournament without replacement" selection operator was utilised. It consists of a series of "tournaments" in which two individuals are selected at random, and the winner is chosen according to a prefixed probability (0.6 in our case), which must be set greater for the fittest individual. Moreover, as the variation without replacement scheme was adopted, individuals cannot be selected more than once. Employed genetic operators are classic Holland's crossover and mutation with probability of 1.0 and 2/44, respectively. In particular, the latter probability implies, on an average, two bits mutated for each individual, as the genotype length was exactly 44 (see Tab. 2). The number of individuals forming the initial population was set to 256, while the number of individuals to be replaced at each GA step was set to 16.

One of the most important aspects of the calibration phase is the definition of the fitness function, i.e. the function which measures the level of agreement between the real event and the simulated one. In order to the model to well describe the overall spatio-temporal dynamics of the system, the simulation should be able to produce outcomes which correspond with the real phenomenon in terms of its relevant characteristics (e.g. areal extension and shape, lava thickness and temperature, invasion velocity) at some specific instants $\{t_1, \ldots t_k\} \in [0, \tau]$, being τ the duration of the real event. Unfortunately, complete mapping of lava characteristics corresponding to instants $t < \tau$ are rarely surveyed by geologists, while only the final shape of invasion is usually available. On the other hand, for risk analysis purposes, the correspondence between real and simulated lava shapes in the instant $t = \tau$ in which the real phenomenon reaches a steady-state condition, is of particular interest. For the above reasons, the fitness function to be maximised was defined in such a way to combine the two objectives as:

$$\Theta = \sqrt{\Omega_\tau \, \Omega_{\tau + \Delta\tau}} \qquad (8)$$

Table 2. The best set of SCIARA-fv parameters as obtained through calibration phase

Parameter	Explored range	Encoding bits	Best value
p_s	$[60, 180]$	8	155.29 s
p_{Tv}	-	-	1373 K
p_{Tsol}	$[1123, 1173]$	8	1165.35 K
p_{adv}	$[0.1, 2.0]$	4	0.7 m
p_{adsol}	$[6.0, 30.0]$	6	12 m
p_{cool}	$[10^{-16}, 10^{-13}]$	16	$2.9 \times 10^{-14} \, m^3 \, K^{-3}$
p_a	-	-	5 m

where $\Omega \in [0, 1]$ is a suitable measure of the spatial agreement between real and simulated lava invasions, with $\Omega = 1$ in case of perfect overlapping, while $\Delta\tau$ is a small surplus time. When $\Theta = 1$, then real and simulated events perfectly overlap with the further condition $\Omega_\tau = \Omega_{\tau+\Delta\tau}$, meaning that the simulation reaches the steady-state exactly at the same time as the real event does. As regards function Ω in (8), it was defined as:

$$\Omega_t = \sqrt{\frac{m(R \cap S)}{m(R \cup S)}} \tag{9}$$

where t is the instant of evaluation, R and S represent the sets of cells affected by the real and simulated event, respectively, while $m(A)$ denotes the measure of the set A. Note that $\Omega \in [0,1]$ and its value is 0 if the real and simulated events are completely disjoint, being $m(R \cap S)=0$, while it is 1 in case of perfect overlap, being $m(R \cap S) = m(R \cup S)$.

4   A Case of Study

At 3.00 AM on July 18-th, 2001, an eruption started from the fracture of Mount Calcarazzi, on the southern flank of Mt Etna (Sicily, Italy), at 2100 m a.s.l. The event was fed by a medium lava flow rate (about $7m^3/s$) and, due to the steep descent of the terrain in that area, pointed southwards creating the main danger for the towns of Nicolosi and Belpasso. After 10 days of activity, it reached its maximum extension, which was almost 6 Km in terms of run-out. Such event was chosen as the reference case for the calibration phase, as it was considered sufficiently representative of Etnean lava flows and even characterised by a relative brief duration. In particular, the second feature allowed to execute the elevated number of simulations required by the GA in a reasonable amount of time. Moreover, to further speedup the experiments execution, a Master-Slave parallel GA version was considered, instead of the sequential one, which simply split the individuals' fitness evaluation over the available "slave" processors, while the GA steps are managed by the "master" one. It represents the simplest example of parallel GA [8]. Accordingly, calibration was performed on a Nec TX7 NUMA machine, composed by 4 quadri-processors Itanium class nodes, with an overall RAM memory of 32 GB and a peak performance of 64 GFLOPS.

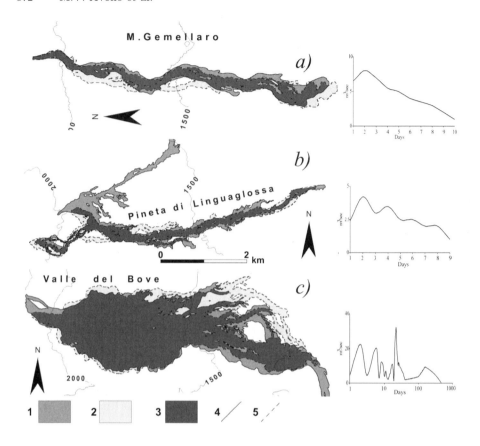

Fig. 1. Comparison between real and simulated events (charts report the lava emission rate). Keys: 1) Area affected by the real event but missed by the simulation; 2) Area incorrectly forecasted as invaded by lava; 3) Area correctly forecasted as invaded by lava; 4) Limits of real event, 5) Limits of simulated event.

On the basis of previous empirical attempts, ranges within which the values of the CA parameters are allowed to vary were individuated in order to define the GA search space (see Tab. 2), and a set of 10 experiments iterated for 100 steps. As regards the fitness function, Ω_τ was evaluated after $\tau = 10$ days (which corresponds to the duration of the real event), while $\Delta\tau$ was set to 3 days. Results of the calibration phase are reported in Tab. 2. Note that parameter p_{Tv} was set to a prefixed value, which corresponds to the typical temperature of Etnean lava flows at vents. Parameter p_a was also prefixed, as it was imposed by the detail of the considered topographic data. The parameter set listed in Tab. 2 allowed to reproduce the considered 2001 Nicolosi Etnean lava flow (see Fig. 1, case a), giving rise to a fitness $\Theta = 0.72$, which corresponds to $\Omega_\tau = 0.74$ in terms of areal comparison. Such values indicate a good performance of the model considering the significant uncertainties which affect both the measured effusion rate and the available topography data (while the GA proved to be able to explore quite well the parameter space, since in a calibration test against the outcomes of the model itself it provided a solution with $\Omega_\tau \approx 0.95$). Besides, as it can be observed in Fig. 1, a fitness value of 0.72 can be considered as satisfactory from a risk assessment and hazard mapping point of view, since the adopted measure is fairly severe in terms of shape concordance.

The validation phase was carried out by testing the obtained parameters to other real cases of study: the 2002 Linguaglossa and the $1991-93$ Valle del Bove events, both regarding Mt Etna (Sicily). The first lasted 9 days, the second 473 days. Results are graphically illustrated in Fig. 1, case b) and c), respectively with the corresponding emission rates. As expected, the best set of parameters (see Tab. 2) permitted a satisfactorily reproduction of the considered phenomena: in quantitative terms, the obtained Ω_τ values were 0.71 and 0.85, respectively.

5   Conclusions

Simulating real lava invasions for hazard management requires models charac- terized by high reliability, robustness and low computational cost. Latest results, reported in this paper, confirm that the macroscopic approach based on CA and evolutionary computation can provide such characteristics. Nevertheless, further improvements could be achieved overcoming some of the simplified assumptions of the presented model. Future research work will be addressed in this direction.

References

1. Crisci, G., Di Gregorio, S., Ranieri, G.: A cellular space model of basaltic lava flow. In: Proceedings International AMSE Conference Modelling & Simulation (Paris, France, Jul.1-3). (1982) 65–67
2. Barca, D., Crisci, G., Di Gregorio, S., Nicoletta, F.: Cellular automata for simulating lava flows: A method and examples of the Etnean eruptions. Transport Theory and Statistical Physics **23** (1994) 195–232
3. Crisci, G., Rongo, R., Di Gregorio, S., Spataro, W.: The simulation model sciara: the 1991 and 2001 lava flows at mount Etna. Journal of Volcanology and Geothermal Research **132** (2004) 253–267
4. Di Gregorio, S., Serra, R.: An empirical method for modelling and simulating some complex macroscopic phenomena by cellular automata. Future Generation Computer Systems **16** (1999) 259–271
5. McBirney, A., Murase, T.: Rheological properties of magmas. Annual Review of Earth Planetary Sciences **12** (1984) 337–357
6. Park, S., Iversen, J.: Dynamics of lava flow: Thickness growth characteristics of steady two-dimensional flows. Geophysical Research Letters **11** (1984) 641–644
7. Spataro, W., D'Ambrosio, D., Rongo, R., Trunfio, G.A.: An evolutionary approach for modelling lava flows through cellular automata. In Sloot, P.M.A., Chopard, B., Hoekstra, A.G., eds.: LNCS - ACRI. (2004) 725–734
8. Cantu-Paz, E.: Efficient and Accurate Parallel Genetic Algorithms. Kluwer Academic Publishers, Norwell, MA, USA (2000)

Acceleration of Preconditioned Krylov Solvers for Bubbly Flow Problems

J.M. Tang and C. Vuik

Delft University of Technology,
Faculty of Electrical Engineering, Mathematics and Computer Science,
Delft Institute of Applied Mathematics,
Mekelweg 4, 2628 CD Delft, The Netherlands
{j.m.tang,c.vuik}@tudelft.nl

Abstract. We consider the linear system which arises from discretization of the pressure Poisson equation with Neumann boundary conditions, coming from bubbly flow problems. In literature, preconditioned Krylov iterative solvers are proposed, but these show slow convergence for relatively large and complex problems. We extend these traditional solvers with the so-called deflation technique, which accelerates the convergence substantially. Several numerical aspects are considered, like the singularity of the coefficient matrix and the varying density field at each time step. We demonstrate theoretically that the resulting deflation method accelerates the convergence of the iterative process. Thereafter, this is also demonstrated numerically for 3-D bubbly flow applications, both with respect to the number of iterations and the computational time.

Keywords: deflation, conjugate gradient method, preconditioning, symmetric positive semi-definite matrices, bubbly flow problems.

1 Introduction

Recently, moving boundary problems have received much attention in literature, due to their applicative relevance in many physical processes. One of the most popular moving boundary problems is modelling bubbly flows, see e.g. [12]. These bubbly flows can be simulated, by solving the well-known Navier-Stokes equations for incompressible flow:

$$\begin{cases} \dfrac{\partial u}{\partial t} + u \cdot \nabla u + \dfrac{1}{\rho}\nabla p = \dfrac{1}{\rho}\nabla \cdot \mu\left(\nabla u + \nabla u^T\right) + g; \\ \nabla \cdot u = 0, \end{cases} \qquad (1)$$

where g represents the gravity and surface tension force, and ρ, p, μ are the density, pressure and viscosity, respectively. Eqs. (1) can be solved using, for instance, the pressure correction method [7]. The most time-consuming part of this method is solving the symmetric and positive semi-definite (SPSD) linear system on each time step, which comes from a second-order finite-difference

Y. Shi et al. (Eds.): ICCS 2007, Part I, LNCS 4487, pp. 874–881, 2007.

discretization of the Poisson equation with possibly discontinuous coefficients and Neumann boundary conditions:

$$\begin{cases} \nabla \cdot \left(\frac{1}{\rho}\nabla p\right) = f_1, & \mathbf{x} \in \Omega, \\ \frac{\partial}{\partial \mathbf{n}}p = f_2, & \mathbf{x} \in \partial\Omega, \end{cases} \tag{2}$$

where $\mathbf{x}$ and $\mathbf{n}$ denote the spatial coordinates and the unit normal vector to the boundary $\partial\Omega$, respectively. In the 3-D case, domain Ω is chosen to be a unit cube. Furthermore, we consider two-phase bubbly flows, so that ρ is piecewise constant with a relatively high contrast:

$$\rho = \begin{cases} \rho_0 = 1, & \mathbf{x} \in \Lambda_0, \\ \rho_1 = 10^{-3}, & \mathbf{x} \in \Lambda_1, \end{cases} \tag{3}$$

where Λ_0 is water, the main fluid of the flow around the air bubbles, and Λ_1 is the region inside the bubbles.

The resulting linear system which has to be solved is

$$Ax = b, \quad A \in \mathbb{R}^{n \times n}, \tag{4}$$

where the singular coefficient matrix A is SPSD and $b \in \text{range}(A)$. In practice, the preconditioned Conjugate Gradient (CG) method [4] is widely used to solve (4), see also References [1,2,3,5]. In this paper, we will restrict ourselves to the Incomplete Cholesky (IC) decomposition [8] as preconditioner, and the resulting method will be denoted as ICCG. In this method,

$$M^{-1}Ax = M^{-1}b, \quad M \text{ is the IC preconditioner,}$$

is solved using CG. ICCG shows good performance for relatively small and easy problems. For complex bubbly flows or for problems with large jumps in the density, this method shows slow convergence, due to the presence of small eigenvalues in the spectrum of $M^{-1}A$, see also [13].

To remedy the bad convergence of ICCG, the deflation technique has been proposed, originally from Nicolaides [11]. The idea of deflation is to project the extremely small eigenvalues of $M^{-1}A$ to zero. This leads to a faster convergence of the iterative process, due to the fact that CG can handle matrices with zero-eigenvalues [6] and the effective condition number becomes more favorable. The resulting method is called Deflated ICCG or shortly DICCG, following [17], and it will be further explained in the next section.

2   DICCG Method

In DICCG, we solve

$$M^{-1}PA\tilde{x} = M^{-1}Pb, \quad P \text{ is the deflation matrix,}$$

using CG, where

$$P := I - AZE^{-1}Z^T, \quad E := Z^TAZ, \quad Z \in \mathbb{R}^{n \times r}, \quad r \ll n. \tag{5}$$

Piecewise-constant deflation vectors are used to approximate the eigenmodes corresponding to the components which caused the slow convergence of ICCG. More technically, deflation subspace matrix $Z = [z_1 \ z_2 \ \cdots \ z_r]$ consists of deflation vectors z_j with

$$z_j(\mathbf{x}) = \begin{cases} 0, & \mathbf{x} \in \Omega \setminus \bar{\Omega}_j; \\ 1, & \mathbf{x} \in \Omega_j, \end{cases}$$

where the domain Ω is divided into non-overlapping subdomains Ω_j, which are chosen to be cubes, assuming that the number of grid points in each spatial direction is the same. Note that, due to the construction of the sparse matrix Z, matrices AZ and E are sparse as well, so that the extra computations with the deflation matrix P are relatively cheap.

3   Application of DICCG to Bubbly Flow Problems

The deflation technique works well for invertible systems and when the deflation vectors are based on the geometry of the problem, see also References [9, 10]. Main questions in this paper are:

- is the deflation method also applicable to linear systems with singular matrices?
- is the deflation method with fixed deflation vectors also applicable to problems, where the position and radius of the bubbles change in every time step?

The second question will be dealt in the next section, where numerical experiments will be presented to show the success of the method for time-dependent bubbly flow problems.

First, we show that DICCG can be used for singular matrices. Due to the construction of matrix Z and the singularity of A, the coarse matrix $E := Z^T A Z$ is also singular. In this case, E^{-1} does not exist. We propose several new variants of deflation matrices P:

(i) invertibility of A is forced resulting in a deflation matrix P_1, i.e., we adapt the last element of A such that the new matrix, denoted as $\widetilde{A}$, is invertible;

(ii) a column of Z is deleted resulting in a deflation matrix P_2, i.e., instead of Z we take $[z_1 \ z_2 \ \cdots \ z_{r-1}]$ as the deflation subspace matrix;

(iii) systems with a singular E are solved iteratively resulting in a deflation matrix P_3, i.e., matrix E^{-1}, as given in Eq. (5), is considered to be a pseudo-inverse.

As a result, Variant (i) and (ii) give a non-singular matrix E and, in addition, the real inverse of E is not required anymore in Variant (iii). Subsequently, we can prove that the three DICCG variants are identical in exact arithmetic, see Theorem 1.

Theorem 1. $P_1 \widetilde{A} = P_2 A = P_3 A.$

Proof. The proof can be found in [15, 16].

We observe that the deflated systems of all variants are identical. From this result, it is easy to show that the preconditioned deflated systems are also the same. Since the variants are equal, any of them can be chosen in the numerical experiments. We will apply the first variant for convenience, and the results and efficiency of this variant will be demonstrated numerically, in the next section.

4   Numerical Experiments

We test the efficiency of the DICCG method for two kind of test problems.

4.1   Test Case 1: Stationary Problem

First, we take a 3-D bubbly flow application with eight air-bubbles in a domain of water, see Figure 1 for the geometry. We apply finite differences on a uniform Cartesian grid with $n = 100^3$, resulting in a very large but sparse linear system $Ax = b$ with SPSD matrix A.

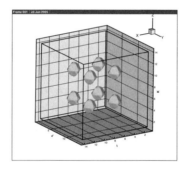

Fig. 1. An example of a bubbly flow problem: eight air-bubbles in a unit domain filled with water

Then, the results of ICCG and DICCG can be found in Table 1, where ϕ denotes the final relative exact residual and DICCG$-r$ denotes DICCG with r deflation vectors. Moreover, we terminate the iterative process, when the relative update residuals are smaller than the stopping tolerance $\epsilon = 10^{-8}$.

From Table 1, one observes that the larger the number of deflation vectors, the less iterations DICCG requires. With respect to the CPU time, there is an optimum, namely for $r = 10^3$. Hence, in the optimal case, DICCG is more than five times faster compared to the original ICCG method, while the accuracy of both methods are comparable!

Similar results also hold for other related test cases. Results of ICCG and DICCG for the problem with 27 bubbles can be found in Table 2. In addition, it appears that the benefit of the deflation method is larger when we increase the number of grid points, n, in the test cases, see also [16].

Table 1. Convergence results of ICCG and DICCG$-r$ solving $Ax = b$ with $n = 100^3$, for the test problem as given in Figure 1

Method	# Iterations	CPU Time (s)	ϕ ($\times 10^{-9}$)
ICCG	291	43.0	1.1
DICCG-2^3	160	29.1	1.1
DICCG-5^3	72	14.2	1.2
DICCG-10^3	**36**	**8.2**	**0.7**
DICCG-20^3	22	27.2	0.9

Table 2. Convergence results of ICCG and DICCG$-r$ solving $Ax = b$ with $n = 100^3$, for the test case with 27 bubbles

Method	# Iterations	CPU Time (sec)	ϕ ($\times 10^{-9}$)
ICCG	310	46.0	1.3
DICCG-2^3	275	50.4	1.3
DICCG-5^3	97	19.0	1.2
DICCG-10^3	**60**	**13.0**	**1.2**
DICCG-20^3	31	29.3	1.2

Finally, for the test case with 27 bubbles, the plots of the residuals during the iterative process of both ICCG and DICCG can be found in Figure 2. Notice that the behavior of the residuals of ICCG are somewhat irregularm due to the presence of the bubbles. For DICCG, we conclude that the larger r, the more linear the residual plot is, so the faster the convergence of the iterative process. Apparently, the eigenvectors associated to the small eigenvalues of $M^{-1}A$ have been well-approximated by the deflation vectors, if r is sufficiently large.

4.2   Test Case 2: Time-Dependent Problem

Next, we present some results from the 3-D simulation of a rising air bubble in water, in order to show that the deflation method is also applicable to real-life problems with varying density fields. We adopt the mass-conserving level-set method [13] for the simulations, but it could be replaced by any operator-splitting method, in general. At each time step, a pressure Poisson equation has to be solved, which is the most time-consuming part of the whole simulation. Therefore, during this section we only concentrate on this part at each time step. We investigate whether DICCG is efficient for all those time steps.

We consider a test problem with a rising air bubble in water without surface tension. The exact material constants and other relevant information can be found in [13, Sect. 8.3.2]. The starting position of the bubble in the domain and the evolution of the movement during the 250 time steps are given in Figure 3.

In [13], the Poisson solver is based on ICCG. Here, we will compare this method to DICCG with $r = 10^3$ deflation vectors, in the case of $n = 100^3$. The results are presented in Figure 4.

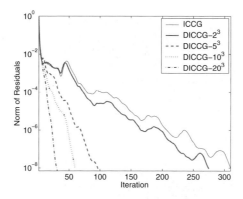

Fig. 2. Residual plots of ICCG and DICCG$-r$, for the test problem with 27 bubbles and various number of deflation vectors r

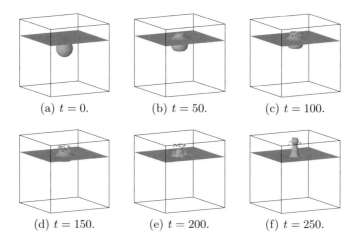

Fig. 3. Evolution of the rising bubble in water without surface tension in the first 250 time steps

From Subfigure 4(a), we notice that the number of iterations is strongly reduced by the deflation method. DICCG requires approximately 60 iterations, while ICCG converges between 200 and 300 iterations at most time steps. Moreover, we observe the erratic behavior of ICCG, whereas DICCG seems to be less sensitive to the geometries during the evolution of the simulation. Also with respect of the CPU time, DICCG shows very good performance, see Subfigure 4(b). At most time steps, ICCG requires 25–45 seconds to converge, whereas DICCG only needs around 11–14 seconds. Moreover, in Figure 4(c), one can find the gain factors, considering both the ratios of the iterations and the CPU time between ICCG and DICCG. From this figure, it can be seen that DICCG needs approximately 4–8 times less iterations, depending on the time step. More important,

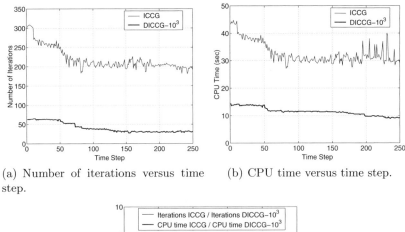

(a) Number of iterations versus time step.

(b) CPU time versus time step.

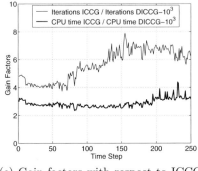

(c) Gain factors with respect to ICCG and DICCG.

Fig. 4. Results of ICCG and DICCG with $r = 10^3$, for the simulation with a rising air bubble in water

DICCG converges approximately 2–4 times faster to the solution compared to ICCG, at all time steps.

In general, we see that, compared to ICCG, DICCG decreases significantly the number of iterations and the computational time as well, which are required for solving the pressure Poisson equation with discontinuous coefficients, in applications of 3-D bubbly flows.

5   Conclusions

A deflation technique has been proposed to accelerate the convergence of standard preconditioned Krylov methods, for solving bubbly flow problems. In literature, this deflation method has already been proven to be efficient, for linear systems with invertible coefficient matrix and not-varying density fields in time. However, in our bubbly flow applications, we deal with linear systems with a singular matrix and varying density fields. In this paper, we have shown, both theoretically and numerically, that the deflation method with fixed subdomain deflation vectors can also be applied to this kind of problems. The method appeared to be robust and very efficient in various numerical experiments, with respect to both the number of iterations and the computational time.

References

1. Benzi, M.: *Preconditioning techniques for large linear systems: A survey*, J. Comp. Phys., **182** (2002), 418–477
2. Gravvanis, G.A.: *Explicit Approximate Inverse Preconditioning Techniques*, Arch. Comput. Meth. Eng., **9** (2002), 371–402.
3. Grote, M.J., Huckle, T.: *Parallel preconditioning with sparse approximate inverses*, SIAM J. Sci. Comput., **18** (1997), 838–853.
4. Hestenes, M.R., Stiefel, E.: *Methods of Conjugate Gradients for Solving Linear Systems*, J. Res. Nat. Bur. Stand., **49** (1952), 409–436.
5. Huckle, T.: *Approximate sparsity patterns for the inverse of a matrix and preconditioning*, Appl. Num. Math., **30** (1999), 291–303.
6. Kaasschieter, E.F.: *Preconditioned Conjugate Gradients for solving singular systems*, J. Comp. Appl. Math., **24** (1988), 265–275.
7. Kan, J.J.I.M. van: *A second-order accurate pressure correction method for viscous incompressible flow*, SIAM J. Sci. Stat. Comp., **7** (1986), 870–891.
8. Meijerink, J.A., Vorst, H.A. van der: *An iterative solution method for linear systems of which the coefficient matrix is a symmetric M-matrix*, Math. Comp., **31** (1977), 148–162.
9. Nabben, R., Vuik, C.: *A comparison of Deflation and Coarse Grid Correction applied to porous media flow*, SIAM J. Numer. Anal., **42** (2004), 1631–1647.
10. Nabben, R., Vuik, C.: *A Comparison of Deflation and the Balancing Preconditioner*, SIAM J. Sci. Comput., **27** (2006), 1742–1759.
11. Nicolaides, R.A.: *Deflation of Conjugate Gradients with applications to boundary value problems*, SIAM J. Matrix Anal. Appl., **24** (1987), 355–365.
12. Van der Pijl, S.P., Segal, A., Vuik, C., Wesseling, P.: *A mass-conserving Level-Set method for modelling of multi-phase flows*, Int. J. Num. Meth. in Fluids, **47**(4) (2005), 339–361.
13. Van der Pijl, S.P.: *Computation of bubbly flows with a mass-conserving level-set method*, PhD thesis, Delft University of Technology, Delft (2005).
14. Sousa, F.S., Mangiavacchi, N., Nonato, L.G., Castelo, A., Tome, M.F., Ferreira, V.G., Cuminato, J.A., McKee, S.: *A Front-Tracking / Front-Capturing Method for the Simulation of 3D Multi-Fluid Flows with Free Surfaces*, J. Comp. Physics, **198** (2004), 469–499.
15. Tang, J.M., Vuik, C.: *On Deflation and Singular Symmetric Positive Semi-Definite Matrices*, J. Comp. Appl. Math., *to appear* (2006).
16. Tang, J.M., Vuik, C.: *An Efficient Deflation Method applied on 2-D and 3-D Bubbly Flow Problems*, Elec. Trans. Num. Anal., *submitted* (2006).
17. Vuik, C., Segal, A., Meijerink, J.A.: *An efficient preconditioned CG method for the solution of a class of layered problems with extreme contrasts in the coefficients*, J. Comp. Phys., **152** (1999), 385–403.

An Efficient Characteristic Method for the Magnetic Induction Equation with Various Resistivity Scales

Jiangguo (James) Liu

Department of Mathematics, Colorado State University, Fort Collins, CO 80523, USA
liu@math.colostate.edu
http://www.math.colostate.edu/~liu/

Abstract. In this paper, we develop an efficient characteristic finite element method (FEM) for solving the magnetic induction equation in magnetohydrodynamics (MHD). We carry out numerical experiments on a two dimensional test case to investigate the influence of resistivity at different scales. In particular, our numerical results exhibit how the topological structure and energy of the magnetic field evolve for different scales of resistivity. Magnetic reconnection can also be observed in the numerical experiments.

Keywords: characteristic method, convection-diffusion, magnetohydrodynamics(MHD), magnetic resistivity, numerical simulation.

1   Introduction

Magnetohydrodynamics (MHD) is the study of the interaction between the magnetic field and the flow of electrically conducting fluids, typically plasmas, liquid metals, or brine. The MHD equations are a combination of the Navier-Stokes equations for fluid motion and the Maxwell equations for electromagnetism coupled through Lorentz body force and Ohmic heating [2,3]:

$$\partial_t \rho + \nabla \cdot (\rho \mathbf{v}) = 0 \quad \text{(Mass conservation)}$$

$$\partial_t (\rho \mathbf{v}) + \nabla \cdot (\rho \mathbf{v} \otimes \mathbf{v} - \mathbf{B} \otimes \mathbf{B}) + \nabla p^* = \mathbf{0} \quad \text{(Momentum conservation)}$$

$$\partial_t E + \nabla \cdot \left[(E + p^*)\mathbf{v} - (\mathbf{v} \cdot \mathbf{B})\mathbf{B} \right] = \varepsilon |\nabla \times \mathbf{B}|^2 \quad \text{(Energy conservation)}$$

$$\partial_t \mathbf{B} = \nabla \times (\mathbf{v} \times \mathbf{B}) + \varepsilon \Delta \mathbf{B} \quad \text{(Magnetic induction equation)}$$

$$\nabla \cdot \mathbf{B} = 0, \quad \text{for all } t \geq 0 \quad \text{(Solenoidal property)}$$

$$(1)$$

where ρ is the mass density (per unit volume of the liquid), $\mathbf{m} = \rho \mathbf{v}$ the momentum density, $\mathbf{v}$ the velocity field, $\mathbf{B}$ the magnetic field, E the total energy density, ε the magnetic resistivity that represents the effect of Ohmic heating, t the time variable, and $\otimes$ stands for the tensor product.

Y. Shi et al. (Eds.): ICCS 2007, Part I, LNCS 4487, pp. 882–889, 2007.

Some estimates suggest that up to 99% of the entire universe is filled by plasmas [2]. Man-made and natural plasmas can be found in plasma TVs, plasma lamps, lightning, aurora, and solar winds (plasmas flowing out from the Sun at a very high speed and in the radially outward direction). Table 1 lists the resistivity of various types of plasmas. Among them, the 5k eV plasma is used in large lab fusion experiments [2].

Table 1. The resistivity of various types of plasmas

Plasma	Resistivity (ohm m)	Plasma	Resistivity (ohm m)
Earth's ionosphere	2×10^{-2}	100 eV plasma	5×10^{-7}
Solar corona	5×10^{-5}	5k eV plasma	1×10^{-9}
Interstellar gas	5×10^{-7}		

The simplest MHD is the ideal MHD model that assumes the fluid has no resistivity and can be treated as a perfect conductor. The ideal MHD equations can be reformulated as a hyperbolic system and then hyperbolic solvers apply. If the resistivity terms are retained, the model is referred as resistive MHD.

Ideal MHD is applicable for a limited time before resistivity becomes too important to ignore. This limited time span varies. It could be hundreds to thousands of years for the Sun, and hence much longer than the actual lifetime of a sunspot, so it is reasonable to ignore the resistivity. However, a meter-sized volume of seawater has a magnetic diffusion time measured in milliseconds. Instabilities existing in a plasma can increase its effective resistivity. The enhanced resistivity usually results from the formation of small scale structures such as current sheets or fine scale magnetic turbulence. In these situations, the ideal MHD model is broken. When resistivity is present, magnetic reconnection may happen, in which a plasma releases stored magnetic energy as waves, bulk mechanical acceleration of material, particle acceleration, or heat. In addition, most numerical methods for ideal MHD inevitably introduce numerical diffusion and finally break down. Therefore, efficient numerical methods for resistive MHD are much needed.

However, resistive MHD presents a challenge due to the widespread time scales, the nonlinear coupling, the fineness of spatial grids for resolving current sheets, etc. To tackle this multi-physics and multi-scale problem, one could adopt an operator decomposition approach. Efficient solvers based on characteristic tracking for MHD kinematics (the influence of the velocity field on the magnetic field) and the fluid equations (the first three equations in (1) can be solved as a hyperbolic system) are developed respectively. Strategies for controlling errors of coupling between the characteristic solvers and the hyperbolic solvers (reflection of the interaction of fluid motion and electromagnetism) are also to be investigated.

2   A Characteristic Finite Element Method for the Magnetic Field Induction Equation

The magnetic induction equation and the solenoidal (divergence-free) property together describe the kinematics of MHD flows, i.e., the influence of the velocity field on the magnetic field. The divergence-free property $\nabla \cdot \mathbf{B} = 0$ should be preserved by numerical methods, globally or locally, in the classical or weak sense. One approach is to express the magnetic field as the curl of a vector potential $\mathbf{A}$, i.e., to define $\mathbf{B} = \nabla \times \mathbf{A}$. This approach is particularly convenient in two dimensions, where $\mathbf{A} = (0, 0, A_3)$ and $\mathbf{B} = (B_1, B_2, 0) = (\partial_y A_3, -\partial_x A_3, 0)$. If the fluid is incompressible ($\nabla \cdot \mathbf{v} = 0$), then the magnetic field equation reduces to a scalar convection-diffusion equation about the magnetic potential A_3:

$$\partial_t A_3 + \nabla \cdot (\mathbf{v} A_3 - \varepsilon \nabla A_3) = 0. \tag{2}$$

Convection-diffusion problems like equation (2) arise also in groundwater contaminant remediation, petroleum reservoir simulation, and many other applications. A general form for this type of problems is

$$\begin{cases} u_t + \nabla \cdot (\mathbf{v} u - \mathbf{D} \nabla u) = f(\mathbf{x}, t), & \mathbf{x} \in \Omega, \ t \in (0, T], \\ u(\mathbf{x}, 0) = u_0(\mathbf{x}), & \mathbf{x} \in \Omega, \\ \text{Suitable boundary conditions}, \end{cases} \tag{3}$$

where $\Omega \subset \mathbb{R}^d (d = 1, 2, 3)$ is a domain with boundary $\Gamma = \partial\Omega$, $u(\mathbf{x}, t)$ is the unknown function, $\mathbf{v}(\mathbf{x}, t)$ a prescribed velocity field, $\mathbf{D}(\mathbf{x}, t)$ a diffusion matrix, and $f(\mathbf{x}, t)$ a source/sink term. Let $\mathbf{n}$ be the unit outward normal vector on Γ. We define the inflow, outflow, and noflow boundaries Γ^I, Γ^O, Γ^N as

$$\Gamma^I = \{\mathbf{x} \in \Gamma : \mathbf{v} \cdot \mathbf{n} < 0\}, \quad \Gamma^O = \{\mathbf{x} \in \Gamma : \mathbf{v} \cdot \mathbf{n} > 0\}, \quad \Gamma^N = \{\mathbf{x} \in \Gamma, \mathbf{v} \cdot \mathbf{n} = 0\}.$$

Dirichlet, Neumann, or Robin boundary conditions are posed as

$$u(\mathbf{x}, t) = g_1^{type}(\mathbf{x}, t), \quad (\mathbf{x}, t) \in \Gamma^{type} \quad \text{(Dirichlet)},$$
$$-\mathbf{D}\nabla u(\mathbf{x}, t) \cdot \mathbf{n} = g_2^{type}(\mathbf{x}, t), \quad (\mathbf{x}, t) \in \Gamma^{type} \quad \text{(Neumann)}, \tag{4}$$
$$(\mathbf{v} u - \mathbf{D}\nabla u)(\mathbf{x}, t) \cdot \mathbf{n} = g_3^{type}(\mathbf{x}, t), \quad (\mathbf{x}, t) \in \Gamma^{type} \quad \text{(Robin)},$$

where $type = I, O, N$ represents an inflow, outflow, or noflow boundary type.

Solutions to convection-dominated ($|\mathbf{v}| \gg |\mathbf{D}|$) convection-diffusion problems exhibit physical features such as steep fluid fronts, current sheets, and shocks, which pose serious challenges to numerical methods for these problems. Standard finite difference or finite element methods produce either excessive nonphysical oscillations or extra numerical diffusion, which smears these physical features. The Eulerian-Lagrangian method developed in [5] (see also the references therein) is a characteristic FEM for these problems. This method naturally incorporates all types of boundary conditions in its formulation, is not subject to the severe restrictions imposed by the Courant-Friedrichs-Lewy (CFL) condition, and generates accurate numerical solutions even if large time steps are used. We briefly review this method in this section.

Let $0 = t_0 < t_1 < \cdots < t_{n-1} < t_n < \cdots < t_N = T$ be a partition of $[0, T]$ with $\Delta t_n = t_n - t_{n-1}$. We multiply equation (3) by a test function $w(\mathbf{x}, t)$ that vanishes outside the space-time strip $\Omega \times (t_{n-1}, t_n]$ and is discontinuous in time at time t_{n-1}. Integration by parts leads to the following weak form

$$
\int_\Omega u(\mathbf{x}, t_n) w(\mathbf{x}, t_n) \, d\mathbf{x} + \int_{t_{n-1}}^{t_n} \int_\Omega (\mathbf{D}\nabla u) \cdot \nabla w \, d\mathbf{x}dt
$$

$$
+ \int_{t_{n-1}}^{t_n} \int_{\partial\Omega} (\mathbf{v}u - \mathbf{D}\nabla u) \cdot \mathbf{n}w \, dSdt - \int_{t_{n-1}}^{t_n} \int_\Omega u(w_t + \mathbf{v} \cdot \nabla w) \, d\mathbf{x}dt \quad (5)
$$

$$
= \int_\Omega u(\mathbf{x}, t_{n-1}) w(\mathbf{x}, t_{n-1}^+) \, d\mathbf{x} + \int_{t_{n-1}}^{t_n} \int_\Omega (fw)(\mathbf{x}, t) \, d\mathbf{x}dt,
$$

where dS is the differential element on $\partial\Omega$ and $w(\mathbf{x}, t_{n-1}^+) := \lim_{t \to t_{n-1}^+} w(\mathbf{x}, t)$ arises from the fact that $w(\mathbf{x}, t)$ is discontinuous in time at time t_{n-1}.

In the Eulerian-Lagrangian method, all test functions are chosen to satisfy the adjoint equation

$$
w_t + \mathbf{v} \cdot \nabla w = 0. \quad (6)
$$

Characteristic tracking is also an important part of the method. One solves the following initial value problem by numerical methods, e.g., the 2nd order Runge-Kutta method, since an exact solution is usually unavailable.

$$
\begin{cases} \dfrac{d\mathbf{y}}{ds} = \mathbf{v}(\mathbf{y}(s; \mathbf{x}, t), s), \\[2mm] \mathbf{y}(s; \mathbf{x}, t)|_{s=t} = \mathbf{x}. \end{cases} \quad (7)
$$

Let Γ_n^I be the restriction of Γ^I on $[t_{n-1}, t_n]$ and similarly for Γ_n^O. Variable time steps $\Delta t^I(\mathbf{x}, t_n)$, $\Delta t^O(\mathbf{y}, t)$ are defined according to backward tracking [5]. Truncation of the diffusion and the source terms (the 2nd terms on the left and the right sides of (5)) leads to the following numerical scheme

$$
\int_\Omega u(\mathbf{x}, t_n) w(\mathbf{x}, t_n) d\mathbf{x} + \int_\Omega \Delta t^I(\mathbf{x}, t_n)(\mathbf{D}\nabla u \cdot \nabla w)(\mathbf{x}, t_n) d\mathbf{x}
$$

$$
+ \int_{\Gamma_n^O} \Delta t^O(\mathbf{y}, t)(\mathbf{D}\nabla u) \cdot \nabla w)(\mathbf{y}, t) \, (\mathbf{v} \cdot \mathbf{n}) \, dS
$$

$$
+ \int_{\Gamma_n^O} (\mathbf{v}u - \mathbf{D}\nabla u) \cdot \mathbf{n} \, w(\mathbf{y}, t) dS + \int_{\Gamma_n^I} (\mathbf{v}u - \mathbf{D}\nabla u) \cdot \mathbf{n} \, w(\mathbf{y}, t) dS \quad (8)
$$

$$
= \int_\Omega u(\mathbf{x}, t_{n-1}) w(\mathbf{x}, t_{n-1}^+) d\mathbf{x} + \int_\Omega \Delta t^I(\mathbf{x}, t_n) f(\mathbf{x}, t_n) w(\mathbf{x}, t_n) d\mathbf{x}
$$

$$
+ \int_{\Gamma_n^O} \Delta t^O(\mathbf{y}, t) f(\mathbf{y}, t) w(\mathbf{y}, t) \, (\mathbf{v} \cdot \mathbf{n}) \, dS.
$$

A finite element implementation seeks the trial function $U_n(\mathbf{x})$ from a conforming finite element space S_h. Each test function $w(\mathbf{x}, t)$ needs to satisfy the adjoint equation (6) and also $w(\mathbf{x}, t_n) \in \mathcal{S}_h$. Then the discretized system has a banded symmetric positive definite coefficient matrix and the linear system can be solved by the conjugate gradient method.

3   Numerical Experiments on the Resistivity Scales

In this section, we perform numerical experiments on a test case that simulates the role of the magnetic field in a convecting plasma. The example was originally proposed in [7] and recently studied in [8]. In this paper, the Eulerian-Lagrangian method discussed in the previous section is used as a unified numerical method for the two dimensional magnetic potential equation (2). By "unified", we mean that the solver can be applied to the equation with or without the resistivity term. Lagrangian $\mathcal{P}_1$ elements are used on a $128 \times 128 \times 2$ structured triangular mesh (a rectangle is divided into two triangles). The second order Gaussian quadrature is employed for all triangle elements. The solver uses a large time step $\Delta t = 0.1$. The 2nd order Runge-Kutta method with 16 micro steps is used for tracking characteristics. We test the example with four different values for resistivity ($\varepsilon = 0$, $5 \times 10^{-7}, 5 \times 10^{-5}, 2 \times 10^{-2}$) and observe the changes in the magnetic field energy and topology.

The spatial domain is $\Omega = [-0.5, 0.5]^2$. The velocity field is prescribed as

$$\mathbf{v}(x, y) = (-\sin(2\pi x)\cos(2\pi y), \cos(2\pi x)\sin(2\pi y)).$$

All four sides of the domain are noflow boundaries. The boundary conditions are

$$(A_3)|_{x=-0.5} = 1, \quad (A_3)|_{x=0.5} = 0, \quad \frac{\partial A_3}{\partial n}\Big|_{y=-0.5} = \frac{\partial A_3}{\partial n}\Big|_{y=0.5} = 0.$$

The initial condition shown in Figure 1 is $(A_3)|_{t=0} = 0.5 - x$, which corresponds to an initial (horizontally) uniform magnetic field.

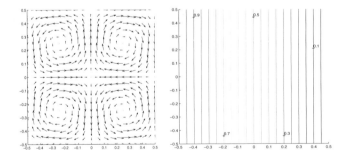

Fig. 1. (Left): the velocity field; (Right): the initial uniform magnetic potential, contour line values from 0 to 1 with an increment 0.05

This model problem exhibits different scales of resistivity. The convection and diffusion processes behave at disparate scales. But the characteristic FEM discussed in Section 2 is an efficient solver for this type of problems. Accordingly, there are two time scales in the numerical simulation. The macro time step for the numerical scheme (8) is related to diffusion and could be relatively large. It is not subject to the CFL condition. But the convection happens at a much faster pace. Small micro time steps are used for solving equations (6) and (7), so that steep fluid fronts can be accurately resolved.

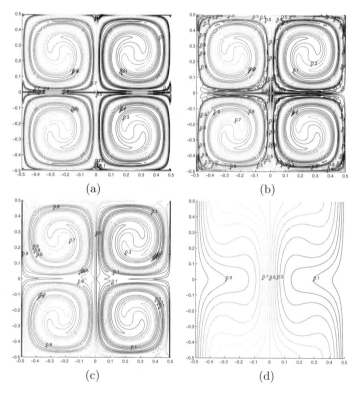

Fig. 2. Topology of the magnetic field at the same time $T = 3$ with different resistivity: (a) $\varepsilon = 0$; (b) $\varepsilon = 5 \times 10^{-7}$; (c) $\varepsilon = 5 \times 10^{-5}$; (d) $\varepsilon = 2 \times 10^{-2}$

Note that contours of the magnetic potential are actually magnetic field lines. When resistivity disappears, the magnetic equation becomes a pure convection equation, the magnetic field is "frozen", In other words, a thin rope-like volume of fluid surrounding a magnetic field line will continue to lie along the line, even though it might be twisted and distorted [2]. The topology of the magnetic field does not change. However, the geometry of the magnetic field changes due to the convection, and as a result, the magnetic energy increase. This can be observed in the left subplot of Figure 3. It is also known from Figure 2(a) that all magnetic field lines are confined to each of the four convection cells.

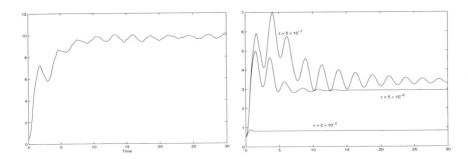

Fig. 3. Evolution of the magnetic field energy over the time period $[0, T] = [0, 30]$. (Left): $\varepsilon = 0$; (Right): $\varepsilon = 5 \times 10^{-7}, 5 \times 10^{-5}, 2 \times 10^{-2}$ (top, middle, bottom, respectively).

Table 2. Maximal magnetic field strength at various time moments

	$\varepsilon = 0$	$\varepsilon = 5 \times 10^{-7}$	$\varepsilon = 5 \times 10^{-5}$	$\varepsilon = 2 \times 10^{-2}$
$T = 0.1$	1.867	1.867	1.865	1.705
$T = 1$	60.837	49.917	40.281	4.839
$T = 5$	66.807	66.940	31.122	4.671
$T = 10$	64.000	65.251	33.406	4.671
$T = 30$	64.000	59.101	33.340	4.671

When resistivity is present, the magnetic field diffuses through the fluid in a manner similar to the diffusion of heat through a solid. The change of the magnetic flux on a closed curve due to the resistivity can produce a nonzero component in the electric field that is parallel to the magnetic field, but this does not imply magnetic reconnection will necessarily happen [2]. As shown in Figure 2, a resistivity of $\varepsilon = 2 \times 10^{-2}$ is too strong, no magnetic reconnection can happen. However, for $\varepsilon = 5 \times 10^{-5}$, magnetic reconnection happens around $T = 3$. Magnetic field lines from different magnetic domains are spliced to one another, changing the overall topology of the magnetic field. An X-type conjunction can be observed in Figure 2(c).

In real plasma physics, an X-type neutral line is believed to form between the Earth's magnetic field and a solar wind [1]. The evolution of the coronal magnetic field due to magnetic reconnection was investigated in [6].

Figure 3 shows the magnetic field energy evolution for all four cases. For $\varepsilon = 2 \times 10^{-2}$, the strong resistivity quickly brings the magnetic field to its steady state. For $\varepsilon = 5 \times 10^{-5}$, the time period for the magnetic field to reach its equilibrium is relatively longer. In general, this is related to the Alfvén time scale of a plasma [2]. All four cases exhibit an energy spike in the early stage, due to the energy infusion from the initial convection-dominance. Table 2 lists the maximal strength of the magnetic fields at several time moments. These results are consistent with those about energy evolution.

4   Concluding Remarks

Other examples on applying the characteristic FEM to resolve the kinematics of 2-dim resistive MHD flows can be found in [4], which contains also a comparison of the numerical solutions of the time-dependent convection-diffusion equation for magnetic induction and the steady-state convection-diffusion equation describing the equilibrium of a magnetic field. While [4] examines the role of velocity fields, this paper focuses on the role of the scale of resistivity.

The divergence-free property is an important physical property that should be respected by numerical methods. Another approach is to use locally divergence-free (LDF) finite elements, especially for the three dimensional magnetic induction equation. Combining LDF finite elements and the discontinuous Galerkin method to solve the ideal MHD equations has been reported in [3]. We are developing numerical methods for the resistive magnetic induction equation based on combination of characteristic tracking and discontinuous or nonconforming LDF finite elements. The results will be reported elsewhere.

This paper focuses on the role of resistivity in the kinematics of MHD, This is a one-way coupling. But MHD is a coupled system about the interaction between fluid motion and electromagnetism. Numerical methods for the fully coupled problems in resistive MHD are the topics of our ongoing research projects.

References

1. Dungey, J.W.: Interplanetary Magnetic Field and the Auroral Zones. Phys. Rev. Lett. **6** (1961) 47–48
2. Gurnett, D.A., Bhattacharjee, A.: Introduction to Plasma Physics with Space and Laboratory Applications, Cambridge University Press. 2005
3. Li, F., Shu, C.W.: Locally Divergence-free Discontinuous Galerkin Methods for MHD equations. J. Sci. Comput. **22/23** (2005) 413–442
4. Liu, J., Tavener, S., and Chen, H.: ELLAM for Resolving the Kinematics of Two Dimensional Resistive Magnetohydrodynamic Flows. Preprint.
5. Wang, H., Dahle, H.K., Ewing, R.E., Espedal, M.S., Sharpley, R.C., Man, S.: An ELLAM Scheme for Advection-Diffusion Equations in Two Dimensions. SIAM J. Sci. Comput. **20** (1999) 2160–2194
6. Wu, S.T., Wang, A.H., Plunkett, S.P., Michels, D.J.: Evolution of Gloabl-scale Coronal Magnetic Field due to Magnetic Reconnection: The Formulation of the Observed Blob Motion in the Coronal Steamer Belt. Astro. Phys. J. **545** (2000) 1101–1115
7. Weiss, N.O.: The Expulsion of Magnetic Flux by Eddies. Proc. Roy. Soc. A. **293** (1966) 310–328
8. Zegeling, P.A.: On Resistive MHD Models with Adaptive Moving Meshes. J. Sci. Comput. **24** (2005) 263–284

Multiscale Discontinuous Galerkin Methods for Modeling Flow and Transport in Porous Media

Shuyu Sun[1] and Jürgen Geiser[2]

[1] Department of Mathematical Sciences, Clemson University
O-221 Martin Hall, Clemson, SC 29634-0975, USA
shuyu@clemson.edu
[2] Department of Mathematics, Humboldt-Universität zu Berlin
Unter den Linden 6, D-10099 Berlin, Germany
geiser@mathematik.hu-berlin.de

Abstract. Multiscale discontinuous Galerkin (DG) methods are established to solve flow and transport problems in porous media. The underlying idea is to construct local DG basis functions at the coarse scale that capture the local properties of the differential operator at the fine scale, and then to solve the DG formulation using the newly constructed local basis functions instead of conventional polynomial functions on the coarse scale elements. Numerical examples are provided for demonstrating their effectiveness.

1 Introduction

Flow and reactive transport in porous media are fundamental processes arising in many diversified fields such as petroleum engineering, groundwater hydrology, environmental engineering, soil mechanics, earth sciences, chemical engineering and biomedical engineering [13,8]. Major challenges for realistic simulations of simultaneous flow, transport and chemical reaction include multiple temporal and spatial scales, long simulation time periods, and multiple coupled nonlinear components. In particular, the critical effect of fine scale processes on coarser scales requires numerical methods to effectively account for the interactions across various scales. Since it is far beyond current computational power to resolve the finest scale directly, one usually has to incorporate the fine scale effects within a coarse-scale approximation by various multiscale numerical schemes [4,5].

DG are finite element methods using discontinuous piecewise polynomial spaces and specialized bilinear forms to weakly impose boundary conditions and interelement continuities [6,7,1,11,3]. The methods have recently become popular in the scientific and engineering communities due to their many appealing features including local mass conservation, small numerical diffusion and little oscillation, and support of variable local approximations and nonconforming meshes. In this paper, we shall establish a two-scale framework for primal DG methods.

Y. Shi et al. (Eds.): ICCS 2007, Part I, LNCS 4487, pp. 890–897, 2007.

2 Discontinuous Galerkin Schemes

2.1 Governing Equations

Let Ω denote a bounded polygonal domain in $\mathbb{R}^d$, ($d = 1, 2$, or 3) and let T denote the final simulation time. The flow problem we consider is single-phase flow in porous media:

$$- \nabla \cdot (\mathbf{K} \nabla p) \equiv \nabla \cdot \mathbf{u} = q, \qquad (x, t) \in \Omega \times (0, T], \tag{1}$$

where the pressure p or the Darcy's velocity is the unknown variable to be solved. The conductivity $\mathbf{K}$ is a given parameter, and it usually varies greatly with geological space. The imposed external total flow rate q is a sum of sources (injection) and sinks (extraction).

The modeling equation for single-species reactive transport in a single flowing phase in porous media involves both spatial and temporal derivatives:

$$\frac{\partial \phi c}{\partial t} + \nabla \cdot (\mathbf{u} c - \mathbf{D}(\mathbf{u}) \nabla c) = qc^* + r(c), \ (x, t) \in \Omega \times (0, T], \tag{2}$$

where the species concentration c is the unknown variable to be solved. The effective porosity ϕ is a given parameter, and it usually varies spatially. The dispersion-diffusion tensor $\mathbf{D}(\mathbf{u})$ is a given function of Darcy velocity $\mathbf{u}$. The notation c^* is the injected concentration c_w if $q > 0$ and is the resident concentration c if $q \leq 0$. The reaction term $r(c)$ is a given function to describe the kinetic biogeochemical reaction.

We write $\partial \Omega = \overline{\Gamma}_D \cup \overline{\Gamma}_N = \overline{\Gamma}_{\text{in}} \cup \overline{\Gamma}_{\text{out}}$, where Γ_D is the Dirichlet boundary and Γ_N is the Neumann boundary for the flow subproblem and $\Gamma_D \cap \Gamma_N = \emptyset$; Γ_{in} is the inflow boundary and Γ_{out} is the outflow/noflow boundary condition. We assume the following boundary conditions:

$$p = p_B, \qquad (x, t) \in \Gamma_D \times (0, T], \tag{3}$$

$$\mathbf{u} \cdot \mathbf{n} = u_B, \qquad (x, t) \in \Gamma_N \times (0, T], \tag{4}$$

$$(\mathbf{u} c - \mathbf{D}(\mathbf{u}) \nabla c) \cdot \mathbf{n} = c_B \mathbf{u} \cdot \mathbf{n}, \qquad (x, t) \in \Gamma_{\text{in}} \times (0, T], \tag{5}$$

$$(-\mathbf{D}(\mathbf{u}) \nabla c) \cdot \mathbf{n} = 0, \qquad (x, t) \in \Gamma_{\text{out}} \times (0, T]. \tag{6}$$

The initial condition is specified by

$$c(x, 0) = c_0(x), \qquad x \in \Omega. \tag{7}$$

Here, c_B is the inflow boundary concentration and c_0 the initial concentration.

2.2 DG Schemes

Let $\mathcal{E}_h$ be a family of non-degenerate and possibly non-conforming partitions of Ω composed of line segments if $d = 1$, triangles or quadrilaterals if $d = 2$, or tetrahedra, prisms or hexahedra if $d = 3$. The set of all interior points (for $d = 1$), edges (for $d = 2$), or faces (for $d = 3$) for $\mathcal{E}_h$ is denoted by Γ_h. The discontinuous finite element space is taken to be $\mathcal{D}_r\left(\mathcal{E}_h\right) := \left\{\phi \in L^2(\Omega) : \left.\phi\right|_E \in \mathbb{P}_r(E), \ E \in \mathcal{E}_h\right\}$, where $\mathbb{P}_r(E)$ denotes the space of polynomials of (total) degree less than or equal to r on E. Let $E_i, E_j \in \mathcal{E}_h$ and $\gamma = \partial E_i \cap \partial E_j \in \Gamma_h$ with $\mathbf{n}_\gamma$ exterior to E_i. We now define the average and jump for a smooth function ϕ by $\{\phi\} := \frac{1}{2}\left(\left.(\phi|_{E_i})\right|_\gamma + \left.(\phi|_{E_j})\right|_\gamma\right)$, $[\phi] := \left.(\phi|_{E_i})\right|_\gamma - \left.(\phi|_{E_j})\right|_\gamma$. Denote the upwind value of the concentration $\left.c^*\right|_\gamma$ as $\left.c\right|_{E_i}$ if $\mathbf{u} \cdot \mathbf{n}_\gamma \geq 0$ and $\left.c\right|_{E_j}$ if $\mathbf{u} \cdot \mathbf{n}_\gamma < 0$. We introduce the bilinear forms $a(p, \psi)$ and $B(c, w; \mathbf{u})$ for the flow and transport problems respectively:

$$a(p, \psi) := \sum_{E \in \mathcal{E}_h} \int_E \mathbf{K}\nabla p \cdot \nabla \psi + J_0^\sigma\,(p, \psi)$$

$$- \sum_{\gamma \in \Gamma_h} \int_\gamma \{\mathbf{K}\nabla p \cdot \mathbf{n}_\gamma\}\,[\psi] - s_{\text{form}} \sum_{\gamma \in \Gamma_h} \int_\gamma \{\mathbf{K}\nabla \psi \cdot \mathbf{n}_\gamma\}\,[p] +$$

$$- \sum_{\gamma \subset \Gamma_D} \int_\gamma \mathbf{K}\nabla p \cdot \mathbf{n}_\gamma \psi - s_{\text{form}} \sum_{\gamma \subset \Gamma_D} \int_\gamma \mathbf{K}\nabla \psi \cdot \mathbf{n}_\gamma p,$$

$$B(c, w; \mathbf{u}) := \sum_{E \in \mathcal{E}_h} \int_E (\mathbf{D}(\mathbf{u})\nabla c - c\mathbf{u}) \cdot \nabla w - \int_\Omega cq^- w$$

$$- \sum_{\gamma \in \Gamma_h} \int_\gamma \{\mathbf{D}(\mathbf{u})\nabla c \cdot \mathbf{n}_\gamma\}\,[w] - s_{\text{form}} \sum_{\gamma \in \Gamma_h} \int_\gamma \{\mathbf{D}(\mathbf{u})\nabla w \cdot \mathbf{n}_\gamma\}\,[c]$$

$$+ \sum_{\gamma \in \Gamma_h} \int_\gamma c^*\mathbf{u} \cdot \mathbf{n}_\gamma\,[w] + \sum_{\gamma \subset \Gamma_{\text{out}}} \int_\gamma c\mathbf{u} \cdot \mathbf{n}_\gamma w + J_0^\sigma\,(c, w),$$

where $s_{\text{form}} = -1$ for NIPG (the Nonsymmetric Interior Penalty Galerkin method [7]) or OBB-DG (the Oden-Babuška-Baumann formulation of DG [6]), $s_{\text{form}} = 1$ for SIPG (the Symmetric Interior Penalty Galerkin method [12,9,11]), and $s_{\text{form}} = 0$ for IIPG (the Incomplete Interior Penalty Galerkin method [9, 2, 11, 10]). For notational convenience, we use the same s_{form} and penalty term for both flow and transport problems, but they can be chosen differently in practice. Here q^+ is the injection source term, and q^- is the extraction source term, i.e., $q^+ := \max(q, 0)$, $q^- := \min(q, 0)$. In addition, we define the interior penalty term as $J_0^\sigma(c, w) := \sum_{\gamma \in \Gamma_h} r^2 \sigma_\gamma \int_\gamma [c]\,[w]\,/h_\gamma$, where σ_γ is the penalty parameter on γ. We have $\sigma_\gamma \equiv 0$ for OBB-DG. The linear functionals for flow and transport problems are defined respectively as

$$l(\psi) := (q, \psi) - s_{\text{form}} \sum_{\gamma \subset \Gamma_D} \int_e K\nabla \psi \cdot \mathbf{n}_\gamma p_B - \sum_{\gamma \subset \Gamma_N} \int_e \psi u_B, \qquad (8)$$

$$L(w; \mathbf{u}, c) := \int_\Omega r\,(c)\,w + \int_\Omega c_w q^+ w - \sum_{\gamma \subset \Gamma_{\text{in}}} \int_\gamma c_B \mathbf{u} \cdot \mathbf{n}_\gamma w. \qquad (9)$$

For the flow problem (1) and (3)-(4), we seek a time-independent solution $P^{DG} \in \mathcal{D}_r\left(\mathcal{E}_h\right)$ satisfying,

$$a\left(P^{DG}, v\right) = l(v), \qquad \forall v \in \mathcal{D}_r\left(\mathcal{E}_h\right). \qquad (10)$$

For the transport problem (2) and (5)-(7), the continuous-in-time DG approximation $C^{DG}(\cdot, t) \in \mathcal{D}_r(\mathcal{E}_h)$ is the solution of the following ODEs:

$$\left(\frac{\partial \phi C^{DG}}{\partial t}, w \right) + B(C^{DG}, w; \mathbf{u}^{DG}) = L(w; \mathbf{u}^{DG}, C^{DG}), \tag{11}$$

$$\forall w \in \mathcal{D}_r(\mathcal{E}_h), \quad t \in (0, T],$$
$$\left(\phi C^{DG}, w \right) = \left(\phi c_0, w \right), \quad \forall w \in \mathcal{D}_r(\mathcal{E}_h), \ t = 0. \tag{12}$$

3 Multiscale Formulation for Discontinuous Galerkin

We now establish a multiscale DG formulation for solving the flow problem in porous media. The same multiscale approach can be applied to the transport problem as well, but is omitted here for brevity. In a fine mesh $\mathcal{E}_h$, we consider an r-order approximation space $\mathcal{D}_r(\mathcal{E}_h)$. The DG solution P_h in the space $\mathcal{D}_r(\mathcal{E}_h)$ can be obtained by solving the following algebraic linear system:

$$a(P_h, v) = l(v), \qquad \forall v \in \mathcal{D}_r(\mathcal{E}_h).$$

Flow in geological media usually involves behaviors across several scales, but direction solution at a very fine scale is far beyond current computational power. Let us assume that we can only afford to solve the flow equation in a coarse space $\mathcal{D}_R(\mathcal{E}_H)$ directly, where $R < r$ and $H > h$. That is, current computational capability only allows us to solve for P_H in the space $\mathcal{D}_R(\mathcal{E}_H)$ from the following algebraic system:

$$a(P_H, v) = l(v), \qquad \forall v \in \mathcal{D}_R(\mathcal{E}_H). \tag{13}$$

We note that the bilinear form $a(\cdot, \cdot)$ contains slightly different terms on different meshes. However, by choosing the penalty parameters properly, we can make the coarse mesh bilinear form identical to the restriction of the fine mesh bilinear form to the coarse space.

Since the conductivity (or permeability) could be highly oscillatory on the fine scale, we would like to incorporate the fine scale effects within a coarse-scale approximation. One natural solution is to decompose the fine space into the coarse space component and its complement: $\mathcal{D}_r(\mathcal{E}_h) = \mathcal{D}_R(\mathcal{E}_H) \oplus V_f$. If we seek $P_H \in \mathcal{D}_R(\mathcal{E}_H)$ and $P_f \in V_f$ such that

$$a(P_H, v_H) = l(v_H) - a(P_f, v_H), \qquad \forall v_H \in \mathcal{D}_R(\mathcal{E}_H), \tag{14}$$
$$a(P_f, v_f) = l(v_f) - a(P_H, v_f), \qquad \forall v_f \in V_f. \tag{15}$$

We can recover the solution P_h at the fine scale by letting $P_h = P_H + P_f$. Giving P_f, (14) requires only computation of a solution in the coarse space, and thus is affordable. However, (15) is still computationally intractable. To improve efficiency, we replace the space V_f by a locally supported space: $V_{f0} := \{v \in V_f : v|_{\partial E} = 0, \forall E \in \mathcal{E}_H\}$. Our multiscale algorithm is to seek $P_H \in \mathcal{D}_R(\mathcal{E}_H)$ and $P_{f0} \in V_{f0}$ such that

$$a(P_H, v_H) = l(v_H) - a(P_{f0}, v_H), \qquad\qquad \forall v_H \in \mathcal{D}_R(\mathcal{E}_H), \qquad (16)$$
$$a(P_{f0}, v_{f0}) = l(v_{f0}) - a(P_H, v_{f0}), \qquad\qquad \forall v_{f0} \in V_{f0}. \qquad (17)$$

We note that (17) is decoupled among the elements in the coarse mesh $\mathcal{E}_H$. To solve (17), it is equivalent to solve N local problems, where N is the number of elements in the coarse mesh $\mathcal{E}_H$. For each element $E \in \mathcal{E}_H$, the local problem is to seek $P_{f0,E} \in V_{f0}|_E$ such that

$$a(P_{f0,E}, v_{f0,E}) = l(v_{f0,E}) - a(P_H, v_{f0,E}), \qquad\qquad \forall v_{f0,E} \in V_{f0}|_E.$$

The final solution we obtain is $P_{\mathrm{MS}} := P_H + P_{f0} = P_H + \sum_{E \in \mathcal{E}_H} P_{f0,E}$.

The two equations (16) and (17) are coupled through the right-hand sides. Though we can solve the two-scale system (16)-(17) iteratively, but a substitution method is more efficient. For each basis function v_H in the coarse space V_H, we solve for $\phi(v_H) \in V_{f0}$ such that

$$a(\phi(v_H), v_{f0}) = l(v_{f0}) - a(v_H, v_{f0}), \qquad\qquad \forall v_{f0} \in V_{f0}.$$

Of course, the solution of $\phi(v_H)$ involves only a local problem within a coarse mesh element. We then construct the multiscale basis by $\phi_{\mathrm{MS}}(v_H) := \phi(v_H) + v_H$. Obviously, we know $a(\phi_{\mathrm{MS}}(v_H), v_{f0}) = l(v_{f0})$, $\forall v_{f0} \in V_{f0}$. We now define a multiscale approximation space $V_{\mathrm{MS}} := \mathrm{span}\{\phi_{\mathrm{MS}}(v_H) : \forall v_H \in \mathcal{D}_R(\mathcal{E}_H)\}$. Now our multiscale scheme (16)-(17) becomes the problem of seeking $P_{\mathrm{MS}} \in V_{\mathrm{MS}}$ such that

$$a(P_{\mathrm{MS}}, v_{\mathrm{MS}}) = l(v_{\mathrm{MS}}), \qquad\qquad \forall v_{\mathrm{MS}} \in V_{\mathrm{MS}}. \qquad (18)$$

Since the multiscale space V_{MS} has the same dimension as the coarse space $\mathcal{D}_R(\mathcal{E}_H)$, the computational time and memory requirement for solving (18) is similar to that for solving (13), provided the multiscale basis functions have been constructed. The construction of multiscale basis involves only solutions of local problems and can be obtained efficiently. In addition, the problem coefficient for the construction of multiscale basis does not have to be stored, which substantially saving memory requirement compared to the direction solution of the flow problem in the fine space.

4   Numerical Results

We consider the problem (1)-(7) over the domain $\Omega = (0,1)^2$. The conductivity tensor $\mathbf{K}$ is a diagonal matrix with $\mathbf{K}_{11} = \mathbf{K}_{22}$, and the distribution of $\mathbf{K}_{11}$ across the domain is plotted in Figure 1. We impose no source/sink term, i.e. $q = 0$. For the flow problem, we impose Dirichlet boundary conditions $p = 1$ and $p = 0$ on the left and right boundaries, respectively. The boundary conditions

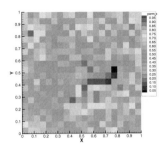

Fig. 1. Conductivity field for the flow problem

on top and bottom boundaries are assumed to be no-flow. For the transport problem, we let the effective porosity to be 1 (for convenience), we impose no reaction and we assume zero diffusion-dispersion. The advection part uses the Darcy velocity computed from the flow problem, with inflow concentration being constant 1 and with initial concentration being 0 everywhere.

NIPG is employed to solve the flow problem. The penalty parameter for NIPG is chosen to be 1. The fine mesh used is a 20x20 uniform rectangular grid and the coarse one is a 5x5 uniform rectangular mesh. The polynomial degree for the flow problem is $R = r = 2$ for both the fine and the coarse spaces. For the transport problem, we only test the Darcy velocity computed from flow for predicting advection behaviors, and thus we use fine mesh for all cases. Advection is solved by OBB-DG with lowest order polynomials (constants) and explicit Euler's temporal integration with uniform time step $\Delta t = 0.001$. The spatial discretization for advection here is equivalent to the upwinding finite volume method.

Figure 2 displays the results of pressure, velocity and concentration. The results from our multiscale algorithm are plotted in the first column. For comparison, the DG results in the coarse and fine meshes are provided in the second and third columns. Comparison indicates the computed pressure from the multiscale approach is very close to the one computed from DG in the fine mesh. Unlike the coarse DG result, the multiscale pressure result captures the fine scale information in a similar fashion as the DG result in the fine mesh; this is especially pronounced in the area around the point (0.8, 0.5). Simulations of advection indicate that the multiscale DG and coarse mesh DG methods produce nonphysical overshoots and undershoots, while the fine mesh DG method possesses no overshoot/undershoot. This is because the velocity produced from the fine mesh DG method is locally mass conservative in the fine mesh but the velocities computed from the multiscale DG and coarse mesh DG methods are not. However, the overshoot/undershoot produced by the multiscale DG is much smaller than that produced by the coarse mesh DG method. We also note that the multiscale approach leads to inaccurate velocity in a few locations along the no-flow boundary, particularly at the point (0.1, 1). This is because we strongly impose zero pressure in the local problem for computing multiscale basis functions. A potential improvement is to collect the local problems that touch the domain no-flow boundary and modify the corresponding boundaries in the local problems by no-flow rather than zero Dirichlet boundary condition.

Compared to the classical multiscale implementation of continuous Galerkin methods [4], the multiscale DG scheme is computational simpler. Unlike continuous Galerkin, the support of coarse basis function in DG sits within a single coarse mesh element, thus the construction of multiscale basis function is also limited to a single coarse mesh element. In addition, the multiscale DG algorithm has more flexibility in the construction of multiscale basis functions. For example, in this paper, we have strongly enforced zero Dirichlet boundary condition for the local problems to construct multiscale basis functions. However, we could also use other boundary conditions in implementation. For example, we could weakly impose homogeneous Dirichlet boundary condition for the local problems, or we could strongly or weakly impose homogeneous Neumann (i.e. no-flow) boundary condition for the local problems.

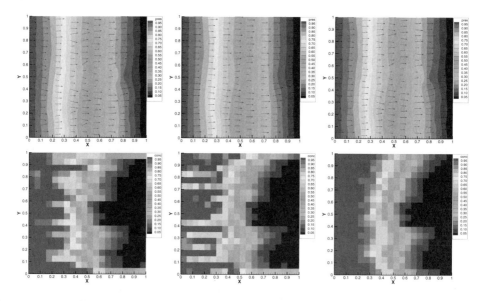

Fig. 2. NIPG results (left column: multiscale DG solution; middle column: coarse mesh DG solution; right column: fine mesh DG solution; top row: pressure and velocity for the flow problem; bottom row: concentration at $t=1$ for the transport problem)

5   Conclusions

The four primal DG schemes, namely, NIPG, OBB-DG, SIPG, and IIPG, have been considered to solve flow and transport problems in porous media. We have systematically establish a theoretical framework for multiscale DG methods. The underlying idea is first to construct local basis functions at the coarse scale that capture the local properties of the differential operator at the fine scale. The second step of the algorithm is to solve the DG formulation using the newly constructed local basis functions instead of conventional polynomial functions on the coarse scale elements. Numerical examples have been provided to demonstrate that the proposed multiscale DG indeed captures many important features of flow and transport at fine scales but yet cost roughly only the computational time of the corresponding coarse scale problem. We have also compared multiscale DG algorithms with multiscale continuous Galerkin methods, and found that the multiscale implementation of DG is probably simpler and more flexible than the multiscale implementation of continuous Galerkin. A future work is to investigate various other options for multiscale DG implementation; in particular, it will be interesting to weakly impose homogeneous Neumann condition for the local problems to construct multiscale basis functions, as this will likely reduce the error of the velocity and introduce little nonphysical oscillation.

References

1. Arnold, D. N.: An interior penalty finite element method with discontinuous elements. SIAM J. Numer. Anal. **19** (1982) 742–760
2. Dawson, C., Sun, S., Wheeler, M. F.: Compatible algorithms for coupled flow and transport. Comput. Meth. Appl. Mech. Eng. **193** (2004) 2565–2580
3. Geiser, J.: Mixed discretisation methods for the discontinuous galerkin method with analytical test-functions. Preprint No. 2006-8 of Humboldt University of Berlin, Department of Mathematics, Germany (2006)
4. Hou, T. Y., Wu, X.: A multiscale finite element method for elliptic problems in composite materials and porous media. J. Comput. Phys. **134** (1997) 169–189
5. Hughes, T. J. R.: Multiscale phenomena: Green's functions, the Dirichlet-to-Neumann formulation, subgrid scale models, bubbles and the origins of stabilized methods. Comput. Methods Appl. Mech. Engrg. **127** (1995) 387–401
6. Oden, J. T., Babuška, I., Baumann, C. E.: A discontinuous hp finite element method for diffusion problems. J. Comput. Phys. **146** (1998) 491–516
7. Rivière, B., Wheeler, M. F., Girault, V.: A priori error estimates for finite element methods based on discontinuous approximation spaces for elliptic problems. SIAM J. Numer. Anal. **39** (2001) 902–931
8. Steefel, C. I., Van Cappellen, P.: Special issue: Reactive transport modeling of natural systems. Journal of Hydrology. **209** (1998) 1–388
9. Sun, S.: Discontinuous Galerkin methods for reactive transport in porous media. PhD thesis, The University of Texas at Austin (2003)
10. Sun, S., Wheeler, M. F.: Discontinuous Galerkin methods for coupled flow and reactive transport problems. Appl. Numer. Math. **52** (2005) 273–298
11. Sun, S., Wheeler, M. F.: Symmetric and nonsymmetric discontinuous Galerkin methods for reactive transport in porous media. SIAM Journal on Numerical Analysis. **43** (2005) 195–219
12. Wheeler ,M. F.: An elliptic collocation-finite element method with interior penalties. SIAM J. Numer. Anal. **15** (1978) 152–161
13. Yeh, G. T., Tripathi, V. S.: A model for simulating transport of reactive multispecies components: model development and demonstration. Water Resources Research. **27** (1991) 3075–3094

Fourier Spectral Solver for the Incompressible Navier-Stokes Equations with Volume-Penalization

G.H. Keetels[1], H.J.H. Clercx[1,2], and G.J.F. van Heijst[1]

[1] Department of Physics,
Eindhoven University of Technology, The Netherlands
[2] Department of Applied Mathematics,
University of Twente, The Netherlands
http://www.fluid.tue.nl

Abstract. In this study we use a fast Fourier spectral technique to simulate the Navier-Stokes equations with no-slip boundary conditions. This is enforced by an immersed boundary technique called volume-penalization. The approach has been justified by analytical proofs of the convergence with respect to the penalization parameter. However, the solution of the penalized Navier-Stokes equations is not smooth on the surface of the penalized volume. Therefore, it is not *a priori* known whether it is possible to actually perform accurate fast Fourier spectral computations. Convergence checks are reported using a recently revived, and unexpectedly difficult dipole-wall collision as a test case. It is found that Gibbs oscillations have a negligible effect on the flow evolution. This allows higher-order recovery of the accuracy on a Fourier basis by means of a post-processing procedure.

Keywords: Immersed boundary, volume-penalization, Fourier spectral methods, dipole-wall collision, vortices.

1 Introduction

Fourier spectral methods are widely used in the CFD community to solve flow problems with periodic boundary conditions. Higher order accuracy can be achieved provided that the solution of the problem is sufficiently smooth. Moreover, these methods are fast, relatively easy to implement even for performing parallel computations. Incorporation of no-slip boundaries is, however, not straightforward. Therefore we use the volume-penalization method of Arquis & Caltagirone [2]. They model an obstacle or domain boundary as a porous object. By decreasing the permeability the penalized Navier-Stokes solution converge towards the Navier-Stokes solution with no-slip boundary conditions (see Ref.[1],[4]). A delicate issue is that a very steep velocity profile appears inside the porous obstacle. This can be a drawback for Fourier spectral methods as Gibbs oscillations might deteriorate the stability and accuracy of the scheme.

Y. Shi et al. (Eds.): ICCS 2007, Part I, LNCS 4487, pp. 898–905, 2007.

On the other hand, the numerical simulations of Kevlahan & Ghidaglia [10] and Schneider [11] have shown that is possible to perform stable Fourier spectral computations with volume-penalization for flow around cylinders.

It is important to extend this analysis and to determine the accuracy of the Fourier spectral computations. An important issue is to fully quantify the role of the Gibbs effect on the flow dynamics: is it possible to recover higher-order accuracy of the Fourier spectral scheme? A very challenging dipole-wall collision experiment is used as a test problem. The Chebyshev spectral and finite differences computations of Clercx & Bruneau [5] have shown that a dipole colliding with a no-slip wall is a serious challenge for CFD methods. In particular, the formation and detachment of very thin boundary layers, containing high-amplitude vorticity, during the collision process and the subsequent formation of small-scale vorticity patches in the near-wall region can possibly deteriorate the accuracy of the flow computation. This dramatically affects the dynamics of the flow after the impact.

2   Fourier Collocation Scheme

In the volume-penalization approach from Arquis & Caltagirone [2] an obstacle with a no-slip boundary is considered as a porous object with infinitely small permeability. The flow domain Ω_f is embedded in a larger domain Ω, such that $\Omega_f = \Omega \setminus \overline{\Omega_s}$, where Ω_s represents the volume of the porous objects. The interaction between the flow and the obstacles is modelled by adding a Darcy drag term to the Navier-Stokes equations locally inside Ω_s. This yields the penalized Navier-Stokes equations

$$\partial_t \mathbf{u} + (\mathbf{u} \cdot \nabla)\mathbf{u} + \nabla p - \nu \Delta \mathbf{u} + \frac{1}{\epsilon} H \mathbf{u} = 0 \qquad \text{in} \qquad \Omega \times [0, T] \,, \qquad (1)$$

where $\mathbf{u} = (u(\mathbf{x}, t), v(\mathbf{x}, t))$ is the Eulerian velocity, $p = p(\mathbf{x}, t)$ the scalar kinetic pressure, ν the kinematic viscosity, ϵ the penalization parameter and the mask function H is defined as

$$H = \begin{cases} 1 & \text{if} \quad \mathbf{x} \in \overline{\Omega}_s \\ 0 & \text{if} \quad \mathbf{x} \in \Omega_f. \end{cases} \qquad (2)$$

Figure 1 shows some examples of different geometries. In this study we use the channel geometry (Fig. 1b) and the square bounded geometry (Fig. 1c). This allows a comparison with classical methods, such as Chebyshev spectral computations, that are well adapted to these geometries. The continuity condition $\nabla \cdot \mathbf{u} = 0$ accompanies the penalized Navier-Stokes equations in Ω. On the domain boundary $\partial\Omega$ we take a periodic boundary condition, such that Fourier spectral methods can be applied. Carbou & Fabrie [4] have rigorously shown that the L_2-norm of the differences in the velocity and velocity gradients of the penalized Navier-Stokes equations and the Navier-Stokes equations with no-slip boundary conditions is proportional to $\sqrt{\epsilon}$. To ensure the C^1 continuity of the velocity, an asymptotically thin boundary layer proportional to $\sqrt{\nu\epsilon}$ appears

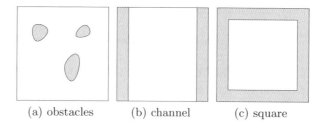

(a) obstacles   (b) channel   (c) square

Fig. 1. Decomposition of a square computational domain Ω into porous objects Ω_s (dashed) and flow domain Ω_f (white)

inside the obstacle. As a consequence, the limit $\epsilon \to 0$ might result in two problems for a numerical approximation. i) The velocity gradient profile becomes extremely steep, which might result in Gibbs oscillations when using Fourier spectral schemes. ii) The validity of the $\sqrt{\epsilon}$ error bound might be affected in case the details of the penalization boundary layer are not resolved. Furthermore, due to the slow convergence rate proportional to $\sqrt{\epsilon}$ the time step has to be decoupled from the penalization parameter in order to achieve a stiffly stable time integration scheme. In this study this is achieved by a collocation approach of the penalized Navier-Stokes equations. A third-order extrapolated backward differentiation scheme (BDF) with exact differentiation of the diffusion term is applied,

$$\alpha_0 \boldsymbol{u}_N^{n+1} + \delta t \boldsymbol{L}_N(\boldsymbol{u}_N^{n+1}) = -\sum_{h=1}^{3}\left(\delta t \beta_h \boldsymbol{G}_N(\boldsymbol{u}_N^{n+1-h}) + \alpha_h \boldsymbol{u}_N^{n+1-h}\right) e^{h\delta t \nu \Delta} \quad (3)$$

$$\tilde{\nabla} \cdot \boldsymbol{u}_N = 0 \quad (4)$$

where $\boldsymbol{G}_N = (\boldsymbol{u}_N \cdot \tilde{\nabla})\boldsymbol{u}_N + \tilde{\nabla}p_N$, $\boldsymbol{L}_N = \frac{1}{\epsilon}H\boldsymbol{u}_N$, $e^{h\delta t \nu \Delta}$ is the semi-group of the heat kernel and δt the time step. A tilde marks that collocation derivatives are used here in order to distinguish from the exact Galerkin derivative. The coefficients α_j and β_j are determined by third-order backward differentiation and extrapolation, respectively. The right-hand side of Eq. (3) is computed in transform space (Ref. [9]). Aliasing is avoided by applying the zero-padding technique introduced by Orszag, generally referred to as the 2/3-rule (see, for details Ref. [3]). Three forward and four backward FFTs are required in total.

3   Results

We consider two set of experiments. The first set concerns a normal dipole-wall collision where the dipole traverses from the center of the domain perpendicular to the wall. For these experiments a channel geometry Fig. 1b is considered. A high-resolution benchmark computation is achieved by using Chebyshev expansion in the direction perpendicular to the wall and Fourier expansion in the

periodic direction. For details of the Chebyshev-Fourier code see Kramer [8]. The
second test problem is an oblique dipole-wall collision experiment in a square
bounded geometry Fig. 1c. The benchmark computation for this case is conducted with Chebyshev expansions in both directions [6]. For details of the initial
conditions and setup of the simulations see Clercx & Bruneau [5]. Fig. 2 show the
result of a normal dipole-wall experiment at $Re = 1000$. The Reynolds number
Re is based on the total kinetic energy of the flow $E(t) = \frac{1}{2} \int_{-1}^{1} \int_{-1}^{1} \mathbf{u}^2(\mathbf{x}, t) dx dy$,
and the half width of the domain $W = 1$. As the dipole impinges the wall Gibbs
oscillations in the vorticity isolines become apparent. This can be related to steep
gradients of the velocity inside the porous obstacle. Note that only one half of the
domain is shown because of the symmetry of a normal dipole-wall collision. The
oscillations are more pronounced near the wall than in the interior of the flow
domain. As the vortex moves into the interior of the flow domain at $t = 0.5$ the
wiggles disappear (see Fig. 2d). This indicates that the observed oscillations do
not have a serious dynamical effect on the evolution of the dipole-wall collision.
Fig 2 b,e show the result of a post-processing or smoothing technique developed
by Tadmor & Tanner [12]. The mollified result in Fig. 2b shows that the Gibbs
oscillations can safely be removed from the vorticity isolines. To determine the
accuracy of the scheme we consider the normalized L_2-error δ_N of the vorticity.
The error is computed in a box $\mathbf{x} \in [-0.99, 0.99]$ that is slightly smaller than

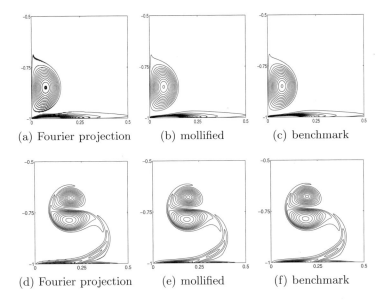

Fig. 2. Vorticity isolines of a normal dipole-wall collision with $Re = 1000$ at $t = 0.3$ (a,b,c) and $t = 0.5$ (d,e,f). 1364×1364 active Fourier modes, $\delta t = 10^{-5}$ and $\epsilon = 2.5 \times 10^{-5}$ are used. Benchmark computation is conducted with 1024 Chebyshev modes and 2048 active Fourier modes and $\delta t = 10^{-5}$. Contour levels are drawn for -270..,-50,-30,-10,10,30,50,..270.

the computational domain $\mathbf{x} \in [-1, 1]$. This is motivated by the fact that the post-processing procedure of Tadmor & Tanner [12] is only second order in the close vicinity of a discontinuity, but is higher-order accurate at a sufficient distance. From Fig. 3 it can be deduced that the truncation error of the Fourier projection shows only first order behavior in case Gibbs oscillations are present around $t = 0.3$. The mollification procedure of Tadmor & Tanner [12] recovers, however, a higher-order accuracy rate of the Fourier spectral scheme. After the first collision around $t = 0.5$ the Fourier projection without mollification shows a higher-order decay rate of the error in the vorticity. Mollification slightly improves the accuracy see Fig. 3b. An important issue is to check if it is possible

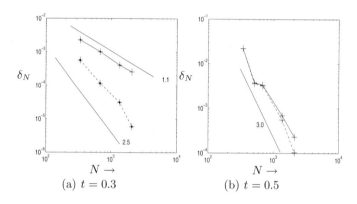

(a) $t = 0.3$ (b) $t = 0.5$

Fig. 3. Truncation error δ_N computed with respect to the highest resolution run with $N = 2730$ active Fourier modes in both directions. The time step $\delta t = 2 \times 10^{-5}$ is fixed for all runs. The error of the Fourier projection (solid) and the mollified result (dashed).

to find a balance between the truncation error and the penalization error. Fig. 4 shows a set of computations where a balance between both error sources is achieved. By choosing an appropriate number of active Fourier modes N and value of the penalization parameter ϵ, the vorticity isolines coincide (see Fig. 4c). By keeping ϵ fixed and increasing N until a saturation level is reached in the error in the vorticity versus the high-resolution benchmark computation one obtains a measure for the penalization error. The decay rate (not shown here) is proportional to $\epsilon^{0.7}$, which is consistent with the theoretical upperbound proportional to $\sqrt{\epsilon}$ derived by Carbou & Fabrie [4].

To increase the complexity of the flow problem we consider an oblique collision in the square bounded geometry Fig. 1c at a higher Reynolds number $Re = 2500$. For this problem it is necessary to perform mollification in both directions. Fig. 5 shows that the Fourier spectral scheme with post-processing is able to compute the isolines of the small-scale structures in this flow problem correctly as well. Some global measures are presented in Fig. 6. The total enstrophy $Z(t)$ is the L_2-norm of the vorticity in the flow domain Ω_f. The angular momentum of the

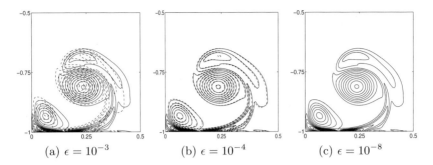

(a) $\epsilon = 10^{-3}$ (b) $\epsilon = 10^{-4}$ (c) $\epsilon = 10^{-8}$

Fig. 4. Contour plots (dashed) of the vorticity after the second collision at $t = 0.8$ for different values of the penalization parameter ϵ with respect to the benchmark simulation (solid). Contour levels are drawn for -270,-250,...,-50,-30,-10,10,30,50,..,250, 270. Number of active Fourier modes $N = 682$ (a), $N = 1364$ (b) and $N = 2730$ (c). Benchmark computation (solid) is conducted with 1024 Chebyshev modes and 2048 active Fourier modes and $\delta t = 10^{-5}$.

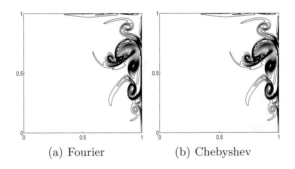

(a) Fourier   (b) Chebyshev

Fig. 5. Vorticity isolines of an oblique dipole-wall collision at $t = 0.6$ obtained with a Fourier spectral method with volume-penalization with 2730 active Fourier modes and $\delta t = 2 \times 10^{-5}$ (a) Benchmark computation conducted with 640 Chebyshev modes in both directions and $\delta t = 1.25 \times 10^{-5}$ (b) Contour levels are drawn for -270,-250,...,-50,-30,-10,10,30,50,..,250, 270

flow is computed with respect to the center of the container. It is found that the maximum error in the total enstrophy is smaller than one percent. Recall that the quality of post-processing procedure of Tadmor & Tanner [12] is only second order in the vicinity of the wall and cannot be applied on the wall itself. Therefore the accuracy in the total enstrophy is slightly limited. Clercx & Bruneau [5] found that is difficult to achieve convergence in the total angular momentum, especially after the second collision. Fig. 6b shows a good agreement between the high-resolution benchmark computation with 640 Chebyshev modes in both directions and the Fourier spectral computation with $N = 2730$ active Fourier modes. Note that the Fourier spectral method corresponds to an equidistant grid, while the Chebyshev method uses a Gauss-Lobatto grid that is strongly

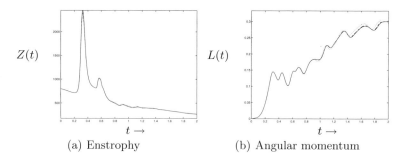

(a) Enstrophy   (b) Angular momentum

Fig. 6. Total enstrophy $Z(t)$ and angular momentum $L(t)$ with respect to the center of the domain versus time. Chebyshev benchmark computation 640 Chebyshev modes and $\delta t = 1.25 \times 10^{-5}$ (solid), Fourier spectral computation with volume-penalization obtained with $N = 1364$ (dots) and $N = 2730$ active Fourier modes (dashed). For both runs $\delta t = 2 \times 10^{-5}$ and penalization parameter $\epsilon = 10^{-8}$ is used.

refined in the corners of the domain (see Ref. [6]). Therefore it is not suprising that the required number of active Fourier modes is larger than the number of Chebyshev modes to achieve mode convergence for this specific problem.

4   Conclusion and Discussion

The numerical results of a normal and an oblique dipole-wall collision demonstrate that it is possible to conduct stable and accurate Fourier spectral computations using volume-penalization. Gibbs oscillations are present in the Fourier projections but are not dynamically active. In addition, the accuracy can be recovered using post-processing. Similar results are reported for Fourier spectral computations of other non-smooth problems (see Ref. [7]). For instance on the simulation of shock development in the 2D Euler equations. In our case it is, however, not necessary to apply any artificial viscosity to stabilize the scheme. The penalization error can be controlled by an appropriate choice of the penalization parameter. The theoretical upperbound of the penalization error obtained by Carbou & Fabrie [4] proportional to $\sqrt{\epsilon}$ is confirmed. It actually scales slightly better proportional to $\epsilon^{0.7}$ for the dipole-wall problem. This is a remarkable result since the asymptotically thin boundary layer that appears inside the obstacle is unresolved in our simulations. This result may be related to the formal expansion of penalized Navier-Stokes solution (see Ref. [4] for details). Only the higher-order terms in the $\sqrt{\epsilon}$ expansion rely on the details of the penalization boundary layer. As a consequence, a $\sqrt{\epsilon}$ accuracy of the penalized Navier-Stokes equations is possible without computing the boundary layer components. In our opinion the combination of Fourier spectral methods and volume-penalization can be useful to pursue DNS of turbulence in complex geometries. Note that this can easily be achieved by a different choice of the mask function.

References

1. Angot, P., Bruneau, C.-H, Fabrie, P.: A penalization method to take into account obstacles in viscous flows, *Numer. Math.* **81** (1999), 497
2. Arquis, E, Caltagirone, J.P.: Sur les conditions hydrodynamique au voisinage d'une interface milieu fluide-milieu poreux: application à la convection naturelle *C. R. Acad. Sci. Paris*, **299**, Série II, 1-4 (1984)
3. Canuto, C., Hussaini, M.Y., Quarteroni,A., Zang, T.A.: Spectral Methods in Fluid Dynamics Springer-Verlag, Berlin (1987).
4. Carbou, G., Fabrie, P.: Boundary layer for a penalization method for viscous incompressible flow, *Adv. Differential Equations* **8** (2003) 1453
5. Clercx, H.J.H., Bruneau, C.-H : The normal and oblique collision of a dipole with a no-slip boundary, *Comput. Fluids* **35** (2006) 245
6. Clercx, H.J.H.: A spectral solver for the Navier-Stokes equations in the velocity-vorticity formulation for flows with two non-periodic direction, *J. Comput. Phys.* **137** (1997) 186.
7. Gottlieb, D., Gottlieb, S.: Spectral methods for discontinuous problems. In Griffiths, D.F., Watson, G.A. (eds.): *Proc. 20th Biennial Conference on Numerical Analysis*, University of Dundee, (2003) 65
8. Kramer, W: Dispersion of tracers in two-dimensional bounded turbulence. *Ph.D. thesis*, Eindhoven University of Technology, The Netherlands (2007)
9. Keetels G.H., D'Ortona, U., Kramer, W., Clercx, H,J,H, Schneider, K., van Heijst, G.J.F: Fourier spectral and wavelet solvers for the incompressible Navier-Stokes equations with volume-penalization: convergence of a dipole-wall collision. submitted to *J. Comput. Phys.*
10. Kevlahan, N.K.R., Ghidaglia J.-M.: Computation of turbulent flow past an array of cylinders using spectral method with Brinkman penalization. *Eur. J. Mech. B-Fluid.* **20** (2001) 333
11. Schneider, K.: The numerical simulation of transient flow behaviour in chemical reactors using a penalization method. *Comput. Fluids* **34** (2005) 1223
12. Tadmor, E., Tanner, J.: Adaptive mollifiers for high resolution recovery of piecewise smooth data from its spectral information. *Found. Comput. Math.* **2** (2002) 155

High Quality Surface Mesh Generation for Multi-physics Bio-medical Simulations

Dominik Szczerba, Robert McGregor, and Gábor Székely

Computer Vision Lab, ETH, CH-8092 Zürich, Switzerland
domi@vision.ee.ethz.ch

Abstract. Manual surface reconstruction is still an everyday practice in applications involving complex irregular domains, necessary for modeling biological systems. Rapid development of biomedical imaging and simulation, however, requires automatic computations involving frequent re-meshing of (r)evolving domains that human-driven generation can simply no longer deliver. This bottleneck hinders the development of many applications of high social importance, like computational physiology or computer aided medicine. While many commercial packages offer mesh generation options, these depend on high quality input, which is rarely available when depending on image segmentation results. We propose a simple approach to automatically recover a high quality surface mesh from low-quality, oversampled and possibly non-consistent inputs that are often obtained via 3-D acquisition systems. As opposed to the majority of the established meshing techniques, our procedure is easy to implement and very robust against damaged or partially incomplete, inconsistent or discontinuous inputs.

1 Introduction

Generating a mesh is a necessary pre-condition when obtaining numerical solutions of partial differential equations. An adequate mesh highly impacts both the accuracy and the efficiency of numerical procedures. Since the inception of the finite element method dating back to the middle of the last century, automatic mesh generation with sufficient quality over an arbitrary domain has remained a central topic of intensive research, without being able to reach a fully satisfying solution up to now. Even though a tremendous number of approaches has been described, at the end none of them offers truly interaction-free, general-purpose processing. From the perspective of half a century's work it seems clear that mesh generation, despite its scientific context and origin, bears every sign of artistic sculpturing that escapes automation due to complex decision making and continuous adaptation during the creation process. Interactive meshing is therefore an everyday practice in applications involving complex static domains. For man-made objects, usually emerging from CAD applications, even a few almost-automatic volume meshing methods are available, like the advancing front technique [1], [2]. Such methods usually work very well for geometries represented as either a constructive solid geometry (CSG), non-uniform rational B-spline (NURB) patches, or by an already existing surface mesh of sufficient quality. The rapid development of biomedical imaging and subsequent simulation, however, requires computations involving frequent re-meshing of largely (r)evolving anatomical domains that human-driven

Y. Shi et al. (Eds.): ICCS 2007, Part I, LNCS 4487, pp. 906–913, 2007.

generation can simply no longer deliver. In these applications the previously listed requirements hardly ever hold: a constructive geometry description of an anatomical domain may not make any sense at all, NURB patches are obviously not available, and surface meshes obtained directly from segmentation are generally of very low quality, oversampled, often with broken topology. This serious bottleneck hinders development on many domains of high social importance like computational physiology or computer aided medicine. The computational pipeline required for such applications can be sketched as follows: data acquisition $\rightarrow$ domain segmentation $\rightarrow$ surface representation $\rightarrow$ volume meshing $\rightarrow$ discretization of governing equations $\rightarrow$ solution. The data, acquired via imaging techniques like MRI, CT, US, laser scanning, etc., becomes input to a segmentation procedure (see e.g. [3]). Quite rudimentary algorithms, like the marching cube method [4] are then used to represent the identified objects by volumetric or surface meshes, which are, however, too low quality to be directly used for numerical simulations. High quality surface reconstruction suitable as input to volume meshing algorithms has therefore generated a lot of interest in computer science.

2   Related Work

Several established mesh simplification algorithms are available [5], [6]. The techniques include merging coplanar facets, decimation, re-tiling, local re-meshing, vertex clustering, energy function optimization, wavelet-based methods and many other. They are very useful in downsizing of oversampled input meshes, such methods do not aim, however, to improve input mesh quality.

Automatic surface mesh generation from constructive solid geometry (CSG) or stereo-lithography (STL) representations is largely used in mechanical engineering (e.g., [7], [8], [9]). To derive a high quality surface mesh for such representations, information about the object's boundary elements (faces, vertices) must be first extracted. This works very well for geometries which can be decomposed into simple primitives, however, the required boundary evaluation using e.g. Bézier patches or algebraic surfaces does not always produce the expected results for inherently irregular and unpredictable physiological domains. Similarly, methods based on feature recognition (e.g., [10], [11]) or surface parameterization (e.g., [12], [13]) suffer from the same fundamental limitation.

A robust and almost automatic surface reconstruction can be achieved using knowledge based methods, as demonstrated by e.g. [14]. The approach to blood vessel meshing relies on medial axis extraction and subsequent insertion of boundary faces along the sweeping axes. Such methods can be attractive for specific cases where *a priori* assumptions about the domain's shape can be made, however, are far from what can be called a versatile approach.

The family of algorithms based on computational geometry (e.g., [15], [16], [17]), Delaunay refinement (e.g., [18], [19]) or optimization principles (e.g., [20]) is demonstrated to often provide high quality outputs, in many cases with guarantees on convergence and for lower limits on element quality. These sophisticated methods are generally very sensitive to the consistency, continuity and differentiability of the input. In practice they often fail on real-life biomedical data due to precondition violations. In addition they are usually quite difficult to implement.

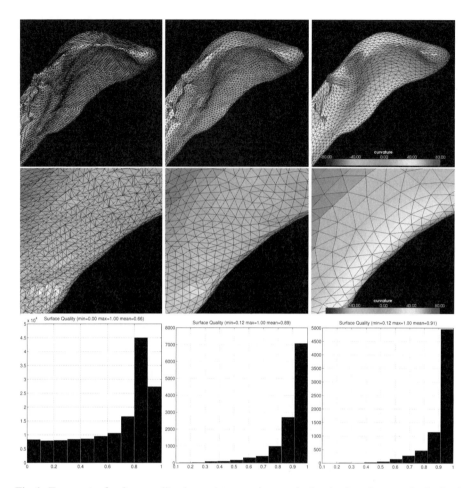

Fig. 1. Fragment of a low quality, inconsistent and excessively sized surface mesh obtained from the segmentation of a human lateral ventricle (left), its high quality uniform (middle) and curvature-adapted reconstruction (right). The upper and middle row shows patches of the reconstructed surface with different magnifications, while histograms of the corresponding triangle quality (normalized radii-ratios) are presented by the lower figures.

The number of approaches based on smoothing is too large to provide a comprehensive summary in this overview. The basic idea behind them is to relax the nodal positions such that the elements eventually become evenly sized. Some more advanced versions attempt to eliminate the shrinking effect (e.g. [21]) by e.g. applying a band pass filter or simply projecting the relaxed nodes back to the original boundary. This often works sufficiently well if a topologically correct input mesh is available for the relaxation. In general, these methods are easy to implement but suffer from loss of high frequency features. In addition, they do not offer adaptive refinement options, which are necessary for efficient numerical solution procedures.

The work reported in [22] demonstrates the ability of mass-spring dumper models to generate high quality tetrahedral meshes by relaxing pre-refined and pre-compressed mass-spring connections, which, while expanding, naturally fill the volume constrained by the boundary. The method is efficient and robust, if a (very expensive) consistent volumetric representation of the input is available. Shortcomings include discontinuous refinement and resulting tetrahedralizations not being of the Delaunay type. This method is similar to ours in that it relies on physical and not on mathematical principles. The important differences are that 1) we use a very cheap, possibly inconsistent input represented as a polygonal mesh and not volumetric data; 2) we allow for smooth refinement of the re-generated mesh and 3) we produce topology conforming to the Delaunay condition.

To complete the survey, mesh-free approaches have also to be mentioned, which eliminate the discretization problems inherent to meshes by fundamentally not relying on them. Even though it sounds very attractive, these techniques are at their infancy while still relying on some local topology. There are also somewhat exotic approaches based on fractal geometries, chaos theories or neural networks that will not be discussed here.

3   Method

We propose an iterative procedure as follows: 1. initialization; 2. removal of outside elements (if any); 3. calculation of reaction forces; 4. time integration; 5. correction at boundaries; 6. convergence check, eventual termination. 7. loop back to 2. The procedure is initialized by a set of points filling the bounding box of the input surface (1). The loop begins by the removal of the elements found outside the object (2). In subsequent steps, a physical equilibrium between the remaining points is sought by enforcing a prescribed distance between points (3,4):

$$M\frac{\partial^2 r}{\partial t^2} + D\frac{\partial r}{\partial t} = F_r,$$

with F_r being the reaction forces and incorporating an internal energy dissipation term regulated by D for stability. Special care is needed next to the object's boundary: the points there must be magnetically snapped to it, allowed only to slide along the surface but not to detach (5). We have tested both mass-spring dumpers and charged particle clouds for the actual implementation and did not observe any qualitative differences. It is only important to note that the structure must be initially preloaded with some potential energy such that expansion forces act already during the early iteration stages. Also, we observed that somewhat better results are achieved using magnetic and not Hookean reaction forces. After some iterations (7), this procedure results in evenly distributed points that very closely follow the surface features (6). Topology provided by a robust Delaunay connectivity algorithm [23] produces a tetrahedral mesh of surprisingly good quality, but not free from some badly conditioned elements (slivers). Their removal in general is far from trivial, but not impossible [24]. In fig. 2 we demonstrate that with an additional effort it is possible to repair and re-use those tetrahedral meshes. A detailed description of how we deal with this issue is, however, beyond the scope of this paper.

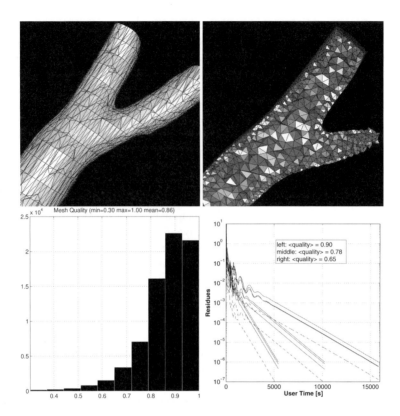

Fig. 2. Meshes generated by the algorithm proposed by this paper. **Top:** a surface mesh of a human abdominal aortic bifurcation resulting directly from the segmentation of the underlying MRI data (**left**) and the boundary layer generated by an adaptive volumetric mesh refinement (**right**). **Bottom, left:** the quality histogram (normalized radii-ratios) for the resulting mesh. **Bottom, right:** the convergence rates of subsequent fluid simulations performed on different meshes. There are 3 simulations for different mesh qualities and there are 4 curves for each of them: three momentum residues (solid lines) and a mass residue (dashed line).

However, often only a surface mesh is needed, e.g., for conversion to NURB patches or simply to generate other volume elements then tetrahedrons. In such cases, a surface mesh can be easily extracted by preserving only the triangles having one adjacent tetrahedron and removing all others (fig 1).

The procedure sketched above results in a high quality uniform mesh closely following features of the input surface. However, to simulate large domains, adaptive refinement is often sought to reduce the solution accuracy in the regions out of interest, and this way decrease the load on the underlying PDE solver by reducing the number of finite elements used. Such adaptation can be achieved by modifying point-point interactions to make the regional equilibrium radius correspond to the locally requested element size instead of being a global constant. The refinement map can be arbitrarily specified by the user or automatically derived from some selected features of the input, like the distance from the boundary or the local curvature. If curvature or any other feature is used, which is only defined on the object's surface, its effect must be propagated into the interior elements e.g. by diffusion:

$$\frac{\partial c}{\partial t} = div(grad(c)) + R_c,$$

with c the surface scalar to be distributed. The reaction term, R_c, is necessary to control the magnitude of the resulting gradient of c inside the volume to ensure smooth distribution of the refinement map. There are two possible choices to integrate this process into the discussed procedure: a *multi-grid* or a *coupled* method. In the first case a desired refinement map is provided on an existing, previously generated, uniform mesh and will be continuously interpolated onto a converging refined mesh. This gives very good control of the adaptive refinement process, but requires additional implementation efforts and computing time. The second possible option is to formulate a *coupled* problem, where the distance/curvature based refinement function is evolving simultaneously with the formation of the adaptive mesh, while using its actual state for the numerical representation of the map. This is much easier to implement, much faster to compute and surprisingly stable, but does not allow one to check the refinement map beforehands. Either of these methods will result in a smoothly refined mesh, with an adaptive element size. Note that the technique does not require any pre-existing expensive voxel-based (e.g. level set) description of the input surface. As a consequence, large inputs with very fine details can be resolved on a standard PC, whereas volumetric storage of all necessary transient information about a modest 512^3 domain results in several gigabytes of memory usage.

This method is nearly automatic with minor manual corrections eventually needed to enforce specific information about complex features to be followed for adaptive meshing. The disadvantage of the technique in the first place is its inferior speed inherent to any methods based on solving differential equations. In addition, we are not aware of either mathematically proven convergence behavior or lower limits on element quality. In practice, however, the algorithm always generated meshes with the desired properties and quality, as demonstrated by the examples on the figures.

4 Conclusions

We have presented an automatic technique for high quality reconstruction of oversampled, low-grade and possibly topologically inconsistent surface meshes that can be an attractive alternative to existing approaches. Our procedure is easy to implement, and is very robust against damaged or partially incomplete, inconsistent or discontinuous input. Compared to smoothing-based approaches our technique does not result in any feature loss and naturally offers refinement options (distance to selected regions, curvature, etc.). The method is automatic, manual input may only be necessary if the map used to govern adaptive refinement cannot be calculated from the input data (including the original, non-segmented acquisition) alone. The obtained high quality triangular meshes can become input to other applications like NURB patching or other volumetric meshing procedures. If tetrahedral meshes are sought, however, the volumetric mesh generated as a by-product of our method could be rectified by a sliver removal procedure.

Fig. 2 demonstrates the effect of volumetric mesh quality on the performance of the subsequent numerical simulations. The human abdominal bifurcation was segmented out of raw image data and three meshes with different average qualities were produced using the presented method. The meshes were subsequently used in numerical flow computations using a momentum-pressure PDE solver. Using tetrahedrons of average quality around 0.9 resulted in 3 times faster convergence when compared to average quality around 0.65. In case when only a surface mesh is sought for further processing, it is crucial to note that its quality will strongly influence the quality of the results, as is the case with procedures like the advancing front method.

Obviously, the presented method is not limited to bio-medical applications. Due to its strong feature preserving nature it can be used in general engineering applications where e.g. sharp edges or point-like singularities need to be preserved. The major disadvantages of our technique are inferior speed and missing theoretical bounds on element quality. However, we did not detect any quality defects on the numerous meshes we have generated up to now.

Acknowledgments

This work is a part of the Swiss National Center of Competence in Research on Computer Aided and Image Guided Medical Interventions (NCCR Co-Me), supported by the Swiss National Science Foundation.

References

1. Schöberl, J.: Netgen an advancing front 2d/3d-mesh generator based on abstract rules. Computing and Visualization in Science **V1**(1) (1997) 41–52
2. Tristano, J., Owen, S., Canann, S.: Advancing front surface mesh generation in parametric space using a reimannian surface definition (1998)
3. Yushkevich, P.A., Piven, J., Cody Hazlett, H., Gimpel Smith, R., Ho, S., Gee, J.C., Gerig, G.: User-guided 3D active contour segmentation of anatomical structures: Significantly improved efficiency and reliability. Neuroimage (2006)
4. Lorensen, W.E., Cline, H.E.: Marching cubes: a high resolution 3d surface construction algorithm. Computer Graphics (ACM) **21**(4) (1987) 163–169
5. Kim, S.J., Kim, C.H., Levin, D.: Surface simplification using a discrete curvature norm. Computers & Graphics **26**(5) (2002) 657–663
6. Balmelli, L., Liebling, T., Vetterli, M.: Computational analysis of mesh simplification using global error. Computational Geometry **25**(3) (2003) 171–196
7. Boender, E., Bronsvoort, W.F., Post, F.H.: Finite-element mesh generation from constructive-solid-geometry models. Computer-Aided Design **26**(5) (1994) 379–392
8. Bechet, E., Cuilliere, J.C., Trochu, F.: Generation of a finite element mesh from stereolithography (stl) files. Computer-Aided Design **34**(1) (2002) 1–17
9. Rypl, D., Bittnar, Z.: Generation of computational surface meshes of stl models. Journal of Computational and Applied Mathematics **192**(1) (2006) 148–151

10. Cuilliere, J.C., Maranzana, R.:   Automatic and a priori refinement of three-dimensional meshes based on feature recognition techniques. Advances in Engineering Software **30**(8) (1999) 563–573
11. Chappuis, C., Rassineux, A., Breitkopf, P., Villon, P.: Improving surface meshing from discrete data by feature recognition. Engineering with Computers **V20**(3) (2004) 202–209
12. Lee, C.K., Hobbs, R.E.: Automatic adaptive finite element mesh generation over rational b-spline surfaces. Computers & Structures **69**(5) (1998) 577–608
13. Hormann, K., Labsik, U., Greiner, G.: Remeshing triangulated surfaces with optimal parameterizations. Computer-Aided Design **33**(11) (2001) 779–788
14. Antiga, L., Ene-Iordache, B., Caverni, L., Paolo Cornalba, G., Remuzzi, A.: Geometric reconstruction for computational mesh generation of arterial bifurcations from ct angiography. Computerized Medical Imaging and Graphics **26**(4) (2002) 227–235
15. Frey, P.J., Borouchaki, H.: Geometric surface mesh optimization. Computing and Visualization in Science **V1**(3) (1998) 113–121
16. Garimella, R.V., Shashkov, M.J., Knupp, P.M.: Triangular and quadrilateral surface mesh quality optimization using local parametrization. Computer Methods in Applied Mechanics and Engineering **193**(9-11) (2004) 913–928
17. Montenegro, R., Escobar, J., Montero, G., Rodríguez, E.: Quality Improvement of Surface Triangulations. (2006)
18. Oudot, S., Rineau, L., Yvinec, M.: Meshing volumes bounded by smooth surfaces. In: Proc. 14th International Meshing Roundtable. (2005) 203–219 meshing the volume bounded by the surfaces.
19. Dey, T., Li, G., Ray, T.: Polygonal Surface Remeshing with Delaunay Refinement. (2006)
20. Escobar, J.M., Rodriguez, E., Montenegro, R., Montero, G., Gonzalez-Yuste, J.M.: Simultaneous untangling and smoothing of tetrahedral meshes. Computer Methods in Applied Mechanics and Engineering **192**(25) (2003) 2775–2787
21. Cebral, J.R., Löhner, R.: From medical images to anatomically accurate finite element grids. International Journal for Numerical Methods in Engineering **51**(8) (2001) 985–1008
22. Marroquim, R., Cavalcanti, P.R., Esperança, C., Velho, L.: Adaptive multi-resolution triangulations based on physical compression. Communications in Numerical Methods in Engineering **21**(10) (2005) 571–580
23. Barber, C.B., Dobkin, D.P., Huhdanpaa, H.: The quickhull algorithm for convex hulls. ACM Transactions on Mathematical Software **22**(4) (1996) 469–483
24. Labelle, F.: Sliver removal by lattice refinement. In: SCG '06: Proceedings of the twenty-second annual symposium on Computational geometry, New York, NY, USA, ACM Press (2006) 347–356

Macro-micro Interlocked Simulation for Multiscale Phenomena

Kanya Kusano, Shigenobu Hirose, Toru Sugiyama, Shinichiro Shima,
Akio Kawano, and Hiroki Hasegawa

The Earth Simulator Center,
Japan Agency for Marine-Earth Science and Technology
3173-25 Showa-machi, Kanazawa-ku Yokohama, Kanagawa 236-0001,
Japan
kusano@jamstec.go.jp
http://www.es.jamstec.go.jp/esc/research/Holistic/index.en.html

Abstract. A new methodology for the simulation of multiscale processes, called Macro-Micro Interlocked (MMI) Simulation, is introduced. The MMI simulation is carried out by the two-way connection of different numerical models, which may handle macroscopic and microscopic dynamics, respectively. The MMI simulation are applied to several multiscale phenomena, for instance, cloud formation, gas detonation, and plasma dynamics. The results indicate that the MMI simulation provide us an effective and prospective framework for multiscale simulation.

Keywords: simulation, macro-micro Interlocked, multiscale, cloud formation, detonation, aurora, plasma, fluid, super-droplet, DSMC.

1 Introduction

Multi-scale phenomena, in which the elementary process of micro-scale and the system evolution of macro-scale are tightly connected to each other, quickly comes up as a crucial issue in any state of the art research fields, such as material science, plasma physics, chemistry, astrophysics, geo-science, bio-science and so on. Computer simulation is usually performed on basis of the theoretical description like the partial differential equation, and thus the applicability of simulation model has to be constrained by the limitation of the basic theory behind. So far, the multi-scale simulations have been developed by the two different manners. The first is the extension of macroscopic simulation, in which the microscopic effects are included as phenomenological parameters, just like the anomalous resistivity caused by micro-scale turbulence in high-temperature plasmas and the bulk cloud parameterization in atmospheric global circulation model. However, since the phenomenological parameterization is not guaranteed to work well in unbeknown circumstances, the applicability of such a way would be restricted particularly as an effective tool for prediction of unknown phenomena.

Y. Shi et al. (Eds.): ICCS 2007, Part I, LNCS 4487, pp. 914–921, 2007.

Another approach is the large-scale microscopic simulation, whereby the macroscopic phenomena are attempted to be built from the elementary block of micro-scale. The large-scale molecular dynamics (MD) simulation is the typical example of that. However, it is computationally so demanding, that the practical application is hardly feasible despite of the incredible capability of the latest super-computers.

Macro-Micro Interlocked (MMI) Simulation is the new type of computation, which is recently introduced by Sato [1]. MMI simulation consists of the macro- and micro-simulation models, which are carried out simultaneously and interconnected to each other. The micro-model is performed under the environmental condition controlled by the macro-model, and it feeds back the microscopic information to the macro-model. The fundamental principle of MMI simulation is in common with that of the heterogeneous multiscale method (HMM) [2]. Furthermore, the way to interconnect macro- and micro-scales is similar to Equation-Free approach [3], although, in MMI simulation, we have to adopt some macro-scale model rather than using a coarse projective method. However, originally MMI simulation has been proposed as a computational algorithm suitable to the hardware of heterogeneous architecture, called MMI simulator, in which different types of architecture fitting respectively to the numerical models of macro- and micro-scales are interconnected. In particular, heterogeneous vector/scalar architecture is the most promising design of the MMI simulator, and we have developed several applications for that.

The objective of this paper is to demonstrate the applications of MMI simulation. For that, we will show the brief results of typical MMI simulations for cumulus cloud formation, gas detonation, and plasma dynamics in the following sections, and the prospects of the MMI simulation will be summarized in Sec.3.

2   Applications

2.1   Cloud Simulation

Although clouds play a crucial role in meteorological phenomena and climate change like global warming, the numerical modeling of cloud is not well established yet. The reason of that is attributed to the fact that the formation of clouds and the development of precipitation are essentially governed by mutiscale-multiphysics processes. The macro-scale processes such as the fluid motion of moist air associated with clouds is called "cloud dynamics", and the micro-scale processes such as the condensation and coalescence of water droplets are called "cloud microphysics". These two processes mutually affect each other. Numerical methods to simulate the interaction between them should be developed to understand and predict cloud-related phenomena.

Cloud dynamics model to describe the fluid motion of atmosphere has been well developed so far. However, it is still difficult to perform the accurate simulation of cloud microphysics, though several simulation methods, such as bulk parameterization and spectral (bin) methods have been proposed.

We recently developed a novel simulation model of cloud microphysics, named Super-Droplet Method (SDM) [4], which is a particle-based cloud microphysics model. The super-droplet is defined as computational particle representing multiple real droplets, which have common properties, *e.g.* position, velocity, cloud condensation nuclear (CCN), and electric charge. The motion of each super-droplet is calculated by the equation of motion or by the assumption that each droplet immediately attains terminal velocity. The condensation and the evaporation of droplets can be directly calculated based on Köhler's theory using the properties of super-droplet and the state variable of atmospheric environment. The coalescence of droplet is handled in stochastic manner.

The cloud dynamics is simultaneously calculated with SDM using the nonhydrostatic model equations,

$$\rho \frac{D\boldsymbol{U}}{Dt} = -\nabla P - (\rho + \rho_w)\boldsymbol{g} + \lambda \nabla^2 \boldsymbol{U}, \tag{1}$$

$$P = \rho R_d T, \tag{2}$$

$$\frac{D\theta}{Dt} = -\frac{L}{c_p \Pi} S_v + \kappa \nabla^2 \theta, \tag{3}$$

$$\frac{D\rho}{Dt} = -\rho \nabla \cdot \rho, \tag{4}$$

$$\frac{Dq_v}{Dt} = S_v + \kappa \nabla^2 q_v, \tag{5}$$

where ρ is the density of moist air, ρ_w the density of liquid water, $\boldsymbol{U}$ the wind velocity, P the pressure, λ the viscosity, R_d the gas constant for dry air, T the temperature, θ the potential temperature, L the latent heat, q_v the mixing ratio of vapor, S_v the source of water vapor, and Π is the Exner function, respectively. The density of droplets

$$\rho_w(\boldsymbol{x}, t) = \sum_i \xi_i m_i(t) w(\boldsymbol{x}, \boldsymbol{x}_i),$$

and the conversion ratio between vapor and liquid

$$S_v(\boldsymbol{x}, t) = -\frac{1}{\rho(\boldsymbol{x}, t)} \frac{\partial \rho_w(\boldsymbol{x}, \boldsymbol{t})}{\partial t},$$

are sent to the fluid model, and used in the source terms, where i is the index of super-droplet, ξ_i the multiplicity of super-droplet, m_i the mass of droplet, and w is the shape function of super droplet.

SDM is the hybrid model, in which the particle-based direct simulation Monte-Carlo (DSMC) model and the fluid model are coupled. By evaluating the integrated squared error for the distribution function in parameter space of droplet properties, we can show that the computational efficiency of SDM is higher than the spectral (bin) method, when more than four properties of cloud droplet are taken into account. (Refer to [4] for further detail.) It implies that the particle-continuum coupled model is more effective for more complicated system, compared to the conventional methods.

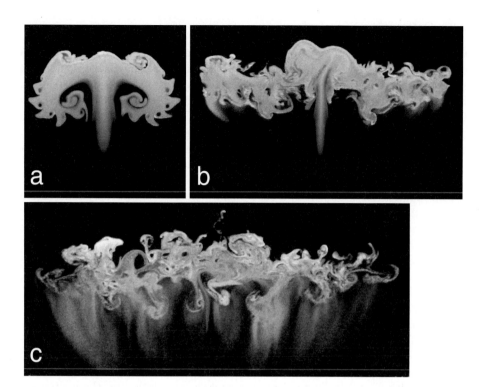

Fig. 1. The distribution of cloud droplets calculated by the coupled simulation of the super-droplet method and non-hydrostatic model. Color represents the size of droplet. (a) At $t = 960$sec after simulation starts, upwelling initiated by warm bubble causes the condensation growth of cloud droplets (white particles). (b) At $t = 1188$sec, droplets bigger than 10 microns in radius (blue particles) further grow due to collision-coalescence. (c) At $t = 1512$sec, droplets (yellow particles), which are bigger than 100 micron, are precipitated as rain-drops. Inside the cloud, turbulence-like structure is strongly driven by cumulus convection.

In order to demonstrate the feasibility of SDM, it has been applied to the simulation of cloud formation and precipitation in maritime cumulus. The simulation system is given by 2-dimensional $x - z$ domain just for simplicity. The initial state consists of un-saturated stratified layer, which is slightly unstable only under 2km in altitude, and number of tiny droplets, which contain soluble substance as Cloud Condensation Nuclear (CCN), are uniformly distributed in the entire domain. As shown in Fig.1, the results indicate that the particle-continuum coupled model may work to simulate the whole process from cloud formation to precipitation without introducing any empirical parameterizations. The new model may provide a powerful tool for the study of cloud-related various problems, although the predictability of SDM should be evaluated more carefully.

2.2   Detonation Simulation

Combustion fluid dynamics is a typical subject of multiscale simulation, in which chemical reaction is mutually interacted with the macroscopic flow dynamics. In the conventional methods, the reaction is treated by the Arrhenius rate equation. However, the reaction rate of the equation must be derived from the distribution function assuming local thermal equilibrium (LTE). Although the assumption could be satisfied in normal condition of fluids, when the local Knudsen number, defined as the ratio of mean free path to characteristic length scale, is larger than 0.01, the assumption of LTE may not be valid. In the case, the flow should be treated as rarefied one and we have to solve the Boltzmann equation.

Especially, detonation, which is sustained by shock wave driven by combustion wave, is the case, because the thickness of detonation front may be comparable to the mean free path. It implies that the Arrhenius rate equation may not be valid on the detonation front.

So, we have developed a novel method for simulation of combustion by connecting a microscopic molecular model and a macroscopic continuum model, those are based on the Boltzmann equation and the Navier-Stokes equation, respectively. We adopted non-steady DSMC method [5] for the molecular model, and the continuum model is carried out by the HLLC method.

Our detonation model is an extension of Hybrid Continuum-Atomistic Simulation, which has been quickly developed by many authors [6,7,8]. The algorithm to connect the molecular model and the continuum model is summarized as follows: (1) Gradient of pressure is monitored during the simulation by the continuum model, and the molecular model is embedded in the region where the steepness exceed some threshold and the continuum model is failed. (2) Interlocking layers are laid around the outer-most region of the molecular domains, where particles are generated to let the distribution function be able to reproduce the macroscopic variables, density, velocity and pressure. Numerical flux on the outer-most boundary of the continuum domain is calculated from the particle motion.

Some test results of the two-dimensional detonation simulation are shown in Fig.2. It represents the distribution of pressure on three different snapshots. The detonation front bounded by dotted line corresponds to the domain, where the particle-based model is adopted. The smooth connection between fluid and particle-based models indicates that the continuum-atomistic simulation is applicable also to the combustion process.

2.3   Plasma Simulation

Plasma inherently forms a multi-scale system, in which there are different characteristic scales related to electron and ions. The key issue in multiscale plasma processes is how macroscopic magnetohydrodynamics (MHD) is related to microscopic particle kinetics. For instance, in magnetic reconnection that is responsible for explosive energy release in high-temperature plasmas, the kinetics might be important especially in the restricted diffusion region, which is formed when anti-parallel magnetic fluxes collide each other [9].

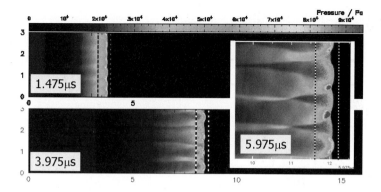

Fig. 2. The pressure distribution of the detonation simulation at three different snapshots. The particle-based domain, which corresponds to the region bounded by dotted lines, tracks the detonation front. The particle-based region at $5.975\mu s$ is zoomed-in on right subset.

So, we have developed the new algorithm for such a multiscale plasma dynamics. Our model is constituted by the connection of the particle-in-cell (PIC) model and the magnetohydrodynamic (MHD) model, the former of which is embedded in the MHD simulation domain[10]. This MHD-PIC interlocked model is resembled to the hybrid continuum-atomistic simulation, which was explained in the previous subsection. However, it should be mentioned that, in plasma simulation, several characteristics described by PIC model are negated in continuum (MHD) model, in contrast to the hydrodynamic simulations. So, in order to make the smooth interlocking between MHD and PIC models, we have to introduce some special filter, which passes only the MHD mode from the PIC domain to the MHD domain.

The filtering process is actualized by the following procedure. First, the MHD simulation is advanced from time t_n to $t_n + \Delta t$ in the whole domain including the PIC domain as the predictor phase (Fig.3a). Second, the PIC simulation is advanced to $t_n + \Delta t$ only in the PIC domain, in which the boundary condition is given by the MHD simulation of the predictor phase. Note that the time step in PIC is much shorter than that of MHD, and the boundary condition in each PIC step has to be interpolated. The electric field and the electric current density of the PIC result, $\boldsymbol{E}_{PIC}$ and $\boldsymbol{J}_{PIC}$, and those of the MHD predictor results, $\boldsymbol{E}_{MHD}$ and $\boldsymbol{J}_{MHD}$, are mixed as following

$$\boldsymbol{E}_* = w(\boldsymbol{x})\boldsymbol{E}_{PIC} + (1 - w(\boldsymbol{x}))\boldsymbol{E}_{MID},$$

$$\boldsymbol{J}_* = w(\boldsymbol{x})\boldsymbol{J}_{PIC} + (1 - w(\boldsymbol{x}))\boldsymbol{J}_{MID},$$

where $w(\boldsymbol{x})$ is the weighting function that smoothly ramped from 0 to 1 on the periphery of the PIC domain, as shown in Fig.3b. Finally, using $\boldsymbol{E}_*$ and $\boldsymbol{J}_*$, the MHD simulation is advanced again from t_n to $t_n + \Delta t$. This is the corrector phase, and the procedure above is repeated until the calculation is finished. In

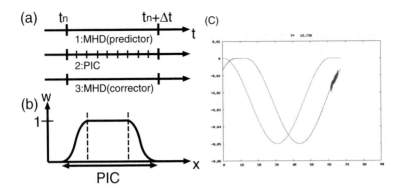

Fig. 3. (a) The algorithm for time advancement of the MMI plasma simulation. (b) The schematic diagram of spatial relation between the weighting function w and the PIC domain. (c) The result of the test simulation for Alfvén wave propagation.

this algorithm, the kinetic process comes out with $\boldsymbol{E}_{PIC}$ and $\boldsymbol{J}_{PIC}$, which are used instead of the generalized Ohm's law in the corrector phase.

Figure 3c shows the magnetic wave form in the test-simulation for the one-dimensional Alfvén wave propagation. Green and blue curves represent the initial state and the result of the MHD model after a while, where red line indicates the result of PIC, which is embedded in a small region. The wave form is scarcely affected by interlocking with the PIC model, as high-frequency modes appear in the PIC domain. It indicates that the particle-continuum connection model is applicable also to plasma simulations.

3   Summary

Three different applications for the MMI simulation are demonstrated. In any examples, the particle-continuum connection technique plays a crucial role. However, it is worthwhile to note that the particle-continuum connection in the cloud simulation is not based on the domain decomposition method unlike the other two applications of plasma and detonation, because cloud is ubiquitous phenomena and we cannot restrict the cloud domain into a limited region.

MMI simulation based on the particle-continuum connection technique might be suitable to MMI simulator of heterogeneous architecture. For instance, vector and scalar architectures fit for the algorithms of continuum and particle models, respectively. In fact, the parallelization with MPI is convenient for the implementation of MMI simulation, because the macro and micro models can be assigned to different ranks, and it could be easily ported also to the heterogeneous architecture system.

Although MMI simulator could be a potential tool for multiscale simulation, at least so far the interlocking algorithm has been developed almost in ad hoc manner. Therefore, it is not yet clear what is the best fit architecture for each

MMI algorithm, and more systematic research for the matching between the algorithm and the architecture of MMI simulator will be required.

Acknowledgments. This work is performed as a part of the program of "Development of Multi-scale Coupled Simulation Algorithm" in the Earth Simulator Center, JAMSTEC.

References

1. Sato, T.: Macro-Micro Interlocked Simulator, Journal of Physics: Conference Series SciDAC 2005 Scientific Discovery through Advanced Computing, Vol.16, (2005) 310–316
2. Weinan E, Engquist, B., Li, X., Ren, W., and Vanden-Eijnden,E.: Heterogeneous Multiscale Methods: A Review, Commun. Comput. Phys. 2, 3 (2007) 367-450.
3. Kevrekidis, I.G., Gear, C.W., Hummer, G.: Equation-free: The computer-aided analysis of complex multiscale systems Aiche J. 50, 7 (2004) 1346-1355.
4. Shima, S., Kusano, K, Kawano, A., Sugiyama, T., and Kawahara, S.: Super-Droplet Method for the simulation of Cloud and Precipitation; a Particle-Based Microphysics Model Coupled with Non-hydrostatic Model. submitted to J. Atm. Sci.
5. Bird, G.A.: Molecular Gas Dynamics and the Direct Simulation of Gas Flows. Clarendon Press, Oxford (1994).
6. Aktas, O., Aluru, N.R.: A Combined Continuum/DSMC Technique for Multiscale Analysis of Microfluidic Filters, J. Comp. Phys. 178, 2 (2002) 342–372.
7. Wijesinghe, H.S., Hadjiconstantinou, N.G.: Discussion of Hybrid Atomistic-Continuum Methods for Multiscale Hydrodynamics, Int. J. Multiscale Computational Engineering 2, 2 (2004) 189–202.
8. Schwartzentruber, T.E., Boyd, I.D.: A Hybrid Particle-Continuum Method Applied to Shock Waves. J. Comp. Phys. 215, 2 (2006) 402-416.
9. Kusano, K., Maeshiro, T., Yokoyama, T., & Sakurai, T.: The Trigger Mechanism of Solar Flares in a Coronal Arcade with Reversed Magnetic Shear. Astrophys. j. 610 (2004) 537–549
10. Sugiyama, T., Kusano, K.: Multiscale Plasma Simulation by the Interlocking of Magnetohydrodynamic Model and Particle-in-Cell Kinetic Model. submitted to J. Comp. Phys.

Towards a Complex Automata Framework for Multi-scale Modeling: Formalism and the Scale Separation Map

Alfons G. Hoekstra[1], Eric Lorenz[1], Jean-Luc Falcone[2], and Bastien Chopard[2]

[1] Section Computational Science, Faculty of Science, University of Amsterdam,
Kruislaan 403, 1098 SJ Amsterdam, The Netherlands
{alfons,lorenz}@science.uva.nl
http://www.science.uva.nl/research/scs
[2] Department of Computer Science, University of Geneva, 24, Rue du Général
Dufour, 1211 Geneva 4, Switzerland
{Bastien.Chopard,Jean-Luc.Falcone}@cui.unige.ch
http://spc.unige.ch/

Abstract. Complex Automata were recently proposed as a paradigm to model multi-scale complex systems. The concept is formalized and the scale separation map is further investigated in relation with its capability to specify the components of Complex Automata. Five classes of scale separation are identified, each potentially giving rise to a specific multi-scale modeling paradigm. A number of canonical examples are briefly discussed.

Keywords: Complex Automata, Cellular Automata, Multi-Scale Modeling, Scale Separation Map.

1 Introduction

Complex Automata (CxA) were recently proposed as a paradigm to model multi-scale complex systems [1]. The key idea is that a multi-scale system can be decomposed into N single-scale Cellular Automata (CA) that mutually interact across the scales. The decomposition is achieved by building a scale map on which each system can be represented as an area according to its spatial and temporal scales. Processes having well separated scales are easily identified as the components of the multi-scale model.

In this contribution we further elaborate on the CxA framework by first providing a formal definition. Next, we investigate in some detail the concept of the scale separation map, by taking into account the temporal and spatial scales that are covered by a single CA. We thus identify 5 classes of scale separation, each potentially giving rise to specific multi-scale modeling paradigms.

To illustrate our framework we discuss a number of canonical examples, demonstrating how to formulate them as a CxA, how the decomposition into a number of single scale CAs can be obtained, and how these systems are characterized by their mutual positions on the scale separation map.

Y. Shi et al. (Eds.): ICCS 2007, Part I, LNCS 4487, pp. 922–930, 2007.

2   Complex Automata Modeling

2.1   Definition

A CxA is a collection of interacting CA. We refer the reader to standard text-books for a complete definition of a CA [2]. Here we shall define a CA as a tuple

$$\mathcal{C} = < A(\Delta x, \Delta t, L, T), S, R, G, F > \tag{1}$$

where A is the spatial domain. It is made of cells of size Δx and it spans a region of size L. The quantity Δt is the time step and T is the number of iterations during which the CA will be run. Therefore, processes with time scales between Δt and T can be represented and spatial scales ranging from Δx to L can be resolved.

We call S the set of all possible states of each cell and R the evolution rule. G is the topology describing the neighborhood relation.

At the boundaries of A, additional information is needed (boundary conditions). We define F as the flux of information exchanged at each iteration between the system and its environment.

From definition 1 we can define a CxA as a graph $\mathcal{X} = (V, E)$ where V is the set of vertices and E the set of edges with the following properties

- Each vertex is a CA $\mathcal{C}_i = < A_i(\Delta x_i, \Delta t_i, L_i, T_i), S_i, R_i, G_i, F_i >$
- each edge E_{ij} is a coupling procedure describing the interaction between $\mathcal{C}_i$ and $\mathcal{C}_j$. In practice, E_{ij} will define how information is exchanged between the two subsystems.

During the initialization phase, this graph is built according to the modeler specifications.

2.2   Execution Model

Each vertex (subsystem) knows its time step with respect to the global time as well as it spatial location within the whole computational domain. This knowledge is in part given by Δx_i and Δt_i

Edges mediate an asynchronous communication between the vertices they connect. The main idea is that data from one CA that need to be known by another is written in the link that connects them, as soon as it is available. Similarly, when the recipient CA needs new input information from its neighbors, it reads it from the link. If this information is not yet available, the CA has to wait. In this way one achieves a decentralized, asynchronous communication model, which is compatible with parallelization or distributed computing.

The format under which this information is exchanged depends on the type of coupling. Yet, clearly, a conceptual model is needed to find a generic way of exchanging information (with little programming from the users), or to have a simple model to program these links. Smart edges (i.e. edges with computing and communication capabilities) should result from this information exchange procedure.

2.3   Genericity of CA's

The fact that each CA (vertex of the CxA) has a common instruction flow (because it is a CA) gives a way to implement generic coupling mechanisms and achieve the above proposed execution model.

A generic CA can be programmed using the propagation-collision paradigm instead of the gather-update paradigm. In this approach, the following steps implement all possible CA's

```
procedure executeCA
   for time t1 to t2, by step dt
      forall cells
          propagation
          computeBoundary
          collision
      end forall
   end for
end procedure
```

Here we assume that all submodels will be cast in this structure, which is possible if the model complies with the CA definition. Clearly, a lattice Boltzmann (LB)[3] model fulfills this requirement. In an agent based model, the same fundamental operations are also performed with cells replaced by agents. The propagation procedure sends the local states of each cell to the neighbors that need it. So, the propagation operator assumes an underlying topology of interconnection. In an agent based model, a special agent can be defined as a centralized information repository.

The data structure is a set of cells or agents which is traversed in any order because all the above operation are, in nature, parallel operations.

The computeBoundary procedure is needed to specify the values of the variable that are defined by the external environment. In the case of a LB fluid simulation, the missing density distributions at the wall that are computed in this procedure.

Collision is the procedure in which the evolution rule is executed for each cell, using the cell values obtained from propagation and computeBoundary.

2.4   Coupling Mechanisms

In the literature, several ways of coupling have been identified. From reference [4], we have

- Sub-Domain coupling (**SDC**), called domain decomposition in [4]
- Hierarchical-Model coupling (**HMC**), called heterogenous modeling method (HMM) in [4].

In the SDC adjacent spatial domains are described by different models on space-time grids of possible different resolution. Note that the two adjacent subdomains can possibly overlap.

In HMC, some parameters or variables of a main model (e.g. the CA rule, collision operators) are first computed locally (i.e for some selected cells) on the fly by a finer scale model.

The above SDC and HMC coupling mechanisms can be easily incorporated in the flow structure of `executeCA`, in agreement with the concept of smart edges connecting the CA's together.

For instance, the `propagation` step produces information that is sent to neighboring spatial cells, possibly those belonging a another CA. In case of SDC (subdomain coupling) the outgoing information from the boundary of one system is exaclty what is needed by the adjacent subdomain. Thus `propagation` has to **write** the adequate information in the edge interconnecting the two sub-models. Similarly, the procedure `computeBoundary` will **read** the missing information from the corresponding edge.

The HMC (hierarchical model coupling) is typically affecting the `collision` procedure by giving the value of a parameter needed by the collision rule and which is computed by a subscale model. For instance, this parameter could be a threshold on the blood shear stress that will lead to blood clotting. Such a threshold would typically result from the own dynamics of the cells in the endothelium.

3   The Scale Separation Map

The Scale Separation Map (SSM) is defined as a two dimensional map with the horizontal axis coding for temporal scales and the vertical axis coding for spatial scales. Each subsystem occupies a certain area on this map. Fig. 1 shows an example of such SSM, in which three subsystems have been identified. Subsystem 1 operates on small spatial scales, and short time scales, process 2 at intermediate scales, and process three at large scales. This could e.g. be processes operating at the micro-, meso-, and macro scale.

Consider two processes A and B with their own specific spatial - and temporal scale, denoted by ξ_i and τ_i respectively ($i \in \{A, B\}$). Assume that A has the largest spatial scale. In case the spatial scales are the same, A has the largest

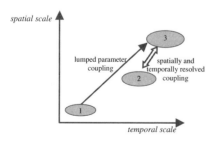

Fig. 1. A scale map, showing three subsystems and their mutual couplings

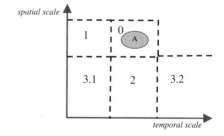

Fig. 2. Interaction regions on the scale map

temporal scale. In other words, $(\xi_B < \xi_A)$ OR $(\xi_B = \xi_A$ AND $\tau_B < \tau_A)$. We can now place A on the scale map and then investigate the different possibilities of placing B on the map relative to A. This will lead to a classification of types of multi-scale coupling, as in Fig. 2.

Depending on where B is, we find the following regions:

Region 0: A and B overlap, so we do not have a scale separation, we are dealing here with a single-scale multi-science model.

Region 1: Here $\xi_B = \xi_A$ AND $\tau_B < \tau_A$, so we observe a separation of time scales at the same spatial scale.

Region 2: Here $\xi_B < \xi_A$ AND $\tau_B = \tau_A$, so we observe a separation in spatial scales, like coarse and fine structures on the same temporal scale.

Region 3: Separation in time - and spatial scales. Region 3.1 is the well-known micro $\Leftrightarrow$ macro coupling, so fast processes on a small spatial scale coupled to slow processes on a large spatial scale. This type of multi-scale model has received most attention in the literature, and the SDC and HMC coupling paradigms explained earlier have mostly been applied in this region. In region 3.2 we have the reversed situation, a slow process on small spatial scales coupled to a fast process on large spatial scales. We believe that this region is very relevant in for instance coupling of biological with physical processes, where the biological process is e.g. the slow response of cells to a faster physical process on a larger scale (e.g. blood flow in arteries).

Note that we do not have to consider other regions of the scale map, because then the role of A and B just reverses, and we fall back to one of the five cases identified above.

Next we address the question of the area that process A and B occupy on the scale map, and from that, how to quantify the regions $0 - 3$ on the scale map. As discussed earlier, a single scale CA is characterized by a spatial discretization Δx and a system size L, where $\Delta x < \xi < L$. The number of CA cells in the full domain is then $N^{(x)} = L/\Delta x$. We introduce $\delta^{(x)}$ and $\eta^{(x)}$ so that the relevant spatial scale ξ is represented by $10^{\delta^{(x)}}$ cells (i.e. $\Delta x = \xi/10^{\delta^{(x)}}$) and the spatial extension of the CA is $10^{\eta^{(x)}}$ times the spatial scale, i.e. $L = \xi 10^{\eta^{(x)}}$, and therefore $N^{(x)} = 10^{\eta^{(x)}+\delta^{(x)}}$. Likewise for the temporal domain, i.e. a single scale CA has a time step Δt and the CA is simulated over a time span T, and we have $\Delta t < \tau < T$. The number of time steps $N^{(t)} = T/\Delta t$. The discretization has been chosen such that the temporal scale is represented by $10^{\delta^{(t)}}$ time steps (i.e. $\Delta t = \tau/10^{\delta^{(t)}}$) and that simulation time of the CA is $10^{\eta^{(t)}}$ times the spatial scale, i.e. $T = \tau 10^{\eta^{(t)}}$ and $N^{(t)} = 10^{\eta^{(t)}+\delta^{(t)}}$.

A process' position on the scale map is now fully determined by the tuple $\{\xi, \delta^{(x)}, \eta^{(x)}; \tau, \delta^{(t)}, \eta^{(t)}\}$, and is drawn in Fig. 3, where the axes are now on a logarithmic scale. On such logarithmic SSM the process is rectangular with

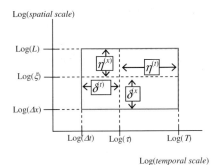

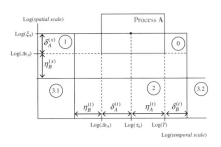

Fig. 3. Position of a process with parameters $\{\xi, \delta^{(x)}, \eta^{(x)}; \tau, \delta^{(t)}, \eta^{(t)}\}$ on the logarithmic scale map

Fig. 4. Interaction regions on the logarithmic scale map, in more detail

area $(\delta^{(t)} + \eta^{(t)}) \times (\delta^{(x)} + \eta^{(x)})$ asymmetrically centered around the point $(\log(\tau), \log(\xi))$.

In the special case that $\delta^{(x)} = \eta^{(x)} = \delta^{(t)} = \eta^{(t)} = 1$ (a reasonable first order assumption) we see that the process is symmetrically centered around $(\log(\tau), \log(\xi))$ and that the size of the box extends 2 decades in each dimension.

In Fig. 4 we show the extension of Fig. 2, where regions $1 - 3$ now have well defined positions and size. Depending on the location of process B, that is the point $(\log(\tau_B), \log(\xi_B))$ on the SMM, and with all information on the spatial and temporal extensions of process A and B, we can unambiguously find in which region of the scale map they are located with respect to each other.

Consider once more region 3. In region 3.1 we find that $L_B < \Delta x_A$ and $T_B < \Delta t_A$. As said earlier, this is the classical micro $\Leftrightarrow$ macro coupling, and in our language this means the full spatio-temporal extend $T_B \times L_B$ of process B is smaller than one single spatio-temporal step $\Delta t_A \times \Delta x_A$ of process A.

Region 3.2 also exhibits separation of time and length scales, but now the situation is quite different. We find that, just like in region 3.1, $L_B < \Delta x_A$. So, the spatial extend of process B is smaller than the grid spacing of process A. However, now we find that $T_A < \Delta t_B$. In other words, the full time scale of process A is smaller then the time step in process B. This will result in other modeling and simulation paradigms than in region 3.1. Typically, the coupling between A and B will involve time averages of the dynamics of the fast process A.

Let us now turn our attention to the regions where there is overlap on the temporal - or spatial scales, or both (regions 0, 1, and 2, in Fig. 4). In all these cases we can argue that we have partial or full overlap of the scales, giving rise to different types of (multi-scale) modeling and simulation. We say that the scales fully overlap if the point $(\log(\tau_B), \log(\xi_B))$ falls within (one of) the scales spanned by process A. On the other hand, there is partial overlap if $(\log(\tau_B), \log(\xi_B))$ falls *outside* (one of) the scales spanned by process A, but the rectangular area of process B still overlaps with (one of) the scales spanned by process A. The region of partial scale overlap can also be considered as a region of gradual scale separation, a boundary region between the scale separated regions 1, 2 and 3 and region 0. Simulations of this kind of multi-scale system would typically involve CxA's with local grid refinements, or multiple time stepping approaches, or a combination of both. We save this more detailed discussion for a future publication.

4   Examples

In this section we briefly indicate how some problems can be expressed as a CxA, with the benefit of code reusability, modeling flexibility and design efficiency.

4.1   Grid Refinement

The problem of grid refinement in LB fluid simulations is important in order to resolve small spatial scales, for instance around boundaries. The two subdomains C and F are the regions where a coarse and fine grid are defined, respectively. Such a situation is described in more detail in [5]. The two regions are updated by applying a LB (i.e a CA) dynamics such as executeCA. In addition, they interact because C and F have a common spatial interface, at the boundary between the two regions. Thus, the coupling is of SCD type. The CxA edge connecting the two submodels consists of transfering the information for the cells at the interface, from one grid to the other. In grid refinement the grid spacing is typically refined by some small number (say 2 or 4), and the timestep is scaled accordingly. So, in the language of the SSM, processes C and F will have partially overlapping spatial and temporal scales.

Typically, the edge will collect the data from grid C, rescale it, then perform a space and time interpolation in order to provide grid F with a data set at scales Δx_F and Δt_F. Similarly, the same edge will also translate the data produced by F to the requirement of grid C.

From the point of view of grids C and F the coupling is achieved simply by writing and reading the specific buffers managed by the edge, thus fully decoupling the model part from its interaction with other system components.

4.2   Time Splitting in Reaction-Diffusion

Time splitting has been introduced in LB models for reaction-diffusion [6] in order to deal with a wide range of reaction and diffusion constants in the same simulation. The idea is to consider diffusion as a first, larger scale model. Reaction acts between the diffusion steps, at each spatial cell, with a finer time resolution than the diffusion process. The reaction process is modeled as an implicit solver for a local differential equation. So we have here a HMC type of coupling. In this case, the subsystem coupling edge simply consists in reading the input chemical concentrations given by the diffusion process and pass them to the reaction solver. The new concentrations are then written back to the cell for the next diffusion step. Depending on the characteristic time scales of the

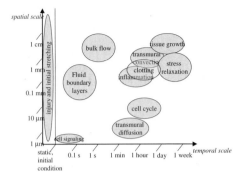

Fig. 5. A SSM for in stent re-stenosis

diffusion (specified by the diffusion constants) and the reactions (specified by the rate constants), the CxA is has fully separated time scales (region 1 on the SMM), or the time scales are partially overlapping.

4.3   In-stent Restenosis

Within the EU funded COAST project [7] we have chosen the treatment and progression of coronary artery disease as a prototypical multi-scale multi-science complex system to be modeled within the CxA framework. We will address the adverse vessel wall remodelling (re-stenosis), which occurs in some patients after placement of a metal frame (stent) within the artery lumen to expand and support the vessel at the site of a stenosis. This restenosis is due to the growth of scar tissue in between the strut of the stent, which tends to block the lumen. A full description of this application is beyond the scope of this manuscript. Here, it serves as an illustrative example of how we plan to apply CxA modeling. In-Stent Restenosis involves a large number of biological and physical processes on many spatial and temporal scales. Fig. 5 shows a simplified SMM for this process, showing relevant physical processes (such as bulk flow, transmural diffusion) and biological processes (such as cell signaling, inflammation and clotting) on their (estimated) characteristic scales. We are currently in the process of defining single scale CA or agent-based models for all processes, and once this is available, we will cast this SMM into the form of Fig. 3 and 4.

5   Discussion and Conclusions

The simulation of multiscale, multiscience complex systems are a central challenge in computationa science. Cellular Automata are a powerful framework to model spatially extended dynamical systems. In order to cope with multiscale and multi-physics, we have introduced Complex Automata as a set of interacting CA's. Together with a description of system components through a scale separation map,

CxA offer a flexible and intuitive framework to solve problems in which several different physical processes at different spatial and temporal scales interact.

Acknowledgments. This research is supported by the European Commission, through the COAST project [7] (EU-FP6-IST-FET Contract 033664)

References

1. Hoekstra, A., Chopard, B., Lawford., P., Hose, R., Krafczyk, M., and Bernsdorf, J.: Introducing Complex Automata for Modelling Multi-Scale Complex Systems. In Proceedings of the European Conference on Complex Systems ECCS'06, ISBN 0-9554123-0-7 (2006).
2. Chopard, B. and Droz, M.: Cellular Automata Modeling of Physical Systems. Cambridge University Press (1998).
3. Succi, S.: The Lattice Boltzmann Equation, for Fluid Dynamics and Beyond. Oxford University Press (2001).
4. E, W., Engquist, B., Li, X, Ren, W., and Vanden-Eijnden, E.: Heterogeneous Multiscale Methods, A Review. Commun. Comput. Phys., 2, 367-450 (2007).
5. Dupuis, A. and Chopard, B.: Theory and applications of an alternative lattice Boltzmann grid refinement algorithm. Phys. Rev. E, 67, 066707 (2003).
6. Alemani, D., Chopard, B., Buffle, J. and Galceran, J.: LBGK method coupled to time splitting technique for solving reaction-diffusion processes in complex systems. Physical Chemistry and Chemical Physics 7, 3331–3341 (2005).
7. http://www.complex-automata.org/

Multilingual Interfaces for Parallel Coupling in Multiphysics and Multiscale Systems

Everest T. Ong[1], J. Walter Larson[2,3], Boyana Norris[2], Robert L. Jacob[2],
Michael Tobis[4], and Michael Steder[4]

[1] Department of Atmospheric and Oceanic Science, University of Wisconsin,
Madison, Wisconsin, USA
{eong,larson,norris,jacob}@mcsanl.gov,
{steder,tobis}@gmail.com
[2] Mathematics and Computer Science Division, Argonne National Laboratory,
Argonne, IL 60439, USA
[3] ANU Supercomputer Facility, The Australian National University,
Canberra ACT 0200, Australia
[4] Department of Geophysical Sciences, University of Chicago, Chicago, IL USA

Abstract. Multiphysics and multiscale simulation systems are emerging as a new grand challenge in computational science, largely because of increased computing power provided by the distributed-memory parallel programming model on commodity clusters. These systems often present a *parallel coupling problem* in their intercomponent data exchanges. Another potential problem in these coupled systems is language interoperability between their various constituent codes. In anticipation of combined parallel coupling/language interoperability challenges, we have created a set of interlanguage bindings for a successful parallel coupling library, the Model Coupling Toolkit. We describe the method used for automatically generating the bindings using the Babel language interoperability tool, and illustrate with short examples how MCT can be used from the C++ and Python languages. We report preliminary performance reports for the MCT interpolation benchmark. We conclude with a discussion of the significance of this work to the rapid prototyping of large parallel coupled systems.

1 Introduction

Multiphysics and multiscale models are emerging or are in active use in many fields, including meteorology and climate, space weather, combustion and reactive flow, fluid-structure interactions, material science, and hydrology. Multiphysics models portray complexity stemming from the mutual interactions among a system's many constituent subsystems. Multiscale models depict complex phenomena originating in interactions between a system's multiple prevalent spatiotemporal scales. In both multiphysics and multiscale models, these interactions are called *couplings*, and thus they are *coupled models*.

Coupled systems are very demanding computer applications and often require high-performance computing for their solutions. Most HPC platforms today are

Y. Shi et al. (Eds.): ICCS 2007, Part I, LNCS 4487, pp. 931–938, 2007.

distributed-memory multiprocessor clusters programmed with a message-passing programming model. This approach creates a new concomitant problem—*the parallel coupling problem* (PCP)[1]. Specifically, implementing couplings between message-passing models that solve their equations of evolution on distributed domains entails the description, transfer, and transformation of *distributed data.* Thus the PCP is an important emerging problem in computational science.

Typical coupled models have purpose-built *ad hoc* solutions to the PCP. For example, numerous coupling packages exist in the climate area alone, including Ocean-Atmosphere-Sea Ice-Surface (OASIS; [2]) coupler, Projet D'Assimilation par Logiciel Multi-methodes (PALM; http://cerfacs.fr/ palm), the Flexible Modeling System (FMS; http://gfdl.noaa.gov/fms/), and the Earth System Modeling Framework (ESMF; http://esmf.ucar.edu). Each of the cited examples has its own custom (and slightly different) solution to the same underlying problem (the PCP), but implemented with numerous domain-specific assumptions (e.g., mesh descriptions based on Arakawa grids from geophysical fluid dynamics). The PCP is sufficiently widespread across many distinct scientific fields that an application-neutral software solution is desirable.

MCT[1] is an open-source package that supports rapid development of parallel coupling interfaces between MPI-based parallel codes. MCT is in active use in numerous applications, most notably as coupling middleware for CCSM[2], and as a prototype coupling layer for the WRF[3]. MCT is highly scalable and performance portable, supporting both vector and commodity microprocessor architectures. Its Fortran-based API is naturally compatible with scientific applications, as Fortran remains the dominant programming language for science. MCT's programming model is minimally invasive, and scientific-programmer-friendly.

Here we report results of our work to broaden the applicability of the MCT programming model from its native Fortran API. Our motivations for extending MCT to other languages are 1) to broaden MCT's applicability from that of a Fortran-based toolkit to a callable framework that allows one to compose parallel coupling mechanisms in multiple programming languages, 2) to allow coupling of separately developed codes implemented in different languages, and 3) to leverage MCT's robust and efficient Fortran compute kernels and build coupling mechanisms in languages better suited to object-oriented programming (OOP). The strategy we have chosen is to create a set of multilingual bindings that can be installed on top of the MCT code base, rather than making them an inextricable part of MCT. This separation of concerns is in keeping with the MCT philosophy of offering a minimally invasive programming model, and will be addressed in some detail in Section 3. We believe that the outcome of the work reported here will enable rapid prototyping of parallel coupled models implemented in different languages, and should be of great interest to both current and would-be multiphysics and multiscale modeling teams.

---

[1] Model Coupling Toolkit, http://www.mcs.anl.gov/mct
[2] The Community Climate System Model, http://www.ccsm.ucar.edu/
[3] Weather Research and Forecasting Model, http://wrf-model.org

2   The Model Coupling Toolkit

MCT [1,3] is a Fortran toolkit that eases the programming of parallel couplings in MPI-based applications. MCT addresses the parallel data processing part of the PCP, while maximizing developer flexibility in choices regarding parallel coupled model architecture. MCT achieves this by providing a set of Fortran modules designed to emulate object oriented classes and methods[4] and a library of routines that perform parallel data transfer and transformation. These classes and methods amount to programming shortcuts that are used *à la carte* to create custom parallel couplings. MCT supports parallel coupling for both serial and parallel compositions—and combinations thereof—and also supports single and multiple executables.

MCT has nine classes for use as parallel coupling building blocks. Three datatypes constitute the MCT data model, encapsulating storage of multi-field integer- and real-valued data (AttrVect), the grids or spatial discretizations on which the data reside (GeneralGrid), and their associated domain decompositions (GlobalSegMap). MCT's data transfer facility's fundamental class is a lightweight component registry (MCTWorld) containing a directory of all components to be coupled and a process rank translation table supporting intercomponent messaging. One-way parallel data transfer message scheduling is encapsulated in the Router class, and this function for parallel data redistribution is embodied in the Rearranger. Data transformations supported directly by MCT are handled by three additional classes. The Accumulator is a set of time-integration registers for state and flux data. MCT supports regridding of data in terms of sparse linear transformations, with user-supplied transform coefficients stored in coordinate (COO) format by the SparseMatrix class. The SparseMatrixPlus class encapsulates matrix element storage and the necessary communications scheduling for parallel matrix-vector multiplication.

MCT has a library of routines that manipulate MCT datatypes to perform parallel coupling, supporting both blocking and non-blocking parallel data transfer and redistribution. MCT's transformation library routines support 1) parallel linear transforms used for intergrid interpolation; 2) time accumulation of flux and state data; 3) computation of spatial integrals required for flux conservation diagnoses; and 4) merging of outputs from multiple models for input to another model.

MCT is invoked through the *use* statement of Fortran90/95. One *uses* MCT modules to gain access to MCT datatype definitions and library interfaces, one *declares* variables of MCT datatypes to express distributed data to be exchanged and transformed, and one *invokes* MCT library routines to perform parallel data transfer and transformation. This is analogous to importing a class, instantiating an object which is a member of that class, and invoking the class methods associated with the object. A simple example of how MCT is used to construct

---

[4] The choice of Fortran as MCT's implementation language was driven by Fortran's continuing dominance as the language of choice in scientific programming. The developers of MCT implemented OOP features manually in Fortran, and the use of the terms class and method follow Decyk et al. [4].

a GlobalSegMap domain decomposition descriptor is shown in the code fragment below. More detailed examples MCT usage can be found in [1,3] and in the example codes bundled in the MCT source distribution, which can be downloaded from the MCT Web site.

```
use m_GlobalSegMap, only : GlobalSegMap, GlobalSegMap_Init => init
implicit none
type(GlobalSegMap) :: AtmGSMap
integer, dimension(:) :: starts, lengths
integer :: myRoot, myComm, myCompID ! MPI communicator, root process
integer :: myCompID                  ! MCT component ID
! initialize segment start and length arrays starts(:) and lengths(:)...
:
! Create and initialize MCT GlobalSegMap
  call GlobalSegMap_init(AtmGSMap, starts, lengths, myRoot, myComm, &
                         myCompID)
```

3   Construction of the Multi-lingual Interfaces for MCT

The MCT API is expressed using Fortran derived types and pointers, complicating considerably the challenge of interfacing MCT to other programming languages. This is due to the lack of a specific standard for array descriptors in Fortran90/95[5]. Thus, interfacing available contemporary Fortran to other languages remains notoriously difficult. One solution is to hard-code wrappers, which can be cumbersome, time-consuming, hard to maintain, and error-prone. The only automatic solution known to the authors is a vendor-by-vendor implementation of array descriptors such as CHASM[5].

Our multilingual interfaces are defined using the Scientific Interface Definition Language (SIDL). These interfaces are processed by a language interoperability tool called Babel[6], which leverages the vendor-specific array descriptors provided by CHASM. Babel currently supports interoperability between C, f77, Fortran90/95, C++, Java, and Python. Babel is used to generate glue code from an iterface description, thus avoiding modification of the original source code. This has the important advantage of separation of concerns; that is, we view language interoperability as a distinct problem from the algorithmics of parallel coupling. Thus MCT's scientist-friendly Fortran-based programming model is untouched, while langauage interoperability is available to those who need it. Our use of Babel enables us to create multilingual MCT bindings and distribute them as a separate package that references MCT. This approach superior to ESMF, whose fundamental types (e.g., ESMF_Array), are implemented in C for interfacing ESMF with possible future C applications, while important functions (such as Regrid) are implemented in Fortran. Language interoperability is *internal* to the ESMF software, not necessarily a user feature.

---

[5] This has been rectified by the BINDC specification in Fortran2003, but this standard is only now beginning to be implemented by compiler vendors.

As mentioned earlier, the SIDL interfaces and classes for MCT are processed by Babel to generate interlanguage "glue" code to bridge the caller/callee language gap. This glue code comprises *skeletons* in the callee's programming language (Fortran in the case of MCT), the *internal object representation* (IOR), which is implemented in C, and *stubs* that are generated in the caller programming language (e.g., C++ and Python). We have inserted calls to the MCT library in the Babel-generated implementation files that initially contain the empty function definitions from the SIDL interfaces (we will refer to them as *.IMPL files*). The IOR and stubs that provide the inter-language glue are generated by Babel automatically and require no modifications by the user. Working MCT bindings and example codes for both C++ and Python can be downloaded from the MCT Web Site. Included with the bindings are the Babel base classes and other core pieces of glue code that must be compiled against a pre-installed MCT, eliminating the need to install Babel (which can be nontrivial).

An example SIDL code block for the MCT AttrVect class and excerpt of the associated .IMPL file for the skeleton code are shown in Figure 1. The directed dotted grey arrows on the figure show the calling path by which an application written in some non-Fortran language accesses MCT by first calling a stub in the application's implementation language, which in turn calls the C IOR, which then calls the Fortran skeleton, and via it MCT.

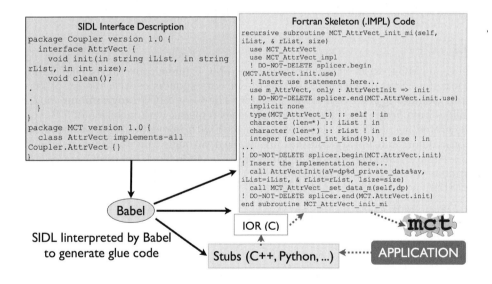

Fig. 1. Schematic for SIDL/Babel-based generation of MCT's multilingual interfaces and calling path from applications in other languages back to MCT

Our multilingual MCT bindings correspond to a *subset* of the MCT API because at this time Babel does not support the use of optional arguments, a Fortran feature that is widely used in MCT. For routines with only one or two optional arguments, we have created static SIDL interfaces for each possible combination of optional arguments. For some of MCT's spatial integration and merging routines, which have in some cases four or more optional arguments, we decided to support only a subset of the possibilities, which will be expanded in the future as needed.

4   The MCT Multilingual Programming Model

MCT's native Fortran programming model described in section 2 consists of module usage, declaration of derived types, and invocation of library routines. The C++ and Python MCT programming models are analogous, but within each language's context. We will illustrate the differences in the respective programming models with a simple example taken from the example code in the MCT Fortran distribution and its C++ and Python counterparts. Below we show code excerpts from the MCT multilingual example applications and the original Fortran. The full example code demonstrates the kind of coupling found in climate models such as CCSM, with focus on atmosphere-ocean interactions via a third component called a coupler. The atmosphere, ocean, and coupler each execute on their own respective pool of MPI processes, making this example a parallel composition. The code fragments in this section are from the "coupler" parts of the respective example codes, and are for the initialization of an MCT GlobalSegMap domain decomposition descriptor.

4.1   C++

In C++, MCT exists as a *namespace* named MCT, and a collection of header files containing class declarations, in one-to-one correspondence with the set of Fortran modules in the original MCT source. The C++ MCT programming model differs from Fortran as follows: module use is replaced with inclusion of a Babel-generated header file; declaration is replaced with invocation of a no-argument constructor; and library routines are invoked through calls to Babel-generated C++ stubs that reference MCT functions. The code block below illustrates creation of an MCT GlobalSegMap in C++. The most subtle usage point is the use of SIDL arrays required by the MCT C++ interfaces.

```
#include "MCT_GlobalSegMap.hh"
...
  // Create SIDL arrays indexed (note starting index is 1 or
  // compatibility with Fortran
  int32_t dim1 = 1;
  int32_t lower1[1] = {1};
  sidl::array<int32_t> start =
      sidl::array<int32_t>::createRow(dim1, lower1, lower1);
  start.set(1, (myrank * localsize) + 1);
...
  // Create an MCT GlobalSegMap domain decomposition descriptor
  MCT::GlobalSegMap AtmGSMap = MCT::GlobalSegMap::_create();
  AtmGSMap.initd0(start, length, 0, comm, compid);
```

4.2   Python

In Python, MCT exists as a *package* named MCT. The Python MCT programming model differs from MCT's native Fortran as follows: module use is replaced with Python package import; declaration is replaced with invocation of a constructor; and library routines are invoked through calls to Babel-generated Python stubs. The code block below illustrates creation of an MCT GlobalSegMap in Python. Babel's support of SIDL arrays in Python is handled by the Numeric package, thus the creation of the arrays start and length as Numeric arrays.

```
import Numeric
from MCT import GlobalSegMap
...

    # Create start and length arrays--only the 'start' array shown here
    start = Numeric.zeros(2,Numeric.Int32)
    start[1] = (myrank*localsize)+1
...

    # Describe decomposition with MCT Global Seg Map
    AtmGSMap = GlobalSegMap.GlobalSegMap()
    AtmGSMap.initd0(start,length, 0,comm, compid)
```

5   Performance

Babel has been used successfully in various projects to generate interlanguage bindings with low performance overheads (e.g., see [7]). We evaluated the performance of the MCT C++ API for MCT's atmosphere-ocean parallel interpolation benchmark. The atmosphere-to-ocean grid operations are: 720 interpolation calls to regrid a bundle of two fields, and two sets of 720 interpolastion calls to regrid six fields. The atmospheric grid is the CCSM 3.0 T340 grid (512 latitudes by 1024 longitudes), and the ocean grid is the POP $0.1°$ grid (3600×2400 grid points). We ran the experiments on the Jazz cluster in the Laboratory Computing Resource Center at Argonne National Laboratory. Jazz is a 350-node cluster of 2.4 GHz Pentium Xeon processors connected by a Myrinet 2000 switch. The compilers used in this study were Absoft 9.0 Fortan and gcc 3.2.3. Timings (in seconds) for this benchamark for both MCT's native Fortran and for the Babel-generated C++ API are summarized in Table 1. The overhead decreases from 1.6% for small numbers of processors to less than one percent for larger numbers of processors, where the amount of work performed by MCT is enough to amortize the cost of executing the interlanguage glue code for each call to MCT.

Table 1. Timings (in seconds) of the MCT A2O Benchmark

Number of Processors	8	16	32
Native Fortran	1985.6	1084.6	556.9
C++ via Babel	2016.5	1085.1	559.6

6   Conclusions

We have created a set of multilingual bindings for the Model Coupling Toolkit. These bindings were created using the Babel language interoperability tool from a SIDL description of the MCT API. The resultant glue code is sufficiently robust to support proof-of-concept example applications, and impose relatively little performance overhead. The Babel-generated glue code and C++ and Python coupling example codes are now publicly available for download at the MCT Web site. The MCT programming model has been expanded beyond its native Fortran, making this robust and well-tested parallel coupling package available for use in coupling MPI-based parallel applications implemented in other languages. This is a first step towards our long-term goal of enabling fast prototyping of large, multilingual parallel coupled systems. The multilingual MCT bindings are also capable of supporting coupling of parallel applications implemeted in multiple programming language. We have employed them in a daring object-oriented Python re-implementation of the CCSM coupler (pyCPL), resulting in a Pythonic CCSM that supports Fortran models interacting via a Python coupler. Promising preliminary peformance results are a matter for further study.

Acknowledgements. This work was supported by the US Department of Energy's Scientific Discovery through Advanced Computing under Contract DE-AC02-06CH11357, by the National Science Foundation under award ATM-0121028, and by the Australian Partnership for Advanced Computing.

References

1. Larson, J., Jacob, R., Ong, E.: The Model Coupling Toolkit: A new Fortran90 toolkit for building multi-physics parallel coupled models. Int. J. High Perf. Comp. App. **19**(3) (2005) 277–292
2. Valcke, S., Caubel, A., Vogelsang, R., Declat, D.: Oasis3 ocean atmosphere sea ice soil user's guide. Technical Report TR/CMGC/04/68, CERFACS, Toulouse, France (2004)
3. Jacob, R., Larson, J., Ong, E.: M×N communication and parallel interpolation in CCSM3 using the Model Coupling Tookit. Int. J. High Perf. Comp. App. **19**(3) (2005) 293–308
4. Decyk, V.K., Norton, C.D., Syzmanski, B.K.: Expressing object-oriented concepts in Fortran90. ACM Fortran Forum **16**(1) (1997) 13–18
5. Rasmussen, C.E., Sottile, M.J., Shende, S.S., Malony, A.D.: Bridging the language gap in scientific computing: The CHASM approach. Concurrency and Computation: Practice and Experience **18**(2) (2006) 151–162
6. Dahlgren, T., Epperly, T., Kumfert, G.: Babel User's Guide. CASC, Lawrence Livermore National Laboratory. version 0.9.0 edn. (January 2004)
7. Kohn, S., Kumfert, G., Painter, J., Ribbens, C.: Divorcing language dependencies from a scientific software library. In: Proc. Tenth SIAM Conference on Parallel Processing in Scientific Computing, Portsmouth, VA (August 2001)

On a New Isothermal Quantum Euler Model: Derivation, Asymptotic Analysis and Simulation

Pierre Degond[1], Samy Gallego[1], and Florian Méhats[2]

[1] MIP (UMR CNRS 5640), Université Paul Sabatier,
118 Route de Narbonne, 31062 Toulouse Cedex 4 (France)
`degond,gallego@mip.ups-tlse.fr`
[2] IRMAR (UMR CNRS 6625), Université de Rennes,
Campus de Beaulieu, 35042 Rennes Cedex (France)
`florian.mehats@univ-rennes1.fr`

Abstract. In the first part of this article, we derive a New Isothermal Quantum Euler model. Starting from the quantum Liouville equation, the system of moments is closed by a density operator which minimizes the quantum free energy. Several simplifications of the model are then written for the special case of irrotational flows. The second part of the paper is devoted to a formal analysis of the asymptotic behavior of the quantum Euler system in two situations: at the semiclassical limit and at the zero-temperature limit. The remarkable fact is that in each case we recover a known model: respectively the isothermal Euler system and the Madelung equations. Finally, we give in the third part some preliminary numerical simulations.

1 Introduction

Various works have been devoted in the past decade to the derivation of quantum hydrodynamic models for semiconductors in order to simulate nanoscale devices such as tunneling diodes or lasers. The interest in such models comes from the fact that they are supposed to describe quantum transport in highly collisional situations and to be computationally less expensive than corresponding quantum microscopic models, such as the Schrödinger equation or the Wigner equation.

To design quantum hydrodynamic models with temperature effects, the route which has been usually followed consists in incorporating to classical fluid models some "quantum" correction terms, based on the Bohm potential [5,6,7]. However, such approaches are not obvious to justify from physical principles. Moreover, quantum corrections involving the Bohm potential produce high order terms in these systems and make their resolution difficult, from the mathematical and numerical points of view.

In 2003, Degond and Ringhofer proposed in [3] a different manner to derive quantum hydrodynamic models, by closing the systems of moments thanks to a quantum entropy minimisation principle. This systematic approach, which is an extension of Levermore's moment method to the framework of quantum mechanics, seems very fruitful, although it is still formal. Following this first

Y. Shi et al. (Eds.): ICCS 2007, Part I, LNCS 4487, pp. 939–946, 2007.
© Springer-Verlag Berlin Heidelberg 2007

paper, an entropic quantum drift-diffusion model has been derived, analyzed and then discretized numerically in [4]. Numerical comparisons of this model with existing quantum transport models in [2] showed that it is a good candidate model for quantum device simulations in diffusive regimes.

It has been possible to rewrite recently in [1] the isothermal quantum hydro-dynamic model in a simpler and differential way. In this article, we remind the methodology used to derive this model.

We also investigate several asymptotic limits for this model: at the semiclassi-cal limit and at the zero-temperature limit. The remarkable fact is that in each case we recover a known model: respectively the isothermal Euler system and the Madelung equations.

Some simplifications of the model for the special case of irrotational flows allow us to present some preliminary numerical results in one dimension.

We finish this introduction by giving some possible applications of the Quan-tum Euler model. We first have in mind the semiconductor industry where engi-neers have first introduced Hydrodynamics models with $\hbar^2$ quantum corrections in order to simulate nanoscale devices such as the so-called Resonant Tunneling Diode. The use of these models have permitted to exhibit interesting features of RTD such as negative resistance or hysteresis. The Quantum Euler model could be also used in quantum chemistry, or other areas of physics such as quantum optics, the study of superfluidity, etc...

2   Derivation of the Model

2.1   The Underlying Quantum Microscopic Model

At the microscopic scale, a quantum system evolving in $\mathbb{R}^3$ and subject to a po-tential $V(x,t)$ can be described by a time-dependent density operator satisfying the quantum Liouville equation:

$$i\hbar \partial_t \varrho = [\mathcal{H}, \rho] + i\hbar \mathcal{Q}(\varrho), \tag{1}$$

where $\hbar$ is the scaled Planck constant and $\mathcal{H}$ is the Hamiltonian:

$$\mathcal{H} = -\frac{\hbar^2}{2}\Delta + V(x,t), \tag{2}$$

and $\mathcal{Q}(\rho)$ is an unspecified collision operator which describes the interaction of the particles with themselves and with their environment and accounts for dissipation mechanisms. By a density operator, we shall always mean a posi-tive, Hermitian, trace-class operator acting on $L^2(\mathbb{R}^3)$. Let $W[\varrho](x,p)$ denote the Wigner transform of ρ:

$$W[\varrho](x,p) = \int \varrho\left(x - \frac{1}{2}\eta, x + \frac{1}{2}\eta\right) e^{\frac{i\eta \cdot p}{\hbar}} d\eta, \tag{3}$$

where $\varrho(x,x')$ is the distribution kernel of ϱ:

$$\varrho\psi = \int \varrho(x,x')\psi(x')\,dx'.$$

The inverse Wigner transform (or Weyl quantization) is given by the following formula:

$$W^{-1}[w]\psi(x) = \int w\left(\frac{x+y}{2}, p\right) \psi(y) e^{\frac{ip(x-y)}{\hbar}} \frac{dp\,dy}{(2\pi\hbar)^3}, \tag{4}$$

and defines $W^{-1}[w]$ as an operator acting on the element ψ of L^2. The Wigner transform W and its inverse W^{-1} are Isometries between $\mathcal{L}^2$ (the space of operators such that the product $\varrho\varrho^\dagger$ is trace-class, where $\varrho^\dagger$ is the Hermitian conjugate of ϱ) and $L^2(\mathbb{R}^{2d})$:

$$\mathrm{Tr}\{\varrho\sigma^\dagger\} = \int W[\varrho](x,p)\overline{W[\sigma](x,p)} \frac{dx\,dp}{(2\pi\hbar)^3}. \tag{5}$$

Taking the Wigner transform of (1), we get the following collisional Wigner equation for $w = W[\varrho]$:

$$\partial_t w + p \cdot \nabla_x w + \Theta^\hbar[V]w = Q(w), \tag{6}$$

with

$$\Theta^\hbar[V]w = -\frac{i}{(2\pi)^3\hbar} \int (V(x+\frac{\hbar}{2}\eta) - V(x-\frac{\hbar}{2}\eta)) \times$$
$$\times w(x,q)\, e^{i\eta\cdot(p-q)}\, dq\, d\eta. \tag{7}$$

and $Q(w)$ is the Wigner transform of $\mathcal{Q}(\varrho)$.

2.2 The Moment Method

If we suppose that the collision operator conserves the mass and momentum, integrations with respect to the variable $p \in \mathbb{R}^3$ enable to obtain equations for the first two moments $n(t,x) = \int w(t,x,p)\frac{dp}{(2\pi\hbar)^3}$ (the mass density) and $n(t,x)u(t,x) = \int p\,w(t,x,p)\frac{dp}{(2\pi\hbar)^3}$ (the current density):

$$\partial_t n + \nabla \cdot nu = 0, \tag{8}$$
$$\partial_t(nu) + \nabla\Pi + n\nabla V = 0. \tag{9}$$

Of course, this system of equations is not closed, since the pressure tensor Π is still expressed in terms of the microscopic Wigner function $w(t,x,p)$:

$$\Pi = \int_{\mathbb{R}^d} p \otimes p\,w\, \frac{dp}{(2\pi\hbar)^3}. \tag{10}$$

The quantum Euler model is thus complete only as soon as the closure assumption is made precise. Hence, by analogy with Levermore's methodology [9] and according to [3], we close the system by replacing the expression (10) of Π by another one in terms of n and nu:

$$\Pi = \int_{\mathbb{R}^d} p \otimes p\,w_{n,nu}^{eq}\, \frac{dp}{(2\pi\hbar)^3}, \tag{11}$$

where $w_{n,nu}^{eq} = W[\varrho_{n,nu}^{eq}]$ is the so-called quantum Maxwellian, depending only on n and nu in a non local (and non explicit) way, as defined in next subsection.

2.3   The Quantum Free Energy Minimization Principle

We intend to describe the effect of the interaction of a quantum system, subject to a potential V, with a thermal bath at temperature T. To this aim, it is convenient to introduce the quantum free energy defined by:

$$G(\varrho) = \operatorname{Tr}\left\{T(\varrho(\log\varrho - 1)) + \mathcal{H}\varrho\right\}. \tag{12}$$

where $\mathcal{H} = -\frac{\hbar^2}{2}\Delta + V$ is the Hamiltonian.

The main assumption concerning the interaction between the system and the thermal bath is that the first two moments n and nu are conserved during these interactions. According to this statement, we now claim the quantum free energy minimization principle (the reader can refer to [3] for details). Let the functions n and nu be given. Consider the following constrained minimization problem:

$$\min\left\{G(\varrho) \text{ such that } \int W[\varrho]\frac{dp}{(2\pi\hbar)^3} = n \text{ and } \int p\,W[\varrho]\frac{dp}{(2\pi\hbar)^3} = nu\right\}. \tag{13}$$

The solution, if it exists, is called the local equilibrium density operator (or quantum Maxwellian) associated to n and nu. Lagrange multiplier theory for the constrained problem (13) (see [3]) shows that there exist a scalar function A and a vector function B, both real valued and defined on $\mathbb{R}^3$, such that this local equilibrium density operator takes necessarily the form:

$$\varrho_{n,nu}^{eq} = \exp\left(-\frac{1}{T}H(A,B)\right), \tag{14}$$

where $H(A,B)$ is the following modified Hamiltonian:

$$H(A,B) = W^{-1}\left[\frac{1}{2}(p-B)^2 + A\right] = \frac{1}{2}(i\hbar\nabla + B)^2 + A. \tag{15}$$

2.4   The Quantum Euler Model

According to the definition of the quantum Maxwellian (14) and to commutation properties with the Hamiltonian, it is possible to simplify the pressure tensor and to express it with respect to n, nu, A and B (see [1]) and we obtain the following system:

$$\partial_t n + \nabla \cdot nu = 0, \tag{16}$$

$$\partial_t(nu) + \nabla(nu \otimes B) + n(\nabla B) \cdot (u - B) + n\nabla(V - A) = 0, \tag{17}$$

where the extensive quantities n and nu and the associated intensive ones A and B are linked by the following constitutive equations:

$$n = W[\varrho_{n,nu}^{eq}] = \sum_{p\in\mathbb{N}} \exp\left(-\frac{\lambda_p}{T}\right)|\psi_p|^2,$$

$$nu = W[p\,\varrho_{n,nu}^{eq}] = \sum_{p\in\mathbb{N}} \exp\left(-\frac{\lambda_p}{T}\right)\mathcal{I}m\left(\hbar\nabla\psi_p\,\overline{\psi_p}\right), \tag{18}$$

and where $(\lambda_p, \psi_p)_{p \in \mathbb{N}}$ denotes the complete set of eigenvalues and normalized eigenvectors of the modified Hamiltonian defined by

$$H(A, B) = \frac{1}{2} \left(i\hbar\nabla + B \right)^2 + A. \tag{19}$$

2.5   Special Case of Irrotational Flows

We conclude this section by dealing with the special case of irrotational flows. Let n, nu, A, B be given according to the definition of the quantum Maxwellian. Assume moreover that u is an irrotational vector field, *i.e.* that there exists $S(x)$ such that $u = \nabla S$. Then B is defined by

$$B = u = \nabla S \tag{20}$$

In this case, the quantum Euler system (16)–(19) can be rewritten as follows:

$$\partial_t n + \nabla \cdot nu = 0, \tag{21}$$
$$\partial_t(nu) + \nabla(nu \otimes u) + n\nabla(V - A) = 0, \tag{22}$$

where n and A are coupled by the constitutive equation:

$$n = \sum_{p \in \mathbb{N}} \exp\left(-\frac{\lambda_p}{T} \right) |\psi_p|^2, \tag{23}$$

and where $(\lambda_p, \psi_p)_{p \in \mathbb{N}}$ denotes the complete set of eigenvalues and normalized eigenvectors of the modified Hamiltonian defined by

$$H(A, 0) = -\frac{\hbar^2}{2}\Delta + A. \tag{24}$$

As we can see, the system is much simpler in this case since it does not depend on B anymore. The only non-local relation links the intensive variable A to the extensive one n.

3   Formal Asymptotics

We investigate here various asymptotic approximations of the quantum Euler system as dimensionless parameters go to zero. Our aim here is to draw some connections between this model and several other existing models.

3.1   Semiclassical Asymptotics

It is possible to perform an $\hbar$ expansion of the "quantum Maxwellian" $\varrho_{n,nu}^{eq}$. By keeping in this expansion the terms up to the order $\mathcal{O}(\hbar^2)$, one gets formally some approximate constitutive equations linking (n, nu) and (A, B):

$$A = T \ln n_0 - T \ln n + \frac{\hbar^2}{6} \frac{\Delta\sqrt{n}}{\sqrt{n}} - \frac{\hbar^2}{24} |\omega|^2 + \mathcal{O}(\hbar^4), \tag{25}$$

$$nB = nu + \frac{\hbar^2}{12} \nabla \times (n\omega) + \mathcal{O}(\hbar^4), \tag{26}$$

where we have denoted $\omega = \nabla \times u$ and n_0 is the effective density of state $n_0 = \left(\frac{T}{2\pi\hbar^2}\right)^{3/2}$.

Then, inserting these equations (25), (26) in (16), (17), one gets an approximate quantum Euler model:

$$\partial_t n + \nabla \cdot nu = 0, \tag{27}$$

$$\partial_t(nu) + \nabla(nu \otimes u) + T\nabla n + n\nabla V - \frac{\hbar^2}{6}n\nabla\left(\frac{\Delta\sqrt{n}}{\sqrt{n}}\right) +$$

$$+\frac{\hbar^2}{12}\omega \times (\nabla \times (n\omega)) + \frac{\hbar^2}{24}n\nabla(|\omega|^2) = \mathcal{O}(\hbar^4). \tag{28}$$

It is readily seen from (27) and (28) that, as $\hbar \to 0$, the quantum Euler system converges formally to the isothermal Euler system. Moreover, if we drop the terms $\mathcal{O}(\hbar^4)$ in (27), (28), the obtained model is the so-called quantum hydrodynamic model [7], up to additional terms depending only on $\omega = \nabla \times u$. For irrotational flows, we recover the quantum hydrodynamic model. These observations were already made in [8].

3.2   The Zero Temperature Limit

An opposite limit to the semiclassical asymptotics is the zero-temperature limit. Assume that the solution $(n^T, n^T u^T, A^T, B^T)$ of the quantum Euler system (16)–(19) with temperature T, admits a non trivial limit (n, nu, A, B) as the dimensionless temperature T goes to zero. Then these functions satisfy the Madelung equations:

$$\partial_t n + \nabla \cdot nu = 0, \tag{29}$$

$$\partial_t u + (u \cdot \nabla)u + \nabla V - \frac{\hbar^2}{2}\nabla\left(\frac{\Delta\sqrt{n}}{\sqrt{n}}\right) = 0, \tag{30}$$

$$A = \frac{\hbar^2}{2}\frac{\Delta\sqrt{n}}{\sqrt{n}} - C, \qquad B = u, \tag{31}$$

where C is a constant and the system is described by a pure-state:

$$\varrho(x, x') = \psi(x)\,\overline{\psi(x')}, \tag{32}$$

with

$$\psi = \sqrt{n}\,e^{iS/\hbar} \quad \text{and} \quad \nabla S = u.$$

This gives a clear link between the quantum Euler model and the fluid dynamic expression of the Schrödinger equation.

4   Numerical Results

Inspired by the previous work realized for the quantum drift-diffusion model [4,2], we have implemented a 1D numerical scheme for solving the quantum

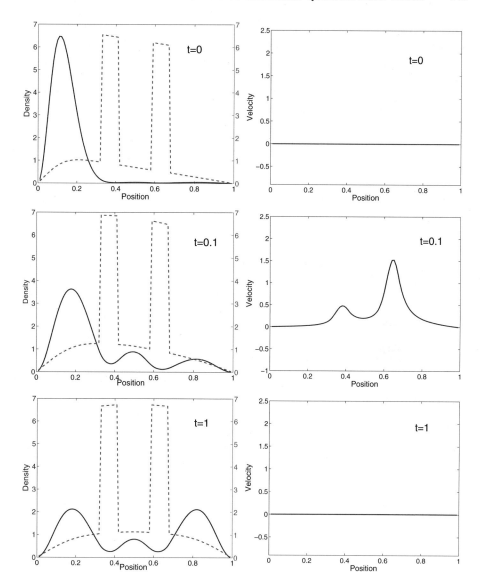

Fig. 1. Numerical solution of the quantum Euler model with relaxation for $t = 0$, $t = 0.1$ and $t = 1$. Left: density $n(t,x)$ (solid line) and total electrical potential $(V^s + V^{ext})(t,x)$ (dashed line) as functions of the position x. Right: velocity $u(t,x)$ as function of the position x.

Euler system with relaxation coupled to the Poisson equation on the domain $\Omega = [0,1]$. In 1D, the velocity is of course irrotational so that the quantum Euler model reads under the reduced form (21)-(24), and the Poisson equation is given by:

$$- \alpha^2 \Delta V^s = n, \qquad (33)$$

with α the scaled Debye length. We have chosen for the parameters $\hbar^2/2T = 0.1$ and $\alpha^2 = 0.1$. We choose an initial density concentrated on the left of the device and take the initial velocity equal to zero. The external potential V^{ext} is chosen to be a double barrier (in fact, the device here is a simplified Resonant Tunneling Diode without doping) and Figure 1 shows the evolution of the electron density $n(x)$ on the left, and the velocity $u(x)$ on the right for $t = 0$, $t = 0.1$ and $t = 1$. We can see electrons going through the barriers by tunneling effect.

5   Conclusion and Perspectives

In this paper, we have written in a simpler and differential way the isothermal version of the quantum hydrodynamic model derived in [3]. We have then written some formal asymptotics of the model as dimensionless parameters tend to zero. It appears formally that the semiclassical limit of the model is the Classical isothermal Euler system while the zero temperature limit of the model gives the Madelung equations. We have also written several simplifications of the model when the velocity is irrotational which allows us to perform numerical simulations on a simple 1D device. These preliminary numerical simulations seem to indicate that the model gives meaningful results in realistic situations. The study of the numerical scheme together with comparisons with other models will be presented in a future article.

References

1. P. Degond, S. Gallego, F. Méhats, *Isothermal quantum hydrodynamics: derivation, asymptotic analysis and simulation*, accepted at SIAM MMS.
2. P. Degond, S. Gallego, F. Méhats, *An entropic Quantum Drift-Diffusion model for electron transport in resonant tunneling diodes*, to appear in J. Comp. Phys.
3. P. Degond, C. Ringhofer, *Quantum Moment Hydrodynamics and the Entropy Principle*, J. Stat. Phys. **112** (2003), no. 3-4, 587–628.
4. S. Gallego, F. Méhats, *Entropic discretization of a quantum drift-diffusion model*, SIAM J. Numer. Anal. **43** (2005), no. 5, 1828–1849.
5. C. Gardner, *The quantum hydrodynamic model for semiconductor devices*, SIAM J. Appl. Math. **54** (1994), no. 2, 409–427.
6. I. Gasser, A. Jüngel, *The quantum hydrodynamic model for semiconductors in thermal equilibrium*, Z. Angew. Math. Phys. **48** (1997), no. 1, 45–59.
7. A. Jüngel, *Quasi-hydrodynamic semiconductor equations*, Progress in Nonlinear Differential Equations, Birkhäuser, 2001.
8. A. Jüngel, D. Matthes, *A derivation of the isothermal quantum hydrodynamic equations using entropy minimization*, Z. Angew. Math. Mech. **85** (2005), 806–814.
9. C. D. Levermore, *Moment closure hierarchies for kinetic theories*, J. Stat. Phys. **83** (1996), 1021–1065.

Grate Furnace Combustion:
A Submodel for the Solid Fuel Layer

H.A.J.A. van Kuijk, R.J.M. Bastiaans, J.A. van Oijen,
and L.P.H. de Goey

Eindhoven University of Technology, P.O. Box 513, 5600 MB,
Eindhoven, The Netherlands
h.a.j.a.v.kuijk@tue.nl

Abstract. The reduction of NO_x-formation in biomass fired grate furnaces requires the development of numerical models. To represent the variety in scales and physical processes playing a role in the conversion, newly developed submodels are required. Here, a submodel for the reverse combustion process in the solid fuel layer on the grate is described. The submodel is shown to give good predictions for the velocity of the combustion front as well as for the spatial profiles of porosity, oxygen mass fraction and temperature. These predictions are essential input for NO_x-calculations.

Keywords: Grate furnace, reverse combustion, biomass.

1 Introduction

A popular way of medium small-scale thermal biomass conversion is the use of grate furnace combustion [1]. The operating conditions and design of such a furnace have to be chosen carefully to meet the NO_x emission limits. Numerical models can support the making of these choices, provided that accurate submodels for the phenomena occurring in the oven are available.

The formation of NO_x in the furnace (cf. [1]) is strongly determined by the combustion process in the furnace. Generally, the conversion processes takes place in two zones. In the primary conversion zone, the solid fuel is gasified on a moving grate. In the secondary conversion zone, additional air is supplied to burn out the resulting gas mixture. Due to the low combustion temperatures in both zones, the main source for the NO is the fuel-N (fuel-bound nitrogen). In the solid fuel layer, the fuel-N is released as so-called N-precursors, mainly NH_3. In the secondary conversion zone, part of these precursors is converted to NO.

The combustion process as well as the NO_x-formation process is highly complex due to the different physical phenomena involved, i.e. transport of heat and mass as well as chemical reactions. Furthermore, a wide range of length scales is involved. The overall combustion process is determined by the conversion of single particles, occurring at a typical length scale of a particle diameter. Multiple particles are converted simultaneously in a reaction wave, propagating

Y. Shi et al. (Eds.): ICCS 2007, Part I, LNCS 4487, pp. 947–954, 2007.

trough the solid fuel layer in the primary conversion zone on a length scale of the bed height. Finally, the gases originating from the solid fuel layer burn out and transfer their heat to the boiler present in the oven at a length scale given by the dimension of the oven.

To represent the variety in scales and physical processes affecting the NO_x-formation process, submodels for (i) the combustion process in the solid fuel layer, (ii) the release of N-precursors from the solid fuel layer (iii) the gas phase combustion process and (iv) the oxidation of the N-precursors are required [2]. Without a submodel for the combustion process in the solid fuel layer, advanced models for the secondary combustion zone (cf. [3]) have to operate with a boundary condition based on measurements just above the solid fuel layer, which strongly restricts their flexibility (cf. [4]).

In developing a model for the solid fuel layer, its resemblance to a fixed bed combustion process can be exploited. In a reference frame traveling at the same velocity as the grate, the combustion process is similar to reverse combustion in a fixed bed [5]. In this mode, air is flowing in the upward direction through the fuel layer, while the reaction front propagates downward. Reverse fixed bed combustion processes can be described with 1-dimensional models.

Existing 1-dimensional numerical models for reverse fixed bed combustion (e.g. [6], [7]) and coal have the disadvantage that due to the great amount of detail in the equations and their time-dependent formulation, they require calculation times that are too long to study changes in furnace design and operating conditions [4]. The complexity may also lead to numerical difficulties (cf. [8]). Application of sensitivity analyses to these models is generally not reported, which raises the question up to which extent their complexity and the resulting calculation times are justified.

The present work addresses the need for a model that is able to describe the reaction front propagating in the solid fuel layer with only the essential processes and at low computational costs. Therefore, a limited set of equations based on a model from the literature [9] is used. The predictions of the model are at the length scale of the bed, while in the model chemistry and transport processes at the level of a single particle are accounted for. At the moment the model presented here is operating with a parameter set for coal combustion, but as the combustion process for wood-like biomass is essentially the same, the model can be adapted for this.

In this work, two important new aspects of the model are presented. Firstly, it is shown that this model can be solved also numerically, while in the past it has been solved analytically only [10]. This opens the possibility to extend the model for increased accuracy and flexibility if necessary. Secondly, it is shown that stationary solutions for the reverse combustion process can be obtained by using a reference frame attached to the combustion front. This enables convenient performance of parametric studies and results in fast calculation times. In addition, with the numerical analysis it can also be shown that the effect of heat and mass and heat transfer limitations in the fuel bed lead to a change in flame structure and thus to the applicability of analytical solution methods. This study is presented elsewhere (cf. [11]).

An important feature of the model is that it can be combined with a submodel for the release rates of N-precursors. These release rates are strongly determined by the conversion rate and temperature, (cf. [12]), two quantities that can be predicted with the model. These quantities can serve as input for a submodel consisting of N-precursor release rates based on measurements in a heated grid reactor (cf. [13]) or TGA (cf. [12]))

It is the purpose of this paper to illustrate the advantages of the numerical implementation of the model and the possibilities it offers to serve as a submodel for grate furnace combustion. An extensive comparison of model results with experiments falls not within the present scope. Suffice it to remark that the model currently predicts conversion rates that have the correct order of magnitude compared to experimental results (eg. [14]) which is good, considering that i) the model is not highly detailed, while fixed bed conversion is a complex process, ii) no adaption of the model constants was performed to fit the model to the measurements, iii) there are large uncertainties in some of the model constants and iv) there are large uncertainties in experimental results reported (compare [10] with [14])

The outline of this paper is as follows. First, the model equations are discussed (Sec. 2). Then, it is shown that the model can predict spatial profiles of the variables and the conversion rate of the solid material (Sec. 3). Finally, the conclusions are presented (Sec. 4).

2 Model Equations and Data

In this section a short review of the model equations and data presented in Ref. [10] is given. The model consists of a set of three equations for the porosity of the fuel bed ϵ, the temperature T and the oxygen mass fraction Y as a function of the spatial coordinate x. In order to obtain this set, the original timedependent model equations have been transformed to a reference frame traveling at the same velocity as the combustion front [9]. This results in:

$$v_s \frac{d}{dx}\left(\rho_s(1-\epsilon)\right) = -R \tag{1}$$

$$v_s \frac{d}{dx}\left(\epsilon\rho_g Y\right) + \frac{d}{dx}\left(\epsilon\rho_g v_g Y\right) - \frac{d}{dx}\left(\rho_g D \frac{dY}{dx}\right) = \nu R \tag{2}$$

$$v_s \frac{d}{dx}\left((1-\epsilon)\rho_s c_{ps} T + \epsilon\rho_g c_{pg} T\right) + \frac{d}{dx}\left(\epsilon\rho_g v_g c_{pg} T\right) - \frac{d}{dx}\left(\Lambda \frac{dT}{dx}\right) = -\Delta H_r R \tag{3}$$

Here, v_s is the velocity of the solid, which is moving due to the change in reference frame. The first terms on the rhs. of each equation are convective fluxes involving v_s. These result from the transformation of time-derivatives in the original reference frame. Other symbols in this set of equations are the reaction source term R, reaction enthalpy ΔH_r, effective coefficients for dispersion D and conduction Λ as well as specific heat c_p and density ρ for gas and solid (subscripts g and s, respectively). The boundary conditions are given at the unburnt side $(u; x \to -\infty)$ and burnt side $(b; x \to \infty)$

$$\epsilon|_u = 0, \qquad Y|_u = Y_u, \qquad T|_u = T_u; \qquad \frac{dY}{dx}\bigg|_b = 0 \qquad \frac{dT}{dx}\bigg|_b = 0. \qquad (4)$$

In this paper, instead of using v_s and v_g, results are presented in terms of the gas mass flux and solid mass flux at the unburnt side, given by

$$m_{gu} = \epsilon_u \rho_{gu} v_{gu}; \qquad m_{su} = (1 - \epsilon_u)\rho_s v_{su}. \qquad (5)$$

Summarizing, the stationary model equations (1,2) and (3) together with the boundary conditions can be used to obtain $\epsilon(x)$, $Y(x)$ and $T(x)$ as well as m_{su} as a a function of gas mass flux m_{gu}.

Expressions for the reaction source term, the diffusion coefficient and the conductivity coefficient are required to solve the model. To obtain an expression for R the heterogeneous surface reaction of coal with oxygen

$$C + \nu O_2 \to (1 + \nu)\, CO_2, \qquad (6)$$

is used. As measurements indicate that the primary product in the reaction zone is CO_2 [10], this is a reasonable assumption. The volumetric reaction rate per unit volume can be expressed by

$$R = \frac{1}{k_r^{-1} + k_m^{-1}} S^{2/3} \rho_g Y. \qquad (7)$$

Here, k_r is an Arrhenius rate constant, k_m a mass transfer coefficient and S the specific surface area of the fuel bed. Introducing the dimensionless mass transfer coefficient,

$$K_m = \frac{k_m}{k_r} \qquad (8)$$

it can be seen that for $K_m \to \infty$, R is limited by kinetics, while $K_m \to 0$ indicates that R is limited by mass transfer effects. In the kinetically limited case, the expression reduces to that given by Field [15]:

$$R = k_r \rho_g Y S. \qquad (9)$$

with k_r given by

$$k_r = A_f T \exp\left(-\frac{E_a}{RT}\right), \qquad (10)$$

in which the pre-exponential factor is indicated with A_f and the activation energy by E_a. The specific surface area in the rate expression can be related to particle properties. For spherical particles of diameter d_p in an SC (simple cubic) structure reacting as a shrinking core, the specific surface area during conversion can then be expressed by [10]

$$S = S_u \left(\frac{1 - \epsilon}{1 - \epsilon_u} \right)^{2/3}.$$ (11)

Finally, expressions for the gas density, the effective conductivity the dispersion coefficient and the mass transfer coefficient are needed. The gas density is given by

$$\rho_g = \rho_{gu} \frac{T_u}{T_b}.$$ (12)

Correlations for Λ, D and k_m are given in [10]. The model equations and additional data are implemented in the one dimensional laminar flame code Chem1d [16].

Table 1. Data used for the calculations for a propagating reaction front in a bed of coal particles [10]

Parameter	Unit	Value
T_u	K	288
ΔH	$J\,g^{-1}$	28.9×10^3
c_{ps}	$J\,kg^{-1}K^{-1}$	1.8
ρ_s	$g\,cm^{-3}$	1220×10^{-3}
ρ_{sa}	$g\,cm^{-3}$	470×10^{-3}
d_p	cm	2.5
$\rho_{gu}(T_u)$	$g\,cm^{-3}$	1.226×10^{-3}
c_{pg}	$J\,kg^{-1}K^{-1}$	1.13
Y_u	−	0.23
A_f	$cm\,K^{-1}s^{-1}$	6×10^4
E_a	$kJ\,mol^{-1}$	150
ν	−	2.67

3   Results

Model results consist of spatial profiles of ϵ, T and the heat release $\dot{q}$. Here, we present solutions for varying m_{gu} (cf. Fig. 1). The latter quantity can be determined by the equation $\dot{q} = -\Delta H_r R$ and is a measure for the magnitude reaction source term. The results show that with increasing m_{gu}, $\dot{q}$ increases. At the same time, the width of the boundary layers in the profiles for T and Y increases. Due to the increased flow of air, the conversion of the solid fuel ϵ_b as well as T_b become higher.

From the spatial flame profiles for different m_{gu}, the conversion ϵ_b as well as T_b can be determined as a function of m_{gu} (cf. Figs. 2b, 2c). By using T_b in the evaluation of K_m (cf. (8)), K_{mb} can be determined (cf. Fig. 2d). Furthermore, m_{su} results from the calculations (cf. Fig. 2a). It can be observed that m_{su} initially increases when m_{gu} increases, until a maximum is reached. Then, m_{su}

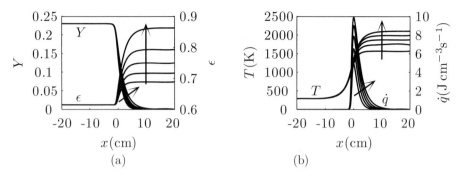

Fig. 1. Spatial flame profiles for varying gas mass flux. Results for Y and ϵ (a) and T and $\dot{q}$ (b). The arrows indicate results obtained with increasing m_{gu} from $m_{gu} = 4.9 \times 10^{-3}$ g cm^{-2}s^{-1} to $m_{gu} = 14.6 \times 10^{-3}$ g cm^{-2}s^{-1} in steps of 2.4×10^{-3} g cm^{-2}s^{-1}.

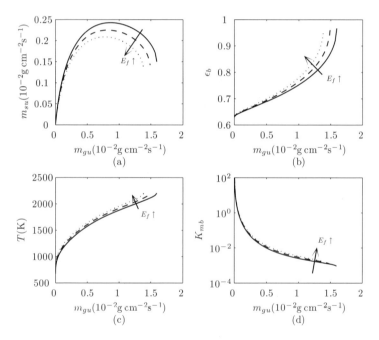

Fig. 2. Flame parameters as a function of gas mass flux for varying activation energy. Results for the solid mass flux (a), the porosity (b), the temperature (c) and dimensionless mass transfer coefficient (d). Solid line: $E_a = 150$ kJ mol^{-1}, dashed line: $E_a = 160$ kJ mol^{-1}, dotted line: $E_a = 170$ kJ mol^{-1}.

decreases until extinction takes place. The shape of the curve of m_{su} as a function of m_{gu} (Fig. 2) was explained in earlier works in terms of the counterbalancing effects of the increased oxygen supply to and and increased convective heat transport out of the reaction zone with increasing m_{gu} (cf. [17]). The temperature T_b and porosity ϵ_b increase continuously with increasing m_{gu}. (Due to limitations of the current numerical implementation, the extinction point $\epsilon_b = 1$ is not reached completely). The dimensionless mass transfer coefficient K_{mb} (i.e. K_m evaluated at T_b) shows that for low m_{gu}, the source term is controlled by kinetics ($K_{mb} > 1$). At $m_{gu} \sim 1.0 \times 10^{-3}$ g cm^{-2}s^{-1} mass transfer effects begin to play a role ($K_{mb} < 1$). Furthermore, it can be seen that all parameters are sensitive to changes in E_a.

A typical calculation time to obtain results for a variation of m_{gu} from zero up to the extinction point is only ten minutes on a PC. Furthermore, with other models for coal beds with an instationary formulation (e.g. [18]), it is no longer necessary to wait until a steady state is reached when a parameter is varied.

4   Conclusions

From the numerical results presented in this paper, it can be concluded that the two most important quantities that determine the release of the N-precursors can be predicted, namely the front velocity and temperature. These quantities can be determined as a function of gas mass flow, the main controlling parameter in a grate furnace. In the calculations, both kinetics and mass transfer limitations can be taken in into account. Furthermore, the effect of fuel parameters (particle size and bed density) that show strong variations in practical combustion processes has been included via the correlations used to determine the model coefficients.

The implementation of a stationary set of equations as is performed here has two important advantages. Firstly, the calculation times are short. This makes the method proposed here suitable for application as a submodel in a grate furnace. Moreover, the the model can be extend further with additional important phenomena and still remain within the limits of reasonable calculation times. In addition, the instationary implementation makes it convenient to perform parametric studies. This makes it achievable to study the effect of additional phenomena included in the model by means of a sensitivity analysis to determine their effect on the velocity of the reaction front, the temperature and the structure of the reaction zone.

First, the model will be extended to describe the additional chemistry necessary for biomass conversion by using the Chem1D modules for the gas phase kinetics. It is also planned to include the effect of other parameters that are shown to have a large impact on the combustion process in practical applications, e.g. the moisture content of the fuel.

References

1. S. van der Loo, H. Koppejan, Handbook of Biomass: Combustion and Cofiring, Twente University Press, 2002.
2. R. J. M. Bastiaans, H. A. J. A. van Kuijk, B. A. Albrecht, J. A. van Oijen, L. P. H. de Goey, Catalysis for Renewables, Wiley-VCH, 2006, Ch. Thermal biomass conversion and NOx emissions.

3. B. A. Albrecht, R. J. M. Bastiaans, J. A. van Oijen, L. P. H. de Goey, NOx emissions modelling in biomass combustion grate furnaces, in: A. Reis, W. J., W. Leuckel (Eds.), Proceedings of the 7th European Conference on Industial Furnaces and Boilers, Porto, Portugal, 2006.

4. R. Scharler, T. Feckl, I. Obernberger, Modification of a Magnussen constant of the Eddy Dissipation Model for biomass grate furnaces by means of hot gas in-situ FT-IR absorption spectroscopy, Progress in Computational Fluid Dynamics 3 (2-4) (2003) 102–111.

5. H. Thunman, B. Leckner, Ignition and propagation of a reaction front in cross-current bed combustion of wet biofuels, Fuel 80 (4) (2001) 473–481.

6. H. Thunman, B. Leckner, Influence of size and density of fuel on combustion in a packed bed, Proceedings of the Combustion Institute 30 (2) (2005) 2939–2946.

7. J. C. Wurzenberger, S. Wallner, H. Raupenstrauch, J. G. Khinast, Thermal conversion of biomass: Comprehensive reactor and particle modeling, AIChE Journal 48 (10) (2002) 2398–2411.

8. P. T. Radulovic, M. U. Ghani, L. D. Smoot, An improved model for fixed bed coal combustion and gasification, Fuel 74 (4) (1995) 582–594.

9. R. Gort, J. J. H. Brouwers, Theoretical analysis of the propagation of a reaction front in a packed bed, Combust. Flame 124 (1-2) (2001) 1–13.

10. R. Gort, On the propagation of a reaction front in a packed bed, Ph.D. thesis, Twente University, Enschede, The Netherlands (1995).

11. H. A. J. A. van Kuijk, R. J. M. Bastiaans, J. A. Oijen, L. P. H. de Goey, Reverse fixed bed combustion: kinetically controlled and mass transfer controlled flame structures. To be published.

12. W. de Jong, A. Pirone, M. A. Wojtowicz, Pyrolysis of miscanthus giganteus and wood pellets: TG-FTIR analysis and reaction kinetics, Fuel 82 (9) (2003) 1139–1147.

13. A. Toland, R. J. M. Bastiaans, A. Holten, L. P. H. Goey, Kinetics of CO release from bark and medium density fibre board pyrolysis, Biomass and Bioenergy. Submitted for publication.

14. P. Nicholls, Underfeed combustion, effect of preheat, and distribution of ash in fuel beds, Bulletin of the United States Bureau of Mines 378 (1934) 1–76.

15. M. A. Field, D. W. Gill, B. B. Morgan, P. G. W. Hawksly, Combustion of Pulverized Coal, BCURA, Leatherhead, 1967.

16. Chem1d, A one-dimensional laminar flame code. www.combustion.tue.nl.

17. D. Lozinski, J. Buckmaster, Quenching of reverse smolder, Combust. Flame 102 (1-2) (1995) 87–100.

18. J. Cooper, W. L. H. Hallett, A numerical model for packed-bed combustion of char particles, Chemical Engineering Science 55 (20) (2000) 4451–4460.

Introduction to the ICCS 2007 Workshop on Dynamic Data Driven Applications Systems

Frederica Darema

National Science Foundation, USA
darema@nsf.gov

Abstract. This is the 5th International Workshop on Dynamic Data Driven Applications Systems (DDDAS), organized in conjunction with ICCS. The DDDAS concept entails the ability to dynamically incorporate data into an executing application simulation, and in reverse, the ability of applications to dynamically steer measurement processes. Such dynamic data inputs can be acquired in real-time on-line or they can be archival data. DDDAS is leading to new capabilities by improving applications modeling and systems management methods, augmenting the analysis and prediction capabilities of simulations, improving the efficiency of simulations and the effectiveness of measurement systems. The scope of the present workshop provides examples of research and technology advances enabled through DDDAS and driven by DDDAS. The papers presented in this workshop represent ongoing multidisciplinary research efforts by an international set of researchers from academe, industry, national and research laboratories.

Keywords: applications, measurements, dynamic runtime, sensors, grids.

1 Introduction

The Dynamic Data Driven Applications Systems (DDDAS) [1,6] concept entails the ability to dynamically incorporate data into an executing application simulation, and in reverse, the ability of applications to dynamically steer measurement processes. Such dynamic data inputs can be acquired in real-time on-line or they can be data retrieved from archival storage. The DDDAS concept offers the promise of improving modeling methods, augmenting the analysis and prediction capabilities of application simulations, improving the efficiency of simulations and the effectiveness of measurement systems. The kind of multidisciplinary advances that can be enabled through the DDDAS concept, as well as technology capabilities required to support DDDAS environments, have been discussed in [1-8, 10-11]; papers presented in the DDDAS/ICCS Workshops Series provide examples of advances enabled through an increasing number of projects. The DDDAS Program [5] announced in 2005, with seeding efforts on DDDAS having started previously (2000 – 2005) through the NSF ITR Program [6], have provided some initial funding support for these efforts. In particular, the DDDAS Program was co-sponsored by multiple NSF Directorates and Offices, and by NOAA and NIH, and in cooperation with Programs in the European

Y. Shi et al. (Eds.): ICCS 2007, Part I, LNCS 4487, pp. 955–962, 2007.

Community and the United Kingdom [5]. The value of DDDAS has been increasingly recognized by the broader community and has been discussed in the reports of "blue ribbon" studies [7, 8] and community driven workshops [10, 11]. DDDAS is a "revolutionary and important" direction [7]. This impact of DDDAS to scientific, engineering, and societal areas is increasingly becoming widely and internationally recognized: the 2006 NSF sponsored DDDAS Workshop [4], drew participation from multiple US agencies, and there is an established cooperation of European programs with the US on DDDAS. Furthermore, at the same time there is an increasing interest by industry on DDDAS, and this is manifested through the participation of industry in a number of the DDDAS research projects.

The need for synergistic multidisciplinary research in enabling DDDAS capabilities has been articulated from the outset and consistently thereafter [1-8]; and specifically, there is need for synergistic multidisciplinary research advances in applications modeling, in mathematical and statistical algorithms, in measurement methods and data management, and in computer systems software, targeted towards creating DDDAS capabilities and environments. Furthermore, in [2] the case is made that the DDDAS concept leads to an infrastructure that goes beyond the traditional notions of computational grids, and provides an impetus for new directions and capabilities in CyberInfractrucure environments [9]. Presently an increasing number of multidisciplinary research efforts on DDDAS is being pursued by an international set of researchers from academe, industry, national and research laboratories, leading to new capabilities in application modeling and measurement capabilities, computer sciences methods, and development of enhanced software systems architectural frameworks supporting DDDAS environments, and driving new directions in CyberInfrastructure [4].

The efforts presented in this workshop provide examples of technologies and new capabilities that are being developed based on the DDDAS concept and in supporting DDDAS environments. The www.cise.nsf.gov/dddas and www.dddas.org sites are source of information on a multitude of projects and other efforts encompassed under the DDDAS rubric. The present workshop is the fifth of a series of international workshops on the DDDAS topic that have been organized as part of ICCS, starting with ICCS'03. The Workshop provides a forum where project investigators update the broader community on the progress in their project. In addition, the discussions that take place, provide an opportunity for leveraging advances made in related projects, and facilitating objectives for creating collaborations and liaisons among complimentary projects, thus leading "systems of systems" through horizontal integration across projects, and creation of "software architectural frameworks" through vertical integration leveraging across projects. In addition to DDDAS/ICCS and the NSF sponsored workshops [1, 6], other workshops and community building activities are organized by the research community [e.g. 10, 11] further broadening the outreach to other researchers and application areas. The papers presented in the present workshop represent a sample of the scope of technical efforts pursued. This introduction serves as an overview to place in context the work presented in the workshop.

2   Overview of Work Presented in This Workshop

The papers presented in this workshop represent a sample of the ongoing research efforts to enable DDDAS capabilities, in a wide set of areas, such as: enhanced civil and mechanical systems; health systems; critical infrastructure systems, such as electrical power grids, transportation and urban water distribution systems, emergency detection, response and disaster mitigation; enhanced methods for severe weather analysis and prediction; air, water, and subsurface chemical contaminant propagation and mitigation of damage; human and social systems; and enhanced computer systems software methods and environments. The papers represent ongoing efforts synergistic, multidisciplinary research advances in applications modeling, applications algorithms, measurement methods, and in systems software. These efforts contribute and build the layers of software architectural frameworks that are being developed to enable and support DDDAS capabilities and requirements. A number of projects have interactions and collaborations with relevant industrial partners, and with regional and other governmental organizations, to utilize existing infrastructure resources and to demonstrate in realistic situations the new capabilities developed. Each paper, together with an overview of the respective project, discusses at depth recent technical advances made in the project.

Papers [12-15] discuss applications related to human health: from personalized pharmaceuticals, to advanced medical intervention and customized medical treatment. In [12] is discussed the onset of a paradigm shift in pharmaceutical and biotechnology research methods, leading towards a future world of personalized medicines, which will be enabled through a foundational framework that will be based on the DDDAS concept.   In [13] new frameworks and architectures for brain-machine interface communication and control capabilities are developed, using mixture models to enhance rehabilitation of disabled patients.   In [14] and [15] are presented new modeling methods and distributed computational frameworks for advanced medical procedures.   Paper [14] discusses an image-guided neurosurgery and grid-enabled software environment by integrating real-time acquisition of intra-operative Magnetic Resonance Imaging (I-MRI) with the preoperative MRI, fMRI, and DT-MRI data.   Paper [15] discusses work on novel advances for cancer tumor treatment by utilizing real-time imaging data to provide dynamic control of a laser heat source targeted to ablate only the tumor-tissue, and at the same time imparting minimal damage to the tumor surrounding area.

In papers [16-22], new applications and methods are enabled in the areas of environmental and natural resource management. Dynamic streaming of data into the application models and architecture and adaptive control of sensor networks, is a common aspect of the work discussed here.   Papers [16] and [17] discuss new capabilities on DDDAS methods for application models driving and establishing system architecture, enabling adaptive management and control of stationary and mobile sensors for monitoring ecosystems such as forests [16] and aqueous environments [17], and addressing issues such reducing errors and uncertainty, dynamic control of the sensor network for power management, and adaptive placement of mobile sensors.   Papers [18-21] address advances through DDDAS in contaminant tracking in subsurface and aqueous environments (e.g. oil spills in [18]), or in the atmosphere ([19-21]). In [19] DDDAS methods are used to perform analysis and accurately predict the path of the contaminant, by effectively solving the forward problem, as well as the inverse problem, so that the location of the contaminant source can be identified. Furthermore the project is developing enhanced reduced-order-modeling approaches to enable such predictions and analysis to be done in near-real time. Papers [20] and [21] use the DDDAS concept to develop enhanced Kalman Filtering methods and an integrated framework for analysis of transport of atmospheric chemical pollutants (like sulfur) in wide geographical regions. In the context of a research effort addressing the development of DDDAS environments for coastal and environmental applications, paper [22] focuses on discussing the advances in a software framework for providing accurate and timely analysis for hurricane events. Paper [23] uses DDDAS approaches for improving analysis and the accuracy of predicting the path of wildfires, through dynamic incorporation, into executing simulations, measurement data dynamically acquired through aerial survey, thermal sensors on the ground, winds and humidity measurements, together with archival data on the terrain and vegetation.

Papers [24-28] discuss work that is aimed to create DDDAS-based enhanced critical infrastructure systems, where simulation models use dynamic data inputs and control the data acquisition process in real-time evolving conditions to provide optimized planning, operation and response and mitigation of the impact of adverse unexpected conditions. Advances made include work reported in [24] on ad-hoc simulation modeling and monitoring data for adaptive management of transportation systems to optimize conditions in urban traffic and to provide decision support for responding to adverse and disruptive events such as needs emergency evacuation of a city area. Work reported in [25] on adaptive monitoring and detection of water contamination in a city system, and determining the source of the contaminant; this work can lead to more safe urban water systems and provide response capabilities in cases of adverse events affecting a city's water system. Papers [26-28] discuss the development of DDDAS environments for optimized management of electrical power-grid systems, to predict, respond, and mitigate, failures such as power demand surge, or disruptive events such as failure of transformers and thus avoid extensive blackouts. The methods include streaming monitoring data in to continually running simulation models, and through simulation control determine adaptively the kinds of further monitoring needed, as well as develop robust methods in the presence of measurement uncertainties.

Papers [29-32] deal with behavioral and situational awareness management systems, which, through DDDAS approaches, seek to develop advanced analysis, prediction, and response capabilities. Paper [29] discusses work where data from continually monitored cell-phone activity is streamed into agent-based simulations, to detect potentially disruptive events, like traffic congestion or an adverse event in a large crowd gathering, or evacuation situations. Work discussed in [30] is aimed at developing methods and frameworks to support large DDDAS "mixed reality" environments of agent-based simulations together with full scale preparedness exercise, by modeling individual and group dynamics, and using on-line monitoring for analyzing and predicting the onset of behaviors and actions, in situations such as the effects and impact of an epidemic disease, or a chemical or radiological release, and guiding responders actions as such events evolve. Paper [31] discusses DDDAS methods to track facial expressions and body movement to detect human behaviors such as deception and predict intent. In [32] is the use of DDDAS approaches to enhance fidelity of simulations for social sciences studies environments, through adaptive data collection processes, to minimize the impact of input data uncertainties.

Papers [33-38] discuss novel methods of weather analysis and forecast, including for situations of severe or extreme weather events. In [33] and [34] are discussed DDDAS-based simulation and measurement methods and software frameworks, for supporting the required cyber-infrastructure for tornadic events, and for improving the accuracy of predicting the tornado path. Paper [33] presents software infrastructure advances needed to support real-time, dynamic and adaptive feed-back loop between dynamic incorporation of monitoring data into the executing simulation models, and in return continuous and simulation-based adaptive control of multiple and heterogeneous on-line measurements, such as radar measurements, winds, atmospheric pressure, temperature, humidity, etc. Work in [34] is developing new algorithmic methods, by applying active learning with support vector machines approaches, to improve the methods for discriminating between conditions that lead to onset of tornados versus those that do not.   In [35] are discussed new advances in adaptive observation strategies through integration of non-linear adaptive weather prediction models, as well as algorithms that learn from past-experience, for adaptive planning and control of the best discrete locations and times for additional measurements under time-varying conditions, in order to minimize the forecast errors for "high-impact" weather in the Pacific.   Paper [36] uses DDDAS to enhance and overcome limitations of traditional data-assimilation approaches; the new methods reported here treat jointly position and amplitude data assimilation, and extend the methodology to multivariate fields, as a generalized solution applicable to meteorological, oceanographic and hydrological applications.   While measurements in ambient environments are very important and used extensively in many modeling efforts (including many of the ones presented in this workshop), there is value also derived from more controlled measurement environments; this is the approach in the efforts discussed in [37] and [38] which use DDDAS-based adaptive modeling and measurement methods for weather forecasting. Paper [37] discusses the experimental infrastructure as a testing observatory for laboratory simulations of planetary flows that functions in real- time. Paper [38] discusses the software methods for more efficient execution of the weather model simulator, including fluids modeling, algorithms and performance.

Papers [39-43] address DDDAS-based methods on structural monitoring, fault-tolerance, and optimized management of systems like buildings, aircraft, wind-turbines, fluid thermal systems, optimized performance in materials. In [39] are presented advances through DDDAS for prediction of "near real-time" structural dynamics and impact, in a structure like an aircraft, through reduced-order modeling methods integrated with on-line sensor data, together with the development of a novel sensor-data compression algorithm and it's application to fast detection of structural damage. The research pursued in [40] applies DDDAS to materials characterization by enabling new methods for determining a priori and as the experiment proceeds the collection of data needed to characterize the material. Paper [41] is employing DDDAS for evaluation of fluid-thermal systems for conditions such of a heated jet injected into a laminar boundary layer. Paper [42] is applying DDDAS to novel capabilities for more effective fault diagnosis is wind turbines, and more efficient operation of such systems. The efforts in [43] are developing new systems methods for supporting DDDAS environments for analysis of structural aspects of sensor instrumented buildings, and specifically new methods for managing distributed communication and coordination of over large self-organized networks of heterogeneous devices, under critical conditions with resource constraints and needs for prioritization of resource allocation.

Papers [44-48] discuss algorithmic and software advances for supporting DDDAS environments. The focus of paper [44] is on novel algorithms to support parallel multi-block numerical models coupled with global stochastic optimization algorithms, dynamic multi-dimensional, scientific datasets, and adaptive runtime management strategies, resulting into a software architectural framework for developing large-scale and complex decision-making DDDAS systems. In [45] methods are developed for dynamic intrusion detection and a software framework to support the requirements of such applications. Work in [46] uses DDDAS to improve on the precision and robustness of signal and speech recognition process. In [47] are discussed methods and a framework to enable verification of dynamically composable and configurable simulations, such as those required in DDDAS environments. Paper [48] discusses the challenges of propagation and compounding of uncertainties and novel methods on increasing confidence in evolving complex simulations in DDDAS environments.

3 Summary

DDDAS has been gaining recognition as an important and revolutionary concept that creates new capabilities in applications and measurement methods and systems. Through efforts that are developing DDDAS environments, we are in a strong path towards accomplishing such objectives and realize the DDDAS promise.

References

1. NSF Workshop, March 2000; www.cise.nsf.gov/dddas
2. F. Darema, Grid Computing and Beyond: The Context of Dynamic Data Driven Applications Systems, Proceedings of the IEEE, Special Issue on Grid Computing, March 2005
3. F. Darema, *Dynamic Data Driven Applications Systems: A New Paradigm for Application Simulations and Measurements*, ICCS'04; F. Darema, Dynamic Data Driven Applications Systems: New Capabilities for Application Simulations and Measurements, ICCS'05; and F. Darema, Introduction to the ICCS2006 Workshop on Dynamic Data Driven Applications Systems
4. NSF Sponsored Workshop on DDDAS-Dynamic Data Driven Applications Systems, Jan 19-20, 2006 [www.cise.nsf.gov/dddas]
5. DDDAS-Dynamic Data Driven Applications Systems Program Solicitation (NSF 05-570); www.cise.nsf.gov/dddas
6. NSF Information Technology Research (ITR) Program (1999-2004)
7. *SBES: Simulation-Based Engineering Science - Revolutionizing Engineering through Simulation,* May 2006; (http://www.ices.utexas.edu/events/SBES_Final_Report.pdf)
8. Defense Modeling, Simulation, and Analysis – Meeting the Challenge, National Reseach Council of the National Academies, Report, 2006,

9. Cyberinfrastructure Report http://www.communitytechnology.org/nsf_ci_report
10. International Mini-Workshop on DDDAS (Dynamic Data Driven Application Systems: *"Optimal measurement and control of distributed parameter systems using mobile actuator and sensor networks"*; Dec. 2006, Utah State University (CSOIS: www.csois.usu.edu)
11. 2006 DDDAS and Environmental Problems Workshop, National Laboratory for Scientific Computing (LNCC), Petropolis, Brasil, July 10-14, 2006
12. *Pharmaceutical Informatics and the Pathway to Personalized Medicines*; Sangtae Kim and Venkat Venkatasubramanian
13. *Towards Real-Time Distributed Signal Modeling for Brain Machine Interfaces;* Jack DiGiovanna, Loris Marchal, Prapaporn Rattanatamrong, Ming Zhao, Shalom Darmanjian, Babak Mahmoudi, Justin Sanchez, José Príncipe, Linda Hermer-Vazquez, Renato Figueiredo and José Fortes
14. *Grid-Enabled Software Environment for Enhanced Dynamic Data-Driven Visualization and Navigation During Image Guided Neurosurgery;* Nikos Chrisochoides, Andriy Fedorov, Andriy Kot
15. *Using Cyber-Infrastructure for Dynamic Data Driven Laser Treatment of Cancer;* C. Bajaj, J. T. Oden, K. R. Diller, J. C. Browne, J. Hazle, I. Babu¡ska, J. Bass, L. Bidaut, L. Demkowicz, A. Elliott, Y. Feng, D. Fuentes, B. Kwon, S. Prudhomme, R. J. Sta_ord, and Y. Zhang
16. *From Data Reverence to Data Relevance: Model-Mediated Wireless Sensing of the Physical Environment;* Paul G. Flikkema, Pankaj K. Agarwal, James S. Clark, Carla Ellis, Alan Gelfand, Kamesh Munagala, and Jun Yang
17. *AMBROSia: an Autonomous Model-Based Reactive Observing System;* David Caron, Abhimanyu Das, Amit Dhariwal, Leana Golubchik, Ramesh Govindan, David Kempe, Carl Oberg, Abhishek Sharma, Beth Stauffer, Gaurav Sukhatme, Bin Zhang
18. *Dynamically Identifying and Tracking Contaminants in Water Bodies;* Craig C. Douglas, Martin J. Cole, Paul Dostert, Yalchin Efendiev, Richard E. Ewing, Gundolf Haase, Jay Hatcher, Mohamed Iskandarani6, Chris R. Johnson, and Robert A. Lodder
19. *Hessian-based model reduction for large-scale data assimilation problems;* Omar Bashir, Omar Ghattas, Judith Hill, Bart van Bloemen Waanders, and Karen Willcox
20. *Localized Ensemble Kalman Dynamic Data Assimilation for Atmospheric Chemistry;* Adrian Sandu, Emil M. Constantinescu, Gregory R. Carmichael, Tianfeng Chai, John H. Seinfeld, and Dacian D¢aescu
21. *Data Assimilation in Multiscale Chemical Transport Models;* Lin Zhang and Adrian Sandu
22. *Building a Dynamic Data Driven Application System for Hurricane Forecasting*; Gabrielle Allen
23. *A Dynamic Data Driven Wildland Fire Model;* Jan Mandel, Jonathan D. Beezley, Lynn S. Bennethum, Soham Chakraborty, Janice L. Coen, Craig C. Douglas, Jay Hatcher, Minjeong Kim, and Anthony Vodacek
24. *Ad Hoc Distributed Simulation of Surface Transportation Systems;* R. M. Fujimoto, R. Guensler, M. Hunter, K. Schwan, H.-K. Kim, B. Seshasayee , J. Sirichoke, W. Suh
25. *Cyberinfrastructure for Contamination Source Characterization in Water Distribution Systems;* Sarat Sreepathi, Kumar Mahinthakumar, Emily Zechman, Ranji Ranjithan , Downey Brill, Xiaosong Ma, and Gregor von Laszewski
26. *Integrated Decision Algorithms for Auto-Steered Electric Transmission System Asset Management;* James McCalley, Vasant Honavar, Sarah Ryan, William Meeker, Daji Qiao, Ron Roberts, Yuan Li, Jyotishman Pathak, Mujing Ye, Yili Hong

Pharmaceutical Informatics and the Pathway to Personalized Medicines

Sangtae Kim[*] and Venkat Venkatasubramanian

Purdue University, West Lafayette, IN 47907, USA
kim55@purdue.edu

Abstract. The present computational era has witnessed the evolution of the field of computational biology into two somewhat disjoint communities: one group focused on the "wriggling of molecules" (e.g. molecular dynamics) and another focused on pattern matching and informatics (e.g. sequence alignment). The separation starts relatively early in the training of the community, as is evident from the relative weighting of calculus vs. computer science courses in the curriculum of undergraduate majors in the various disciplines of computational biology. But at the dawn of the petascale era in computational science, exciting advances in healthcare can be envisioned at a new niche in the computational biology ecosystem – a niche formed by the reunion of the separated branches of computational biology. We consider some illustrative examples from pharmaceutical informatics of data-intensive informatics approaches applied to "objects" that are not simply numbers or static molecular structures but instead are (CPU-intensive) molecular simulations. These results suggest a promising pathway for applying petascale informatics to shift the paradigm in pharmaceutical and biotechnology R&D to a future world of personalized medicines. The early examples from pharmaceutical informatics also indicate that the foundational framework will be based on the concept of Dynamic Data Driven Application Systems (DDDAS).

---

[*] Conference speaker.

Y. Shi et al. (Eds.): ICCS 2007, Part I, LNCS 4487, p. 963, 2007.
© Springer-Verlag Berlin Heidelberg 2007

Towards Real-Time Distributed Signal Modeling for Brain-Machine Interfaces

Jack DiGiovanna[1], Loris Marchal[2], Prapaporn Rattanatamrong[2], Ming Zhao[2], Shalom Darmanjian[2], Babak Mahmoudi[1], Justin C. Sanchez[3], José C. Príncipe[2], Linda Hermer-Vazquez[4], Renato Figueiredo[2], and José A.B. Fortes[2]

[1] Dep. of Biomedical Engineering, [2] Dep. of Electrical and Computer Engineering, [3] Dep. of Pediatrics, [4] Dep. of Psychology
University of Florida, Gainesville, Florida, USA
{fortes, renato, lindahv, principe, jcs77}@ufl.edu

Abstract. New architectures for Brain-Machine Interface communication and control use mixture models for expanding rehabilitation capabilities of disabled patients. Here we present and test a dynamic data-driven (BMI) Brain-Machine Interface architecture that relies on multiple pairs of forward-inverse models to predict, control, and learn the trajectories of a robotic arm in a real-time closed-loop system. A method of window-RLS was used to compute the forward-inverse model pairs in real-time and a model switching mechanism based on reinforcement learning was used to test the ability to map neural activity to elementary behaviors. The architectures were tested with *in vivo* data and implemented using remote computing resources.

Keywords: Brain-Machine Interface, forward-inverse models.

1 Introduction

Among other applications, Brain-machine interfaces (BMIs) for motor control can potentially enable patients with spinal injuries to regain autonomy as well as create new and revolutionary forms of man-machine interaction. Research is providing an increased understanding of how brains control and learn movement, how neuronal signals can be detected and processed, and how computers can be used to control artificial limbs in real time. This paper reports progress of our work in these fronts.

In [2] we described a BMI architecture that relies on pairs of forward-inverse models to predict, control and learn the trajectories of a robotic arm in a real-time closed-loop system. Section 2 extends it into a dynamically data-driven adaptive system (DDAS) where the choice and learning of computational models is determined by predicted model responsibilities and observed trajectory errors. The experimental setup is described in Section 3. The real-time computation and learning of the models require the matrix operations and adaptive algorithms discussed in Section 4. Experimental results and conclusions are reported in Sections 5 and 6.

Y. Shi et al. (Eds.): ICCS 2007, Part I, LNCS 4487, pp. 964–971, 2007.
© Springer-Verlag Berlin Heidelberg 2007

2   A Mixed-Model DDDBMI Architecture

The basis of our design methodology for BMIs is rooted in a model-based approach to motor control proposed by Kawato and Wolpert [1, 3]. The major difference is that for BMIs the inputs are a mixture of neuronal outputs from motor cortex and feedback signals which are processed by computer hardware to generate an output to a prosthetic arm (Fig 1). The architecture uses the concept called the Multiple Paired Forward-Inverse Model (MPFIM) which consists of multiple pairs of models, each comprising a forward model (for movement planning) and an inverse model (for movement execution). Individual *model-pairs* or combinations of model-pairs control motion on the basis of feedback data (visual or proprioceptive).

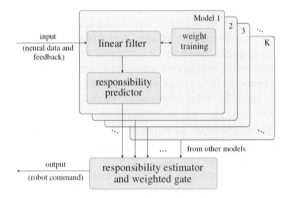

Fig. 1. Multiple paired models and the responsibility predictor for the DDDAS based BMI

In the DDDAS framework, the feedforward models are implemented with conventional optimal input-output mappers (adaptive FIR filters) using supervised learning [4]. Therefore, we include both the neural signal and musculoskeletal transformation of those signals to prosthetic joint actuations (lookup table) in our inverse model. The MPFIM architecture requires a Responsibility Predictor/Estimator (RP) to serve as the gating mechanism to select which local 'expert' (forward-inverse model pair) is most appropriate based on the current system inputs. In the original architecture, the RP uses 'contextual signal' such as 'sensory information and cognitive plans' to determine the responsibility of each local expert at the current time based on the state of the motor control system and environment [1]. Research suggests that the a representation of the state of the motor control system is present in the motor cortex [5]. Following this view, this work estimates the firing rates from the forelimb area of primary motor cortex [6] to create a contextual (state) signal.

The RP is trained to map neural state to the appropriate forward-inverse model pair. That model pair generates an elementary action (or moveme analogous to phoneme) for the prosthetic limb. To determine the model-pairs for each neural state, we use a modified version of reinforcement learning (RL), which provides a learning mechanism that is similar to those of a biological motor system [7]. Indeed, the learner is not told what actions to take but must discover which actions yield the most reward by trying them [8], very similar to operant conditioning in psychology.

Conceptually, RL teaches an agent a policy to interact with their environment in a fashion that maximizes reward over time [8]. RL requires 3 signals to try to learn the optimal policy: the 'state' of the environment, the 'reward' provided by the environment, and the 'action' taken by the agent. From the BMI perspective, 'reward' is achieved by reaching a targeted position. For RL, a positive reward signal is generated for task completion (i.e. reach-and-grasp); to encourage efficiency, the reward signal is negative for actions that do not immediately complete the task.

3 Experimental Design

To investigate and exploit the distributed representation of neurons in the rat motor cortex and implement them in a novel computational framework, we have developed a novel experimental design involving a "reach and grasp" task where a cue is given to initiate controlled movement of a robotic arm from a rest position to one of two targets. This paradigm is designed to be comparable to a real situation where the patient's neural signal and rewards are known but kinematic variables are unavailable. We developed a "testbed" behavioral task that rats could perform stereotypically across trials, that could be relatively easily modeled using our grid computing approach, that would facilitate rats' switching from arm control to brain control, and finally, that could be modified to introduce brain control of the MPFIM without the availability of a desired response, as is conventionally done in neurorobotic research. The animal training paradigm (see Fig. 2) was designed to gradually shape the rats into the desired reaching behavior. Neural firing rates from a rat were recorded using a Tucker-Davis (Alachua, Florida) Pentusa system during a go no-go behavioral task as described in [9]. The task was defined such that the animal initiates all trials (nose poke breaking an IR beam). The animal must press a lever in the behavioral box while a LED cue is present.

Additionally, a robotic arm moves in a 3-dimensional grid to target levers within the rat's field of vision [10] but in a discrete area (outside the cage) that the rat cannot reach. To earn a reward, the rat and robotic arm must press and hold their respective levers for 250 ms simultaneously. Performance of this task must exceed 80% before

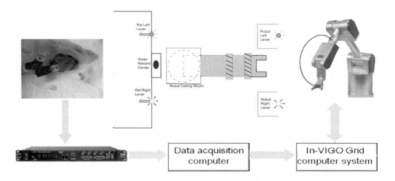

Fig. 2. Animal Experimental Paradigm. The rat is in the cage on the left side. In brain control, the rat levers are not extended, the robotic arm is the only way to earn rewards.

electrode implantation or the animal was excluded from the study. This criterion ensures that the animal is discriminating between cues rather than guessing. Video analysis confirmed (> 95% trials) that the animal presses with the paw contralateral to the lever; ispilateral neural firings reflect this difference in behavior. In this reach-and-grasp task all data is time synchronized to the Pentusa's clock (neural signal, reward times, and lever position) and sent to remote computers for analysis. However, we do not record the kinematic variables of the animal's arms.

4 Computational Framework

We now turn to the question of how to train the forward-inverse models pairs and the gating mechanism (i.e. the individual likelihood models and the RP) of Fig. 1.

4.1 Weight Training for Linear Models

The output of an optimal linear filter of order p is computed by $Y = W^T \times X$ where X is the vector of the last p input data and W is the matrix of weights. To train the weights, we are given the past values of X and Y on a large time window (also called the training window). A common method is then to find the weights W that gives the smallest sum of square errors on Y when applied on the X values (least squares method). When all data are available at once, the weights can be computed by $W = R_X^{-1} \times r_{XY}$ where R_X is the auto-correlation matrix of signal X, and r_{XY} the cross-correlation matrix between X and Y. Computing the auto-correlation matrix and its inverse requires a huge amount of computation, especially when the training window is large. This is the reason why we focus on recursive updates to compute the optimal weights, specifically the recursive least square (RLS) algorithm [11].

The principle of the recursive least square method is to obtain rank one updates of both for R_X^{-1} and r_{XY} at each time a new data vector is received. Many variants exist such as the sliding window-RLS [14] or the QR-RLS [15], which uses a QR factorization of the auto-correlation matrix. Because of the requirement of large training windows (up to 10,000 samples), we adopt the sliding window-RLS algorithm, which is based on the Woodbury formula [13]. We implemented a general algorithm, providing the following features: (1) multidimensional inputs and outputs and separated modules for auto- and cross-correlation recursion, so that the same signal can be used to train a lot of different models at low cost; (2) strict sliding window, the matrices are updated with new data and "downdated" with old ones; (3) ability to use a forgetting factor to increase the importance of recent data.

This algorithm is specifically designed for online training, by updating its internal data structure at each data received. It works on a special set of weights, called "training weights" and does not interfere with current model computations. When the user asks for it, or at predefined time-steps, the weights of the model computation are updated and replaced by the training weights. The RLS algorithm could also be used on offline data, to initialize a set of weights on data recorded in a past experiment. However, the RLS algorithm is not efficient for this purpose, so we also provide an implementation of the block least square (BLS) algorithm for offline training. Fig. 3 shows the raw neural data along with the lever press and a linear model output.

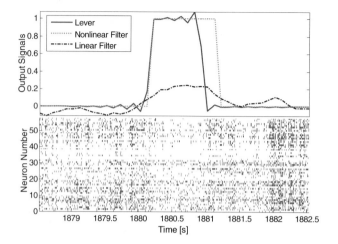

Fig. 3. Representative lever press, linear, and non-linear filter output for this paradigm [top]. Raster plot of recorded neural firing during the lever press [bottom].

4.2 Inverse Model Computation

In the initial testing of this DDDBMI architecture, the set of possible desired trajectories is restricted to six movemes (or simplified models of moving forward, back, up, down, left, and right) of equal length. The starting position of the robot prosthetic is also constant for all trials. This allows us to compute a lookup table of necessary joint actuations for any possible prosthetic orientation – desired trajectory pair in our workspace. This lookup table is only a placeholder for future inverse models where the initial positions may not be known a priori and the set of movemes may be modified through training.

These future inverse models would require an optimization using a biomechanical model of the prosthetic limb to be controlled. Such an optimization averages 24.5 function evaluations using Matlab's fmincon[1]. Computing a single inverse model (performed every 100 ms) requires, on average, 784 multiplications, 1054 additions, and 343 trigonometric function evaluations.

4.3 RL Computation

Here we train the switching mechanism through an application of RL. In the classical embodiment of RL, a model of the animal's brain would be considered an 'agent' which learns (through RL or other methods) to achieve a goal or reward. Here the states of the environment and rewards are used to drive new actions (i.e. which model pair to select). In a BMI experimental paradigm though, one has access to not only the real animal brain but can observe the spatio-temporal activation of brain states (indirectly related to the environment) as the animal seeks a goal or reward. Thus,

---

[1] fmincon is a function provided by the MATLAB Optimization Toolbox for constrained nonlinear optimization; it finds a constrained minimum of a scalar function of several variables starting at an initial estimate.

from the BMI point of view, neural activity is external to, and can not be directly modified by, the agent (algorithm); it must be considered part of the environment [8]. The BMI RL agent must use information in the neural signal to create movement commands for the robot; therefore the agent must learn the state-to-action mapping.

We devised a novel architecture where the animal's neural firing is part of the environment and defines the state, actions occur in a discrete physical space (a separate portion of the environment), and the RL algorithm serves as the agent. The animal can see the actions of the robot in the environment; however, actions do not directly influence the state of the rat brain. This is an important shift in RL architecture because states are decoupled from actions

5 Results

Presently, linear models are being studied; the software must support the following requirements. With about 100 neurons and a time window of 10 steps to compute a five-dimension output, each linear model roughly use 5000 weights. Thus the computing power needed for one model is small; however, we want to use hundreds of models and need to distribute this computation on the available resources. Additionally, each linear model is constantly training its weights as data is received, to improve its behavior. This training has to be done in parallel with the computation of the output, and may represent a huge amount of computation. The software for signal processing is composed of communicating processes, currently in MPI. One global process receives data from the Brain Institute and broadcasts them to a number of processes, each of them responsible for one model. For each model, a thread, running in background, is constantly training the weights. All the computation-intensive operations are implemented using function calls to BLAS (gotoBLAS) and LAPACK libraries. The timing results for a close-loop experiment on a 512MB Pentium 3 1.16GHz computing node using one model are the following:

Gathering data from the animal's brain	3.4 ms
Overall communication between both laboratories	0.87 ms
Computing the output of one model	0.20 ms
Inter-process communication and data management	23 ms
Robot Control	0.028 ms
Weight training	31.8 ms
From collection of brain data to robot control	28 ms

This shows that one closed-loop using only one model takes much less than the bound of 100ms, even when training the weights of the model on the same machine.

The MPFIM architecture tested here contains two fixed movemes (actions) and two target locations; thus, we can model the movements of the prosthetic in the environment to generate the reward signal for RL. The 'value' (responsibility in the MPFIM) of each moveme (forward-inverse model pair) is approximated by an artificial neural network (ANN) based on the animal's brain state. Watkin's $Q(\lambda)$

learning [8] was used to train the ANNs and decide which moveme is most appropriate given the current neural state. *In vivo* testing of the experimental paradigm and MPFIM architecture was performed in one dimensional space with two possible movemes (-1 or +1,i.e. left or right robotic movements). In this form of RL, we performed value function approximation using both single and multilayer perceptrons to test the performance differences for achieving the target. To earn a reward, the BMI needed to reach a target of -7 or +7 starting from position 0. There were 46 training trials (23 trials each for left and right targets) and 16 testing trials (8 targets for each direction). All trials contained 10 samples of neural data (1 s) which terminated in a water reward in the *in vivo* animal recordings. Three different ANN topologies were used to approximate the value of the neural state: single layer perceptron (SLP), multilayer perceptron (MLP-nonlinear output), and a multilayer perceptron with linear outputs.

After training the value function, the MPFIM architecture was tested on the 16 novel trials not used for training. Performance was evaluated as the percent of trials that the robot reached the targets. For this test, the null hypothesis is that the neural data contains no information about the movement intent of the animal and the MPFIM will learn regardless of neural state. To test this hypothesis, a surrogate data set was constructed by randomizing the temporal sequence of neural activity collected from all of the neurons. The results for real neural data and surrogate neural data are presented in table 1. The real neural data can hit the cued target between 81% and 94% of the trials. In contrast, the null hypothesis was shown to be incorrect through the performance of the surrogate data (hitting the cued target 31% of the time). Although we only present results from a simplified environment, we show that the RL does generalize to novel data and presents a viable technique for switching in the RP.

The training and testing performance of this architecture can also be measured in terms of its computational complexity. Temporal difference error [8] was used to train the networks where each error signal is applied to a history of past states (inputs). Therefore, the average number of weight updates per sample depends on the length of trials that the algorithm has to find a reward. For a 1s trial length, the algorithm averages 4.5 weight updates per step (100 ms). In animal experiments, we allow 6.5 s for the animal to earn a reward. This trial length averages 33 weight updates per step. For the best performing case (SLP), the weight training consists of multiplications on the order of $(4M+1)*A$ and adds on the order of $(M+1)*A$ for each update. Here M corresponds to the number of input neurons (ranging from 10s to 100s) and A is the number of actions. This example indicates that future architectures utilizing online training will require high-performance computing resources.

Table 1. MPFIM performance in 1D environment

	SLP	MLP(NO)	MLP(LO)
Real Data	93.7%	81.2%	81.2%
Surrogates	31.2%	31.2%	31.2%

[note: MLPs have 3 hidden PEs]

6 Conclusions

The DDAS BMI architecture can serve as a general-purpose neural–to-motor mapping framework with modular forward and inverse models with a semi-supervised gating mechanism. This architecture is well prepared to deal with the uncertainties in the nature and generation of motor commands. We showed how the capabilities of linear and nonlinear feedforward models can be implemented in real-time, simultaneously allowing one to determine the most appropriate functional form of the motor commands. Distributed processing provided the necessary computational power for performing *in vivo* experimentation on the timescales of the behaving animal. Implementation of the reinforcement learning for mapping to the elemental constructs of movement is a first step in the development of semi-supervised, goal-directed BMI techniques for paralyzed patients who are unable to generate a desired movement response. Future work will focus on scaling up the biomechanical modeling and behavioral response of the animal to 3-D robot control.

Acknowledgments

This work is supported in part by the NSF under Grant No. CNS-0540304 and by the BellSouth Foundation.

References

1. Kawato,M.,Wolpert, D. M.: Multiple paired forward inverse models for motor control. Neural Networks. vol. 11(1998) 1317-1329
2. Fortes, J., Figueiredo, R., Hermer-Vazquez, L., Principe, J., Sanchez, J.: A New Architecture for Deriving Dynamic Brain-Machine Interfaces. In: ICCS-DDDAS(2006)
3. Erhan Oztop, D. W., Kawato,M.: Mental state inference using visual control parameters Cognitive Brain Research.vol. 22(2004)129-151
4. Carmena, J., et al.: Learning to control a brain–machine interface for reaching and grasping by primates. PLoS Biology. vol. 1(2003) 1-16
5. Doya, K.:What are the computations of the cerebellum, the basal ganglia and the cerebral cortex?. Neural Networks.vol. 12(1999) 961-974
6. Nicolelis, M. ,et al.:Reconstructing the engram: simultaneous, multisite, many single neuron recordings. Neuron. vol. 18(1997)529-537
7. Bower, G. :Theories of Learning, 5th ed. Englewood Cliffs: Prentice-Hall, Inc.(1981)
8. Sutton, R., Barto,A.: Reinforcement learning: an introduction. MIT Press (1998)
9. DiGiovanna, J., Sanchez, J., Principe, J.: Improved Linear BMI Systems via Population Averaging.In: IEEE Conf. Eng. in Medicine and Biology Society (2006)
10. Whishaw, I.:*The behavior of the laboratory rat*. NY: Oxford University Press (2005).
11. Haykin, S.: *Adaptive Filter Theory*: Prentice Hall (2002)
12. Golub, G. H. and Van Loan, C. F.: Matrix computations. Johns Hopkins (1989).
13. Liu, H. and He, Z.: A sliding-exponential window rls adaptive filtering algorithm: properties and applications. Signal Process., 45(3):357-368 (1995).
14. Sakai, H. and Nakaoka, H.: A fast sliding window QRD-RLS algorithm. Signal Process., 78(3):309-319 (1999).

Using Cyber-Infrastructure for Dynamic Data Driven Laser Treatment of Cancer

C. Bajaj[1], J.T. Oden[1], K.R. Diller[2], J.C. Browne[1], J. Hazle[3], I. Babuška[1],
J. Bass[1], L. Bidaut[3], L. Demkowicz[1], A. Elliott[3], Y. Feng[1], D. Fuentes[1],
B. Kwon[1], S. Prudhomme[1], R.J. Stafford[3], and Y. Zhang[1]

[1] Institute for Computational Engineering and Sciences,
[2] Department of Biomedical Engineering,
The University of Texas at Austin, Austin TX 78712, USA
{oden,babuska,bass,leszek,feng,fuentes,serge,jessica}@ices.utexas.edu,
kdiller@mail.utexas.edu, {bajaj,browne,juneim}@cs.utexas.edu
[3] University of Texas M.D. Anderson Cancer Center,
Department of Diagnostic Radiology, Houston TX 77030, USA
{jhazle,jstafford,Andrew.Elliott,Luc.Bidaut}@di.mdacc.tmc.edu
http://dddas.ices.utexas.edu

Abstract. Hyperthermia based cancer treatments are used to increase the susceptibility of cancerous tissue to subsequent radiation or chemotherapy treatments, and in the case in which a tumor exists as a well-defined region, higher intensity heat sources may be used to ablate the tissue. Utilizing the guidance of real-time treatment data while applying a laser heat source has the potential to provide unprecedented control over the outcome of the treatment process [6,12]. The goals of this work are to provide a working snapshot of the current system architecture developed to provide a real-time finite element solution of the problems of calibration, optimal heat source control, and goal-oriented error estimation applied the equations of bioheat transfer and demonstrate that current finite element technology, parallel computer architecture, peer-to-peer data transfer infrastructure, and thermal imaging modalities are capable of inducing a precise computer controlled temperature field within the biological domain.

1 Introduction

Thermal therapies delivered under various treatment modalities permit a minimally invasive and effective cancer treatment that eradicates the disease, maintains functionality of infected organs, and minimizes complications and relapse. The physical basis for thermal therapies is that exposing cells to temperatures outside their natural environment for certain periods of time can damage and even destroy the cells. However, one of the limiting factors in all forms of thermal therapies, including cryotherapy, microwave, radio-frequency, ultrasound, and laser, is the ability to control the energy deposition to prevent damage to adjacent healthy tissue [13].

Y. Shi et al. (Eds.): ICCS 2007, Part I, LNCS 4487, pp. 972–979, 2007.

Current imaging technology allows the imaging of the geometry of tissue and an overlaying temperature field using MRI and MRTI (MR Temperature Imaging) technology. MRTI has the the ability to provide fast, quantitative temperature imaging in a variety of tissues, and the capability of providing biologically relevant information regarding the extent of injury immediately following a thermal therapy [4]. Image guidance [12,15] has the potential to facilitate unprecedented control

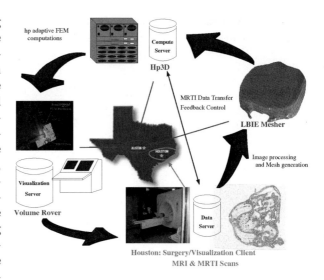

Fig. 1. Schematic of the peer to peer communication architecture used to control the laser treatment process. Feedback control is achieved through the continual interaction of the data, compute, and visualization modules.

over bioheat transfer by providing real time treatment monitoring through temperature feedback during treatment delivery. A similar idea using ultrasound guided cryotherapy has been studied and shows good results [13].

The ultimate goal of this work is to deliver a computational model of bioheat transfer that employs real-time, patient specific data and provides real-time high fidelity predictions to be used concomitantly by the surgeon in the laser treatment process. The model employs an adaptive hp-finite element approximation of the nonlinear parabolic Pennes equation and uses adjoint-based algorithms for inverse analysis, model calibration, and adaptive control of cell damage. The target diseases of this research are localized adenocarcinomas of the breast, prostate, cerebrum, and other tissues in which a well-defined tumor may form. The algorithms developed also provide a potentially viable option to treat other parts of the anatomy in patients with more advanced and aggressive forms of cancer who have reached their limit of radiation and chemotherapy treatment.

2   Software Architecture

A schematic of the software architecture embedded in the control loop is shown in Figure 1. Figure 2 illustrates the main software modules and communication methods between software modules. Multiple client-server applications utilizing a remote procedure calling protocol connect the actual laboratory at M.D. Anderson Cancer Center in Houston, TX to the computing and visualization center in Austin, TX. Prior to treatment, the LBIE Mesher[1] uses MRI data to generate a finite element mesh of the patient-specific biological domain. Goal-oriented estimation and adaption is used to optimize the mesh to a particular quantity of interest [9]. The tool then proceeds to solve an optimal control problem, wherein the laser parameters (location of optical fiber, laser power, etc.) are controlled to eliminate/sensitize cancer cells, minimize damage to healthy cells, and control Heat Shock Protein (HSP) expression. Upon initiation of the treatment process, the compute server employs real-time MRI data to co-register the computa-

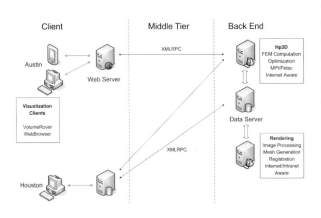

Fig. 2. Three Tier Cyber-software architecture. Computation, data transfer, and visualization are done on the backend compute nodes. A middle tier of XMLRPC connections connects the backend to the visualization clients in both Austin & Houston.

tional domain and MRTI data is used to calibrate the bioheat transfer model to the biological tissue values of the patient. As the data server, in Houston, delivers new data intermittently to the client, in Austin, computation is compared to the measurements of the real-time treatment and an appropriate course of action is chosen according to the differences seen. A parallel computing paradigm built from the Petsc [2] software infrastructure is used to meet the demands of rapid calibration and adapting the computational mesh and models to control approximation and modeling error. Volume Rover[1] [1] is used to achieve

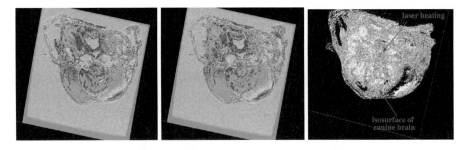

Fig. 3. Selected Slices of Canine MRI Brain Data, used for mesh generation. MRTI thermal data and Iso-surface visualization of Canine MRI Brain Data illustrate the location of heating [8].

---

[1] Software available at: http://cvcweb.ices.utexas.edu/cvc

efficient visualization of the volumetric MRI and thermal MRTI images simultaneously with the finite element prediction. From a computational point of view, the orchestration of a successful laser treatment is to solve the problems of co-registration, calibration, optimal control, and mesh refinement invisibly to the surgeon, and merely provide the surgeon with an interface to the optimal laser parameters and visualization of the computational prediction of the treatment treatment.

3   Image Segmentation, Meshing, and MRTI-Registration

Figure 4 shows a quality hexahedral mesh obtained for finite element simulations from a set of MRI data (256x256x34 voxels) of a canine brain, Figure 3. The field view of the MRI images was 200mm x 200mm with each image spaced 1mm apart. First, the image processing techniques, available in Volume Rover [1], were used to improve the quality of imaging data. Contrast enhancing techniques improved the contrast and anisotropic and bilateral diffusion [3] removed noise. Two dimensional segmentation was performed via a manual tracing of boundaries on each image slice, and the stack of contours tiled to form an initial water-tight triangulated surface. Three dimensional segmentation [1] could not be used because of the anisotropy in the imaging data. After the geometric model was obtained, geometric flow smoothed the geometric model and a geometric

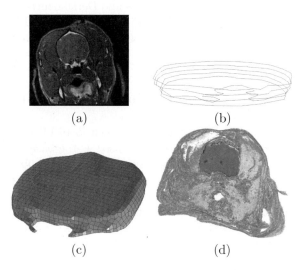

(a)   (b)

(c)   (d)

Fig. 4. Canine brain data. (a) Segmentation of the canine brain boundaries from the transverse 34-slice stack of 256x256 MRI data. A single slice is shown in gray-scale intensities, with the segmented boundary in red. (b) stack of 2D contours obtained from segmentation. (c) 8820 hexahedral mesh elements, Jacobian quality > .05. (d) combined volume visualization of the 256x256x32 MRI data of a canine head, with the embedded subset of hexahedral finite element mesh of the segmented canine brain.

volumetric map using the signed distance function method was created. The hexahedral mesh was generated using an octree-based isocontouring method. Geometric flow [17], pillowing [7] and the optimization method were used to improve the mesh quality. The constructed hexahedral mesh has two important properties: good aspect ratios and there is at most one face lying on the boundary for each element.

The day of treatment, a FFT-based technique is used to register the finite element mesh to the current position of the patient. The registration software has been rigorously tested against a suite of validation experiments using phantom materials. The phantom materials are fabricated with two materials of contrasting image density in which an inner smaller object is placed asymmetrically within the larger object. The materials are composed of 2 % agar gel and at least three 2 mm nylon beads are introduced as fiducials. The suite of data consists of several 3D images of incremental translational and rotational rigid body motions of the phantom material as well as images of incremental deformation of the phantom material. The data is provided for the image registration community from the DDDAS project webpage[2].

The final image processing step is to overlay the MRTI thermal data onto the finite element mesh. A median and Deriche filter are used to remove the inherent noise from the MRTI data, Figure 5. The filtered MRTI data is interpolated onto the finite element solution space. The order of interpolation is determined by the order of the mesh.

4   Calibration, Optimal Control, and Error Estimation

Pennes model [10] has been shown [5,16,14] to provide very accurate prediction of bioheat transfer and is used as the basis of the finite element prediction. The control paradigm involves three major problems: calibration of the Pennes bioheat transfer model to patient specific MRTI data, optimal positioning and power supply of the laser heat source, and computing goal oriented error estimates. During the laser treatment process, all three problems are solved in tandem by separate groups of processors communicating amongst each other as needed. The variational form of the governing Pennes bioheat transfer model is as follows:

> Given a set of model, β, and laser, η, parameters,
> Find $u(\mathbf{x}, t) \in \mathcal{V} \equiv H^1\left([0, T], H^1(\Omega)\right)$ s.t.
> $$B(u, \beta; v) = F(\eta; v) \qquad \forall v \in \mathcal{V}$$

where the explicit functional dependence on the model parameters, β, and laser parameters, $\eta = (P(t), \mathbf{x}_0)$, are expressed as follows

$$B(u, \beta; v) = \int_0^T \int_\Omega \left[\rho c_p \frac{\partial u}{\partial t} v + k(u, \beta) \nabla u \cdot \nabla v + \omega(u, \beta) c_{blood}(u - u_a) v \right] dx dt$$

$$+ \int_0^T \int_{\partial \Omega_C} hu \, v \, dA dt + \int_\Omega u(\mathbf{x}, 0) \, v(\mathbf{x}, 0) \, dx$$

---

[2] Project Website: dddas.ices.utexas.edu

$$F(\eta; v) = \int_0^T \int_\Omega 3P(t)\mu_a\mu_{tr} \frac{\exp(-\mu_{eff}\|\mathbf{x} - \mathbf{x}_0\|)}{4\pi\|\mathbf{x} - \mathbf{x}_0\|} v\, dx dt$$

$$+ \int_0^T \int_{\partial\Omega_C} h u_\infty\, v\, dAdt - \int_0^T \int_{\partial\Omega_N} \mathcal{G}\, v\, dAdt + \int_\Omega u^0\, v(\mathbf{x}, 0)\, dx$$

$$\mu_{tr} = \mu_a + \mu_s(1 - \gamma) \qquad \mu_{eff} = \sqrt{3\mu_a\mu_{tr}}$$

Here $k\left[\frac{J}{s \cdot m \cdot K}\right]$ and $\omega\left[\frac{kg}{s\, m^3}\right]$ are bounded functions of u, c_p and c_{blood} are the specific heats, u_a the arterial temperature, ρ is the density, and h is the coefficient of cooling. P is the laser power, μ_a, μ_s are laser coefficients related to laser wavelength and give probability of absorption of photons by tissue, γ is the anisotropy factor, and $\mathbf{x}_0$ is the position of laser photon source. Constitutive model data and details of the optimization process are given in [8,11].

5   Data Transfer, Visualization, and Current Results

Conventional data transfer methods and software rendering visualization tools pose a major bottleneck in developing a laser treatment paradigm in which high performance computers control the bioheat data transferred from a remote site. The data transfer problem is addressed through the use of client-server applications that use a remote procedure calling protocol to transfer data directly between physical memory instead of incurring the overhead of a writing to disk and transferring data. Volume Rover [1] is able to achieve high performance interactive visualization through the use of modern programmable graphics hardware to provide combined geometry and volume rendering displays, Figure 4. Software rendering is limited by the memory and processor.

Computational time used to advance the Pennes model equations forward in time is not a bottleneck. Computations are done at the Texas Advanced Computing Center on a Dual-Core Linux Cluster. Each node of the cluster contains two Xeon Intel Duo-Core 64-bit processors (4 cores in all) on a single board, as an SMP unit. The core frequency is 2.66GHz and supports 4 floating-point operations per clock period. Each node contains 8GB of memory. The average execution times of a representative 10 second simulation is approximately 1 second, meaning that in a real time 10 second span Pennes model can predict out to more than a minute. Equivalently, in a 10 second time span, roughly 10 corrections can be made to calibrate the model coefficients or optimize the laser parameters.

The typical time duration of a laser treatment is about five minutes. During a five minute span, one set of MRTI data is acquired every 6 seconds. The size of each set of MRTI data is $\approx$330kB (256x256x5 voxels). Computations comparing the predictions of Pennes model to experimental MRTI taken from a canine brain show very good agreement, Figure 5. A manual craniotomy of a canine skull was preformed to allow insertion of an interstitial laser fiber. A finite element mesh of the biological domain generated from the MRI data is shown in Figure 4. The mesh consists of 8820 linear elements with a total of

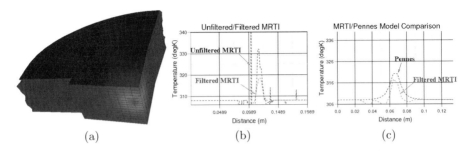

Fig. 5. (a) Contours of Pennes model prediction overlayed onto the finite element mesh. (b),(c) Simultaneous cutline comparison of Pennes model prediction, Filtered MRTI data, and Unfiltered MRTI data. Cutline taken through laser source.

9872 degrees of freedom. MRTI thermal imaging data was acquired in the form of five two dimensional 256x256 pixel images every six seconds for 120 time steps. The spacing between images was 3.5mm. The MRTI data was filtered then projected onto the finite element mesh. Figure 5 shows a cutline comparison between the MRTI data and the predictions of Pennes model. It is observed that the results delivered by the computational Pennes model slightly over diffuses the heat profile peaks compared to measured values. However, at early times the maximum temperature value is within 5% of the MRTI value.

6   Conclusions

Results indicate that reliable finite element model simulations of hyperthermia treatments can be computed, visualized, and provide feedback in the same time span that the actual therapy takes place. Combining these prediction capabilities with an understanding of HSP kinetics and damage mechanisms at the cellular and tissue levels due to thermal stress will provide a powerful methodology for planning and optimizing the delivery of hyperthermia therapy for cancer treatments.

The entire closed control loop in currently being tested on agar and ex-vivo tissue samples in preparation for the first real time computer guided laser therapy, which is anticipated within the upcoming year. The culmination of adaptive hp-finite element technology implemented on parallel computer architectures, modern data transfer and visualization infrastructure, thermal imaging modalities, and cellular damage mechanisms to provide cancer treatment tool will be a significant achievement in the field of computational science.

Acknowledgments. The research in this paper was supported in part by the National Science Foundation under grants CNS-0540033, IIS-0325550, and NIH Contracts P20RR0206475, GM074258. The authors also acknowledge the important support of DDDAS research by Dr. Frederica Darema of NSF.

References

1. C. Bajaj, Z. Yu, and M. Aue. Volumetric feature extraction and visualization of tomographic molecular imaging. *Journal of Structural Biology*, 144(1-2):132–143, October 2003.

2. Satish Balay, William D. Gropp, Lois C. McInnes, and Barry F. Smith. Petsc users manual. Technical Report ANL-95/11 - Revision 2.1.5, Argonne National Laboratory, 2003.

3. W. Jiang, M. Baker, Q. Wu, C. Bajaj, and W. Chiu. Applications of bilateral denoising filter in biological electron microscopy. *Journal of Structural Biology*, 144:Issues 1-2:114–122, 2003.

4. M. Kangasniemi et al. Dynamic gadolinium uptake in thermally treated canine brain tissue and experimental cerebral tumors. *Invest. Radiol.*, 38(2):102–107, 2003.

5. J. Liu, L. Zhu, and L. Xu. Studies on the three-dimensional temperature transients in the canine prostate during transurethral microwave thermal therapy. *J. Biomech. Engr*, 122:372–378, 2000.

6. R. J. McNichols et al. MR thermometry-based feedback control of laser interstitial thermal therapy at 980 nm. *Lasers Surg. Med.*, 34(1):48–55, 2004.

7. S. A. Mitchell and T. J. Tautges. Pillowing doublets: refining a mesh to ensure that faces share at most one edge. In *Proc. 4th International Meshing Roundtable*, pages pages 231–240, 1995.

8. J. T. Oden, K. R. Diller, C. Bajaj, J. C. Browne, J. Hazle, I. Babuška, J. Bass, L. Demkowicz, Y. Feng, D. Fuentes, S. Prudhomme, M. N. Rylander, R. J. Stafford, and Y. Zhang. Dynamic data-driven finite element models for laser treatment of prostate cancer. *Num. Meth. PDE,*, accepted.

9. J. T. Oden and S. Prudhomme. Goal-oriented error estimation and adaptivity for the finite element method. *Computers and Mathematics with Applications*, 41(5–6):735–756, 2001.

10. H. H. Pennes. Analysis of tissue and arterial blood temperatures in the resting forearm. *J. Appl. Physiol.*, 1:93–122, 1948.

11. M. N. Rylander, Y. Feng, J. Zhang, J. Bass, Stafford R. J., J. Hazle, and K. Diller. Optimizing hsp expression in prostate cancer laser therapy through predictive computational models. *J. Biomed Optics*, 11:4:041113, 2006.

12. R. Salomir et al. Hyperthermia by MR-guided focused ultrasound: accurate temperature control based on fast MRI and a physical model of local energy deposition and heat conduction. *Magn. Reson. Med.*, 43(3):342–347, 2000.

13. K. Shinohara. Thermal ablation of prostate diseases: advantages and limitations. *Int. J. Hyperthermia*, 20(7):679–697, 2004.

14. J.W. Valvano and et al. An isolated rat liver model for the evaluation of thermal techniques to measure perfusion. *ASME J. Biomech. Eng.*, 106:187–191, 1984.

15. F. C. Vimeux et al. Real-time control of focused ultrasound heating based on rapid MR thermometry. *Invest. Radiol.*, 34(3):190–193, 1999.

16. L. Xu, M.M. Chen, K.R. Holmes, and H. Arkin. The evaluation of the pennes, the chen-holmes, the weinbaum-jiji bioheat transfer models in the pig kidney vortex. *ASME HTD*, 189:15–21, 1991.

17. Y. Zhang, C. Bajaj, and G. Xu. Surface smoothing and quality improvement of quadrilateral/hexahedral meshes with geometric flow. In *Proceedings of 14th International Meshing Roundtable*, volume 2, pages 449–468., 2005.

Grid-Enabled Software Environment for Enhanced Dynamic Data-Driven Visualization and Navigation During Image-Guided Neurosurgery[*]

Nikos Chrisochoides[1], Andriy Fedorov[1], Andriy Kot[1], Neculai Archip[2],
Daniel Goldberg-Zimring[2], Dan Kacher[2], Stephen Whalen[2], Ron Kikinis[2],
Ferenc Jolesz[2], Olivier Clatz[3], Simon K. Warfield[3],
Peter M. Black[4], and Alexandra Golby[4]

[1] College of William & Mary, Williamsburg, VA, USA
[2] Department of Radiology, Brigham and Women's Hospital, Boston, MA, USA
[3] Department of Radiology, Children's Hospital, Boston, MA, USA
[4] Department of Neurosurgery, Brigham and Women's Hospital, Boston, MA, USA

Abstract. In this paper we present our experience with an Image
Guided Neurosurgery Grid-enabled Software Environment (IGNS-GSE)
which integrates real-time acquisition of intraoperative Magnetic Resonance Imaging (IMRI) with the preoperative MRI, fMRI, and DT-MRI
data. We describe our distributed implementation of a non-rigid image
registration method which can be executed over the Grid. Previously,
non-rigid registration algorithms which use landmark tracking across the
entire brain volume were considered not practical because of the high
computational demands. The IGNS-GSE, for the first time ever in clinical practice, alleviated this restriction. We show that we can compute
and present enhanced MR images to neurosurgeons during the tumor
resection within minutes after IMRI acquisition. For the last 12 months
this software system is used routinely (on average once a month) for clinical studies at Brigham and Women's Hospital in Boston, MA. Based on
the analysis of the registration results, we also present future directions
which will take advantage of the vast resources of the Grid to improve
the accuracy of the method in places of the brain where precision is critical for the neurosurgeons.

Keywords: Grid, Data-Driven Visualization.

1 Introduction

Cancer is one of the top causes of death in the USA and around the world.
Medical imaging, and Magnetic Resonance Imaging (MRI) in particular, provide great help in diagnosing the disease. In brain cancer cases, MRI provides

---

[*] This research was supported in part by NSF NGS-0203974, NSF ACI-0312980, NSF
ITR-0426558, NSF EIA-9972853.

Y. Shi et al. (Eds.): ICCS 2007, Part I, LNCS 4487, pp. 980–987, 2007.

extensive information which can help to locate the tumor and plan the resection strategy. However, deformation and shift of brain structures is unavoidable during open brain surgery. This creates discrepancies as compared to the preoperative imaging during the operation.

It is possible to detect the brain shift during the surgery. One of the means to do this is IMRI. IMRI provides sparse dynamic measurements, which can be used to align (register) the preoperative data accordingly. In this way, high-quality, multimodal preoperative imaging can be used during the surgery.

However, registration is a computationally-intensive task, and it cannot be initiated before IMRI becomes available. Local computing resource available at a particular hospital may not allow to perform this computation in time. The goal of our research is to use geographically distributed computing resources to expedite the completion of this computation. In the on-going collaboration between Brigham and Women's Hospital (BWH) in Boston, MA, and College of William and Mary (CWM) in Williamsburg, VA, we are studying how widely available commodity clusters and Grid resources can facilitate the timely delivery of registration results. We leverage our work from the state-of-the art registration method [1], and extensive experience with distributed processing and dynamic load balancing [2,3].

We have designed a robust distributed implementation of the registration method, which meets the following requirements concerning (1) execution speed, (2) reliability, (3) ease-of-use and (4) portability of the registration code. We evaluate this prototype implementation in a geographically-distributed environment, and outline how IGNS computations can benefit from large-scale computing resources like *TeraGrid*.

2   Near-Real-Time Non-rigid Registration

2.1   Registration Algorithm

The registration method was first presented in [1], and subsequently evaluated in [4]. The computation consists of preoperative and intraoperative components. Intraoperative processing starts with the acquisition of the first intraoperative scan. However, the *time-critical* part of the intraoperative computation is initiated when a scan showing shift of the brain is available. The high-level timeline of this process is shown in Fig. 1.

Here we briefly describe the three main steps of the algorithm:

1. the patient-specific tetrahedral mesh model is generated from the segmented intra-cranial cavity (ICC). The ICC segmentation [4] is prepared based on pre-operative imaging data. As the first intraoperative scan is available, pre-operative data is rigidly (i.e., using translation and rotation) aligned with the patient's head position;
2. with the acquisition of the intraoperative image showing brain deformation, sparse displacement field is estimated from the intra-operative scan using blockmatching [1]. Its computation is based on the minimization of

the correlation coefficient between regions, or blocks, of the pre-operative (aka floating) image and the real-time intra-operative (aka fixed) image;

3. the FEM model of the intra-cranial cavity with linear elastic constitutive equation is initialized with the mesh and sparse displacement field as the initial condition. An iterative hybrid method is used to discard the outlier matches.

Steps 2 and 3 are time critical and should be performed as the surgeons are waiting. In the context of the application we define the *response time* as the time between the acquisition of the intra-operative scan of the deformed tissue and the final visualization of the registered preoperative data on the console in the operating room. These steps performed intraoperatively form the Dynamic Data-Driven Application System (DDDAS [1]) steered by the IMRI-acquired data. Our broad objective is to minimize the perceived (end-to-end) response time of the DDDAS component.

2.2 Implementation Objectives

We have completed an evaluation of the initial PVM implementation [1] of the de-scribed registration approach. The evaluation was done on retrospective datasets obtained during past image-guided neurosurgeries. We identified the following problems:

1. The execution time of the original non-rigid registration code is data-dependent and varies between 30 and 50 minutes, when computed on a high-end 4 CPU workstation. The scalability of the code is very poor due to work-load imbalances.
2. The code is designed as a single monolithic step (since it was not evaluated in the intraoperative mode) to run and a single failure at any point requires to restart the registration from the beginning.
3. The original code is not intuitive to use, a number of implementation-specific parameters are required to be set in the command line. This makes it cum-bersome and error-prone to use during neurosurgery. The possible critical delays are exacerbated in the case the code has to run remotely on larger Clusters of Workstations (CoWs).
4. The original code is implemented in PVM which is not supported by many sites due to widespread use of MPI standard for message passing.

Based on the evaluation of the original code, the following implementation objectives were identified:

High-performance. Develop an efficient and portable software environment for parallel and distributed implementation of real-time non-rigid registra-tion method for both small scale parallel machines and large scale geographi-cally distributed CoWs. The implementation should be able to work on both dedicated, and time-shared resources.

---

[1] The notion of DDDAS was first coined and advocated by Dr.Darema, see http://dddas.org.

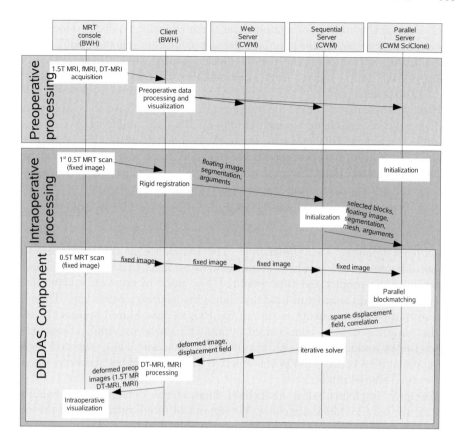

Fig. 1. Timeline of the image processing steps during IGNS (the client is running at BWH, and the server is using multiple clusters at CWM, for fault-tolerance purposes)

Quality-of-service (QoS). Provide functionality not only to sustain failure but also to dynamically replace/reallocate faulty resources with new ones during the real-time data acquisition and computation.

Ease-of-use. Develop a GUI which automatically will handle exceptions (e.g., faults, resource management, and network outages).

We have developed an implementation which addresses the aforementioned objectives [5]. Next we briefly highlight some of the implementation details.

2.3   Implementation Details

Multi-level Distributed Block Matching. In order to find a match for a given block, we need the block center coordinates, and the areas of the fixed and floating images bounded by the block matching window [1]. The fixed and floating images are loaded on each of the processors during the initialization step, as shown in Fig. 1. The total workload is maintained in a *work-pool* data structure.

Each item of the work-pool contains the three coordinates of the block center (total number of blocks for a typical dataset is around 100,000), and the best match found for that block (in case the block was processed; otherwise that field is empty). We use the master-worker computational model to distribute the work among the processors.

However, because of the scarce resource availability we have to be able to deal with computational clusters which belong to different administrative domains. In order to handle this scenario, we use hierarchical multi-level organization of the computation with master-worker model. We use a separate master node within each cluster. Each master maintains a replica of the global work-pool, and is responsible for distributing the work according to the requests of the nodes within the assigned cluster, and communicating the execution progress to the other master(s).

Multi-level Dynamic Load Balancing. The imbalance of the processing time across different nodes involved in the computation is caused by our inability or difficulty to predict the time required per block of data on a given architecture. The main sources of load imbalance are *platform-dependent*. These are caused by the heterogeneous nature of the PEs we use. More importantly, some of the resources may be time-shared by multiple users and applications, which affect the processing time in an unpredictable manner. The (weighted-) static work assignment of any kind is not effective when some of the resources operate in the time-shared mode.

We have implemented a multi-level hierarchical dynamic load balancing scheme for parallel block matching. We use initial rough estimation of the combined computational power of each cluster involved in the computation (based on CPU clock speed) for the weighted partitioning of the work-pool and initial assignment of work. However, this is a rough "guess" estimation, which is adjusted at runtime using a combination of master/worker and work-stealing [6,7] methods. Each master has a copy of the global work-pool, which are identical in the beginning of the computation. The portion of the work-pool assigned to a specific cluster is partitioned in meta-blocks (a sequence of blocks), which are passed to the cluster nodes using the master-worker model. As soon as all the matches for a meta-block are computed, they are communicated back to the master, and a new meta-block is requested. In case the portion of the work-pool assigned to a master is processed, the master continues with the "remote" portions of work (i.e., those, initially assigned to other clusters). As soon as the processing of a "remote" meta-block is complete, it is communicated to all the other master nodes to prevent duplicated computation.

Multi-level Fault Tolerance. Our implementation is completely decoupled, which provides the first level of fault tolerance, i.e., if the failure takes place at any of the stages, we can seamlessly restart just the failed phase of the algorithm and recover the computation. The second level of fault tolerance concerns with the parallel block matching phase. It is well-known that the vulnerability of parallel computations to hardware failures increases as we scale the size of the system.

We would like to have a robust system which in case of failure would be able to continue the parallel block matching without recomputing results obtained before the failure. This functionality is greatly facilitated by maintaining the previously described work-pool data-structure which is maintained on by the master nodes.

The work-pool data-structure is replicated on the separate file-systems of these clusters, and has a tuple for each of the block centers. A tuple can be either empty, if the corresponding block has not been processed, or otherwise it contains the three components of the best match for a given block. The work-pool is synchronized periodically between the two clusters, and within each cluster it is updated by the PEs involved. As long as one of the clusters involved in the computation remains operational, we will be able to sustain the failure of the other computational side and deliver the registration result.

Ease-of-Use. The implementation consists of the client and server components. The client is running at the hospital site, and is based on a Web-service, which makes it highly portable and easy to deploy. On the server side, the input data and arguments are transferred to the participating sites. Currently, we have a single server responsible for this task. The computation proceeds using the participating available remote sites to provide the necessary performance and fault-tolerance.

Table 1. Execution time (sec) of the intra-surgery part of the implemented web-service at various stages of development

Setup	ID						
	1	2	3	4	5	6	7
High-end workstation, using original PVM implementation	1558	1850	2090	2882	2317	2302	3130
SciClone (240 procs), no load-balancing	745	639	595	617	570	550.4	1153
SciClone (240 procs) and CS lab(29 procs), dynamic 2-level load-balancing and fault-tolerance	**30**	**40**	**42**	**37**	**34**	**33**	**35**

3   Initial Evaluation Results

Our preliminary results use seven image datasets acquired at BWH. The computations for two of these seven registration computations were accomplished during the course of surgery (at the College of William and Mary), while the rest of the computations were done retrospectively. All of the intra-operative computations utilized *SciClone* (a heterogeneous cluster of workstations located at CWM, reserved in advance for the registration computation) and the workstations of the student lab (time-shared mode). The details of the hardware configuration can be found in [5]. Data transfer between the networks of CWM

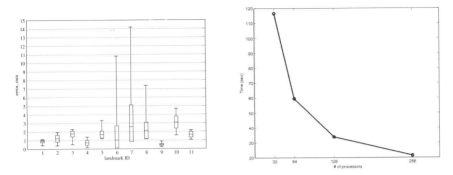

Fig. 2. Registration accuracy dependence on proper parameter selection (left). Excellent scalability of the code on the NCSA site of *TeraGrid* (right) enables intraoperative search for optimal parameters.

and BWH (subnet of Harvard University) are facilitated by the Internet2 backbone network with the slowest link having bandwidth of 2.5 Gbps.

The initial evaluation results are summarized in Table 1. We were able to reduce the total response time to 2 minutes (4 minutes, including the time to transfer the data). We showed, that dynamic load balancing is highly effective in time-shared environment. Modular structure of the implemented code greatly assisted in the overall usability and reliability of the code. The fault-tolerance mechanisms implemented are absolutely essential and introduce a mere 5-10% increase in the execution time. We have also evaluated our implementation on the *Mercury* nodes of the NCSA *TeraGrid* site [8]. The 64-bit homogeneous platform available at NCSA allows for high sustained computational power and improved scalability of the code (see Fig. 2).

4   Discussion

The registration algorithm we implemented has a number of parameters whose values can potentially affect the accuracy of the results. The evaluation of all parameters is computationally demanding (the parameter space of the algorithm has high dimensionality), which requires vast computational resources that are available only over the Grid. Based on the preliminary analysis, registration accuracy is dependent on the parameter selection. Fig. 2 shows the spread of registration precision at expert-identified anatomical landmarks. Given the distributed resources available within *TeraGrid*, we should be able to compute in parallel registration results which use different parameter settings. As the multiple registrations become available, the surgeon will specify the area of interest within the brain, and the registration image which gives the best *effective* accuracy in that particular region will be selected.

The implemented framework proved to be very effective during on-going clinical study on nonrigid registration. However, more work needs to be done to make the framework portable and easy to deploy on an arbitrary platform. Once this is complete, the registration can be provided as a ubiquitously-available Web service. The concerns about resource allocation and scheduling on a shared resource like *TeraGrid* are of high importance. The presented research utilized time-shared resources together with a large cluster operating in dedicated mode. However, we are currently investigating other opportunities, e.g., SPRUCE and urgent computing [9] on *TeraGrid*.

Acknowledgments. This work was performed in part using computational facilities at the College of William and Mary which were enabled by grants from Sun Microsystems, the National Science Foundation, and Virginia's Commonwealth Technology Research Fund. We thank SciClone administrator Tom Crockett for his continuous support and personal attention to this project. We acknowledge support from a research grant from CIMIT, grant RG 3478A2/2 from the NMSS, and by NIH grants R21 MH067054, R01 RR021885, P41 RR013218, U41 RR019703, R03 EB006515 and P01 CA067165.

References

1. Clatz, O., Delingette, O., Talos, I.F., Golby, A., Kikinis, R., Jolesz, F., Ayache, N., Warfield, S.K.: Robust non-rigid registration to capture brain shift from intra-operative MRI. IEEE Trans. Med. Imag. **24**(11) (2005) 1417–1427
2. Barker, K., Chernikov, A., Chrisochoides, N., Pingali, K.: A load balancing framework for adaptive and asynchronous applications. IEEE TPDS **15**(2) (February 2004) 183–192
3. Fedorov, A., Chrisochoides, N.: Location management in object-based distributed computing. In: Proc. of IEEE Cluster'04. (2004) 299–308
4. Archip, N., Clatz, O., Whalen, S., Kacher, D., Fedorov, A., Kot, A., Chrisochoides, N., Jolesz, F., Golby, A., Black, P.M., Warfield, S.K.: Non-rigid alignment of preoperative MRI, fMRI, and DT-MRI with intra-operative MRI for enhanced visualization and navigation in image-guided neurosurgery. NeuroImage (2007) (in press).
5. Chrisochoides, N., Fedorov, A., Kot, A., Archip, N., Black, P., Clatz, O., Golby, A., Kikinis, R., Warfield, S.K.: Toward real-time image guided neurosurgery using distributed and Grid computing. In: Proc. of IEEE/ACM SC06. (2006)
6. Blumofe, R., Joerg, C., Kuszmaul, B., Leiserson, C., Randall, K., Zhou, Y.: Cilk: An efficient multithreaded runtime system. In: Proceedings of the 5th Symposium on Principles and Practice of Parallel Programming. (1995) 55–69
7. Wu, I.: Multilist Scheduling: A New Parallel Programming Model. PhD thesis, School of Comp. Sci., Carnegie Mellon University, Pittsburg, PA 15213 (July 1993)
8. TeraGrid Project: TeraGrid Home page (2006) `http://teragrid.org/`, accessed 23 April 2006.
9. Beckman, P., Nadella, S., Trebon, N., Beschastnikh, I.: SPRUCE: A system for supporting urgent high-performance computing. In: Proc. of WoCo9: Grid-based Problem Solving Environments. (2006)

From Data Reverence to Data Relevance: Model-Mediated Wireless Sensing of the Physical Environment

Paul G. Flikkema[1], Pankaj K. Agarwal[2], James S. Clark[2], Carla Ellis[2], Alan Gelfand[2], Kamesh Munagala[2], and Jun Yang[2]

[1] Northern Arizona University, Flagstaff AZ 86001 USA
[2] Duke University, Durham, NC USA

Abstract. Wireless sensor networks can be viewed as the integration of three subsystems: a low-impact *in situ* data acquisition and collection system, a system for inference of process models from observed data and *a priori* information, and a system that controls the observation and collection. Each of these systems is connected by feedforward and feedback signals from the others; moreover, each subsystem is formed from behavioral components that are distributed among the sensors and out-of-network computational resources. Crucially, the overall performance of the system is constrained by the costs of energy, time, and computational complexity. We are addressing these design issues in the context of monitoring forest environments with the objective of inferring ecosystem process models. We describe here our framework of treating data and models jointly, and its application to soil moisture processes.

Keywords: Data Reverence,Data Relevance, Wireless Sensing.

1 Introduction

All empirical science is based on measurements. We become familiar with these quantitative observations from an early age, and one indication of our comfort level with them is the catchphrase "ground truth". Yet one characteristic of the leading edge of discovery is the poor or unknown quality of measurements, since the instrumentation technology and the science progress simultaneously, taking turns pulling each other forward in incremental steps.

Wireless sensor networking is a new instrument technology for monitoring of a vast range of environmental and ecological variables, and is a particularly appropriate example of the interleaving of experiment and theory. There are major ecological research questions that must be treated across diverse scales of space and time, including the understanding of biodiversity and the effects on it of human activity, the dynamics of invasive species (Tilman 2003), and identification of the web of feedbacks between ecosystems and global climate change. Wireless sensor networks have great potential to provide the data to help answer these questions, but they are a new type of instrumentation with

Y. Shi et al. (Eds.): ICCS 2007, Part I, LNCS 4487, pp. 988–994, 2007.

substantial constraints: the usual problems of transducer noise, nonlinearities, calibration, and sensitivity to temperature and aging are compounded by numerous potential sensor and network failure modes and intrinsically unreliable multihop data transmissions.

Moreover, the entire measurement and networking enterprise is severely constrained by limited power and energy. There is substantial redundancy in data collected within wireless networks (Fig. 1). Yet the capacity to collect dense data when it provides valuable information is one of the key motivations for the technology. Clearly, there is need to control the measurement process with model-based evaluation of potential observations.

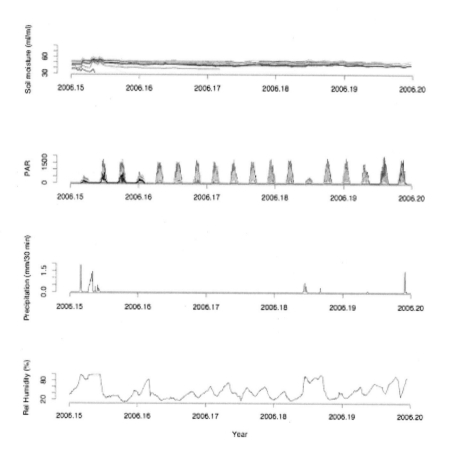

Fig. 1. Examples of four variables measured in the Duke Forest wireless sensor network showing different levels of redundancy at different scales

Measurements without underlying data and process models are of limited use in this endeavor. Indeed, most useful measurements, even when noiseless and unbiased, are still based on an underlying model, as in processing for sensors and satellite imagery (Clark et al. 2007). These often-implicit or even forgotten models have limitations that are bound up in their associated data. For example, a fundamental operation in environmental monitoring is sampling in space and time, and one approach is to estimate temporal and spatial bandwidths to establish sampling rates. However, the usual frame of reference in this case is the classic Shannon sampling theorem, and the requirement of finite bandwidth in turn forces definition of the signal over all time, a clear model-based limitation.

The phenomena of interest in environmental monitoring are highly time-varying and non-stationary, and laced with measurement and model uncertainty. These factors are the key motivation for the application of the dynamic distributed data application systems (DDDAS) paradigm to wireless sensor networks (Flikkema et al. 2006). DDDA systems are characterized by the coupling of the concurrent processes of data collection and model inference, with feedbacks from one used to refine the other. The fact that resources—energetic and economic—are limited in wireless sensor networks is in some sense an opportunity. Rather than accept measurements as the gold standard, we should embrace the fact that both measurements and models can be rife with uncertainty, and then tackle the challenge of tracking and managing that uncertainty through all phases of the project: transducer and network design; data acquisition, transfer, and storage; model inference; and analysis and interpretation.

2   Dynamic Control of Network Activity

Looking at two extreme cases of data models—strong spatial correlation combined with weak local correlation and vice versa—can shed some light on the trade-offs in designing algorithms that steer network activity. First, consider the case when the monitored process is temporally white but spatially coherent. This could be due to an abrupt global (network-wide) change, such as the onset of a rainstorm in the monitoring of soil moisture. In this case, we need snapshots at the natural temporal sampling rate, but only from a few sensor nodes. Data of the needed fidelity can then be obtained using decentralized protocols, such as randomized protocols that are simple and robust (Flikkema 2006). Here, the fusion center or central server broadcasts a cue to the nodes in terms of activity probabilities. The polar opposite is when there is strong temporal coherence but the measurements are statistically independent in the spatial domain. One example of this is sunfleck processes in a forest stand with varying canopy density. Since most sensor nodes should report their measurements, but infrequently, localized temporal coding schemes can work well.

Our overall effort goes beyond data models to the steering of network activity driven by ecosystem process models, motivated by the fact that even though a measured process may have intrinsically strong dynamics (or high bandwidth), it may be driving an ecosystem process that is a low-pass filter, so that the original data stream is strongly redundant with respect to the model of interest. Our approach is to move toward higher-level modeling that reveals what data is important.

A common criticism might arise here: what if the model is wrong? First, given the imprecision and unreliability of data, there is no a priori reason to favor data. For example, we often reject outliers in data preprocessing, which relies on an implicit "reasonableness" model. Yet an outlier could be vital information. Thus any scheme must dynamically allocate confidence between the model and the incoming data. By using a Bayesian approach, this allocation can be made in a principled, quantitative manner (Clark 2005, Clark 2007, MacKay 2003).

Any algorithm that uses model-steered sampling and reporting (rather than resorting to fixed-rate sampling at the some maximum rate) will make errors with a non-zero probability. To mitigate these errors, our strategy is based concurrent execution of the same models in the fusion center as in individual sensors. Using this knowledge, the fusion center can estimate unreported measurements; the reliability of these estimates is determined by the allowed departure of the predicted value from the true value known by the sensing node. The fusion center can also run more sophisticated simulation-based models that would be infeasible in the resource-constrained sensors, and use the results to broadcast model-parameter updates.

Clearly, a missing report could be due to a communication or processing error rather than a decision by the sensor. By carefully controlling redundancy within the Bayesian inference framework, which incorporates models for both dynamic reporting and failure statistics (Silberstein et al. 2007), it become possible to infer not only data and process models, but node- and network-level failure modes as well. Finally, in our experiments, each sensor node archives its locally acquired data in non-volatile memory, allowing collection of reference data sets for analysis.

3   Example: Soil Moisture Processes

Soil moisture is a critical ecosystem variable since it places a limit on the rate of photosynthesis and hence plant growth. It is a process parameterized by soil type and local topography and driven by precipitation, surface runoff, evapotranspiration, and subsurface drainage processes. Because it is highly non-linear, it is much more accessible to Bayesian approaches than ad hoc inverse-modeling techniques. Bayesian techniques permit integration of process noise that characterizes our level of confidence in the model. In practice, it may more productive to use a simple model with fewer state variables and process noise instead of a model of higher dimension with poorly known sensitivity to parameter variations.

Once the model is obtained (for example, using training data either from archival data or a "shake-out" interval in the field), the inferred parameters can then be distributed to the sensor nodes. The nodes then use the model as a predictor for new measurements based on the past. The observation is transmitted only when the discrepancy between the measurement and the predicted value exceeds a threshold (again known to both the sensor and the fusion center). Finally, the model(s) at the fusion center are used to recover the unreported changes.

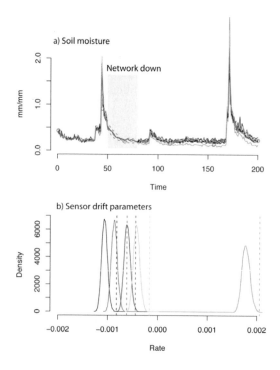

Fig. 2. a) Simulated soil moisture data (solid line) and simulated observations (colored lines) from sensors that drift. b) Posterior estimates of parameter drift become part of the model that is used to determine which observations to collect (Fig. 3).

In the simulation results shown in Figure 2 (Clark et al. 2007), the underlying soil moisture is shown as a solid black line. Here we use a purely temporal model in each sensor node. Five sensors are shown in different colors, with calibration data as red dots. To emphasize that the approach does not depend on a flawless network, we assume that all sensors are down for a period of time. The 95% predictive intervals for soil moisture (dashed lines) show that, despite sensor drift and even complete network failure, soil moisture can be accurately predicted. For this particular example, the estimates of drift parameters are somewhat biased (Figure 2b), but these parameters are of limited interest, and have limited impact on predictive capacity (Figure 2a). The impact on reporting rate and associated energy usage is substantial as well (Clark et al. 2007).

Our strategy is to incorporate dynamic reporting starting with simple, local models in an existing wireless sensor network architecture (Yang 2005). As shown by the soil moisture example, even purely temporal models can have a significant impact. From a research standpoint, it will be useful to first determine the effectiveness of dynamic reporting driven by a local change-based model where a node reports an observation only if it has changed from the previously reported observation by a specified absolute amount. This is simple to implement and requires a negligible increase in processing time and energy. In general, local

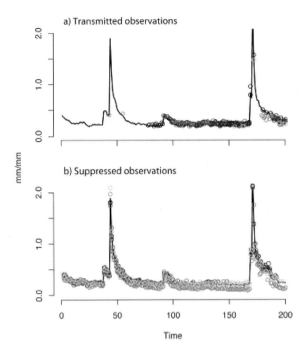

Fig. 3. A simple process model, with field capacity and wilting point, together with a data model that accommodated parameter drift (Fig. 2b) allows for transmission of only a fraction of the data (solid dots in (a)). Far more of the measurements are suppressed (b), because they can be predicted.

models have the advantage of not relying on collaboration with other sensor nodes and its associated energy cost of communication.

What about the problem of applying one model for data collection and another for modeling those data in the future? It is important to select models for data collection that emphasize data predictability, rather than specific parameters to be estimated. For example, wilting point and field capacity are factors that make soil moisture highly predictable, their effects being evident in Figures 1 and 2. By combining a process model that includes just a few parameters that describe the effect of field capacity and wilting point and a data model that includes sensor error, the full time series can be reconstructed based on a relatively small number of observations (Figure 3)(Clark et al. 2007).

4   Looking Ahead

Researchers tend to make an observation, find the most likely value, and then treat it as deterministic in all subsequent work, with uncertainty captured only in process modeling. We have tried to make the case here for a more holistic approach that captures uncertainty in both data and models, and uses a framework to monitor and manage that uncertainty. As wireless sensor network deployments become larger and more numerous, researchers in ecology and the environmental sciences will become inundated with massive, unwieldy datasets filled with numerous flaws and artifacts. Our belief is that much of this data may be redundant, and that many of the blemishes may be irrelevant from the perspective of inferring predictive models of complex, multidimensional ecosystems processes. Since the datasets will consume a great deal of time and effort to document, characterize, and manage, we think that that the time for model-mediated sensing has arrived.

References

1. NEON: Addressing the Nation's Environmental Challenges. Committee on the National Ecological Observatory Network (G. David Tilman, Chair), National Research Council. 2002. ISBN: 0-309-09078-4.
2. Clark, J.S. Why environmental scientists are becoming Bayesians. Ecol. Lett. 8:2-14, 2005.
3. Clark, J.S. *Models for Ecological Data: An Introduction.* Princeton University Press, 2007.
4. Clark, J.S., Agarwal, P., Bell, D., Ellis, C., Flikkema, P., Gelfand, A., Katul, G., Munagala, K., Puggioni, G., Silberstein, A., and Yang, J. Getting what we need from wireless sensor networks: a role for inferential ecosystem models. 2007 (in preparation).
5. Flikkema, P. The precision and energetic cost of snapshot estimates in wireless sensor networks. Proc. IEEE Symposium on Computing and Communications (ISCC 2006), Pula-Cagliari, Italy, June 2006.
6. Flikkema, P., Agarwal, P., Clark, J.S., Ellis, C., Gelfand, A., Munagala, K., and Yang, J. Model-driven dynamic control of embedded wireless sensor networks. Workshop on Dynamic Data Driven Application Systems, International Conference on Computational Science (ICCS 2006), Reading, UK, May 2006.
7. MacKay, D.J.C. *Information Theory, Inference, and Learning Algorithms.* Cambridge University Press, 2003.
8. Silberstein, A., Braynard, R., Filpus, G., Puggioni, G., Gelfand, A., Munagala, K., and Yang, J. Data-driven processing in sensor networks. Proc. 3rd Biennial Conference on Innovative Data Systems Research (CIDR '07), Asilomar, California, USA, January 2007.
9. Yang, Z., et al. WiSARDNet: A system solution for high performance in situ environmental monitoring. Second International Workshop on Networked Sensor Systems (INSS 2005), San Diego, 2005.

AMBROSia: An Autonomous Model-Based Reactive Observing System*

David Caron, Abhimanyu Das, Amit Dhariwal, Leana Golubchik**,
Ramesh Govindan, David Kempe, Carl Oberg, Abhishek Sharma, Beth Stauffer,
Gaurav Sukhatme, and Bin Zhang

University of Southern California, Los Angeles, CA 90089
**leana@cs.usc.edu

Abstract. Observing systems facilitate scientific studies by instrumenting the real world and collecting corresponding measurements, with the aim of detecting and tracking phenomena of interest. Our AMBROSia project focuses on a class of observing systems which are *embedded* into the environment, consist of *stationary and mobile* sensors, and *react* to collected observations by reconfiguring the system and adapting which observations are collected next. In this paper, we report on recent research directions and corresponding results in the context of AMBROSia.

1   Introduction

Observing systems facilitate scientific studies by instrumenting the real world and collecting measurements, with the aim of detecting and tracking phenomena of interest. Our work focuses Reactive Observing Systems (ROS), i.e., those that are (1) *embedded* into the environment, (2) consist of *stationary and mobile* sensors, and (3) *react* to collected observations by reconfiguring the system and adapting which observations are collected next. The goal of ROS is to help scientists verify or falsify hypotheses with useful samples taken by the stationary and mobile units, as well as to analyze data autonomously to discover interesting trends or alarming conditions. We explore ROS in the context of a marine biology application, where the system monitors, e.g., water temperature and light as well as concentrations of micro-organisms and algae in a body of water.

Current technology (and its realistic near future prediction) precludes sampling all possibly relevant data: bandwidth limitations between stationary sensors make it impossible to collect all sensed data, and time & storage capacity constraints for mobile entities curtail the number and locations of samples they can take. To make good use of limited resources, we are developing a framework capable of optimizing and controlling the set of samples to be taken at any given time, taking into consideration the

---

* This research has been funded by the NSF DDDAS 0540420 grant. It has also been funded in part by the NSF Center for Embedded Networked Sensing Cooperative Agreement CCR-0120778, the National Oceanic and Atmospheric Administration Grant NA05NOS47812228, and the NSF EIA-0121141 grant.

** Contact author.

Y. Shi et al. (Eds.): ICCS 2007, Part I, LNCS 4487, pp. 995–1001, 2007.

application's objectives and system resource constraints. We refer to this framework as AMBROSia (Autonomous Model-Based Reactive Observing System). In [7] we give an overview of the AMBROSia framework as well as the experimental system setting and the corresponding marine biology application. In this paper we report on our recent research directions and corresponding results, in the context of AMBROSia.

As already noted, one of the core functionalities of AMBROSia is the selection of samples which (1) can be retrieved at reasonably low energy cost, and (2) yield as much information as possible about the system. The second property in particular will change dynamically: in reaction to past measurements, different observations may be more or less useful in the future. At any given time, the system must select the most informative samples to retrieve based on the model at that point. We briefly outline a mathematical formulation of this problem and results to date in Section 2.

In ROS accurate measurements are useful to scientists seeking a better understanding of the environment. However, it may not be feasible to move the static sensor nodes after deployment. In such cases, mobile robots could be used to augment the static sensor network, hence forming a robotic sensor network. In such networks, an important question to ask is how to coordinate the mobile robots and the static nodes such that estimation errors are minimized. Our recent efforts on addressing this question are briefly outlined in Section 3.

The successful use of ROS, as envisioned in AMBROSia, partly depends on the system's ability to ensure the collected data's quality. However, various sensor network measurement studies have reported transient faults in sensor readings. Thus, another important goal in AMBROSia is automated high-confidence fault detection, classification, and data rectification. As a first step towards that goal, we explore and characterize several qualitatively different classes of fault detection methods, which are briefly outlined in Section 4.

Our concluding remarks are given in Section 5.

2   A Mathematical Formulation of Sample Selection

Mathematically, our sample selection problem can be modeled naturally as a subset selection problem for regression: Based on the small number of measurements X_i taken, a random variable Z (such as average temperature, chlorophyll concentration, growth of algae, etc.) is to be estimated as accurately as possible. Different measurements X_i, X_j may be partially correlated, and thus partially redundant, a fact that should be deduced from past models. In a pristine and abstract form, the problem can thus be modeled as follows:

We are given a covariance matrix C between the random variables X_i, and a vector $\mathbf{b}$ describing covariances between measurements X_i and the quantity Z to be predicted (C and $\mathbf{b}$ are estimated based on the model). In order to keep the energy sampling cost small, the goal is to find a small set S (of size at most k) so as to minimize the *mean squared prediction error* [4,8] $\mathrm{Err}(Z, S) := E[(Z - \sum_{i \in S} \alpha_i X_i)^2]$, where the α_i are the optimal regression coefficients specifically for the set S selected.

The selection problem thus gives rise to the well-known *subset selection problem for regression* [10], which has traditionally had many applications in medical and social studies, where the set S is interpreted as a good predictor of Z. Finding the best set S of size k is NP-hard, and certain approximation hardness results are known [2,11]. However, despite its tremendous importance to statistical sciences, very little was known in terms of approximation algorithms until recent results by Gilbert et al. [6] and Tropp [17] established approximation guarantees for the very special case of nearly independent X_i variables.

In ongoing work, we are investigating several more general cases of the subset selection problem for regression, in particular with applications to selecting samples to draw in sensor network environments. Over the past year, we have obtained the following key results (which are currently under submission [1]):

Theorem 1. *If the pairwise covariances between the X_i are small (at most $1/6k$, if k variables can be selected), then the frequently used Forward Regression heuristic is a provably good approximation.*

The quality of approximation is characterized precisely in [1], but omitted here due to space constraints. This result improves on the ones of [6,17], in that it analyzes a more commonly used algorithm, and obtains somewhat improved bounds. The next theorem extends the result to a significantly wider class of covariance matrices, where several pairs can have higher covariances.

Theorem 2. *If the pairs of variables X_i with high covariance (exceeding $\Omega(1/4k)$) form a tree, then a provably good approximation can be obtained in polynomial time using rounding and dynamic programming.*

While this result significantly extends the cases that can be approximated, it is not directly relevant to measuring physical phenomena. Hence, we also study the case of sensors embedded in a metric space, where the covariance between sensors' readings is a monotone decreasing function of their distance. The general version of this problem is the subject of ongoing work, but [1] contains a promising initial finding:

Theorem 3. *If the sensors are embedded on a line (in one dimension), and the covariance decreases roughly exponentially in the distance, then a provably good approximation can be obtained in polynomial time.*

The algorithm is again based on rounding and a different dynamic program, and makes use of some remarkable properties of matrix inverses for this type of covariance matrix. At the moment, we are working on extending these results to more general metrics (in particular, two-dimensional Euclidean metrics), and different dependencies of covariances on the distance.

3   Scalar Field Estimation

Sensor networks provide new tools for observing and monitoring the environment. In aquatic environments, accurately measuring quantities such as temperature, chlorophyll, salinity, and concentration of various nutrients is useful to scientists seeking a better understanding of aquatic ecosystems, as well as government officials charged with ensuring public safety via appropriate hazard warning and remediation measures.

Broadly speaking, these quantities of interest are scalar fields. Each is characterized by a single scalar quantity which varies spatiotemporally. Intuitively, the more the readings near the location where a field estimate is desired, the less the reconstruction error. In other words, the spatial distribution of the measurements (the *samples*) affects the estimation error. In many cases, it may not be feasible to move the static sensor nodes after deployment. In such cases, one or more mobile robots could be used to augment the static sensor network, hence forming a sensor-actuator network or a robotic sensor network.

The problem of adaptive sampling: An immediate question to ask is how to coordinate the mobile robots and the static nodes such that the error associated with the estimation on the scalar field is minimized subject to the constraint that the energy available to the mobile robot(s) is bounded. Specifically, if each static node makes a measurement in its vicinity, and the total energy available to the mobile robot is known, what path should the mobile robot take to minimize the mean square integrated error associated with the reconstruction of the entire field? Here we assume that the energy consumed by communications and sensing is negligible compared to the energy consumed in moving the mobile robot. We also assume that the mobile robot can communicate with all the static nodes and acquire sensor readings from them. Finally, we focus on reconstructing phenomena which do not change temporally(or change very slowly compared to the time it takes the mobile robot to complete a tour of the environment).

The domain: We develop a general solution to the above problem and test it on a particular set up designed to monitor an aquatic environment. The experimental set up is a systems of anchored buoys (the static nodes), and a robotic boat (the mobile robot) capable of measuring temperature and chlorophyll concentrations. This testbed is part of the NAMOS (Networked Aquatic Microbial Observing System) project (http://robotics.usc.edu/~namos), which is used in studies of microbial communities in freshwater and marine environments [3,15].

Contributions: We propose an adaptive sampling algorithm for a mobile sensor network consisting of a set of static nodes and a mobile robot tasked to reconstruct a scalar field. Our algorithm is based on local linear regression [13,5]. Sensor readings from static nodes (a set of buoys) are sent to the mobile robot (a boat) and used to estimate the Hessian Matrix of the scalar field (the surface temperature of a lake), which is directly related to the estimation error. Based on this information, a path planner generates a path for the boat such that the resulting integrated mean square error (IMSE) of the field reconstruction is minimized subject to the constraint that the boat has a finite amount of energy which it can expend on the traverse. Data from extensive (several km) traverses in the field as well as simulations, validate the performance of our algorithm.

We are currently working on how to determine the appropriate resolution to discretize the sensed field. One interesting observation from the simulations and experiments is that when the initial available energy is increased, the estimation errors decrease rapidly and level off instead of decreasing to zero. Theoretically, when the energy available to the mobile node increases, more sensor readings can be taken and hence the estimation errors should keep decreasing. By examining the path generated by the adaptive sampling algorithm, we found that when the initial energy is enough for the mobile node to go through all the 'important' locations, increasing the initial energy does not have much effect on the estimation error. We plan to investigate advanced path planning strategies and alternative sampling design strategies in future work.

4   Faults in Sensor Data

With the maturation of sensor network software, we are increasingly seeing longer-term deployments of wireless sensor networks in real world settings. As a result, research attention is now turning towards drawing meaningful scientific inferences from the collected data [16]. Before sensor networks can become effective replacements for existing scientific instruments, it is important to ensure the quality of the collected data. Already, several deployments have observed faulty sensor readings caused by incorrect hardware design or improper calibration, or by low battery levels [12,16].

Given these observations, and the realization that it will be impossible to always deploy a perfectly calibrated network of sensors, an important research direction for the future will be automated detection, classification, and root-cause analysis of sensor faults, as well as techniques that can automatically scrub collected sensor data to ensure high quality. A first step in this direction is an understanding of the prevalence of faulty sensor readings in existing real-world deployments.

We focus on a small set of sensor faults that have been observed in real deployments: single-sample spikes in sensor readings (SHORT faults), longer duration noisy readings (NOISE faults), and anomalous constant offset readings (CONSTANT faults). Given these fault models, our work makes the following two contributions.

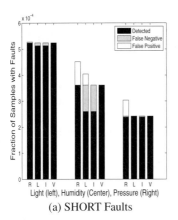
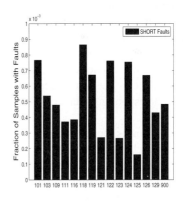

(a) SHORT Faults	(b) SHORT Faults: Light Sensor

Fig. 1. GDI data set

Detection Methods. We have explored three qualitatively different techniques for automatically detecting such faults from a trace of sensor readings. Rule-based methods leverage domain knowledge to develop heuristic rules for detecting and identifying faults. Linear Least-Squares Estimation (LLSE) based methods predict "normal" sensor behavior by leveraging sensor correlation, flagging deviations from the normal as sensor faults. Finally, learning-based methods (based on Hidden Markov Models) are trained to statistically detect and identify classes of faults.

Our findings indicate that these methods sit at different points on the accuracy/robustness spectrum. While rule-based methods can detect and classify faults, they can be sensitive to the choice of parameters. By contrast, the LLSE method is a bit more robust to parameter choices but relies on spatial correlations and cannot classify faults. Finally, our learning method (based on Hidden Markov Models) is cumbersome, partly because it requires training, but it can fairly accurately detect and classify faults. We also explored hybrid detection techniques, which combine these three methods in ways that can be used to reduce false positives or false negatives, whichever is more important for the application. These results are omitted for brevity and the interested reader is referred to [14].

Evaluation on Real-World Datasets. We applied our detection methods to real-world data sets. Here, we present results from the Great Duck Island (GDI) data set [9], where we examine the fraction of faulty samples in a sensor trace.

The predominant fault in the readings was of the type SHORT. We applied the SHORT rule, the LLSE method, and Hybrid(I) (a hybrid detection technique) to detect SHORT faults in light, humidity and pressure sensor readings. Figure 1(a) shows the overall prevalence (computed by aggregating results from all the 15 nodes) of SHORT faults for different sensors in the GDI data set. (On the x-axis of this figure, the SHORT rule's label is **R**, LLSE's label is **L**, and Hybrid(I)'s label is **I**.) The Hybrid (I) technique eliminates any false positives reported by the SHORT rule or the LLSE method. The intensity of SHORT faults was high enough to detect them by visual inspection of the entire sensor readings timeseries. This ground-truth is included for reference in the figure under the label **V**. It is evident from the figure that SHORT faults are relatively infrequent. They are most prevalent in the light sensor readings (approximately 1 fault every 2000 samples). Figure 1(b) shows the distribution of SHORT faults in light sensor readings across various nodes. (Here, node numbers are indicated on the x-axis.) SHORT faults do not exhibit any discernible pattern in the prevalence of these faults across different sensor nodes; the same holds for other sensors, but we have omitted the corresponding graphs for brevity.For results on other data sets, please refer to [14].

Our study informs the research on ensuring data quality. Even though we find that faults are relatively rare, they are not negligibly so, and careful attention needs to be paid to engineering the deployment and to analyzing the data. Furthermore, our detection methods could be used as part of an online fault diagnosis system, i.e., where corrective steps could be taken during the data collection process based on the diagnostic system's results.

5   Concluding Remarks

Overall, our vision for AMBROSia is that it will facilitate observation, detection, and tracking of scientific phenomena that were previously only partially (or not at all) observable and/or understood. In this paper we outlined results corresponding to some of our recent steps towards achieving this vision.

References

1. A. Das and D. Kempe. Algorithms for subset selection in regression, 2006. Submitted to STOC 2007.
2. G. Davis, S. Mallat, and M. Avellaneda. Greedy adaptive approximation. *Journal of Constructive Approximation*, 13:57–98, 1997.
3. Amit Dhariwal, Bin Zhang, Carl Oberg, Beth Stauffer, Aristides Requicha, David Caron, and Gaurav S. Sukhatme. Networked aquatic microbial observing system. In *the Proceedings of the IEEE International Conference of Robotics and Automation (ICRA)*, May 2006.
4. G. Diekhoff. *Statistics for the Social and Behavioral Sciences*. Wm. C. Brown Publishers, 2002.
5. Jianqing Fan. Local linear regression smoothers and their minimax efficiencies. *The Annals of Statistics*, 21(1):196–216, 1993.
6. A. Gilbert, S. Muthukrishnan, and M. Strauss. Approximation of functions over redundant dictionaries using coherence. In *Proc. ACM-SIAM Symposiun on Discrete Algorithms*, 2003.
7. Leana Golubchik, David Caron, Abhimanyu Das, Amit Dhariwal, Ramesh Govindan, David Kempe, Carl Oberg, Abhishek Sharma, Beth Stauffer, Gaurav Sukhatme, and Bin Zhang. A Generic Multi-scale Modeling Framework for Reactive Observing Systems: an Overview. In *Proceedings of the Dynamic Data Driven Application Systems Workshop held with ICCS*, 2006.
8. R. A. Johnson and D. W. Wichern. *Applied Multivariate Statistical Analysis*. Prentice Hall, 2002.
9. Alan Mainwaring, Joseph Polastre, Robert Szewczyk, and David Cullerand John Anderson. Wireless Sensor Networks for Habitat Monitoring . In *the ACM International Workshop on Wireless Sensor Networks and Applications. WSNA '02*, 2002.
10. A. Miller. *Subset Selection in Regression*. Chapman and Hall, second edition, 2002.
11. B. Natarajan. Sparse approximation solutions to linear systems. *SIAM Journal on Computing*, 24:227–234, 1995.
12. N. Ramanathan, L. Balzano, M. Burt, D. Estrin, E. Kohler, T. Harmon, C. Harvey, J. Jay, S. Rothenberg, and M. Srivastava. Rapid Deployment with Confidence: Calibration and Fault Detection in Environmental Sensor Networks. Technical Report 62, CENS, April 2006.
13. D. Ruppert and M. P. Wand. Multivariate locally weighted least squares regression. *The Annals of Statistics*, 22(3):1346–1370, 1994.
14. A. Sharma, L. Golubchik, and R. Govindan. On the Prevalence of Sensor Faults in Real World Deployments. Technical Report 07-888, Computer Science, University of Southern California, 2007.
15. Gaurav S. Sukhatme, Amit Dahriwal, Bin Zhang, Carl Oberg, Beth Stauffer, and David Caron. The design and development of a wireless robotic networked aquatic microbial observing system. *Environmental Engineering Science*, 2007.
16. Gilman Tolle, Joseph Polastre, Robert Szewczyk, David Culler, Neil Turner, Kevin Tu, Stephen Burgess, Todd Dawson, Phil Buonadonna, David Gay, and Wei Hong. A Macroscope in the Redwoods. In *SenSys '05: Proceedings of the 2nd international conference on Embedded networked sensor systems*, pages 51–63, New York, NY, USA, 2005. ACM Press.
17. J. Tropp. *Topics in Sparse Approximation*. PhD thesis, University of Texas, Austin, 2004.

Dynamically Identifying and Tracking Contaminants in Water Bodies

Craig C. Douglas[1,2], Martin J. Cole[3], Paul Dostert[4], Yalchin Efendiev[4],
Richard E. Ewing[4], Gundolf Haase[5], Jay Hatcher[1], Mohamed Iskandarani[6],
Chris R. Johnson[3], and Robert A. Lodder[7]

[1] University of Kentucky, Department of Computer Science, 773 Anderson Hall,
Lexington, KY 40506-0046, USA
hatch22@fastmail.us
[2] Yale University, Department of Computer Science, P.O. Box 208285
New Haven, CT 06520-8285, USA
douglas-craig@cs.yale.edu
[3] University of Utah, Scientific Computing and Imaging Institute, Salt Lake City, UT
84112, USA
{crj,mjc}@cs.utah.edu
[4] Texas A&M University, Institute for Scientific Computation, 612 Blocker, 3404
TAMU, College Station, TX 77843-3404, USA
richard_ewing@tamu.edu, {dostert,efendiev}@math.tamu.edu
[5] Karl-Franzens University of Graz, Mathematics and Computational Sciences,
A-8010 Graz, Austria
gundolf.haase@uni-graz.at
[6] University of Miami, Rosenstiel School of Marine and Atmospheric Science, 4600
Rickenbacker Causeway, Miami, FL 33149-1098, USA
mohamed.iskandarani@rsmas.miami.edu
[7] University of Kentucky, Department of Chemistry, Lexington, KY, 40506-0055, USA
lodder@contactincontext.org

Abstract. We present an overview of an ongoing project to build a
DDDAS for identifying and tracking chemicals in water. The project involves a new class of intelligent sensor, building a library to optically
identify molecules, communication techniques for moving objects, and
a problem solving environment. We are developing an innovative environment so that we can create a symbiotic relationship between computational models for contaminant identification and tracking in water
bodies and a new instrument, the Solid-State Spectral Imager (SSSI), to
gather hydrological and geological data and to perform chemical analyses. The SSSI is both small and light and can scan ranges of up to about
10 meters. It can easily be used with remote sensing applications.

1 Introduction

In this paper, we describe an intelligent sensor and how we are using it to create
a dynamic data-driven application system (DDDAS) to identify and track contaminants in water bodies. This DDDAS has applications to tracking polluters,

Y. Shi et al. (Eds.): ICCS 2007, Part I, LNCS 4487, pp. 1002–1009, 2007.

finding sunken vehicles, and ensuring that drinking water supplies are safe. This paper is a sequel to [1].

In Sec. 2, we discuss the SSSI. In Sec. 3, we discuss the problem solving environment that we have created to handle data to and from SSSI's in the field. In Sec. 4, we discuss In Sec. 5, we state some conclusions.

2   The SSSI

Using a laser-diode array, photodetectors, and on board processing, the SSSI combines innovative spectroscopic integrated sensing and processing with a hyperspace data analysis algorithm [2]. The array performs like a small network of individual sensors. Each laser-diode is individually controlled by a programmable on board computational device that is an integral part of the SSSI and the DDDAS.

Ultraviolet, visible, and near-infrared laser diodes illuminate target points using a precomputed sequence, and a photodetector records the amount of reflected light. For each point illuminated, the resulting reflectance data is processed to separate the contribution of each wavelength of light and classify the substances present. An optional radioactivity monitor can enhance the SSSI's identification abilities.

The full scale SSSI implementation will have 25 lasers in discrete wavelengths between 300 nm and 2400 nm with 5 rows of each wavelength, consume less than 4 Watts, and weigh less than 600 grams. For water monitoring in the open ocean, imaging capability is unnecessary. A single row of diodes with one diode at each frequency is adequate. Hence, power consumption of the optical system can be reduced to approximately one watt.

Several prototype implementations of SSSI have been developed and are being tested at the University of Kentucky. These use an array of LEDs instead of lasers.

The SSSI combines near-infrared, visible, and ultraviolet spectroscopy with a statistical classification algorithm to detect and identify contaminants in water. Nearly all organic compounds have a near-IR spectrum that can be measured. Near-infrared spectra consist of overtones and combinations of fundamental mid-infrared bands, which makes near-infrared spectra a powerful tool for identifying organic compounds while still permitting some penetration of light into samples [3].

The SSSI uses one of two techniques for encoding sequences of light pulses in order to increase the signal to noise ratio: Walsh-Hadamard or Complementary Randomized Integrated Sensing and Processing (CRISP).

In a Walsh-Hadamard sequence multiple laser diodes illuminate the target at the same time, increasing the number of photons received at the photo detector. The Walsh-Hadamard sequence can be demultiplexed to individual wavelength responses with a matrix-vector multiply [4]. Two benefits of generating encoding sequences by this method include equivalent numbers of on and off states for each sequence and a constant number of diodes in the on state at each resolution point of a data acquisition period.

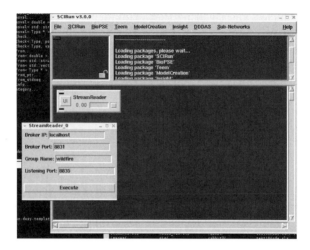

Fig. 1. SCIRun screen with telemetry module in forefront

CRISP encoding uses orthogonal pseudorandom codes with unequal numbers
of on and off states. The duty cycle of each code is different, and the codes are
selected to deliver the highest duty cycles at the wavelengths where the most
light is needed and lowest duty cycle where the least light is needed to make the
sum of all of the transmitted (or reflected) light from the samples proportional
to the analyte concentration of interest.

3   Problem Solving Environment SCIRun

SCIRun version 3.0 [5,6], a scientific problem solving enironment, was released in
late 2006. It includes a telemetry module based on [7], which provides a robust
and secure set of Java tools for data transmission that assumes that a known
broker exists to coordinate sensor data collection and use by applications. Each
tool has a command line driven form plus a graphical front end that makes it so
easy that even the authors can use the tools.

In addition there is a Grid based tool that can be used to play back already collected data. We used Apple's XGrid environment [8] (any Grid environment will work, however) since if someone sits down and uses one of the
computers in the Grid, the sensors handled by that computer disappear from
the network until the computer is idle again for a small period of time. This
gives us the opportunity to develop fault tolerant methods for unreliable sensor
networks.

The clients (sensors or applications) can come and go on the Internet, change
IP addresses, collect historical data, or just new data (letting the missed data
fall on the floor). The tools were designed with disaster management [9] in mind
and stresses ease of use when the user is under duress and must get things right
immediately.

A new Socket class was added to SCIRun, which encapsulates the socket traffic and is used to connect and transfer data from the server. The client handshakes with the server, which is informed of an ip:port where the client can be reached, and then listens on that port. Periodically, or as the server has new data available, the server sends data to the listening client.

The configuration for SCIRun was augmented to include libgeotiff [10]. SCIRun then links against this client and has its API available within the modules. This API can be used to extract the extra information embedded in the tiff tags in various supported formats. For example, position and scale information can be extracted so that the images can be placed correctly.

To allow controller interfaces to be built for the SSSI, a simulation of the device has been written in Matlab. This simulation follows the structure of the firmware code and provides the same basic interface as the firmware device. Data files are used in place of the SSSI's serial communication channel to simulate data exchange in software. Matlab programs are also provided to generate sample data files to aid in the development of Hadamard-Walsh and CRISP encodings for various SSSI configurations. The simulation also provides insight into the SSSI's firmware by emulating the use of oversampling and averaging to increase data precision and demonstrating how the data is internally collected and processed. The simulation can be used for the development of interfaces to the SSSI while optimization and refinement of the SSSI firmware continues.

SCIRun has a Matlab module so that we can pipe data to and from the SSSI emulator. As a result, we can tie together the data transfer and SSSI components easily into a system for training new users and to develop virtual sensor networks before deployment of a real sensor network in the field.

4   Accurate Predictions

The initial deployment of the sensor network and model will focus on estuarine regions where water quality monitoring is critical for human health and environmental monitoring. The authors will capitalize on an existing configuration of the model to the Hudson-Raritan Estuary to illustrate the model's capabilities (see [1] for details). We will consider passive tracer driven by external sources:

$$\frac{\partial C(x,t)}{\partial t} - L(C(x,t)) = S(x,t), \; C(x,0) = C^0(x) \; x \in \Omega,$$

where C is the concentration of contaminant, S is a source term and L is linear operator for passive scalar (advection-diffusion-reaction). L involves the velocity field which is obtained via the forward model based on the two-dimensional Spectral Element Ocean Model (SEOM-2D). This model solves the shallow water equations and the details can be found in our previous paper [1]. We have developed the spectral element discretization which relies on relatively high degree (5-8th) polynomials to approximate the solution within flow equations. The main features of the spectral element method are: geometric flexibility due to its unstructured grids, its dual paths to convergence: exponential by increasing polynomial degree or algebraic via increasing the number of elements, dense computational kernels with sparse inter-element synchronization, and excellent scalability on parallel machines.

We now present our methodology for obtaining improved predictions based on sensor data. For simplicity, our example is restricted to synthetic velocity fields. Sensor data is used to improve the predictions by updating the solution at previous time steps which is used for forecasting. This procedure consists of updating the solution and source term history conditioned to observations and reduces the computational errors associated with incorrect initial/boundary data, source terms, etc., and improves the predictions [11,12,13]. We assume that the source term can be decomposed into pulses at different time steps (recording times) and various locations. We represent time pulses by $\delta_k(x,t)$ which corresponds to contaminant source at the location $x = x_k$.

We seek the initial condition as a linear combination of some basis functions $C^0(x) \approx \tilde{C}^0(x) = \sum_{i=1}^{N_D} \lambda_i \varphi_i^0(x)$. We solve for each i,

$$\frac{\partial \varphi_i}{\partial t} - L(\varphi_i) = 0, \ \varphi_i(x,0) = \varphi_i^0(x).$$

Thus, an approximation to the solution of $\frac{\partial C}{\partial t} - L(C) = 0, \ C(x,0) = C^0(x)$ is given by $\tilde{C}(x,t) = \sum_{i=1}^{N_D} \lambda_i \varphi_i(x,t)$. To seek the source terms, we consider the following basis problems

$$\frac{\partial \psi_k}{\partial t} - L(\psi_k) = \delta_k(x,t), \ \psi_k(x,0) = 0$$

for ψ and each k. Here, $\delta_k(x,t)$ represents unit source terms that can be used to approximate the actual source term. In general, $\delta_k(x,t)$ have larger support both in space and time in order to achieve accurate predictions. We denote the solution to this equation as $\{\psi_k(x,t)\}_{k=1}^{N_c}$ for each k. Then the solution to our original problem with both the source term and initial condition is given by

$$\tilde{C}(x,t) = \sum_{i=1}^{N_D} \lambda_i \varphi_i(x,t) + \sum_{k=1}^{N_c} \alpha_k \psi_k(x,t).$$

Thus, our goal is to minimize

$$F(\alpha,\lambda) = \sum_{j=1}^{N_s} \left[\left(\sum_{k=1}^{N_c} \alpha_k \psi_k(x_j,t) + \sum_{k=1}^{N_D} \lambda_k \varphi_k(x_j,t) - \gamma_j(t) \right)^2 \right] + \\ \sum_{k=1}^{N_c} \tilde{\kappa}_k \left(\alpha_k - \tilde{\beta}_k \right)^2 + \sum_{k=1}^{N_D} \hat{\kappa}_k \left(\lambda_k - \hat{\beta}_k \right)^2, \quad (1)$$

where N_s denotes the number of sensors. If we denote $N = N_c + N_d$, $\mu = [\alpha_1, \cdots, \alpha_{N_c}, \lambda_1, \cdots, \lambda_{N_D}]$, $\eta(x,t) = [\psi_1, \cdots, \psi_{N_c}, \varphi_1, \cdots, \varphi_{N_D}]$,

$\beta = \left[\tilde{\beta}_1, \cdots, \tilde{\beta}_{N_c}, \hat{\beta}_1, \cdots, \hat{\beta}_{N_D} \right]$, and $\kappa = [\tilde{\kappa}_1, \cdots, \tilde{\kappa}_{N_c}, \hat{\kappa}_1, \cdots, \hat{\kappa}_{N_D}]$ then we want to minimize

$$F(\mu) = \sum_{j=1}^{N_s} \left[\left(\sum_{k=1}^{N} \mu_k \eta_k(x_j, t) - \gamma_j(t) \right)^2 \right] + \sum_{k=1}^{N} \kappa_k (\mu_k - \beta_k)^2.$$

This leads to solving the least squares problem $A\mu = R$ where

$$A_{mn} = \sum_{j=1}^{N} \eta_m(x_j, t) \eta_n(x_j, t) + \delta_{mn} \kappa_m,$$

and

$$R_m = \sum_{j=1}^{N} \eta_m(x_j, t) \gamma_j(t) + \kappa_m \beta_m.$$

We can only record sensor values at some discrete time steps $t = \{t_j\}_{j=1}^{N_t}$. We want to use the sensor values at $t = t_1$ to establish an estimate for μ, then use each successive set of sensor values to refine this estimate. After each step, we update and then solve using the next sensor value.

Next, we present a representative numerical result. We consider contaminant transport on a flat surface, a unit dimensionless square, with convective velocity in the direction $(1, 1)$. The source term is taken to be 0.25 in $[0.1, 0.3] \times [0.1, 0.3]$ for the time interval from $t = 0$ to $t = 0.05$. Initial condition is assumed to have the support over the entire domain. We derive the initial condition (solution at previous time step) by solving the original contaminant transport problem with some source terms assuming some prior contaminant history.

To get our observation data for simulations, we run the forward problem and sample sensor data at every 0.05 seconds for 1.0 seconds. We sample at the following five locations: $(0.5, 0.5)$, $(0.25, 0.25)$, $(0.25, 0.75)$, $(0.75, 0.25)$, and $(0.75, 0.75)$.

When reconstructing, we assume that there is a subdomain $\Omega_c \subset \Omega$ where our initial condition and source terms are contained. We assume that the source term and initial condition can be represented as a linear combinations of basis functions defined on Ω_c. For this particular model, we assume the subdomain is $[0, 0.4] \times [0, 0.4]$ and we have piecewise constant basis functions. Furthermore, we assume that the source term in our reconstruction is nonzero for the same time interval as $S(x, t)$. Thus we assume the source basis functions are nonzero for only $t \in [0, 0.05]$.

To reconstruct, we run the forward simulation for a 4×4 grid of piecewise constant basis functions on $[0, 0.4] \times [0, 0.4]$ for both the initial condition and the source term. We then reconstruct the coefficients for the initial condition and source term using the approach proposed earlier. The following plot shows a comparison between the original surface (in green) and the reconstructed surface (in red). The plots are for $t = 0.1$, 0.2, 0.4 and 0.6. We observe that the recovery at initial times is not very accurate. This is due to the fact that we have not

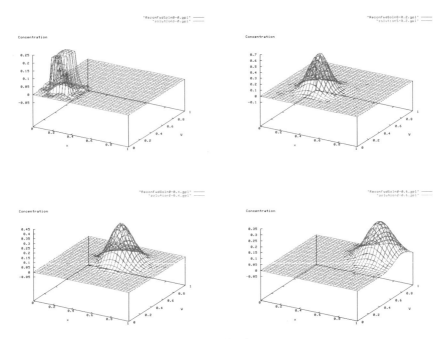

Fig. 2. Comparison between reconstructed (red) solution and exact solution at $t = 0.1$ (upper left), $t = 0.2$ (upper right), $t = 0.4$ (lower left), and $t = 0.6$ (lower right)

collected sufficient sensor data. As the time progresses, the prediction results improve. We observe that at $t = 0.6$, we have nearly exact prediction of the contaminant transport.

To account for the uncertainties associated with sensor measurements, we consider an update of initial condition and source terms, within a Bayesian framework. The posterior distribution is set up based on measurement errors and prior information. This posterior distribution is complicated and involves the solutions of partial differential equations. We developed an approach that combines least squares with a Bayesian approach, such as Metropolis-Hasting Markov chain Monte Carlo (MCMC) [14], that gives a high acceptance rate. In particular, we can prove that rigorous sampling can be achieved by sampling the sensor data from the known distribution, thus obtaining various realizations of the initial data. Our approach has similarities with the Ensemble Kalman Filter approach, which can also be adapted in our problem. We have performed numerical studies and these results will be reported elsewhere.

5   Conclusions

In the last year, we have made strides in creating our DDDAS. We have developed software that makes sending data from locations that go on and off the Internet and possibly change IP addresses rather easy to work with. This is a stand alone package that runs on any devices that support Java. It has also been integrated into newly released version 3.0 of SCIRun and is in use by other groups, including surgeons while operating on patients. We have also developed software that simulates the behavior of the SSSI and are porting the relevant parts so that it can be loaded into the SSSI to get real sensor data. We have developed algorithms that allow us to achieve accurate predictions in the presence of errors/uncertainties in dynamic source terms as well as other external conditions. We have tested our methodology in both deterministic and stochastic environments and have presented some simplistic examples in this paper.

References

1. Douglas, C.C., Harris, J.C., Iskandarani, M., Johnson, C.R., Lodder, R.A., Parker, S.G., Cole, M.J., Ewing, R.E., Efendiev, Y., Lazarov, R., Qin, G.: Dynamic contaminant identification in water. In: Computational Science - ICCS 2006: 6th International Conference, Reading, UK, May 28-31, 2006, Proceedings, Part III, Heidelberg, Springer-Verlag (2006) 393–400
2. Lowell, A., Ho, K.S., Lodder, R.A.: Hyperspectral imaging of endolithic biofilms using a robotic probe. Contact in Context 1 (2002) 1–10
3. Dempsey, R.J., Davis, D.G., R. G. Buice, J., Lodder, R.A.: Biological and medical applications of near-infrared spectrometry. Appl. Spectrosc. 50 (1996) 18A–34A
4. Silva, H.E.B.D., Pasquini, C.: Dual-beam near-infrared Hadamard. Spectrophotometer Appl. Spectrosc. 55 (2001) 715–721
5. Johnson, C.R., Parker, S., Weinstein, D., Heffernan, S.: Component-based problem solving environments for large-scale scientific computing. Concur. Comput.: Practice and Experience 14 (2002) 1337–1349
6. SCIRun: A Scientific Computing Problem Solving Environment, Scientific Computing and Imaging Institute (SCI). http://software.sci.utah.edu/scirun.html (2007)
7. Li, W.: A dynamic data-driven application system (dddas) tool for dynamic reconfigurable point-to-point data communication. Master's thesis, University of Kentucky Computer Science Department, Lexington, KY
8. Apple OS X 10.4 XGrid Features, Apple, inc. http://www.apple.com/acg/xgrid (2007)
9. Douglas, C.C., Beezley, J.D., Coen, J., Li, D., Li, W., Mandel, A.K., Mandel, J., Qin, G., Vodacek, A.: Demonstrating the validity of a wildfire DDDAS. In: Computational Science - ICCS 2006: 6th International Conference, Reading, UK, May 28-31, 2006, Proceedings, Part III, Heidelberg, Springer-Verlag (2006) 522–529
10. GeoTiff. http://www.remotesensing.org/geotiff/geotiff.html (2007)
11. Douglas, C.C., Efendiev, Y., Ewing, R.E., Ginten, V., Lazarov, R.: Dynamic data driven simulations in stochastic environments. Computing 77 (2006) 321–333
12. Douglas, C.C., Efendiev, Y., Ewing, R.E., Ginting, V., Lazarov, R., Cole, M.J., Jones, G.: Least squares approach for initial data recovery in dynamic data-driven applications simulations. Comp. Vis. in Science (2007) in press.
13. Douglas, C.C., Efendiev, Y., Ewing, R.E., Ginting, V., Lazarov, R., Cole, M.J., Jones, G.: Dynamic data-driven application simulations. interpolation and update. In: Environmental Security Air, Water and Soil Quality Modelling for Risk and Impact Assessment. NATO Securtity through Science Series C, New York, Springer-Verlag (2006)
14. Robert, C., Casella, G.: Monte Carlo Statistical Methods. Springer-Verlag, New York (1999)

Hessian-Based Model Reduction for Large-Scale Data Assimilation Problems[*]

Omar Bashir[1], Omar Ghattas[2], Judith Hill[3],
Bart van Bloemen Waanders[3], and Karen Willcox[1]

[1] Massachusetts Institute of Technology, Cambridge MA 02139, USA
bashir@mit.edu,kwillcox@mit.edu
[2] The University of Texas at Austin, Austin TX 78712
omar@ices.utexas.edu
[3] Sandia National Laboratories, Albuquerque NM 87185
jhill@sandia.gov,bartv@sandia.gov

Abstract. Assimilation of spatially- and temporally-distributed state observations into simulations of dynamical systems stemming from discretized PDEs leads to inverse problems with high-dimensional control spaces in the form of discretized initial conditions. Solution of such inverse problems in "real-time" is often intractable. This motivates the construction of reduced-order models that can be used as surrogates of the high-fidelity simulations during inverse solution. For the surrogates to be useful, they must be able to approximate the observable quantities over a wide range of initial conditions. Construction of the reduced models entails sampling the initial condition space to generate an appropriate training set, which is an intractable proposition for high dimensional initial condition spaces unless the problem structure can be exploited. Here, we present a method that extracts the dominant spectrum of the input-output map (i.e. the Hessian of the least squares optimization problem) at low cost, and uses the principal eigenvectors as sample points. We demonstrate the efficacy of the reduction methodology on a large-scale contaminant transport problem.

Keywords: Model reduction; data assimilation; inverse problem; Hessian matrix; optimization.

1 Introduction

One important component of Dynamic Data Driven Application Systems (DDDAS) is the continuous assimilation of sensor data into an ongoing simulation. This inverse problem can be formulated as an optimal control problem,

[*] Partially supported by the National Science Foundation under DDDAS grants CNS-0540372 and CNS-0540186, the Air Force Office of Scientific Research, and the Computer Science Research Institute at Sandia National Laboratories. Sandia is a multiprogram laboratory operated by Sandia Corporation, a Lockheed-Martin Company, for the US Department of Energy under Contract DE-AC04-94AL85000.

Y. Shi et al. (Eds.): ICCS 2007, Part I, LNCS 4487, pp. 1010–1017, 2007.

in which the controls are the initial conditions, the constraints are the state equations describing the dynamics of the system, and the objective is the difference between the state observations and those predicted by the state equations, measured in some appropriate norm.

When the physical system being simulated is governed by partial differential equations in three spatial dimensions and time, the forward problem alone (i.e. solution of the PDEs for a given initial condition) may requires many hours of supercomputer time. The inverse problem, which requires repeated solution of the forward problem, may then be out of reach in situations where rapid assimilation of the data is required. In particular, when the simulation is used as a basis for forecasting or decision-making, a *reduced model* that can execute much more rapidly than the high-fidelity PDE simulation is then needed. A crucial requirement for the reduced model is that it be able to replicate the output quantities of interest (i.e. the observables) of the PDE simulation over a wide range of initial conditions, so that it may serve as a surrogate of the high fidelity PDE simulation during inversion.

One popular method for generating a reduced model is through a projection basis (for example, by proper orthogonal decomposition in conjunction with the method of snapshots). To build such a reduced order model, one typically constructs a training set by sampling the space of (discretized) initial conditions. When this space is high-dimensional, the problem of adequately sampling it quickly becomes intractable. Fortunately, for many *ill-posed* inverse problems, many components of the initial condition space have minimal or no effect on the output observables. This is particularly true when the observations are sparse. In this case, it is likely that an effective reduced model can be generated with few sample points. The question is how to locate these sample points.

Here, we consider the case of a linear forward problem, and propose that the sample points be associated with dominant eigenvectors of the Hessian matrix of the misfit function. This matrix maps inputs (initial conditions) to outputs (observables), and its dominant eigenvectors represent initial condition components that are most identifiable from observable data. Thus, one expects these eigenvectors to serve as good sample points for constructing the reduced model.

In Section 2, we describe the model reduction framework we consider, and in Section 3 justify the choice of the dominant eigenvectors of the Hessian by relating it to solution of a certain greedy optimization problem to locate the best sample points. Section 4 illustrates the methodology via application to a data assimilation inverse problem involving transport of an atmospheric contaminant.

2   Reduced-Order Dynamical Systems

Consider the general linear initial-value problem

$$x(k+1) = Ax(k), \qquad k = 0, 1, \ldots, T-1, \tag{1}$$

$$y(k) = Cx(k), \qquad k = 0, 1, \ldots, T, \tag{2}$$

$$x(0) = x_0, \tag{3}$$

where $x(k) \in \mathbb{R}^N$ is the system state at time t_k, the vector x_0 contains the specified initial state, and we consider a time horizon from $t = 0$ to $t = t_T$. The vector $y(k) \in \mathbb{R}^Q$ contains the Q system outputs at time t_k. In general, we are interested in systems of the form (1)–(3) that result from spatial and temporal discretization of PDEs. In this case, the dimension of the system, N, is very large and the matrices $A \in \mathbb{R}^{N \times N}$ and $C \in \mathbb{R}^{Q \times N}$ result from the chosen spatial and temporal discretization methods.

A reduced-order model of (1)–(3) can be derived by assuming that the state $x(k)$ is represented as a linear combination of n basis vectors,

$$\hat{x}(k) = V x_r(k), \tag{4}$$

where $\hat{x}(k)$ is the reduced model approximation of the state $x(k)$ and $n \ll N$. The projection matrix $V \in \mathbb{R}^{N \times n}$ contains as columns the orthonormal basis vectors V_i, i.e., $V = [V_1 \ V_2 \ \cdots \ V_n]$, and the reduced-order state $x_r(k) \in \mathbb{R}^n$ contains the corresponding modal amplitudes for time t_k. Using the representation (4) together with a Galerkin projection of the discrete-time system (1)–(3) onto the space spanned by the basis V yields the reduced-order model with state x_r and output y_r,

$$x_r(k+1) = A_r x_r(k), \qquad k = 0, 1, \ldots, T-1, \tag{5}$$
$$y_r(k) = C_r x_r(k), \qquad k = 0, 1, \ldots, T, \tag{6}$$
$$x_r(0) = V^T x_0, \tag{7}$$

where $A_r = V^T A V$ and $C_r = CV$.

For convenience of notation, we write the discrete-time system (1)–(3) in matrix form as

$$\mathbf{A}\mathbf{x} = \mathbf{F}x_0, \qquad \mathbf{y} = \mathbf{C}\mathbf{x}, \tag{8}$$

where $\mathbf{x} = \left[x(0)^T \ x(1)^T \ \ldots \ x(T)^T \right]^T$, $\mathbf{y} = \left[y(0)^T \ y(1)^T \ \ldots \ y(T)^T \right]^T$, and the matrices $\mathbf{A}$, $\mathbf{F}$, and $\mathbf{C}$ are appropriately defined functions of A and C. Similarly, the reduced-order model (5)–(7) can be written in matrix form as

$$\mathbf{A}_r \mathbf{x}_r = \mathbf{F}_r x_0, \qquad \mathbf{y}_r = \mathbf{C}_r \mathbf{x}_r, \tag{9}$$

where $\mathbf{x}_r$, $\mathbf{y}_r$, $\mathbf{A}_r$, and $\mathbf{C}_r$ are defined analogously to $\mathbf{x}$, $\mathbf{y}$, $\mathbf{A}$, and $\mathbf{C}$ but with the appropriate reduced-order quantities, and $\mathbf{F}_r = [V \ 0 \ \ldots \ 0]^T$.

In many cases, we are interested in rapid identification of initial conditions from sparse measurements of the states over a time horizon; we thus require a reduced-order model that will provide accurate outputs for any initial condition contained in some set $\mathcal{X}_0$. Using the projection framework described above, the task therefore becomes one of choosing an appropriate basis V so that the error between full-order output $\mathbf{y}$ and the reduced-order output $\mathbf{y}_r$ is small for all initial conditions of interest.

3   Hessian-Based Model Reduction

To determine the reduced model, we must identify a set of initial conditions to be sampled. At each selected initial condition, a forward simulation is performed to generate a set of states, commonly referred to as snapshots, from which the reduced basis is formed. The key question is then how to identify important initial conditions that should be sampled. Our approach is motivated by the greedy algorithm of [5], which proposed an adaptive approach to determine the parameter locations at which samples are drawn to form a reduced basis. The greedy algorithm adaptively selects these snapshots by finding the location in parameter–time space where the error between the full-order and reduced-order models is maximal, updating the basis with information gathered from this sample location, forming a new reduced model, and repeating the process.

In the case of the initial-condition problem, the greedy approach amounts to sampling at the initial condition $x_0^* \in \mathcal{X}_0$ that *maximizes* the error between the full and reduced-order outputs. For this formulation, the only restriction that we place on the set $\mathcal{X}_0$ is that it contain vectors of unit length. This prevents unboundedness in the optimization problem, since otherwise the error in the reduced system could be made arbitrarily large.

The key step in the greedy sampling approach is thus finding the worst-case initial condition x_0^*, which can be achieved by solving the optimization problem,

$$x_0^* = \arg\max_{x_0 \in \mathcal{X}_0} \ (\mathbf{y} - \mathbf{y}_r)^T (\mathbf{y} - \mathbf{y}_r) \tag{10}$$

$$\text{where} \quad \mathbf{A}\mathbf{x} = \mathbf{F}x_0, \tag{11}$$

$$\mathbf{y} = \mathbf{C}\mathbf{x}, \tag{12}$$

$$\mathbf{A}_r\mathbf{x}_r = \mathbf{F}_r x_0, \tag{13}$$

$$\mathbf{y}_r = \mathbf{C}_r\mathbf{x}_r. \tag{14}$$

Equations (10)-(14) define a large-scale optimization problem, which includes the full-scale dynamics as constraints. The linearity of the state equations can be exploited to eliminate the full-order and reduced-order states and yield an equivalent unconstrained optimization problem,

$$x_0^* = \arg\max_{x_0 \in \mathcal{X}_0} \ x_0^T H^e x_0, \tag{15}$$

where

$$H^e = \left(\mathbf{C}\mathbf{A}^{-1}\mathbf{F} - \mathbf{C}_r\mathbf{A}_r^{-1}\mathbf{F}_r \right)^T \left(\mathbf{C}\mathbf{A}^{-1}\mathbf{F} - \mathbf{C}_r\mathbf{A}_r^{-1}\mathbf{F}_r \right). \tag{16}$$

It can be seen that (15) is a quadratic unconstrained optimization problem with Hessian matrix $H^e \in \mathbb{R}^{N \times N}$. From (16), it can be seen that H^e is a symmetric positive semidefinite matrix. Since we are considering initial conditions of unit norm, the solution x_0^* maximizes the Rayleigh quotient; therefore, the solution of (15) is given by the eigenvector corresponding to the largest eigenvalue of H^e.

This eigenvector is the initial condition for which the error in reduced model output prediction is largest.

Rather than constructing a reduced model at every greedy iteration, and determining the dominant eigenvector of the resulting error Hessian H_e, an efficient one-shot algorithm can be constructed by computing the dominant eigenmodes of the Hessian matrix

$$H = \left(\mathbf{CA}^{-1}\mathbf{F}\right)^T \left(\mathbf{CA}^{-1}\mathbf{F}\right). \tag{17}$$

Here, $H \in \mathbb{R}^{N \times N}$ is the Hessian matrix of the full-scale system, and does not depend on the reduced-order model. As before, H is a symmetric positive semi-definite matrix. It can be shown that, under certain assumptions, the eigenvectors of H with largest eigenvalues approximately solve the sequence of problems defined by (10)–(14) [3].

These ideas motivate the following basis-construction algorithm for the initial condition problem. We use the dominant eigenvectors of the Hessian matrix H to identify the initial-condition vectors that have the most significant contributions to the outputs of interest. These vectors are in turn used to initialize the full-scale discrete-time system to generate a set of state snapshots that are used to form the reduced basis (using, for example, the proper orthogonal decomposition).

4   Application: Model Reduction for 3D Contaminant Transport in an Urban Canyon

We demonstrate our model reduction method by applying it to a 3D airborne contaminant transport problem for which a solution is needed in real time. Intentional or unintentional chemical, biological, and radiological (CBR) contamination events are important national security concerns. In particular, if contamination occurs in or near a populated area, predictive tools are needed to rapidly and accurately forecast the contaminant spread to provide decision support for emergency response efforts. Urban areas are geometrically complex and require detailed spatial discretization to resolve the relevant flow and transport, making prediction in real-time difficult. Reduced-order models can play an important role in facilitating real-time turn-around, in particular on laptops in the field. However, it is essential that these reduced models be faithful over a wide range of initial conditions, since in principle any initial condition can be realized. Once a suitable reduced-order model has been generated, it can serve as a surrogate for the full model within an inversion framework to identify the initial conditions given sensor data (the full-scale case is discussed in [1]).

To illustrate the generation of a reduced-order model that is accurate for arbitrary high-dimensional initial conditions, we consider a three-dimensional urban canyon geometry occupying a (dimensionless) $15 \times 15 \times 15$ domain. Figure 1 shows the domain and buildings, along with locations of six sensors, all placed at a height of 1.5. Contaminant transport is modeled by the advection-dispersion equation,

$$\frac{\partial w}{\partial t} + \boldsymbol{v} \cdot \nabla w - \kappa \nabla^2 w = 0 \quad \text{in } \Omega \times (0, t_f), \tag{18}$$

$$w = 0 \quad \text{on } \Gamma_D \times (0, t_f), \tag{19}$$

$$\frac{\partial w}{\partial n} = 0 \quad \text{on } \Gamma_N \times (0, t_f), \tag{20}$$

$$w = w_0 \quad \text{in } \Omega \text{ for } t = 0, \tag{21}$$

where w is the contaminant concentration, $\boldsymbol{v}$ is the velocity vector field, κ is the diffusivity, t_f is the time horizon of interest, and w_0 is the given initial condition. Γ_D and Γ_N are respectively the portions of the domain boundary over which Dirichlet and Neumann boundary conditions are applied. Eq. (18) is discretized in space using an SUPG finite element method with linear tetrahedra, while the implicit Crank-Nicolson method is used to discretize in time. Homogeneous Dirichlet boundary conditions are specified for the concentration on the inflow boundary, $\bar{x} = 0$, and the ground, $\bar{z} = 0$. Homogeneous Neumann boundary conditions are specified for the concentration on all other boundaries.

The velocity field, $\boldsymbol{v}$, required in (18) is computed by solving the steady laminar incompressible Navier-Stokes equations, also discretized with SUPG-stabilized linear tetrahedra. No-slip conditions, i.e. $\boldsymbol{v} = 0$, are imposed on the building faces and the ground $\bar{z} = 0$. The velocity at the inflow boundary $\bar{x} = 0$ is taken as known and specified in the normal direction as

$$v_x(z) = v_{\max} \left(\frac{z}{z_{\max}} \right)^{0.5},$$

with $v_{\max} = 3.0$ and $z_{\max} = 15$, and zero tangentially. On the outflow boundary $\bar{x} = 15$, a traction-free (Neumann) condition is applied. On all other boundaries ($\bar{y} = 0$, $\bar{y} = 15$, $\bar{z} = 15$), we impose a combination of no flow normal to the boundary and traction-free tangent to the boundary. The spatial mesh for the full-scale system contains 68,921 nodes and 64,000 tetrahedral elements. For both basis creation and testing, a final non-dimensional time $t_f = 20.0$ is used, and discretized over 200 timesteps. The Peclet number based on the maximum inflow velocity and domain dimension is Pe=900. The PETSc library [2] is used for all implementation.

Figure 2 illustrates a sample forward solution. The test initial condition used in this simulation, meant to represent the system state just after a contaminant release event, was constructed using a Gaussian function with a peak magnitude of 100 centered at a height of 1.5. For comparison with the full system, a reduced model was constructed based on the dominant Hessian eigenvector algorithm discussed in the previous section, with $p = 31$ eigenvector initial conditions and $n = 137$ reduced basis vectors (these numbers were determined based on eigenvalue decay rates). Eigenvectors were computed using the Arnoldi eigensolver within the SLEPc package [4], which is built on PETSc. Figure 3 shows a comparison of the full and reduced time history of concentration at each output

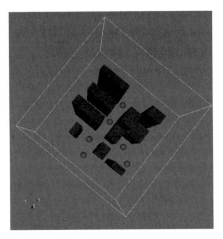

Fig. 1. Building geometry and locations of outputs for the 3-D urban canyon problem

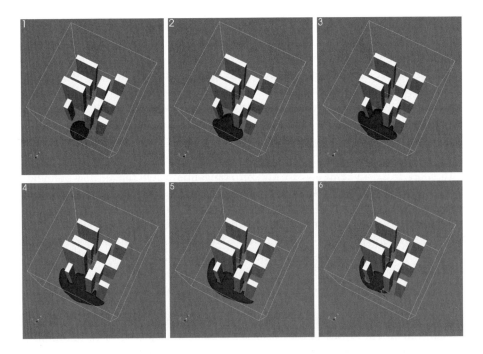

Fig. 2. Transport of contaminant concentration through urban canyon at six instants in time, beginning with the initial condition shown in upper left

location. There is no discernible difference between the two. The figure demonstrates that a reduced system of size $n = 137$, which is solved in a matter of seconds on a desktop, can accurately replicate the outputs of the full-scale system of size $N = 65,600$. We emphasize that the (offline) construction of the

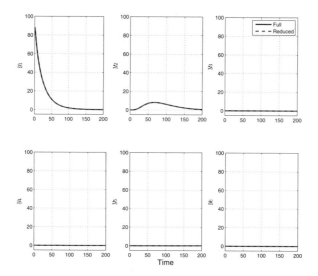

Fig. 3. Full (65,600 states) and reduced (137 states) model contaminant predictions at the six sensor locations for urban canyon example

reduced-order model targets only the specified outputs, and otherwise has no knowledge of the initial conditions used in the test of Figure 3.

References

1. V. Akçelik, G. Biros, A. Draganescu, O. Ghattas, J. Hill, and B. van Bloemen Waanders. Dynamic data-driven inversion for terascale simulations: Real-time identification of airborne contaminants. In *Proceedings of SC2005, Seattle, WA*, 2005.
2. S. Balay, K. Buschelman, V. Eijkhout, W. Gropp, D. Kaushik, M. Knepley, L. McInnes, B. Smith, and H. Zhang. PETSc users manual. Technical Report ANL-95/11 - Revision 2.1.5, Argonne National Laboratory, 2004.
3. O. Bashir. Hessian-based model reduction with applications to initial condition inverse problems. Master's thesis, MIT, 2007.
4. V. Hernandez, J. Roman, and V. Vidal. SLEPc: A scalable and flexible toolkit for the solution of eigenvalue problems. *ACM Transactions on Mathematical Software*, 31(3):351–362, sep 2005.
5. K. Veroy, C. Prud'homme, D. Rovas, and A. Patera. *A posteriori* error bounds for reduced-basis approximation of parametrized noncoercive and nonlinear elliptic partial differential equations. AIAA Paper 2003-3847, Proceedings of the 16th AIAA Computational Fluid Dynamics Conference, Orlando, FL, 2003.

Localized Ensemble Kalman Dynamic Data Assimilation for Atmospheric Chemistry*

Adrian Sandu[1], Emil M. Constantinescu[1], Gregory R. Carmichael[2],
Tianfeng Chai[2], John H. Seinfeld[3], and Dacian Dăescu[4]

[1] Department of Computer Science, Virginia Polytechnic Institute and State
University, Blacksburg, VA 24061.
{asandu, emconsta}@cs.vt.edu
[2] Center for Global and Regional Environmental Research, The University of Iowa,
Iowa City, 52242-1297.
{gcarmich,tchai}@cgrer.uiowa.edu
[3] Department of Chemical Engineering, California Institute of Technology, Pasadena,
CA 91125.
seinfeld@caltech.edu
[4] Department of Mathematics and Statistics, Portland State University.
daescu@pdx.edu

Abstract. The task of providing an optimal analysis of the state of
the atmosphere requires the development of dynamic data-driven systems (DDDAS) that efficiently integrate the observational data and the
models. Data assimilation, the dynamic incorporation of additional data
into an executing application, is an essential DDDAS concept with wide
applicability. In this paper we discuss practical aspects of nonlinear ensemble Kalman data assimilation applied to atmospheric chemical transport models. We highlight the challenges encountered in this approach
such as filter divergence and spurious corrections, and propose solutions
to overcome them, such as background covariance inflation and filter
localization. The predictability is further improved by including model
parameters in the assimilation process. Results for a large scale simulation of air pollution in North-East United States illustrate the potential
of nonlinear ensemble techniques to assimilate chemical observations.

1 Introduction

Our ability to anticipate and manage changes in atmospheric pollutant concentrations relies on an accurate representation of the chemical state of the
atmosphere. As our fundamental understanding of atmospheric chemistry advances, novel data assimilation tools are needed to integrate observational data
and models together to provide the best estimate of the evolving chemical state
of the atmosphere. The ability to dynamically incorporate additional data into
an executing application is a fundamental DDDAS concept (http://www.cise.
nsf.gov/dddas.) We refer to this process as data assimilation. Data assimilation
has proved vital for meteorological forecasting.

* This work was supported by the National Science Foundation through the award
NSF ITR AP&IM 0205198 managed by Dr. Frederica Darema.

In this paper we focus on the particular challenges that arise in the application of nonlinear ensemble filter data assimilation to atmospheric chemical transport models (CTMs). Atmospheric CTMs solve the mass-balance equations for concentrations of trace species to determine the fate of pollutants in the atmosphere [16]. The CTM operator, $\mathcal{M}$, will be denoted compactly as

$$c_i = \mathcal{M}_{t_{i-1} \to t_i}\left(c_{i-1},\, u_{i-1},\, c_{i-1}^{\text{in}},\, Q_{i-1}\right), \tag{1}$$

where c represents the modeled species concentration, c^{in} the inflow numerical boundary conditions, u the wind fields, Q the surface emission rates, and the subscript denotes the time index. In our numerical experiments, we use the Sulfur Transport Eulerian Model (STEM) [16], a state-of-the-art atmospheric CTM.

Kalman filters [12] provide a stochastic approach to the data assimilation problem. The filtering theory is described in Jazwinski [10] and the applications to atmospheric modeling in [13]. The computational burden associated with the filtering process has prevented the implementation of the full Kalman filter for large-scale models. Ensemble Kalman filters (EnKF) [2,5] may be used to facilitate the practical implementation as shown by van Loon et al. [18]. There are two major difficulties that arise in EnKF data assimilation applied to CTMs: (1) CTMs have stiff components [15] that cause the filter to *diverge* [7] due to the lack of ensemble spread and (2) the ensemble size is typically small in order to be computationally tractable and this leads to filter *spurious corrections* due to sampling errors. Kalman filter data assimilation has been discussed for DDDAS in another context by Jun and Bernstein [11].

This paper addresses the following issues: (1) *Background covariance inflation* is investigated in order to avoid *filter divergence*, (2) *localization* is used to prevent spurious filter corrections caused by small ensembles, and (3) *parameters are assimilated together with the model states* in order to reduce the model errors and improve the forecast. The paper is organized as follows. Section 2 presents the ensemble Kalman data assimilation technique, Section 3 illustrates the use of the tools in a data assimilation test, and Section 4 summarizes our results.

2   Ensemble Kalman Filter Data Assimilation

Consider a nonlinear model $c_i = \mathcal{M}_{t_0 \to t_i}(c_0)$ that advances the state from the initial time t_0 to future times t_i $(i \geq 1)$. The model state c_i at t_i $(i \geq 0)$ is an approximation of "true" state of the system c_i^t at t_i (more exactly c_i^t is the system state projected onto the model space space). Observations y_i are available at times t_i and are corrupted by measurement and representativeness errors ε_i (assumed Gaussian with mean zero and covariance $\mathbb{R}_i$), $y_i = \mathcal{H}_i\left(c_i^t\right) + \varepsilon_i$. Here $\mathcal{H}_i$ is an operator that maps the model state to observations.

The data assimilation problem is to find an optimal estimate of the state using both the information from the model (c_i) and from the observations (y_i).

The (ensemble) Kalman filter estimates the true state c^t using the information from the current best estimate c^f (the "forecast" or the background state) and the observations y. The optimal estimate c^a (the "analysis" state) is obtained as a linear combination of the forecast and observations that minimize the variance of the analysis (P^a)

$$c^a = c^f + P^f H^T (H P^f H^T + \mathbb{R})^{-1} (y - \mathcal{H}(c^f)) = c^f + K (y - \mathcal{H}(c^f)) . \quad (2)$$

The forecast covariance P^f is estimated from an ensemble of runs (which produces an ensemble of E model states $c^f(e)$, $e = 1, \cdots, E$). The analysis formula (2) is applied to each member to obtain an analyzed ensemble. The model advances the solution from t_{i-1} to t_i, then the filter formula is used to incorporate the observations at t_i. The filter can be described as

$$c_i^f(e) = \mathcal{M}(c_{i-1}^a(e)), \ c_i^a(e) = c_i^f(e) + K_i \left(y_i - \mathcal{H}_i(c_i^f(e)) \right). \quad (3)$$

The results presented in this paper are obtained with the practical EnKF implementation discussed by Evensen [5].

2.1 The Localization of EnKF (LEnKF)

The practical Kalman filter implementation employs a small ensemble of Monte Carlo simulations in order to approximate the background covariance (P^f). In its initial formulation, EnKF may suffer from spurious correlations caused by sub-sampling errors in the background covariance estimates. This allows for observations to incorrectly impact remote model states. The filter *localization* introduces a restriction on the correction magnitude based on its remoteness.

One way to impose localization in EnKF is to apply a decorrelation function ρ, that decreases with distance, to the background covariance. Following [8], the EnKF relation (2) with some simplifying assumptions becomes

$$c_i^a = c_i^f + \rho(D^c) \circ P_i^f H_i^T \left(\rho(D^y) \circ \left(H_i P_i^f H_i^T \right) + \mathbb{R}_i \right)^{-1} \left(y_i - \mathcal{H}_i(c_i^f) \right), \quad (4)$$

where $D^{\{c,y\}}$ are distance matrices with positive elements ($d_{i,j} \geq 0$), and $0 \leq \rho(d_{i,j}) \leq 1$, $\rho(0) = 1$, $\forall i, j$. The decorrelation function ρ is applied to the distance matrix and produces a decorrelation matrix (decreasing with the distance). The operation 'o' denotes the Schur product that applies elementwise $\rho(D)$ to the projected covariance matrices $P^f H^T$ and $H P^f H^T$, respectively. Here, D^y is calculated as the distance among the observation sites, and D^c contains the distance from each state variable to each observation site.

We considered a Gaussian distribution for the decorrelation function, ρ. Since our model generally has an anisotropic horizontal–vertical flow, we consider the two correlation components (and factors, δ) separately:

$$\rho \left(D^h, D^v \right) = \exp \left[- \left(D^h/\delta^h \right)^2 - \left(D^v/\delta^v \right)^2 \right], \quad (5)$$

where D^h, D^v, δ^h, δ^v are the horizontal and vertical components. The horizontal correlation-distance relationship is determined through the NMC method [14]. The horizontal NMC determined correlations were fitted with a Gaussian distribution, $\delta^h = 270$ km. The vertical correlation was chosen as $\delta^v = 5$ grid points.

2.2   Preventing Filter Divergence

The "textbook application" of EnKF [5] may lead to filter divergence [7]: EnKF shows a decreasing ability to correct ensemble states toward the observations. This is due to an underestimation of the model error covariance magnitude during the integration. The filter becomes "too confident" in the model and "ignores" the observations in the analysis process. The solution is to increase the covariance of the ensemble and therefore decrease the filter's confidence in the model. The following are several ways to "inflate" the ensemble covariance.

The first method is the *additive inflation* [4], where the model errors are simulated by adding uncorrelated noise (denoted by η) to the model (η_-) or analysis (η_+) results. This increases the diagonal entries of the ensemble covariance. Since the correlation of the model errors is to a large extent unknown, white noise is typically chosen. With the notation (3), $c_i^f(e) = \mathcal{M}\big(c_{i-1}^a(e) + \eta_-(e)\big) + \eta_+(e)$. The second method is the *multiplicative inflation* [1], where each member's deviation from the ensemble mean is multiplied by a constant ($\gamma > 1$). This increases each entry of the ensemble covariance by that constant squared (γ^2). The ensemble can be inflated before ($\gamma_{\{-\}}$) or after ($\gamma_{\{+\}}$) filtering: $c_i^{\{f/a\}}(e) \leftarrow \langle c_i^{\{f/a\}} \rangle + \gamma_{\{-/+\}}$, where $\langle \cdot \rangle$ denotes the ensemble average.

A third possibility for covariance inflation is through perturbations applied to key model parameters, and we refer to it as *model-specific inflation*. This approach focuses on sources of uncertainty that are specific to each model (for instance in CTMs: boundary conditions, emissions, and meteorological fields). With the notation (3) and considering p as a set of model parameters, the model-specific inflation can be written as $c_i^f(e) = \mathcal{M}\big(c_{i-1}^a(e), \alpha_{i-1}(e)\, p_{i-1}\big)$, where $\alpha(e)$ are random perturbation factors of the model parameters.

2.3   Inflation Localization

The traditional approach to covariance inflation increases the spread of the ensemble equally throughout the computational domain. In the LEnKF framework, the corrections are restricted to a region that is rich in observations. These states are corrected and their variance is reduced, while the remote states (i.e., the states that are relatively far from the observations' locations) maintain their initial variation which is potentially reduced only by the model evolution. The spread of the ensemble at the remote states may be increased to unreasonably large values through successive inflation steps. And thus, the covariance inflation needs to be restricted in order to avoid the over-inflation of the remote states.

A sensible inflation restriction can be based on the localization operator, $\rho(D)$, which is applied in the same way as for the covariance localization. The localized multiplicative inflation factor, γ_ℓ, is given by

$$\gamma_\ell(i,j,k) = \max\{\rho(D^c(i,j,k))\}(\gamma - 1) + 1, \qquad (6)$$

where γ is the (non-localized) multiplicative inflation factor and i, j, k refer to the spatial coordinates. In this way, the localized inflation increases the ensemble spread only in the information-rich regions where filter divergence can occur.

3   Numerical Results

The test case is a real-life simulation of air pollution in North–Eastern U.S. in July 2004 as shown in Figure 1.a (the dash-dotted line delimits the domain). The observations used for data assimilation are the ground-level ozone (O_3) measurements taken during the ICARTT [9,17] campaign in 2004 (which also includes the initial concentrations, meteorological fields, boundary values, and emission rates). Figure 1.a shows the location of the ground stations (340 in total) that measured ozone concentrations and an ozonesonde (not used in the assimilation process). The computational domain covers $1500 \times 1320 \times 20$ Km with a horizontal resolution of 60×60 Km and a variable vertical resolution.

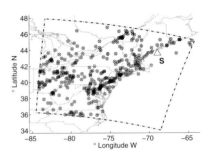

The simulations are started at 0 GMT July 20th with a four hour initialization step ([-4,0] hours). The "best guess" of the state of the atmosphere at 0 GMT July 20th is used to initialize the deterministic solution. The ensemble members are formed by adding a set of unbiased perturbations to the best guess, and then evolving each member to 4 GMT July 20th. The perturbation is formed according to an AR model [3] making it flow dependent. The 24 hours assimilation window starts at 4 GMT July 20th (denoted by [1,24] hours). Observations are available at each integer hour in this window, i.e., at 1, 2, ..., 24 hours (Figure 1.a). EnKF adjusts the concentration

Fig. 1. Ground measuring stations in support of the ICARTT campaign (340 in total) and the ozonesonde (S) launch location

fields of 66 "control" chemical species in each grid point of the domain every hour using (2). The ensemble size was chosen to be 50 members (a typical size in NWP). A 24 hour forecast window is also considered to start at 4 GMT July 21st (denoted by [24,48] hours).

The performance of each data assimilation experiment is measured by the R^2 correlation factor (correlation^2) between the observation and the model solution. The R^2 correlation results between the observations and model values for all the numerical experiments are shown in Table 1. The deterministic (best guess) solution yields an R^2 of 0.24 in the analysis and 0.28 in the forecast windows. In Table 1 we also show the results for a 4D-Var experiment.

Figure 2.a shows the O_3 concentration measured at a Washington DC station and predicted by the EnKF and LEnKF with model-specific inflation. Figure 2.b shows the ozone concentration profile measured by the ozonesonde for the EnKF and LEnKF with additive inflation. Two effects are clear for the "textbook" EnKF. The filter diverges after about 12 hours (2.a), and spurious corrections are made at higher altitudes (2.b), as the distance from the observation (ground) sites increases. The vertical profile in Figure 2.b shows great improvement in the analyzed solution of LEnKF. The results in Table 1 confirm the benefits of localization by dramatically improving the analysis and forecast fit.

Table 1. The R^2 measure of model-observations match in the assimilation and forecast windows for EnKF, 4D-Var, and LEnKF. (Multiplicative inflation: $\gamma_- \leq 4$, $\gamma_+ \leq 4$; Model-specific inflation: 10% emissions, 10% boundaries, 3% wind).

Method & Details	R^2 analysis	R^2 forecast
Deterministic solution, no assimilation	0.24	0.28
EnKF, "textbook application"	0.38	0.30
4D-Var - 50 iterations	0.52	0.29
LEnKF, model-specific inflation	0.88	0.32
LEnKF, multiplicative inflation	0.82	0.32
LEnKF, additive inflation	0.92	0.31
LEnKF with parameter assimilation, and multiplicative localized inflation	0.89	0.41

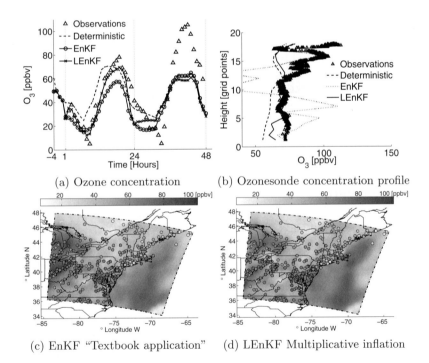

(a) Ozone concentration

(b) Ozonesonde concentration profile

(c) EnKF "Textbook application"

(d) LEnKF Multiplicative inflation

Fig. 2. Ozone concentration (a) measured at a Washington DC station (ICARTT ID: 510590030) and predicted by EnKF ("textbook") and LEnKF with model-specific inflation, and (b) measured by the ozonesonde for EnKF and LEnKF. Ground level ozone concentration field (c,d) at 14 EDT in the forecast window measured by the ICARTT stations (shown in color coded filled circles) and predicted EnKF and LEnKF.

3.1   Joint State and Parameter Assimilation

In regional CTMs the influence of the initial conditions is rapidly diminishing with time, and the concentration fields are "driven" by emissions and by lateral boundary conditions. Since both of them are generally poorly known, it is of considerable interest to improve their values using information from observations. In this setting we have to solve a joint state-parameter assimilation problem [6].

The emission rates and lateral boundary conditions are multiplied by specific correction coefficients, α. These correction coefficients are appended to the model state. The LEnKF data assimilation is then carried out with the augmented model state. With the notation (1), LEnKF is applied to

$$\left[c_i^f \ \alpha_i^{\{1,2\}} \right]^T = \left[\mathcal{M}_{t_{i-1} \to t_i} \left(c_{i-1}^a, u_{i-1}, \alpha_{i-1}^{\{1\}} c_{i-1}^{in}, \alpha_{i-1}^{\{2\}} Q_{i-1} \right) \ \alpha_{i-1} \right]^T.$$

For α, we consider a different correction for each species and each gridpoint. The initial ensemble of correction factors is an independent set of normal variables and the localization is done in the same way as in the state-only case.

The R^2 after LEnKF data assimilation for combined state and emission correction coefficients (presented in Table 1) show improvements in both the forecast and the analysis windows. Figures 2.(c,d) show the ground level ozone field concentration at 14 EDT in the forecast window measured by the ICARTT stations, EnKF with state corrections and LEnKF with joint state-parameter corrections. In the LEnKF case under consideration the addition of the correction parameters to the assimilation process improves the assimilated solution (especially on the inflow boundary (West)).

4   Conclusions

This paper discusses some of the challenges associated with the application of nonlinear ensemble filtering data assimilation to atmospheric CTMs. Three aspects are analyzed in this study: *filter divergence* - CTMs tend to dampen perturbations; *spurious corrections* - small ensemble size cause wrong increments, and *model parametrization errors* - without correcting model errors in the analysis, correcting the state only does not help in improving the forecast accuracy.

Experiments showed that the filter diverges quickly. The influence of the initial conditions fades in time as the fields are largely determined by emissions and by lateral boundary conditions. Consequently, the initial spread of the ensemble is diminished in time. Moreover, stiff systems (like chemistry) are stable - small perturbations are damped out quickly in time. In order to prevent filter divergence, the spread of the ensemble needs to be explicitly increased. We investigated three approaches to ensemble covariance inflation among which model-specific inflation is the most intuitive. The "localization" of EnKF is needed in order to avoid the spurious corrections noticed in the "textbook" application. The correlation distances are approximated using the NMC method. Furthermore, covariance localization prevents over-inflation of the states that are remote from observation. LEnKF increased both the accuracy of the analysis and forecast at the observation sites and at distant locations (from the observations).

Since the solution of a regional CTM is largely influenced by uncertain lateral boundary conditions and by uncertain emissions it is of great importance to adjust these parameters through data assimilation. The assimilation of emissions and boundary conditions visibly improves the quality of the analysis.

References

1. J.L. Anderson. An ensemble adjustment Kalman filter for data assimilation. *Mon. Wea. Rev*, 129:2884–2903, 2001.
2. G. Burgers, P.J. van Leeuwen, and G. Evensen. Analysis scheme in the ensemble Kalman Filter . *Mon. Wea. Rev*, 126:1719–1724, 1998.
3. E.M. Constantinescu, T. Chai, A. Sandu, and G.R. Carmichael. Autoregressive models of background errors for chemical data assimilation. *To appear in J. Geophys. Res.*, 2006.
4. M. Corazza, E. Kalnay, and D. Patil. Use of the breeding technique to estimate the shape of the analysis "errors of the day". *Nonl. Pr. Geophys.*, 10:233–243, 2002.
5. G. Evensen. The ensemble Kalman filter: theoretical formulation and practical implementation. *Ocean Dynamics*, 53, 2003.
6. G. Evensen. The combined parameter and state estimation problem. *Submitted to Ocean Dynamics*, 2005.
7. P.L. Houtekamer and H.L. Mitchell. Data assimilation using an ensemble Kalman filter technique . *Mon. Wea. Rev*, 126:796–811, 1998.
8. P.L. Houtekamer and H.L. Mitchell. A sequential ensemble Kalman filter for atmospheric data assimilation . *Mon. Wea. Rev*, 129:123–137, 2001.
9. ICARTT. ICARTT home page:http://www.al.noaa.gov/ICARTT .
10. A.H. Jazwinski. *Stochastic Processes and Filtering Theory*. Academic Press, 1970.
11. B.-E. Jun and D.S. Bernstein Least-correlation estimates for errors-in-variables models . *Int'l J. Adaptive Control and Signal Processing*, 20(7):337–351, 2006.
12. R.E. Kalman. A new approach to linear filtering and prediction problems. *Trans. ASME, Ser. D: J. Basic Eng.*, 83:95–108, 1960.
13. R. Menard, S.E. Cohn, L.-P. Chang, and P.M. Lyster. Stratospheric assimilation of chemical tracer observations using a Kalman filter. Part I: Formulation. *Mon. Wea. Rev*, 128:2654–2671, 2000.
14. D.F. Parrish and J.C. Derber. The national meteorological center's spectral statistical-interpolation analysis system. *Mon. Wea. Rev*, (120):1747–1763, 1992.
15. A. Sandu, J.G. Blom, E. Spee, J.G. Verwer, F.A. Potra, and G.R. Carmichael. Benchmarking stiff ODE solvers for atmospheric chemistry equations II - Rosenbrock solvers. *Atm. Env.*, 31:3,459–3,472, 1997.
16. A. Sandu, D. Daescu, G.R. Carmichael, and T. Chai. Adjoint sensitivity analysis of regional air quality models. *J. of Comp. Phy.*, 204:222–252, 2005.
17. Y. Tang et al. The influence of lateral and top boundary conditions on regional air quality prediction: a multi-scale study coupling regional and global chemical transport models. *Submitted to J. Geophys. Res.*, 2006.
18. M. van Loon, P.J.H. Builtjes, and A.J. Segers. Data assimilation of ozone in the atmospheric transport chemistry model LOTOS. *Env. Model. and Soft.*, 15:603–609, 2000.

Data Assimilation in Multiscale Chemical Transport Models

Lin Zhang and Adrian Sandu

Department of Computer Science, Virginia Polytechnic Institute and State University,
Blacksburg, VA 24061, US
{lin83, asandu}@vt.edu

Abstract. In this paper we discuss variational data assimilation using the STEM atmospheric Chemical Transport Model. STEM is a multiscale model and can perform air quality simulations and predictions over spatial and temporal scales of different orders of magnitude. To improve the accuracy of model predictions we construct a dynamic data driven application system (DDDAS) by integrating data assimilation techniques with STEM. We illustrate the improvements in STEM field predictions before and after data assimilation. We also compare three popular optimization methods for data assimilation and conclude that L-BFGS method is the best for our model because it requires fewer model runs to recover optimal initial conditions.

Keywords: STEM, Chemical Transport Model, Data Assimilation.

1 Introduction

The development of modern industry has brought about much pollutant to the world, which has deep influence to people's life. To analyze and control the air quality, large comprehensive models are indispensable. STEM(Sulfur Transport Eulerian Model) [1] is a chemical transport model used to simulate the air quality evolutions and make predictions. A large variety of species in the air are changing at different time and space magnitude, which requires STEM to be a multiscale system to fully simulate these changes in the atmosphere.

The dynamic incorporation of additional data into an executing application is an essential DDDAS concept with wide applicability (http://www.cise.nsf.gov/dddas). In this paper we focus on data assimilation, the process in which measurements are used to constrain model predictions; the information from measurements can be used to obtain better initial conditions, better boundary conditions, enhanced emission estimates, etc. Data assimilation is essential in weather/climate analysis and forecast activities and is employed here in the context of atmospheric chemistry and transport models.

Kalman filter technique [2] gives a stochastic approach to the data assimilation problem, while variational methods (3D-Var, 4D-Var) provide an optimal control approach. Early applications of the four-dimensional variational (4D-Var) data assimilation were presented by Fisher and Lary [3] for a stratospheric photochemical box model with trajectories. Khattatov et al. [4] implemented both the 4D-Var and a

Y. Shi et al. (Eds.): ICCS 2007, Part I, LNCS 4487, pp. 1026–1033, 2007.

Kalman filter method using a similar model. In the past few years variational methods have been successfully used in data assimilation for comprehensive three-dimensional atmospheric chemistry models (Elbern and Schmidt [5], Errera and Fonteyn [6]). Wang et al. [7] provide a review work of data assimilation applications to atmospheric chemistry. As our STEM model is time dependent in 3D space, 4D-Var is the appropriate approach to data assimilation.

The paper is organized as follows. Second section introduces the STEM Chemical Transport Model. Theory and importance of 4D-Var data assimilation are presented in the third section, followed by some results of data assimilation using the STEM in the forth section. In section five, we briefly describe L-BFGS(Limited-memory Broyden Fletcher Goldfarb Shanno method), Nonlinear Conjugate Gradients and Hessian Free Newton methods and assess their performances in the STEM model. Summary and conclusions are given in section six.

2 The STEM Chemical Transport Model

The STEM is a regional atmospheric Chemical Transport Model(CTM). Taking emissions, meteorology(wind, temperature, humidity, precipitation etc.) and a set of chemical initial and boundary conditions as input, it simulates the pollutants behavior in the selected domain. In the following we give the mathematical description of the Chemical Transport Model [8].

2.1 Chemical Transport Model in STEM

We denote u the wind filed vector, K the turbulent diffusivity tensor, ρ the air density in $molecues/cm^3$. Let v_i^{dep} be the deposition velocity of species i, Q_i the rate of surface emissions, and E_i the rate of elevated emissions for this species. The rate of chemical transformations f_i depends on absolute concentration values; the rate at which mole-fraction concentrations change is then $f_i(\rho c)/\rho$.

Consider a domain Ω which covers a region of the atmosphere with the boundary $\partial\Omega$. At each time moment the boundary of the domain is partitioned into $\partial\Omega = \Gamma^{IN} \cup \Gamma^{OUT} \cup \Gamma^{GR}$, where Γ^{GR} is the ground level portion of the boundary; Γ^{IN} is the inflow part of lateral or top boundary and Γ^{OUT} the outflow part.

The evolution of concentrations in time is described by the material balance equations

$$\frac{\partial c_i}{\partial t} = -u \cdot \nabla c_i + \frac{1}{\rho}\nabla\cdot(\rho K\nabla c_i) + \frac{1}{\rho}f_i(\rho c) + E_i, \quad t^0 \le t \le T \tag{1}$$

$$c_i(t^0, x) = c_i^0(x), \tag{2}$$

$$c_i(t, x) = c_i^{IN}(t, x) \quad for \quad x \in \Gamma^{IN}, \tag{3}$$

$$K\frac{\partial c_i}{\partial n} = 0 \quad for \quad x \in \Gamma^{OUT}, \tag{4}$$

$$K \frac{\partial c_i}{\partial n} = V_i^{dep} c_i - Q_i \quad for \quad x \in \Gamma^{GR}, \quad for \quad all \quad 1 \le i \le s. \tag{5}$$

The system (1)-(5) builds up the *forward (direct) model*. To simplify the presentation, in this paper we consider the initial state c^0 as parameters of the model. It is known that this does not restrict the generality of the formulation.

An infinitesimal perturbation δc^0 in the parameters will result in perturbations $\delta c_i(t)$ of the concentration fields. These perturbations are solutions of the *tangent linear model*. In the direct sensitivity analysis approach we can solve the model (1)-(5) together with the tangent linear model forward in time.

2.2 Continuous Adjoint Model in STEM

Consider a scalar response functional defined in terms of the model solution $c(t)$

$$J(c^0) = \int_0^T \int_\Omega g(c(t, x)) dx dt \tag{6}$$

The response depends implicitly on the parameters c^0 via the dependence of $c(t)$ on c^0. The continuous adjoint model is defined as the adjoint of the tangent linear model. By imposing the Lagrange identity and after a careful integration by parts one arrives at the following equations that govern the evolution of the adjoint variables:

$$\frac{\partial \lambda_i}{\partial t} = -\nabla \cdot (u \lambda_i) - \nabla \cdot (\rho K \nabla \frac{\lambda_i}{\rho}) - (F^T (\rho c) \lambda)_i - \phi_i \quad T \ge t \ge t^0 \tag{7}$$

$$\lambda_i(T, x) = \lambda_i^F(x) \tag{8}$$

$$\lambda_i(t, x) = 0 \quad for \quad x \in \Gamma^{IN} \tag{9}$$

$$\lambda_i u + \rho K \frac{\partial(\lambda_i / \rho)}{\partial n} \quad for \quad x \in \Gamma^{OUT} \tag{10}$$

$$\rho K \frac{\partial(\lambda_i / \rho)}{\partial n} = V_i^{dep} \lambda_i \quad for \quad x \in \Gamma^{GR}, \quad for \quad all \quad 1 \le i \le s \tag{11}$$

Where

$$\phi_i(t, x) = \frac{\partial g(c_1, \dots, c_n)}{\partial c_i}(t, x), \quad \lambda_i^F(x) = 0, \tag{12}$$

and $\lambda_i(t, x)$ are the adjoint variables associated with the concentrations $c_i(t, x)$, $1 \le i \le s$. In the above $F = \partial f / \partial c$ is the Jacobian of the chemical rate function f. To obtain the ground boundary condition we use the fact that $u \cdot n = 0$ at ground level. We refer to (7)-(11) as the (continuous) adjoint system of the tangent linear model. In the context of optimal control where the minimization of the functional (6) is required, the adjoint variables can be interpreted as Lagrange multipliers.

The adjoint system (7)-(11) depends on the states of the forward model (i.e. on the concentration fields through the nonlinear chemical term $F(\rho c)$ and possibly through the forcing term ϕ for nonlinear functionals. Note that the adjoint initial condition is posed at the final time T such that the forward model must be first solved forward in time, the state $c(t,x)$ saved for all t, then the adjoint model could be integrated backwards in time from T down to t^0.

2.3   Properties of STEM

The model uses the SAPRC-99(Statewide Air Pollution Research Center's chemical mechanism) [9] and KPP(the Kinetic PreProcessor) [10], to determine chemical reactions. KPP implements integration of chemical mechanism using implicit Rosenbrock and Runga-Kutta method in both forward and adjoint model.

The STEM model runs multiscale simulations in both time and space. From the time aspect of view, it ranges from 10^{-6} seconds for fast chemical reactions to days' simulation measured in hours. Fast chemical reactions are referred to atomic level reactions, such as O, OH radical activities, while long term simulation usually accounts for atmospheric species transportation in large range. When it comes to spatial scales, STEM is able to simulate in range measured in meters, such as emissions of pollutants like NO, NO_2, CO_2, Volatile Organic Compounds (VOC) and particles from vehicles. Besides, continental scales as large as thousands of kilometers are usually used in STEM for air quality simulation. So far STEM has been employed for simulations over U.S., Asia and Europe.

For this paper, we use STEM to run on a horizontal resolution of 60Km by 60Km, with 21 vertical levels defined in the Regional Atmospheric Modeling System's sigma-z coordinate system. The domain covers northeast of U.S, ranging from 68°W to 85°W and from 36°N to 48°N. The simulations are carried out from 8am EDT to 8pm EDT on July 20, 2004, and dynamical time step is 15 minutes.

3   4D-Var Data Assimilation in STEM

Data assimilation is the process by which measurements are used to constrain the model predictions; the information from measurements can be used to obtain better initial conditions, better boundary conditions, enhanced emission estimates, etc. Data assimilation combines information from three different sources: the physical and chemical laws of evolution (encapsulated in the model), the reality (as captured by the observations), and the current best estimate of the distribution of tracers in the atmosphere.

In this paper we focus on obtaining optimized initial conditions which are essential in forward model integration. 4D-Var data assimilation can be used to STEM and is expected to improve air quality forecasting. In practice, directly measuring the parameters of the atmospheric conditions in large range is difficult because of sampling, technical and resource requirements. And due to the complexity of the chemical transport model, the number of possible state variables is extremely large. However, data acquisition for field and parameter estimates via data assimilation is feasible, even though enough observations are still need to fulfill data assimilation.

Figure 1 shows the employment of data assimilation in multiscale systems. The atmospheric chemical and transport processes take place in a variable range of time and space. The observations for data assimilation can come from local stations, planes, and satellites, measured from second to weeks in time and from *nm* to *km* in space. Data assimilation helps models to improve environmental policy decisions.

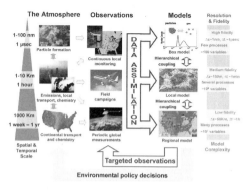

Fig. 1. Atmospheric Chemical and Transport Processes take place at multiple temporal and spatial scales

We applied 4D-Var data assimilation to obtain the optimal initial conditions using STEM by minimizing a cost function that measures the misfit between model predictions and observations, as well as the deviation of the solution from the background state. The cost function is formulated as

$$\min J\left(c^0\right) = \tfrac{1}{2}\left(c^0 - c^B\right)^T B^{-1}\left(c^0 - c^B\right) + \tfrac{1}{2}\sum_{k=0}^{N}\left(H_k c^k - c_{obs}^k\right)^T R_k^{-1}\left(H_k c^k - c_{obs}^k\right) \qquad (13)$$

and our goal is to minimize the cost function J. In the above formula, c^B represents the 'a priori' estimate (background) of the initial values and B is associated covariance matrix of the estimated background error. H_K is an observation operator and c_{obs}^k is the real observations depending on time k. The covariance matrix R_k^{-1} accounts for observations and representativeness errors.

The 4D-Var data assimilation is implemented by an iterative optimization procedure: each iteration requires STEM to run a forward integration to obtain the value of the cost function and an adjoint model to evaluate the gradient. Since model states are as high as 10^6 in our air quality simulation problem, it is prohibitive to evaluate Hessian of the cost function. Therefore, we choose three optimization methods that only require the values and the gradients of the cost function.

4   Results for Data Assimilation

We performed data assimilation to optimize initial conditions. The simulation interval is from 8am EDT to 8pm EDT on July 20, 2004. Figure 2 shows the simulation domain and three selected AIRNOW stations. AIRNOW stations provide hourly observations of ground level ozone throughout the entire month of July 2004. To show the change of ozone in this time interval at one location, we choose three out of all AirNow stations. The ozone time series at these three stations are illustrated in Figure 3. From the figure, we can find that the assimilated lines are closer to observations than non-assimilated lines, which indicates the improvement in model predictions after data assimilation. This is also confirmed by the scatter and quantile-quantile plots of Figure 4, which indicate that the correlation coefficient between model predictions and observations increases considerably from $R^2 = 0.15$ for the original model to $R^2 = 0.68$ after assimilation.

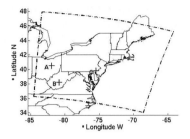

Fig. 2. Three selected stations where O3 time series are considered

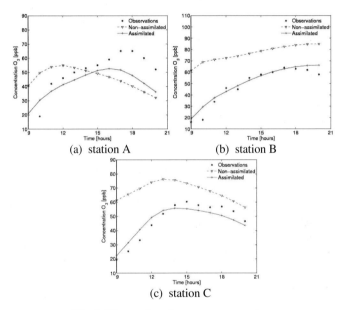

(a) station A   (b) station B

(c) station C

Fig. 3. Time series of ozone concentrations

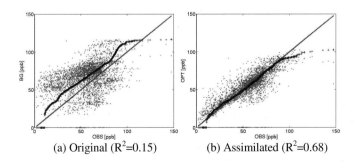

(a) Original (R^2=0.15)   (b) Assimilated (R^2=0.68)

Fig. 4. Scatter plot and quantile-quantile plot of model-observations agreement

5   Assessment of Three Optimization Methods

We applied three optimization methods: L-BFGS, Nonlinear Conjugate Gradient and Hessian Free Newton for data assimilation in STEM and assess the performances of them. The principle of L-BFGS [11] is to approximate Hessian matrix G by a symmetric positive definite matrix H, and update H at each step using the information from the m most recent iterations. Nonlinear Conjugate Gradient method is an iterative method and generates a set of search directions $\{p_0, p_1, ..., p_m\}$ conjugating with each other for $i \neq j$. The Fletcher-Reeves Conjugate Gradient method is used in this paper. Hessian Free Newton is an inexact Newton method in which the Hessian matrix is not available, and we use automatic differentiation or finite differences to approximate the products of the Hessian times a vector [12].

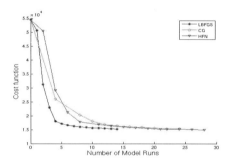

Fig. 5. Decrease of the cost function vs. the number of model runs for three methods

These methods are tested respectively to optimize initial concentrations. All the optimizations start at the same cost function of around 54800 and converge at about 16000 within 15 iterations, which proves that they are all able to solve data assimilation in STEM model. The difference lies in the number of model runs when they converge. For every model run, the STEM calls forward model and adjoint model to evaluate value and gradient of the cost function for optimization subroutine, so the more model runs, the more time needed in optimization. Figure 5 shows performances of these methods in terms of model runs they required. It is obvious that L-BFGS converges the fastest. We can conclude that of the three optimization methods L-BFGS is the best for data assimilation in STEM.

6   Conclusions

STEM is a multiscale atmospheric chemical transport model and has been used for air quality simulation regionally. Model Predictions can be improved by the technique of data assimilation. Data assimilation allows combining the information from both observations and STEM forward and adjoint models to obtain best estimates of the three-dimensional distribution of tracers. Reanalyzed fields can be used to initialize air quality forecast runs and have the potential to improve air quality predictions of the STEM model. Therefore, STEM is closely correlated to data assimilation. In this paper, we perform 4D-Var data assimilation over northeast of U.S. using the STEM model to optimize initial conditions. Both data and figures show the great improvement for simulation of STEM after data assimilation. Besides, we assess performance of three optimization methods that implement data assimilation in the STEM model. The results imply that L-BFGS best fits the STEM model of the three methods. Future work will focus on implementing second order adjoint model in STEM to provide Hessian of the cost function. In this way we can utilize some more accurate optimization methods for data assimilation.

Acknowledgemets. This work was supported by the Houston Advanced Research Center (HARC) through the award H59/2005 managed by Dr. Jay Olaguer and by the National Science Foundation through the award NSF ITR AP&IM 0205198 managed by Dr. Frederica Darema.

References

1. Carmichael, G.R., Peters, L.K., Saylor R.D.: The STEM-II regional scale acid deposition and photochemical oxidant model - I. An overview of model development and applications. Atmospheric Environment 25A: 2077-2090, 1990.
2. Kalman, R.E.: A new approach to linear filtering and prediction problems. Trans. ASME, Ser. D: J. Basic Eng., 83:95-108, 1960.
3. Fisher, M. and Lary, D.J.: Lagrangian four-dimensional variational data assimilation of chemical species. Q.J.R. Meteorology, 121:1681-1704, 1995.
4. Khattatov, B.V., Gille, J. C., Lyjak, L.V., Brasseur, G. P., Dvortsov, V. L., Roche, A. E., and Walters, J.: Assimilation of photochemically active species and a case analysis of UARS data. Journal of Geophysical Research, 104:18715-18737, 1999.
5. Elbern, H. and Schmidt, H.: A 4D-Var chemistry data assimilation scheme for Eulerian chemistry transport modeling. Journal of Geophysical Research, 104(5):18583-18589, 1999.
6. Errera, Q. and Fonteyn, D.: Four-dimensional variational chemical assimilation of CRISTA stratospheric measurements. Journal of Geophysical Research, 106(D11):12253-12265, 2001.
7. Wang, K.Y., Lary, D.J., Shallcross, D.E., Hall, S.M., and Pyle, J.A.: A review on the use of the adjoint method in four-dimensional atmospheric-chemistry data assimilation. Q.J.R. Meteorol. Soc., 127(576(Part B)):2181-2204, 2001.
8. Sandu, A., Daescu, D.N., Carmichael, G.R. and Chai, T.: Adjoint Sensitivity Analysis of Regional Air Quality Models. Journal of Computational Physics, Vol. 204: 222-252, 2005.
9. Carter, W.P.L.: Documentation of the SPARC-99 chemical mechanism for VOC reactivity assessment final report to California Air Resources Board. Technical Report. University of California at Riverside, 2000.
10. Damian, V., Sandu, A., Damian, M., Potra, F. and Carmichael, G.R.:The Kinetic PreProcessor KPP - A Software Environment for Solving Chemical Kinetic. Computers and Chemical Engineering, Vol. 26, No. 11: 1567-1579, 2002.
11. Liu, D.C. and Nocedal, J.: On the limited memory BFGS method for large-scale optimization. Math. Programming 45: 503-528, 1989.
12. Morales, J.L., and Nocedal, J.: Enriched Methods for Large-Scale Unconstrained Optimization. Computational Optimization and Applications, 21: 143-154, 2002.

Building a Dynamic Data Driven Application System for Hurricane Forecasting

Gabrielle Allen

Center for Computation & Technology and Department of Computer Science,
Louisiana State University, Baton Rouge, LA 70803
gallen@cct.lsu.edu
http://www.cct.lsu.edu

Abstract. The Louisiana Coastal Area presents an array of rich and
urgent scientific problems that require new computational approaches.
These problems are interconnected with common components: hurricane
activity is aggravated by ongoing wetland erosion; water circulation models are used in hurricane forecasts, ecological planning and emergency response; environmental sensors provide information for models of different
processes with varying spatial and time scales. This has prompted programs to build an integrated, comprehensive, computational framework
for meteorological, coastal, and ecological models. Dynamic and adaptive
capabilities are crucially important for such a framework, providing the
ability to integrate coupled models with real-time sensor information,
or to enable deadline based scenarios and emergency decision control
systems. This paper describes the ongoing development of a Dynamic
Data Driven Application System for coastal and environmental applications (DynaCode), highlighting the challenges of providing accurate and
timely forecasts for hurricane events.

Keywords: Dynamic data driven application systems, DDDAS, hurricane forecasting, event driven computing, priority computing, coastal
modeling, computational frameworks.

1 Introduction

The economically important Louisiana Coastal Area (LCA) is one of the world's
most environmentally damaged ecosystems. In the past century nearly one-third
of its wetlands have been lost and it is predicted that with no action by 2050 only
one-third of the wetlands will remain. Beyond economic loss, LCA erosion has
devastating effects on its inhabitants, especially in New Orleans whose location
makes it extremely vulnerable to hurricanes and tropical storms. On 29th August
2005 Hurricane Katrina hit New Orleans, with storm surge and flooding resulting
in a tragic loss of life and destruction of property and infrastructure. Soon after,
Hurricane Rita caused similar devastation in the much less populated area of
southwest Louisiana. In both cases entire communities were destroyed.

To effectively model the LCA region, a new comprehensive and dynamic
approach is needed including the development of an integrated framework for

Y. Shi et al. (Eds.): ICCS 2007, Part I, LNCS 4487, pp. 1034–1041, 2007.

coastal and environmental modeling capable of simulating all relevant interacting processes from erosion to storm surge to ecosystem biodiversity, handling multiple time (hours to years) and length (meters to kilometers) scales. This framework needs the ability to dynamically couple models and invoke algorithms based on streamed sensor or satellite data, locate appropriate data and resources, and create necessary workflows on demand, all in real-time. Such a system would enable restoration strategies, improve ecological forecasting, sensor placement, control of water diversion for salinity, or predict/control harmful algal blooms, and support sea rescue and oil spill response. In extreme situations, such as approaching hurricanes, results from multiple coupled ensemble models, dynamically compared with observations, could greatly improve emergency warnings.

These desired capabilities are included in the emerging field of Dynamic Data Driven Application Systems (DDDAS), which describes new complex, and inherently multidisciplinary, application scenarios where simulations can dynamically ingest and respond to real-time data from measuring devices, experimental equipment, or other simulations. In these scenarios, simulation codes are in turn also able to control these varied inputs, providing for advanced control loops integrated with simulation codes. Implementing these scenarios requires advances in simulation codes, algorithms, computer systems and measuring devices.

This paper describes work in the NSF funded DynaCode project to create a general DDDAS toolkit with applications in coastal and environmental modeling; a futuristic scenario (Sec. 2) provides general needs (Sec. 3) for components (Sec. 4). The rest of this section describes ongoing coastal research and development programs aligned with DynaCode, forming a scientific research foundation:

Lake Pontchartrain Forecast System. During Hurricane Katrina storm surge water from Lake Pontchartrain flooded New Orleans via breaches in outfall canals. The Army Corp of Engineers plans to close Interim Gated Structures at canal mouths during future storms, but this takes several hours, cannot occur in strong winds, and must be delayed as long as possible for storm rain water drainage.

The Lake Pontchartrain Forecast System (LPFS), developed by UNC and LSU, provides timely information to the Army Corp to aid in decision making for gate closing. LPFS is activated if a National Hurricane Center advisory places a storm track within 271 nautical miles of the canals, and an ensemble of storm surge (ADCIRC) runs is automatically deployed across the Louisiana Optical Network Initiative (LONI, http://www.loni.org) where mechanisms are in place to ensure they complete within two hours and results are provided to the Corp.

Louisiana CLEAR Program. The Coastal Louisiana Ecosystem Assessment and Restoration (CLEAR) program is developing ecological and predictive models to connect ecosystem needs with engineering design. CLEAR has developed a modeling tool to evaluate restoration alternatives using a combination of modules that predict physical processes, geomorphic features, and ecological succession. In addition, simulation models are being developed to provide an ecosystem forecasting system for the Mississippi Delta. This system will address questions such as what will happen to the Mississippi River Deltaic Plain under different scenarios of restoration alternatives, and what will be the benefits to society?

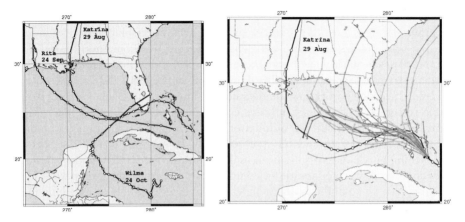

Fig. 1. Timely forecasts of the effects of hurricanes and tropical storms is imperative for emergency planning. The paths and intensity of the devastating hurricanes Katrina, Rita and Wilma [left] during 2005, as with other storms, are forecast from five days before expected landfall using different numerical and statistical models [right]. Model validity depends on factors such as the storm properties, location and environment.

SURA Coastal Ocean Observing and Prediction (SCOOP): The SCOOP Program [1] (http://scoop.sura.org) involves a diverse collaboration of coastal modelers and computer scientists working with government agencies to create an open integrated network of distributed sensors, data and computer models. SCOOP is developing a broad community-oriented cyberinfrastructure to support coastal research activities, for which three key scenarios involving distributed coastal modeling drive infrastructure development: 24/7 operational activities where various coastal hydrodynamic models (with very different algorithms, implementations, data formats, etc) are run on a daily basis, driven by winds from different atmospheric models; retrospective modeling where researchers can investigate different models, historical data sets, analysis tools etc; and most relevant for DDDAS, hurricane forecasting. Here, severe storm events initiate automated model workflows triggered by National Hurricane Center advisories, high resolution wind fields are generated which then initiate ensembles of hydrodynamic models. The resulting data fields are distributed to the SCOOP partners for visualization and analysis, and are placed in a highly available archive [2].

2   Data Driven Hurricane Forecast Scenario

When advisories from the National Hurricane Center indicate that a storm may make landfall in a region impacting Louisiana, government officials, based on information provided by model predictions (Fig. 1) and balancing economic and social factors, must decide whether to evacuate New Orleans and surrounding towns and areas. Such advisories are provided every six hours, starting from some five days before the storm is predicted to make landfall. Evacuation notices for large cities like New Orleans need to be given 72 hours in advance.

Here we outline a complex DDDAS scenario which provides hurricane predictions using ensemble modeling: *A suddenly strengthening tropical depression tracked by satellite changes direction, worrying officials. An alert is issued to state researchers and an advanced autonomic modeling system begins the complex process of predicting and validating the hurricane path. Realtime data from sensor networks on buoys, drilling platforms, and aircraft, across the Gulf of Mexico, together with satellite imagery, provide varied resolution data on ocean temperature, current, wave height, wind direction and temperature. This data is fed continuously into a ensemble modeling tool which, using various optimization techniques from a standard toolkit and taking into account resource information, automatically and dynamically task farms dozens of simulations, monitored in real-time. Each simulation represents a complex workflow, with closely coupled models for atmospheric winds, ocean currents, surface waves and storm surges. The different models and algorithms within them, are dynamically chosen depending on physical conditions and required output sensitivity. Data assimilation methods are applied to observational data for boundary conditions and improved input data. Validation methods compare data between different ensemble runs and live monitoring data, with data tracking providing additional information for dynamic decisions. Studying ensemble data from remotely monitored simulations, researchers steer computations to ignore faulty or missing input data. Known sensitivity to uncertain sensor data is propagated through the coupled ensemble models quantifying uncertainty. Sophisticated comparison with current satellite data is made with synthesized data from ensemble models to determine in real-time which models/components are most reliable, and a final high resolution model is run to predict 72 hours in advance the detailed location and severity of the storm surge. Louisiana's Office of Emergency Preparedness disseminates interactive maps of the projected storm surge and initiates contingency plans including impending evacuations and road closures.*

3   System Requirements

Such scenarios require technical advances across simulation codes, algorithms, computer systems and measuring devices. Here we focus on different technical issues related to the various components of a DDDAS simulation toolkit:

- **Data Sources & Data Management.** Data from varied sources must be integrated with models, e.g. wind fields from observational sources or computer models. Such data has different uncertainties, and improving the quality of input data, on demand, can lower forecast uncertainty. The ability to dynamically create customized ensembles of wind fields is needed, validated and improved with sensor data, for specified regions, complete with uncertainty functions propagated through models. Sensor data from observing systems must be available for real-time verification and data assimilation. Services for finding, transporting and translating data must scale to complex workflows of coupled interacting models. Emergency and real-time computing scenarios demand highly available data sources and data transport that is fault tolerant with guaranteed quality of service. Leveraging new optical networks requires mechanisms to dynamically reserve and provision networks and data scheduling capabilities. Metadata describing the huge amounts of distributed data is also crucial. and must include provenance information.

- **Model-Model Coupling and Ensembles.** Cascades of coupled circulation, wave and transport models are needed. Beyond defining interfaces, methods are needed to track uncertainties, create and optimize distributed workflows as the storm approaches, and invoke models preferentially, based on algorithm performance and features indicated by input data.

Cascades of models, with multiple components at each stage, lead to potentially hundreds of combinations where it is not known *a priori* which combinations give the best results. Automated and configurable ensemble modeling across grid resources, with continuous validation of results against observations and models, is critical for dynamically refining predictions. Algorithms are needed for dynamic configuration and creation of ensembles to provide predictions with a specifiable, required accuracy. In designing ensembles, the system must consider the availability and "cost" of resources, which may also depend on the threat and urgency e.g. with a Category 5 Hurricane requiring a higher quality of service than a Category 3 Hurricane.

- **Steering.** Automated steering is needed to adjust models to physical properties and the system being modeled, e.g. one could steer sensor inputs for improved accuracy. The remote steering of model codes, e.g. to change output parameters to provide verification data, or to initiating the reading of new improved data, will require advances to the model software. Beyond the technical capabilities for steering parameters (which often requires the involvement of domain experts), the steering mechanism must require authentication, with changes logged to ensure reproducibility.

- **Visualization and Notification.** Detailed visualizations, integrating multiple data and simulation sources showing the predicted effect of a storm are important for scientific understanding and public awareness (e.g. of the urgency of evacuation, or the benefits of building raised houses). Interactive and collaborative 3-D visualization for scientific insight will stress high speed networks, real-time algorithms and advanced clients. New visualizations of verification analysis and real-time sensor information are needed. Notification mechanisms to automatically inform scientists, administrators and emergency responders must be robust and configurable. Automated systems require human intervention and confirmation at different points, and the system should allow for mechanisms requiring authenticated response with intelligent fallback mechanisms.

- **Priority and Deadline Based Scheduling.** Dealing with unpredictable events, and deploying multiple models concurrently with data streams, provides new scheduling and reservation requirements: priority, deadline-based, and co-scheduling. Computational resources must be available on demand, with guaranteed deadlines for results; multiple resources must be scheduled simultaneously and/or in sequence. Resources go beyond traditional computers, including archival data, file systems, networks and visualization devices.

Policies need to be adopted at computing centers that enable event-driven computing and data streaming; computational resources of various kinds need to be available on demand, with policies reflecting the job priority.

4   DynaCode Components

In the DynaCode project this functionality is being developed by adapting and extending existing software packages, and building on the SCOOP and LPFS scenarios. Collectively, the packages described below form the basis for a "DDDAS Toolkit", designed for generic DDDAS applications, with specific drivers including the hurricane forecast scenario:

- **Cactus Framework.** Cactus [3], a portable, modular software environment for developing HPC applications, has already been used to prototype DDDAS-style applications. Cactus has numerous existing capabilities relevant for DDDAS, including extreme portability, flexible and configurable I/O, an inbuilt parameter steering API and robust checkpoint/restart capabilities. Cactus V5, currently under development, will include additional crucial features, including the ability to expose individual 'thorn' (component) methods with web service interfaces, and to interoperate with other framework architectures. Support for the creation and archiving of provenance data including general computational and domain specific information is being added, along with automated archiving of simulation data.
- **User Interfaces.** An integrated, secure, web user interface developed with the GridSphere portal framework builds on existing relevant portlets [4]. New portlets include Threat Level and Notification, Ensemble Monitoring, Ensemble Track Visulization and Resource Status. A MacOSX "widget" (Fig. 2, left) displays hurricane track information and system status.
- **Data Management.** Through the SCOOP project a highly reliable coastal data archive [2] has been implemented at LSU, with 7TB of local storage and 7TB of remote storage (SDSC SRB) for historical data. This archive was designed to ingest and provide model (surge, wave, wind) and observational (sensor, satellite) data, and is integrated with a SCOOP catalogue service at UAH. To support dynamic scenarios a general trigger mechanism was added to the archive which can be configured to perform arbitrary tasks on arrival of certain files. This mechanism is used to drive ensemble configuration and deployment, notification and other components. DynaCode is partnered with the NSF funded PetaShare project which is developing new technologies for distributed data sharing and management, in particular data-aware storage systems and data-aware schedulers.
- **Ensembles.** Ensembles for DynaCode scenarios are currently deployed across distributed resources with a management component executed on the archive machine. The system is designed to support rapid experimentation rather than complex workflows, and provides a completely data-driven architecture with the ensemble runs being triggered by the arrival of input wind

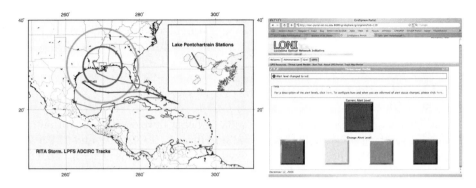

Fig. 2. [Left] The LPFS system is activated if any ensemble member places the hurricane track within 271 nautical miles of the canal mouths (inside the inner circle). [Right] A threat level system allows the threat level to be changed by trusted applications or scientists. This triggers the notification of system administrators, customers and scientists and the setting of policies on compute resources for appropriately prioritized jobs. The diagram shows a portal interface to the threat level system.

files. As ensemble runs complete, results are fed to a visualization service and also archived in the ROAR Archive [2]. Metadata relating to the run and the result set is fed into the catalog developed at UAH via a service interface.

- **Monitoring.** The workflow requires highly reliable monitoring to detect failures and prompt corrective action. Monitoring information (e.g. data transfers/job status) is registered by the various workflow components and can be viewed via portal interfaces. A spool mechanism is used to deliver monitoring information via log files to a remote service, providing high reliability and flexibility. This ensures the DynaCode workflow will not fail due to unavailability of the monitoring system. Also, the workflow executes faster than a system where monitoring information is transported synchronously.

- **Notification.** A general notification mechanism sends messages via different mechanisms to configurable role-based groups. The system currently supports email, instant messaging, and SMS text messages, and is configurable via a GridSphere portlet. The portlet behaves as a messaging server that receives updates from e.g. the workflow system and relays messages to subscribers. Subscribers can belong to different groups that determine the information content of messages they receive, allowing messages to be customized for e.g. system administrators, scientists or emergency responders.

- **Priority Scheduling & Threat Levels.** Accurate forecasts of hurricane events, involving large ensembles, need to be completed quickly and reliably with specific deadlines. To provide on-demand resources the DynaCode workflow makes use of policies as well as software. On the large scale resources of LONI and CCT, the queues have been configured so that it is possible to preempt currently running jobs and free compute resources at extremely short notice. Large queues are reserved for codes that can checkpoint and restart. These queues share compute nodes with preemptive queues that pre-empt jobs in the 'checkpoint' queues when they receive jobs to run. Software such as SPRUCE (http://spruce.teragrid.org/) is being used to provide el-evated priority and preemptive capabilities to jobs that hold special tokens reducing user management burden from system administrators.

A "Threat Level" service has been developed; trusted applications or users can set a global threat level to red, amber, yellow or green (using web service or portal interfaces), depending on the perceived threat and urgency (Fig. 2). Changes to the threat level triggers notification to different role groups, and is being integrated with the priority scheduling system and policies.

– **Co-allocation.** DynaCode is partnering with the NSF Enlightened Com-puting project which is developing application-enabling middleware for optical networks. The HARC co-allocator, developed through the En-lightened-DynaCode collaboration, can already allocate reservations on com-pute resources and optical networks, and is being brought into production use on the LONI network to support DynaCode and other projects.

Acknowledgments. The author acknowledges contributions from colleagues in the SCOOP, CLEAR and LPFS projects and collaborators at LSU. This work is part of the NSF DynaCode project (0540374), with additional funding from the SURA Coastal Ocean Observing and Prediction (SCOOP) Program (including ONR Award N00014-04-1-0721, NOAA Award NA04NOS4730254). Computa-tional resources and expertise from LONI and CCT are gratefully acknowledged.

References

1. Bogden, P., Allen, G., Stone, G., Bintz, J., Graber, H., Graves, S., Luettich, R., Reed, D., Sheng, P., Wang, H., Zhao, W.: The Southeastern University Research Associa-tion Coastal Ocean Observing and Prediction Program: Integrating Marine Science and Information Technology. In: Proceedings of the OCEANS 2005 MTS/IEEE Conference, Sept 18-23, 2005, Washington, D.C. (2005)
2. MacLaren, J., Allen, G., Dekate, C., Huang, D., Hutanu, A., Zhang, C.: Shelter from the Storm: Building a Safe Archive in a Hostile World. In: Proceedings of the The Second International Workshop on Grid Computing and its Application to Data Analysis (GADA'05), Agia Napa, Cyprus, Springer Verlag (2005)
3. Goodale, T., Allen, G., Lanfermann, G., Massó, J., Radke, T., Seidel, E., Shalf, J.: The Cactus framework and toolkit: Design and applications. In: High Perfor-mance Computing for Computational Science - VECPAR 2002, 5th International Conference, Porto, Portugal, June 26-28, 2002, Berlin, Springer (2003) 197–227
4. Zhang, C., Dekate, C., Allen, G., Kelley, I., MacLaren, J.: An Application Portal for Collaborative Coastal Modeling. Concurrency Computat.: Pract. Exper. **18** (2006)

A Dynamic Data Driven Wildland Fire Model

Jan Mandel[1,2], Jonathan D. Beezley[1,2], Lynn S. Bennethum[1],
Soham Chakraborty[3], Janice L. Coen[2], Craig C. Douglas[3,5], Jay Hatcher[3],
Minjeong Kim[1], and Anthony Vodacek[4]

[1] University of Colorado at Denver and Health Sciences Center, Denver, CO
80217-3364, USA
[2] National Center for Atmospheric Research, Boulder, CO 80307-3000, USA
[3] University of Kentucky, Lexington, KY 40506-0045, USA
[4] Rochester Institute of Technology, Rochester, NY 14623-5603, USA
[5] Yale University, New Haven, CT 06520-8285, USA

Abstract. We present an overview of an ongoing project to build
DDDAS to use all available data for a short term wildfire prediction.
The project involves new data assimilation methods to inject data into
a running simulation, a physics based model coupled with weather prediction, on-site data acquisition using sensors that can survive a passing
fire, and on-line visualization using Google Earth.

1   Introduction

DDDAS for a short-term wildland fire prediction is a challenging problem. Techniques standard in geophysical applications generally do not work because the
nonlinearity of fire models are much stronger than those of the atmosphere or
the ocean; data is incomplete; and it is not clear which model is the best for
physical representation and faster than real time speed.

2   Fire Model

The goal in wildland fire modeling is to predict the behavior of a complex system
involving many processes and uncertain data by a physical model that reproduces
important dynamic behaviors. Our overall approach is to create a mathematical
model at the scales at which the dominant behaviors of the system occur and
the data exist.

Perhaps the simplest PDE based wildland fire model [2] is of the form

$$\frac{dT}{dt} = \nabla \cdot (k\nabla T) + \overrightarrow{v} \cdot \nabla T + A\left(Sr\left(T\right) - C_0\left(T - T_a\right)\right), \tag{1}$$

$$\frac{dS}{dt} = -C_S Sr\left(T\right), \tag{2}$$

where the reaction rate is a modified Arrhenius rate,

$$r(T) = e^{-B/(T-T_a)}. \tag{3}$$

Y. Shi et al. (Eds.): ICCS 2007, Part I, LNCS 4487, pp. 1042–1049, 2007.

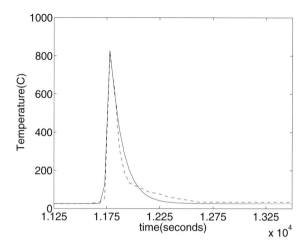

Fig. 1. Measured time-temperature profile (dotted line) in a wildland fire at a fixed sensor location, digitized from [1], and computed profile (solid line) [2]. Coefficients of the model were identified by optimization to match to measured profile.

Eq. (1) represents a 2D balance of energy in a fire layer of some unspecified finite vertical thickness, (2) represents the balance of fuel, T is the temperature of the fire layer, $r(T) \in [0,1]$ is the reaction rate, assumed to depend only on the temperature, $S \in [0,1]$ is the fuel supply mass fraction (the relative amount of fuel remaining), k is the diffusion coefficient, A is the temperature rise per second at the maximum burning rate with full initial fuel load and no cooling present, B is the proportionality coefficient in the modified Arrhenius law, C_0 is the scaled coefficient of the heat transfer to the environment, C_S is the fuel relative disappearance rate, T_a is the ambient temperature, and $\overrightarrow{v}$ is the given wind speed from the atmosphere.

Physics-based models of the form similar to (1-3) are known [3,4,5]. The system (1-2) with the reaction rate (3) is formally the same as combustion equations of premixed fuel, with cooling added. Our interpretation is that the equations are a rough approximation of the aggregated behavior of small-scale random combustion. We are going to add variables and equations only as needed to match the observed fire behavior. Possible future extension include pyrolysis, multiple fuels, crown fire, and moisture (treated as a fuel with negative heat content).

The reaction rate (3) has been modified to be exactly zero at ambient temperature (according to chemical kinetics, the reaction rate should be zero only at absolute zero temperature); consequently equations (1-3) admit traveling wave solutions. The temperature in the traveling wave has a sharp leading edge, followed by an exponentially decaying cool-down trailing edge (Fig. 1). The wave speed can be found numerically [5], but no mathematical proof of the existence of traveling waves and their speed for system (1-3) seems to be known.

Since weather is a very significant influence on fire and fire in turn has influence on weather, coupling of the fire model with the Weather Research Forecast Model

(WRF) [6,7] is in progress [8]. It is now well established that fire dynamics can be understood only by modeling the atmosphere and the fire together, e.g., [9].

3   Coefficient Identification

The reason for writing the coefficients of model (1-3) in that particular form is that it is basically hopeless to use physical coefficients and expect a reasonable solution. The fuel is heterogeneous, the fire blends into the atmosphere, and it is unclear what exactly are, e.g., the diffusion and the heat loss coefficients. Instead, we consider information that can be reasonably obtained. For example, consider the heat balance term at full fuel load ($S = 1$),

$$f(T) = r(T) - C_0(T - T_a).$$

Generally, there are three temperatures which cause $f(T)$ to be zero (where the heat created by the reaction is balanced by the cooling): T_a, arising from our modification of the Arrhenius rate (otherwise there is one zero just above T_a because some reaction is always present); the autoignition temperature; and finally the "high temperature regime" [10], which is the hypothetical steady burning temperature assuming fuel is constantly replenished. Substituting reasonable values for the latter two temperatures allows us to determine reasonable approximate values of the coefficients B and C_0. Assuming zero wind, we can then determine the remaining coefficients in 1D dynamically from a measured temperature profile in an actual wildland fire [1]. Reasonable values of the remaining coefficients are obtained by (i) passing to a nondimensional form and matching nondimensional characteristics of the temperature profile, such as the ratio of the widths of the leading and the trailing edge and (ii) using the traveling wave speed and intensity to match the scale. Once reasonable starting values are known, the coefficients can be further refined by optimization to match the measured temperature profile (Fig. 1). See [2] for details. Identification of coefficients in the presence of wind in the model is in progress [11].

4   Numerical Solution

The coefficients identified from a 1D temperature profile were used in a 2D model, discretized by standard central finite differences with upwinding of the convection term. Implicit time stepping by the trapezoidal method (Crank-Nicholson) was used for stability in the reaction term and to avoid excessively small time steps. The large sparse system of linear equations in every time step was solved by Newton's method with GMRES as the linear solver, preconditioned by elimination of the fuel variables (which are decoupled in space) and FFT. One advantage of this approach is that even after more fuels are added, the fuel variables can still be eliminated at every node independently, and the resulting system has the same sparsity structure as the heat equation alone. Typically, mesh steps of $1m$ – $5m$ and time steps of $1s$ – $5s$ are required to resolve the moving fire front and to get a reasonable simulation [11].

5   Data Assimilation

Data assimilation is how the data is actually injected into a running model. We have chosen a probabilistic approach to data assimilation, which is common in geosciences [12]. The ensemble filter method used here is one possible approach. This method estimates the system state from all data available by approximating the probability distribution of the state by a sample, called an ensemble. The ensemble members are advanced in time until a given *analysis time*. At the analysis time, the probability density, now called the *prior* or the *forecast*, is modified by accounting for the data, which are considered to be observed at that time. This is done by multiplying the prior density by data likelihood and normalizing (the Bayesian update). The new probability density is called the *posterior* or the *analysis* density. The data likelihood is a function of state. It is equal to the density of the probability that the data could have been obtained for the given state. It can be found from the probability density of data error, which is assumed to be known (every measurement must be accompanied by an error estimate or it is meaningless), and an *observation function* (sometimes called the *forward operator* in an inverse problem context). The argument of the observation function is a model state and its value is what the correct value of the data should be for that state.

The Ensemble Kalman Filter (EnKF) [13,14] works by forming the analysis ensemble as *linear combinations of the forecast ensemble*, and the Bayesian update is implemented by linear algebra operating on the matrix created from ensemble members as columns.

In a wildland fire application, EnKF produces nonphysical states that result in instability and model breakdown [2,15]. The linear combination of several states with a fire in slightly different locations cannot match data with a fire in another location. In a futile attempt to match the data, EnKF makes the analysis ensemble out of crazy linear combinations of the forecast ensemble. These can be fatal immediately. Even if they are not, the analysis ensemble tends to bear little resemblance to the data. This can be ameliorated to some degree by penalization of nonphysical states [2,15] but not well enough to produce a reliable method: the filter is capable of only a small adjustment of the fire location, and its parameters have to be finely tuned for acceptable results, especially the data variance is required to be artificially large.

For this reason, we have developed a new method that updates the location of firelines directly, called the *morphing ensemble filter* [16,17], which is based on techniques borrowed from registration in image processing. Given two images as pixel values on the same domain, the registration problem is to find a mapping of the domain that turns one of the given images into the other with a *residual* (the remaining difference) that is as small as possible. We write this mapping as the identity plus a *registration mapping*. The registration mapping should be as small and as smooth as possible. Hence, one natural way to solve the automatic registration problem is by optimization, and we use a method based on [18]. Once the registration mapping and the residual are found, we can construct intermediate images between the two images by applying the registration mapping

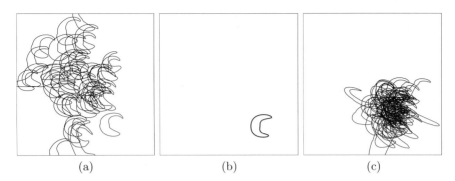

(a)   (b)   (c)

Fig. 2. Data assimilation by the morphing ensemble filter [16]. Contours are at $800K$, indicating the location of the fireline. The reaction zone is approximately inside the moonshaped curve. This fireline shape is due to the wind. The forecast ensemble (a) was obtained by a perturbation of an initial solution. The fire in the simulated data (b) is intentionally far away from the fire in the initial solution. The data for the EnKF engine itself consisted of the transformed state obtained by registration of the image. The analysis (c) shows the adjusted location of the fire in the analysis ensemble members.

and adding the residual, both multiplied by some number between zero and one. Generalizing this observation, we can choose one fixed state with a fire in some location, and register all ensemble members. The registration mappings and the residuals then constitute a transformed ensemble. *The morphing ensemble filter consists of applying the EnKF to the transformed ensemble and thus using intermediate states rather than linear combinations.* This results in an adjustment of both the intensity and the position of the fire. Now the fireline in the data can be anywhere in the domain, and the data variance can be small without causing filter divergence.

Fire data in the form of an image can be assimilated by registering the image and then using the transformed image as the observation. The observation function is then only a pointwise transformation of the model state to gridded values and interpolation. In the general case, the transformed observation function is defined by composition of functions, and so it is a highly nonlinear function of the registration mapping. So, assimilation of weather station data and sensor data will be more complicated and can result in strongly non-Gaussian multimodal probability distributions. For example, a sensor can read low temperature either because the fire did not get there yet or because it has already passed, so the analysis density after assimilating one sensor reading would be concentrated around these two possibilities.

Another view is that the morphing EnKF works by transforming the state so that after the transformation, the probability distribution is closer to Gaussian. We have also developed another new technique, called predictor-corrector filters [17], to deal with the remaining non-Gaussianity.

6 Data Acquisition

Currently, we are using simple simulated data, as described in the previous section. In the future, data will come from fixed sensors that measure temperature, radiation, and local weather conditions. The fixed Autonomous Environmental Sensors (AESs), positioned so as to provide weather conditions near a fire, are mounted at various heights above the ground on a pole with a ground spike (Fig. 3a). This type of system will survive burnovers by low intensity fires. The temperature and radiation measurements provide a direct indication of the fire front passage and the radiation measurement can also be used to determine the intensity of the fire [20]. The sensors transmit data and can be reprogrammed by radio.

Data will come also from images taken by sensors on either satellites or airplanes. Three wavelength infrared images can then be processed using a variety of algorithms to extract which pixels contain a signal from fire and to determine the energy radiated by the fire and even the normal velocity of the fireline (Fig. 3b).

The data is related to the model by an observation function. Currently, we are using simple simulated observation functions, as described in the previous section. Our software framework supports using multiple types of observation functions (image, weather station, etc.). Each data item must carry information about which observation function is to be used and any requisite parameters (coordinates, scaling factors, etc.) in metadata.

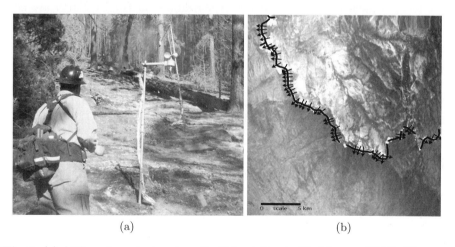

(a)   (b)

Fig. 3. (a) AES package deployed on the Tripod Fire in Washington (2006). The unit shown has been burned over by the fire but is still functional. The datalogger package is buried a small distance beneath the surface to shield it from the effects of the fire. (b) Airborne image processed to extract the location and propagation vector of the fireline (reproduced from [19] by permission).

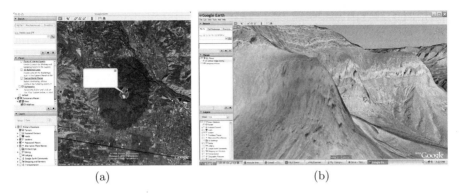

(a)   (b)

Fig. 4. The Google Earth Fire Layering software tool. (a) Closeup on a single fire (b) 3D view.

7   Visualization

Our primary tool for visualization is storing gridded data from the model arrays into files for visualization in Matlab. Our Google Earth Fire visualization system (Fig. 4) greatly simplifies map and image visualization and will be used for model output in the future. The user can control the viewing perspective, zooming into specific sites, and selecting the time frame of the visualization within the parameters of the current available simulation. Google Earth is quickly becoming a de-facto standard and wildland fire visualizations in Google Earth are now available from several commercial and government sources, e.g., [21,22].

Acknowledgment

This research has been supported by the National Science Foundation under grants CNS 0325314, 0324989, 0324988, 0324876, and 0324910.

References

1. Kremens, R., Faulring, J., Hardy, C.C.: Measurement of the time-temperature and emissivity history of the burn scar for remote sensing applications. Paper J1G.5, Proceedings of the 2nd Fire Ecology Congress, Orlando FL, American Meteorological Society (2003)
2. Mandel, J., Bennethum, L.S., Beezley, J.D., Coen, J.L., Douglas, C.C., Franca, L.P., Kim, M., Vodacek, A.: A wildfire model with data assimilation. CCM Report 233, http://www.math.cudenver.edu/ccm/reports/rep233.pdf (2006)
3. Asensio, M.I., Ferragut, L.: On a wildland fire model with radiation. International Journal for Numerical Methods Engineering **54** (2002) 137–157
4. Grishin, A.M.: General mathematical model for forest fires and its applications. Combustion Explosion and Shock Waves **32** (1996) 503–519

5. Weber, R.O., Mercer, G.N., Sidhu, H.S., Gray, B.F.: Combustion waves for gases ($Le = 1$) and solids ($Le \to \infty$). Proceedings of the Royal Society of London Series A **453** (1997) 1105–1118
6. Michalakes, J., Dudhia, J., Gill, D., Klemp, J., Skamarock, W.: Design of a next-generation weather research and forecast model. Towards Teracomputing: proceedings of the Eighth Workshop on the Use of Parallel Processors in Meteorology, European Center for Medium Range Weather Forecasting, Reading, U.K., November 16-20, 1998. ANL/MCS preprint number ANL/MCS-P735-1198 (1998)
7. WRF Working Group: Weather Research Forecasting (WRF) Model. http://www.wrf-model.org (2005)
8. Beezley, J.D.: Data assimilation in coupled weather-fire models. Ph.D. Thesis, in preparation (2008)
9. Coen, J.L.: Simulation of the Big Elk Fire using using coupled atmosphere-fire modeling. International Journal of Wildland Fire **14** (2005) 49–59
10. Frank-Kamenetskii, D.A.: Diffusion and heat exchange in chemical kinetics. Princeton University Press (1955)
11. Kim, M.: Numerical modeling of wildland fires. Ph.D. Thesis, in preparation (2007)
12. Kalnay, E.: Atmospheric Modeling, Data Assimilation and Predictability. Cambridge University Press (2003)
13. Evensen, G.: Sequential data assimilation with nonlinear quasi-geostrophic model using Monte Carlo methods to forecast error statistics. Journal of Geophysical Research **99 (C5)** (1994) 143–162
14. Houtekamer, P., Mitchell, H.L.: Data assimilation using an ensemble Kalman filter technique. Monthly Weather Review **126** (1998) 796–811
15. Johns, C.J., Mandel, J.: A two-stage ensemble Kalman filter for smooth data assimilation. Environmental and Ecological Statistics, in print. CCM Report 221, http://www.math.cudenver.edu/ccm/reports/rep221.pdf (2005)
16. Beezley, J.D., Mandel, J.: Morphing ensemble Kalman filters. CCM Report 240, http://www.math.cudenver.edu/ccm/reports/rep240.pdf (2007)
17. Mandel, J., Beezley, J.D.: Predictor-corrector and morphing ensemble filters for the assimilation of sparse data into high dimensional nonlinear systems. 11th Symposium on Integrated Observing and Assimilation Systems for the Atmosphere, Oceans, and Land Surface (IOAS-AOLS), CD-ROM, Paper 4.12, 87th American Meterological Society Annual Meeting, San Antonio, TX, January 2007.
18. Gao, P., Sederberg, T.W.: A work minimization approach to image morphing. The Visual Computer **14** (1998) 390–400
19. Ononye, A.E., Vodacek, A., Saber, E.: Automated extraction of fire line parameters from multispectral infrared images. Remote Sensing of Environment (to appear)
20. Wooster, M.J., Zhukov, B., Oertel, D.: Fire radiative energy for quantitative study of biomass burning: derivation from the BIRD experimental satellite and comparison to MODIS fire products. Remote Sensing of Environment **86** (2003) 83–107
21. NorthTree Fire International: Mobile Mapping/Rapid Assessment Services. http://www.northtreefire.com/gis/ (2007)
22. USGS: RMGSC - Web Mapping Applications for the Natural Sciences. http://rockyitr.cr.usgs.gov/rmgsc/apps/Main/geiDownloads.html (2007)

Ad Hoc Distributed Simulation of Surface Transportation Systems

R.M. Fujimoto, R. Guensler, M. Hunter, K. Schwan, H.-K. Kim,
B. Seshasayee, J. Sirichoke, and W. Suh

Georgia Institute of Technology, Atlanta, GA 30332 USA
{fujimoto@cc, randall.guensler@ce, michael.hunter@ce,
schwan@cc}.gatech.edu

Abstract. Current research in applying the Dynamic Data Driven Application Systems (DDDAS) concept to monitor and manage surface transportation systems in day-to-day and emergency scenarios is described. This work is focused in four, tightly coupled areas. First, a novel approach to predicting future system states termed ad hoc distributed simulations has been developed and is under investigation. Second, on-line simulation models that can incorporate real-time data and perform rollback operations for optimistic ad hoc distributed simulations are being developed and configured with data corresponding to the Atlanta metropolitan area. Third, research in the analysis of real-time data is being used to define approaches for transportation system data collection that can drive distributed on-line simulations. Finally, research in data dissemination approaches is examining effective means to distribute information in mobile distributed systems to support the ad hoc distributed simulation concept.

Keywords: surface transportation systems, ad hoc distributed simulations, rollback operations.

1 Introduction

The Vehicle-Infrastructure Integration (VII) initiative by government agencies and private companies is deploying a variety of roadside and mobile sensing platforms capable of collecting and transmitting transportation data [1-3]. With the ongoing deployment of vehicle and roadside sensor networks, transportation planners and engineers have the opportunity to explore new approaches to managing surface transportation systems, offering the potential to allow the creation of more robust, efficient transportation infrastructures than was possible previously. Effective and efficient system management will require real-time determinations as to which data should be monitored, and at what resolutions. Distributed simulations offer the ability to predict future system states for use in optimizing system behaviors both in day-to-day traffic conditions as well as in times of emergency, e.g., under evacuation scenarios. Data collection, data processing, data analysis, and simulations performed by system agents (sub-network monitoring systems, base stations, vehicles, etc.) will lessen communication bandwidth requirements and harness surplus computing

Y. Shi et al. (Eds.): ICCS 2007, Part I, LNCS 4487, pp. 1050–1057, 2007.

capacity. Middleware to manage the distributed network, synchronize data and results among autonomous agents, and resolve simulation output conflicts between agents using disparate data sets become critical activities in such a system. Dynamic, data-driven application systems (DDDAS) offer the potential to yield improved efficiencies in the system that can reduce traffic delays and congestion, pollution, and ultimately, save lives during times of crisis.

We are addressing this challenge through a distributed computing and simulation approach that exploits in-vehicle computing and communication capabilities, coupled with infrastructure-based deployments of sensors and computing equipment. Specifically, we envision a system architecture that includes in-vehicle computing systems, roadside compute servers (e.g., embedded in traffic signal controllers) as well as servers residing in traffic management centers (TMCs). The remaining sections provide an overview of specific research examining key elements of this system. The next section describes a concept called ad hoc distributed simulations that are used to project future system states.

2   Ad Hoc Distributed Simulations

Consider a collection of in-vehicle simulations that are interconnected via wireless links and (possibly) wired network infrastructure. Individually, each simulation only models a portion of the traffic network – that which is of immediate interest to the "owner" of the simulator, Figure 1.

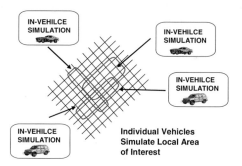

Fig. 1. In-Vehicle Simulation

Collectively, these simulations could be used to create a kind of distributed simulation system with the ability to make forecasts concerning the entire transportation infrastructure as a whole. One can envision combining in-vehicle simulators with simulations running within the roadside infrastructure, e.g., within traffic signal controller cabinets, simulations combining sub-regions of the transportation network, and simulations running in traffic management centers to create a large-scale model of a city's transportation system, as shown in Figure 2.

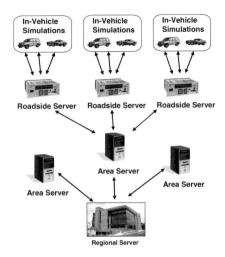

Fig. 2. Ad Hoc Distributed Simulation Structure

We term a collection of autonomous, interacting simulations tied together in this fashion an *ad hoc distributed simulation*. Like a conventional distributed simulation, each simulator within an ad hoc distributed simulation models a portion of the overall system, and simulators exchange time stamped state information to collectively create a model of the overall system. However, in a conventional distributed simulation the system being modeled is designed in a *top-down* fashion. Specifically, the system is neatly partitioned into non-overlapping elements, e.g., geographic regions, and a process is assigned to model each element. By contrast, an ad hoc distributed simulation is created in a *bottom-up* fashion with no overarching intelligence governing the partitioning and mapping of the system to processes. Rather, the distributed simulation is constructed in an "ad hoc" fashion, in much the same way an arbitrary collection of mobile radios join together to form an ad hoc wireless network. The elements of the physical system modeled by different simulators in an ad hoc distributed simulation may overlap, leading to possibly many duplicate models of portions of the system, as seen in Figure 1. Other parts of the system may not be modeled at all. For example, an in-vehicle transportation simulator may be only modeling the portion of the road network along the vehicle's intended path to reach its destination. Thus, ad hoc distributed simulations differ in important fundamental ways from conventional distributed simulations.

Ad hoc distributed simulations are on-line simulation programs, meaning they are able to capture the current state of the system through measurement, and then execute forward as rapidly as possible to project a future state of the system. By assumption, each of the simulators making up the distributed simulation can simulate some portion of the system faster than real time.

Ad hoc distributed simulations require a synchronization protocol to coordinate interactions among other simulations. For this purpose we have developed an optimistic (rollback-based) synchronization protocol designed for use in these systems. Each in-vehicle simulator utilizes information concerning traffic conditions and predictions of future system states (e.g., road flow rates) to complete its simulation. If this information changes beyond certain parameterized limits, the simulator rolls back, and corrects its previously computed results. Based on this protocol, a prototype ad hoc distributed simulation system has been developed using both a custom-developed cellular automata traffic simulator as well as a commercial simulation tool called VISSIM, described next. Further details of this work are presented in [4].

3   Transportation Simulation Models

An ad hoc transportation simulation based on a cellular automata model was developed for our initial investigations. The simulation consists of agents modeling vehicles, traffic signal controllers and traffic lights. The simulation operates in a timestep fashion. At every timestep, each agent decides what operation to perform next based on its own characteristics and the state of the system at the end of previous interval. Each vehicle agent includes characteristics such as origin, destination, maximum speed of the vehicle, and driver characteristics such as aggressiveness. Each vehicle has complete knowledge of the road topography around it. At the end of each timestep each vehicle agent makes decisions whether to move, stop, accelerate, or decelerate based on the following four rules:

- *Acceleration*: if the velocity v of a vehicle is lower than max velocity and if the distance to the next car ahead is larger than v + gap, the speed is increased: $v = v$ + acceleration.
- *Slowing down* (due to other cars): If a vehicle at site X sees the next vehicle at site $X + j$ (with $j < v$), it reduces its speed to j: $v = j$
- *Randomization*: The velocity a vehicle, if greater than zero, is decreased with probability p_{dec}: $v = v$ - deceleration.
- *Car motion*: each vehicle is advanced v cells.

Similarly, the traffic controller may also change its state, specifically the stop-go flag, at the end of each time step.

We conducted experiments in the management of 20 client simulations using the cellular automata simulator covering a 10 intersection corridor. Under steady conditions, the distributed simulation client provided a system representation similar to that of a replicated trial experiment of the entire network, demonstrating that the ad hoc approach offers potential for accurately predicting future systems states Further, when a spike in the traffic flow was introduced into the network, the distributed clients again successfully modeled the traffic flows; however there was a short delay (up to approximately four minutes) in the upstream client transmitting the increased flows to the downstream clients.

While the cellular automata simulation model allowed for the development of an understanding of the behavior of the ad hoc distributed approach it is desirable that future experimentation be conducted with a significantly more detailed, robust transportation simulation model. Further, the ability to adapt existing commercial simulation software for use in ad hoc distributed simulations would significantly improve the likelihood that the technology would be used. An initial investigation into implementing the ad hoc strategy using the off-the-shelf transportation simulation model VISSIM was conducted. VISSIM is widely used by private firms and public agencies for transportation system analysis. VISSIM is a discrete, stochastic, time step based microscopic simulation model developed to model urban traffic (freeway and arterial) and public transit operations. The model is capable of simulating a diverse set transportation network features, such as facility types, traffic control mechanism, vehicle and driver types, etc. Individual vehicles are modeled in VISSIM using a psycho-physical driver behavior model developed by Wiedemann. The underlying concept of the model is the assumption that a driver can be in one of four driving modes: free driving, approaching, following, or braking [5]. VISSIM version 4.10 was utilized for this investigation.

Investigation into the use of VISSIM has proven very hopeful. Utilizing the VISSIM COM interface [6] it is possible to access many of the VISSIM object, methods, and properties necessary to implement simulation roll backs and automate VISSIM clients disbursed among workstations connected by a local area network. The initial investigation discussed in the section focuses on the ability to implement a roll back mechanism in VISSIM. Through the VISSIM COM interface it is possible at anytime t_i during a simulation run to save the current state of the simulation model. At any later time t_j, where $j > i$, it is possible to stop the simulation run, load the simulation state from time t_i, update the simulation attributes indicated by the roll back (i.e. arrival rate on some link), and restart the simulation from time t_i. An initial experiment was conducted using VISSIM demonstrating that the roll back algorithm was successfully implemented. This simulator was driven by traces of simulation-generated traffic flow data, and demonstrated to accurately predict future states (as represented from the trace data). Further details of these experiments are presented in [4].

4 Real-Time Dynamic Data Analysis

The precision of the real-time data varies depending on the level of data aggregation. For example, minute-by-minute data are more precise than hourly average data. We examined the creation of an accurate estimate of the evolving state of a transportation system using real-time roadway data aggregated at various update intervals.

Using the VISSIM model described in the previous section, a simulation of the transportation network in the vicinity of the Georgia Institute of Technology in Atlanta, Georgia was utilized to represent the real world, with flow data from the Georgia Tech model provided to a smaller simulation of two intersections within the network. The "real-world" flow data (i.e. flow data measured from the large scale Georgia Tech model) was aggregated in different intervals and used to dynamically drive the two-intersection model. The desire of this study was to explore how well the small-scale simulation model is able to reflect the real world scenario when fed data at different aggregation levels. This work explored congested and non-congested traffic demand at data five different aggregation time intervals: 1 sec., 10 sec., 30 sec., 60 sec., and 300 sec.

For the non-congested conditions, there existed minor differences in the average value of the considered performance metrics (arrival time and delay) and the performance metric difference values for the tested scenarios. However, there was a clear trend of increasing the root mean square error (RMSE) as the aggregation interval increased. Varying the upstream origin of the arrival streams also tended to influence the RMSE values more than the average values. From these results it can be seen that under non-congested conditions, the average of performance metrics alone are likely not good indicators of the ability of a data driven simulation to reflect real world operations. Measures of variation such as RMSE should also be considered.

Unlike the non-congested conditions, the average values of the performance metrics in congested conditions were considerably different for the large real world simulation than the local simulation. The RMSE values also were significantly greater that those in the non-congested scenarios. There is also not a clear trend of the local simulation providing an improved reflection of the large simulation when given the smaller aggregation intervals. For the tested scenarios, the impact of congestion dominated the impact of a selected aggregation interval and upstream arrival pattern. The use of outflow constraints significantly improved the local model performance. These constraints helped capture the impact on the local simulation of congestion that occurs outside the local model boundaries. Where the boundaries of the congested region fall outside of the local simulation it becomes readily apparent that both the inflow and outflow parameters of the simulation must be dynamically driven to achieve a reasonable reflection of the real world conditions.

In the deployment of in-vehicle simulations these experiments highlight the need for realistic field measured inflow and outflow data streams, which are not currently widely available. However, as sensor technologies have advanced, the amount of available real-time field data is increasing dramatically. The quantity of available real-time data is expected to continue to climb at an ever-increasing rate. This tidal wave of real-time data creates the possibility of a wide variety of data driven transportation applications. This effort has begun to examine some of the innumerable potential uses of this data. Further details of these results are described in [7].

5 Data Dissemination

The proposed DDDAS system relies on mobile, wireless ad hoc networks to interconnect sensors, simulations, and servers. The experiments reported earlier were conducted over a local area network, and thus do not take into account the performance of the wireless network. However, the performance of the network infrastructure can play an important role in determining the overall effectiveness of the ad hoc distributed simulation approach. Thus, a separate effort has been devoted to examining the question of the impact of network performance on the results produced by the ad hoc distributed simulation.

Clearly, a single hop wireless link between the vehicles and servers severely limits the wireless coverage area, motivating the use of a multihop network. However, standard routing protocols designed for end-to-end communication in mobile ad hoc networks cause each node to maintain the state of their neighboring nodes, for efficient routing decisions. Such a solution does not scale to vehicular networks, where the nodes are highly mobile, and route maintenance becomes expensive. Other routing protocols such as those discussed in [8,9] attempt to address this problem using flooding or optimistic forwarding techniques.

Vehicle ad hoc network (VANET) data dissemination protocols are typically designed to address data sharing among the vehicles, and related applications such as multimedia, control data, etc. Distributed simulations, however, define a different data transfer model, and consequently, the solutions designed for data sharing applications may not perform well when transposed to the demands of simulations. Additionally, data transfers in VANETs is inherently unreliable, and drops and delays in message delivery can be highly dependent on factors such as traffic density and wireless activity, thus having a strong impact on the simulation itself.

To address this challenge, a data dissemination framework for addressing the routing demands of a distributed simulation in the VANET environment has been developed. Our framework uses a combination of geographic routing and controlled flooding to deliver messages, with no organization enforced among the vehicles. The design parameters of the framework are currently under study in order to assess how this impacts the overall accuracy and reliability of simulation results.

6 Future Research

Research in ad hoc distributed simulations and their application to the management of surface transportation systems is in its infancy, and many questions remain to be addressed. Can such a distributed simulation make sufficiently accurate predictions of future system states to be useful? Can they incorporate new information and revise projections more rapidly and/or effectively than conventional approaches, e.g., global, centralized simulations? Does networking the individual simulators result in better predictions than simply leaving the individual simulators as independent simulations that do not exchange future state information? How would an ad hoc distributed simulation be organized and operate, and what type of synchronization mechanism is required among them? Will the underlying networking infrastructure required to support this concept provide adequate performance? These are a few of the questions that remain to be answered.

A broader question concerns the applicability of the ad hoc simulation approach to other domains. While our research is focused on transportation systems, one can imagine simulations like those discussed here might be used in other on-line simulation applications such as management of communication networks or other critical system infrastructures.

Acknowledgement

The research described in this paper was supported under NSF Grant CNS-0540160, and is gratefully acknowledged.

References

1. Werner, J.: Details of the VII Initiative 'Work in Progress' Provided at Public Meeting(2005)
2. Bechler, M., Franz, W.J., Wolf, L.:Mobile Internet Access in FleetNet. In KiVS (2003)
3. Werner, J.: USDOT Outlines the New VII Initiative at the 2004 TRB Annual Meeting(2004)
4. Fujimoto, R. M., Hunter, M., Sirichoke, J. Palekar, M. Kim, H.-K., Suh, W. :Ad Hoc Distribtued Simulations. Principles of Advanced and Distributed Simulation(2007)
5. PTV, VISSIM User Manual 4.10. 2005, PTV Planung Transport Verkehr AG: Karlsruhe, Germany(2005).
6. PTV, VISSIM COM, User Manual for the VISSIM COM Interface, VISSIM 4.10-12. 2006, PTV Planung Transport Verkehr AG: Karlsruhe, Germany(2006).
7. Hunter, M. P., Fujimoto, R. M. , Suh, W., Kim, H. K. :An Investigation of Real-Time Dynamic Data Driven Transportation Simulation. In: Winter Simulation Conference(2006)
8. Rahman, W. Olesinski, Gburzynski, P.: Controlled Flooding in wireless ad-hoc networks. In: Proc. of IWWAN(2004)
9. Wu, H., Fujimoto, R., Guensler, R., Hunter, M.: MDDV: A Mobility-Centric Data Dissemination Algorithm for Vehicul ar Networks. In: Proceedings of the VANET Conference(2004)

Cyberinfrastructure for Contamination Source Characterization in Water Distribution Systems

Sarat Sreepathi[1], Kumar Mahinthakumar[1], Emily Zechman[1], Ranji Ranjithan[1], Downey Brill[1], Xiaosong Ma[1], and Gregor von Laszewski[2]

[1] North Carolina State University, Raleigh, NC, USA
{sarat_s, gmkumar, emzechma, ranji, brill, xma}@ncsu.edu
[2] University of Chicago, Chicago, IL, USA
gregor@mcs.anl.gov

Abstract. This paper describes a preliminary cyberinfrastructure for contaminant characterization in water distribution systems and its deployment on the grid. The cyberinfrastructure consists of the application, middleware and hardware resources. The application core consists of various optimization modules and a simulation module. This paper focuses on the development of specific middleware components of the cyberinfrastructure that enables efficient seamless execution of the application core in a grid environment. The components developed in this research include: (i) a coarse-grained parallel wrapper for the simulation module that includes additional features for persistent execution, (ii) a seamless job submission interface, and (iii) a graphical real time application monitoring tool. The performance of the cyberinfrastructure is evaluated on a local cluster and the TeraGrid.

1 Introduction

Urban water distribution systems (WDSs) are vulnerable to accidental and intentional contamination incidents that could result in adverse human health and safety impacts. Identifying the source and extent of contamination ("source characterization problem") is usually the first step in devising an appropriate response strategy in a contamination incident. This paper develops and tests a preliminary grid cyberinfrastructure for solving this problem as part of a larger multidisciplinary DDDAS [1] project that is developing algorithms and associated middleware tools leading to a full fledged cyberinfrastructure for threat management in WDSs [2].

The source characterization problem involves finding the contaminant source location (typically a node in a water distribution system) and its temporal mass loading history ("release history") from observed concentrations at several sensor locations in the network. The release history includes start time of the contaminant release in the WDS, duration of release, and the contaminant mass loading during this time. Assuming that we have a "forward simulation model" that can simulate concentrations at various sensor locations in the WDS for given source characteristics, the source characterization problem, which is an "inverse problem", can be formulated as an optimization problem with the goal of finding a source that can minimize the difference between the simulated and observed concentrations at the

Y. Shi et al. (Eds.): ICCS 2007, Part I, LNCS 4487, pp. 1058–1065, 2007.

sensor nodes. This approach is commonly termed "simulation-optimization" as the optimization algorithm drives a simulation model to solve the problem. Population based search methods such as evolutionary algorithms (EAs) are popular methods to solve this problem owing to their exploratory nature, ease of formulation, flexibility in handling different types of decision variables, and inherent parallelism. Despite their many advantages, EAs can be computationally intensive as they may require a large number of forward simulations to solve an inverse problem such as source characterization. As the larger DDDAS project relies heavily on EA based methods [3][4] for solving source characterization and sensor placement problems, an end-to-end cyberinfrastructure is needed to couple the optimization engine to the simulation engine, launch the simulations seamlessly on the grid, and track the solution progress in real-time. Thus the primary objective of this research is to develop a prototype of this grid cyberinfrastructure.

1.1 Related Work

Existing grid workflow systems such as CoG Kit [5] and Kepler [6] support pre-processing, post-processing, staging data/programs, and archival of results for a generic application on the grid. However, they do not provide custom solution to an application that requires frequent runtime interactions among its components (i.e., optimization and simulation components) at a finer granularity. They also require that the execution time of the core component (e.g., simulation) to be significantly large in order to amortize the overhead induced by the workflow system. In the WDS application, a single simulation instance can take anywhere from several milliseconds to several minutes depending on the network. If we need a system that can cater to any problem then it would not be feasible to use existing workflow systems (for smaller problems) without a significant performance penalty. To address this, a custom framework is developed in this research that can not only aggregate a large number of small computational tasks but also allows for persistent execution of these tasks during interactions with the optimization component in a batch environment. Existing workflow systems also do not provide support for real time monitoring of simulation-optimization runs from the perspective of a WDS application. Hence a real time visualization tool has been developed to inform the quantitative progress of the application to domain scientists.

2 Architecture

The high level architecture of the preliminary cyberinfrastructure developed in this paper is shown in Fig 1. The optimization toolkit (which is a collection of optimization methods) interacts with the simulation component (parallel EPANET) through the middleware. The middleware also communicates with the grid resources for resource allocation and program execution.

Typically the user invokes a script that launches the optimization toolkit and the visualization engine from a client workstation. The optimization toolkit then receives observed data from the sensors (or reads a file that has been synthetically generated) and then calls the middleware interface to invoke the simulation engine. The middleware interface then surveys the available resources and launches the simulation engine on the available resources through batch submission scripts or interactive commands. The middleware also transmits the sets of decision variables (e.g., variables representing source characteristics) generated by the optimization engine to the simulation engine via files. The simulation engine calculates the fitness values corresponding to the sets of decision variables sent by the optimization engine. These are then transmitted back to the optimization and visualization engines via files. The optimization engine processes this data and sends new sets of decision variables back to the simulation engine for the next iteration of the algorithm. The simulation engine maintains a persistent state until all the iterations are completed.

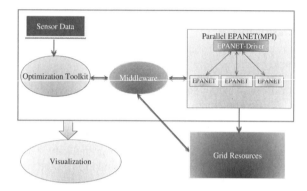

Fig. 1. Basic Architecture of the Cyberinfrastructure

The following subsections provide a brief description of the component developments involved in this cyberinfrastructure. Readers interested in additional details should refer to [7]. Subsequent discussions assume that the optimization engine uses EA based methods and the problem solved is source identification. However, the basic architecture is designed to handle any optimization method that relies on multiple simulation evaluations and any WDS simulation-optimization problem.

2.1 Simulation Model Enhancements

The simulation engine, EPANET [8] is an extended period hydraulic and water-quality simulation toolkit developed at EPA. It is originally developed for the Windows platform and provides a C language library with a well defined API [8]. The original EPANET was ported to Linux environments and customized to solve simulation-optimization optimization problems by building a "wrapper" around it. For testing purposes, limited amount of customization was built into the wrapper to solve source identification problems. The wrapper uses a file-based communication system to interoperate with existing EA based optimization tools developed in diverse development platforms such as Java [3] and Matlab [4]. It also aggregates the EPANET simulations into a single parallel execution for multiple sets of source characteristics to amortize the startup costs and minimize redundant computation.

Parallelization

The parallel version of the wrapper is developed using MPI and referred to as **'pepanet'**. The middleware scripts are designed to invoke multiple 'pepanet' instantiations depending on resource availability. Within each MPI program, the master process reads the base EPANET input file (WDS network information, boundary conditions etc.) and an input file generated by the optimization toolkit that contains the source characteristics (i.e., decision variables). The master process then populates data structures for storing simulation parameters as well as the multiple sets of contamination source parameters via MPI calls. The contamination source parameter sets are then divided among all the processes equally ensuring static load balancing. Each process then simulates its assigned group of contamination sources successively. At the completion of assigned simulations, the master process collects results from all the member processes and writes it to an output file to be processed by the optimization toolkit.

Persistency

The evolutionary computing based optimization methods that are currently in use within the project exhibit the following behavior. The optimization method submits some evaluations to be computed (generation), waits for the results and then generates the next set of evaluations that need to be computed. If the simulation program were to be separately invoked every generation, it needs to wait in a batch environment for acquiring the requisite computational resources.

But if the pepanet wrapper is made persistent, the program needs to wait in the queue just once when it is first started. Hence pepanet was enhanced to remain persistent across generations. In addition to amortizing the startup costs, the persistent wrapper significantly reduces the wait time in the job scheduling system. The persistent wrapper achieves this by eliminating some redundant computation across generations. One all evaluations are completed for a given generation (or evaluation set) the wrapper maintains a wait state by "polling periodically" for a sequence of input files whose filenames follow a pattern. The polling frequency can be tuned to improve performance (see section 3). This pattern for the input and output file names can be specified as command line arguments facilitating flexibility in placement of the files as well as standardization of the directory structure for easy archival.

2.2   Job Submission Middleware

Consider the scenario when the optimization toolkit is running on a client workstation and the simulation code is running at a remote supercomputing center. Communication between the optimization and simulation programs is difficult due to the security restrictions placed at current supercomputing centers. The compute nodes on the supercomputers cannot be directly reached from an external network. The job submission interfaces also differ from site to site.

In light of these obstacles, a middleware framework based on Python has been developed to facilitate the interaction between the optimization and simulation components and to appropriately allocate resources. The middleware component utilizes public key cryptography to authenticate to the remote supercomputing center from the client site. The middleware then transfers the file generated by the optimization component to the simulation component on the remote site using available file transfer protocols. It then waits for the computations to be completed at the remote sites and then fetches the output file back to the client site. This process is repeated until the termination signal is received from the client side (in the event of solution convergence or reaching iteration limit). The middleware script also polls for resource availability on the remote sites to allocate appropriate number of processors to minimize queue wait time by effectively utilizing the backfill window of the resource scheduler. When more than one supercomputer site is involved, the middleware divides the simulations proportionally among the sites based on processor availability and processor speed. A simple static allocation protocol is currently employed.

2.3   Real-Time Visualization

The current visualization toolkit is geared toward the source identification problem and was developed with the following goals in mind: (i) Visualize the water distribution system map and the locations where the optimization method is currently searching for contamination sources, (ii) Visualize how the search is progressing from one stage (generation) of the optimization algorithm to the next to facilitate understanding of the convergence pattern of the optimization method. The tool has been developed using Python, Tkinter and Gnuplot. Fig 2 shows a screenshot of the visualization tool after the optimization method found the contamination source for an example problem instance. It shows the map of the water distribution system marking the "true" source (as it is known in the hypothetical test case) and the estimated source found by the optimization method. It also provides a plot comparing the release history of the true source and the estimated source. A multi-threaded implementation enables the user to interact with the tool's graphical interface while the files are being processed in the backend.

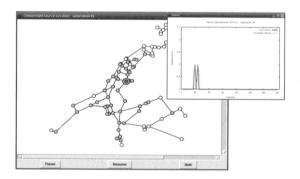

Fig. 2. Visualization Tool Interface showing the Water Distribution System Map and Concentration profile for the true (red) and estimated (green) sources

3   Performance Results

Performance results are obtained for solving a test source identification problem involving a single source. The sensor data is synthetically generated using a hypothetical source. An evolutionary algorithm (EA) is used for solving this problem [3]. The following platforms are used for evaluating the performance of the cyberinfrastructure: (i) Neptune, a 11 node Opteron Cluster at NCSU consisting of 22 2.2 GHz AMD Opteron(248) processors and a GigE Interconnect, and (ii) Teragrid Linux Cluster at NCSA consisting of 887 1.3-1.5 GHz Intel Itanium 2 nodes and a Myrinet interconnect.

Teragrid results are confined to simulations deployed on a single cluster but with the optimization component residing on a remote client site. Additional results including preliminary multi-cluster Teragrid results are available in [7]. The cyberinfrastructure has also been demonstrated on SURAgrid resources [9]. For timing purposes, the number of generations in the EA is fixed at 100 generations even though convergence is usually achieved for the test problem well before the 100th generation. The population size was varied from 600 to 6000 but the results in this paper are restricted to the larger population size.

Timers were placed within the main launch script, middleware scripts, optimization toolkit and the simulation engine to quantify the total time, optimization time, simulation time, and overhead due to file movements. Additional timers were placed within the simulation engine to break down the time spent in waiting or "wait time" (includes the optimization time and all overheads) and time spent in calculations. Preliminary tests revealed that the waiting time within the simulation code was exceedingly high when the optimization toolkit and root process of the wrapper (simulation component) are on different nodes of the cluster. When both are placed on the same compute node, wait time reduced by a factor of more than 15 to acceptable values. Additional investigation indicated that these were due to file system issues. Further optimization of the polling frequency within the simulation engine improved wait time by an additional factor of 2. Once these optimizations were performed, the wait time predominantly consisted of optimization calculations, since the overhead is relatively negligible. Fig 3 illustrates the parallel performance of the application after these optimizations on the Neptune cluster up to 16 processors. As expected, the computation time within the simulation engine (parallel component) scales nearly linearly while the overhead and the optimization time (serial component) remain more or less constant.

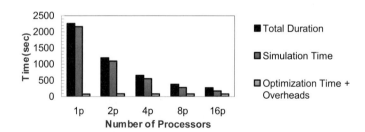

Fig. 3. Performance of the application up to 16 processors on Neptune cluster

For timing studies on the Teragrid, the optimization engine was launched on Neptune and the simulations on the Teragrid cluster. Again, preliminary performance tests indicated that the wait times were significantly impacted by file system issues. The performance improved by a factor of three when the application is moved from a NFS file system to a faster GPFS file system. Further improvement in wait time (about 15%) was achieved by consolidating the network communications within the middleware. Fig 4 shows the speedup behavior on the Teragrid cluster for up to 16 processors. While the computation time speeds up nearly linearly the wait time is much higher when compared to the timings of Neptune (Fig 3). Additional breakdown on the wait time indicated that 14% was spent in optimization and the remaining 86% in overheads including file transfer costs. The increased file transfer time is not unexpected as the transfers occur over a shared network between the geographically distributed optimization and simulation components. Comparison of simulation times with Neptune indicates that the Teragrid processors are also about three times slower.

It is noted that the WDS problem solved is relatively small (about 20 seconds (Neptune) or 1 minute (TeraGrid) per 6000 simulations using 1 processor) thus making the overheads appear relatively high. As the overhead will remain approximately constant with increased problem sizes we expect the results to improve considerably for larger problems. Furthermore, several enhancements are planned to minimize overheads (see next section).

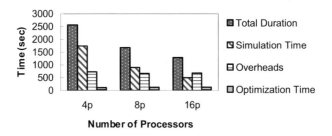

Fig. 4. Performance of the application up to 16 processors on NCSA Teragrid Linux Cluster

4 Conclusions and Future Work

An end-to-end solution for solving WDS contamination source characterization problems in grid environments has been developed. This involved coarse-grained parallelization of simulation module, middleware for seamless grid deployment and a visualization tool for real-time monitoring of application's progress. Various performance optimizations such as improving processor placements, minimizing file system overheads, eliminating redundant computations, amortizing queue wait times, and multi-threading visualization were carried out to improve turnaround time. Even with these optimizations, the file movement overheads were significant when the client and server sites were geographically distributed as in the case of Teragrid. Several future improvements are planned including optimization algorithm changes that can allow for overlapping file movements and optimization calculations with simulation calculations, localizing optimization calculations on remote sites by partitioning techniques, and minimizing file transfer overhead using grid communication libraries.

Acknowledgments. This work is supported by National Science Foundation (NSF) under Grant Nos. CMS-0540316, ANI-0540076, CMS-0540289, CMS-0540177, and NMI-0330545. Any opinions, findings and conclusions or recommendations expressed in this material are those of the authors and do not necessarily reflect the views of the NSF.

References

1. Darema, F. Introduction to the ICCS 2006 Workshop on Dynamic data driven applications systems. *Lecture Notes in Computer Science 3993*, pages 375-383, 2006.
2. Mahinthakumar, G., G. von Laszewski, S. Ranjithan, E. D. Brill, J. Uber, K. W. Harrison, S. Sreepathi, and E. M. Zechman, An Adaptive Cyberinfrastructure for Threat Management in Urban Water Distribution Systems, *Lecture Notes in Computer Science, Springer-Verlag*, pp. 401-408, 2006.(International Conference on Computational Science (3) 2006: 401-408)
3. Zechman, E. M. and S. Ranjithan, "Evolutionary Computation-based Methods for Characterizing Contaminant Sources in a Water Distribution System," *Journal of Water Resources Planning and Management,* (submitted)
4. Liu, L., E. M. Zechman, E. D. Brill, Jr., G. Mahinthakumar, S. Ranjithan, and J. Uber "Adaptive Contamination Source Identification in Water Distribution Systems Using an Evolutionary Algorithm-based Dynamic Optimization Procedure," *Water Distribution Systems Analysis Symposium*, Cincinnati, OH, August 2006
5. CoG Kit Project Website, http://www.cogkit.org
6. Kepler Project Website, http://www.kepler-project.org
7. Sreepathi, S., Cyberinfrastructure for Contamination Source Characterization in Water Distribution Systems, Master's Thesis, North Carolina State University, December 2006.
8. Rossman, L.A.. The EPANET programmer's toolkit. In Proceedings of Water Resources Planning and Management Division Annual Specialty Conference, ASCE, Tempe, AZ, 1999.
9. Sreepathi, S., Simulation-Optimization for Threat Management in Urban Water Systems, Demo, Fall 2006 Internet2 Meeting, December 2006.

Integrated Decision Algorithms for Auto-steered Electric Transmission System Asset Management

James McCalley, Vasant Honavar, Sarah Ryan, William Meeker, Daji Qiao,
Ron Roberts, Yuan Li, Jyotishman Pathak, Mujing Ye, and Yili Hong

Iowa State University, Ames, IA 50011, US
{jdm, honavar, smryan, wqmeeker, daji, rroberts, tua, jpathak,
mye, hong}@iastate.edu

Abstract. Electric power transmission systems are comprised of a large number of physical assets, including transmission lines, power transformers, and circuit breakers, that are capital-intensive, highly distributed, and may fail. Managing these assets under resource constraints requires equipment health monitoring integrated with system level decision-making to optimize a number of various operational, maintenance, and investment-related objectives. Industry processes to these ends have evolved ad-hoc over the years, and no systematic structures exist to coordinate the various decision problems. In this paper, we describe our progress in building a prototype structure for this purpose together with a software-hardware environment to deploy and test it. We particularly focus on the decision algorithms and the Benders approach we have taken to solve them in an integrated fashion.

Keywords: asset management, Benders decomposition, condition monitoring, decision algorithms, electric transmission, optimization, service-oriented architecture, software-hardware.

1 Introduction

There are three interconnected electric power transmission grids in North America: the eastern grid, the western grid, and Texas. Within each grid, power supplied must equal power consumed at any instant of time; also, power flows in any one circuit depend on the topology and conditions throughout the network. This interdependency means that should any one element fail, repercussions are seen throughout the interconnection, affecting system economic and engineering performance. Overall management requires *decision* in regards to how to operate, how to maintain, and how to reinforce and expand the system, with objectives being risk minimization and social welfare maximization. The three decision problems share a common dependence on equipment health or propensity to fail; in addition, their solutions heavily influence future equipment health. As a result, they are coupled, and optimality requires solution as a single problem. However, because network size (number of nodes and branches) together with number of failure states is so large, such a problem, if solved using traditional optimization methods, is intractable. In addition, the three decision problems differ significantly in decision-horizon, with operational decisions

Y. Shi et al. (Eds.): ICCS 2007, Part I, LNCS 4487, pp. 1066–1073, 2007.

implemented within minutes to a week, maintenance decisions within weeks to a couple of years, and investment decisions within 2-10 years. Therefore, excepting the common dependence and effect on equipment health, the coupling is sequential, with solution to latter-stage problem depending on solution to former-stage problems. Because of this, the industry has solved them separately, with the coupling represented in a very approximate fashion via human communication mechanisms. We conjecture that resulting solutions are not only suboptimal, but they are not even very good solutions, a conjecture which motivates the work reported here.

A previous paper [1] described an initial design for a hardware-software prototype capable of auto-steering information-decision cycles inherent to managing operations, maintenance, and planning of the high-voltage electric power transmission systems. Section 2 of this paper describes a refined version of this overall design together with progress in implementing it. Section 3 summarizes the various optimization problems, providing problem statements when solved individually. Section 4 provides a new formulation, based on Benders decomposition, for a subgroup of problems, an approach that we eventually intend to apply to the entire set. Section 5 concludes.

2   Overall Design and Recent Progress

Figure 1 illustrates design of our prototype system for auto-steering information-decision processes for electric transmission system asset management. This section overviews intended implementation and recent progress of the 5 different layers.

Layer 1, The power system: The prototype centers on a continuously running model of the Iowa power system using network data provided by a local utility company using a commercial-grade operator training simulator (OTS). The OTS is provided by ArevaT&D (www.areva-td.com) and comprises the same energy management software system used by many major transmission control centers all over the world. The dataset on which this software system runs is the same dataset used by the utility company at their control center. This presents information security requirements that must be satisfied in our lab, since the data represents a critical national infrastructure. The work to implement this is intensive and is being supported under a cost-sharing arrangement between ArevaT&D, ISU, and the utility company.

Layer 2, Condition sensors: Transformers are the most expensive single transmission asset, with typical costs between $1-5M. The utility company has over 600 of them some of which have well exceeded their ~40 year design life. All units undergo yearly dissolved gas-in-oil analysis (DGA) which, similar to a human blood test, provides information useful for problem diagnosis and prediction. We have obtained this data for all units and are using it to perform life prediction of the units. In addition, we are installing a real-time DGA monitor (www.kelman.co.uk) in one of the largest and oldest units and have been working on methods of transforming this data into health indicators that can be used in our decision algorithms.

Layer 3, Data communication and integration: The transformer monitor is equipped with a cellular modem provided by Cannon (www.cannontech.com) that communicates the real-time data to our lab. A federated data integration system has been designed to provide efficient, dependable, and secure mechanisms for interfacing Layer 4 data transformation algorithms with the data resources [2].

Layer 4, Data processing and transformation: The data available for equipment health prediction includes transformer monitoring and test data and weather/vegetation data which is useful for estimating probabilistic failure indices of transformers and overhead transmission lines [3].

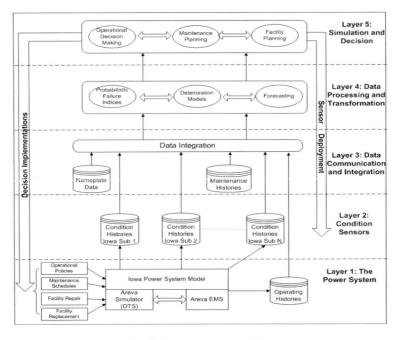

Fig. 1. Prototype system design

Layer 5, Simulation and decision: This layer utilizes probabilistic failure indices from layer 4 together with short and long-term system forecasts to drive integrated stochastic simulation and decision models. Resulting operational policies, maintenance schedules, and facility expansion plans are implemented on the power system (as represented by the ArevaT&D simulator). The decision models are also used to discover the value of additional information. This valuation will be used to drive the deployment of new sensors and redeployment of existing sensors, impacting Layer 2. The integration of decision models is further described in Section 3.

A service-oriented architecture (SOA) is used for this software system. This framework, PSAM-s, for Power System Asset Management employs a Web services-based SOA . The core of the framework is the PSAM-s engine comprised of multiple services responsible for enabling interaction between users and other services that offer specific functionality. These services are categorized into internal services (part of the PSAM-s engine) and external services. The internal services include submission, execution, brokering, monitoring, and storage. The external services include data provision and information processing. These services and their overall architecture are illustrated in Fig. 2; additional description is provided in [4].

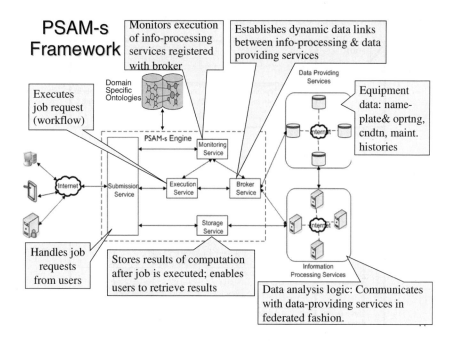

Fig. 2. A Service-Oriented Architecture

3   Layer 5: Simulation, Decision and Information Valuation

There are 6 basic risk-economy decision problems associated with power systems operation, maintenance, and planning, as illustrated in Table 1. The table illustrates the sequential coupling between the various problems in terms of information that is passed from one to the other. Information required to solve a problem is in its diagonal block and in the blocks left of that diagonal. Output information from solving a problem is below its diagonal block and represents input information for the lower-level problems. We briefly summarize each of these problems in what follows.

Operations: There are three operational sub-problems [5, 6].

- *Unit commitment (UC)*: Given an hourly total load forecast over the next day or week, identify the hourly sequence of generation commitments (which generators are interconnected to the grid?) to maximize welfare (minimize costs) subject to the requirement that load must be satisfied, and also subject to physical limits on each generator associated with supply capability, start-up, and shut-down times.
- *Optimal power flow (OPF)*: Given the unit commitment solution together with load requirements at each bus, and the network topology, determine the allocation of load to each generator and each generator's voltage set point to maximize social welfare, subject to Kirchoff's laws governing electricity behavior (encapsulated in a set of nonlinear algebraic "power flow" equations) together with constraints on branch flows, node voltages, and generator supply capability.

Table 1. Summary of Power System Risk-Economy Decision Problems

From \ To		Operations T=1-168 hrs			Maintenance T=1-5 yrs		Planning T=5-10 yrs
		Unit commit (UC)	Optimal power flow (OPF)	Security Assessmnt (SA)	Shortterm maint	Longterm maint	Investment planning
Operations	Unit commit (UC)	Total load					
Operations	Optimal power flow (OPF)	Units committed	Bus loads, topology				
Operations	Security Assessment (SA)	Units committed	Operating condition	Weather, failure data inst. cndtn data			
Maintenance	Shortterm maint	Units committed	Operating condition	Operating (risk) history	Maint effcts, failure data, cdt history, resources		
Maintenance	Longterm maint	Units committed	Operating condition	Operating (risk) history	ST maint schedule, ST eqp deter rate	Cost of capital, failure data cdt history	
Planning	Investment planning	Units committed	Operating condition	Operating (risk) history	ST maint schedule, ST eqp. deter rate	LT maint schedule, LT eqp. deter rate	Cost of capital, failure data, cdt history

- *Security assessment (SA)*: Given the operating condition (which is economically optimal), find the best tradeoff between minimizing supply costs and minimizing risk associated with potential failures in the network. Presently, the industry solves this problem by imposing hard constraints on risk (or conditions associated with risk), thus obtaining a single objective optimization problem, but it is amendable to multiobjective formulation.

Maintenance: There are two maintenance-related sub-problems [7, 8].

- *Short-term maintenance*: Given a forecasted future operating sequence over an interval corresponding to a budget period (e.g., 1 year), together with a set of candidate maintenance tasks, select and schedule those maintenance tasks which most effectively reduce cumulative future risk, subject to resource (budget and labor) and scheduling constraints.
- *Long-term maintenance*: For each class of components, given a future loading forecast, determine an inspection, maintenance, and replacement schedule to maximize its operational reliability and its residual life at minimum cost. This multiobjective problem is typically addressed with the single objective to maximize residual life subject to constraints on operational reliability and cost.

Planning [9, 10]: Given a set of forecasted future load growths and corresponding operating scenarios, determine a network expansion plan that minimizes investment costs, energy production costs, and risk associated with potential failures in the network, subject to Kirchoff's laws together with constraints on branch flows, node voltages, and generator physical supply capabilities. This problem is often solved by minimizing investment and production costs while imposing constraints on risk.

4 Benders Decomposition and Illustration

Benders decomposition is an appropriate method for problems that are sequentially nested such that solution to latter-stage problems depends on solution to former-stage problems. Mixed integer problems can be posed in this way as can stochastic programming problems. The operational problem described in Section 3, consisting of the sequence of UC, OPF, and SA, is both. To illustrate concepts, consider:

$$Min : z = c(x) + d(y) \tag{1}$$

$$Problem\ P \qquad s.t. \quad A(x) \qquad \geq b \tag{1a}$$

$$E(x) + F(y) \geq h \tag{1b}$$

This problem can be represented as a two-stage decision problem [11]:

Stage 1 (Master Problem): Decide on a feasible x^* only considering (1a);

$$Min : z = c(x) + \alpha'(x) \tag{2}$$

$$s.t. \quad A(x) \qquad \geq b \tag{2a}$$

where $\alpha'(x)$ is a guess of stage 2 regarding stage 1 decision variable x, to be updated by stage 2.

Stage 2 (Subproblem): Decide on a feasible y^* considering (1b) given x^* from stage 1.

$$\alpha(x^*) = Min \quad d(y) \tag{3}$$

$$s.t. \quad F(y) \geq h - E(x^*) \tag{3a}$$

The partition theorem for mixed-integer programming problems [12] provides an optimality rule on which Benders decomposition is based. If we obtain optimal solution (z^*, x^*) in the first stage and then obtain optimal solution y^* in the second stage, if $c(x^*) + d(y^*) = z^*$, then (y^*, x^*) is the optimal solution for *Problem P*. The interaction between stages 1 and 2 is shown in Fig. 3.

The procedure of Benders decomposition is a learning process (try-fail-try-inaccurate-try-...-solved). In the left part of Fig. 3, when the stage 1 problem is solved, the optimal value is then sent to stage 2. Stage 2 problem has two steps: 1)

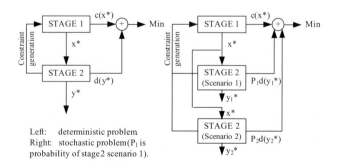

Left: deterministic problem
Right: stochastic problem (P_1 is probability of stage 2 scenario 1).

Fig. 3. Benders decomposition (modified from [11])

Check if the optimal solution from stage 1 is feasible. If it is not feasible, the stage 2 problem sends feasibility cuts back to stage 1 to be repeated under the additional constraints found in stage 2 to be in violation. 2) Check if the optimal guess of stage 2 from stage 1 is accurate enough. If it is not, a new estimation of $\alpha'(x)$ is sent to stage 1. If the optimal rule is met, the problem is solved. This process is easily expanded to the stochastic programming case, as illustrated in the right part of Fig. 3 where the optimal value from stage 1 is sent to stage 2, which has multiple scenarios. The process is exactly the same as the deterministic case, except that all constraint cuts and the optimal value from stage 2 are weighted by the probability of the scenario.

A 6-bus test system, Fig. 4, is used to illustrate. Generators are located at buses 1, 2, 6; loads at buses 3, 4, 5. Possible contingencies considered include any failure of a single circuit. Detailed data for the system are provided in [5]. Figure 5 plots total cost of supply against time for a 24 hour period for two different scenarios: "average" uses contingency probabilities under normal weather, and "10*average" uses contingency probabilities under stormy weather. We observe in Fig. 5 the increased cost required to reduce the additional risk due to the stormy weather. Although the UC solution is the same in the two cases illustrated in Fig 5, it changes if the contingency probabilities are zero, an extreme situation which in fact corresponds to the way UC is solved in practice where UC and SA are solved separately. This is evidence that better solutions do in fact result when the different problems are solved together.

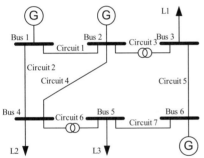

Fig. 4. 6-bus test system   **Fig. 5.** Effect of contingency

5 Conclusions

Aging, capital intensive equipment comprise electric power grids; their availability largely determines the economic efficiency of today's electricity markets on which a nation's economic health depends; their failure results in increased energy cost, at best, and widespread blackouts, at worst. The balance between economy and reliability, or risk, is maintained via solution to a series of optimization problems in operations, maintenance, and planning, problems that traditionally are solved separately. Yet, these problems are coupled, and so solving them together necessarily improves on the composite solution. In this paper, we described a hardware-software system designed to address this issue, and we reported on our progress in developing this system, including acquisition of a real-time transformer monitor and of a commercial-grade power system simulator together with corresponding data modeling the Iowa power system. We also designed a service-oriented architecture to guide development of our software system. Finally, we implemented an optimization framework based on Benders decomposition to efficiently solve our sequential series of decision problems. This framework is promising; we expect it to be an integral part of our power system asset management prototype as we continue to move forward in its development.

Acknowledgments

The work described in this paper is funded by the National Science Foundation under grant NSF CNS0540293.

References

1. J. McCalley, V. Honavar, S. Ryan, W. Meeker, R. Roberts, D. Qiao and Y. Li, "Auto-steered Information-Decision Processes for Electric System Asset Management," in Computational Science - ICCS 2006, 6th International Conference, Reading, UK, May 28-31, 2006, Proceedings, Part III, Series: Lecture Notes in Computer Science , Vol. 3993, V. Alexandrov, G. van Albada, P. Sloot, J. Dongarra, (Eds.), 2006.
2. J. Pathak, Y. Jiang, V. Honavar, J. McCalley, "Condition Data Aggregation with Application to Failure Rate Calculation of Power Transformers," Proc. of the Hawaii International Conference on System Sciences, Jan 4-7, 2006, Poipu Kauai, Hawaii.
3. F. Xiao, J. McCalley, Y. Ou, J. Adams, S. Myers, "Contingency Probability Estimation Using Weather and Geographical Data for On-Line Security Assessment," Proc. of the 9th Int. Conf. on Probabilistic Methods Applied to Pwr Sys, June 11-15, 2006, Stockholm, Sweden.
4. J. Pathak, Y. Li, V. Honavar, J. McCalley, "A Service-Oriented Architecture for Electric Power Transmission System Asset Management," 2nd International Workshop on Engineering Service-Oriented Applications: Design and Composition, Dec. 4, 2007, Chicago, Ill.
5. Y. Li, J. McCalley, S. Ryan, "Risk-Based Unit Commitment," to appear in Proc. of the 2007 IEEE PES General Meeting, June, 2007, Tampa Fl.
6. F. Xiao, J. McCalley, "Risk-Based Multi-Objective Optimization for Transmission Loading Relief Strategies," to appear, Proc. of the 2007 IEEE PES Gen Meeting, June, 2007, Tampa Fl.
7. J. McCalley, V. Honavar, M. Kezunovic, C. Singh, Y. Jiang, J. Pathak, S. Natti, J. Panida, "Automated Integration of Condition Monitoring with an Optimized Maintenance Scheduler for Circuit Breakers and Power Transformers," Final report to the Power Systems Engineering Research Center (PSerc), Dec., 2005.
8. Y. Jiang, J. McCalley, T. Van Voorhis, "Risk-based Maintenance Optimization for Transmission Equipment," IEEE Trans on Pwr Sys, Vol 21, I 3, Aug. 2006, pp. 1191 – 1200.
9. J. McCalley, R. Kumar, O. Volij, V. Ajjarapu, H. Liu, L. Jin, W. Zhang, Models for Transmission Expansion Planning Based on Reconfigurable Capacitor Switching," Chapter 3 in "Electric Power Networks, Efficiency, and Security," John Wiley and Sons, 2006.
10. M. Ye, S. Ryan, and J. McCalley, "Transmission Expansion Planning with Transformer Replacement," Proc. of 2007 Industrial Engr. Research Conf, Nashville, Tn, May 20-22, 2007.
11. S. Granville et al., "Mathematical decomposition techniques for power system expansion planning," Report 2473-6 of the Electric Power Research Institute, February 1988.
12. J. Benders, "Partitioning procedures for solving mixed-variables programming problems," Numerische Mathematik 4: 238–252, 1962.

DDDAS for Autonomic Interconnected Systems: The National Energy Infrastructure

C. Hoffmann[1], E. Swain[2], Y. Xu[2], T. Downar[2], L. Tsoukalas[2],
P. Top[3], M. Senel[3], M. Bell[3], E. Coyle[3], B. Loop[5], D. Aliprantis[3],
O. Wasynczuk[3], and S. Meliopoulos[4]

[1] Computer Sciences, Purdue University, West Lafayette, Indiana 47907, USA
[2] Nuclear Engineering, Purdue University, West Lafayette, Indiana 47907, USA
[3] Electrical and computer engineering, Purdue University,
West Lafayette, Indiana 47907, USA
[4] Electrical and computer engineering, Georgia Institute of Technology,
Atlanta, Georgia 30332, USA
[5] PC Krause and Associates, West Lafayette, Indiana 47906, USA

Abstract. The most critical element of the nation's energy infrastructure is our electricity generation, transmission, and distribution system known as the "power grid." Computer simulation is an effective tool that can be used to identify vulnerabilities and predict the system response for various contingencies. However, because the power grid is a very large-scale nonlinear system, such studies are presently conducted "open loop" using predicted loading conditions months in advance and, due to uncertainties in model parameters, the results do not provide grid operators with accurate "real time" information that can be used to avoid major blackouts such as were experienced on the East Coast in August of 2003. However, the paradigm of Dynamic Data-Driven Applications Systems (DDDAS) provides a fundamentally new framework to rethink the problem of power grid simulation. In DDDAS, simulations and field data become a symbiotic feedback control system and this is refreshingly different from conventional power grid simulation approaches in which data inputs are generally fixed when the simulation is launched. The objective of the research described herein was to utilize the paradigm of DDDAS to develop a marriage between sensing, visualization, and modelling for large-scale simulation with an immediate impact on the power grid. Our research has focused on methodological innovations and advances in sensor systems, mathematical algorithms, and power grid simulation, security, and visualization approaches necessary to achieve a meaningful large-scale real-time simulation that can have a significant impact on reducing the likelihood of major blackouts.

1 Introduction

The August 2003 blackout cascaded throughout a multi-state area of the U.S. and parts of Canada within a few minutes. Hindsight reveals the blackout was the result of the system operating too close to the point where synchronous

Y. Shi et al. (Eds.): ICCS 2007, Part I, LNCS 4487, pp. 1074–1082, 2007.
© Springer-Verlag Berlin Heidelberg 2007

operation of the generators could be lost (transient stability limit). In this situation, the inability of operators and system controllers to quickly respond to unexpected anomalies in the system resulting in the blackout. An August, 2004 IEEE Spectrum article entitled "The Unruly Power Grid" [1] emphasizes the need for accurate system modelling and fast computation in order to provide operators and system controllers with timely on-line information regarding the dynamic security of the entire electric grid, thereby reducing the likelihood of future cascading system failures.

One of the main difficulties in controlling complex, distributed systems such as the electric power grid is its enormous scale and complexity. Presently, transient stability studies are conducted open loop based upon predicted load profiles months in advance. As such, the results are only approximations of what might happen based upon the assumptions, load estimates, and parameters assumed in the computer studies. Distributed dynamic estimation based on dynamic GPS-synchronized data has the potential of alleviating the shortcomings of the present state of the art. The simulation of the evolution of the system state forward in time using dynamic and synchronized measurements of key system parameters appears to be within reach. Fast, accurate, data-driven computation can revolutionize operating procedures and open the door to new and challenging research in the operation and control of the electric power grid.

There are several key areas that must be addressed to achieve the overall research objective: faster-than-real-time simulation, distributed dynamic sensing, the development of behavioral models for situations where physics-based models are impracticable, and visualization. Faster-than-real-time simulation alone is insufficient – the simulation must be accurate and track the true operating conditions (topology, parameters, and state) of the actual system. An integrated sensing network is essential to ensure that the initial state of the simulated system matches the present state of the actual system, and to continually monitor and, if necessary, adjust the parameters of the system model. Moreover, it has been recognized that the dynamic characteristic of loads can influence the voltage stability of the system. Due to their sheer number and the uncertainties involved, it does not appear possible to model distributed loads and generators, such as wind turbines and photovoltaic sources, using deterministic physics-based models. Neural networks may be used in cases where physics-based models do not exist or are too complex to participate in a faster-than-real-time simulation. Finally, even if the goal of an accurate faster-than-real time predictive simulation is achieved, the number of potential contingencies to be evaluated at any instant of time will be intractable. Therefore, methods of identifying vulnerabilities using visualization and the development of suitable stability metrics including heuristic approaches are sought.

The research performed at Purdue University, the Georgia Institute of Technology, and PC Krause and Associates has addressed each of these critical areas. The purpose of this paper is to describe the research conducted heretofore and key results obtained to date.

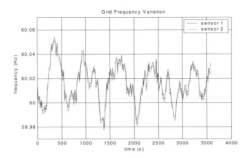

Fig. 1. Grid frequency variations at two locations in the power grid separated by 650 mi. NTP and post-processing techniques were used to synchronize time stamps of the data to within $10\,\mu s$.

2   Distributed Dynamic Sensing

Presently, sufficient measurements are taken so that, with the aid of a state estimator, the voltage magnitudes, phase angles, and real and reactive power flows can be monitored and updated throughout the grid on a 30-90 second cycle. A current project, supported by the Department of Energy (DOE), has as its goal to instrument every major bus with a GPS-synchronized phasor measurement unit. While this would provide system operators with more up-to-date top-level information, a concurrent bottom-up approach can provide even more information, reduce uncertainty, and provide operators and researchers alike with more accurate models, an assessment of the true state of the system, and diagnostic information in the event of a major event such as a blackout. The large scale of the power grid means that many sensors in various locations will be required for monitoring purposes. The number of sensors required for effective monitoring means they must be inexpensive, while maintaining the required accuracy in measurements. A custom designed sensor board including data acquisition hardware and a time stamping system has been developed. This sensor uses the Network Time Protocol (NTP) and WWVB, a US-based time signal that, because of its long wavelength, propagates far into buildings (unlike GPS). This new board will provide the required accuracy in measurement and time stamping at a very reasonable cost. The goal is synchronization of sensors deployed throughout the grid to within $10\,\mu s$ or better because a one degree error in a phase estimate corresponds to a 46-μs delay.

Since November 2004, six proof-of-concept sensors were deployed in various locations in the U.S. These sensors have been in operation and have been collecting data since that time. The sensors communicate through the Internet, and regularly send the collected data to a central server. A significant amount of effort has gone into developing an automated system wherein the sensor data is automatically processed and archived in a manner that can be easily retrieved via a web interface. In the future, this will allow additional sensors to be added with only minimal effort. A typical example is shown in Fig. 1, which depicts

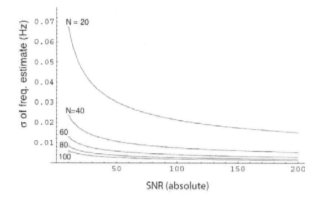

Fig. 2. Cramer–Rao lower bounds on the variance of the error for estimates of the signal frequency based on 1200 samples per second (20 per period) of the 60 Hz power signal with additive white Gaussian noise

frequency estimates from two proof-of-concept sensors (one in West Lafayette, Indiana, and the other in Maurice, Iowa) over the period of an hour. The data shows the synchronization of the relative frequency changes in the electric grid even in geographically distant sensor locations. The technique used to estimate the frequency of the power signal is equivalent to the maximum likelihood estimate for large SNR's, which is typically the case with our measurements. We have derived the Cramer–Rao lower bounds on the variance of the error of this (unbiased) estimator. The results are shown in Fig. 2.

3   Distributed Heterogeneous Simulation

Recent advancements in Distributed Heterogeneous Simulation (DHS) have enabled the simulation of the transient performance of power systems at a scope, level-of-detail, and speed that has been heretofore unachievable. DHS is a technology of interconnecting personal computers, workstations, and/or mainframes to simulate large-scale dynamic systems. This technology provides an increase in simulation speed that approaches the cube of the number of interconnected computers rather than the traditional linear increase. Moreover, with DHS, different simulation languages may be interconnected locally or remotely, translation to a common language is unnecessary, legacy code can be used directly, and intellectual property rights and proprietary designs are completely protected. DHS has been successfully applied to the analysis, design, and optimization of energy systems of ships and submarines [2], aircraft [3], land-based vehicles [4], the terrestrial power grid [5], and was recently featured in an article entitled "Distributed Simulation" in the November 2004 issue of Aerospace Engineering [6]. DHS appears to be an enabling technology for providing faster-than-real-time detailed simulations for automated on-line control to enhance survivability of large-scale systems during emergencies and to provide guidance to ensure safe and stable system reconfiguration. A DHS of the Indiana/Ohio terrestrial electric grid has been developed by PC Krause and Associates with cooperation from the Midwest Independent System Operator (MISO), which may be used as a real-life test bed for DDDAS research.

4 Visualization

The computational and sensory tasks generate data that must be scrutinized to assess the quality of simulation, of the models used to predict state, and to assess implications for the power grid going forward. We envision accurate visual representations of extrapolations of state evolution, damage assessment, and control action consequences to guide designers and operators to make timely and well-informed decisions. The overriding concern is to ensure a smooth functioning of the power grid, thus it is critical to provide operators with situation awareness that lets them monitor the state of the grid rapidly and develop contingency plans if necessary. An easily understood visual presentation of state variables, line outages, transients, phase angles and more, is a key element for designers and operators to be in a position to perform optimally.

The Final Report on the August 14th Blackout in the United States and Canada by the U.S.–Canada Power System Outage Task Force [7] states clearly that the absence of accurate real-time visualization of the state of the electric grid at FirstEnergy (FE) was an important shortcoming that led to the blackout of August 2003. FE was simply in the dark and could not recognize the deterioration leading to the wide-spread power failure. Our mission, therefore, is to assist grid system operators to monitor, predict, and take corrective action.

The most common representation of the transmission grid is the one-line diagram, with power flow indicated by a digital label for each arc of the graph. Recently, the diagrams have been animated [8], allowing insight into power flow. Flow has also been visualized using dynamically sized pie charts that grow superlinearly as a line approaches overload. The one-line diagram is impractical for visualizing complex power systems: buses overlap with each other and with lines to which they are not connected, and lines intersect. Although considerable research in information visualization has been devoted to graph drawing [9], contouring is a simpler and more effective method for power grids. Contouring is a classic visualization technique for continuous data, which can be applied to discrete data by interpolation. Techniques have been developed for contouring bus data (voltage, frequency, phase, local marginal prices), and of line data (load, power transfer distribution factors) [8, 10, 11]. Phase-angle data as measured directly via GPS-based sensors or semi-measured/calculated from existing sensors/state estimators may be used as an indicator of transient stability margin or via "energy" functions.

We have developed tools for visualizing phase angle data that allow the operators to detect early and correct quickly conditions that can lead to outages [12]. To provide real-time assessment of power grid vulnerability, large amounts of measured data must be analysed rapidly and presented timely for corrective response. Prior work in visualization of power grid vulnerability handles only

Fig. 3. Visualization of Midwestern power grid

the case of static security assessment; a power flow is run on all contingencies in a set and the numerical data is visualized with decorated one-line diagrams [13]. Strategically, it is an advantage to use preventative techniques that lead to fast apprehension of the visually presented data and thus to situation awareness. Displays created based on the theory of preattentive processing (from the scientific field of human cognitive psychology) can provide such a visualization environment [14]. These displays would use a combination of colors and glyphs to provide underlying information, as depicted in Fig. 3.

5   Neural Distributed Agents

Neural networks (NNs) provide a powerful means of simplifying simulations of subsystems that are difficult or impossible to simulate using deterministic physics-based models, such as the composite loads in the power grid. The ability of a neural distributed agent [15] to reduce the computational burden and to help achieve real-time performance has been demonstrated. Instead of directly computing the state of the coupled systems, the states of each subsystem are pre-computed and functionalized to the state variables using NN methods. The state of the coupled system for any given condition is then constructed by interpolating the pre-calculated data. The approach used involved decomposition of the system into several subsystems and utilizing a NN to simulate the most computationally intensive subsystems. The NN produces pre-computed outcomes based on inputs to respective subsystems. The trained NN then interpolates to obtain a solution for each iteration step. Periodic updates of the NN subsystem solution will ensure that the overall solution remains consistent with the physics-based DHS. NN-based models further facilitate real-time simulation by reducing computational complexity of subsystem models. Similar concepts have been used for real-time simulation of the dispersion model of chemical/biological agents in a 3-D environment [16].

This concept was demonstrated using a small closed electrical system consisting of a generator connected to transmission lines. The generator subsystem was created in Simulink [17] and provided the template for training the NN. The generator subsystem was then replaced by a NN configuration with the goal of reducing by at least two orders of magnitude the time required to simulate the computationally intensive physics-based generator equations used in the Simulink subsystem. The current phase of the project restricts the NN to simulating only steady-state electrical generator response. The generator subsystem physics model provided corresponding I/O pairs by varying parameters in the transmission lines. These pairs were then used to train the NN such that the subsystem output would be pre-computed prior to any simulations using the subsystem. More than 160 system variations were obtained that produced a stable steady-state system. To accurately train the network, the generator subsystem data obtained from the original model was split into a training data set and a testing data set. The training data set was used to create a feed-forward back propagation NN. The output of the NN consisted of six electrical current values. The electrical current consisted of the magnitude and phase angle for each of the three phases of the system. The NN input consisted of six components of the electrical voltage, similar to the components of the current. The trained networks were tested using the testing data set consisting of 20 data points that were used to compare the results of the completed NN with the results produced by the original generator subsystem model. The mean square error between the outputs of the physics and NN models were used to measure the validity of the NN training.

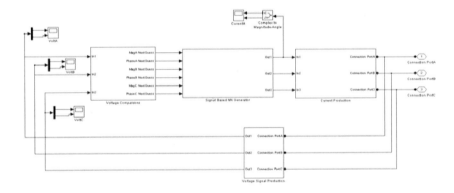

Fig. 4. Neural network test system for electrical generator

The generator NN was then integrated into Simulink as shown in Fig. 4. This system interacts with the generator load, which is connected through the connection ports on the right side of the figure. A Matlab script was created that allows the Simulink model to be called iteratively until a solution is reached. The neural network-based system showed very good agreement with the original Simulink model. Using a convergence tolerance of 0.0004 for the relative iteration change of the voltage values, the maximum relative error in the voltage values was 0.00056 and the maximum relative error in the current values was 0.0022. Execution time comparisons showed a dramatic time reduction when using the NN. The Simulink model required about 114,000 iterations and 27 minutes CPU time to simulate 30 seconds real time on a 531-MHz computer. The NN-based system was able to determine solutions for over half the cases in less than 15 seconds on a 2.16-GHz computer. All cases were completed in less than 120 seconds. Further reductions in the execution time are anticipated, as well as the ability to perform transient analysis.

6 Summary

The research conducted at Purdue University, the Georgia Institute of Technology, and PC Krause and Associates address the critical areas needed to achieve the overall goal of an accurate real-time predictive simulation of the terrestrial electric power grid: faster-than-real-time simulation, distributed dynamic sensing, the development of neural-network-based behavioral models for situations where physics-based models are impracticable, and visualization for situation awareness and identification of vulnerabilities. DDDAS provides a framework for the integration of the results from these coupled areas of research to form a powerful new tool for assessing and enhancing power grid security.

References

1. Fairley, P.: The unruly power grid. IEEE Spectrum (August 2004)
2. Lucas, C.E., Walters, E.A., Jatskevich, J., Wasynczuk, O., Lamm, P.T., Neeves, T.E.: Distributed heterogeneous simulation: A new paradigm for simulating integrated naval power systems. WSEAS/IASME Trans. (July 2004)
3. Lucas, C.E., Wasynczuk, O., Lamm, P.T.: Cross-platform distributed heterogeneous simulation of a MEA power system. In: Proc. SPIE Defense and Security Symposium. (March 2005)
4. PC Krause and Associates: Virtual prototyping vehicle electrical system, management design tool. Technical report, SBIR Phase I Contract W56HZV-04-C-0126 (September 2004)
5. Jatskevich, J., Wasynczuk, O., Mohd Noor, N., Walters, E.A., Lucas, C.E., Lamm, P.T.: Distributed simulation of electric power systems. In: Proc. 14th Power Systems Computation Conf., Sevilla, Spain (June 2002)
6. Graham, S., Wong, I., Chen, W., Lazarevic, A., Cleek, K., Walters, E., Lucas, C., Wasynczuk, O., Lamm, P.: Distributed simulation. Aerospace Engineering (November 2004) 24–27
7. U.S.–Canada Power System Outage Task Force: Final report on the August 14th blackout in the United States and Canada, https://reports.energy.gov. Technical report (April 2004)
8. Overbye, T.J., Weber, J.D.: New methods for the visualization of electric power system information. In: Proc. IEEE Symposium on Information Visualization. (2000)
9. Wong, N., Carpendale, S., Greenberg, S.: Edgelens: An interactive method for managing edge congestion in graphs. In: Proc. IEEE Symposium on Information Visualization. (2003)

10. Anderson, M.D. et al: Advanced graphics zoom in on operations. IEEE Computer Applications in Power (2003)
11. Weber, J.D., Overbye, T.J.: Power system visualization through contour plots. In: Proc. North American Power Symposium, Laramie, WY (1997)
12. Solomon, A.: Visualization strategies for monitoring power system security. Master's thesis, Purdue University, West Lafayette, IN (December 2005)
13. Mahadev, P., Christie, R.: Envisioning power system data: Vulnerability and severity representations for static security assessment. IEEE Trans. Power Systems (4) (November 1994)
14. Hoffmann, C., Kim, Y.: Visualization and animation of situation awareness http://www.cs.purdue.edu/homes/cmh/distribution/Army/Kim/overview.html.
15. Tsoukalas, L., Gao, R., Fieno, T., Wang, X.: Anticipatory regulation of complex power systems. In: Proc. European Workshop on Intelligent Forecasting, Diagnosis, and Control, Santorini, Greece (June 2001)
16. Boris, J.P. et al: CT-Analyst: Fast and accurate CBR emergency assessment. In: Proc. First Joint Conf. on Battle Management for Nuclear, Chemical, Biological and Radiological Defense, Williamsburg, VA (November 2002)
17. The Mathworks: Simulink. http://www.mathworks.com/products/simulink/

Implementing Virtual Buffer for Electric Power Grids

Rong Gao and Lefteri H. Tsoukalas

Applied Intelligent Systems Lab, School of Nuclear Enigineering, Purdue University,
400 Central Drive, West Lafaytte, IN, USA
{gao, tsoukala}@ecn.purdue.edu

Abstract. The electric power grid is a vital network for every aspect of our life. The lack of buffer between generation and consumption makes the power grid unstable and fragile. While large scale power storage is not technically and economically feasible at present stage, we argue that a virtual buffer could be effectively implemented through a demand side management strategy built upon the concept of dynamic data driven paradigm. An intelligent scheduling module implemented inside a smart meter enables customers to access electricity safely and economically. The managed use of electricity acts effectively as a buffer to provide the much needed stability for the power grid. Pioneering efforts intending to implement these concepts have been conducted by groups such as the Consortium for the Intelligent Management of Electric Grid. Progresses have been made and reported in this paper.

Keywords: Electric Power Grid, Artificial Intelligence, Dynamic Data Driven.

1 Introduction

The electric power grid is a vital network for every aspect of our life. The grid is so complex that it experiences the same instability problem as other complex systems do [1]. The system can ran into an unstable point without notice until a large scale collapse spreads, such as the major northeast blackout of 2003. The fragile nature of the electric power grid, to a great extend, has to do with the fact that electricity can not be stored in the grid. Electricity is consumed when it is generated. The lack of cushion between generation and consumption constitutes a major unstable factor for the power system.

On the other hands, it is helpful to study another example of complex system, the Internet. The Internet connects billions of users in the world. The degrees of connectivity and irregularity are enormous. Yet surprisingly high stability is achieved considering the openness provide by the Internet. Users have the maximum freedom to use the resources with little restriction, something that is hard to imagine for the electric power grid. The key behind this impressive performance is the protocols, a set of rules that very player agrees to. Protocols regulate the creation, transfer and use of the resources. The Internet protocols roots on the assumption that very one has equal access to the resources on the network. This principle of fairness is the guideline for resolving conflict originated from the competition of resources. The protocols are the reason why the Internet is able to maintain functional under malignant environment.

Y. Shi et al. (Eds.): ICCS 2007, Part I, LNCS 4487, pp. 1083–1089, 2007.

The protocols are feasible and enforceable because of one big assumption that the resource (information) can be stored somewhere in the network. This assumption is the cornerstone for many key concepts in the protocols, such as routing and conflict resolving. In the Internet, resources are generated, stored, and consumed. The lag between generation and consumption ranges from milliseconds to seconds or minutes depending on the traffic in the network. The delay has proven an important stabling factor for the network.

Having had a successful case at hand, it is natural to ask whether we can duplicate the same kind of success in the electric power grid. As the initial attempts, we can try to develop protocols similar to the ones used in the Internet to regulate the generation, transportation, and consumption of the electricity. This is certainly unrealistic for the current power grid. However, as the discussion and research on smart power meter have been going for a while, we have the reason to believe that in the near future every household will have one of these intelligent device installed. The intelligent meter is connected to the network and is able to communicate with the outside world such as the electricity supplier. The meter will have the computing capability, making possible the implementation of a set of fairy complicated protocols. With these smart meters at our disposal, there is only one thing between us and the Internet-type stable electric power grid - we need the means to store electricity. There have been lots of efforts being spent on electricity storage. Some successes have been reported [2]. But the solution for large scale storage that can make a big impact to the power grid is still not technically and economically feasible. In this paper, we argue that through the utilization of information technologies, we can implement a virtual buffer between the generation and consumption without the need to physically store the electricity in the grid.

2 Virtual Buffer for Electric Power Grid

As we know, the lack of buffer between generation and consumption is the source of problem for many complex systems such as the electric power grid. Even though we don't have a feasible way to physically store large amount of electricity in the network so far, we are able to mimic the effect of storage through a virtual buffer.

A virtual buffer between generation and consumption can be created through demand side management strategy which is build upon the practices of dynamic data driven paradigms. With the emergence of intelligent meters, it is possible to dynamically schedule the use of electricity of every customer. This dynamic scheduling will create a sheet of virtual buffer for generation and consumption as we argue here. Under the new paradigm, the consumption of electricity of every customer is intelligently managed. Customers don't power up their electricity-hungry machines at will. Rather, they make smart decisions after balancing the costs and benefits. For example, some non-urgent activities, such as laundry, can be scheduled to sometime during the day when electricity is abundant and cheap. The costs of the electricity are determined by the supply-to-demand ratio and the capacity of the network to transfer the resources. This managed use of resources is analogous to the access control widely used in the Internet. A buffer between generation and consumption is therefore created, virtually. No physic laws are broke. The electricity is still actually consumed when generated. However, from the customer point of view, with dynamic consumption scheduling the resources (electricity) are created and then stored somewhere in the power grid before they are used. The analogy is shown in Fig. 1. The virtual buffer can greatly increase the stability of the power grid.

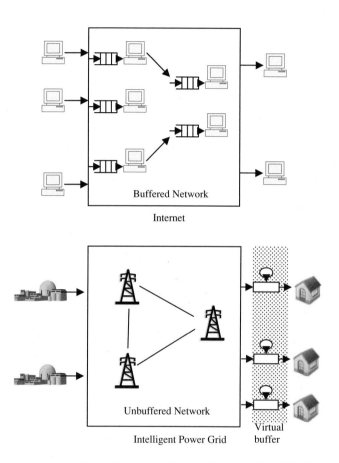

Fig. 1. Internet and Power Grid with Virtual Buffer

Dynamic scheduling has to be carried out by software agents or intelligent agents. The intelligent agents will act on behalf of their clients, making reasonable decision based the analysis of the situation. One of the most important analysis powers an agent has to possess is the anticipation capability. The intelligent agent needs to predict its client's future consumption pattern to make scheduling possible. In other words, load forecasting capability is the central piece for such a system. A pioneering effort, conducted by CIMEG, has recognized this need and advanced many techniques to make that possible.

3 CIMEG and TELOS

In 1999, EPRI and DOD funded the *Consortium for the Intelligent Management of the Electric Power Grid (CIMEG)* to develop intelligent approaches to defend power systems against potential threats [3]. CIMEG was led by Purdue University and included partners from The University of Tennessee, Fisk University, TVA and ComEd (now Exelon). CIMEG advanced an anticipatory control paradigm with which power systems can act proactively based on early perceptions of potential threats. It uses a bottom-up approach to circumvent the technical difficulty of defining the global health of power system at the top level, ash shown in Fig. 2. The concept of Local Area Grid (LAG) is extremely important in CIMEG. A LAG is a demand-based autonomous entity consisting of an appropriated mixture of different customers, charged with the necessary authority to maintain its own health by regulating or curtailing the power consumption of its members. Once all LAGs have achieved security, the whole grid, which is constructed by a number of LAGs in a hierarchical manner, achieves security as well. To pursue the health of a LAG, intelligent agents are used. Intelligent agents monitor every load within the LAG, forecasting the power consumption of each individual load and taking anticipatory actions to prevent potential cascade of faults. A prototypical system called the Transmission Entities with Learning-capabilities and On-line Self-healing (TELOS) has been developed and implemented in the service area of Exelon-Chicago and Argonne National Laboratory [4].

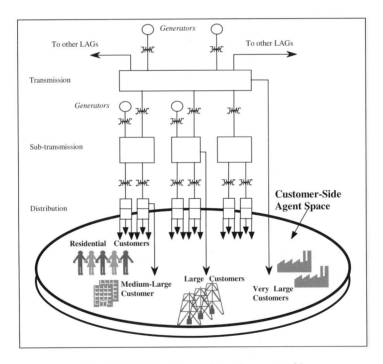

Fig. 2. CIMEG's picture of electric power grid

CIMEG's mission is to create a platform for managing the power grid intelligently. In CIMEG's vision, customers play more active role than in current power system infrastructure. Lots of solicitation and negotiations are involved, as shown in Fig. 3. The customer, who is represented by an intelligent meter in Fig. 3, predicts its need for the electricity and place order in the market. The amount of the order is influenced by the market price of the electricity, which is further determined by the difference of the demand and supply and also the capacity of the network. Economic models with price elasticity are used in the process [5]. The active interaction between customers and suppliers create a virtual buffer between consumption and generation as discussed earlier.

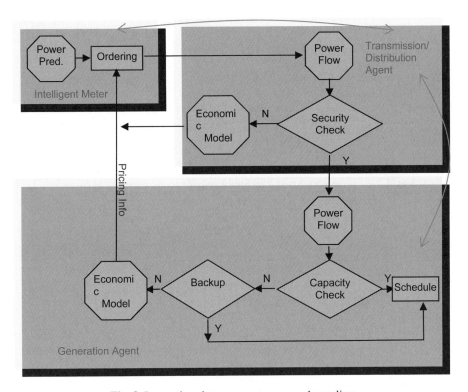

Fig. 3. Interactions between customers and suppliers

One of the key assumptions taken by CIMEG is that the demand of customers can be predicted to a useful degree of accuracy. Among many existing prediction algorithms, neural networks outperform others because of their superb capabilities in handling non-linear behaviors [6].

TELOS achieves improved forecasting accuracy by performing load forecast calculations at the customer site. The customer is capable of tracking local weather conditions as well as the local load demand. TELOS forecasts are based on neuro-fuzzy algorithms, which are highly adaptive to changing conditions. Preliminary results show that simple feed forward neural networks are well suited to predicting load at the customer scale. Fig. 4 shows neural-network-based load prediction for a large commercial customer.

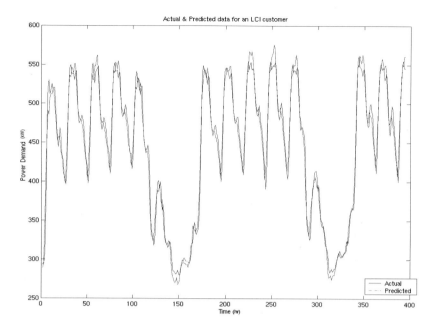

Fig. 4. Neural-network-based load forecasting for actual commercial customer

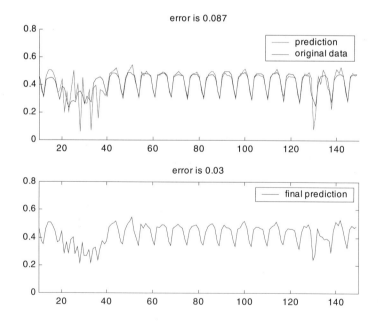

Fig. 5. The prediction results with (bottom) and without (upper) compensations

Neural networks are well qualified for handling steady state cases, wherever typical behaviors have been seen and generalized. Some unexpected events such as favorite sport programs seriously cause enormous error in prediction. CIMEG has developed a fuzzy logic module to process such situation. This module, called PROTREND, quickly detects any deviation from the steady state and generates a compensation for the final prediction. The results in Fig. 5 show a consumption pattern over a winter break. PROTREND effectively compensates a large portion of errors and makes the final prediction results more reliable.

4 Conclusions and Future Work

A regulation between generation and consumption is vital for the stability of a complex system with limited resources. An effective way for regulation is to create buffers where resources can stored when not needed and released when requested. We have shown that an effective form of buffer can be simulated with the intelligent management of the consumption when no means of storage is available. CIMEG has made some promising results, demonstrating the feasibility of intelligent management based on the anticipation of consumption. Accurate load forecasting is possible. However, CIMEG's demo was implemented on a very small system. Large scale demonstrations are necessary, both on simulation and theoretical analysis.

Acknowledgments

This project is partly supported by the National Science Foundation under contract no 0540342-CNS.

References

1. Amin, M.:Toward Self-Healing Energy Infrastructure Systems. In: IEEE Compouter Applications in Power, Volume 14, Number 1, 220-28 (2001)
2. Ribeiro, P. F., et al: Energy Storage Systems for Advanced Power Application.In: Proceedings of the IEEE, Vol 89, Issue 12, p1744-1756,(2001)
3. CIMEG, Intelligent Management of the Power Grid: An Anticipatory, Multi-Agent, High Performance Computing Approach, EPRI, Palo Alto, CA, and Applied Intelligent Systems Lab, School of Nuclear Engineering, Purdue University, West Lafayette, IN, 2004
4. Gao, R., Tsoukalas, L. H.:Anticipatory Paradigm for Modern Power System Protection. In:ISAP, Lemnos, Greece (2003)
5. Gao, R., Tsoukalas, L. H.: Short-Term Elasticities via Intelligent Tools for Modern Power Systems,.In: MedPower'02, Athens, Greece(2002)
6. Tsoukalas, L., Uhrig, R. : Fuzzy and Neural Approaches in Engineering. New York: Wiley, (1997)

Enhanced Situational Awareness: Application of DDDAS Concepts to Emergency and Disaster Management[*]

Gregory R. Madey[1], Albert-László Barabási[2], Nitesh V. Chawla[1],
Marta Gonzalez[2], David Hachen[3], Brett Lantz[3], Alec Pawling[1],
Timothy Schoenharl[1], Gábor Szabó[2], Pu Wang[2], and Ping Yan[1]

[1] Department of Computer Science & Engineering
University of Notre Dame, Notre Dame, IN 46556, USA
{gmadey, nchawla, apawling, tschoenh, pyan}@nd.edu
[2] Department of Physics
University of Notre Dame, Notre Dame, IN. 46556, USA
{alb, m.c.gonzalez, gabor.szabo, pwang2}@nd.edu
[3] Department of Sociology
University of Notre Dame, Notre Dame, IN 46556, USA
{dhachen, blantz}@nd.edu

Abstract. We describe a prototype emergency and disaster information system designed and implemented using DDDAS concepts. The system is designed to use real-time cell phone calling data from a geographical region, including calling activity – who calls whom, call duration, services in use, and cell phone location information – to provide enhanced situational awareness for managers in emergency operations centers (EOCs) during disaster events. Powered-on cell phones maintain contact with one or more within-range cell towers so as to receive incoming calls. Thus, location data about all phones in an area are available, either directly from GPS equipped phones, or by cell tower, cell sector, distance from tower and triangulation methods. This permits the cell phones of a geographical region to serve as an ad hoc mobile sensor net, measuring the movement and calling patterns of the population. A prototype system, WIPER, serves as a test bed to research open DDDAS design issues, including dynamic validation of simulations, algorithms to interpret high volume data streams, ensembles of simulations, runtime execution, middleware services, and experimentation frameworks [1].

1 Introduction

For disaster and emergency response managers to perform effectively during an event, some of their greatest needs include quality communication capabilities and high levels of situational awareness [2-4]. Reports from on-scene coordinators, first responders, public safety officials, the news media, and the affected population can

---

[*] The material presented in this paper is based in part upon work supported by the National Science Foundation, the DDDAS Program, under Grant No. CNS-050312.

Y. Shi et al. (Eds.): ICCS 2007, Part I, LNCS 4487, pp. 1090–1097, 2007.

provide managers with point data about an emergency, but those on-scene reports are often inaccurate, conflicting and incomplete with gaps in geographical and temporal coverage. Additionally, those reports must be fused into a coherent picture of the entire affected area to enable emergency managers to effectively respond.

The prototype Wireless Phone Based Emergency Response System (WIPER) is designed using the concepts of Dynamic Data Driven Application Systems (DDDAS) with the goal of enhancing situational awareness of disaster and emergency response managers [5, 6]. The subsequent sections review open research issues with the DDDAS concept and design and development features of the WIPER prototype that address some of those open issues. Those features include an analysis of the local and the global structure of a large mobile communication network, several algorithms for detecting anomalies in streaming data, agent-based simulations for classification and prediction, a distributed system design using web services, and the design and implementation of the WIPER DSS interface.

2 Dynamic Data Driven Application Systems

Dynamic data driven application systems (DDDAS) were explored in detail in two NSF workshops: one in early 2000 [7] and one in 2006 [1]. The first workshop concluded that the DDDAS concept offered the potential of greatly improving the accuracy and efficiency of models and simulations. The workshop final report identified more research in the areas of 1) dynamic, data driven application technologies, 2) adaptive algorithms for injecting and steering real-time data into running simulations, and 3) systems software that supports applications in dynamic environments. At following conferences, initial research and applications exploring these research areas were reported [8]. A fourth area of research important to the DDDAS concept emerged, that of measurement systems; the dynamic steering of the data collection needed by the simulations may require improvements in measurement, observation and instrumentation methods. In 2004, Darema described the DDDAS concept as:

> Dynamic Data Driven Application Systems (DDDAS) entails the ability to incorporate additional data into an executing application - these data can be archival or collected on-line; and in reverse, the ability of applications to dynamically steer the measurement process. The paradigm offers the promise of improving modeling methods, and augmenting the analysis and prediction capabilities of application simulations and the effectiveness of measurement systems. This presents the potential to transform the way science and engineering are done, and induce a major impact in the way many functions in our society are conducted, such as manufacturing, commerce, hazard management, and medicine [8].

The second major NSF workshop on DDDAS (2006) highlighted progress to date, and research yet to be conducted on open issues in the DDDAS concept [1, 9]. The prototype WIPER system explores all four research areas relevant to the DDDAS concept [5, 6] and many of the open DDDAS research issues. Those open DDDAS research areas include: 1) *Dynamic and continuous validation of models, algorithms, systems, and system of systems*: WIPER uses an ensemble of agent-based simulations to test hypothesis about the emergency event, dynamically validating those simulations against streaming data, 2) *Dynamic data driven, on demand scaling and resolution* – the WIPER simulations request detailed data from the data source, providing higher resolution where needed to support dynamic validation, 3) *Data format: collections taken across different instrument types can range in format and units* – the WIPER data source includes multiple types of data, including location and movement, cell call data, service type, e.g., voice, data, SMS, 4) *System Software, especially runtime execution, and middleware service and computational infrastructure* – WIPER is employing a novel application of web services (SOA), messaging, and asynchronous Javascript and XML (AJAX) to integrate heterogeneous services and user displays, 5) *Mathematical and Statistical Algorithms, especially advances in computational simulation tools for modeling, and system and integration of related algorithms* – WIPER includes new algorithms for monitoring streaming data [10, 11] and new insight from mobile phone call data that will be incorporated into planned algorithms for anomaly detection [12, 13].

3   Data

The WIPER system uses both actual call and location data provided by a cellular carrier and synthetic data to simulate emergencies. The actual data is used to calibrate the generation of synthetic data and to help design the anomaly detection algorithms. An anomaly, a possible indication of an emergency, triggers the execution of an ensemble of simulations, each dynamically validated against new streaming data. All data is anonymized to protect privacy. During development, testing and evaluation,

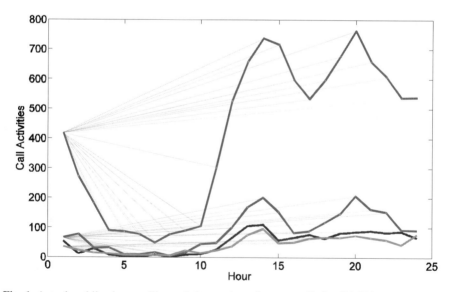

Fig. 1. Actual mobile phone calling activity per hour from a small city (20,000 population) for a 24 hour period starting at midnight. The four data series are associated with 4 different cell towers (all cell sectors combined for each tower). This data is used to calibrate synthetic data and to validate anomaly detection algorithms.

the data is stored in a database with software modules streaming the data to simulate the real-time data streams the system would see if deployed. The data that the WIPER system uses is aggregate in nature, does not track individual cell phones by actual user ID, and does not include the content of phone calls or messages. An example of the call activity over a 24 hour period or a small city with 4 cell towers is shown in Fig. 1. The synthetic data is validated against such data.

4 Design

The design and operation of the WIPER DDDAS prototype is shown schematically in Fig. 2. The system has five components: 1) Data Source, 2) Historical Data Storage, 3) Anomaly Detection and Alert System (DAS), 4) Simulation and Prediction System (SPS), and 5) Decision Support (DSS) with Web Interface. Additional details about the operation of each component can be found elsewhere [5, 6]. The Simulation and Prediction System is triggered by the detection of an anomaly and an alert signal from the DAS. An ensemble of agent-based simulations evaluates possible explanations for the anomaly and alert. The dynamically validated simulation is used to classify the event and to predict its evolution. This and other WIPER provided information is displayed on a web-based console to be used by emergency managers for enhance situational awareness as shown in Fig. 3.

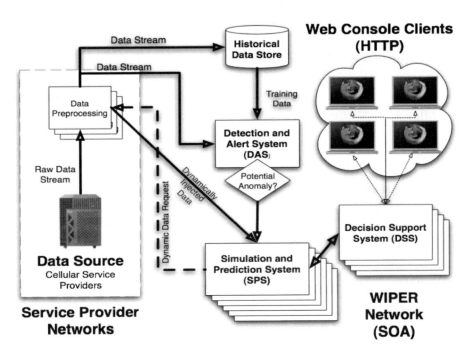

Fig. 2. Architecture of the WIPER DDDAS system: Either synthetic data or actual data from the cell phone providers is used for training, real time anomaly detection and dynamic validation of the simulation and prediction system

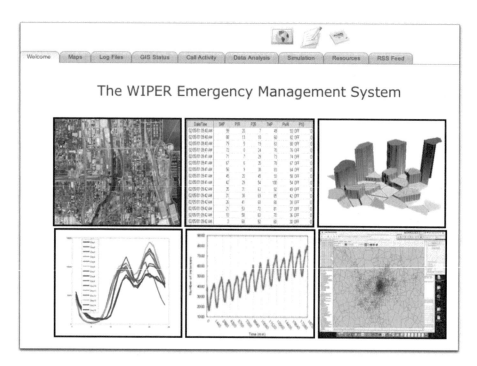

Fig. 3. The WIPER DSS console is web based, using asynchronous Javascript and XML (AJAX). Emergency managers can view streaming call activity statistics from the area of an incident, maps, weather information, GIS visualizations, and animations derived from agent-based simulation used to predict the evolution of an incident.

5   Data, Algorithms and Analysis

The DDDAS concept is characterized by dynamic data, the first two D's in DDDAS. In the WIPER system the dynamic data is cell phone activity: temporal calling patterns, statistics, locations, movement, and the evolving social networks of callers. In order to improve the detection and classification of anomalies and to dynamically validate predictive simulations, the WIPER project has conducted to date, five different, but complementary investigations. Each contributes understanding and methods for anomaly detection on streaming cell phone activity data.

5.1   Structure and Tie Strengths in Mobile Communication Networks

Cell phone calling activity databases provide detailed records of human communication patterns, offering novel avenues to map and explore the structure of social and communication networks potentially useful in anomaly detection. This investigation examined the communication patterns of several million of mobile phone users, allowing the simultaneous study of the local and the global structure of a society-wide communication network. We observed a coupling between interaction strengths and the network's local structure, with the counterintuitive consequence that social networks are robust to the removal of the strong ties, but fall apart following a phase transition if the weak ties are removed [13].

5.2 Anomaly Detection Using a Markov Modulated Poisson Process

Cell phone call activity records the behavior of individuals, which reflects underlying human activities across time. Therefore, they appear to have multi-level periodicity, such as weekly, daily, hourly, etc. Simple stochastic models that rely on aggregate statistics are not able to differentiate between normal daily variations and legitimate anomalous (and potentially crisis) events. In this investigation we developed a framework for unsupervised learning using Markov modulated Poisson process (MMPP) to model the data sequence and use the posterior distribution to calculate the probability of the existence of anomalies over time [11].

5.3 Mapping and Visualization

The cell phone calling data currently includes user locations and activity at a cell-sized level of resolution. The size of a wireless cell can vary widely and depends on many factors, but these can be generalized in a simple way using a Voronoi diagram. A Voronoi lattice is a tiling of polygons in the plane constructed in the following manner: Given a set of points P (in our case, a set of towers) construct a polygon around each point in P such that for all points in the polygon around p_0, the point is closer to p_0 than to any other point in P. Thus we can construct a tiling of a GIS space into cells around our towers. We build a 3D image based on the activity at the site of interest as shown in Fig. 4. This 3D view gives a good representation of the comparative activity levels in the cells [6, 14].

5.4 Hybrid Clustering Algorithm for Outlier Detection on Streaming Data

We developed a hybrid clustering algorithm that combines k-means clustering, the leader algorithm, and statistical process control. Our results indicate that the quality of the clusters produced by our algorithm are comparable to, or better than, those produced by the expectation maximization algorithm using sum squared error as an evaluation metric. We also compared the outlier set discovered by our algorithm with the outliers discovered using one nearest neighbor. While our clustering algorithm produced a number of significant false positive and false negatives, most of the outlier detected by our hybrid algorithm (with proper parameter settings) were in fact outliers. We believe that our approach has promise for clustering and outlier detection on streaming data in the WIPER Detection and Alert System [10].

5.5 Quantitative Social Group Dynamics on a Large Scale

Interactions between individuals result in complex community structures, capturing highly connected circles of friends, families, or professional cliques in social networks. Although most empirical studies have focused on snapshots of these communities, because of frequent changes in the activity and communication patterns of individuals, the associated social and communication networks are subject to constant evolution. Our knowledge of the mechanisms governing the underlying community dynamics is limited, but is essential for exploiting the real time streaming cell phone data for emergency management. We have developed a new algorithm based on a clique percolation technique, that allows, for the first time, to investigate in detail the time dependence of overlapping communities on a large scale and to uncover basic relationships of the statistical features of community evolution. The focus of this investigation on the networks formed by the calls between mobile phone users, observing that these communities are subject to a number of elementary evolutionary steps ranging from community formation to breakup and merging, representing new dimensions in their quantitative interpretation. We found that large groups persist longer if they are capable of dynamically altering their membership, suggesting that an ability to change the composition results in better adaptability and a longer lifetime for social groups. Remarkably, the behavior of small groups displays the opposite, the condition for stability being that their composition remains unchanged [12].

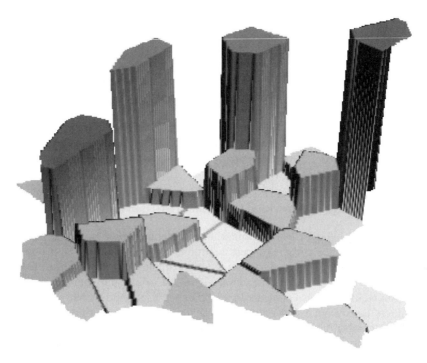

Fig. 4. A 3D view of activity in mobile phone cells. Each polygon represents the spatial area serviced by one tower. Cell color and height are proportional to the number of active cell phone users in that cell for a unit of time.

6 Summary

WIPER is designed for real-time monitoring of normal social and geographical communication and activity patterns of millions of cell phone users, recognizing unusual human agglomerations, potential emergencies and traffic jams. WIPER uses streams of high-resolution data in the physical vicinity of a communication or traffic anomaly, and dynamically injects them into agent-based simulation systems to classify and predict the unfolding of the emergency in real time. The agent-based simulation systems dynamically steer local data collection in the vicinity of the anomaly. Distributed data collection, monitoring, analysis, simulation and decision support modules are integrated to generate traffic forecasts and emergency alerts for engineering, public safety and emergency response personnel for improved situational awareness.

References

[1] NSF, "DDDAS Workshop Report," http://www.dddas.org/nsf-workshop2006/-wkshp_report.pdf, 2006.

[2] J. Harrald and T. Jefferson, "Shared Situational Awareness in Emergency Management Mitigation and Response," in Proceedings of the 40th Annual Hawaii International Conference on Systems Sciences: Computer Society Press, 2007.

[3] Naval Aviation Schools Command, "Situational Awareness," http://wwwnt.cnet.navy.mil/crm/crm/stand_mat/seven_skills/SA.asp, 2007.

[4] R. B. Dilmanghani, B. S. Manoj, and R. R. Rao, "Emergency Communication Challenges and Privacy," in Proceedings of the 3rd International ISCRAM Conference, B. Van de Walle and M. Turoff, Eds. Newark, NJ, 2006.

[5] G. Madey, G. Szábo, and A.-L. Barabási, "WIPER: The integrated wireless phone based emergency response system," Proceedings of the International Conference on Computational Science, Lecture Notes in Computer Science, vol. 3993, pp. 417-424, 2006.

[6] T. Schoenharl, R. Bravo, and G. Madey, "WIPER: Leveraging the Cell Phone Network for Emergency Response," International Journal of Intelligent Control and Systems, (forthcoming) 2007.

[7] NSF, "Workshop on Dynamic Data Driven Application Systems," www.cise.nsf.gov/dddas, 2000.

[8] F. Darema, "Dynamic Data Driven Application Systems: A New Paradigm for Application Simulations and Measurements," in ICCS'04, Krakow, Poland, 2004.

[9] C. C. Douglas, "DDDAS: Virtual Proceedings," http://www.dddas.org/virtual_proceeings.html, 2006.

[10] A. Pawling, N. V. Chawla, and G. Madey, "Anomaly Detection in a Mobile Communication Network," Proceedings of the NAACSOS, 2006.

[11] Y. Ping, T. Schoenharl, A. Pawling, and G. Madey, "Anomaly detection in the WIPER system using a Markov modulated Poisson distribution," Working Paper, Notre Dame, IN: Computer Science & Engineering, University of Notre Dame, 2007.

[12] G. Palla, A.-L. Barabási, and T. Viscsek, "Quantitative social group dynamics on a large scale," Nature (forthcoming), 2007.

[13] J.-P. Onnela, J. Saramäki, J. Hyvönen, G. Szábo, D. Lazer, K. Kaski, J. Kertész, and A.-L. Barabási, "Structure and tie strengths in mobile communication networks," PNAS (forthcoming), 2007.

[14] T. Schoenharl, G. Madey, G. Szábo, and A.-L. Barabási, "WIPER: A Multi-Agent System for Emergency Response," in Proceedings of the 3rd International ISCRAM Conference, B. Van de Walle and M. Turoff, Eds. Newark, NJ, 2006.

AIMSS: An Architecture for Data Driven Simulations in the Social Sciences

Catriona Kennedy[1], Georgios Theodoropoulos[1], Volker Sorge[1],
Edward Ferrari[2], Peter Lee[2], and Chris Skelcher[2]

[1] School of Computer Science, University of Birmingham, UK
[2] School of Public Policy, University of Birmingham, UK
C.M.Kennedy,G.K.Theodoropoulos@cs.bham.ac.uk

Abstract. This paper presents a prototype implementation of an intelligent assistance architecture for data-driven simulation specialising in qualitative data in the social sciences. The assistant architecture semi-automates an iterative sequence in which an initial simulation is interpreted and compared with real-world observations. The simulation is then adapted so that it more closely fits the observations, while at the same time the data collection may be adjusted to reduce uncertainty. For our prototype, we have developed a simplified agent-based simulation as part of a social science case study involving decisions about housing. Real-world data on the behaviour of actual households is also available. The automation of the data-driven modelling process requires content interpretation of both the simulation and the corresponding real-world data. The paper discusses the use of Association Rule Mining to produce general logical statements about the simulation and data content and the applicability of logical consistency checking to detect observations that refute the simulation predictions.

Keywords: Architecture,Data Driven Simulations, Social Sciences.

1 Introduction: Intelligent Assistance for Model Development

In earlier work[1] we proposed a conceptual architecture for the intelligent management of a data driven simulation system. In that architecture, a software "assistant" agent should compare simulation predictions with data content and adapt the simulation as necessary. Similarly, it should adjust the data collection depending on simulation predictions. In this paper, we present a proof-of-concept prototype that is being developed as part of the AIMSS project[1] (Adaptive Intelligent Model-building for the Social Sciences). This is an exploratory implementation of the conceptual architecture.

A key issue the AIMSS project is trying to address is "evidence based model development": this can be understood as an iterative process involving the following stages:

1. Formulate initial model and run simulation;
2. Once the simulation has stabilised, inspect it visually and determine whether it makes interesting predictions which need to be tested;

---

[1] http://www.cs.bham.ac.uk/research/projects/aimss/

Y. Shi et al. (Eds.): ICCS 2007, Part I, LNCS 4487, pp. 1098–1105, 2007.

3. Collect the relevant data and analyse it;
4. Determine if the simulation predictions are supported by the data;
5. If the data does not support the predictions, determine whether the model should be revised. Experiment with variations of the original simulation and return to Step 2.

The goal of the AIMSS project is to investigate the role of DDDAS in the automation of this process for the social sciences. The project is focusing on qualitative data and agent-based models.

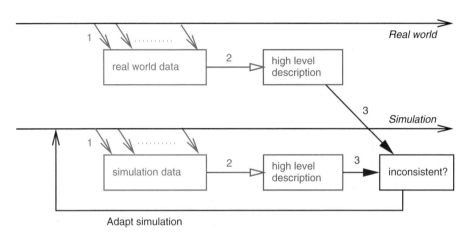

Fig. 1. Data driven adaptation of a simulation

A schematic diagram of the AIMSS concept of data-driven adaptation is shown in Figure 1. We can think of the simulation as running in parallel with events in the real world, although in a social science scenario, there are two important differences:

1. The simulation does not physically run in parallel with the real world. Instead, the real world data is usually historical. However, the data input could be reconstructed to behave like a stream of data being collected in parallel).
2. The simulation is abstract and does not correspond to a particular observed system. This means that the variable values read from the data cannot be directly absorbed into the simulation as might be typical for a physical model. Instead, the simulation represents a sequence of events in a typical observed system. The data may be collected from multiple real world instances of the general class of systems represented by the simulation. (For example, the simulation could be about a typical supermarket, while the data is collected from multiple real supermarkets).

The figure can be divided into two processes: the first is a process of interpreting both the simulation predictions and the data content and determining whether they are consistent (arrows 1, 2 and 3 in the figure). The second involves adapting the simulation in the event of an inconsistency. In the current version of the prototype we have focused on the first process. This requires the following automated capabilities:

- Interpretation of simulation states (at regular intervals or on demand): due to the abstract and qualitative nature of the simulation, this is not just about reading the current variable values, but about generating a *high level description* summarising patterns or trends.
- Interpretation of real world data: the same methods are required as for the simulation interpretation, except that the data is often more detailed and is usually noisy. Therefore pre-processing is required, which often involves the integration of data from multiple sources and the generation of higher level datasets that correspond to the simulation events.
- Consistency checking to determine whether the simulation states mostly agree with descriptions of data content.
- Re-direction and focusing of data collection in response to evaluation of simulation states or uncertainty in the data comparison.

It is expected that the simulation and data interpretation will be complemented by human visualisation and similarly the consistency/compatibility checking may be overriden by the "common sense" judgements of a user. Visualisation and compact summarisation of the data are important technologies here.

2   Social Science Case Study

As an example case study, we are modelling agents in a housing scenario, focusing on the circumstances and needs of those moving to the social rented sector, and emphasising qualitative measures such as an agent's perception of whether its needs are met.

We have implemented the agent-based simulation using RePast[2]. The environment for the agents is an abstract "housing space" that may be divided into subspaces. One example scenario is where the space is divided into 4 "regions" (R1-4): R1: expensive, small city centre apartments; R2: inexpensive cramped city towerblocks in a high crime area; R3: Modest suburb; R4: Wealthy suburb (large expensive houses with large gardens). At initialisation, homes are allocated randomly to regions with largest number in inner city and city centre. Households are allocated randomly to regions initially with varying densities. A household is represented by a single agent, even if it contains more than one member. Precise densities and other attributes of each region (such as crime level etc.) can be specified as parameters.

The simulation is a sequence of steps in which agents decide whether they want to move based on a prioritised sequence of rules. These rules are simplified assumptions about decisions to move. Each rule has the form:

if (condition i not satisfied) *then* look for homes satisfying conditions $1, .., i$

where i is the position in a priority list which is indexed from 1 to n. For example, if condition 1 is "affordability", this is the first condition to be checked. If the current

---

[2] http://repast.sourceforge.net/

home is not affordable, the agent must move immediately and will only consider affordability when selecting other homes (as it is under pressure to move and has limited choice). The agent will only consider conditions further down the list if the conditions earlier in the list are already satsified by its current home. For example, if the agent is relatively wealthy and its current housing is good, it may consider the pollution level of the neighbourhood as being too high. Before moving into a new home, it will take into account the pollution level of the new area, along with all other conditions that were satisfied in the previous home (i.e. it must be affordable etc.). We have used a default scenario where the conditions are prioritised in the following order: "affordability", "crime level", "living space", "condition of home", "services in neighbourhood" and "pollution level". Those agents that find available homes move into them, but only a limited number are vacant (depending on selected parameters). An agent becomes "unhappy" if it cannot move when it wants to (e.g. because its income is too low).

Clearly, the above scenario is extremely simplified. The decision rules do not depend on the actions of neighbours. Since this is a proof-of-concept about intelligent management of a simulation, the actual simulation model itself is a minor part of the prototype. According to the incremental prototyping methodology, we expect that this can be gradually scaled up by successively adding more realistic simulations. More details on the simulation are in [2].

2.1 Data Sources

For the feasibility study, we are using a database of moves into the social rented sector for the whole of the UK for one year. This is known as CORE (Continuous Recording) dataset. Each CORE record include fields such as household details (age, sex, economic status of each person, total income), new tenancy (type of property, number of rooms, location), previous location and stated reason for move (affordability, overcrowding etc).

The simulation is a sequence of moves from one house to another for a typical housing scenario over a period of time measured in "cycles". The CORE data contains *actual* moves that were recorded in a particular area (England).

3 Interpretation and Consistency Checking

The first process of Figure 1 is the generation of datasets from both the simulation and the real world. To define the structure of the data, an ontology is required to specify the entities and attributes in the simulation, and to define the state changes. There are actually two components required to define a simulation:

1. Static entities and the relations of the model. For example, households and homes exist and a household can move from one region to another; a household has a set of *needs* that must be satisfied;
2. Dynamic behaviour: the decision rules for the agent as well as probabilistic rules for dynamic changes in the environment and household status (ageing, having children, changes in income etc.) The way in which these entities are *initialised* should also be stated as part of the model, as this requires domain knowledge (e.g. initial densities of population and houses etc.) For more detailed models, this becomes increasingly non-trivial, see e.g. [3]).

In the AIMSS prototype, both these components are specified in XML. Later we will consider the use of OWL[3]. The XML specification also includes the agent rules. These specifications are machine-readable and may potentially be modified autonomously. We are building on existing work on this area [4].

The entities and attributes are used to define the structure of data to be sampled from the simulation as well as the structure of a high level dataset to be derived from the pre-processing of the raw data. At the end of this process, we have two datasets, one records a sequence of simulated house moves, the other contains a sequence of actual moves.

3.1   Data Mining: Recognising General Patterns

The second stage in Figure 1 is the generation of high level descriptions. These are general statements about the developments in the simulation and in the real world. For this purpose, we are investigating data mining tools.

We have done some initial experimentation with Association Rule Mining using the Apriori algorithm [5], which is available in the WEKA Machine Learning package [6]. Association rules are a set of "if ... then" statements showing frequently occuring associations between combinations of "attribute = value" pairs. This algorithm is suited to large databases containing qualitative data which is often produced in social science research. Furthermore, it is "unsupervised" in the sense that predefined classes are not given. This allows the discovery of unexpected relationships.

An association rule produced by Apriori has the following form:

if (a_1 and a_2 and ... and a_n) s_1 then (c_1 and c_2 and .. c_m) s_2 conf(c)

where $a_1, ... a_n$ are *antecedents* and $c_1, ..., c_m$ are *consequents* of the rule. Both antecedents and consequents have the form "attribute = value". s_1 and s_2 are known as the *support* values and c is the *confidence*. The support value s_1 is the number of occurrences (records) in the dataset containing all the antecedents on the left side. s_2 is the number of occurrences of both the right and left sides together. Only those collections of items with a specified minimum support are considered as candidates for construction of association rules. The confidence is s_2/s_1. It is effectively the accuracy of the rule in predicting the consequences, given the antecedents. An example minimum confidence may be 0.9.

The higher the support and confidence of a rule, the more it represents a regular pattern in the dataset. If these measures are relatively low, then any inconsistency would be less "strong" than it would be for rules with high confidence and high support. The values of attributes are mutually exclusive. They are either strings or nominal labels for discrete intervals in the case of numeric data.

The following are some example rules that were mined from the simulation data using the agent rules above and environmental parameters guided by domain specialists

```
S1: if (incomeLevel=1 and moveReason=affordability) 283
then newHomeCost=1 283 conf(1)
```

---

[3] Ontology Web Language: http://www.w3.org/2004/OWL/

This specifies that if the income level is in the lowest bracket and the reason for moving was affordability then the rent to be paid for the new home is in the lowest bracket. The following is an example from the CORE data:

```
D1: if (moveReason=affordability and incomeLevel=1) 102
then newHomeCost=2 98 conf(0.96)
```

This has a similar form to S1 above, except that the new home cost is in the second lowest bracket instead of the lowest.

3.2   Consistency-Checking

Assuming that the CORE data is "typical" if sampled for a minimum time period (e.g. a year), the simulation can also be sampled for a minimum number of cycles beginning after a stabilisation period. The simulation-generated rule above is an example prediction. To test it, we can apply consistency checking to see if there is a rule that was discovered from the data that contradicts it. This would indicate that the available data does not support the current model. Some existing work on postprocessing of association rules includes contradiction checking. For example, [7] uses an "unexpectedness" definition of a rule, given previous beliefs. These methods may be applied to an AIMSS type architecture, where the "beliefs" are the predictions of a simulation.

Efficient algorithms for general consistency checking are available, e.g. [8]. We are currently investigating the application of such algorithms to our work and have so far detected simple inconsistencies of the type between S1 and D1 above.

4   Towards Dynamic Reconfiguration and Adaptation

Work is ongoing to develop mechanisms to dynamically adjust the data collection and the simulation. Data mining often has to be fine-tuned so that the analysis is focused on the most relevant attributes. The rules generated from the simulation should contain useful predictions to be tested and the rules generated from the data have to make statements about the same entities mentioned in the prediction. Data mining parameters may be adjusted, e.g. by selecting attributes associated with predicted negative or positive outcomes.

The consistency checking may still be inconclusive because there is insufficient data to support or refute the prediction. In this case the ontology should contain pointers to additional data sources, and these may be suggested to the user before data access is attempted.

The dynamic adjustment of data collection from the simulation is also important so that focusing on particular events is possible (as is the case for the real world). Currently the data generated from the simulation is limited and only includes house moves that have actually taken place. This can be extended so that data can be sampled from the simulation which represents different viewpoints (e.g. it may be a series of spatial snapshots or it focus on the dynamic changes in the environment instead of actions of agents).

4.1 Adaptation and Model Revision

The ontology and behaviour model may be adapted, since they are represented in a machine-readable and modifiable form. Possible forms of adaptation include the following:

- Modify the initial values of parameters (such as e.g. initial density of homes in a particular kind of region) or the probabilities used to determine the initial values or to determine when and how they should change.
- Add new attributes or extend the range of values for existing attributes as a result of machine learning applied to the raw data.
- Modify agent behaviour rules or add new ones;
- Modify the order of execution of behaviour rules.

Note that behaviour rules are intended to give a causal explanation, while association rules merely show correlations. Furthermore, association rules may represent complex emergent properties of simple behaviour rules.

Populations of strings of behaviour rules may be subjected to an evolutionary algorithm (such as genetic algorithms [9]) to evolve a simulation that is most consistent with the reality in terms of behaviour. Behaviour models that are most "fit" can be regarded as good explanations of the observed data. However, domain experts would have to interact with the system to filter out unlikely behaviours that still fit the available data.

4.2 Limitations of the Current Approach

One limitation of the current prototype is that the pre-processing of the raw data is too much determined by artificial boundaries. For example, Association Rule Mining requires that numeric values are first divided into discrete intervals (e.g. "high", "medium", "low" for income and house prices). The problem of artificial divisions can be addressed by the use of clustering [10] to generate more natural classes, which can then be used as discrete attribute values for an Association Rule miner. Conceptual Clustering [11] addresses the need for clusters to relate to existing concepts. Instead of just relying on one method, a combination of different pattern recognition and machine learning methods should be applied to the different datasets.

Another limitation of the approach we have taken is that the model-building process is determined by a single interpretation of the data (i.e. one ontology). In future work we plan to capture multiple ways of describing the events to be modelled by involving representatives of different social groups (stakeholders) in the initial model-building process. Multiple ontologies can lead to multiple ways of generating data from a simulation (or possibly even multiple simulations). Analysis of simulation predictions and real world observations is then not dependent on a single interpretation. Therefore the fault-tolerance of the system can be enhanced.

5 Conclusion

As part of the AIMSS project we have developed a simple prototype demonstrating some of the features required for the assistant agent architecture presented in an earlier study. Although this prototype is far too simple to be used operationally to generate real-world models, it serves as a proof-of-concept and can be used as a research tool by social scientists to help with exploratory model building and testing. Future work will involve the development of more realistic simulations, as well as the use of a wider range of data analysis and machine learning tools.

Acknowledgements

This research is supported by the Economic and Social Research Council as an e-Social Science feasibility study.

References

1. Kennedy, C., Theodoropoulos, G.: Intelligent Management of Data Driven Simulations to Support Model Building in the Social Sciences. In: Workshop on Dynamic Data-Driven Applications Simulation at ICCS 2006, LNCS 3993, Reading, UK, Springer-Verlag (May 2006) 562–569
2. Kennedy, C., Theodoropoulos, G., Ferrari, E., Lee, P., Skelcher, C.: Towards an Automated Approach to Dynamic Interpretation of Simulations. In: Proceedings of the Asia Modelling Symposium 2007, Phuket, Thailand (March 2007)
3. Birkin, M., Turner, A., Wu, B.: A Synthetic Demographic Model of the UK Population: Methods, Progress and Problems. In: Second International Conference on e-Social Science, Manchester, UK (June 2006)
4. Brogan, D., Reynolds, P., Bartholet, R., Carnahan, J., Loitiere, Y.: Semi-Automated Simulation Transformation for DDDAS. In: Workshop on Dynamic Data Driven Application Systems at the International Conference on Computational Science (ICCS 2005), LNCS 3515,, Atlanta, USA, Springer-Verlag (May 2005) 721–728
5. Agrawal, R., Srikant, R.: Fast Algorithms for Mining Association Rules in Large Databases. In: Proceedings of the International Conference on Very Large Databases, Santiage, Chile: Morgan Kaufmann, Los Altos, CA (1994) 478–499
6. Witten, I.H., Frank, E.: Data Mining: Practical Machine Learning Tools and Techniques. Elsevier, San Fransisco, California (2005)
7. Padmanabhan, B., Tuzhilin, A.: A Belief-Driven Method for Discovering Unexpected Patterns. In: Knowledge Discovery and Data Mining. (1998) 94–100
8. Moskewicz, M., Madigan, C., Zhao, Y., Zhang, L., Malik, S.: Chaff: Engineering an Efficient SAT Solver. In: Design Automation Conference (DAC 2001), Las Vegas (June 2001)
9. Mitchell, M.: An Introduction to Genetic Algorithms. MIT Press (1998)
10. Jain, A.K., Murty, M.N., Flynn, P.J.: Data Clustering: A Review. ACM Computing Surveys **31**(3) (September 1999)
11. Michalski, R.S., Stepp, R.E.: Learning from Observation: Conceptual Clustering. In Michalski, R.S., Carbonell, J.G., Mitchell, T.M., eds.: *Machine Learning: An artificial intelligence approach*. Morgan Kauffmann, Palo Alto, CA:Tioga (1983) 331–363

Bio-terror Preparedness Exercise in a Mixed Reality Environment*

Alok Chaturvedi, Chih-Hui Hsieh, Tejas Bhatt, and Adam Santone

Purdue Homeland Security Institute, Krannert School of Management, 403 West
State Street, Purdue University, West Lafayette, IN 47907-2014
{alok, hsiehc, tejas, santone}@purdue.edu

Abstract. The paper presents a dynamic data-driven mixed reality environment to complement a full-scale bio-terror preparedness exercise. The environment consists of a simulation of the virtual geographic locations involved in the exercise scenario, along with an artificially intelligent agent-based population. The crisis scenario, like the epidemiology of a disease or the plume of a chemical spill or radiological explosion, is then simulated in the virtual environment. The public health impact, the economic impact and the public approval rating impact is then calculated based on the sequence of events defined in the scenario, and the actions and decisions made during the full-scale exercise. The decisions made in the live exercise influence the outcome of the simulation, and the outcomes of the simulation influence the decisions being made during the exercise. The mixed reality environment provides the long-term and large-scale impact of the decisions made during the full-scale exercise.

1 Introduction

The Purdue Homeland Security Institute (PHSI) created a Dynamic Data-Driven Mixed Reality Environment to support a full-scale bio-terror preparedness exercise. In a mixed reality environment certain aspects of the scenario are conducted in the live exercise, while others are simulated. Actions and outcomes in the live exercise influence the simulated population, and the actions and outcomes of the simulation affect the lessons learned. The simulation modeled the public health aspect of the virtual population, as well as the economy of the virtual geographies. The artificial population would also voice a public opinion, giving a measure of support for the decisions and actions the government is taking on their behalf. The simulation provided the capability to analyze the impact of the crisis event as well as the government response.

With such powerful capabilities, there are numerous advantages to using the simulation to augment the live exercise. The simulation allows us to scale the scenario to a much larger geographical area than possible with just a live exercise, thereby allowing key decision makers to keep the bigger picture in mind.

---

* This research was partially supported by the National Science Foundation's DDDAS program grant # CNS-0325846 and the Indiana State 21st Century Research and Technology award #1110030618.

Y. Shi et al. (Eds.): ICCS 2007, Part I, LNCS 4487, pp. 1106–1113, 2007.

The simulation can execute in faster-than-real-time, allowing the participants to analyze the long-term impacts of their actions in a matter of minutes. The simulation also provides the ability to move forward and backward in virtual time, to analyze possible future implications of current actions, or to go back and retry the response to achieve better results.

In the future we hope to allow the participants to have greater interaction with the simulation. The participants would receive continuously updated statistics from the simulation and the live exercise. This will allow them to make more strategic decisions on the scenario. With hands-on simulation training provided, or technical support staff taking actions on behalf of the participants, more time can be spent analyzing the results of the simulation than dealing with the simulation itself. The simulation is intended to be used as a tool for discussion of critical issues and problems in a response and recovery scenario. With the live exercise and the simulation connecting in real-time, the accuracy of the simulation will greatly improve, thereby providing more meaningful information to the key players.

2   Computational Modeling

The computational modeling is based on the Synthetic Environments for Analysis and Simulation (SEAS) platform. SEAS provides a framework that is unbiased to any one specific scenario, model, or system and can be used to represent fundamental human behavior theories without restrictions of what can be modeled common in modern simulation efforts. The enabling technology leverages recent advances in agent-based distributed computing to decouple control as well as data flow. SEAS is built from a basis of millions of agents operating within a synthetic environment. Agents emulate the attributes and interactions of individuals, organizations, institutions, infrastructure, and geographical decompositions. Agents join together to form networks from which evolve the various cultures of the global population. Intricate relationships among political, military, economic, social, information and infrastructure (PMESII) factors emerge across diverse granularities. Statistics calculated from the simulation are then used to provide measurable evaluations of strategies in support of decision making.

The fundamental agent categories in SEAS are the individuals, organizations, institutions, and infrastructure (IOIIG). The population agents of these fundamental types will form higher order constructs in a fractal-like manner, meaning sufficient detail exists at multiple levels of focus, from world constructs to individuals. Higher order constructs include political systems (type of government, political parties/factions), militaries (soldiers, institutions, branches of service), economic systems (formal banking networks and black-market structures), social systems (tribes, religious groups, neighborhoods) and information systems (print, broadcast, internet). Agents representing individuals are used to model the populace in the synthetic environment. Individual agents are categorized into citizen and leader agents. An individual's well being is based on a model consisting of eight fundamental needs: basic, political, financial, security, religious, educational, health, and freedom of movement. The desire and perceived level of each of the well being categories are populated taking into account the socio-economic class of the individual the agent represents.

Citizen agents are constructed as a proportional representation of the societal makeup of a real nation. A citizen agent consists of a set of fundamental constructs: traits, well being, sensors, goals, and actions. The traits of citizen agents, such as race, ethnicity, income, education, religion, gender, and nationalism, are configured according to statistics gathered from real world studies. Dynamic traits, such as religious and political orientations, emotional arousal, location, health, and well being, result during simulation according to models that operate on the citizen agents and interactions they have with other agents. The traits and well being determine the goals of a citizen agent. Each citizen agent "senses" its environment, taking into account messages from leaders the citizen has built a relationship with, media the citizen subscribes to, and other members in the citizen's social network. Each citizen agent's state and goals can change as a result of interactions the citizen has with its environment. A citizen agent can react to its environment by autonomously choosing from its repertoire of actions. Additionally, a citizen agent's set of possible actions can change during the course of the simulation, such as when a citizen agent resorts to violence. Traits, well-being, sensors, and actions together determine the behavior of the citizen agent.

Clusters of citizen and leader agents form organizations. Citizen agents voluntarily join organizations due to affinity in perspective between the citizens and the organization. An organization agent's behavior is based on a foundation consisting of the desires of the organization's leaders and members. Organizational leadership constantly seeks maintenance and growth of the organizational membership by providing tangible and intangible benefits, and citizens subscribe based on a perceived level of benefit that is received from the organization. Additionally, through inter-organization networks, attitudes and resources may be shared among organizations. Through these internal and external interactions, organizations cause significant changes in perception and attitude change and become core protagonists of activism in the model. In turn, an organization exercises its power through the control over its resources and its ability to procure and maintain its resource base.

Institution agents are represented as 'governmental entities' such as the army, police, legislature, courts, executive, bureaucracy, and political parties-entities that are able to formulate policies that have legal binding, and have more discretionary resources. SEAS models institutions as structures that are products of individual choices or preferences, being constrained by the institutional structures (i.e. an interactive process). Institutions are like formal organizations with an additional power to influence the behaviors of members and non-members.

Media agents also play a significant role in providing information to other agents in the form of reports on well-being and attitudes. Media organizations consist of television, radio, newspapers, and magazines. The media make choices of what information to cover, who to cover, what statements to report, what story elements to emphasize and how to report them. Incidents are framed on well-being components, and formalized in a media report. Media is able to set the agenda for domestic policies as well as foreign policy issues. Citizens subscribe to media organizations based on their ideological bend. Media organizations act primarily to frame the issues for their audiences in such a way that they increase their viewer-ship as well as their influence.

Agents interact with the environment and respond, i.e., take action, to exogenous variables that may be specified by human agents or players in the environment as well as inputs from other agents. This is implemented with the introduction of inputs and outputs that each agent possesses. Inputs consist of general "environmental sensors" as well as particular "incoming message sensors." The incoming message sensors are singled out because of the importance ascribed to each agent to be able to query messages from the environment discriminately. The agent also possesses ports characterized collectively as "external actions" that allow the agent to submit its actions or messages to the environment. Finally, the agent possesses an internal set of rules classified as "internal actions" that result in the agents "external actions" on the basis of the sensor inputs as well as the traits/attributes and intelligence structure of each agent.

2.1 Virtual Geographies

The simulation will consist of a fictitious community that contains all the relevant features (hospitals, railways, airports, lakes, rivers, schools, business districts) from any Indiana community. This fictitious community can be customized to mimic a real community within Indiana. The virtual geography may be divided into high population density residential areas, low population density residential areas, commercial areas as well as uninhabitable areas. There can be various level of granularity for different communities as needed for the scenario (from international, national, state, district, county, city to city block levels).

2.2 Computational Epidemiology of Synthetic Population

The virtual community will have a virtual population represented by artificial agents. An agent is able to represent the activity of a human through a combination of learned variables and interactions. Research has shown that these agents act as the vertices of a societal network, and that their interactions comprise the edges of the network [Wasserman, 1994]. Like living beings, each agent has different interactions and experiences, and thus acts differently when faced with a situation. And while these evolving differences are essential for a useful simulation, certain predefined traits are also necessary. As an example, though all students in a class may be exposed to a flu virus, certain members will be more susceptible, and case severity will differ among those who contract the illness. For this reason, parameters must be assigned that define the susceptibility of an agent to a given pathogen. The high number of relevant attributes for each agent serves to differentiate each agent from its peers. But as the artificial agents grow in complexity, they must also grow in number, in order to maintain the characteristics of the society they seek to describe. Once the society has been sufficiently populated, the artificial agents begin to interact with and learn from each other, forming an environment well suited for analysis and interaction by human agents.

In addition to these behaviors, each agent is endowed with certain characteristics that help to differentiate the population. These attributes help to model the variability in human response to a situation. As an example, a wealthier individual may be more likely to leave a high-risk area, if only because of the financial independence he or she enjoys. The following is a partial list of characteristics that serve to differentiate one artificial agent from another: Age, Sex, Income, Education, and Health.

The decision-making process for an artificial agent is a simplified version of the decision making process of rational humans. When faced with a decision, rational humans consider numerous variables. Using a combination of intuition, experience, and logic, one selects the alternative that leads to a certain goal - usually happiness. And while different decisions vary in magnitude, the underlying cognitive model remains relatively constant. As such, while different physical or psychological needs take precedence in different situations, the human decision-making process can be modeled by considering each need in a hierarchical manner. To illustrate, scholarship has shown that, when presented with a threatening environment, the primary focus of a living being shifts to ensuring its own survival. The list that follows partially describes the variables that an artificial agent considers before making a decision: Security, Information Level, Health, Basic Necessities, Mobility and Freedom, Financial Capability, and Global Economy.

In the SEAS environment, as in the real world, reproductive rates and propagation vary according to the type of disease. Similarly, variables such as population density, agent mobility, social structure, and way of life interact to determine the proliferation of the disease. The government officials, or human agents, interact with the system and control propagation via such means as vaccination, treatment, or agent isolation. The options available to the human agents are the same as in real life, and the effectiveness of these interactions is modeled using statistically verified historical information [Longini, et. al 2000].

2.3 Public Opinion Model

While the safety of the artificial agents takes highest precedence, government officials must consider the overall spirit of the population when making decisions. To illustrate, though safety may be maximized by quarantining a city in every instance of potential attack [Kaplan, Craft & Wein, 2002], such restrictive measures may not be tolerated by the population. To enhance the level of learning they can achieve through the simulation, the human agents must consider the impact on public sentiment that each of their decisions may have. As in real life, each artificial agent determines his or her happiness level using a combination of variables: Current health status - must be alive in order to hold an opinion; perceived security; information level; basic necessities; and freedom of mobility.

2.4   Economic Impact Model

The simulation projects the long term economic impact of the crisis scenario. The economic impact of the crisis scenario as well as the government response is modeled based on the following criteria: Loss from impact on public health, cost of response, loss of productivity, loss of business, and loss from impact on public opinion.

3   Application: Indiana District 3 Full-Scale Exercise 2007

For the Indiana District 3 full-scale exercise in January 2007, the crisis scenario involved the intentional release of aerosolized anthrax at a fictitious district-wide festival. The following section describes the district-wide results of the simulation in more detail.

3.1   Detailed Public Health Statistics for District 3

As shown in Fig. 1, the scenario assumed that 22,186 people were initially exposed to the anthrax. As the epidemiology of anthrax initiates, 17,693 people who were not exposed to enough anthrax spores move into the recovered health status. The remaining 4,493 people started showing symptoms and falling sick, with a brief period of reduction in symptoms, followed by high fever and an 80% chance of shock and death. The remaining 892 people eventually recovered. Note: This graph is for the no-intervention case (assuming no mass prophylaxis was started).

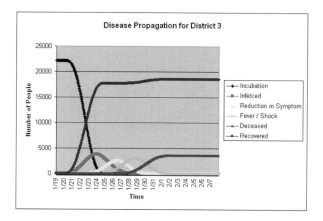

Fig. 1. Disease progression within the population of District 3 over a period of 3 weeks

3.2   Economic Impact for District 3

Even with a strong public health response, the District would have to deal with a tremendous economic impact (close to 400 million dollars in the long term) due to the crisis situation, as seen from Fig. 2.

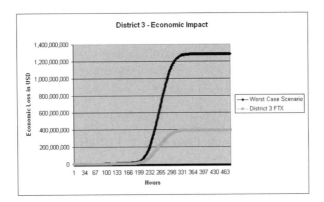

Fig. 2. The economic loss is approximately 900 million dollars less than worst case

3.3   Public Opinion Impact for District 3

In the worst case scenario, the population became aware of the crisis when people started to die from the anthrax exposure, as seen in Fig. 3. Hence, public opinion drops at a later date than when the government announces plans for mass prophylaxis. Even though public opinion dropped sooner, it did not go as low as the worst case scenario, due to the proactive and efficient government response.

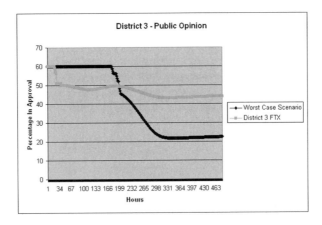

Fig. 3. The public opinion is 21.47% more in approval of the government than the worst case scenario due to the government response in curbing the situation

4   Conclusion

The quick response of the local agencies participating in the exercise resulted in fewer casualties in all counties within District 3. The counties were quick to determine the shortage of regimens available to them locally, and in requesting additional regimens from the state. All the counties would have been able to complete the simulated mass prophylaxis of the exposed population within the timeline guided by the CDC (within 48 hours after exposure) based on their survey responses. This created a dramatic difference (saving 2,505 lives) in the public health statistics of District 3 as compared to the worst case.

While the worst case economic loss would have been around 1.3 billion dollars for District 3, the estimated economic loss for the District 3 Exercise was only 392 million due to the government response and public health actions. Enormous long term economic loss could have crippled the entire district, if the crisis was not handled properly. This situation was avoided during the District 3 Exercise due to positive and efficient government actions.

Initial drop in public opinion was due to the inability of the government to prevent the terror attack from taking place, however, in the long term, the government response and gain of control over the situation, stabilized the public opinion. Based on the public opinion results, it would take some time before the public opinion would come back to normal levels - it likely would take aggressive media campaigns and public service announcements, by the public information officers as well as the elected officials, to mitigate the general state of panic and fear of such an attack happening again.

References

1. Data from CDC website (http://www.bt.cdc.gov/agent/anthrax/anthrax-hcp-factsheet.asp)
2. Data from FDA website (http://www.fda.gov/CBER/vaccine/anthrax.htm)
3. Reducing Mortality from Anthrax Bioterrorism: Strategies for Stockpiling and Dispensing Medical and Pharmaceutical Supplies. Dena M. Bravata. Table 1
4. Systematic Review: A Century of Inhalational Anthrax Cases from 1900 to 2005 Jon-Erik K. Holty P.g: 275
5. Center for Terrorism Risk Management Policy (http://www.rand.org)
6. The Roper Center for Public Opinion Research at the University of Connecticut
7. E.H. Kaplan, D.L. Craft and L.M. Wein, Emergency response to a smallpox attack: The case for mass vaccination, PNAS, 2002
8. I.M. Longini, M. Elizabeth Halloran, A. Nizam, et al., Estimation of the efficacy of life, attenuated influenza vaccine from a two-year, multi-center vaccine trial: implications for influenza epidemic control, Vaccine, 18 (2000) 1902-1909

Dynamic Tracking of Facial Expressions Using Adaptive, Overlapping Subspaces

Dimitris Metaxas, Atul Kanaujia, and Zhiguo Li

Department of Computer Science, Rutgers University
{dnm,kanaujia,zhli}@cs.rutgers.edu

Abstract. We present a *Dynamic Data Driven Application System (DDDAS)* to track 2D shapes across large pose variations by learning non-linear shape manifold as overlapping, piecewise linear subspaces. The learned subspaces adaptively adjust to the subject by tracking the shapes independently using Kanade Lucas Tomasi(KLT) point tracker. The novelty of our approach is that the tracking of feature points is used to generate independent training examples for updating the learned shape manifold and the appearance model. We use landmark based shape analysis to train a Gaussian mixture model over the aligned shapes and learn a Point Distribution Model(PDM) for each of the mixture components. The target 2D shape is searched by first maximizing the mixture probability density for the local feature intensity profiles along the normal followed by constraining the global shape using the most probable PDM cluster. The feature shapes are robustly tracked across multiple frames by dynamically switching between the PDMs. The tracked 2D facial features are used deform the 3D face mask.The main advantage of the 3D deformable face models is the reduced dimensionality. The smaller number of degree of freedom makes the system more robust and enables capturing subtle facial expressions as change of only a few parameters. We demonstrate the results on tracking facial features and provide several empirical results to validate our approach. Our framework runs close to real time at 25 frames per second.

1   Introduction

Tracking deformable shapes across multiple viewpoints is an active area of research and has many applications in biometrics, facial expressions analysis and synthesis for deception, security and human-computer interaction applications. Accurate reconstruction and tracking of 3D objects require well defined delineation of the object boundaries across multiple views.

Landmark based deformable models like Active Shape Models(ASM)[1]have proved effective for object shape interpretation in 2D images and have lead to advanced tools for statistical shape analysis. ASM detects features in the image by combining prior shape information with the observed image data. A major limitation of ASM is that it ignores the non-linear geometry of the shape manifold. Aspect changes of 3D objects causes shapes to vary non-linearly on a hyper-spherical manifold.

Y. Shi et al. (Eds.): ICCS 2007, Part I, LNCS 4487, pp. 1114–1121, 2007.

A generic shape model that would fit any facial expression is difficult to train, due to numerous possible faces and relative feature locations. In this work we present a generic framework to learn non-linear shape space as overlapping piecewise linear subspaces and then dynamically adapting the shape and appearance model to the Face of the subject. We do this by accurately tracking facial features across large head rotations and re-training the model specific to the subject using the unseen shapes generated from KLT tracking. We use the Point Distribution Models(PDM) to represent the facial feature shapes and use ASM to detect them in the 2D image. Our generic framework enables large scale automated training of different shapes from multiple viewpoints. The shape model is composed of the *Principal Components* that account for most of the variations arising in the data set. Our *Dynamic Data Driven framework* continuously collects different shapes by tracking feature points independently and adjusts the principal components basis to customize it for the subject.

2   Related Work

A large segment of research in the past decade has focused on incorporating non-linear statistical models for learning shape manifold. Murase et. al. [2] showed that pose from multiple viewpoint when projected onto eigenspaces generates a 2D hypersphere manifold. Gong et. al [3] used non-linear projections onto the eigenspace to track and estimate pose from multiple viewpoints. Romdhani et al. [4] proposed an ASM based on Kernel PCA to learn shape variation of face due to yaw. Several prominent work exist on facial feature registration and tracking, use appearance based models(AAM)[5,6]. [5] uses multiple independent 2D AAM models to learn correspondences between features of different viewpoints. The most notable work in improving ASM to learn non-linearities in the training data is by Cootes et. al[7] in which large variation is shapes is captured by parametric Gaussian mixture density, learned in the principal subspace. Unlike [5], our framework does not require explicit modeling of head pose angles. Although we use multivariate gaussian mixture model to learn initial clusters of the shape distribution, our subspaces are obtained by explicitly overlapping the clusters.

3   Learning Shape Manifold

An Active Shape Model(ASM) is a landmark based model that tries to learn a statistical distribution over variations in shapes for a given class of objects. Changes in viewpoint causes the object shapes to lie on a hyper-sphere and cannot be accurately modeled using linear statistical tools.

Face shape variation across multiple aspects is different across human subjects. It is therefore inaccurate to use a static model to track facial features for different subjects. Our approach to dynamically specialize the learned shape manifold to a human subject provides an elegant solution to this problem. However tracking shapes across multiple aspects requires modeling and synthesis of paths between the source and target shapes lying on a non-linear manifold. In our framework non-linear region is approximated as a combination of multiple smaller linear subregions. For the first frame, we search the shape subspace iteratively by searching along the normals of the landmark points and simultaneously constraining it to lie on the shape manifold. The path between the source shape and the target shape is traversed by searching across multiple subspaces that constitute the non-linear shape surface. For the subsequent frames, we track the facial features independent of the prior shape model. The tracked shapes are used to learn Principal Components of the shape and appearance models that capture the variations specific to the human subject face. As a pre-requisite for shape analysis, all the 2D planar shapes are aligned to the common co-ordinate system using Generalized Procrustes Analysis[8]. The tangent space approximation $\mathbf{T_s}$ projects the shapes on a hyper-plane normal to the mean vector and passing through it. Tangent space is a linear approximation of the general shape space so that the Procrustes distance can be approximated as euclidean distance between the planar shapes. The cluster analysis of shape is done in the global tangent space. We assume a generative multivariate Gaussian mixture distribution for both the global shapes and the intensity profile models(IPMs). The conditional density of the shape $\mathbf{S_i}$ belonging to an N-class model $p(\mathbf{S_i}|\text{Cluster}) =$

$$\sum_{j=1}^{N} \gamma_j (2\pi)^{-(\frac{N}{2})} \|\mathbf{C_j}\|^{-1/2} \exp\{-\frac{1}{2}(S_i - (\mu_j + P_j b_j))^T \mathbf{C_j}^{-1}(S_i - (\mu_j + P_j b_j))\} \quad (1)$$

We also assume a diagonal covariance matrix $\mathbf{C_j}$. γ_j are the cluster weights and (μ_j, P_j, b_j) are the mean, eigen matrix and eigen coefficients respectively for the principle subspace defined for each cluster. The clustering can be achieved by the EM algorithm with variance flooring to ensure sufficient overlapping between the clusters. For each of the N clusters we learn a locally linear PDM using PCA and using the eigenvectors to capture significant variance in the cluster(98%). The intensity profiles for the landmark points also exhibit large variation when trained over multiple head poses. The change in face aspects causes the profiles to vary considerably for the feature points that are occluded. The multivariate Gaussian mixture distribution(1) is learned for the local intensity profiles model(IPM) in order to capture variations that cannot be learned using a single PCA model.

Overlapping Between Clusters: It is important that the adjacent clusters overlap sufficiently to ensure switching between subspaces during image search and tracking. We can ensure subspace overlap by using boundary points between adjacent clusters to learn the subspace for both the clusters. These points can be obtained as nearest to the cluster center but not belonging to that cluster.

4   Image Search in the Clustered Shape Space

Conventional ASM uses an Alternating Optimization(AO) technique to fit the shape by searching for the best matched profile along the normal followed by constraining the shape to lie within the learned subspace. The initial average

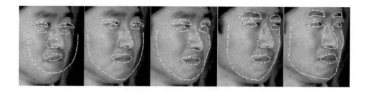

Fig. 1. Iterative search across multiple clusters to fit the face. The frames correspond to iteration 1(Cluster 1), iter. 3(Cluster 5), iter. 17(Cluster 7), iter. 23(Cluster 6) and final fit at iter. 33(Cluster 6) for level 4 of the Gaussian pyramid.

shape is assumed to be in a region near to the target object. We use robust Viola-Jones face detector to extract a bounding box around the face and use its dimensions to initialize the search shape. The face detector has 99% detection rate for faces with off-plane and in-plane rotation angles $\pm30^\circ$. We assign the nearest Cluster$_i$ to the average shape based on the mahalanobis distance between the average shape and the cluster centers in the global tangent space. The image search is initiated at the top most level of the pyramid by searching IPM along normals and maximizing the mixture probability density (1) of the intensity gradient along the profile. The model update step shifts the shape to the current cluster subspace by truncating the eigen coefficients to lie within the allowable variance as $\pm2\sqrt{\lambda_i}$. The shape is re-assigned the nearest cluster based on the mahalanobis distance and the shape coefficients are re-computed if the current subspace is different from the previous.

The truncation function to regularize the shapes usually generates discontinuous shape estimates. We use the truncation approach, due to its low computational requirement and faster convergence. The above steps are performed iteratively and converges irrespective of the initial cluster of the average shape.

5   Dynamic Data Driven Tracking Framework

We track the features independent of the ASM by Sum of Squared Intensity Difference(SSID) tracker across consecutive frames[9]. The SSID tracker is a method for registering two images and computes the displacement of the feature by minimizing the intensity matching cost, computed over a fixed sized window around the feature. Over a small inter-frame motion, a linear translation model can be accurately assumed. For an intensity surface at image location $\mathbf{I}(\mathbf{x_i}, \mathbf{y_i}, \mathbf{t_k})$, the tracker estimates the displacement vector $\mathbf{d} = (\delta\mathbf{x_i}, \delta\mathbf{y_i})$ from new image $\mathbf{I}(\mathbf{x_i} + \delta\mathbf{x}, \mathbf{y_i} + \delta\mathbf{y}, \mathbf{t_{k+1}})$ by minimizing the residual error over a window $\mathcal{W}$ around $(\mathbf{x_i}, \mathbf{y_i})$ [9].

$$\int_{\mathcal{W}} [\mathbf{I}(\mathbf{x_i} + \delta\mathbf{x}, \mathbf{y_i} + \delta\mathbf{y}, \mathbf{t_{k+1}}) - \mathbf{g}.\mathbf{d} - \mathbf{I}(\mathbf{x_i}, \mathbf{y_i}, \mathbf{t_k})] \, d\mathcal{W} \qquad (2)$$

The inter-frame image warping model assumes that for small displacements of intensity surface of image window $\mathcal{W}$, the horizontal and vertical displacement of the surface at a point $(\mathbf{x_i}, \mathbf{y_i})$ is a function of gradient vector $\mathbf{g}$ at that point.

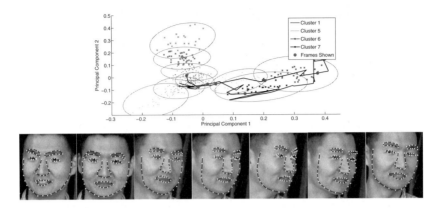

Fig. 2. *(Best Viewed in Color)*Tracking the shapes across right head rotation.**(Top)** The cluster projections on 2D space using 2 principal modes(for visualization)and the bounded by hyper-ellipsoid subspace. The right head rotation causes the shape to vary across the clusters. The red circles corresponds to the frames 1, 49, 68, 76, 114, 262 and 281. The entire tracking path lies within the subspace spanned by the hyper-ellipsoids.**(Bottom)** The images of the tracking result for the frames shown as red markers in the plot.

The tracking framework generates a number of new shapes not seen during the training for ASM and hence provides independent data for our dynamic data driven application systems. Both the appearance (IPMs) and the shape models are composed of *Principal Vector* basis that are dynamically updated as we obtain new shapes and IPMs for the landmark points. For the shape $\mathcal{X}_{i+1}$ at time step $(i + 1)$, the covariance matrix $\mathbf{C_i}$, is updated as

$$\mathbf{C_{i+1}} = ((N + i) - \frac{K}{N + i}) * \mathbf{C_i} + \frac{K}{N + i} * \mathcal{X}_{i+1}{}^T \mathcal{X}_{i+1} \tag{3}$$

where N is the number of training examples and i is the current tracked frame. The updated covariance matrix $\mathbf{C_{i+1}}$ is diagonalized using power method to obtain new set of basis vectors. The subspace corresponding to these basis vectors encapsulates the unseen shape. The sequence of independent shapes and IPMs for the landmarks are used to update the current and neighboring subspaces, and the magnitude of updates can be controlled by the predefined learning rate K. The number of PCA basis vectors(eigenvectors) may also vary as a result of updation and specialization of the shape and the appearance model. Fig. 3 illustrates the applicability of our adaptive learning methodology to extreme facial expressions of surprise, fear, joy and disgust (not present in training images). For every frame we align the new shape $\mathbf{Y_t}$ to the global average shape $\overline{\mathbf{X}}_{\mathbf{init}}$ and re-assign it to the nearest Cluster$_i$ based on mahalanobis distance. Finally after every alternate frame we ensure that the shape $\mathbf{Y_t}$ obtained from tracking is a plausible shape by constraining the shape to lie on the shape manifold of the current cluster. Fig. 2 shows the path (projection on 2 principal components) of a shape(and

Fig. 3. 2D Tracking for extreme facial expressions

the corresponding cluster) for a tracking sequence when the subject rotates the head from frontal to full right profile view and back. The entire path remains within the plausible shape manifold spanned by the 9 hyper-ellipsoid subspaces.

6   Deformable Model Based 3D Face Tracking

Deformable model based 3D face tracking is the process of estimation, over time, of the value of face deformation parameters (also known as the state vector of the system) based on image forces computed from face image sequences. Our objective is to build a dynamically coupled system that can recover both the rigid motion and deformations of a human face, without the use of manual labels or special equipment. The main advantage of deformable face models is the reduced dimensionality. The smaller number of degree of freedom makes the system more robust and efficient, and it also makes post-processing tasks, such as facial expression analysis, more convenient based on recovered parameters. However, the accuracy and reliability of a deformable model tracking application is strongly dependent on accurate tracking of image features, which act as 2D image force for 3D model reconstruction. Low level feature tracking algorithms, such as optical flows, often suffer from occlusion, unrealistic assumptions etc. On the other hand, model based 2D feature extraction method, such as active shape model, has been shown to be less prone to image noises and can deal with occlusions. In this paper, we take advantage of the coupling of the 3D deformable model and 2D active shape model for accurate 3D face tracking. On the one hand, 3D deformable model can get more reliable 2D image force from the 2D active shape model. On the other hand, 2D active shape model will benefit from the good initialization provided by the 3D deformable model, and thus improve accuracy and speed of 2D active shape model. The coupled system can handle large rotations and occlusions. A 3D deformable model is parameterized by a set of parameters $\mathbf{q}$. Changes in $\mathbf{q}$ causes geometric deformations of the model. A particular point on the surface is denoted by $\mathbf{x}(\mathbf{q}; \mathbf{u})$ with $\mathbf{u} \in \Omega$. The goal of a shape and motion estimation process is to recover parameter $\mathbf{q}$ from face image sequences. To distinguish between shape estimation and motion tracking, the parameters $\mathbf{q}$ can be divided into two parts: static parameter $\mathbf{q}_s$, which describes the unchanging features of a particular face, and dynamic parameter $\mathbf{q}_m$, which describes the global (rotation and translation of the head) and local deformations (facial expressions) of an observed face during tracking. The deformations can also be divided into two parts: $\mathbf{T}_s$ for shape and $\mathbf{T}_m$ for motion (expression), such

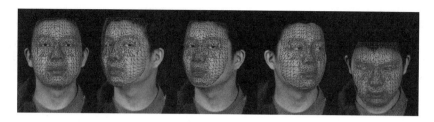

Fig. 4. 3D tracking results of deformable mask with large off-plane head rotations

that $\mathbf{x}(\mathbf{q}; \mathbf{u}) = \mathbf{T}_m(\mathbf{q}_m; \mathbf{T}_s(\mathbf{q}_s; s(\mathbf{u})))$ The kinematics of the model is $\dot{\mathbf{x}}(\mathbf{u}) = \mathbf{L}(\mathbf{q}; \mathbf{u})\dot{\mathbf{q}}$, where $\mathbf{L} = \frac{\partial \mathbf{x}}{\partial \mathbf{q}}$ is the model Jacobian. Considering the face images under a perspective camera with focal length f, the point $\mathbf{x}(\mathbf{u}) = (x, y, z)^T$ projects to the image point $\mathbf{x}_p(\mathbf{u}) = \frac{f}{z}(x, y)^T$. The kinematics of the new model is given by:

$$\dot{\mathbf{x}}_p(\mathbf{u}) = \frac{\partial \mathbf{x}_p}{\partial \mathbf{x}} \dot{\mathbf{x}}(\mathbf{u}) = (\frac{\partial \mathbf{x}_p}{\partial \mathbf{x}} \mathbf{L}(\mathbf{q}; \mathbf{u}))\dot{\mathbf{q}} = \mathbf{L}_p(\mathbf{q}; \mathbf{u})\dot{\mathbf{q}} \tag{4}$$

where the projection Jacobian matrix is

$$\frac{\partial \mathbf{x}_p}{\partial \mathbf{x}} = \begin{bmatrix} f/z & 0 & -fx/z^2 \\ 0 & f/z & -fy/z^2 \end{bmatrix} \tag{5}$$

which converts the 2D image forces to 3D forces. Estimation of the model parameters $\mathbf{q}$ is based on first order Lagrangian dynamics [10], $\dot{\mathbf{q}} = \mathbf{f}_q$ Where the generalized forces $\mathbf{f}_q$ are identified by the displacements between the actual projected model points and the identified corresponding 2D image features, which in this paper are the 2D active shape model points. They are computed as:

$$\mathbf{f}_q = \sum_j (\mathbf{L}_p(\mathbf{u}_j)^T \mathbf{f}_{image}(\mathbf{u}_j)) \tag{6}$$

Given an adequate model initialization, these forces will align features on the model with image features, thereby determining the object parameters. The dynamic system is solved by integrating over time, using standard differential equation integration techniques:

$$\mathbf{q}(t+1) = \mathbf{q}(t) + \dot{\mathbf{q}}(t)\Delta t \tag{7}$$

Goldenstein *et. al* showed in [11] that the image forces $\mathbf{f}_{image}$ and generalized forces $\mathbf{f}_q$ in these equations can be replaced with affine forms that represent probability distributions, and furthermore that with sufficiently many image forces, the generalized force converges to a Gaussian distribution. In this paper, we take advantage of this property by integrating the contributions of ASMs with other cues, so as to achieve robust tracking even when ASM methods and standard 3D deformable model tracking methods provide unreliable results by themselves.

7   Conclusion

In this work we have presented a real time DDDAS framework for detecting and tracking deformable shapes across non-linear variations arising due to aspect changes. Detailed analysis and empirical results have been presented about issues related to the modeling non-linear shape manifolds using piecewise linear models. The shape and appearance model updates itself using new shapes obtained from tracking the feature points. The tracked 2D features are used to deform the 3D face mask and summarize the facial expressions using only a few parameters. This framework has many application in face-based deception analysis and we are in the process of performing many tests based on relevant data.

Acknowledgement

This work has been supported in part by the National Science Foundation under the following two grants NSF-ITR-0428231 and NSF-ITR-0313184.

Patent Pending

The current technology is protected by patenting and trade marking office, *"System and Method for Tracking Facial Features,"*, Atul Kanaujia and Dimitris Metaxas, Rutgers Docket 07-015, Provisional Patent #60874, 451 filed December, 12 2006. No part of this technology may be reproduced or displayed in any form without the prior written permission of the authors.

References

1. Cootes, T.: An Introduction to Active Shape Models. Oxford University Press (2000)
2. Murase, H., Nayar, S.: Learning and recognition of 3D Objects from appearance. IJCV (1995)
3. Gong, S., Ong, E.J., McKenna, S.: Learning to associate faces across views in vector space of similarities to prototypes. BMVC (1998)
4. Romdhani, S., Gong, S., Psarrou, A.: A Multi-View Nonlinear Active Shape Model Using Kernel PCA. BMVC (1999)
5. Cootes, T., Wheeler, G., Walker, K., Taylor, C.: View-Based Active Appearance Models. BMVC (2001)
6. Edwards, G.J., Taylor, C.J., Cootes, T.F.: Learning to Identify and Track Faces in Image Sequences. BMVC (1997)
7. Cootes, T., Taylor, C.: A mixture model for representing shape variation. BMVC (1997)
8. Goodall, C.: Procrustes methods in the statistical analysis of shape. Journal of the Royal Statistical Society (1991)
9. Tomasi, C., Kanade, T.: Detection and Tracking of Point Features. Technical Report CMU-CS-91-132 (1997)
10. Metaxas, D.: Physics-Based Deformable Models: Applications to Computer Vision, Graphics and Medical Imaging. Kluwer Academic Publishers (1996)
11. Goldenstein, S., Vogler, C., Metaxas, D.: Statistical Cue Integration in DAG Deformable Models. PAMI (2003)

Realization of Dynamically Adaptive Weather Analysis and Forecasting in LEAD: Four Years Down the Road

Lavanya Ramakrishnan, Yogesh Simmhan, and Beth Plale

School of Informatics, Indiana University, Bloomington, IN 47045,
{laramakr, ysimmhan, plale}@cs.indiana.edu

Abstract. Linked Environments for Atmospheric Discovery (LEAD) is a large-scale cyberinfrastructure effort in support of mesoscale meteorology. One of the primary goals of the infrastructure is support for real-time dynamic, adaptive response to severe weather. In this paper we revisit the conception of dynamic adaptivity as appeared in our 2005 DDDAS workshop paper, and discuss changes since the original conceptualization, and lessons learned in working with a complex service oriented architecture in support of data driven science.

Keywords: Weather Analysis and Forecasting.

1 Introduction

Linked Environments for Atmospheric Discovery (LEAD)[2] [1] is a large-scale cyberinfrastructure effort in support of mesoscale meteorology. This is accomplished through middleware that facilitates adaptive utilization of distributed resources, sensors and workflows, driven by an adaptive service-oriented architecture (SOA). As an SOA, LEAD encapsulates both application and middleware functionality into services. These services include both atomic application tasks as well as resource and instrument monitoring agents that drive the workflow. The project is broad, with significant effort expended on important efforts such as education and outreach.

LEAD was conceived in the early 2000's in response to the then state-of-the-art in meteorology forecasting. Forecasts were issued on a static, cyclic schedule, independent of current weather conditions. But important technology and science factors were converging to make it possible to transform weather forecasting by making forecast initiation automatic and responsive to the weather. The grid computing community was focused on web services as a scalable, interoperable architecture paradigm [12]. Research was occurring on one-pass data-mining algorithms for mesoscale phenomena [3]. The CASA Engineering Research Center

---

[1] Funded by National Science Foundation under Cooperative Agreements: ATM-0331594 (OU), ATM-0331591 (CO State), ATM-0331574 (Millersville), ATM-0331480 (IU), ATM-0331579 (UAH), ATM03-31586 (Howard), ATM-0331587 (UCAR), and ATM-0331578 (UIUC).

[4] was building small, inexpensive high resolution Doppler radars. Finally, large-scale computational grids, such as TeraGrid, began to emerge as a community resource for large-scale distributed computations.

In this paper we revisit the concept of dynamic adaptivity as was presented in our DDDAS workshop paper of 2005 [1], a conceptualization that has grown and matured. We discuss the facets of the model as they exist today, and touch on lessons learned in working with a complex SOA in support of data driven science.

2   System Model

Creating a cyberinfrastructure that supports dynamic, adaptive responses to current weather conditions requires several facets of dynamism. The service framework must be able to respond to weather conditions by detecting the condition then directing and allocating resources to collect more information about the weather and generate forecasts. Events also occur as execution or run-time phenomena: problems in ingesting data, in network failure, in the resource availability, and in inadequate model progress for instance. Adaptivity is driven by several key requirements:

User-Initiated Workflows. A typical mode of usage of the LEAD system is user-initiated workflow through a portal (also known as "science gateway") where a user composes a workflow or selects a pre-composed workflow and configures the computational components, and data selection criteria. In this scenario, the system needs mechanisms to procure resources and enable workflow execution, provide recovery mechanisms from persistent and transient service failures, adapt to resource availability, and recover from resource failures during workflow execution.

Priorities of Workflows. The LEAD cyberinfrastructure simultaneously supports science research and educational use, so workflow prioritization must be supported. Consider the case of an educational LEAD workshop where resources have been reserved through out-of-band mechanisms for advanced reservation. Resource allocation needs to be based on existing load on the machines, resource availability, the user priorities and workflow load. The bounded set of resources available to the workshop might need to be proportionally shared among the workflow users. If a severe weather event were to occur during the workshop, resources might need to be reallocated and conflicting events might need some arbitration.

Dynamic Weather Events. We consider the case of dynamic weather event detection with data mining. Users have the freedom to specify dynamic mining criteria from the portal, and use the triggers from detected weather phenomena as the basis for automated forecast initiation. This freedom creates resource arbitration issues. Multiple weather events and their severity might factor into assigning priorities between users for appropriate allocation of limited available resources, for instance.

Advanced User Workflow Alternatives. An advanced user has a workflow and provides a set of constraints (e.g., a time deadline) and the work to be done. Tradeoffs may have to be made to arbitrate resources. For instance, the user might be willing to sacrifice forecast resolution to get early results which might then define the rest of the workflow.

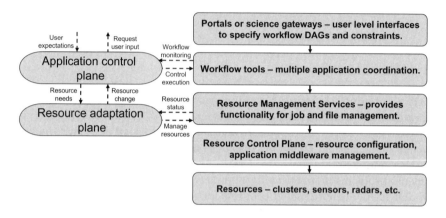

Fig. 1. Service and resource stack is controlled by application control plane interacting at workflow level; resource adaptation plane effects changes to underlying layers. Stream mining is a user-level abstraction, so executes as a node in a workflow.

The conceptualization of the system as reported in the 2005 DDDAS workshop paper [1] casts an adaptive infrastructure as an adaptation system that mirrors the forecast control flow. While a workflow executes services in the generation of a forecast, the adaptive system is busy monitoring the behavior of the system, the application, and the external environment. Through pushing events to a single software bus, the adaptive system interacts with the workflow system to enact appropriate responses to events.

The model eventually adopted is somewhat more sophisticated. As shown in Figure 1, the external-facing execution is one of users interacting through the portal to run workflows. The workflows consume resources, and access to the resources is mediated by a resource control plane [11]. The adaptive components are mostly hidden from the user. A critical component of the application control plane is monitoring workflow execution (Section 4). The resource adaptation plane manages changes in resource allocation, in consultation with the application control plane, and in response to a number of external stimuli (Section 5).

At the core of the LEAD architecture is a pair of scalable publish-subscribe event notification systems. One is a high-bandwidth event streaming bus designed to handle large amounts of distributed data traffic from instruments and other remote sources [5]. The second bus handles communication between the service components of the system. While not as fast as a specialized bus, it does not need to be. Its role is to be the conduit for the notifications related to the response triggers and all events associated with the workflow enactment as well as the overall state of the system [7]. This event bus is based on the WS-Eventing standard endorsed by Microsoft, IBM and others and is a very simple XML message channel.

3   Dynamic Data Mining

Dynamic weather event responsiveness is achieved by means of the Calder stream processing engine (SPE) developed at Indiana University [5] to provide on-the-wire continuous query processing access, filtering, and transforming of data in data streams. Functionality includes access to a large suite of clustering data mining algorithms for detecting mesoscale weather conditions developed at the University of Alabama Huntsville [3]. The SPE model is a view of data streams as a single coherent data repository (a "stream store") of indefinite streams, with provisioning for issuing SQL-like, continuous queries to access streams. The layer that transports stream data is a binary publish-subscribe system. Sensors and instruments are currently added to the stream network by a manual process of installing a point-of-presence in front of the instrument that converts events from the native format to the system's XML format and pub-sub communication protocol.

As an example of the SPE in LEAD, suppose a user wishes to keep an eye on the storm front moving into Chicago later that evening. He/she logs into the portal, and configures an agent to observe NEXRAD Level II radar data streams for severe weather developing over the Chicago region. The request is in the form of a continuous query. The query executes the mining algorithm repeatedly. Data mining will result in a response trigger, such as "concentration of high reflectivity found centered at lat=x, lon=y". The Calder service communicates with other LEAD components using the WS-Eventing notification system. It uses an internal event channel for the transfer of data streams. The query execution engine subscribes to channels that stream observational data as events that arrive as bzipped binary data chunks and are broken open to extract metadata that is then stored as an XML event.

4   System Monitoring

The dynamic nature of the workflows and data products in LEAD necessitates runtime monitoring to capture the workflow execution trace including invocation of services, creation of data products, and use of computational, storage, and network resources. There are two key drivers to our monitoring: to gather a near real-time view of the system to detect and pinpoint problems, and for building an archive of process and data provenance to assist in resource usage prediction and intelligent allocation for future workflow runs. Orthogonally, there are three types of monitoring that are done: application resource usage monitoring, application fault monitoring, and quality of service monitoring. In this section we describe the monitoring requirements and detail the use of the Karma provenance system [6] in monitoring the LEAD system.

Monitoring Resource Usage. Resource usage monitoring provides information on resource behavior, predicted time for workflow completion, and helps guide the creation of new "soft" resources as a side-effect of workflow execution. Taking a top-down view, resource usage starts with workflows that are launched through the workflow engine. The workflows, which behave as services consume resources from the workflow engine, itself a service. Various services that form part of the workflow represent the next level of resources used. Services launch application instances that run on computational nodes consuming compute resources. Data transfer between applications consumes network bandwidth while staging files consumes storage resources. In addition to knowledge about the available resource set present in the system that might be available through a network monitoring tool and prior resource reservations, real-time resource usage information will give an estimate of the resources available for allocation less those that might be prone to faults. In case of data products, creation of replicas as part of the workflow run makes a new resource (data replica) available to the system. Similar "soft" resources are transient services that are created for one workflow but may be reused by others. Resource usage information also allows us to extrapolate into the future behavior aiding resource allocation decisions.

Monitoring Application Faults. Dynamic systems have faults that take place in the applications plane and need appropriate action such as restarting the application from the last checkpoint, rerunning or redeploying applications. Faults may take place at different levels and it is possible that a service failure was related to a hardware resource failure. Hence sufficient correlation has to be present to link resources used across levels. Faults may take place in service creation because of insufficient hardware resources or permissions, during staging because of missing external data files or network failure, or during application execution due to missing software libraries.

Monitoring Quality of Service. Monitoring also aids in ensuring a minimum quality of service guarantee for applications. Workflow execution progress is tracked and used to estimate completion time. If the job cannot finish within the window stated, it may be necessary to preempt a lower priority workflow. The quality of data is determined through real-time monitoring and the maintenance of a quality model [8]. Data quality is a function of, among other attributes, the objective quality of service for accessing or transferring the data as well as the subjective quality of the data from a user's perspective, which may be configured for specific application needs.

LEAD uses the Karma provenance system for workflow instrumentation and services to generate real-time monitoring events and for storing them for future mining. Karma defines an information model [6] built upon a process-oriented view of workflows, and a data-oriented view of product generation and consumption. Activities describe the creation and termination of workflows and services, invocation of services and their responses (or faults), data transferred, consumed and produced by applications, and computational resources used by applications. The activities help identify the level at which the activity took place (workflow, service, application) and the time through causal ordering, along with attributes that describe specific activities. For example, the data produced activity generated by a service would describe the application that generated the data, the workflow it was part of and the stage in the workflow, the unique ID for the data along with the specific URL for that replica, and the timestamp of creation.

The activities are published as notifications using the publish-subscribe system that implements the WS-Eventing specification [7]. The Karma provenance service subscribes to all workflow related notifications and builds a global view of the workflow execution by stitching the activities together. One of the views exported by the provenance service through its querying API is the current state of a workflow in the form of an execution trace. This provides information about the various services and applications used by the workflow, the data and compute resources used, and the progress of the workflow execution. More fine-grained views can also be extracted that details the trace of a single service or application invocation. The information collected can be mined to address the needs postulated earlier. In addition, the activity notifications can also be used directly to monitor the system in real-time.

5   Adaptation for Performability

In today's grid and workflow systems, where nature of the applications or workflow is known apriori, resource management, workflow planning and adaptation techniques are based on performance characteristics of the application [9]. The LEAD workflows, in addition to having dynamic characteristics, also have very tight constraints in terms of time deadlines, etc. In these types of workflows, it is important to consider the reliability and timely availability of the underlying resources in conjunction with the performance of the workflow. Our goal is to adapt for performability, a term originally defined by J. Meyer [10]. Performability is used as a composite measure of performance and dependability, which is the measure of the system's performance in the event of failures and availability. The bottom-up performability evaluation of grid resources and the top-down user expectations, workflow constraints or needs of the application guides the adaptation in the application control plane and the resource adaptation planes. Adaptation might include procuring additional resources than originally anticipated, changing resources and/or services for scheduled workflows, reaction to failures, fault-tolerance strategies, and so on.

To meet the performability guarantees of the workflow, we propose a two-way communication in our adaptation framework, between the resource adaptation plane and the application control plane (see Figure 1). The application control plane interacts with the resource control plane to inquire about resource status and availability, select resources for workflow execution, and guide resource recruitment decisions. In turn, the application control plane needs information from the resource layer about resource status and availability and, during execution, about failures or changes in performance and reliability.

The adaptive system has workflow planner and controller components. The workflow planner applies user constraints and choices in conjunction with resource information to develop an "online" execution and adaptation plan. The workflow controller is the global monitoring agent that controls the run-time execution. The workflow planner comes up with an annotated plan to the original user-specified DAG that is then used by the workflow controller to monitor and orchestrate the progress of the workflow. The LEAD workflows have a unique set of requirements that drive different kinds of interaction between the workflow planner and the resource control plane. As discussed in section 2, the LEAD system is used for educational workshops. In this scenario, the workflow planner might need to distribute the bounded set of available resources among the workshop participants. It is possible during the course of the execution, additional resources become available which could be used by the existing workflows. When notified of such availability the workflow planning step can reconfigure the workflows to take advantage of the additional resources. In this scenario, the workflow adaptation will be completely transparent to the end user.

The goal of the workflow controller is to control the schedule and the adaptation tasks of the workflow engine. The workflow controller can potentially receive millions of adaptation events sometimes requiring conflicting actions and hence it uses an arbitration policy. For example if a weather event occurs during an educational workshop, resources will need to be reallocated and the other workflows paused till the higher priority workflows are serviced. The workflow controller uses pre-determined policies to determine the level of adaptation and dynamism it can respond to without additional intervention. Some adaptation decisions might require external human intervention, for example, if all TeraGrid machines go down at the same time. This multi-level adaptation framework enhances existing grid middleware allowing resource and user workflows to interact to enable a flexible, adaptive, resilient environment that can change to resource variability in conjunction with changing user requirements.

6   Conclusion

The LEAD adaptation framework provides a strong foundation for exploring the effect of complex next-generation workflow characteristics, such as hierarchical workflows, uncertainties in execution path, on resource coordination. Four years into the LEAD project we are considerably closer to realizing the goal of a fully adaptive system. Architecting a system in a multidisciplinary, collaborative academic setting is benefited by the modular nature of a service-oriented architecture. The loosely coupled solutions need only minimally interact. And as the infrastructure begins to take on large numbers of external and educational users, most notably to serve the Nationally Collegiate Forecasting contest issues of reliability and long queue delays become the most immediate and pressing of issues, issues for which the adaptation framework described here is well suited.

Acknowledgements. The authors thank the remaining LEAD team, including PIs Kelvin Droegemeier (Oklahoma Univ.), Mohan Ramamurthy (Unidata), Dennis Gannon (Indiana Univ.), Sara Graves (Univ. of Alabama Huntsville), Daniel A. Reed (UNC Chapel Hill), and Bob Wilhelmson (NCSA). Author Lavanya Ramakrishnan conducted some of the work while a researcher at UNC Chapel Hill.

References

1. B. Plale, D. Gannon, D. Reed, S. Graves, K. Droegemeier, B. Wilhelmson, M. Ramamurthy. Towards Dynamically Adaptive Weather Analysis and Forecasting in LEAD. In *ICCS workshop on Dynamic Data Driven Applications* and *LNCS*, 3515, pp. 624-631, 2005.
2. K. K. Droegemeier, D. Gannon, D. Reed, B. Plale, J. Alameda, T. Baltzer, K. Brewster, R. Clark, B. Domenico, S. Graves, E. Joseph, D. Murray, R. Ramachandran, M. Ramamurthy, L. Ramakrishnan, J. A. Rushing, D. Weber, R. Wilhelmson, A. Wilson, M. Xue and S. Yalda. Service-Oriented Environments for Dynamically Interacting with Mesoscale Weather. *Computing in Science and Engineering*, 7(6), pp. 12-29, 2005.
3. X. Li, R. Ramachandran, J. Rushing, S. Graves, Kevin Kelleher, S. Lakshmivarahan, and Jason Levit. Mining NEXRAD Radar Data: An investigative study. In *American Meteorology Society annual meeting*, 2004.
4. K.K. Droegemeier, J. Kurose, D. McLaughlin, B. Philips, M. Preston, S. Sekelsky, J. Brotzge, V. Chandresakar. Distributed collaborative adaptive sensing for hazardous weather detection, tracking, and predicting. In *International Conference on Computational Science (ICCS)*, 2004.
5. Y. Liu, N. Vijayakumar, B. Plale. Stream Processing in Data-driven Computational Science. In *7th IEEE/ACM International Conference on Grid Computing (Grid'06)*, 2006.
6. Y. L. Simmhan, B. Plale, and D. Gannon. A Framework for Collecting Provenance in Data-Centric Scientific Workflows. In *International Conference on Web Services (ICWS)*, 2006.
7. Y. Huang, A. Slominski, C. Herath and D. Gannon. WS-Messenger: A Web Services-based Messaging System for Service-Oriented Grid Computing. In *Cluster Computing and the Grid (CCGrid)*, 2006
8. Y. L. Simmhan, B. Plale, and D. Gannon. Towards a Quality Model for Effective Data Selection in Collaboratories. In *IEEE Workshop on Workflow and Data Flow for Scientific Applications (SciFlow)*, 2006.
9. A. Mandal, K. Kennedy, C. Koelbel, G. Marin. G, J. Mellor-Crummey, B. Liu and L. Johnsson. "Scheduling Strategies for Mapping Application Workflows onto the Grid". In *IEEE International Symposium on High Performance Distributed Computing*, 2005.
10. J. Meyer, On Evaluating the Performability of Degradable Computing Systems. In *IEEE Transactions Computers*, 1980.
11. L. Ramakrishnan, L. Grit, A. Iamnitchi, D. Irwin, A. Yumerefendi and J. Chase, "Toward a Doctrine of Containment: Grid Hosting with Adaptive Resource Control," In *ACM/IEEE SC 2006 Conference (SC'06)*, 2006.
12. I. Foster, C. Kesselman (eds), "The Grid 2: Blueprint for a New Computing Infrastructure," *Morgan Kaufmann Publishers Inc*, 2003.

Active Learning with Support Vector Machines for Tornado Prediction

Theodore B. Trafalis[1], Indra Adrianto[1], and Michael B. Richman[2]

[1] School of Industrial Engineering, University of Oklahoma, 202 West Boyd St, Room 124,
Norman, OK 73019, USA
ttrafalis@ou.edu, adrianto@ou.edu
[2] School of Meteorology, University of Oklahoma, 120 David L. Boren Blvd, Suite 5900,
Norman, OK 73072, USA
mrichman@ou.edu

Abstract. In this paper, active learning with support vector machines (SVMs) is applied to the problem of tornado prediction. This method is used to predict which storm-scale circulations yield tornadoes based on the radar derived Mesocyclone Detection Algorithm (MDA) and near-storm environment (NSE) attributes. The main goal of active learning is to choose the instances or data points that are important or have influence to our model to be labeled and included in the training set. We compare this method to passive learning with SVMs where the next instances to be included to the training set are randomly selected. The preliminary results show that active learning can achieve high performance and significantly reduce the size of training set.

Keywords: Active learning, support vector machines, tornado prediction, machine learning, weather forecasting.

1 Introduction

Most conventional learning methods use static data in the training set to construct a model or classifier. The ability of learning methods to update the model dynamically, using new incoming data, is important. One method that has this ability is active learning. The objective of active learning for classification is to choose the instances or data points to be labeled and included in the training set. In many machine learning tasks, collecting data and/or labeling data to create a training set is costly and time-consuming. Rather than selecting and labeling data randomly, it is better if we can label the data that are important or have influence to our model or classifier.

In tornado prediction, labeling data is considered costly and time consuming since we need to verify which storm-scale circulations produce tornadoes in the ground. The tornado events can be verified from facts in the ground including photographs, videos, damage surveys, and eyewitness reports. Based on tornado verification, we then determine and label which circulations produce tornadoes or not. Therefore, applying active learning for tornado prediction to minimize the need for the instances and use the most informative instances in the training set in order to update the classifier would be beneficial.

Y. Shi et al. (Eds.): ICCS 2007, Part I, LNCS 4487, pp. 1130–1137, 2007.

In the literature, the Mesocyclone Detection Algorithm (MDA) attributes [1] derived from Doppler radar velocity data have been used to detect tornado circulations. Marzban and Stumpf [1] applied artificial neural networks (ANNs) to classify MDA detections as tornadic or non-tornadic circulations. Additionally, Lakshmanan et al. [2] used ANNs and added the near-storm environment (NSE) data into the original MDA data set and determined that the skill improved marginally. Application of support vector machines (SVMs) using the same data set used by Marzban and Stumpf [1] has been investigated by Trafalis et al. [3]. Trafalis et al. [3] compared SVMs with other classification methods, such as ANNs and radial basis function networks, concluding that SVMs provided better performance in tornado detection. Moreover, a study by Adrianto et al. [4] revealed that the addition of NSE data into the MDA data can improve performance of the classifiers significantly. However, those experiments in the literature were conducted using static data.

In this paper, we investigated the application of active learning with SVMs for tornado prediction using the MDA and NSE data. We also compared this method to passive learning with SVMs using these data where the next instances to be added to the training set are randomly selected.

2   Data and Analysis

The original data set was comprised of 23 attributes taken from the MDA algorithm [1]. These attributes measure radar-derived velocity parameters that describe various aspects of the mesocyclone. Subsequently, 59 attributes from the NSE data [2] were incorporated to this data set. The NSE data described the pre-storm environment of the atmosphere on a broader scale than the MDA data, as the MDA attributes are radar-based. Information on wind speed, direction, wind shear, humidity lapse rate and the predisposition of the atmosphere to accelerate air rapidly upward over specific heights were measured by the NSE data. Therefore, the MDA+NSE data consist of 82 attributes.

3   Methodology

3.1   Support Vector Machines

The SVM algorithm was developed by Vapnik and has proliferated into a powerful method in machine learning [5-7]. This algorithm has been used in real-world applications and is well known for its superior practical results. In binary classification problems, the SVM algorithm constructs a hyperplane that separates a set of training vectors into two classes (Fig. 1). The objective of SVMs (the primal problem) is to maximize the margin of separation and to minimize the misclassification error. The SVM formulation can be written as follows [8]:

$$\min \frac{1}{2}\|\mathbf{w}\|^2 + C\sum_{i=1}^{l}\xi_i$$

$$\text{subject to } y_i\left(\langle \mathbf{w}\cdot\mathbf{x}_i\rangle + b\right) \geq 1 - \xi_i, \ \xi_i \geq 0, \ i = 1,...,l$$

(1)

where w is the weight vector perpendicular to the separating hyperplane, b is the bias of the separating hyperplane, ξ_i is a slack variable, and C is a user-specified parameter which represents a trade off between generalization and misclassification. Using Lagrange multipliers α, the SVM dual formulation becomes [8]:

$$\max Q(\alpha) = \sum_{i=1}^{l} \alpha_i - \frac{1}{2} \sum_{i=1}^{l} \sum_{j=i}^{l} \alpha_i \alpha_j y_i y_j \mathbf{x}_i \mathbf{x}_j$$

$$\text{subject to } \sum_{i=1}^{l} \alpha_i y_i = 0, \ \ 0 \le \alpha_i \le C, \ \ i = 1,...,l$$

(2)

The optimal solution of Eq. (1) is given by $\mathbf{w} = \sum_{i=1}^{l} \alpha_i y_i \mathbf{x}_i$ where $\alpha = (\alpha_1,...,\alpha_l)$ is the optimal solution of the optimization problem in Eq. (2). The decision function is defined as:

$$g(\mathbf{x}) = \text{sign}(f(\mathbf{x})), \text{ where } f(\mathbf{x}) = \langle \mathbf{w} \cdot \mathbf{x} \rangle + b$$

(3)

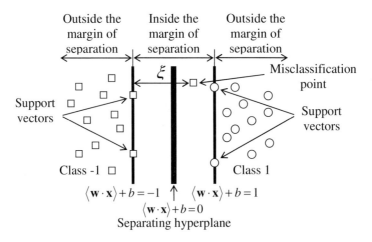

Fig. 1. Illustration of support vector machines

For solving nonlinear problems, the SVM algorithm maps the input vector $\mathbf{x}$ into a higher-dimensional feature space through some nonlinear mapping Φ and constructs an optimal separating hyperplane [7]. Suppose we map the vector $\mathbf{x}$ into a vector in the feature space $(\Phi_1(\mathbf{x}),...,\Phi_n(\mathbf{x}),...)$, then an inner product in feature space has an equivalent representation defined through a kernel function K as $K(\mathbf{x}_1,\mathbf{x}_2) = \langle \Phi(\mathbf{x}_1) \cdot \Phi(\mathbf{x}_2) \rangle$ [8]. Therefore, we can introduce the inner-product kernel as $K(\mathbf{x}_i,\mathbf{x}_j) = \langle \Phi(\mathbf{x}_i) \cdot \Phi(\mathbf{x}_j) \rangle$ and substitute the dot-product $\langle \mathbf{x}_i \cdot \mathbf{x}_j \rangle$ in the dual problem in Eq. (2) with this kernel function. The kernel function used in this study is the radial basis function (RBF) with $K(\mathbf{x}_i,\mathbf{x}_j) = \exp\left(-\gamma \|\mathbf{x}_i - \mathbf{x}_j\|^2\right)$ where γ is the parameter that controls the width of RBF.

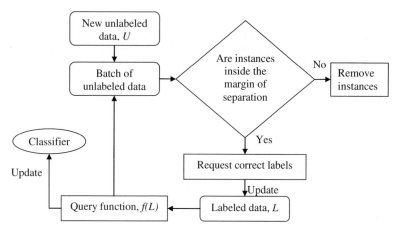

Fig. 2. Active learning with SVMs scheme

3.2 Active Learning with SVMs

Several active learning algorithms with SVMs have been proposed by Campbell et al. [9], Schohn and Cohn [10], and Tong and Koller [11]. Campbell et al. [9] suggested that the generalization performance of a learning machine can be improved significantly with active learning. Using SVMs, the basic idea of the active learning algorithms is to choose the unlabeled instance for the next query closest to the separating hyperplane in the feature space which is the instance with the smallest margin [9-11]. In this paper, we choose the instances that are inside the margin of separation to be labeled and included in the training set. Since the separating hyperplane lies in the middle of the margin of separation, these instances will have an effect on the solution. Thus, the instances outside the margin of separation will be removed.

Suppose we are given an unlabeled pool U and a set of labeled data L. The first step is to find a query function $f(L)$ where, given a set of labeled data L, determine which instances in U to query next. This idea is called the pool-based active learning. Scheme of active learning can be found in Fig. 2.

3.3 Measuring the Quality of the Forecasts for Tornado Prediction

In order to measure the performance of a tornado prediction classifier, it is important to compute scalar forecast evaluation scores such as the Critical Success Index (CSI), Probability of Detection (POD), False Alarm Ratio (FAR), Bias, and Heidke Skill Score (HSS), based on a "confusion" matrix or contingency table (Table I). Those skill scores are defined as: CSI = a/(a+b+c), POD = a/(a+c), FAR = b/(a+b), Bias = (a+b)/(a+c), and HSS = 2(ad-bc)/[(a+c)(c+d)+(a+b)(b+d)].

It is important not to rely solely on a forecast evaluation statistic incorporating cell d from the confusion matrix, as tornadoes are rare events with many correct nulls. This is important as there is little usefulness in forecasting "no" tornadoes every day. Indeed, the claim of skill associated with such forecasts including correct nulls for rare events has a notorious history in meteorology [13].The CSI measures the accuracy of a solution equal to the total number of correct event forecasts (hits) divided by the total number of tornado forecasts plus the number of misses (hits + false alarms + misses) [12]. It has a range of 0 to 1, where 1 is a perfect value. The POD calculates the fraction of observed events that are correctly forecast. It has a perfect score of 1 and a range is 0 to 1 [14]. The FAR measures the ratio of false alarms to the number of "yes" forecasts. It has a perfect score of 0 with its range of 0 to 1 [14]. The Bias computes the total number of event forecasts (hits + false alarms) divided by the total number of observed events. It shows whether the forecast system is under-forecast (Bias < 1) or overforecast (Bias > 1) events with a range of 0 to +∞ and perfect score of 1 [14]. The HSS [15] is commonly used in forecasting since it considers all elements in the confusion matrix. It measures the relative increase in forecast accuracy over some reference forecast. In the present formulation, the reference forecast is a random guess. A skill value > 0 is more accurate than the reference. It has a perfect score of 1 and a range of -1 to 1.

Table 1. Confusion matrix

		Observation	
		Yes	**No**
	Yes	hit	false alarm
Forecast		a	b
	No	miss	correct null
		c	d

4 Experiments

The data were divided into two sets: training and testing. In the training set, we had 382 tornadic instances and 1128 non-tornadic instances. In order to perform online setting experiments, the training instances were arranged in time order. The testing set consisted of 387 tornadic instances and 11872 non-tornadic instances. For both active and passive learning experiments, the initial training set was the first 10 instances consisted of 5 tornadic instances and 5 non-tornadic instances. At each iteration, new data were injected in a batch of several instances. Two different batch sizes, 75 and 150 instances, were used for comparison. In passive learning with SVMs, all incoming data were labeled and included in the training set. Conversely, active learning with SVMs only chooses the instances from each batch which are most informative for the classifier. Therefore, the classifier was updated dynamically at each iteration. The performance of the classifier can be measured by computing the scalar skill scores (Section 3.3) on the testing set. The radial basis function kernel with $\gamma = 0.01$ and $C = 10$ was used in these experiments. The experiments were performed in the Matlab environment using LIBSVM toolbox [16].

Before training a classifier, the data set needs to be normalized. We normalized the training set so that each attribute has the mean of 0 and the standard deviation of 1. Then, we used the mean and standard deviation from each attribute in the training set to normalize each attribute in the testing set.

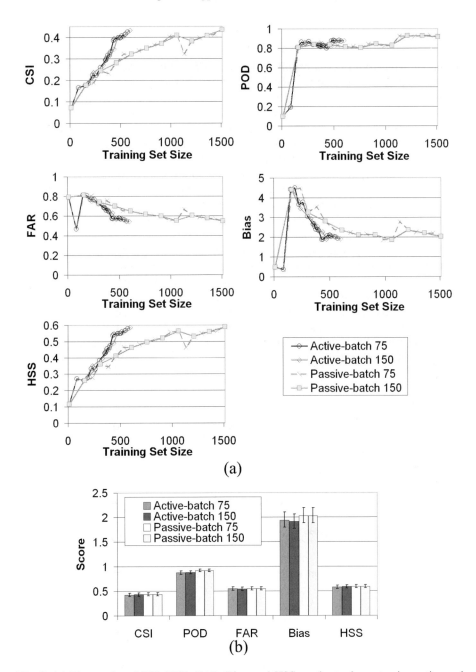

Fig. 3. (a) The results of CSI, POD, FAR, Bias, and HSS on the testing set using active and passive learning at all iterations. (b) The last iteration results with 95% confidence intervals on the testing set.

5 Results

It can bee seen from Fig. 3a for all skill scores, CSI, POD, FAR, Bias, and HSS, active learning achieved relatively the same scores as passive learning using less training instances. From the FAR diagram (Fig. 3a), we noticed that at early iteration the active and passive learning FAR with the batch size of 75 dropped suddenly. It happened because the forecast system was underforecast (Bias < 1) at that stage. Ultimately, every method produced overforecasting. Furthermore, Fig. 3b showed the last iteration results with 95% confidence intervals after conducting bootstrap resampling with 1000 replications [17]. The 95% confidence intervals between active and passive learning results with the batch sizes of 75 and 150 overlapped each other for each skill score, so the differences were not statistically significant. These results indicated that active learning possessed similar performance compared to passive learning using the MDA and NSE data set.

The results in Fig. 4 showed that active learning significantly reduced the training set size to attain relatively the same skill scores as passive learning. Using the batch size of 75 instances, only 571 labeled instances were required in active learning whereas in passive learning 1510 labeled instances were needed (Fig. 4a). This experiment reveals that about 62.6% reduction was realized by active learning. Using the batch size of 150 instances, active learning can reduce the training set size by 60.5% since it only needed 596 labeled instances whereas passive learning required 1510 labeled instances (Fig. 4b).

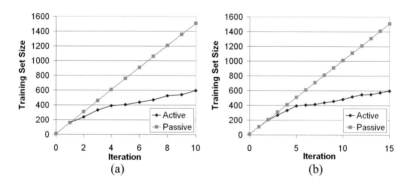

Fig. 4. Diagrams of training set size vs. iteration for the batch sizes of (a) 75 and (b) 150 instances

6 Conclusions

In this paper, active learning with SVMs was used to discriminate between mesocyclones that do not become tornadic from those that do form tornadoes. The preliminary results showed that active learning can significantly reduce the size of training set and achieve relatively similar skill scores compared to passive learning. Since labeling new data is considered costly and time consuming in tornado prediction, active learning would be beneficial in order to update the classifier dynamically.

Acknowledgments. Funding for this research was provided under the National Science Foundation Grant EIA-0205628 and NOAA Grant NA17RJ1227.

References

1. Marzban, C., Stumpf, G.: A neural network for tornado prediction based on Doppler radar-derived attributes. J. Appl. Meteorol. 35 (1996) 617-626
2. Lakshmanan, V., Stumpf, G., Witt, A.: A neural network for detecting and diagnosing tornadic circulations using the mesocyclone detection and near storm environment algorithms. In: 21st International Conference on Information Processing Systems, San Diego, CA, Amer. Meteor. Soc. (2005) CD–ROM J5.2
3. Trafalis, T.B., Ince, H., Richman M.B. Tornado detection with support vector machines. In: Sloot PM et al. (eds). Computational Science-ICCS (2003) 202-211
4. Adrianto, I., Trafalis, T.B., Richman, M.B., Lakshmivarahan, S., Park, J.: Machine learning classifiers for tornado detection: sensitivity analysis on tornado data sets. In: Dagli C. Buczak, A., Enke, D., Embrechts, M., Ersoy, O. (eds.): Intelligent Engineering Systems Through Artificial Neural Networks, Vol. 16. ASME Press (2006) 679-684
5. Boser, B.E., Guyon, I.M., Vapnik, V.N.: A training algorithm for optimal margin classifiers. In: Haussler D (ed): 5th Annual ACM Workshop on COLT. ACM Press, Pittsburgh, PA (1992) 144-152
6. Vapnik, V.N.: The Nature of Statistical Learning Theory. Springer Verlag, New York (1995)
7. Vapnik, V.N.: Statistical Learning Theory. Springer Verlag, New York (1998)
8. Haykin S.: Neural Networks: A Comprehensive Foundation. 2nd edn. Prentice Hall, New Jersey (1999)
9. Campbell, C., Cristianini, N., Smola, A.: Query learning with large margin classifiers. In: Proceedings of ICML-2000, 17th International Conference on Machine Learning. (2000)111-118
10. Schohn, G., Cohn, D.: Less is more: Active learning with support vector machines. In: ICML Proceedings of ICML-2000, 17th International Conference on Machine Learning, (2000) 839-846
11. Tong, S., Koller, D.: Support vector machine active learning with applications to text classification. J. Mach. Learn. Res. 2 (2001) 45-66
12. Donaldson, R., Dyer, R., Krauss, M.: An objective evaluator of techniques for predicting severe weather events. In: 9th Conference on Severe Local Storms, Norman, OK, Amer. Meteor. Soc. (1975) 321-326
13. Murphy, A.H.: The Finley affair: a signal event in the history of forecast verifications. Weather Forecast. 11 (1996) 3-20
14. Wilks, D.: Statistical Methods in Atmospheric Sciences. Academic Press, San Diego, CA (1995)
15. Heidke P.: Berechnung des erfolges und der gute der windstarkvorhersagen im sturmwarnungsdienst, Geogr. Ann. 8 (1926) 301-349
16. Chang, C., Lin, C.: LIBSVM: a library for support vector machines. Software available at <http://www.csie.ntu.edu.tw/~cjlin/libsvm> (2001)
17. Efron, B., Tibshirani, R.J.: An Introduction to the Bootstrap. Chapman & Hall, New York (1993)

Adaptive Observation Strategies for Forecast Error Minimization

Nicholas Roy[1], Han-Lim Choi[2], Daniel Gombos[3], James Hansen[4], Jonathan How[2], and Sooho Park[1]

[1] Computer Science and Artificial Intelligence Lab
Massachusetts Institute of Technology
Cambridge, MA 02139
[2] Aerospace Controls Lab
Massachusetts Institute of Technology
Cambridge, MA 02139
[3] Department of Earth and Planetary Sciences
Massachusetts Institute of Technology
Cambridge, MA 02139
[4] Marine Meteorology Division
Naval Research Laboratory
Monterey, CA 93943

Abstract. Using a scenario of multiple mobile observing platforms (UAVs) measuring weather variables in distributed regions of the Pacific, we are developing algorithms that will lead to improved forecasting of high-impact weather events. We combine technologies from the nonlinear weather prediction and planning/control communities to create a close link between model predictions and observed measurements, choosing future measurements that minimize the expected forecast error under time-varying conditions.

We have approached the problem on three fronts. We have developed an information-theoretic algorithm for selecting environment measurements in a computationally effective way. This algorithm determines the best discrete locations and times to take additional measurement for reducing the forecast uncertainty in the region of interest while considering the mobility of the sensor platforms. Our second algorithm learns to use past experience in predicting good routes to travel between measurements. Experiments show that these approaches work well on idealized models of weather patterns.

1 Introduction

Recent advances in numerical weather prediction (NWP) models have greatly improved the computational tractability of long-range prediction accuracy. However, the inherent sensitivity of these models to their initial conditions has further increased the need for accurate and precise measurements of the environmental conditions. Deploying an extensive mobile observation network is likely to be costly, and measurements of the current conditions may produce different results in terms of improving forecast performance [1,2]. These facts have led to the development of observation strategies where additional sensors are deployed to achieve the best performance according to some

Y. Shi et al. (Eds.): ICCS 2007, Part I, LNCS 4487, pp. 1138–1146, 2007.

measures such as expected forecast error reduction and uncertainty reduction [3]. One method for augmenting a fixed sensor network is through the use of "adaptive" or "targeted" observations where mobile observing platforms are directed to areas where observations are expected to maximally reduce forecast error under some norm (see, for example, NOAA's Winter Storm Reconnaissance Program [4]). The hypothesis is that these directed measurements provide better inputs to the weather forecasting system than random or gridded use of the observing assets.

This paper describes an adaptive observation strategy that integrates nonlinear weather prediction, planning and control to create a close link between model predictions and observed measurements, choosing future measurements that minimize the expected forecast error under time-varying conditions. The main result will be a new framework for coordinating a team of mobile observing assets that provides more efficient measurement strategies and a more accurate means of capturing spatial correlations in the system dynamics, which will have broad applicability to measurement and prediction in other domains. We first describe the specific non-linear weather prediction model used to develop our adaptive observation strategy, and then describe a global targeting algorithm and a local path planner that together choose measurements to minimize the expected forecast error.

2   Models of Non-linear Weather Prediction

While there exist large-scale realistic models of weather prediction such as the Navy's Coupled Ocean Atmosphere Prediction System (COAMPS), our attention will be restricted to reduced models in order to allow computationally tractable experiments with different adaptive measurement strategies. The Lorenz-2003 model is an extended model of the Lorenz-95 model [1] to address multi-scale feature of the weather dynamics in addition to the basic aspects of the weather motion such as energy dissipation, advection, and external forcing. In this paper, the original one-dimensional model is extended to two-dimensions representing the mid-latitude region $(20 - 70 \deg)$ of the northern hemisphere. The system equations are

$$\dot{y}_{ij} = - \xi_{i-2\alpha,j}\xi_{i-\alpha,j} + \frac{1}{2\lfloor \alpha/2 \rfloor + 1} \sum_{k=-\lfloor \alpha/2 \rfloor}^{k=+\lfloor \alpha/2 \rfloor} \xi_{i-\alpha+k,j} y_{i+k,j}$$

$$- \mu \eta_{i,j-2\beta}\eta_{i,j-\beta} + \frac{\mu}{2\lfloor \beta/2 \rfloor + 1} \sum_{k=-\lfloor \beta/2 \rfloor}^{k=+\lfloor \beta/2 \rfloor} \eta_{i,j-\beta+k} y_{i,i+k} \qquad (1)$$

$$- y_{ij} + F$$

where

$$\xi_{ij} = \frac{1}{2\lfloor \alpha/2 \rfloor + 1} \sum_{k=-\lfloor \alpha/2 \rfloor}^{k=+\lfloor \alpha/2 \rfloor} y_{i+k,j}, \quad \eta_{ij} = \frac{1}{2\lfloor \beta/2 \rfloor + 1} \sum_{k=-\lfloor \beta/2 \rfloor}^{k=+\lfloor \alpha/2 \rfloor} y_{i,j+k}, \qquad (2)$$

where $i = 1, \ldots, L_{on}, j = 1, \ldots, L_{at}$. The subscript i denotes the west-to-eastern grid index, while j denotes the south-to-north grid index. The dynamics of the (i, j)-th grid point depends on its longitudinal 2α-interval neighbors (and latitudinal 2β) through the advection terms, on itself by the dissipation term, and on the external forcing ($F = 8$ in this work). When $\alpha = \beta = 1$, this model reduces to the two-dimension Lorenz-95 model [3]. The length-scale of this model is proportional to the inverse of α and β in each direction: for instance, the grid size for $\alpha = \beta = 2$ amounts to 347 km × 347 km. The time-scale is such that 0.05 time units are equivalent to 6 hours in real-time.

2.1 State Estimation

A standard approach to state estimation and prediction is to use a Monte Carlo (ensemble) approximation to the extended Kalman Filter, in which each ensemble member presents an initial state estimate of the weather system. These ensembles are propagated (for a set forecast time) through the underlying weather dynamics and the estimate (i.e., the mean value of these ensembles) is refined by measurements (i.e., updates) that are available through the sensor network. The particular approximation used in this work is the sequential ensemble square root filter [5] (EnSRF). In the EnSRF, the propagation of the mean state estimate and covariance matrix amounts to a nonlinear integration of ensemble members, improving the filtering of non-linearities compared to standard EKF techniques and mitigating the computational burden of maintaining a large covariance matrix [6,5]. The ensemble mean corresponds to the state estimate, and the covariance information can be obtained from the perturbation ensemble,

$$\mathbf{P} = \tilde{\mathbf{X}}\tilde{\mathbf{X}}^T/(L_E - 1), \qquad \tilde{\mathbf{X}} \in \mathbb{R}^{L_S \times L_E} \tag{3}$$

where L_S is the number of state variables and L_E is the ensemble size. $\tilde{\mathbf{X}}$ is the perturbation ensemble defined as

$$\tilde{\mathbf{X}} = \eta \left(\mathbf{X} - \bar{\mathbf{x}} \times \mathbf{1}^T \right) \tag{4}$$

where $\mathbf{X}$ is the ensemble matrix, a row concatenation of each ensemble member, and $\bar{\mathbf{x}}$ is the ensemble mean, the row average of the ensemble matrix. $\eta (\geq 1)$ is the covariance inflation factor introduced to avoid underestimation of the covariance by finite ensemble size. The propagation step for EnSRF is the integration

$$\mathbf{X}^f(t + \Delta t) = \int_t^{t+\Delta t} \dot{\mathbf{X}} dt, \qquad \mathbf{X}(t) = \mathbf{X}^a(t), \tag{5}$$

with $\mathbf{X}^f$ and $\mathbf{X}^a$ denoting the forecast and analysis ensemble, respectively. The measurement update step for the EnSRF is

$$\bar{\mathbf{x}}^a = \bar{\mathbf{x}}^f + \mathbf{K}(\mathbf{y} - \mathbf{H}\bar{\mathbf{x}}^f) \tag{6}$$

$$\tilde{\mathbf{X}}^a = (\mathbf{I} - \mathbf{KH})\tilde{\mathbf{X}}^f \tag{7}$$

where $\mathbf{y}$ denotes the observation vector and $\mathbf{H}$ is the linearized observation matrix. $\mathbf{K}$ denotes the appropriate Kalman gain, which can be obtained by solving a nonlinear matrix equation stated in terms of $\mathbf{X}$[5]. The sequential update process avoids solving a nonlinear matrix equation and provides a faster method for determining $\mathbf{K}$. The ensemble update by the m-th observation is

$$\tilde{\mathbf{X}}^{m+1} = \tilde{\mathbf{X}}^m - \alpha_m \beta_m \mathbf{p}_i^m \tilde{\xi}_i^m, \tag{8}$$

$$\alpha_m = 1 / \left(1 + \sqrt{\beta_m R_i} \right), \beta_m = 1 / (\mathbf{P}_{ii}^m + R_i) \tag{9}$$

where measurement is taken for i-th state variable. $\mathbf{p}_i^m, \tilde{\xi}_i^m$, and $\mathbf{P}_{ii}^m$ are the i-th column, the i-th row, and the (i,i) element of the prior perturbation matrix $\mathbf{P}^m$ respectively. α_m is the factor for compensating the mismatch of the serial update and the batch update, while $\beta_m \mathbf{p}_l^m$ amounts to the Kalman gain.

Figure 1(a) shows an example true state of a Lorenz model (top) over the 36×9 state variables. The bottom frame shows the estimated state at the same time. This estimate is computed from an EnSRF using 200 ensemble members. Observations are taken at 66 fixed (routine) locations represented by blue circles; note that there are regions where routine observations are sparse, representing areas such as open ocean where regular measurements are hard to acquire. Figure 1(b) (top) shows the squared analysis error between true state and ensemble estimates from the upper figure, that is, the actual forecast error. The lower panel shows the ensemble variance, that is, the expected squared forecast error. Note that the expected and true error are largely correlated; using 200 ensemble members was enough to estimate the true model with reasonable error as shown in the figures.

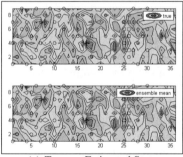
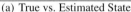
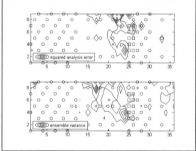

(a) True vs. Estimated State   (b) Performance Analysis

Fig. 1. (a) Top panel: the true state of the Lorenz system, where the intensity correlates with the state value. Lower panel: The estimated state of the system, using 200 ensemble members. (b) Top panel: the actual forecast error. Lower panel: the ensemble variance.

For the purposes of forecast error, we are typically interested in improving the forecast accuracy for some small region such as the coast of California, rather than the entire Pacific. A verification region is specified as $\mathbf{X}[v]$ and verification time t_v in our experiments, as shown by the red squares in Figure 1. Our goal is therefore to choose measurements of $\mathbf{X}$ at time t to minimize the forecast error at $\mathbf{X}[v]$ at time t_v.

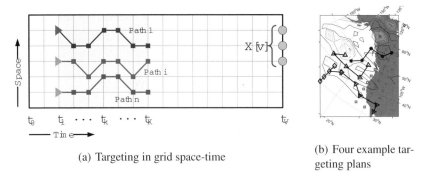

(a) Targeting in grid space-time

(b) Four example targeting plans

Fig. 2. (a) Multi-UAV targeting in the grid space-time. (b) Targeting of four sensor platforms for the purpose of reducing the uncertainty of 3-day forecast over the west coast of North America.

3   A Targeting Algorithm for Multiple Mobile Sensor Platforms

The targeting problem is how to assign multiple sensor platforms (e.g. UAVs) to positions in the finite grid space-time (Figure 2a) in order to reduce the expected forecast uncertainty in the region of interest $X[v]$. We define the targeting problem as selecting n paths consisting of K (size of the targeting time window) points that maximize the information gain at $X[v]$ of the measurements taken along the selected paths.

A new, computationally efficient *backward selection* algorithm forms the backbone of the targeting approach. To address the computation resulting from the expense of determining the impact of each measurement choice on the uncertainty reduction in the verification site, the backward selection algorithm exploits the commutativity of mutual information. This enables the contribution of each measurement choice to be computed by propagating information backwards from the verification space/time to the search space/time. This significantly reduces the number of times that computationally expensive covariance updates must be performed. In addition, the proposed targeting algorithm employs a branch-and-bound search technique to reduce computation required to calculate payoffs for suboptimal candidates, utilizing a simple cost-to-go heuristics based on the diagonal assumption of the covariance matrix that provides an approximate upper bound of the actual information gain. The suggested heuristic does not guarantee an optimal solution; nevertheless, in practice it results in a substantial reduction in computation time while incurring minimal loss of optimality, which can be improved by relaxing a bounding constraint.

Figure 2(b) depicts an illustrative solution of the four-agent (black $\Diamond$, $\triangle$, $\triangleright$, $*$) targeting problem for enhancing the 3-day forecast of the west coast of North America (red $\square$); the time interval between the marks is three hours.

The computation time of the targeting algorithm grows exponentially with the number of sensor platforms and the size of the targeting window increase, in spite of the reduction in computational cost. Thus, further approximations that decompose the computation and decision making into different topologies and choices on the planning horizon will be explored. These have been shown to avoid the combinatorial explosion of the computation time, and the performance of the approximation scheme turns out to depend highly on the communication topology between agents. The existence of inter-agent sharing of the up-to-date covariance information has also been shown to be essential to achieve performance.

4   Trajectory Learning

Given a series of desired target locations for additional measurements, an appropriate motion trajectory must be chosen between each pair of locations. Rather than directly optimizing the trajectory based on the current state of the weather system, the system will learn to predict the best trajectory that minimizes the forecast by examining past example trajectories. The advantage to this approach is that, once the predictive model is learned, each prediction can be made extremely quickly and adapted in real time as additional measurements are taken along the trajectory. The second advantage is that by careful selecting the learning technique, a large number of factors can be considered in both the weather system and the objective function, essentially optimizing against a number of different objectives, again without incurring a large computational penalty.

The problem of learning a model that minimizes the predicted forecast error is that of reinforcement learning, in which an agent takes actions and receives some reward signal. The goal of the agent is to maximize over its lifetime the expected received reward (or minimize the received cost) by learning to associate actions that maximize reward in different states. Reinforcement learning algorithms allow the agent to learn a policy $\pi : x \rightarrow a$, mapping state x to action a in order to maximize the reward.

In the weather domain, our cost function is the norm of the forecast error at the verification state variables $(\mathbf{X}[v])$ at the verification time t_v, so that the optimal policy π^* is

$$\pi^*(\mathbf{X}) = \operatorname*{argmin}_{\pi \in \Pi} E_{\mathbf{X}_{t_v}[v]} \left[\left\| (\tilde{\mathbf{X}}_{t_v}[v] | h(\pi), \mathbf{X}) - \mathbf{X}_{t_v}[v] \right\| \right] \tag{10}$$

If our set of actions is chosen to be a class of paths through space, such as polynomial splines interpolating the target points, then the policy attempts to choose the best spline to minimize our expected forecast error. Notice that this policy maps the current state $\mathbf{X}$ to the action a; however, the policy does not have access to the current weather state but only the current estimate of the weather given by the EnSRF. The learner therefore computes the policy that chooses actions based on the current estimate given by the mean $\tilde{\mathbf{X}}$ and covariance Σ of the ensemble.

In order to find the optimal policy π^*, a conventional reinforcement learning algorithm spends time trying different trajectories under different examples of weather conditions, and modelling how each trajectory predicts a different forecast error. The learning problem then becomes one of predicting, for a given EnSRF estimate $\tilde{\mathbf{X}}$ and Σ, the expected forecast error $\xi \in \mathbb{R}$ for each possible trajectory $a \in A$:

$$(\tilde{\mathbf{X}}, \Sigma) \times A \rightarrow \mathbb{R}. \tag{11}$$

Once this functional relationship is established, the controller simply examines the predicted error ξ for each action a given the current state estimate and chooses the action with the least error.

With access to a weather simulator such as the Lorenz model, we can simplify this learning problem by turning our reinforcement learning problem into a "supervised" learning problem, where the goal of the learner is not to predict the forecast error ξ of each possible trajectory conditioned on the current weather estimate, but rather to predict the best trajectory, a converting the regression problem of equation (11) into a classification problem, that is,

$$(\tilde{\mathbf{X}}, \Sigma) \to A. \tag{12}$$

Although regression and classification are closely linked (and one can often be written in terms of another), we can take advantage of some well-understood classification algorithms for computing policies. The classification algorithm used is the multi-class Support Vector Machine [7], which assigns a label (i.e., our optimal action) to each initial condition. The SVM is a good choice to learn our policy for two reasons: firstly, the SVM allows us to learn a classifier over the continuous state space $(\tilde{\mathbf{X}}, \Sigma)$. Secondly, the SVM is generally an efficient learner of large input spaces with a small number of samples; the SVM uses a technique known as the "kernel trick" [7] to perform classification by projecting each instance to a high-dimensional, non-linear space in which the inputs are linearly separable according to their class label.

4.1 Experimental Results

Training data for the Lorenz model was created by randomly generating initial conditions of the model, creating a set of ensemble members from random perturbations to the initial conditions and then propagating the model. Figure 3(a) shows a plot of 40 initial conditions used as training data created by running the model forward for several days and sampling a new initial condition every 6 hours, re-initializing the model every 5 days. Each row corresponds to a different training datum $\mathbf{X}_t$, and each column corresponds to a state variable X^i. While the data are fairly random, the learner can take advantage of the considerable temporal correlation; notice the clear discontinuity in the middle of the data where the model was re-initialized.

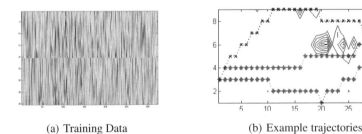

(a) Training Data   (b) Example trajectories

Fig. 3. (a) A plot of 40 training instances $\mathbf{X}$. Each column corresponds to a state variable X^i and each row is a different $\mathbf{X}_t$. (b) Three example trajectories from our action space A. Our action space consisted of 5 trajectories that span the region from the same start and end locations.

Each training instance was labelled with the corresponding optimal trajectory. We restricted the learner to a limited set of 5 actions or candidate trajectories, although this constraint will be relaxed in future work. All candidate trajectories started from the same mid-point of the left side of the region and ended at the same mid-point of the right side of the region. Three examples of the five trajectories are shown in Figure 3(b); the trajectories were chosen to be maximal distinct through the area of sparse routine observations in the centre of the region. From the initial condition, the model and ensemble were propagated for each trajectory. During this propagation, routine observations were taken every 5 time units and then a forecast was generated by propagating the ensemble for time equivalent to 2 and 4 days, without taking additional observations. The forecast error was then calculated by the difference between the ensemble estimate and the true value of the variables of the verification region. Each initial condition was labelled with the trajectory that minimized the resultant forecast error.

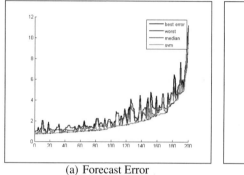
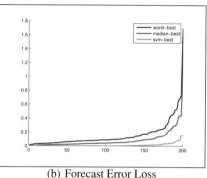

(a) Forecast Error   (b) Forecast Error Loss

Fig. 4. (a) Forecast error at the verification region after 2 days for the 200 largest losses in test data, sorted from least to greatest. (b) Forecast error loss, where the loss is taken with respect to the forecast error of the best trajectory.

Figure 4(a) shows the forecast error of best, median, worst and SVM trajectory for the 200 most difficult (highest forecast error) initial conditions in terms of the forecast error in the verification region. Notice that the forecast error of the SVM trajectory tracks the best trajectory relatively closely, indicating good performance. Figure 4(b) is an explicit comparison between the worst, median and SVM trajectories compared to the best trajectory for the same 200 most difficult training instances. Again, the SVM has relatively little loss (as measured by the difference between the forecast error of the SVM and the forecast error of the best trajectory) for many of these difficult cases.

In training the learner, two different kernel (non-linear projections in the SVM) were tested, specifically polynomial and radial basis function (RBF) kernels. Using cross-validation and a well-studied search method to identify the best kernel fit and size, a surprising result was that a low-order polynomial kernel resulted in the most accurate prediction of good trajectories. A second surprising result is that in testing different combinations of input data, such as filter mean alone, compared to filter mean and filter covariance, the filter covariance had relatively little effect on the SVM performance. This effect may be related to the restricted action class, but further investigation is warranted.

5 Conclusion

The spatio-temporal character of the data and chaotic behavior of the weather model makes adaptive observation problem challenging in the weather domain. We have described two adaptive observation techniques, including a targeting algorithm and a learning path planning algorithm. In the future, we plan to extend these results using the Navy's Coupled Ocean Atmosphere Prediction System (COAMPS), a full-scale regional weather research and forecasting model.

References

1. Lorenz, E.N., Emanuel, K.A.: Optimal sites for supplementary weather observations: Simulation with a small model. Journal of the Atmospheric Sciences **55**(3) (1998) 399–414
2. Morss, R., Emanuel, K., Snyder, C.: Idealized adaptive observation strategies for improving numerical weather prediction. Journal of the Atmospheric Sciences **58**(2) (2001)
3. Choi, H.L., How, J., Hansen, J.: Ensemble-based adaptive targeting of mobile sensor networks. In: Proc. of the American Control Conference (ACC). (To appear. 2007)
4. : http://www.aoc.noaa.gov/article_winterstorm.htm. Available online (last accessed June 2005)
5. Whitaker, J., Hamill, H.: Ensemble data assimilation without perturbed observations. Monthly Weather Review **130**(7) (2002) 1913–1924
6. Evensen, G., van Leeuwen, P.: Assimilation of altimeter data for the agulhas current using the ensemble kalman filter with a quasigeostrophic model. Monthly Weather Review **123**(1) (1996) 85–96
7. Cristianini, N., Shawe-Taylor, J.: An Introduction to Support Vector Machines. Cambridge University Press, Cambridge, UK (2000)

Two Extensions of Data Assimilation by Field Alignment[*]

Sai Ravela

Earth, Atmospheric and Planetary Sciences
Massachusetts Institute of Technology
ravela@mit.edu

Abstract. Classical formulations of data-assimilation perform poorly when fore-cast locations of weather systems are displaced from their observations. They compensate position errors by adjusting amplitudes, which can produce unacceptably "distorted" states. Motivated by cyclones, in earlier work we show a new method for handling position and amplitude errors using a single variational objective. The solution could be used with either ensemble or deterministic methods. In this paper, extension of this work in two directions is reported. First, the methodology is extended to multivariate fields commonly used in models, thus making this method readily applicable. Second, an application of this methodology to rainfall modeling is presented.

1 Introduction

Environmental data assimilation is the methodology for combining imperfect model predictions with uncertain data in a way that acknowledges their respective uncertainties. It plays a fundamental role in DDDAS. However, data assimilation can only work when the estimation process properly represents all sources of error. The difficulties created by improperly represented error are particularly apparent in mesoscale meteorological phenomena such as thunderstorms, squall-lines, hurricanes, precipitation, and fronts. Errors in mesoscale models can arise in many ways but they often manifest themselves as errors in the position. We typically cannot attribute position error to a single or even a small number of sources and it is likely that the position errors are the aggregate result of errors in parameter values, initial conditions, boundary conditions and others. In the context of cyclones operational forecasters resort to ad hoc procedures such as *bogussing* [4] . A more sophisticated alternative is to use data assimilation methods. Unfortunately, sequential [10], ensemble-based [9] and variational [12,3] state estimation methods used in data assimilation applications adjust amplitudes to deal with position error. Adjusting amplitudes doesn't really fix position error, and instead, can produce unacceptably distorted estimates.

In earlier work [16], we show how the values predicted at model grid points can be adjusted in amplitude and moved in space in order to achieve a better fit to observations. The solution is general and applies equally well to meteorological, oceanographic, and

---

[*] This material is supported NSF CNS 0540259.

Y. Shi et al. (Eds.): ICCS 2007, Part I, LNCS 4487, pp. 1147–1154, 2007.

hydrological applications. It involves solving two equations, in sequence, described as follows:

Let $X = X(\mathbf{r}) = \{X[\underline{r}_1^T] \ldots X[\underline{r}_m^T]\}$ be the model-state vector defined over a spatially discretized computational grid Ω, and $\mathbf{r}^T = \{\underline{r}_i = (x_i, y_i)^T, i \in \Omega\}$ be the position indices. Similarly, let $\mathbf{q}$ be a *vector* of displacements. That is, $\mathbf{q}^T = \{\underline{q}_i = (\Delta x_i, \Delta y_i)^T, i \in \Omega\}$, and $X(\mathbf{r} - \mathbf{q})$ represents *displacement* of X by $\mathbf{q}$. The displacement field $\mathbf{q}$ is real-valued, so $X(\mathbf{r} - \mathbf{q})$ must be evaluated by interpolation if necessary.

We wish to find $(X, \mathbf{q})$ that has the maximum a posteriori probability in the distribution $P(X, \mathbf{q}|Y)$, where Y is the observation vector. Using Bayes rule we obtain $P(X, \mathbf{q}|Y) \propto P(Y|X, \mathbf{q})P(X|\mathbf{q})P(\mathbf{q})$. Assume a linear observation model with uncorrelated noise in space and time, the component densities to be Gaussian and the displacement field solution is smooth and non-divergent. Then, the following Euler-Lagrange equations are obtained and solved sequentially.

1) Alignment. Define $\mathbf{p} = \mathbf{r} - \mathbf{q}$ and the alignment equation is then written at each grid node i as:

$$w_1 \nabla^2 \underline{q}_i + w_2 \nabla(\nabla \cdot \underline{q}_i) + \left[\nabla X^T |_{\mathbf{p}} H^T R^{-1} \left(H \left[X \left(\mathbf{p} \right) \right] - Y \right) \right]_i = 0 \qquad (1)$$

Equation 1 introduces a forcing based on the residual between fields. The constraints on the displacement field allow the forcing to propagate to a consistent solution. Equation 1 is non-linear, and is solved iteratively.

2) Amplitude Adjustment: The aligned field $X(\hat{\mathbf{p}})$ is used in the second step for a classical Kalman update:

$$\hat{X}(\hat{\mathbf{p}}) = X(\hat{\mathbf{p}}) + B_{\hat{Q}} H^T (H B_{\hat{Q}} H^T + R)^{-1} (Y - H \, X(\hat{\mathbf{p}})) \qquad (2)$$

The covariance $B_{\hat{Q}}$ is computed after alignment. It can be estimated using ensembles (when each has been aligned), or any method that would otherwise be used on the field X.

2   Multivariate Formulation

To be applicable in practice, the field alignment algorithm must be extended to the multivariate case, including vector fields. As an example, consider 2D fields (3D-fields can be handled analogously), say two components of velocity and pressure. Now partition the state X and observations Y into component fields, P, U, V for pressure. To align X to Y, we constrain individual fields to have the same displacements. Displacing vector fields, however, involves the Jacobian of the deformation (when the wind field rotates, both the coordinate and wind-field components rotate). Therefore, if ψ is a scalar function undergoing a deformation $\psi(\mathbf{r} - \mathbf{q})$, then the gradient vector field undergoes a transformation that is expressed as $(\nabla \mathbf{q})^T \nabla \psi|_{\mathbf{r} - \mathbf{q}}$. We introduce, therefore, for the wind velocity components, the variables $\tilde{U}, \tilde{V}$, defined as

$$\begin{pmatrix} \tilde{U}_i \\ \tilde{V}_i \end{pmatrix} = (\nabla \underline{q}_i)^T \begin{pmatrix} U_i \\ V_i \end{pmatrix} \qquad (3)$$

The field alignment equation now looks as follows:

$$w_1 \nabla^2 \underline{q}_i + w_2 \nabla(\nabla \cdot \underline{q}_i) + \left\{ [\nabla P^T|_{\mathbf{r}-\mathbf{q}} H_P{}^T R_P^{-1} \left(H_P P (\mathbf{r} - \mathbf{q}) - Y_P \right) \right\}_i$$
$$+ \left\{ [\nabla \tilde{U}^T|_{\mathbf{r}-\mathbf{q}} H_U{}^T R_U^{-1} \left(H_U \tilde{U} (\mathbf{r} - \mathbf{q}) - Y_U \right) \right\}_i$$
$$+ \left\{ [\nabla \tilde{V}^T|_{\mathbf{r}-\mathbf{q}} H_V{}^T R_V^{-1} \left(H_V \tilde{V} (\mathbf{r} - \mathbf{q}) - Y_V \right) \right\}_i = 0$$

Here R is the observation noise covariance, H_P, H_U, H_V are the observation operators, and Y_U, Y_V, Y_P are the component fields of the observation vector. We demonstrate the use of the 2-step method on multivariate fields next.

Example: An example is now demonstrated using pressure and velocity. Figures 1 contain "filled" iso-contour plots of state fields. The first guess fields shown down the first column is rotated by 40° from truth (shown in second column). Measurements are made at every 5th pixel on truth, marked in white dots or black crosses in third column. There is an amplitude error of a multiplicative factor 1.1, the observational noise, 30% of peak amplitude in pressure, and 7% in velocity components, i.i.d. The background error covariance (modeled isotropic, see [16] for flow dependent results) in this example is substantially more uncertain than the observational uncertainty.

The fourth column of Figure 1 depicts the analysis of 3DVAR-only for pressure and velocity components. The pressure and velocity fields are clearly "smeared". In contrast, the rightmost column of Figure 1 depicts a much better analysis when field alignment is used as the preprocessing step to 3DVAR. Note that the first guess and observations fed to 3DVAR-only and field alignment algorithm is the same.

In Figure 2, using the nonlinear balance equation, we compare the analyzed and diagnosed pressure fields to study how well-balanced alignment is. Result: the alignment followed by 3DVAR preserves the balance to a far greater degree than 3DVAR alone, see Figure 2.

Figure 3 compares analysis and balance errors for various cases of sparsity. The x-axis of the left panel depicts sparsity; 1 implies every location was observed, 10 implies every 10th location was observed. The y-axis represents normalized error, normalized by the maximum pressure differential. The left panel's bar charts contain filled bars comparing the analyzed pressure vs. truth, using open bars for 3DVAR and closed ones for field alignment followed by 3DVAR. We can see that as the observations get more sparse, 3DVAR performance degrades more sharply than a field alignment preprocessed version.

The right panel of Figure 3 compares the analyzed and diagnosed pressure, with the x-axis representing sparsity and the y-axis, normalized error. The differences rise sharply for 3DVAR-only and, in fact, after about a sparsity of 6 pixels, 3DVAR-only breaks down. The analysis does not compensate for position error and this is clearly seen in the analysis errors shown in the right panel corresponding to this case. Therefore, although the diagnosed and analyzed pressure fields in 3DVAR-only find themselves in good agreement, they are quite far away from the truth! In contrast, compensating position error using field alignment yields analysis and balance errors that are much smaller. They do grow, but much more slowly as function of observational sparsity.

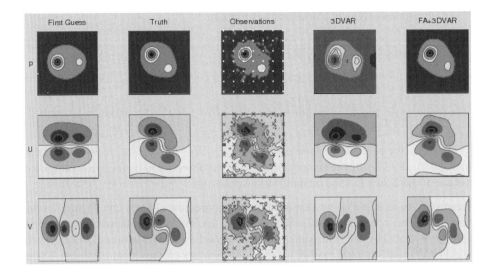

Fig. 1. The left column of panels is the first guess, the second column truth, the third shows observations, taken at indicated locations, the fourth shows 3DVAR-only analysis and the rightmost shows field alignment followed by 3DVAR with identical truth, first guess and observations. The top row corresponds to pressure, the second row to U component of velocity and the third column is the V component.

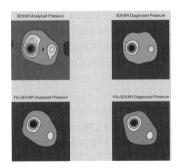

Fig. 2. This figure depicts a comparison of balance between 3DVAR and our approach. The analyzed and diagnosed 3DVAR pressures (top two panels) are substantially different than the corresponding pressure fields using 3DVAR after alignment.

3   Application to Velocimetry

In contrast to hurricanes, we apply the methodology to rainfall modeling. Rainfall models broadly fall into two categories. The first is a meteorological or the quantitative precipitation forecasting model, such as Mesoscale Model (MM5) [2], the step-mountain Eta coordinate model [1], and the Regional Atmospheric Modeling System (RAMS) [5], etc. The second type is the spatiotemporal stochastic rainfall model. It aims to summarize the spatial and temporal characteristics of rainfall by a small set of

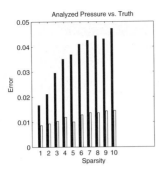
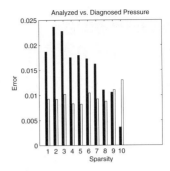

Fig. 3. The x-axis of these graphs represents sparsity. The y-axis of the left panel shows the normalized error between the analyzed pressure and truth, and the right panel shows the normalized error between analyzed and diagnosed pressure. The filled bars depict the 3DVAR-only case, and the open bars are for field alignment followed by 3DVAR.

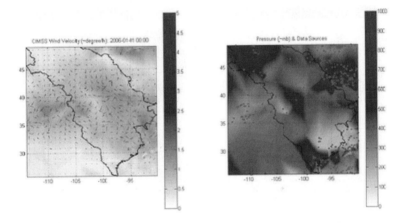

Fig. 4. CIMSS Winds derived from GOES data at 2006-04-06-09Z (left) and pressure (right). The velocity vectors are sparse and contain significant divergence.

parameters [6]. This type of model usually simulates the birth and decay of rain-cells and evolve them through space and time using simple physical descriptions. Despite significant differences among these rainfall models, the concept of propagating rainfall through space and time are relatively similar.

The major ingredient required to advect rainfall is a velocity field. Large spatial-scale (synoptic) winds are inappropriate for this purpose for a variety of reasons. Ironically, synoptic observations can be sparse to be used directly and although synoptic-scale wind analyses produced from them (and models) do produce dense spatial estimates, such estimates often do not contain variability at the meso-scales of interest. The motion of mesoscale convective activity is a natural source for velocimetry. Indeed, there exist products that deduce "winds" by estimating the motion of temperature, vapor and other fields evolving in time [7,8].

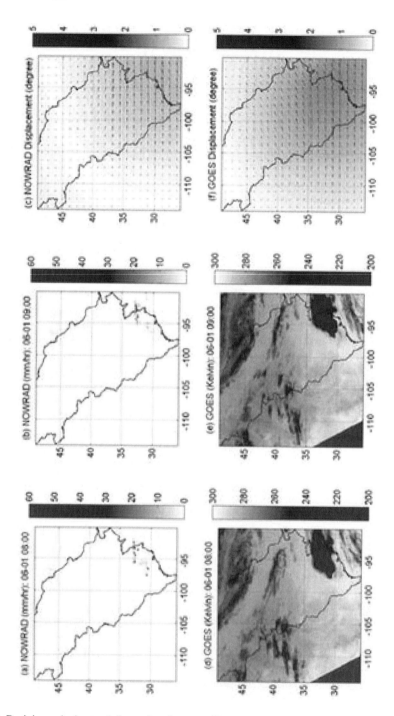

Fig. 5. Deriving velocimetry information from satellite observations, Nexrad (top), GOES (bottom). See text for more information.

In this paper, we present an algorithm for velocimetry from observed motion from satellite observations such as GOES, AMSU, TRMM, or radar data such as NOWRAD. This is obtained by the direct application of equation 1 to two time separated images. This approach provides marked improvement over other methods in conventional use. In contrast to correlation based approaches used for deriving velocity from GOES imagery, the displacement fields are dense, quality control is implicit, and higher-order and small-scale deformations can be easily handled. In contrast with optic-flow algorithms [13,11], we can produce solutions at large separations of mesoscale features, at large time-steps or where the deformation is rapidly evolving. A detailed discussion is presented in [15].

Example: The performance of this algorithm is illustrated in a velocimetry computation. To compare, we use CIMSS wind-data satellite data [8], depicted in Figure 4 obtained from CIMSS analysis on 2006-06-04 at 09Z. CIMSS wind-data is shown over the US great plains, and were obtained from the sounder. The red dots indicate the original location of the data. The left subplot shows wind speed (in degree/hr). The right ones show pressure, and the location of raw measurements in red.

It can be seen in the map in Figure 4 that the operational method to produce winds generate sparse vectors and, further, has substantial divergence. Considering the length-scales, this isn't turbulence and wind vectors are more likely the result of weak quality control. A more detailed discussion is presented in [15].

In contrast, our method produces dense flow fields, and quality control is implicit from regularization constraints. Figure 5(a,b) shows a pair of NOWRAD images at 2006-06-01-0800Z and 2006-06-01-0900Z respectively, and the computed flow field in Figure 5(c). Similarly, Figure 5(d,e,f) show the GOES images and velocity from the same time frame over the deep convective rainfall region in the Great Plains example. The velocities are in good agreement with CIMSS derived winds where magnitudes are concerned, but the flow-fields are smooth and visual confirmation of the alignment provides convincing evidence that they are correct.

4   Discussion and Conclusions

The joint position-amplitude assimilation approach is applicable to fields with coherent structures. Thus, problems in reservoir modeling, convection, rainfall modeling, tracking the ocean and atmosphere will benefit. The solution to the assimilation objective can be computed efficiently in two steps: diffeomorphic alignment followed by amplitude adjustment. This solution allows ready use with existing methods, making it an attractive option for operational practice.

The alignment formulation does not require features to be identified. This is a significant advantage in sparse observations when features cannot be clearly delineated. The alignment formulation can be extended easily to multivariate fields and can be used for a variety of velocimetry problems including particle image velocimetry, velocity from tracer-transport, and velocity from GOES and other satellite data. In relation to GOES, our approach implicitly provides quality control in terms of smoothness, and produces dense displacement fields.

To complete this body of work on position-amplitude estimation, we are conducting research in the following directions:

1. We recently demonstrated [14] that an ensemble filter can be developed when both observations and states have position and amplitude error. This situation occurs in the context of rainfall models, where both satellite derived rain cells and model forecast cells contain position and amplitude error.
2. The position-amplitude smoother: We develop an optimal fixed-interval and fixed-lag ensemble smoother [17]. Our results show that fixed-interval ensemble smoothing is linear in the interval and fixed-lag is independent of lag length. We are extending this smoother to the position-amplitude problem.
3. New constraints: The smoothness constraint has been observed to provide weak control in certain problems. In ongoing work, we have reformulated the alignment problem using a spectral constraint on the deformation field.

References

1. T. L. Black. The new nmc moesoscale eta model: Description and forecast examples. *Weather and Forecasting*, 9(2):265–278, 1994.
2. F. Chen and J. Dudhia. Coupling an advanced land surface-hydrology model with the penn state-ncar mm5 modeling system. part i: Model implementation and sensitivity. *Monthly Weather Review*, 129(4):569–585, 2001.
3. P. Courtier. Variational methods. *J. Meteor. Soc. Japan*, 75, 1997.
4. C. Davis and S. Low-Nam. The ncar-afwa tropical cyclone bogussing scheme. *Technical Memorandum, Air Force Weather Agency (AFWA), Omaha, NE, [http://www.mmm.ucar.edu/mm5/mm5v3/tc-report.pdf]*, 2001.
5. A. Orlandi et al. Rainfall assimilation in rams by means of the kuo parameterisation inversion: Method and preliminary results. *Journal of Hydrology*, 288(1-2):20–35, 2004.
6. C. Onof et al. Rainfall modelling using poisson-cluster processes: A review of developments. *Stochastic Environmental Research and Risk Assessment*, 2000.
7. C. S. Velden et al. Upper-tropospheric winds derived from geostationary satellite water vapor observations. *Bulletin of the American Meteorological Society*, 78(2):173–195, 1997.
8. C. Velden et al. Recent innovations in deriving tropospheric winds from meteorological satellites. *Bulletin of the American Meteorological Society*, 86(2):205–223, 2005.
9. G. Evensen. The ensemble kalman filter: Theoretical formulation and practical implementation. *Ocean Dynamics*, 53:342–367, 2003.
10. A. Gelb. *Applied Optimal Estimation*. MIT Press, 1974.
11. D. J. Heeger. Optical flow from spatiotemporal filters. *International Journal of Computer Vision*, pages 279–302, 1988.
12. A. C. Lorenc. Analysis method for numerical weather predictin. *Q. J. R. Meteorol. Soc.*, 112:1177–1194, 1986.
13. H.-H Nagel. Displacement vectors derived from second order intensity variations in image sequences. *Computer Vision, Graphics and Image Processing*, 21:85–117, 1983.
14. S. Ravela and V. Chatdarong. How do we deal with position errors in observations and forecasts? In *European Geophysical Union Annual Congress*, 2006.
15. S. Ravela and V. Chatdarong. Rainfall advection using velocimetry by multiresolution viscous alignment. Technical report, arXiv, physics/0604158, April 2006.
16. S. Ravela, K. Emanuel, and D. McLaughlin. Data assimilation by field alignment. *Physica D (Article in Press)*, doi:10.1016/j.physd.2006.09.035, 2006.
17. S. Ravela and D. McLaughlin. Fast ensemble smoothing. *Ocean Dynamics(Article in Press)*, DOI 10.1007/s10236-006-0098-6, 2007.

A Realtime Observatory for Laboratory Simulation of Planetary Circulation*

S. Ravela, J. Marshall, C. Hill, A. Wong, and S. Stransky

Earth, Atmospheric and Planetary Sciences
Massachusetts Institute of Technology
ravela@mit.edu

Abstract. We present a physical, laboratory-scale analog of large-scale atmospheric circulation and develop an observatory for it. By combining observations
of a hydro-dynamically unstable flow with a 3D numerical fluid model, we obtain
a real-time estimate of the state of the evolving fluid which is better than either
model or observations alone. To the best of our knowledge this is the first such
observatory for laboratory simulations of planetary flows that functions in real
time. New algorithms in modeling, parameter and state estimation, and observation targeting can be rapidly validated, thus making weather and climate application accessible to computational scientists. Properties of the fluid that cannot be
directly observed can be effectively studied by a constrained model, thus facilitating scientific inquiry.

1 Introduction

Predicting planetary circulation is fundamental for forecasting weather and for studies of climate change. Predictions are typically made using general circulation models
(GCMs), which implement the discretized governing equations. It is well-known that
the prediction problem is hard [4]. Models typically have erroneous parameters and parameterizations, uncertain initial and boundary conditions, and their numerical schemes
are approximate. Thus not only will the error between physical truth and simulation
evolve in a complex manner, but the PDF of the evolving model state's uncertainty is
unlikely to retain the true state within it. A way forward is to constrain the model with
observations of the physical system. This leads to a variety of inference problems such
as estimating initial and boundary conditions and model parameters to compensate for
model inadequacies and inherent limits to predictability. Constraining models with observations on a planetary scale is a logistical nightmare for most researchers. Bringing
the real world into an appropriate laboratory testbed allows one to perform repeatable
experiments, and so explore and accelerate acceptance of new methods.

A well-known analog of planetary fluid-flow is a thermally-driven unstable rotating
flow [2,3]. In this experiment a rotating annulus with a cold center (core) and warm periphery (exterior) develops a circulation that is dynamically similar to the mid-latitude
circulation in the atmosphere (see Figure 1). We have built an observatory for this laboratory experiment with the following components: Sensors to take measurements of

---

* This material is supported NSF CNS 0540248.

Y. Shi et al. (Eds.): ICCS 2007, Part I, LNCS 4487, pp. 1155–1162, 2007.

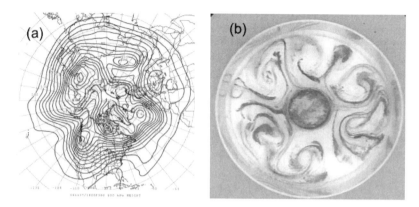

Fig. 1. Image (a) shows the 500hPa heights for 11/27/06:1800Z over the northern hemisphere centered at the north pole. Winds flow along the contours. Image (b) shows a tracer (dye) in a laboratory analog. The tank is spinning and the camera is in the rotating frame. Tracer droplets initially inserted at the periphery (red dye, warm region) and around the central chilled can (green dye, cold region) has evolved to form this pattern. The laboratory analog and the planetary system are dynamically akin to one-another. We study the state-estimation problem for planetary flows using the laboratory analog.

the evolving physical system, a numerical model trying to forecast the system, and an algorithm to combine model and observations. The challenges in building such a system are rather similar to the large-scale problem, in at least four ways. Nonlinearity: The laboratory analog is nonlinear and the numerical model is the same used in planetary simulations. Dimensionality: The size of the state of the numerical model is of the same order as planetary simulations. Uncertainty: The initial conditions are unknown, and the model is imperfect relative to the physical system. Realtime: Forecasts must be produced in better than realtime. This corresponds to a time of order ten seconds in our laboratory system within which a forecast-observe- estimate cycle must be completed.

In this report, we discuss the realtime system and focus on the problem of estimating initial conditions, or state. This estimation problem is posed as one of filtering and we demonstrate a two-stage assimilation scheme that allows realtime model-state estimates.

2   The Observatory

The observatory, illustrated in Figure 2, has a physical and computational component. The physical component consists of a perspex annulus of inner radius 8cm and outer radius of 23cm, filled with 15cm of water and situated rigidly on a rotating table. A robotic arm by its side moves a mirror up and down to position a horizontal sheet of laser light at any depth of the fluid. Neutrally buoyant fluorescent particles are embedded in water and respond to incident laser illumination. They appear as a plane of textured dots in the 12-bit quantized, $1K \times 1K$ images (see Figure 4) of an Imperx camera. These images are transported out of the rotating frame using a fiber-optic rotary joint (FORJ or slip-ring). The actual configuration of these elements is shown in a photograph of our rig in Figure 3.

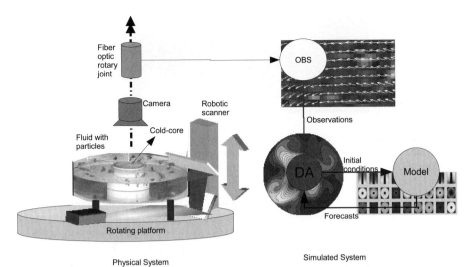

Fig. 2. The laboratory observatory consists of a physical system: a rotating table on which a tank, camera and control system for illumination are mounted. The computational part consists of a measurement system for velocimetry, a numerical model, and an assimilation system. Please see text for description.

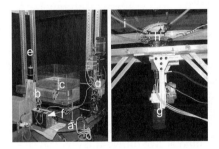

Fig. 3. The apparatus consists of (a) the rotating platform, (b) the motorized mirror, (c) the tank, (d) electronics, (e) a rig on which a camera is mounted, (g). Laser light comes from direction (f) and bounces off two mirrors before entering the tank. The fiber optic rotary joint (FORJ) (h) allows images to leave rotating frame and is held stably by bungee chords (i).

The computational aspects of the observatory are also shown in Figure 2. A server acquires particle images and ships them to two processors that compute optic-flow in parallel (Figure 2, labeled (OBS)). Flow vectors are passed to an assimilation program (Figure 2, labeled (DA)) that combines them with forecasts to estimate new states. These estimates become new initial conditions for the models. We now go on to discuss individual components of this system.

2.1   Physical Simulation and Visual Observation

We homogenize the fluid with neutrally buoyant particles and spin the rotating platform, typically with a period of six seconds. After twenty minutes or so the fluid entrains itself

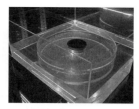

Fig. 4. The rotating annulus is illuminated by a laser light sheet shown on the left. The camera in the rotating frame sees embedded particles shown on the right. Notice the shadow due to the chiller in the middle. The square tank is used to prevent the laser from bending at the annulus interface.

to the rotation and enters into solid body rotation. The inner core is then cooled using a chiller. Within minutes the water near the core cools and becomes dense. It sinks to the bottom to be replenished by warm waters from the periphery of the annulus, thus setting up a circulation. At high enough rotation rates eddies form; flowlines bend forming all sorts of interesting structures much like the atmosphere; see Figure 1.

Once cooling commences, we turn off the lights and turn on the continuous wave 1W 532nm laser, which emits a horizontal sheet of light that doubles back through two periscoped mirrors to illuminate a sheet of the fluid volume (see Figure 4). An imaging system in the rotating frame observes the developing particle optic-flow using a camera looking down at the annulus.

The ultra-small pliolite particles move with the flow. We see the horizontal component and compute optical flow from image pairs acquired 125-250ms apart using LaVision's DaVis software. Flow is computed in 32×32 windows with a 16 pixel uniform pitch across the image. It takes one second to acquire and compute the flow of a single 1Kx 1K image pair. An example is shown in Figure 5.

Observations are gathered over several levels, repeatedly. The mirror moves to a preprogrammed level, the system captures images, flow is computed, and the mirror moves to the next preprogrammed level and so on, scanning the fluid volume in layers. We typically observe the fluid at five different layers and so observations of the whole fluid are available every 5 seconds and used to constrain the numerical model of the laboratory experiment.

2.2 Numerical Model

We use the MIT General Circulation Model developed by Marshall et al. [6,5] to numerically simulate the circulation. The MIT-GCM is freely available software and can be configured for a variety of simulations of ocean or atmosphere dynamics.

We use the MIT-GCM to solve the primitive equations for an incompressible Boussinesq fluid in hydrostatic balance. Density variations are assumed to arise from changes in temperature. The domain is three-dimensional and represented in cylindrical coordinates, as shown in Figure 6(a), the natural geometry for representing an annulus. In experiments shown here, the domain is divided into 23 bins in radius (1cm/bin), 120 bins in orientation ($3°$ bins). The vertical coordinate is discretized non uniformly using 15 levels and covering 15cm of physical fluid height, as shown in Figure 6(b). The fluid is modeled as having a free slip upper boundaries and a linear implicit free surface. The

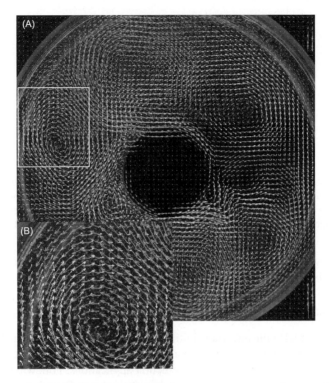

Fig. 5. A snapshot of our interface showing model velocity vectors (yellow), and observed velocities (green) at some depth. The model vectors macroscopically resemble the observations, though the details are different, since the model began from a different initial condition and has other errors. Observations are used to constrain model states, see section 2.1.

lateral and bottom boundaries are modeled as no-slip. The temperature at the outer core is constant and at the inner core is set to be decreasing with a profile interpolated from sparse measurements in a separate experiment (see Figure 6(b)). The bottom boundary has a no heat-flux condition. We launched the model from a random initial temperature field. A 2D slice is shown in Figure 6(c).

The MIT-GCM discretizes variables on an Arakawa C-grid [1]. Momentum is advected using a second-order Adams Bashforth technique. Temperature is advected using an upwind-biased direct space-time technique with a Sweby flux-limiter [7]. The treatment of vertical transport and diffusion is implicit. The 2D elliptic equation for the surface pressure is solved using conjugate gradients. In Figure 5, the model velocities are overlaid on the observed velocities after suitably registering the model geometry to the physical tank and interpolating. Despite the obvious uncertainty in initial conditions and other approximations, the model preserves the gross character of flow observed in the physical fluid, but at any instant the model-state differs from the observations, as expected.

The model performs in better than realtime. On an Altix350, using one processor, we barely make it[1], but on 4 processors we are faster by a factor of 1.5. The reason for this

---

[1] In ongoing work with Leiserson et al. we seek to speedup using multicore processors.

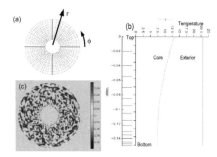

Fig. 6. (a) The computational domain is represented in cylindrical coordinates. (b) Depth is discretized with variable resolution, to resolve the bottom-boundary finely. The lateral boundary conditions were obtained by interpolating sparse temperature measurements taken in a separate run and the bottom boundary is no flux. (c) shows a random initial condition field for a layer.

performance is the non-uniform discretization of the domain using nonuniform vertical levels, which is also sufficient to resolve the flow.

2.3   State Estimation

An imperfect model with uncertain initial conditions can be constrained through a variety of inference problems. In this paper, we estimate initial conditions, or state estimation, which is useful in weather prediction applications. Following well-known methodology, when the distributions in question are assumed Gaussian, the estimate of state X_t at time t is the minimum of the following quadratic objective

$$J(X_t) = (X_t - X_t^f)^T B_t^{-1}(X_t - X_t^f) + \\ (Y_t - h(X_t))^T R^{-1}(Y_t - h(X_t)) \tag{1}$$

Here, X_t^f is the forecast at time t, B is the forecast error-covariance, h is the observation operator and R is the observation error-covariance.

We use a two-stage approach. In the first stage, forecast and measured velocities at each of the five observed levels are separately assimilated to produce velocity estimates. Thermal wind is then used to map the implied temperature fields at these corresponding levels. Linear models estimated between model-variables in the vertical are finally used to estimate velocity and temperature at all model layers. Velocities at individual layers are assimilated using an isotropic background error covariance. In effect, we assume that the model does not have sufficient skill at startup and the first stage, therefore, is designed to nudge the model to develop a flow similar to the observations. Because the domain is decomposed into independent 2D assimilations and many independent 1D regressions, this stage is very fast.

After running the first step for a few iterations, the second stage is triggered. Here we use an ensemble of model-states to represent the forecast uncertainty and thus use the ensemble Kalman filter to compute model-state estimates. It is impractical to do large ensemble simulations. So we use time samples of the MIT-GCM's state and

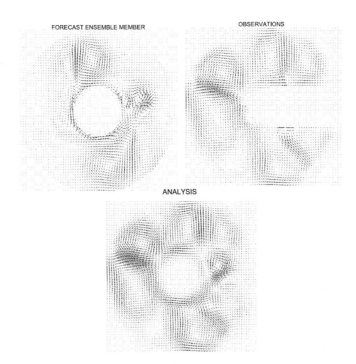

Fig. 7. The forecast velocity field at a time $t = 10min$ (left), observations at this layer (right) and estimated velocity (bottom). The shadow precludes measurements in a large area and produces noisy vectors at the shadow boundary.

perturb snapshots azimuthally with a mean perturbation of 0° and standard deviation 9° to statistically represent the variability of model forecasts. In this way the ensemble captures the dominant modes with very few numerical simulations. This method produces effective state estimates very efficiently.

The model is spun up from a random initial condition and forecasts 10 seconds ahead in approximately 6 seconds. Ensemble members are produced by time sampling every model second and perturbed to construct 40 member forecast ensembles. Planar velocity observations of 5 layers of the model are taken in parallel. In current implementation, assimilation is performed off-line though its performance is well within realtime. In under 2 seconds, the models and observations are fused to produce new estimates. The new estimated state becomes a new initial condition and the model launches new forecasts.

In Figure 7 the planar velocity of a forecast simulation, observations and estimates at the middle of the tank is shown 10 minutes into an experiment. As can be seen, the observations are noisy and incomplete due to the shadow (see Figure 4). The estimate is consistent with the observations and fills-in missing portions using the forecast. The error between the observations and estimates is substantially reduced. Please note that all 5 levels of observations are used and the entire state is estimated, though not depicted here for lack of space.

3   Conclusions

The laboratory analog of the mid-latitude circulation is a robust experiment, and the numerical model is freely available. Thus the analog serves as a new, easy-to-use, testbed. We have built a realtime observatory that to the best of our knowledge has not been reported before. Our hope is that the datasets generated here would find useful application to other researchers to apply their algorithms. Realtime performance is achieved through parallelism (observations), domain-reduction (model) and an efficient method to generate samples and compute updates (estimation).

A successful observatory also opens a number of exciting possibilities. Once the numerical model faithfully tracks the physical system, properties of the fluid that cannot easily be observed (surface height, pressure fields, vertical velocities etc.) can be studied using the model. Tracer transport can be studied using numerical surrogates. Macroscopic properties such as effective diffusivity can be studied via the model. For weather prediction, the relative merits of different state estimation algorithms, characterizations of model error, strategies for where to observe, etc etc, can all be studied and results reported on the laboratory system will be credible.

Of particular interest is the role of the laboratory analog for DDDAS. By building the infrastructure in the first year of this work, we can take on DDDAS aspects of this research in the second. In particular, we are interested in using the model-state uncertainty to optimize the number of observed sites, and locations where state updates are computed. In this way we expect to steer the observation process and use the observations to steer the estimation process.

Acknowledgment

Jean-Michel Campin and Ryan Abernathy's help in this work is gratefully acknowledged.

References

1. A. Arakawa and V. Lamb. Computational design of the basic dynamical processes of the ucla general circulation model. *Methods in Computational Physics, Academic Press*, 17:174–267, 1977.
2. R. Hide and P. J. Mason. Sloping convection in a rotating fluid. *Advanced Physics*, 24:47–100, 1975.
3. C. Lee. *Basic Instability and Transition to Chaos in a Rapidly Rotating Annulus on a Beta-Plane*. PhD thesis, University of California, Berkeley, 1993.
4. E. N. Lorenz. Deterministic nonperiodic flow. *J. Atmos. Sci.*, 20:130–141, 1963.
5. J. Marshall, A. Adcroft, C. Hill, L. Perelman, and C. Heisey. A finite-volume, incompressible navier stokes model for studies of the ocean on parallel computers. *J. Geophysical Res*, 102(C3):5753–5766, 1997.
6. J. Marshall, C. Hill, L. Perelman, and A. Adcroft. Hydrostatic, quasi-hydrostatic and nonhydrostatic ocean modeling. *Journal of Geophysical Research*, 102(C3):5733–5752, 1997.
7. P. K. Sweby. High resolution schemes using flux-limiters for hyperbolic conservation laws. *SIAM Journal of Numerical Analysis*, 21:995–1011, 1984.

Planet-in-a-Bottle: A Numerical Fluid-Laboratory System*

Chris Hill[1], Bradley C. Kuszmaul[1], Charles E. Leiserson[2], and John Marshall[1]

[1] MIT Department of Earth and Planetary Sciences, Cambridge, MA 02139, USA
[2] MIT Computer Science and Artificial Intelligence Laboratory, Cambridge, MA
02139, USA

Abstract. Humanity's understanding of the Earth's weather and climate depends critically on accurate forecasting and state-estimation technology. It is not clear how to build an effective dynamic data-driven application system (DDDAS) in which computer models of the planet and observations of the actual conditions interact, however. We are designing and building a laboratory-scale dynamic data-driven application system (DDDAS), called *Planet-in-a-Bottle*, as a practical and inexpensive step toward a planet-scale DDDAS for weather forecasting and climate model. The Planet-in-a-Bottle DDDAS consists of two interacting parts: a fluid lab experiment and a numerical simulator. The system employs *data assimilation* in which actual observations are fed into the simulator to keep the models on track with reality, and employs *sensitivity-driven observations* in which the simulator targets the real-time deployment of sensors to particular geographical regions and times for maximal effect, and refines the mesh to better predict the future course of the fluid experiment. In addition, the feedback loop between targeting of both the observational system and mesh refinement will be mediated, if desired, by human control.

1 Introduction

The forecasting and state-estimation systems now in place for understanding the Earth's weather and climate consists of myriad sensors, both in-situ and remote, observing the oceans and atmosphere. Numerical fluid simulations, employing thousands of processors, are devoted to modeling the planet. Separately, neither the observations nor the computer models accurately provide a complete picture of reality. The observations only measure in a few places, and the computer simulations diverge over time, if not constrained by observations. To provide a better picture of reality, application systems have been developed that incorporate *data assimilation* in which observations of actual conditions in the fluid are fed into computer models to keep the models on track with the true state of the fluid. Using the numerical simulation to target the real-time deployment of sensors to particular geographical regions and times for maximal effect, however, is only a

---

* This research was supported by NSF Grant 0540248.

Y. Shi et al. (Eds.): ICCS 2007, Part I, LNCS 4487, pp. 1163–1170, 2007.

research area and not yet operational. Under the NSF DDDAS program, we are designing and building a laboratory analogue dynamic data-driven application system (DDDAS) that includes both data assimilation and *sensitivity-driven observations* to explore how one might proceed on the planetary scale.

In meteorology, attempts to target adaptive observations to "sensitive" parts of the atmosphere is already an area of active research [11]. Observation-targeting field experiments, such as the Fronts and Atlantic Storm-Track EXperiment (FASTEX) and the NORth Pacific EXperiment (NORPEX), have demonstrated that by using, for example, objective adjoint techniques, it is possible, in advance, to identify regions of the atmosphere where forecast-error growth in numerical forecast models is maximally sensitive to the error in the initial conditions [3]. The analysis sensitivity field is then used to identify promising targets for the deployment of additional observations for numerical weather prediction. Such endeavors, although valuable, are enormously expensive and hence rare, and the experimental results are often not repeatable.

In oceanography, the consortium for Estimating the Circulation and Climate of the Ocean (ECCO [13]) is an operational state-estimation system focusing on the ocean. This consortium is spearheaded by some of the authors of the present project but also involves scientists at the Jet Propulsion Laboratory (JPL), the Scripps Institute of Oceanography (SIO), as well as other MIT researchers. Ours has been a broad-based attack — we have coded new models of the atmosphere and ocean from the start with data assimilation in mind [9,10,8,1]. Both forward and adjoint models (maintained through automatic differentiation [7]) have been developed. We have also collaborated with computer scientists in the targeting of parallel computers [12] and in software engineering [5].

Unfortunately, deploying a DDDAS for ECCO or weather forecasting is currently unrealistic. Scientific methodology is uncertain, the cost would be substantial, and the technical hurdles to scaling to a global, high-resolution system, especially in the area of software, are immense.

We therefore are designing and building a laboratory-scale DDDAS, which we call *Planet-in-a-Bottle* (or *Bottle*, for short), a practical and inexpensive step towards a planet-scale DDDAS. The Bottle DDDAS will emulate many of the large-scale challenges of meteorological and oceanographic state-estimation and forecasting but provide a controlled setting to allow systematic engineering strategies to be employed to devise more efficient and accurate techniques. The DDDAS will consist of two interacting parts: a fluid lab experiment and a numerical simulator. Observations taken from the laboratory experiment will feed into the simulator, allowing the simulator to refine the irregular mesh underlying the simulation and better predict the future course of the fluid experiment. Conversely, results from the numerical simulation will feed back to the physical system to target observations of the fluid, achieving a two-way interplay between computational model and observations. In addition, the feedback loop between targeting of both the observational system and mesh refinement will be mediated, if desired, by human control.

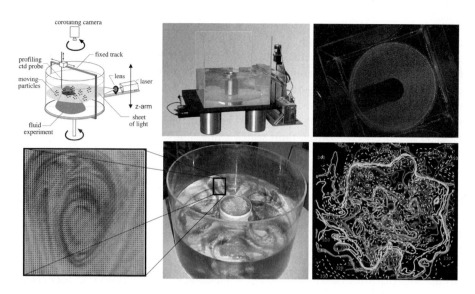

Fig. 1. The Planet-in-a-Bottle laboratory setup. The top row shows, on the left, a schematic of the tank-laser assembly, in the middle, the apparatus itself and, on the right, the laser sheet illuminating a horizontal plane in the fluid for the purpose of particle tracking via PIV. Note that the whole apparatus is mounted on a large rotating table. The bottom row shows, on the left, dye streaks due to circulation in the laboratory tank with superimposed velocity vectors from PIV, in the middle, eddies and swirls set up in the tank due to differential heating and rotation and, on the right, a snapshot of the temperature of the tropopause in the atmosphere (at a height of roughly 10km) showing weather systems. The fundamental fluid mechanics causing eddies and swirls in the laboratory tank in the middle is the same as that causing the weather patterns on the right.

The Planet-in-a-Bottle DDDAS will provide an understanding of how to build a planet-scale DDDAS. By investigating the issues of building a climate modeling and weather forecasting DDDAS in a laboratory setting, we will be able to make progress at much lower cost, in a controlled environment, and in a environment that is accessible to students and other researchers. The MITgcm [10, 9, 6, 2] software that we use is exactly the same software used in planet-scale initiatives, such as ECCO, and so our research promises to be directly applicable to planet-scale DDDAS's of the future. MITgcm is a CFD engine, built by investigators from this project, designed from the start with parallel computation in mind, which in particular runs on low-cost Linux clusters of processors.

When this project started, both the Bottle lab and the Bottle simulator already existed (see Figure 1), but they had not been coupled into a DDDAS system. In particular, the simulator was far too slow (it ran at about 100 times real time), and it lacked sufficient accuracy. Data assimilation of observations occured off-line after the fluid experiment has been run, and the feedback loop that would allow the simulator to target observations in the fluid is not yet available. Building the Planet-in-a-Bottle DDDAS required substantial research that we broke into two research thrusts: dynamic data-driven science for fluid modeling, and algorithms and performance.

The first research thrust involves developing the dynamic data-driven science for fluid modeling in the context of MITgcm running in the Bottle environment. We are investigating *adjoint models*, which are a general and efficient representation of model sensitivity to any and all of the parameters defining a model. We are studying how adjoint models can be used to target observations to yield key properties of the fluid which cannot be directly measured, but only inferred by the synthesis of model and data. We are also studying their use in guiding the refinement of irregular meshes to obtain maximal detail in the Bottle simulation. We estimate that the computational demands for a practical implementation will be 5–10 times that of an ordinary forward calculation. The implementation also challenges the software-engineering infrastructure we have built around MITgcm, which today is based on MPI message passing and data-parallel computations.

The second thrust of our research focuses on algorithmic and performance issues. The previous Bottle simulation is far too slow to be usable in a DDDAS environment. We believe that the performance of the simulation can be improved substantially by basing the simulation on adaptive, irregular meshes, rather than the static, regular meshes now in use, because many fewer meshpoints are required for a given solution. Unfortunately, the overheads of irregular structures can negate their advantages if they are not implemented efficiently. For example, a naive implementation of irregular meshes may not use the memory hierarchy effectively, and a poor partitioning of an irregular mesh may lead to poor load balancing or high communication costs. We are investigating and applying the algorithmic technology of *decomposition trees* to provide provably good memory layouts and partitions of the irregular meshes that arise from the fluid simulation.

2   The Laboratory Abstraction: Planet-in-a-Bottle

The Bottle laboratory consists of the classic *annulus experiment* [4], a rotating tank of water across which is maintained a lateral temperature gradient, cold in the middle (representing the earth's pole), warm on the outside (representing the equator). The apparatus is shown in Figure 1. Differential heating of the rotating fluid induces eddies and geostrophic turbulence which transfer heat radially from "equator to pole," and hence the "Planet-in-a-Bottle" analogy. This class of experiment has been a cornerstone of geophysical fluid dynamics and serves as a paradigm for the fluid dynamics and hydrodynamical instability processes that underlie our weather and key processes that maintain the pole-equator temperature gradient of the planet.

We use a video camera to capture the flow in a chosen horizontal plane by introducing neutrally buoyant pliolite particles into the fluid and illuminating them with a laser sheet. The top right panel of Figure 1 shows the laser illuminating pliolite particles at a particular depth. The captured images are recorded at a rate of 30 frames per second with a resolution of 1024×780 pixels. The video images are fed to an image-processing system, which calculates the velocity field of the particles using PIV (particle imaging velocimetry). In addition, an array of sensors in the fluid record the temperature. The resulting data is fed into the Bottle simulator. The Bottle simulator consists of a computer system running an ensemble of 30 simulation kernels, each of which runs the MITgcm, a computational fluid dynamics (CFD) code that we have developed.

As illustrated in Figure 2, the simulator process divides time into a series of epochs, each with a duration about the same as one rotation of the laboratory experiment (representing one day, or in the laboratory, typically on the order of 10 seconds). At the beginning of each epoch, the simulator initializes each of the 30 kernels with an ensemble Kalman filter derived estimate of the current state of the fluid based on the observed fluid state and the simulation state from the preceeding epoch. The simulator perturbs the initial state of each kernel in a slightly different manner. The 30 kernels then run forward until the end of the epoch, at which point the results are combined with the next frame of observations to initialize the next epoch.

The DDDAS we are building will enhance the current data assimilation with real-time, sensitivity-driven observation. The assimilation cycle will be preceeded by a forecast for the upcoming epoch. At the end of the forecast, using the adjoint of the forward model, a sensitivity analysis will determine which locations in the fluid most affect the outcomes in which we are interested. The simulator will then direct the motor controlling the laser sheet to those particular heights in the fluid and direct the camera to zoom in on the particular regions. A dynamically driven set of observations will then be made by the camera, the image-processor will calculate the velocity field of the particles, and the simulator will compute a new estimate of the fluid state. In addition, the Planet-in-a-Bottle system will dynamically refine the mesh used by the Kalman filter simulation kernels so that more computational effort is spent in those regions to which the outcome is most sensitive. This forecast and assimilation process will be repeated at each epoch.

This is a direct analogue of the kind of large scale planetary estimates described in, for example, [13]. Everything about the laboratory system — the assimilation algorithms, the simulation codes, the observation control system, the real-time constraints, and the scalable but finite compute resources is practically identical to technologies in operational use in oceanography and meteorology today. Moreover — and this is at the heart of the present project — it provides an ideal opportunity to investigate real-time data-driven applications because we can readily (i) use the adjoint of our forward model to target observations of an initial state to optimise the outcome (ii) refine the numerical grid in regions of particular interest or adjoint sensitivity (iii) experiment with intervention by putting humans in the loop, directly controlling the observations and/or grid refinement process. Although the physical scale of the laboratory experiment is small, it requires us to bring up the end-to-end combined physical-computational system and consider many issues encountered in the operational NWP and

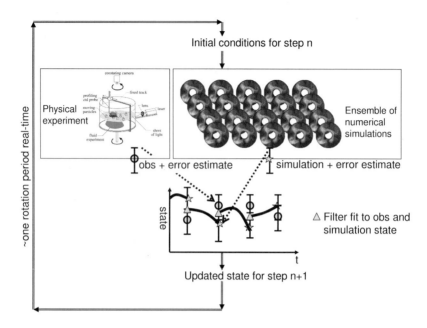

Fig. 2. Schematic of the ensemble Kalman filter. The interval between Kalman filter updates is one rotation period of the fluid experiment (typically $\approx$ 10s). The physical system (on the left) is observed with noisy and incomplete observations. An ensemble (on the right) of simulation kernels is run, each from a perturbed initial state, producing simulations with an error estimate. A filter combines the observations with the ensembles to yield a new updated state for the next epoch.

large-scale state-estimation communities. Notably, the MITgcm software we currently use for numerical simulation of the laboratory fluid is also in use in planetary scale initiatives. Consequently, innovations in computational technologies tested in the laboratory will directly map to large-scale, real-world problems.

3   Progress

Our team has made significant progress on our Bottle DDDAS in the first year. We have finished developing about 70% of the physical infrastructure, 80% of the computational infrastructure, and have developed a first-version of the end-to-end system demonstrating real-time simulation, measurement and estimation processes.

We have developed robust protocols for physical simulation, instrumented the system to acquire and process data at a high-bandwidth. Currently the system can process about 60 MBytes/sec of raw observational and model data to produce state estimates in real-time. The system includes a camera and a fiber-optic rotary joint, a laser light-sheet apparatus, a robotic arm to servo the light sheet to illuminate a fluid plane, rotating fluid homogenized with neutrally buoyant particles. A thermal control subsystem is presently being integrated to produce climatological forcing signals of temperature gradient (and hence heat flux).

Observations are produced using a particle image velocimetry method, procedures for which were perfected in the first year. A distributed computing infrastructure has been developed for generating velocity measurements. We now frequently use this subsystem for gathering observations.

We have made refinements to the MIT-GCM using a nonuniform domain decomposition, and this is coupled with a 3rd order accurate advection scheme. The model functions in realtime on an Altix 350.

Our goal has been to develop a high-performance application that is also superbly interactive. The observation subsystem is designed to be easily reconfigurable, but the details of the distributed computation are entirely hidden from the user. Similarly, easy to configure model interface is being developed to allow the user to dynamically reconfigure the model parameters whilst leaving the actual implementation of computation out of the picture. Both observations and models perform in realtime, thus the effect of parameter changes are observed in realtime, and this is the basis for user-perceived interactivity. Data-assimilation is implemented using matlab; a language most researchers know. We are incorporating the StarP system and thus the implementation of distributed computation is hidden from the user. Students and researchers alike can use, almost verbatim, their code on the large-scale problem and change algorithms rapidly. We believe that this architecture strikes the right balance between performance and interaction, and we hope to see its benefits as we incorporate it in research and the classroom in the next two years.

We have demonstrated that data-assimilation inference from data and models can also be performed in real-time. We have developed a two-stage approach that automatically switches assimilation from a weak prior (low model skill) mode to a strong prior mode. Both these function in real-time. The latter mode uses an ensemble to construct subspace approximations of the forecast uncertainty. The ensemble is generated robustly by combining perturbations in boundary conditions with time-snapshots of model evolution. This new approach entirely diminishes the computational bottleneck on the model performance because only few model integrations are required, and surrogate information of the state's uncertainty is gleaned from exemplars constructed from the temporal evolution of the model-state. A forty member ensemble (typical in meteorological use) thus requires around 4 separate model simulations; an easy feat to accomplish.

Work on two other ensemble-based assimilation methods has also progressed. First, a fast ensemble smoother (Ocean Dynamics, to appear) shows that fixed-interval smoothing is $O(n)$ in time, and fixed-lag smoothing is independent of the lag length. Second, we have developed a scale-space ensemble filter that combines graphical, multiscale models with spectral estimation to produce rapid estimates of the analysis state. This method is superior to several popular ensemble methods and performs in $O(n \log n)$ of state size n, and is highly parallelizable.

References

1. A. Adcroft, J-M Campin, C. Hill, and J. Marshall. Implementation of an atmosphere-ocean general circulation model on the expanded spherical cube. *Mon. Wea. Rev.*, pages 2845–2863, 2004.

2. A. Adcroft, C. Hill, and J. Marshall. Representation of topography by shaved cells in a height coordinate ocean model. *Mon. Wea. Rev.*, pages 2293–2315, 1997.

3. N. Baker and R. Daley. Observation and background adjoint sensitivity in the adaptive observation-targeting problem. *Q.J.R. Meteorol. Soc*, 126(565):1431–1454, 2000.

4. M. Bastin and P. Read. A laboratory study of baroclinic waves and turbulence in an internally heated rotating fluid annulus with sloping endwalls. *J. Fluid Mechanics*, 339:173–198, 1997.

5. C. Hill, C. DeLuca, Balaji, M. Suarez, and A. DaSilva. The architecture of the Earth System Modeling Framework. *Computing in Science and Engineering*, 6(4):18–28, 2004.

6. C. Hill and J. Marshall. Application of a parallel Navier-Stokes model to ocean circulation. In *Proceedings of Parallel Computational Fluid Dynamics: Implementations and Results Using Parallel Computers*, pages 545–552, 1995.

7. J. Marotzke, R. Giering, K. Q. Zhang, D. Stammer, C. Hill, and T. Lee. Construction of the adjoint MIT ocean general circulation model and application to Atlantic heat transport sensitivity. *J. Geo. Res.*, 104(C12):29,529–29,547, 1999.

8. J. Marshall, A. Adcroft, J-M. Campin, C. Hill, and A. White. Atmosphere-ocean modeling exploiting fluid isomorphisms. *Monthly Weather Review*, 132(12):2882–2894, 2004.

9. J. Marshall, A. Adcroft, C. Hill, L. Perelman, and C. Heisey. A finite-volume, incompressible Navier Stokes model for studies of the ocean on parallel computers. *J. Geophys. Res.*, 102, C3:5,753–5,766, 1997.

10. J. Marshall, C. Hill, L. Perelman, and A. Adcroft. Hydrostatic, quasi-hydrostatic and nonhydrostatic ocean modeling. *J. Geophys. Res.*, 102, C3:5,733–5,752, 1997.

11. T.N. Palmer, R. Gelaro, J. Barkmeijer, and R. Buizza. Vectors, metrics, and adaptive observations. *J. Atmos. Sci*, 55(4):633–653, 1998.

12. A. Shaw, Arvind, K.-C. Cho, C. Hill, R. P. Johnson, and J. Marshall. A comparison of implicitly parallel multi-threaded and data-parallel implementations of an ocean model based on the Navier-Stokes equations. *J. of Parallel and Distributed Computing*, 48(1):1–51, 1998.

13. D. Stammer, C. Wunsch, R. Giering, C. Eckert, P. Heimbach, J. Marotzke, A. Adcroft, C. Hill, and J. Marshall. Volume, heat, and freshwater transports of the global ocean circulation 1993–2000, estimated from a general circulation model constrained by World Ocean Circulation Experiment (WOCE) data. *J. Geophys. Res.*, 108(C1):3007–3029, 2003.

Compressed Sensing and Time-Parallel Reduced-Order Modeling for Structural Health Monitoring Using a DDDAS

J. Cortial[1], C. Farhat[1,2], L.J. Guibas[3], and M. Rajashekhar[3]

[1] Institute for Computational and Mathematical Engineering
[2] Department of Mechanical Engineering
[3] Department of Computer Science
Stanford University
Stanford, CA 94305, U.S.A
jcortial@stanford.edu,cfarhat@stanford.edu,
guibas@cs.stanford.edu,manj@stanford.edu

Abstract. This paper discusses recent progress achieved in two areas related to the development of a Dynamic Data Driven Applications System (DDDAS) for structural and material health monitoring and critical event prediction. The first area concerns the development and demonstration of a sensor data compression algorithm and its application to the detection of structural damage. The second area concerns the prediction in near real-time of the transient dynamics of a structural system using a nonlinear reduced-order model and a time-parallel ODE (Ordinary Differential Equation) solver.

1 Introduction

The overall and long-term goal of our effort is to enable and promote active health monitoring, failure prediction, aging assessment, informed crisis management, and decision support for complex and degrading structural engineering systems based on dynamic-data-driven concepts. Our approach involves the development a Dynamic Data Driven Applications System (DDDAS) that demonstrates numerically, as much as possible, its suitability for structural health monitoring and critical event prediction. An outline of this approach, its objectives, and preliminary architectural concepts to support it can be found in [1].

The main objective of this paper is to describe progress achieved in two areas of activity that are directly related to drastically reducing the overall sensing and computational cost involved in the utilization of such a DDDAS. For this purpose, our efforts draw experiences and technologies from our previous research on the design of a data-driven environment for multiphysics applications (DDEMA) [2,3].

The first area of progress pertains to the development and implementation of an efficient data compression scheme for sensor networks that addresses the issue of limited communication bandwidth. Unlike other data compression algorithms,

Y. Shi et al. (Eds.): ICCS 2007, Part I, LNCS 4487, pp. 1171–1179, 2007.

this scheme does not require the knowledge of the compressing transform of the input signal. It constitutes a first step towards the efficient coupling between a given sensor network and a given computational model.

The second area of progress pertains to the development and implementation of a numerical simulator for the prediction in near real-time of the transient dynamics of a structural system. The new aspect of this effort is the generalization to nonlinear second-order hyperbolic problems of the PITA (Parallel Implicit Time-integration Algorithms) framework developed in [4,5] for linear time-dependent problems.

Preliminary verification and performance assessment examples that furthermore illustrate the basic concepts behind both methodologies outlined above are also provided in this paper.

2   An Efficient Data Compression Scheme for Sensor Networks

In the context of a DDDAS for health monitoring and critical event prediction, the sensor network collects the monitored data of the structure and sends it to the numerical simulator. This simulator is equipped with full- and reduced-order computational models that are validated for healthy states of the structure. Differences between sensed and computed (possibly in near real-time) data, or sensed data and data retrieved in real-time from a computational data base in the simulator, form a special class of sparse signals that can be used to locate, assess, and possibly predict the evolution of structural damage as well as dynamically update the computational models [6,7]. Hence, the sensor network acts in general as the source of the data used for guiding and driving numerical simulations and/or updating the underlying computational models.

Communication bandwidth is a very limited resource in sensor networks. For this reason, the trivial approach where all sensor readings are sent to the simulator is unlikely to be deployed. Compressed sensing offers a venue to a different approach that avoids expensive communication between the sensor network and the numerical simulator. Such an alternative approach is not only more practical for health monitoring in general, but also essential for critical event prediction. In this section, we discuss our chosen data compression scheme whose theory was developed in [8]. Unlike many other data compression algorithms, this scheme does not require any information about the compressing transform of the input signal. It only assumes that the signal is compressible — that is, it has a sparse representation in some orthogonal basis. In health monitoring applications, this assumption typically holds given that as mentioned above, a signal usually consists of the difference between a measured information and a predicted one. Since a sparse signal can be reconstructed from the knowledge of a few random linear projections [8], collecting only these few random projections constitutes a feasible mechanism for data compression and provides the sought-after saving in communication time.

2.1   Compressed Sensing Overview

Let $x_0 \in \mathbb{R}^m$ denote a signal of interest that has a sparse representation in some orthogonal basis, and $\phi \in \mathbb{R}^{n \times m}$, with $n < m$, denote the random matrix generated by the "uniform spherical ensemble" technique described in [8]. Consider the vector $y = \phi x_0$. This matrix-vector product represents the result of n random linear projections — or samples — of the original signal x_0. From the knowledge of y, one can reconstruct an approximation $\tilde{x}_0$ of x_0 as follows

$$\tilde{x}_0 = \arg\min_x ||x||_1 \text{ subject to } y = \phi x \tag{1}$$

The reconstruction technique summarized by the above equation involves essentially linear, non-adaptive measurements followed by a nonlinear approximate reconstruction. In [8], it is proved that despite its undersampling ($n < m$), this technique leads to an accurate reconstruction $\tilde{x}_0$ when $n = O(N \log(m))$, where N denotes the number of largest transform coefficients of x_0 — that is, the number of transform coefficients of x_0 that would be needed to build a reasonable approximation of this signal. For example, for a k-sparse system — that is, a system with only k values different from zero or larger in absolute value than a certain minimum threshold — $N = k$.

2.2   Implementation

In our proposed sensing approach, we avoid expensive communication between the sensor network and the simulator by communicating to the latter only summaries of the original sensor readings. These summaries contain however enough information to allow the extraction of the deviant sensor readings and localize the sensors announcing deviant information. Here, "deviant" refers to a measurement that is significantly different from an expected value obtained by simulation and therefore is indicative, for example, of structural damage.

More specifically, our approach to compressed sensing is composed of two main steps that are described next.

Extraction of Compressed Summaries. Let $s_0 \in \mathbb{R}^m$ and $p_0 \in \mathbb{R}^m$ denote the measured and numerically predicted readings at m sensor nodes, respectively. We compute n ($n << m$) random linear projections of these readings as follows

$$y_s = \phi s_0, \ y_p = \phi p_0 \tag{2}$$

The n compressed measurements of sensor readings represented by y_s are then communicated over the network to the simulator. Since $n << m$, this leads to a reduction in communication time.

Decoding of a Compressed Representation. At the simulator, we compute the difference between the sensor readings and corresponding simulator predictions

$$\delta y = y_s - y_p = \phi s_0 - \phi p_0 = \phi(s_0 - p_0) \tag{3}$$

For a computational model that is validated for a healthy state of the structure, the simulator predictions should be in principle in reasonable agreement with the sensor readings most of the time. Therefore, the signal $(s_0 - p_0)$ is in principle a sparse and therefore compressible one that can be accurately reconstructed by solving the minimization problem underlying Eq. (1) after substituting x by $(s - p)$ and y by δy from Eq. (3). Then, as demonstrated in Section 2.4, the reconstructed value $(\tilde{s}_0 - \tilde{p}_0)$ of $(s_0 - p_0)$ can be exploited, after noise removal, to localize the deviant sensor readings, predict the location of damage and possibly characterize it.

2.3   Summary

The compressed summaries outlined above behave like an *error detecting code*. They allow the system to detect events of interest, which facilitates automatic detection. They also contain enough information to allow the localization of the sensor nodes contributing to deviant readings. Finally, compressed sensing allows faster data acquisition by lowering the communication burden.

2.4   Preliminary Verification

To illustrate the compressed sensing methodology described above, we consider here the case of an F-16 fighter jet that is suddenly hit by an explosive bullet in one of its wings. For simplicity, we ignore the damage formation process in order to focus on the compressed sensing issues. Hence, we simulate here the inflicted damage by suddenly introducing in the Computational Structural Dynamics (CSD) and Computational Fluid Dynamics (CFD) models a realistic hole that alters the stiffness, mass, and aerodynamic properties of the aircraft (see Fig. 1). We equip the simulator with validated, full- and reduced-order, aeroelastic computational models of the F-16 such as those presented in [9,10,11]. We assume that 9131 sensor nodes are deployed over the entire fighter jet. We note

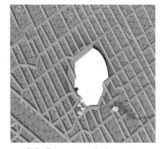

(a) CFD (wireframe) & CSM (shaded)   (b) Structural damage

Fig. 1. CFD and CSM models of an F-16 fighter with a hole in the port side wing

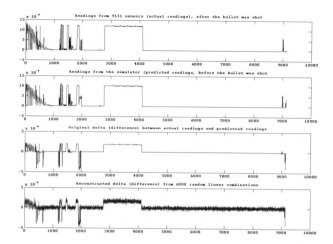

Fig. 2. In all plots, the x-axis represents the sensor nodes and the y-axis represents the sensed, predicted, difference, and reconstructed difference values of the vertical component of the displacement field at the sensor nodes

that this massive instrumentation hypothesis is currently unrealistic because of sensor weight issues. We are currently addressing this aspect of the problem by conducting various trade-off studies. We simulate the data measured by the sensor nodes after damage infliction using the AERO simulator [9,10] and compare it to the pre-damage data obtained by numerical predictions using the validated computational models.

Fig. 2 reports our findings using the compressed sensing approach with $m = 6000$ — that is, when collecting 6000 instead of 9131 readings. The peak values of the reconstructed difference signal shown in the bottom most diagram of this figure correlate well with the locations of the sensor nodes that are closest to the region hit by the explosive bullet. This demonstrates the potential of compressed sensing for locating structural damage. Using such a massive instrumentation also allows the in-flight estimation of the size of the damage, among other properties.

3   A Time-Parallel Computational Framework for Faster Numerical Predictions

The transient dynamics of a complex structural system are typically modeled by a set of Partial Differential Equations (PDEs) that are semi-discretized by the finite element method. The solution of the resulting semi-discrete systems of equations is typically organized around two loops: (a) an inner-loop which evaluates some quantities related to the state of the structure at a specific time-instance, and (b) an outer one which advances the state of the structure from a given time-instance to the next one. Hence, the computational cost associated with the time-integration of the semi-discrete systems governing structural dynamics problems can be divided into two main components: (a) the CPU time incurred by the inner-loop — that is, by the evaluation of the displacement field and/or related quantities at a given time-instance, and (b) that incurred by the outer-loop — that is, by time-advancing the complete state of the structure. To a large extent, parallel algorithms for the solution of transient dynamics problems have mainly addressed the reduction of the CPU time associated with the inner-loop. Such parallel algorithms can be described as space-parallel algorithms. By comparison, the parallelization of the outer-loop of a structural dynamics time-integrator — or time-parallelism — has received little attention in the literature. This is because, in general, time-parallelism where the solution is computed simultaneously at different time-instances is harder to achieve than space-parallelism, due to the inherently sequential nature of the time-integration process. However, time-parallelism can be of paramount importance to fast computations, for example, when space-parallelism is unfeasible, cannot exploit all the processors available on a given massively parallel system, or when predicting structural dynamics responses in near-real-time requires exploiting both space- and time-parallelisms.

Usually, space-parallelism is unfeasible when the number of degrees of freedom (dofs) is too small to allow amortization of the interprocessor communication overhead. Reduced-order computational models, which can be a critical component of an on-board DDDAS, examplify this scenario as they often involve only a few dofs. Yet, these and other computational models with relatively few dofs can be (or feel like they are) CPU intensive when time-integration is performed over a relatively large time-interval, and/or real-time predictions are desired, for example, for on-line critical event prediction or control purposes. Predicting in near-real-time the time-evolution of the variables governed by such models calls for time-parallelism. Finally, when considering off-line transient structural dynamics computations on massively parallel computers with thousands of processors such as the Department of Energy's ASCI machines, space-parallelism alone is not always the optimal strategy for reducing the time-to-solution [4]. Indeed, in many cases such as low-frequency applications, the mesh resolution dictated by accuracy considerations can be accommodated by a fraction of the total number of available processors. It is well-known that in such cases, because of hardware-related issues such as memory bandwidth effects and interprocessor communication costs, the solution time typically increases when the number of processors N_p is increased beyond a certain critical value, N_p^{cr}, while the mesh size is maintained constant. In this sense, the smaller is N_p^{cr} compared to the maximum number of available processors, N_p^{max}, the less optimal is the space-parallel implementation strategy, and the more adequate becomes an approach that combines both space- and time-parallelisms in order to reduce as much as possible the solution time by exploiting all N_p^{max} processors [4].

During the last decade, at least three approaches have been proposed for time-parallelizing time-integrators (for example, see [4] for a brief review). Among these approaches, the second-generation PITA [5] stands out as the only computational approach that has demonstrated a potential for accelerating the solution of second-order hyperbolic problems such as those encountered in structural dynamics applications. However, this computational framework was developed in [5] for linear problems only. Under this DDDAS research effort, we have recently generalized it to nonlinear dynamics problems such as those encountered in failure prediction and aging assessment.

3.1 Nonlinear Structural Dynamics Computational Models

Nonlinear structural dynamics computational models can often be written as

$$M\ddot{u} + f^{int}(\dot{u}, u) = f^{ext}(t), \ u(t_0) = u_0, \ \dot{u}(t_0) = \dot{u}_0 \tag{4}$$

where M, u, and $\dot{u}$ denote the mass matrix, and displacement and velocity vectors, respectively, f^{int} denotes the vector of internal forces which typically vary nonlinearly with u and $\dot{u}$ and f^{ext} denotes the vector of external forces. In the context of a DDDAS, Eq. (4) represents a full-order nonlinear computational model when discussing off-line simulations and a reduced-order model (ROM) when discussing on-line simulations.

3.2 Nonlinear PITA

The nonlinear PITA framework for the time-parallel solution of Eq. (4) is an iterative one. It can be described as follows. The time-domain of interest is partitioned into time-slices whose boundary points are treated as a coarse time-grid. In a pre-processing step, the solution is approximated on this coarse time-grid to provide a seed — that is, an initial condition — for each time-slice, using a sequential but computationally inexpensive numerical procedure. Next, the following computational steps are performed. First, a preferred time-integrator is applied independently and therefore concurrently in each time-slice to advance the solution from the starting point of this time-slice to its end point. This introduces jumps in the computed solution at the interior points of the coarse time-grid. Then, a Newton-based corrective step is performed on the original time-grid to improve the accuracy of the seeds and reduce the magnitude of these jumps. This two-step sequence of computations is repeated until convergence is reached — that is, until the solution jumps at the interior points of the coarse time-grid are eliminated. The computational burden associated with the corrective Newton step is alleviated by the introduction of a carefully constructed projector, so that the sequential computational complexity of each iteration of a nonlinear PITA is only slightly larger to that of its underlying (or preferred), time-sequential, time-integration algorithm. Therefore, the asymptotic parallel speedup of the nonlinear PITA using N_p processors is roughly equal to N_p/N_{it} where N_{it} denotes the number of iterations for convergence. This result reveals that the nonlinear PITA is mostly interesting when N_{it} can be kept as small as possible, say of the order of 2 to 4.

3.3   Preliminary Performance Assessment

To illustrate the potential of the recently developed nonlinear PITA for near real-time simulations such as those envisioned on an on-line DDDAS, we consider here the solution of Eq. (4) in the context of a reduced-order model of the F-16 aircraft shown in Fig. 1 and a cyclic (fatigue) load. We choose to equip PITA with the midpoint rule as the underlying time-integrator and decompose the time-domain of interest in 20 time-slices. The computational results reported in Fig. 3 reveal that for the considered problem, the nonlinear PITA converges in 3 iterations, which highlights its potential for supporting an on-line DDDAS for health monitoring.

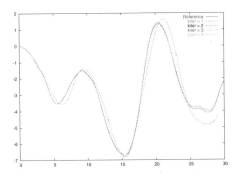

Fig. 3. Convergence of the nonlinear PITA equipped with the midpoint rule and applied to the solution of Eq. (4) in the context of an on-line reduced-order model

Acknowledgement. The authors acknowledge the support by the National Science Foundation under Grant CNS-0540419.

References

1. Farhat, C., Michopoulos, J. G., Chang, F. K., Guibas, L. J., Lew, A. J., Towards a Dynamic Data Driven System for Structural and Material Health Monitoring. International Conference on Computational Science, **3**, (2006), 456–464.
2. Michopoulos, J., Tsompanopoulou, P., Houstis, E., Rice, J., Farhat, C., Lesoinne, M., Lechenault, F., DDEMA: A Data Driven Environment for Multiphysics Applications. In: Proceedings of International Conference of Computational Science - ICCS'03, Sloot, P.M.A., et al. (Eds.) Melbourne Australia, June 2-4, **LNCS 2660**, Part IV, Springer-Verlag, Haidelberg, (2003), 309–318.
3. Michopoulos, J., Tsompanopoulou, P., Houstis, E., Farhat, C., Lesoinne, M., Rice, J., Joshi, A., On a Data Driven Environment for Multiphysics Applications. Fut. Generation Comp. Sys., **21**:6, (2005), 953–968.
4. Farhat C., Chandesris M., Time-Decomposed Parallel Time-Integrators: Theory and Feasability Studies for Fluid, Structure, and Fluid-Structure Applications. Internat. J. Numer. Meths. Engrg, **58**:9, (2003), 1397–1434.

5. Farhat, C., Cortial, J., Dastillung, C., Bavestrello, H., Time-Parallel Implicit Integrators for the Near-Real-Time Prediction of Linear Structural Dynamic Responses. Internat. J. Numer. Meths. Engrg, **67**:5, (2006), 697–724.

6. Farhat, C., Hemez, F., Updating Finite Element Dynamic Models Using an Element-by-Element Sensitivity Methodology. AIAA J., **31**:9, (1993), 1702–1711.

7. Doebling, S.W., Hemez, F.M., Peterson, L.D., Farhat, C., Improved Damage Location Accuracy Using Strain Energy-Based Mode Selection Criteria. AIAA J., **35**:4, (1997), 693–699.

8. Donoho, D., Compressed Sensing. IEEE Transactions on Information Theory, **52**:(4), (2006), 1289–1306.

9. Farhat, C., Geuzaine, P., Brown, G., Application of a Three-Field Nonlinear Fluid-Structure Formulation to the Prediction of the Aeroelastic Parameters of an F-16 Fighter. Comput. & Fluids, **32**, (2003), 3–29.

10. Geuzaine, P., Brown, G., Harris, C., Farhat, C., Aeroelastic Dynamic Analysis of a Full F-16 Configuration for Various Flight Conditions. AIAA J., **41**, (2003), 363–371.

11. Lieu, T., Farhat, C., Lesoinne, M., Reduced-Order Fluid/Structure Modeling of a Complete Aircraft Configuration, Comput. Meths. Appl. Mech. Engrg, **195**, (2006), 5730–5742.

Multi-level Coupling of Dynamic Data-Driven Experimentation with Material Identification

John G. Michopoulos[1] and Tomonari Furukawa[2]

[1] Computational Multiphysics Systems Laboratory
Special Projects Group, Code 6390.2
Center for Computational Material Science
Naval Research Laboratory, USA
john.michopoulos@nrl.navy.mil
[2] ARC Centre of Excellence in Autonomous Systems
School of Mechanical and Manufacturing Engineering, J17,
The University of New South Wales, Sydney, NSW 2052, Australia
t.furukawa@unsw.edu.au

Abstract. We describe a dynamic data-driven methodology that is capable of simultaneously determining both the parameters of a constitutive model associated with the response of a composite material, and the optimum experimental design that leads to the corresponding material characterization. The optimum design of experiments may contain two parts. One involving the identification of the parameters that are tunable prior to performing the experiment such as specimen characteristics and a priori loading path. The other is involving the parameters characterizing the experiment during the experiment itself such as the directionality of the loading path for the case of multi-axial loading machine. A multi-level coupled design optimization methodology is developed and applied to demonstrate the concept. Essential to the process is the development of objective functions that express the quality of the experimental procedure in terms of the uniqueness and distinguishability associated with the inverse solution of the constitutive model determination. The examples provided are based on the determination of the linear constitutive response of a laminate composite material.

1 Introduction

The advent of dynamic data driven application systems (DDDAS) and associated technologies, as they have been evolving rapidly during the past five years [1] has stimulated us to explore various areas beyond the realm of how data can be used to dynamically drive computational model formation or drive the associated simulations. The opportunity to explore the feasibility of designing experiments -before and during the time of their execution- necessary for the collection of data needed for characterizing a material system, defines the focus and goal of this paper as a natural extension of our ongoing DDDAS efforts [2,3].

Utilization of data-driven design optimization practices to determine constitutive behavior parameters of materials under mechanical loadings has been

Y. Shi et al. (Eds.): ICCS 2007, Part I, LNCS 4487, pp. 1180–1188, 2007.

traditionally based on experimental procedures with fixed architectures and no regard to how experimental designs may affect the quality of the material parameter estimation processes. However, the advent of multi-degree of freedom mechatronic systems capable of multidimensional mechanical loading [4,5,6], has introduced the potential of multiple designs for experimental processes to acquire the necessary behavioral data.

In the present paper we are proposing a multi-level design optimization methodology that intertwines three successive and dynamic design optimization subprocesses and is based to information theoretic grounds. One is responsible for the traditional material identification associated with the linear or nonlinear constitutive behavior; the other is responsible for the characterization of the proper loading path followed by a multidimensional loading frame online with the previous state of material characterization. The third is responsible for characterizing aspects of the experiment prior to conducting the experiment (i.e. offline). These aspects can be the geometrical characteristics of the shape of the specimen, lamination characteristics of it as well as the loading path itself as well.

Here we introduce the idea of a meta-objective functions that are constructed to express the performance of the constitutive model characterization optimization subprocess employed in the material identification level. Thus, the experimental design both in the offline and online senses, is generated dynamically as data are being acquired in a fashion that optimizes the performance of the optimization employed for the material parameter estimation.

The paper continues with a section that defines the proposed methodology. Subsequently, an application related to characterizing the elastic response of a composite material is described where the performance of the characterization process is defined in terms of the uniqueness and distinguishability of the obtained parameter set as solution of a singular value decomposition (SVD) problem. An application example is described and the paper concludes with a short discussion of the findings and the plans.

2   Multi-level Approach

The approach followed employs a process that involves multi-level design optimization that relates three stages or levels of intervention. The first stage involves the material model parameter characterization that occurs at has associated with it the master time marching loop. The second stage involves the determination of parameters that characterize the experiment during its exercise and therefore it is a real-time activity (online) relative to the experimental process intended for data acquisition. Its time marching loop is in parallel to that of the material identification one of stage one. The third stage is concerned with determining the parameters describing the experiment prior to actually performing the experiment. This stage offline stage has a time marching loop that is completely independent of the previous two.

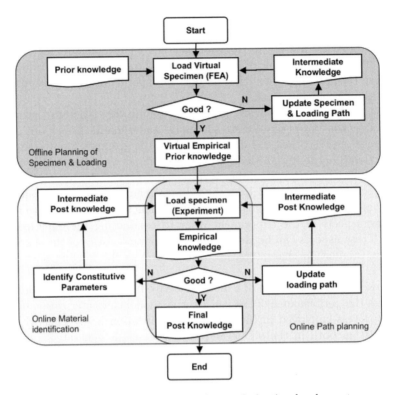

Fig. 1. Flow Chart of the interacting optimization levels or stages

A schematic representation of the linking of these stages is reflected by the block diagram in Fig. 1 where the flow chart of the interacting levels of design optimization are depicted.

As shown in Fig. 1 the online material identification (OMI) level contains the task of identifying the parameters associated with the behavior model in general, and the material constitutive model in particular. The branching node implies the existence of an optimizer that effectively implements the minimization of the difference between the behavior of the analytical model of the system that is under determination and the physical system itself as it is measured by the experimental frame tasked to expose it to the widest possible variety of excitation conditions. Then, a performance specification for the OMI is used to form the objectives of another optimizer in the level of the online path-planning level (OPP). The result of this optimization is the determination of the parameters describing the evolution of the loading path. Similarly, but separately in time, a performance specification for the OMI is used to form the objectives of the last optimizer in the level of the offline planning of specimen and loading (OPSL). From an information theoretic perspective levels OMI and OPP are using empirical knowledge derived from measured behavior of the respective systems and are producing final knowledge (identified behaviors) while they are building it iteratively trough successive stages of intermediate post knowledge. In the OPSL level prior knowledge based entirely on a priori considerations (like the initial geometric specification of the specimen) is used to feed the process and produce virtual empirical prior knowledge through various stages of intermediate knowledge.

3 Composite Material System Application

A linear anisotropic material with four moduli expressing its constitutive characteristics is considered as the system to be determined in level OMI. We have already demonstrated [7,8] that this problem can be reduced to the following linear (with respect to the unknown parameter vector) relation

$$\mathbf{G}\left(\theta\right)\mathbf{q}_M = \mathbf{w}_k, \tag{1}$$

where $\mathbf{G}\left(\theta\right) = [\hat{\mathbf{g}}_1\left(\theta\right), ..., \hat{\mathbf{g}}_n\left(\theta\right)]^T$ is a $m \times 4$-dimensional array that contributes the finite element approximation of the internal energy stored in the system from an increment of strain from point $K-1$ to point K in a manner that does not contain the material moduli since that is contributed by the 4x1 array $\mathbf{q}_M$ of the unknown parameters on the left side of Eq. (1). In the right side the m-dimensional array $\mathbf{w}_K$ contains the scaled external work that is applied as excitation into the system for all loading increments.

In Eq. (1) the right side represents the measured output of the system, while the left hand side represent the corresponding changing inside the system due to all possible excitation inputs. Its solution can be achieved by using any of the three methods available for implementation of least squares approximation [9]. Here we will focus on using SVD for the purpose of determining the parameters of the material model associated with level-OMI. We will also neglect additional computational performance criteria and focus only on potential measures of performance of the SVD implementation from an algorithmic perspective.

The solution containing the identified parameters according to Eq. (1) can be written in the form [7,8].

$$\mathbf{q}_M = \mathbf{G}\left(\theta\right)^+ \mathbf{w}_k \tag{2}$$

where

$$\mathbf{G}\left(\theta\right)^+ = \left\{\mathbf{G}\left(\theta\right)^T \mathbf{G}\left(\theta\right)\right\}^{-1} \mathbf{G}\left(\theta\right)^T \tag{3}$$

is the pseudoinverse of $\mathbf{G}\left(\theta\right)$ and it exists uniquely only when $n \geq 4$ or when the system of linear equations represented by Eq. (2) is over-determined.

4 Performance Measures

What determines the quality of solving Eq. (1) is now reduced to determining the quality of applying Eq. (2) and therefore the quality of the process associated with establishing the pseudoinverse array defined by Eq. (3).

We have identified [6,7] that the concepts of "uniqueness" and "distinguisha-bility" of the obtained solution, can be used as performance metrics for the quality of the determination of the parameter column array $\mathbf{q}_M$.

The uniqueness of the solution depends not only on the size of $\mathbf{G}(\theta)$ but also on whether $\mathbf{G}(\theta)^T \mathbf{G}(\theta)$ in Eq. (3) has and inverse matrix (or this matrix is fully ranked ($r = 4$)). In order to guarantee the uniqueness and further prepare for designing optimal experiments via OPP or OPSL optimization, the proposed technique obtains the singular values of the matrix as a result of singular value decomposition (SVD) [10] as they express the factorization of in the form:

$$\mathbf{G}(\theta) = \mathbf{USV}^T \tag{4}$$

where $\mathbf{U} \in \Re^{n \times n}$ and $\mathbf{V} \in \Re^{4 \times 4}$ are orthogonal to each other and $\mathbf{S} \in \Re^{n \times 4}$ is a diagonal array with real, non-negative singular values $s_i, \forall i \in \{1, ..., 4\}$. These singular values can be used to define the two measures characterizing parameter identification, one being the distinguishability and the other being the uniqueness of the solution and are described as follows.

Distinguishability can be defined as the property of the obtained solution to provide the largest possible variation of the measured response of two sys-tems when their material parameters are very close to each other. It has been demonstrated that when two materials systems exhibit a small difference in their properties, then the difference of the values of their corresponding responses (ob-served experimentally) depend linearly on S [10]. Thus, any expression of the combined effect of the elements of S as it increases, has the ability to distinguish two materials that are seemingly close to each other from a properties perspec-tive, by producing exaggerated energy responses that are scaled values of these property variations.

We have defined distinguishability as the product of all singular values, and uniqueness as the inverse of the condition number of $\mathbf{G}(\theta)^+$ according to

$$F^d \equiv \prod_{i=1}^{n} s_i \quad \text{and} \quad F^u \equiv \frac{1}{c} = \frac{s_{\min}}{s_{\max}} \leq 1 \tag{5}$$

where $s_{\min}, s_{\max}$ are the minimum and maximum of the four singular values involved, respectively.

Uniqueness has been defined as the measure of whether $\mathbf{G}(\theta)^T \mathbf{G}(\theta)$ in Eq. (3) has an inverse matrix or not, is equivalent to the existence of $(\mathbf{S}^T \mathbf{S})^{-1}$ as shown by substituting Eq. (4) into Eq. (3) that yields

$$\mathbf{G}(\theta)^+ = \mathbf{V} \{\mathbf{S}^T \mathbf{S}\}^{-1} \mathbf{S}^T \mathbf{U}^T. \tag{6}$$

The necessary and sufficient condition for this to occur is that $|\mathbf{S}^T \mathbf{S}| = s_1^2 s_2^2 s_3^2 s_4^2 \neq 0$ i.e., $s_i \neq 0, \forall i \in \{1, ..., 4\}$. It is important to underscore here that distinguisha-bility increases as any of the s_i increases, while uniqueness increases as the condition number decreases to unity.

5   Loading Path Optimization

If in addition to determining the material parameters we require that this is achieved such as distinguishability and uniqueness are as high as possible, then we have defined the goals of the optimization involved in levels OPP and OPSL. The design variables at OPP level have to therefore be connected with what is controllable in an experimental setup used to acquire experimental data used for identifying the material parameters at OMI. Such parameters can be those that define the evolution of the loading path, such as total number of increments, loading path increment magnitude, and loading path increment orientation. For the case of a displacement controlled two degree of freedom (2-DoF) testing machine used for experimentation the parameter vector to be identified per loading increment could be could be formed by the measure of displacement increment and the angle denoting the change orientation of the loading path between to successive increments defined according to

$$\Delta u_K = \|\mathbf{u}_{K+1} - \mathbf{u}_K\|, \quad \Lambda_{K,K+1} = \tan^{-1}\left(\frac{u_{y|K+1} - u_{y|K}}{u_{x|K+1} - u_{x|K}}\right), \tag{7}$$

with the total boundary displacement vector defined by its components along the two axes according to the usual definition $\mathbf{u}_k \equiv [u_x, u_y]_k = [u_{x|k}, u_{y|k}]$. Since increased uniqueness and distinguishability express a sense of reliability of the SVD process used for determining the material parameters in level OMI, we can define a vectorial objective function that needs to be maximized for maximum reliability and is formed from two components $J^d_{K,K+1}, J^u_{K,K+1}$ that represent the corresponding increments of distinguishability and uniqueness along an increment of loading [10]. However, the solution of the two objective functions problem is not given by a single point but by a space satisfying the Pareto-optimality, which is often referred to as Pareto-optimal front [10]. To avoid the computational cost associated with this approach the problem is reformulated in a manner that only one scalar objective function is constructed according to the following generalized form

$$J_{K,K+1} \equiv (1 - \mu)J^d_{K,K+K_n} + \mu J^u_{K,K+K_n} \rightarrow \max_{\Lambda_{K,K+K_n}} \tag{8}$$

where each objective function is given by the scaled increment:

$$J^\alpha_{K,K+K_n} \equiv \frac{\bar{F}^\alpha_{0,K+K_n} - F^\alpha_K}{F^\alpha_K}, \quad \forall \alpha \in \{d, u\} \tag{9}$$

and $\mu \in [0,1]$ is a weighting factor that controls the bias towards one or the other component of the objective function. The formulation expressed by Eqs. (8,9) avoids the computational tedious derivation of the Pareto-front. This scalar objective function can be extended beyond the algorithmic efficiency measures utilized up to now to capture computational efficiency metrics such as speed of computation.

6   Numerical Application

To demonstrate the proposed concepts the material selected for generating the necessary simulated experimental data is a typical laminate constructed from an epoxy resin/fiber laminae system of type AS4/3506-1 with a balanced +/- 30 degrees stacking sequence.

All subsequent computational results have been produced by the implementation of the analysis presented earlier within MATLAB [11].

Figure 2 shows the resulting laminate material parameters evolution as a function of the number of the experimental points from the optimization conducted at level OMI.

The solution for the optimization at level OPP or OPSL is expressed in terms if the loading path defined on the $u_x - u_y$ plane as shown in Fig. 3.

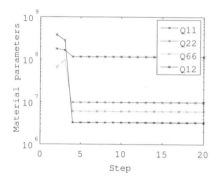
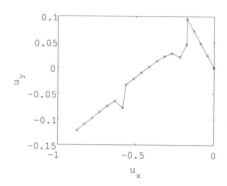

Fig. 2. Fig. 1 Evolution of determined material moduli from stage OMI

Fig. 3. Fig. 2 Evolution of loading path from stage OPP

7   Conclusions

The preliminary framework of a multi-level methodology was proposed; this methodology can succeed in both the determination of the parameters characterizing the response of a system, as well as characterizing the design parameters of an experiment required to collect data necessary for the systemic characterization.

The approach was applied in the context of an anisotropic material system. The systemic constitutive response of a linear anisotropic behavior to be identified was selected to be the elastic system fully defined from its five elastic moduli. These were the design variables for the OMI level of optimization. The experimental model considered for the OPP level of optimization was chosen to represent the description of a loading path in a 2-dimensional loading space. Implied here is the existence of a 2-degree of freedom loading frame, capable of applying such a loading path and measure both the path and reaction mechanical load characteristics for each increment.

To achieve a definition of the objective function at OPP level the quantities of distinguishability and uniqueness were introduced as performance metrics of the design optimization process at level-2 that quantifies the performance of the SVD process employed. To achieve this, a two-component meta-objective function was constructed to be maximized. Maximization of this dual objective function leads to determines dynamically (for the OPP case) and in prior time (for the OPSL case) the experimental design specification in terms of loading path direction parameter.

Numerical simulation of the entire process was performed to demonstrate its feasibility. We demonstrated that the material moduli unknowns can be determined simultaneously to the loading path characteristics needed to design the necessary experiment.

Various extensions of this work will be attempted in the future, while simulation as the activity of exercising the determined model will also be added to complete the essential triad (dynamic and simultaneous physical model identification, design of experiments and design of simulation) of activities associated with a DDDAS [1].

Acknowledgement. The first author acknowledges the support by the National Science Foundation under grant 0540419 and the support of NRL's 6.1 core-funding program. Both authors acknowledge also the support by the Office of Naval Research.

References

1. Darema,F., Introduction to the ICCS2006 Workshop on Dynamic Data Driven Applications Systems. International Conference on Computational Science (3), (2006), 375–383
2. Michopoulos, J., Tsompanopoulou, P., Houstis, E., Farhat, C., Lesoinne, M., Rice, J., Joshi, A., On a Data Driven Environment for Multiphysics Applications, Fut. Generation Comp. Sys., **21**(6), (2005), 953–968.
3. Farhat, C., Michopoulos, J. G., Chang, F. K., Guibas, L. J., Lew, A. J., Towards a Dynamic Data Driven System for Structural and Material Health Monitoring. International Conference on Computational Science, (3) (2006), 456-464.
4. Mast, P. W., Michopoulos, J. G., Thomas, R. W., Badaliance, R., and Wolock I., Characterization of strain-induced damage in composites based on the dissipated energy density: Part I - basic scheme and formulation. Int. Jnl. of Theor. and Applied Fract. Mech., **22**, (1995), 71-96.
5. Michopoulos,J., Computational and Mechatronic Automation of Multiphysics Research for Structural and Material Systems, in "Recent advances in Composite Materials", Kluwer Academic publishing, (2003), 9–21.
6. Michopoulos, J., Mechatronically automated characterization of material constitutive respone. in: Proc. of the 6th World Congress on Computational Mechanics (WCCM-VI), (2004), 486-491.
7. Michopoulos, J.G., Furukawa, T., Kelly, D.W., A Continuum Approach for Identifying Elastic Moduli of Composites, Proceedings of the 16th European Conference of Fracture, Alexandroupolis, Greece, July 3-7, 2006 July 2006, Springer 2006. L, 1416 (2006).

8. Michopoulos, J.G., Furukawa, T., Effect of Loading Path and Specimen Shape on Inverse Identification of Elastic Properties of Composites, Proceedings of IDETC/CIE 2006 ASME 2006, Paper DETC-2006-99724, CD-Rom ISBN: 0-7918-3784-X, (2006).

9. Lawson C.L., R.J. Hanson R.J., Solving Least Squares Problems, Prentice-Hall, Englewood Cliffs, NJ, (1974). Reprinted by SIAM Publications, Philadelphia, PA, (1996).

10. Michopoulos, J.G., Furukawa, T., Design of Multiaxial Tests for Characterizing Anisotropic Materials, Int. J. Numer. Meth. Engng., In print.

11. The Mathworks, Matlab , http://www.mathworks.com

Evaluation of Fluid-Thermal Systems by Dynamic Data Driven Application Systems - Part II

D. Knight, Q. Ma, T. Rossman, and Y. Jaluria

Dept of Mechanical and Aerospace Engineering
Rutgers - The State University of New Jersey
New Brunswick, NJ 08903
knight@soemail.rutgers.edu

Abstract. A Dynamic Data Driven Application Systems (DDDAS) methodology is developed for evaluation of fluid-thermal systems wherein a complete specification of the boundary conditions is not known *a priori* and experimental diagnostics are restricted to a limited region of the flowfield. The Closed Loop formulation of DDDAS is used whereby experiment and simulation are synergized in an iterative manner to determine the unknown boundary conditions, thereby enabling a full simulation (and hence, evaluation) of the fluid-thermal system. In this DDDAS methodology, the experiment directs the simulation and vice-versa. The DDDAS methodology is applied to a heated jet injected into a laminar boundary layer where the jet temperature and velocity are not known *a priori* for the simulations. The DDDAS methodology accurately determines the unknown jet temperature and velocity.

1 Introduction

In a wide range of fluid-thermal systems, there is typically limited access to the flow domain for experimental measurements of the flowfield (*e.g.*, pressure, species concentration, temperature and velocity). Examples of such systems include combustors, furnaces and reactors. For example, an optical fiber drawing furnace typically has an infrared sensor to monitor the temperature of the heating element at a single location [1]. Consequently, the necessary boundary conditions (*e.g.*, inflow, outflow, solid boundary, etc) for computational simulations (using a Computational Fluid Dynamics [CFD] code such as Fluent$^{©}$) are not completely known, and therefore a simulation cannot be performed.

The objective of this research is the development of a Dynamic Data Driven Applications Systems methodology that synergizes experiment and simulation in fluid-thermal systems to determine the unknown boundary conditions, thereby enabling a complete simulation of the fluid-thermal system.

2 Dynamic Data Driven Application Systems

The Dynamic Data Driven Applications Systems (DDDAS) concept was described in the DDDAS Workshop held at the National Science Foundation in

Y. Shi et al. (Eds.): ICCS 2007, Part I, LNCS 4487, pp. 1189–1196, 2007.

March 2000 [2] and further described by Darema [3] (see also the DDDAS web-page [4]). DDDAS is a unique approach to engineering and scientific research wherein experiment and simulation interact in a synergistic, symbiotic manner. There are two different implementations of DDDAS: Open Loop and Closed Loop. In Open Loop, experimental data is streamed into the simulation (or vice-versa) to achieve greater accuracy, detail and/or robustness. An example is the Dynamic Data Driven Wildfire Modeling methodology of Mandel et al [5] and Douglas et al [6]. In Closed Loop, experiment and simulation interact in an iterative manner, i.e., the experiment guides the simulation and the simulation in turn guides the experiment [3]. An example is the Dynamic Data Driven Optimization Methodology (DDDOM) developed by Knight et al [7,8].

3   Description of Research

3.1   Objective

The objective is the development of a DDDAS methodology for synergizing experiment and simulation to evaluate fluid-thermal systems wherein the experimental measurements are restricted in region and scope, and the a priori boundary conditions for simulation are incomplete. The Closed Loop DDDAS concept is used wherein the experiment directs the simulation and vice-versa in an iterative manner. The strategy is to approximate the unknown boundary conditions by minimizing the error in the prediction of the measured data (i.e., the experiment driving the simulation), and to identify needed subsequent experimental measurements to reduce the error (i.e., the simulation driving the experiment) and also subsequent additional simulations to reduce the error (i.e., the experiment driving the simulation).

3.2   Configuration

We consider a rectangular jet injected perpendicular to an incompressible laminar boundary layer. The configuration is shown in Fig. 1. The inflow is an equilibrium laminar boundary layer in air defined by the specified freestream conditions (velocity U_∞, static pressure p_∞ and static temperature T_∞) and boundary layer thickness δ_∞ (i.e., the laminar boundary layer thickness that would exist at the location of the jet exit in the absence of the jet). The jet is defined by the jet average velocity U_j, static pressure p_j, and static temperature T_j. The computational domain ABCDE is shown. For the simulations, the freestream conditions (velocity U_∞, static pressure p_∞, static temperature T_∞ and boundary layer thickness δ_∞) are assumed known. By analogy to the optical fiber furnace, the jet average velocity U_j and static temperature T_j are assumed unknown (insofar as the simulations are concerned), while the jet exit pressure p_j is assumed known. The objective is the determination of the jet average velocity U_j and static temperature T_j based upon a Closed Loop DDDAS methodology. The range of values for U_j and T_j provided to the DDDAS methdology are indicated in Table 1 (see Section 3.7).

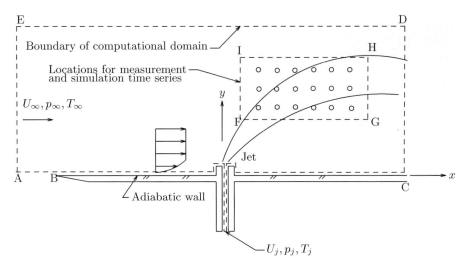

Fig. 1. Flow configuration

3.3 Experiments

The experiments were performed in the Rutgers Low Speed Wind Tunnel. The experimental configuration is shown in Fig. 2a. A two-dimensional slot jet protrudes from a flat plate with the jet centerline at a distance of 188 mm from the leading edge of the plate. The flat plate has a sharp leading edge and is raised 50 mm above the wind tunnel floor to avoid the tunnel floor boundary layer. The jet slot width is 3.2 mm and the spanwise depth is 545 mm. The jet fences protrude 8.8 mm above the flat plate and each fence is 6.4 mm wide in the x-direction. The jet exit temperature was measured by a small bead thermocouple at the jet exit and was observed to vary less than 2% across the jet exit. Additional details are presented in Knight *et al* [9].

A diode laser system was used to measure the time-varying absorbance across the flowfield in the spanwise direction at selected locations. The measured laser absorbance is related to the local thermodynamic conditions and gas concentrations through the spectral absorption coefficient and Beer's law

$$A = 1 - \frac{I}{I_o} = 1 - \exp\left(-k_v L\right)$$

where $k_v = S\phi P$, S is the temperature dependent linestrength (cm^2-cm^{-1}), ϕ is the lineshape function (1/cm^{-1}), P is the partial pressure of the absorbing species given in terms of number density (cm^{-3}), and L is the path length (cm). The spectrally-dependent absorbance was converted from the time domain to the wavelength domain using laser calibration tuning curves. The current injection versus wavelength tuning was determined using an optical spectrum analyzer prior to performing the experiments. The spectrally-dependent absorbance (k_n) can be integrated versus wavelength to remove the effect of the lineshape

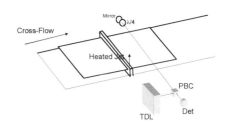

(a) Wind tunnel model   (b) Diode laser configuration

Fig. 2. Experiment

function. The integrated absorbance (k) only depends on the linestrength and number density of absorbers. The temperature dependency of both quantities is known, and therefore the path-averaged temperature can be obtained from the absorption data. Further details are presented in Ma *et al* [10].

3.4   Simulations

The two-dimensional, laminar, unsteady Navier-Stokes equations were solved using Fluent©. The fluid is air. The Incompressible Ideal Gas Law was used together with the Boussinesq approximation. The molecular dynamic viscosity was modeled using Sutherland's Law. The simulations are second-order accurate in space and time. The spatial reconstruction is 3rd order MUSCL, and the temporal integration is Implicit Dual-Time Stepping. Twenty inner time steps at a specified inner Courant number of five are used per outer (physical) time step. A constant outer timestep $\Delta t = 4 \cdot 10^{-4}$ sec was used for all simulations corresponding to an outer Courant number $CFL = \Delta t U_\infty / \Delta x_{min} = 4$ to 8 for the range 4 m/s $\leq U_\infty \leq$ 8 m/s considered in this study. Further details are presented in Knight *et al* [9]. Table 1 summarizes the range of parameters for the simulations.

3.5   Validation

A validation study was performed to assess the accuracy of the simulations by comparison with experiment. Details are presented in Knight *et al* [9]. An experiment was performed with $U_\infty = 4$ m/s, $T_\infty = 299$ K, $p_\infty = 101$ kPa, $U_j = 8.11$ m/s, $T_j = 398$ K and $p_j = 101$ kPa. The mean temperature was measured at three locations in the flowfield downstream of the jet using the diode laser absorbance method (Section 3.3) and thermocouple. The freestream and jet pressure, temperature and velocity conditions were provided to the simulation using Fluent©. The computed mean temperature at the three locations within the flowfield agreed with the experimental mean temperature within 10 K. This represents also the uncertainty in the experimental measurement.

3.6   Response Surface Models

The energy equation is decoupled from the mass and momentum equations (neglecting variations in density and buoyancy effects), and thus the static temperature behaves as a passive scalar. The temperature field must therefore scale as $T(x, y, t) - T_\infty = (T_j - T_\infty)f(x, y, t; U_j, U_\infty)$. Therefore, a quadratic Response Surface Model (RSM) for the time mean static temperature $T_m(x, y)$ may be constructed at a fixed position (x, y) in the flowfield according to

$$T_m(x, y) - T_\infty = (T_j - T_\infty) \left[\beta_o(x, y) + \beta_1(x, y) \left(\frac{U_j}{U_\infty} \right) + \beta_2(x, y) \left(\frac{U_j}{U_\infty} \right)^2 \right]$$

The coefficients $\beta_i(x, y)$ are obtained from simulations performed for a fixed value $T_j - T_\infty$ (selected from the range indicated in Table 1) and a set of U_j selected from the range indicated in Table 1 For this study, the values $U_j = 4$, 6 and 8 m/s were selected. The freestream conditions $(U_\infty, T_\infty, p_\infty)$ and jet pressure p_j are fixed as indicated in Table 1.

3.7   DDDAS Methodology

The Closed Loop DDDAS methodology integrates experiment and simulation in a synergistic, iterative manner to achieve a complete evaluation of the fluid-thermal system. There are five steps:

1. Select Monitor Locations for Simulations
   A set S_s of monitor locations for the simulations is selected. At each monitor location, a time series of the static temperature is obtained in every simulation. The number of monitor locations can be arbitrarily large and is limited only by available disk storage for the time series data. A total of eighteen monitor locations were selected (Table 2).
2. Generate Response Surface Model Based on Simulations for Fixed ΔT_j^i
   A fixed value of $\Delta T_j^i = T_j^i - T_\infty$ $(i = 1, 2, \ldots)$ is chosen from within the range of values indicated in Table 1. Simulations are performed for a set of values U_j for the fixed ΔT_j^i and time series at each of the monitor locations in S_s is recorded. A Response Surface Model for the mean static temperature T_m at each monitor location is generated based upon the assumed value of ΔT_j^i and selected values of U_j (see Section 3.6).
3. Select Monitor Locations for Experiments
   A subset S_e^k of the monitor locations S_s is selected for the kth experiment $(k = 1, 2, \ldots)$. Each experiment is performed at the same U_j and T_j; however, the values of U_j and T_j are not known for any of the simulations. Note that the size of the set S_e^k is small compared to S_s due to the substantial amount of time required for the experimental measurements.

4. Estimate Experimental Values for $T_j - T_\infty$ and U_j

The experimental mean temperatures are compared with the Response Surface Models to estimate the value of $T_j - T_\infty$ and U_j in the experiment (see below).

5. Determine New Measurement Locations

The Response Surface Model is used to select the next set of monitor locations S_e^{k+1} from S_s. Step No. 4 is repeated to provide a revised estimate of $T_j - T_\infty$ and U_j. Based upon the estimated value of $T_j - T_\infty$, the procedure repeats from Step 2 until convergence of the predicted values for U_j and T_j.

Table 1. Flow Conditions

Parameter	Value
U_∞ (m/s)	4.0
T_∞ (K)	290.
p_∞ (kPa)	101.8
U_j (m/s)	4.0 to 8.0
T_j (K)	350 to 450
p_j (kPa)	101.8

Table 2. Location of Monitors

No.	x (cm)	y (cm)	No.	x (cm)	y (cm)	No.	x (cm)	y (cm)
1	1.2	2.0	7	1.2	3.0	13	1.2	4.0
2	3.2	2.0	8	3.2	3.0	14	3.2	4.0
3	5.2	2.0	9	5.2	3.0	15	5.2	4.0
4	7.2	2.0	10	7.2	3.0	16	7.2	4.0
5	9.2	2.0	11	9.2	3.0	17	9.2	4.0
6	11.2	2.0	12	11.2	3.0	18	11.2	4.0

The estimate of the experimental values $T_j - T_\infty$ and U_j at each step in the procedure is obtained as follows. The square error between the experimental mean temperature and the Response Surface Model for each possible subset of l locations within S_e^k is computed as

$$E = \sum_l \left\{ \Delta T_{m_e} - \Delta T_j \left[\beta_0(x,y) + \beta_1(x,y) \left(\frac{U_j}{U_\infty} \right) + \beta_2(x,y) \left(\frac{U_j}{U_\infty} \right)^2 \right] \right\}^2$$

where $\Delta T_j = T_j - T_\infty$, $\Delta T_{m_e} = T_{m_e} - T_\infty$, and the sum is over l locations within S_e^k (the minimum number for l is 2). For example, assume S_e^k contains six locations and let $l = 2$. For each possible set of two locations from S_e^k, the values of ΔT_j and U_j that minimize E are determined. This yields fifteen triplets $(\Delta T_j, U_j, E)$. For a given value of l, the predicted values of ΔT_j and U_j, denoted by ΔT_j^l and U_j^l, are taken to be the triplet with the minimum E (i.e., the values of ΔT_j and U_j with the smallest square error). The procedure is repeated for all values of l from $l = 2$ to $n = $ size S_e^k. The estimate for the experimental value of $T_j - T_\infty$ is the average of these values $T_j - T_\infty = (n-1)^{-1} \sum_{l=2}^{l=n} \Delta T_j^l$ and similarly for U_j.

3.8   Results

The Closed Loop DDDAS methodology was applied to determine the experimental $T_j - T_\infty$ and U_j. A total of eighteen monitor locations were selected (Step 1). Response Surface Models were generated for all monitor locations for an assumed value $\Delta T_j = 66$ K (Step 2). Based upon these models, six locations (Nos.

3, 9, 10, 14, 15 and 16) from Table 2 were selected for the experiment (Step 3). Note that the experimental U_j and $T_j - T_\infty$ were selected by the experimentalists (Q. Ma and T. Rossman) but not communicated to the person performing the Closed Loop DDDAS Method (D. Knight) until the DDDAS method was converged. Using the experimental mean temperature measurements at the six locations, the estimated values $\Delta T_j = 110 \pm 16$ K and $U_j = 7.3 \pm 1$ m/s were obtained using the Response Surface Models (Step 4). An additional set of locations for experiments was defined based upon the Response Surface Models (Nos. 2, 4, 5 and 17) (Step 5). A revised estimate $\Delta T_j = 120 \pm 16$ K and $U_j = 7.1 \pm 1$ m/s were obtained using the Response Surface Models (Step 4). A revised $T_j - T_\infty = 115$ K was selected for creation of the Response Surface Models (Step No. 2) recognizing that the value originally used ($T_j - T_\infty = 66$ K) was significantly below the value predicted by the Response Surface Models. Steps 4 and 5 were repeated using the new Response Surface Models yielding the estimate $T_j - T_\infty = 105 \pm 13$ K and $U_j = 7.1 \pm 1$ m/s. The actual experimental values are $T_j - T_\infty = 107 \pm 10$ K and $U_j = 8.0$ m/s. The predicted values for $T_j - T_\infty$ and U_j thus agree with the experimental measurements to within the experimental uncertainty (Section 3.5), thereby validating the Closed Loop DDDAS methdology.

Table 3. Results of DDDAS Method

Iteration	Predicted $T_j - T_\infty$ (K)	U_j (m/s)
1	110 ± 16	7.3 ± 1
2	120 ± 16	7.1 ± 1
3	105 ± 13	7.1 ± 1
Exp	107 ± 10	8

4   Conclusions

A methodology for evaluation of fluid-thermal systems is developed based upon the Dynamic Data Driven Application Systems approach. The methodology is intended for fluid-thermal systems where complete specification of the boundary conditions is not known *a priori* and experimental measurements are restricted to a subregion of the fluid-thermal domain. The methodology synergizes experiment and simulation in a closed-loop, iterative manner to achieve a full evaluation of the fluid-thermal system. Results are presented for the configuration of a heated jet injected into a laminar boundary layer where the jet temperature is not known *a priori*. The DDDAS methodology accurately predicts the unknown jet temperature and jet velocity.

Acknowledgments

The research is sponsored by the US National Science Foundation under grant CNS-0539152 (1 Oct 2005 - 30 Sept 06). The program manager is Dr. Frederica Darema.

References

1. S. Roy Choudhury and Y. Jaluria. Practical Aspects in the Drawing of an Optical Fiber. *Journal of Materials Research*, 13:483–493, 1998.
2. Dynamic Data Driven Applications Systems (DDDAS) Website. National Science Foundation, http://www.nsf.gov/cise/cns/darema/dddas/index.jsp.
3. F. Darema. Dynamic Data Driven Applications Systems: A New Paradigm for Application Simulations and Measurements. In *Fourth International Conference on Computational Science*, pages 662–669, Berlin, 2004. Springer-Verlag.
4. Dynamic Data Driven Applications Systems (DDDAS) Homepage. http://www.dddas.org/.
5. J. Mandel, M. Chen, L. Franca, C. Johns, A. Pulhalskii, J. Coen, C. Douglas, R. Kremens, A. Vodacek and W. Zhao. A Note on Dynamic Data Driven Wildfire Modeling. In *Fourth International Conference on Computational Science*, pages 725–731, Berlin, 2004. Springer-Verlag.
6. C. Douglas, J. Beezley, J. Coen, D. Li, W. Li, A. Mandel, J. Mandel, G. Qin and A. Vodacek. Demonstrating the Validity of a Wildfire DDDAS. In *Sixth International Conference on Computational Science*, pages 522–529, Berlin, 2006. Springer-Verlag.
7. D. Knight, G. Elliott, Y. Jaluria, N. Langrana and K. Rasheed. Automated Optimal Design Using Concurrent Integrated Experiment and Simulation. AIAA Paper No. 2002-5636, 2002.
8. H. Zhao, T. Icoz, Y. Jaluria and D. Knight. Application of Data Driven Design Optimization Methodology to a Multi-Objective Design Optimization Problem. To appear, *Journal of Engineering Design*, 2007.
9. D. Knight, Q. Ma, T. Rossman and Y. Jaluria. Assessment of Fluid-Thermal Systems by Dynamic Data Driven Application Systems. In *International Conference on Modeling and Optimization of Structures, Processes and Systems*, University of Kwazulu-Natal, Durban, South Africa, January 2007.
10. Q. Ma, Y. Luo, T. Rossman, D. Knight and Y. Jaluria. Diode Laser Measurements for DDDAS: Flowfield Reconstruction Using Dynamic Experimental and Numerical Data. AIAA Paper No. 2006-2974, 2006.

Dynamic Data-Driven Fault Diagnosis of Wind Turbine Systems

Yu Ding[1], Eunshin Byon[1], Chiwoo Park[1], Jiong Tang[2], Yi Lu[2], and Xin Wang[2]

[1] Texas A&M University, College Station, TX 77843
yuding@iemial.tamu.edu
[2] University of Connecticut, Storrs, CT 06269
jtang@engr.uconn.edu

Abstract. In this multi-university collaborative research, we will develop a framework for the dynamic data-driven fault diagnosis of wind turbines which aims at making the wind energy a competitive alternative in the energy market. This new methodology is fundamentally different from the current practice whose performance is limited due to the non-dynamic and non-robust nature in the modeling approaches and in the data collection and processing strategies. The new methodology consists of robust data pre-processing modules, interrelated, multi-level models that describe different details of the system behaviors, and a dynamic strategy that allows for measurements to be adaptively taken according to specific physical conditions and the associated risk level. This paper summarizes the latest progresses in the research.

1 Introduction

Wind turbines convert the kinetic energy of wind into the electrical energy, which provides a pollution-free source of electricity. Today wind power is considered the fastest growing energy source around the world. The key issue for all renewable energy utilizations is the cost and the marketability [1]. In the US, Class 6 sites (with average wind speeds of 6.7 m/s at 10 m height) can in theory market electricity at prices of 3 to 4c/kWh, which, together with the tax credit (1.7c/kWh), allows wind energy to compete with traditional energy sources. As more sites are developed, easily accessible new Class 6 sites are becoming less available. Emphasis has now been shifted to Class 4 wind sites (5.8 m/s at 10 m height), which cover vast areas of the Great Plains from northern Texas to the US-Canada border. Class 4 sites represent almost 20 times the developable wind resource of Class 6 sites. Currently the electricity at Class 4 sites can be generated at costs in the range of 5 to 6 c/kWh. In order to position the wind energy as an attractive option, the Federal Wind Energy Program (FWEP) has set a goal to reduce the cost of electricity generated at Class 4 sites to 3c/kWh. One major hurdle to achieve this goal is the high cost for maintaining the wind turbines at remote areas, given the labors, time, and heavy-duty equipment involved. The FWEP's goal can be achieved only after the number of false alarms, the failure-caused down time, and the actual maintenance cost of wind turbines can be significantly reduced.

Clearly, a reliable, robust fault detection and diagnosis system plays a critical role in making the wind power more marketable. In a modern wind turbine, the most expensive and fault-prone components are the gearbox and the blades, especially for

Y. Shi et al. (Eds.): ICCS 2007, Part I, LNCS 4487, pp. 1197–1204, 2007.

those low speed wind turbines operating at Class 4 sites. While researchers have pursued fault diagnosis for similar structures for decades, diagnosis of wind turbine components is unique and poses new challenge. First, these low speed wind turbines constantly operate under non-stationary conditions and involve very complicated gearboxes with multiple stages of gears to speed up a hundred times or more. Second, in order to increase the unit power conversion efficiency, the blades used are significantly longer, up to 70m in length as compared to the 20m ones currently installed in high speed wind turbines. For these blades with such enormous size and made of composite materials, traditional modal-information-based fault diagnosis is difficult to be implemented on-line and is insensitive to the dominant failure mode, namely the delamination within the layered composites [2].

The most challenging aspect of wind turbine fault diagnosis comes from the high requirements placed on the accuracy and credibility, owing to the cost-sensitive nature related to the FWEP's goal. The currently available signal processing and diagnosis methods are deterministic in nature, leading to a large number of false alarms [3]. In view of these challenges and the urgent need, in this project we develop a new framework for the effective and robust diagnosis of wind turbine systems using the dynamic data-driven methodology. This methodology, as illustrated in Fig. 1, consists of robust data pre-processing modules for highly sensitive feature extraction, interrelated models that describe different details of the system behaviors at multiple levels, and a dynamic strategy that allows the measurements to be adaptively taken according to the specific physical conditions and the associated risk level. In this paper, we summarize the progresses made in the research.

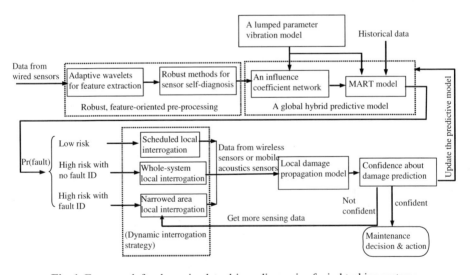

Fig. 1. Framework for dynamic, data-driven diagnosis of wind turbine systems

2   Signal Pre-processing and Local-Level Detection

At local level, we are concerned about the health conditions of the gearbox and the blades. For the monitoring of the gearbox, vibratory signals during the wind turbine operation collected by accelerometers will be used. For the monitoring of the blades, we mainly rely on an active damage interrogation scheme that uses embedded piezoelectric actuators to generate Lamb wave propagation in the blades. The wave propagation anomaly will indicate damage occurrence such as delamination.

The common feature of the aforementioned signals that are used for health monitoring is that they are non-stationary either due to the wind turbine operation, or due to the nature of the wave excitation. For such signals, it is better to choose time-frequency representations, among which wavelet analysis is particularly useful, to extract their critical features. Different from traditional Fourier transform-based frequency analysis, wavelet transforms lead to the flexibility in using narrow windows for the analysis of high-frequency content and wide windows for low frequencies. Newland [4] developed harmonic wavelet and its generalized form, whose wavelet levels represent non-overlapping frequency bands that can help interpret frequency contents and link detection results with physical meanings. Liu improved the generalized harmonic wavelet transform by applying an entropy-based algorithm for best basis selection, in the sense that it gives the sparsest representation of the signal [5]. Since the time-frequency interpretation is adaptive to specific signals, finding the common wavelet bases for a set of samples (signals) in the presence of noise requires extra improvement, which is the focus of our current study.

From the Fourier transform in the frequency domain

$$W_{mnk}(\omega) = \begin{cases} \dfrac{1}{(n-m)2\pi} e^{-i\omega\frac{k}{n-m}} & m2\pi \leq \omega \leq n2\pi \\ 0 & \text{otherwise} \end{cases} \tag{1}$$

Newland derived the family of generalized harmonic wavelets [4]

$$w_{mnk}(t) = w_{mn}(t - \frac{k}{n-m}) = \frac{\exp\left[in2\pi(t - \dfrac{k}{n-m})\right] - \exp\left[im2\pi(t - \dfrac{k}{n-m})\right]}{(n-m)i2\pi t} \tag{2}$$

where m and n are the level parameters, $0 \leq m \leq n$, and integer k denotes the translation parameter within the level (m, n). The advantage of harmonic wavelets is that signal analysis is restricted to specific frequency bands with known physical meanings, and these bands are represented by corresponding wavelet levels. A discrete algorithm is developed to calculate coefficients by computing the inverse discrete Fourier transform (IDFT) of successive blocks (each corresponds to a level m, n) of the Fourier coefficients of the input signal.

Liu treated each selection $\{(m_0, n_0), (m_1, n_1), \ldots, (m_{L-1}, n_{L-1})\}$ as a partition of $\Omega = \{0, 1, \ldots, N_f\}$ and developed a Shannon entropy-based algorithm to search a partition tree for the best partition (in the sense that the signal can be represented most sparsely) [5]. For any sequence $\mathbf{x} = \{x_j\}$, the Shannon entropy is defined as

$$H(\mathbf{x}) = -\sum_j p_j \log p_j \tag{3}$$

where $p_j = |x_j|^2 / \|\mathbf{x}\|^2$, and $p_j \log p_j$ is set as 0 if $p_j = 0$. The entropy above is a measure of the sparsity and therefore we expect smaller entropy for a better partition of wavelet coefficients. We incorporate this technique to the pre-processing of Lamb wave-based damage detection in laboratory beam-type specimens. The iteration procedure for a sequence of 16 elements using a binary partition tree is illustrated in Fig. 2 (Phase II). For the initial partition, every single Fourier coefficient forms an 'initial subgroup', whose the Shannon entropy can be called initial entropy. Then in Step 1, every 2 successive Fourier coefficients form a 'Step 1 subgroup', whose entropy is calculated and compared with the corresponding sum of initial entropies. For the example in Fig. 2, the sum entropy of the 'initial subgroups' 1 and 2 is smaller than that of the 'Step 1 subgroup' 1. The former subgroups, instead of the latter, are therefore kept after the selection in Step 1. After the entire iteration, the features of the input signal will be highlighted by projecting samples onto those best basis functions.

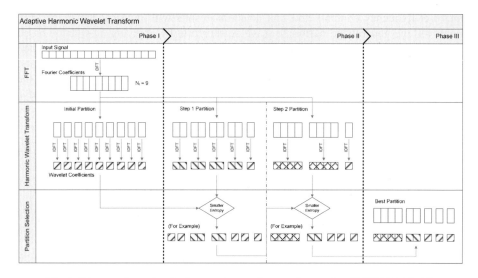

Fig. 2. Discrete-time adaptive harmonic wavelet transform using FFT/IFFT (N=16)

The above strategy of adaptive harmonic wavelet transform (AHWT) is signal-dependent. Multiple signals from the same gearbox or blade may, however, lead to different wavelet basis partition simply due to the existence of noise. In order to build a baseline dataset, the detection algorithm requires a common wavelet basis set for all the samples. We hence extend the adaptive harmonic wavelet transform for multivariate applications. Each time series sample $s_l = \{s_l(r), \ r = 0,1,\ldots,N-1\}$ collected from the accelerometer or piezoelectric sensor is associated with a set of basis functions $\{w_{mnk}\}_l$, then all the L samples are projected on $\{w_{mnk}\}_l$ to yield a matrix of wavelet coefficients $[\mathbf{a_1} \ldots \mathbf{a_L}]_l$. The total Shannon entropy can be defined as $H([\mathbf{a_1} \ldots \mathbf{a_L}]_l) = H(\mathbf{a_1})+\ldots+ H(\mathbf{a_L})$, where $l=1,\ldots,L$ (the number of samples). We select the common wavelet basis set $\{w_{mnk}\}_u$ such that the wavelet coefficients $[\mathbf{a_1} \ldots \mathbf{a_L}]_u$ have the smallest total Shannon entropy, i.e.,

$$u = \arg \min_l H([\mathbf{a_1} \ldots \mathbf{a_L}]_l) \tag{4}$$

We call $[\mathbf{a_1} \ldots \mathbf{a_L}]_u$ the baseline dataset; similarly, the on-line collected response signal can be processed by applying the above procedure.

With the AHWT as basis for feature extraction, we may then use several statistical analysis tools for local-level detection.   Principal component analysis (PCA) transforms a number of correlated variables into a smaller number of uncorrelated new variables called principal components.   Consider a block of baseline data and let $\mathbf{C}$ be the corresponding covariance matrix.   PCA yields an orthogonal (eigenvector) matrix $\mathbf{V} = [\mathbf{v_1} \ldots \mathbf{v_K}]$ matrix and a diagonal (eigenvalue) matrix $\mathbf{D} = \mathrm{diag}(\lambda_1,\ldots,\lambda_K)$ such that $\mathbf{CV} = \mathbf{VD}$. The eigenvalues are arranged in descending order $\lambda_j \geq \lambda_{j+1}$. Introducing the effective rank rk of $\mathbf{X}$ [6], we can discard the eigenvectors associated with $\lambda_{rk+1},\ldots, \lambda_K$ and form a modified eigenvector matrix $\mathbf{V_m}$. Here we choose rk as the smallest number so that the accumulative energy is above a certain threshold $ET\%$,

$$rk_0 = \min rk \;\; s.t. \;\; \sum_{j=1}^{rk}\lambda_j > ET\% \sum_{j=1}^{K}\lambda_j \tag{5}$$

given that the eigenvalues represent the distribution of the original energy among each of the eigenvectors. The local-level detection is facilitated by the Hotelling's T^2 analysis [7].   First, baseline data are sub-grouped and used to establish an upper control limit UCL^1 under a certain confidence level $100(1-\alpha)\%$, where α indicates the error probability $(0<\alpha<1)$.   Then in phase II, a distinction is made between the baseline and the online sensor data using a modified upper control limit UCL^2. If any calculated T^2 value exceeds the phase II upper control limit, we may conclude, with

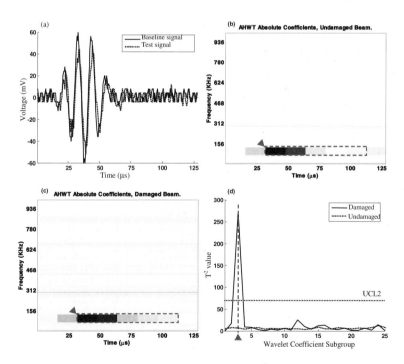

Fig. 3. Piezoelectric interrogation and wave signals: local-level data-driven signal processing

confidence level 100(1-α)%, that the structure is in its damaged state. Fig. 3 shows an example of successful detection of crack damage in a laboratory beam structure using piezoelectric-based damage interrogation.

3 Dynamic Interrogation and Decision-Making

We intend to develop a "risk-based dynamic decision making" policy, considering the loss that may be caused by undetected component failures and the cost of doing unnecessary maintenance work for normally operating components. In this dynamic procedure, we classify the sensor feedbacks and the subsequent actions into three levels: (1) the normal level, and thus operate as is, when no sign indicates any serious problems; (2) the alert level and then put the monitoring into the intensive care mode; (3) the alarm level, and thus dispatch the maintenance crew for on-site repairing. Under intensive care, one need update the sensor measurements more frequently, use advanced mobile sensors for obtaining additional information, and invoke computationally intensive modules for predicting failure modes with reduced uncertainty. So the intensive care mode will incur additional cost but will not be as expensive as dispatching the crew. The decision will be made dynamically by weighing a group of factors.

Factors need to be considered including the costs and risks associated with different actions to be taken (e.g., dispatch a maintenance crew or not), weather conditions now and in the near future which affects both the wind power generation, the severity of any existing mechanical problem, and the feasibility of a repairing mission, the loss caused by the disruption to power generation as the result of a major maintenance or as the result of an utter failure. A hidden Markov model [8] is being used to model the wind turbine's health status. Weather is an important condition for wind turbine operations because wind is the primary source for producing wind power [9]. Non-stationarity and irregularity in wind conditions induce the fatigues causing many component failures. The weather condition also affect the feasibility and cost of a repairing mission since doing a repair under a severe weather

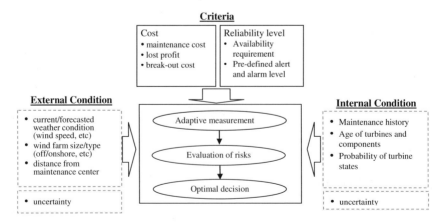

Fig. 4. Factors considered in the dynamic decision making process

condition will be more costly than doing so under a fair condition. Shutting down the wind turbine for repair will also incur losses in power production so one would be better off to schedule such an action during a low wind period than a high wind period. In addition to the weather condition, a wind farm type such as where the wind farm is located (for instance, offshore will need special care and vessel), the distance between the wind farm and maintenance center are also the external factors in our decision making process.

As for the internal conditions related to the wind turbine, the probabilities about the turbine state, which is derived from sensor signals, are the most important input factor. Other factors include the age of the turbine and the repair history and so on. Our main objective in this decision making process is to minimize the risk caused by the information uncertainties [10, 11] from the external and internal factors. The criteria and influential factors are summarized in Fig. 4. Next, we present a simple example to explain the dynamic decision making process.

In the following example, we simplify the above consideration by assuming that the external conditions remain constant. We assume that a wind turbine could be in two health states: survival or failure and the sensor feedback can be classified into three categories: that is, $O_t \in \{o_1, o_2, o_3\}$ or $O_t' \in \{o_1', o_2', o_3'\}$, where O_t' is obtained from an advanced sensor under intensive care, and o_i indicates an escalated risk of turbine failure as i increases. Similar treatments were also used prior studies [8].

Fig. 5 shows a possible trajectory of sensory data and the corresponding scenario of decision rules at each epoch. The wind turbine operates normally during $t_1 \sim t_3$. At t_4, o_2 is observed from a regular sensor, implying that an intensive care is needed. When invoking the advanced sensor, we got an o_1' so we operate wind turbine as usual. Afterwards, o_2 and o_2' are observed both at t_5 and t_6. Then we calculate the risk based on both sensor feedbacks and found the failure risk is not high enough at t_5 so we remain in the intensive care. But the failure risk crosses the alarm level at t_6 so we need to dispatch the crew for maintenance. After the repair, the wind turbine returns to the normal level. At t_{10}, an alarm signal is detected so we need to dispatch the crew right away without going through an intensive care period.

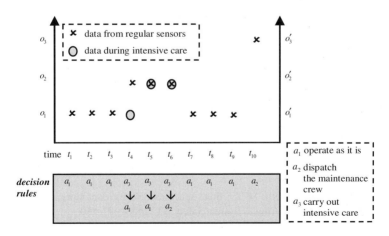

Fig. 5. A trajectory of sensor signals and decision rules

4 Summary

This paper summarizes the recent progresses made in the dynamic data-driven fault diagnosis of wind turbine systems. The health monitoring of wind turbine systems is inherently challenging due to the high requirements placed on the accuracy and reliability of the monitoring system. To fundamentally solve the relevant issues, a series of tools and strategies have been developed and explored, which include robust data-driven local signal processing and preliminary detection algorithms, redundancy analysis for sensor network reliability and robustness, and global strategy for dynamic interrogation and decision making.

Acknowledgements

The authors gratefully acknowledge the support of the National Science Foundation through the grant number CMMI-0540132 and CMMI-0540278.

References

1. National Wind Technology Center, Wind Program Multi-Year Technical Plan, http://www.nrel.gov/wind_meetings/2003_imp_meeting/pdfs/mytp_nov_2003.pdf.
2. Zou, Y., Tong, L., and Steven, G.P.: Vibration-based model-dependent damage (delamination) identification and health monitoring for composite structures – a review. Journal of Sound and Vibration, **230** (2000) 357-378.
3. Tumer, I.R., and Huff, E.M.: Analysis of triaxial vibration data for health monitoring of helicopter gearboxes. ASME Journal of Vibration and Acoustics. **125** (2003) 120 – 128.
4. Newland, D.E.: Harmonic wavelet analysis. Proceedings: Mathematical and Physical Sciences. **443** (1993) 203-225.
5. Liu, B.: Adaptive harmonic wavelet transform with applications in vibration analysis. Journal of Sound and Vibration. **262** (2003) 45-64.
6. Konstantinides, K., and Yao, K.: Statistical analysis of effective singular value in matrix rank determination. IEEE Transactions on Acoustics, Speech & Signal Processing. **36** (1988) 757-763.
7. Johnson, R.A., and Wichern, D.W.: Applied Multivariate Statistical Analysis, 5th Edition. Prentice Hall (2002).
8. Baruah P. and Chinnam, R. B.: HMMs for diagnosis and prognosis in machining process. International Journal of Production Research. **43** (2005) 1275-1293.
9. European Commission: Advanced maintenance and repair for offshore wind farms using fault prediction and condition monitoring techniques. Project report of European Commission under Contract NNE5/2001/710 (2005).
10. Gebraeel, N.: Sensory-updated residual life distributions for components with exponential degradation patterns. IEEE Transactions on Automation Science & Engineering. **3** (2006) 382-393.
11. Pedregal, D.J. and Carnero, M. C.: State space models for condition monitoring: a case study. Reliability Engineering & System Safety. **91** (2006) 171-180.

Building Verifiable Sensing Applications Through Temporal Logic Specification

Asad Awan, Ahmed Sameh, Suresh Jagannathan, and Ananth Grama

Department of Computer Sciences, Purdue University, W. Lafayette, IN 47907

Abstract. Sensing is at the core of virtually every DDDAS application. Sensing applications typically involve distributed communication and coordination over large self-organized networks of heterogeneous devices with severe resource constraints. As a consequence, developers must explicitly deal with low-level details, making programming time-consuming and error-prone. To reduce this burden, current sensor network programming languages espouse a model that relies on packaged reusable components to implement relevant pieces of a distributed communication infrastructure. Unfortunately, programmers are often forced to understand the mechanisms used by these implementations in order to optimize resource utilization and performance, and to ensure application requirements are met. To address these issues, we propose a novel and high-level programming model that directly exposes control over sensor network behavior using temporal logic specifications, in conjunction with a set of system state abstractions to specify, generate, and automatically validate resource and communication behavior for sensor network applications. TLA+ (the temporal logic of actions) is used as the underlying specification language to express global state abstractions as well as user invariants. We develop a synthesis engine that utilizes TLC (a temporal logic model-checker) to generate detailed actions so that user-provided behavioral properties can be satisfied, guaranteeing program correctness. The synthesis engine generates specifications in TLA+, which are compiled down to sensor node primitive actions. We illustrate our model using a detailed experimental evaluation on our structural sensing and control testbed. The proposed framework is integrated into the COSMOS macroprogramming environment, which is extensively used to develop sensing and control applications at the Bowen Lab for Structural Engineering at Purdue.

1 Introduction

Sensor networks are integral to most DDDAS applications. They are often composed of large numbers of low-cost motes with wireless connectivity and varying sensing capabilities. Motes are characterized by limited resources including memory, CPU, network capacity, and energy budget. As DDDAS applications become widespread, there is increasing realization of the complexity associated with building robust sensor network applications. Much of this complexity stems from the need to implement reliable distributed coordination in dynamic environments, under severe performance and resource constraints. Driven by these underlying constraints, sensor network programming often involves low-level system details and communication mechanisms. Consequently, even simple applications pose challenging implementation problems, motivating high-level

Y. Shi et al. (Eds.): ICCS 2007, Part I, LNCS 4487, pp. 1205–1212, 2007.

programming models and abstractions. With this overarching goal, we have developed a comprehensive sensor macroprogramming environment, COSMOS, that supports programmability, device heterogeneity, scalable performance, and robustness. Our macroprogramming environment integrates node and network operating system kernels with a language and compilation infrastructure that has been validated on a range of applications with varying performance requirements.

A critical component of our macroprogramming environment is foundational support for automatic synthesis and verification of distributed coordination mechanisms from formal specifications. While conventional programming paradigms leverage high-level abstractions for defining communication and synchronization protocols, these abstractions often do not provide mechanisms for enforcing requirements that maximize resource utilization and flexibility. In contrast, we present a high-level programming model for sensor network programming in which program behavior is expressed as invariants that directly capture performance and resource constraints. To this end, the paper makes two basic contributions – (i) it demonstrates the use of temporal logic for defining specifications that capture a large class of behaviors that arise in sensor programs; and (ii) it presents techniques for automatically generating efficient communication and other coordination actions from these specifications.

Recent work in sensor networks has resulted in a large number of communication protocol implementations available as reusable components [2]. Use of these components requires the programmer to interface local data sensing and processing at the nodes with available communication library components. These component implementations represent varying domain specific optimizations and protocol designs. For example, tree routing based aggregation, as used in TinyDB [6], allows collection of network data at the root node, while ring-gradient based aggregation, used in Synopsis Diffusion [7] supports similar functionality, but provides significantly enhanced resilience through the use of multi-path routing. This comes at the cost of possible data duplication and out-of-order data delivery. On the other hand, the Trickle [5] protocol provides dissemination of data from a single node to all the nodes in the network, which represents complementary functionality. Interestingly, a single application may require both aggregation/collection and dissemination on different distributed dataflow paths with varying constraints on robustness, in-ordered delivery, and latency.

Conventional sensor network application development, consequently, requires a developer to select and compose coordination mechanisms from available primitives to satisfy application specifications. For the gradient versus tree routing example above, factors that impact selection include (i) acceptable resource overhead to achieve the required level of resilience, and (ii) whether the aggregation functions are duplicate and order insensitive. The first consideration represents a tradeoff of resource consumption and resilience, while the second factor deals with cross-component conflicts, specified in terms of requirements on distributed dataflow semantics. Due to application needs and resource constraints, the developer is forced to account for low-level implementation details of communication mechanisms–making program development time and effort intensive. This paper addresses the problem by providing powerful formalisms and automated techniques for generating the necessary implementations that are guaranteed to satisfy user constraints.

2   Overview of Proposed Approach

We target three key aspects of sensor network programming – correctness, program-mability, and performance. We do this by enabling the user to provide a high-level specification of aggregate system behavior for the application, and provide a pro-gram synthesis and compilation infrastructure to automatically generate efficient ex-ecutables from these specifications. Specifications are expressed in temporal logic; formulae in this logic define invariants over system state. For example, the formula ($\Box$ *RoutingTableSize* < *150Bytes*) expresses a constraint on a global system abstraction (*RoutingTableSize*) that must hold over all state transitions. Notably, this specification provides a separation of concerns: any routing implementation that meets this constraint is feasible under this invariant. We support the following abstractions:

(A1) **Member node groups abstractions.** These abstractions allow description of data source, forwarding and processing, and sink nodes that participate in distributed coordination. For example, to express the fact that all nodes are data sources, and that there exists a unique root in the network, we specify:

$$\Box((\forall s \in Senders) \Rightarrow (s \in Nodes) \tag{1}$$
$$\wedge (\exists r \in Receivers.r \in Root)$$
$$\wedge (\mid Root \mid = 1))$$

Here, *Senders* defines a set of data source nodes, *Receivers* defines a set of sinks, and *Root* defines a singleton set containing a distinguished receiver.

(A2) **Routing resource consumption abstractions.** These allow specification of control requirements on the overhead and performance of data routing and hop-by-hop forwarding. For example, to express the fact that there is no routing re-dundancy, we specify:

$$\Box(\forall s, r_1, r_2 \in Nodes. \tag{2}$$
$$Comm(s, r_1) \wedge Comm(s, r_2) \Rightarrow r_1 = r_2)$$

Comm defines a binary relation on nodes that expresses a communication relation among senders and receivers in the network.

(A3) **Link abstractions.** These abstractions provide control over properties of com-munication links (e.g., radio energy consumption, packet batching for congestion control, etc.). For example, the following formula expresses the constraint that the radio power on a mote should be turned off if the size of its input buffer is less than some threshold, and that the power should eventually be turned on if the buffer is non-empty (because of data inserted from the sensor end). Further-more, the maximum size of the buffer should never exceed a specified constant (MaxLen). Observe there are several implementations that could satisfy this spec-ification; one might resume power as soon as the threshold level is exceeded, another might periodically service the buffer, but may choose to drop packets to enforce the invariant on buffer size.

$$\Box(\forall n \in Nodes. \tag{3}$$
$$(\mid Buffer(n) \mid < Threshold \Rightarrow Radio(n) = \text{down})$$
$$\wedge (\mid Buffer(n) \mid > 0 \Rightarrow \Diamond Radio(n) = \text{up})$$
$$\wedge (\mid Buffer(n) \mid < MaxLen))$$

(A4) **Data management buffers.** These allow specification of properties relating to distributed data coordination and processing, and the control of associated resources (such as memory, service delays, etc.). For example, the following formula captures a time-ordered precedence among data received by nodes. Specifically, it obligates a data packet p_1 with an earlier timestamp than data packet p_2 (as defined by the *Epoch* function) to be either serviced or evicted before p_2. (We use $a \rightsquigarrow b$ to denote $a \Rightarrow \Diamond b$.)

$$\Box(\forall p_1, p_2 \in Data. \forall n \in Nodes. p_1, p_2 \in Buffer(n) \tag{4}$$
$$Epoch(p_1) < Epoch(p_2) \rightsquigarrow$$
$$(p_1 \notin Buffer(n) \wedge p_2 \in Buffer(n)))$$

Observe that these specifications capture salient aspects of system behavior with respect to application requirements without fixing a specific set of implementation choices. Indeed, in practice, some of the specifications above may possibly have several implementations. We derive a feasible implementation from the composition of *all* provided specifications. This is complicated by the fact that an implementation choice made in response to one specification may conflict with other specifications. For example, we could satisfy the specification in Equation 1 using either a tree or gradient routing implementation. In the case of gradient routing, though, packet duplication conflicts with the invariant in Equation 2.

We automatically synthesize a valid implementation that satisfies user-defined specifications, based on invariants provided by library components. Unfortunately, not all user-provided specifications are directly supported by implementations. For example, the specification in Equation 4 imposes a precedence relation on received data. However, no library implementations in existing sensor network systems actually provide such functionality because data management is often tightly coupled to application semantics. Thus, mapping this specification to an implementation requires us to inject additional details.

The user does not specify how the given invariants and properties are met. The actions required to meet the user specified invariants are automatically generated by a *synthesis engine*. Since user specifications may be incomplete, it is possible that not all desired invariants can be satisfied based on user-provided specifications. Invariant violations that arise because of incomplete specifications are reported by the TLC model checker. Using these violations the synthesis engine generates refined specifications based on knowledge of the abstractions exposed by the system. The synthesis process terminates when a provably correct (i.e., model checked) specification is achieved. Finally, the automatically-generated specifications (MCAutoGen) together with the interface description provided by the user module are compiled to node primitives. The compiler infrastructure uses a combination of two approaches to generate native code. First, based on the user constraints over system state variables, it locates components from an annotated library that provide the required mechanisms to meet user requirements. Inter-component dependencies and conflicts are automatically resolved. Second, based on the detailed actions generated by the synthesis engine it automatically generates node primitives from TLA+ specifications.

Table 1. Invariants for each of the distributed coordination tasks of the seismic sensing application

Sys Abs	MaxFlow	HiResFlow	CtrlFlow	Total
Member	RcvrID=Root		SndrID=Root	2
Routing	FwdLinks=1, RtableSize≤150B MaintenanceInterval=10000ms		FwdLinks=Bcast RtableSize≤10B	5
Link	-		-	0
MsgType	[TimeSeq]	[NodeID, SeqNum]	[]	3
In-Net Buf	MLenInv, MOrderInv, MAggregateMax	HLenInv, HOrderInv	-	5
Rcv Buf	MLenInv, MOrderInv, MAggregateMax	HLenInv, HFlushInv, HOrderInv	-	6
Timer	MFBServiceRate, MRBServiceRate,	HFBServiceRate, HRBServiceRate,	CFBServiceRate, CRBServiceRate,	6

3   Evaluation

Our application testbed is a seismic-sensing application, currently operational at the Bowen Lab for monitoring a three story concrete structure. The application uses tens of nodes, to sense acceleration (using on-board accelerometer sensors) and displacement. Seismic activity generates high resolution data from the network, which is used to extract and study the frequency response. Our implementation of this application is as follows: time-windowed max acceleration value is extracted from the sensor network using in-network aggregation. This data is fed to a user provided controller function, which detects seismic activity and disseminates a trigger message to all nodes. The controller function runs on a unique node (root node) in the network. In response to the trigger, all nodes generate high-resolution data, which is forwarded to the root node, where a user provided FFT function evaluates the frequency response of the structure.

The application specification includes three distributed communication and coordination tasks: (i) MaxFlow, which involves collection and in-network aggregation of max sensor readings, (ii) HiResFlow, which involves collection of high resolution sensor readings, and (iii) CtrlFlow, which involves dissemination of controller trigger messages to all nodes in the network. The invariants on state abstractions for each of these flows are shown in Table 1. Since the max values and high resolution data flows from all nodes in the network to the root node, the user can share the system abstractions including member, routing, and link abstractions. On the other hand, CtrlFlow involves data dissemination from the root to all nodes and needs to be independently controlled. For MaxFlow and HiResFlow the application specifies that forwarding should be duplicate sensitive, and provides routing table size and topology maintenance resource constraints. On the other hand, for dissemination, the application requires that duplicates are necessary and that routing memory resource consumption should be low. For CtrlFlow the user does not require any constraint on buffering, except that its service rate should be high (CXXServiceRate). The user specifies buffer length and packet ordering for both the buffers of MaxFlow and \verbHiResFlow+. In addition, for MaxFlow in-network aggregation is specified. For HiResFlow the application

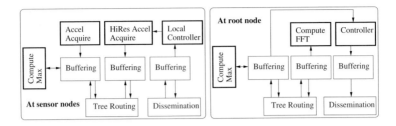

Fig. 1. Interaction of components in the synthesized seismic sensing application. Separate program fragments are generated for the sensor nodes and the root node. User functions are represented by thick border boxes.

specifies (using HFlushInv) that sequence numbers for data from each node should be consecutive (because several consecutive packets are required for FFT to generate a reasonable frequency analysis). For the entire application only 27 invariants are used, in all. The interactions of the components of the generated program are illustrated in Figure 1. The user provides only six functions, which implement data acquisition and processing.

Performance. We evaluate the performance of the synthesized program on a 10 node (plus one root), wireless sensor network testbed. We use Mica2 motes with MTS310 multi sensor boards [1] as sensor nodes. Mica2 has an ATmega128 micro-controller running at 7.37MHz, with a 128KB program ROM, 4KB data RAM, and Chipcon CC1000 radio (with a maximum through put of 12.3kbps). The MTS310 board supports a wide variety of sensors including a 2-D accelerometer, which is used in our experiments. The root node is a workstation with Intel Pentium 4, 1.70GHz and 512MB RAM, running Linux 2.6.17.7 kernel. The root is interfaced to the Mica2 radio network using a MIB510 [1] board connected to the serial port.

The key performance concerns in our application include (i) reliability of network communication, and (ii) congestion control, due to the heavy traffic when high resolution data is communicated to the root node. To evaluate the ability of our programming

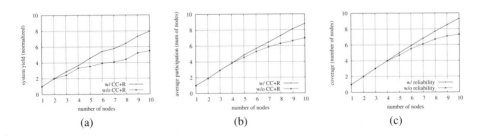

Fig. 2. Congestion control and communication reliability components are inserted into the synthesized application due to additional user invariants. The plots show that the use of these components increases system performance substantially.

model to meet user requirements we execute the application with and without the following invariants:

$$Inv \triangleq \wedge MH\_link.ReliabilityOvh \leq 1 \tag{5}$$
$$MH\_link.CongestionBatching \leq 4$$
$$Ctrl_link.ReliabilityOvh \leq 1$$

These invariants imply that the communication of max values and high resolution data (MH_XX) should use reliability, with a maximum overhead of one packet, and congestion triggered buffering of upto three packets. Similarly, reliability, with a maximum overhead of one packet, should be provided for controller messages. The performance results, for varying network size, for the synthesized application, with and without the above invariants, are shown in Figure 2. The evaluation includes the following. (a) System yield for high resolution data (normalized to the yield of a single node), where yield measures the amount of useful data received and processed at the root node. (b) Average number of nodes whose data was used in each max value calculation. Note that one max value is generated per epoch. (c) Coverage, in terms of the number of nodes that received the dissemination trigger message in the first refresh interval. Note that the dissemination protocol uses periodic refreshes to account for network dynamics. However, the response time increases with the number of nodes that receive the trigger in the first refresh interval.

The plots in Figure 2 show that by using congestion control and reliable forwarding, approximately 40% more yield of high resolution data, and 30% higher participation in max aggregation is achieved, when the network has 10 nodes. Similarly, using reliability for trigger dissemination achieves about 14% higher coverage in a network of 10 nodes. These results show that using simple high-level invariants, the user can substantially affect system performance without having to deal with low-level programming details. Furthermore, the invariants allow constraining the resource overheads involved in achieving desired performance.

3.1 Specifying Resource-Quality Tradeoffs

Most sensor network applications utilize in-network data processing. Nodes in a wireless sensor network dynamically form self-organized data routing topologies. Hence, the number of nodes that feed data to a forwarding node, where processing takes place, varies dynamically. Given limited buffer memory, applications should adaptively manage processing to cater to surges in input data rate or risk dropping messages due to overflowing buffers. However, developing such robust applications is difficult. Our language enables users to provide multiple data processing functions, which are selectively triggered based on buffer state, thus, allowing load conditioning. The user provides processing intensive functions (e.g., input signal convolution, FFT, or statistical analysis of input data), and their less processing-intensive counterparts that yield lower quality approximations. Following is a simple load conditioning invariant example.

$$LoadCondInv \triangleq \text{IF } Len(fb) < Threshold \tag{6}$$
$$\text{THEN } UseFunc(\text{"DataFFT"}, fb, fb[End])$$
$$\text{ELSE } UseFunc(\text{"AproxFFT"}, fb, fb[End])$$

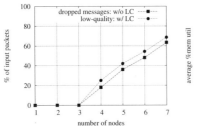 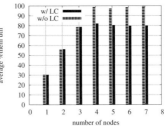

Fig. 3. Memory vs quality tradeoff. Load conditioning prevents lost data by using a faster computation, which yields approximate results, to maintain memory threshold.

Driven by resource limitations, users tradeoff quality with resource consumption. The plots in Figure 3 illustrates an evaluation of a system with and without the above invariant. Without load conditioning, input data is dropped when the input rate increases (as the number of data source nodes increase). With load conditioning the memory threshold is maintained, while high load conditions trigger low quality approximations. For example, with data from 4 sources, approximately 25% of output data is low quality while 75% is high quality.

4   Concluding Remarks

In this paper, we presented a powerful and novel formalism for development of verifiable sensing applications based on temporal logic specifications. We also present an automatic synthesis engine, that uses a model checker, TLC, along with low-level system abstractions, to generate code guaranteed to satisfy programmer specified invariants.

References

1. Crossbow Inc. http://www.xbow.com/wireless_home.aspx.
2. Cheng Tien Ee, Rodrigo Fonseca, Sukun Kim, Daekyeong Moon, Arsalan Tavakoli, David Culler, Scott Shenker, and Ion Stoica. A modular network layer for sensornets. In *Proc. OSDI'06*, November 2006.
3. Jason Hill, Robert Szewczyk, Alec Woo, Seth Hollar, David Culler, and Kristofer Pister. System architecture directions for networked sensors. In *Proc. of ASPLOS-IX*, November 2000.
4. Leslie Lamport. *Specifying Systems: The TLA+ Language and Tools for Hardware and Software Engineers*. Pearson Education, Inc, 2002.
5. Philip Levis, Neil Patel, David Culler, and Scott Shenker. Trickle: A self-regulating algorithm for code propagation and maintenance in wireless sensor networks. In *Proc. of NSDI '04*, March 2004.
6. Samuel Madden, Michael Franklin, Joseph Hellerstein, and Wei Hong. TinyDB: an acquisitional query processing system for sensor networks. *ACM Transactions on Database Systems*, 30(1):122–173, March 2005.
7. S. Nath, P. B. Gibbons, S. Seshan, and Z. Anderson. Synopsis diffusion for robust aggregation in sensor networks. In *Proc. of SenSys '04*, November 2004.

Dynamic Data-Driven Systems Approach for Simulation Based Optimizations*

Tahsin Kurc[1], Xi Zhang[1], Manish Parashar[2], Hector Klie[3], Mary F. Wheeler[3], Umit Catalyurek[1], and Joel Saltz[1]

[1] Dept. of Biomedical Informatics, The Ohio State University, Ohio, USA
{kurc,xizhang,umit,jsaltz}@bmi.osu.edu
[2] TASSL, Dept. of Electrical & Computer Engineering, Rutgers,
The State University of New Jersey, New Jersey, USA
parashar@caip.rutgers.edu
[3] CSM, ICES, The University of Texas at Austin, Texas, USA
{klie,mfw}@ices.utexas.edu

Abstract. This paper reviews recent developments in our project that are focused on dynamic data-driven methods for efficient and reliable simulation based optimization, which may be suitable for a wide range of different application problems. The emphasis in this paper is on the coupling of parallel multiblock predictive models with optimization, the development of autonomic execution engines for distributing the associated computations, and deployment of systems capable of handling large datasets. The integration of these components results in a powerful framework for developing large-scale and complex decision-making systems for dynamic data-driven applications.

1 Introduction

The problem of optimal decision making and reliable parameter estimation on physical domains is generally one of the most challenging tasks in many engineering and scientific applications, such as biomedicine, structural mechanics, energy and environmental engineering. In these applications, the feasibility of the tasks depend heavily on how simulations and optimization are effectively combined [1,2,3,4,5,6,7,8,9,10]. The overall goal is both to generate good estimates of optimal parameter values and to reliably predict end results. The objective function of the optimization process can be viewed as a performance measure that depends on an array of controlled variables (e.g., the number and locations of sensors), which define the decision parameters, and a vector of uncontrolled conditions (e.g., electromagnetic properties of the subject sample in magnetic resonance imaging, subsurface properties in reservoir simulations, boundary

* The research presented in this paper is supported in part by the National Science Foundation Grants ACI-9984357, EIA-0103674, EIA-0120934, ANI-0335244, CNS- 0305495, CNS-0426354, IIS-0430826, ACI-9619020 (UC Subcontract 10152408), ANI-0330612, EIA-0121177, SBR-9873326, EIA-0121523, ACI-0203846, ACI-0130437, CCF-0342615, CNS-0406386, CNS-0426241, ACI-9982087, CNS-0305495, NPACI 10181410, Lawrence Livermore National Laboratory under Grant B517095 (UC Subcontract 10184497), Ohio Board of Regents BRTTC BRTT02-0003, and DOE DE-FG03-99ER2537.

Y. Shi et al. (Eds.): ICCS 2007, Part I, LNCS 4487, pp. 1213–1221, 2007.

conditions), which represent the characteristics of the physical domain. Finding a solution to the objective function requires a systematic search of the parameter space and evaluation of corresponding scenarios within the physical domain. An exhaustive search of the space is often unfeasible, since the space can consist of thousands to millions of data points and requires the evaluation of a very large number of potential scenarios corresponding to these points. Moreover, the values of uncontrolled variables are not known precisely, introducing a high level of uncertainty into the problem. A possible approach is to couple optimization algorithms with simulations and experimental measurements to enable a systematic evaluation of the scenarios. This approach, together with the experienced judgment of specialists, can allow for better assessment of uncertainty and can significantly reduce the risk in decision-making [11,12]. However, such an approach is characterized by dynamic interactions between complex numerical models, optimization methods, and data. Major obstacles to using this approach include the large computational times required by the complex simulations, the challenges of integrating dynamic information into the optimization process, and the requirements of managing and processing large volumes of dynamically updated datasets (obtained either from simulations or sensors).

In this paper, we provide an overview of the problems in optimization in domains with large search space and the approaches we have devised to address these issues. Our efforts have resulted in (1) algorithms to support parallel multi-block numerical models coupled with global stochastic optimization algorithms, (2) execution engines that implement adaptive runtime management strategies to enable efficient execution of optimization and simulation processes in distributed and dynamic computational environments, and (3) systems to handle very large and potentially dynamic multi-dimensional, scientific datasets on large scale storage systems. This paper is structured as follows. The next section provides further motivations through the description of different simulation based optimization scenarios. Section 3 describes support for multiblock simulations. Data management and processing solutions are presented in Section 4. Section 5 concludes the paper..

2   Optimization Challenges in Large-Scale Domains

In making decisions based on a modeling of the physical domain under study, there are at least two key goals to consider: (1) the design and deployment of man-made objects to optimize a desired response, and (2) the reproduction of the behavior of the physical phenomenon by matching the numerical model response with the field measurements. The first goal deals with forecasting the behavior of the model under a given set of conditions. The objective here is to find the optimal values of the operational parameters. The second goal implies an estimation and understanding of the parameters (state and control variables) describing the model (e.g., porosity, permeability, pressures, temperatures, geometry), and analysis and validation of the corresponding numerical model. The experimental data obtained from sensors can be compared with predictions obtained through a numerical model with the objective of reducing mismatch between the computed and observed data.

Consider optimization in oil reservoir management [1] in the context of the first goal described above. The number and locations of wells in an oil reservoir have significant impact on the productivity and environmental effects of the reservoir (i.e., optimum economic revenue, minimum bypassed hydrocarbon, minimal environmental hazards/impact). If the oil extraction wells are not placed carefully, large volumes of bypassed oil may remain in the field. The amount of water that needs to be injected (in order to drive the oil toward extraction wells) and disposed of also depends on the number and locations of injection wells as well as the extraction wells. For example, assume the objective function is formulated as maximizing profit. Here, the decision parameters are the number of injection and extraction wells and their locations. The uncontrolled conditions are the geological and fluid properties of the reservoir and the economic parameters (e.g., the cost of pumping water). Even with a fixed number of wells, finding a solution for this objective function requires the evaluation of a large number of possible configurations. For each placement of the wells, the reservoir model has to be evaluated for many time steps in order to calculate required parameters (i.e., the effective volume of extracted oil and the net value). Moreover, the lack of complete information about these properties requires the use of stochastic approaches in the generation of equally probable scenarios (or realization) using Monte Carlo simulation or in the solution of stochastic PDEs [13].

Another example is high-field Magnetic Resonance Imaging (MRI). High-field MRI devices (e.g., 7 Tesla systems) offer high signal-to-noise ratios, better contrast, greater shift dispersion, and thus the ability to obtain better, higher resolution images [14]. However, a problem in high-field MRI is the non-uniform detection of signals across the subject (sample) being imaged. That is, the brightness in the final image of the sample varies across the spatial domain; some regions are brighter (finer details can be seen), whereas other regions are darker, depending on the location and voltage strength of the coils in the device. A challenging design and operational problem, therefore, is to determine the location of the coils and the amount and duration of voltage values per coil to achieve optimal distribution of brightness across the subject volume. This involves simulating signal distribution within the volume for a given set of design/operational parameters (i.e., coil locations, coil voltage values, and the duration of voltage at each coil) [15], and searching the space of design and operational parameters for optimal values. To speed up the execution of simulations, multi-resolution Grid techniques can be used [16].

From a computational viewpoint, these applications perform function evaluations by solving sets of coupled nonlinear PDEs in three dimensions, for multiple times steps and for different sets of parameters. Furthermore, the simulation and optimization components require the synthesis and querying of data to search the parameter space. The datasets are generated using simulations or obtained from sensors. They are large, multi-dimensional, multi-scale, and may be generated and stored at disparate locations. In the following sections, we present techniques and tools to address these problems.

3   Distributed Multi-block Simulations

The ability to define different regions with differing degree of granularity provides a valuable means for focusing computational resources to critical areas, specially since simulating high fidelity descriptions of the entire system may be computationally too expensive. Furthermore, incorporating automatic model reduction into solution procedures provides an additional means for increasing computational efficiency by lumping parameters, and simplifying basic principles. From the conceptual and computational standpoint, different models and interactions may take place in the same domain at different spatial and temporal scales. In order to deal with the accurate and efficient solution of these problems, the spatial physical domain may be decomposed (i.e., decoupled) into different blocks or subdomains under the assumption that different algorithms, physical models or scales are solved. In order to preserved the integrity of the overall solution, continuity of fluxes and pressures are imposed across each subdomain.

One way to efficiently establish this spatial coupling/decoupling among subdomains is through mortar spaces [17,18,19,20,21]. These spaces allows imposing physically driven matching conditions on block interfaces in a numerically stable and accurate way. Some of the computational advantages of the multi-block approach include: (1) multi-physics, different physical processes/mathematical models in different parts of the domain may be coupled in a single simulation; (2) multi-numerics, different numerical techniques may be employed on different subdomains; (3) multi-scale resolution and adaptivity, highly refined regions may be coupled with more coarsely discretized regions and dynamic grid adaptations may be performed locally on each block; and (4) multi-domains, highly irregular domains may be described as unions of more regular and locally discretized subdomains with the possibility of having interfaces with non-matching grids.

A key challenge in parallel/distributed multi-block formulations are the dynamic and complex communication and coordination patterns resulting from the multi-physics, multi-numerics, multi-scale and multi-domain couplings. These communication and coordination patterns depend on state of the phenomenon being modeled. Moreover, they are determined by the specific numerical formulation, domain decomposition and sub-domain refinement algorithms used, which, in most practical cases, is known only at runtime. Implementing these communication and coordination patterns using commonly used parallel programming frameworks is non-trivial.

Seine/MACE [22,23], developed as part of this effort, provides a virtual shared space abstraction to support interactions in parallel multi-block simulations. Seine builds on two key observations: (a) formulations of the targeted simulations are based on geometric multi-dimensional domains (e.g., a grid or a mesh) and (b) interactions in these applications are typically between entities that are geometrically close in this domain (e.g., neighboring cells, nodes or elements). Rather than implementing a general and global associative space, Seine defines geometry-based transient interaction spaces, which are dynamically created at runtime, and each of which is localized to a sub-region of the application geometric domain. The interaction space can then be used to share objects between nodes whose computational sub-domains geometrically intersect with that region. To share an object using the interaction space, nodes do not have to know of, or synchronize with each other at the application layer. Sharing objects in the Seine model is similar to that in a tuple space model. Furthermore, multiple shared spaces can exist simultaneously in the application domain. An experimental evaluation demonstrates its scalability and low operational overheads, as well as its ability to effectively support distributed multiblock simulations [23].

4 Data Management and Processing

Management, querying, and processing of data generated and referenced during optimization present several challenges. In this section, we describe these challenges and the techniques and tools we have developed to address them.

A dataset generated by an optimization run consists of the values of the input and output parameters (i.e., the values of controlled variables and uncontrolled conditions, and the output of the objective function) along with the output from simulations of a numerical model of the physical domain. User-defined metadata also is needed to describe a given optimization run (e.g., the names of the optimization methods used, the id of the optimization run, the name of the simulation model used). By maintaining these data types and datasets, a large-scale knowledge base can be created and used to speed up the execution of optimization runs and to carry out post-optimization analyses. In the simplest case, the knowledge base can be queried during optimization to see if a given step, or a subset of numerical simulations at that step, have already been evaluated, potentially in a previously executed or concurrently executing optimization run. In this case, the query can be formulated to search the knowledge base based on the metadata to check if a given optimization step has already been evaluated and its output stored in the system. Similarly, during post-optimization analyses, a user may want to compare and correlate results obtained from one optimization run with results from another set of optimization runs. However, querying of metadata and information discovery in large, decentralized, and dynamic environments is a challenging problem. To address this challenge, we have developed Squid [24], which is a decentralized distributed information discovery system that supports complex content-based queries including ranges and wildcards. It guarantees that all existing data elements that match a query will be found with bounded costs in terms of the number of messages and the nodes involved. A key innovation is a locality preserving indexing scheme that effectively maps the multidimensional information space to physical nodes. This indexing scheme makes use of Space Filling Curves (SFC) [25]. Squid defines two basic operations: "publish" and "query". In the publishing process, the keywords describing the content of the data element and the SFC-mapping are used to construct the index for the data element, and this index is used to store the element at the appropriate node in the overlay. In the simple querying process, the query is translated into the corresponding region in the multi-dimensional information space and the relevant clusters of the SFC-based index space, and the appropriate nodes in the overlay are queried.

Another challenge is the large volumes of data and dynamically updated multidimensional datasets that are generated by simulations or field sensors. At each step of an optimization run, one or more simulations (using different values for the uncontrolled conditions) may need to be executed over many simulation time steps on a large mesh modeling the physical domain. Even a single optimization run may generate multiple terabytes of data. During an optimization run, simulations can be executed on distributed collections of compute systems. The environment can be heterogeneous, consisting

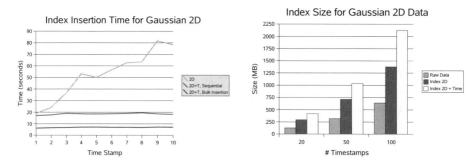

Fig. 1. (a) Index insertion time. (b) Index size.

of systems with different storage and computation capacities. Moreover, datasets can be dynamically updated, from new simulations or from sensors and monitors in the field; data from sensors can be used to validate simulation results as well as to refine numerical models. When datasets are updated and new data elements are added, indexes as well as the organization of the dataset on disk and in memory need to be updated. For example, assume the dataset is initially partitioned into a set of chunks so that every chunk contains an approximately equal number of elements and also contains elements that are close to each other in the multi-dimensional space of the domain. When new data elements are added, one approach is to create new chunks from these data elements. One problem with this approach is that it may result in a very large number of chunks, increasing the cost of indexing. Moreover, if the coordinates of new data elements are spread across the multi-dimensional space, the bounding boxes of the resulting chunks may be large. This may introduce I/O overheads; a query may intersect these chunks even though the chunks may not have any data elements required for the query. Alternately, existing data chunks can be reorganized to store the new data elements. The advantage of this approach is that it reduces the number of chunks and is also more likely to result in chunks with tighter bounding boxes. However, it will require reorganization of the dataset on disk, resulting in I/O overheads during updates. The indexes also need to be be updated when new data is added. A simple approach would be to update indexes as new data arrives. This would require frequent updates to the index. Another possible approach would be to aggregate data updates, organize them, and do a bulk update to the index. We have developed the STORM framework to address the storage and querying of such very large datasets. STORM [26,27] is a service-based middleware that is designed to provide basic database support for very large datasets where data is stored in flat files. Such datasets commonly arise in simulation studies and in biomedical imagery. STORM supports execution of SQL-style SELECT queries on datasets stored on distributed storage systems. These services provide support for indexing, data extraction, execution of user-defined filters, and transferring of data from distributed storage nodes to client nodes in parallel.

Within the STORM framework, we are also investigating runtime support to efficiently execute range and sampling queries against dynamic multi-dimensional datasets. Experimental results on data organization and index updates are presented in Figure 1. Figure 1(a) compares the performance of different strategies for inserting incoming data tuples into R-tree indexes [28] (to speed up range queries) in terms of insertion time. In these experiments, we used synthetically generated datasets. Data objects are generated within a 2D normalized $[1 \times 1]$ space using a Gaussian distribution for their coordinates at each time stamp. The size of each data tuple is fixed at 64 bytes and we generated data for 500 time stamps. We fixed the number of data objects to be $100,000$. For the index, we have three options. First, we only use the data attribute as the index attribute, second we use the time stamp as an index attribute with sequential insertion, and third we use the time stamp as an index attribute with bulk insertion. The figure shows that the insertion time for the index with only the data attribute as index attribute increase almost linearly as the size of the index increases. On the other hand, the insertion time for the index with time as an index attribute is almost constant, is independent of the size of the original index, and only depends on the size of the inserted data. Furthermore, bulk insertion achieves the best insertion performance, as expected. We should note that by using time stamp as an indexing attribute, we effectively increase the attribute space and the overall storage space for the index file will likely increase. Figure 1(b) shows that the overall size for the index with time as an index attribute is larger than the index with only data attribute as index attributes. This is a space-efficiency tradeoff, and as disk storage capacity continues to grow at very fast rate and price per GB storage continues to fall, we believe that the insertion efficiency would be the deciding factor in most real-world scenarios.

5   Conclusions

Coupling optimization algorithms with simulations and experimental measurements represents an effective approach for optimal decision making and reliable parameter estimation in engineering and scientific applications, and can allow for better assessment of uncertainty and significantly reduce the risk in decision-making. However, the scale, dynamism and heterogeneity of computation and data involved in the approach present significant challenges and require effective computational and data management support. Specifically, achieving high levels of performance and accuracy on multiphysics and multiscale simulation based optimizations require the development of sophisticated tools for distributed computation and large scale data management. Further, dynamic data-driven simulations require the timely orchestration of several components that may deal with different models, to respond adequately to disparate data streams and levels of information. Uncertainty is unavoidable in data and models, and may govern the entire dynamic data-driven process if it is not assessed and managed in an adaptive manner. This paper reported on models, algorithms and middleware level solutions developed by the authors to address these challenges. These solutions have been shown to effectively enable reliable optimizations and decision-making processes for oil reservoir management, contaminant management and other related applications.

References

1. Bangerth, W., Klie, H., Parashar, M., Mantosian, V., Wheeler, M.F.: An autonomic reservoir framework for the stochastic optimization of well placement. Cluster Computing **8**(4) (2005) 255–269

2. Parashar, M., Klie, H., Catalyurek, U., Kurc, T., Bangerth, W., Matossian, V., Saltz, J., Wheeler, M.F.: Application of Grid-enabled technologies for solving optimization problems in data-driven reservoir studies. Future Generation of Computer Systems **21** (2005) 19–26

3. Eldred, M., Giunta, A., van Bloemen Waanders, B.: Multilevel parallel optimization using massively parallel structural dynamics. Structural and Multidisciplinary Optimization **27** (1-2) (2004) 97–109

4. Eldred, M., Giunta, A., van Bloemen Waanders, B., Wojtkiewicz Jr., S., Hart, W., Alleva, M.: DAKOTA, A Multilevel Parallel Object-Oriented Framework for Design Optimization, Parameter Estimation, Uncertainty Quantification, and Sensitivity Analysis. Version 3.0 Users Manual. Sandia Technical Report SAND2001-3796. (2002)

5. Stori, J., King, P.W.C.: Integration of process simulation in machining parameter optimization. Journal of Manufacturing Science and Engineering (USA) **121**(1) (1999) 134–143

6. Fu, M.C.: Optimization via simulation: A review. Annals of Operations Research (Historical Archive) **53**(1) (1994) 199–247

7. Gehlsen, B., Page, B.: A framework for distributed simulation optimization. In: Proceedings of the 2001 Winter Simulation Conference. (2001) 508–514

8. Schneider, P., Huck, E., Reitz, S., Parodat, S., Schneider, A., Schwarz, P.: A modular approach for simulation-based optimization of mems. In: SPIE Proceedings Series Volume 4228: Design, Modeling, and Simulation in Microelectronics. (2000) 71–82

9. Mendes, P., Kell, D.B.: Non-linear optimization of biochemical pathways: applications to metabolic engineering and parameter estimation. Bioinformatics **14**(10) (1998) 869–883

10. Klie, H., Bangerth, W., Gai, X., Wheeler, M., Stoffa, P., Sen, M., Parashar, M., Catalyurek, U., Saltz, J., Kurc, T.: Models, methods and middleware for Grid-enabled multiphysics oil reservoir management. Engineering with Computers **22** (2006) 349–370

11. Spall, J.C.: Introduction to stochastic search and optimization: Estimation, simulation and control. John Wiley & Sons, Inc., Publication, New Jersey (2003)

12. Yeten, B., Durlofsky, L.J., Aziz, K.: Optimization of nonconventional well type, location, and trajectory. SPE Journal **8**(3) (2003) 200–210 SPE 86880.

13. Zhang, D.: Stochastic Methods for Flow in Porous Media. Academic Press, Inc (2002)

14. Hoult, D., Richard, R.: The signal to noise ratio of the nuclear magnetic resonance experiment. Journal of Magnetic Resonance **24** (1976) 71–85

15. Han, Y., Wright, S.: Analysis of rf penetration effects in mri using finite-difference time-domain method. Proc SMRM 12th Annu. Meeting, New York (1993)

16. Zivanovic, S.S., Yee, K.S., Mei, K.K.: A subgridding method for the time-domain finite-difference method to solve maxwell's equations. IEEE Trans. Microwave Theory Tech. **39** (1991) 471–479

17. Arbogast, T., Cowsar, L.C., Wheeler, M.F., Yotov, I.: Mixed finite element methods on non-matching multiblock grids. SIAM J. Numer. Anal. **37** (2000) 1295–1315

18. Li, J., Wheeler, M.F.: Uniform convergence and superconvergence of mixed finite element methods on anisotropically refined grids. SIAM Journal on Numerical Analysis **38**(3) (2000) 770–798

19. Peszyńska, M., Wheeler, M.F., Yotov, I.: Mortar upscaling for multiphase flow in porous media. Comput. Geosci. **6**(1) (2002) 73–100

20. Wheeler, M.F., Peszynska, M.: Computational engineering and science methodologies for modeling and simulation of subsurface applications. (Advances in Water Resources) in print.

21. Wheeler, M.F., Yotov, I.: Physical and computational domain decompositions for modeling subsurface flows. In Mandel, J., et al., eds.: Tenth International Conference on Domain Decomposition Methods, Contemporary Mathematics, vol 218, American Mathematical Society (1998) 217–228

22. Zhang, L., Parashar, M.: A dynamic geometry-based shared space interaction framework for parallel scientific applications. In: Proceedings of the 11th Annual International Conference on High Performance Computing (HiPC 2004). Volume 3296., Bangalore, India, LNCS, Springer-Verlag (2004) 189–199

23. Zhang, L., Parashar, M.: Seine: A dynamic geometry-based shared space interaction framework for parallel scientific applications. Concurrency and Computations: Practice and Experience **18**(15) (2006) 1951–1973

24. Schmidt, C., Parashar, M.: Enabling flexible queries with guarantees in p2p systems. IEEE Network Computing, Special Issue on Information Dissemination on the Web **8**(3) (2004) 19–26

25. Sagan, H.: Space-Filling Curves. Springer-Verlag (1994)

26. Narayanan, S., Kurc, T., Catalyurek, U., Saltz, J.: Database support for data-driven scientific applications in the grid. Parallel Processing Letters **13**(2) (2003) 245–271

27. Narayanan, S., Kurc, T.M., Catalyurek, U.V., Saltz, J.H.: Servicing seismic and oil reservoir simulation data through grid data services. In: Proceedings of VLDB Workshop Data Management in Grid 2005 (VLDB DMG'05). (2005) 98–109

28. Guttman, A.: R-Trees: A dynamic index structure for spatial searching. In: Proceedings of the 1984 ACM-SIGMOD Conference. (1984) 47–57

DDDAS/ITR: A Data Mining and Exploration Middleware for Grid and Distributed Computing

Jon B. Weissman, Vipin Kumar, Varun Chandola, Eric Eilertson, Levent Ertoz, Gyorgy Simon, Seonho Kim, and Jinoh Kim

Dept. of Computer Science and Engineering,
University of Minnesota, Twin Cities
jon@cs.umn.edu

Abstract. We describe our project that marries data mining together with Grid computing. Specifically, we focus on one data mining application - the Minnesota Intrusion Detection System (MINDS), which uses a suite of data mining based algorithms to address different aspects of cyber security including malicious activities such as denial-of-service (DoS) traffic, worms, policy violations and inside abuse. MINDS has shown great operational success in detecting network intrusions in several real deployments. In sophisticated distributed cyber attacks using a multitude of wide-area nodes, combining the results of several MINDS instances can enable additional early-alert cyber security. We also describe a Grid service system that can deploy and manage multiple MINDS instances across a wide-area network.

Keywords: Data mining, Grid Computing.

1 Introduction

MINDS contains various modules for collecting and analyzing massive amounts of network traffic (Figure 1). Typical analyses include behavioral anomaly detection, summarization, scan detection and profiling. Additionally, the system has modules for feature extraction and filtering out attacks for which good signatures have been learned [3]. Each of these modules will be individually described in the subsequent sections. Independently, each of these modules provides key insights into the network. When combined, which MINDS does automatically, these modules have a multiplicative effect on analysis. As shown in the figure, MINDS system involves a network analyst who provides feedback to each of the modules based on their performance to fine tune them for more accurate analysis.

While the anomaly detection and scan detection modules aim at detecting actual attacks and other abnormal activities in the network traffic, the profiling module detects the dominant modes of traffic to provide an effective profile of the network to the analyst. The summarization module aims at providing a concise representation of the network traffic and is typically applied to the output of the anomaly detection module to allow the analyst to investigate the anomalous traffic in very few screen-shots.

MINDS is deployed at the University of Minnesota, where several hundred million network flows are recorded from a network of more than 40,000 computers every day.

Y. Shi et al. (Eds.): ICCS 2007, Part I, LNCS 4487, pp. 1222–1229, 2007.

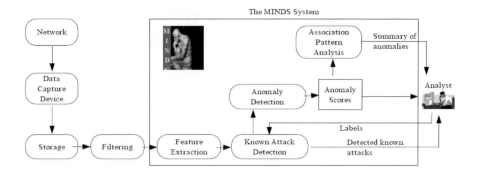

Fig. 1. Minds Architecture

MINDS is also part of the Interrogator architecture at the US Army Research Lab - Center for Intrusion Monitoring and Protection (ARL-CIMP), where analysts collect and analyze network traffic from dozens of Department of Defense sites [4].

1.1 Anomaly Detection

The core of MINDS is a behavioral anomaly detection module that is based upon a novel data-driven technique for calculating the distance/similarity between points in a high dimensional space. A key advantage of this technique is that it makes it possible to meaningfully calculate similarity between records that have a mixture of categorical and numerical attributes (such as network traffic records) based upon the nature of the data. Unlike other anomaly detection methods extensively investigated by the intrusion detection community, this new framework does not suffer from a high number of false alarms. In fact, ARL-CIMP considers MINDS to have the first effective anomaly detection scheme for intrusion detection. A key strength of this technique is its ability to find behavioral anomalies. Some real examples from its use in the DoD network are identification of streaming video from a DoD office to a computer in a foreign country and identification of a back door on a hacked computer. To the best of our knowledge, no other existing anomaly detection technique is capable of finding such complex behavior anomalies while maintaining very low false alarm rate. A multi-threaded parallel formulation of the anomaly detection module allows analysis of network traffic from many sensors in near real time at the ARL-CIMP.

1.2 Detecting Distributed Attacks

Another interesting aspect of the problem of Intrusion Detection is that often times the attacks are launched from multiple locations. In fact, individual attackers often control a large number of machines and may use different machines to launch a different step of the whole attack. Moreover the targets of the attack could be distributed across multiple sites. Thus an intrusion detection system running at one site may not have enough information to detect the attack by itself. Rapid detection of such distributed cyber attacks requires an inter-connected system of IDSs that can ingest network traffic data in near real-time, detect anomalous connections, communicate their results to other IDSs, and incorporate the information from other IDSs to enhance the anomaly scores of such threats. Such a system consists of several autonomous IDSs that share their knowledge bases with each other for swift detection of malicious, large-scale cyber attacks. We illustrate the distributed aspect of this problem with the following example. Figure 2 shows the 2-dimensional global IP space such that every IP allocated in the world is represented in some block. The black region represents the unallocated IP space. Figure 3 shows a graphical illustration of suspicious connections originating from outside (box on the right) to machines inside the University of Minnesota's IP space (box on the left) in a typical time window of 10 minutes. Each red dot in the right box represents a connection made by that machine to an internal machine on port 80 that is suspicious. In this case, this means that the internal machine being contacted does not have a web-server running, thus making the external machines that are attempting to make connections to port 80, to be suspected attackers. The right box indicates that most of these potential attackers are clustered in specific address blocks of the Internet. A close examination shows that most of the dense areas belong to network blocks located in cable/AOL users in USA or blocks allocated to Asia and Latin America. There are totally 999 unique sources involved on the outside trying to contact 1126 destinations inside the U of M IP network space. The total number of involved flows is 1516 which means that most of the external sources made just one suspicious connection to inside. It is hard to tag a source as malicious based on just one connection. If multiple sites running the same analysis across the IP space report the same external source as suspicious, it would make the classification much more accurate.

The ideal scenario would be that we bring in the data collected at these different sites at one place and then analyze it. But this is not feasible due to following reasons – firstly, the data is naturally distributed and is more suited for a distributed analysis;

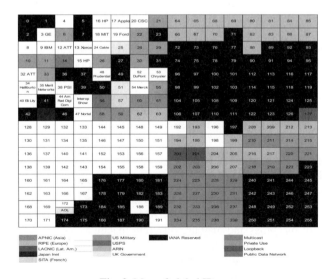

Fig. 2. Map of global IP space

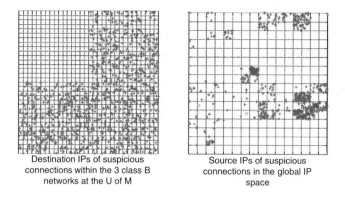

Destination IPs of suspicious
connections within the 3 class B
networks at the U of M

Source IPs of suspicious
connections in the global IP
space

Fig. 3. Suspicious traffic on port 80

secondly, the cost of merging huge amounts of data and running analysis at one site is also very high and finally there are privacy, security and trust issues that arise in sharing the network data between different organizations. Thus what is really required is a distributed framework in which these different sites can independently analyze their data and then share the high-level patterns and results while honoring the privacy of data from individual sites. The implementation of such a system would require handling distributed data, privacy issues and the use of data mining tools, and would be much easier if a middleware provided these functions. Development and implementation of such a system (see Figure 4) is currently in progress as part of an NSF funded collaborative project between University of Minnesota, University of Florida and University of Illinois, Chicago.

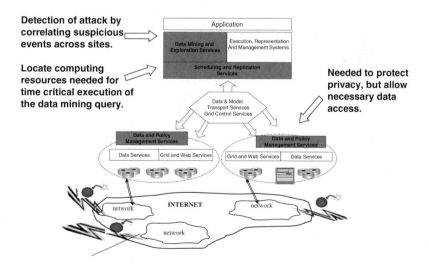

Fig. 4. Distributed network intrusion detection system

2 Grid Middleware

To support distributed data mining for network intrusion (as envisioned in Figure 4), we have developed a framework that leverages Grid technology. The MINDS system is transformed into a Grid service and deployed on distributed service containers. By separating the process of training set generation (data pre-processing) from the given input data, the data can be decomposed into multiple fragments and distributed among different MINDS Grid services for parallel analysis. The MINDS Grid service front-end is an entry point for the MINDS service requests. The scheduling module inside the front-end sets up a plan regarding where to pre-process data for training set generation if necessary, how to decompose the data, where to send the decomposed data (that is, where to run the MINDS Grid service back-end), whether to deploy new MINDS back-end Grid services if necessary, where to aggregate the result datasets, and finally where to store. The MINDS Grid service front-end may coordinate with different Grid services such as storage service, and with various middleware-level services such as replication service, data management service (e.g., GridFTP [6] for data transfer), and security infrastructure for authentication and authorization. For regulated data access within the data mining community, a community level security authority (CSA, Community Security Authority) is required. The authority expresses policies regarding four main principals in the community – users (and user groups), resource providers (storage provider and compute provider), data (and data groups), and applications - and relationships between them. The CSA uses a catalogue service to manage the catalogue of raw input data, processed data (alert and summary), and replicated datasets, and it maintains a database containing policies.

The system consists of three main component services: MINDS Service, Storage Service, and Community Security Authority (Figure 5). In order to address the first requirement - exploitation of geographically and organizationally distributed computing resources to solve data-intensive data mining problem, we designed the MINDS service as a composite service of a front-end and multiple back-ends. The MINDS Grid service front-end supports planning, scheduling, and resource allocation for MINDS anomaly analysis. MINDS Grid service front-end service should adapt dynamically changing environment to make efficient decisions. We have developed runtime middleware frameworks that can be plugged in the front-end: a dynamic service hosting framework [7], a resource management middleware for dynamic resource allocation and job scheduling [8]. In collaboration with these runtime middleware frameworks, the front-end can set up a plan for data pre-processing (training set generation), input data decomposition, and parallel anomaly analysis, and aggregation of analysis results from MINDS back-end services, and finally store the aggregated analysis result into one or more storage services. Each MINDS back-end service does the actual MINDS anomaly analysis by taking one or more decomposed input data fragments and a training data. The MINDS front-end service coordinates with different Grid services such as storage service, Grid Security Infrastructure, and other various middleware-level services such as catalogue service, replica management service, data management service (e.g., GridFTP for data transfer) on top of Globus Toolkit infrastructure. To ensure the privacy and sensitivity of data, every communication between clients and Grid services are encrypted based on TLS

(Transport Level Security)-based communication and PKI (Public Key Infrastructure)-based authentication and authorization.

2.1 Prototype Implementation

We have developed a Java-based prototype of the MINDS grid service consisting of a front-end service and multiple back-end service using GT4.0. Through a user interface, user-custom configuration files and the location of input network flow data are submitted to the MINDS front-end service. The front-end sets up a plan and runs analysis in parallel on multiple back-end services. On completion of analysis, each back-end service returns the result to the front-end and the front-end aggregates the results and calls selected storage service to store the analysis result. The MINDS grid service is packaged as a GAR (Grid Application aRchive) to be deployed into Globus container or is packaged as a Web Application aRchive file (WAR) to be deployed into Tomcat service container.

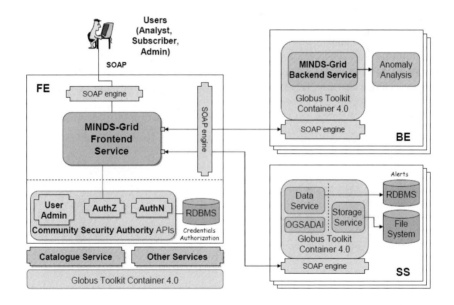

Fig. 5. MINDS Grid System Architecture

3   Performance Evaluation

We measure performance of the system on a testbed built on four Linux systems (2x 652 MHz Intel Pentium III processor with 2GB memory running ubuntu 6.06) connected by 100MB Ethernet and a Windows system (1.8 GHz Intel Pentium Mobile processor with 1GB memory running Microsoft Windows XP) connected by Wireless LAN 802.11b. We deployed MINDS Grid service front-end on a Linux machine, three MINDS Grid service back-ends on other three Linux machines, and finally two different versions of storage services on a Windows machine. In all experiments, at least 10 trials are conducted and the measurements are averaged with error bars with a 95% confidence interval. The experiments consist of three parts. First, we measure the performance of each functional unit varying input data size. The functional units are 1) connection setup time, 2) application running time (MINDS analysis), 3) alert storing time, and 4) alert retrieval time. Second experiment is to evaluate the efficiency of distributed MINDS analysis. We use three different decomposition factors (1 to 3) in this experiment.

3.1 Performance Evaluation Functional Units

We decompose the workflow into multiple functional units and measure the time delay of each functional unit to understand which functional units are dominant. With regard to input data, three different input data sets (1,000 records, 10,000 records, and 50,000 records) are used. As shown in figure 6 (top), the overall execution time linearly increases as the input data size increases. Figure 6 (bottom) shows detailed performance of each functional unit in a logarithmic scale. First of all, there is no difference in the connection setup time among different input size as expected. Secondly, MINDS analysis and Retrieval are dominant operations in the sense that the execution time of each operation increases sharply as the input size increases. On the other hand, the input size does not impact that much on the execution time of store operation. Hence, we can see Analysis and Retrieval operations can be bottlenecks in the overall performance.

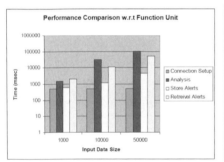 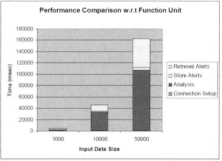

Fig. 6. Performance of Functional Units

4   Conclusion

We described the two components of this project. First, MINDS was presented which uses a suite of data mining based algorithms to address different aspects of cyber security including malicious activities such as denial-of-service (DoS) traffic, worms, policy violations and inside abuse. MINDS has shown great operational success in detecting network intrusions in several real deployments. We also presented a security enabled Grid system that supports distributed data mining, exploration and sharing using MINDS. The system addresses issues pertaining to the three main requirements of distributed data mining on Grid: 1) exploiting of geographically and organizationally distributed computing resources to solve data-intensive data mining problems, 2) ensuring the security and privacy of sensitive data, and 3) supporting seamless data/computing resource sharing. We designed a system architecture are built on a layered Grid system stack that address 1 and 3. For 2, we developed a community security authority (CSA) which supports secure communication between entities, authentication, and authorized access control. We leveraged existing technologies such as TLS, PKI, and grid security infrastructure to support secure communication, user authentication, and mutual authentication between software clients and servers. Two access schemes - RPBAC and SCBAC were developed to effectively regulate various security related activities and access in the MINDS Grid VO.

References

[1] L. Ertoz, M. Steinbach, V. Kumar: *A New Shared Nearest Neighbor Clustering Algorithm and its Applications*, 2nd SIAM International Conference on Data Mining (2002)

[2] V. Chandola and V. Kumar: *Summarization – Compressing Data into an Informative Representation*, Technical Report, TR 05-024, Dept. of Computer Science, Univ of Minnesota, Minneapolis, USA, (2005)

[3] L. Ertoz, E. Eilertson, A. Lazarevic, P. Tan, J. Srivastava, V.Kumar, and P. Dokas: The MINDS - Minnesota Intrusion Detection System, *"Next Generation Data Mining,* MIT Press" (2004)

[4] E. Eilertson, L. Ertoz, V. Kumar, and K. Long: Minds -- a new approach to the information security process, *In the 24th Army Science Conference.* US Army (2004).

[5] W. W. Cohen: Fast effective rule induction, In *International Conference on Machine Learning* (ICML) (1995)

[6] Globus GT4: www.globus.org (2006)

[7] J.B. Weissman, S. Kim, and D. England, A Dynamic Grid Service Architecture: *IEEE International Symposium on Cluster Computing and the Grid (CCGrid2005)*, Cardiff, UK (2005)

[8] B. Lee and J.B. Weissman: ``Adaptive Resource Selection for Grid-Enabled Network Services", *2nd IEEE International Symposium on Network Computing and Applications* (2003)

A Combined Hardware/Software Optimization Framework for Signal Representation and Recognition

Melina Demertzi[1], Pedro Diniz[2], Mary W. Hall[1],
Anna C. Gilbert[3], and Yi Wang[3],⋆

[1] USC/Information Sciences Institute
4676 Admiralty Way, Suite 1001
Marina del Rey, CA 90292 USA
[2] Instituto Superior Técnico
Technical University of Lisbon
Tagus Park, 2780 Porto Salvo
PORTUGAL
[3] The University of Michigan
Ann Arbor, MI 48109 USA

Abstract. This paper describes a signal recognition system that is jointly optimized from mathematical representation, algorithm design and final implementation. The goal is to exploit signal properties to jointly optimize a computation, beginning with first principles (mathematical representation) and completed with implementation. We use a BestBasis algorithm to search a large collection of orthogonal transforms derived from the Walsh-Hadamard transform to find a series of transforms which best discriminate among signal classes.The implementation exploits the structure of these matrices to compress the matrix representation, and in the process of multiplying the signal by the transform, reuse the results of prior computation and parallelize the implementation in hardware. Through this joint optimization, this dynamic, data-driven system is able to yield much more highly optimized results than if the optimizations were performed statically and in isolation. We provide results taken from applying this system to real input signals of spoken digits, and perform the initial analyses to demonstrate the properties of the transform matrices lead to optimized solutions.

1 Introduction

Signal and speech recognition require precision and robustness in the detection process. Not only do different peoples' vocal patterns have different spectral representations, but also within an utterance, phonemes have a variety of spectral signatures. All of these spectral patterns change over time, possibly even abruptly with a single utterance. For these reasons, speech recognition is a challenging task. When the desired detection system includes hard constraints on

---

⋆ This work supported by NSF CNS DDDAS-SMRP: 0540407.

Y. Shi et al. (Eds.): ICCS 2007, Part I, LNCS 4487, pp. 1230–1237, 2007.

hardware and software resources, as in embedded platforms, using compute-intensive recognition systems based solely upon the spectral analysis from the Fast-Fourier Transform (FFT) might exceed available compute resources or power/energy budgets.

In this paper, we describe an approach to speech recognition that is highly optimized according to the expected signals or phonemes, through a *joint optimization of mathematical representation, and hardware or software implementation*. The overall goal of the project is to exploit signal properties in signal processing applications to jointly optimize a computation, beginning from first principles with the mathematical representation, and carried through to the implementation in hardware and software. It is data-driven, in that the signal properties guide the optimization strategy, and it is dynamic, in that the algorithm execution will ultimately be tailored to the input signal. The optimization strategy begins with the mathematical derivation of a transform most appropriate for a training set of phonemes. From this mathematical representation, we exploit properties of the transform structure to significantly reduce the computations that are performed and the storage and memory bandwidth requirements, as compared to FFTs or straightforward implementations of multiplying the signal by the transform. At the implementation level, we exploit the ability to derive customized hardware implementations using field-programmable gate-arrays (FPGAs) and/or the corresponding software implementation.

Overall, the richness of the design space of mathematical transformations and hardware/software implementations makes the development of good solutions accessible to a handful of highly trained specialists only. A common approach relies on optimizing the individual components of the system. We believe a better solution lies in starting with a high-level representation of the problem that can be optimized from first principles. Thus, while this paper focuses on speech recognition of digits, this work can be thought of as a starting point for a more general methodology for deriving highly optimized signal recognition systems. Similar strategies are used in optimizing signal processing in SPIRAL [5,3], which optimizes linear transforms from high-level expression of signal processing formulas, but this paper is looking at a different transform and specifically targeting a hardware implementation.

To illustrate the concept of joint mathematical, hardware or software optimization for voice recognition, we consider the problem of recognizing spoken digits "0" through "9", where "0" is represented by both "oh" and "zero". A common approach to this digit recognition problem relies on using training samples to derive custom transform matrices associated with one or more digits that are then used to distinguish signals from other digits, as shown in Figure 1. In this figure, we see a process where a training set of signals representing digits are analyzed offline to derive, in this case, a transform associated with each spoken digit. The BestBasis algorithm derives a series of transforms which will best discriminate among the digit classes. Each transform encodes a choice of feature vectors or set for the signal classes. In the on-line processing phase, these

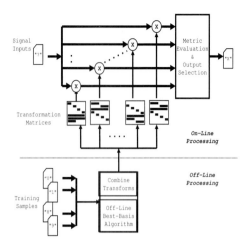

Fig. 1. Overview of speech recognition approach

transforms are applied to the input signals, followed by the metric evaluation and output selection step of a classifier to determine which digit has been spoken.

This paper is organized as follows. In the next section we describe the mathematical foundation of our specific domain and in Section 3 we describe a set of optimization opportunities in the context of the hardware implementation of a specific set of transformations. Section 4 describes some preliminary results, followed by a conclusion.

2   Mathematical Foundations

2.1   Walsh Wavelet Packets

The Walsh-Hadamard transform H_n of size $2^n \times 2^n$ is an orthogonal transform closely related to the Fourier transform. Because the Walsh-Hadamard transform is an orthonormal basis, each row of the matrix H_n corresponds to a basis vector. Let W_n^ℓ denote the ℓth row vector of the Hadamard matrix H_n. We use these vectors to build a redundant or overcomplete collection of vectors of size $2^L \geq 2^n$. To form a Walsh wavelet packet vector, we choose some W_n^ℓ of length $2^n \leq 2^L$ and we "insert" it in a vector of zeros (of length 2^L) at position $2^n k$ for some integer $k = 0, \ldots, 2^{L-n} - 1$. The collection of all such vectors $W_{n,k}^\ell$ for $n = 0, \ldots, L-1$, $k = 0, \ldots, 2^{L-n} - 1$, and $\ell = 0, \ldots, 2^n - 1$ is the full collection of Walsh wavelet packets of length 2^L. This collection defines a matrix with dimensions $L2^L \times 2^L$. If we apply the matrix to a vector, we obtain more wavelet packet coefficients than points in the vector.

Because the collection of Walsh wavelet packets is overcomplete, there are multiple ways in which to represent a vector as a linear combination of wavelet packets. This collection even includes an exponential-sized family of orthonormal bases, which means that there are exponentially many different transforms to choose from, including those with good discriminatory quality, good compressive ability, etc. We can choose our basis in which to represent our vector depending on different design criteria.

In Figure 2, we illustrate two different matrices which instantiate two different orthonormal transforms of a vector of length $2^L = 8$. Observe that the matrices have different sparsity structure; that is, the matrix on the left has many fewer nonzero entries than the matrix on the right. The notation for the Walsh wavelet

Fig. 2. Two matrices which instantiate two different orthonormal Walsh wavelet packet transformations of a vector of length $2^L = 8$

packet vectors encodes the structure of the nonzero blocks we see in the rows of the matrices in Figure 2. We denote each vector by $W^\ell_{n,k}$ where n determines the extent or width of the non-zero block, k determines which column the block begins in (column $2^n k$), and ℓ determines the sign pattern on the block. Because we have an orthonormal transform, each pattern is normalized by $2^{-n/2}$. Observe that both the block length and the initial column are powers of 2.

2.2   Best-Basis Algorithm

Although the collection of orthonormal transforms is exponentially large, the BestBasis algorithm searches this collection efficiently because of the implicit binary structure of this collection. We obtain the Walsh wavelet packet coefficients of the signal by recursively applying the H_1 matrix to the signal. This recursive application generates a binary tree of wavelet packet coefficients.

We index the nodes B^ℓ_n of the binary tree with n for the scale or length of the Walsh wavelet packet and ℓ for the sign pattern or frequency. Within each node B^ℓ_n, we have the coefficients corresponding to the vectors $W^\ell_{n,k}$. An orthonormal basis in the exponential family is a maximal antichain in the binary tree and there are $O(2^L)$ such bases. This decomposition allows us to formulate a dynamic programming algorithm to efficiently optimize over all admissible bases. See [1] for the details of the BestBasis algorithm. In our case, the optimization criterion is the discriminability of the basis and we refer to [4] for details of the discrimination criterion.

3   Exploiting Transformation Structure in Hardware

As described in section 2, the basic Walsh-Hadamard transform and its derived Walsh wavelet packet transforms have a very specific structure that enables various hardware-oriented improvements, when computing the dot-product of a transformation matrix with an input signal vector. We now describe some of these opportunities.

3.1   Compressed Representation

The transformation matrices described in the previous section have a structure that lends itself to a compressed data representation. Each row of the matrix has a single non-zero block that can be represented as a coefficient that is a specific power-of-two and a non-zero block of only $+1$ and -1 coefficients.

Consider a 4096×4096 matrix, which in 32-bit single precision would be represented in its binary form as a more than 60 Mbyte object. Given the above structure, we can greatly compress this representation to reduce bandwidth and storage requirements of the matrix. Each row can be represented with the following set of information:

1. n, where 2^n is the length of the block, and $2^{-n/2}$ represents the coefficient of the block.
2. $index$, the position in the row representing the start of the block.
3. $b(index, index + 2^{n-1})$ a set of 2^n bits, with each bit representing the sign of the corresponding element in the row (0 for 1, 1 for -1).

In the worst case, where the entire block is non-zero, each row of the matrix takes 12 bits to represent n (limited to 4096), 12 bits to represent $index$, and 4096 bits to represent b, or 515 bytes per row and no more than 2 MBytes for the entire matrix. We see in the next section that most rows of the matrices contain many zeros, so we can compress the matrices much more than this.

3.2   Optimized Implementation for a Row

To compute a single coefficient, we multiply a single row of the matrix with the input vector. Because of the structure of any single given row, we need only perform a multiplication by b, the bit pattern, and a floating point number $2^{-n/2}$. A single multiplication by a coefficient $2^{-n/2}$ is done relatively inexpensively in floating-point representation. Furthermore, the bit patterns can be implemented efficiently as well. For example, the specific non-zero block $b(0, 4)$ can be implemented fairly efficiently using a custom three-adder hardware module in two steps. During the first step, all the values are added using three adders and making use of a arithmetic negation unit. In a second step the interim result is multiplied by the $2^{-4/2} = 2^{-2}$ value. Figure 3 (left) illustrates this computation using a dedicated or custom hardware implementation.

An alternative implementation that exploits the fact that the block $b(0, 4)$ can be decomposed into two sub-blocks with the same coefficients but with opposite sign value is illustrated in Figure 3 (right). Here the computation is done in three steps but only two hardware adders are required. During the first step the values corresponding to the two $+1$ coefficients are added and saved in a register. In a second step the same adders are used to add the values corresponding to

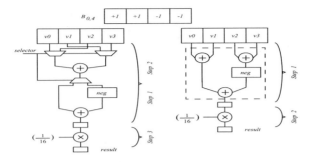

Fig. 3. A Custom Dot-Product Unit for $b(0, 4)$

the two -1 coefficients. The ouput is complemented to account for the negative sign of the coefficient. Finally in a third step, two partial results are added and multiplied by the 2^{-2} value. We can see that a faster implementation is possible if we double the number of adders and perform all four additions in a single step. A clear space-time trade-off can also be exploited in the implementation of these operations.

We make a final observation that the multiplication by the power of two can frequently be eliminated (depending on whether n is even or odd) and replaced with a shift operation. We will see in the next section that the number of distinct powers of two is small so it is straightforward to anticipate likely values for this shift prior to hardware design.

3.3   Computation Reuse

Because of the specific representation of non-zero blocks, an implementation can share partial results across rows. If two rows have non-zero blocks that overlap across the columns and there is a subset of the unitary coefficients with the same or opposite sign, then the partial sums when computing the matrix vector product can be reused. Figure 4 depicts such opportunity for reuse. These opportunities require that the actual hardware design be more sophisticated as it needs to track opportunities for partial reuse, and index and save partial results in internal memory or registers.

In addition to the reuse opportunities within one specific transform (a single matrix-vector multiplication), there are also opportunities for data and computation reuse and parallel execution when we apply multiple transforms of the same input signal vector.

3.4   Adaptive Transformation Application

We have assumed that the digit recognition procedure is a simple sequential decision process where at each step we use a transformation built to identify a specific digit. This is a sequential process that can be improved by organizing the decision tree as a binary tree or even as a single node from which emanate multiple branches corresponding to the decision of each digit. The problem of a

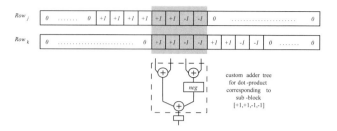

Fig. 4. Partial Computation Reuse Example

recognition in a single step or as a single transformation is that the reliability of the transformation in disambiguating all the digits might be low. On the other hand testing individually each digit is time consuming. This suggests an adaptive strategy where the implementation would start rather optimistically with a single transformation. This would allow it to identify a set of possible candidates for the input digit, say d of the 10 digits. Next the implementation would use a specific precomputed transformation for those d or simply iterate sequential over individual transformations for those d digits to clearly identify them.

4   Preliminary Results

We now present preliminary results derived from real transformation matrices obtained from training samples. These were clean spoken digits from the training portion of the TIDIGITS corpus [2]. There are three male and three female speakers. Each speaker utters the digits "zero" through "nine" and "oh" in isolation. To derive the transformation matrices, we perform the BestBasis algorithm from Section 2. Each sample included a 4096 timed sample thus leading to matrices of 4096×4096 values and to input vectors that are 4096 elements long.

As expected, the number of non-zero values for each of the transformation matrices is fairly low. Overall, for the male subjects the number of non-zeros is 15% whereas for the females this number is only 10%. For the male subjects the number of non-zeros is higher for the higher-valued digits whereas for the female subjects the reverse is true.

For this same set of matrices we determine the size of the non-zero blocks across all rows of the various matrices. The block sizes 64, 1024 and 2048 occur in every matrix, whereas other matrices may lack one of the other block sizes. For brevity, we omit these histograms.

We have developed a 4×4 custom hardware adder-tree unit depicted in Figure 3 using Xilinx ISE 7.1i tool set and targeting the Virtex-II-Pro (xc4vfx140) device. This design uses $1,714$ slices (2% of the device capacity) and attains a clock rate of 199.3 MHz. This unit is fully pipelined with an initiation interval of 12 clock cycles and overall latency of 30 clock cycles. The multiplier introduces an additional 5 clock cycle latency as it corresponds to a set of shift and normalization operations.

We then compare the performance of this custom unit with a base implementation with one single-precision adder and multiplication unit for a sample set of input signals and matrices from both female and male subjects for a wide range of input digits. For these input signals and input matrices we simulate the performance of the matrix-vector dot-product calculation and compare the performance relative to a hardware implementation using a single adder and multiplication unit. The relative performance results reveal that on average the speedup using this custom unit across all matrices from female subjects is 3.88 and is 3.91 for the male subjects respectively. We attribute this slight difference to the fact that for male subjects there are several matrices that have blocks of unit size which do not require an addition operation and have trivial multiplication. Irrespective of this minor performance difference, the speedups are nearly ideal for a custom unit performing four concurrent addition operations.

5   Conclusion

In this work, we explore the opportunities for data-driven signal processing optimization using transforms derived from the Walsh-Hadamard transform and the BestBasis selection algorithm. We are currently exploring a wide range of optimization opportunities for these transforms and the specific problem of digit recognition to highlight the opportunities for cross domain joint optimization between the mathematical, software and hardware implementation domains. This approach is data-driven, in that we use the signal properties to guide optimization, and dynamic, in that we will ultimately use features of incoming signals to optimize execution. We provide preliminary results taken from applying this system to real input signals of spoken digits, and perform the initial analyses to demonstrate that exploiting the properties of the transform matrices leads to optimized solutions. The preliminary evidence leads us to believe that this general concept is also applicable to other signal domains, not just digit recognition, particularly in resource-constrained environments.

References

1. R. R. Coifman and M. V. Wickerhauser. Entropy-based algorithms for best basis selection. *IEEE Transactions on Information Theory*, 38(2):713–718, 1992.
2. D. Ellis. Sound examples http://www.ee.columbia.edu/~dpwe/sounds/tidigits.
3. M. Puschel, J. Moura, J. Johnson, D. Padua, M. Veloso, B. Singer, J. Xiong, F. Franchetti, A. Gacic, Y. Voronenko, K. Chen, R. W. Johnson, and N. Rizzolo. Spiral: Code generation for dsp transforms. *Proceedings of the IEEE special issue on Program Generation, Optimization, and Adaptation*, 93(2):232–275, 2005.
4. N. Saito and R. R. Coifman. Local discriminant bases. *Mathematical Imaging: Wavelet Applications in Signal and Image Processing, Proc. SPIE*, 2303, 1994.
5. J. Xiong, J. Johnson, R. Johnson, and D. Padua. Spl: A language and compiler for dsp algorithms. In *Proc. of the ACM 2001 Conference on Programming Language Design and Implementation*, 2001.

Validating Evolving Simulations in COERCE

Paul F. Reynolds Jr., Michael Spiegel, Xinyu Liu, and Ross Gore

Computer Science Department
University of Virginia
{reynolds, ms6ep, xl3t, rjg7v}@cs.virginia.edu

Abstract. We seek to increase user confidence in simulations as they are adapted to meet new requirements. Our approach includes formal representation of uncertainty, lightweight validation, and novel techniques for exploring emergent behavior. Uncertainty representation, using formalisms such as Dempster-Shafer theory, can capture designer insight about uncertainty, enabling formal analysis and improving communication with decision and policy makers. Lightweight validation employs targeted program analysis and automated regression testing to maintain user confidence as adaptations occur. Emergent behavior validation exploits the semi-automatic adaptation capability of COERCE to make exploration of such behavior efficient and productive. We describe our research on these three technologies and their impact on validating dynamically evolving simulations.

1 Introduction

Uncertainty pervades model assumptions, and so frequently model designers must base decisions on little more than informed guesses. This condition presents both an opportunity and a risk. Opportunity arises from the rich set of outcomes a model can produce while exercising reasonable alternatives created by uncertainty. Risk results from not knowing which assumptions, and combinations of assumptions, reflect truth. Exploiting the opportunity while managing the risk is our ultimate goal: we seek to limit the consequences of uncertainty while providing an opportunity to adapt simulations to meet new requirements. We place high priority on maintaining user confidence in the correctness of a simulation as adaptation proceeds.

No formal approach to representing uncertainty in model descriptions or simulation languages exists. Therefore our investigation into uncertainty representation has begun with a clean slate and a broad opportunity. Our approach is to explore uncertainty representation for all aspects of model uncertainty, and not just those that best serve the needs we find in simulation adaptation. However, our investigation of uncertainty representation methods treats support of adaptation as a high priority. Our goal is to enable formal representation of potential model inputs, model outcomes, and related likelihoods –probabilities and plausibilities– for the purpose of reducing risk. The representation formalism must support automated and semi-automated analysis, and improvement of communication among model designers and implementers and related policy and decision-making personnel and processes. We discuss our progress on uncertainty representation in section 2.

Simulation adaptation, both before execution (static) and during execution (dynamic), is a high payoff capability that can save mechanical, human and monetary

Y. Shi et al. (Eds.): ICCS 2007, Part I, LNCS 4487, pp. 1238–1245, 2007.

resources.   Our adaptation technology, COERCE [1], exploits expert knowledge provided at design time for enhancing simulation adaptability both before and during execution. COERCE increases the potential for simulations to adapt to changes occurring in, for example, DDDAS environments that combine simulation and live data. COERCE employs software annotations, which we call "flexible points," to capture expert knowledge about model assumptions and alternatives. Flexible point alternatives often reflect a significant degree of uncertainty. Treatment of uncertainty, which can be daunting enough for models as defined in a traditional sense, can become considerably more complex when interactions among alternatives for model assumptions and design decisions – flexible points – are also considered. Thus arises the need for safeguards for ensuring user confidence in a simulation as is it progresses through the COERCE adaptation process

Factors that influence the efficacy of user confidence safeguards include management of uncertainty, cost of meeting user confidence goals, and technical feasibility of tools for supporting desired guarantees.   Current software validation methods, which would seem to offer the best hope for maintaining user confidence, do not apply well to simulation adaptation. As a rule they cost too much and require more information than is generally available. Similarly, technologies for validating emergent behaviors in simulations are nascent [2]. We view efficient validation and emergent behavior explanation as essential safeguards during simulation adaptation. Thus, we have concluded that we must create methods for maintaining user confidence as a simulation is adapted, and we must provide support for exploring and validating emergent behaviors, all in the presence of model uncertainty, as adaptation proceeds. Our approaches for addressing user confidence issues include lightweight validation and semi-automatic emergent behavior exploration We discuss our approaches to lightweight validation and emergent behavior exploration in sections 3 and 4, resp.

2   Representations of Uncertainty

We are designing a language for the formal representation of uncertainty in modeling and simulation, for quantitative risk assessment. Modeling under uncertainty has been of paramount importance in the public and private sector for the past half century, as quantitative methods of analysis have been developed to take advantage of computational resources. Our work brings together the fields of computer science and risk analysis. Some prominent public policy examples of uncertainty analysis in simulation include the technical studies for the Yucca Mountain nuclear waste repository [3], the assessment reports of the Intergovernmental Panel on Climate Change [4], and the guidelines from the Office of Budget and Management that recommend formal quantitative uncertainty analysis for major rules involving annual economic effects of $1 billion or more [5].

Our uncertainty representation design effort possesses two primary goals. The first is representation of continuous and discrete random variables as first-class citizens in a programming language. We aim to employ multiple mathematical frameworks for the representation of random variables. Each mathematical framework displays a tradeoff of relative expressive power for ease of use. In following subsections, we will show how three mathematical frameworks can be appropriate when varying degrees of information are available. Probability theory suffers from three primary weaknesses when representing uncertainty [6]. First, a precise probability value must be assigned to each element in the set of possible outcomes. It may not be possible to assign exact values or even assign reasonable approximations when little information is available. Second, probability theory imposes Laplace's principle of insufficient reason when no information is available. When n mutually exclusive possible outcomes are indistinguishable except for their names, they must each be assigned a probability of 1/n. Third, conflicting evidence cannot be represented in traditional probability theory. By assigning probabilities to individual elements, we can express neither incompatibility nor a cooperative effect between multiple sources of information.

Our second design goal is the capacity to specify calibration techniques for uncertainty representations in all three mathematical frameworks. Modeling under uncertainty implies the absence of perfect information, but often partial information exists in the form of observations on the model's expected behavior. Simulation practitioners expect to make the best possible use of the information available to them. A Bayesian engine is able to support the calibration of probability theory, possibility theory, and probability mass functions, and we consider such an approach.

2.1 Imprecise Probabilities

Several different mathematical systems can be used to perform uncertainty analysis. We will focus on probability theory, probability boxes, and the Dempster-Shafer theory of evidence. Probability theory is the most traditional representation of uncertainty and the one most familiar to non-mathematicians. The use of probability theory attempts to provide a quantitative analysis to answer the following three questions: (1) what can go wrong, (2) how likely is it that will happen, and (3) if it does happen, what are the consequences? [7]. Probability as a representation of subjective belief is common in quantitative risk analysis. Safety assessments must deal with rare events and thus it is difficult to assess the relative frequencies of these events [8].

2.2 Probability Boxes

Probability boxes define upper and lower boundaries for the probabilities of a set of events [9]. These boundaries (represented by $\overline{P}(X)$ and $\underline{P}(X)$) can provide information not available using traditional probability theory.   A gambler's interpretation of $\underline{P}(X)$ is that it represents the highest price s/he is willing to pay in order to receive one dollar if X occurs, or receive nothing if X does not occur. Similarly, $\overline{P}(X)$ represents the infimum selling price of an event, which is the lowest price that s/he is willing to receive in order to sell one dollar if X occurs. Probability boxes are the upper and lower distribution functions ($\overline{F}$ and $\underline{F}$) of an event X where $\underline{F}(x) = \underline{P}(X \leq x)$ and $\overline{F}(x) = \overline{P}(X \leq x)$. Upper and lower distribution functions allow an analyst to make no assumptions about the shape of the true probability distribution function. A series of coherency axioms ensure that $\underline{F}(x) \leq F(x) \leq \overline{F}(x)$ for all real numbers x. Probability boxes enable some separation of epistemic uncertainty and aleatory uncertainty [10], [11]. Under classical probability theory, the principle of indifference dictates one should select a uniform distribution when presented with a lack of information concerning the shape of that distribution. Traditional probabilistic analysis eliminates the epistemic uncertainty of the model and can result in misleading risk assessment calculations.

Regan et al. [12] investigated EPA calculations for Ecological Soil Screening Levels (Eco-SSLs) in Superfund ecological risk assessments. The study compared deterministic calculations of Eco-SSLs with a Monte Carlo approach and a probability bounds approach. Results show that Eco-SSL estimates using conservative deterministic methods were greater than estimates using probability bounds methods by two to three orders of magnitude. Median-based deterministic calculations resulted in estimates approximately one order of magnitude greater than conservative deterministic methods. Estimates based on Monte Carlo simulation generally fell between conservative and median-based deterministic estimates. The Monte Carlo simulation fails to produce a conservative estimate due to a combination of assumptions about dependencies between variables, and assumptions about the shape of the probability distribution curves. The authors "believe that probability bounds analysis is most useful as a tool for identifying the extent of uncertainty in model application and can assist in reducing this uncertainty."

2.3 Dempster-Shafer Theory of Evidence

In Dempster-Shafer theory, the concept of imprecise probabilities is extended to account for both non-specificity and discord of available evidence [13]. Probability boxes account for non-specificity by propagating lower and upper bounds without specifying the shape of the distribution. But probability boxes require that all the available evidence concludes in one non-overlapping interval. Dempster-Shafer theory allows a decision maker to reason about several candidate probability intervals for a random process, even when they conflict with one another. Dempster-Shafer theory is formulated in terms of a function known as the basic probability assignment. If Ω is the set of all possible outcomes and 2^{Ω} its power set, then a basic probability assignment is defined as $m(A) : 2^{\Omega} \rightarrow [0, 1]$ such that: $m(\emptyset) = 0$ and $\sum_{A \in 2^{\Omega}} m(A) = 1$.

Helton et al. [14] present a Dempster-Shafer risk analysis of a hypothetical safety system that is exposed to fire. The safety system consists of one weak link (WL) component and one strong link (SL) component that are both exposed to thermal heating. Both components will ultimately fail at sufficiently high temperatures. The weak link component is designed to fail safe during accidents and render the system inoperational. The strong link component is designed to be robust and resistant to extreme environments. Risk analysis is performed to assess the likelihood that the WL component will fail first. A time-dependent thermal response curve is used to model the high temperature scenario. The model contains 11 uncertain parameters such as initial temperatures, maximum temperatures, thermal constants, frequency responses, and expected values and standard deviations of normal distributions.

Dempster-Shafer theory enables the expression of several forms of partial information concerning uncertain parameters. For example, peak amplitude of the WL temperature transient (T) was measured in laboratory environments. Three measurement techniques resulted in different recorded intervals for the parameter: $T_1 = -500 \pm 40°C$, $T_2 = -1000 \pm 60°C$, $T_3 = -1800 \pm 80°C$. All three sources are considered equally credible, yet the intervals give conflicting information. Equal credibility can be expressed by assigning $m(T_1) = m(T_2) = m(T_3) = 1/3$. Evidence theory can also express nested probability structures. The thermal heating time constant (H) of the WL temperature transient is expressed with the following confidence intervals: $H_1 = 0.27 \leq H \leq 0.30$ min^{-1} with 30% confidence, $H_2 = 0.25 \leq H \leq 0.35$ min^{-1} with 50% confidence, and $H_3 = 0.20 \leq H \leq 0.40$ min^{-1} with 100% confidence. H_1, H_2, and H_3 are nested intervals that can be interpreted as plausibility measurements on H. Calculating backwards from the plausibility measurements yields the basic probability assignments $m(H_1) = 0.3$, $m(H_2) = 0.2$, and $m(H_3) = 0.5$.

Formal representations of uncertainty can be treated as sources of information regarding both model flexibility and bounds on model correctness. In the following sections we report on our approach to ensuring user confidence in model correctness as alternatives are explored in the COERCE adaptation process, and we report on exploiting the flexibility that representations of uncertainties present for validating emergent behaviors in simulation execution.

3 Lightweight Validation

Automated lightweight validation is meant to maintain user confidence that a simulation adaptation is proceeding in accordance with expectations. The approach is triggered when a user explores alternatives that uncertainty and model assumptions present, and then requires assurance that certain *correctness properties* have been maintained. A subset of requirements –correctness properties– most important to the user is identified, thus reducing analysis cost. This approach reflects a lightweight cost-efficient analysis that builds confidence, as a replacement for traditionally expensive, full validation or even regression testing methods. The concept is designed and parts have been tested as we discuss more below.

While COERCE improves the cost-effectiveness of simulation adaptation, there is a risk of introducing errors into the simulation during adaptation. When changes are made through the adaptation process, there may be inconsistencies and even conflicts among the changes. Therefore, the results of the adaptations must be validated or verified. Complete validation using statistical methods [15] is generally too expensive to be applied at each step of the adaptation process and should be applied only after a user believes that completed adaptations will not be reversed.

Our work has focused on using abstraction methods to improve the cost-effectiveness of validation [16]. We have identified two uses of abstraction: guiding optimization and checking coercion. We have explored extensions of existing abstraction methods such as program slicing, data approximation, behavior reduction and decomposition, [17, 18], and we explored an abstraction based on partial traces [19]. Our study of the adaptation of an abstract bicyclist simulation demonstrated the benefit of using abstraction methods. With a data approximation method, it took three hours to filter out 12.4% of invalid combinations of flexible point values. With partial trace abstraction it took five minutes to filter 31.6% of invalid combinations. With a control abstraction method, our results showed that previously determined optimal flexible point bindings did not extrapolate to similar but different paths for the bicyclist. Of the abstraction tools we have explored, fully-automated program slicing is one of the most promising. We have developed a prototype program slicer based on the program analysis tool, SOOT [20]. The slicer can perform inter-procedural slicing of Java programs using SOOT's program analysis framework.

Future work will employ *impact analysis*. Software impact analysis concerns estimating portions of software that can be affected if a proposed software change is made. The approach we envision will operate as follows. First, a subset of requirements, represented as correctness properties, is identified to help maintain user confidence in changes brought about by an adaptation, and to reduce analysis cost. Once the correctness properties and changes are known, impact analysis is performed to extract an impact set within the software. Built on this impact set, *automated* test case generation techniques are employed to generate test cases targeted specifically at the changes. Then, regression testing is employed using the test cases generated to validate correctness properties. If faults are detected, the adaptation process resumes until another solution is found. One contribution of our work beyond simulation adaptation will be our new methods for automated generation of test cases.

4 Validating Emergent Behaviors

Simulation behavior is emergent if it is unexpected and stems from interactions of underlying model components. Emergent behavior can be beneficial for the insight it provides. Emergent behavior can be harmful if it reflects an error in model construction. Because models often include a great deal of uncertainty, it is important that users have tools available for establishing the validity of emergent behaviors.

Validation of emergent behaviors requires an exploration capability that extends a model beyond its original intended use, so that users can test hypotheses about the characteristics of emergent behaviors. Need for a model extension capability requires adaptation which COERCE supports [1, 19]. We call our adaptation-based exploration process *Explanation Exploration.* Explanation Exploration (EE) allows a user to observe characteristics of emergent behavior as a simulated phenomenon is semi-automatically driven towards *conditions of interest*, as we explain further below.

Multiple advantages arise as a result of using COERCE: 1) COERCE flexible points enable capture of a broader range of model abstraction alternatives (both structural and parametric [19]) than a typical parameterized approach supports, 2) because COERCE employs semi-automated search methods, users can efficiently explore questions they might not have otherwise investigated, and 3) users can explore relationships between simulation behaviors they understand, but do not necessarily know how to induce, directly or indirectly, and emergent behaviors.

What constitutes a behavior can vary. Of importance to us is how a user relates choices about flexible points and knowledge of uncertainty to behaviors, and how behaviors are related to each other. An emergent behavior, E, occurs when some subset of observable simulation behaviors exhibits a pattern of unexpected behavior(s) across a set of simulation trials. An emergent behavior in a sailboat simulation may be: "the velocity of the sailboat is sometimes greater than the true wind speed when the sailboat's orientation is near perpendicular to the true wind direction." Given an emergent behavior, a user must establish if expectations regarding simulation behaviors need to be modified to include the emergent behavior. Alternatively the user may decide the emergent behavior is an error and not valid. EE facilitates this decision process. The user generally needs to formulate hypotheses about the relationship between alternatives (arising from flexible points or uncertainties) and variations of E, manifested as a function of bindings chosen for the flexible points.

A user will identify either a direct coupling between a set of alternatives (e.g. flexible points) and emergent behaviors, or an indirect coupling. Informally a direct coupling hypothesis is:

Direct Coupling Hypothesis: Within selected sets of bindings for a selected set of flexible points, predictable behavior E_{dc} related to E will be manifested in accordance with user expectations.

Exploration of direct couplings can be conducted in a straight-forward manner: alternatives for flexible point bindings can be tested and impact on emergent behaviors can be analyzed directly. Indirect couplings pose more interesting challenges because a user may not be able to hypothesize a direct link between alternatives (e.g. flexible points), bindings for those flexible points and expectations about an emergent behavior. However, it may be possible to identify instrumentable conditions within the simulation that can be related directly to emergent behavior expectations. If the user can then identify flexible points that relate directly to the intermediate conditions then a composition of the direct relationships yields a direct relationship between the flexible points and the emergent behavior. However, it is often the case that the user does not know how to make the intermediate conditions occur directly. If s/he can offer possible relevant sets of flexible points and bindings then a hypothesis may be testable with the support of search methods, such as COERCE. Informally an indirect coupling hypothesis is:

Indirect Coupling Hypothesis: For a range of allowable sets of bindings for a range of allowable flexible points, there are cases when intermediate condition C arises. When C arises, behaviors E_{ic} related to emergent behaviors E will be manifested in accordance with user expectations.

Because the user does not know which specific flexible point sets or bindings will cause condition C to arise, search will be employed. The relationship between C and E_{ic} is conjecture on the part of the user, to be established by the outcome of testing the indirect coupling hypothesis.

5 Summary

Model uncertainty presents a significant challenge to model and simulation designers, implementers and users and the policy and decision makers who often depend on their product. Model adaptation to satisfy new requirements, whether of a static (before execution time) nature, or a DDDAS dynamically adapting nature, compounds the challenge. Without methods for harnessing uncertainty and managing user confidence, particularly when model adaptation takes place, simulation designers will continue to face stiff challenges to the validity of their models. Here we have presented our approach to formal representation of uncertainty in model descriptions and simulations and the methods we have designed (and partially implemented) for maintaining user confidence in a model. Our analysis is not complete, but our objectives are clear and our designs are mature, as reflected in the work presented here.

Acknowledgments. The authors gratefully acknowledge the support of the NSF under grant 0426971.

References

1. Waziruddin, S., Brogan, D.C., Reynolds, P.F.: Coercion through optimization: A classification of optimization techniques. In: Proceedings of the Fall Simulation Interoperability Workshop. (2004)
2. Davis, P.K. "New Paradigms and Challenges", *Proceedings of 2005 Winter Simulation Conference*, IEEE, Piscataway, NJ, 2005, pp. 293-302.
3. OCRWM. Yucca mountain science and engineering report REV 1. DOE/RW-0539-1. U.S. Department of Energy, Office of Civilian Radioactive Waste Management, Las Vegas, Nevada, 2002.
4. IPCC. Ipcc fourth assessment report climate change. Intergovernmental Panel on Climate Change, 2007.
5. OMB. Circular a-4, regulatory analysis. Office of Management and Budget, September 2003.
6. Sentz, K and Ferson, S. Combination of evidence in dempster-shafer theory. Sandia National Laboratories, 2002.
7. Kaplan, S. and Garrick, B. On the quantitative definition of risk. Risk Analysis, 1(1):11–27, 1981.
8. Apostolakis, G.. The concept of probability in safety assessment of technological systems. Science, 250:1359–1364, December 1990.
9. Walley, P.. Statistical Reasoning with Imprecise Probabilities. Chapman/Hall, 1991.
10. Ferson, S. and Ginzburg, L.. Different methods are needed to propagate ignorance and variability. Reliability Engineering and System Safety, 54:133–144, 1996.
11. Ferson, S. and Hajagos, J. Arithmetic with uncertain numbers: rigorous and best possible answers. In: Reliability Engineering and System Safety, 85:135–152, 2004.
12. Regan, H., Sample, B. and Ferson, S.. Comparison of deterministic and probabilistic calculation of ecological soil screening levels. Environmental Toxicoloy and Chemistry, 21(4):882–890, 2002.
13. Shafer, G.. A mathematical theory of evidence. Princeton University Press, 1976.
14. Helton, J., Oberkampf, W. and Johnson, J.. Competing failure risk analysis using evidence theory. Risk Analysis, 25(4):973–995, 2005.
15. Easterling, R. and Berger, J. Statistical foundations for the validation of computer models. In: V&V State of the Art: Proceedings of Foundations '02. 2002.
16. Liu, X., Reynolds, P. and Brogan, D.. Using abstraction in the verification of simulation coercion. In: Proceedings of the 2006 Conference on Principles of Advanced and Distributed Simulation (PADS). 119-128, 2006.
17. Holzmann, G.. The SPIN Model Checker: Primer and Reference Manual. Addison-Wesley, Boston, Massachusetts, 2004.
18. Lynch, N., Segala, R., and Vaandrager, F. Hybrid, I/O Automata, In: Technical Report: MIT-LCS-TR-827d, MIT Lab. for Computer Science, Jan. 2003.
19. Carnahan, J., Reynolds, P., and Brogan, D. Language constructs for identifying flexible points in coercible simulations. In: Proceedings of the 2004 Fall Simulation Interoperability Workshop. Simulation Interoperability Standards Organization, Orlando, Florida. September 2004
20. Vallee-Rai, R., Co, P., Gagnon, E., Hendren, L., Lam, P., and Sundaresan, V.,. Soot - a Java bytecode optimization framework. In: Proceedings of the 1999 Conference of the Centre for Advanced Studies on Collaborative Research, 1999.

Equivalent Semantic Translation from Parallel DEVS Models to Time Automata

Shoupeng Han and Kedi Huang

College of Mechaeronics Engineering and Automation,
National University of Defense Technology, 410073, Changsha Hunan, China
hspself@163.com

Abstract. Dynamic reconfigurable simulation based on Discrete Event System Specification (DEVS) requires efficient verification of simulation models. Traditional verification method of DEVS model is based on I/O test in which a DEVS model is regarded as a black box or a grey box. This method is low efficient and insufficient because input samples are often limited. This paper proposes a formal method which can translate Parallel DEVS model into a restrict kind of Timed Automata (TA) with equivalent behaviors. By this translation, a formal verification problem of Parallel DEVS model can be changed into the formal verification of according timed automata.

Keywords: Discrete Event System Specification (DEVS), Timed Transition System (TTS), Semantic Equivalence, Timed Automata (TA).

1 Introduction

One of the intended consequences of utilizing simulations in dynamic, data-driven application system (DDDAS) [1] is that the simulations will adjust to new data as it arrives. These adjustments will be difficult because of the unpredictable nature of the world and because simulations are so carefully tuned to model specific operating conditions. Accommodating new data may require adapting or replacing numerical methods, simulation parameters, or the analytical scientific models from which the simulation is derived. These all require simulation modules in DDDAS to be dynamically reconfigurable. Furthermore, the substitutions must be verified before they are injected into the running simulation so that they can behave required properties. So, how to verify simulation system and simulation component more efficiently is a critical problem for dynamic reconfigurable simulation. Formal verification is a kind of method which can search the whole state space in acceptable time. In contrast to traditional verification method based on I/O test, it has many advantages, such as the verification process can be automated and the resources are more limited etc. This paper will propose a formal verification method for models formalized with discrete event system specification (DEVS) [2] [3] which is a representative formalism in modeling time varying reactive discrete systems. This method is based on trace equivalence which can change the verification problem of DEVS model to the verification of TA [4] with equivalent behaviors.

Y. Shi et al. (Eds.): ICCS 2007, Part I, LNCS 4487, pp. 1246–1253, 2007.

2   Formal Semantic for Parallel DEVS Model

Though Parallel DEVS has a well-defined syntax, its behavior is given as a kind of abstract simulator informally. In order to verify Parallel DEVS models formally, a formal behavior semantic of it must be defined at first [5] [6]. In this paper, Timed Transition System (TTS) is selected as the semantic base of DEVS models' behaviors.

2.1   Parallel DEVS [2]

Parallel DEVS formalism consists of two parts, atomic and coupled models. An atomic model is a structure:

$$M = \ <X,\ Y,\ S,\ s_0,\ \delta_{int},\ \delta_{ext},\ \delta_{con},\ \lambda,\ ta\ > \tag{1}$$

With

$X = \{(p,v) \mid p \in IPorts,\ v \in X_p\}$ is the set of input events, $IPorts$ is the set of input ports, and X_p is the range of values of port p ;

$Y = \{(p,v) \mid p \in OPorts,\ v \in Y_p\}$ is the set of output events, $OPorts$ is the set of output ports, and Y_p is the range of values of port p ;

S is the set of states, s_0 is the initial state;

$\delta_{int} : S \rightarrow S$ is the internal transition function;

$\delta_{ext} : Q \times X^b \rightarrow S$ is the extern transition function, X^b is a set of bags over elements in X, $Q = \{(s,e) \mid s \in S, 0 < e < ta(s)\}$, e is the elapsed time since last state transition;

$\delta_{con} : S \times X^b \rightarrow S$ is the confluent transition function, subject to $\delta_{con}(s,\phi) = \delta_{int}(s)$;

$ta : S \rightarrow R_{0,\infty}^+$ is the time advance function, $R_{0,\infty}^+$ refers to nonnegative real value;

Atomic models may be coupled in the DEVS formalism to form a coupled model. A coupled model tells how to connect several component models together to form a new model. A coupled model is defined as follows:

$$N =< X,Y,D,\{M_i \mid i \in D\},\{I_i\},\{Z_{i,j}\} > \tag{2}$$

where,

X is a set of input events;

Y is a set of output events;

D is a set of components names;

For each i in D , M_i is a component model; I_i is the set of influencees for i;

For each j in I_i , $Z_{i,j}$ is the i-to-j output translation function. It includes three cases: (1) $Z_{i,j} : X \rightarrow X_j$, if $i = N$; (2) $Z_{i,j} : Y_j \rightarrow Y$, if $j = N$; (3) $Z_{i,j} : Y_i \rightarrow X_j$, if $i \neq N$ and $j \neq N$. Above three functions are also called *EIC*, *EOC* and *IC*。

2.2 Timed Transition System (TTS)

A Timed Transition system T_t is represented as a 5-tuple:

$$T_t =< S, init, \Sigma, D, T > \qquad (3)$$

where

S is a possibly infinite set of states;

$init$ is an initial state;

$\Sigma = Act \cup R_0^+$ is the alphabet, where Act is a set of discrete actions;

D is a set of discrete transitions, noted $s \xrightarrow{x} s'$, where $x \in \Sigma$ and $s, s' \in S$, asserting that "from state s the system can instantaneously move to state s' via the occurrence of the event x";

T is a set of time-passage transitions, noted $s \xrightarrow{d} s'$, where $s, s' \in S$, asserting that "from state s the system can move to state s' during a positive amount of time d in which no discrete events occur". Time-passage transition is assumed to satisfy two axioms:

Axiom 1: If $s \xrightarrow{d} s'$ and $s' \xrightarrow{d'} s''$, then $s \xrightarrow{d+d'} s''$;

Axiom 2: Each time-passage step $s \xrightarrow{d} s'$ has a trajectory.

The trajectory used usually is I-trajectory.   A I-trajectory is defined as a function $\omega : I \rightarrow S$, where I is a closed interval of real value beginning with 0. ω also satisfy following property: for $\forall d, d' \in I$, if $d < d'$, then $\omega(d) \xrightarrow{d'-d} \omega(d')$. We often use $\omega.ltime$ to represent the supremum of I, $\omega.fstate$ to denote the first state $\omega(0)$ in I, and $\omega.lstate$ to denote the last state $\omega(\omega.ltime)$ in I. So, there is a $[0, d]$-trajectory with transition $s \xrightarrow{d} s'$, where $\omega.fstate = s$ and $\omega.lstate = s'$.

Timed execution fragment: Given a TTS, its timed execution fragment is a finite altering sequence $\gamma = \omega_0 a_1 \omega_1 a_2 \omega_2 \cdots a_n \omega_n$, where ω_i is a I-trajectory, a_i is a discrete event and $\omega_i.lstate \xrightarrow{a_{i+1}} \omega_{i+1}.fstate$. The length and initial state of γ is noted as $\gamma.ltime$ and $\gamma.ltime$, where $\gamma.ltime = \Sigma_i \omega_i.ltime$ and $\gamma.fstate = \omega_0.fstate$. All possible timed execution fragments of T_t is described as a set $execs(T_t)$.

Timed trace: Every timed execution fragment $\gamma = \omega_0 a_1 \omega_1 a_2 \omega_2 \cdots a_n \omega_n$ has an according timed trace noted $trace(\gamma)$, which is defined as an altering sequence consists of pairs noted $(a_i, \omega_i.ltime)$, where the orders of a_i in $trace(\gamma)$ is just the same as they occur in γ. All possible timed traces of T_t is described as $traces(T_t)$.

2.3 Semantic of DEVS Model Based on TTS

Behavior semantic of atomic Parallel DEVS model can be described by timed execution fragment and time trace. For a Parallel DEVS model defined in equation (1), its execution fragment is a finite altering sequence $\gamma = \omega_0 a_1 \omega_1 a_2 \omega_2 \cdots a_n \omega_n$ which includes two kinds of transitions:

Time-passage transition: each ω_i in γ is a map from interval $I_i = [0, t_i]$ to global state space of D. For $\forall j, j' \in I_i \mid j < j'$, if $\omega_i(j) = (s, e)$, then $\omega_i(j') = (s, e + j' - j)$.

Discrete event transition: discrete events in parallel can be divided into two categories: input and output events. According to the execution of DEVS, when they occur concurrently, output events must be dealt first. As shown in abstract simulator of DEVS, when an atomic model is going to send output, it will receive a event "$*$" first, "$*$"event and output event can been seen as a pair because they must be concatenated simultaneity; if there is no input event when output event occurs, the component will receive an empty event ϕ. If $(s, e) = \omega_{i-1}(\sup(I_{i-1}))$ and $(s', e') = \omega_i(\inf(I_i))$, one of following conditions should be satisfied:

$$a_i = *, e = ta(s),\ (s', e') = (s, e)\ \text{and}\ \lambda(s) \in Y \tag{3.1}$$

$$a_i \in X, \delta_{ext}(s, e, a_i) = s',\ e' = 0\ \text{and}\ e < ta(s) \tag{3.2}$$

$$a_i \in X, \delta_{con}(s, a_i) = s',\ e' = 0\ \text{and}\ e = ta(s) \tag{3.3}$$

$$a_i = \phi, \delta_{int}(s) = s',\ e' = 0\ \text{and}\ e = ta(s) \tag{3.4}$$

If two timed systems have the same timed trace set, they have the same timed behavior set obviously. The behavior semantic of Parallel DEVS model can be described by a TTS which has the same timed trace set of the DEVS model. For a DEVS model $M = < X, Y, S, s_0, \delta_{int}, \delta_{ext}, \delta_{con}, \lambda, ta >$, there is an translation noted T_{ts} upon it, which can translate M into a semantic equivalent TTS:

$$T_{ts}(M) = < S_M, init_M, \Sigma_M, D_M, T_M > \tag{4}$$

with:

$S_M = Q_M = \{(s, e) \mid s \in S, 0 \le e \le ta(s)\}$ is the state set, it is equal to the global state space of M;

$init_M = (s_0, 0)$ is the initial state;

$\Sigma_M = (X \cup \{\phi, *\}) \cup Y \cup R_0^+$ is the alphabet;

D_M is the discrete event transition, for $\forall x \in ((X \cup \{\phi\}) \cup Y)$, it is defined as:

$D_M = \{(s, e) \xrightarrow{x} (s', 0) \mid (s, e) \text{ and } s' \text{ satisfy the condition described in equation(3)}\}$

T_M is the time-passage transition, for $\forall d \in R_0^+$, it is defined as:

$$T_M = \{(s, e) \xrightarrow{d} (s, e') \mid (s, e) \in Q_M, e' = e + d, 0 \le e' \le ta(s)\}$$

Obviously, $T_{ts}(M)$ has the equivalent semantic with M.

3 Semantic of Coupled Parallel DEVS Model Based on TA

3.1 Timed Automata

A timed automata consists of a finite automata augmented with a finite set of clock variables, and transitions with clock constraints, additionally the locations can contain local invariant conditions.

For a set of clocks noted X, clock constraint over it is defined by: $\varphi := x \sim c \,|\, x - y \sim c \,|\, \neg\varphi \,|\, \varphi_1 \wedge \varphi_2$, where x and y are clocks in X, c is a nonnegative real value integer constant, φ_1 and φ_2 are clock constraints. The set of all constraints over X is noted $\Phi(X)$.

A clock assignment over a set X of clocks is a function $v : X \to R_{0,\infty}^+$, it assign each clock a real value. If a clock constraint φ is true for all clocks under a clock assignment v, we call v satisfy φ, noted $v \in \varphi$. For $Y \subseteq X$, $[Y \mapsto t]v$ means each clock $x \in Y$ is assigned a value t, while each clock in $X - Y$ satisfy v.

A timed automata is defined as a 6-tuple: $A =< L, l_0, \Sigma, C, I, E >$, where L is a finite set of location; l_0 is an initial location; Σ is a finite alphabet; C is a finite set of clock; $I : L \to \Phi(X)$ is a map, it assign each $l \in L$ a clock constraint in $\Phi(X)$; $E \subseteq L \times \Sigma \times \Phi(C) \times 2^C \times L$ is a set of transition, $< l, a, \varphi, \lambda, l' >$ means a transition from l to l' when an action a occurs and φ is satisfied by all clocks, $\lambda \subseteq C$ is the set of clocks needed to be reset when the transition occurs. $< l, a, \varphi, \lambda, l' >$ is also noted $l \xrightarrow{a, \varphi, \lambda} l'$.

For a timed automata $A =< L, l_0, \Sigma, C, I, E >$, its semantic is defined as a TTS:

$$T_t =< S_A, init_A, \Sigma_A, D_A, T_A >$$

where

S_A is a set of state which consists of a pair (l, v), where $l \in L$ and $v \in I(l)$;

$init_A = (l_0, v_0)$, $v_0 \in I(l_0)$ is a clock assignment, and for each $x \in C$, $v_0(x) = 0$;

$\Sigma_A = \Sigma \cup R_{0,\infty}^+$ is the alphabet;

T_A is a set of time-passage transition, for a state (l, v) and a nonnegative real value $d \geq 0$, if each $0 \leq d' \leq d$ satisfies $(v + d') \in I(l)$, then $(l, v) \xrightarrow{d} (l, v + d)$;

D_A is a set of discrete event transition, for a state (l, v) and a transition $< l, a, \varphi, \lambda, l' >$, if $v \in \varphi$, then $(l, v) \xrightarrow{a} (l', [\lambda \mapsto 0]v)$ and $[\lambda \mapsto 0]v \in I(l')$.

3.2   From Atomic Finite DEVS to TA

TA has more expressiveness than DEVS model to model timed systems. In this section, we will seek to find a translation function T_{aut} which can translate a Parallel DEVS model M to a trace equivalent TA noted $T_{aut}(M)$. As shown in Fig.1, TTS is selected as the common semantic base for these two models. Because infinite DEVS model cannot be formal verified for its infinite global state space, we only consider finite Parallel DEVS here, the finiteness of DEVS model include two aspects: first, the number of states is finite; second, the number of external transitions should be also finite, i.e. each external transition function should be a finite piecewise function.

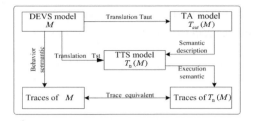

Fig. 1. Illustration of semantic equivalence between DEVS model M and its according TA model $T_{aut}(M)$. TTS model is the middle base for this equivalence.

For an atomic DEVS model $M = < X, Y, S, s_0, \delta_{int}, \delta_{ext}, \delta_{con}, \lambda, ta >$, there is an equivalent semantic translation noted T_{aut} upon it, which can translate a DEVS model into a semantic equivalent TA (shown in Fig.2):

$$T_{aut}(M) =< L_M, l_0(M), \Sigma_M, C_M, I_M, E_M > \tag{5}$$

with

$L_M = S$ is a set of location;

$l_0(M) = s_0$ is an initial location;

$\Sigma_M = \{X \bigcup \{\phi, *\}\} \bigcup Y$ is an alphabet;

C_M is a clock set, because there is a time variable e in atomic DEVS model, so there is also a clock in C_M, noted $C_M = \{e\}$;

$I_M : L_M \rightarrow \Phi(C_M)$ is a map, for $\forall l \in L_M$, $I_M(l): 0 \leq e < ta(l)$;

$E_M \subseteq L_M \times \Sigma_M \times \Phi(C_M) \times 2^{C_M} \times L_M$ is the set of transition, for $(l \xrightarrow{a,\varphi,\kappa} l') \in E_M$, it includes three cases:

(1) If $a = *$, then $(\varphi : e = ta(l)) \wedge (\kappa = \{\}) \wedge (l' = l))$; $\tag{6.1}$

(2) If $a \in X$, then $(\varphi : 0 < e < ta(l)) \wedge (\kappa = \{e\}) \wedge (l' = \delta_{ext}((l,e),a))$ or $\tag{6.2}$
$(\varphi : e = ta(l)) \wedge (\kappa = \{e\}) \wedge (l' = \delta_{con}(l,a))$;

(3) If $a = \phi$, then $(\varphi := e = ta(l)) \wedge (\kappa = \{e\}) \wedge (l' = \delta_{int}(l))$; $\tag{6.3}$

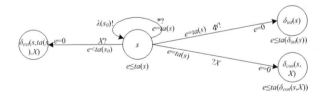

Fig. 2. Illustration of the translation from DEVS model M to its according TA model $T_{aut}(M)$. Where "!" refers to output event, "?" refers to received event, "*" refers to an internal event sent by the coordinator of M, which will be discussed in section3.3.

It is easy to find that the semantic of $T_{aut}(M)$ is just also the TTS $T_{ts}(M)$ defined in equation (4). Thereby, the semantic of $T_{aut}(M)$ is equal to the semantic of M.

3.3   From Coupled DEVS Model to Composition of TAs

Given two TAs $A_1 =< L_1, l_1^0, \Sigma_1, C_1, I_1, E_1 >$, $A_2 =< L_2, l_2^0, \Sigma_2, C_2, I_2, E_2 >$ and two sets of clock C_1 and C_2 don't intersect, the parallel composition of them is a TA noted $A_1 \parallel A_2 =< L_1 \times L_2, l_1^0 \times l_2^0, \Sigma_1 \cup \Sigma_2, C_1 \cup C_2, I, E >$, where $I(l_1, l_2) = I(l_1) \wedge I(l_2)$ and transition E is defined as follows:

(1) For $\forall a \in \Sigma_1 \cap \Sigma_2$, if transition $< l_1, a, \varphi_1, \lambda_1, l_1' >$ and $< l_2, a, \varphi_2, \lambda_2, l_2' >$ exist in E_1, E_2 respectively, then $< (l_1, l_2), a, \varphi_1 \wedge \varphi_2, \lambda_1 \cup \lambda_2, (l_1', l_2') >$ is included in E ;

(2) For $\forall a \in \Sigma_1 \setminus \Sigma_2$, transition $< l_1, a, \varphi_1, \lambda_1, l_1' >$ in E_1 and $l_2 \in L_2$, transition $< (l_1, l_2), a, \varphi, \lambda_1, (l_1', l_2) >$ is included in E ;

(3) For $\forall a \in \Sigma_2 \setminus \Sigma_1$, transition $< l_2, a, \varphi_2, \lambda_2, l_2' >$ in E_2 and $l_1 \in L_1$, transition $< (l_1, l_2), a, \varphi_2, \lambda_1, (l_1, l_2') >$ is included in E ;

The composition of DEVS model is completed by a coordinator, which has the same interface as an atomic DEVS model and it can also be included in another coordinator. The communication among DEVS models is a weak synchronization relationship which is controlled by the coordinator: when time elapsed equal to the time advancement, state transition of child model must occurs in parallel after all of them have sent the output. This weak synchronization can be depicted by channels in TA, where a channel is signed by "?" and "!" event with the same event name. In this way, the coupled model of a coordinator and its child models have just the same behavior as parallel composition of TAs (shown in Fig.3.) according with them.

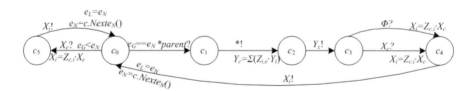

Fig. 3. Illustration of the translation from coordinator of coupled DEVS model to its according TA model. Where "c" is the coordinator discussed here; "!" refers to output event, "?" refers to received event, "*parent*" refers to an internal event for driven output sent by the parent of the coordinator; "*" is the internal event for driving output the coordinator send to its child models; "X_c", "Y_c" refer to input and output event the coordinator interact with outside environment; "X_i", "Y_i" refers to the input and output events the coordinator send to and receive from its child models, "X_i"; "e_L" is the time of last event; "e_N" is the time of next output event scheduled, function $Next_{e_N}()$ refers the function to update the value of e_N when transition has just occurred. In many kinds of TAs (such as TA defined in UPPAAL[7]), global variables are allowed in TA, so X_i, Y_i, X_c, and Y_c can all be defined as integer value or integer vectors, which will improve the efficiency. "e_G" is the global virtual time controlled by the root coordinator.

3.4 Semantic of the Whole DEVS Model

A virtual simulation system is a closed system; the execution of the whole simulation model is controlled by a time management unit which is due to schedule the time advancement of the whole system. In DEVS based simulation, the time advancement is managed by a special coordinator named root coordinator, which has no interaction with outside environment and only be responsible for advancing global virtual time e_G and translating the output of one of its child model to the input of another child model. It is a reduced coordinator and its according TA is showed in Fig.4.

Fig. 4. Illustration of the translation from root coordinator of the whole DEVS simulation model to its according TA model. Where "r" refers to the root coordinator discussed here; "*Root*" refers to an internal event for driven output of its child models; "e_G" is the global virtual time; function $Next e_G()$ is the function to update the value of e_G when a simulation step has just passed. Internal event between it and its child models noted "X_j" and "Y_w" are also be defined as integer value or integer vectors like Fig.3.

4 Conclusion

Trace equivalence is laid as the basis of the translation from Parallel model to Timed Automata model. This is an observable equivalence because trace can be observed from outside. This kind of equivalence can be used in DEVS component based simulation where the things we concerned most are not the internal detail of the components but the behaviors of them. Based on this equivalent translation, formal DEVS model verification can be realized easily using existing model checking methods and tools upon timed automata.

References

1. Darema, F.: Dynamic Data Driven Application systems: A New Paradigm for Application Simulations and Measurements. Lecture Notes in Computer Science, Vol. 3038. Springer-Verlag, Heidelberg (2004) 662–669
2. Zeigler, B.P., Praehofer, H., Kim, T.G.: Theory of Modeling and Simulation, 2nd edn. Academic Press, New York (2000)
3. Hu, X., Zeigler, B.P., Mittal, S.: Variable Structure in DEVS Component-Based Modeling and Simulation. SIMULATION: Transaction of the Society for Modeling and Simulation International, Vol.81. (2005)91–102
4. Alur, R., Dill, D.L.: A theory of timed automata. Theoretical Computer Science, Vol. 126. (1994) 183–235
5. Labiche, D.: Towards the Verification and Validation of DEVS Models. In: Proceedings of the Open International Conference on Modeling & Simulation. (2005)295–305
6. Dacharry, H.P., Giambiasi, N.: Formal Verification with timed automata and DEVS models. In: Proceedings of sixth Argentine Symposium on Software Engineering. (2005)251–265
7. Larsen, K.G., Pettersson, P.: Uppaal in a Nutshell. Int. Journal on Software Tools for Technology Transfer, Vol. 1. (1997)134–152

Author Index

Printing: Mercedes-Druck, Berlin
Binding: Stein + Lehmann, Berlin

Lecture Notes in Computer Science

For information about Vols. 1–4393

please contact your bookseller or Springer

Vol. 4446: C. Cotta, J. van Hemert (Eds.), Evolutionary Computation in Combinatorial Optimization. XII, 241 pages. 2007.

Vol. 4445: M. Ebner, M. O'Neill, A. Ekárt, L. Vanneschi, A.I. Esparcia-Alcázar (Eds.), Genetic Programming. XI, 382 pages. 2007.

Vol. 4444: T. Reps, M. Sagiv, J. Bauer (Eds.), Program Analysis and Compilation, Theory and Practice. X, 361 pages. 2007.

Vol. 4443: R. Kotagiri, P.R. Krishna, M. Mohania, E. Nantajeewarawat (Eds.), Advances in Databases: Concepts, Systems and Applications. XXI, 1126 pages. 2007.

Vol. 4440: B. Liblit, Cooperative Bug Isolation. XV, 101 pages. 2007.

Vol. 4439: W. Abramowicz (Ed.), Business Information Systems. XV, 654 pages. 2007.

Vol. 4438: L. Maicher, A. Sigel, L.M. Garshol (Eds.), Leveraging the Semantics of Topic Maps. X, 257 pages. 2007. (Sublibrary LNAI).

Vol. 4433: E. Şahin, W.M. Spears, A.F.T. Winfield (Eds.), Swarm Robotics. XII, 221 pages. 2007.

Vol. 4432: B. Beliczynski, A. Dzielinski, M. Iwanowski, B. Ribeiro (Eds.), Adaptive and Natural Computing Algorithms, Part II. XXVI, 761 pages. 2007.

Vol. 4431: B. Beliczynski, A. Dzielinski, M. Iwanowski, B. Ribeiro (Eds.), Adaptive and Natural Computing Algorithms, Part I. XXV, 851 pages. 2007.

Vol. 4430: C.C. Yang, D. Zeng, M. Chau, K. Chang, Q. Yang, X. Cheng, J. Wang, F.-Y. Wang, H. Chen (Eds.), Intelligence and Security Informatics. XII, 330 pages. 2007.

Vol. 4429: R. Lu, J.H. Siekmann, C. Ullrich (Eds.), Cognitive Systems. X, 161 pages. 2007. (Sublibrary LNAI).

Vol. 4427: S. Uhlig, K. Papagiannaki, O. Bonaventure (Eds.), Passive and Active Network Measurement. XI, 274 pages. 2007.

Vol. 4426: Z.-H. Zhou, H. Li, Q. Yang (Eds.), Advances in Knowledge Discovery and Data Mining. XXV, 1161 pages. 2007. (Sublibrary LNAI).

Vol. 4425: G. Amati, C. Carpineto, G. Romano (Eds.), Advances in Information Retrieval. XIX, 759 pages. 2007.

Vol. 4424: O. Grumberg, M. Huth (Eds.), Tools and Algorithms for the Construction and Analysis of Systems. XX, 738 pages. 2007.

Vol. 4423: H. Seidl (Ed.), Foundations of Software Science and Computational Structures. XVI, 379 pages. 2007.

Vol. 4422: M.B. Dwyer, A. Lopes (Eds.), Fundamental Approaches to Software Engineering. XV, 440 pages. 2007.

Vol. 4421: R. De Nicola (Ed.), Programming Languages and Systems. XVII, 538 pages. 2007.

Vol. 4420: S. Krishnamurthi, M. Odersky (Eds.), Compiler Construction. XIV, 233 pages. 2007.

Vol. 4419: P.C. Diniz, E. Marques, K. Bertels, M.M. Fernandes, J.M.P. Cardoso (Eds.), Reconfigurable Computing: Architectures, Tools and Applications. XIV, 391 pages. 2007.

Vol. 4418: A. Gagalowicz, W. Philips (Eds.), Computer Vision/Computer Graphics Collaboration Techniques. XV, 620 pages. 2007.

Vol. 4416: A. Bemporad, A. Bicchi, G. Buttazzo (Eds.), Hybrid Systems: Computation and Control. XVII, 797 pages. 2007.

Vol. 4415: P. Lukowicz, L. Thiele, G. Tröster (Eds.), Architecture of Computing Systems - ARCS 2007. X, 297 pages. 2007.

Vol. 4414: S. Hochreiter, R. Wagner (Eds.), Bioinformatics Research and Development. XVI, 482 pages. 2007. (Sublibrary LNBI).

Vol. 4412: F. Stajano, H.J. Kim, J.-S. Chae, S.-D. Kim (Eds.), Ubiquitous Convergence Technology. XI, 302 pages. 2007.

Vol. 4411: R.H. Bordini, M. Dastani, J. Dix, A.E.F. Seghrouchni (Eds.), Programming Multi-Agent Systems. XIV, 249 pages. 2007. (Sublibrary LNAI).

Vol. 4410: A. Branco (Ed.), Anaphora: Analysis, Algorithms and Applications. X, 191 pages. 2007. (Sublibrary LNAI).

Vol. 4409: J.L. Fiadeiro, P.-Y. Schobbens (Eds.), Recent Trends in Algebraic Development Techniques. VII, 171 pages. 2007.

Vol. 4407: G. Puebla (Ed.), Logic-Based Program Synthesis and Transformation. VIII, 237 pages. 2007.

Vol. 4406: W. De Meuter (Ed.), Advances in Smalltalk. VII, 157 pages. 2007.

Vol. 4405: L. Padgham, F. Zambonelli (Eds.), Agent-Oriented Software Engineering VII. XII, 225 pages. 2007.

Vol. 4403: S. Obayashi, K. Deb, C. Poloni, T. Hiroyasu, T. Murata (Eds.), Evolutionary Multi-Criterion Optimization. XIX, 954 pages. 2007.

Vol. 4401: N. Guelfi, D. Buchs (Eds.), Rapid Integration of Software Engineering Techniques. IX, 177 pages. 2007.

Vol. 4400: J.F. Peters, A. Skowron, V.W. Marek, E. Orłowska, R. Słowiński, W. Ziarko (Eds.), Transactions on Rough Sets VII, Part II. X, 381 pages. 2007.

Vol. 4399: T. Kovacs, X. Llorà, K. Takadama, P.L. Lanzi, W. Stolzmann, S.W. Wilson (Eds.), Learning Classifier Systems. XII, 345 pages. 2007. (Sublibrary LNAI).

Vol. 4398: S. Marchand-Maillet, E. Bruno, A. Nürnberger, M. Detyniecki (Eds.), Adaptive Multimedia Retrieval: User, Context, and Feedback. XI, 269 pages. 2007.

Vol. 4397: C. Stephanidis, M. Pieper (Eds.), Universal Access in Ambient Intelligence Environments. XV, 467 pages. 2007.

Vol. 4396: J. García-Vidal, L. Cerdà-Alabern (Eds.), Wireless Systems and Mobility in Next Generation Internet. IX, 271 pages. 2007.

Vol. 4395: M. Daydé, J.M.L.M. Palma, Á.L.G.A. Coutinho, E. Pacitti, J.C. Lopes (Eds.), High Performance Computing for Computational Science - VECPAR 2006. XXIV, 721 pages. 2007.

Vol. 4394: A. Gelbukh (Ed.), Computational Linguistics and Intelligent Text Processing. XVI, 648 pages. 2007.

Lecture Notes in Computer Science 4487

Commenced Publication in 1973
Founding and Former Series Editors:
Gerhard Goos, Juris Hartmanis, and Jan van Leeuwen